HEAVY EQUIPMENT POWER TRAINS AND SYSTEMS

Timothy W. Dell, Ph.D.

Second Edition

Publisher
The Goodheart-Willcox Company, Inc.
Tinley Park, IL
www.g-w.com

Copyright © 2024
by
The Goodheart-Willcox Company, Inc.

Previous editions copyright 2019

All rights reserved. No part of this work may be reproduced, stored, or transmitted
in any form or by any electronic or mechanical means, including information storage
and retrieval systems, without the prior written permission of
The Goodheart-Willcox Company, Inc.

Library of Congress Control Number: 2022941727

ISBN 978-1-68584-445-5

1 2 3 4 5 6 7 8 9 – 24 – 27 26 25 24 23 22

The Goodheart-Willcox Company, Inc. Brand Disclaimer: Brand names, company names, and illustrations for products and services included in this text are provided for educational purposes only and do not represent or imply endorsement or recommendation by the author or the publisher.

The Goodheart-Willcox Company, Inc. Safety Notice: The reader is expressly advised to carefully read, understand, and apply all safety precautions and warnings described in this book or that might also be indicated in undertaking the activities and exercises described herein to minimize risk of personal injury or injury to others. Common sense and good judgment should also be exercised and applied to help avoid all potential hazards. The reader should always refer to the appropriate manufacturer's technical information, directions, and recommendations; then proceed with care to follow specific equipment operating instructions. The reader should understand these notices and cautions are not exhaustive.

The publisher makes no warranty or representation whatsoever, either expressed or implied, including but not limited to equipment, procedures, and applications described or referred to herein, their quality, performance, merchantability, or fitness for a particular purpose. The publisher assumes no responsibility for any changes, errors, or omissions in this book. The publisher specifically disclaims any liability whatsoever, including any direct, indirect, incidental, consequential, special, or exemplary damages resulting, in whole or in part, from the reader's use or reliance upon the information, instructions, procedures, warnings, cautions, applications, or other matter contained in this book. The publisher assumes no responsibility for the activities of the reader.

The Goodheart-Willcox Company, Inc. Internet Disclaimer: The Internet resources and listings in this Goodheart-Willcox Publisher product are provided solely as a convenience to you. These resources and listings were reviewed at the time of publication to provide you with accurate, safe, and appropriate information. Goodheart-Willcox Publisher has no control over the referenced websites and, due to the dynamic nature of the Internet, is not responsible or liable for the content, products, or performance of links to other websites or resources. Goodheart-Willcox Publisher makes no representation, either expressed or implied, regarding the content of these websites, and such references do not constitute an endorsement or recommendation of the information or content presented. It is your responsibility to take all protective measures to guard against inappropriate content, viruses, or other destructive elements.

Image Credits. Front cover: tanger/Shutterstock.com, Dalton Dinglestad/Shutterstock.com

Heavy Equipment Power Trains and Systems was developed to educate students who are planning a career in the off-highway industry, especially construction and agricultural equipment. However, this text will empower students working in other heavy equipment power train fields, like forestry or mining, and is appropriate for students who are pursuing a certificate, associate's degree, bachelor's degree, or master's degree.

The author and publisher proudly support the mission of the AED Foundation, which encourages continuous learning, provides educational opportunities for today's employees, and improves the availability and quality of tomorrow's equipment industry technicians. The contents of this text have been carefully correlated to the Power Trains section and key areas of the Safety and Hydraulics/Hydrostatics sections of AED's "Standards for Construction Equipment Technology."

This textbook includes traditional power train content such as safety, overhead lifting, belts, chains, gearing, manual transmissions, planetary transmissions, countershaft transmissions, powershift and automatic transmissions, torque converters, hydrostatic drives, brake systems, axles, differentials, final drives, suspensions, tires, undercarriages, track steering, wheeled steering systems, and electric drives. In addition, the book is especially unique in providing an in-depth explanation of late-model continuously variable transmissions.

The text provides fundamental instruction on hydrostatic transmissions with the goal of having students fully comprehend the systems rather than needing to rely on discrete (or limited) troubleshooting charts. Students will also gain instruction in hydraulic diagnostic principles such as how to tap into the system, what tools are available, how to properly use those tools, and how to perform the tests safely.

Heavy Equipment Power Trains and Systems provides students the necessary foundational building blocks, equipping them for a bright future in off-highway power train technology. The book includes over 1000 images comprised of multicolor line art, cross-sectional drawings, photographs, and 3-D images. The book includes review questions, case studies, and helpful techniques for diagnosing power trains.

About the Author

Timothy W. Dell, Ph.D. is a Professor of Automotive Technology at Pittsburg State University and serves as the department's Diesel and Heavy Equipment Coordinator. Dr. Dell received his doctoral degree in curriculum and instruction from Kansas State University, a master of science degree in technology education from Pittsburg State University, and a bachelor of science degree in automotive technology with an emphasis in diesel and heavy equipment from Pittsburg State University.

He began his career working for Case IH in their Technical Service Group specializing in combine diagnostics. He has served on John Deere's Agricultural National Service Training Advisory Board and has been the adviser of Pittsburg State University's Caterpillar ThinkBIGGER four-year degree since its inception. He currently teaches Automotive Electricity and Electronic Systems, Fluid Power, Automotive Automatic Transmissions, Advanced Hydraulic Systems, and Construction Equipment Systems. He has also taught Off-Highway Systems. In addition, he teaches three- and four-day workshops to industry representatives and educators on the topics of hydraulic systems and heavy equipment power train systems for the Kansas Center for Career and Technical Education. Dr. Dell has served as the automotive department chair for four years but returned to the classroom full-time to pursue his passion for curriculum development and teaching. He is also the author of *Hydraulic Systems for Mobile Equipment*.

Reviewers

The author and publisher wish to thank the following industry and teaching professionals for their valuable input into the development of *Heavy Equipment Power Trains and Systems*.

James Aakre (retired)
North Dakota State College of Science
Wahpeton, ND

David Cook (retired)
Illinois Central College
East Peoria, IL

Steven Don
Montana State University-Northern
Havre, MT

Charles Ferguson
Lincoln College of Technology
Nashville, TN

Donald Flewelling
Central Arizona College
Coolidge, AZ

Ed Frederick (retired)
State Technical College
Linn, MO

Tom Grothous
University of Northwestern Ohio
Lima, OH

Jerry Hansen
Tooele Technical College
Tooele, UT

Gerald Holmes
Idaho State University
Pocatello, ID

Jeff Klehr
Central Lakes College
Staples, MN

Rene Legault
British Columbia Institute of Technology
Delta, BC, Canada

James Mack
Berks Career and Technology Center
Oley, PA

James Maxwell
Lincoln College of Technology
Nashville, TN

Darcy Moss
Grande Prairie Regional College
Fairview, AB

Chauncey Pennington
Pittsburg State University
Pittsburg, KS

Jeremy Riley
Lake Area Technical College
Watertown, SD

Derrick Russel
Highland Community College
Baileyville, KS

Gary Shore
British Columbia Institute of Technology
Delta, BC, Canada

Chris Thompson
Alexandria Technical & Community College
Alexandria, MN

Joe Valora
Elizabethtown Community and Technical College
Elizabeth, KY

Gary Wenter
Reedley College
Reedley, CA

Acknowledgments

This book has been in progress for numerous years. It is difficult to remember everyone who has participated in the development of the book. The book would not have been possible without the generous support of many individuals. I would like to thank my wife Bertha for her amazing patience and continual support, my son Calvin for his numerous hours in developing 3-D hydrostatic transmission graphics, my family for their encouragement, my students for their valuable input and assistance, and all of the reviewers for their time and expertise in reviewing the chapters.

Goodheart-Willcox Publisher and I would also like to thank the following companies, organizations, and individuals for their contribution of resource material, images, or other support in the development of *Heavy Equipment Power Trains and Systems*.

- AGCO—Matthew Keller, Sherwood Wheeler, Dallas Grothusen, Joe Henry, and Stacy Hofferbert
- Auburn Gear—Gary Grogg
- B & B Hydraulics—Bill Speakman, Dennis Rayl, Ray Miller, and Kyler Ridgeway
- Bosch Rexroth—Günter Luckhardt, Doug Wilson, David Eckerd, Glen Turner, Rüdiger Weiss, Dario Alfredo Caputo, Eduard Engel and David Gingery
- Boyd Cat—Ray Genet, Nick Hill, and Jacob Reynolds
- Caterpillar—Andrew Henry, Simon Bishop (retired), Zack Palmquist, Todd Cole, Bobby Kellum, Pete Holman (retired), Steve Hitch (retired), Kelley Maxwell, Nick Johnson, Travis McClung, and Brent Dyche
- CNH—Kelly Burgess, Ted Polzer, Russell Skewes, Derek Lee, Pete Steiner, Scott McElroy, Daniel Mitchell, Vang Moua, Cody Garrett, Kim Moulds, Roger Lewno, Ed Wojcik, Lane Robert, David North (retired), and Russel Schuchaskie (retired)
- Fabick Caterpillar—Mackenzie McDaniel
- Flaherty Farms—Calvin and Brian Flaherty
- Foley Caterpillar—Chris Scharrer, Shannon Dudley, Jarrod Haas, Cory Forshee, Tommy Phelps, Aaron Monhollon, Anthony Dahl, Jon Robinson, Gregg Haas (retired), Jeff Smarsh, Jack Frederick, and Jeremy Thoennes
- Ft. Scott Community College—Kent Aikin
- General Electric—Brent Wood
- Hayes Manufacturing, Inc.—Zeb Sieting
- Hitachi—Glenn Blackburn
- Illinois Central College—David Cook (retired) and Colin Campbell
- Industrial Sealing & Lubrication, Inc.—David Consiglio
- J & M Clutch and Converter—Terry McWhirter
- Jim Radell Construction—Dennis Shouse
- John Deere—Neil Miller, Glen Oetken, John Bowman, William Eck, Ben Hertle, Ross Graham, Matt Davied, Chris Bennet, and Mark Colvin
- Joseph Industries—Samuel Osinsky
- Joy Global—Ken Gould (retired)
- Kiewit Construction—Marlin Klotz, Scott Morris, Josh Overmeyer, and Bill Biesterfeld
- Kirby-Smith Machinery—Brian DeVore
- Kunshek Chat & Coal Company—Scott Kunshek and Bob Krumby (retired)
- Lang Diesel Incorporated—John Stewart and Ethan Meier
- Lake Area Technical College—Corey Mushitz, Jacob Beutler
- Lincoln Land Community College—Jeff Gardner
- Michael Farms—Bryan Bell and Jim Michael

Minnesota State Community and Technical College—Dick Weber
Murphy Tractor—Clark Rutledge
Oehme Farms—Dustin Oehme
Payne's Machine Shop Inc.—Jim Payne and Andy Rogers
Palmer Johnson Power Systems—Mike Kopetsky
Pittsburg State University—Chauncey Pennington, Bob Schroer, Philip McNew, David Oldham, Mike Elder, Scott Norman, Rion Huffman, Christel Benson
Red International Communications—Kerrie Kashani
Ron Morey Excavation—Ron Morey

New to This Edition

Some notable updates in this edition of *Heavy Equipment Power Trains and Systems* include additional information about torque multipliers and pressure taps in Chapter 2, drills in Chapter 3, and centrifugal clutches in Chapter 8. Chapter 9 has updates on Caterpillar planetary transmissions including the new elevated dozer four-speed transmission, and an introduction to fracking transmissions. The Allison TC10 transmission was relocated to Chapter 10 due to its countershaft design. Chapter 10 also includes the description and clutch apply chart for the John Deere e23 transmission. Chapter 11 includes a case study for diagnosing warped clutches in non-electronic transmissions. Chapter 12 contains updates to three-stage torque converters and torque dividers, and new content on a multiple-disc stator clutch and dozers with lockup clutches. Chapter 18 has a new vibration analysis case study. Chapter 19 includes a New Holland telehandler parking brake case study. Chapter 23 lists several styles of Caterpillar track chains. Chapter 24 has new content on agricultural tractors with four rubber tracks. Chapter 25 includes new content on equipment with multi-steer axles, steering amplifiers, and Danfoss MultiAxis steering.

The chapters with significant updates include:

- Chapter 13: new content on Danfoss single-servo pumps, Rexroth DA speed-sensing-pumps, Danfoss pressure-limiting controls, John Deere X9 overspeed limit control, and improvements to Eaton IPOR controls.
- Chapter 14: new content on how to adjust Bosch Rexroth DA pumps, shaft runout, a combine case study, and a dual-path skid steer.
- Chapter 15: new content on mechanical variators, John Deere X9 pro-drive twin HST motor transmission, and Caterpillar wheel loaders with twin HST transmissions.
- Chapter 26: includes some of the most extensive enhancements. It has been reorganized around the main topics of alternators, six types of electric-drive motors (including synchronous switched reluctance motors), electric-drive mining trucks, electric-drive wheel loaders (including extensive new material), electric-drive dozers, electric and hybrid electric excavators, and electric/hybrid safety and service.

Credentialing Partners and Support

Goodheart-Willcox appreciates the value of industry credentials, certifications, and accreditation. We are pleased to partner with leading organizations to support students and programs in achieving credentials. Integrating industry-recognized credentialing into a career and technical education (CTE) program provides many benefits for the student and for the institution. By achieving third-party certificates, students gain confidence, have proof of a measurable level of knowledge and skills, and earn a valuable achievement to include in their résumés. For educators and administrators, industry-recognized credentials and accreditation validate learning, enhance the credibility of programs, and provide valuable data to measure student performance and help guide continuous program improvement.

The AED and ASE Education Foundation Connections

Goodheart-Willcox is pleased to partner with the *Associated Equipment Distributors (AED)* and *ASE Education Foundation* by correlating **Heavy Equipment Power Trains and Systems** to both the *AED Standards for Construction Equipment* and the *ASE Education Foundation Medium/Heavy Duty Truck* task list. These standards were created in concert with industry and subject matter experts to match real-world job skills and marketplace demands.

To see how **Heavy Equipment Power Trains and Systems** correlates to credentialing and certification standards, visit the Correlations tab at www.g-w.com/heavy-equipment-power-trains-systems-2024

Features of the Textbook

The instructional design of this textbook includes student-focused learning tools to help you succeed. This visual guide highlights these features.

Chapter Opening Materials

Each chapter opener contains a list of learning objectives. **Objectives** clearly identify the knowledge and skills to be gained when the chapter is completed.

Additional Features

Additional features are used throughout the body of each chapter to further learning and knowledge. **Warnings** alert you to potentially dangerous materials and practices. **Cautions** alert you to practices that could potentially damage equipment or instruments. **Notes** are tips that help you develop critical thinking, diagnostic and troubleshooting skills needed in the workplace today. **Step-by-Step Procedures** are presented throughout the textbook to provide clear instructions for hands-on service activities. **Pro Tips** provide advice and guidance that is especially applicable for on-the-job situations. **Case Studies** describe real-life situations encountered by technicians in the field to help you understand what you can anticipate and expect in the workplace.

Illustrations

Illustrations have been designed to clearly and simply communicate the specific topic. Numerous illustrations have been replaced or updated for this edition. Photographic images have been updated to show the latest equipment.

End-of-Chapter Content

End-of-chapter material provides an opportunity for review and application of concepts. A concise **Summary** provides an additional review tool and reinforces key learning objectives. This helps you focus on important concepts presented in the text. **Know and Understand** questions enable you to demonstrate knowledge, identification, and comprehension of chapter material. **Apply and Analyze** questions extend learning and develop your abilities to use learned material in new situations and to break down material into its component parts. **Critical Thinking** questions develop higher-order thinking and problem-solving, personal, and workplace skills.

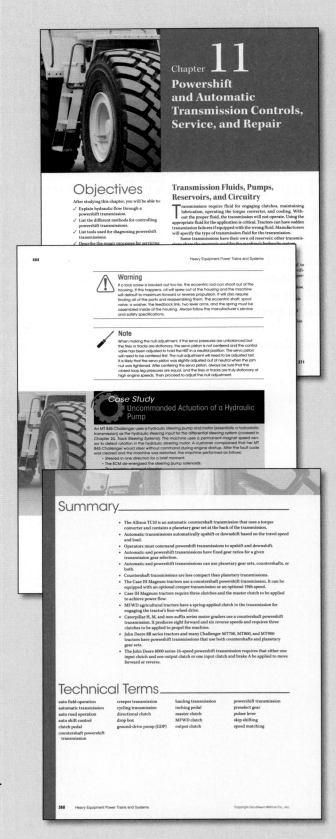

TOOLS FOR STUDENT AND INSTRUCTOR SUCCESS

Student Tools

Student Text

Heavy Equipment Power Trains and Systems is a comprehensive text that focuses on the theory, diagnosis, and service of power train and related systems in heavy equipment used in construction, mining, forestry, and agriculture.

Lab Workbook

- Hands-on practice includes questions and activities.
- Jobs offer students opportunities to perform various hands-on tasks like those they will be required to perform in the industry.

G-W Digital Companion

For digital users, e-flash cards and vocabulary exercises allow interaction with content to create opportunities to increase achievement.

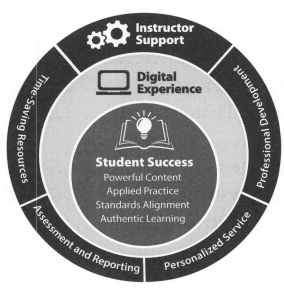

Instructor Tools

LMS Integration

Integrate Goodheart-Willcox content within your Learning Management System for a seamless user experience for both you and your students. EduHub LMS–ready content in Common Cartridge® format facilitates single sign-on integration and gives you control of student enrollment and data. With a Common Cartridge integration, you can access the LMS features and tools you are accustomed to using and G-W course resources in one convenient location—your LMS.

G-W Common Cartridge provides a complete learning package for you and your students. The included digital resources help your students remain engaged and learn effectively:

- **Digital Textbook**
- Online **Lab Workbook** content
- **Drill and Practice** vocabulary activities

When you incorporate G-W content into your courses via Common Cartridge, you have the flexibility to customize and structure the content to meet the educational needs of your students. You may also choose to add your own content to the course.

For instructors, the Common Cartridge includes the Online Instructor Resources. QTI® question banks are available within the Online Instructor Resources for import into your LMS. These prebuilt assessments help you measure student knowledge and track results in your LMS gradebook. Questions and tests can be customized to meet your assessment needs.

Online Instructor Resources

- The **Instructor Resources** provide instructors with time-saving preparation tools such as answer keys, editable lesson plans, and other teaching aids.
- **Instructor's Presentations for PowerPoint®** are fully customizable, richly illustrated slides that help you teach and visually reinforce the key concepts from each chapter.
- Administer and manage assessments to meet your classroom needs using **Assessment Software with Question Banks**, which include hundreds of matching, completion, multiple choice, and short answer questions to assess student knowledge of the content in each chapter.

See www.g-w.com/heavy-equipment-power-trains-systems-2024 for a list of all available resources.

Professional Development

- Expert content specialists
- Research-based pedagogy and instructional practices
- Options for virtual and in-person Professional Development

1	Shop Safety and Practices	1
2	Tools and Fasteners	31
3	Construction Equipment Identification	63
4	Agricultural Equipment Identification	107
5	Lifting	159
6	Belt and Chain Drives	187
7	Manual Transmissions	215
8	Clutches and Planetary Controls	253
9	Planetary Gear Set Theory	283
10	Powershift and Automatic Transmission Theory	327
11	Powershift and Automatic Transmission Controls, Service, and Repair	371
12	Hydrodynamic Drives	411
13	Hydrostatic Drives	435
14	Hydrostatic Drive Service and Diagnostics	479
15	Continuously Variable Transmissions	511
16	Differentials	563
17	Final Drives	587
18	Axles and Driveshafts	615
19	Hydraulic Brake Systems	643
20	Air Brake Systems	689
21	Suspension Systems	731
22	Tires, Rims, and Ballasting	765
23	Undercarriages	797
24	Track Steering Systems	823
25	Wheeled Steering Systems	849
26	Electric and Hybrid Drive Systems	883

Chapter 1
Shop Safety and Practices.... 1
- Personal Protective Equipment (PPE) 2
- Emergency Preparedness 3
- Fluid Hazards 7
- Pneumatic Hazards................................ 9
- Safe Practices.................................... 11
- Job Hazard Analysis (JHA) 14
- Tracking Accidents 14
- Machine Safety 15
- Machine Operation Safety 19
- Welding Safety................................... 21
- Oxygen and Acetylene 22

Chapter 2
Tools and Fasteners....... 31
- General Hand Tool Safety 31
- Hand Tools 32
- Power Tools...................................... 40
- Additional Tools and Equipment................. 41
- Measuring Tools.................................. 49
- Fasteners 53

Chapter 3
Construction Equipment Identification............ 63
- Overview of Construction Equipment 63
- Excavators....................................... 64
- Wheel Loaders................................... 70
- Track Loaders 75
- Dozers... 76
- Motor Graders................................... 79
- Scrapers... 82
- Haul Trucks 85
- On-Highway Dump Trucks....................... 87
- Skid Steers...................................... 89
- Forklifts and Telehandlers 92
- Compactors...................................... 95
- Trenchers.. 97
- Drills.. 99

Chapter 4
Agricultural Equipment Identification............ 107
- Overview of Agricultural Equipment............ 107
- Tillage Tools.................................... 108
- Planting and Seeding Equipment................ 112
- Mowers, Cutters, and Conditioners............. 119
- Rakes... 122
- Hay Balers 123
- Forage Harvesters 127
- Combines....................................... 130
- Tractors 138
- Sprayers 147
- Fertilizer Applications 151

Chapter 5
Lifting 159
- Lifting by Hand................................. 159
- Lifting Equipment 160
- Lifting Principles 176
- Planning a Lift.................................. 182

Chapter 6
Belt and Chain Drives 187
- Relationship of Speed and Torque 188
- Belts and Pulleys................................ 188
- Belts.. 189
- Pulleys 199
- Chains and Sprockets........................... 200
- Types of Chain Drives.......................... 201
- Checking Chains and Sprockets for Wear........ 206
- Chain and Sprocket Ratios 210

Chapter 7
Manual Transmissions... 215
- Types of Gearing................................ 215
- Gear Ratios..................................... 219
- Advantages and Disadvantages of Gearing....... 221
- Types of Manual Transmissions................. 221
- Shift Levers, Rails, and Cams................... 241
- Bearings.. 242

Chapter 8
Clutches and Planetary Controls............. 253
- Traction Clutches................................. 253
- Traction Clutch Service and Adjustment......... 259
- Multiple-Disc Clutches........................... 263
- One-Way Clutches................................. 270
- Electromagnetic Clutches and Brakes............ 274
- Bands... 276
- Slip Clutch....................................... 277
- Centrifugal Clutches.............................. 278
- Mechanical Schematics............................ 278

Chapter 9
Planetary Gear Set Theory............... 283
- Simple Planetary Gear Set........................ 283
- Multiple Planetary Gear Sets..................... 289
- Simplified Transmissions......................... 289

Chapter 10
Powershift and Automatic Transmission Theory..... 327
- On-Highway Countershaft Multiple Disc Clutch Transmissions....................... 327
- Off-Highway Transmission Applications.......... 329
- Countershaft Powershift Transmissions.......... 332
- Combination Planetary Countershaft Powershift Transmission.................... 348

Chapter 11
Powershift and Automatic Transmission Controls, Service, and Repair....... 371
- Transmission Fluids, Pumps, Reservoirs, and Circuitry................................ 371
- Transmission Controls............................ 377
- Transmission Diagnostics......................... 391
- Multiple-Disc Clutch Service and Repair......... 397
- Transmission Removal and Installation.......... 403

Chapter 12
Hydrodynamic Drives.... 411
- Hydrostatic and Hydrodynamic Drives........... 411
- Torque Converters................................ 412
- Dynamic Retarders............................... 429
- Contamination Control........................... 430

Chapter 13
Hydrostatic Drives....... 435
- Hydrodynamic Drives............................. 435
- Attributes of Hydrostatic Drives................. 435
- Applications of Hydrostatic Drives............... 437
- Hydrostatic Charge Pumps and Main Piston Pumps.......................... 444
- Hydrostatic Drive Operation and Oil Flow....... 453
- Variable Hydrostatic Drive Motor Applications... 461
- Additional Hydrostatic Valving.................. 466

Chapter 14
Hydrostatic Drive Service and Diagnostics......... 479
- Centering Adjustments........................... 479
- Opening a Danfoss Series-90 Manual Bypass Valve............................... 485
- Bosch Rexroth DA Valve Plate Adjustment....... 486
- Shaft Run-Out.................................... 488
- Startup (Commissioning) of a Hydrostatic Drive. 489
- Hydrostatic Drive Diagnostics by Symptom...... 491
- Do Not Attempt to Isolate a Hydrostatic Pump... 504
- Using a Hydrostatic Transmission Test Stand..... 504

Chapter 15
Continuously Variable Transmissions........... 511
- Mechanical Variators............................. 511
- Mechanical Input Plus Hydraulic Variator CVT.... 512
- Hydrostatic Drives............................... 555

Chapter 16
Differentials.............. 563
- Introduction to Differentials..................... 563
- Types of Differentials............................ 567
- Diagnosis... 576
- Repairing Differentials........................... 577

Chapter 17
Final Drives.............. 587
- Introduction to Final Drives..................... 587
- Final Drives Integrated into the Axle Housing.... 588
- Final Drives Not Integrated into an Axle Housing...................................... 594
- Servicing and Repairing Final Drives............. 602
- Failure Analysis.................................. 605

Chapter 18
Axles and Driveshafts.... 615
- Axles 615
- Duo-Cone Seal..................................... 619
- Drivelines 621
- Power Take-Off (PTO) Driveshaft 636

Chapter 19
Hydraulic Brake Systems 643
- Brake Types 643
- Hydraulic Principles 645
- Hydraulic Brake Controls 647
- Hydraulic Slack Adjusters....................... 655
- Types of Mechanical and Hydraulic Brakes....... 655
- Accumulators 669
- Cooling Internal Wet Disc Brakes 679
- Towing 681
- Adjusting Internal Wet Disc Brakes 683

Chapter 20
Air Brake Systems 689
- Air Brake Supply System Components............ 689
- Dual Brake Circuit................................ 702
- Off-Highway Single Circuits..................... 708
- Air Brake Valves 710
- Foundation Brakes................................ 714
- Air-Over-Hydraulic Brake Actuator 724
- Air Brakes versus Hydraulic Brakes 725

Chapter 21
Suspension Systems...... 731
- Construction Suspension Systems................ 732
- Agricultural Suspension Systems 741
- Cab Suspension Systems 759

Chapter 22
Tires, Rims, and Ballasting 765
- Tire Safety... 765
- Types of Tires..................................... 766
- Tire Nomenclature................................ 775
- Wheel Slip 780
- Inspecting Tires 781
- Ton-Mile per Hour Value 782
- Wheels and Rims 783
- Changing a Tire on a Multi-Piece Rim........... 785
- Adjusting Agricultural Tractor Wheel Spacing ... 787
- Weighting and Ballasting Agricultural Tractors ... 790

Chapter 23
Undercarriages.......... 797
- Undercarriage Classifications.................... 798
- Undercarriage Components..................... 799
- Track Tension 807
- Splitting the Track................................ 809
- Undercarriage Inspection....................... 809
- Undercarriage Wear 815
- Undercarriage Operating Tips 816
- Caterpillar SystemOne Undercarriage........... 816
- Rubber Track Systems 817

Chapter 24
Track Steering Systems ... 823
- Twin-Track Turning Radiuses.................... 823
- Common Track Steering System Designs......... 824
- Two-Speed Planetary Steering Systems.......... 840
- Independent-Geared Lever-Operated Track Steering Systems 841
- Tractors Equipped with Four-Track Undercarriages............................. 843

Chapter 25
Wheeled Steering Systems 849
- Types of Wheeled Steering 849
- Steering Control Units 856
- Steering Amplifiers 863
- Traditional Steering Priority Valves 864
- Diagnosing Hydraulic Steering Systems 866
- Electronic Steering Features 867
- Caterpillar Command Control Steering 874
- Motor Graders.................................... 876

Chapter 26
Electric and Hybrid Drive Systems 883
- Alternators.. 884
- Traction Motors 888
- Electric-Drive Mining Trucks 895
- Electric-Drive Wheel Loaders 899
- Electric-Drive Dozers 906
- Electric and Hybrid Electric Excavators 908
- Electric/Hybrid Safety and Service 910

Appendix................. 915
- Fastener Information 915
- Equipment Sizes 919
- Mechanical Systems 924

Glossary.................. 930

Index 967

Feature Contents

Case Studies

Job Experience Provides Emergency Preparedness.............................4
Treating Fluid-Injection Injuries 10
Working On-Site..11
Worn Offset Chain Drive... 204
Gear Ratios... 220
Determining the Cause of Fluid Loss 372
Transmission Input Shaft Speed... 384
Identifying a Dragging or Warped Clutch 393
Failed Forward Directional (High and Low) Clutches 401
Centering the Servo Piston.. 482
Uncommanded Actuation of a Hydraulic Pump 484
Overheating Hydraulic Oil When Driving Under Light Loads 493
Charge Pressure Will Not Drop 30 PSI in Forward or Reverse 496
Weak HST 500
Plugged DCV Orifice... 502
Diagnosing a Caterpillar D5H Dozer with Vibration Problems 632
Parking Brake Adjustment 682

Chapter 1
Shop Safety and Practices

Objectives

After studying this chapter, you will be able to:
- ✓ List the types of PPE used for working on heavy equipment.
- ✓ Explain steps (actions) for being prepared for emergencies.
- ✓ Describe the different types of hazards associated with working with fluids and pneumatics.
- ✓ Demonstrate safe methods for working on heavy equipment systems.
- ✓ Explain the purpose of a job hazard analysis.
- ✓ Explain the different classifications of workplace accidents.
- ✓ List multiple safety factors related to heavy equipment.
- ✓ List risks associated with operating heavy equipment.
- ✓ Lists the risks associated with welding.
- ✓ List the risks associated working with oxygen and acetylene gases.

Working around and with heavy equipment exposes personnel to the potential for injury on a daily basis. It is the employer's responsibility to provide a workplace that is free from hazards that could cause physical harm and to ensure employees comply with industry safety standards. These same safety standards also require employees to comply with all safety standards, rules, and regulations. This chapter reviews many basic safe practices and the personal protective equipment, emergency preparedness, basic first aid, and machine safety that are essential to heavy equipment technicians. See **Figure 1-1**.

Goodheart-Willcox Publisher

Figure 1-1. The undercarriages were removed from this Challenger rubber track tractor. The undercarriages weigh several thousand pounds. The 40,000-pound tractor also had to be lifted. Excellent planning must take place to ensure that the task can be completed safely.

Personal Protective Equipment (PPE)

A   Photo Melon/Shutterstock.com

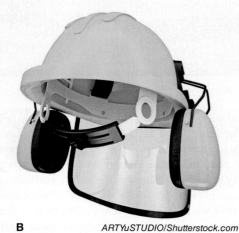

B   ARTYuSTUDIO/Shutterstock.com

Figure 1-2. Eye and face protection includes safety glasses and face shields, many of which are designed for specific tasks. A—Safety glasses. B—Hard hat, earmuffs, and face shield.

Depending on the country in which you are working, you must adhere to the government's regulating authority for safe workplaces. The *Occupational Safety and Health Administration (OSHA)* is the United States federal agency that is responsible for ensuring that employees have a safe work environment. OSHA's regulations carry the force of law, and companies can be fined for failing to follow them. Technicians working at mining sites often must follow a stricter set of rules. The *Mine Safety and Health Administration (MSHA)* is the United States federal agency responsible for ensuring mine site safety.

OSHA regulations require employers to provide workers with *personal protective equipment (PPE)*, consisting of equipment and clothing that is designed to protect the employees from potential injuries or illnesses. Common PPE includes eye protection, gloves, hard hats, boots, and hearing protection. Construction sites may require safety vests. The employer is also responsible for training the workers for proper use of PPE.

Eye Protection

Eye protection is one of the most important pieces of PPE to be worn by technicians. Safety glasses should be equipped with side shields to prevent eye injury from flying debris and fluid being sprayed from multiple angles. Regular sunglasses and prescription eyeglasses are not approved eye PPE because they do not have impact-resistant lenses or side shields. Personnel with prescription eyeglasses may acquire prescription safety glasses or wear approved PPE, such as safety goggles or a face shield, over the traditional eyeglasses. See **Figure 1-2**.

Clothing

PPE includes task-appropriate clothing. Many companies require their technicians to wear long pants, safety vests, and prohibit long hair and any loose-fitting clothing, as these could become caught in operating components. Some technicians wear coveralls. Technicians may choose to wear nitrile gloves to avoid prolonged exposure to oils and chemicals. However, gloves do not protect personnel from fluid injection injuries, which will be discussed later in this chapter.

 Warning

In order to fully concentrate on the job, technicians should not listen to music on headphones.

Hearing Protection

Diesel-powered, off-highway equipment can produce harmful noise, which can cause hearing loss with prolonged exposure. Noise generated by running machinery is one of the negative attributes of working on heavy equipment. Some systems, such as hydrostatic transmissions, generate considerably more noise than electric drive or manual transmissions when the machine is driving under a heavy load.

Hearing protection is required by OSHA when the workplace noise level reaches certain levels. The maximum permissible noise level without hearing protection ranges from 85 decibels to 140 decibels, depending on the frequency of the noise. Many employers adopt more stringent safety guidelines, requiring employees to wear hearing protection at lower noise levels. Hearing protection can consist of earplugs or earmuffs. See **Figure 1-3**.

Hard Hats

Technicians are frequently tasked with traveling to construction job sites and mine sites. Both OSHA and MSHA require hard hats to be worn on the job site to provide protection from falling objects. Technicians have been denied access or evicted from job sites for failing to follow OSHA and MSHA hard hat regulations.

There are two types of hard hats, Type I and Type II. Type I hard hats reduce the force from only the top of the head and Type II hard hats reduce the force of impact from the top as well as the side of the head. Companies may also use a hard hat color code to easily identify personnel on the worksite. See **Figure 1-4**.

Foot Protection

Most companies require their technicians to wear boots on the job for foot protection. OSHA regulations specify that employee's feet be protected from falling objects, rolling objects, and sole piercings. MSHA's protective footwear regulation does not specifically require steel-toe boots, but most mine sites require their employees to wear steel-toe boots.

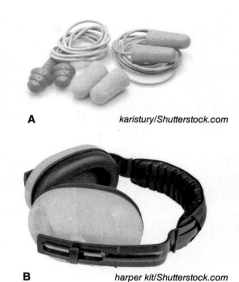

A — *karistury/Shutterstock.com*
B — *harper kit/Shutterstock.com*

Figure 1-3. Hearing protection devices have noise reduction ratings (NRR) that indicate the amount of protection they provide from various decibel levels. A—Earplugs. B—Earmuffs.

Emergency Preparedness

Due to the sheer size and power of the equipment, heavy equipment technicians work in an environment that inherently has risks. Technicians must be prepared for emergencies. Before stepping into the shop, technicians should know where safety equipment is located and how to use it. Technicians who disregard this rule are placing themselves and those around them at risk.

Hard Hat Industrial Classes	
Class G hard hats	Provide impact and penetration resistance along with limited voltage protection (up to 2200 volts).
Class E hard hats	Provide the highest level of protection against electrical hazards, with high-voltage shock and burn protection (up to 20,000 volts). They also provide protection from impact and penetration hazards by flying/falling objects.
Class C hard hats	Provide lightweight comfort and impact protection but offer no protection from electrical hazards.

Optional Color Code				
Yellow	**Blue**	**Gray**	**White**	**Green**
General laborers	Electrical workers	Site visitors	Supervisors	New or probationary employees
Earthmoving equipment operators	Technical advisers	General laborers	Visitors	Inspectors
			Engineers	
			Architects	

Goodheart-Willcox Publisher

Figure 1-4. Hard hat classes are based on the level of protection they provide from impact and electrical hazards. Although color coding is not standardized nor required, many companies choose to use specific colors for people who are working on or visiting a site.

Case Study
Job Experience Provides Emergency Preparedness

A good example of emergency preparedness was reported by an instructor after his students experienced a fire while working in the shop. The instructor learned about the incident after a prepared student had quickly extinguished the fire. The student had been in the US Navy and worked aboard a submarine. Starting in boot camp, sailors learn that they have two jobs. The first job is a firefighter. The other job, such as electronics technician, welder, or avionics technician, truly is secondary to firefighting. This sailor had been properly educated to know where the fire extinguishing equipment was located and how to use it. A person in the middle of the ocean does not have the luxury of calling the fire department. Heavy equipment technicians can also be tasked with working in remote locations and must be prepared for the worst-case scenario with little or no help from others.

Fire Suppression

Fire extinguishers are rated based on the class(es) of fire they put out. Class A fires are fueled by combustible solids, such as wood, paper, or cardboard. Class B fires are fueled by combustible gases, oils, and greases. Class C fires are electrical fires. Class D fires are caused by the ignition of combustible metals. Some fire extinguishers are rated to extinguish more than one type of fire. For example, ABC fire extinguishers can extinguish Class A, Class B, and Class C fires. The ABC fire extinguisher is the type most commonly used by heavy equipment technicians. See **Figure 1-5**.

Warning
Using the wrong class of fire extinguisher on a fire can make the situation worse. For example, if an extinguisher rated for only Class A fires is used on a Class B fire, it could spread the fire. The same extinguisher used on a Class C fire could result in electric shock.

Each fire extinguisher must be inspected to ensure that it is fully charged. If the extinguisher is charged and ready for use, the extinguisher's gauge needle will point to the green section of the gauge. Fire extinguishers must also have an inspection tag indicating the most recent inspection date. The tag lists the date and the initials of the inspector. See **Figure 1-6A**.

Warning
Fire extinguishers expire and will not operate as designed after their expiration date.

While working in a shop, technicians need to know the location of fire extinguishers, fire exits, and fire alarm switches, **Figure 1-6B**. Responsible employers train their employees in safe practices and may test personnel on their emergency preparedness.

Onboard Systems

A machine may be equipped with a rubber grommet or port that serves as a fire extinguisher receptacle. See **Figure 1-7**. In the event of a fire, the fire extinguisher hose is pressed through the center of the port, and the extinguisher is operated. The port allows

Chapter 1 | Shop Safety and Practices

Fire Classifications

Class	Description	Requires	New Symbol	Old Symbol
A	Ordinary Combustibles (Materials such as wood, paper, and textiles)	Cooling/quenching	🔥	A
B	Flammable Liquids (Liquids such as grease, gasoline, oils, and paints)	Blanketing or smothering	🔥	B
C	Electrical Equipment (Wiring, computers, switches and any other energized electrical equipment)	A nonconducting agent	🔥	C
D	Combustible Metals (Flammable metals such as magnesium and lithium)	Blanketing or smothering	🔥	D

Fire Extinguishers

Type	Description		Typically Approved for Use On	Not for Use On
Pressurized Water	Water under pressure		A	B C D
Carbon Dioxide (CO_2)	Carbon dioxide (CO_2) gas under pressure		B C	A D
Foam	Aqueous film-forming foam (AFFF) or film-forming fluoroprotein (FFFP)		A B	C D
Dry Chemical, Multipurpose Type	Typically contains ammonium phosphate		A B C	D
Dry Chemical, BC Type	May contain sodium bicarbonate or potassium bicarbonate		B C	A D
Dry Powder	May contain sodium chloride, sodium carbonate, copper, or graphite		D	A B C

Goodheart-Willcox Publisher

Figure 1-5. Fire and fire extinguisher classification charts.

Goodheart-Willcox Publisher

Figure 1-6. Fire extinguishers must be inspected regularly. A—The gauge indicates that this extinguisher is fully charged and ready for use. The tag lists the date the extinguisher was inspected as well as the inspector's initials. B—Fire extinguishers and fire alarms are often located near an entryway. The fire extinguisher shown here is rated for Class A, B, and C fires.

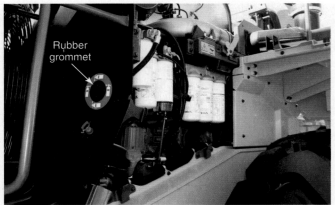

Figure 1-7. The rubber grommet on this Volvo wheel loader is a receptacle through which a fire extinguisher's nozzle can be inserted before the extinguisher is operated. This design allows the engine compartment to remain closed to limit the intake of outside air. In this photo the engine compartment door has been opened, exposing the engine's filters.

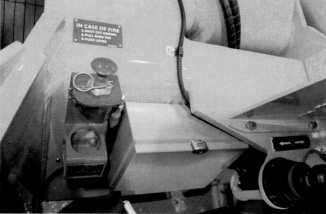

Figure 1-8. A—This D10T Caterpillar dozer has a factory-installed, onboard fire suppression system.
B—This underground mining loader has a button that is used to manually deploy an onboard fire suppression system.

the agent to be deployed with the engine compartment closed, which prevents a rush of outside air (containing oxygen) that could cause the fire to flare. Always review and follow the manufacturer's literature for using the extinguishing port.

The most expensive off-highway machines and underground mining machines commonly use onboard fire-suppression systems. Extra care must be taken when servicing machines with onboard systems. These systems use inert gases, such as nitrogen, and chemical agents to smother a fire. See **Figure 1-8A**.

Onboard systems may be automatically triggered by a heat-sensitive wire or sensor or manually actuated by an operator, **Figure 1-8B**. Inadvertently deploying a system can be costly as well as dangerous for the technician. The deployment of the extinguishing agent could eliminate the technician's oxygen supply if he or she is working in a confined area, such as an engine compartment.

Combustibles in the Shop

Technicians frequently use tools that discharge high amounts of heat, such as welders, plasma cutters, and torches. Excessive heat must be kept away from hydraulic cylinders, hoses, steel lines, accumulators, and other hydraulic components. In addition to components on machines, heat must be kept away from fluids and containers that are not part of the machine, such as chemicals, fuels, lubricants, and aerosol cans. In addition, all combustible fluids and chemicals should be properly stored in a flammable safety storage cabinet when not in use.

Hydraulic Oil

Although hydraulic oil is not highly volatile, it can ignite if it is heated to its flash point. Typical hydraulic oil flash points range from 338°F to 590°F (170°C to 310°C). Unfortunately, technicians have lost their lives due to machine fires. A diesel engine's exhaust, especially the turbocharger, is a source of heat that can cause oil from a ruptured hose to quickly ignite.

Safety Data Sheet (SDS)

Technicians work with a wide variety of chemicals and products, such as hydraulic oils, greases, engine oils, and cleaners. *Safety data sheets (SDS)* are printed materials that provide end users important information regarding products. An SDS includes information on the effects of skin or eye exposure, ingestion, and inhalation as well as the actions that should be taken for each type of exposure. Employees must know the location of the data sheets and be able to quickly access that information in case of an emergency. The categories of information provided in an SDS are listed in **Figure 1-9**.

Product and company identification	Physical and chemical properties
Emergency phone number	Stability and reactivity
Composition information on ingredients	Toxicological information
Hazards identification	Ecological information
First-aid measures	Disposal considerations
Firefighting measures	Transport information
Accidental release measures	Regulatory information
Handling and storage	Other information specific to the chemical/material
Exposure controls and personal protection	

A — *Nattawit Khomsanit/Shutterstock.com* B — *Goodheart-Willcox Publisher*

Figure 1-9. A—An unobstructed view of the location of safety data sheets in a shop. It is helpful to review the information included in the SDS before working with hazardous materials. B—All safety data sheets include critical information for each of these categories.

First-Aid Stations

Knowing the locations of a first-aid kit, an eyewash station, and a safety shower will also assist a technician in being prepared for an emergency. Technicians working in a new environment should familiarize themselves with the workplace and note the location of all first-aid and safety equipment. See **Figure 1-10**. It is also recommended that a basic first-aid kit be kept on heavy equipment machines.

First Aid

Unfortunately, accidents do occur. Many heavy equipment personnel work long distances from metropolitan areas. Technicians can save lives by receiving first-aid and cardiopulmonary resuscitation (CPR) training. *First aid* involves treating an injured person at the job site, where the injury occurred, to help sustain their life until medical personnel can arrive. *Cardiopulmonary resuscitation (CPR)* is the use of manual chest compressions and breathing into the patient's mouth when an individual's heart stops beating or he or she quits breathing. Many shops and job sites have an *automated external defibrillator (AED)* that can be used in the event a person's heart stops beating. Today's defibrillators provide audible instruction on how to install and properly use the AED. However, personnel should be prepared by becoming CPR certified, which includes learning the proper use of AEDs.

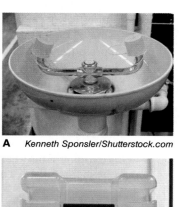

A — *Kenneth Sponsler/Shutterstock.com*

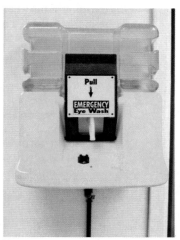

B — *ten43/Shutterstock.com*   C — *jirapong/Shutterstock.com*

Figure 1-10. Know the location of eyewash stations and safety showers. A—This eyewash station is tied into the building's plumbing. B—This station is self-contained and can be installed where there is no plumbing. C—Many shops are equipped with a safety shower.

Fluid Hazards

Heavy equipment machines use pressurized hydraulic and fuel systems that pose serious risks. Technicians are at risk of injuries, such as burns and fluid injection, any time they work with pressurized fluids. *Fluid injection* occurs when pressurized fluid penetrates the skin, most commonly caused by a burst hose. Untrained and careless technicians are at risk of receiving serious burns, having a limb amputated or worse yet, losing their lives.

Fluid-Injection Injuries

Hydraulic system pressures can exceed 7000 psi (482 bar) and some fuel system pressures can exceed 35,000 psi (2413 bar). It has been medically noted that it only takes a pressure of 7 atmospheres (or 100 psi) for fluid to puncture the skin. It has also been reported that a high-pressure oil leak can cause fluid to spray at a velocity of 300 meters per second (671 mph).

Although fluid-injection injuries are rare, the injection injury itself is a very small pinhole to the skin, so small that it is easily overlooked. Some patients have reported that the injury did not initially cause intense pain, leading many patients to delay seeking medical attention. If medical attention is delayed, swelling and pain will increase.

Patients need to quickly seek the care of a surgeon. The two immediate treatments consist of surgical decompression and debridement (the removal of damaged tissue). The severity of the injury is affected by multiple factors: the quantity of fluid injected, the pressure and velocity of the injected fluid, the toxicity of the fluid, and the amount of time before medical attention is obtained.

Patients with injection injuries can mistake the cause of their injury to be something as small as a nick to the hand. This oversight allows the fluid to cause further damage to the skin, which can lead to gangrene if not properly treated. Forty percent of all fluid injection injuries result in some form of amputation, and amputation is required nearly 100% of the time if the patient does not receive prompt medical care. Injuries resulting from system pressures of 7000 psi (482 bar) and higher result in amputation nearly 100% of the time. Even if the limb can be saved, patients typically lose some or all function of the limb.

For these reasons technicians should never use their hands for trying to locate hard-to-find leaks. Some manufacturers recommend using cardboard for locating a pinhole leak. The most common cause for injection injuries is a ruptured hose. Technicians should avoid handling pressurized fluid conductors because they are unable to predict when a hose might burst.

Pinhole leaks emit a fine mist making it difficult to identify which hose is leaking. Placing cardboard between two hoses can help identify the leaking hose and save time and money by indicating the correct hose to replace.

 Warning

Numerous service manuals specify unsafe test procedures, such as checking flow by holding a hose in a 5-gallon bucket, putting a technician at risk for fluid injection and burn injuries. As a result, it is possible for veteran supervisors to unknowingly recommend unsafe practices. Always refer to the most current and safest test procedures. The following is a list of safe practices that should always be followed when working with hydraulic systems. The list is not all-inclusive and your employer or instructor may have additional precautions that you must follow.

- *Never* connect or disconnect plumbing to a system under pressure.
- Do *not* handle hoses, gauges or components that are under pressure. It takes only 100 psi to cause a fluid injection injury and common hydraulic operating temperatures will cause severe burns if you come in contact with the fluid.
- Shut off and depressurize systems *before* working on them.
- Use flowmeters to measure flow.
- Use pressure gauges to measure pressure.
- *Never* apply heat to a fluid line or fluid component, including accumulators.
- *Never* use a makeshift device to load a hydraulic actuator.

Technicians, operators, and customers frequently take shortcuts that endanger themselves and others. The old saying is that hindsight is 20/20. If a person only knew when something was going to cause harm, they would have taken preventive measures. Many personnel work with machinery without a healthy respect of the potential risks and with little expectation that something can go wrong. As a result, they can become complacent, rush through procedures, and take shortcuts, putting themselves and others at serious risk.

Note

Many manufacturers put fluorescent dye in the oil and use a black light to locate hard-to-find leaks.

Because of the rarity of fluid injections, it is necessary for technicians to be prepared in the event of an injury. The International Fluid Power Society (IFPS) provides its members a reminder card that can be carried on their person to remind them of the five things to share with the emergency room personnel:

- Type of fluid.
- Quantity of fluid injected.
- The fluid pressure.
- How far the fluid injury has spread.
- The amount of time since the injury.

Technicians should also have fast and easy access to the fluid's SDS so that the data sheet can be provided to the surgeon as well.

Burns

During a hot summer day, hydraulic operating temperatures can exceed 200°F (93°C). Malfunctioning hydraulic systems can overheat, causing the oil temperatures to exceed 300°F (149°C). In the event of a hose failure, a technician can receive serious burns. Burns are categorized as first-degree, second-degree, third-degree, and fourth-degree depending on their severity.

- First-degree burns are the least severe and only affect the outer layer of skin. These burns are often treated with cool running water. These burns appear red and may cause swelling.
- A second-degree burn affects the two outer layers of the skin and must be treated by medical personnel to prevent infection and reduce the victim's pain. The burn will look red, splotchy, and blistered and may cause disfigurement.
- Third-degree burns penetrate through the first two layers of the skin and reach the inner hypodermis layer and require *immediate* medical attention. The victim's skin is usually charred black or dry and white. Do *not* apply an ointment or ice.
- Fourth-degree burns are even more severe, resulting in damage to deeper tissue, nerves, muscle, and bones. They require expert medical treatment. Patients lose feeling in the burn area due to the nerve damage.

Warning

In the event of a fire, remember to "stop, drop, and roll." If helping someone on fire, use a blanket to smother the fire and call 911.

Pneumatic Hazards

Heavy equipment technicians work with compressed air systems, known as *pneumatic systems*. See **Figure 1-11**. Some machines are equipped with air compressors that provide air pressure for suspension or brake systems. Shops are also equipped with air compressors that supply air pressure for pneumatic tools. Although compressed air systems are essential tools and may not appear hazardous, careless use of compressed air poses dangers to personnel including eye and lung injuries.

Jumjang/Shutterstock.com

Figure 1-11. Repair shops use large air compressors for powering tools.

- Eye injuries may be caused by particles and other flying debris that has been stirred up by compressed air when it is used to clean an area. In addition, as little as 12 psi can force an eye out of its socket.
- Lung injuries may occur when fine dust particles and other debris are stirred up and inhaled. Direct inhalation of compressed air can cause the lungs, intestines, or stomach to burst.

Case Study
Treating Fluid-Injection Injuries

A technician was working on the header float system of a self-propelled windrower. The system had a relief pressure of 2100 psi (145 bar), but a pressure sensor was reading 3800 psi (262 bar). To determine if the pressure sensor was malfunctioning, a diagnostic test port was going to be installed in the circuit to directly measure the circuit's pressure.

The technician shut off the machine and followed the service manual's procedure for depleting the pressure in the circuit. Note that the circuit did have an accumulator. The circuit was bled by manually pressing a bypass valve multiple times.

A wrench was used to crack the fitting on a 1/4" hydraulic hose. Approximately a half gallon of oil leaked from the cracked hydraulic line. After oil quit draining from the hose, the technician began to remove the hose by hand. Keep in mind that the oil had quit draining and the attached hose end was quite loose, with no tension on the fitting.

The technician used his hand to back off the remaining threads on the loose fitting, and this is when things went awry. A tremendous amount of fluid under high pressure blew out of the hose, injecting fluid into the technician's fingers. One finger had approximately a half square inch of skin removed by the force of the hydraulic fluid. An inch-long blister immediately formed on his middle finger and the technician was completely covered in oil.

He covered his bleeding fingers and traveled 40 minutes to the hospital. He chose the hospital that was 40 minutes away because it was a little larger facility. He assumed the doctors at the larger facility would have more experience with this type of injury.

When he arrived at the hospital, oil was still oozing out of his fingers. The doctor soaked his hand and treated the wound as a common hand injury. The technician was unsettled by the lack of concern shown by the doctor. The technician attempted to give the doctor the number for the manufacturer's 24-hour medical hotline so she could consult with them regarding the injury. The emergency room doctor advised the technician that she had gone to medical school, and that the problem was just a common hand injury. After soaking and wrapping the fingers, the doctor sent the technician home, stating that he might feel some tingling, numbness, and soreness.

The technician still felt uneasy about the course of treatment he had received and called the medical hotline. The hotline attendant advised the technician regarding hydraulic fluid injuries, what symptoms might occur, and what information to provide the medical personnel, which included the oil's SDS.

Approximately two hours after he left the emergency room, his finger and the blister on it began to swell quite large, his fingers tingled, his arm went numb, and he suddenly felt as if he had the flu, causing him to vomit violently.

The technician called the manufacturer's medical hotline again, and was advised to go to a different hospital. The hotline attendant spoke to the emergency room's physician's assistant who then called the state's university hospital and consulted with a hand surgeon.

The medical staff was unable to find a puncture wound on the fingers. They took an X-ray of the hand to investigate the extent of the damage. They lanced the blister on the finger and drained four cubic centimeters of oil from the technician's finger. The hand immediately began to feel better. They brushed the wound to clean away the remaining hydraulic oil. Unfortunately, the medical staff was advised to not administer local anesthesia because it would interfere with the treatment.

- Internal bodily injury, such as an *embolism*, may occur if air bubbles penetrate the skin and enter the bloodstream. An embolism can block a blood vessel and cause a stroke or heart attack and result in death.

Cleaning with Compressed Air

Compressed air can be very useful for cleaning dust and other debris from hard-to-reach places or around intricate machinery. In the United States, OSHA requires that compressed air be less than 30 psi if it is being used for cleaning and that proper PPE be used for protection. In some Canadian locations, it is against the law to use compressed air for cleaning certain items, such as benches, machinery, and clothing. Tool manufacturers offer OSHA-approved air nozzles that limit the pressure to less than 30 psi.

Alternatives to Compressed Air

Some employers do not allow the use of compressed air for cleaning due to the hazards presented and instead use vacuum cleaners with proper filtration. Another alternative for cleaning up materials that do not pose an inhalation risk is to sweep.

Safe Practices

Accidents can be costly to both the employee and the company. An injured technician may require substantial recovery time, which can cause personal hardship and cost the company thousands of dollars in worker compensation. The machine availability may also be reduced while the technician is recovering. Responsible employers enforce a safety-first environment and properly train employees before allowing them to work on any machinery. To further encourage safe practices, many companies offer safety bonuses to employees or shops for working consecutive weeks or months without a reportable incident or injury.

Know the System

It is important to gain an understanding of a system before working on a machine. Although heavy equipment machines have many similarities, they also have many differences. A technician can reduce the chance of damage and injury by reviewing the manufacturer's manual and studying the system components before beginning a job.

Case Study
Working On-Site

A customer requests some assistance with a tractor. The tractor's transmission clutch was replaced, and after the new clutch was installed, the hydraulic three-point hitch began malfunctioning. The hydraulic hitch is now jerky, sluggish, and sometimes will not lift a bale of hay. The technician has no familiarity with the tractor. If the technician is limited to taking only one item to the tractor, what should the one-and-only item be?

Inexperienced technicians often recommend taking a service manual, a pressure gauge, a flowmeter, a bucket of oil, or even an experienced technician. However, it is surprising that inexperienced technicians very seldom mention the single most important item to bring, *safety glasses*! In this real case scenario, the hydraulic system spewed oil in the face of the technician. Fortunately, the one-and-only item that was brought to the tractor was the pair of safety glasses that technician was wearing. Technicians must work with the expectancy that the hydraulic system could fail and must be prepared for when that failure occurs.

Note

Technicians working on-site must be aware of potential dangers specific to the site. For example, the silage (wet chopped crop that ferments) kept in farm silos is kept from spoiling by a lack of oxygen. Due to the fermentation process and enclosed area, toxic gases, such as nitrogen dioxide and carbon dioxide, may build up in the work area and create breathing hazards that require respiration equipment for protection.

Before Beginning a Job

Before a technician begins a job, he or she must ensure the worksite, whether it is on-site or in the shop, is safe and the machine that will be worked on has been disabled. One manufacturer recommends that the machine or implement be lowered to the ground, the engine shut off, and the ignition key removed. Most manufacturers recommend disconnecting the battery's negative cable before completing any substantial work on the machine. This practice ensures that someone will not crank the engine while a technician is in a dangerous position. This practice also prevents machine and equipment damage. For example, disabling the machine might prevent a pump failure by not allowing an engine to crank while the reservoir is empty. A technician may also install a safety tag in the cab that clearly states "Do not operate." The tag should include the name of the technician, the date, and time.

Lock-Out, Tag-Out (LOTO)

If a technician is diagnosing or repairing an electrically powered machine on-site, such as in a mine, he or she must follow a lock-out, tag-out procedure. The machine's power must be shut off and a lock and tag must be installed to prevent power from being restored to the machine. The tag may include the date and time as well as the technician's name. After the required service is completed, the lock and tag are removed. See **Figure 1-12**.

Safety Shields and Guards

The agricultural, construction, and mining industries have all spent tremendous amounts of energy to develop safe practices to ensure people are protected from rotating shafts, belts, gears, pumps, motors, and other moving or heated components. Manufacturers also install safety guards and panels on machinery to protect personnel from injury. Unfortunately, people are injured or killed each year while working carelessly around machines from which the safety guards have been removed. If the safety guards must be removed to access a component, the technician must exercise extreme caution and replace the safety guard before performing running tests. The evening television news and morning newspapers are an unfortunate place to be reminded about the consequences of not working safely around machines or disregarding safety shields and guards. Many technicians personally know someone who has been injured while working on a machine without safety guards or shields. Do not allow yourself to become a statistic by taking unnecessary shortcuts.

Rob Byron/Shutterstock.com

Figure 1-12. Typical tag and lock used for lock-out procedures.

Note

Guards on European agricultural equipment require a tool for removal, such as a screwdriver to unlatch a cam lock.

Power Take-Off (PTO)

One component that is especially important to have guarded is the agricultural *power take-off (PTO)*. A PTO provides a mechanical power source to drive implements, such as a baler, mower, grinder, or posthole digger. PTO shafts rotate at high speeds with a high torque and can be very dangerous. PTOs are essential to agricultural work but are also one of the primary causes of injury and death in the agricultural industry. PTOs have factory-installed safety guards and shields to minimize the risk of entanglement in the shaft. Do *not* operate machinery connected to a PTO if the safety guards are not in place. See **Figure 1-13**.

The PTO's shaft contains a lock on the coupler. It must also be fully functional to prevent the shaft from coming uncoupled during operation. Chapter 4, *Agricultural Equipment Identification*, provides more information on agricultural PTOs.

Exhaust Ventilation

An engine's exhaust fumes contain carbon monoxide, which can cause sickness or death. Whenever a machine is running in the shop, the shop's exhaust ventilation system must be turned on and connected to the machine's exhaust pipe. Exhaust ventilation systems vary and each technician working in a shop should be trained to properly connect and use the shop's system. See **Figure 1-14**. Some shops have additional exhaust fans to aid in exhausting fumes and heat generated by running engines. Regular inspection and maintenance should be performed on the system to ensure it is working properly.

Overhead Shop Doors

Overhead shop doors with high clearance are used to allow machines to enter and exit the shop. These doors should be *fully opened* or *fully closed* at all times to prevent a driver from hitting and damaging the door and machine and potentially injuring the driver or nearby personnel.

Goodheart-Willcox Publisher

Figure 1-13. The PTO should have a telescoping cover that keeps objects from getting tangled in the PTO shaft. The cover is normally chained so that it remains stationary while the driveshaft spins.

Goodheart-Willcox Publisher

Figure 1-14. This shop uses a retractable overhead duct system that is attached to a high-volume fan that draws the exhaust fumes out of the shop.

Fall Protection

One common regulation that technicians must follow is the need for fall protection. *Fall protection* is achieved with a full-body harness and a shock-absorbing lanyard that is securely attached to a secure anchor point. In the event that a technician slips or falls, the harness prevents the technician from falling to the ground. See **Figure 1-15**. Fall protection is required when working at elevated heights. The minimum height is dependent on the industry but most require protection for employees working at any height above 4′, 5′, or 6′.

> **Note**
>
> The full-body harness and shock-absorbing lanyard is sometimes called a *fall-arrest system*. Dealerships and repair shops are often designed with fall protection in mind and provide secure places for workers to attach their fall protection harness.

Job Hazard Analysis (JHA)

Some employers require employees to conduct a *job hazard analysis (JHA)* before performing any task. The JHA helps determine potential risks related to the job, tools, and surrounding environment in order to reduce these risks. The JHA form provides space for the technician to write the tasks that will be performed and the potential hazards that will be encountered during the job. Information that may be recorded on a JHA includes the following:

- The name of the employee conducting the JHA.
- A description of the task to be completed.
- The level of risk: low, moderate, high.
- A sequential plan for completing the job.
- The proper PPE required for the task.
- The types of risks on the job (falls, contact with electrical or gas utility lines, working alone, pinch points, and slip hazards).
- List of risks related to manual labor, such as lifting, repeated motion, vibration, and transporting components or tools.
- List of potential hazards and actions that will be taken to prevent the potential injuries.
- Rating of the chances of injury and the plan to lower the odds.
- A reviewer's name and signature.

Tracking Accidents

As a means of ensuring companies maintain safe workplaces, OSHA has strict guidelines for recording and reporting occupational injuries and illness that must be followed for complete compliance. OSHA requires companies to report *recordable accidents* or *incidents* when the accident results in any of the following:

- Fatality.
- Loss of consciousness.
- Illness or injury that requires extended care by a physician and/or hospitalization.

A NShu/Shutterstock.com

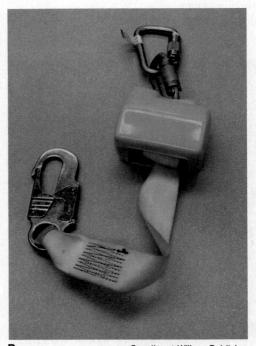
B Goodheart-Willcox Publisher

Figure 1-15. Fall protection allows technicians to work at elevated heights and prevents injuries caused by falls. A—A technician is strapped into a full-body harness. B—A shock-absorbing lanyard is used to secure the harness to an anchor point on the machine.

- Absence from future workdays.
- The employee's work being limited.

In addition to complying with OSHA mandates, many employers keep more detailed records to record both recordable and nonrecordable accidents. An employer may define a **nonrecordable accident** as one in which the employee sees a physician but receives only minor care, such as a temporary bandage. The terms *loss-time accident* and *non-loss time accident* may be used to indicate whether the accident resulted in the employee's absence for a period of time. Employers can use these detailed records to help improve their safety program through careful examination of causes and employee preparedness.

Machine Safety

Most accidents that occur when working with heavy machinery can be attributed to personnel performing unsafe acts. These **unsafe acts** occur when workers willingly choose to take unnecessary risks. It takes training and a conscientious effort to maintain a safe machine work environment. With proper training and discipline, accidents can be greatly reduced or prevented, enabling personnel to remain safe and on the job. The following sections describe many of the safe practices heavy equipment technicians should know and use while on the job.

Wheel Chocks

Wheel chocks are wedge-shaped blocks that are inserted in the front and rear of a machine's tire when the machine is not in service, being unloaded, or being serviced. See **Figure 1-16A**. The wheel chocks prevent the vehicle or trailer from moving in the event that the brake mechanism fails. OSHA requires that the parking brake be set and the rear tires of any commercial motor vehicle (CMV) be chocked before being unloaded. MSHA requires machines to be chocked if they are parked on a surface that is not level. Many employers require the use of wheel chocks on *all* machines, including service trucks. Wheel chocks are typically stored on the machine for easy access. See **Figure 1-16B**.

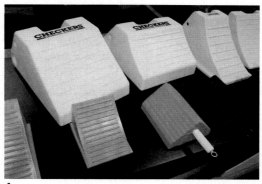

A *Goodheart-Willcox Publisher*

 Caution
Only use legitimate wheel chocks designed to prevent the machine from rolling. Rocks or any other makeshift devices are *not* approved wheel chocks.

Entering and Exiting Machines

Technicians frequently enter and exit machines during repair or maintenance procedures. Most machines require technicians to climb steps or a stepladder to enter the operator's cab. While climbing up or down the stairs/steps, personnel must always face toward the stairs and use three points of contact. The **three points of contact** can be two hands and one foot or two feet and one hand. See

B *NelliGal/Shutterstock.com*

Figure 1-16. A—Wheel chocks are made of different types of material, such as plastic, wood, rubber, or steel. B—The front bumper of this haul truck has a place to hang the truck's wheel chocks.

Figure 1-17. Using three points of contact helps minimize the risks that are associated with entering and exiting the machine, such as slick steps or having a boot becoming lodged in the crevice of a step. Falls from ladders or stairs can cause serious injury, and personnel should never climb facing away from the steps or using only two points of contact while clinging onto tools, supplies, files, or a book.

Implement Locks

A technician may use an implement lock to hold an implement in place while accessing a component that would otherwise be blocked. For example, a tractor's loader might need to be raised to gain access to a front engine component that is blocked when the loader is in the lowered position. An ***implement lock*** typically consists of a steel brace that is inserted over a hydraulic cylinder's rod. The steel brace, locked between the machine frame and the cylinder, prevents the hydraulic cylinder from moving and locks the implement in place. See **Figure 1-18**. Implement locks are also included on most cabs and on other implements, such as on the header or feeder lift cylinder on a combine. See **Figure 1-19**.

Warning

Never work below an unsupported component, implement, or machine. A technician who takes shortcuts and fails to properly secure a machine or component may be severely injured or even lose his or her life. Working below a suspended load that is being held only by fluid pressure is essentially betting your life on the strength of a hydraulic hose or a hydraulic-cylinder's seal. If the hose or seal ruptures, the implement will lower and potentially crush anyone below it.

Note

Many construction machines use an implement hydraulic lockout lever or switch. See **Figure 1-20**. When the hydraulic lockout has been actuated, it typically blocks pilot oil pressure from the directional control valves to prevent the valves from operating.

Scott A. Frangos/Shutterstock.com

Figure 1-17. This operator is using three points of contact, two feet and one hand.

Goodheart-Willcox Publisher

Figure 1-18. Loaders commonly have an implement lock. A—Notice the loader frame is in a lowered position and the red implement lock is in its stored position. B—The implement lock has been placed over the loader's lift cylinder rod and is preventing the loader from lowering.

Figure 1-19. The loader's cab can be raised to service the machine. A—The red cab lock is used to prevent the operator's cab from lowering. B—The red cab lock is placed in a horizontal position when it is not bracing the lifted cab.

Figure 1-20. The excavator's hydraulic lockout lever has been lifted to the lockout position. The hydraulic controls are disabled in this position.

Articulation Steering Locks

Several mobile machines steer by means of a center articulation joint that allows the front and rear frame to pivot in the middle of the machine. Examples of machines that use articulated steering are four-wheel drive agricultural tractors, wheel loaders, motor graders, and haul trucks. As the steering wheel is turned, it causes the steering cylinders to pivot the machine's articulation joint to steer the tractor. It is very dangerous for personnel to be near a tractor's articulation joint when the tractor is running. One bump of the steering wheel could cause the tractor to steer, causing a fatality or serious injury. Articulated tractors have steering locks that lock the front and rear frame together to prevent the tractor from articulating. See **Figure 1-21**.

Figure 1-21. The red brace is the articulation steering lock. It is shown in its stored position. When the left pin is removed, the brace can be rotated and aligned with the hole on the right side frame. When the brace is locked in place across the joint, the loader will not articulate.

Dump Bed Locks

Dump trucks also use locks to hold the bed in a raised position to provide safe access to the powertrain. The locks are designed to prevent the hydraulic bed dump cylinders from retracting. The rigid-frame dump truck in **Figure 1-22** uses two pins to lock the dump bed in a raised position.

Rollover Protective Structure (ROPS)

Most mobile machines are equipped with a *rollover protective structure (ROPS)*. The ROPS is a safety device designed to prevent the machine from crushing the operator if the machine rolls over. To be effective, a machine's ROPS must be used in conjunction with the machine's seat belt. Fatalities often occur when the unsecured operator panics during a rollover and attempts to jump out of the machine or falls out of the seat and into the machine's path.

Figure 1-22. This Caterpillar 777G dump truck has two lock pins that are used to lock the bed in a raised position. When the bed is fully raised, the pins are inserted through the two holes above the suspension cylinders to lock the bed in a raised position.

Some structures are as simple as a U-shaped bar bolted to the machine's frame, **Figure 1-23A**. This design is also known as a two-post ROPS because of the two attaching posts. Some large riding lawn mowers, compact utility tractors, and compactors have a foldable ROPS that allows the tractor to be driven into a low-clearance storage facility. See **Figure 1-23B**. Tractors operated on steep slopes or near cliffs have a higher risk for rolling over and may be equipped with a full ROPS that extends past the operator's cab to the front of the tractor. See **Figure 1-23C**. Full ROPS are also used in forestry applications. In rare cases, machines designed to operate only on flat surfaces at low travel speeds may not be equipped with an ROPS, **Figure 1-23D**.

Many ROPS are designed as an integral part of a machine's cab. It is critical that the ROPS not be modified in any way. Modifications made by drilling, cutting, shortening, lengthening, or welding will compromise the system's integrity.

> **Warning**
>
> An ROPS may be removed from a machine to allow highway bridge clearance during transport. The ROPS should not be removed before the machine is loaded on a trailer and it should be installed before unloading the machine from the trailer to prevent injuries or fatalities from a rollover that occurs during loading or unloading.

Figure 1-23. Different types of rollover protective structures are used on different machines. A—This compact utility tractor's ROPS consists of a simple U-shaped bar. Large bolts fasten the ROPS to each side of the tractor's frame. B—A foldable ROPS enables a machine to enter areas with low clearance. The ROPS on this compactor is folded. C—This dozer is equipped with a full ROPS that extends from the operator's cab to the front of the machine. D—Pavers are examples of machines that have little risk of rolling over because they work on flat surfaces and travel at slow speeds and typically do not have an ROPS.

Seat Belts

As mentioned, an ROPS will not protect an operator unless the operator is wearing a seat belt or operator restraint system. OSHA and MSHA require the use of seat belts and specify that seat belts follow the Society of Automotive Engineers (SAE) seat belt regulations (J386 for off-road machines and J1194 for agricultural wheeled tractors). The seat belt must be clearly marked with the year it was made, the manufacturer's name, and the model number. The seat belt should be replaced if it is worn, frayed, cracked, or any part of the restraint is not working. Some manufacturers also specify that seat belts more than five years old or that have been on the machine more than three years should be replaced.

Note

One large contractor places signs in their machines and trucks that state "seat belts are a condition of employment," meaning that "no seat belt" equals "no job."

Falling Object Protective Structure (FOPS)

Machines used on sites or for jobs where the operator is at risk from falling objects are typically equipped with both an ROPS and a ***falling object protective structure (FOPS)***. An FOPS is commonly used on rigid-frame haul trucks. See **Figure 1-24**. The cabs on haul trucks are at risk from falling objects when excavators, loaders, and shovels load the trucks. The FOPS and ROPS may be designed into the same structure.

Stay out of Compartments

Machinery and surrounding compartments impose great risk to personnel. In a combine for example, the separator housing can have multiple large drive belts and the grain tank can have spinning augers. Personnel must always stay out of compartments while the machine is operating or they risk severe injury or death.

Machine Operation Safety

Technicians may have to operate or move machinery during a service or repair job and must be fully aware of safe practices associated with machine operation. The following material includes some basic safety practices to be used when operating heavy equipment.

Know the Machine's Capacities and Limitations

One key component to machinery safety is knowing the machine's capacities, including the machine's operating weight and its workload limitations. For example, if the machine's operating weight is 40,000 lb (20 tons) it should not be transported over a bridge that is rated at 16 tons (32,000 lb). Another example would be driving a tractor with a full loader. It is very easy to tip a tractor when operating with a full load in the bucket and traveling at moderate speeds. Operating loads and static tipping loads for loaders are explained in Chapter 3, *Construction Equipment Identification*.

Jose Luis Stephens/Shutterstock.com

Figure 1-24. The truck bed serves as the FOPS in a rigid haul truck application.

Machine Blind Spots

Many mobile machines, such as skid steers, haul trucks, and combines, have poor rear visibility. The lack of rear visibility creates a blind spot which makes it difficult to back up, especially in close quarters. This blind spot also creates risk for any person who is located at the rear of the machine. For this reason, all personnel working in close proximity to the machine must ensure that the machine's operator is aware of their presence and can clearly see them as he or she backs up the machine. A spotter may be used to guide the operator. A *spotter* is a person that is in direct radio communication with the machine operator.

Overhead Power Lines

Overhead power lines exert extremely high voltage and pose serious risks to heavy equipment operating in their vicinity. Most overhead power lines have no protective insulation and any machine contact can cause serious damage to the machine and severe injury or death to personnel. Some mobile machines, such as cranes, telehandlers, forklifts and personnel lifts, are at a greater risk of contacting power lines due to their height. Agricultural machines, such as combines or cotton pickers, may also be tall enough to make contact with low overhead power lines. Extreme care should be taken when operating machines near power lines. Spotters should be used to help guide the operator.

Starting a Machine

All heavy equipment machines should be started only when the operator is seated in the driver's seat with the transmission in neutral and the parking brake applied. Oftentimes, an operator working in the field may attempt to start a machine by using a tool to jump across the starter solenoid's terminals while standing on the ground. This unsafe practice is more common with farmers who are often working miles from their workshop and do not have access to the proper tools. Unfortunately, people are often severely injured or killed when the engine starts and the tractor, which may have been left in gear, begins moving and runs them over. Even if the engine does not start, it is possible that the starter is strong enough to propel the tractor while it is cranking and cause injury. Some manufacturers offer kits that cover the starter's solenoid to prevent personnel from attempting to start the machine from the ground. **Figure 1-25A** shows an older starter solenoid with the connectors exposed and **Figure 1-25B** shows a late model tractor with the starter wires covered.

It is also important to ensure that the engine is shut off, the parking brake is applied, and the implements are lowered to the ground before exiting the machine. Some manufacturers have additional safeguards listed in the service literature that are to be used when turning off and exiting a machine.

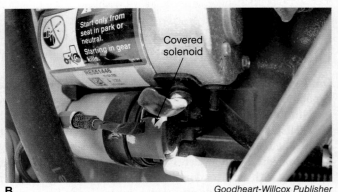

Goodheart-Willcox Publisher

Figure 1-25. Shorting the terminals on a starter solenoid will engage the starter. A—Old tractor starters often have exposed solenoid connections. B—On late-model tractors, the connections are covered to prevent personnel from bypassing the starter switch. Warning stickers are also used to remind personnel of the risk involved with starting a tractor at the starter's solenoid.

No Riders

Personnel should never allow riders or passengers on machines that are equipped with only one operator seat. The second person could easily fall and get hurt or accidentally actuate controls due to the confined space in the cab. Additionally, the operator restraint and ROPS in a single-seat cab is designed for a single operator's protection.

Some larger agricultural tractors and combines are equipped with a small second seat in the operator's cab. The seat is sometimes called a buddy seat. See **Figure 1-26**. Manufacturers often stipulate that the seat is only there for the purpose of training operators how to properly operate the machine.

Towing Large Implements

Manufacturers often specify a minimum amount of tractor horsepower needed to tow implements. The minimum amount of horsepower specified is typically based solely on the size and weight of the machine needed to safely tow the implement. Towing implements with an undersized tractor will damage the tractor and increase the risk of an accident and injury. *Never* tow implements with an undersized tractor.

Goodheart-Willcox Publisher

Figure 1-26. Combines and large agricultural tractors may be equipped with a second seat, often called a buddy seat.

Welding Safety

Heavy equipment technicians are often required to weld broken or damaged components. ***Welding*** is the process of using heat to fuse two pieces of metal. Welding poses a wide range of safety and health risks, including exposure to toxic fumes, burns, eye damage, electric shock, and noise. While many important safety practices are included in this chapter, it is essential for technicians to read and comply with owner's manuals, product safety labels, and all applicable industry standards. Technicians must also read and understand *Safety in Welding, Cutting, and Allied Processes* (ANSI Z49.1). This publication is the industry standard for establishing safe welding practices.

Welding PPE

Proper welding PPE is required to avoid injury from the risks associated with welding. Although there are some variations in PPE depending on the type of welding and the location of the work, all welding requires a helmet and face shield, gloves, jacket or apron, flame-resistant pants, and boots. See **Figure 1-27**. Some types of welding also require a respirator or hearing protection.

- Welding helmets are equipped with a protective lens to protect the welder's eyes from intense ultraviolet (UV) and infrared (IR) radiation emitted during welding. The lenses are rated from 2 to 14, and the rating needed is based on the type of welding and amperage being used. The helmet also protects the welder's face from hot spatter.
- Welding gloves are heat-resistant and have a flame-retardant lining. The gloves protect hands and forearms from heat, electrical shock, UV and IR radiation, hot spatter, and abrasion. Thicker gloves can have better insulation, but are less flexible. The gloves must be regularly inspected for tears and holes.

zilber42/Shutterstock.com

Figure 1-27. It is imperative for technicians to wear the proper PPE when welding.

- Welding jackets and aprons protect the body from heat, fire, UV and IR radiation, and welding spatter.
- A welder's pants must be flame-resistant. The pants must have no tears, holes, or cuffs. Welding chaps may also be worn over pants for additional protection.
- Steel-toe boots with rubber soles are commonly worn in a welding environment. Some welders place covers over their boots to further protect their feet.
- The welding environment is often noisy and may require ear protection. Earplugs may be worn comfortably with a welding helmet.
- Respirators can be worn to protect from inhalation dangers, such as airborne particulates, toxic fumes, and smoke. The type of respirator that should be used varies by situation and the type of welding being performed.

Welding jobs often need to be done on-site and care must be taken to ensure the area is safe and any additional safety precautions needed are taken. Welders may also have to work in awkward positions to reach damaged components, and additional care must be taken to ensure the welding can be performed safely.

Figure 1-28. Flammables must be stored in a flammable safety storage cabinet.

Figure 1-29. The acetylene working pressure gauge redlines at 15 psi.

Oxygen and Acetylene

Many heavy equipment technicians choose to weld with an oxyacetylene system because it is very portable and highly versatile. The torch can be used for heating, soldering, brazing, welding, and cutting steel. It may also be used to remove oxidation from fasteners and expand metal for an easier press fit. Oxyacetylene welding requires a mixture of oxygen and acetylene. When the mixture of oxygen and acetylene burns, the resulting flame can reach a temperature of 5700°F. Welding temperatures are high enough to melt the base metal. Technicians who may need to perform welding repairs should be properly trained to transport, use, and store welding equipment before they attempt any welding repairs.

 Warning

Inspect your work area for combustible materials, such as aerosol cans, oils, fuel, and chemicals, before welding. When not in use, all combustibles must be stored in a steel flammable material storage cabinet that meets local fire code. See **Figure 1-28**.

Acetylene Gas Safety

Acetylene gas is a very unstable and highly flammable gas. An acetylene cylinder's internal design contains a porous mass that is saturated with liquid acetone. The acetone is used to absorb the acetylene gas to improve its stability. Acetylene must only be stored in these specially designed cylinders.

When an acetylene regulator is being adjusted, the working pressure must never exceed 15 psi. The acetylene regulator working pressure gauge should have a red warning area on the scale beyond 15 psi. See **Figure 1-29**. Never use acetylene gas at working pressures above 15 psi. Acetylene cylinders are charged up to 250 psi, but the acetylene cylinder has a porous mass that is saturated with liquid acetone to keep the acetylene stable.

Acetylene gas should not be drawn out of the cylinder at a rate of more than one seventh of the cylinder volume per hour. For example, if the acetylene cylinder's volume equaled 210 cubic feet, the maximum withdraw rate must not exceed 30 cubic feet per hour; otherwise acetone will begin to be withdrawn from the cylinder. The acetylene gas is the fuel source for the torch.

Oxygen Gas Safety

The second gas used in oxyacetylene torches is compressed oxygen. Pure oxygen is not, by itself, a flammable gas. However, it is an oxidizing gas that accelerates the burning of combustibles, causing them to burn faster and hotter. This accelerant characteristic is one of the safety concerns for working with compressed oxygen. Due to the explosive potential of oil exposed to oxygen, never allow oil to mix with compressed oxygen. Oxygen is heavier than atmospheric air and can temporarily pool on the ground while it is mixing back into the atmosphere. This can accelerate the combustion of any flammable materials in the oxygen-rich area.

Oxygen cylinders store compressed oxygen at pressures up to 2100 psi. See **Figure 1-30**. The cylinders must be handled with care. If the cylinder valves are broken, the cylinders can become lethal projectiles. If the cylinder valve is opened carelessly, the pressurized gas presents the same hazards as a pneumatic system.

Goodheart-Willcox Publisher

Figure 1-30. Oxygen cylinders can be charged up to 2100 psi. The gauge on the right shows the cylinder pressure of the gas. The gauge on the left shows the working pressure of the gas.

Safety Devices

Oxyacetylene torches can **backfire**, which is the result of the flame traveling back through the torch tip, resulting in a single loud pop which extinguishes the torch's flame. Backfires can occur if the gas pressure is too low, if the torch tip contacts the heated metal, or if the tip is obstructed. If the popping noise turns into a whistle, then the backfire has become a **flashback**, which is the flame traveling back through the torch head. If the flashback were allowed to continue traveling backward, it could cause the hoses and cylinders to violently explode. Flashback travels at twice the speed of sound, which does not allow a person enough time to shut off the cylinders. Some oxyacetylene torches use two check valves to prevent the reverse flow of gas and/or two flashback arrestors to quench a flashback, preventing the flame from traveling backward. See **Figure 1-31**.

Check Valve

The **check valve** contains a spring that presses against a poppet-style check valve, which stops the reverse flow of gas. The check valves are often installed on the torch before the hoses, one for oxygen and one for acetylene. The check valves cannot stop a flame from traveling backward, but they do prevent reverse flow of gas.

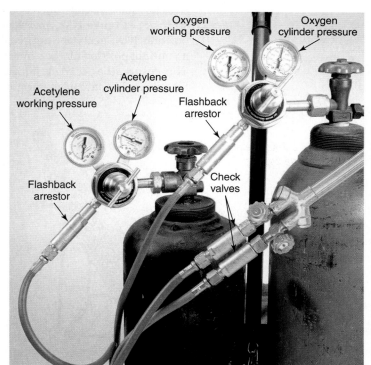

Thermadyne Industries, Inc.

Figure 1-31. This image shows oxygen and acetylene regulators properly attached to the cylinders.

Flashback Arrestor

A *flashback arrestor* is a safety device designed to quench a flame, preventing a flame from traveling back into the hose or regulator. Flashback arrestors are installed on a welding outfit either in place of or in combination with check valves. They are normally physically longer than the check valves. Like check valves, flashback arrestors prevent the reverse gas flow. However, they also have an element that extinguishes any flame that reaches it. A flashback arrestor is placed between each hose and the torch or between each hose and regulator. Today, flashback arrestors are commonly included with the torch. An old torch might have neither check valves nor flashback arrestors.

Cylinders

Gas cylinders may also be equipped with safety devices to prevent the rupture of the cylinders if the cylinder pressure increases. Cylinder pressures can rise if they are exposed to heat, such as a fire.

Oxygen and nitrogen cylinders are equipped with a **burst** or **rupture disk**, which is designed to rupture at a specified pressure. This allows the gas to be emitted out of the tank at a controlled rate, reducing the pressure in the cylinder and preventing it from exploding. The disk will not reclose once the cylinder's pressure drops below the rupture setting. The disk acts as a pressure fuse.

Acetylene cylinders may be equipped with a *fusible plug* filled with a metal alloy that will melt at a specific temperature. Once the metal alloy melts, it allows the cylinder to release the acetylene gas at a controlled rate. Again, this safety feature releases built-up pressure to prevent the cylinder from exploding.

Cylinder Storage and Transport

When the gas cylinders are not in use, they must be stored properly. Removable protective caps should be installed and the cylinders must be secured with restraining chains or straps. See **Figure 1-32**. Acetylene cylinders must always be stored and used in an upright position. If the acetylene tank is placed in a horizontal position, it must be allowed to sit in an upright position for the same length of time it was lying horizontal before it can be used. This ensures the acetylene and acetone have not separated. Preferably, the tank should sit for at least 24 hours in an upright position before its use.

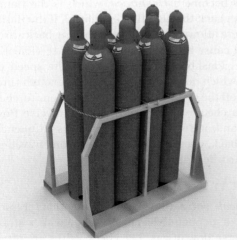

A — narin phapnam/Shutterstock.com B — topae/Shutterstock.com

Figure 1-32. Gas cylinders must be properly secured when not in use. A—This oxygen cylinder is secured to its cart with a chain, and the protective cap is installed. B—These acetylene cylinders are secured to a steel pallet with chains. The cylinders have their protective caps installed as well.

Colors

Colored warning labels are used on compressed gas cylinders to identify the contents. Agencies, such as OSHA, the Department of Transportation (DOT), and the National Fire Protection Agency (NFPA) have established standard requirements for the colors and icons used. For example, the DOT placard for acetylene cylinders uses a red diamond label with a white flame. Oxygen cylinders are labeled with a yellow diamond label with a flame over a circle, indicating that oxygen is an oxidizing gas. The actual color of gas cylinders is not standardized. The acetylene torch hose is red and the oxygen hose is green.

Additional Oxyacetylene Gas Safety Tips

- For any given task, use only equipment designed for that task and the pressure ranges that will be used. This applies to regulators, hoses, cylinders, and torches.
- Wear appropriate PPE, including flame- and spark-resistant cotton clothing, eye and face protection, and gloves.
- Keep all PPE and welding equipment free of oil and grease.
- Never transfer acetylene into another storage cylinder.
- Work in well-ventilated areas to prevent the buildup of carbon monoxide and other dangerous fumes.

Electrical Shock

Although oxyacetylene does not require electricity, other types of welding use electricity to power welders (welding machines). The human body will easily conduct electricity when it is placed in its path and may result in a person's *electrocution* (death from electrical shock). To minimize the risk of electrocution, personnel can take preventive measures.

- Wear dry clothing and PPE.
- Ensure that you are dry and insulated from the metal and the ground.
- Do not allow your skin or anything wet to touch the electrode or its metal.
- Do not stand on wet surfaces.
- Do not touch the metal parts.
- Ensure cable insulation is in good condition.
- Shut off the welder when it is not in use.

Although electrocution can occur anywhere, technicians must take additional precautions when working on-site instead of in the shop. Technicians should evaluate the site and note any wet areas, power lines, and other electrical equipment in use before they begin working on a machine.

Summary

- Technicians must be equipped and trained for proper use of personal protective equipment (PPE).
- Technicians must receive instruction on safe work practices.
- Safety data sheets (SDS) must be on hand for all chemicals and hazardous materials being used in the shop.
- Heavy equipment technicians often work alone in remote locations and need to be prepared for emergencies.
- All shop personnel should be aware of the location of the shop's first-aid stations.
- Pressurized fluids pose risks to personnel and machinery in the form of burns, fluid injection injuries, and machine fires.
- Compressed air systems pose risk to personnel in the form of eye, lung, and internal injuries.
- All safety guards and shields should be properly installed on machines to ensure the operator's safety while the machine is running.
- Exhaust ventilation systems must be on and properly connected when a machine is running in the shop.
- Fall protection will prevent a technician from falling to the ground when working at elevated heights.
- A job hazard analysis (JHA) should be performed before a job is begun.
- A recordable accident occurs when an injury results in a fatality, medical care or hospitalization is required, the employee must be absent from future workdays, or the employee's work is limited.
- Recordable accidents include incidents that result in fatalities, loss of consciousness, having to see a physician, missing work, and having work limitations. An example of a nonrecordable accident could involve seeing a physician, but receiving only a temporary bandage.
- Wheel chocks should be used when a machine is being unloaded or serviced and when it is parked on a slope.
- When entering or exiting a machine, always use three points of contact and face the machine.
- Never work below an unsupported implement. Always use implement locks to secure the implement.
- Articulated tractors have steering locks that lock the front and rear frames together to prevent the tractor from articulating.
- ROPS and seat belts protect operators in the event of a machine rollover.
- FOPS protect operators from falling objects.
- All heavy equipment machines should be started only when the operator is seated in the driver's seat with the transmission in neutral and the parking brake applied.
- Before towing an implement, ensure the tractor is large enough to handle the size of the implement.
- Welding poses a wide range of safety and health risks, including exposure to toxic fumes, burns, eye damage, electric shock, and noise. Proper PPE must always be worn.
- Oxyacetylene welding uses oxygen and unstable, highly flammable acetylene. The welding process generates heat that is high enough to melt metal.

Technical Terms

- acetylene gas
- automated external defibrillator (AED)
- backfire
- burst disk
- cardiopulmonary resuscitation (CPR)
- check valve
- electrocution
- embolism
- falling object protective structure (FOPS)
- fall protection
- first aid
- flashback
- flashback arrestor
- fluid injection
- fusible plug
- implement lock
- job hazard analysis (JHA)
- Mine Safety and Health Administration (MSHA)
- nonrecordable accident
- Occupational Safety and Health Administration (OSHA)
- personal protective equipment (PPE)
- pneumatic system
- power take-off (PTO)
- recordable accident
- recordable incident
- rollover protective structure (ROPS)
- rupture disk
- safety data sheets (SDS)
- spotter
- three points of contact
- unsafe acts
- welding
- wheel chocks

Review Questions

Answer the following questions using the information provided in this chapter.

Know and Understand

1. Of the following, _____ are approved PPE.
   A. contact lenses
   B. safety glasses
   C. sunglasses
   D. traditional prescription glasses

2. All of the following are examples of PPE, *EXCEPT*:
   A. safety glasses.
   B. hard hats.
   C. earplugs.
   D. neck braces.

3. All of the following present additional hazards when exposed to heat, *EXCEPT*:
   A. hydraulic cylinders.
   B. solder.
   C. gas cylinders.
   D. accumulators.

4. A technician's arm breaks out in a rash after rebuilding a hydraulic cylinder. Where should the technician look for information related to this problem?
   A. Service manual.
   B. Operator's manual.
   C. Training manual.
   D. Safety data sheet (SDS).

5. A hydraulic system operating at high system pressure can spray fluid up to _____.
   A. 60 mph
   B. 300 mph
   C. 670 mph
   D. 2000 meters per second

6. In the event of a fluid injection in a technician's hand, all of the following will affect the chances of potential amputation, *EXCEPT*:
   A. delay before receiving medical attention.
   B. quantity of fluid.
   C. the type of fluid.
   D. weather conditions.

7. All of the following are used for fall protection, EXCEPT:
   A. netting.
   B. an anchor point.
   C. a body harness.
   D. a shock-absorbing lanyard.

8. Which of the following is a form completed by employees as a way to analyze risks prior to starting the job?
   A. FHA.
   B. JHA.
   C. SHA.
   D. NHA.

9. An accident occurred on the job. All of the following will result in a recordable accident, EXCEPT:
   A. receiving prescription medication.
   B. receiving stitches.
   C. receiving a temporary bandage.
   D. having to be reassigned to a light-duty job task.

10. How many points of contact should be maintained when entering or exiting a machine?
    A. One.
    B. Two.
    C. Three.
    D. Four.

11. A technician is exiting a tractor. Which way should he or she face while exiting?
    A. Toward the machine.
    B. Away from the machine.
    C. Depends on the type of machine.
    D. Does not matter.

12. Which of the following actions must be performed when a cab or implement must be lifted to service a component?
    A. Technicians are never permitted to work under a lifted cab or implement.
    B. Use a hydraulic cylinder to support the cab or implement.
    C. Use the manufacturer's safety support lock.
    D. The implement must be dismantled.

13. An implement lockout switch or lever normally prevents implement operation by eliminating _____.
    A. hydraulic pilot oil
    B. solenoid operational relay
    C. engine from cranking
    D. hydraulic oil from returning to the reservoir

14. All of the following machines are likely to use an ROPS, EXCEPT:
    A. a dozer operating on a steep incline.
    B. a dozer operating in a forestry application.
    C. an agricultural tractor operating on a steep slope.
    D. an asphalt paver laying pavement.

15. A haul truck bed can also serve as an _____.
    A. FOPS
    B. ROPS
    C. Both A and B.
    D. Neither A nor B.

16. When is it okay to start a tractor by jumping across a tractor's starter's solenoid terminals?
    A. If the transmission is in gear.
    B. When the park brake is released.
    C. When the service brake is released.
    D. Never. The machine should only be started from the operator's seat.

Apply and Analyze

17. OSHA stands for Occupational Safety and Health _____.
18. The _____ is required to provide PPE for employees.
19. The acronym *PPE* stands for personal protective _____.
20. According to some manufacturers, when checking for high-pressure hydraulic leaks, a technician should use a piece of _____.
21. The minimum fluid pressure that can cause oil to penetrate a human's skin is _____ psi.
22. A Class _____ fire extinguisher is used to extinguish oil or grease fires.
23. A Class _____ fire extinguisher is used to extinguish electrical fires.
24. The most common cause of fluid injection injuries is a burst _____.
25. A(n) _____ degree burn affects only the outer layer of skin.
26. The acronym *ROPS* stands for rollover protective _____.
27. Some manufacturers recommend that seat belts that have been on a machine more than _____ years be replaced.
28. Welding helmets are equipped with a protective lens to protect the welder's eyes from intense _____ and infrared radiation emitted during welding.
29. Acetylene must never be used at a working pressure above _____ psi.

Critical Thinking

30. Some manufacturers require a technician to run a machine and make adjustments, such as adjusting the null adjustment on a hydrostatic transmission, while measuring pressures. What are some steps a technician can take to avoid being injured while measuring flows, pressures, and making adjustments to a machine?

Bannafarsai_Stock/Shutterstock.com
Personnel should wear bright clothing or safety vests in addition to all other required PPE when working around heavy equipment.

Chapter 2
Tools and Fasteners

Objectives

After studying this chapter, you will be able to:
- ✓ Describe the safe use of hand tools.
- ✓ List common hand tools used by heavy equipment technicians.
- ✓ List common power tools used by heavy equipment technicians.
- ✓ Describe additional tools used by heavy equipment technicians.
- ✓ Explain how to use measuring tools.
- ✓ Identify and describe different types of fasteners.

Hand and power tools are essential to a heavy equipment technician's ability to properly service, diagnose, and repair machines. Many technicians must acquire their own tools and often invest thousands of dollars to build a complete set of quality tools. It is essential to properly store and maintain hand and power tools to ensure a long service life. See **Figure 2-1**. Well-organized tools also save time and allow a technician to complete tasks more efficiently. Additionally, technicians must also have a safe and efficient means of transporting tools to a job site. This chapter reviews many of the hand and power tools a heavy equipment technician will use in the course of his or her work.

General Hand Tool Safety

Although each shop will have its own safety and operational rules, there are general hand tool safety rules you should always follow:
- Wear eye protection at all times. Use safety glasses with side shields to fully protect your eyes.
- Use the proper personal protective equipment (PPE) for the job.

seksan kingwatcharapong/Shutterstock.com

Figure 2-1. Tools that are well-maintained and organized help technicians work safely and efficiently.

- Remove all jewelry, including rings, necklaces, watches, and bracelets, before entering the work area to prevent entanglement with tools and machinery.
- Use tools only for the job for which they are intended and only in the way you have been trained to operate or use them. Using the wrong tool for a job may result in damage and injury.
- Always follow the safety and maintenance procedures outlined by the manufacturer.
- Use only tools that are in good condition. Inspect the tools prior to use and remove them from service if they are cracked, bent, in need of repair, or show fatigue.
- Properly maintain and regularly inspect hand tools to keep them in safe working order.
- Keep work areas clean and clutter-free at all times.

Hand Tools

Many hand tools are used by heavy equipment technicians. Some may be specific to the industry while many are used in a variety of professions. Tools are designed for specific tasks, such as turning nuts and bolts, driving or removing screws, measuring, drilling holes, and prying objects.

Warning

When working under a machine and using tools overhead, keep your face positioned away from the tool to prevent items, such as a loose socket, nut, or bolt, from hitting you in the face. To avoid injury, do *not* pull the wrench toward your face.

Wrenches

One of the most common hand tools used by technicians is the wrench. Wrenches are available in SAE or metric sizes that coincide with the measurement of nuts and bolts. Most technicians must own two sets of wrenches, one SAE and one metric. Wrenches also vary in shape and are designed for specific tasks. Most heavy equipment technicians use several types of wrenches including the open-end wrench, combination wrench, ratcheting wrench, and flare-nut wrench.

Open-End Wrench

Open-end wrenches are used for loosening, turning, and tightening fasteners, such as bolts, nuts, and fittings. They are not used for applying a specific amount of torque. An open-end wrench does not fully grip the entire circumference of a bolt or nut and therefore is not used for applying high torque values. See **Figure 2-2A**.

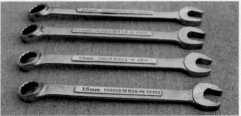

Goodheart-Willcox Publisher

Figure 2-2. A—Open-end wrenches are used for loosening and tightening line fittings. B—Box-end wrenches may be flat or offset. They are designed to completely surround the fastener. Offset wrenches are shown here. C—Combination wrenches have an open-end and a box-end.

Note

Service literature specifies the torque that must be used for tightening hardware. ***Torque*** is leveraged force being applied through an arcing motion. It factors in the force applied and the mechanical advantage of leverage. For example, if a person applied 30 lb of force at the end of a 1′ long wrench, 30 foot-pounds of torque would be exerted through the wrench. If the same 30-lb force was applied to a 2′ long wrench, 60 foot-pounds of torque would be exerted. Torque applied to a lever equals force times the distance between the point where the force is applied and the lever's fulcrum. Torque can be measured in:

- Inch-pounds (in-lb).
- Foot-pounds (ft-lb) (for fastener tightness specifications).
- Pound-feet (lb-ft) (for vehicle, axle, or engine torque).
- Newton meters (N·m).

Box-End Wrench

Box-end wrenches surround the entire circumference of the bolt head and the wrench is therefore less likely to round the head of the bolt or nut. It provides a better grip than open-end wrenches. Like open-end wrenches, box-end wrenches are thin and can be used in areas with little space between the face of the bolt and an adjacent component. It is time-consuming to tighten or loosen fasteners with a box-end wrench because the wrench must be repositioned after each turn. A box-end wrench that is angled is referred to as an ***offset wrench***. An offset wrench does not interfere with hardware in the same plane as the nut being attached. See **Figure 2-2B**.

A ***combination wrench*** contains an open-end and a box-end on the same wrench. Typically, the open- and box-end are the same size. See **Figure 2-2C**.

Flare-Nut Wrench

A ***flare-nut wrench***, also known as a ***line wrench***, is designed for loosening and tightening fittings that attach lines, such as fuel lines. A traditional open-end wrench will not grip all six points on the fitting. A flare-nut wrench resembles a box-end wrench, but has a gap in the box end that allows the wrench to slip over the tubing and grip all six sides of the fitting. See **Figure 2-3**.

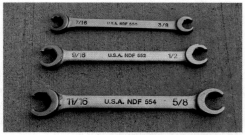

Goodheart-Willcox Publisher

Figure 2-3. Flare-nut wrenches are used to tighten and loosen fuel fittings, brake fittings, and other fittings that secure lines and tubing.

 Warning

Pulling a wrench toward you will reduce the risk of injuring fingers and knuckles if a fastener breaks loose. If you are pushing the wrench when a fastener breaks loose, your hand may slam into a neighboring component, such as a steel frame or cast-iron housing. If the wrench must be pushed while you are working close to another component, push the wrench with an open palm. Using your palm will reduce the risk of hand injuries because the wrench, rather than your hand, will hit the neighboring component if the fastener breaks loose. See **Figure 2-4**.

Ratchet and Socket

A ***ratchet*** is a device in which a toothed wheel, known as an anvil, is located inside a bar. The anvil is engaged by a toothed pawl that permits motion in only one direction and free wheels in the opposite direction. A ***pawl*** is a piece of curved metal, sometimes in the shape of a lever, that meshes with the teeth on the anvil. The ratchet may contain only one reversible pawl or two individual pawls, a left and right pawl. A center switch allows a technician to reverse the ratchet's direction of torque.

A ***socket*** is a short steel tube that attaches to the ratchet handle's quick-release, square drive coupler. Like a box-end wrench, a socket surrounds the entire circumference of the bolt head. The combination of a ratchet handle and socket is often referred to as a *socket driver*, *socket wrench*, or *ratcheting wrench*. A ***ratcheting wrench*** is actually a wrench with a built-in ratcheting mechanism. See **Figure 2-5**.

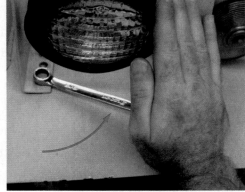

Goodheart-Willcox Publisher

Figure 2-4. When pushing a wrench toward another component, use an open palm to reduce the chances of injuring your fingers and knuckles.

Goodheart-Willcox Publisher

Figure 2-5. Ratcheting wrenches vary in design and size. A and B—Combination wrenches with open and box ends. C and D—Offset (angled) ratcheting wrenches with box ends.

Flex Handle

A *flex handle*, also known as a *breaker bar*, is a stronger handle that should be used in place of a ratchet for loosening stubborn fasteners. The flex handle is longer than a traditional ratchet handle. It has a pivot that allows the handle to be repositioned by 180° without removing the socket from the fastener. The pivot is stronger than the ratcheting mechanism in a ratchet handle. Because of its increased strength and length, the flex handle is able to apply greater torque than a ratchet.

Speed Handle

A *speed handle*, also known as a speed wrench, is an alternative to power tools that is used for quickly installing cap screws or nuts. The wrench is a rod with a C-shaped center section, a rotating handle on one end, and a long shaft on the other end. The long shaft ends in a square head drive that attaches to sockets. The tool is helpful for quickly installing numerous fasteners. The speed handle gives technicians a better feel for the fastener's threads and level of torque compared to power tools. The disadvantage of the speed handle is that it is only capable of applying low torque to a fastener. See **Figure 2-6**.

Drive Size

Ratchet drives vary in size to accommodate different socket drive sizes. A *ratchet drive size* is the size of the square that fits the opening on sockets or other attachments. Common ratchet drive sizes are 1/4″, 3/8″, 1/2″, and 3/4″. See **Figure 2-7**. Air ratchets, which will be discussed later, are also available with 1″ drives.

Socket drive sizes match ratchet drive sizes, but not all socket sizes are available for all drive sizes. Drive adapters can be installed on the ratchet if the ratchet's drive size is too large or too small to use with the desired socket. Reducing adapters reduce the drive by one size. For example, a reducing adapter would be used to couple a 3/8″ drive ratchet to a 1/4″ drive socket. An expander adapter increases the drive by one size. For example, an expander adapter would allow a socket with a 3/4″ drive to be driven by a ratchet with a 1/2″ drive.

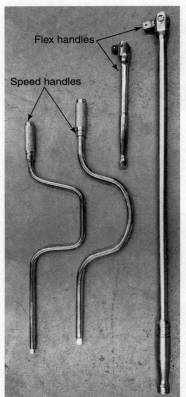

Goodheart-Willcox Publisher

Figure 2-6. Flex handles are used to loosen extremely tight or stubborn hardware. Speed handles are used for quickly installing cap screws and nuts.

 Warning

Standard chrome sockets should not be used in place of impact sockets designed specifically for air ratchets/impact wrenches. Standard sockets are not designed to handle the high torque of an air ratchet or impact wrench and may shatter and become a safety hazard.

Sockets

Individual sockets have two sizes, the *socket drive size* (attaches to ratchet) and the fastener's head size (opening of the socket). See **Figure 2-8**. Sockets are also identified with points, the most common being 6-point and 12-point sockets. It is important to use the correct socket size and the most appropriate number of points to prevent stripping the head of the fastener. The number of points determines the amount of surface contact there is between the socket and the fastener.

- A 6-point socket is similar to a 6-point box-end wrench in that they both provide the most surface contact and are less likely to slip when removing stubborn hex-head (6-sided) fasteners.
- A 12-point socket is used on 12-point fasteners. A 12-point socket should only be used on a 6-point fastener when the fastener's location makes it difficult to rotate the 6-point socket onto the fastener.
- The less common 8-point socket may be used to tighten or loosen square-head fasteners.

Chapter 2 | Tools and Fasteners

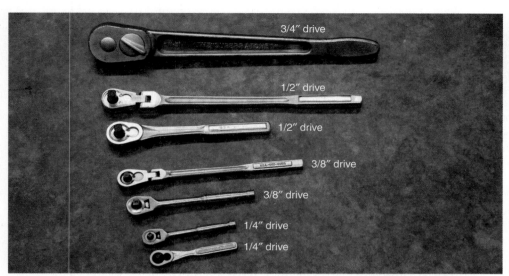

Figure 2-7. Ratchet drives range in size from 1/4″ to 3/4″.

Sockets may be traditional thin-walled chrome or thick-walled black. Chrome sockets are used with hand ratchets, and black sockets are designed to be used with air impact drivers, but can be used with hand ratchets as well. Depending on their length, sockets are classified as shallow, semi-deep, or deep. Shallow sockets are the shortest and deep sockets are the longest.

Sockets are typically sold in sets and may include various handles and extensions as well as a wide range of socket sizes. A socket set may also include both US customary and metric sizes.

An open-end *crowfoot socket* resembles an open-end or flare-nut wrench. It has a square drive designed for use with a ratchet handle, breaker bar, or torque wrench. The cylindrical design allows access to otherwise hard-to-reach areas. See **Figure 2-9**.

Torque Wrenches

Torque wrenches are calibrated tools used to apply a specified amount of torque to a fastener through an attached socket. Using a torque wrench will prevent over-tightening, which could cause thread damage, broken fasteners, or part distortion. Common torque wrenches are the beam, dial, and clicker.

Figure 2-8. Common socket drive sizes are 1/4″, 3/8″, 1/2″, and 3/4″. The sockets can be shallow, medium, or deep. Some sockets have a universal swivel that helps remove fasteners in hard-to-reach locations.

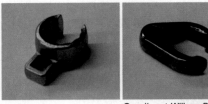

Figure 2-9. Crowfoot sockets allow technicians to use a torque wrench to tighten a line fitting.

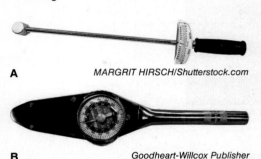

A

B

Figure 2-10. A—Beam torque wrenches are used to measure the amount of torque it takes to turn a fastener, gear, or shaft. B—The dial on a dial torque wrench indicates the amount of torque being applied to the fastener.

A

B

Figure 2-11. A—A clicker torque wrench pivots (clicks) when the torque setting has been reached. B—The scale on the handle allows the technician to set the wrench for the desired torque.

Beam Torque Wrench

A beam torque wrench uses a beam that moves along a numerical scale to display the amount of torque being applied. The beam torque wrench is used for measuring rolling torque. For example, after installing a pinion gear on a differential, the sequence might require a technician to measure the amount of torque it takes to turn the pinion gear. A beam torque wrench is useful for measuring this torque. See **Figure 2-10A**.

Dial Torque Wrench

A dial torque wrench has a round dial that indicates the amount of torque being applied to the fastener. Dial torque wrenches are also used for checking rolling torque. See **Figure 2-10B**.

Clicker Torque Wrench

The clicker torque wrench allows the technician to preset torque with an adjustable handle and lock collar. To set the desired torque, the collar is unlocked and the handle is rotated clockwise to increase or counterclockwise to decrease the torque setting. The torque wrench may have a traditional scale etched on the wrench, which must be read manually, or a digital scale with a quick and easy-to-read display. When a clicker torque wrench is used to tighten a fastener, the wrench clicks when the applied torque reaches the torque setting, **Figure 2-11**. However, do not rely solely on hearing a click when applying low torque values with a clicker wrench because the click may not be audible. Sometimes the click must be felt. If you do not hear the click and are not paying attention, you can easily overtighten a fastener and cause damage to the fastener or component.

Pro Tip

When finished with a torque wrench, rotate the handle counterclockwise and set it to the *lowest setting on the numerical scale*. Keep in mind that the lowest setting is *not* fully backed off but is the lowest tension. Frequently leaving tension on the torque wrench for long periods of time will require the wrench to be recalibrated.

Torque Multipliers

Heavy equipment uses large components that may require tremendous amounts of torque to tighten or remove hardware. A *torque multiplier* is a gear box with a 1/2″, 3/4″, or 1″ drive size. A torque multiplier is used to multiply the torque manually applied by the technician. They come in a variety of shapes and sizes and are especially useful if power tools are unavailable or space limitations will not allow the use of a breaker or extension bar. Torque multipliers are available with different gear ratios. For example, the torque multiplier in **Figure 2-12** has an 18.5:1 multiplication ratio with a maximum output limit of 3200 pounds of force. If a technician applies 54 lb-ft to the gearbox, it will deliver 1000 lb-ft of output torque.

The multiplier has an input drive and an output drive. A ratchet or wrench is attached to the input drive, and a socket is attached to the output. As a technician applies torque to the input drive, the torque multiplier attempts to rotate in the opposite direction and needs to be held stationary. The multiplier can have three different anchor configurations: hand-bar, plate, or a compact design. The hand-bar type is a common style that allows a technician to apply input torque with the input wrench while holding the reaction hand-bar. The plate style has mounting bolts that allows the torque multiplier to be bolted directly to the component.

The compact multiplier has a small reaction fixture that rests against a solid object. Be sure to follow the service literature for safe operation.

One popular compact torque multiplier is designed for removing seized wheel lug nuts. It has an anchor arm that forks off to the side and rests on a neighboring lug nut. The arm provides the steady anchor necessary for the tool to multiply the necessary torque for removing the seized lug nut. The multiplier comes with a hand crank wrench that acts like an input speed handle for driving the torque multiplier.

Some torque multipliers have an anti-wind-up ratchet feature that keeps the multiplier loaded while it is being operated. This feature can make it difficult to remove the tool. The tool has a button similar to small ratchets that allows the tool to reverse its direction of rotation. If the tool is stuck on the fastener and difficult to remove, first apply torque in the same direction to load the torque multiplier. Then, switch the direction of rotation on the torque multiplier by pressing the CW/CCW button. Next, rotate the input wrench in the opposite direction to unload the tension on the socket. This sequence should loosen the tool so it can be easily removed from the fastener.

Goodheart-Willcox Publisher

Figure 2-12. This hand-bar torque multiplier has an 18.5:1 ratio.

Screwdrivers

Screwdrivers are used to remove and attach screws. The most common screwdriver tips are flat, Phillips, and Torx. Screwdrivers are *not* chisels, punches, or pry bars and should only be used to tighten and loosen screws. See **Figure 2-13**.

Punches

A *punch* is used with a hammer to align components or to drive pins. Punches are made of steel or brass. See **Figure 2-14A**. A brass punch is used when special care is required to prevent harming a metal surface, such as when driving out a bearing's race.

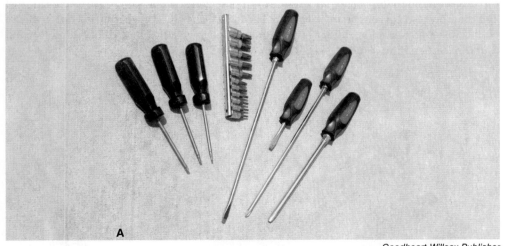

Goodheart-Willcox Publisher

Figure 2-13. A—Screwdrivers are offered in a wide range of lengths. B—Different types of fastener sockets or slots require the use of different screwdriver tips. Some screwdrivers have a socket at the end for interchangeable tips.

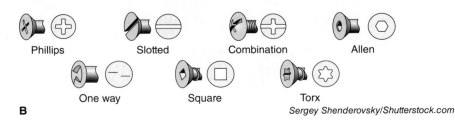

Sergey Shenderovsky/Shutterstock.com

Figure 2-14. A—Punches are used for driving a roll pin (pin punch), aligning a hole (drift punch), or making a dimple to start drilling a hole (center punch). B—Mushrooming is the dangerous flattening and distortion of the head of a punch. A mushroomed head should be ground to remove the formed edges to reduce the risk of injury.

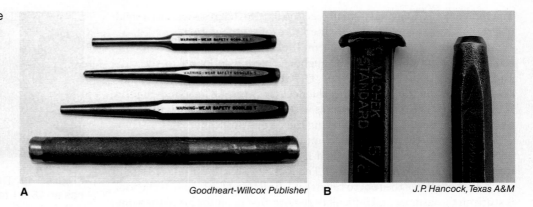

A Goodheart-Willcox Publisher   B J.P. Hancock, Texas A&M

- A *drive-pin punch* or *pin punch* has a straight shank and is used to drive pins in and out of a bore. (A pin is used to secure collars, gears, and shafts to another component.)
- A *center punch* is used to make a dimple on a metal surface, which can be used as a starting point for drilling a hole.
- A *drift punch* has a tapered shaft and is used to align a hole into which a bolt or other type of fastener may be inserted.

 Warning

When the end of a punch becomes mushroomed from repeated use (see **Figure 2-14B**), the mushroomed material must be removed to prevent metal shards from flying off the edge the next time the punch is struck with a hammer.

Chisels

Chisels are hand tools used for cutting and chipping metal surfaces. They are made of steel and their beveled tips or cutting edges are hardened. The chisel type is based on the shape of the chisel tip. Common chisel types include cold, cape, half-round, diamond-point, and round-nose. See **Figure 2-15**.

Figure 2-15. Chisels use hardened edges to cut and chip metal surfaces.

Goodheart-Willcox Publisher

Hammers and Mallets

Hammers are used to drive chisels and punches and may also be used for assembling or disassembling components. There are many types of hammers and mallets designed for specific tasks, and you should always use the proper hammer or mallet for the task at hand. For example, a ball-peen hammer is commonly used in heavy equipment work for most jobs, but a mallet would be used when there is a risk of damaging a component. See **Figure 2-16**.

Pliers

Technicians use different types of pliers designed specifically for gripping, holding, cutting, and assembling components. Common types include slip-joint, adjustable, diagonal, needle-nose, and snap-ring pliers. See **Figure 2-17**.

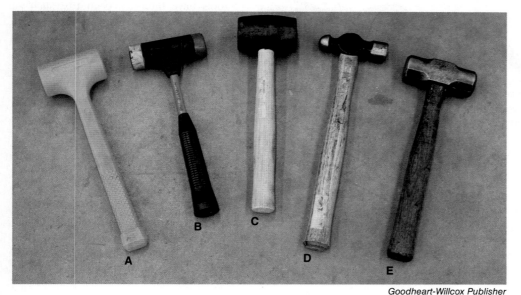

Goodheart-Willcox Publisher

Figure 2-16. Mallets and hammers are designed for specific tasks and offered in a variety of sizes.
A—Dead-blow hammer.
B—Soft-faced mallet.
C—Rubber mallet.
D—Ball-peen hammer.
E—Sledgehammer.

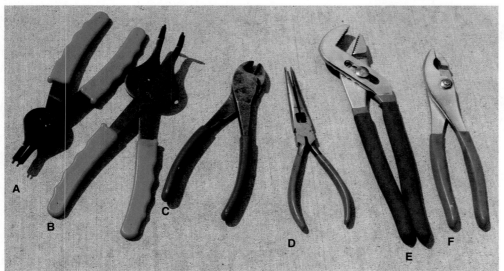

Goodheart-Willcox Publisher

Figure 2-17. Use the correct type of pliers to avoid tool and material damage or injury.
A and B—Snap-ring pliers.
C—Diagonal cutters.
D—Needle-nose pliers.
E—Multi-groove adjustable pliers.
F—Slip-joint pliers.

Power Tools

In addition to hand tools, heavy equipment technicians use power tools for speed and increased power. Power tools may use pneumatic, electric, or hydraulic power and may be stationary or portable.

Electric Tools

Electric power tools may be portable, battery-operated tools or they may plug into an electrical outlet. Drills, lights, grinders, soldering irons, chop saws, and welders are examples of common electric power tools, **Figure 2-18**. Electric power tools have built-in safeguards, such as double insulation, to help prevent shock or electrocution. A handheld electric tool is considered double-insulated if the wire is insulated and the housing is made of non-conducting material (case-insulated). To decrease the risk of electrocution, power cords should be inspected regularly for damage and fraying. Special care must be taken when using electric tools around water or in a damp environment because there is an increased risk of electrical shock.

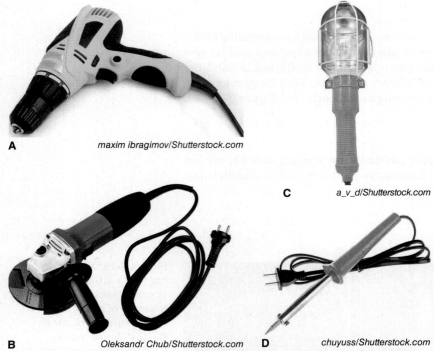

Figure 2-18. Drills, lights, grinders, and soldering irons are common electric power tools. A—Electric drill. B—Electric grinder. C—Drop light. D—Soldering iron.

Pneumatic Tools

Air ratchets, air drills, die grinders, and air chisels are a few examples of pneumatic or air-powered tools. The tools vary in size and power. For example, air ratchets are offered in 3/8″, 1/2″, 3/4″, and 1″ drives. See **Figure 2-19**. The tools are typically powered by shop compressed air. Service trucks are often equipped with an onboard air compressor for operating air tools.

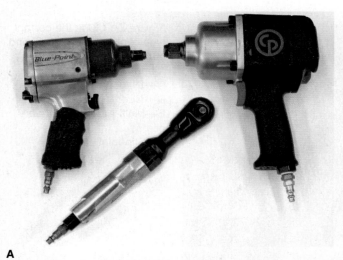

Figure 2-19. Examples of 3/8″, 1/2″, 3/4″, and 1″ drive air ratchets. A—The smaller-sized impact wrenches and air ratchets are used for lower torque applications. B—The 1″ impact is significantly heavier and normally has a handle on the rear or top of the wrench. It is used for heavy-duty torque applications.

Warning

Never attempt to hold the nut with your hand when attaching hardware with an air ratchet. The speed and force exerted through the air ratchet may cause the nut to quickly spin and slice open your fingers or hand.

Hydraulic Tools

Hydraulic power tools provide the force required to press, pull, and install components on heavy machinery. The most common hydraulic power tool is the hydraulic press. See **Figure 2-20**. A hydraulic press may be manually operated with a lever, use air-over-hydraulics (air power to drive a hydraulic cylinder), or use electric-over-hydraulics (electric-powered hydraulic pump). Hydraulic presses may be stationary or portable.

Hydraulic presses produce thousands of pounds of force that can easily break or fracture components. *Always* wear eye protection and a face shield while operating a hydraulic press to protect your eyes and face from flying debris. Care should also be taken to avoid pinch points between the press and component.

Goodheart-Willcox Publisher

Figure 2-20. Hydraulic presses are used for pressing bearings onto shafts, gears, and housings.

Additional Tools and Equipment

In addition to hand and power tools, technicians use a variety of other tools and equipment for diagnosing, servicing, and repairing heavy equipment. These tools and equipment are used for tasks such as fluid pressure testing, cleaning, metal cutting, and media blasting.

Fluid Pressure Testing Equipment

Maintaining accurate pressure settings is essential for maximizing machine productivity. Many types of pressure testing equipment are used to measure and monitor pressure in heavy equipment systems, including:
- Diagnostic pressure taps.
- Traditional pressure gauges.
- Differential pressure gauges.
- Quadrigages.
- Electronic digital pressure meters.
- Snubbers.

Diagnostic Pressure Taps

To measure pressure on older systems, technicians threaded pressure gauges into a system port or installed a T-fitting in the line. Today, standardized diagnostic pressure taps are installed in machine fluid pressure systems by the manufacturer. A ***diagnostic pressure tap*** is a device placed in fluid pressure systems to create a port through which safe and easy pressure measurements may be taken.

The mobile equipment industry commonly uses two types of diagnostic pressure taps. One type, a quick coupler, allows a technician to quickly snap a gauge onto the male pressure tap. See **Figure 2-21A**. These pressure taps are a standardized SAE J1502 interchange, and are found on numerous makes of machines. The quick coupler works similarly to air tool quick couplers and require the outside collar to be slid rearward in order for the coupler to be removed from the pressure tap. However, one difference between the air tool couplers and the pressure taps is that the pressure tap's collar does not have to be slid rearward during installation. The female coupler must simply be pushed onto the male tip for the assembly to snap together as a sealed connection.

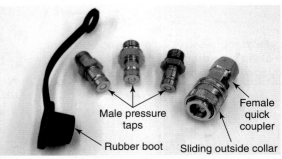

Goodheart-Willcox Publisher

Figure 2-21. Pressure taps are used to attach gauges to measure pressures. A—Male diagnostic pressure taps come in a variety of thread types and sizes. The taps are commonly placed in fluid pressure systems to speed the process of measuring pressures. The female quick coupler is threaded onto the pressure gauge. B—A photo of a hydraulic pressure test kit. It uses a threaded connection. The fitting has a small bore in its center. The hose has a central probe that is inserted inside the fitting's bore, which will open the check valve. The caps are often chained to the fittings.

 Warning

Never remove a pressure gauge from a pressurized system or fluid will spray from the line, causing burns or a fluid-injection injury. If the gauge is registering pressure, do *not* remove the gauge until the system pressure has been depleted.

A second style of pressure tap uses a threaded connection instead of a quick coupler design. See **Figure 2-21B**. Hydrotechnik call the series of fittings Minimess test points. Thompkins calls them Pro-Test test-points.

The advantage of the quick-connect and threaded permanent test port designs over older threaded service ports is that the pressure taps can remain in the system, preventing contamination that occurs with threaded service ports. The disadvantage is that the tap can seep fluid or allow dirt or other contaminants to enter the system. To reduce the potential of contamination, a rubber boot or dust cap can be used to cover the tap. John Deere labels the pressure taps "DR" for diagnostic receptacles.

Permanent pressure taps provide another advantage. Many gauges have pipe threads and tend to leak with repeated removal and installation. The design of the pressure taps ensures a leak-free connection for the ports in the hydraulic system and for gauges equipped with a quick coupler.

Additional permanent diagnostic pressure taps may be installed in a system. The taps can be purchased from the original equipment manufacturer (OEM) or from a supplier.

The pressure taps are available in different designs:
- National pipe taper (NPT).
- O-ring boss (ORB).
- O-ring face seal (ORF).
- Joint Industry Council 37° flare fitting (JIC).

Some common pressure taps are 1/4″ NPT and 1/8″ NPT threads. The pressure taps also use ORB, ORF, and JIC fitting designs. Charts are available to help locate part numbers for pressure taps. See the *Appendix* for a *Pressure Tap Part Number Chart*.

Different systems require specific types of fittings and may use varying combinations of washers and O-rings to seal connections. Always refer to the OEM service literature before installing pressure taps to ensure the proper materials and taps are used.

Traditional Pressure Gauge

The most common and least expensive pressure-measuring device is a traditional mechanical pressure gauge. See **Figure 2-22**. Mechanical gauges are designed specifically

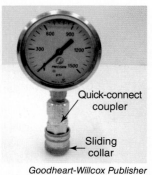

Goodheart-Willcox Publisher

Figure 2-22. A quick-connect coupler has been attached to this liquid-filled mechanical pressure gauge. The gauge will measure up to 1500 psi.

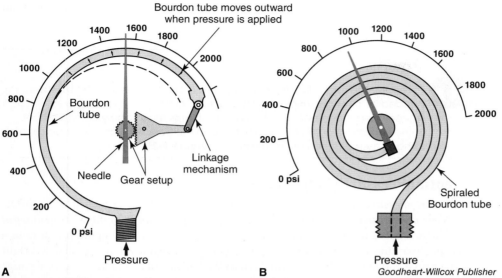

Figure 2-23. Several varieties of Bourdon-tube pressure gauges are available. A—A C-shaped Bourdon-tube pressure gauge uses a tube that will flex as hydraulic pressure is applied to the base of the gauge. B—A spiral Bourdon-tube pressure gauge uses a flattened tube spiraled around the center gauge needle.

Figure 2-24. A helical Bourdon-tube pressure gauge has a helical coil in the middle of the tube.

to measure vacuum, low pressure, medium pressure, or high pressure. A mechanical gauge can easily be damaged if installed in the wrong type of system. For example, a low-pressure gauge will be damaged if it is installed in a high-pressure hydraulic system. The physical sizes of the gauges vary, with the most common being approximately 2.5″ in diameter.

Mechanical gauges may be dry or filled with a liquid, such as glycerin. The liquid reduces needle bouncing caused by vibrations and pulsations that may cause the needle to break. At the top of a liquid-filled gauge is a plug or vent that caps the port that was used to fill the gauge with liquid.

If a gauge with a closable vent reads pressure while no pressure is applied, the vent should be opened to allow the glycerin pressure to equalize with atmospheric pressure. This allows the gauge's needle to return to the zero position prior to use. If the vent has been cut open and cannot be resealed, orient the gauge with the vent in the vertical position to prevent the glycerin from leaking out. If a plugged, liquid-filled gauge reads a pressure while no pressure is applied, the plug can be lifted slightly to allow the needle to return to zero. The glycerin can cause the gauge to discolor when exposed to sunlight for long periods of time, making the gauge difficult to read.

Bourdon-Tube Pressure Gauge

A ***Bourdon-tube pressure gauge*** is the most common type of mechanical pressure gauge used by hydraulic technicians. The gauge was invented by Eugene Bourdon in 1849. Bourdon tube gauges can have one of several configurations:

- C-shaped Bourdon-tube pressure gauge—pressure gauge in which the C-shaped tube flexes and straightens as hydraulic pressure is applied to the base of the gauge. Mechanical linkage attached to the sealed tip of the tube transfers tube deflection to the needle. See **Figure 2-23A**.
- Spiral Bourdon-tube pressure gauge—pressure gauge in which fluid is applied to a long spiral-shaped tube. The end of the tube is connected to a needle. As fluid pressure changes, the spiraled tube contracts or expands, resulting in deflection of the needle. See **Figure 2-23B**.
- Helical Bourdon-tube pressure gauge—pressure gauge that uses a tube with a short, helical coil in the middle of the tube. See **Figure 2-24**. The tube is connected to linkage that transmits the amount of pressure to the gauge's needle.

Pressure gauges used to measure low pressures use a bellows or a diaphragm instead of a Bourdon tube.

Pro Tip

To reduce the risk of fluid-injection injuries, a pressure gauge can be suspended in a hanger, minimizing the need to handle the live gauge. A gauge hanger can be constructed from a truck mud flap by cutting a hole in the flap and placing the gauge through the hole. An S hook can be used to hang the holder on the machine. See **Figure 2-25**.

Goodheart-Willcox Publisher

Figure 2-25. A gauge holder minimizes the need to handle a live gauge.

Differential Pressure Gauge

A *differential pressure gauge* allows a technician to simultaneously measure the difference between two pressures. A differential pressure gauge is helpful for measuring *margin pressure*, which is the difference between pump outlet pressure and signal pressure in a load-sensing hydraulic system. To use the gauge, one pressure line is connected to the right side of the gauge and another line is connected to the left side of the gauge. A second gauge is mounted on the left to monitor the highest system pressure. See **Figure 2-26**.

Quadrigage and Tetra Gauge

Another style of pressure gauge is the MICO Quadrigage, which is similar to the Caterpillar Tetra Gauge. Both gauges have three pressure gauges and allow testing a range of pressures. The gauge's maximum pressure varies based on the gauge model.

The gauge in **Figure 2-27** can measure pressures ranging from a vacuum to as high as 6000 psi (414 bar). This type of gauge is useful when multiple ports will be used to test system pressures and the technician is uncertain which port is the high-, medium-, or low-pressure port. The technician could safely connect the gauge to any of the three ports. The gauge uses an internal cutoff that will prevent high pressure from damaging the two low-pressure gauges. For example, if the gauge was measuring a pressure below 72 psi, only the low-pressure gauge would register a pressure. Once the pressure climbs above 72 psi, the pressure would also register on the medium-pressure scale. The system pressure is read from only the gauge reporting the highest pressure.

OTC Electronic Digital Pressure Meter

The greatest benefit of this tool is safety. A technician measuring 10,000 psi (689 bar) can safely handle a low-voltage pressure transducer and meter rather than a live pressure gauge. The OTC meter has three color-coded pressure transducers designed to measure a specific pressure range:

- The blue transducer is used to measure a pressure range from a vacuum (negative pressure) up to 500 psi (34 bar).
- The red transducer is used to measure pressures up to 5000 psi (345 bar).
- The orange transducer measures pressures up to 10,000 psi (689 bar).

Goodheart-Willcox Publisher

Figure 2-26. A differential pressure gauge is used to simultaneously measure the difference between two pressures.

Goodheart-Willcox Publisher

Figure 2-27. A MICO Quadrigage or a Caterpillar Tetra Gauge allows a technician to tap into a test coupler that can have pressures ranging from a vacuum to high pressure. This gauge will accurately display pressures ranging from a vacuum to 6000 psi.

Caution

The pressure transducers may cost 5 to 6 times the price of a traditional liquid-filled gauge. Placing a low-pressure transducer in a high-pressure system will ruin the transducer. The meter must also be set on the appropriate range to ensure accurate readings. See **Figure 2-28**.

Pro Tip

Install a quick-connect coupler on the transducer to ensure a long, leak-free life.

The OTC electronic digital pressure meter (OEM 1600) reads pressure ranges from vacuum to 10,000 psi and has features that measure temperature, monitor both negative and positive pressure changes, safely provide minimum/maximum system pressures, and determine differential pressure. It also has four pressure ports.

Temperature

A temperature probe, consisting of a thermocouple, is used for measuring the temperature of component surfaces, compartments, and fluids. The thermocouple is plugged into the meter and the meter's dial is turned to the Fahrenheit or Celsius position. The meter's temperature ranges are –40°F to 2500°F and –40°C to 1350°C.

Four Pressure Ports

The pressure meter can be attached simultaneously to four different ports. However, the meter does not simultaneously display all four pressures. The sensor select button allows the user to view each reading separately.

Delta Zero

This feature allows the technician to monitor both negative and positive pressure changes. When the delta zero button is pressed, *Delta Zero* is displayed on the screen and the existing system pressure reading on the screen becomes zero. As the pressure changes, the meter displays a positive or negative reading. For example, if delta zero is selected and the pressure drops 50 psi (3 bar), the meter displays –50 psi.

Minimum/Maximum (MIN/MAX)

This feature allows the technician to concentrate on operating the machine while obtaining readings for the lowest (MIN) and highest (MAX) system pressures. Once the machine is running and the meter is measuring a fluid pressure, the operator presses the MIN/MAX button. After the machine has run for a period of time, the operator presses the MIN/MAX button again and MIN is displayed on the screen with the recorded minimum operating pressure. Pressing the button once more displays MAX on the screen as well as the recorded maximum pressure.

Differential Pressure

This feature displays the difference in two pressures that are being measured simultaneously. This feature is used in load-sensing hydraulic systems to measure margin pressure. The steps for using this feature are as follows:

1. First, connect two of the meter's electrical cords to the meter's receptacles, Δ1 and Δ2. Connect the other end of each cord to a pressure transducer. Connect each pressure transducer to one of the machine's pressure ports.

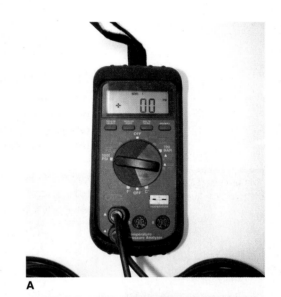

A

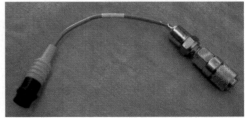

B *Goodheart-Willcox Publisher*

Figure 2-28. The OTC electronic pressure meter offers a safe method for measuring very high pressures. A—For added convenience and safety, the meter has 20′ electrical leads that connect the transducers to the meter, allowing the technician to distance himself or herself from the line. B—The pressure transducers use a pipe thread fitting, which is not designed for repeated use. Installing a quick-connect coupler to the transducer will minimize wear.

2. Choose the appropriate pressure range on the left side of the meter and then zero that sensor by pressing the sensor zero button. Repeat this step for the second sensor.
3. Turn the dial to select the appropriate colored delta symbol (Δ) based on the pressure value being measured.
4. The meter will display the difference between the pressures.

The meter will display the difference of the pressure of sensor Δ2 in respect to sensor Δ1. For example, if sensor Δ2 is measuring pump outlet pressure of 2500 psi (172 bar) and sensor Δ1 is measuring a signal pressure of 2350 psi (162 bar), the meter would read +150 psi because the pressure at sensor Δ2 is 150 psi (10 bar) more than at sensor Δ1.

The opposite would occur if the sensors were reversed. For example, if sensor Δ1 measured a pump outlet pressure of 3000 psi (207 bar) and sensor Δ2 measured a signal pressure of 2900 psi (200 bar), the meter would read –100 psi (–7 bar) because the pressure at sensor Δ2 is 100 psi (7 bar) less than at sensor Δ1.

Cleaning Equipment

Machines and their components often need to be cleaned before a technician can perform an inspection, service, or repair. Dirt and grime can be removed from machines and large components with high-pressure washers, and parts washers can be used to clean smaller components. Both types of cleaning equipment pose risks and require training and PPE to ensure operator safety.

High-Pressure Washers

A *high-pressure washer* uses a pump to build water pressure that is used to spray a stream of water at high velocities. See **Figure 2-29**. The pump may be powered by an electric motor or a gasoline or diesel engine. Most pressure washers have the option of mixing soap with the water and some commercial models have a burner to heat the water.

franco lucato/Shutterstock.com
Figure 2-29. High-pressure washers are used for cleaning shop floors and heavy machinery.

Pressure washer guns use different angled tips: 40°, 25°, 15°, and 0°. The 0° tip creates the highest water velocities. Some high-pressure washing systems can build water pressures 5000 psi or higher. Those pressures combined with narrow angled tips can cause the fluid to slice off fingers, pierce PPE, or inject water, detergent, chemicals, and debris into your skin. To prevent personal injury, operators must never point a pressure washer at a person or a part of your body. Operators should also wear proper PPE, including the following:

- Hard hat.
- Full-face shield and safety glasses.
- Rubber gloves.
- Reinforced rubber boots with slip-resistant soles.
- Water-repellent and cut-resistant suit.

 Warning
The amount of thrust exerted by the high-velocity water can cause the spray gun to kick back or fall from the operator's hands. The spray gun must always be held with two hands to ensure safe handling.

Parts Washers

A *parts washer* has a basin similar to a large sink, a lid, a drain, and an electric pump that recirculates the solvent. The pump is used to deliver cleaning fluid through a nozzle at low pressure and low flow. An *aqueous parts washer* uses a safe, biodegradable, water-based solution. A *solvent parts washer* uses a flammable solvent that efficiently removes grease, oil, and dirt. Technicians brush the component as the solvent flows over the part. See **Figure 2-30**. When the solvent has reached the end of its useful life, it is typically deemed a hazardous waste due to contamination from the parts it has cleaned. Hazardous waste requires proper disposal, and repair shops hire companies to periodically service and drain their solvent tanks to ensure they are in compliance with government regulations.

In addition to being flammable, the solvent is a skin-irritant and harmful if ingested, inhaled, or if it comes in contact with eyes. Safe use of a parts washer includes the use of proper PPE and safety practices, including the following:

- Safety glasses and full-face shield.
- Protective gloves and slip-resistant footwear.
- Close-fitting clothing, secured long or loose hair, and no jewelry.
- Only operating the washer in a well-ventilated room.
- Keeping flames or other potential ignition sources, such as smoking, grinding, and welding, away from the washer.
- Having easy accessibility to the solvent's SDS.
- Properly wiring and grounding the electrical service to the washer.
- Keeping the lid on the washer when it is not in use.

Kalabi Yau/Shutterstock.com

Figure 2-30. Parts washers use a pump, nozzle, and sprayer to spray solvent or a cleaning solution on a part as the technician scrubs it with a brush.

 Warning
Note that parts washers using combustible solvents must be equipped with an automatic closing lid in the event of a fire.

 Pro Tip
Gloves help protect hands and fingers from cuts, scrapes, and bruises. When working on oily components, nitrile gloves can protect the skin from exposure to oil, grease, soot, and other by-products and chemicals.

Media Blasting Equipment

Technicians use media blasting equipment to remove paint and rust from machines and their components. Media blasting equipment uses high-velocity air to spray the equipment with an abrasive media, such as silica sand, steel shot, crushed glass, or plastic beads. More environmentally friendly media, such as sodium bicarbonate or ground walnuts, may also be used.

Large, permanently mounted machines are used outdoors to clean large components, such as a tractor frame. Smaller, self-contained, cabinet-type media blasters are used to clean smaller components. See **Figure 2-31**. Cabinet-type media blasters have a viewing window and two gloved openings. These features allow the technician to safely observe and manipulate the part inside the cabinet.

A *Goodheart-Willcox Publisher* B *Milos Stojanovic/Shutterstock.com*

Figure 2-31. Media blasters can be portable or cabinet models. A—This media blaster has a cabinet with built-in gloves. It is used for cleaning small components. B—Air supplied respirators should be used with portable media blasters. Clean air is provided through the hose at the back of the helmet.

The high air velocity and flying abrasives pose serious health risks to the operator. Technicians must receive sufficient training and wear PPE to ensure personal safety. Safe practices that must be followed include the following:

- Adequate hearing protection must always be worn. The media blasting equipment and the abrasive striking the equipment generate a great deal of noise.
- Respiratory PPE should provide a clean air source and adequate filtration. The air-borne particulates from the material being removed and the abrasives being used can cause lung damage if inhaled.
- The high-velocity air and airborne particulates could also damage the operator's eyes. If the respiratory equipment does not include a full helmet with eye protection, the operator must wear a full-face shield and safety goggles as well as head protection.
- The high-velocity air poses risks of skin injection, and the operator must wear protective clothing, leather gloves that cover the forearms, and boots. The abrasive material will also cut unprotected skin.
- There should be designated areas for the operator to change and shower and for storing all contaminated clothing.

Metal Cutting and Welding Equipment

Heavy equipment technicians often perform repairs that require cutting and welding metal. Many heavy equipment technicians choose to weld with an oxyacetylene system because it is very portable and highly versatile. By changing torch tips, the torch can be used for heating, soldering, brazing, welding, and cutting steel. See **Figure 2-32**. Torches are commonly used to heat corroded fasteners to make them easier to remove and to expand metal for an easier press fit.

Figure 2-32. Oxyacetylene equipment includes a torch with an assortment of cutting and welding tips, torch handle, an oxygen regulator, acetylene regulator, hoses, goggles, and a spark striker.

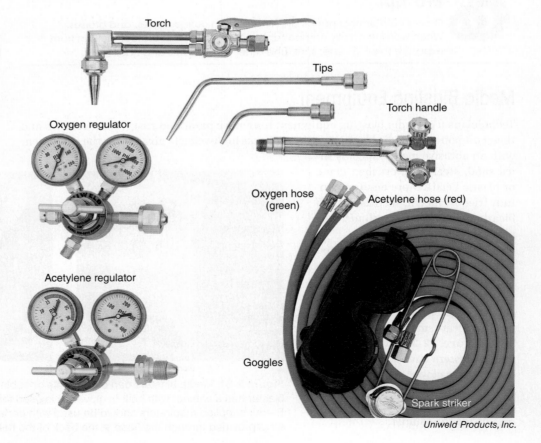

Uniweld Products, Inc.

An *oxyacetylene torch* uses a combination of oxygen and acetylene gases to weld, heat, and cut metal. As discussed in Chapter 1, *Shop Safety and Practices*, acetylene is a very unstable gas that must be handled with great care, including the proper storage, maintenance, and transportation of cylinders. The oxygen cylinders used in this type of system contain compressed oxygen at pressures up to 2100 psi and must also be handled properly. Technicians must receive proper training before using an oxyacetylene system or any type of welding equipment. It is also necessary to wear the appropriate PPE when welding to prevent burns and eye injuries.

Measuring Tools

Heavy equipment technicians must know how to use different measuring tools to take precise and accurate measurements. In addition to common tools, such as standard tape measures and rules, technicians use micrometers, dial calipers, dial indicators, and feeler gauges. Measurements are taken using either US customary (inches) or metric units.

Micrometers

Micrometers are precision instruments that can measure to one-thousandth of an inch (0.001″) or one-hundredth of a millimeter (0.01 mm). Technicians may use several types of micrometers, including outside, inside, and depth micrometers.

- An *outside micrometer* is used to measure external dimensions, diameters, or thicknesses.
- An *inside micrometer* is used for internal measurements of cylinders or other part openings.
- A *depth micrometer* is used to measure the depth of an opening.

The basic micrometer components are the sleeve (also known as a barrel), rotating thimble, and ratchet. See **Figure 2-33**. The micrometer's thimble is rotated until the spindle is near the component being measured. The ratchet should be used for the final adjustment. The ratchet prevents overtightening of the thimble which would lead to false readings and distortion of the micrometer's calibration.

Pro Tip

Always check the calibration of the micrometer before taking any measurement. A small outside micrometer (for example 0″–1″ micrometer) can be checked by threading the thimble until the spindle bottoms out against the anvil. The zero line on the thimble must align with the zero line on the sleeve. If the marks do not align, use a calibration spanner wrench to turn the sleeve until the marks are aligned. A gauge bar is used to check accuracy on larger micrometers. For example, a 1″ gauge bar would be used to check the accuracy of a 1″–2″ micrometer.

Reading a Metric Micrometer

Outside micrometers measure external dimensions, diameters, or thicknesses. The sleeve of the metric outside micrometer in **Figure 2-34** is marked with numbers in increments of 5, which represent millimeters (mm). Each of the smaller lines between the numbers represents 1 mm. The lines below the centerline of the sleeve represent half millimeters (0.5 mm). The numbers on the thimble are also marked in increments of 5 and range

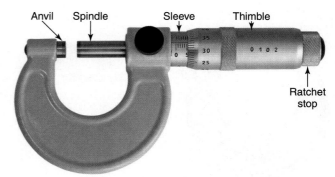

aarrows/Shutterstock.com

Figure 2-33. This outside micrometer uses a metric scale. The graduations on the sleeve are in 1 mm and 0.5 mm increments. The numbers on the thimble represent increments of one-hundredth of a millimeter.

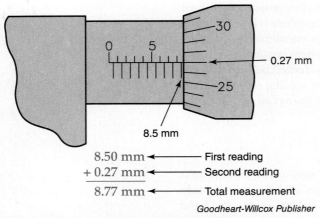

Figure 2-34. Reading a metric micrometer.

from 0 to 50. The small lines between each number represent one-hundredth of a millimeter.

Each revolution of the thimble equals a total of 0.5 mm. The first revolution will cause the thimble to align with the 0.5 mm mark. Another full revolution of the thimble will cause the thimble to align with the first 1 mm mark. Every two complete revolutions of the thimble equal 1 mm.

To read the micrometer, first determine how many millimeter marks are displayed on the sleeve. In **Figure 2-34**, the reading is 8 mm. Next, determine if any half millimeter marks are showing on the sleeve. In this case, there is one half millimeter mark appearing after the 8 mm mark. The next step is to read the graduated scale on the thimble. Remember that these numbers represent one-hundredth of a millimeter; therefore, the thimble reading in **Figure 2-34** is 0.27 mm. The total reading is 8.77 mm.

Pro Tip

If a specification is given in units of inches and a metric micrometer is the only tool available, the metric measurement must be converted. One inch is equal to 25.4 mm. The 8.77 mm can be divided by 25.4 to obtain a standard measurement of 0.3452″.

Reading a Standard Micrometer

The sleeve of the standard 1″ outside micrometer in **Figure 2-35A** is marked with numbers in increments of 0.100″. The single line between each number represents 0.050″ and the lines below the centerline are increments of 0.025″. The numbers on the thimble are also marked in increments of 5, with each line representing one-thousandth of an inch (0.001″). One revolution of the thimble equals 0.025″.

To read the standard micrometer in **Figure 2-35B**, first read the numbers on the sleeve. The *2* on the sleeve represents 0.200″. Below the centerline of the sleeve, there are three lines visible between the edge of the thimble and the *2* line on the sleeve. This represents 0.075″. The thimble is set on two lines above the number *10*, which represents 0.012″. The total measurement is determined by adding the three individual readings together. In this case, the total measurement equals 0.287″.

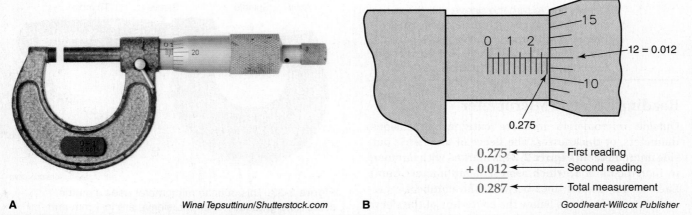

Figure 2-35. Reading a standard US customary micrometer. A—The lines on the thimble of a standard micrometer each represent one-hundredth of an inch. The lines on the sleeve represent 0.025″. B—Reading a standard micrometer.

Pro Tip

To convert the measurement to millimeters, multiply the total reading by 25.4:

0.287″ × 25.4 = 7.289 mm

Inside Micrometers

Inside micrometers are used to measure internal openings, such as the inside diameter of a bearing race. The micrometer consists of an anvil, thimble, short handle, lock screw, and a place to insert an interchangeable rod. See **Figure 2-36**. The length of the interchangeable rod selected is based on the diameter being measured. A set of rods is typically included with the micrometer.

Reading an inside micrometer is similar to reading an outside micrometer. However, an inside micrometer might read only in measurements from 0 to 0.5″ and require the use of short 0.5″ spacer rods rather than relying solely on rods in graduations of 1″.

Fouad A. Saad/Shutterstock.com

Figure 2-36. This inside micrometer measures in metric units. One revolution of the thimble of an inside metric micrometer equals 0.5 mm.

Depth Micrometers

Depth micrometers are used to measure the depth of an opening. Depth micrometers are similar to inside micrometers in that they also use interchangeable rods. See **Figure 2-37A**. Reading a depth micrometer is similar to reading an outside micrometer with two exceptions. First, the numbers on the sleeve of a depth micrometer are reversed from those on the sleeve of an outside micrometer. Second, when reading the value on the sleeve, you read the markings "hidden" under the thimble rather than reading the values visible on the sleeve.

For example, in **Figure 2-37B**, 9 is the first visible number on the sleeve, so 8 is the first number hidden by the sleeve. Use the 8 for your measurement. This indicates that the depth being measured is more than 0.800″ but less than 0.900″. Each line on the sleeve represents 0.025″. Two lines are visible, so one line (0.025″) is hidden by the thimble. The total depth measurement would be calculated as follows:

+ 0.800″ (because the 8 is concealed and the 9 is visible)
+ 0.800″ (because the 8 is concealed and the 9 is visible)
+ 0.010″ (because that is the number represented on the thimble)
= 0.835″ (total depth of measurement)

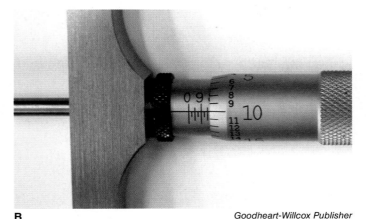

A B *Goodheart-Willcox Publisher*

Figure 2-37. Depth micrometers require choosing an appropriate length rod based on the depth being measured. A—A depth micrometer set includes rods of various lengths. B—This depth micrometer reading is 4.835″.

Figure 2-38. This standard inch dial calipers is being used to measure the outside diameter of a rod with a diameter of 0.375″.

When reading a depth micrometer, you must first determine the length of the inserted rod. For the example above, if a 4″–5″ rod had been inserted, the depth measurement would have been 4.835″.

Dial Calipers

Dial calipers are precision tools used to take four different measurements: inside, outside, depth, and step. A caliper has a graduated numerical scale (0–100) on the beam that runs the full length of the caliper. Each number represents one-tenth of an inch (0.1″), or, more precisely, 100 one-thousandths of an inch (0.100″). In **Figure 2-38**, the caliper is being used to measure the outside diameter of a rod. The *3* showing on the beam represents 0.300″. The dial (which also has a scale of 0–100) reading is 75, which equals 0.075″. The total measurement would be 0.375″. **Figure 2-39A** shows the depth rod at the end of the dial caliper, which is used for measuring the depth of a component.

A component with a step can have the length of the step measured by laying the back of the caliper flat on the step, with the caliper jaws opened wide and the top edge of the caliper perfectly aligned with the edge of the step. Next, slide the jaws of the caliper closed until the bottom inside measurement jaw of the caliper contacts the face of the step. The measurement is then read normally. **Figure 2-39B** shows using a dial caliper to measure the width of a vise jaw pad.

 Pro Tip
Always clean the jaws and check the calibration before using dial calipers. If the dial does not read zero with the jaws closed, adjust the dial to zero.

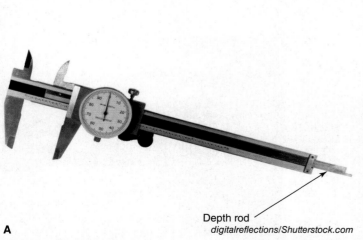

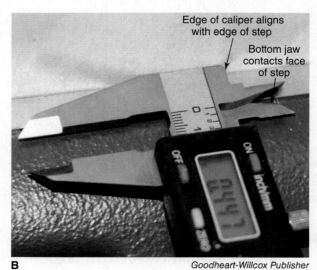

Figure 2-39. Dial calipers can be used to measure depth and step. A—The depth rod protrudes from the end of the caliper. The rod is used to measure the depth of a component. B—This dial caliper is being used to demonstrate how to measure the thickness of a vise jaw using the step method.

Dial Indicator

A *dial indicator* is a precision measuring device used to make a variety of measurements, including runout and shaft endplay. Dial indicators must be temporarily mounted to a fixture, such as the magnetic base illustrated in **Figure 2-40A**. The magnetic base has a switch that can be turned on to secure the tool and off to enable easy removal. An adjustable, snake-like base can be used to clamp the indicator on the flange of a component when working with nonferrous metals. See **Figure 2-40B**.

A standard dial indicator measures in thousandths of an inch (0.001″). The outside dial is rotated to set the needle to zero prior to taking a measurement.

Feeler Gauge

A *feeler gauge* is a measuring device with a series of thin blades that are used to measure small distances between two components, such as the distance between a pressure plate and a snap ring in a clutch (clutch pack disengaged clearance). A feeler gauge is often used when adjusting engine valves. Multiple blades can be stacked to establish the desired spacing or take a measurement. See **Figure 2-41**. Feeler gauges may use US customary (thousandths of an inch) or metric units (millimeters).

Fasteners

Manufacturers use a wide variety of threaded fasteners and hardware to attach and secure components. Fasteners vary by size, strength, and thread type. The most common fasteners are bolts, studs, nuts, and screws.

Caution

Fasteners are designed for specific applications. Failing to use the proper fastener can result in damage to the materials and failure of the fastener. Always refer to manufacturer specifications when replacing original fasteners.

Bolts

A *bolt* is a fastener with external threads on part of its shaft and a head on the opposite end. The shank is the nonthreaded portion of the shaft, located between the bolt head and the threads. The bolt length equals the distance from the end of the bolt to the face of the head. Bolt heads are square, round with a socket, 12-point, or hexagonal, with the hexagonal being the most common. See **Figure 2-42**. Bolts are typically used in combination with nuts. As the nut is drawn closer to the bolt head, a clamping action results that holds parts together.

A

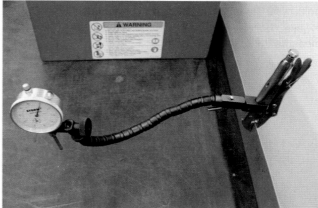

B   *Goodheart-Willcox Publisher*

Figure 2-40. Dial indicators are used to measure endplay, runout, and backlash. A—Dial indicators can be used with a magnetic base which can be turned off, making it easier to remove and reposition. They can also be mounted with a flexible fixture. B—A flexible snake-like fixture can be coupled with a pair of locking pliers.

Four Oaks/Shutterstock.com

Figure 2-41. Feeler gauges have different size (thickness) blades. The blades can be stacked to measure clearances larger than the thickest single blade. In such cases, the thicknesses of the stacked blades are added together to determine the clearance.

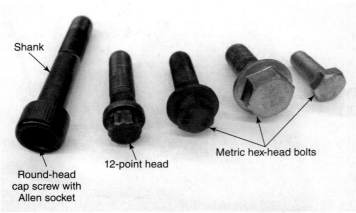

Figure 2-42. A bolt has a head, shank, and threaded end. Metric bolts have the numeric grade marked on the bolt head. Cap screws are similar to bolts but have tighter tolerances. They may have Allen sockets or hex heads.

Grades and Classifications

Bolts are classified by diameter, length, number of threads, and grade.

- The diameter is typically measured on the outside of the threads.
- The length is measured from the bottom of the head to the end of the fastener.
- The number of threads on the shaft is specified by TPI (threads per inch for US customary fasteners) and thread pitch (distance between threads) for metric fasteners. Thread pitch or TPI can be measured with a thread pitch gauge. See **Figure 2-43**.
- The *grade* is a measurement of the bolt's tensile and yield strengths. A bolt's *tensile strength* is the amount of stress that can be placed on the bolt before it stretches and breaks. *Yield strength* is the point at which the bolt will deform plastically and will not return to its original condition.

Grade Markings

The Society of Automotive Engineers (SAE) provides grades for US customary bolts. The number of marks and the location of the marks on top of a standard US customary bolt indicate its grade. On metric bolts, the numeric grade is marked on the bolt head. See **Figure 2-44**. Refer to the *Appendix* for *Metric Bolt Sizes*, *US Customary Bolt Sizes*, *SAE Bolt Grade Strengths*, *Metric Bolt Grade Strengths*, and *Nut Grade Markings*.

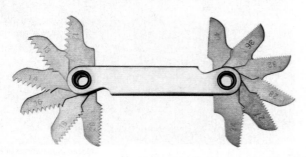

Figure 2-43. Thread pitch gauges are helpful for identifying the bolt's pitch.

Pro Tip
A bolt can be replaced with a higher grade bolt, but should *never* be replaced with a lower grade bolt.

Figure 2-44. Bolts have markings on the head that identify their grade. A—The number of marks on the head of a US customary bolt indicates the bolt's grade. B— Metric bolts are identified with numbers. These bolts are grades 8.8 and 10.9.

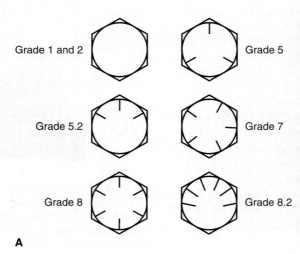

A

B

>
> **Note**
>
> Most bolts and nuts use traditional right-handed threads, which require the nut to be turned clockwise (right-hand) to be tightened and counterclockwise to be loosened. Occasionally left-handed threads will be used, which require the opposite rotation, counterclockwise (left-hand) to be tightened and clockwise to be loosened. When the bolt is positioned with the threads facing upward, a right-handed threaded bolt will have the threads slanting upward to the right. A left-handed threaded bolt will have the threads slanting upward to the left.

Studs

A *stud* is a headless rod with threads on both ends. See **Figure 2-45**. Manufacturers' service materials will specify whether a stud or bolt should be used.

Nuts

Nuts are hex-shaped fasteners with internal threads designed to mate with the threads on bolts and studs. The nuts are used to clamp, attach, and fasten components. The most commonly used is the hex nut. Other types of nuts include acorn, castle, jam, wing nut, lock nut, and flange nut. See **Figure 2-46**.

- Acorn nuts have a slightly domed top that covers the end of the bolt.
- Castle and slotted nuts have a slotted end that is used with a cotter pin to prevent loosening.
- A jam nut is a thinner nut sometimes used as a second nut to "jam" or lock a primary nut in place.
- Wing nuts are used in applications where the nut must be removed or installed frequently. It is designed to easily be turned by hand but may also be turned with slip-joint pliers.
- A lock or prevailing torque nut typically uses nylon to prevent the bolt from loosening. Nylon locking nuts must not be reused.
- A flange nut has a flange located at the bottom that serves as a washer.

Goodheart-Willcox Publisher

Figure 2-45. A stud is a rod with threads on both ends.

Washers

A *washer* is a thin piece of metal that is placed under a bolt head or nut to distribute the clamping force of the fastener. Three types of washers are flat, locking, and star. See **Figure 2-47**. Flat washers spread the clamping force placed on the nut. Lock washers are

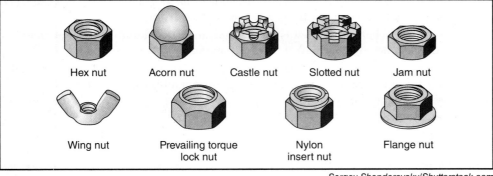

Sergey Shenderovsky/Shutterstock.com

Figure 2-46. The most common type of nut is the hex nut that is often used with a locking washer. Nuts may also have a locking nylon ring or a castle shape designed for use with a cotter key.

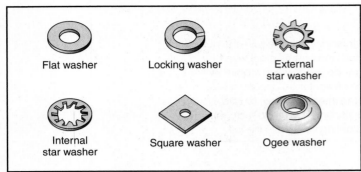

Sergey Shenderovsky/Shutterstock.com

Figure 2-47. Washers are made in a variety of shapes and sizes and from different types of material.

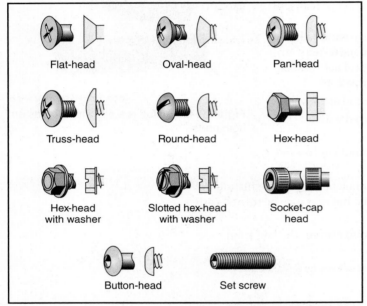

Sergey Shenderovsky/Shutterstock.com

Figure 2-48. Screw heads range in style and size.

Goodheart-Willcox Publisher

Figure 2-49. These round-head cap screws have a barrel-shaped head in place of the traditional hexagon head.

used to reduce the chances of the fastener loosening. Star washers can have internal or external teeth that grip the metal as they are twisted. Star washers are used on sheet metal. Some nuts have an attached washer which speeds the installation process.

Screws

A *screw* is a threaded fastener that is designed to be used in a threaded bore, or to cut threads into the material into which it is installed. Screws are tightened by applying torque to the head of the fastener. They are made from different materials in many shapes and sizes and with different types of threads. The shaft may be threaded along its entire length or it may be only partially threaded and have a smooth section, similar to a bolt. Sheet metal screws have threads the full length of the screw. The threads taper to a sharp point at the end of the screw. They are designed to form their own threads in the material as they are tightened. Common types of screws are cap screws, flat-head screws, pan-head screws, and set screws. See **Figure 2-48**.

- A cap screw resembles a traditional bolt except for the washer attached below the head. The washer does not extend past the head. Cap screws may have a traditional hex head or a round head with a hex socket in the center. Round head cap screws are often used when the fastener must recess inside the bore of a component. Cap screws are used in applications with closer tolerances than those in which bolts are used. Another difference between bolts and cap screws is that cap screws are tightened to apply the fastener's torque. If a bolt is used, it is paired with a nut, which is tightened to apply the fastener's torque. See **Figure 2-49**.
- Flat-head screws have a tapered head that fits in the countersunk hole of a component. When fully installed, the top of the screw remains flat or flush with the top of the component surface.
- A round-head screw has a head that resembles one-half of a sphere. The head has a slot or socket so it can be turned with a matching driver.
- The head of a pan-head screw is somewhat similar to a round-head screw, but the head is thinner and may have a flatter appearance.
- A set screw is a small, fully threaded screw that has no shoulder or head. The end of the screw is recessed and, depending on the opening, can be turned with a screwdriver or Allen wrench.

Torx Fasteners

Torx fasteners have heads with a 6-point, star-shaped recess that provides more area for the screwdriver tip to engage than the recess in a Phillips-head or a flat-tip screw. Some installations require repeated removal and reinstallation of a screw, which may eventually cause the head to strip. A Torx-head screw is less likely to strip than a Phillips-head or flat-tip screw.

Retainers

Retainers, in the form of pins and snap rings, are also used to hold components in place. There are three types of pins: clevis pin, cotter pin, and roll pin. The pins and snap rings are available in a variety of sizes.

- A *clevis pin* has a straight shank and shouldered head with a drilled hole through the end of the clevis. See **Figure 2-50A**.
- A *cotter pin* is a flexible teardrop-shaped pin with two legs. The legs are spread apart after the pin is inserted through the bolt and/or castle nut to prevent the fastener from loosening. Cotter pins must not be reused. See **Figure 2-50B**.
- A *roll pin* is a small, short tube with a slot that runs the length of the pin. As the pin is inserted into the component, the pin compresses to hold itself in place (in its bore). See **Figure 2-50C**.
- Snap rings are frequently used to retain components. Snap rings can be classified as external, internal, E-clip, or C-clip. See **Figure 2-50D**.

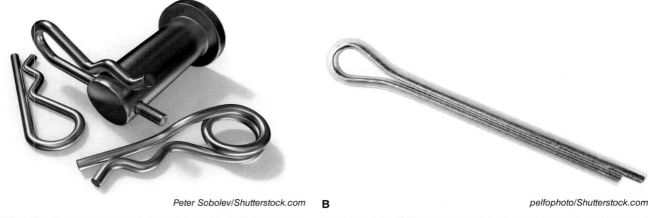

A *Peter Sobolev/Shutterstock.com* B *pelfophoto/Shutterstock.com*

C *Goodheart-Willcox Publisher* D *Goodheart-Willcox Publisher*

Figure 2-50. A variety of retainers are used to hold components in place. A—A clevis pin consists of a smooth shank pin with a shouldered head and a drilled passage on the end to hold a cotter pin or a spring clip. B—A cotter pin is a soft metal pin with two bendable legs. After the pin is slid through a hole in a shaft or bolt, the ends of the pin are separated and bent outward to hold the pin in place. C—Roll pins are slotted tubes. D—Retaining clip styles include E-clips, C-clips, outside clips, and inside clips.

Summary

- When working with hand tools, be sure to use proper PPE, remove jewelry, use the tools for their intended purpose, and follow the manufacturer's safety procedures
- Wrenches are used to loosen and tighten fasteners.
- Ratchets and speed handles speed the process of loosening and tightening fasteners.
- Torque wrenches are required for applying a specified torque to a fastener.
- Torque multipliers multiply the input torque and are used to apply large amounts of torque to a fastener.
- Electric power tools are powered by battery or shop electricity.
- Air tools, such as drills, grinders, ratchets, and chisels, are powered by air compressors.
- Hydraulic power tools provide the force required to press, pull, and install components on heavy machinery.
- Diagnostic pressure taps provide a safe and easy method for measuring system pressures.
- The Bourdon tube pressure gauge is the most common mechanical pressure gauge.
- Multipurpose gauges, such as a MICO Quadrigage, can be designed to measure pressures ranging from a vacuum up to 6000 psi.
- Electronic pressure meters provide the safest means for measuring high system pressures.
- High-pressure washers use pressures as high as 5000 psi (or higher) to clean heavy machinery and components.
- Media blasting equipment uses high-velocity air and an abrasive media to remove paint and rust from machines and components.
- Micrometers are offered in three different styles: inside, outside, and depth.
- Micrometers can measure in millimeters or thousandths of an inch.
- Dial calipers can be used to measure inside diameter, outside diameter, depth, and step.
- Feeler gauges are used to measure small distances between two components, such as the distance between a pressure plate and a snap ring in a clutch.
- Bolts are offered in metric or US customary sizes.
- Bolts are classified by diameter, length, number of threads, and grade (strength).
- Nuts are hex-shaped fasteners with internal threads designed to mate with the threads on bolts and studs.
- Washers are used under a nut or bolt head to distribute the clamping force of the fastener.
- Screws are threaded fasteners that are made with different materials in many shapes and sizes and with different types of threads.
- Retainers, such as clevis pins, cotter pins, roll pins, and snap rings, are used to help keep components in place.

Technical Terms

- bolt
- Bourdon-tube pressure gauge
- box-end wrench
- breaker bar
- chisel
- clevis pin
- combination wrench
- cotter pin
- crowfoot socket
- diagnostic pressure tap
- dial calipers
- dial indicator
- differential pressure gauge
- feeler gauge
- flare-nut wrench
- flex handle
- grade
- high-pressure washer
- line wrench
- margin pressure
- micrometer
- nut
- offset wrench
- open-end wrench
- oxyacetylene torch
- parts washer
- pawl
- punch
- ratchet
- ratchet drive size
- ratcheting wrench
- roll pin
- screw
- socket
- socket drive size
- speed handle
- stud
- torque
- torque multiplier
- torque wrench
- washer

Review Questions

Answer the following questions using the information provided in this chapter.

Know and Understand

1. Using a _____ speeds the process of loosening and tightening fasteners.
   A. ratchet
   B. box-end wrench
   C. torque multiplier
   D. ball-peen hammer

2. All of the following are common ratchet drive sizes, EXCEPT:
   A. 3/8".
   B. 1/2".
   C. 5/8".
   D. 3/4".

3. The _____ is the most commonly used device for measuring pressure.
   A. bellows pressure gauge
   B. Bourdon-tube pressure gauge
   C. quadrigage
   D. electronic pressure meter

4. There are _____ gauges in the Caterpillar Tetra Gauge assembly.
   A. two
   B. three
   C. four
   D. five

5. The primary advantage of using an electronic pressure meter is _____.
   A. that it costs less than a comparable Bourdon-tube gauge
   B. that it measures atmospheric pressure
   C. technician safety
   D. that it creates flow and pressure pulsations in a hydraulic system

6. Which feature on the OTC electronic pressure meter is used to monitor both positive and negative pressure changes?
   A. MIN/MAX.
   B. Delta zero mode.
   C. Differential pressure.
   D. Temperature.

7. Which feature on the OTC electronic pressure meter is used to record the highest and lowest system pressures for a given test port?
   A. MIN/MAX.
   B. Delta zero mode.
   C. Differential pressure.
   D. Temperature.

8. Some pressure washers may exert fluid pressures as high as _____.
   A. 500 psi
   B. 1500 psi
   C. 2500 psi
   D. 5000 psi
9. Which of the following may require the use of a respirator?
   A. Measuring fluid pressure.
   B. Media blasting.
   C. Fastener torquing.
   D. High-pressure washing.
10. What is a common numerical scale found on the thimble of a metric outside micrometer?
    A. 0 to 10
    B. 0 to 25
    C. 0 to 50
    D. 0 to 100
11. Which of the following requires selecting the appropriate-sized rod to be placed inside the tool before taking a measurement?
    A. Outside micrometer.
    B. Depth micrometer.
    C. Dial caliper.
    D. Feeler gauge.
12. On which of the following are the numbers on the sleeve in reverse order?
    A. Outside micrometer.
    B. Inside micrometer.
    C. Depth micrometer.
    D. Dial caliper.
13. A dial caliper is used to measure all of the following, EXCEPT:
    A. inside diameter.
    B. outside diameter.
    C. depth.
    D. runout.
14. Which of the following is a common numerical scale found on the dial of a dial caliper?
    A. 0–10.
    B. 0–25.
    C. 0–50.
    D. 0–100.
15. Bolts are specified by all of the following, EXCEPT:
    A. diameter and length.
    B. pitch.
    C. grade.
    D. color.
16. The strength of SAE bolts are signified by _____.
    A. color
    B. the lines on the bolt head
    C. a numerical decimal on the bolt head
    D. the type of shank
17. The strength of metric bolts are signified by _____.
    A. color
    B. the lines on the bolt head
    C. a number stamped on the bolt head
    D. the type of shank
18. A _____ is a headless rod with threads on both ends.
    A. bolt
    B. screw
    C. stud
    D. nut

Analyze and Apply

19. Safety glasses should include _____ shields.
20. _____ wrenches are primarily used for loosening and tightening fittings that attach lines, such as fuel lines.
21. Diagonal cutters, slip-joint, and needle-nose are types of _____.
22. What is the maximum number of pressure readings an OTC electronic pressure meter can display on its screen at one time?
23. A handheld electric tool is considered double-insulated if the wire is insulated and the housing is made of _____ material.
24. _____ is used to fill liquid-filled pressure gauges.
25. A(n) _____ can be used to measure runout and shaft endplay.
26. A(n) _____ uses a series of thin blades to measure small clearances.
27. Bolt strength is specified as _____.

Critical Thinking

28. Explain how to use a torque multiplier to remove a seized nut.

29. A 0–1″ outside micrometer like the one in Figure 2-35B has the number 1 visible on the thimble as well as the single half line visible between the 1 and 2. There are no small lines appearing at the bottom to the right side of the half line. The thimble is aligned to the numeral 5. What is the measurement?

30. Describe how to properly use a feeler gauge to measure the clearance between a snap-ring and a neighboring component if the clearance is larger than the thickest leaf.

Vietnam Stock Images/Shutterstock.com

Special impact sockets must be used with impact wrenches. Impact sockets are tempered for increased toughness, have thicker walls than their hand socket counterparts, and have a black oxide finish rather than a chrome finish. If a standard hand socket is used with an impact wrench, it can shatter and seriously injure the user.

Chapter 3
Construction Equipment Identification

Objectives

After studying this chapter, you will be able to:
- ✓ Identify the common types of construction equipment and mining equipment.
- ✓ Explain construction machinery controls and performance characteristics.
- ✓ Describe multiple applications for different types of construction equipment.
- ✓ List a range of sizes of each type of construction machine.

Mobile equipment manufacturers produce a wide variety of machines for use in industry. A few of the industries with specialized machinery include construction, agriculture, forestry, and mining.

Overview of Construction Equipment

Construction equipment is very diverse in design and function. The machinery can be used for building a wide variety of structures:
- Roads, highways, and bridges (civil construction).
- Homes, condominiums, and neighborhoods (residential construction).
- Malls, schools, manufacturing plants, and skyscrapers (commercial construction).
- Power plants and oil refineries (energy construction).

A traditional construction equipment dealer offers contractors support in the form of sales, parts, and service for general construction equipment. General construction equipment includes the following types of machines:
- Excavators.
- Wheel loaders.
- Track loaders.
- Dozers.
- Motor graders.
- Scrapers.
- Haul trucks.
- Skid steers and compact track loaders.
- Forklifts and telehandlers.
- Compactors.
- Trenchers.
- Drills (horizontal drilling machines and rock drills).

Excavators

Excavators are machines that are designed to dig or scoop up material, transfer it to another location, and dump it. The name excavator, or backhoe, is derived from the way the equipment operates. The backhoe operates by drawing back its bucket like a garden hoe. See **Figure 3-1**.

Controls

Excavator operations are controlled through a pair of joysticks. The joystick positions for different commands are shown in **Figure 3-2**. The following excavator's movements can be controlled by the joysticks:

- House swing, or slew.
- Boom raise or lower.
- Stick (also known as the arm or dipper) extend or retract (also known as crowd).
- Bucket extend or curl.

Excavator performance is often specified in cycle times and breakout force. *Cycle time* is the amount of time it takes for a bucket, stick, or boom to actuate—extend, retract, or both fully extend and retract—or for the excavator's house to swing, or *slew*. Cycle times can be specified for just one function, such as the bucket, or can be a combined function, such as stick crowd and bucket curl. *Breakout force* is the amount of force exerted at the bucket teeth as it is digging. An excavator has three breakout forces, specified as bucket curling force, stick crowding force, and a combination of bucket curling and stick crowding forces. The combination of bucket curl and stick crowd provides the maximum breakout force. Manufacturer performance manuals provide this data, as well as numerous other parameters, such as the following:

Goodheart-Willcox Publisher

Figure 3-1. A traditional track-type excavator provides a stable base, allowing the excavator to travel across soft, loose, and soupy soils. They are not designed to travel at high speeds or for long distances.

Chapter 3 | Construction Equipment Identification

ISO Excavator Control Pattern

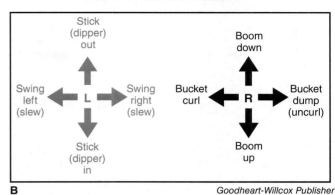

Figure 3-2. Excavators use two joysticks for actuating most of the machine's functions. A—In an ISO control pattern, the left joystick operates the boom. B—In a backhoe control pattern, the left joystick operates the stick (dipper/arm). Some machines allow the operator to switch between control patterns as desired.

- Maximum reach at ground level—how far bucket's teeth will extend by extending the boom, stick, and bucket forward on the ground.
- Maximum digging depth—how deep of a hole can be dug.
- Dump height—the maximum height at which the bucket can be unloaded.
- Maximum travel speed.

Excavators can be fitted with different types of attachments, such as a thumb, hammer, shear, rotary tree shredder, ripper, crane arm or jib, extendable stick, second boom, large cleanout bucket, rock bucket, rotating bucket, or a demolition clamping claw/grapple. See **Figure 3-3**.

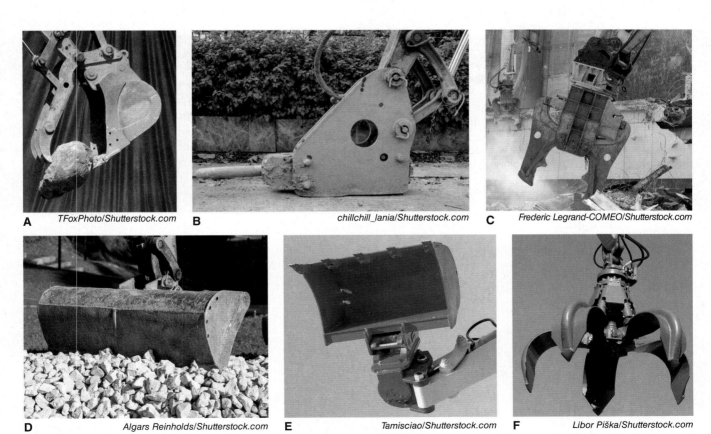

Figure 3-3. An excavator can be equipped with a variety of attachments. A—Thumb. B—Hammer. C—Shear. D—Large cleanout bucket. E—Rotating bucket. F—Grapple.

> **Note**
> Excavators sometimes are equipped with a **quick coupler** that allows the operator to quickly uncouple the bucket and change attachments. The coupler can be mechanically or hydraulically actuated. See **Figure 3-4**.

Goodheart-Willcox Publisher

Figure 3-4. Couplers allow operators to quickly uncouple and reattach work tools and buckets.

Excavator Applications

Excavators are used for building roads, loading trucks, general excavation, demolition, trenching, laying pipe, maneuvering scrap or material, cleaning ponds or waterways (known as dredging), forestry, quarrying, and mining. Excavators come in a variety of different designs to perform these tasks.

The design variations can include the type of propulsion system the excavator uses to move from location to location. Some excavators, like the one shown in **Figure 3-1**, are equipped with tracks. The tracks distribute the weight of the machine over a relatively large surface area. This gives the excavator better traction than wheels would in loose or soggy soil. Wheeled excavators are designed to operate on firm, stable ground. See **Figure 3-5**. Wheels allow an excavator to move faster than a track drive does, but they tend to sink into soft soil. Some wheeled excavators have a two-piece boom, improving their ability to work in tight quarters. Some two-piece booms are offset, allowing them to articulate in the middle and crowd the bucket next to a wall.

The designs of the digging implements themselves are varied to make the equipment better suited to perform certain tasks. For example, telescoping boom excavators are good for working in low clearance areas, removing concrete and asphalt, finishing a grade, and laying rock on a sloped side hill, **Figure 3-6**. The booms can rotate, allowing their attachments to be moved into different positions. These excavators can be mounted on a machine with tracks or wheels, a truck chassis, or even a railroad chassis.

A long-reach excavator is equipped with a longer boom and stick than a conventional excavator. This allows them to work in a larger area without moving. This is especially useful for dredging operations, where the excavator may be confined to a barge or the shore. See **Figure 3-7**. It is also useful during the demolition of tall structures.

Loader backhoes are one of the most universal machines used on construction sites. The backhoe can be used to excavate, trench, or remove sidewalks and driveways. The loader is commonly used to load dirt or rock and can be fitted with pallet forks to serve as a forklift. See **Figure 3-8**.

CapturePB/Shutterstock.com

Figure 3-5. Wheeled excavators are used in applications that require the excavator to travel farther and faster between job sites. This wheeled excavator has a two-piece boom.

Excitator/Shutterstock.com

Figure 3-6. This telescoping boom excavator is being used to dig a trench.

jack_photo/Shutterstock.com

Figure 3-7. Long-reach excavators are used for dredging and sometimes for demolition of tall structures.

Goodheart-Willcox Publisher

Figure 3-8. This loader backhoe is configured with an extendable stick. Operators use this feature to make a smooth, flat bottom trench by extending the stick and carefully drawing the bucket and stick back toward the machine. The backhoe is also equipped with sideshift, allowing it to be shifted to the rear right or left edge of the machine.

Some excavators have an integral propulsion and hydraulic system, while others must be attached to another piece of equipment in order to function. These excavator implements consist of the boom, stick, hydraulic actuators, and controls, but rely on another piece of equipment for their propulsion and hydraulic source. Common examples of this type of excavator include a three-point–hitch excavator and a skid-steer–mounted excavator, **Figure 3-9**.

 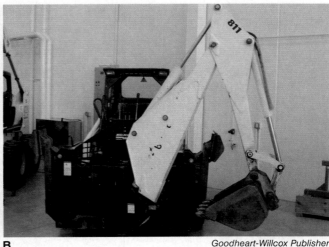

Figure 3-9. Some excavators are mounted to other equipment. A—A three-point–hitch backhoe can be mounted to an agricultural tractor. The backhoe is pinned to the tractor's three-point hitch. B—Skid steers can be equipped with an excavator attachment.

Sizes

Excavators can range in size from a rubber-track mini or compact excavator to a 4500 hp hydraulic shovel. The mini and small excavators commonly use rubber-belted tracks. Mid-size and larger use traditional steel tracks.

Smaller excavators can be equipped with an independent boom swing. Using the cab swing and the boom swing allows the operator to excavate flush beside a wall, such as a basement wall. Some mini excavators can gain access to a backyard through a traditional entry gate, which is only 32″ wide. See **Figure 3-10**. A mini excavator can also be equipped with a dozer blade and hydraulic tread adjustment providing narrow or wider track spacing. Manufacturers recommend adjusting the track width while the machine is moving forward or rearward.

Machine weight is a contributing factor in the load that the excavator can handle. A heavier excavator can lift a heavier load or be designed with a greater reach than a lighter excavator. Large excavators have heavy counterweights, weighing several tons, placed on the back of the machine. Some excavators have controls built into the machine so that the excavator can install and remove its own counterweight. See **Figure 3-11**. See the *Appendix* for *Excavator Sizes*.

Figure 3-10. Mini and compact excavators use rubber tracks. A—This compact excavator is equipped with an independent boom swing. B—This mini excavator is capable of fitting into very narrow work areas.

A **B** *Goodheart-Willcox Publisher*

Figure 3-11. Excavators have large counterweights attached to the rear of the machine. A—This excavator has its counterweight removed. Notice the hydraulic cylinder and chain sprocket that are used to hoist the counterweight back onto the excavator. B—This counterweight has been removed from the excavator.

Forestry Excavation Equipment

The forestry industry utilizes specialized forestry excavators, also called **knuckle boom loaders,** that act as a crane for moving and loading felled trees. Not all knuckle boom loaders are placed on a traditional twin track excavator platforms. Some are placed on a chassis with oversized logging tires. Others are located on the back of a truck.

Another forestry machine built on an excavator platform is a **feller buncher.** A cutting head attachment is placed at the end of the boom. The feller buncher can gather multiple trees in the attachment. A hydraulic cutterhead cuts through the trees, and the boom then places the felled trees next to other felled trees.

Mining Excavation Equipment

Mining sites can use four different types of excavation machines, **Figure 3-12:**
- Hydraulic front-loading shovels.
- Wire rope shovels.
- Drag lines.
- Giant wheeled excavators.

Hydraulic shovels can be equipped with a front-loading shovel bucket or the traditional backhoe bucket. They resemble an excavator but are much larger, **Figure 3-12A**. The largest hydraulic shovels can weigh more than 1000 tons and have a bucket with a volume close to 70 yd^3.

Wire rope shovels, also known as rope shovels, can be powered by electricity or a diesel engine. They use large wire ropes for operating a boom and front-loading bucket. Rope shovels can be designed larger than hydraulic shovels. The largest rope shovel buckets have a capacity of more than 100 tons. See **Figure 3-12B**.

Draglines are powered by electricity. They load the bucket by using wire ropes to drag the bucket into the dirt, **Figure 3-12C**. Draglines can use a crawler-type undercarriage consisting of two oval-shaped tracks or be fitted with two large walking platforms. A walking platform system consists of two long feet that are mounted to movable legs that allow the two platforms to slowly move one step at a time.

Giant mining excavators have large shovel buckets on a rotating wheel. These machines provide the largest mining excavation capacity. The buckets dump the material on a conveyor belt. See **Figure 3-12D**.

Additional Mining Equipment

The mining industry uses unique machines for underground mining as well as massive machines for aboveground strip mining. Types of underground mining include room and pillar and longwall. **Room and pillar mining** consists of mining the deposits while leaving

Figure 3-12. Mining operations are performed with a variety of excavating equipment. A—Hydraulic shovel. B—Wire rope shovel. C—Drag lines. D—Giant bucket-wheel excavator.

a grid pattern of square pillars to support the roof. The *longwall mining* process uses roof supports to allow access for a shearing machine with a rotating cutter-head to cut into the face of a deposit, such as a coal seam. See **Figure 3-13**. The cutter-head advances back and forth across the length of the longwall. A conveyor delivers the deposits out of the mine. During the mining process, the roof supports and equipment move forward into the coal seam, allowing a controlled collapse of the roof behind the roof supports. The collapsing roof is necessary to relieve the geological pressure after the deposits are removed. If the roof does not collapse, it indicates geological irregularities that place miners at risk.

Surface mining removes surface soil to expose deposits of the material being mined. It has the largest negative effect on the landscape but is able to recover the highest percentage of deposits. Longwall mining can remove the second largest percentage of deposits. Room and pillar mining removes the lowest percentage of mining deposits due to the grid of pillars left in place.

Mining equipment is used to mine a wide variety of metals and deposits, such as coal, copper, gold, gypsum, lithium, phosphate, potash, salt, silver, trona, and zinc. In addition to excavators, mines use enormous haul trucks (electric and mechanical powered), dozers, loaders, drills, conveyors, and large motor graders.

Wheel Loaders

Wheel loaders are commonly used for loading trucks and hoppers and, if equipped with pallet forks, maneuvering materials. Wheel loaders, also called front-end loaders, are often equipped with four equally sized tires and are propelled with all-wheel drive. The transmission can be hydrostatic drive, power shift, or electric drive, which are discussed later in the book.

Loader backhoes are the exception and are often equipped with large rear tires and smaller front tires. Loader backhoes can be two-wheel drive or have the option of four-wheel drive.

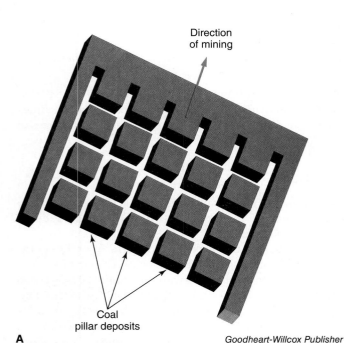

Figure 3-13. Three common mining methods can be used to extract deposits. A—Room and pillar mining leaves behind a grid of square pillars. B—Longwall mining removes sediments along a long wall and requires the use of roof supports. C—Surface mining digs an open pit to gain access to deposits.

Most wheel loaders articulate to steer the tractor. Depending on the model, articulated wheel loaders can articulate up to 43° to provide steering. A few loaders are designed with rigid frames and have front and rear steerable axles. For example, JCB manufactures a loader backhoe with four-wheel steer.

Controls

The loader's common hydraulic controls are:
- Loader lift.
- Bucket curl.
- Loader auxiliary (for a grapple or broom).
- Steering (right or left).

Loader Performance

As with excavators, breakout force and cycle times are two key measures of a loader's performance. The method of measuring a loader's breakout force is standardized by the Society of Automotive Engineers (SAE). The force can be measured by lifting the loader or can be measured by curling (tilting) the bucket. The machine is placed on a hard surface, the transmission is placed in neutral, and the brakes are released. The bucket is positioned parallel to the ground and a force is measured four inches behind the bucket's cutting edge. The force is measured in pounds, kilonewtons, or kilograms. Note that the rear of the machine must be allowed to lift and cannot be latched to the ground.

Loader performance is also measured with a variety of cycle times:
- Raise cycle time—the amount of time it takes to lift the loader frame from the ground to the fully lifted position.

- Lower cycle time—the amount of time it takes to lower the loader from the fully lifted position to the fully lowered position.
- Dump cycle time—the amount of time it takes to hydraulically tilt the bucket from the fully curled position (rolled back position) to the fully dumped position.
- Rack back cycle time—the amount of time it takes to curl the bucket from the dumped position to the fully curled back position.
- Total cycle time—the sum of the four cycle times: raise, lower, dump, and rack back.

Mining wheel loaders can be equipped with a cycle timer that records the machine's cycle times. Personnel can later download the cycle times and evaluate them for ways to improve operator productivity.

Wheel Loader Applications

Wheel loaders can be equipped with a variety of different sizes of buckets or other attachments. Common loader attachments are pallet forks, broom, four-in-one bucket, side dump bucket, roll out bucket, lifting jib boom, and grapples.

Four-in-one buckets use hydraulics to split the bucket, opening the front face of the bucket. It can be used as a clam shell for loading loose material or for increasing the loader's dump height, since the bucket does not have to be tipped forward to dump. The bucket can also be used as a dozer blade (front fully raised) or a scraper (front partially raised). This type of bucket is also known as a *clam shell bucket* or *multipurpose bucket*. See **Figure 3-14**.

Side dump buckets allow the wheel loader to load the bucket in a narrow row and remain in its position while dumping the bucket to the side. The side dump bucket has one end (side) that is open, which allows the bucket to dump out of its side. The bucket might be equipped with a cushioned cylinder to reduce shock loads during operation. The use of a side dump bucket can reduce cycle times and increase the machine's productivity because the loader does not have to drive to the face of the dump truck bed. The loader can remain beside the haul truck and still be able to dump its load.

A *roll out bucket* increases the loader's dump height by rolling the bucket forward, away from the loader's frame.

Wheel Loader Types

The two most common loader frame designs are parallel linkage and Z-bar linkage. *Parallel linkage* loaders have the bucket curl linkage placed in parallel to the lift arms. Parallel linkage is also known as eight bar because it consists of eight moving bars or levers. It is used on Caterpillar integrated tool carrier (IT) wheel loaders. See **Figure 3-15**. In the *Z-bar linkage* loaders,

Goodheart-Willcox Publisher

Figure 3-14. Four-in-one buckets have the front half of the bucket hydraulically actuated so that it can be opened and closed.

Goodheart-Willcox Publisher

Figure 3-15. This Caterpillar integrated tool carrier (IT) uses parallel loader lift linkages.

the linkage that controls bucket curl forms the shape of a Z. See **Figure 3-16**.

Wheel loaders can be equipped with *self-leveling* controls that hold the work tool at the desired pitch as the loader frame is lifted and lowered. The operator does not have to readjust the position of the bucket while the loader is moving. Loader buckets can be kept level by means of mechanical linkage, hydraulic controls, or electro-hydraulic controls.

Loader frames can be equipped with quick couplers, enabling the bucket or work tool to be quickly removed and reinstalled. Some couplers require manual decoupling while others are hydraulically operated. Refer to **Figure 3-17**. Changing attachments on loaders without couplers requires pulling the attachment's pins, which takes longer than swapping attachments with a quick coupler.

A *static tipping load* is the amount of force that a loader must exert to lift the machine's rear tires off the ground. Manufacturers specify two static tipping loads on wheel loaders: *straight static tipping load* and *fully articulated static tipping load*. The straight static tipping load is the amount of load required to tip the loader when the loader is steered straight ahead, with zero degrees of articulation. The fully articulated static tipping load is the amount of load required to tip the loader when the tractor is fully steered to the right or left. It takes less load to tip a fully steered loader than one positioned straight ahead. Half (50%) of a wheel loader's static tipping load equals the wheel loader's *operating load*, as defined by the Society of Automotive Engineers (SAE). A loader's operating load accounts for dynamic loads placed on a loader while it is moving, which could potentially cause the loader to tip or roll over.

Goodheart-Willcox Publisher

Figure 3-16. Compared to parallel linkage, Z-bar linkage can improve operator visibility. Some manufacturers state it also provides better breakout force throughout the bucket's entire range of motion.

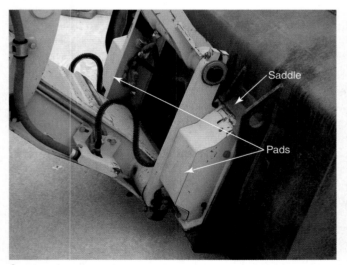 

A B

Goodheart-Willcox Publisher

Figure 3-17. Loaders are often equipped with a quick coupler, making it easy and fast to switch attachments. A—This quick coupler has a rectangular pad on both the left and right sides of the loader frame. The coupler is inserted into a rectangular saddle in the attachment (bucket). A pin is placed through the coupler, holding the bucket to the loader frame. B—This quick coupler requires the attachment to have a U-shaped saddle. The loader frame contains two upper pins, and they are inserted into the attachment's frame. As the loader is lifted, the bottom portion of the coupler will align and can then be pinned to the attachment. Normally, the pins are hydraulically operated.

Warning

A loader's operating load must be respected. The faster a loader is traveling and the more weight a loader contains, the easier it is for the loader to tip and roll. Always wear seat belts and travel slowly when carrying heavy loads.

The *skip loader* is a unique type of wheel loader. It is also known as a landscape tractor or an industrial loader. The machine resembles an agricultural utility tractor with a rear box blade. See **Figure 3-18**. It is based on a loader backhoe chassis but has a hydraulic box blade attached to a rear three-point hitch with hydraulic top and tilt function in place of a backhoe. The box blade's hydraulic controls include lift, lower, tilt to one side or optional both sides, tip forward and rearward. The box blade also has a multiple shank ripper that can be hydraulically lifted and lowered. The tractor is often an open cab design but can have an enclosed cab. The tractor can also be equipped with an agricultural style rear 540 rpm PTO. The loader requires factory rear counterweights due to the box blade being lighter than a traditional backhoe. The seat swivels, allowing the operator to gain a better rear view of the box blade. The machine provides a smoother ride than skid steers and can get in tighter spaces than a motor grader.

Sizes

Wheel loaders can range from compact all the way up to the Komatsu WE2350-2 mining loader that can be equipped with a 70 yd^3 bucket. A loader's bucket is rated in struck and heaped capacities. The *struck capacity* is the volume of material that it takes to fill a bucket in a level position after the excess material has been removed. The material is removed (or struck) by placing a straightedge across the bucket's cutting edge and the rear of the bucket. The *heaped capacity* is equivalent to the struck capacity plus the additional material that would be heaped on top of the bucket when the top plane of the bucket is parallel to the ground. See **Figure 3-19**. The *angle of repose* describes the slope of the bucket's dirt in relationship to the bucket's top plane. Manufacturers use a slope of 2:1 (or a 30° angle of repose) when calculating a machine's rated heaped capacity. However, in the field, the angle of repose will differ based on the type of material and the moisture content of the material.

Wheel loaders are available in a wide variety of sizes. See the *Appendix* for *Wheel Loader Sizes*.

Large mining loaders are available with four different lift options: standard lift, high lift, extended lift, and super-high lift.

The different loader heights provide mines with options for more lifting force and less height, or more height and less lifting force. A loader with an extended lifting height can load and place the dirt into the center of the dump bed of a taller truck. However, the loaders with a higher lift also have reduced lift capacity. Even though the bucket capacity is reduced for higher lifting capabilities, the bucket volume can be large if the material is not dense, for example coal.

Goodheart-Willcox Publisher

Figure 3-18. A skip loader has a box blade attached to a three-point hitch at the back of the machine.

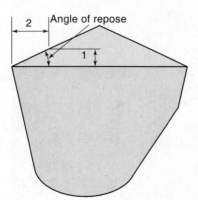

Goodheart-Willcox Publisher

Figure 3-19. The heaped bucket capacity of a wheel loader equals the amount of material the bucket can hold at a 2:1 angle of repose, which is a 30° angle from horizontal.

Note

In quarry and mining applications, loaders are matched to the size of the haul truck. A loader that is appropriately sized to match the haul truck will fully load the truck with three or four buckets, also described as three or four passes. Large mine trucks are commonly loaded with shovels that can fully load the truck in four passes or fewer.

Track Loaders

Track loaders are easily confused with dozers. Both can be configured with low oval shaped crawler undercarriages and have large structures in front. However, the structure at the front of a track loader is not a blade, but a bucket, and the bucket has a higher dump height than a dozer's blade. Track loaders are designed to operate on more stable ground than a dozer but can traverse across loose soils better than wheel loaders. They use a rigid frame, as compared to some dozers that use pivoting undercarriages, which will be explained in Chapter 23, *Undercarriages*.

Controls

Like wheel loaders, track loader's common hydraulic controls are:
- Loader lift.
- Bucket curl.
- Loader auxiliary (for a grapple or broom).

The propulsion and steering of track loaders are different from that of wheel loaders. Wheel loaders normally use articulated steering and a traditional powershift or hydrostatic transmission. Modern track loaders use dual-path hydrostatic transmissions for both propulsion and steering. A ***dual-path hydrostatic transmission*** consists of one hydraulic pump and one hydraulic motor to drive each track. Older track loaders used a complex series of countershafts to provide the loader's propulsion and steering. An example of an older track loader steering and propulsion system is shown in Chapter 24, *Track Steering Systems*.

Track Loader Applications

Track loaders are used in a wide variety of applications, including the following:
- Excavating basements.
- Demolition work.
- Landfill operations.
- Waste handling.
- Excavation work in areas with limited overhead clearance, such as under overhead power lines, under a low-lying bridge, or in a tunnel.
- Steel track loaders can be used for loading hot metal slag in steel mills where wheel loaders cannot be used.

The SAE rated operating load for track loaders is 35% of the static tipping load. A track loader's static tipping load is the amount of force that a loader exerts to lift the rear of the machine's tracks off the ground.

Track Loader Types

Manufacturers offer two types of track loaders: rubber-track loaders and traditional steel-track loaders, also called crawler loaders. Rubber-track loaders can be labeled as mini skid steers, compact track loaders (CTLs), or multi-terrain loaders (MTLs). Rubber-track loaders are very similar to skid steers. They use the same type of controls and normally have dual-path hydrostatic drive transmissions. Skid steers use four rubber tires for propulsion and track loaders use two rubber belts. Chapter 17, *Final Drives*, will explain the differences in the final drives that are used in skid steers and rubber-track loaders. Rubber-track loaders are built by numerous manufacturers. Refer to **Figure 3-20**.

Goodheart-Willcox Publisher

Figure 3-20. This Bobcat T650 rubber-track loader is rated at 74 hp (55 kW) and has an operating weight of 9113 pounds.

Figure 3-21. This Ditch Witch SK 750 rubber-track loader is powered by a Kubota 24.8 hp diesel engine and weighs 2890 pounds.

Some track loaders are so small that the operator stands on the back of the machine or walks behind the loader instead of sitting in a traditional operator's cab. The loaders allow operators to easily travel through a homeowner's walk-through gate, which allows operators to gain access to a backyard without tearing down a fence. Refer to **Figure 3-21**.

A limited number of manufacturers (Caterpillar, John Deere, Liebherr, and Zoomlion) continue to manufacture traditional steel track loaders. See **Figure 3-22**. Like wheel loaders, track loaders can also be configured with a parallel loader linkage or a Z-bar linkage.

Sizes

Track loaders are available in a variety of weights and horsepower. They range in size from small walk-behind loaders to large 300 hp steel track loaders. See the *Appendix* for *Track Loader Sizes*.

Dozers

Dozers are equipped with a heavy-duty blade, which is commonly used for pushing things like another tractor (scraper), trees, or excavating dirt. The dozer can be used for ripping, rough leveling, finish grading, and winching. See **Figure 3-23**.

Controls

The most common type of dozer controls are propulsion, steering, and blade controls. Some dozers, such as hydrostatically propelled dozers, integrate propulsion with steering. Other dozers, such as powershift-propelled dozers, separate the propulsion and steering. These dozers use differential steering or steering clutches and brakes.

The three most common types of dozer steering systems are:
- Dual-path hydrostatic drive.
- Steering clutch and brake.
- Differential steer.

Figure 3-22. Traditional track loaders are configured with an oval undercarriage, giving the machine a low center of gravity.

Figure 3-23. This Caterpillar elevated sprocket dozer is equipped with a full rollover protective structure (ROPS) and a ripper. The blade's push arms are located outside the machine's tracks.

Chapter 24, *Track Steering Systems*, explains several different types of dozer steering systems. Chapter 9, *Planetary Gear Set Theory*, details a traditional Caterpillar dozer three-speed powershift transmission that is used with steering clutch and brake systems or with differential steering systems. Although dozers appear to be quite simple, they can use different types of controls.

Blade Lift

Dozers have ***push arms*** that connect the tractor to the dozer blade and can be lifted and lowered. They transfer the force of the tractor to the blade. Traditional dozers use heavy-duty push arms on the outside of the undercarriage. See **Figure 3-23**.

Smaller dozers used in finish grading applications often have a narrow U-shaped push arm frame inside the undercarriage. See **Figure 3-24**. Most dozers use two cylinders for lifting the push arms, but some dozers can be configured with just one lift cylinder, for example the Caterpillar D7E electric drive dozer. See **Figure 3-25**.

Figure 3-24. Hydraulic cylinders on this Caterpillar D5H dozer can angle the blade either by pushing the left side of the blade forward and pulling the right side rearward or by pushing the right side forward and pulling the left side rearward. The dozer also has hydraulic cylinders that can lift and tilt the blade.

Pitch

Some dozers are equipped with two hydraulic cylinders between the push arms and points high on the blade, **Figure 3-25**. These cylinders allow the operator to control the blade's ***pitch*** (angle of the blade as it is rotated around a horizontal axis) by simultaneously retracting or extending the pitch cylinders. There are three angles of blade pitch. When the cylinders are fully extended, the blade is in the dump position. When the cylinders are fully retracted, the blade is in the carry position. In the dig position, the blade is tipped forward but not fully forward like the dump position. Other dozers have a mechanical adjustment, a threaded turnbuckle for example, that provides a few degrees of pitch variability. See **Figure 3-26**. The way a dozer controls blade pitch is similar to the way a loader dumps and curls its bucket, but with much less arc.

Tilt

Some dozer blades can be actuated downward to the left or to the right. This is referred to as ***blade tilt***. Blade tilt can be a mechanical adjustment or a hydraulic adjustment. If the dozer is equipped with two hydraulic pitch cylinders, one cylinder is extended and the other is retracted to tilt the blade. See **Figure 3-25**. If the dozer in **Figure 3-25** had a hydraulic cylinder between the one push arm and the blade and a rigid arm in place of the other pitch cylinder, the sole hydraulic cylinder would actually be a tilt cylinder and not a pitch cylinder. Other dozer blades have a single hydraulic tilt cylinder horizontally located behind the blade,

Figure 3-25. This Caterpillar D7E electric drive dozer uses a single hydraulic cylinder to lift the dozer's blade.

Figure 3-26. A mechanical turnbuckle is sometimes used to manually adjust the blade's pitch.

as shown in **Figure 3-24**. This tilt cylinder attaches to a central pedestal on the push arm frame and to an outer edge of the blade. Extending or retracting the cylinder changes the tilt of the blade.

Angle

The dozer in **Figure 3-24** has two cylinders located on top of the U-shaped push arm frame. These cylinders provide an angle function, which allows the operator to push either the right or left side of the blade forward. A dozer with a blade that can hydraulically lift, angle, and tilt is known as a *PAT dozer* for Power Angle Tilt. PAT dozers are used for finish grading work, backfilling, and spreading. PAT dozers are sometimes referred to as having a *six-way blade* because the dozer blade can move in six separate directions:

- Lift.
- Lower.
- Angle forward on the right.
- Angle forward on the left.
- Tilt right.
- Tilt left.

Rear Implements

Dozers can also be equipped with rear implements, such as a hydraulically controlled winch, drawbar, or ripper.

Applications

Dozers are used for a number of different applications. Dozing applications include clearing fields, rough leveling, and finish grade work. The following are examples of specific dozer applications:

- Slot dozing—cutting a slot the width of the machine and capturing the dirt in the slot. When a second pass is made, the side walls of the slot help trap the material cut by the blade, allowing the operator to move more material than if the dozer was not working inside a slot. The maximum slot depth should be no higher than the top of the blade.
- Straight dozing—making a straight and level cut and then carrying away the material to dump later. The material is not trapped in a slot.
- Side cutting—using the left or right end (side) of the dozer blade to cut into a hill or berm. The dozer blade can be set so that it simultaneously rough levels the material that is side cut.
- Back filling—pushing material dumped from a truck into a ravine to be filled.
- V-trenching—tilting the blade and dozing a V-shaped trench.

A variety of dozer blades are available to perform the aforementioned applications:

- Straight blade (S blade)—type of blade used in heavy dozing applications with soils that are difficult to cut.
- Universal blade (U blade)—a blade with ends that extend forward, allowing the blade to hold more material. This type of blade used with lighter materials that are easier to push, such as waste material, coal, or loose material.
- Semi-U blade—a blade with edges that extend forward more than those on an S blade but less than those on a U blade.
- Cushion blade—a blade that has a wear-resistant liner plate above the blade's cutting edge that spans the length and width of the blade. The cutting edge is reinforced. The cushion blade is used for pushing scrapers. Overall, it is a narrow-width blade.
- Carry dozer blade (CD blade)—a blade with ends that extend forward similarly to those on a U blade, but farther. This type of blade is used on a Caterpillar D11T CD (carry dozer), which is commonly used in coal applications.

Additional dozer applications include ripping, pulling an implement, and pushing a scraper. Ripper designs can vary from having multiple lightweight shanks for light-duty applications to having a single heavy-duty shank for difficult ripping applications. Although it is rare, dozers can use a drawbar to pull an implement. Pusher dozers are used for pushing open bowl scrapers. The dozer only pushes the scraper while the scraper is making its cut. The dozer is especially helpful if the soil conditions are slick, if the soil is difficult to cut, or if the scraper is making a deep cut.

Types

Dozers can be classified as wheel dozers and track dozers. Wheel dozers are available for high-horsepower applications, such as mining operations. They should be used only when footing conditions are stable. Wheeled dozers have lower operating and ownership costs than track-type dozers, and their travel speeds are close to three times as fast as those of track-type dozers. Wheel dozers normally use articulated steering. See **Figure 3-27.**

The majority of track-type dozers have steel tracks, but a dozer can also have rubber tracks. The John Deere 764 articulated four-track high-speed dozer is an example of a dozer that uses rubber tracks. See **Figure 3-28**.

Sizes

Dozers are manufactured in a wide variety of sizes, ranging from small dozers up to mining dozers. See the *Appendix* for *Dozer Sizes*.

Motor Graders

Motor graders, sometimes called blades or graders, are commonly used for snow removal, maintaining rock roads, or finish work such as preparing roadways and parking lots for paving. A motor grader uses a moldboard-shaped blade that is mounted to

abutyrin/Shutterstock.com

Figure 3-27. Wheel dozers, like traditional track dozers, use two push arms for securing the blade to the dozer. Wheel dozers provide much faster travel speeds than traditional track-type dozers.

Goodheart-Willcox Publisher

Figure 3-28. John Deere's 764 high-speed dozer uses four rubber tracks. It provides both high travel speeds and good traction.

a circle. The circle consists of an internal-toothed ring gear that is rotated by a gearbox, allowing the operator to rotate the position of the blade. The circle is attached to the front portion of the motor grader's frame by a T- or V-shaped drawbar. See **Figure 3-29**. The drawbar is attached to the frame by a ball socket, which allows the circle drawbar assembly to be hydraulically actuated (lifted and lowered, or shifted side-to-side). In addition to the moldboard blade, a motor grader can also be equipped with a dozer blade on the front of the grader. See **Figure 3-30**.

Figure 3-29. This motor grader uses a V-shaped drawbar that attaches the circle to the front of the motor grader's frame. Notice the scarifier that is located directly behind the front axle.

Figure 3-30. Front-mounted dozer blades are sometimes installed on graders. The dozer blade is useful for working in tight quarters, such as in a parking lot, because it allows the operator to move dirt out of the corners. If the grader is not equipped with a dozer blade, a second tractor, such as a loader, would be required to finish the parking lot's corners.

Controls

A motor grader is one of the most hydraulically complex machines used in the construction industry. Operators must be highly skilled and trained to maximize a motor grader's functions. Common motor grader hydraulic functions include the following:

- Blade right lift or lower.
- Blade left lift or lower.
- Circle turn left and right.
- Circle shift to the left.
- Circle shift to the right.
- Blade shift right or left.
- Blade tip forward or rearward.
- Wheel lean right or left.
- Front wheel steer.
- Articulation steer.
- Scarifier lift or lower.
- Ripper lift or lower.

In older motor graders, the operator stands while operating the tractor and actuates a series of levers, also known as an antler rack, and a steering wheel. See **Figure 3-31**. In modern motor graders, an operator remains seated in a comfortable seat with arm rests to perform the same functions using a pair of joysticks. Each joystick has multiple controls integrated into it, **Figure 3-32**. Caterpillar's late-model GC-series motor grader has an antler rack and steering wheel that can be positioned toward the operator to allow the operator to remain seated.

Many motor graders use a hydraulic system with post-spool compensation, also known as *flow sharing*. A flow-sharing hydraulic system, when the operator requests more oil than the pump is capable of delivering, the hydraulic system proportions the available hydraulic pump flow to each of the commanded hydraulic functions based on the amount of oil requested for each function.

Applications

Motor graders are used in a variety of applications, including the following:

- Maintenance of roads, parking lots, and runways.
- Leveling roads.
- Crowning roads.
- Spreading fill.
- Forming ditches.
- Forming banks.
- Removing snow.
- Scarifying and ripping hard soils and asphalt.

Some motor grader applications can be performed at faster travel speeds than other applications. For example, maintaining a rock road can be performed at higher road speeds than performing a finish grade prior to paving a parking lot.

Figure 3-31. Some motor graders use a series of levers that actuate the machine's hydraulic control valves.

A

B

Figure 3-32. Caterpillar M series motor graders use a left and right joystick, which replaces the old manual lever operated controls and steering wheel. A—The left joystick controls the forward and reverse propulsion, powershift transmission upshift and downshift, left side blade lift and lower, wheel lean, articulation steer, articulation center, and steer right or left. B—The right joystick controls right blade lift and lower, blade shift left and right, blade tip forward and rearward, and circle shift right and left.

Types

Older, smaller motor graders can have a fixed frame with only front-wheel steer. Modern motor graders have both front-wheel steer and articulation steer. The combination of both steering functions enables the grader to turn in a tighter turning radius. If the front axle is steered in the opposite direction of the articulation, the grader's front frame can be placed on one side of a windrow while the blade can direct and smooth the windrow without either the front or rear wheels having to cross or ride on top of the windrow. Chapter 25, *Wheeled Steering Systems,* further explains combination steering.

The motor grader normally has rear-wheel drive, powered by a powershift transmission. Each side of the tractor is driven by a chain drive system that connects the differential to the two drive wheels on that side. Because the chain drives on each side of the grader each consist of two drive sprockets, two chains, and two driven sprockets, they are often called the **tandems**. This type of drive system is further explained in Chapter 17, *Final Drives.* A lockable differential delivers power to the left and right tandems. The differential allows the inside drive wheels to be driven at a slower speed than the outside wheels. Differentials will be explained in Chapter 16, *Differentials.*

Some motor graders use hydrostatic drives to power the front wheels, enabling the grader to have six-wheel drive. Each of the front hydraulically driven wheels can be powered independently from the other. As a result, when the vehicle is turning, the inside drive wheel can be driven at a slower speed than the outside wheel to prevent skidding.

Sizes

Manufacturers of motor graders include Case, Caterpillar, John Deere, Komatsu, LeeBoy, Mauldin, NorAm, and Volvo. Motor graders are available in a wide variety of sizes, ranging from small up to mining motor graders. See the *Appendix* for *Motor Grader Sizes.*

> **Note**
> In quarry and mining applications, a haul road should be approximately four to five times the width of the haul truck. Mining motor graders have a blade that is 24 feet wide to allow them to prepare a haul road with a minimal number of passes.

Scrapers

Scrapers are a class of construction equipment designed to scrape up material, hold it so it can be transported to a different location, and then dump and level the materials. Scrapers contain a box assembly with rigid side walls that form a bowl. A cutting edge is installed on the front lower edge of the bowl. **Figure 3-33** shows an ***open bowl scraper*** design, which has a movable ***apron*** that forms the front wall of the bowl when it is lowered. As the operator hydraulically lifts the apron, it rotates upward, opening the front of the bowl. The operator then lowers the bowl assembly so the scraper can cut material as the bowl is pulled forward by the draft arms. As the scraper is pulled forward, the blade cuts a swath of material the width of the bowl and scoops it into the bowl. Operators can control the depth of the cut by partially opening or closing the apron and lowering or raising the bowl. When the bowl is fully dropped and the apron is opened an equal distance, the scraper cuts at its maximum depth. If the apron is closed slightly and the bowl is raised slightly, the scraper will scrape only several inches of material.

Scrapers can be towed by another tractor or can be powered by their own engine. They are the only type of construction machine that can self-load material, haul it to a faraway location, and then self-unload. Note that many open bowl scrapers are loaded with the assistance of another machine, which will be discussed later in this chapter.

Open bowl scrapers have a hydraulically actuated ***ejector*** at the rear of the scraper's bowl. When the operator is ready to unload, the apron is first fully lifted, and then the ejector is pushed forward by a telescoping hydraulic cylinder. The scraper continues to travel as it is unloaded, and the cutting edge levels the dumped material.

Note

If an inexperienced operator forgets to retract the ejector and begins loading the bowl, the dirt may flow over the top of the ejector and fall behind it. If this happens, the material must be manually removed using a shovel.

Controls

The following are the most commonly used functions in an open bowl scraper:
- Bowl lift and lower.
- Apron lift and lower.
- Ejector extension or retraction.

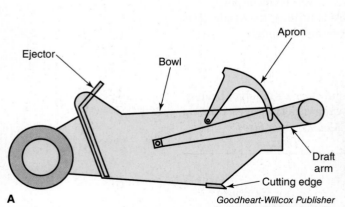
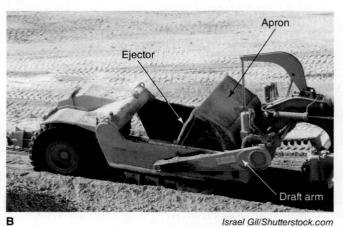

Figure 3-33. Open bowl scrapers contain an apron that forms the front wall of the bowl, a rear ejector, a cutting edge, and draft arms. A—The basic parts of an open bowl scraper. B—This open bowl scraper is unloading. Note that the ejector is all the way forward and the apron is fully open.

If the scraper is a tractor-type scraper, it will have an automatic transmission. Tractor scrapers have a high-arching gooseneck hitch that allows the tractor to make sharp turns, up to 90°.

Applications

Scrapers are used in applications where a lot of material needs to be moved a long distance. Scrapers are frequently used for hauling dirt while building highways or creating large man-made lakes and reservoirs. Scrapers can be used in large residential construction projects, such as preparing land for a new housing subdivision, and for preparing large commercial job sites. Scrapers are also used in mines for hauling coal. Coal bowl scrapers have a much larger capacity because coal is much less dense than traditional soils.

Types

Scrapers can be classified as open bowl or elevating. As previously mentioned, open bowl scrapers have a movable front wall, known as an apron. Some open bowl scrapers are configured with a vertical auger in the center of the bowl. The auger helps lift the material into the bowl. See **Figure 3-34**. The ejector in this design has flared corners, similar to wings. The ejector corners extend forward so that, when the ejector is extended, the corners take up the space between the auger and the side walls. The ejector extends only partially and stops when it reaches the auger. The auger is driven while the scraper is unloading.

Elevating scrapers have an elevator assembly that consists of steel flighting mounted on chains. A hydraulic motor drives the elevator, causing the elevator to direct material into the bowl of the scraper. See **Figure 3-35**.

When elevating scrapers unload, the ejector is extended halfway forward until it approaches the elevator. The floor is also retracted and the elevator is reversed to unload the bowl.

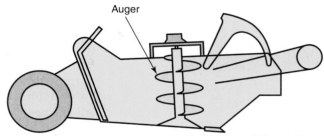

Goodheart-Willcox Publisher

Figure 3-34. A single vertical auger is sometimes placed in an open bowl scraper. The auger helps load the bowl by lifting material vertically and distributing the material throughout the bowl.

 Caution
Pusher dozers should not be used to assist loading of elevating scrapers, as the dozer can harm the elevator's chain and flighting.

Scrapers are classified as either tractor scrapers or towed scrapers. Tractor scrapers have their own engine and power train for propulsion and operation. Towed scrapers are pulled with other machines, often large four-wheel-drive agricultural tractors. See **Figure 3-36**. The large agricultural tractors commonly pull two or three towable scrapers in series. Small towable scrapers can be pulled by row crop agricultural tractors, also known as mechanical front-wheel-drive (MFWD) tractors.

Billy Gadbury/Shutterstock.com

Figure 3-35. Elevating scrapers use chain-driven bars, known as flighting, to load the scraper. Normally, the elevator is hydraulically driven.

Figure 3-36. A towable scraper is normally pulled by the drawbar of an agricultural tractor. Multiple towable scrapers can be pulled by one tractor.

Note

Although most towable scrapers are open bowl, Miskin manufactures a towable elevating scraper. Miskin also manufactures an *air-assisted scraper*. Air directed through the floor of this towable scraper loosens the dirt, helping the scraper load more dirt.

There are several different classes of tractor scrapers:
- Single engine.
- Twin engine (also known as tandem scrapers).
- Push-pull.

Single engine scrapers have the engine in the front frame. The most common type of single engine scraper is the elevating scraper.

Twin-engine scrapers have an engine, transmission, axle, and differential at the front and rear of the machine. Tandem scrapers often use two separate throttle pedals for controlling the two engine speeds. The front engine is called the tractor engine and the rear engine is called the scraper engine.

Both the front and rear engines deliver power to their dedicated torque converter, transmission, and differential axle assembly. The tractor engine has more horsepower than the scraper engine. The tractor engine drives the hydraulic pumps to control the steering, bowl operations, cushion hitch, and push-pull bail controls. If the machine uses air brakes, the compressor is driven by the tractor's engine.

A Caterpillar 637K tractor transmission has eight forward speeds and one reverse speed, while the scraper transmission has only four forward speeds and one reverse speed. Both power trains are used for scraping and climbing steep inclines. The tractor's eight-speed transmission and higher-horsepower engine provide fast road speeds, up to 34.7 mph (55.8 km/h). The tractor's transmission can be shifted manually or automatically, while the scraper's transmission can only be shifted automatically and is controlled by the tractor-scraper control system.

Warning

If a tandem scraper turns a sharp corner and the operator accidentally accelerates the rear engine by itself, it could cause the tractor to roll.

Push-pull scrapers are designed to work in pairs. The machine has a hydraulically actuated bail at the front of the machine and a push plate and hook at the rear. Two or more scrapers connect to each other through the bail and hooks. See **Figure 3-37**. As the scrapers enter the cut, one scraper will load while the other scraper will push. After the first scraper is loaded, it will pull the rear scraper as it loads.

Both towed scrapers and tractor scrapers may have a *cushioned hitch*. A cushioned hitch is a hitch design that improves the life of the hitch by reducing the shock loads placed on it. The use of a cushioned hitch also reduces operator fatigue.

A cushioned hitch consists of one double-acting hitch-leveling hydraulic cylinder, two nitrogen gas accumulators, a hitch-leveling valve, and an electronic control module. During transport, the two accumulators cushion the cylinder, dampening the hitch. The operator locks out the cushioned hitch while the scraper is loading the bowl to ensure that the cutting edge remains in a fixed position. Once the bowl is loaded and raised, the cushioned hitch can be reactivated to provide a smoother ride.

Byron W. Moore/Shutterstock.com

Figure 3-37. Push-pull tractor scrapers have a push plate and hook at the rear of the machine and a bail at the front. The bail latches into the hook of another scraper.

Sizes

The smallest types of scrapers are towable, with bowl capacities ranging from 5.5 yd^3 to 26 yd^3. Larger scrapers are self-propelled, with bowl capacities up to 44 yd^3. See the *Appendix* for *Tractor Scraper Sizes*.

Haul Trucks

Like scrapers, haul trucks are off-road vehicles used for hauling material long distances, often 2000 feet or more. However, unlike scrapers, haul trucks are not capable of self loading and must be loaded by a loader, excavator, or shovel. See **Figure 3-38**.

 Note

Many mining haul trucks have a dump bed that extends over the operator's cab, serving as the FOPS, as explained in Chapter 1, *Shop Safety and Practices*.

Gingerss/Shutterstock.com

Figure 3-38. Haul trucks are commonly used in mining applications and are often loaded with rope shovels.

Figure 3-39. This mining haul truck's transmission shift lever is to the left of the truck's dump control lever, which is also known as a hoist lever.

Controls

Like traditional automotive trucks, haul trucks have an automatic transmission, front steerable axle, rear drive axle, and parking and service brakes. A mining haul truck's transmission shift control lever is shown in **Figure 3-39**. Haul truck transmissions can have six to nine forward speeds and two to three reverse speeds.

The truck's dump bed, also known as a *hoist*, is hydraulically raised and lowered, allowing the load to be dumped. The bed is normally lifted by two telescoping hydraulic cylinders. Typically, the hydraulic cylinders can be commanded to perform three functions: lift, lower, and hold.

Note

Articulated dump trucks can be equipped with an ejector similar to that in an open bowl scraper. The ejector enables the truck to dump without raising the bed.

Caterpillar mining trucks use an electronic feature called a *body-up reverse neutralizer*. It prevents the operator from backing the truck up while the bed is raised. Any time the operator shifts the transmission to reverse with the bed raised, the transmission will automatically be shifted to neutral.

Some haul trucks incorporate the use of a retarder. *Retarders* are devices that help slow the machine and reduce brake wear. A retarder may consist of a rotor inside a fluid-filled housing. The rotor is driven by the axle and must move through the fluid in the retarder housing as it rotates. The resistance provided by the fluid is used to slow the truck. Some manufacturers use the electric drive or service brakes to retard the truck. Retarders are explained in Chapter 12, *Hydrodynamic Drives*.

Applications

Haul trucks are found on most residential and commercial construction job sites and are used for hauling rock, dirt, clay, and other soils. Mines use haul trucks to haul away the *overburden*, or *spoil* (material that is located above the mined deposits), and the mining deposits.

Trucks can be equipped with a variety of different types of beds, depending on the application. Three examples of mining beds include:

- Coal body—the largest volume bed placed on a truck. It has a large volume because it is used to carry low-density materials, like coal.
- Taconite body—the smallest volume bed placed on a truck. It has a small volume because it is used to carry high-density materials, like iron ore.
- Oil sand body—the truck's exhaust is plumbed through the bed so the oil sand stays warm. This allows the product to flow more freely while being dumped.

Beds may be equipped with *liners*, which are steel plates that help protect the walls or floors of the bed. A liner placed in the center of the floor of the bed is helpful when a shovel loads the truck. Liners on the walls are useful if the truck is being loaded with a loader, which can cause wear to the walls of the bed.

Types

The two most common types of haul trucks are the rigid frame and articulating truck.

Rigid-Frame Haul Trucks

Rigid-frame haul trucks have a solid, unarticulated frame and usually have a single set of dual rear wheels and a pair of front-steer wheels. Many rigid-frame mining trucks have *rock ejectors*, which typically consist of steel bars fastened to the bed of the truck and extending down through the gap between the dual tires. See **Figure 3-40**. The ejectors prevent large rocks from getting wedged between the dual tires. Because of the size and travel speeds of the trucks, a large wedged rock could dislodge and be launched several hundred yards with lethal consequences. Properly maintaining haul roads and training operators to steer away from large rocks helps minimize the need for rock ejectors.

Mining trucks can use a mechanical drive, consisting of a planetary-gear automatic transmission, or can be electrically driven. Electric drive trucks can be powered solely by an onboard diesel-engine-driven generator that sends electrical current to electric drive motors, **Figure 3-41**, or they can receive additional assistance through overhead power lines. This system of providing additional electric power to the truck through overhead power lines is called *trolley assist*. Trucks that must travel long uphill grades can benefit from trolley assist. It increases the truck's travel speed and horsepower, decreases fuel consumption, and lengthens the life of the engine. When the truck connects to the power lines, the engine is reduced to an idle. Trucks that use trolley assist must be specifically designed by the manufacturer for those applications.

Andrey B Bannov/Shutterstock.com

Figure 3-40. Rock ejectors prevent rocks from becoming wedged between the dual tires.

Manufacturers of rigid haul trucks include Caterpillar, Hitachi, Komatsu, Liebherr, and Volvo.

 Note

BelAZ produces a 500-ton payload rigid frame dump truck that uses two 2300-hp diesel engines, which each drive an electric generator. The truck is all-wheel drive and has dual wheels at both the rear drive axle and the front steering drive axle. It is manufactured in the European country of Belarus.

Goodheart-Willcox Publisher

Figure 3-41. This General Electric (GE) wheel motor is driven by alternating current (AC). It is used on 240-ton haul trucks.

Articulated Dump Trucks

Articulated dump trucks (ADTs) commonly use two rear drive axles equipped with single wheels and are frequently equipped with a front drive axle, resulting in six wheels that are all powered (6×6). See **Figure 3-42**. ADTs use articulation steering, which is explained in Chapter 25, *Wheeled Steering Systems*. The truck can also be equipped with automatic traction control and differential locks to limit wheel slip. ADTs have travel speeds up to 33 mph. Manufacturers of articulating dump trucks include Bell, Caterpillar, John Deere, Komatsu, Rokbak, and Volvo.

Crawler Dump Beds

Terramac and Morooka produce track-type dump trucks. The rubber tracks provide greater traction than wheeled trucks in soft, loose, or soggy terrain. Some models combine a rotating house and dump bed above a crawler frame. See **Figure 3-43**.

Sizes

Haul trucks are normally rated in three categories: rated capacity, struck volume, and heaped capacity. See the *Appendix* for *Haul Truck Sizes*. Note that the US Customary unit of weight is a short ton, equaling 2000 pounds. A metric ton equals 2204 pounds.

On-Highway Dump Trucks

Some construction companies use on-highway dump trucks to transport materials from a worksite to a dumping site. On-highway dump trucks are available in a variety of configurations to meet the demands of different jobs, **Figure 3-44**. Examples include the following:
- Conventional 10-wheel dump truck, also called an end-dump.

Figure 3-42. Articulated dump trucks are often equipped with six-wheel drive and use articulated steering.

Figure 3-43. This Morooka crawler has a rotating house equipped with a dump bed.

Figure 3-44. On-highway dump trucks are also commonly used on construction sites. A—10-wheel dump trucks are common in residential construction. B—Tractor-trailer vertical-dump trucks are used for hauling more material longer distances. Notice the truck requires lots of overhead clearance in order to dump the trailer. C—Tractor-trailer side-dump trucks can dump in areas with very little overhead clearance.

- Tractor-trailer vertical-dump trailer.
- Tractor-trailer side-dump trailer.
- Tractor-trailer belly-dump trailer.
- Tractor-trailer belt-dump trailer.

Skid Steers

Skid steer loaders (SSLs) are compact self-propelled loaders, and sometimes called uniloaders. Skid steers use dual-path hydrostatic transmissions to steer and propel the loader. Each hydraulic motor is responsible for driving two tires on one side of the loader. See **Figure 3-45**.

Because both wheels on each side of the skid steer are driven by the same hydraulic motor, they rotate at the same speed in the same direction. The term skid steer is derived from the way the machine steers. Since the wheels on the loader do not pivot, they must skid left or right as the vehicle turns. For example, if the right drive wheels are rotating forward and the left drive wheels are rotating backward, the front wheels will skid to the left and the rear wheels will skid to the right. This will cause the loader to turn to the left.

Manufacturers design the loader's weight distribution to make it easier for the machine to pivot on the front wheels when the bucket is loaded and on the rear wheels when the bucket is empty. For example, one manufacturer designs the skid steer to have 70% of the machine's weight placed on the rear wheels when the loader's bucket is empty. This causes the machine to pivot on the rear wheels when steering with an empty bucket. When the bucket is loaded, the front wheels have 70% of the machine's weight, causing the machine to pivot on the front wheels. See **Figure 3-46**.

Skid steers are some of the most difficult machines to repair or service because they are very compact, with little clearance between the components. Some manufacturers attempt to alleviate these problems by keeping service in mind as they design new models.

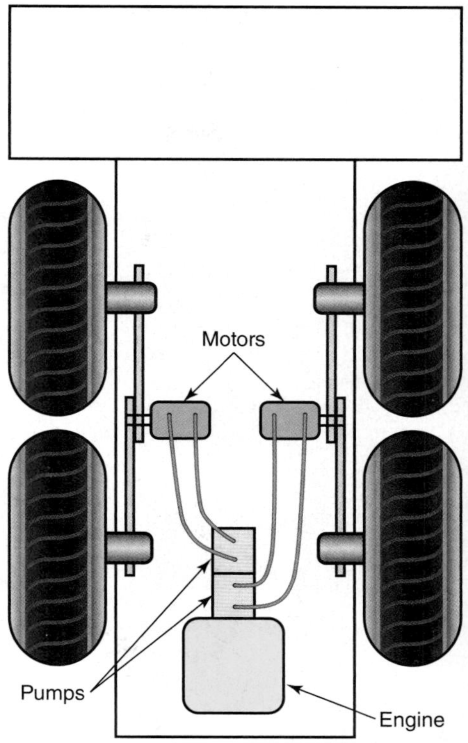

Figure 3-45. Skid steers use one hydraulic pump and one hydraulic motor to drive one side of the loader.

Controls

Older skid steers have two propulsion levers, one on the right side of the operator and one on the left side of the operator. When both levers are moved forward an equal distance, the loader moves forward. When the levers are both moved rearward an equal distance, the loader moves backward. When one lever is moved forward and the other is moved rearward,

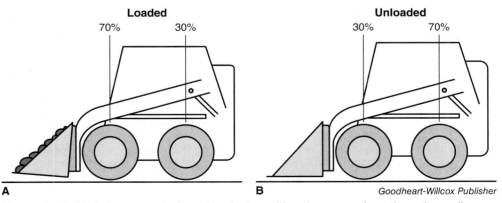

Figure 3-46. Skid steers are designed to pivot on either the rear or front tires, depending on whether the bucket is empty or loaded. A—Weight distribution on a loaded skid steer. B—Weight distribution on an unloaded skid steer.

the loader steers toward the side of the rearward lever. The distance the lever moves determines the speed of the hydraulic motor, which causes the machine to move or turn faster.

Older skid steers use foot controls to operate the loader, bucket, and auxiliary hydraulics:

- Left foot pedal tipped forward lowers the loader.
- Left foot pedal tipped rearward lifts the loader.
- Right foot pedal tipped forward dumps the bucket.
- Right foot pedal tipped rearward curls the bucket.
- Center foot pedal controls the auxiliary hydraulics.

Late-model skid steers use joystick controls for operating the loader, bucket, propulsion, and steering. See **Figure 3-47**. Some machines are equipped with enhanced electronic controls. For example, Bobcat offers their selectable joystick control (SJC), which allows the operator to electronically control four programmable features:

- Speed management—allows the engine speed to be set independently from the loader's travel speed. Some hydraulic attachments, such as trenchers, planers, mowers, and snowblowers, might require a higher engine speed, but the operator might want a slower travel speed.
- Horsepower management—prevents the engine from stalling when the loader is pushing into a difficult load.
- Drive response—establishes how fast the loader's propulsion will react based on how far the joystick is actuated.
- Steering drift compensation—enables the loader to send more oil to either the left or right hydraulic motor, which is useful when using offset attachments. Offset attachments place a larger force on one side of the machine.

Note

Some late-model skid steers can be operated with remote controls, which can reduce risks to operators by keeping them out of hazardous areas.

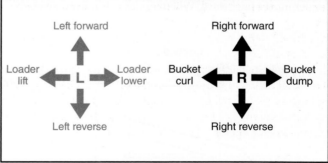

Figure 3-47. Skid steer controls use left and right side joysticks. The joysticks are detented to a neutral position. When the operator releases the levers, the machine quits moving and the hydraulic functions stop. A—The ISO skid steer pattern uses the left joystick to control propulsion and steering and the right joystick to control the loader. B—The H pattern uses both left and right joysticks for propulsion and steering. The left joystick controls the loader lift and lower. The right joystick controls the bucket curl.

Applications

Skid steers are used in a wide variety of applications, including residential construction, commercial construction, and farm and ranch applications. Although skid steers are used for smaller jobs than excavators, some large skid steers have been used to excavate basements.

Skid steers have more attachments available than any other mobile machine. Examples of skid steer attachments include augers, backhoes, blades, buckets, bale spears, brooms, cement mixers, hydraulic hammers, log grapples, mowers, pallet forks, planers, rakes, scrap shears, snowblowers, tree spades, tree shears, trenchers, tillers, and vibratory rollers.

Types

Most skid steers are equipped with four drive wheels. Some manufacturers offer a set of field-attachable steel tracks. As mentioned earlier in the chapter, rubber-track loaders are very similar to skid steers and have similar controls. However, the differences between the machines extend beyond the fact that one is equipped with tires and the other is equipped with tracks. Rubber-track loaders use a final drive gear set and track drive sprocket at the rear of the machine. Skid steers have a centrally located hydraulic motor that operates two drive chains to drive each side's wheels. Chapter 17, *Final Drives*, will explain both types of final drives.

Bobcat manufactures a unique all-wheel steer loader that is different from traditional skid steers in several important ways. Unlike the wheels on a traditional skid steer, both the front and rear wheels on an all-wheel steer loader can pivot to steer the machine. Also, an all-wheel steer loader does not use drive chains. Instead, each drive tire is independently and hydraulically controlled. The operator can select the traditional skid steer mode or the all-wheel steer mode. When in the all-wheel steer mode, the loader pivots the rear tires in the opposite direction of the front tires in order to provide a sharp turning radius. Although the all-wheel steering mode does not steer the loader as sharply as the traditional skid steer mode, it does significantly reduce tire wear and is much less destructive to lawns and turf.

Skid steers are available with two different lift arm designs, radius lift path and vertical lift path. As the term implies, a ***radial lift*** design causes the lift arms to move through a pronounced arc as they are lifted and lowered. The radial lift design is good for digging and provides an extended reach at mid-level heights. It is good for loading and unloading flat-bed trucks, backfilling, and dumping material over a wall. A ***vertical lift*** design arranges the hydraulic cylinders and lift arms so the arms move in a very shallow arc. This gives the loader more reach when the load is fully raised. The load is kept closer to the machine, making the vertical lift design a good choice for loading tall trucks.

The majority of skid steers use dual lift arms. JCB and Volvo offer skid steers with a single lift arm. The single-lift-arm design improves safety and visibility and allows the operator to enter the machine without having to crawl over the bucket and loader frame. See **Figure 3-48**.

Sizes

Skid steers vary in size, power, and lifting capacity. They range from small walk-behind models that lift less than a thousand pounds to large models capable of lifting more than a ton and a half. See the *Appendix* for *Skid Steer Sizes*.

Art Konovalov/Shutterstock.com

Figure 3-48. JCB and Volvo manufacture a single-lift-arm skid steer, which improves visibility and safety and makes it easier to enter the machine.

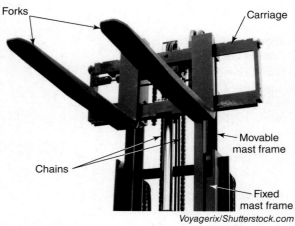

Figure 3-49. A traditional forklift uses a sliding mast frame, a pair of chains, a carriage, and a pair of sliding forks.

Forklifts and Telehandlers

Forklifts and telehandlers are frequently used by construction personnel for lifting and positioning materials and components. Most service repair facilities have at least one forklift. A forklift, also known as a fork truck, lifts materials using a ***carriage assembly***, which has two forks that are allowed to slide horizontally across the carriage. The forks have spring-loaded safety pins that lock the forks in position on the carriage. See **Figure 3-49.**

The forklift mast is comprised of two or more C-shaped frames that extend and retract within each other. One of the mast frames is fixed to the forklift chassis. The other mast frames are free to slide vertically. As the forklift's mast is lifted, one or two hydraulic cylinders lift the moveable frame up through the fixed frame. The forklift carriage is raised and lowered via a pair of chains. One end of each chain is fastened to the fixed mast. The chains loop around the sheaves mounted on the movable mast frame and the opposite end of the chains are attached to the mast's carriage assembly.

As the forklift's mast lift cylinders are extended, the movable mast frame begins to extend out of the fixed mast frame. As the movable mast frame extends, it pulls the chains, which, in turn, lift the carriage.

A ***telehandler***, also known as a rough-terrain forklift, operates similarly to a hydraulic telescoping crane boom. The boom can be hydraulically lifted and hydraulically extended. See **Figure 3-50**. A telehandler's boom is similar to a forklift's mast because the boom is extended and retracted with chains, one set of chains to extend the boom and one set to retract the boom. It takes more operator skill to properly position the forks because as the boom is lifted, it also has to be simultaneously extended. On a traditional forklift, the mast is lifted in a single plane.

Controls

A traditional forklift has, at a minimum, a hydraulic mast raise and lower function. Many forklifts allow the operator to side shift the mast to the left and right, and possibly tilt the mast forward or rearward.

Some forklifts have additional hydraulic functions, such as hydraulically widening or narrowing the space between the forks, or hydraulically pushing the carriage forward, away from the mast, or retracting the carriage back to the mast.

Some forklifts have a traditional transmission and a rear hydraulically steerable axle with a differential. Other forklifts steer by driving one drive wheel forward and the other rearward. These forklifts have one or two rear caster wheels, similar to an agricultural swather. Swathers are discussed in Chapter 4, *Agricultural Equipment Identification.*

A telehandler uses hydraulic controls to lift and lower the boom, extend and retract the boom, lower and lift the outriggers (also known as stabilizers), tilt the carriage forward or rearward, side shift the carriage to the left or right, and steer the machine.

The telehandler's boom includes a hydraulic ***compensation cylinder*** that is responsible for hydraulically self-leveling the carriage (forks or front attachment). The compensation cylinder is mechanically pinned to the boom. If the operator is only lifting or lowering the boom (and not tilting the carriage), as the boom is raised and lowered, it causes the compensation cylinder to move. The oil flow from the compensation cylinder travels through the leveling valve and moderates the amount of oil traveling to the carriage tilt cylinder. This keeps the carriage in the last commanded position, which is often a level position. This function allows the operator to raise and lower the boom without having to continuously readjust the tilt of the carriage.

Telehandlers are normally all-wheel drive and four-wheel steer. Steering can be performed by the front axle or rear axle alone, or as a coordinated steer, also known as

indykb/Shutterstock.com

Figure 3-50. A telehandler uses a lift cylinder to lift the boom. The boom is hydraulically extended by a telescoping hydraulic cylinder.

four-wheel steer. With four-wheel steer, the front and rear axles steer in opposite directions to provide a tighter turn. The wheels can also be steered with both axles steering in the same direction, also known as four-wheel *crab steer*. See **Figure 3-51**. Crab steer causes the machine to move laterally while staying pointed straight ahead. Axles are further explained in Chapter 18, *Axles and Driveshafts*, and wheeled steering systems are explained in Chapter 25, *Wheeled Steering Systems*.

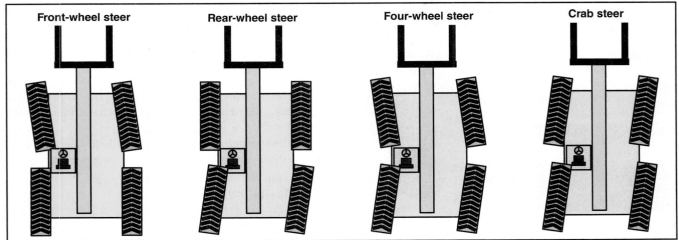

Goodheart-Willcox Publisher

Figure 3-51. Two-wheel steer is performed by the front or rear axle alone and results in a large turning radius. Four-wheel steer allows the telehandler to turn in a tight radius. Crab steer allows the telehandler to stay oriented straight ahead while moving at an angle. It is accomplished by angling front and rear wheels in the same direction.

Applications

Traditional forklifts normally use only pallet forks. However, some telehandlers can have other work tools, such as hay grapple, hay spear, or a loader bucket, attached in place of traditional pallet forks.

Types

Traditional forklifts typically have propane or diesel engines. Forklifts can also use compressed-natural-gas or gasoline engines.

Industrial plants commonly use electric battery-powered forklifts. The electric truck batteries are normally charged daily. These heavy-duty ***deep-cycle batteries*** allow the battery to be drained during the shift and recharged after the shift is complete. Repeatedly deeply discharging and recharging a conventional battery would dramatically shorten its life span. Deep-cycle batteries are specifically designed to be deeply discharged and recharged frequently.

Forklifts used in warehouses and loading docks are often small compact trucks with hard tires designed for use on solid flooring. **Figure 3-52** shows a compact high-lift forklift being used in a warehouse application. Forklifts can also range in size up to large rough-terrain forklifts with masts that can extend 20′ or more.

Most traditional forklifts use single-acting lift cylinders, which do not use hydraulic fluid pressure to retract the cylinder. Instead, the mast and carriage are lowered by gravity alone. However, some home repair stores that deliver materials to job sites use truck-mounted forklifts with double-acting hydraulic masts. To prepare the forklifts for transport, the forks are placed inside rails in the back of a delivery truck. When the mast is hydraulically lowered, the entire fork truck is hydraulically lifted up to the bed and secured so that it can be hauled with the load of materials. See **Figure 3-53**.

Sizes

Forklifts range in sizes from compact, which lift as little as 2000 pounds, to massive trucks that lift tens of thousands of pounds. Smaller forklifts are normally designed to work on hard surfaces inside shops and buildings. The large forklifts can be designed to operate on hard surfaces or in worksites with soft soils.

Forklifts are normally rated in at least three different capacities. The following are average ranges for these three capacities:

- Lift capacity: 2000–20,000 pounds
- Lift height: 9–21 feet
- Fork length: 4–7 feet

Penka Todorova Vitkova/Shutterstock.com

Figure 3-52. In warehouses, compact high-lift forklifts allow personnel to place pallets on high shelves.

Goodheart-Willcox Publisher

Figure 3-53. Home repair stores use forklifts with double-acting lift cylinders, which will pull the forklift up to the truck bed when the lift cylinder is hydraulically retracted. This truck-mounted forklift has three wheels, but they can also be designed with four wheels.

Compactors

Compactors, also known as rollers, are used for compacting soils, aggregate, asphalt, and garbage. Most compactors are self-propelled by means of a hydrostatic transmission. However, landfill compactors use a power train similar to that in a large four-wheel-drive wheel loader. This type of power train uses a torque converter and power shift transmission to deliver power to the front and rear axle, providing all-time four-drum drive. Other compactors are pulled by a tractor's drawbar or can be driven as an attachment on a machine, such as a skid steer loader or an excavator. See **Figure 3–54**.

Controls

Common compactor controls consist of propulsion, articulated steering, vibratory roller, and dozer blade. Propulsion is normally provided by means of a hydraulic motor. Compactors can have a variety of tire and wheel arrangements. If a compactor has two rollers (smooth or pad), both rollers are hydraulically driven. Some compactors use a single roller and two rubber drive tires that are hydraulically driven. Other compactors have multiple pneumatic tires that serve as the roller. These machines are known as pneumatic compactors. See **Figure 3-55**.

A
Michael Zysman/Shutterstock.com

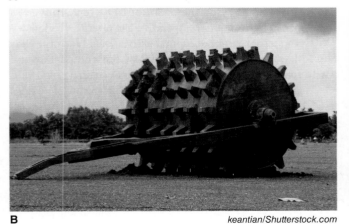
B *keantian/Shutterstock.com*

C *Byron W. Moore/Shutterstock.com*

Figure 3-54. Compactors can be self-propelled, towed, or attachments on another piece of equipment. A—This landfill compactor has a dozer blade and compactor drums in place of wheels. B—Pull-type pad foot compactors are pulled behind large tractors. C—This compactor is an attachment on an excavator.

Figure 3-55. Wheel and drum configurations among compactors vary. A—This tandem-roller compactor is equipped with two smooth rollers. B—Single roller compactors use drive tires in place of a second roller. C—Compactors can use pneumatic tires for compacting.

Most compactors use articulation steering. The articulation joint will steer the compactor to the right and to the left.

A *vibratory compactor* uses a hydraulic motor to drive a shaft that contains an eccentric weight. As the hydraulic motor is rotated, the eccentric weight exerts a centrifugal force within the roller, causing it to vibrate and compact the material. The centrifugal force can be measured in thousands of pounds or kilonewtons.

The vibratory compactor has two adjustments: frequency and amplitude. The frequency changes the quantity of vibrations for a given time period. The compactor's different frequencies are specified in one of two units:

- Hertz (Hz)—the number of vibrations per second.
- VPM—the number of vibrations per minute.

The operator can also vary the vibratory roller's amplitude, resulting in compaction depth that is often specified in thousands of an inch or decimal millimeter. Sometimes the amplitude is varied by reversing the direction of rotation of the vibratory compactor.

Vibratory rollers emit standard vibration or oscillatory vibration. Standard vibration acts like a hammer and directs the vertical vibrations straight to the ground directly below the roller. An oscillatory vibrating compactor contains two vibratory weights inside a single drum. The vibration is oscillated horizontally to the rear and to the front of the drum, which kneads the asphalt rather than hammers the asphalt. Oscillatory vibration is better for paving thinner asphalt, such as 1.5" thick and when paving bridges because it emits less stress to the bridge deck.

Many vibratory rollers have an operator station that can be pivoted allowing the operator to be repositioned in relationship to the roller to optimize the operator's view.

Compactors can be equipped with a dozer blade that has four functions: lift, lower, hold, and float. Soil compactors and waste compactors are two examples that use dozer blades.

Asphalt compactors are equipped with a row of nozzles used for spraying water onto the roller to prevent the material from sticking to the drum. See **Figure 3-56**.

Figure 3-56. Asphalt compactors use spray nozzles to wet the drum to prevent material from sticking to the roller.

Applications

Compactors are used for compacting multiple types of materials including rock, sand, gravel, silt, clay, asphalt, and garbage. The compactors are sometimes labeled based on the application:

- Landfill compactor.
- Soil compactor.
- Asphalt compactor.

Roller Designs

Compactor rollers are offered in a variety of different designs. Compactors can have smooth vibratory drums for tasks like compacting and smoothing granular materials like asphalt. Other compactors may be equipped with padfoot drums for compressing fine, cohesive materials like dirt. A sheep's foot roller is used for compacting sublayers of materials, such as silt and clay, while fluffing the top layer. See **Figure 3-57**. Pneumatic tire rollers are often for final smoothing of asphalt. The soft pneumatic tires conform to the surface and are able to smooth the transitions between surface irregularities better than steel drums.

A   B *Goodheart-Willcox Publisher*

Figure 3-57. Some drums have protrusions to focus the force of the roller over a smaller surface area. A—Padfoot rollers have tapered pads on the drum. As the roller rotates, each pad compacts the soil. Manufacturers vary the placement of the padfeet. B—The early style of sheep's foot compactor consisted of pads located on top of pedestals that were attached to the drum. However, today many people call a traditional padfoot roller a sheep's foot compactor.

 Note

Some compactors can be quickly converted from smooth rollers to a padfoot roller design. The padfoot drum consists of two half shells that are bolted to the smooth drum. Sometimes the scraper bar has to be removed first. See **Figure 3-58**.

Two other tools used for site compaction are a high energy pull-type impact compactor and a loaded dump truck. An impact compactor consists of a trailing lopsided implement that is pulled across the soil. Because the compactor is lopsided, it does not roll evenly causing it to slam into the soil as it dragged across the soil. The roller can have a 3-sided triangular shape, a 4-sided rectangular shape, or a 5-sided pentagon shape. See **Figure 3-59**. The impact compactor can be pulled across a field faster than a vibratory roller. However, the compactor will severely shake the operator and tractor.

A fully loaded 10-wheel dump truck can also be used to compact the soil. The operator will often crosshatch the soil by driving the loaded dump truck over the soil in a crossing pattern.

Sizes

Compactors range in size from a small compactor that the operator must walk behind to a large landfill compactor that weighs more than 120,000 pounds. See the *Appendix* for *Compactor Sizes*.

Goodheart-Willcox Publisher

Figure 3-58. Padfeet can be added to a smooth roller with the use of two half shells. When the padfoot design is no longer needed, the half shells can be removed from the smooth roller.

Trenchers

Several different types of machines are used to excavate trenches. The trenches are required for laying utilities like fiber optics, sewer lines, gas lines, water lines, and drainage tile. Some trenches are designed to form concrete footings.

Types

The four common types of trenchers are the traditional excavator/backhoe, chain, rock wheel, and plow. Trenchers can be equipped with four drive wheels or drive tracks. The tracks can be rubber belts or steel tracks. Small walk-behind trenchers can be equipped with only two drive wheels, four wheels, or twin tracks. A trenching attachment can also be added to a skid steer.

One of the most common types of trencher is the chain-drive trencher. See **Figure 3-60**. The chain can be quite narrow to excavate

Goodheart-Willcox Publisher

Figure 3-59. A high-energy inertia compactor is pulled behind a tractor. This one is a four-sided design.

Figure 3-60. A chain-drive trencher is a machine with a chain mounted on a blade, similar to a large chain saw, at the rear of the machine. A—This small chain drive trencher is the type used to bury cable or small pipe. The auger at the front of the chain lifts the loose dirt to both sides of the trench. B—Pipeline trenchers use wide belts for digging wide trenches.

Figure 3-61. A rock wheel trencher is used to cut through rock and concrete.

Figure 3-62. A vibrating plow trencher completes both tasks of digging a deep narrow trench and burying cables as it moves through the field.

a thin trench for burying cable or small pipe. The chain can also be wide for digging a trench for pipeline or long foundations. The advantage of a wide-belt chain trencher is that it trenches a consistent flat-bottom trench.

Rock wheel trenchers have hardened teeth mounted to a wheel. They are used to cut through rock and concrete. See **Figure 3-61**. The carbide teeth have a short life. Depending on the application, the teeth can require replacement in as little as 15 minutes or as long as 40 hours of service. The carbide teeth must be free to rotate on their axis or they will fail prematurely.

A vibratory plow trencher is designed to pull a deep vibrating plow through soils while simultaneously burying cable. The trencher normally has a boom and spindle mounted on the front of the trencher. The spindle allows cable to unwind off a spool while the machine moves through the field and digs the trench. See **Figure 3-62**.

Controls

Trenchers commonly use hydrostatic transmissions for propulsion to drive tires or tracks, but can be equipped with a power-shift transmission. If the trencher has a rigid frame with tires, it normally has four-wheel steering. Articulated trenchers use hydraulic articulation to steer.

In addition to hydraulic steering and propulsion, trenchers can be equipped with several additional hydraulic controls. For example, trenchers may be equipped with a small dozer blade (lift, lower, and hold), a backhoe (swing, boom, stick and bucket), and outriggers, also known as ***stabilizers***, that provide a stable footing for operation. See **Figure 3-63**.

Applications

Trenchers are used for laying residential and commercial utilities, such as gas lines, water lines, sewer lines, electrical lines, and fiber optic cables. Pipeline trenchers can dig trenches up to 48″ wide and 18′ deep to bury large water lines, gas lines, and oil lines.

Robert J. Beyers II/Shutterstock.com

Figure 3-63. This trencher is an articulated machine with a chain trencher on the back and dozer blade, backhoe, and stabilizers mounted on the front of the machine.

Sizes

Trenchers vary in size from a small walk-behind chain trencher to a large pipeline trencher that weighs over 200,000 pounds. See the *Appendix* for *Trencher Sizes*.

Drills

Drills are used for installing utility cables, drilling rock, and drilling for resources like water, oil or gas. Two common types of drills are rock drills and horizontal directional drills.

Rock Drills

Rock drills are designed for drilling through rock formations so explosives can be inserted to blast the rock. Rock drills are used in quarries and for building highways. The rock drill can be designed as a standalone machine that serves the sole purpose of drilling rock, or it can be added as an attachment to an excavator. The drill uses a drill head to drill through the rock and uses additional drill rod to increase the drilling depth. The drill head can be angled in any position between vertical upward, horizontal, and vertical downward positions. However once that angle is set, the drill head will drill in that fixed direction, and cannot change direction.

Horizontal Directional Drills

For cable-laying operations, a drill is sometimes used instead of a trencher. See **Figure 3-64**. Most drills used for laying cable are capable of steering the drill head in different direction while drilling, and are therefore called ***horizontal directional drills*** (HDDs).

The drill is designed to push a series of long pipes, called drill rods, underground several hundred feet, creating a pilot bore. The drill rod is advanced, and a drill bit on the end of the drill rod cuts through the soil and roots to create the bore. Drill fluid, consisting of

Figure 3-64. A horizontal directional drill pushes a bit underground. A—The water line supplies water for the drilling fluid. The machine automatically attaches the drill rod stored on the side of the machine to lengthen the drill's capabilities. B—The front of the drill has two long anchoring shafts that are hydraulically driven several feet into the ground so that the drill can remain stationary while it is pushing and pulling the pipe. In this image, the far shaft has been inserted and the shaft on the near side is being inserted into the ground.

water, clay, and chemicals, is sprayed through the front of the bit to lubricate the bit as it is pushed or pulled through the ground. Drilling fluids can vary depending on the soil conditions, for example sticky soil or porous soil. Bentonite clay is commonly added to the fluid mixture. The fluid needs to be biodegrable as well.

The operator guides the drill bit back to the surface when it reaches its target. A technician removes the drill bit and attaches multiple conduits and a reamer to the drill rod, which is then pulled back to the drilling machine. See **Figure 3-65**. The reamer typically enlarges the pilot bore to 1.5 times the diameter of the conduit being installed.

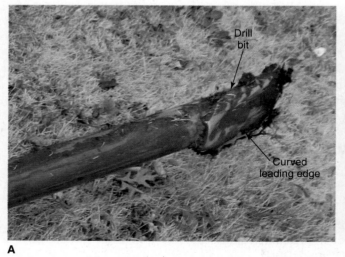

Figure 3-65. Horizontal directional drills are designed to pull multiple conduits underground so that fiber optic cables can be inserted in the conduit. A—The horizontal directional drill has exited the ground. B—The drill bit is removed and a large reamer is threaded onto the drill rod. A bridle and swivel are inserted onto the reamer. The reamer is rotated and water is pushed through the reamer while the drill pulls the orange conduit underground back to the drilling machine. The swivel allows the reamer to rotate without rotating the conduit.

The drill then pulls the conduit back through the ground to the drilling machine. Telephone, internet, or other wire cables are inserted in the conduits. Compressed air is used to blow the cables through the conduits.

The machine has a carriage assembly that is moved forward and rearward by a rack and pinion gear system. See **Figure 3-66**. A drill bit is placed on the end of the drill rod. The carriage will push the drill rod, guiding it underground. When the first drill rod is driven as far as possible, the carriage releases from the first drill rod, returns to its starting position, threads another 10-foot section of drill rod into the first, and continues drilling. This process continues until the drill bit arrives at its target. Another worker uses a handheld electronic locator and a two-way radio to communicate with the drill operator, who guides the underground drill. The handheld locator is used to determine the depth and the pitch of the drill bit. See **Figure 3-67**.

Notice in **Figure 3-65A** that the drill bit has an overall gradual curve, with one edge curved more than the other. As the drill bit is pushed through the earth, it tends to move in the direction of the curvature. The operator looks at the machine's monitor to know the orientation of the bit. To change the direction of the bore hole, the operator can rotate the bit so its curvature points in a different direction. The position of the bit will guide the drill rod as it is pushed underground. If the operator wants the bit to move straight ahead, the bit is spun continuously, which allows the bit to be pushed straight forward. The reamer bit shaft has a slight bend which allows the operator to control the direction of the bit by rotating the bit shaft.

A rock horizontal drill uses a different configuration. Its pipe has another smaller pipe located in its center (one pipe inside another). The internal pipe is attached to the bit and driven to bore through solid rock. This internal rotating pipe is protected by the outside pipe. The outer pipe has a slight bend that allows the operator to steer the bit. To change direction, the operator rotates the outer pipe so the bit points in the desired direction of travel. If the operator wants the bit to move straight ahead, the outer pipe is continuously rotated.

During the drilling operation, drilling fluid is continuously pumped into the bore through the pipe. The drill fluid consists of water and an additive called bentonite. The drilling fluid cools the bit, lubricates the drill pipe, flushes cut material from the bore, and helps form a wall by sealing the bored passageway. As more fluid is pumped into the bore, the used fluid returns to the drill's entry point through the gap between the pipe and the bore. When it returns, it is filtered to remove sediment and is pumped back into the bore in a continuous cycle.

Goodheart-Willcox Publisher

Figure 3-66. Hydraulic motors move the carriage and spin the pipe.

Goodheart-Willcox Publisher

Figure 3-67. Handheld locators are electronic devices used to determine the location, pitch, and depth of the drill bit. The locator screen displays the temperature, depth, and steered position of the bit.

Summary

- Excavators use a boom, stick and bucket for demolition, trenching, laying pipe, maneuvering scrap, cleaning ponds or waterways (dredging).
- Wheel loaders are all four-wheel-drive articulated tractors used for loading trucks and hoppers and maneuvering materials. Loaders use a parallel linkage or a Z-bar linkage.
- Track loaders are similar to dozers, but use a bucket instead of a blade. They also have a higher dump height than a dozer.
- Dozers are used for pushing scrapers and trees, excavating dirt, ripping, rough leveling, and finish grading. Dozers might be configured with an S, U, semi-U, or cushion blade.
- Motor graders are commonly used for snow removal, maintaining rock roads, and finish work. The blade is attached to a circle that is hydraulically rotated, lifted, and shifted.
- Motor graders traditionally have more hydraulic functions than most other construction machines.
- Scrapers use a cutting edge to cut materials, which are then automatically scooped into a bowl.
- Open bowl scrapers use an apron, which forms a movable front wall of the bowl, and an ejector, which serves as a movable back wall of the bowl.
- Elevating scrapers use chains and flighting for loading the bowl.
- Smaller off-road haul trucks use articulated steering.
- Rigid-frame haul truck payloads range in size from 40 tons to 500 tons.
- Haul trucks commonly use some type of retarding mechanism to allow the machine to travel safely at higher speeds and reduce the wear on service brakes.
- Skid steers use dual-path hydrostatic transmissions that drive a pair of chains for propelling the right and left side drive wheels. The loader skids in order to steer.
- A forklift's mast consists of a sliding frame that raises and lowers a pair of chains that are connected to a carriage assembly.
- Telehandlers resemble a telescoping hydraulic crane. The boom is raised and lowered and hydraulically extended.
- Compactors are designed to compact soils, aggregate, asphalt, and landfill materials.
- Compactors can be designed with a smooth drum, a vibratory roller, or a padfoot roller.
- Compactors can be equipped with a single drum, tandem drums, or four drums.
- Trenches are formed with an excavator, chain trencher, rock wheel trencher, or plow.
- Rock drills bore holes in rock formations so that explosives can be used to blast the rock formations.
- Horizontal directional drills are used to pull conduit underground so that cables can later be inserted inside the conduit.

Technical Terms

- air-assisted scraper
- angle of repose
- apron
- articulated dump truck (ADT)
- blade tilt
- body-up reverse neutralizer
- breakout force
- carriage assembly
- compensation cylinder
- crab steer
- cushioned hitch
- cycle time
- deep-cycle batteries
- dual-path hydrostatic transmission
- ejector
- elevating scraper
- feller buncher
- flow sharing
- four-in-one bucket
- heaped capacity
- hoist
- horizontal directional drill
- knuckle boom loader
- liners
- longwall mining
- open bowl scraper
- operating load
- parallel linkage
- PAT dozer
- pitch
- push arms
- quick coupler
- radial lift
- retarder
- rigid-frame haul truck
- rock drill
- rock ejectors
- roll out bucket
- room and pillar mining
- self-leveling
- side dump bucket
- six-way blade
- skid steer
- skip loader
- slew
- stabilizers
- static tipping load
- struck capacity
- surface mining
- tandems
- telehandler
- trolley assist
- vertical lift
- vibratory compactor
- Z-bar linkage

Review Questions

Answer the following questions using the information provided in this chapter.

Know and Understand

1. All of the following are examples of excavator performance parameters, *EXCEPT*:
   A. breakout force.
   B. cycle time.
   C. maximum reach.
   D. maximum drawbar pull.

2. All of the following are used for excavation in the mining industry, *EXCEPT*:
   A. hydraulic and wire rope shovels.
   B. drag lines.
   C. giant wheeled excavators.
   D. long reach excavators.

3. Which machine is normally all-wheel drive, uses articulated steering, and uses four tires that are all the same size?
   A. Motor grader.
   B. Loader backhoe.
   C. Wheel loader.
   D. Skid steer.

4. Which term describes the loader feature that keeps the bucket at the same pitch as the loader frame is raised and lowered?
   A. Self-leveling.
   B. Non-adjustable leveling.
   C. Pitch leveling.
   D. Infinite leveling.

5. Which of the following loader bucket capacities is based on a 2:1 angle of repose?
   A. Struck capacity.
   B. Heaped capacity.
   C. Loaded capacity.
   D. Shallow capacity.

6. What type of large mining machine is available with standard lift, high lift, extended lift, or super high lift?
   A. Dozers.
   B. Wheel loaders.
   C. Haul trucks.
   D. Shovels.

7. When matching a loader to a haul truck, what is the recommended number of passes it should take to load the truck?
   A. 1 to 2.
   B. 3 to 4.
   C. 5 to 6.
   D. 7 to 8.

8. Which of the following control a dozer's dig, carry, and dump functions?
   A. Lift cylinders.
   B. Pitch cylinders.
   C. Angle cylinders.
   D. Tilt cylinders.

9. A motor grader's blade is mounted to a _____.
   A. triangle
   B. square
   C. circle
   D. rectangle

10. All of the following blade functions on a motor grader can be hydraulically actuated, *EXCEPT*:
    A. shifted right or left.
    B. tipped forward or rearward.
    C. lifted right or left.
    D. reversed 180°.

11. On a scraper, what is the name of the front wall of the scraper than can be lifted or lowered in an arcing motion?
    A. Apron.
    B. Ejector.
    C. Bowl.
    D. Elevator.

12. Rock ejectors prevent rocks from _____.
    A. wedging between the drive tires
    B. entering the engine intake
    C. damaging the hydraulic reservoir
    D. accidentally being loaded into the dump bed

13. Technician A states that end dump tractor trailers are good for low-clearance areas. Technician B states that side dumps and ejector haul trucks are good for low clearance areas. Who is correct?
    A. Technician A.
    B. Technician B.
    C. Both A and B.
    D. Neither A nor B.

14. Skid steers use what system configuration to power the drive wheels?
    A. A bevel gear set.
    B. A planetary gear set.
    C. A bull-and-pinion gear set.
    D. Chain drives.

15. Skid steers normally use what type of power transmission?
    A. Syncroshift.
    B. Powershift.
    C. Electric drive.
    D. Dual-path hydrostatic drive.

16. The forks on a forklift attach to _____.
    A. the carriage
    B. cylinders
    C. the carrier
    D. the car body

17. All of the following are used to dig trenches, *EXCEPT*:
    A. a loader bucket.
    B. a plow.
    C. a chain.
    D. a rock wheel.

18. A _____ mechanism is used to move the carriage on a horizontal drill.
    A. spiral bevel gear
    B. rack-and-pinion
    C. bull-and-pinion
    D. planetary gear set

Apply and Analyze

19. _____ manufactures a telescoping boom excavator.
20. A(n) _____-type of wheel-loader bucket can act like a clam shell.
21. The _____-type of loader linkage provides improved breakout forces throughout its entire range of motion.
22. A wheel loader's operating load is _____ percent of its static tipping load.
23. A track loader's operating load is _____ percent of its static tipping load.
24. A PAT dozer is equipped with a(n) _____-way blade.
25. On a scraper, the moveable rear wall is called the _____.
26. _____ dozing allows the operator to trap material between the side walls of the previous cut and helps the dozer push more material.
27. Skid steers equipped with a _____ loader lift arm design is best for loading tall trucks.
28. A(n) _____ is the only type of construction machine that is capable of self-loading, hauling the material long distances, and unloading by itself.

Critical Thinking

29. What type of machine has the largest size variation from the smallest all the way up to a mining machine and list the sizes?
30. Explain the difference between a rock drill and a horizontal directional drill.

James Mattil/Shutterstock.com

Expos give manufacturers an opportunity to introduce prospective customers to their latest models and innovations.

Chapter 4
Agricultural Equipment Identification

Objectives

After studying this chapter, you will be able to:

✓ Identify the common types of agricultural equipment.
✓ Describe the different types of tillage equipment.
✓ Describe the different types of seeding equipment.
✓ Explain the different types of mowers.
✓ List the different types of rakes.
✓ Describe the different types of hay balers.
✓ Explain the difference in forage harvesters.
✓ Describe the different functions of a combine harvester.
✓ Identify the different types of agricultural tractors.
✓ Describe the types of agricultural sprayers.
✓ Describe the types of fertilizer applicators.

A heavy equipment technician working for an agricultural equipment dealer will work on a wide range of equipment with many types of systems. Heavy equipment is used in the agricultural industry for tasks such as soil preparation, planting and fertilizing crops, spraying pesticides, and harvesting and transporting crops. Many of these tasks are performed with agricultural implements that use power from the tractor to function. Some implements use the tractor's hydraulic system. Other implements have their own hydraulic pump, typically driven by the tractor's PTO and use hydraulic power to operate. Still other implements are purely mechanical and are driven off the tractor's PTO. Today's implements also interact with the tractor's ECM to perform more efficiently.

Overview of Agricultural Equipment

A traditional agricultural equipment dealer supports farmers by providing sales, parts, and service on a wide range of agricultural equipment. Agricultural equipment includes the following types of machines and implements:

- Tillage tools.
- Planting and seeding equipment.
- Mowers, cutters, and conditioners.
- Rakes.
- Hay balers.
- Forage harvesters.
- Combines.
- Tractors.
- Spreaders and sprayers.
- Specialty equipment.

Tillage Tools

Tilling is the act of cutting, agitating, stirring, and overturning the soil in preparation for seeding and to control weeds. It is typically performed with a tractor and a tilling implement. The type of equipment used varies based on the geographic region, soil type, and the desired result. For example, a farmer needing to till deeper than a few inches would use primary tillage equipment. ***Primary tillage*** aggressively tills the soil. The tools are designed to till the soil at deeper levels, ranging anywhere from 6″ to 2′. A farmer who wants to till soil at only a minimal depth would use secondary tillage equipment. ***Secondary tillage*** is less aggressive and works the soil at much shallower depths, usually less than 6″. Sometimes a secondary tillage tool is used after a field has been tilled with a primary tillage tool.

General types of tillage tools include the following types of implements:

- Moldboard plow.
- Disc harrow.
- Field cultivator.
- Chisel plow.
- V-ripper (sub soiler).
- Spike-toothed harrow.
- Rod weeder.
- Combination tool.
- Rotary hoe.
- Finish harrow.
- Wide-sweep blade plow.
- Rotary tiller.

Moldboard Plow

The ***moldboard plow*** is a tillage tool that slices into and inverts the soil, burying the crop residue on the first pass. After one pass of a moldboard plow, 90% to 100% of the crop residue has been inverted and buried. However, the moldboard plow leaves behind an uneven field with large mounds or clods of soils. For this reason, the field must be re-plowed with another tool, such as a disc harrow, before it can be planted. See **Figure 4-1**.

Most late-model moldboard plows are reversible and can invert the soil to the left or to the right as needed. The plow contains two complete sets of plow blades. The blades in the second set are curved in the opposite direction of those in the first set. Only one set of blades engage the soil at a time. To reverse the direction the soil turns, the second set of blades is rotated down, into the working position. This also moves the first row of blades up and out of the way. As the second set of blades moves through the soil, the soil is turned over in the opposite direction.

Disc Harrow

A ***disc harrow*** consists of a group of discs, often called a ***gang***. See **Figure 4-2A**. The individual discs are free to rotate on bearings as they are pulled through the soil. The gang can be designed to pivot or lift against spring tension as it encounters rough soil or obstacles in the field. Examples of disc harrows include single discs, double discs, offset discs, and double offset discs. See **Figure 4-2B**.

Taina Sohlman/Shutterstock.com

Figure 4-1. A moldboard plow inverts the soil, burying the crop residue.

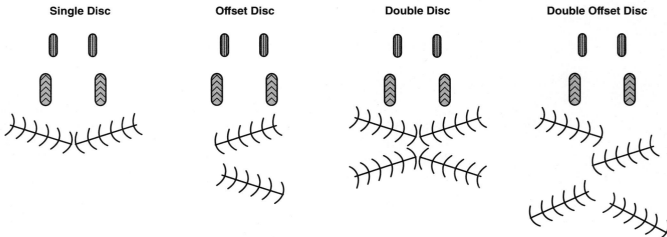

Figure 4-2. Disc harrows are offered in a variety of sizes and styles. The implement contains several discs. A—Most disc harrows have four or more gangs of discs. B—Examples of disc harrow styles are single, offset, double, and double offset. Notice the single and offset have two gangs of discs, and the double discs have four gangs of discs.

The discs can range in size, weight, shape, and design. An individual disc can range from 16″ in diameter and weighing 40 lb up to 42″ in diameter and weighing up to 300 lb. Discs can have a dished shape or cone shape. They can have material cut out from the outside circumference, making them more aggressive for cutting through crop residue. Heavy concrete ballasting can be added to the top of the disc harrow to give it greater cutting action. Large harrows are sometimes pulled through rough terrain by dozers or large four-wheel-drive tractors.

Disc harrows can be lightweight for secondary tillage or heavy duty for primary tillage. They are designed to invert the soil and smooth the seed bed. Large disc harrows, such as double or double offset discs, invert 35% to 65% of the crop residue below the soil on a single pass.

Field Cultivator

Field cultivators use multiple rows of **shovels**, also called **sweeps**, to break up the soil. They are designed to operate at shallower depths than moldboard plows or disc harrows and are used to prepare the soil for planting (secondary tillage). See **Figure 4-3**. Each shovel is mounted to a spring steel shank that is connected to a trip mechanism. The trip mechanism allows the shovel and shank assembly to release and pivot upward if it hits an obstacle in the field. This prevents damage to the shank and shovel. Most trip mechanisms are designed to automatically reset after the cultivator passes over the obstacle; others must be reset manually.

Shovels can range from 8″ to 18″ wide. The shovels can be designed with varying angles of pitch cast into them, such as 5° or 7°. The placement of the shovels affects the flow of debris through the implement. If the cultivator becomes plugged or has bent shovels or shanks, it can force the implement to pull to the side and cause the tillage tool to miss the soil. This deflection of the tillage tool is known as an *askewness angle*.

A field cultivator does not invert the soil like a moldboard plow or a disc harrow. It leaves most of the previous year's crop residue on top of the surface, which helps maintain moisture and reduces soil erosion.

Chisel Plow

A **chisel plow** looks very similar to a field cultivator because it also has multiple spring-loaded steel shanks. However, a chisel plow is used for primary tillage and works deeper below the surface than a field cultivator. It does not contain wide sweeps, but instead uses V-shaped or twisted spikes. See **Figure 4-4**.

V-Ripper

A **V-ripper**, also known as a **sub-soiler**, uses heavy-duty rippers to cut deep into the soil. They can break through deep layers of soil, allowing the crop to grow further below the surface and improving moisture penetration. The rippers are mounted solidly to the implement's tool bar. They are much stronger than the shanks found on cultivators or chisel plows. See **Figure 4-5**.

Spike-Toothed Harrow

A **spike-toothed harrow** is a secondary tillage tool that can be pulled with a tractor's drawbar, attached to another secondary tillage tool, or attached to a tractor's three-point hitch. It has a mat of spikes that are dragged across the soil's surface to smooth it. See **Figure 4-6**.

Goodheart-Willcox Publisher

Figure 4-3. Field cultivators use multiple rows of shovels mounted to spring steel shanks. The shovels shatter the root structure below the soil's surface. The trip mechanisms allow the shanks to lift out of the ground if they encounter an obstacle.

Goodheart-Willcox Publisher

Figure 4-4. A chisel plow looks similar to a field cultivator but has narrow spikes that operate at lower soil depths.

Goodheart-Willcox Publisher

Figure 4-5. A V-ripper is a type of sub-soil tillage tool. They open the soil to improve root and water penetration.

Rod Weeder

A *rod weeder* has a long rod that is dragged under the soil's surface to uproot vegetation. The rod can be fixed or mounted inside bearings. The bearings are located in multiple arms that resemble a vertical arm or ripper. The rod is rotated at speeds up to 150 rpm by a ground-driven wheel or hydraulic motor. As the rod rotates, it grabs, wraps up, and pulls out any roots it encounters.

Combination Tool

A *combination tool* is a single implement consisting of multiple tillage tools. The combination tool may include any combination of disc harrows, field cultivators, chisel plows, finish harrows, and spike-toothed harrows. See **Figure 4-7.**

Goodheart-Willcox Publisher

Figure 4-6. A spike-toothed harrow has multiple spikes attached to bars and is used for tilling the soil.

Rotary Hoe

A *rotary hoe* consists of several spiked wheels that are dragged across the top of the soil. The hoe is a good cultivator for reducing or eliminating weeds. It is sometimes used to help improve the plant emergence or to minimize soil crusting, which can occur after heavy rains. The implement is pulled at high field speeds, up to 12 mph.

Goodheart-Willcox Publisher

Figure 4-7. Many tillage implements consist of multiple tools on a single implement. This Great Plains TC5323 is called a turbo chisel plow. It has 22″ wavy-edged discs, called coulter blades, in front of chisel plows. It has chopper wheels behind the chisel plows to break dirt clods.

Finish Harrow

A *finish harrow* uses a rotary bladed spool to pulverize the soil, which prepares an excellent seedbed for a planter. The tool can be placed in front of a spiked-toothed harrow or after a secondary tillage tool. See **Figure 4-8**.

Wide-Sweep Blade Plow

A *wide-sweep blade plow* is similar to a field cultivator except that it uses a single row of very wide sweeps. The sweeps can be up to 72″ wide. They are used to slice through the roots. The plow leaves most of the crop residue on top of the soil's surface.

Rotary Tiller

A *rotary tiller* is normally a three-point-hitch–mounted implement used to till soil. It is driven by a tractor's power take-off (PTO). The tiller has a rotating shaft with multiple knives that pulverize the soil. Rotary tillers are commonly used for tilling gardens. See **Figure 4-9**. PTOs and three-point hitches are covered later in this chapter.

Planting and Seeding Equipment

Several different types of planters and seeders are used for planting crops. Regardless of the type of seeder, it must perform several functions:

- Open the furrow.
- Meter the seed.
- Place the seed.
- Close the furrow.
- Firm the seedbed.

Goodheart-Willcox Publisher

Figure 4-8. Finish harrows pulverize the seedbed, which makes it very smooth.

Seeders can be categorized into three categories:
- Row crop planters.
- Drills and air seeders.
- Broadcast seeders.

Planters

A *row crop planter* is a type of seeder that controls the planting based on a seed population of a set number of seeds per acre or hectare. Row crop planters typically plant with wider spacing between rows than drills do. An example range of row spacing with row crop planters is 15″ to 40″. The space between the planted rows can be later cultivated to minimize weeds. The number of rows a planter will seed varies from a few to dozens of rows. A typical very small planter might plant four rows at a time, and a very large row planter might plant 48 rows at a time. For example, the John Deere DB120 48Row30 row crop planter has a 120′-wide tool bar to accommodate the 48 rows spaced 30″ apart.

Figure 4-9. Rotary tillers are three-point-hitch–mounted implements that are powered by a PTO. They are commonly used for tilling gardens.

The planter's number of rows and spacing between rows are chosen for compatibility with the machine harvesting the crop, such as a 12-row 20″-spaced corn header, or an 8-row 40″-spaced cotton picker. For example, a 12-row corn header could harvest fields that were planted by 12-, 24-, 36-, or 48-row planters and an 8-row cotton picker could harvest fields that were planted with 8-, 16-, or 24-row planters.

Opening the Furrow

A trench, known as a furrow, must be opened to plant seeds. Typically, two rubber gauge wheels and two steel disc openers work together to cut a furrow of the desired depth. See **Figure 4-10**. The gauge wheels ride on top of the soil while the dual steel discs ride between and below the gauge wheels, cutting the furrow into the soil. The depth of the disc openers in relation to the *gauge wheels* determines the depth of the furrow. Using the depth control lever to move the disc openers further below the gauge wheels will result in a deeper furrow. Using the depth control lever to raise the disc openers in relation to the gauge wheels will decrease the depth of the furrow.

Other types of trench openers are shovels, found on air seeders, and shoes or runners, which were used on older planters.

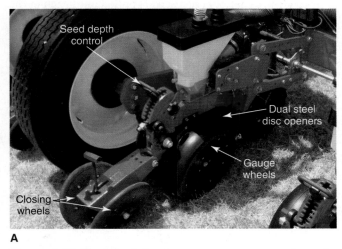
A

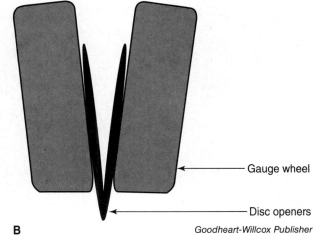
B

Figure 4-10. Dual steel disc openers are a common type of opener used on planters. A—This White brand row crop planter is manufactured by AGCO. It uses the traditional dual disc openers. B—This simplified front view illustrates the relationship of the gauge wheel position to the disc openers.

Figure 4-11. Vacuum planters use a hydraulic motor to drive a fan that develops vacuum, which is distributed through a manifold to each row unit.

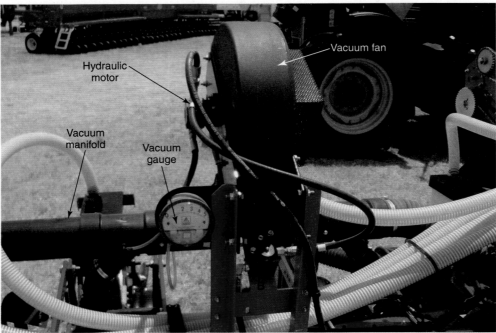

Metering

One of the most common types of seed metering systems used in row crop planters is the vacuum plate. A hydraulic motor spins a fan, which creates a vacuum that is applied to each row unit through a manifold. A planter's vacuum gauge is shown in **Figure 4-11.**

The seeds drop into each row unit through a hopper, which can be an individual hopper located at each row unit or can be distributed through a single large hopper. John Deere has a Central Commodity System (CCS), which uses a large central hopper to supply seed to smaller hoppers located at each row unit.

The vacuum pulls the seeds into the dimples of a vertically positioned seed disc. The speed of the rotating seed disc is critical to the population rate. A *singulator* is located inside each row unit's seed meter. Its purpose is to ensure that only one seed is pulled onto each of the disc's individual seed holes. See **Figure 4-12**.

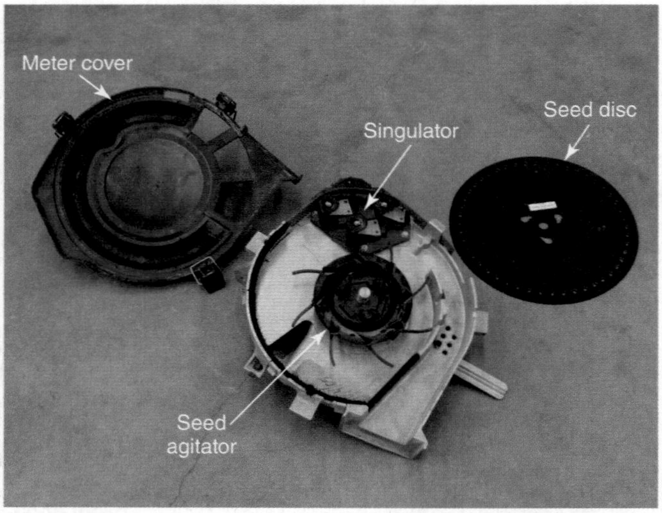

Older planters operated the metering mechanisms through a ground-driven chain drive system. Late-model planters often use a hydraulic motor or electric motor to drive the metering mechanism. The motor can infinitely vary the speed of the meters, allowing the planter to vary the population based on a prescription programmed into the planter's monitor. John Deere uses a 56-volt DC electric motor to drive and vary the seed meter on their ExactEmerge planters.

Other seed metering devices include seed-plate planters, which were common decades ago, and the Case IH Early Riser positive air drum seeder. The seed-plate planters dropped seeds onto a rotating plate that metered the seed. The Case IH Early Riser planter uses a hydraulic motor to build a positive air pressure, which holds seeds against a rotating drum. A seed release wheel rotates on the outside circumference of the drum. As the

Figure 4-12. This Case IH Advanced Seed Meter (ASM) is a vacuum meter. The vacuum is applied through the cover and pulls the seed onto the seed disc. The singulator ensures that only one seed is held in each seed disc hole.

release wheel rolls across the dimpled drum, it causes the seed to fall out of the drum's dimple. The seeds fall into the discharge manifold and are routed to each row unit through a manifold tube. A planter can also use positive air pressure to hold the seed against a vertical rotating disc. This metering system is similar to those in vacuum planters, except positive pressure is used in place of vacuum.

Placing the Seed

Seeds can be placed into the furrow by dropping them through a traditional gravity-drop seed tube or can be placed by a mechanical conveyor that more accurately controls the placement of the seed. Top performing planters place the seeds at a proper depth and have excellent *singulation*. The term *singulation* refers to placing single seeds, with consistent spacing between seeds. The goal is for the seeds to emerge in an even, picket-fence spacing, eliminating double seeds or skips of seeds.

New advances in planter technology include mechanisms that convey the seed to the open trench, which improves singulation over the gravity-drop seed tube design. John Deere's ExactEmerge uses a seed brush conveyor that is driven by an electric drive motor at a speed proportional to the planter's travel speed. The planter can travel at a higher ground speed, for example 10 mph, without having large losses due to skipping seeds (*skips*) or placing two seeds next to each other (*doubles*). Electronic controls that include a seed sensor for each row unit allow the operator to monitor seed spacing in real time and make adjustments to the planter to maximize the seed placement. See **Figure 4-13**.

Planters can use different methods for exerting *down force* on the row units. The force is exerted through hydraulic actuators or air bags at each row unit, **Figure 4-14**. The down force helps the row units cut through crop residue and cut seed trenches in difficult soils. Too much down force causes additional wear and tear on the row units and wheel slip in older mechanical drive planters. Too little down force causes a shallow seed trench.

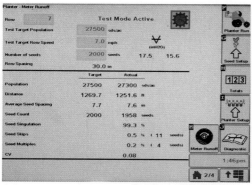

Goodheart-Willcox Publisher

Figure 4-13. A John Deere ExactEmerge planter has an onboard monitor that can display statistics about seed distribution. Notice that for this test run on row number seven, the singulation percentage is 99.3.

Goodheart-Willcox Publisher

Figure 4-14. An air bag is used to apply down force on this planter's row unit.

Down force margin equals the difference between the amount of planter down force plus the weight of the row unit minus the amount of down force that is required to cut through soil. For example, if the planter has 150 lb of down force applied to the row unit, and the row unit weighs 100 lb (totaling 250 lb) and if the soil requires 200 lb of down force to cut through the soil, then the down force margin equals 50 lb. Some positive down force margin is desirable, such as 50 lb to 100 lb.

Planters and GPS

Today's global positioning satellite (GPS) systems allow farmers to enter a planter prescription that varies seed population in different areas of the field. A planter in which the seed distribution rates can be adjusted is called a *variable rate planter*. These planters can optimize a field's yield by varying the seed population to match the terrain and growing conditions in different areas of the field. As previously mentioned, a hydraulic or electric motor varies the seed meter speed, which varies the seed population. A planter prescription is developed using GPS location, past yield maps, grid soil sampling, and other data, such as moisture maps.

An electronic control module (ECM) can also control a planter's clutches to provide automated row shutoffs. By enabling a planter to shut off its individual rows, or a group (or section) of rows, the ECM provides *swath control*. This feature enables the planter to avoid planting on top of previously planted rows, which occurs when planting in odd-shaped fields, such as triangular fields. The feature can save on seed and fertilizer that would have been overlapped on previously planted rows.

Planter clutches can be electrically or electrical-pneumatically operated. The clutches vary in design and operation. Some shut off the mechanical drive to the seed meter. Others work by eliminating the row unit's vacuum, thereby eliminating the ability to plant seed.

Note
A hybrid seed is a seed that has been developed using two different plants. The seed producer develops different types of hybrid seeds to maximize crop production. Hybrid seeds can be classified as a defensive hybrid, which is designed to be hardy enough to grow in poor soil conditions, or an offensive hybrid, which is designed to improve yields in good soil conditions.

A *multi-hybrid planter* has the ability to change the type of seed it is planting based on the location in the field. In areas that produce smaller yields, the planter can plant defensive hybrid seeds, which will increase crop performance. The planter can plant offensive hybrids in areas that normally produce high yields, which will also increase crop performance. The multi-hybrid planters require multiple seed hoppers for the different hybrid seeds.

Closing the Furrow
Planters can use shovels, knives, discs, wheels, or chains to cover the seed bed. The planter in **Figure 4-10A** uses two closing wheels to close the furrow.

Firming the Seedbed
A seedbed needs to be firmed in order for the seed to properly germinate. Some planters have a press wheel that serves the sole purpose of firming the seedbed. Other planters use firming wheels that both close the furrow and firm the seedbed.

Chemicals
Planters also dispense chemicals, such as fertilizers, herbicides, and insecticides, during planting operations. The chemicals can be liquid or granular and are used to boost yields, deter weeds, and deter insects. The chemicals require their own reservoir, metering system, and drive mechanism.

Markers
Non-GPS-assisted planters use *markers* as a guide for row positioning. Markers are discs positioned on each end of the planter. As the planter is pulled through the field, one of the markers will dig a furrow that is used as a reference point for the driver to straddle on the next pass. See **Figure 4-15**.

Planter Sensors
Two common types of sensors used on row crop planters are seed tube sensors and hopper level sensors. *Seed tube sensors* alert the operator if a seed tube becomes plugged and no longer allows seed to drop into the trench. *Hopper level sensors* alert the operator when the seed hoppers become low and need to be refilled.

Frames
A planter can also be classified by the method the tractor uses to connect and pull it through the field:
- Drawbar (drawn planters)—the largest type of planter. See **Figure 4-16A**. Large drawn planters normally can fold in a forward or vertical position for transport.
- Three-point-hitch mounted (integral)—are lifted and lowered by the tractor's hitch, **Figure 4-16B.**
- Stackable (or unit planter)—a variation of three-point-hitch–mounted planters. The outside rows can be stacked above the center of the planter to provide a narrow transport width, **Figure 4-16C.**

Figure 4-15. A marker is used to mark the location for the tractor to straddle as it makes its next pass in the field.

Figure 4-16. Planters can be classified as drawn, three-point mounted, or a stackable unit planter. A—Drawbar planter. B—Three-point-hitch–mounted planter. C—Stackable unit planter.

Drills

Another common type of seeding implement is the *drill*, which meters seed by volume, meaning pounds per acre (kilograms per hectare), or bushel per acre (quintal per hectare). Drill seeding is not specified in seeds per acre because drills meter such large volumes of seeds. Drills plant in much narrower rows than a row crop planter, typically spaced 5″ to 12″ apart. Examples of crops that are drilled are small grains like oats, wheat, barley, rice, and sometimes soybeans. See **Figure 4-17**.

Grain drills are normally ground driven through end wheels, wing wheels, or press wheels. A press wheel grain drill uses the group of press wheels to both firm the seedbed and drive the seed metering device. See **Figure 4-18**.

Goodheart-Willcox Publisher

Figure 4-17. This Landoll grain drill has 64 openers. The two outside wings fold forward for transport. It uses the wing wheels to drive the seed meter.

Clayton Thacker/Shutterstock.com

Figure 4-18. A press wheel drill uses press wheels to drive the drill's meter.

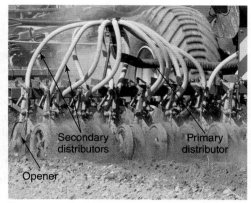

Figure 4-19. Air seeders are used for seeding large volumes of small grains. A—Seed and chemicals are stored in the seed cart. B—Primary and secondary distributors deliver the product to the openers.

Air seeders also plant small grains by volume. They can perform two functions simultaneously, tilling and planting in a single pass through the field. An air seeder integrates a chisel plow and a large seed cart that contains both the seed and chemicals. The cart uses a fan to blow air across an orifice, creating a low pressure area above the orifice that sucks up seed and chemicals. The product (seed and chemicals) is metered and blown through a network of distribution tubes, sometimes known as the primary and secondary distributors. The primary distributors receive the product from the cart and send it to the secondary distributors. The secondary distributors send the product through smaller tubes to each of the chisel plow openers, where it is released. See **Figure 4-19**.

Broadcast Spreaders

A ***broadcast spreader*** uses a paddle-type spinner that slings seed, lime, fertilizer, or chemicals in a circular fashion. See **Figure 4-20**. Broadcast spreaders used as seeders require another tool to cover the seed. They are used for planting non-row crops.

Mowers, Cutters, and Conditioners

Several different agricultural machines are designed to mow hay, forage, and grain crops. Examples include:
- Mowers.
- Swathers.
- Mower conditioners.
- Harvesting headers:
- Combine straight cut.
- Forage harvester header.

Figure 4-20. Broadcast spreaders sling the chemicals, fertilizer, or seeds in a circular pattern.

Mowers

Mowers are designed to cut and lay down hay crops to dry in the field. Some mowers use a *sickle bar knife drive* for cutting crops. The knife drive has triangular knife sections riveted or bolted to the cutter bar, which is cycled back and forth through cutter guards. As the knife is cycled back and forth, it causes a shearing action between the knife and the guards, which cuts the crop's stalks. See **Figure 4-21A**.

Manufacturers use different types of drives to cause the reciprocating (back and forth) action of a sickle bar, including pitman drives, wobble boxes, and planetary-gear boxes. In a *pitman arm drive*, a linkage called a pitman arm is connected to a pulley cam assembly. Since the pitman arm is connected off center, it reciprocates as the pulley turns. This reciprocating motion is translated through the pivot arm to the knife. See **Figure 4-21B**.

In rotary mowers, a mechanical gearbox is driven by a PTO. The gearbox rotates several oval discs. The discs are sometimes called turtles because of their shape. The discs have two attached knives that spin at speeds up to 200 mph. Because the blades always spin in the same direction, only one side is used for cutting. The knives can be flipped to extend their useful life. Rotary mowers can operate at higher field speeds than a traditional sickle bar mower.

Rotary mowers use a horizontal gearbox that drives the rotating discs. On many rotary mowers, the gearbox consists of several individual modules. This modular design makes the mower cheaper and easier to repair because only the bad module must be replaced, rather than the entire gearbox assembly.

Each rotary disc is driven via a *shear hub* that is designed to shear in the event a knife hits an obstruction. The sacrificial hubs are normally easy and cheap to replace compared to replacing the mower's gearbox. See **Figure 4-22**.

Swathers

A *swather*, or *windrower*, performs two functions: it mows the crop and gathers the crop into a single swath, also known as a windrow. A swather can be self-propelled, containing its own engine and transmission, or it can be pulled and powered by a tractor. See **Figure 4-23**.

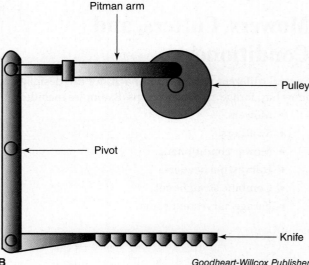

Figure 4-21. Sickle bars can be used on mowers, combine headers, and forage harvester headers. A—Sickle bar mowers slide a knife through guards, causing the sickle to shear the crop stalks like scissors. B—Pitman arm drives are one method creating the reciprocating motion of the knife drive.

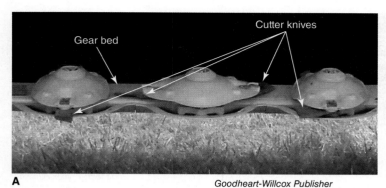 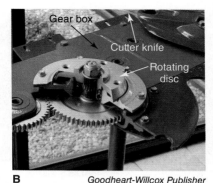

Figure 4-22. Rotary mowers cut standing crops. A—Rotary mowers have a horizontal gearbox, also known as a gearbed, that drives several discs with swinging cutter knives. B—The knife drive gearbox is made of multiple modules that drive the rotary discs and knives. This cutaway illustrates the spur gears used to drive one rotary disc module.

Figure 4-23. A self-propelled swather is also known as a windrower. It uses a straight cut header similar to that on a combine. It cuts the crop and gathers it into a single windrow that can later be picked up by a combine or baler. Swathers can also be pulled by a tractor.

Mower Conditioners

A *mower conditioner (MOCO)* is a swather that contains two long, horizontal rollers that are designed to crimp or crush the hay crop. This conditioning process cracks the stems, causing the hay to dry faster and be ready to bale sooner than hay that has been conventionally mowed and field dried. The conditioner rollers can be rubber, urethane, or steel. If the conditioner uses two smooth rolls, they are called *crushing rollers*. If the conditioner uses one or two rollers with lugs, the rollers are known as *crimping rollers*. See **Figure 4-24A** and **Figure 4-24B.**

A mower conditioner can be used as a swather to windrow grain crops. However, the conditioner *must* be disabled or it will shatter the grain, resulting in costly grain loss. Some mower conditioners use an impeller instead of rollers to condition the hay. The impeller has numerous swinging tines (or fingers) that rub the crop against the conditioner's hood. See **Figure 4-24C.**

Figure 4-24. Mower conditioners use either rollers or impellers to crimp or crush the hay to speed the drying process. A—Rubber crimper rollers. B—Steel crimper rollers. C—Impeller.

> **Note**
> Harvesters also cut crops as part of the harvesting operation. These machines will be discussed later in the chapter.

Rakes

A *rake* is used to windrow hay. They can be used to rake hay that has been cut and left to dry in the field, rake two small windrows into a single larger windrow, or to turn over hay that has become wet. Common types of rakes are wheel rakes, parallel bar rakes, and rotary rakes.

Wheel Rake

A *wheel rake* has several thin wheels with tines around their circumferences. As the rake is pulled through the field, the wheels rotate (ground driven), which lifts the hay into windrows. See **Figure 4-25**.

Figure 4-25. Wheel rakes are driven by the ground as they are pulled across the field.

Parallel Rake

Parallel rakes have several rotating bars with tines. The rotating bars form a rotating reel. The rake can be driven by the ground, a PTO, or a hydraulic motor. Single or multiple parallel rakes are pulled by a tractor. See **Figure 4-26**.

Rotary Rake

A *rotary rake* has one or more rotating vertical spindles, known as rotors. Each rotor has several horizontal tine arms extending outward from it. A number of vertical tines are attached to the end of each tine arm. As the rotor spins and the tine arms rotate, the tines gather the hay and roll it over to a side curtain, which forms the windrow. The rotors are driven by the tractor's PTO. See **Figure 4-27**.

Chapter 4　Agricultural Equipment Identification

TFoxFoto/Shutterstock.com
Figure 4-26. Parallel rakes contain parallel bars with tines that rotate and lift the hay into a windrow.

dvoevnore/Shutterstock.com
Figure 4-27. A rotary rake rolls the hay to the side curtain, which forms the windrow.

Tedders

Tedders have several small PTO-driven rotors with tines that are designed to pick up hay and turn it over. Tedders are used to speed the drying of freshly cut hay or hay that needs further drying, such as cut hay soaked by a heavy rain. See **Figure 4-28**. Hay that is turned with a tedder will be ready for baling sooner than hay that is left to dry undisturbed.

Hay Balers

Hay balers are used to pick up a swath of hay and pack it into a bale that can be stored and fed to livestock. Manufacturers produce both square and round balers.

Dar1930/Shutterstock.com
Figure 4-28. Tedders lift the hay and turn it over to speed the drying process.

Square Balers

Square balers form rectangular bales. The bales are held together with either baling twine or baling wire. **Figure 4-29** lists examples of square bale sizes.

Conventional Square Baler Operation

A conventional small square baler is pulled behind and to the right side of the tractor. It uses a pickup assembly that lifts the windrow of hay off the ground. See **Figure 4-30**. The pickup delivers the hay to a rotating auger. The auger works in conjunction with reciprocating feeder teeth to deliver the hay into the baler's bale chamber.

A PTO-driven flywheel propels a crank arm that drives a *plunger*. The plunger reciprocates inside the chamber. The plunger has a knife that cuts the hay as it is pushed into the bale chamber, which forms a bale flake. The plunger packs the flakes into a firm bale. The feeder teeth must be timed so that they are out of the chamber when the plunger is stroked in the chamber, otherwise the plunger will break the teeth. As the bale flakes are compressed into the bale, the bale advances through the baler. As the bale advances, strands of baling twine are fed out and strung across the top and bottom of the bale.

Small Square Baler								
Model	Baler Chamber Size	Bale Length	Bale Weight Example	Number of Knotters	Minimum PTO HP Required	Plunger Speed	PTO Speed	Baler Weight
Hesston 1836	14″ × 18″	12″–52″	70 lb	Two	45 hp	92 strokes/min	540 rpm	2700 lb
Hesston 1842	16″ × 18″	12″–52″	90 lb	Two	50 hp	100 strokes/min	540 rpm	4375 lb
Hesston 1844N	15.75″ × 22″	12″–52″	145 lb	Three	80 hp	100 strokes/min	1000 rpm	8000 lb
Large Square Baler								
Model	Baler Chamber Size	Bale Length	Bale Weight Example	Number of Knotters	Minimum PTO HP Required	Plunger Speed	PTO Speed	Baler Weight
Hesston 2250	31.5″ × 34.4″	9′	1200 lb	Four	150 hp	47 strokes/min	1000 rpm	15,505–18,420 lb
Hesston 2270	34.4″ × 47.2″	9′	1500 lb	Six	170 hp	47 strokes/min	1000 rpm	19,701–23,611 lb
Hesston 2290	47.2″ × 50.2″	9′	2000 lb	Six	180 hp	33 strokes/min	1000 rpm	23,197–24,321 lb

Goodheart-Willcox Publisher

Figure 4-29. Small and large square balers vary in chamber size, weights, and required PTO shaft speed and horsepower.

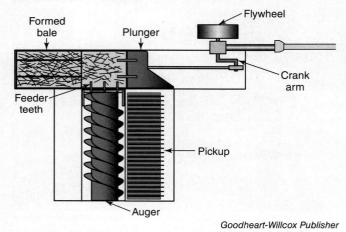

Goodheart-Willcox Publisher

Figure 4-30. This top view of a conventional square baler reveals the plunger that reciprocates in the bale chamber.

The baler has a wheel that rides along the top of the bale. As the wheel rotates, it measures the length of the bale. When the bale reaches a set length, needles move up through the bottom of the baler and deliver the twine to the knotters. The knotters will then cut and tie the two ends of the twine into a knot. See **Figure 4-31**.

Inline Square Baler

Manufacturers produce small and large inline square balers. They are similar to conventional square balers except inline balers are pulled directly behind the tractor, making them easier to transport by road. However, because of their design, the tractor must straddle the windrow of hay in the field. Large inline square balers use much larger components and operate at half the plunger speed compared to small balers.

The inline baler's pickup delivers the crop to a *pre-charge chamber*. See **Figure 4-32**. As the pickup

Goodheart-Willcox Publisher

Figure 4-31. This knotter assembly is on display at the AGCO Hesston, KS manufacturing facility where the balers are manufactured. The large square balers use six knotters to tie six twines on the bale.

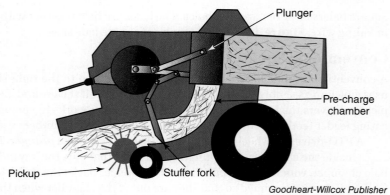

Goodheart-Willcox Publisher

Figure 4-32. A side view of an inline square baler reveals a pre-charge chamber, where the bale flakes are formed. The stuffer fork loads the flake into the bale chamber, where it is compressed by the plunger.

delivers hay into the pre-charge chamber, a flake is formed. A stuffer fork stuffs the flake of hay into the bale chamber. The ***stuffer fork*** places the flake in the bale chamber when the plunger is retracted and out of its way. When the stuffer fork retracts, the plunger moves forward, compressing the flake into the bale. From that point on, the operation of the inline baler is very similar to that of a conventional square baler.

Note

Banana-shaped bales will be produced in an inline baler if the flakes have less hay or if the stuffer fork cannot lift the flake all the way into the bale chamber.

Round Balers

Round balers vary in design and size. Modern round balers are offered in two different designs:
- Bale is formed and carried inside a variable bale chamber.
- Bale is formed and carried inside a fixed bale chamber.

The most common round baler forms and carries the bale inside a variable bale chamber. ***Variable bale chambers*** give farmers the option of changing the bale's diameter, for example 3′ to 6′, while being able to maintain a consistent bale density. Variable bale chambers produce solid core bales that, when viewed from the side, have a consistent spiral shape.

Fixed bale chamber balers can produce only fixed-diameter bales. As the hay is directed to the chamber, there is no method of initially holding pressure against the newly forming bale. The hay remains loose and tumbles until the bale grows in size enough to reach the belts. When the bale is large enough to touch the belts, the belts can apply pressure to the growing bale. This tends to pack the outside of the bale more densely than the inside. As a result, the baler produces a bale with a soft, star-shaped core and an inconsistent density.

Note

Lundell previously manufactured a round baler that formed and rolled the bale along the ground as the tractor pulled the baler. It is no longer in production.

Round Baler Widths

The width of a round baler determines the length of bale it produces. Manufacturers offer balers with three-foot, four-foot, five-foot, and six-foot widths.

Variable Chamber Round Baler Operation

A round baler, like the inline square baler, is pulled directly behind the tractor, requiring the tractor to straddle the windrow of hay. The baler's pickup delivers the hay into the baler's stuffer. See **Figure 4-33**. The rotary stuffer contains a left-hand and a right-hand auger on each end. The stuffer's augers center the hay to the width of the baler's chamber. The stuffer has numerous tines. As the stuffer rotates, the tines stuff the crop into the baler's chamber. A baler's stuffer can also have knives that slice the hay, which helps pack the hay into a tight bale and makes it easier to break apart when feeding livestock. Knife-equipped balers can have an adjustment that allows the operator to choose the number of knives that are slicing the hay entering the bale chamber.

Goodheart-Willcox Publisher

Figure 4-33. A baler's pickup will lift the windrow of hay and place it into the stuffer or bale slicer.

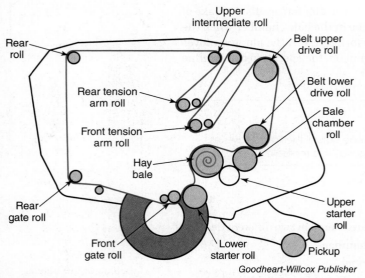

Figure 4-34. A variable bale chamber is designed to have the belts place pressure on the bale from its earliest formation stage.

Round balers have starter rolls that cause the incoming hay to tumble and start the shape of a rolling bale. See **Figure 4-34.** The majority of round balers use multiple long rubber belts that press against the bale. The belts guide, shape, and drive the bale during its development.

> **Note**
>
> Krone Fortima round balers and old New Holland balers do not use rubber belts, but instead use chains and steel bars to form the bale.

In a variable chamber baler, the tension rollers hold the belts against the newly forming bale. As the bale grows in diameter, the tension rollers lift, allowing the belts to follow the bale as it grows in diameter. The steady tension on the bale results in a consistently dense bale.

Old balers had a bale diameter indicator on the baler, which required the operator to continuously look over his or her shoulder to monitor the baler. The operator had to manually initiate bale wrapping when the bale reached the desired diameter. Late-model round balers have an electronic bale monitor inside the tractor's cab. The monitor allows the operator to know that the bale has reached its full diameter.

> **Note**
>
> Round balers require the operator to stop the tractor so the baler can wrap or tie the bale. Once the bale has been tied (or net-wrapped), the bale has been ejected out of the baler, and the rear door has been closed, the operator can begin baling hay again. Square balers allow the operator to continuously move forward with the baling operation without stopping to tie and eject the bale.

When the bale reaches its set diameter, late-model balers automatically wrap the bale. After the bale is wrapped, the following sequence of events occurs:

- The de-clutcher disengages the baler's drive.
- The baler's tailgate lifts, causing the bale to roll out of the baler.
- A C-shaped frame assembly, known as a kicker, pushes the bale away from the baler.
- The kicker retracts.
- The tailgate closes.
- The baler's clutch begins driving the baler, ready to form another bale.

Twine versus Net Wrap

If the baler uses twine, more time is required to wrap the bale because multiple passes of the twine must be wrapped around the bale. If the baler uses net wrap, a mesh sheet material, the wrapping process is sped up dramatically. With net wrap, the bale only has to be spun 1.5 to 3 revolutions, as compared to 15 or more revolutions for twine. However, net wrap is significantly more expensive than twine.

Bale Monitors

In-cab bale monitors enable farmers to remain facing forward while baling hay instead of having to frequently look back at the bale indicator on the baler. Many bale monitors display two or three columns of bars that indicate the shape and diameter of the bale. As the baler is operating, the bale indicators show how the bale is growing and if one side is larger than the other side. See **Figure 4-35**.

Bale monitors can list the number of bales that were baled per field or the total number of bales that the baler has baled. They also provide the operator several different functions for varying the operation of the baler.

Controls

Baler operation can be adjusted electronically or mechanically, depending on the machine design. A few examples of baler controls are bale density, bale diameter, choice of twine or net wrap, spacing between twine, twine proximity from the edge of the bale, and the number of knives in the bale slicer.

A denser bale will hold more hay and be much heavier than a less dense bale. The amount of pressure placed on the baler's tension arms directly controls bale density. Common methods for varying the pressure on the tensioner's arms include coiled springs, air pressure supplied to air bags, and hydraulic pressure applied to a tension cylinder. A pressure gauge is sometimes mounted on the front of the baler for quick viewing of tensioner pressure.

Some balers are equipped to use either twine or net wrap, depending on the operator's preference. Some old balers can use only twine. When the space between the twine is widened, the bale wraps and ties quicker and uses less twine; however, it increases the chances of hay blowing away during transport.

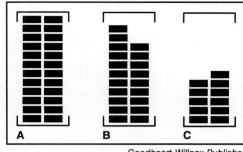

Goodheart-Willcox Publisher

Figure 4-35. The bars of a bale monitor indicate how full the bale is on the right and left sides. A—All 24 bars illuminated indicating the bale is full and squarely formed. B—The left side of the bale is larger than the right side; the operator should weave to the right to even the bale formation. C—The right side of the bale is larger than the left side; the operator should weave to the left to even the bale formation.

Note

Some late-model balers are equipped with automatic lubrication systems, which help ensure the numerous grease points get properly lubricated. Older machines can have aftermarket automatic lubrication systems added.

Forage Harvesters

Forage harvesters cut, chop, and blow the crop, known as *silage*, into a wagon. The whole plant is cut at the ground level, allowing the forage harvester to harvest the stalk and grain of the plant. Forage is cut at high levels of moisture, higher than any other crop. The silage's moisture content can range from 50% to 70%, depending on the storage technique. See **Figure 4-36**. Silage is commonly fed to dairy cattle. Examples of crops harvested for silage are corn, milo, and cereal grain crops. Corn is the most popular silage crop.

Silage can be baled or stored in silos or bunkers. For storage, silage must be packed tightly to eliminate airspace, otherwise the oxygen will rot the silage. If the silage is stored in bunkers, large four-wheel-drive tractors with dozer blades are used to pile and pack the silage. The bunker is then covered with plastic to keep air out. If silage is stored in a silo, it is loaded from the top so gravity will pack the silage to eliminate airspace. Baled silage is wrapped tight in plastic wrap to eliminate oxygen.

Forage harvesters can be self-propelled or pulled and powered by a tractor. See **Figure 4-37**. Self-propelled forage harvesters are the highest-horsepower machines used in the agricultural industry. Pull-type forage harvesters are powered by the tractor's PTO.

oticki/Shutterstock.com

Figure 4-36. Forage harvesters chop the entire plant. The silage is green because it has a high moisture content.

Figure 4-37. Forage harvesters can be self-propelled or pull type. A—Self-propelled forage harvesters require high-horsepower engines. B—Pull-type forage harvesters are pulled and powered by a tractor.

> **Note**
> The engine horsepower for some self-propelled forage harvesters will be stated in PS, for Pferdestarke, which is the German word for horsepower.
> 1 PS = 0.986 US horsepower.

The harvesters can use a few different types of headers: straight cut header, non-directional corn header, row crop header, or a windrow pickup header. See **Figure 4-38**. The row crop header must harvest the crop in the specific row spacing. Some large headers used on self-propelled machines can fold, making it easier to transport the machine.

Corn headers can be equipped with row guidance sensors that are used in conjunction with *auto-guidance systems*. A row guidance sensor senses the location of corn stalks as they enter the header. If the machine moves too far to the right or to the left of the rows of corn stalks, the sensors provide an input to the electronic control module informing it that the machine is steering off course. Many late-model agricultural machines, such as forage harvesters, combines, tractors, and cotton harvesters use electronic global positioning satellite (GPS) systems that can automatically steer the machine as it is working in the field. The machine uses an antenna to receive a signal from multiple satellites to get a position of the machine's location. A controller is used to electronically control the machine's steering system to guide the machine through the field.

Figure 4-38. Forage harvester headers. A—A non-directional corn header allows the harvester to harvest both parallel to the corn rows or across them (perpendicular). B—A windrow pickup header that picks up and feeds the silage into the forage harvester. Both self-propelled and pull-type forage harvesters can use a windrow pickup header.

Forage Harvester Operation

Forage harvesters perform the following functions:
- Cut and guide crop into the machine (header).
- Feed crop into the machine and chop it (cutter head).
- Process crop (crack the kernels).
- Throw or blow the silage into the wagon (cutter head or fan).

The functions of a forage harvester depend on the type of header installed. With a pickup header, crop that has previously been cut, windrowed, and dried is picked up and fed into the chopper. With all other types of headers, the crop is cut and fed into the harvester at the same time.

The heart of a forage harvester is the chopper. It contains a rotating cutter head equipped with knives. A *shear bar* is positioned so that the knives cut the crop as the cutter head rotates past the shear bar. See **Figure 4-39**.

The speed of the feed rollers determines the length of the cut crop. Decreasing the speed of the feed rollers, while maintaining the same cutter head speed, will shorten the length of the cut crop.

Forage harvesters will either cut and throw the silage or cut and blow the silage. On *cut-and-throw forage harvesters*, the high-speed cutter head performs two functions. It cuts the silage and throws the silage so that it can travel through the spout into the wagon. A *cut-and-blow forage harvester,* has a fan downstream from the cutter head that blows the chopped crop through the spout and into the wagon.

Many forage harvesters use a metal detection system to protect the machine. For example, the lower feed roll can contain a large magnet that induces a small current if any metal moves past the feed roll. This induced current signals the controller, which automatically disengages the chopper to protect the machine.

The cutter head's knives will dull from everyday use. Forage harvesters have provisions for sharpening the cutter head knives and adjusting the position of the shear bar. On some machines, both adjustments can be performed automatically. Other harvesters require the tasks to be performed manually. The cutter head knives are sharpened by spinning the cutter head backward while a sharpening stone is stroked laterally back and forth across the rotating cutter head. A shear bar adjustment is also made while the cutter head is being spun backward. The shear bar is slowly advanced toward the cutter head until a slight ticking sound is heard. Then, the shear bar is slightly retracted and tightened in place. A pull-type forage harvester contains two PTO input shafts. One is the normal harvester input shaft, and the other rotates the cutter head in reverse and is used for sharpening the blades and adjusting the shear bar.

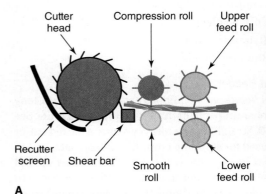
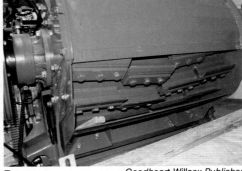

Goodheart-Willcox Publisher

Figure 4-39. Forage harvesters use cutter heads to chop the crop. A—The feed rolls deliver the crop to the cutter head. The cutter head cuts the crop as it spins past the shear bar. B—This cutter head is from a John Deere self-propelled forage harvester.

Warning

Although all agricultural equipment is dangerous, a self-propelled forage harvester poses some of the greatest risks as it operates in the field. The chopper has tremendous horsepower and will chop whatever is fed into the machine. Because the operator must pay constant attention to the location of the wagon to load it evenly, the operator is constantly turning his or her head to the side, which increases the risk of running into unseen objects in the field.

Some forage harvesters contain a kernel processor that is responsible for cracking the corn kernels. After the cutter head chops the crop, it is fed through the center of the kernel processor, which consists of two steel rollers rotating at different speeds. Cracking the kernels ensures that the cattle will receive the full nutritional benefit of the crop.

Combines

The term combine was coined when manufacturers combined multiple harvesting functions into one machine. Prior to the introduction of combines, farmers would *reap* (cut and windrow the crop), *thresh* (rub the crop so the grain breaks out of its hulls), and *winnow* the crop (blow away the chaff and keep the grain). A combine performs these three functions and also transports and unloads the grain.

Combine Functions

A combine harvester is designed to perform the following tasks:
- Cut or strip the standing crop.
- Feed the crop into the thresher.
- Thresh the crop.
- Separate the grain from the chaff.
- Deliver the clean grain to the grain tank.
- Spread the straw and chaff.
- Unload the clean grain into a grain cart or semi-truck hopper.

Cutting or Stripping

The combine's header cuts or strips the crop and feeds the crop into the combine. Combines can operate several different types of headers:
- Straight cut header.
- Windrow pickup header.
- Corn header.
- Stripper header.
- Row crop header.

Straight Cut Header

Straight cut headers are equipped with a rotating reel, long auger or rubber draper belts, and a reciprocating knife. See **Figure 4-40**. Straight cut headers can have a locked cutter bar that is used for standing crops like wheat, oats, barley, and rice. Flex headers allow the cutter bar to flex up and down, enabling the header to cut crops that grow close to the ground, like soybeans. Some flex headers can be locked rigid when they are used for harvesting standing crops.

James W. Copeland/Shutterstock.com

Figure 4-40. Straight cut combine headers directly cut and feed the crop into the combine's feeder house. They can use a long auger or a draper belt.

The width of straight cut headers can range widely, for example from 10′ to 60′ wide. A rotating reel lifts and holds the crop against the header's guards so the knife can cut the crop's stems. After the crop has been cut, the reel delivers the crop into the header's auger or draper belts. The reel can be equipped with plastic or steel tines (a pickup reel), or have long lightweight bars (a bat reel). The reel can have the following adjustments:

- Pitch or angle of the tines.
- Speed, including changing the speed in relationship to the combine's field speed.
- Lift and lower.
- Push fore and aft.

The header can be equipped with a long auger that delivers the crop to the center. Toward the center of the auger, the flighting is replaced with fingers. The center fingers lift the crop into the feeder house. The fingers are timed so that they will not interfere with the feeder house drive. See **Figure 4-41A**. Straight cut headers can use long rubber belts, called *draper belts*, in place of the auger. See **Figure 4-41B**.

Note

A straight cut header is sometimes called a table. The longest straight cut headers will use two reels instead of one long reel. The longest headers also normally use draper belts in place of a long auger.

Windrow Pickup Header

Some crops are first cut and swathed into a windrow so the crop can dry before harvest. A *windrow pickup header* contains a rubber belt with tines that are designed to pick up the windrowed crop and feed it into header's auger, which feeds the crop into the combine's feeder house. See **Figure 4-42**. These headers are used to harvest small grains like wheat, oats, barley, and canola.

A

B

Goodheart-Willcox Publisher

Figure 4-41. Straight cut headers are equipped with either a long auger or rubber draper belts. A—A straight cut header equipped with an auger that has fingers located in the middle, between the auger's flighting. The fingers are timed so that they are retracted when they are facing the feeder house. B—This MacDon brand header uses rubber draper belts to gather the crop to the center of the header.

Sergei Butorin/Shutterstock.com

Figure 4-42. Windrow pickup headers pick up a crop that has been previously cut, swathed, and laid in a windrow on the ground.

Figure 4-43. Two row units on the underside of a corn header are shown here.

Corn Header

Corn headers have individual row units that are designed to pull the corn plant's stock downward allowing the ears to be stripped as they hit the header's stripper plates, located above the stalk rolls. See **Figure 4-43**. The stalk rolls pull the stalk down, the stripper plates remove the ears from the stalk, and the gathering chains deliver the ears to the header's auger. See **Figure 4-44**. Each row unit contains a gearbox that drives both the stalk rolls and the gathering chains. The gathering chain sprockets are driven by worm gears and the stalk rolls are driven with bevel gears.

Corn headers are built to harvest a specific number of rows of corn (for example, 6, 8, 12, 16, or 18 rows) that are spaced a specific distance from each other (for example, 20″, 22″, or 30″).

Some manufacturers build a corn-chopping header that has lawnmower-type blades below the stalk rolls. These rotating blades chop the stalks as the corn is harvested. Chopping the stalks makes it easier to plant the field in the following season. A corn header can also be designed to fold, as shown in **Figure 4-45**, making it easier to transport.

Stripper Header

Shellbourne Reynolds manufactures a *stripper header* with a long rotating rotor with comb-like teeth that strip the grain off the crop's stems, leaving the stems standing in the field. The rotor spins at fast speeds, for example 600 rpm. The header dramatically reduces the amount of crop residue that is fed into the feeder house. The stripper header is used for harvesting cereal crops like wheat, barley, flax, and oats. It is also used for harvesting rice and grass seed. See **Figure 4-46**.

Row Crop Header

Years ago, John Deere built row crop headers. Like a corn header, it harvested crops planted between widely spaced rows and did not use a reel. It was designed to

Figure 4-44. These gathering chain assemblies have been removed and set on a workbench.

 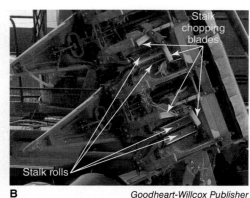

Figure 4-45. Some large corn headers are designed to fold for transport. A—With the header folded, the combine is narrow enough to travel down roads. B—A close-up reveals that the stalk rolls on this header have sharp edges that chop the stalks as they are pulled through. The header also has rotating blades that further chop the stalks.

Figure 4-46. Stripper headers use high-speed rotors equipped with comb-like teeth that strip the gain off the crop's stems. A—Front view of the stripper header. B—The comb-like teeth on a stripper header rotor.

harvest row crops other than corn, such as sunflowers and soybeans. Today, sunflowers and soybeans are usually harvested with straight cut headers.

Feeding

A *feeder house* connects a combine's header to the combine and feeds the crop from the header into the combine. It uses a front drum and a rear beater that are rotated by two, three, or four chains. Feeder slats or bars are attached to the chains. The slats deliver the crop from the header to the combine's thresher. See **Figure 4-47**. The feeder house, also known as the throat, can be very small or large.

Two different types of drums are used, standard and stone retarder. A stone retarder drum has deeply recessed openings where the chains ride, allowing the feeder slats to ride closer to the drum. This reduces the chance of feeding rocks into the feeder house.

 Note
Claas offers optional feeder belts in place of feeder chains, which reduces noise. Belts also last twice as long as the chains.

Feeder houses can also include a rock trap. See **Figure 4-48**. It consists of a tray-like box below the beater and

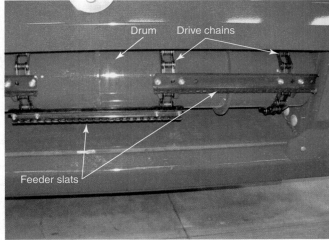

Figure 4-47. Feeder houses contain a drum at the opening of the feeder.

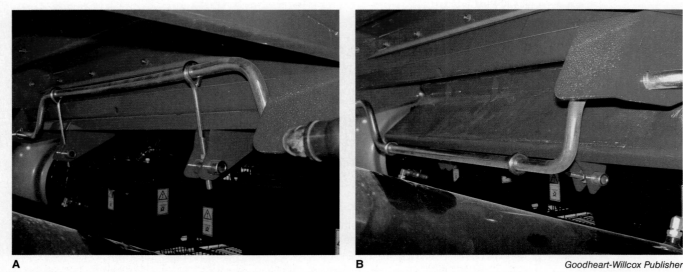

Figure 4-48. Rock traps are located at the rear of a feeder house and provide a place for rocks to accumulate, reducing the chance that a rock is injected into the combine's cylinder or rotor. A—The rock trap door is closed. B—The rock trap door is open.

provides a place for rocks to fall and collect during harvesting. The operator must check the rock trap daily by opening the rock trap door. The door must be closed before the machine is operated.

New Holland has an *Advanced Stone Protection (ASP)* system that uses acoustic sensors to detect the presence of stones. If a stone is detected, the system automatically stops the header and feeder, ejects the stone, sounds an alarm, and displays a message on a monitor. The operator can reset the ASP by reversing the feeder and lifting the header.

Threshing

Combines can be classified as conventional or rotary. The classification is based on where the majority of threshing occurs, a cylinder or rotor(s).

Conventional Cylinder

A *conventional combine* uses a rotating cylinder, positioned parallel to the header and feeder drum, as the *primary* means of threshing the crop. The cylinder contains rasp bars that rotate the grain. A set of wire shells, known as *concaves*, are located under the cylinder and allow the threshed grain to fall into the conveyance augers. See **Figure 4-49.**

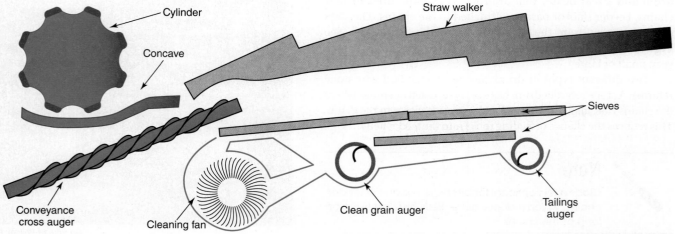

Figure 4-49. Conventional combines use a horizontal cylinder that must thresh the crop on its single pass past the cylinder. It also rethreshes any crop that is sent back to the cylinder from the tailings auger and elevator.

Conventional combines are the oldest type of combine. Because the crop only passes through the cylinder once, the cylinder and concave must be aggressive to properly thresh the crop. Manufacturers use wide feeder houses to spread the crop across the entire width of the cylinder.

Claas Lexion offers a conventional combine that has a small cylinder in front of the main threshing cylinder. This smaller cylinder rotates at 80% of the speed of the main threshing cylinder. It separates approximately 30% of the grain from the crop before the crop enters the main threshing cylinder. Claas calls this design the Accelerated Pre-Separation (APS) system. See **Figure 4-50**. Claas APS has been improved. The threshing cylinder diameter increased from 24" with eight rasp bars to 30" with ten rasp bars. The diameter of the rear impeller was increased from 15" to 24". It is now called APS SYNFLOW HYBRID.

Many *rotary combines* use one rotating rotor as the *primary* means of threshing the crop, **Figure 4-51**. The rotor has rasp bars that rotate the crop around the rotor, churning the crop. This action loosens and shakes the grain off the stalks. A set of concaves surrounding the rotor allows the threshed grain to fall into the conveyance augers, **Figure 4-52**.

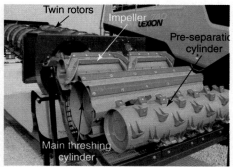

Goodheart-Willcox Publisher

Figure 4-50. The Claas Lexion combine uses an accelerated pre-separation system that has a pre-separation cylinder in front of the main cylinder and the impeller. The twin rotors behind the APS system are used mainly for separation rather than threshing.

Note
Some conventional combines are equipped with twin rotors. However, these rotors are not used as the primary method of threshing the crop, but are used for separating the crop. The John Deere CTS (Cylinder Tine Separation) combine, Caterpillar Challenger combine, Claas Lexion combine, and the Massey Ferguson Delta combine have used this design.

Single Rotor Rotary

Rotary combines are gentler on the crop than conventional combines because the crop makes multiple passes around the rotor rather than a single pass. The grain is released from the stems due to repeated crop-on-crop contact rather than an aggressive metal-on-crop contact during a single pass, as in a conventional combine.

The single rotor used in Case IH, John Deere, AGCO Challenger, Massey Ferguson and Versatile combines is axial (perpendicular to the header) and aligned with the feeder house. The axial rotor requires changing the lateral crop mat flow from the feeder house into a rotary motion inside the rotor. Most manufacturers use a corkscrew-shaped rotor, also known as a bullet rotor (John Deere) or an AFX rotor (Case IH). See **Figure 4-53**. AGCO's

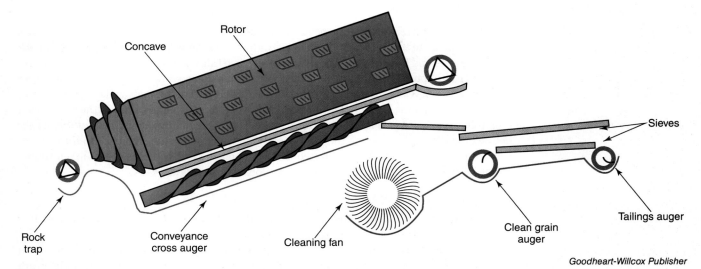

Goodheart-Willcox Publisher

Figure 4-51. On many rotary combines, the combine rotor is axial and aligned with the feeder house.

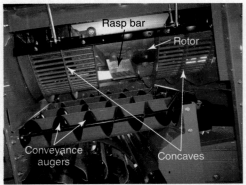

Figure 4-52. The John Deere STS combine uses an axial rotor. Notice in this photo that the center concave has been removed to reveal a rotor rasp bar. The grain falls through the concaves into the conveyance augers.

Figure 4-53. The John Deere STS combine has an axial rotor. The front beater is not a cylinder because it does not have a concave located below it. The beater feeds the crop into the rotor's transition area. Deere's STS rotor uses a rotor cage that increases in diameter in three stages, known as the tri-stream crop flow.

Gleaner brand combine places the rotor in a transverse position, parallel to the header. Gleaner advertises that their combine uses a natural crop flow because the lateral crop mat flow from the feeder house directly feeds one end of the transverse positioned rotor. The feeder house on a Gleaner combine is offset to the right of the machine, allowing the feeder house to directly feed the right side of the rotor, in the full width of the crop mat, directly into the rotor. Like axial rotors, the transverse rotor has vanes to convey the crop from one end of the rotor to the other.

The Versatile combine is the most unique single rotor combine. Instead of using a stationary concave that is opened and closed, the concave spins in the opposite direction of the rotor. It is called RCR for rotating concave rotary. It is marketed for use in cereal crops.

Twin Rotor Rotary

New Holland combines for decades have used twin axial rotors. The twin rotors are used to thresh and separate the crop. In 2021, John Deere introduced the X9 1000 and 1100 combines. These machines also use twin rotors for threshing and separating the crop. They are the largest combines offered by John Deere. The single rotor S-series Deere combines are also still being manufactured.

Separating

After the crop is threshed, the grain needs to be winnowed (separated from the chaff). Combines have a cleaning fan that blows air through *sieves* that look like several finned combs. Refer back to **Figure 4-49** and **Figure 4-51**. The sieves are adjusted from closed to open based on the type of crop and conditions. The cleaning fan directs air through the sieves, blowing away the chaff over the sieves. The sieves are driven forward and rearward as part of the sieve shaker system. Clean grain falls through the sieves, but unthreshed heads of grain cannot. As a result, the unthreshed crop rolls over the edge of the bottom sieve and falls into the tailings auger. The *tailings auger* delivers the unthreshed crop to the tailings elevator, which sends the crop back to the cylinder or rotor to be rethreshed. An *elevator* consists of a housing that contains chain-driven rubber paddles that move grain from an auger to a new location. See **Figure 4-54**.

Delivering the Clean Grain to the Grain Tank

As clean grain falls through the sieves, it is collected by the *clean grain auger*, which is located at the bottom of the combine. It collects the clean grain and delivers it to the clean grain elevator. The clean grain elevator lifts the grain to the inclined clean grain auger, which carries the grain the rest of the way to the top of the grain tank and dumps it into the grain tank. See **Figure 4-55**.

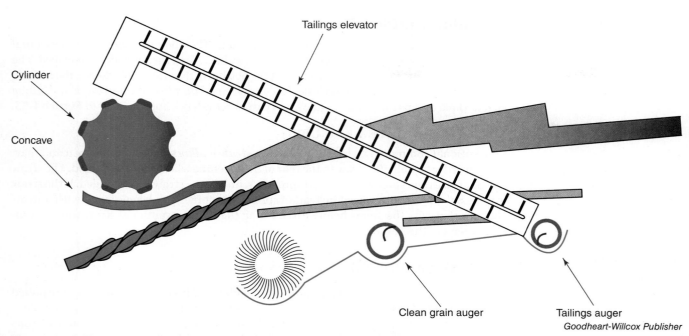

Figure 4-54. The tailings elevator delivers tailings to the cylinder to be rethreshed.

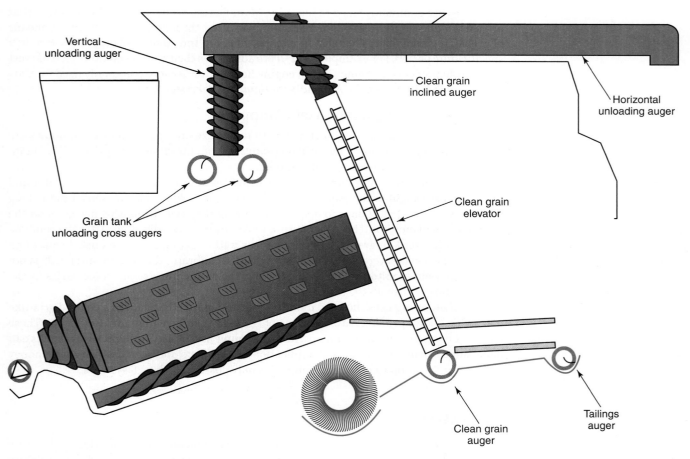

Figure 4-55. The clean grain auger delivers the grain to the clean grain elevator. The elevator sends grain to the grain tank via an inclined auger.

Unloading the Clean Grain

Once the grain tank is full and ready for unloading, the operator swings the horizontal unloading auger out of its seat, also known as a saddle. Next, the unloader is engaged. The unloading cross augers, which are located at the bottom of the grain tank, deliver the grain to the unloading vertical auger. The vertical auger delivers the grain to the horizontal unloading auger, which dumps the grain into the grain cart or semi-truck hopper. Refer to **Figure 4-55**.

Spreading the Straw and Chaff

Conventional combines can be equipped with *straw walkers* that move forward and rearward in order to walk the straw to the rear of the machine where it is discharged. The straw and chaff are then spread using chaff spreaders, unless the straw is to be baled. If the straw is to be baled, it is windrowed instead of spread. Combines can be equipped with a straw chopper that cuts the stems before discharging them. A conventional straw walker combine produces the best quality of straw for baling.

Types of Combines

Years ago, manufacturers produced pull-type combines. These small combines were pulled and powered by large agricultural tractors. Late-model combines are self-propelled. They can also be equipped with hillside leveling kits that enable the machines to harvest crops planted on steep hillsides while keeping the shaker system level.

Sizes

Combines are commonly sized in "classes." Manufacturers currently build machines that are class 5 and larger. Years ago, manufacturers used throughput as the primary means of classifying combines. The term *throughput* refers to the amount of crop harvested in a specific time period. For example, John Deere advertises their class 11 combine can harvest 7,200 bushels of corn per hour. Today, engine horsepower is more commonly used for classifying combine size. See the Appendix for *Combine Harvester Sizes*.

Yield Monitoring versus Yield Mapping

Late-model combines use GPS technology to maximize crop performance. Combines are equipped with yield monitoring, yield mapping, and/or auto-guidance systems. Yield monitoring is often confused with yield mapping.

Yield monitoring monitors grain flow using a sensor in the clean grain elevator and measures the weight of the crop. A controller uses the yield measuring sensor data along with the machine travel data to display the momentary yield (in bushel per acre) on the machine's yield monitor. The yield monitor must be properly calibrated. After the combine harvests a full tank of grain, the grain is measured to get an accurate bushel count and that information is entered into the monitor, which calibrates the yield monitor. GPS is not needed for yield monitoring as it only displays the momentary yield, which varies as the combine moves through the field.

Yield mapping uses the yield monitor data and places that data in a colored grid-like map. This allows producers to see areas of the field that have high and low yields. These maps are compared with other maps, like soil sample and moisture maps, so producers can make adjustments to the rates of application of fertilizers and seeds, which saves money and maximizes crop yields. Yield mapping is only possible with GPS systems.

Tractors

Agricultural tractors are one of the most heavily used machines on today's farms. Tractors can be used for tilling or cultivating, planting and seeding, cutting hay and forage, baling hay, packing silage, spreading fertilizer, grinding grain, spraying pesticides, and pulling carts and wagons.

Controls and Features

Depending on the purpose of the tractor, it might contain a three-point hitch, power take-off (PTO), and/or a loader.

Three-Point Hitch

A *three-point hitch* uses three arms to attach an implement to the tractor. To raise the implement, the tractor's hydraulic system operates hydraulic cylinders that cause a rockshaft to rotate. The rockshaft is splined to the two lift arms. As the rockshaft rotates, it causes two lift arms to lift. These arms are attached to the draft arms by lift links. A third arm, known as the center link, and the two draft arms attach to the implement. As the rockshaft rotates, it causes the tractor to lift the three-point hitch. See **Figure 4-56**. Examples of three-point-hitch–mounted implements are mowers, tillage tools, and planters.

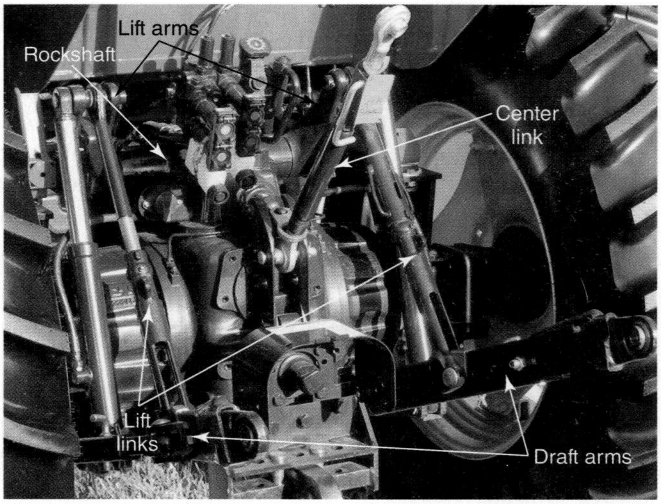

Figure 4-56. A three-point hitch has multiple arms. The tractor's hydraulic system lifts the hitch, causing the attached implement to lift.

Tractors can be equipped with a class 0, 1, 2, 3, or 4 hitch. A hitch's category determines the diameter of the hitch pin and the spacing between the two lower draft arms. See **Figure 4-57**. A sub-compact utility tractor uses either a class 0 or class 1 hitch, while high-horsepower tractors are equipped with class 4 hitches.

Some larger tractors have a hitch switch mounted at the rear of the tractor. Such a switch makes it easier to connect to an implement because the operator is outside the cab and has a better view of the implement. See **Figure 4-58**.

Many tractors attach a *quick-hitch* to the three-point hitch. The quick-hitch acts like a coupler assembly and makes connecting the tractor to the implement much quicker. The implement must match the quick-hitch. In **Figure 4-59**, the top of the quick-hitch has a U-shaped saddle that slides into the implement. The quick-hitch has two spring-loaded latches (one on each of the lower hooks) that keep the implement pins seated in the hitch hooks as the hitch is raised and lowered. The implement is detached by actuating the two levers on top of the quick-hitch (which releases the latches) and lowering the tractor's hitch. Quick-hitches are made for the different sizes of hitches.

Some implements connect to the tractor through only the bottom portion of the quick hitch coupler or the hitch's two draft arms. Notice in **Figure 4-60** that the implement is missing a center link connection.

Category	Center Link Pin Diameter	Draft Arm Pin Link Diameter	Spacing between Draft Arms	Recommended Drawbar Horsepower
0	5/8″ (17 mm)	5/8″ (17 mm)	20″ (500 mm)	<20 hp (<15 kW)
1	3/4″ (19 mm)	7/8″ (22.4 mm)	26″ (718 mm)	20–45 hp (15–35 kW)
2	1″ (25.4 mm)	1 1/8″ (28.7 mm)	32″ (870 mm)	40–100 hp (30–75 kW)
3	1 1/4″ (31.75 mm)	1 7/16″ (37.4 mm)	38″ (1010 mm)	80–225 hp (60–168 kW)
4	1 3/4″ (45 mm)	2″ (51 mm)	46″ (1220 mm)	180 hp (135 kW) and higher

Goodheart-Willcox Publisher

Figure 4-57. Three-point hitches are categorized based on the hitch pin diameter and the space between the draft arms.

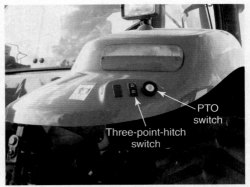

Figure 4-58. The right rear fender of this tractor has a three-point-hitch switch and a PTO switch.

Figure 4-59. Quick-hitches are installed on the tractor's three-point hitch to make connecting a tractor's hitch to an implement much quicker.

Power Take-Offs

Power take-offs (PTOs) consist of a splined stub shaft that protrudes from the front, middle, or rear of the tractor. One end of a matching splined driveshaft slips over the stub shaft and the other end is connected to an implement. When the PTO is activated, it rotates the driveshaft to provide power to the implement. Tractors use one of three types of PTO shafts.

- 540-rpm PTO, 1 3/8″-diameter shaft with 6 splines.
- 1000-rpm PTO, 1 3/8″-diameter shaft with 21 splines.
- 1000-rpm PTO, 1 3/4″-diameter shaft with 20 splines.

Some tractors have a PTO that can rotate at either 540 rpm or 1000 rpm. One such design uses two different PTO stub shafts. See **Figure 4-61**. Another style has a single shaft that is removed from the tractor, flipped, and re-inserted into the tractor to change output speed. In one position, the shaft rotates at 540 rpm, and in the other position, it rotates at 1000 rpm. In the 1000-rpm configuration, power flows directly from the PTO clutch, through the shaft, to the implement. In the 540-rpm configuration, the PTO clutch diverts power through a set of gears that lowers the shaft output speed to 540 rpm.

The most common location for the PTO is at the rear of the tractor. The PTO shaft rotates clockwise when viewed from the rear of the tractor. PTOs are used to power mowers, balers, pumps, grain grinders, elevator augers, snowblowers, and rotary tillers.

Mid-mounted PTOs are located in the middle of the tractor and are used to power belly mowers and snowblowers. Front-mounted PTOs commonly work in conjunction with a three-point-hitch front-mounted implement.

>
> **Note**
> Regardless whether the PTO is 540 rpm or 1000 rpm, the proper PTO speed is obtained when the engine is operating at high rpm. The tractor's tachometer often shows the engine speed at which the tractor delivers the rated PTO speed. See **Figure 4-62**.

Figure 4-60. Some implements connect through the hitch's two lower draft arms, and do not use the third link (center link).

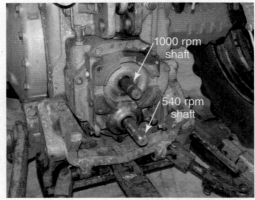

Figure 4-61. This tractor has a 1 3/8″ 1000-rpm PTO shaft on top and a 1 3/8″ 540-rpm on the bottom. This option allows the tractor to drive a 540-rpm implement or a 1000-rpm implement.

Chapter 4   Agricultural Equipment Identification

Figure 4-62. Engine tachometers commonly display the specific engine speed that is necessary to achieve the rated PTO rpm. In this example, when the engine runs at 3150 rpm, the PTO spins at 540 rpm.

PTOs can be classified as one of four different types: transmission-driven, continuous-running, independent, and ground-speed. A *transmission-driven PTO* is powered after the transmission clutch. When the operator disengages the transmission clutch, the transmission and the PTO are both disengaged. Although this PTO is unpopular, it can be found on some smaller tractors with manual transmissions, such as the Kubota B2320, B2620, and B2920. See **Figure 4-63A**.

A *continuous-running PTO* is engaged and disengaged with a two-stage, clutch that is operated by one clutch pedal. When the operator presses the clutch pedal halfway down, the transmission is disengaged. When the operator presses the clutch pedal completely to the floor, both the transmission and the PTO are disengaged. Continuous-running PTOs are rarely used in modern tractors. An example is the John Deere 4005 compact utility tractor. See **Figure 4-63B**.

A continuous-running PTO meets the definition of a *live PTO*, which is a PTO that can continue to operate even if the tractor quits moving. A transmission-driven PTO is not a live PTO, because when the clutch pedal is pressed, the tractor and the PTO both stop.

An *independent PTO* uses a clutch that is independent of the transmission. It receives input directly from the engine. It can be controlled separately from the tractor's transmission. Late-model tractors commonly use electricity to control a solenoid that allows hydraulic fluid to apply the PTO clutch. It is a live PTO that is driven unrelated to ground speed or transmission engagement. This is by far the most common PTO used in late-model tractors. See **Figure 4-63C**.

A *ground-speed PTO*, also known as ground-drive PTO, is powered only when the tractor is moving forward or reverse. Think of it as being connected to the rear axle. See **Figure 4-63D**. If the tractor is stationary, the PTO is not rotating, and if the tractor is traveling at higher speeds, the PTO is rotating at higher speeds.

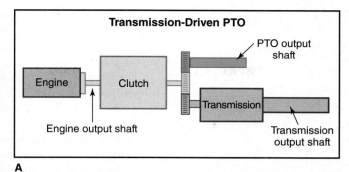

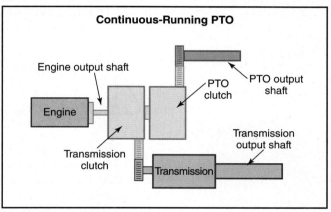

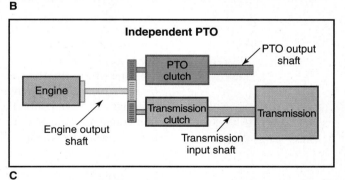

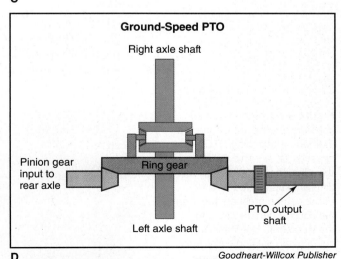

Figure 4-63. PTOs can be classified as four different types. A—A transmission-driven PTO. B—A continuous-running PTO. C—An independent PTO. D—A ground-speed PTO.

A ground-speed PTO is not a live PTO. It can be used to power a PTO-driven rake or used to power a drive-axle on trailers or implements. Ground-speed PTOs are rare compared to independent PTOs. A Fendt 300-series Vario can have the option of a ground-speed PTO.

The tractor's PTO stub shaft can be connected to the implement's driveshaft a few different ways. One design uses a spring-loaded pin. The pin must be pressed and then the implement's driveshaft is slid onto the tractor's stub shaft. The pin is then released. The driveshaft should be pulled rearward to ensure that the lock pin is seated and the shaft is locked in place. See **Figure 4-64A**.

Another method of securing a driveshaft is a spring-loaded collar. The collar must be slid backward, toward the opposite end of the driveshaft, then the driveshaft is slid onto the tractor's PTO stub shaft. After releasing the collar, pull the driveshaft rearward to ensure that it is locked in place. See **Figure 4-64B**.

Warning

Be sure to properly use PTO shaft guards. Place covers over the tractor's PTO stub shaft when it is not in use. Never attempt to couple a driveshaft to the PTO when a tractor is running. Always shut off the engine. If the PTO accidentally engaged, it could cause serious injuries or a fatality.

PTOs are protected by shear pins or slip clutches. A *shear pin* is simply a low-grade sacrificial bolt, like a grade-2 bolt, that couples the PTO driveshaft to the implement. If the implement, for example a mower, hits a hard object, like a tree stump, the bolt shears into two pieces. This prevents excessive stress on parts of the mechanical drive, such as gearboxes or shafts. Many people have made the mistake of installing a higher-grade bolt because they were tired of replacing the shear pin. As a result, the machines and equipment suffered damage, such as a failed gearbox, because the bolt would not shear. See **Figure 4-65A**.

One alternative to a shear pin is a *slip clutch*. It uses springs to sandwich a friction disc between two separate drive discs. If the implement encounters an obstacle, the slip clutch is designed to overcome the spring tension, and will slip rather than causing driveline or implement damage. Manufacturers often recommend slipping the clutch at the beginning of each season by backing off the spring tension and then running the drive to slip the friction disc. This ensures that the clutch is not rusted and seized. See **Figure 4-65B**.

Loaders

Many agricultural tractors are equipped with front-end loaders, which greatly increase their utility. See **Figure 4-66.** Loaders are used for loading dirt, manure, hay, rock, pallets, and many other types of loads. Common attachments include pallet forks, buckets, and bale spears. Agricultural tractor loader frames use parallel linkage (which is explained

Figure 4-64. Spring-loaded couplers are the most common method for attaching PTO driveshafts. A—This PTO driveshaft contains two different spring-loaded pins. Either can be pressed and then the coupler can be slid onto the stub shaft. B—This coupler contains an outside collar that must be slid back before the driveshaft can be slid onto the tractor's stub shaft.

Goodheart-Willcox Publisher

Chapter 4 | Agricultural Equipment Identification

Figure 4-65. Shear pins and slip clutches are used to protect PTO drivelines. A—A shear pin consists of a low-grade bolt that is designed to fail in the event that the implement drive system contacts an obstruction. B—In other cases, a slip clutch is used in place of a shear pin to protect PTO-driven equipment.

Goodheart-Willcox Publisher

Figure 4-66. Loaders are one of the most common attachments added to agricultural tractors.

Taina Sohlman/Shutterstock.com

in Chapter 3, *Construction Equipment Identification*) and some are capable of self-leveling. Many loaders use a coupling mechanism to enable the fast removal and installation of loader attachments. A universal skid steer quick coupler is used on some agricultural loaders.

Drawbars

Nearly every tractor is equipped with a drawbar that is used for pulling implements and trailers. Normally, the drawbar can be extended or retracted to change its length or inverted to change its height. Some drawbars can be moved laterally. A swinging drawbar can swing in a large arc, which is sometimes used when pulling heavy implements or multiple implements. However, the drawbar must be locked in the middle position when pulling PTO-operated equipment. See **Figure 4-67**.

Goodheart-Willcox Publisher

Figure 4-67. Drawbars can be allowed to swing from side to side when pulling heavy implements, as long as the implement does not require PTO power.

Caution

When a tractor is pulling an implement with a PTO shaft through a ditch, if the drawbar is adjusted too short, the implement can press the PTO shaft into the tractor, breaking the tractor's PTO housing. Be sure to follow the service literature.

Note

Some agricultural tractors do not have a three-point hitch, PTO, or loader. One example is a large articulated agricultural tractor used only for pulling a large tillage tool or construction scrapers.

Types and Sizes

Agricultural tractors vary in size from small garden tractors, known as sub-compact tractors, to large articulated tractors. They can be classified in the following categories:
- Sub-compact tractor.
- Compact utility tractor.
- Utility tractor.
- Row crop tractor.
- Articulated four-wheel-drive tractor.
- Rubber-track tractors.

Sub-Compact Tractor

A sub-compact tractor is the smallest type of tractor available that can still be fitted with a loader, three-point hitch, and PTO. They are commonly used for mowing large yards, normally with a belly mower. An example of their engine horsepower range is 18–25 hp. See **Figure 4-68**. These tractors are commonly equipped with hydrostatic drive transmissions, but can be equipped with a manual transmission. The tractor can be two-wheel drive or four-wheel drive.

Compact Utility Tractor

Compact utility tractors offer the same features as sub-compact tractors, but with more horsepower and more size. They are commonly used to maintain larger residential properties. An example range of their engine horsepower is 25 hp–60 hp. See **Figure 4-69**. Some

Henk Jacobs/Shutterstock.com

Figure 4-68. Subcompact utility tractors are the smallest tractors sold by agricultural manufacturers. They can be equipped with a loader, three-point hitch, and a PTO.

logoboom/Shutterstock.com

Figure 4-69. This compact utility tractor is used to maintain a vineyard.

manufacturers offer a few different sizes of compact utility tractors. For example, John Deere offers compact utility tractors in three different categories, based on sizes. The compact utility tractors are also used for commercial applications, such as spraying, garden tilling, brush cutting, dirt work, and snow removal. Compact utility tractors are equipped with either a hydrostatic transmission or a manual transmission. The tractor can be two-wheel drive or four-wheel drive. Many compact tractors have loaders.

Utility Tractor

Utility tractors are used by small crop, dairy, and poultry farms. Horsepower for this category of tractor typically ranges from 45 hp–140 hp. The tractors normally have a manual transmission or a power shift transmission. Utility tractors may be equipped with two-wheel drive or four-wheel drive. They are used for feeding cattle, cleaning barns, and operating hay balers. See **Figure 4-70**. Many utility tractors have loaders, but some applications do not require a loader.

Row Crop Tractor

Row crop tractors have wheels or rubber tracks that can be spaced to match the spacing between crop rows. They are used for tilling fields, cultivating, and planting crops. They can also be used for other functions like spraying, pulling grain carts, and hay baling. They commonly use a powershift or a continuously variable transmission (CVT). The tractor can be equipped with rubber tracks, single, dual, or triple wheels on the rear axle and rubber tracks, single or dual wheels on the front. See **Figure 4-71**. The horsepower range for row crop tractors is typically 140 hp–500 hp.

Utility and row crop tractors are frequently equipped with four-wheel drive, but can be limited to two-wheel drive. Utility and row crop tractors equipped with four-wheel drive can also be called mechanical front-wheel-drive (MFWD) tractors. These tractors have a clutch located inside the transmission that is used to power the front axle. The clutch is normally spring engaged and hydraulically released.

High-Clearance Tractor

High-clearance tractors are a variation of utility and row crop tractors that are designed to have extra ground clearance. See **Figure 4-72**. The tractors are used for spraying and cultivating row crops. They use a special powertrain that provides increased ground clearance.

Four-Wheel-Drive Tractor

Four-wheel-drive tractors have full-time four-wheel drive. Unlike other agricultural tractors, they are equipped with tires or rubber tracks that are all the same size. They can be

Maria Jeffs/Shutterstock.com

Figure 4-70. These two utility tractors are two-wheel drive.

Goodheart-Willcox Publisher

Figure 4-71. This row crop tractor is equipped with front and rear dual wheels. Notice the wheel weights being used on the rear wheels.

Richard Thornton/Shutterstock.com

Figure 4-72. High-clearance tractors are similar to utility and row crop tractors except they have more clearance underneath the tractor.

equipped with single wheels, dual wheels, triple wheels, or rubber tracks when the owner desires low ground pressures. See **Figure 4-73**. They are normally built with an articulating frame that pivots up to 40° when turning. Such tractors are called *articulated agricultural tractors*. Some older four-wheel-drive tractors had a rigid frame and used both front and rear steerable axles. Case IH offers a four-wheel-drive tractor with articulation steer and front axle steer. It will be explained in Chapter 25, *Wheeled Steering Systems*. Four-wheel-drive tractors commonly have a manual or powershift transmission. Their engine horsepower ranges from 360 hp–690 hp. They are used for tilling, planting, pulling grain carts, and pulling scrapers.

Twin-Rubber-Track Tractors

Agricultural tractors can be equipped with twin rubber tracks. They can be sized in the row crop range or four-wheel-drive range. They use a differential steering system, which will be explained in Chapter 24, *Track Steering Systems*. Rubber-track tractors have excellent traction and produce low ground pressures. See **Figure 4-74**.

Note

Tractors can be equipped with auto-guidance systems. As the tractor pulls larger and larger implements, it becomes more challenging for an operator to manually steer the tractor efficiently. If the tractor is steered by hand, it can have too much implement overlap as it makes passes through the field or it can miss points in the field. With today's auto-guidance systems, the tractor can electronically steer itself, eliminating the steering inefficiencies and reducing operator fatigue.

Goodheart-Willcox Publisher

Figure 4-73. Four-wheel-drive tractors use all the same size tires and wheels (or rubber tracks). This tractor is equipped with dual wheels.

Goodheart-Willcox Publisher

Figure 4-74. This Challenger rubber-track tractor uses narrow tracks that can be spaced far apart for farming row crops.

Sprayers

The agricultural industry uses *sprayers*, also known as *applicators*, for applying liquid chemicals to fields. The types of chemicals applied include the following:
- Fertilizers—Chemical nutrients to enrich the soil to increase plant growth.
- Pesticides—Chemicals to control pests. Pesticides can include herbicides and insecticides.
- Herbicides—Chemicals to control weeds.
- Insecticides—Chemicals to control insects.
- Fungicides—Chemicals to control fungi.

Note

Late-model sprayers, such as the John Deere R4030, R4038, R4044 and R4045, are capable of applying liquid or dry fertilizer. To convert the sprayer for dry fertilizer application, a wet solution skid is removed and a dry spinner spreader is installed. This option expands the usefulness of the machine, which would otherwise be sitting unused after the crop has been harvested.

Types

Two types of large agricultural sprayers are available, classified by the method used to pressurize and spray the liquids:
- Hydraulic.
- Centrifugal.

A hydraulic sprayer uses a positive-displacement pump to draw liquid from the chemical tank and push it through nozzles, which atomize the liquid into droplets as it is ejected from the boom. The *positive-displacement pump* (either a piston pump or a roller pump) delivers flow proportional to its speed. Because it designed to displace a set amount of liquid, it will lose practically no flow when pressure increases. As a result, positive-displacement pumps require the use of a pressure relief valve, or the system will burst if it encounters too much resistance. Hydraulic sprayers are also sometimes called high-volume sprayers.

Centrifugal sprayers are the most common type of agricultural sprayer. A centrifugal sprayer operates similarly to a hydraulic sprayer, but the pump is a non-positive-displacement pump. *Non-positive-displacement pumps* do not need a pressure relief valve because they reduce flow when the pumps encounter resistance. If resistance is excessive, flow will stop completely. Two types of non-positive-displacement pumps are commonly used in sprayers, turbine and centrifugal. A diaphragm pump can also be used, and is sometimes labeled a semi-positive-displacement pump.

Sprayers can also be classified based on their method of transportation, such as compact sprayers mounted on a tractor, pull-type sprayers, large self-propelled sprayers, and aircraft sprayers known as crop dusters. See **Figure 4-75**. Many self-propelled sprayers have the boom mounted to the rear of the machine, but they can also be located at the front of the machine.

Functions, Controls, and Specifications

The common functions employed by agricultural sprayers include the following:
- Boom raise and lower.
- Boom lateral tilt.
- Boom horizontal fold.
- Boom vertical fold.

A
oticki/Shutterstock.com

B
Valentin Valkov/Shutterstock.com

C
Stockr/Shutterstock.com

D
Rocky33/Shutterstock.com

Figure 4-75. Types of agricultural sprayers. A—Tractor-mounted sprayers are often mounted on the three-point hitch. B—Pull-type sprayers are pulled behind agricultural tractors. C—Self-propelled sprayers may have the boom mounted on the back, as shown here, or on the front of the machine. D—Crop dusters have the spraying nozzles mounted below the wings.

- Applicator sprayer system.
- Flow and pressure.
- Applicator control valves.
- Chemical agitator (mixer).
- Steering.
- Propulsion.

Boom Raise and Lower

Booms can be lifted and lowered to optimize the application of the chemical. The recommended position of the sprayer's boom is 18″ to 24″ above the crop. Most late-model sprayers are equipped with an automatic boom height control that uses ultrasonic sensors.

Boom Lateral Tilt

The sprayer booms are often equipped with a hydraulic cylinder that allows the boom to be tilted. For example, the boom can be tilted when operating the sprayer on a slope.

Boom Fold

Large booms can be equipped to fold both vertically and horizontally. When the sprayer has arrived at a field to spray, an example boom sequence is the boom is lifted, horizontally unfolded, vertically unfolded, and then lowered. See **Figure 4-76**.

The wing ends have spring latches that allow the wing to pivot if it hits an obstruction or the ground. Once the wing has cleared the obstruction, the latching mechanism will automatically re-latch the wing in place.

Applicator Sprayer Pump

As mentioned, a sprayer has either a positive-displacement or a non-positive-displacement pump. The speed of either type of pump will affect the flow rate produced. The flow can vary from 0 to 120 gallons per minute (gpm). Flow rate is also affected by changes in system pressure. The operating pressure of a centrifugal sprayer typically ranges from 5 psi–80 psi.

Nozzles

Sprayers use electronically controlled solenoid valves for controlling the application of the liquid. Older sprayers used one electronic solenoid valve for each boom section, which contained multiple spray nozzles. Late-model sprayers can use one solenoid valve per nozzle and can directly control each individual spray nozzle. The rate of fluid flow is measured in gallons per acre (gpa).

Nozzles vary in orifice size and spray angle, for example 60° to 140° spray patterns. The desired spray pattern and droplet size varies with the chemical being applied. Operating pressure determines the size of the sprayed droplets. A higher pressure produces smaller droplets. Droplets that are very small form a mist. Misting causes the applied chemicals to drift, which is a problem for pre- and post-emergent herbicides. Overlapping passes should also be avoided when applying chemicals. Any areas of overlap will receive a double dose of chemical.

All of the following will affect the drift, overlap, and spray pattern of chemicals:
- Nozzle orifice size.
- Nozzle angle.
- Height above plant.
- Travel speed.
- Fluid pressure applied to nozzle.

Nozzles are chosen based on how fast the sprayer is traveling in the field and the rate of application (gpa). For example, if a customer desires a faster travel speed, the operating pressure might increase from 45 psi (3 bar) to 75 psi (5 bar). The nozzle must have a large enough orifice to produce the desired flow under those conditions.

TFoxFoto/Shutterstock.com

Figure 4-76. This sprayer has the boom horizontally unfolded but still requires the boom to be vertically unfolded.

Agitation

Sprayers use an agitation valve to divert a portion of the sprayer pump's flow back to the chemical tank to ensure that the chemicals' composition is thoroughly mixed and consistent.

Steering

Self-propelled sprayers can be configured with two-wheel steering or four-wheel steering. See **Figure 4-77**.

Propulsion

Most sprayers are four-wheel drive and use hydrostatic transmissions for propelling the machine. Some older sprayers were two-wheel-drive and mechanically propelled.

Suspension

Some late-model sprayers can travel as fast as 15 mph in the field. If the sprayer is converted to a dry chemical applicator, it can travel as fast as 25 mph in the field and carry up to 10 tons of fertilizer. Some sprayers have a road speed up to 34 mph. Due to these fast speeds, suspension systems are critical on sprayers, perhaps more than on any other agricultural machine. Depending on the manufacturer, the sprayer suspension can be one of three types:

- Hydraulic.
- Air bag.
- Coiled spring.

Some sprayers have independent four-wheel suspension. See **Figure 4-78**. AGCO's RoGator sprayers (RG900B, RG1100B, and RG1300B) have a flex frame that allows a wheel to flex up to 14″ (independent from the other wheels) and still provide ground contact.

Figure 4-77. This John Deere sprayer is equipped with front-wheel steer.

Goodheart-Willcox Publisher

Markers versus GPS

When using wide boom sprayers, it is difficult to know where to apply the next swath of chemicals, especially if the boom is 120′ wide. Some sprayers have foam markers that apply a small line of foam as the sprayer travels through the field. The foam marks a line that the operator can straddle with the sprayer's wheels as it makes the next pass through the field. The marker helps the operator ensure the entire field is sprayed and avoid overlapping chemicals.

Other sprayers use global positioning satellite (GPS) systems instead of foam markers. The GPS system can be used with lightbar or auto guidance technology. The lightbar technology uses GPS, controllers, and a *lightbar*. A lightbar is a display panel with a series of indicator lights to the left and right of center that tells the operator which way to steer. Auto-guidance systems are integrated with the machine's controls so that the electronic control module literally steers the sprayer as it travels through the field. Sprayers with auto guidance can also use swath control to shut off portions of the boom when the sprayer is approaching portions of the field that have already been sprayed.

Goodheart-Willcox Publisher

Figure 4-78. Each wheel on this Case IH sprayer uses a hydraulic shock absorber and coiled spring for suspension.

Sizes

Sprayers are available in a variety of different sizes, ranging from a compact sprayer that can be attached to a utility vehicle or pulled behind a small tractor to a self-propelled machine with 120′-long boom. Sprayers are elevated so that the sprayer can clear the crops as it is operating in the field.

Sprayers can be sized based on boom size and tank volume. Engine horsepower should also be considered when classifying self-propelled sprayers:

Pull-type sprayer
- Boom size: 10′–60′ wide.
- Tank volume: 45 to 1000 gallons.

Self-propelled sprayers
- Boom size: 60′–120′ wide.
- Tank volume: 500 to 1300 gallons.
- Engine horsepower: 130 to 350 engine horsepower.

Fertilizer Applications

Agricultural producers apply *fertilizer* to the soil. Fertilizer is a chemical that promotes plant growth. Fertilizers can be applied in a liquid, gas, or dry state. The fertilizer is injected below the soil or on top of the soil.

Two methods are used for injecting fertilizer below the soil. The first is using a coulter blade to open a furrow, a nozzle that disperses the fertilizer in the furrow, and two closing wheels that close the furrow. See **Figure 4-79A**. The other method uses a knife (similar to a ripper) and a nozzle located behind the knife. See **Figure 4-79B**. As the implement is pulled through the soil, the knife creates an opening for the nozzle to inject the fertilizer. GPS can also be used to control different rates of fertilizer, including varying different sections or row units across the implement.

A

B

Goodheart-Willcox Publisher

Figure 4-79. Liquid fertilizer is injected below the soil using one of two methods. A—A coulter opens a furrow, while a nozzle injects the fertilizer. Two wheels close the furrow. B—Knives and nozzles inject fertilizer below the soil.

One of the most common application methods used by agricultural producers is *side dressing*, which is applying liquid fertilizer below the soil between the rows of crops so that the crop has access to the nutrients at its side.

Anhydrous Ammonia

One common example of a liquid fertilizer is anhydrous ammonia, **Figure 4-80**. At normal temperatures and atmospheric pressure, anhydrous ammonia is a gas. The chemical is compressed into a liquid form so that it can be stored in a tank.

An anhydrous ammonia tank, sometimes called a nurse cart, is pulled in conjunction with the implement. The tank is often connected through a breakaway shut-off valve, which is designed to shut off flow if the tank becomes detached from the implement. Small anhydrous ammonia applicators have a safety cord (rope) strung into the cab, which the operator can pull to shut off all flow of the anhydrous ammonia in an emergency. Larger implements have an electronic control switch that shuts off the flow of anhydrous ammonia.

The anhydrous ammonia applicator controls the liquid with electronic control valves. The liquid flows through a cooler and manifolds, and is distributed through hoses to knives that inject the fertilizer below the soil. See **Figure 4-81**.

After the liquid leaves the manifolds, it begins to turn to a gas. When the gas contacts the soil, it combines with the soil. The hoses between the manifolds and the nozzles must be all the same length so that gas is equally distributed to each of the nozzles. The nozzles located closer to the manifolds will have the hoses curled to accommodate the extra length of hose. Anhydrous ammonia can be knived or coultered into the soil.

Anhydrous ammonia is a very dangerous fertilizer. If it is allowed to touch your skin, it will immediately freeze the skin tissue, causing severe burns. Personnel must use proper PPE, including goggles, rubber gloves, long-sleeve shirt, and a respirator to protect their skin, eyes, and lungs from the ammonia vapor. Most anhydrous ammonia applicators have an onboard water tank so workers can rinse off if they come in contact with the anhydrous ammonia. See **Figure 4-82**.

Charles Brutlag/Shutterstock.com

Figure 4-80. Farmers use anhydrous ammonia to apply nitrogen to their fields. The nitrogen is stored under pressure in a tank.

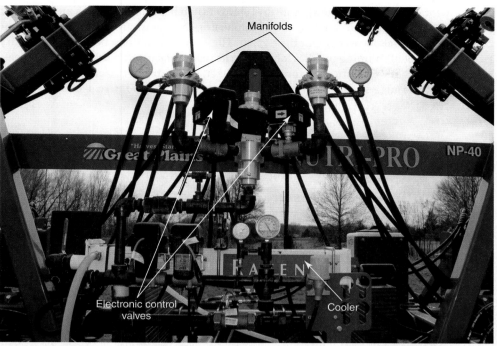

Figure 4-81. An anhydrous ammonia applicator has a cooler, electronic control valves, and manifolds to distribute the anhydrous ammonia to a coulter unit or knife.

Liquid Fertilizer

Implements can also contain a liquid storage tank used to store liquid fertilizer. These implements can also use coulters or knives to inject the liquid fertilizer into the soil.

Liquid Manure

Liquid manure is another example of a fertilizer that is injected into the soil. See **Figure 4-83**. The implement uses a centrifugal pump or positive-displacement pump to deliver the manure to a distribution manifold that distributes the liquid to a row of knives that inject the liquid below the soil. For precision, the applicators measure one of two variables to track the rate of application, the flow rate, or the weight of fertilizer as the tractor is moving through the field.

Figure 4-82. Anhydrous ammonia applicators often have an onboard water tank to allow workers to rinse themselves in case they come in contact with the anhydrous ammonia.

Figure 4-83. Producers can use liquid manure as a fertilizer. A large manure tank and applicator is pulled behind the tractor.

As mentioned earlier, some agricultural sprayers can be reconfigured as an applicator that spreads dry fertilizer. Many fertilizer applicators have a rotating chain conveyor or rubber conveyor belt located under the hopper. The conveyor chain or belt delivers the dry fertilizer to a set of spinning paddles located at the rear of the machine. The paddle wheels rotate to spread dry manure or granulized fertilizer. Other dry fertilizer applicators have hoppers and use augers to meter the fertilizer to the ground. See **Figure 4-84**.

A
Charles Brutlag/Shutterstock.com

B
Goodheart-Willcox Publisher

Figure 4-84. Dry fertilizer is commonly spread on top of the soil. A—This applicator is spreading dry fertilizer to a field. B—This John Deere applicator contains a hydraulically driven belt located under the hopper. The belt delivers the fertilizer to hydraulically driven paddles that spread the fertilizer across the soil.

Summary

- Examples of agricultural equipment include tillage tools, planters, drills, mowers, rakes, balers, forage harvesters, combines, tractors, sprayers, and fertilizer applicators.
- Primary tillage tools operate at depths of six inches and deeper.
- Secondary tillage tools operate at depths less than six inches.
- Planters and drills must open the furrow, meter the seed, place the seed, close the furrow, and firm the seedbed.
- Seeders are categorized as row crop planters, drills, air seeders, and broadcast seeders.
- Planters plant on wider rows and in population units of seeds per acre.
- GPS-aided planters can vary the seed population and provide swath control.
- Drills plant on narrow rows and meter the seed by volume.
- Drills seed small grains like wheat, barley, oats, and rice.
- Air seeders plant small seeds by volume.
- Broadcast seeders use paddle-type spinners to sling the seed.
- Reciprocating knives can be driven by a pitman arm, wobble box drive, or a planetary gearbox drive.
- Rotary mowers spin several oval-shaped discs that contain knives.
- Swathers mow and gather the crop into a windrow. They are self-propelled or pull-type.
- Rakes are used to windrow hay or turn over wet hay. The different types of hay rakes are wheel, parallel bar, and rotary.
- Tedders are used to turn over hay.
- Types of hay balers include large round fixed chamber, large round variable chamber, small square conventional baler, small square inline baler, and large inline square baler.
- Variable chamber round balers can vary the bale diameter. They produce solid core bales with consistent densities.
- Forage harvesters chop the entire plant at high moisture levels, for example, 50 to 70%.
- Forage harvesters can be self-propelled or pull type, and can be equipped with a straight cut header, row crop header, windrow pick-up header, or a non-directional corn head.
- Forage harvesters can cut and feed, chop, crack corn kernels, and throw or blow the silage into the wagon.
- A combine harvester is designed to cut or strip the crop, feed the crop into the thresher, thresh the crop, separate the grain from the chaff, deliver the clean grain to the grain tank, spread the straw and chaff, and unload the clean grain into a grain cart or semi-truck hopper.
- Types of combine headers include straight cut headers, windrow pickup headers, corn headers, stripper headers, and row crop headers.
- Conventional combines use a cylinder as the primary means of threshing grain.
- Rotary combines use a rotor as the primary means of threshing grain.
- Tractors might be equipped with a loader, three-point hitch, or a PTO.
- Three-point hitches are categorized in sizes (0, 1, 2, 3, and 4) based on pin diameter and distance between pins.
- Tractor PTO stub shafts are available in three different sizes and spline configurations (1 3/8″ six splines, 1 3/8″ 21 splines, and 1 3/4″ 20 splines). Six-spline shafts rotate at 540 rpm and 20- and 21-spline shafts rotate at 1000 rpm.

- PTOs can be classified as transmission-driven, continuous-running, independent, and ground-speed.
- Large four-wheel-drive articulated tractors always power all four wheels and the wheels or tracks are all the same size.
- Rubber-track tractors provide improved traction and low ground pressures. The tractors can be manufactured with two or four tracks.
- Examples of sprayers include tractor mounted, pull-type sprayers, self-propelled, and crop dusters.
- A coulter or a knife is used to inject fertilizer below the soil.

Technical Terms

Advanced Stone Protection (ASP)
air seeder
articulated agricultural tractor
auto-guidance system
broadcast spreader
chisel plow
clean grain auger
combination tool
concaves
continuous-running PTO
conventional combine
corn header
cut-and-blow forage harvester
cut-and-throw forage harvester
disc harrow
doubles
down force
draper belts
drill
elevator
feeder house
fertilizer
field cultivator
finish harrow
fixed bale chamber
forage harvester
gang
gauge wheels
ground-speed PTO
hay baler
high-clearance tractor
hopper level sensor
independent PTO
lightbar
live PTO
marker
moldboard plow
mower
mower conditioner (MOCO)
multi-hybrid planter
non-positive-displacement pump
parallel rake
pitman arm drive
plunger
positive-displacement pump
pre-charge chamber
primary tillage
quick-hitch
rake
reap
rod weeder
rotary combine
rotary hoe
rotary rake
rotary tiller
row crop planter
row crop tractor
secondary tillage
seed tube sensor
shear bar
shear hub
shear pin
shovel
sickle bar knife drive
side dressing
sieve
silage
singulation
singulator
skips
spike-toothed harrow
sprayer
straight cut header
straw walkers
stripper header
stuffer fork
sub-soiler
swath control
swather
sweep
tailings auger
tedder
three-point hitch
thresh
tilling
transmission-driven PTO
variable bale chamber
variable rate planter
V-ripper
wheel rake
wide-sweep blade plow
windrow pickup header
windrower
winnow
yield mapping
yield monitoring

Review Questions

Answer the following questions using the information provided in this chapter.

Know and Understand

1. All of the following are common examples of agricultural machines, *EXCEPT*:
   A. combine.
   B. baler.
   C. backhoe.
   D. planter.

2. Which of the following has shovels, or sweeps?
   A. Chisel plow.
   B. Disc harrow.
   C. Field cultivator.
   D. Moldboard plow.

3. Sickle bars can be used in all of the following, *EXCEPT*:
   A. combine corn header.
   B. combine straight cut headers.
   C. forage harvesters straight cut headers.
   D. swathers.

4. Which baler has a pre-charge chamber?
   A. Large round baler with a fixed bale chamber.
   B. Large round baler with a variable bale chamber.
   C. Small inline square baler.
   D. Small conventional square baler.

5. Silage can be stored in all of the following, *EXCEPT*:
   A. a bunker.
   B. a harvester tank.
   C. a silo.
   D. a wrapped bale.

6. Forage harvesters can use all of the following headers, *EXCEPT*:
   A. a non-directional corn header.
   B. a straight-cut header.
   C. a stripper header.
   D. a windrow-pickup header.

7. All of the following tasks are combine harvester functions, *EXCEPT*:
   A. crack the kernel.
   B. cut or strip the crop.
   C. feed the crop.
   D. thresh the crop.

8. Straight cut combine headers with a locked cutter bar can harvest all the following crops, *EXCEPT*:
   A. barley.
   B. rice.
   C. soybeans.
   D. wheat.

9. Which of the following combine headers uses a reel?
   A. Corn header.
   B. Row crop header.
   C. Straight cut header.
   D. Stripper header.

10. Which type of combine produces the best quality of straw?
    A. Conventional straw walker.
    B. Twin rotor.
    C. Axial rotor.
    D. None of the above.

11. Below the cylinder or rotor, combines have a set of wire shells known as _____.
    A. convexs
    B. concaves
    C. sieves
    D. straw walkers

12. In a three-point-hitch, the _____ is in direct contact with the lift cylinders and rotates to lift the hitch.
    A. center link.
    B. draft arm.
    C. lift arm.
    D. rockshaft.

13. What is the largest hitch category?
    A. One.
    B. Three.
    C. Four.
    D. Six.

14. All of the following are examples of PTO stub shafts, *EXCEPT*:
    A. 1 3/8″ 540 rpm 6 splines.
    B. 1 3/8″ 1000 rpm 21 splines.
    C. 1 3/8″ 1000 rpm 6 splines.
    D. 1 3/4″ 1000 rpm 20 splines.

15. What is the most common type of PTO in late model tractors?
    A. Transmission-driven PTO.
    B. Continuous-running PTO.
    C. Independent PTO.
    D. Ground-speed PTO.
16. Slip clutches should _____.
    A. be slipped at the beginning of each season
    B. never slip
    C. slip at 540 rpm
    D. slip at 1000 rpm
17. A shear pin is essentially a _____.
    A. low-grade bolt
    B. high-grade bolt
    C. stainless steel cotter pin
    D. stainless steel clevis pin
18. The operating pressure of a centrifugal sprayer typically ranges from _____ psi.
    A. 0 to 120
    B. 5 to 80
    C. 90 to 120
    D. 120 to 180

Apply and Analyze

19. The term _____ is another name for a group of discs.
20. Singulation describes consistent _____ spacing.
21. A(n) _____ seeder uses a chisel plow and a seed cart.
22. A large round baler with a(n) _____ bale chamber produces a solid core, spiraled bale.
23. _____ is the most common silage crop.
24. Decreasing the speed of a forage harvester's feed rollers while maintaining the same cutter head speed will _____ the length of the cut crop.
25. A(n) _____ combine header is designed to match the spacing between the rows.
26. A transverse rotary combine is manufactured by _____.
27. A(n) _____ describes a housing on a combine that contains chain-driven rubber paddles that are used to move grain from an auger to a new location.
28. One of the most common fertilizer application methods used by agricultural producers is _____ dressing.

Critical Thinking

29. Explain the difference between planters and drills.
30. Describe the crop flow through a combine starting at the header and ending inside the grain tank.

Chapter 5
Lifting

Objectives

After studying this chapter, you will be able to:
- ✓ List proper methods for manually lifting components.
- ✓ List the different types of hoists used by technicians.
- ✓ Identify the different types of cranes.
- ✓ List the different wire rope classifications.
- ✓ Explain the differences in synthetic slings.
- ✓ Explain the different types of chain sling classifications.
- ✓ Identify the different types of fittings used with lifting slings.
- ✓ List common mistakes made while lifting components and machines.
- ✓ Compute sling load angle factors, horizontal sling angles, and included angles placed on slings.
- ✓ Demonstrate the three different types of lifting hitches.
- ✓ Compute sling loads while factoring the center of gravity.

Many heavy equipment components cannot be serviced or repaired without the assistance of lifts, cranes, rigging, or jacks. Using proper lifting procedures, following manufacturers' specifications, and adhering to industry standards will help prevent damage to machines and components and injury to personnel. Technicians should receive training for using, maintaining, and inspecting lifting equipment.

Lifting by Hand

Technicians frequently lift heavy components by hand. Using improper lifting techniques often results in personal injury. An injury may require substantial recovery time, which can cause personal hardship and cost a company thousands of dollars in worker compensation. Hernias and back injuries are common injuries resulting from trying to lift too much or lifting with poor techniques. To prevent these types of injuries, keep the following safe lifting practices in mind.

- Know your physical limitations and do not attempt to lift too much. Ask for help or divide the load into smaller, more manageable portions.
- Do *not* attempt to lift heavy loads positioned away from your body.
- Bend your hips and knees to squat down to the load. Keep the load close to your body and lift with your legs by straightening them.
- Do *not* lift with your back. Bending at the waist and lifting an object will injure your back.
- Keep your back in a straight and vertical position.
- If you must turn, do *not* twist your back. Turn your whole body by using your feet and legs.
- When lowering the load, do not attempt to set the load down away from your body. Set the load down close to your body and, if needed, slide the load away from you.

Lifting Equipment

Technicians use a variety of tools, components, and machines for lifting heavy components. It is very important to use the proper equipment for a given task. It is also vital for the technician to use proper lifting points on the machine to prevent damage and injury. To ensure safe lifting practices are used, all technicians must receive training before using lifting equipment. Jacks, hoists, slings, and cranes are lifting tools commonly used by heavy equipment technicians.

Jacks

One of the most popular lifting devices is a jack. Jacks can be mechanical or hydraulic. The less common mechanical jack uses a threaded rod or lever to raise and lower the jack. Mechanical jacks typically have lower lifting capacities (weight and height) than hydraulic jacks and limited use in the heavy equipment industry. The more popular hydraulic jack has more lifting capacity and requires less force to lift heavy loads. Two types of hydraulic jacks are the bottle jack and the floor jack, **Figure 5-1**. They may be manually operated or, for heavier applications, operated with shop air. Both types are portable but require a firm, stable base.

Bottle jacks and floor jacks both have a range of lifting capacities. Bottle jack ratings range from 4 ton to 50 ton. The height limit also varies. Some models include a threaded adjustment at the end of the ram for additional height. Floor jacks also have a wide range of lifting capacities. Floor jacks have long handles and are mounted on rollers for easy maneuverability. The clearance and height limit vary between models.

Pro Tip

Manufacturers' service literature often includes the weight of the machine and its individual components as well as the location of lifting points.

A B C

Goodheart-Willcox Publisher

Figure 5-1. A variety of jacks are used in the shop. A—Manually operated bottle jacks are used in the field and shop. These jacks have a small footprint and require a firm, stable base. B—Floor jacks are primarily used in the shop. This floor jack is air-powered. C—This floor jack is manually operated and has a five-ton capacity.

Jack Stands

Jack stands are adjustable or fixed-height steel supports. Heavy equipment technicians use jack stands to support components and machines after they have been lifted. See **Figure 5-2**. The stands range in size and weight capacities. Jack stands should be inspected for damage and faults before each use.

Figure 5-2. Heavy-equipment repair shops use a variety of heavy-duty jack stands.

 Warning

Never work under a suspended component or machine that is not properly supported by jack stands. A suspended component could drop due to a failed O-ring, wire rope, or chain link and injure or kill the person working underneath.

Splitting Stands

Splitting stands are commonly used by agricultural equipment technicians. The stands have rollers and are mounted on the front and rear halves of the tractor. Each stand supports half of the tractor to provide access to components, such as the transmission clutch. See **Figure 5-3**.

Blocking and Cribbing

Blocking and cribbing describes a variety of procedures used to stabilize heavy equipment during service or maintenance. *Blocking* refers to the actions taken or methods used to prevent a component or machine from moving. In addition to wheel chocks, implement locks, and other built-in safety devices, technicians use jack stands, component stands, and wooden blocks for blocking. See **Figure 5-4**.

Figure 5-3. Agricultural tractors sometimes must be split to gain access to the transmission or other components. Rollers allow the front and rear half of the tractor to be split and supported.

 Warning

All machines and components must be safely supported to avoid damage and injury. Most manufacturers provide detailed instructions for blocking and cribbing procedures that should be used with a specific piece of equipment. Technicians may also receive training on safe blocking and cribbing practices through manufacturer and government training programs.

Cribbing is a means of temporarily supporting a load with hardwood or engineered plastic cribbing blocks stacked in alternating directions. See **Figure 5-5**. Cribbing has more support and contact area than jacks and better distributes the load, especially on softer surfaces. Although cribbing is sometimes used in shops, it is commonly used to support a machine or component whenever a hard surface is unavailable. Cribbing is one form of blocking.

Figure 5-4. Blocking consists of supporting implements, components, or machines. A dozer transmission is blocked during the transmission removal process.

Goodheart-Willcox Publisher

Figure 5-5. Cribbing is made with alternating wood or engineered plastic blocks. A—Wooden blocks supporting the master clutch of an agricultural four-wheel-drive tractor transmission. B—Engineered plastic cribbing blocks.

 Warning

Wood blocking should always be made of hardwood, such as hickory or oak. *Never* use a softwood for blocking a component. Wood blocking must be straight and uniform, free of defects, kept clean, and never painted or stained due to the increased risk of slippage.

Manufacturers and shops are replacing their wooden blocks with engineered plastic cribbing blocks. The lightweight plastic blocks provide higher load capacities than wood blocks and come with extended warranties. They also do not splinter, split, or crack as easily as wooden blocks, will not rot, and are more resistant to oil and chemicals. The blocks are also designed to interlock, providing a more stable base than wooden blocks.

Cranes

A crane is a hoisting structure or machine used to lift machines and components. Cranes can be classified as telescoping, lattice boom, jib, bridge, gantry, and tower, **Figure 5-6**.

- A hydraulic telescoping crane has an extendable boom which can be retracted to accommodate transport to each job site. Hydraulic telescoping cranes are designed as rough-terrain (RT), all-terrain (AT), or truck-mounted.
- Lattice boom cranes have a boom made with multiple *lacings* (steel tubes) that are attached to four parallel steel tubes referred to as *chords*. In addition to servicing and diagnosing cranes, technicians must also assemble and disassemble the cranes for transport to job sites. They are some of the tallest cranes used in the construction industry.
- A jib crane has a vertical post with a horizontal I-beam that can rotate up to 360°. The jib crane is one of the most popular types of cranes used in service repair shops.
- Truck-mounted cranes can be easily and quickly driven to job sites.
- A bridge crane uses an I-beam that travels the length of a shop on top of two cross beams. These cranes can have capacities large enough to lift entire machines off the ground. Bridge cranes are also referred to as overhead cranes.
- A gantry crane has an I-beam attached to two A-frame ends. Gantry cranes can be mounted on rollers. Shipping ports often use large gantry cranes for loading shipping containers onto truck trailers.
- Tower cranes are assembled on site and used to erect commercial buildings.

Service repair shops typically use jib, bridge, and gantry cranes. Some shops use several smaller jib cranes over a single large bridge crane to help with the shop's work flow. The additional jib cranes allow more than one technician access to a crane at the same time.

Chain Hoists

A chain hoist, or chain fall, provides a mechanical advantage for lifting loads. Chain hoists can be manually operated with a pull chain or mechanical lever or electrically or pneumatically driven. See **Figure 5-7**. In many shops, the chain hoist is mounted on a trolley that rolls along the boom of a bridge or jib crane. Chain hoists are rated by their maximum weight-lifting capacity.

Technicians should slowly lift and closely observe the component being lifted to ensure the part is completely detached from the machine and properly attached to the chain. For

Chapter 5 Lifting

Figure 5-6. Cranes come in many different configurations. A—A telescoping crane. B—Lattice boom crane. C/D—Jib crane. E—Truck-mounted crane. F—Bridge crane. G—Gantry crane. H—Tower crane.

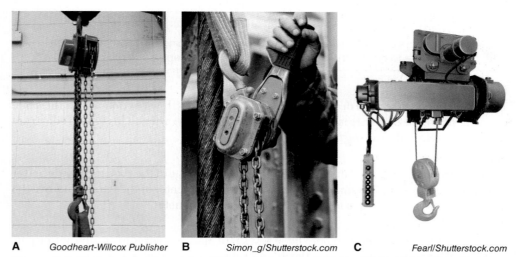

Figure 5-7. Chain hoists are used to lift heavy loads. A—This mechanically operated chain hoist uses a separate pull chain for lifting and lowering the hoist's load chain. B—This chain hoist uses a mechanical lever to lift and lower the chain. C—An electrically operated hoist is used to lift heavier loads. They are sometimes placed on jib and bridge cranes.

example, when separating power train components, a technician will be able to lower the hoist if a bolt is still attached before damage occurs to the hoist or machine.

Components are attached to the hoist with a large hook and a sling. Slings are made with different materials and configurations and may be categorized as wire rope, synthetic web, synthetic rope, and chain slings. Slings have maximum capacity ratings and should be inspected for faults or damage before each use.

Wire Rope

Wire rope is used for lifting heavy components and machines. Wire rope is also used in other heavy equipment applications, such as a crane's hoist and a dozer's winch. Wire rope is classified by the following characteristics:

- Type of core.
- Number of strands.
- Number of wires in a strand.
- Strand wire arrangement.
- Type of rope lay.
- Steel grade.

Wire Rope Core

Wire ropes have a fiber core or an independent wire rope core (IWRC). A fiber core is more flexible but weaker than an IWRC. The type of core selected is based on the intended use and the environment in which it will be used. Technicians must refer to industry standards and manufacturer recommendations when selecting wire rope.

Wire Rope Strands

Wire ropes are classified based on the number of strands in the rope and the approximate number of wires in each strand. See **Figure 5-8**. Note that a 6×7 wire rope has 6 strands and 3 to 14 wires per strand.

Strand Wire Arrangement

The strands in a wire rope can be set in different arrangements, each with its own advantages and disadvantages. Four popular types of strand arrangements are simple, Seale, Warrington, and filler.

Note

Always consult a wire rope supplier when replacing the rope on a winch or crane to ensure the correct type is used as the replacement.

Simple Wire Rope
The term *simple wire rope* describes a rope constructed with wires that are all the same size. The size of the core wire is also the same size as the outer wires. This type of construction may be referred to as *single layer wire rope*.

Seale Wire Rope
A *Seale wire rope* has larger-diameter wires in the outer edge of the rope to provide better resistance to abrasion. The smaller-diameter wires in the center of the strand improve the rope's flexibility. The wires in each layer have the same diameter.

Warrington Wire Rope
The *Warrington wire rope* uses both large and small diameter wires in an alternating pattern. The Warrington wire rope is a flexible and abrasive-resistant wire rope.

Filler Wire Rope
Filler wire rope is constructed with two layers of wire with the same diameter around a center. The inner layer has half as many wires as the outer layer. Small-diameter wires fill

Wire Rope Classification	Number of Strands	Wires in a Strand	Max. Allowable Outer Wires
6 × 7	6	3–14	9
6 × 19	6	15–26	12
6 × 37	6	27–49	18
6 × 61	6	50–74	24
6 × 91	6	75–109	30
7 × 19	7	15–26	12
7 × 37	7	27–49	18
8 × 7	8	3–14	9
8 × 19	8	15–26	12
8 × 37	8	27–49	18
8 × 61	8	50–74	24
19 × 7	17–19	6–9	8
19 × 19	17–19	15–26	12
36 × 7	26–36	6–9	8
35 × 19	26–36	15–26	12

A

Goodheart-Willcox Publisher

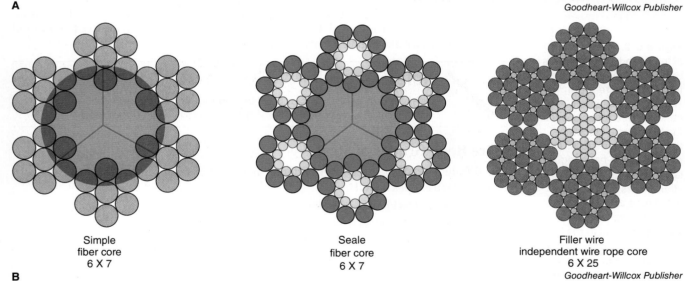

Simple
fiber core
6 X 7

Seale
fiber core
6 X 7

Filler wire
independent wire rope core
6 X 25

Goodheart-Willcox Publisher

B

Figure 5-8. Different types of wire ropes are constructed differently. A—A common wire rope has 6 or 8 strands in a single rope. Notice that the last number is an approximation, meaning that 7 indicates the strand has 3 to 14 wires in a single strand. B—Notice the simple wire rope is made from wires of the same diameter. Seale rope uses large-diameter wire in the outer layer and smaller wire in the inner layer. Filler wire rope has small-diameter wires between the outside and inner layers.

the spaces between the outer and inner layers of wires. The filler wire rope is stronger and has less chance of fatigue than other wire ropes.

 Note

Ropes with larger wires are resistant to abrasion but are less flexible than ropes made with smaller wires.

Wire Rope Lay

Wire rope lay is the way the wires are laid to form strands and the direction the strands are laid around the core. Wire ropes commonly have six or more strands that are wound around a fiber or wire core. The direction the strands are wound around the core determines whether the rope is a left-lay or a right-lay rope. The direction of the winding is viewed by looking at the end of the wire rope and following the travel of the wire strands spiraling away from you. In a *left-hand-lay wire rope*, the strands spiral in a counterclocwise pattern. In a *right-hand-lay wire rope*, the strands spiral in a clockwise direction.

The lay of the rope can be further classified based on whether the individual wires that make up the strand spiral in the same or opposite direction as the strand. In a *regular-lay wire rope*, the wires spiral in the *opposite* direction of the strands. Notice the wires within a single strand appear to run *parallel* to each other and the rope's axis. See **Figure 5-9**. Regular-lay ropes are less likely to kink or unwind, making them easier to handle than Lang-lay ropes. They also resist crushing.

The wires and strands in a *Lang-lay wire rope* spiral in the *same* direction. This type of lay has more surface area and resistance to wear from abrasion. The wires in the strand are at an angle to both the strand and the axis of the rope. Lang-lay ropes have a higher degree of flexibility and fatigue resistance but are more likely to kink, unravel, or be crushed.

An *alternate-lay wire rope*, also known as a reverse lay, has alternating Lang-lay strands and regular-lay strands. The wires in adjacent strands appear to converge, creating a V pattern, similar to a tractor tire tread. The alternate-lay design combines the flexibility and wear resistance of the Lang-lay strands with the strength and crush resistance of the regular-lay strands.

When describing the lay of a rope, both the orientation of the strands and the orientation of strand wires are included. For example, the wire rope shown in **Figure 5-9D** would be described as a right-hand alternate-lay wire rope.

A

B

C

D *Cableworks, Inc.*

Figure 5-9. The direction of the strands and the wires within the strands determine the lay of a wire rope. A—The strands in left-hand-lay wire ropes spiral to the left, counterclockwise, around the rope's axis. This rope has a left-hand regular lay. Note the individual strand wires are parallel to the rope axis. B—The strands in right-hand-lay wire ropes spiral to the right, clockwise, around the rope's axis. This is a right-hand regular-lay rope. Again, note the strand wires are parallel to the rope axis. C—Notice that in a Lang-lay wire rope, the wires in the strands are diagonal to the rope axis. D—In alternate-lay rope, the wires in adjacent strands spiral in opposite directions. This creates a V pattern between the wires of adjacent strands.

 Note
Wire ropes may also be classified as rotation-resistant or nonrotating. The arrangement is designed to resist spin or rotation while under load.

Wire Rope Grades of Steel

The different grades of steel have different tensile strengths. The types of steel used to construct wire ropes include:
- Plow steel (seldom used today).
- Improved plow steel (IPS).
- Extra improved plow steel (EIPS/XIPS).
- Extra, extra improved plow steel (EEIPS/XXIPS).

Inspecting Wire Ropes

Rope manufacturers and machine manufacturers provide specifications for the maximum number of breaks of wire that are allowable in a wire strand and for a given rope lay. To ensure the strength and integrity of wire ropes, they must be inspected regularly for breaks, corrosion, kinks, protruding cores, and bird caging. The term *bird cage* is used to describe the damage caused when a wire rope has been overstressed and does not return to its original state. See **Figure 5-10**. The rope lay length and outer diameter should also be measured to check for stretching and wear.

 Caution
Wire ropes removed from service due to excessive wear or damage must be destroyed to prevent accidental use.

Figure 5-10. Wire rope should be periodically inspected for damage. A—Two strands are completely severed, exposing the rope's core. B—This wire rope has been overstressed, resulting in a separation of the strands (bird cage).

Rope Lay

A rope will lose strength if it has been stretched. Measuring the rope's lay length is a means of determining whether the rope has been stretched. One *rope lay* or *lay length* is the linear distance (measured on the length of the wire rope) that it takes for one strand to make an entire helical revolution around the wire rope. See **Figure 5-11A**. As a rule of thumb, one rope lay length is equal to approximately 7 to 8 diameters of the rope.

Rope Diameter

The outside diameter of wire rope is measured to determine if, due to abrasion, the rope is too worn for safe use or if the internal core has failed. Notice that the measurement in **Figure 5-11B** is made on the high points of two opposing strands. A six-strand rope must be measured across all three pairs of opposing strands. An eight-strand rope must be measured across all four pairs of opposing strands.

Wire Rope Slings

OHSA defines a *wire rope sling* as "an assembly that connects the load to the material handling equipment." The manner in which a wire rope sling is oriented and attached varies, depending on the end terminations and the type of hitch being used. The most common wire rope sling is a rope with an eyelet at each end. The lifting equipment's wire rope or chain is attached to the wire rope sling's end using fittings. See **Figure 5-12**.

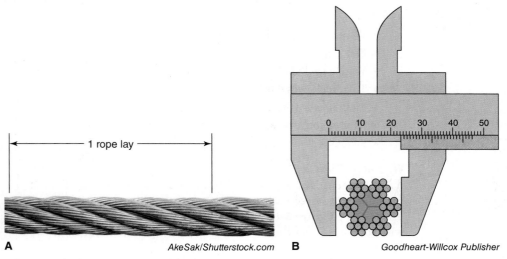

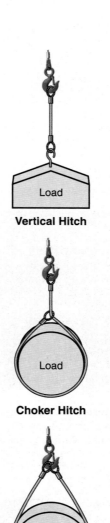

Figure 5-11. A rope lay is the linear distance it takes for one rope strand to wrap helically around the wire rope one complete turn. A—On this six-strand rope, the strand being measured to determine lay is highlighted. B—The diameter of the rope must be measured across the outside edge (crown) of two opposing strands. Do not measure the diameter across the parallel flat edges of four strands.

Figure 5-12. Wire rope slings are used in different hitch configurations that best accommodate each load.

Working Load Limit (WLL)

The *working load limit (WLL)* of a sling is the maximum load a sling can safely lift. Sling lift capacities are given in pounds or tons. A wire rope sling's lift capacity is affected by the following factors:

- Steel grade.
- Rope diameter.
- Strand arrangement.
- Type of core (fiber or steel).
- Type of fitting.
- Horizontal sling angle.

Manufacturers use proof testing to determine the WLL of a sling. *Proof testing* is a testing method in which weight is added to the sling until it is slightly above the WLL or until it fails. The sling's *ultimate strength* is the average load that causes the sling to fail. A sling's ultimate strength value is divided by a safety factor, also known as a design factor, to determine the working load limit. For example, if a wire rope sling with a safety factor of 5 breaks at 20,000 lb of force, the rope's WLL is one-fifth of that value (4000 lb).

Warning
Wire rope slings must not be used if the rope has 10 or more broken wires in a lay of rope or 5 broken wires in one strand in a single lay of rope.

Note
A wire rope sling's WLL is printed on the sling's label or tag. See **Figure 5-13**. The wire rope should be removed from service if the label is missing.

Goodheart-Willcox Publisher

Figure 5-13. Tags attached to wire rope slings provide the WLL for each of the three hitches (vertical, basket, and choker).

Wire Rope End Terminations

Many types of fittings are used to terminate the ends of wire ropes. The end terminations determine the *termination efficiency*, which is the value factored into the rope's WLL based on the termination method. The wire rope terminations and their respective efficiencies are as follows:

- Wire rope clips (80%).
- Fist grips (80%).
- Wedge sockets (80%).
- Turnback eyes (90%).
- Flemish eyes (90%).
- Spelter sockets (100%).
- Swage sockets (100%).

Wire Rope Clips

A *wire rope clip* is a clamp used to fix the loose end of the loop back to the wire rope and create an eyelet. Wire rope clips are typically made with a U-bolt and a saddle. See **Figure 5-14**. Wire rope clips are not used to form eyelets in wire rope slings but they are used in other applications, such as forming an eye on a winch.

The wire rope ends are labeled the live end and the dead end. The *live end* or *live line* is the rope end suspending the weight of the load. The *dead end* or *dead line* is the opposite wire rope end. The dead end is also known as the *free end*. The clip's saddle must be placed on the live end because the U-bolt crushes into the wire rope. A popular phrase

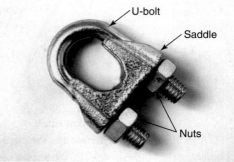

suchinan/Shutterstock.com

Figure 5-14. The saddle of the wire rope clip is secured to the U-bolt with two nuts. Multiple clips may be used to secure the wire rope.

to help you remember this restriction is "Never saddle a dead horse." In **Figure 5-15**, the clip's saddle has been placed on the wrong end of the wire rope, the dead end.

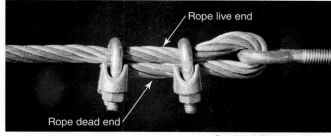

Sergii Votit/Shutterstock.com

Figure 5-15. The illustration shows wire rope saddles incorrectly placed on the dead end of the wire. Saddles belong on the live end.

Note
Manufacturers provide specifications for attaching wire rope clips to wire ropes, such as the number of clips, spacing between the clips, amount of torque to apply, and the amount of wire rope to turn back from the eye.

Fist Grip Clip
A *fist grip clip* is a variation of a wire rope clip that uses a second saddle instead of a U-bolt. This configuration prevents the wire rope from being crushed and ensures the live end of the wire will be saddled. See **Figure 5-16**.

Wedge Sockets
A *wedge socket* is used to secure the end of a crane's wire rope to the crane's load block or overhaul ball. See **Figure 5-17**. Wedge sockets are a quick and secure means of terminating the crane's wire rope end. Wedge sockets are not used to form eyelets in wire rope slings.

To assemble a wedge socket, the wire rope is threaded through the socket, looped, and threaded back through the socket. The wedge is placed inside the eye formed by the wire rope. The rope is wedged tightly in the socket when a load is applied. See **Figure 5-18**. The tail, or dead end, of the wire rope is secured with a wire rope clip. On some devices, the clip's U-bolt can be threaded through a hole in the wedge for additional security.

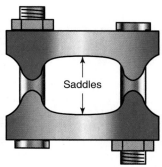

Goodheart-Willcox Publisher

Figure 5-16. A fist grip does not use a U-bolt. The fist grip uses two saddles, which eliminates the chance that a U-bolt will crush the live end of the rope.

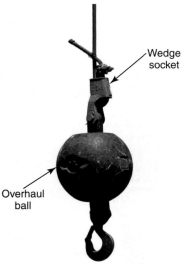

J.D.S./Shutterstock.com

Figure 5-17. Wedge sockets are used for quickly terminating the end of a crane's wire rope to a load block or overhaul ball. The socket is pinned to the crane's load block or overhaul ball.

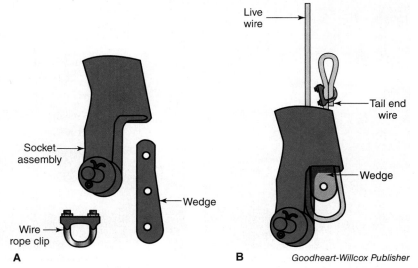

Goodheart-Willcox Publisher

Figure 5-18. In a wedge socket termination, the wire rope is gripped between the socket and wedge. A—A wedge socket termination consists of a socket assembly, a wedge, and a wire rope clip. B—Notice how the wire is looped through the socket assembly. The loop has been enlarged to illustrate how the wire is routed through the socket.

Turnback Eye

A *turnback eye* is formed by passing the wire rope through one side of a swage sleeve, around a thimble, and back through the other side of the swage sleeve. A specialized press is then used to compress the sleeve around the live and dead ends of the rope, securing them. See **Figure 5-19A**. A turnback eye is also known as a *foldback eye*.

Flemish Eye

A *Flemish eye* is formed by initially unwinding the end of a wire rope into two equal parts. The two parts are then looped and woven together to form an eyelet. See **Figure 5-19B**. A swage sleeve is slid over the splice and compressed around it to secure the splice.

Caution

Wire rope clips, fist grips, and wedge sockets are *not* used to form wire rope slings.

Spelter Socket

The *spelter socket* uses a polyester-based compound resin to secure the wire rope in the socket. The liquid resin is poured into the socket containing the wire rope and allowed to cure. The hardened resin secures the wire rope and provides 100% termination efficiency (based on the wire rope's WLL). See **Figure 5-20**.

Swaged Socket

Swaged sockets are formed by placing the fitting and wire rope into a die that is hydraulically pressed through a tapered ring. The swaging process secures the wire rope to the socket, providing 100% termination efficiency. See **Figure 5-21**.

Wire Rope Sling Advantages and Disadvantages

Wire rope slings are heavy, difficult to handle, susceptible to rust, and can cause surface damage to the load. Wire rope slings will not fray as easily as synthetic slings, but pose

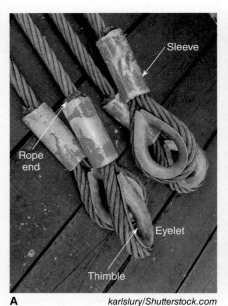

Figure 5-19. Turnback eyes and Flemish eyes are two common types of terminations. A—A turnback eye is formed by placing the two wire ends beside each other. The two wire ends are not woven together as compared to the Flemish eye. B—A Flemish eye is formed by weaving the two parts of a single rope together to form an eyelet. This wire rope sling uses a Flemish eye with a thimble.

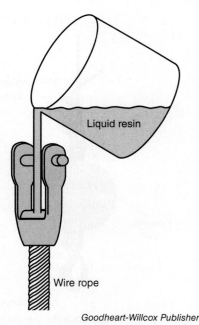

Figure 5-20. Spelter sockets are formed with a compound resin that secures the wire rope in the socket. Spelter sockets have 100% termination efficiency.

more of a physical hazard when the wires do fray. The wire rope slings are also more likely to twist, causing kinks in the wire and damage to the sling. One major advantage of wire rope slings is that they can be used in a wide range of temperatures (–60°F to 400°F [–51°C to 204°C] for steel-core and –60°F to 180°F [–51°C to 82°C] for fiber-core).

Synthetic Slings

Synthetic slings are made of polyester or nylon and may have a flat-strap profile (web slings) or a round profile. The main differences between polyester and nylon slings are the tendency to stretch, level of acid resistance, surface softness, and abrasion resistance. See **Figure 5-22**.

Web Slings

Web slings have a flat-strap profile and are available in a wide range of sizes (width, length, and strength) and designs. See **Figure 5-23**. Web slings have colored-yarn warning cores that show when the outer surface (jacket)

Goodheart-Willcox Publisher

Figure 5-21. These crane guylines use sockets that are swaged onto the wire rope ends to provide 100% termination efficiency.

	Polyester	Nylon
Percentage of stretch at maximum WLL	3% (untreated)	6% (untreated)
	7% (treated)	10% (treated)
Has less reaction to the following chemicals	Acids	Ethers and alcohols

Goodheart-Willcox Publisher

Figure 5-22. Polyester and nylon web slings vary in the amount they stretch as well as their reaction to chemicals.

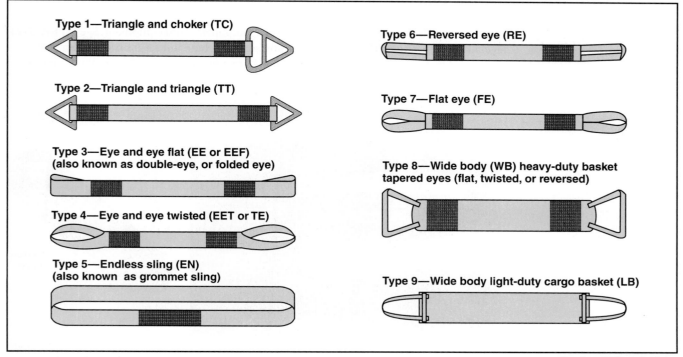

Goodheart-Willcox Publisher

Figure 5-23. Synthetic web slings are manufactured in a variety of styles.

is worn or cut. The sling should be taken out of service and destroyed if the warning core is visible.

Web Sling Ratings

Web slings are rated by their load capacity and given different classifications. According to OSHA standards, web-sling capacity is rated based on the type of hitch, material strength, design factor, angle of loading, diameter of curvature over which the sling is used, and fabrication efficiency. Always refer to the manufacturer's literature and government standards to determine which type of web sling is acceptable for a given task.

Polyester Endless Round Slings

Small polyester yarn fibers are twisted together to form a round, endless sling. Round slings are very flexible and lightweight and manufactured to allow minimal stretching. See **Figure 5-24**. The fibers are protected with a seamless cover.

Inspecting Synthetic Slings

To ensure safe use, synthetic slings must be used according to the manufacturer's specifications and properly maintained. Synthetic slings should be inspected regularly for damage, such as burns, melting, holes, tears, cuts, worn stitching, excessive abrasive wear, knots, brittle or stiff areas, and cracked or broken fittings. If the sling's warning core is exposed or if the sling exhibits damage, the sling should be removed from service and destroyed.

Synthetic Sling Advantages and Disadvantages

Polyester and nylon slings are easy to handle, lightweight, and do not rust. They are also useful for lifting components with surfaces that would be damaged by wire rope slings and chain slings. Web slings have a wider surface area, which provides a better grip on the load. One disadvantage of synthetic slings is that they are more easily damaged than wire rope or chain slings. Polyester and nylon slings also have a much narrower operating temperature range (–40°F to 194°F [–40°C to 116°C]). The slings are also more susceptible to environmental damage and must be stored away from ultraviolet light and harmful chemicals.

Chain Slings

Technicians also use chain slings for hoisting components and equipment. See **Figure 5-25**. A chain sling uses a collector ring, hook, and series of chain links. Chain slings may have one or more legs. Chain slings with multiple legs form a *bridle*. **Figure 5-25B** illustrates a two-leg chain bridle.

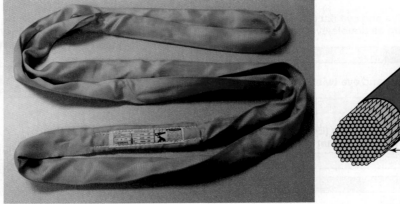
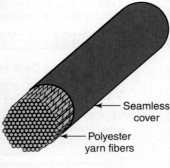

Figure 5-24. Round slings are endless loops with no eyelets. The polyester fibers are wound into a round sling and the ends of the fibers are tied together to create the loop.

Chain Sling Ratings

Chain slings have load capacity ratings (WLL) and grades based on the chain's ultimate breaking strength. Chain grades 8 (80) or 10 (100) are used for chain slings. Other important characteristics that affect a chain sling's rating include the chain diameter, number of legs, type of master link, and the type of bottom attachments. The WLL is listed on the sling's tag.

> **Note**
> Chain slings have a lower safety factor (4) than wire rope and synthetic slings (5).

Inspecting Chain Slings

Chain slings must be inspected for damage and wear, such as cracks, nicks, and gouges. See **Figure 5-26**. Manufacturers and government standards outline specifications for maximum allowable wear. Wear can be determined by measuring the cross section at the link ends. The overall chain length must also be measured to determine if it has been stretched beyond its specifications. The sling's original overall length is included on the sling's tag.

Chain Sling Advantages and Disadvantages

Chain slings can be used in a wider temperature range than polyester and nylon slings. For example, a grade 8 chain sling's operating range is −40°F to 400°F (−40°C to 96°C). Chain slings are more flexible than wire rope slings. However, they are only as strong as their weakest link and the chain is susceptible to rust.

Sling Fittings

Sling fittings are used to attach and maneuver loads. Common sling fittings are master links, collector rings, hooks, shackles, eyebolts, swivels, and thimbles. These fittings are not used exclusively with chain slings and may be used in other applications.

Master Links

A *master link* is the top ring that collects or holds bridle legs. See **Figure 5-27**. The master link is often attached to a hoist's hook. The master link of a double bridle sling is proof tested at a rate of four times the WLL of a single leg. The master link of a three- or four-leg bridle sling is proof tested at a rate of six times the WLL of a single leg.

Collector Rings

Collector rings are also used to collect multiple slings or fittings in a single ring. Collector rings may be oval or pear-shaped. See **Figure 5-28**. Shackles can also be used as collector rings.

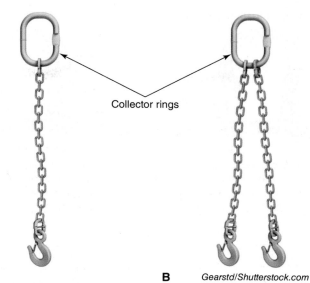

Gearstd/Shutterstock.com

Figure 5-25. Chain slings can also be used for hoisting. A—A chain sling has a collector ring attached to the top and a hook is attached to the bottom of the sling. B—This chain sling is a two-leg bridle.

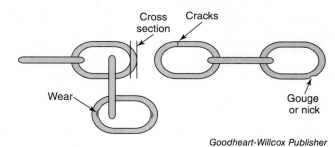

Goodheart-Willcox Publisher

Figure 5-26. Chain slings need to be inspected for cracks, nicks, gouges, and wear.

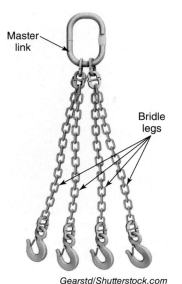

Gearstd/Shutterstock.com

Figure 5-27. Master links hold the bridle legs.

Goodheart-Willcox Publisher

Figure 5-28. Collector rings are used to collect the bridle legs into a single point.

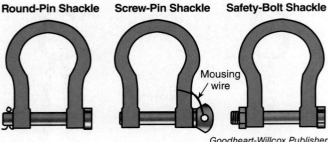

Figure 5-29. Shackles can have three different designs: round pin, screw pin, and safety bolt. Notice the screw pin shackle has been moused. Mousing is used when the shackle is being used in a long term application.

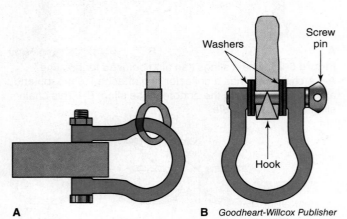

Figure 5-30. Shackles are often used to attach slings and hooks. A—If a shackle is used in a side-loaded application, the WLL is reduced as much as 50%. B—If a shackle is placed in a hook, hardened washers should be used to center the hook to prevent the shackle from tilting, which would place a side load on the shackle and hook.

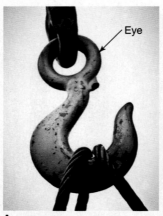
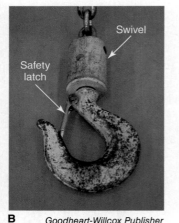

Figure 5-31. Hooks are available in several designs. A—This eye hook is being used to lift two wire rope slings. If the hook does not have a latch and if it is used in a long term application, it should be *moused* to prevent the slings from sliding out of the hook. B—Swivel hooks are commonly used on hoists to allow the load to be rotated. This hook has a movable safety latch.

Shackles

A *shackle* is a U-shaped device that uses a pin or bolt to attach slings to the loads and hooks. Common designs include the screw-pin, round-pin, and safety-type (also known as a bolt-type). See **Figure 5-29**. The round-pin shackle is secured only by a cotter key and therefore can only be used in a straight vertical pull. The screw-pin shackle in **Figure 5-29** has been *moused* with a wire to prevent the pin from loosening. The screw-pin and safety-bolt shackle can be side loaded, but the WLL of the shackle will be reduced. See **Figure 5-30A**.

If a shackle's pin is placed in the valley of the hook, the shackle will slide and cause the load to shift. Hardened washers should be used to keep the shackle centered in the hook. See **Figure 5-30B**.

> **Warning**
>
> When using a shackle with a hoist's wire rope, the wire must *not* be allowed to run across the shackle's pin or it may cause the pin to unthread from the shackle.

Hooks

A hook is used to attach the hoist to the load's slings. The most common designs are the eye hook or swivel hook. See **Figure 5-31**. Some hooks have a safety latch to prevent the sling's end from sliding off. Some hooks have quick-check marks to indicate sling positioning limits. See **Figure 5-32**. Slings must not be positioned outside

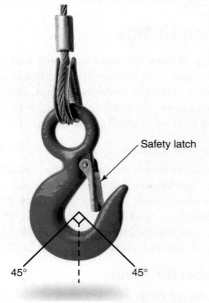

Figure 5-32. Cross angle indicator marks can be placed on the hook as a method for checking the hook's included angle.

the sling's 90° included angle. The *included angle* is the angle formed by the sling directly beneath the hook. The maximum allowable included angle for a hook is 90°.

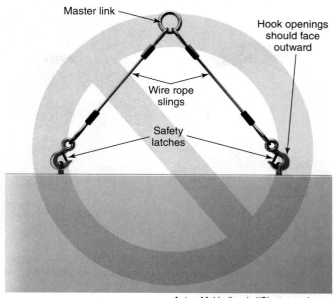

Figure 5-33. Do *not* point the hooks inward as they are in this figure. The hooks should face outward to minimize the risk of the sling or bolt slipping out.

 Caution
The hooks must face outward when used with slings to hoist a load. See **Figure 5-33**.

Eyebolts

Eyebolts are classified as shouldered and non-shouldered. See **Figure 5-34**. The shoulder provides more stability to the base of the eye. Eyebolts without a shoulder should be used for vertical loading only. Shouldered eyebolts can be side loaded (angular loading); however, side loading will greatly reduce the WLL. If a shouldered eyebolt is used in a side-loaded application, the eyelet must be positioned so the load pulls in the same plane as the eyelet, not perpendicular to the eyelet. Because an eyebolt has a shaft with a set number of threads, it may be in the wrong position when fully tightened. To correct this, hardened washers can be placed under the eyebolt so that it is in the correct position when it is fully tightened. See **Figure 5-35**.

Hoist Swivel Rings

A hoist swivel ring is a fitting that is bolted to the load and allows the load to be rotated. See **Figure 5-36**. The ring can pivot in a 180° range of motion and the swivel rotates 360°. The swivel ring's WLL is not reduced when the sling is angled.

 Pro Tip
Sling Savers™ made by Crosby evenly distribute the load across a web sling's eyelet. See **Figure 5-37**.

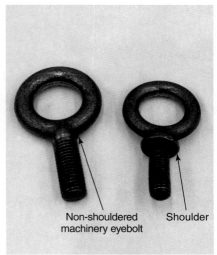

Figure 5-34. These machinery eyebolts can be identified by their fully threaded shanks.

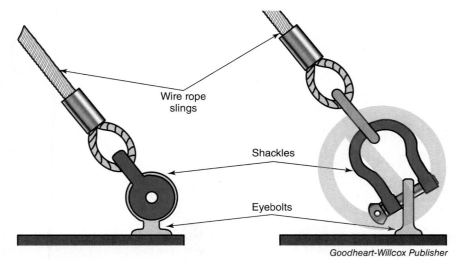

Figure 5-35. Side-loaded eyebolts need to lift in the same plane as the eyebolt's eyelet. Hardened washers are used to correctly position the eyebolt in the direction of the sling.

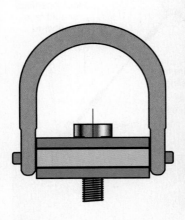

Figure 5-36. Hoist swivel rings are similar in function to shouldered eyebolts, but can swivel to allow the load to be rotated during lifting.

Goodheart-Willcox Publisher

Goodheart-Willcox Publisher

Figure 5-37. The Crosby Sling Saver is placed in the eyelet of a synthetic web sling.

Goodheart-Willcox Publisher

Figure 5-38. Lifting links use one or two chain links and an angled link.

Lifting Links

Lifting links, also known as lifting brackets, are another fitting used by technicians in place of eyebolts. The fitting contains one or two chain links attached to an angled link. See **Figure 5-38.**

Lifting Principles

Mistakes are easily made when lifting heavy components. It is still possible for equipment to be damaged or for personnel to be injured when using a sling or fitting with an adequate WLL rating if fundamental lifting principles are not followed.

Horizontal Sling Angle

Perhaps one of the most important but often overlooked principles in overhead lifting is the horizontal sling angle. The term *horizontal sling angle* describes the angle formed between a horizontal plane and the lifting sling. See **Figure 5-39.** A load value does not

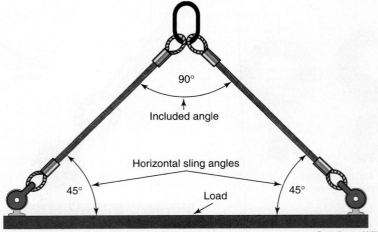

Goodheart-Willcox Publisher

Figure 5-39. The triangle formed by the lifting slings results in two horizontal sling angles and one included angle.

need to be factored in the lift if the sling is making a straight 90° lift. However, if a sling is tilted at an angle less than 90° (perpendicular) to the load, a load multiplier known as the *load angle factor* must be multiplied by the sling's overall load. **Figure 5-40** lists examples of horizontal sling angles and the corresponding load angle factors.

Notice that a horizontal sling angle of 30° requires doubling the sling's load because the load angle factor equals 2.0. This means that if each sling is responsible for lifting 5000 lb, the 30° sling angle will cause the load on the sling, fittings, and component being lifted to rise to 10,000 lb. See **Figure 5-41**.

When a lift is made with two slings that are the same length, the slings form an *isosceles* triangle, which results in two horizontal sling angles that are the same value. The sum of all three angles (two horizontal sling angles and the included angle) equals 180°. In **Figure 5-41**, the sum of the two horizontal sling angles is 60° and the included angle is 120°. The included angle is formed at the top of the lifting triangle by the angle of the two lifting slings.

If the exact horizontal sling angle is unknown, the load angle factor can be determined using the formula: length (L) ÷ height (H) = load factor. See **Figure 5-42**.

Horizontal Sling Angle	Load Angle Factor
90°	1.000
60°	1.155
50°	1.305
45°	1.414
30°	2.000

Goodheart-Willcox Publisher

Figure 5-40. Horizontal sling angles less than 90° require multiplying a load angle factor with the sling's load.

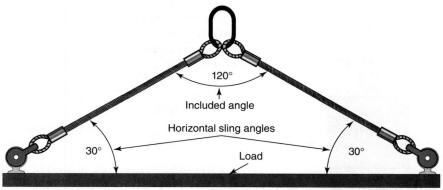

Goodheart-Willcox Publisher

Figure 5-41. This lifting triangle has two horizontal sling angles of 30° and an included angle of 120°.

Warning

OSHA regulations mandate that horizontal sling angles be 30° or greater. Horizontal sling angles less than 30° are unsafe because the load angle factor increases greatly at angles less than 30°. For example, a horizontal sling angle of 10° has a load angle factor of 6, and a horizontal sling angle of 5° has a load angle factor of 12. This significant increase in the load angle factor can cause slings to become overloaded and fail. The 30° horizontal sling angle rule also results in a maximum included angle of no more than 120°. Note that hooks can only be used with included angles of 90° or less. Only shackles and collector rings can be used with 90°–120° included angles.

Pro Tip

If the slings can easily be repositioned and are the same length, then an equilateral triangle can be formed when attaching the slings to the load. (An equilateral triangle has three equal lengths.) The two sling lengths are the same as the horizontal length. The three angles (two horizontal sling angles and the included angle) are each 60°. As explained earlier, a 60° horizontal sling angle has a 1.155 load angle factor.

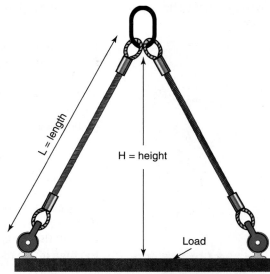

Goodheart-Willcox Publisher

Figure 5-42. Finding the height and length of the lifting triangle allows a technician to compute the load factor.

Hitches

A *hitch* is the manner in which a sling is configured for lifting. The most common lifting hitches are the vertical inline hitch, choker hitch, and basket hitch. The hitch affects the sling's lift capacity and its ability to secure the load.

Vertical Inline Hitch

A vertical inline hitch hangs straight from the hook to the load. No multipliers are required to adjust the sling's lift capacity. See **Figure 5-43**. For example, a 3-ton wire rope sling can lift 6000 lb in a single inline vertical lift.

Choker Hitch

A *choker hitch* is formed with one end of the sling passing under the load and through a fitting or eyelet on the other end of the sling. See **Figure 5-44**. The choker hitch has a reduced lifting capacity due to the horizontal sling angle formed by the hitch. The WLL of a properly formed choker hitch is 85% of the WLL of a vertical inline hitch. The capacity is further reduced if the choker hitch is pushed down toward the load.

Note

The hook must point away from the load when it (the hook) is used to form a choker hitch.

A *double-wrap choker hitch* is formed by wrapping the sling completely around the load once before forming the choker hitch. See **Figure 5-45A**. The hitch is used to secure loose materials, such as several pieces of pipe. The hitch provides 360° of sling contact on the load. A *double-choker hitch* has two separate slings choked on the load. The double-choker hitch is used to lift long loads. See **Figure 5-45B**.

Basket Hitch

A *basket hitch* is formed by trapping the load and placing both ends of the sling into the hoist's hook or collector ring. See **Figure 5-46A**. A perfectly formed basket hitch will have a WLL that is twice the capacity of a vertical inline hitch. A *double-wrap basket hitch* is formed by fully wrapping a single sling around the load. It is used to trap the loose materials in the same fashion as the double-wrap choker hitch. A double-wrap basket hitch can be used for longer loads. A spreader bar is required to maintain the straight vertical angle on both legs of a basket hitch. See **Figure 5-46B**.

Inverted Basket Hitch

An *inverted basket hitch* is an unstable hitch made with a single sling that is loosely placed over a hook with the two ends of the sling attached to the load. The inverted basket hitch in **Figure 5-47** reveals two common mistakes made by inexperienced technicians. Notice the sling is free to slide within the hook. The load could easily shift and cause the sling to slide off the hook. The second mistake is that the horizontal sling angle is too low, causing too high of a load on the sling and fittings.

Center of Gravity

In addition to determining the proper horizontal sling angle, the technician must also determine the load's center of gravity (COG). The *center of gravity (COG)* is the balancing point of the load. Once the COG is determined, the technician can determine the types and lengths of slings that will be needed to securely lift the load.

The image in **Figure 5-48** shows two vertical hitches lifting an oddly shaped load. The black-and-white circle is the COG. The COG in the image is 2′ from the right side of the load

Goodheart-Willcox Publisher

Figure 5-43. Inline vertical hitches do not require a load angle multiplier. Their load is based on the weight of the load plus the weight of the lifting hardware.

Gearstd/Shutterstock.com

Figure 5-44. A choker hitch snugly traps the load.

Figure 5-45. There are different variations of choker hitches for different lifting scenarios. A—This forklift's mast has been laid on the ground so that the lift cylinders can be removed. A double-wrap choker hitch was used to provide 360° sling contact around the slick, oily surface of the hydraulic cylinder. B—Two choker hitches are used for lifting longer loads.

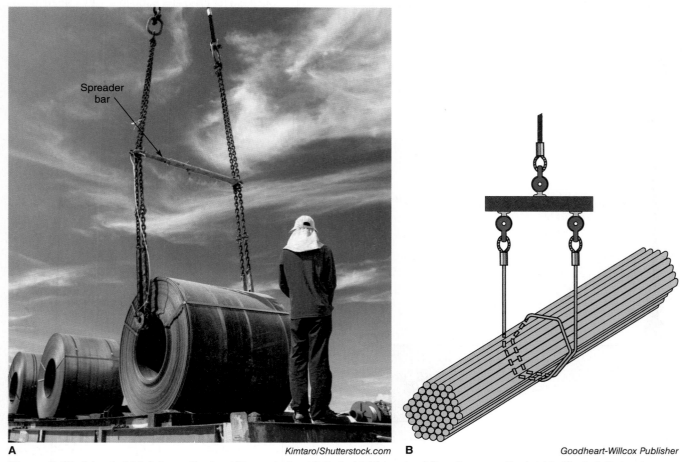

Figure 5-46. A basket hitch traps the load like a choker hitch, but both ends of the sling are attached to the hoist's hook or collector ring. A—This basket hitch is trapping the cylindrical load. Notice both ends of the sling are in a straight vertical position. The spreader bar maintains the vertical position of the slings. B—A double-wrap basket hitch is used with longer loads. In this example, the double-wrap basket hitch is also retaining the loose bundle of material.

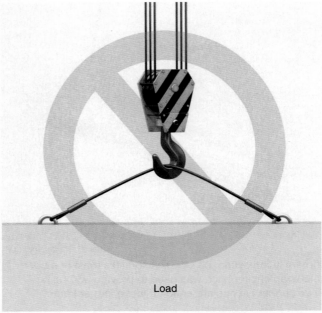

Figure 5-47. An inverted basket hitch is a hazardous lift because the sling can easily slide through the hoist's hook.

and 8′ from the left side. The sling closest to the COG is responsible for more of the load's weight. To determine how much weight each hitch is lifting, use the following formula:

Sling A load = Distance from sling B to COG ÷ overall length of load

Sling B load = Distance from sling A to COG ÷ overall length of load

Calculations for **Figure 5-48**:

Sling A load = 2 ÷ 10 = 20%
10,000 lb × .20 = 2000 lb

Sling B load = 8 ÷ 10 = 80%
10,000 lb × .80 = 8000 lb

As explained in **Figure 5-42**, the load angle factor must be entered into the equation when there is a horizontal sling angle. See **Figure 5-49**.

Sling A load angle factor = 35 ÷ 26 = 1.35
Total load for sling A = 2000 lb × 1.35 = 2700 lb

Sling B load angle factor = 31 ÷ 26 = 1.19
Total load for sling B = 8000 lb × 1.19 = 9520 lb

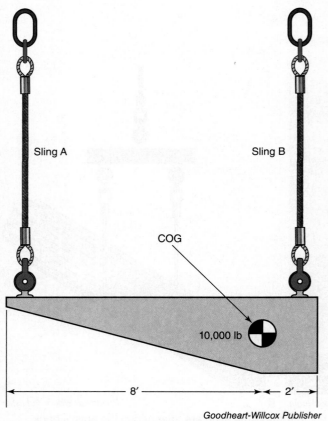

Figure 5-48. The center of gravity is the point where the load is balanced. The COG is not always located in the center of the load.

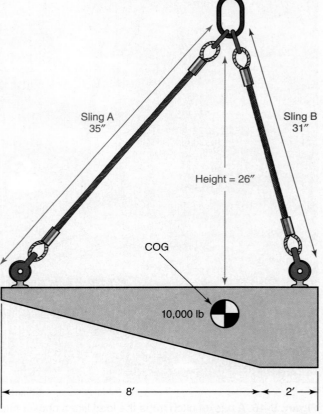

Figure 5-49. The load factor and the COG are different for sling A and sling B.

Trial-and-Error Method

The least desirable method for finding the COG is the trial-and-error method. To begin, the load is lifted with two slings of different lengths. A plumb bob (a weight on a string) is hung from the hoist's master link. A line is drawn on the load to indicate where the plumb bob fell. See **Figure 5-50**. The load is then lowered to the ground and the two slings are swapped. The process is repeated and a second line is drawn on the load to indicate the plumb bob's location. The COG is the point at which the two plumb bob lines intersect.

Warning

Using a trial-and-error method to determine a load's COG can be a dangerous practice. If a trial-and-error method is used, extreme caution should be taken to ensure the slings and fittings are not overloaded, the load is not damaged, and personnel are not injured.

Pro Tip

The COG is always located directly below the main hoist's hook.

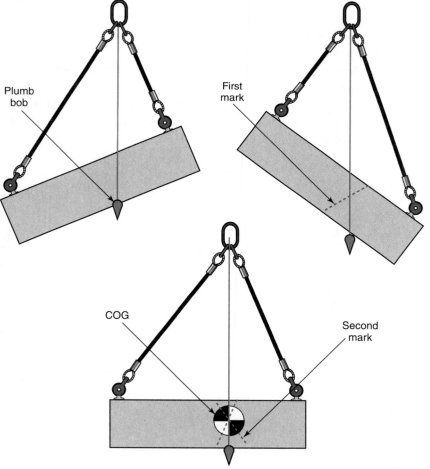

Goodheart-Willcox Publisher

Figure 5-50. Two different length slings and a plumb bob are used in the trial-and-error method. This is the least desirable method for determining a load's COG.

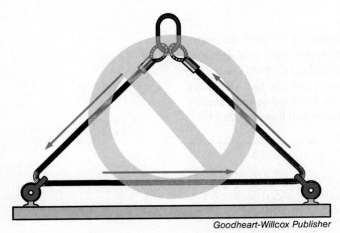

Figure 5-51. Reeving is an improper rigging method in which a single sling is run through both lifting eyes.

Bridle Lifts

When making a lift with a multiple-leg bridle, such as a three- or four-leg bridle, do not make the mistake of calculating the overall WLL as the sum of the bridle legs, even if the legs are the same length, have the same horizontal angle, and the legs share an equal amount of the COG. The load can easily shift when lifted and moved. This shifting may cause the bulk of the weight to become unequally distributed, leaving it (the bulk) on only one or two legs. A four-leg bridle total capacity can only be based on three of the bridle legs.

Reeving

Many technicians use an improper lifting technique called reeving. *Reeving* is attaching a single sling to a load by threading the sling through both lifting eyes and attaching both ends of the sling to the lifting hook. See **Figure 5-51**. Reeving multiplies the side-crushing load placed on the lifting eyes.

Planning a Lift

Planning a lift before you attach and move a load ensures safe lifting and moving of heavy components. Keep the following in mind when planning a lift:

- Inspect the load for anything that might cause a hazard.
- Determine the full weight of the load.
- Determine how many people will be needed to load and execute the lift.
- Determine the type of equipment needed to lift and move the load.
- If the load will be moved to another area, determine the path it will take.
- Determine what type of support equipment will be needed and have it in place before you move the load.

Summary

- Using improper lifting techniques often results in personal injury.
- Hydraulic and mechanical jacks are used for lifting components or machines.
- Blocking refers to the actions taken or methods used to prevent a component or machine from moving.
- Cribbing is a means of temporarily supporting a load with hardwood or engineered plastic blocks stacked in alternating directions.
- Service repair shops are often equipped with gantry, jib, and overhead cranes.
- Chain hoists may be manually operated with a pull chain or mechanical lever or electrically or pneumatically driven.
- Wire ropes are classified by the type of core, number of strands, number of wires in a strand, strand wire arrangement, type of rope lay, and steel grade.
- Wire ropes are inspected before each use for breaks, kinks, protruding cores, and bird caging. The rope lay length and outside diameter are measured to determine wear.
- The working load limit of a wire rope is affected by the rope's grade of steel, diameter, wire arrangement, core type, termination fittings, type of hitch, and horizontal sling angle.
- A sling safety factor is a value that is divided into the sling's ultimate strength value and used to determine the sling's working load limit.
- A sling's ultimate strength is found by testing multiple slings to determine the average amount of force that causes the sling to fail.
- Wire ropes can be terminated using wire rope clips, fist grips, wedge sockets, turnback eyes, Flemish eyes, spelter sockets, and swage sockets. Wire rope clips, fist grips, and wedge sockets cannot be used to form slings.
- The saddle of a wire rope clip must be placed on the live end.
- Synthetic slings are made of polyester or nylon and may have a flat-strap profile (web slings) or a round profile.
- Polyester and nylon web slings vary in the amount they stretch as well as their reaction to chemicals.
- Polyester and nylon slings have colored-yarn warning cores that are exposed when the outer surface is worn or cut.
- A polyester endless round sling is made with small polyester yarn fibers twisted together and connected to form an endless loop.
- Synthetic slings are easy to handle, lightweight, do not rust, and are useful for lifting components with surfaces that would be damaged by wire rope slings and chain slings.
- A chain sling uses a collector ring, hook, and series of chain links.
- A chain sling with multiple legs is called a bridle.
- Chain slings must be inspected for wear, cracks, nicks, gouges, and stretching.
- Fittings used with chain slings include master links, collector rings, hooks, shackles, eyebolts, swivels, and thimbles.
- Screw-pin and safety-bolt shackles are the only type of shackles that can be side loaded, but their WLL is greatly reduced when they are side loaded.
- Non-shouldered eyebolts must not be side loaded.
- The WLL of a side-loaded shouldered eyebolt is greatly reduced.
- The WLL of a swivel ring is not reduced when the sling is angled.
- The horizontal sling angle is formed between the horizontal plane of the component being lifted and the lifting sling.
- Any time a sling is tilted at an angle less than a straight 90° perpendicular angle, a load multiplier, known as a load angle factor, must be multiplied by the sling's overall load.
- A horizontal sling angle of 30° doubles the load on the sling.

- OSHA regulations mandate that sling angles be 30° or greater, which also results in a maximum included angle of no more than 120°.
- The included angle is the top angle of the lifting triangle formed by the lifting slings, located directly below the hook or collector ring.
- A choker hitch is formed with one end of the sling passing under the load and through a fitting or eyelet on the other end of the sling.
- A double-wrap choker hitch is formed by wrapping the sling completely around the load once before forming the choker hitch.
- A double-choker hitch is two separate slings choked on a load.
- A basket hitch is formed by trapping the load and placing both ends of the sling into the hoist's hook or collector ring.
- A double-wrap basket hitch is formed by fully wrapping a single sling around the load.
- An inverted basket hitch is an unstable hitch made with a single sling that is loosely placed over a hook with the two ends of the sling attached to the load.
- The center of gravity (COG) is the balancing point of the load and is located directly below the main hoist's hook.

Technical Terms

- alternate-lay wire rope
- basket hitch
- blocking
- bridle
- center of gravity (COG)
- choker hitch
- collector rings
- cribbing
- dead end
- dead line
- double-choker hitch
- double-wrap basket hitch
- double-wrap choker hitch
- filler wire rope
- fist grip clip
- Flemish eye
- foldback eye
- free end
- hitch
- horizontal sling angle
- included angle
- inverted basket hitch
- Lang-lay wire rope
- lay length
- left-hand-lay wire rope
- live end
- live line
- load angle factor
- master link
- moused
- proof testing
- reeving
- regular-lay wire rope
- right-hand-lay wire rope
- rope lay
- Seale wire rope
- shackle
- simple wire rope
- spelter socket
- splitting stands
- swaged socket
- termination efficiency
- turnback eye
- ultimate strength
- Warrington wire rope
- wedge socket
- wire rope clip
- wire rope lay
- wire rope sling
- working load limit (WLL)

Review Questions

Answer the following questions using the information provided in this chapter.

Know and Understand

1. When lifting components by hand, all of the following should be performed, *EXCEPT*:
   A. know your physical limitations.
   B. keep the load close to your body.
   C. do not attempt to lift heavy loads away from your body.
   D. bend your back.

2. A lattice boom crane is constructed with multiple steel tubes referred to as _____.
   A. lacings
   B. chocks
   C. strings
   D. ropes

3. Left-hand-lay wire ropes always have _____.
   A. strands spiraling counterclockwise
   B. strands spiraling clockwise
   C. wires spiraling counterclockwise
   D. wires spiraling clockwise

4. The wires and strands in a _____ wire rope spiral in the same direction.
   A. Lang-lay
   B. regular-lay
   C. left-hand-lay
   D. right-hand-lay

5. The term _____ describes the maximum load a sling can safely lift.
   A. design factor
   B. proof test
   C. ultimate strength
   D. working load limit

6. The term _____ describes the process of loading a sling to determine the amount of weight that causes the sling to fail.
   A. design factor
   B. proof testing
   C. ultimate strength
   D. working load limit

7. The number _____ is the safety factor of a wire rope sling.
   A. three
   B. four
   C. five
   D. six

8. All of the following describe synthetic slings, *EXCEPT*:
   A. susceptible to environmental damage.
   B. flexible.
   C. covered.
   D. heavy.

9. A _____ sling is made with small yarn fibers that are twisted together and formed into a loop.
   A. polyester endless round
   B. nylon web
   C. polyester web
   D. synthetic chain

10. Chain slings are made of what type of grade of steel?
    A. Grade 8
    B. Grade 10
    C. Both A and B.
    D. Neither A nor B.

11. A _____ shackle can only be used in a straight vertical pull.
    A. round-pin
    B. screw-pin
    C. safety-bolt
    D. Both A and B.

12. The maximum allowable included angle of a hook is _____.
    A. 30°
    B. 60°
    C. 90°
    D. 180°

13. A double-leg hook bridle must be installed with the hooks facing _____.
    A. inward
    B. outward
    C. It does not matter.
    D. It depends on the COG.

14. Which of the following horizontal sling angles has a load angle factor of 2.0?
    A. 30°.
    B. 45°.
    C. 60°.
    D. 90°.
15. A sling is 36″ long. The sling angle is unknown, but the sling's overall height is 24″. What is the sling's load factor?
    A. 0.667.
    B. 1.2.
    C. 1.5.
    D. 2.0.
16. OSHA regulations mandate that horizontal sling angles be no less than _____.
    A. 30°
    B. 45°
    C. 60°
    D. 80°
17. A _____ hitch secures loose materials by providing a 360° sling contact.
    A. choker
    B. double-choker
    C. basket
    D. double-wrap choker

Apply and Analyze

18. Splitting jack stands are commonly used when replacing a transmission _____.
19. The term _____ is used to describe the actions taken to prevent a component, attachment, or machine from moving.
20. The term _____ is used to describe the process of temporarily supporting a load with hardwood or engineered plastic blocks stacked in alternating directions.
21. Rough-terrain, all-terrain, and truck-mounted cranes are examples of _____ cranes.
22. For a shop with a high volume of work, it is preferred to have multiple _____ cranes.
23. The term rope _____ is used to describe the linear distance it takes for one strand to make an entire helical revolution around the wire rope.
24. Chain slings have a safety factor of _____.
25. The wire rope clip's saddle must be placed on the _____ end of the wire.
26. A(n) _____ hitch traps the load and places both sling eyelets in the hoist's hook.
27. A(n) _____ hitch is formed with one end of the sling passing under the load and through a fitting or eyelet on the other end of the sling.
28. The center of _____ is the balancing point of the load.

Critical Thinking

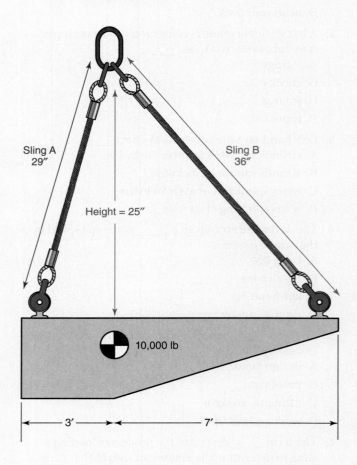

29. Using the illustration, determine the horizontal sling factor for sling B.
30. Using the illustration, determine the total load placed on sling B.

Chapter 6
Belt and Chain Drives

Objectives

After studying this chapter, you will be able to:

- ✓ Explain the relationship of speed and torque for a fixed amount of horsepower.
- ✓ Identify types of belt drives and their applications.
- ✓ Describe pulleys that are used in belt drives.
- ✓ Identify types of sprockets and chain drives and their applications.
- ✓ Explain how to check sprocket and chain wear.
- ✓ Calculate chain drive ratios.

Heavy equipment power systems must be able to change speed, torque, and direction. *Torque* is force applied through an arcing motion. Torque applied to a lever equals force times the distance between the point where the force is applied and the lever's fulcrum, **Figure 6-1**.

In mobile equipment, the following systems are used to change speed, change torque, and reverse power flow:
- Belts and pulleys.
- Chains and sprockets.
- Gears (mechanical transmissions).
  - Spur.
  - Helical.
  - Bevel.
  - Worm.
  - Planetary.
- Electric drives.
- Fluid drives.
  - Torque converters.
  - Hydrostatic transmissions.
- Automatic transmissions.
- Continuously variable transmissions.

This chapter focuses on power transmission through belts and pulleys and chains and sprockets. Later chapters cover other power transmission methods.

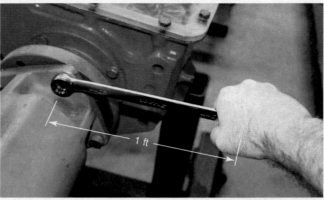

Figure 6-1. Torque applied to the end of a lever equals the force times the length of the lever.

Relationship of Speed and Torque

In any type of mechanical power transmission system, if the input drive (sprocket, pulley, or gear) has a large diameter, and the output drive has a small diameter, speed increases and torque decreases. If the input drive has a small diameter and if the output drive has a large diameter, speed decreases and torque increases. See **Figure 6-2**.

The change of speed and torque has an inverse relationship, **Figure 6-3**. For a fixed amount of horsepower, an increase in either speed or torque causes a decrease in the other. The only way to increase both speed and torque is to increase horsepower.

Belts and Pulleys

A variety of belt drives transmit power in mobile machinery. Drive belts can be used on engines for driving the alternator, water pump, and air-conditioning compressor. Belts are used on machines for driving shafts, gearboxes, pumps, rollers, rotors, drums, and conveyor belts. Many off-highway machines use belts in higher-horsepower applications. For example, some combines have high-horsepower drive belts that transfer 60 hp or more.

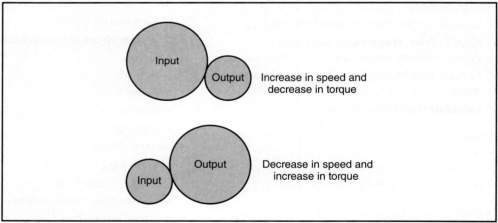

Figure 6-2. If a large-diameter gear, pulley, or sprocket is driving a smaller-diameter component, an increase in speed and decrease in torque results. If a small-diameter gear, pulley, or sprocket drives a larger-diameter component, there is a decrease in speed and increase in torque.

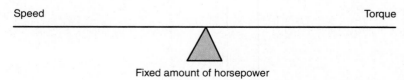

Figure 6-3. For a fixed amount of horsepower, speed can increase only when torque is decreased. Torque can increase only when speed is decreased.

Chapter 6 | Belt and Chain Drives

The earliest belts were made of rope. Belt materials evolved to leather and fabric and then to neoprene rubber. Manufacturers now make belts of synthetic materials because they are stronger, wear less, and last longer.

Belts

Belts can be classified as follows:
- Round belts.
- Flat belts.
- Timing belts.
- V-belts.
- Variable speed belts.
- V-ribbed belts.

Round Belts

A *round belt*, also known as an O-ring belt, has a round cross-sectional area that is used for driving a pulley. Round belts, sometimes called *endless round belts*, are used in serpentine drives and other drives that require the belt to turn a quarter turn or twist. These belts are used in light load applications.

Flat Belts

A *flat belt* has the appearance of a continuous flat strap. These belts are driven with a crowned pulley, which has a slight bevel taper in the center. The crown keeps the belt centered on the pulley, **Figure 6-4**.

Flat belts are rarely found in modern heavy equipment. In the past, they were used on old farm tractors for powering implements, and they continue to be used as conveyors in mining applications. See **Figure 6-5**. Today's agricultural tractors use PTO shafts instead of flat auxiliary drive belts.

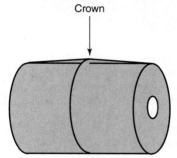

Goodheart-Willcox Publisher

Figure 6-4. Flat belts are driven with a pulley that has a taper, creating a crown. The pulley's crown keeps the belt centered on the pulley.

Round Baler Belts

Round hay balers are one of the few machines that still have flat belts, **Figure 6-6**. Multiple long flat belts are used to form the round bale of hay. The belts have one, two, or three layers (plies). These belts can be made of different materials, including rubber, nylon, and polyester.

Alain Lauga/Shutterstock.com

Figure 6-5. Old tractors used flat belts as an auxiliary power source, similar to today's PTOs.

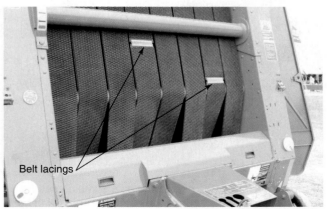

Goodheart-Willcox Publisher

Figure 6-6. A round baler equipped with eight flat belts. Notice the chevron-shaped pattern on the belts.

Equipment manufacturers need a belt that resists puncturing, tearing, or fraying from debris; resists stretching; is gentle on the crop; self-cleans; and lasts a long time. Worn hay baler belts make it difficult to start forming the initial hay bale in challenging crop conditions. Manufacturers engineer belts in a variety of pattern designs to optimize the baler's ability to produce bales in various crop conditions.

Baler belts vary in size. The width can be narrow or wide. The belts are also available in various lengths (circumferences).

Many belts are connected together with a splice, also known as *belt lacings*. See **Figure 6-7A**. The *splice* consists of a group of metal loops that are pressed onto each end of a baler's belt using a belt splice press tool. The belt's two ends are interlaced together and secured with a metal pin to form a hinge. The ends of the pin are bent to prevent the pin from walking out of the splice.

If a belt becomes torn from debris, such as rocks and sticks, technicians cut and remove a section of the belt and splice in the necessary length of new belt. Some belt splices require skiving, which refers to removing an outer layer of the belt in the splice area so the splice plates can fit over the end of the belt. The process requires using a knife to score a line the width of the belt lacing. Then, an electric belt-skiving tool, similar to a handheld electric oscillating multi-function tool, removes a thin layer from the end of the belt. See **Figure 6-7B**.

Goodheart-Willcox Publisher

Figure 6-7. Two ends of a belt can be spliced together. A—Rivets are used to attach the two splices to the belt because they provide strength and durability. The wire pin shown joins this belt's two splices. The uninstalled splice in the middle is an extra service life splice. B—Some belts require the top layer of the belt to be removed from the splice areas prior to splicing. This is known as skiving.

Timing Belts

Timing belts, also known as *synchronous belts*, are flat belts with cogs or teeth. The pulleys used with these belts have cogs or teeth that mesh with those on the belt. These belts are used as a traditional belt drive or for synchronous applications that require two or more pulleys to be synchronized or timed in relationship with each other.

Timing belts are commonly used in low-horsepower engines, such as automotive engines. They drive the engine's camshaft(s) while keeping the camshaft(s) synchronized with the engine's crankshaft. Another example is the timing belt on a Honey Bee combine header. See **Figure 6-8**. The timing belt drives the header's drum and draper belts. Timing belts offer the advantage of no slip, less load placed on the pulley's bearing, less frictional drag, and less bulk compared to traditional V-belts.

V-Belts

A *V-belt* has a V wedge–shaped cross section. **Figure 6-9** shows how to measure the width and depth of a V-belt. The belt's shape is engineered to match the pulley's wedge-shaped pocket. The belt propels the pulley using a wedging action. The pulleys are also called *sheaves*.

Goodheart-Willcox Publisher

Figure 6-8. Honey Bee manufactures a combine header that uses timing belts to drive the header's drum (wide belt) and to drive the small draper belt.

Belt speed is given in feet per minute (fpm) or meters per minute. V-belts normally operate at speeds between 1000 fpm to 4500 fpm, but can be designed to operate up to 10,000 fpm.

V-belts are also known as *friction belts*. The belt is sometimes called a conventional V-belt, classical V-belt, or multiple V-belt, because multiple V-belts can be used to drive a single pulley. V-belt applications include engine cooling fan drives, combines, and lawn mower deck drives.

Compared to chains, V-belts offer the following advantages:

- Do not require lubrication.
- Are quiet and smooth.
- Can be used in a wide range of horsepower applications.
- Can be used in a wide speed range.
- Act like a fuse when overloaded by either slipping or breaking.
- Dampen pulsations.

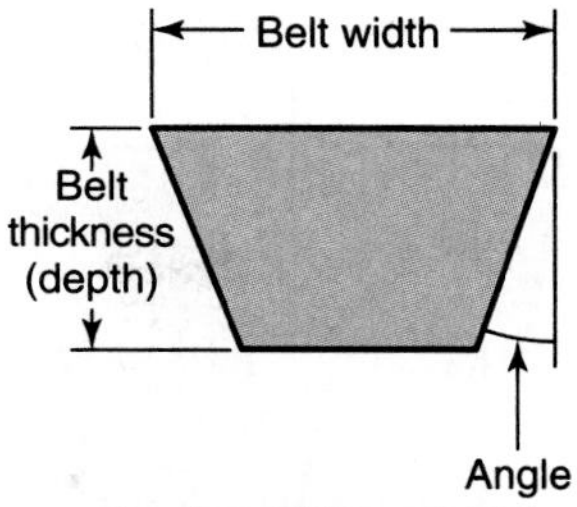

Figure 6-9. V-belts have a V wedge–shaped cross section.

V-Belt Construction

A core of cords made of twisted synthetic fibers in the center of a V-belt gives the belt its strength. Synthetic rubber located below the core cords is called the *compression section*. This section compresses the belt as it is driven by the sheave and supports the polyester core cords. Located directly above the core cords is the tension section, which must be able to stretch as the belt is driven by the sheave. A V-belt cover may be located above the tension section to provide wear resistance for the belt. See **Figure 6-10**. Not all V-belts have a cover.

Some V-belts have Kevlar core cords, which provide higher tensile strength. These belts are sometimes referred to as *aramid* belts. The term *aramid* stands for aromatic polyamide fibers, which are strong, lightweight, man-made materials. These belts are especially helpful if the back of the V-belt has to ride along an idler, which can cause shock loads on the belt.

V-belts are commonly constructed of neoprene, polyurethane, rubber, or urethane. Various designs are available, including conventional V, double V, and cogged V. Conventional V-belts are specified with a prefix letter—A, B, C, D, or E—that signifies the width of the top of the belt. See **Figure 6-11**. *Belt length* is the overall length of the belt if it is cut and measured from one end to the other end.

V-Belt Gauge

Belt manufacturers provide V-shaped plastic gauges called ***V-belt gauges***. The V-belt is placed inside the gauge. The size of the belt is determined by comparing the position of the

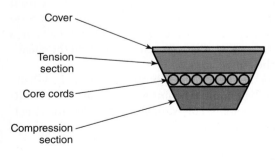

Figure 6-10. V-belts consist of a tension section, core cords, and a compression section. Some V-belts have a cover.

Conventional V-Belt Sizes			
Belt Size	**Belt-Width**	**Thickness (Depth)**	**Examples of Belt Lengths**
A	1/2"	11/32"	13–237"
B	5/8"	7/16"	19–627"
C	7/8"	9/16"	36–551"
D	1 1/4"	3/4"	81–600"
E	1 1/2"	1"	144–660"

Figure 6-11. Conventional V-belts are sized using five letter prefixes. The letter indicates the belt width and thickness. The belt width is the distance across the top of the V-belt. The thickness is the height of the V-belt. The length is the belt's outside circumference.

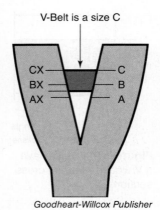

Figure 6-12. A V-belt gauge has a V-shaped slot. A V-belt is placed in the slot. The top of the belt aligns with one of the gauge's indicator bars, which indicates the size of the V-belt. This V-belt is a size C.

Figure 6-13. Two cogged V-belts drive the engine fan in this dozer.

top surface of the belt to the markings on the gauge, **Figure 6-12**. A V-belt gauge helps the technician determine what type of belt to order when the old belt size is unknown.

Heavy-Duty V-Belt

Heavy-duty V-belts are identified by the prefixes HA, HB, HC, HD, and HE. The sizes are the same as the conventional V-belts, except they are heavier-duty belts. Heavy-duty belts are also known as *agricultural V-belts*.

Light-Duty V-Belt

Light-duty V-belts are designed for lighter duty applications. These belts are identified by a two-character prefix, such as 2L, 3L, 4L, or 5L. The number indicates the top width of the belt in units of 1/8″. For example, a 2L is a light-duty belt that has a top width of 2/8″, which is a 1/4″ width belt. These belts are also called *fractional-horsepower (FHP) V-belts*.

Narrow V-Belt

Narrow V-belts have a narrower width than conventional V-belts. The prefix designation is similar to the light-duty V-belt, except the letter *V* designates it as a narrow belt. Narrow V-belt prefixes are 3V, 5V, and 8V, representing 3/8″, 5/8″ and 1″ belts.

Cogged V-Belt

Cogged V-belts, also called *corrugated V-belts*, have corrugations along the bottom of the belt that resemble ribs. The corrugations make the belt more flexible and increase its surface area, which improves its ability to dissipate heat. These factors increase the belt life by 25% to 50%. See **Figure 6-13**. Cogged V-belts can be used with a smaller-diameter sheave, providing more room on the machine. The belt has a slightly higher operating temperature limit than the traditional 140°F (60°C) for a typical V-belt, because the cogged belt runs cooler than the traditional V-belt. Cogged belt prefixes are AX, BX, and CX. Cogged V-belts are also available in narrow sizes. These belts use the narrow V-belt designations followed by an X. For example, a 3VX belt is a 3/8″ narrow cogged V-belt.

 Note

Belts are rated at 1/15th of their ultimate strength. For example, a belt that was tested and failed at 1500 lb is designed to operate at a 100 lb load. Belts also have a maximum speed rating in fpm.

Joined Belts

A drive mechanism can use multiple V-belts that are joined with a top layer of material known as a *tie-band*. The V-belts drive using the same wedging action as individual V-belts. See **Figure 6-14**. This belt is called a ***joined belt*** or ***banded V-belt***. Joined belts are used in applications with long spans or on pulsating drives to reduce slip that would otherwise occur with multiple single drive belts. In these applications, several individual V-belts (in a matched set) are more likely to slip. However, pulley alignment and inspecting the sheave for wear are more important than in other belt applications.

To prevent the tie-band from contacting the sheave edges (bottoming-out), the V-belts are designed to ride higher in the pulley. The belt number begins with a numeral that indicates the number of belts joined into the belt assembly. For example, a 3/BX identifies a belt with three cogged V-belts joined into one belt assembly.

Double V-Belt

A ***double V-belt*** has the appearance of two V-belts placed onto each to form a hexagon. See **Figure 6-15**. Both sides of the belt are used to drive pulleys. The belt can be looped so it drives two pulleys located beside each other in the opposite direction. The belt prefix uses double letters, such as AA, BB, CC, and DD, to indicate the belt's width.

 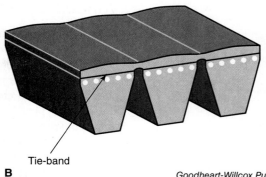

A *Chromatic Studio/Shutterstock.com*   B *Goodheart-Willcox Publisher*

Figure 6-14. Joined belts are also known as banded V-belts. A—This joined belt consists of two V-belts joined together by an outer layer called a tie-band. B—This cross section of a joined belt shows three V-belts joined together by a tie band.

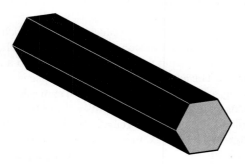

Goodheart-Willcox Publisher

Figure 6-15. A double V-belt is designed to drive on both sides of the belt.

V-Belt Tension

Belts must be properly tensioned. A loose belt causes slip and wear and can overheat or even break as it slips and then grips. A tight belt causes the belt to overheat, places additional load on the sheaves and bearings, and stretches the belt.

Belt tension is measured by applying a specific amount of force to the belt and measuring how much it deflects. See **Figure 6-16**. Many manufacturers provide special tools for measuring belt tension.

A   B *Goodheart-Willcox Publisher*

Figure 6-16. Proper V-belt tension is critical. A—The amount of belt deflection being measured. B—The engine's alternator is mounted on a slotted bracket, which allows the alternator to be moved to add tension to the belt.

Variable Speed Belt

A *variable speed belt* is a wider-width belt with a thin cross section. Variable speed belts are used in applications where a variable-diameter pulley (or sheave) varies the input or output speed. This type of system is known as a *variable-diameter pulley (VDP) system*. A variable-diameter pulley consists of two pulley halves that can be spread apart or moved closer together to change the effective diameter of the pulley and provide a change of speed. In a typical VDP system, one pulley (speed pulley) is responsible for changing input or output speed. The other pulley (tension pulley) maintains the belt's tension.

The speed pulley can be the input (drive) pulley or the output (driven) pulley. There are two speed pulley designs:
- A fixed sheave half and a movable sheave half.
- Two movable halves of a sheave.

If the speed pulley is on the input side, a narrower width provides a taller pulley, resulting in a faster belt speed. A wider pulley results in a shorter pulley diameter, which slows belt speed. If the speed pulley is on the output side, increasing the pulley diameter decreases the rotational speed of the output shaft and decreasing the pulley diameter increases the rotational speed of the output shaft.

The tension pulley must be adjustable because the belt, which has a fixed circumference (overall length), must maintain its tension as the speed pulley changes the drive speed. A combine's rotor belt drive has a tension pulley that is defaulted to a narrow width by means of a torque-sensing assembly. The assembly contains a torque-sensing cam (with two notched ramps), a large coiled spring, and a cam follower (with two cam rollers known as *followers*). The torque-sensing spring holds tension on the tension pulley. When the variable speed belt attempts to widen the tension pulley, the cam twists, causing the tension pulley to provide a stronger grip on the pulley. See **Figure 6-17**. For example, consider that the tension pulley is the output (driven) pulley. If the speed pulley is narrowed to create a taller input pulley, the force of the belt will cause the tension pulley to widen, overcoming the spring tension on the tension pulley.

Variable speed belts have been used in a variety of equipment, including industrial manufacturing machines, lawn mowers, all-terrain vehicles, snowmobiles, farm utility vehicles, and combines. Combines have used variable speed belts to power the cleaning fan, rotor, and separator. In most agricultural equipment with variable-speed belt drives, an electric motor or hydraulic oil controls the width of the variable-diameter pulley. Old combines used mechanical threaded adjustments for varying pulley diameter.

Combine Cleaning Fan VDP

A combine cleaning fan has an electric motor, similar to a windshield wiper motor. The motor varies the width of the output sheave that drives the cleaning fan's shaft. See **Figure 6-18**. As the electric motor spins in one direction, a gear and shaft rotate. This rotation forces a lever to move, causing the pulley to spread apart, decreasing the pulley's effective diameter. When the electric motor is reversed, the pulley closes, increasing the pulley's effective diameter. The cleaning fan's input pulley (drive) uses springs to hold tension on the pulley.

Combine Rotor VDP

Combines can also use a VDP to drive heavier-load applications, such as a rotor or feeder shaft. In these applications, hydraulic fluid pressure adjusts the width of the speed pulley. John Deere STS combines have a hydraulically adjustable VDP that varies the rotor speed. See **Figure 6-19**.

Power is delivered to the VDP through a right-angle gearbox. The speed (input) pulley is closed using a single-acting hydraulic cylinder. The narrowing of the speed pulley causes the belt to ride higher in the pulley and belt speed to increase. When oil is drained from the cylinder, pressure is released from the speed pulley and spring tension on the tension (output) pulley causes the belt to widen the speed pulley.

The rotor drive output pulley is defaulted to a narrow position by means of spring tension. See **Figure 6-20**. As the input speed pulley changes its diameter, the output pulley

Chapter 6 | Belt and Chain Drives

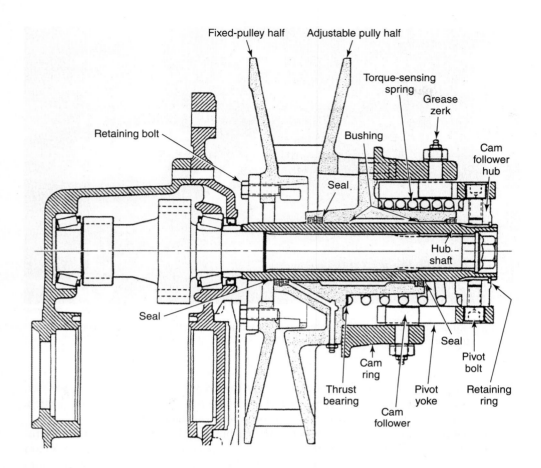

Case New Holland

Figure 6-17. A torque-sensing unit. This unit is used on a Case IH 2188 combine variable rotor belt drive. The top portion of the image illustrates the sheaves spread wide. The bottom portion of the image illustrates the sheaves in a narrow position.

Goodheart-Willcox Publisher

Figure 6-18. On this John Deere STS combine, an electric motor drives the gear shaft. As the gear shaft rotates, it forces the lever to move. The pulley then changes diameters, resulting in a speed change to the cleaning fan.

Goodheart-Willcox Publisher

Figure 6-19. This John Deere STS combine uses a variable-speed belt to drive the rotor. The speed pulley receives power from the right-angle gearbox, and fluid pressure is used to vary the width of the speed pulley.

Figure 6-20. A John Deere STS combine uses a variable speed belt to drive the rotor. The tension pulley drives a two-speed gearbox.

diameter changes based on the belt load. Because the drive pulley has a narrow diameter, the belt forces the driven pulley to widen and overcome the driven pulley spring tension. The driven pulley drives a two-speed gearbox, which drives the rotor.

John Deere also uses a VDP system to vary the header PTO shaft speed so the operator can vary the speed of the header's stalk rolls. See **Figure 6-21**. The output pulley's cam mechanism is a torque-sensing mechanism. As the belt begins to slip, the cam places more tension on the tension pulley. Late-model large-frame John Deere combines use a five-speed gearbox instead of a variable speed belt to drive the header.

Figure 6-21. The shields on this John Deere 9770 STS combine have been removed to expose the drive belt. Notice the large spring on the tension pulley. When the header is installed, the header's PTO shaft will connect to the header PTO stub shaft. The other end of the stub shaft protrudes on the other side of the feeder house to drive the other side of the header. As the speed pulley narrows, the belt speeds up, and the tension pulley widens. This increases the speed at which the header PTO stub shaft rotates.

V-Ribbed Belts

A *V-ribbed belt* is a flat belt formed with multiple V-shaped ribs, tensile cords, and a belt cover. See **Figure 6-22**. It is thinner than a traditional V-belt. Unlike V-belts that transmit power through wedging action, V-ribbed belts drive based on friction, similar to flat belts. V-ribbed belts require 20% more tension than a V-belt. The V-shaped ribs help keep the belt in alignment.

V-ribbed belts are offered in five sizes—PH, PJ, PK, PL, and PM. The PM-sized V-ribbed belts can drive up to 1000 hp. See **Figure 6-23**. One of the most common V-ribbed belt applications is a *serpentine belt*, which is named based on the way the belt snakes around the pulleys in the system. The belt can also be called a *multi-V belt* or *multi-rib belt*. See **Figure 6-24**. The belt is mated with grooved pulleys. Because a V-ribbed belt is much thinner than a V-belt, it is more flexible than a V-shaped belt. For a given amount of horsepower, the belt can be used with a smaller pulley.

Serpentine belts are used as the accessory drive belt. In small horsepower applications, the accessory belt drive system (ABDS) drives the water pump, alternator, and air conditioner's compressor. A disadvantage is that failure of the ABDS belt causes a loss of all the accessories.

For years, an automatic belt tensioner has been used to automatically maintain the belt's tension. Tensioners, like belts, have a service life and should be replaced at a normal service interval. Belts become glazed as they wear, overheating the pulleys and bearings in the accessory drive. Whenever a serpentine belt becomes glazed, replace the belt, idler, and automatic tensioner. Most likely, the belt became glazed due to the faulty tensioner. If the belt only is replaced, the new belt will soon become glazed again.

Although all drive belt systems require the pulleys to be located in the same plane, serpentine belts are more sensitive to this requirement than V-belts. Use a straightedge to determine if the pulleys are located in the same plane. Belt alignment is discussed later in this chapter.

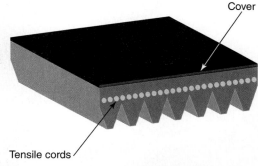

Goodheart-Willcox Publisher

Figure 6-22. V-ribbed belts are thin, flat belts with V-ribs at the bottom of the belt.

Serpentine Belt Gauges

In the late 1990s, belt manufacturers started making serpentine belts with ethylene propylene diene monomer (EPDM) rather than neoprene. Damage to neoprene rubber belts was easier to detect because they would easily crack or lose material from the belt. However, EPDM belts gradually wear over long periods of time and are difficult to visually diagnose, even after high usage. When they wear, the belts bottom-out in the pulleys, resulting in belt slip and heating of the belt.

Belt manufacturers, like Gates and DAYCO, offer different types of tools that measure serpentine belt wear. The Gates serpentine belt gauge tool is placed in the groove of the belt, **Figure 6-25**. The gauge can be used with the belt removed or installed. The height of the gauge should be higher than the tips of the belt ribs, otherwise the belt is worn. If the gauge sits at different heights when different grooves are checked, the pulleys are misaligned.

V-Ribbed Belt Size	Rib Spacing	Belt Height
PH	0.063″ (1.6 mm)	0.118″ (3.0 mm)
PJ	0.092″ (2.3 mm)	0.160″ (4.0 mm)
PK	0.140″ (3.6 mm)	0.200″ (5.0 mm)
PL	0.185″ (4.7 mm)	0.270″ (6.9 mm)
PM	0.370″ (9.4 mm)	0.500″ (12.7 mm)

Goodheart-Willcox Publisher

Figure 6-23. V-ribbed belts have five different sizes defined by the letters PH, PJ, PK, PL, and PM

marekusz/Shutterstock.com

Figure 6-24. Serpentine belts are thin and have multiple V-shaped ribs.

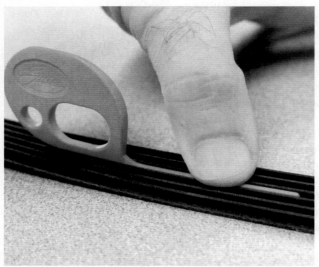

Figure 6-25. A Gates serpentine belt wear gauge is placed inside the belt's grooves. If the gauge sits below the edges of the belt grooves, the belt is worn and needs to be replaced.

Stretch Fit V-Ribbed Belt

In the early 2000s, automotive manufacturers began offering a *stretch fit V-ribbed belt*. This belt does not use a tensioner and must be stretched over the pulleys during installation. Gates calls the belts Stretch Fit belts because the belt is stretched during installation. DAYCO calls the belts *ela belts*, which stands for elastic belts. The drive uses a shorter overall belt length.

Be sure to follow the manufacturer's service literature for installing the belt. Manufacturers have two common procedures for installing a stretch fit V-ribbed belt. One uses a zip-tie, while the other uses a small tool. First, cut the old belt with snips to remove it.

In the zip-tie method, place the belt over the drive pulley, and then place the belt over the driven pulley. The belt will not fully fit over the driven pulley. Place a zip-tie around the belt and through a hole in the driven pulley at the point where the belt is attempting to be guided onto the pulley. Pull the zip-tie tight. Next, rotate the drive pulley until the belt slips onto the driven pulley. Stop rotating the drive pulley after the belt has slipped into the driven pulley. Cut off the zip-tie and run the engine to check the operation of the belt.

The other installation method requires a small tool that resembles a small wheel with a protruding arm. Some special tools have three different threaded holes so they can be used with a wide variety of pulley designs.

Place the tool on top of the belt, **Figure 6-26**. A cap screw is threaded through the tool and extends into a hole in the dished portion of the pulley, acting like a guide bolt. The diameter of the hole in the pulley is larger than the diameter of the cap screw, so the special tool's cap screw does not thread into the hole. Rather, the cap screw acts as an indexing pin and prevents the special tool from rotating as the belt is installed. As the tool is placed on top of the belt and pulley, its only purpose is to guide the belt as the technician manually rotates the pulley's cap screw. As the pulley is rotated, the belt stretches and falls into place on the pulley. Do not use a screwdriver to force the belt onto the pulley. Depending on the pulley design, you may need to place the tool on the front side or the back side (rear face) of the pulley.

Figure 6-26. Stretch fit V-ribbed belt installation method using a special tool.

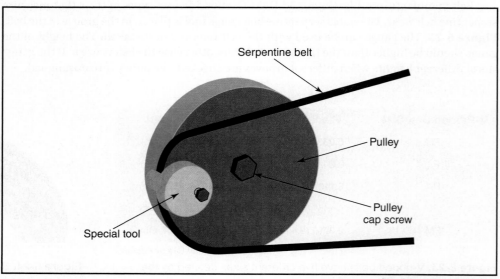

>
> **Note**
> A stretch fit V-ribbed belt is *not* interchangeable with a traditional serpentine belt.

Pulleys

Belt pulleys are manufactured to work in conjunction with the belts. Pulleys are specified in ***pulley pitch diameter***, which is the diameter of the pulley where the V-belt's core cords ride on the pulley. The pitch diameter does not equal the inside diameter or the outside diameter.

The pulleys can be fastened to shafts through splines, key stock, or a Woodruff key. The key stock or key forms the spline that couples the pulley to the shaft. A setscrew, cap screw, jam nut, or snap ring can be used to lock the pulley in place.

Pulley Wear

Pulleys, like belts, wear over a period of time. When replacing a belt, the pulleys should also be checked for wear. V-belts drive by wedging the belt into the side of the pulley. If the belt or pulley is worn, allowing the belt to ride in the bottom of the pulley, the belt will slip.

V-belt sheave gauges are used to measure pulleys for wear. The gauges are specific to the type of belt and the size of the belt and pulley. See **Figure 6-27**. Check for wear between the gauge and the wall of the pulley. No daylight should appear between the gauge and the sheave side wall.

Belt and Pulley Alignment

Worn shafts, pulleys, and bearings can cause belt misalignment, leading to slip, belt wear, and pulley wear. The belt can even fall off the pulley. Check pulley position to see if the pulleys are twisted or out of alignment with the belt. Use a straightedge to see if the pulleys are running in the same plane. See **Figure 6-28**.

Combines use large belts to drive the harvester's rotor. A belt that is out of alignment could cause the rotor to slip, potentially *slugging* the machine (plugging the rotor). A gauge tool is sometimes used to check the rotor belt for proper alignment.

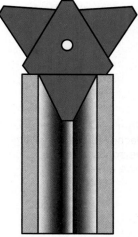

Goodheart-Willcox Publisher

Figure 6-27. A Gates V-belt sheave gauge has three different sizes on one plastic wear gauge. The gauge size is chosen based on the pulley's outside diameter.

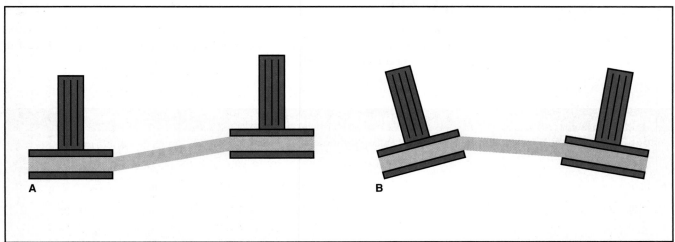

Goodheart-Willcox Publisher

Figure 6-28. Even if a belt's pulleys are running parallel with one another, the two pulleys can be located outside of the same plane. A—The pulleys are parallel, but not in the same plane. B—The pulleys are not in alignment.

Belt and Pulley Ratios

Belt drive ratios are computed using the pulleys' pitch diameters. A drive that uses the same size pulleys for input and output does not modify speed or torque. It simply transmits the power from the input to the output. A drive that has a smaller-diameter input pulley and a larger-diameter output pulley increases torque and decreases speed. A drive with a larger-diameter input pulley and a smaller-diameter output pulley increases speed and decreases torque.

The output ratio is computed by dividing the effective diameter of the output pulley by the effective diameter of the input pulley. The resulting change in output speed is found by dividing the input speed by the output ratio. See **Figure 6-29**. The resulting change in torque is found by multiplying the input torque by the output ratio.

Chains and Sprockets

Chains are fairly economical to build and repair. They do not slip and, therefore, can maintain a set ratio. Compared to belts, chains are stronger and can endure higher temperatures. As a result, the chain drive used for a given amount of horsepower can be smaller than an equivalent belt drive. However, chains stretch over a period of time and require a tensioner to remove the chain's slack. In addition, chains are noisy and require lubrication.

Chains and sprockets are used in heavy equipment to do the following:
- Convey product (move material).
- Time components.
- Lift weight.
- Transfer power.

Conveyor Chain

A *conveyor chain* is used to move material. These chains are common in agricultural machines. The chains used in combine elevators, feeder houses, and corn headers are equipped with rubber flaps (elevators), metal slats (feeder), and flighting (corn header) to move material. Conveyor chain is also called *agricultural chain* or *conveyance chain*. See **Figure 6-30**. This chain type is identified by standardized numbers, such as CA550, CA555, and CA557.

Timing Chains

Timing chains are used to synchronize the rotation of two or more shafts. They serve the same purpose as timing belts. Timing chains are commonly used in engines to keep the camshaft(s) and crankshaft properly timed.

Belt Drive Ratios				
Input Speed	Input Pulley Diameter	Output Pulley Diameter	Ratio	Output Speed
1000 rpm	6"	12"	12 ÷ 6 = 2:1	1000 ÷ 2 = 500 rpm
1000 rpm	9"	12"	12 ÷ 9 = 1.33:1	1000 ÷ 1.33 = 750 rpm
1000 rpm	10"	7"	7 ÷ 10 = 0.7:1	1000 ÷ .7 = 1428 rpm
1000 rpm	8"	8"	8 ÷ 8 = 1:1	1000 ÷ 1 = 1000 rpm

Goodheart-Willcox Publisher

Figure 6-29. The pulleys' effective diameters determine the drive's output ratio. The drive ratio and input speed determine the output speed.

Lifting Chains

Chains and sprockets are sometimes used in heavy equipment to lift heavy objects. Two examples are the chains used on forklift masts and the chain assembly that lifts the counterweight on an excavator. A forklift's mast chain is used every time the forklift lifts a load. In comparison, an excavator counterweight chain is used only when a counterweight is added or removed from the equipment. The chain assembly shown in **Figure 6-31** is used to lift an excavator's counterweight into position so it can be reinstalled on the excavator.

Types of Chain Drives

Non-conveyor type chains are used to transmit power, rather than move material, and are known as *precision chains*. The following six designs are examples of precision chains used in heavy equipment:

- Roller.
- Rollerless.
- Silent.
- Detachable link.
- Pintle.
- Block.

Goodheart-Willcox Publisher

Figure 6-30. This chain is part of the tailings elevator on a John Deere combine. The tailings auger delivers the unthreshed crop to the elevator, and the elevator delivers the crop back to the rotor to be rethreshed.

Roller Chain

Roller chains are formed from a series of connected link assemblies made up from a combination of pins, bushings, rollers, and link plates. The rollers are located around the chain's bushings and ride in the valleys between the teeth of a sprocket. As the chain revolves around the sprocket, the rollers rotate and pull on the bushings, pins, and chain links behind them. In this way, power is transferred from the drive sprocket through the chain to the driven sprocket. Roller chains are the most common chain used in heavy equipment. See **Figure 6-32**. A roller chain consists of the following components:

- Two inner link plates.
- Two outer link plates (side bars).
- Two bushings (sleeves).

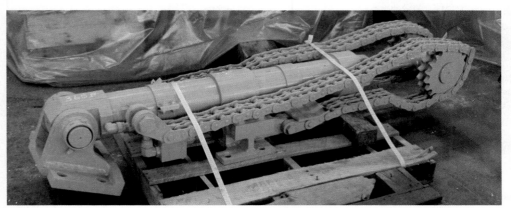

Goodheart-Willcox Publisher

Figure 6-31. This chain assembly has been removed from the rear of an excavator and placed on a pallet.

Figure 6-32. Roller chains have pins, bushings, rollers, link plates, and side bars.

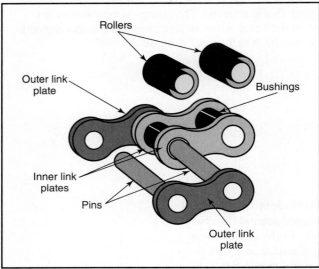

Figure 6-33. Bushings are pressed into the holes located in the two inner links.

Figure 6-34. Sprockets mesh with the roller chain. They are coupled to shafts.

- Two rollers.
- Two pins.

A press is used to insert two bushings into the holes located in the two inner link plates. See **Figure 6-33**. Two rollers are slid over the two bushings and can rotate freely on the bushings. Two pins, also known as *rivets*, are inserted through the center of the bushings. The two pins are pressed into the holes located in the two outer link plates (also known as *side bars*). The outer links connect a pair of adjacent roller links.

Shortening a roller chain requires removing a pair of inner link plates and a pair of outer link plates, which might result in a chain that is too short. Half chain links can be purchased to alleviate this problem. The half chain link resembles a section of an offset chain, which is described later in this chapter.

A *sprocket* is a toothed wheel that meshes with a chain. The sprocket teeth are sized to match the chain's pitch, the roller's diameter, and the width between the links. See **Figure 6-34**.

The American National Standards Institute (ANSI) provides safety and industry standards for US organizations. The chain size indicates the chain's pitch, roller width, and roller diameter. See the *Appendix* for *ANSI Roller Chain Sizes*. A chain's *pitch* is the distance from the center of one pin to the center of the next pin. See **Figure 6-35**.

Sprockets

Chain drives have drive sprockets, driven sprockets, and idler sprockets. The drive sprocket is typically splined to a shaft and is responsible for driving the chain. The driven sprocket is the output member of the chain drive and delivers power to a gear shaft or another sprocket. Idler sprockets can be used for a variety of functions, such as removing slack in long runs of chain, routing the chain around other components, maintaining the chain's tension, or preventing the chain from bouncing when the chain drive is reversed.

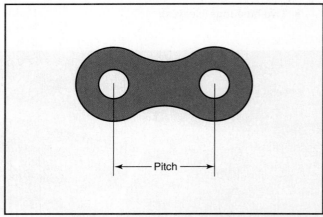

Figure 6-35. The pitch of a chain equals the distance from the center of one pin to the center of the next pin.

Sprockets vary in number of teeth, pitch, thickness, circumference, and style. Sprockets are sized using the ANSI standard chain number. The sprocket's pitch matches the chain's pitch and is specified in 1/8″ units of measurement. See the *Appendix* for *ANSI Roller Chain Sizes*.

A sprocket has a specified ***pitch circumference*** that equals the sprocket's pitch multiplied by the number of teeth on the sprocket. See **Figure 6-36**. If the sprocket is a #80 (with a 1″ pitch) and has 10 teeth, then the sprocket's pitch circumference is 10″.

The ***pitch diameter*** is the diameter across the sprocket's pitch circumference. It can be computed by using the diameter formula (note that C = circumference):

$$\begin{aligned} \text{Diameter} &= C \div \pi \\ &= 10 \div 3.1415 \\ &= 3.18″ \end{aligned}$$

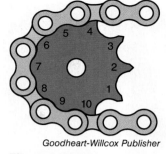

Goodheart-Willcox Publisher

Figure 6-36. This sprocket is a #80, which has a 1″ pitch. Because the sprocket has 10 teeth, the pitch circumference is 10″.

Sprockets are normally designed with one of four hub styles:
- Type A—no hub (plain plate sprocket).
- Type B—hub protruding on one side of the sprocket.
- Type C—hub protruding on both sides of the sprocket.
- Type D—removable hub.

Multiple Strands

Roller chain is offered with single or multiple strands, **Figure 6-37**. In ***multiple-strand chains***, longer pins connect the additional strands. One sprocket assembly consists of several sprockets to mate with the multiple strands.

Agricultural applications for roller chain include combines, balers, older planters, and forage harvesters. The two most common construction equipment applications are driving the wheels on skid steers and motor graders.

Offset Chain

An ***offset chain*** uses only one pair of link plates for each section of chain; there are no separate inner and outer link plates. One end of each link plate has a large hole that receives the bushing. The other end has a small hole that receives the pin. See **Figure 6-38**. The roller is placed around the bushing, and then the pin inserted through the bushing and neighboring links.

Krasi/Shutterstock.com

Figure 6-37. Roller chain is offered in single strands or multiple strands. This image illustrates single-, double-, and triple-strand roller chain.

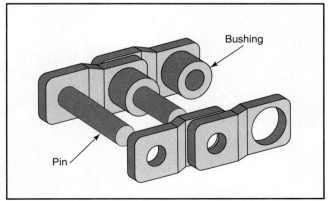

Goodheart-Willcox Publisher

Figure 6-38. Offset links do not have inner and outer links. The narrow end of each link fits inside the wide end of the adjoining link.

Case Study
Worn Offset Chain Drive

Some technicians have experienced a worn offset chain drive that results in a chain that is too long and jumps the sprocket. Usually, this requires replacing both the chain and the sprockets. However, some technicians need to find a temporary solution to keep a machine running for the moment. They remove a single link assembly as a temporary solution to remove the slack. In a traditional roller chain, a longer section of chain would need to be removed due to the need to remove a pair of inner link plates and a pair of outer link plates.

Rollerless Chain

A *rollerless chain* has the same appearance as a traditional roller chain, except it is missing the rollers. Unlike a roller chain, the inside links are a single, solid unit. As in a roller chain, the two neighboring inside links are joined using outside link plates and pins. See **Figure 6-39**. These chains are used for the following applications:

- Slower speed applications.
- Conveyor applications.
- Applications in which chain wear is not a concern.
- Lifting and hoist applications.

Silent Chain

A *silent chain*, as its name implies, is a quieter chain. Each link in this type of chain consists of several link plates assembled with pins. The chain has no rollers. See **Figure 6-40**. It is sometimes referred to as a *multilink* chain. Standard automotive timing chains are typically this type of chain. If additional durability is needed, a roller timing chain is used instead of a silent chain.

Goodheart-Willcox Publisher

Figure 6-39. A rollerless chain resembles a traditional roller chain except it does not have rollers. This chain is propelling a conveyor on an agricultural spreader.

Jeffery B. Banke/Shutterstock.com

Figure 6-40. Silent chains are commonly used as timing chains in automotive applications.

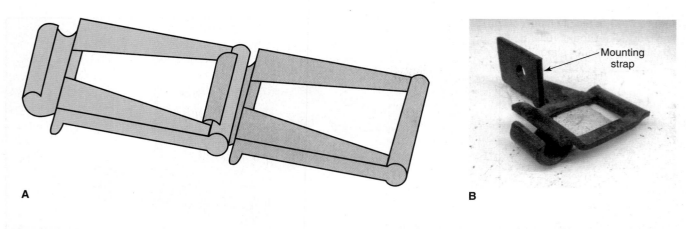

Figure 6-41. A detachable link chain allows the chain to be easily removed. A—One end of the link has a saddle that receives the other side of the next link. B—This detachable link has a mounting strap. C—Detachable link chains inside an agricultural spreader. D—The links of this detachable chain are formed from bent rod.

Detachable Link Chain

A *detachable link chain* is used for low-speed and low-torque applications. It is designed to easily disassemble and assemble. The chain can be driven in only one direction. See **Figure 6-41**.

Pintle Chain

Pintle chains are formed from a series of tapered C-shaped links. The narrow end of each link has a barrel. The opposite end of the link has two small holes that receive pins. The pins pass through the small holes in the wide end of one link and through the barrel on the adjacent link, joining the links. See **Figure 6-42**.

Pintle chain drives are used in low-speed, low-load conveyor applications. They withstand weather and the elements better than traditional roller chain drives. The pitch lengths can be as long as 6″.

Block Chain

In a block chain, blocks replace the inner links, rollers, and bushings. These chains are used in conveyor applications. See **Figure 6-43**.

Figure 6-42. Pintle links are connected together by a pin that passes through the wide end of one link and the narrow end (barrel) of the adjacent link. A conveyor chain on an agricultural spreader is shown here.

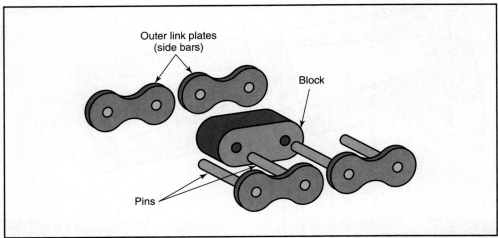

Figure 6-43. A block chain has blocks in place of rollers, bushings, and inner plates.

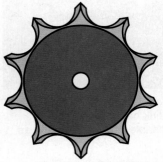

Goodheart-Willcox Publisher

Figure 6-44. Sprockets wear over a period of time. The profile of a new sprocket is shown in gray, and the profile of a worn sprocket is shown in tan. Tooth wear results in less area on the tooth profile. As the teeth wear, they sharpen to a point. Repeated contact with the chain eventually causes wear pockets to form on the sides of the teeth.

Checking Chains and Sprockets for Wear

Both chains and sprockets need to be inspected for wear. If both are worn, but only the chain is replaced, the new chain's life may be reduced by 50%. The sprocket tends to cost less than the chain, so it is good practice to replace the sprocket at the same time as the chain. The chain has already been removed, making it easier and quicker to replace the sprocket.

Worn sprockets cause drive problems, including the possibility of the chain slipping off the sprocket. See **Figure 6-44**. The teeth deform, reducing the area of the tooth profile. The spots where the chain contacts the sprocket wear, forming wear pockets. If the sprockets are driven in both directions, the wear causes the teeth to become pointed. Chains that drive in one direction cause sprocket teeth to wear more on one side, creating a hook-like appearance. The width of the tooth should be measured with a dial caliper and compared to the service literature specification.

 Note

Tsubaki manufactures sprockets with brass wear bars inserted in the face of the sprocket teeth. If the brass wear bar becomes visible, then the sprocket is past its wear limits and needs to be replaced. These sprockets are only available in sizes larger than #100.

Chain Elongation

As the chains pass through the sprockets, the pins and bushings rotate against each other. When the chain drive is loaded, the chain wears, causing the chain to lengthen over a period of time. This is referred to as *chain elongation*.

A worn (lengthened) chain causes chain and sprocket engagement problems. Good preventive maintenance includes scheduling the machine for sprocket and chain replacement in order to prevent unscheduled downtime and ensure that the machine is available for production.

Note

The increased length of chain is sometimes called *chain stretch*, but the increase occurs due to wear and not from stretching links or rollers. Therefore, *chain elongation* is the more common term used by manufacturers and engineers.

Tools used for measuring chain elongation are a calculator, dial caliper, tape measure, and chain wear gauge. Chain measurements must be taken with the chain under load. Some chains, called bi-directional chains, do not make a full 360° rotation. Measure these chains where the sprocket engages the chain and is under load, because that is where the majority of wear occurs.

Measuring Chain Elongation

The following are the steps in measuring chain elongation:

1. Measure with a dial caliper across several rollers (for example, the outside edges of a span of eight rollers, **Figure 6-45**). Record the measurement.
2. Using the inside jaws of the dial caliper, measure the inside distance between the same rollers. Record the measurement.
3. Calculate the average of the two measurements ([outside and inside]/2).
4. Divide the number of pitches (equal to the number of pins measured minus one) into that calculated average. The resulting number is the chain's measured average pitch.
5. The manufacturer's specified pitch is then subtracted from the measured average pitch. That resulting number is the amount of elongation per pitch.
6. To find the percentage of elongation, divide the measured average pitch by the manufacturer's specified pitch, subtract 1, and multiply by 100.
7. To find the elongation per foot, multiply the elongation per pitch by the number of pitches per foot. The result is elongation per foot.

Chain Wear Specifications

Chain wear specifications vary between manufacturers. Some manufacturers recommend replacement of a chain after it has reached 2% elongation, while others recommend replacing the chain after 3% elongation.

For example, a technician is measuring wear across eight rollers of a #80 chain that has a 1.143″ pitch and a maximum allowable elongation of 2%. If the outside dimension measures 8.785″ and the inside dimension measures 7.545″, the average of the two dimensions is 8.165″. To find the percentage of elongation, 8.165″ is divided by 8.00″ (7 pitches × 1.143″) for a result of 1.02. Next, 1 is subtracted from the total and the difference is multiplied by 100, resulting in .02, or 2%. Since the maximum allowable elongation is 2%, this chain would need to be replaced.

A manufacturer may also provide an overall length to measure across a specified number of pins. In this scenario, a tape measure is often used to measure the chain's length across the specified span. The service literature lists a length (distance) for a good chain and the minimum length that would require the chain to be replaced.

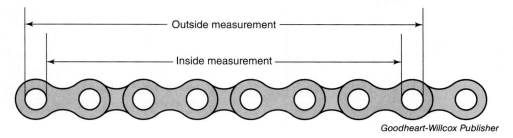

Figure 6-45. Measuring the outside and inside distances across a span of eight rollers.

Goodheart-Willcox Publisher

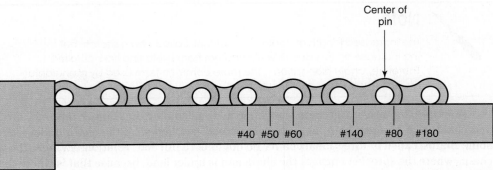

Figure 6-46. A chain wear gauge is placed against the edge of a pin on a #80 chain. The appropriate chain number is referenced on the gauge. If the pin at the end of the specified span has not yet reached the appropriate index mark, the chain elongation is still acceptable. Since the specified pin has not reached the #80 index mark on the wear gauge, the wear on this chain is still within acceptable limits.

Chain Wear Gauge

Chain wear can be measured with a chain wear gauge. There are different types of gauges available, and the methods of using the gauges vary with the type of gauge. One common type of gauge works by measuring the distance between a set number of pins. On this type of wear gauge, each index mark indicates the maximum allowable length for a specific size of chain. See **Figure 6-46**. The edge of the wear gauge is placed against a chain pin. If the center of the pin at the end of measured span aligns with or is located past the index mark for that size chain, the chain is worn and needs to be replaced.

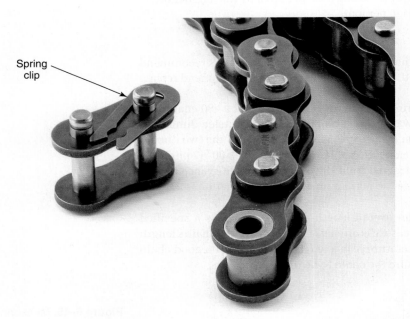

Figure 6-47. A master link is used to join two ends of a chain. A master link can be easily identified by the spring clip that prevents the master link assembly from coming apart in service. The clip should be oriented so the closed end points in the direction of chain travel.

Installing a New Chain

Most chains have a master link for easy chain removal. The *master link* has a spring clip that slides into two recessed slots on the master link pins. See **Figure 6-47**. To split the chain so it can be removed from the drive, the spring clip is removed and then the top link plate is removed. The master link can then be removed, splitting the chain.

A bulk chain purchased for chain replacement may need to be shortened for the specific application. This will require removal of links. A *chain breaker* is a tool used to drive the pins out of the links and split the chain. The master link is then used to connect the two ends of the chain. Chain breakers fasten to the chain, and a mechanical press pushes the pins out of the chain. The breaker can use a long manual lever, a short threaded lever, compressed air, or a hydraulic hand pump.

Lubricating Chains

When driving under load, the chain wears as the pins articulate (rotate) inside the bushings. Proper lubrication helps reduce this wear. Pins and bushings can also corrode, rust, or even seize due to lack of lubrication.

Chain lubrication is critical to maximize the life of a chain drive. Some chains, such as those used in final drives on motor graders and skid steers, ride in a bath lubrication, providing constant lubrication. Some applications have automatic oilers, while other applications require manual lubrication. Manual lubrication requirements vary depending on the machine and application. One common example is to lubricate a roller chain after every 10 hours of operation.

Lubricant

The lubricant must be able to penetrate the clearances between the pins and bushings. Society of Automotive Engineers (SAE) 30-weight oil is often recommended for operating temperatures ranging from 41°F to 104°F (5°C to 40°C). The lubricant should be lighter weight oil if the equipment is operating in colder environments and heavier oil if the equipment is operating in warmer climates.

The lubricant is often applied to the top of the slack side of the chain (usually the lower strand), between the inner and outer plates. The chain should be warm. After lubrication, the chain should be operated to allow the oil to penetrate the clearances between the pins and bushings. See **Figure 6-48**.

Warning

To avoid potential injury, never lubricate a chain while it is operating.

Lubricant Methods

Chains can be lubricated manually or with a bath or spray. Manual chain lubrication can be applied with a brush or dripped from an oil can (known as drip lubrication). Lubricant can be pumped through a zerk (grease fitting) with a hand pump. In bath lubrication, the lower chain strand is directed through an oil bath sump. A disc can also be used with an oil sump. The disc is spun in conjunction with sprocket. The disc, called a *slinger*, is used to sling the oil to lubricate the chain. Spray lubrication involves directly spraying oil onto the chain.

Goodheart-Willcox Publisher

Figure 6-48. Apply lubricating oil to the pins and bushings on the top of the slack side. Direct the oil between the inner and outer plates so it penetrates the clearance between the pins and bushings.

Aftermarket Chain Oilers

Aftermarket chain oilers are sometimes installed on agricultural equipment, such as combines, hay balers, forage harvesters, and mower conditioners. The oiler has an oil reservoir and an oiler pump that delivers oil to the chain. The oiler pump is operated by tapping into a double-acting hydraulic cylinder, such as an unloader swing cylinder on a combine or a tailgate cylinder on a hay baler. The same pressure that actuates the double-acting cylinder is used to operate the oil pump. An adjustment on the pump allows the rate of lubrication to be varied from maximum to no lubrication.

Chain and Sprocket Ratios

The number of teeth on the drive sprocket compared to the number of teeth on the driven sprockets determines the chain drive ratio. A small-diameter input sprocket (fewer teeth) driving a large-diameter output sprocket (more teeth) results in a speed decrease and a torque multiplication. For example, in **Figure 6-49**, if sprocket A is the input sprocket and sprocket B is the output sprocket, the drive ratio would be calculated as follows:

$$\text{Output (driven)} \div \text{Input (drive)} = \text{Drive ratio}$$
$$18 \div 10 = 1.8:1$$

This speed reduction ratio requires input sprocket A to rotate 1.8 revolutions in order to achieve one full output revolution of sprocket B. The input torque would be multiplied by a factor of 1.8 and the output speed would be decreased. For example, if the input rpm is 100, the output rpm would equal 100 ÷ 1.8, or 55 rpm. Notice that the ratio is close to 2, and the output rpm was cut almost in half.

If sprocket B became the input and sprocket A became the output, then the drive ratio would be calculated as follows:

$$\text{Output (driven)} \div \text{Input (drive)} = \text{Drive ratio}$$
$$10 \div 18 = 0.55:1$$

This speed increase ratio causes the output sprocket to rotate one full revolution before the input sprocket completes a full revolution. If the input rpm is 100 rpm, the output rpm would be calculated as follows:

$$100 \div 0.55 = 181 \text{ rpm}$$

Notice that the ratio is close to 0.5 and the output rpm is nearly doubled. The input torque is decreased by a factor of 0.55.

If the two sprockets had the same number of teeth, the output speed and torque would remain the same as the input.

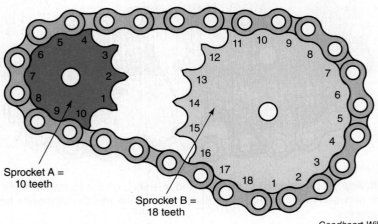

Figure 6-49. Roller chain ratios can be computed using the number of teeth on the sprockets.

Goodheart-Willcox Publisher

Summary

- For a fixed amount of horsepower, an increase in either speed or torque causes a decrease in the other.
- Drive mechanisms that transfer power in heavy equipment can be belts and pulleys or chains and sprockets.
- Belts and pulleys and chains and sprockets can increase speed/decrease torque, decrease speed/increase torque, or transmit power without changing speed and torque.
- Belts are offered in a variety of styles, including round, flat, timing, V-belt, V-ribbed, and variable speed.
- Timing belts or chains and sprockets are used to keep shafts in synchronization.
- Belts must be properly tensioned to prevent slippage, wear, overheating, and stretching of the belt.
- Belt wear can be checked with gauges.
- A variable-diameter pulley (VDP) system is a belt-and-pulley drive system that includes one or more pulleys that can change their effective diameters to change the drive ratio.
- Pulleys should be checked for wear and correct alignment.
- Chains and sprockets are used in heavy equipment to move material, time components, lift weight, and transfer power.
- Chains are offered in a variety of styles—roller, offset roller, rollerless, silent, detachable link, pintle, and block.
- Chains lengthen as they wear, requiring them to be measured for elongation.
- When a worn chain is replaced, it is practical to replace the sprocket at the same time.
- Chains must be lubricated to reduce wear.
- The number of teeth on the input sprocket compared to the output sprocket determines a chain drive's ratio. The number of output sprocket teeth is divided by the number of input sprocket teeth.

Technical Terms

banded V-belt
belt length
chain breaker
chain elongation
cogged V-belt
conveyor chain
detachable link chain
double V-belt
flat belt
friction belt
joined belt
master link
multiple strand chain
offset chain
pintle chain
pitch
pitch circumference
pitch diameter
precision chain
pulley pitch diameter
roller chain
rollerless chain
round belt
serpentine belt
sheave
silent chain
splice
sprocket
stretch fit V-ribbed belt
timing belt
timing chain
torque
variable speed belt
variable diameter pulley (VDP) system
V-belt
V-belt gauge
V-ribbed belt

Review Questions

Answer the following questions using the information provided in this chapter.

Know and Understand

1. The term _____ is defined as force applied through an arcing motion.
   A. horsepower
   B. pressure
   C. speed
   D. torque

2. For a fixed amount of horsepower, if the speed is decreased, torque will _____.
   A. stay the same
   B. decrease in proportion to the decrease in speed
   C. increase
   D. drop to zero

3. Belt speed is specified in _____.
   A. fpm
   B. fps
   C. kph
   D. mph

4. _____ give a V-belt its strength.
   A. Compression sections
   B. Core cords
   C. Covers
   D. Tension sections

5. Which type of V-belt cord does not have as high of a tensile strength as the others?
   A. Aramid.
   B. Aromatic polyamide fibers.
   C. Kevlar.
   D. Polyester.

6. Which of the following is used as a prefix to indicate the width of a conventional V-belt?
   A. Letter.
   B. Decimal number.
   C. Fractional number.
   D. Roman numeral.

7. Which group of V-belts has a narrow belt width?
   A. 2L, 3L, 4L, and 5L.
   B. 3V, 5V, and 8V.
   C. AX, BX, CX.
   D. HA, HB, HC, HD, and HE.

8. A V-belt pulley's pitch diameter equals the _____.
   A. pulley's inside diameter
   B. pulley's outside diameter
   C. diameter where the V-belt's core cords ride on the pulley
   D. average of the pulley's inside and outside diameter

9. A 12″ input pulley is responsible for driving an 8″ output pulley. If the input pulley is spinning at 200 rpm, what is the output pulley's rpm?
   A. 66 rpm.
   B. 133 rpm.
   C. 300 rpm.
   D. 606 rpm.

10. Pulleys can be fastened to shafts through all of the following, EXCEPT:
    A. splines.
    B. key stock.
    C. woodruff key.
    D. rubber belt.

11. All of the following are common functions of a chain drive, EXCEPT:
    A. drive shafts.
    B. lift components.
    C. start engines.
    D. synchronize shafts.

12. All of the following are chain drive types, EXCEPT:
    A. detachable.
    B. noisy.
    C. pintle.
    D. roller.

13. Chain pitch is defined as the _____.
    A. angle of attach that the link approaches the sprocket
    B. distance between the center of two pins
    C. distance between the center of the drive sprocket and driven sprocket
    D. speed of the drive sprocket

14. A roller chain size indicates all of the following, EXCEPT:
    A. diameter of the roller.
    B. overall length of the link plates.
    C. the chain's pitch.
    D. distance between link plates.

15. When is it a good time to replace sprockets?
    A. Every other chain replacement.
    B. When the teeth have become sharply pointed.
    C. When the chain becomes noisy.
    D. When the chain is replaced.

16. Which type of chain does not have outer link plates?
    A. Block.
    B. Offset.
    C. Roller.
    D. Rollerless.

Apply and Analyze

17. A small input pulley driving a large output pulley results in _____ speed.
18. A(n) _____ belt is used in applications that require the belt to twist or turn.
19. A(n) _____ belt is driven with a crowned pulley.
20. A(n) _____ belt is used to keep shafts in synchronization.
21. Late-model serpentine belts are made of _____.
22. A(n) _____ belt has a hexagonal shape.
23. A(n) _____ belt is composed of a group of belts connected by a top layer of material.
24. Pulleys are specified in pulley pitch _____.
25. The _____ part of a roller chain rides in the sprocket's valley.
26. A sprocket's pitch _____ is defined as the pitch times the number of teeth.

Critical Thinking

27. An input sprocket has 12 teeth and is rotating at 50 rpm. The output sprocket has 18 teeth. What is the output speed?
28. An input sprocket has 22 teeth and is rotating at 30 rpm. The output sprocket has 14 teeth. What is the output speed?

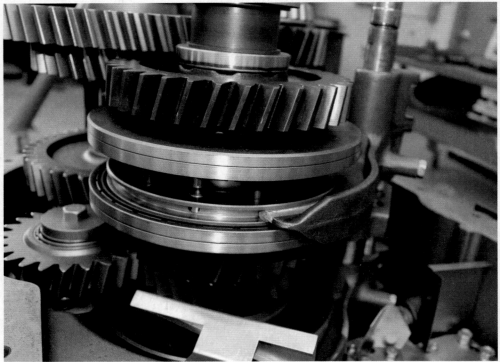

Goodheart-Willcox Publisher

This is a close-up view of a fixed-pin synchronizer used in the 12-speed synchroshift manual transmission of a Case IH 4WD Steiger tractor. The transmission has four synchronizers, two on the speed shaft and two on the range shaft. The synchronizers on the speed shaft provide three forward gears and reverse. The synchronizers on the range shaft provide four speed ranges for a total of twelve forward speeds and three reverse speeds.

Chapter 7
Manual Transmissions

Objectives

After studying this chapter, you will be able to:
- ✓ List the types of gearing used in mobile equipment.
- ✓ Compute gear ratios.
- ✓ Explain gear advantages and disadvantages.
- ✓ Describe the three types of manual transmissions.
- ✓ List the components used for shifting manual transmissions.
- ✓ List the types of bearings used in mobile equipment.

Heavy equipment uses numerous types of gear systems for the following purposes:

- Increasing torque/decreasing speed.
- Increasing speed/decreasing torque.
- Providing direct drive.
- Reversing power flow.
- Changing direction of power flow by 90°.

Internal combustion engines require a transmission that serves four of these purposes: changing speed, changing torque, changing direction of power flow, and providing direct drive. The engine alone does not have the ability to propel a machine unaided by a transmission. Nor can engine rotation be reversed to allow the machine to reverse its direction of travel.

In addition to propulsion, heavy equipment uses gearing for other power transmission purposes, such as driving drums, rotors, belts, augers, rollers, conveyors, reels, and shafts.

Types of Gearing

Many types of gears are used to transmit power in off-highway machinery, including the following:
- Spur.
- Helical.
- Herringbone.
- Ring and pinion.
- Planetary.
- Bevel.
- Rack and pinion.
- Worm gear.

Copyright Goodheart-Willcox Co., Inc.

Spur Gear

A *spur gear* is an external-toothed gear that has straight cut teeth. When two spur gears are meshed with each other, the external-toothed gears will rotate in opposite directions. See **Figure 7-1**.

Straight-cut toothed gears are simple and easy to manufacture. However, compared to helical gears, they are noisy and distribute power across only one tooth at a time, limiting the amount of torque they can transfer.

An imaginary circle that passes through the center of each tooth on a gear is called the *pitch circumference*, or pitch circle. The bottom cavity between the teeth is called the *root*. The space between the top of one gear tooth and the bottom of the mating gear tooth is called the clearance. The sides of spur gear teeth have an involute curvature. This design is used so that the meshed gear teeth roll against each other rather than slide. The rolling action minimizes wear. See **Figure 7-2**.

Two meshing externally toothed gears always rotate in opposite directions from one another. In a transmission, power is typically transmitted from a gear on the input shaft to a gear on the countershaft. This reverses the direction of rotation so that the countershaft rotates in the opposite direction of the input shaft. Another gear on the countershaft is meshed with a gear on the transmission's output shaft. As a result, the output shaft rotates in the same direction as the input shaft, and in the opposite direction as the countershaft.

Transmission designers add an additional gear to transmissions, known as a *reverse idler*, that can be positioned between a gear on the countershaft and a gear on the output shaft. When this gear is in position, power is transferred from the countershaft to the output shaft using three gears instead of two. As a result, the output shaft rotates in the same direction as the countershaft and in the opposite direction as the transmission's input shaft. In **Figure 7-3**, Gear A is the countershaft gear, Gear B is the reverse idler gear, and Gear C is the output shaft gear. A reverse idler or a reverse idler shaft is often used in manual transmissions for reversing the machine's propulsion.

Tewlyx/Shutterstock.com

Figure 7-1. Spur gears have external teeth around the gear's perimeter. Two meshing spur gears will rotate in opposite directions.

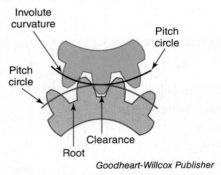

Goodheart-Willcox Publisher

Figure 7-2. An imaginary circle that runs through the center of the gear's teeth is known as the gear's pitch circle.

Helical Gears

Helical gears, like spur gears, have external teeth around the gear's perimeter. However, the gear's teeth are cut at an angle to the gear's centerline. They offer the advantage of quieter operation and have more tooth contact area due to the angled teeth, resulting in a stronger gear. However, helical gears have the disadvantage of creating a side thrust load, which attempts to force the gear and shaft forward or rearward. See **Figure 7-4**. Helical gears are commonly used in numerous mobile equipment applications.

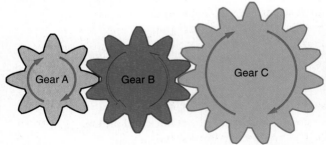

Goodheart-Willcox Publisher

Figure 7-3. If Gear A and Gear C were meshing directly with each other, they would rotate in opposite directions. A reverse idler gear (Gear B) causes Gear C to rotate in the same direction as Gear A.

Tewlyx/Shutterstock.com

Figure 7-4. Helical gear teeth are at an angle to the gear's centerline, which creates a thrust load.

Herringbone Gears

A ***herringbone gear*** has a chevron-shaped tooth pattern, resembling two helical tooth patterns (with opposite angles) that are joined. See **Figure 7-5**. Herringbone gears are quiet and do not produce a side thrust load because the load created by the angle at the front of the tooth is offset by the load on the back of the tooth, which has the opposite angle. These gears are more expensive to manufacture than spur and helical gears. Although a herringbone gear is sometimes mistakenly called a double-helix gear, a double-helix gear has a space between the two rows of helical teeth. The two rows of teeth on a herringbone gear are joined. A double helical gear is cheaper to manufacture than a herringbone gear. Herringbone gears are used on draglines for the hoist and drag, and are found in the turbines of marine ships. They can also be used inside a planetary gear set.

Tewlyx/Shutterstock.com

Figure 7-5. Herringbone gears use chevron-shaped teeth.

External- and Internal-Toothed Gears

A small external-toothed gear can be meshed with a large internal-toothed gear to form a ***ring-and-pinion gear set***. See **Figure 7-6**. The small external-toothed gear is the pinion gear. The large internal-toothed gear is called a ***ring gear*** or ***annulus gear***. Meshing internal- and external-toothed gears rotate in the same direction. These gears are commonly used in planetary gear sets.

Planetary Gear Sets

A ***planetary gear set***, also known as an ***epicyclic gear set***, consists of a sun gear that is surrounded by and meshes with a set of planetary gears that are attached to a planetary carrier. A ring gear surrounds and meshes with the planetary pinions. See **Figure 7-7**. Chapter 9, *Planetary Gear Set Theory*, explains the fundamentals of planetary gear sets.

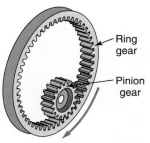

Tewlyx/Shutterstock.com

Figure 7-6. An internal-toothed ring gear and an external-toothed pinion gear will rotate in the same direction. Cranes and excavators use this gearing to swing the upper house of the machine.

Bull-and-Pinion Gear Sets

If a small external-toothed gear is driving a larger external-toothed gear, the gear set is sometimes called a ***bull-and-pinion***. The larger gear is the ***bull gear*** and the smaller gear is the ***pinion gear***. Low track dozers and track loaders sometimes use bull-and-pinion gear sets for the final drive gearing.

Rack-and-Pinion Gear Sets

A ***rack-and-pinion gear set*** uses an external-toothed spur or helical gear to drive an external-toothed rack. The gear set changes rotary motion into linear motion. As the pinion gear is rotated, the rack moves linearly. Rack-and-pinion gears are used in automotive steering systems, the trap door openers on the bottom hoppers on grain semitrailers, to move the drill-head carriage assembly on a horizontal directional drill, and to adjust the wheel spacing on John Deere agricultural tractor bar axles. See **Figure 7-8**.

Bevel Gears

Bevel gears are gears that have straight cut teeth or helical cut teeth on the angled face of the gear. They are used to change the direction of power flow. See **Figure 7-9**. The two gear shafts are positioned at right angles (perpendicular to each other). Bevel gears are used in multiple heavy equipment applications, such as a combine unloading auger drives, combine conveyance augers, combine corn head stalk rolls, differentials, and dozer transmissions, **Figure 7-10**.

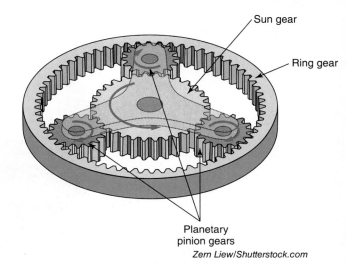

Zern Liew/Shutterstock.com

Figure 7-7. A planetary gear set has a centrally located sun gear that meshes with a set of planetary pinion gears that are pinned to a carrier and surrounded by a ring gear.

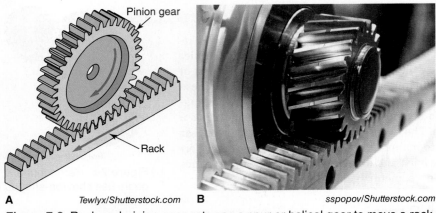

Figure 7-8. Rack-and-pinion gear sets use a spur or helical gear to move a rack linearly. A—This rack-and-pinion gear set has a spur pinion gear and straight cut rack. B—This gear set's pinion gear and rack have helical cut teeth.

Figure 7-9. Straight cut bevel gears redirect power flow by 90°.

Figure 7-10. These straight cut bevel gears are used to change direction of a machine's shaft.

A *spiral bevel gear set* has a small bevel gear with helical teeth that mesh with the helical teeth on a large beveled ring gear. See **Figure 7-11**. The gear set distributes the load across a larger tooth area, making it stronger than a straight-toothed bevel gear set. It is also quieter.

The small pinion gear is often driven by a driveshaft that is powered by the transmission output shaft. The ring gear is attached to the differential carrier. Differentials are explained in Chapter 16, *Differentials*. This gear set is also known as a ring-and-pinion gear set. It provides torque multiplication (speed reduction) and redirects power flow by 90°. If the pinion gear is positioned below the center axis of the ring gear, it is known as a *hypoid gear set*. If the pinion gear is placed above the center axis of the ring gear, it is called an *amboid gear set*. See **Figure 7-12**.

Worm Gears

Worm drive gear sets use gears with helical-cut teeth resembling the threads on a large screw to drive another gear. The driven gear can be a spur gear, rack gear, or sector gear (a gear that resembles a portion of a spur gear) with straight- or helical-cut teeth. This drive changes the direction of power flow. As with bevel gear sets, the two gear shafts are located perpendicular to each other. See **Figure 7-13**.

Figure 7-11. Spiral bevel gear sets are commonly used in rear axles in conjunction with the differential, but can also be used in other gear box applications. A—Spiral bevel gear set. B—Example of a spiral bevel gear set in a gear box mining application.

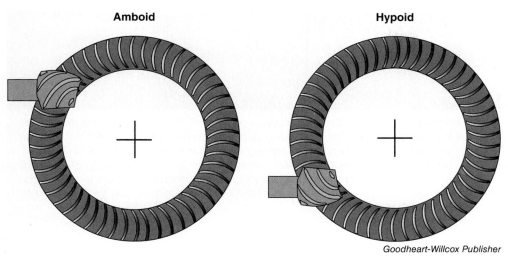

Figure 7-12. Spiral bevel gear sets are commonly used in conjunction with differentials. If the pinion is located above the center axis, it is known as an amboid gear set. If the pinion is located below the center axis of the ring gear, it is known as a hypoid gear set.

A motor grader's circle gearbox uses a worm gear drive to rotate the blade. Worm gears are combined with a rack or sector gear for limited-rotation applications, such as actuating the concaves on a combine. See **Figure 7-14**. For continuous drive applications, such as driving the gathering chains on a corn header, worm gears are paired with spur gears.

Gear Ratios

Gears are one type of simple machine that can change speed, torque, and direction of power flow. A small input gear driving a larger output gear will reduce speed and increase torque. The larger gear has more teeth than the smaller gear. The gear ratio of the gear set can be computed by dividing the number of teeth on the driven gear by the number of teeth on the driving gear.

Figure 7-13. Worm gears change the direction of rotation by 90°.

Torque Multiplication/Speed Reduction

In **Figure 7-15**, Gear A has eight teeth and Gear B has 10 teeth. If Gear A is driving Gear B, the gear ratio results in torque multiplication and speed reduction.

$$10 \div 8 = 1.25:1$$

Gear A will need to rotate 1.25 revolutions in order for Gear B to make one complete revolution. The gear ratio also equals the torque multiplication factor. For example, if Gear A was delivering 200 lb-ft of torque, the torque for Gear B can be computed by multiplying the input torque by the gear ratio.

$$200 \text{ lb-ft} \times 1.25 = 250 \text{ lb-ft of output torque}$$

The output speed can be obtained by multiplying the input speed by the number of teeth on the input gear and dividing that product by the number of teeth on the output gear. For example, if Gear A was rotating 100 rpm, the output of Gear B would equal:

$$(100 \text{ rpm} \times 8) \div 10 = 80 \text{ rpm}$$

Figure 7-14. This combine concave drive uses a worm gear to rotate a shaft. The shaft has threaded rods that attach to the rotor's concaves. As the shaft is rotated, the threaded rods lift and lower the concaves.

Case Study: Gear Ratios

Some heavy equipment operates with deep speed reductions and large torque multiplication due to large gear ratios. For example, a dozer's final drives can have 20:1 gear ratios. The final drive output torque would be 20 times higher than the input torque. A child could spin the input of a 20:1 final drive with 40 pound-feet of torque, and a strong adult would not be able to prevent the final drive output from spinning because the input torque would be multiplied 20 times (40 × 20 = 800 lb-ft) at the output. Final drives will be explained in Chapter 17, *Final Drives*.

Notice that the input gear has 20% fewer teeth than the output gear, resulting in a 20% speed reduction.

Two methods can be used to compute the gear ratio of three gears meshed in series. In **Figure 7-15**, Gear A is driving Gear B, which drives Gear C. The simplest method is to simply use the number of teeth on Gear A and Gear C.

$$15 \div 8 = 1.875:1$$

The more complex method would be to compute ratio of Gear A and B, as well as the ratio of Gear B and C, and then multiplying those two ratios.

$$10 \div 8 = 1.25$$
$$15 \div 10 = 1.5$$
$$1.25 \times 1.5 = 1.875:1$$

From this example you can see that Gear B has no effect on the gear ratio between Gear A and Gear C. A larger or a smaller gear could be used in place of Gear B, and the gear ratio between Gear A and Gear C would remain the same.

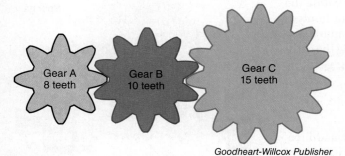

Goodheart-Willcox Publisher

Figure 7-15. Gear ratios are computed using the number of teeth on each gear.

Overdrive

The term overdrive refers to gear ratios that are smaller than 1. Overdrive gear ratios increase speed and decrease torque. If the larger Gear C was the input gear and if the smaller Gear A was the output gear, the gear ratio would be an overdrive ratio:

$$8 \div 15 = 0.533:1 \text{ ratio}$$

If input Gear C was driving at 100 rpm, then output speed of Gear A would equal:

$$(100 \times 15) \div 8 = 187.5 \text{ rpm}$$

Notice that the input gear has nearly twice the teeth (15 teeth) as the output gear (8 teeth), resulting in a speed that is almost twice as fast.

However, the torque could be cut nearly in half. If the input torque equaled 200 lb-ft, it would result in:

$$200 \times 0.533 = 106 \text{ lb-ft}$$

Direct Drive

If the input gear has the same number of teeth as the output gear, the gear ratio is 1:1. This is referred to as a direct drive gear ratio. With a direct drive gear ratio, the output gear rotates with the same speed and torque as the input gear.

Advantages and Disadvantages of Gearing

One advantage of gear drives over belt drives is that gear drives will not slip. Also, they will not stretch like chain drives do, and therefore do not require a tensioner. They can also handle heavier loads than belts and chains. However, they are noisy and more expensive to manufacture.

Types of Manual Transmissions

Heavy equipment can be equipped with a *manual transmission*, which requires the operator to press a clutch pedal and move a gearshift lever to change gears. The three types of manual transmissions are sliding gear, collar shift, and synchronizer shift.

Sliding Gear Transmissions

A *sliding gear transmission* uses straight cut gears placed on parallel shafts. The inside diameter of one or more gears are splined to a transmission shaft, allowing them to slide to different positions along that shaft. When the operator moves the gearshift lever, the splined gears slide into and out of mesh with mating gears on the parallel shaft. Those mating gears are permanently affixed on a common shaft, and the entire unit is referred to as a spool gear. These transmissions are not designed to be shifted while the equipment is moving. The transmission speed is chosen while the machine is stationary, otherwise the operator will grind the gears, causing noise, wear, chipping, or potentially fracturing gear teeth.

Years ago, the sliding gear transmission was used as a standalone means of delivering engine power to the final drive in agricultural tractors, backhoes, and dozers. Today, sliding gear transmissions are commonly used in conjunction with hydrostatic transmissions. The hydrostatic transmission provides the change of direction, either forward or reverse, as well as a limited range of speed change. Sliding gear transmissions that are used in conjunction with hydrostatic drives normally have two, three, or four speeds. A three-speed combine sliding gear transmission provides the operator a low field speed (1st gear), mid-range field speed (2nd gear), and a road speed (3rd gear).

While the equipment is sitting stationary, the operator first selects the sliding gear transmission gear range (1st, 2nd, or 3rd). The operator next actuates the hydrostatic transmission, to control the machine's direction (forward or reverse) and travel speed within the range provided by the selected transmission gear. Hydrostatic transmissions are explained in Chapter 13, *Hydrostatic Drives*, and Chapter 14, *Hydrostatic Drive Service and Diagnostics*.

Case IH Three-Speed Sliding-Gear Combine Transmission

The Case IH three-speed combine transmission in **Figure 7-16** uses a sliding gear design to provide three speeds and neutral. This transmission is used in 2100-, 2300-, 2500-, 88-, and 130-series combines. The operator controls a hydrostatic pump that drives a hydrostatic single-speed or two-speed motor. The motor provides the input into the three-speed transmission.

Notice the sliding gears appear to be part of a parallel twin countershaft. However, each countershaft is actually comprised of two separate shafts that spin independently of each other. The input shaft spins freely inside the change drive gear's shaft. The change driven gear's shaft spins freely inside the spool gear assembly.

The motor provides input into the input shaft that contains a 26-tooth input gear. In 1st and 2nd gear, the input gear drives the spool gear assembly, which contains three gears: a 59-tooth driven gear, a 39-tooth 2nd gear drive gear and a 27-tooth 1st gear drive gear.

In 3rd gear, the 2nd & 3rd gear sliding gear (located on the shaft with the change drive gear) slides to the left, causing the input gear to couple with the change drive gear shaft. This causes power flow to transfer directly from the hydrostatic motor to the change drive gear's shaft, bypassing the spool gear assembly.

Based on the operator's input, the hydrostatic motor's output shaft turns the three-speed transmission's 26-tooth input gear forward or backward at the commanded speed.

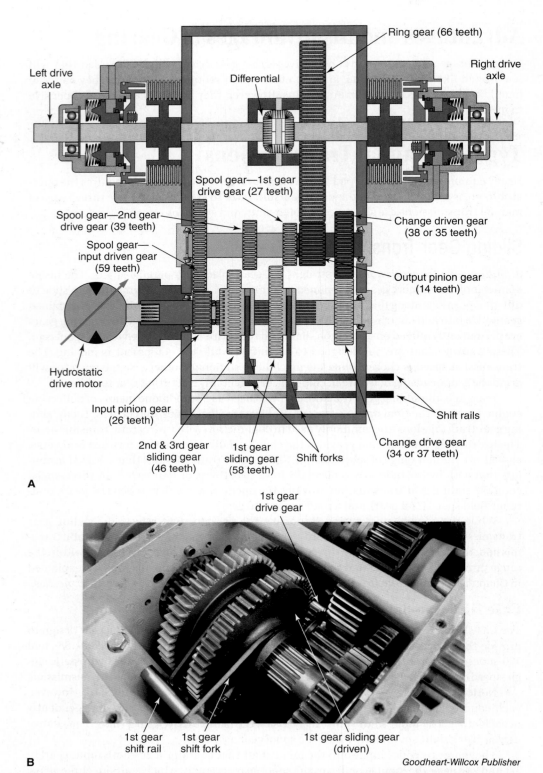

Figure 7-16. This Case IH combine transmission is a three-speed sliding gear transmission that uses two shift rails to slide two gears. Depending on the positions of the sliding gears, the transmission provides 1st gear, 2nd gear, 3rd gear, or neutral. A—The transmission has an input gear that drives a spool gear assembly. The spool gear includes the 1st gear and 2nd gear drive gears. The 2nd & 3rd gear sliding gear and 1st gear sliding gear deliver power to the change drive gear. The change driven gear drives the 14-tooth-output pinion gear, which drives the 66 tooth-ring gear. B—Notice a fork is mounted on a shift rail and is used to slide the gear in and out of mesh.

If 1st or 2nd gear is selected, the 26-tooth input gear drives the 59-tooth gear on the spool gear at a ratio of 2.27:1:

$$59 \div 26 = 2.27:1 \text{ ratio.}$$

In addition to the 59-tooth gear, the spool gear also includes the drive gears for 1st gear (27 teeth) and 2nd gear (39 teeth). If the operator chooses 1st or 2nd gear, the 1st gear sliding gear (58 teeth) or 2nd & 3rd gear sliding gear (46 teeth) slides into mesh with its corresponding drive gear on the spool gear, resulting in one of the following ratios:

$$\text{1st gear: } 58 \div 27 = 2.15:1 \text{ ratio}$$
$$\text{2nd gear: } 46 \div 39 = 1.18 \text{ ratio}$$

3rd gear is achieved by sliding the 2nd & 3rd gear sliding gear over to the transmission input gear. Note that there are short teeth, referred to as dog teeth, on the outer edge of the input gear that match a splined recess in the side of the 2nd & 3rd gear sliding gear. When the 2nd & 3rd gear sliding gear moves to the left, the dog teeth on the input gear seat in the recess in the 2nd & 3rd gear sliding gear, locking the two gears together. Since the 2nd & 3rd gear sliding gear is splined to the change gear shaft, the change gear shaft becomes connected directly to the input gear, resulting in a 1:1 direct drive ratio.

Depending on the gear selection, the 2nd & 3rd gear sliding gear or the 1st gear sliding gear delivers power to the change drive gear. The change drive gear is in constant mesh with the change driven gear. This transmission can have one of two different sets of change gears. The set of change gears used in a given transmission is based on the intended use of the combine and the gear ratio desired:

- Corn grain combine—a 37-tooth change drive gear and 35-tooth change driven gear are used, providing a 0.95:1 ratio.
- Rice combine—a 34-tooth change drive gear and 38-toothed change driven gear are used, providing a 1.12:1 ratio.

The 37/35 change gear transmission is designed for flat land harvesting and provides higher speeds and less torque multiplication. The 34/38 change gear transmission is designed for hilly conditions and provides lower travel speed with higher torque multiplication.

The driven change gear delivers power to the 14-tooth output pinion gear. The pinion gear drives the 66-tooth ring gear attached to the differential carrier, providing a 4.71:1 gear ratio. The matrix in **Figure 7-17** lists the transmission's overall gear ratios based on the type of change gears and the speed selection (1st, 2nd, or 3rd).

Power Flow through a Sliding Gear Transmission

Before the transmission is shifted into any gear, the hydrostatic drive should be commanded to the neutral position and the equipment should be stationary. This will stop rotation of all transmission gears, allowing the sliding gears to be repositioned without grinding.

Transmission Gear Selection	26-Tooth Input Gear to 59-Tooth Spool Gear	Trans Drive Gear to Driven Gear	37/35 Change Gears	34/38 Change Gears	14-Tooth Pinion Gear to 66-Tooth Ring	Total Ratio for 37/35 Transmission	Total Ratio for 34/38 Transmission
1st Gear	2.27:1	2.15:1	0.95:1	1.12:1	4.71:1	21.84:1	25.75:1
2nd Gear	2.27:1	1.18:1	0.95:1	1.12:1	4.71:1	11.99:1	14.13:1
3rd Gear	1:1	1:1	0.95:1	1.12:1	4.71:1	4.47:1	5.7:1

Goodheart-Willcox Publisher

Figure 7-17. Combine three-speed sliding gear transmissions provide a range of ratios from 4.47:1 to 25.75:1, depending on the gear selected and the type of change gears in the transmission.

1st Gear

Figure 7-18. Power flow through 1st gear.

2nd Gear

Figure 7-19. Power flow through 2nd gear.

3rd Gear

Figure 7-20. Power flow through 3rd gear.

1st Gear

If the transmission is shifted into 1st gear, the 1st gear sliding gear is moved into mesh with the spool gear's 1st gear drive gear. Shift linkages simultaneously move the 2nd & 3rd gear sliding gear out of mesh. When the hydraulic drive is moved out of the neutral, the hydraulic motor rotates the transmission input gear, which drives the spool gear's 59-tooth input-driven gear. The spool gear's 1st gear drive gear drives the 1st gear sliding gear. The 1st gear sliding gear is splined to the change gear shaft, which causes the change drive gear to deliver power to the change driven gear. Since the change driven gear and the output pinion gear share a shaft, power is transmitted through the output pinion gear to the ring gear. See **Figure 7-18**. This causes the differential carrier to rotate. The differential delivers power to the right and left output shafts. The output shafts can be allowed to rotate with the differential carrier or they can be held by the right and left parking or service brakes.

2nd Gear

If the transmission is shifted into 2nd gear, the 2nd & 3rd gear sliding gear is moved into mesh with the spool gear's 2nd gear drive gear as the 1st gear sliding gear is moved out of mesh. The hydrostatic motor turns the transmission input gear, which drives the spool gear's input driven gear. The spool gear's 2nd gear drive gear drives the 2nd & 3rd gear sliding gear. The 2nd & 3rd gear sliding gear is splined to the change gear shaft, which causes the change drive gear to deliver power to the change driven gear. The change driven gear delivers power to the output pinion gear, which drives the ring gear. See **Figure 7-19**. The ring gear is directly connected to the differential carrier, which delivers power to the right and left output shafts.

3rd Gear

If the transmission is shifted into 3rd gear, the 2nd & 3rd gear sliding gear is moved into mesh with the transmission input gear. At the same time, the 1st gear sliding gear is moved out of mesh. The hydraulic motor turns the transmission input gear. The 2nd & 3rd gear sliding gear is directly coupled to the input drive gear and splined to the change gear shaft. As a result, power flow is transmitted through the input gear directly to the change drive gear. The change drive gear delivers power to the change driven gear, which delivers power to the output pinion gear. The output pinion gear drives the ring gear, which causes the differential carrier to rotate. See **Figure 7-20**. The differential carrier delivers power to the right and left output shafts.

Sliding gear transmissions are not designed to shift gears while the machine is moving. If the operator attempts to shift gears while the machine is moving, the sliding gear will clash and grind with the spool gear. Even if the machine is at rest, gear clash will occur if the input gear is turning. The sliding gear transmissions in old backhoes and dozers used a flywheel clutch to disengage the engine from the transmission and a brake to stop the input shaft so the sliding gear could mesh with its mating gear.

Collar Shift Transmission

Like sliding gear transmissions, *collar shift transmissions* also use external-toothed gears (spur or helical) and parallel shafts. However, the drive and driven gears in a collar shift transmission are always in mesh.

One transmission shaft, for example either the main shaft or countershaft, has gears directly splined to the shaft. The other transmission shaft uses one of two types of methods for coupling the gear to the shaft.

One method uses a *shift collar* that is splined to the transmission shaft. When the collar is moved into contact with the gear, external splines on the collar slide into internal splines in the gear, locking the gear to the shaft. See **Figure 7-21**. In semitruck applications, the shift collar is known as a *sliding clutch*.

The other style of collar shift transmission has range drive gears that are permanently coupled to the countershaft and in constant mesh with driven range gears. These driven range gears are not splined to the pinion shaft. Each of the driven range gears has a hub with dog teeth. Next to each output range gear is a *shift gear* that is splined to the transmission's shaft. The shift gears have exterior splines that match the dog teeth on the range gear hubs. Each shift gear also has a moveable shift collar with internal splines that match the exterior splines on the shift gear and dog teeth on the range gear. The shift collar can be moved so it covers the dog teeth on the driven range gear hub and half the splines on the shift gear. This couples the shift gear and pinion shaft to the driven range gear

John Deere 4200 Compact Utility Tractor

The example in **Figure 7-22** is the range transmission in a John Deere 4200 compact utility tractor. A *range transmission* consists of a gearbox that provides multiple speed ranges, in this example A range (low), B range (medium), and C range (high). The shift gears are splined to the pinion shaft. When the shift collars are centered over the shift gears,

Figure 7-21. A fork is used to slide the shift collar to engage the gear. This shift collar, also known as a sliding clutch, is used in on-highway semitruck applications.

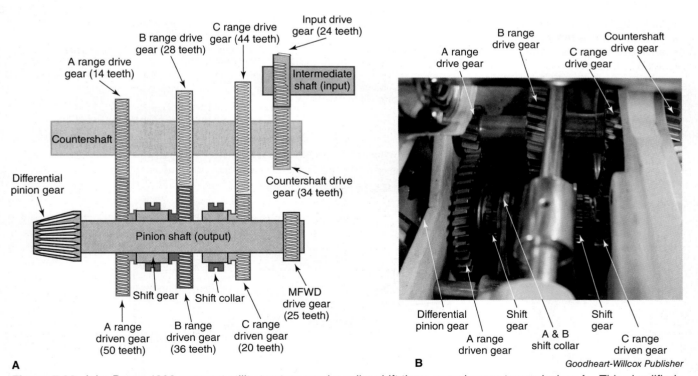

Figure 7-22. John Deere 4200 compact utility tractors used a collar shift three-speed range transmission. A—This simplified drawing depicts a collar shift range transmission in a John Deere 4200 compact utility tractor. When the operator selects A, B, or C, a shift collar couples the shift gear to a range driven gear. B—The range transmission receives power from the speed transmission via the intermediate shaft. The shift gears in this image are in the neutral position.

as they are in **Figure 7-22**, the shift gears and pinion shaft do not rotate with the output gears. If a shift collar is moved over so it covers the dog teeth on the hub of an output range gear, the shift gear and output pinion shaft become coupled to that output range gear and rotate at the same speed as the selected output gear. The remaining two output range gears rotate at different speeds without affecting the pinion shaft. Power flow through the various gear ranges is shown in **Figure 7-23** through **Figure 7-25**.

Both the sliding gear transmission and the collar shift transmission are unsynchronized manual transmissions. On-highway trucks that use collar shift transmissions require the driver to match the engine speed to the transmission speed in order to shift gears while moving. Collar shift transmissions in agricultural tractors are shifted the same way as sliding gear transmissions, requiring the tractor to be stationary. If the machine is moving or the input gear is still turning, clashing and grinding will occur between the shift collar's internal teeth and the external splines on the range gear.

The John Deere three-speed range collar shift transmission illustrated in **Figure 7-22** through **Figure 7-25** is used in conjunction with a synchronized three-speed transmission to provide a total of nine forward speeds and three reverse speeds. The engine delivers power to a mechanically actuated clutch, which sends power to a three-speed synchronized transmission that allows the operator to choose three forward gears or reverse. The three-speed synchronized transmission sends power to the three-speed range collar shift transmission.

With the engine running and the tractor stationary, the operator presses the foot-operated clutch pedal that mechanically decouples the engine from the transmission, allowing the operator to engage a transmission speed and a transmission range gear. With a speed and gear selected, as the operator slowly releases the foot-operated clutch pedal, the tractor will begin to move. If the operator would like to change speeds while the tractor is moving, the clutch pedal is depressed and the operator can choose a different synchronized speed. However, the collar shift ranges must be selected while the tractor is stationary and the clutch pedal is depressed.

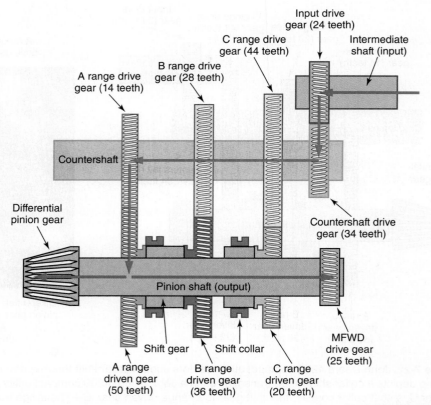

Figure 7-23. When the tractor is shifted into the A range, the A range gears provide a slow speed 3.57:1 ratio.

Goodheart-Willcox Publisher

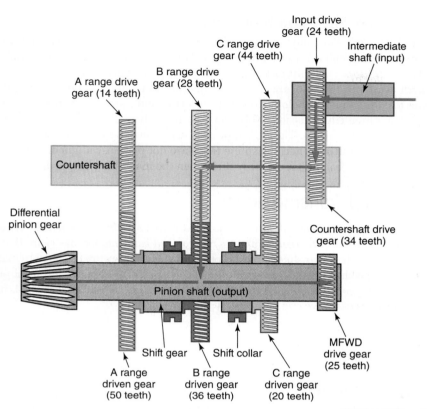

Figure 7-24. When the tractor is shifted into the B range, the B range gears provide a medium speed 1.29:1 ratio.

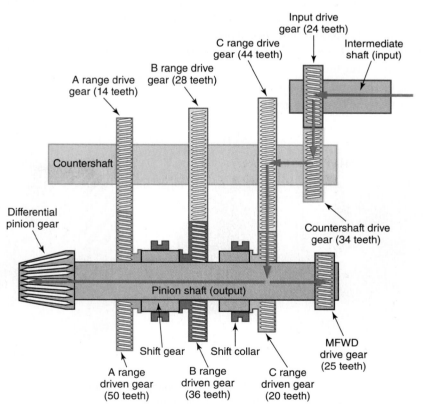

Figure 7-25. When the tractor is shifted into the C range, the C range gears provide a fast speed 0.45:1 ratio.

Case IH Collar-Shift Combine Transmissions

The Case IH 10-, 20-, 230-, and 240-series flagship combines use an electric motor to shift a four-speed sliding-collar range transmission. The electric motor actuates a cam, which moves two shift forks. One shift fork controls first and second gear; the other controls third and fourth gear. The 150, 240, and 250 mid-range series combines and 25-series flagship combines use the same technology except the electric motor and cam control a two-speed sliding-collar range transmission. Note that the parts books incorrectly label the sliding sleeves as synchronizers. The shift fork does use bronze wear blocks but it allows for smooth electrical shifting when the spline teeth are not perfectly aligned. The transmission is a collar-shift design that can only shift when the machine is stopped.

Synchronized Transmissions

Mobile off-highway equipment can also be equipped with a ***synchronized transmission***, also known as a ***synchromesh transmission***. A synchronized transmission has gears that are in constant mesh, like the collar shift transmission. The synchronizers are designed to match the engine speed (input shaft) to the transmission speed (output shaft). The synchronizer prevents grinding gears and gears clashing, while allowing the gears to be shifted as the tractor is moving.

Synchronizers can have one of the following designs:
- Block (cone).
- Pin.
- Friction plate.

Block Synchronizers

Block synchronizers, also known as ***cone synchronizers***, have a hub with internal splines that spline to the transmission's shaft. Although the synchronizers can be located on the input shaft of the transmission, it is more common for the synchronizers to be placed on the output shaft.

The ***synchronizer hub*** also has external splines that align with the synchronizer's sliding sleeve. Keep in mind that the hub is fixed to the output shaft and does not slide, but rotates at the transmission shaft speed (output shaft speed). See **Figure 7-26**.

The ***sleeve***, also known as the slider, has internal splines that align with the following:
- Splines on the hub.
- Dog teeth on the blocker ring.
- Dog teeth on the speed gear.

Goodheart-Willcox Publisher

Figure 7-26. Block synchronizers have a hub that is splined to the transmission shaft. A—As the sleeve is shifted, it causes the blocker ring to mesh with the cone on the speed gear. B—The blocker ring is designed to couple with the cone (shoulder) on the speed gear.

The end of a shift fork sits inside a large groove around the outside of the sleeve. Because the synchronizer hub is splined to the transmission shaft and also splined to the sleeve, the sleeve rotates at the same speed as the transmission shaft and synchronizer hub. The shift fork rides in the sleeve's groove as the sleeve rotates.

The hub is fitted with three *inserts*, also known as *keys*. The inserts are placed in the three slots on the hub and are held in position with one or two snap rings inside the hub. Each insert has a center ridge (raised notch) on top. The ridge sits inside an internal groove or pocket in the sleeve, also called the neutral detent. See **Figure 7-27** and **Figure 7-28**.

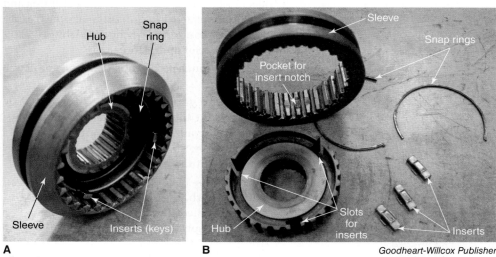

Figure 7-27. Hubs are fitted with three inserts. A—The inserts are held in the three spaces on a hub by one or two snap rings inside the edge of the synchronizer hub. B—The hub, sleeve, inserts, and snap rings have been disassembled to show the components individually. The sleeve has a notch for the insert ridge.

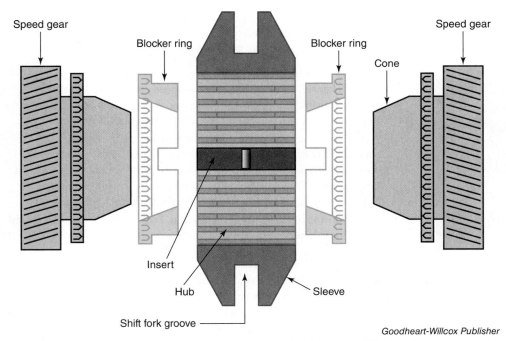

Figure 7-28. A cone synchronizer assembly has a hub, sleeve, two blocker rings, two speed gears, and three inserts.

The ***blocker rings***, also known as ***synchronizer rings***, are made of a soft, wearable material, such as brass. The blocker ring has three notches along its perimeter that the inserts fit into. This contact between the inserts and the blocker ring causes the blocker ring to rotate at the same speed as the synchronizer hub, sleeve, and transmission shaft. The inside bore of the blocker rings is tapered to match the cone on the speed gear. The tapered interface between the blocker ring and the speed gear cone acts like a clutch. When the blocker ring is initially pushed into the speed gear cone, a limited contact area between the two tapered surfaces allows slippage. However, as the blocker ring is pushed further, the area of contact increases and slippage between the blocker ring and speed gear decreases. As a result, the rotational speeds of the synchronizer hub and speed gear equalize. At this point, the sleeve can slide over the dog teeth on the blocker ring and speed gear, coupling the speed gear to the transmission output shaft. Both the blocker ring and the speed gear have sharply pointed dog teeth that make them easier to index with the sleeve's splines during a synchronized shift.

Synchronizer sequence of operation:

1. The operator presses the clutch pedal to disengage engine power from the transmission.
2. Operator chooses the next gear by moving the shift lever toward the desired gear. As the gearshift lever is moved, the shift fork slides the synchronizer's sleeve toward the selected speed gear. Simultaneously, the shift fork for the previously selected gear moves away from that gear, disengaging it.
3. The sleeve pushes the three inserts deeper into their notches in the blocker ring. When the inserts bottom out in their notches, additional movement of the sleeve pushes the blocking ring toward the speed gear.

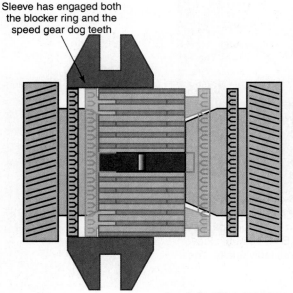

Goodheart-Willcox Publisher

Figure 7-29. In this side view, the sleeve has shifted fully to the left, allowing the sleeve to couple with the blocker ring dog teeth and the speed gear dog teeth.

Note

When the transmission is in a neutral position, the inserts (keys) hold the sleeve centered on the hub because the insert ridges are located inside the sleeve's internal groove or pocket. As the transmission is shifted into gear, the sleeve slides toward a blocker ring and speed gear. The sleeve holds pressure on the inserts (keys), which pushes the blocker ring into the speed-gear's cone.

4. The blocker ring begins to contact the speed gear's cone (shoulder). The blocker ring initially blocks the engagement of gearshift lever because the two transmission speeds (input and output) are not yet synchronized.
5. As the synchronizer hub's rotational speed and the speed gear's rotational speed equalize, the blocker ring aligns (indexes) with the speed gear's dog teeth, allowing the sleeve to fully engage the speed gear. See **Figure 7-29**. This couples the speed gear to transmission shaft. Simultaneously, the gearshift lever "pops" into the desired position.
6. The operator releases the clutch pedal. Engine speed changes based on the new load.

Goodheart-Willcox Publisher

Figure 7-30. This synchronizer is used in a Caterpillar TH330B telehandler. Spring-loaded check balls between the sleeve and the hub hold the synchronizer in the neutral position. The speed gear ring mates to the speed gear. The speed gear ring's cone mates with the blocker ring.

Spring-Detented Block Synchronizers

Some block synchronizers, such as the example in **Figure 7-30**, use internal springs and check balls rather than the insert ridge for their detent. The spring sits below the check ball, pressing force radially between the insert and the hub. The sleeve has a seat for the check ball. The detent spring holds the sleeve in a neutral position. When the synchronizer is shifted into gear, the sleeve slides past the spring-loaded check ball.

Block synchronizers are found in low- and mid-range horsepower applications, including compact utility tractors, smaller MFWD agricultural tractors, loader backhoes, and telehandlers.

Block Synchronizer Inspection

Before disassembling a synchronized transmission, mark the synchronizer components to ensure they can be properly reassembled. Some hubs and sleeves are directional and should be installed in a specific direction.

Synchronizer blocker rings wear over a period of time. Excessive blocker ring wear results in the hub and sleeve clashing and grinding. When a ring can no longer grip the speed gear's tapered cone, the hub and sleeve will not be able to match speed. Synchronizer wear is measured with a feeler gauge. The measurement is normally made during a transmission overhaul, with the synchronizers removed. The speed gear and blocker ring are set flat on a workbench. See **Figure 7-31**. The blocker ring is lightly pressed toward the speed gear while a feeler gauge is inserted between the blocker ring and the speed gear to measure the gap. Equal pressure must be applied to both sides of the blocker ring to keep it from tilting, which would result in an uneven gap.

Goodheart-Willcox Publisher

Figure 7-31. Blocker ring gap is measured during disassembly by evenly applying pressure to the top of the blocker ring while using a feeler gauge to measure the clearance between the blocker ring and the speed gear.

 Note

A new blocker ring will have the largest gap between the ring and the speed gear. As the blocker ring wears, the gap narrows. An excessively worn blocker ring will cause grinding.

If the blocker ring gap is to be measured while the synchronizers are still installed in the transmission, pressure must be applied to the blocker ring, while measuring the gap. See **Figure 7-32**. Pressure can be applied by pushing the sleeve (by hand or with the shift fork or lever) toward the blocker ring, causing the inserts to press against the blocker ring. An alternative method is to slide the sleeve out of the way (engaged into the opposing gear), which allows you to use your fingers to hold the blocker ring against the speed gear. However, this method makes it very difficult to apply equal pressure to the blocker ring. The best method for measuring the blocker ring gap is with the synchronizer assembly removed from the transmission.

The blocker ring should also be inspected for cracks and to ensure the dog teeth are not excessively worn. The dog teeth on the speed gear and the blocker ring should be sharp. If they are worn, it will be more difficult for the sleeve to index with the dog teeth during a shift.

Additional synchronizer wear points are the speed gear's dog teeth, the space where the inserts are located on the synchronizer hub, and the inserts. The splines on the hub and the sleeve must also be checked for wear.

Goodheart-Willcox Publisher

Figure 7-32. A feeler gauge is used to measure the space between the blocker ring and the speed gear to determine the amount of wear on the blocker ring. Pressure must first be used to push the blocker ring into the speed gear without engaging the sleeve onto the speed gear.

Pin Synchronizers

Pin synchronizers perform the same basic function as block synchronizers. They equalize input and output transmission speeds so gears can be engaged without grinding. However, rather than using inserts to move a blocker ring into contact with a speed gear, they use pins to move a collar into contact with a range gear. The collar performs the blocking function of the synchronizer. Pin synchronizers have one of two types of designs, fixed pin or split pin.

Fixed Pin

One example of a ***fixed pin synchronizer*** is used in the rear auxiliary transmission in semi-trucks that provides high and low ranges for the main five-speed transmission, resulting in ten available forward speeds. Speeds one through five are obtained when the range is in low range positions low 1, low 2, low 3, low 4, and low 5. 6th speed is achieved by shifting the range into high and moving the gear selector back to the same position as 1st gear (high 1). High range provides speeds six through ten: 6th (high 1), 7th (high 2), 8th (high 3), 9th (high 4), and 10th (high 5).

Note

Although fixed pin synchronizers are more common in on-highway trucks, they have also been used in off-highway equipment. Case IH 4WD Steiger tractors were offered with a 12-speed and 24-speed synchroshift transmission with fixed pin synchronizers. These transmissions have four synchronizers. Two syncrhonizers are located on the range shaft (A & B and C & D), and two synchronizers are located on the speed shaft (1st & Reverse and 2nd & 3rd). The 12-speed transmission becomes a 24-speed transmission with the addition of a high-low multiple disc clutch gear assembly.

The synchronizer has a low range synchronizer collar and a high range synchronizer collar. Both the high range and the low range collars have friction material. On the high range collar, the friction material is on the inside edge of the collar. On the low range collar, the friction material is on the outside edge of the collar. Both collars also have three pins, each with a large-diameter portion and a small-diameter portion, with a tapered shoulder between. The pins are inserted through a set of chamfered holes in the synchronizer's sliding clutch gear and into matching holes in the other collar. See **Figure 7-33**. Note that the matching holes are slightly elongated. This allows the collars to rotate slightly in relation to each other.

The sliding clutch gear has internal and external splines. The internal splines couple to the transmission's output shaft. As a result, the sliding clutch gear is always rotating at output shaft speed. The external splines match splines in the high range and low range gears and are the means by which the gears can be coupled to the transmission shaft. See **Figure 7-34**.

When the synchronizer is in the neutral position, the high and low range gears are not coupled to the transmission shaft and are free to rotate at different speeds. When the sliding clutch gear is moved to one side or the other, its external splines mesh with either the low range gear or the high range gear, coupling that gear to the transmission shaft.

The sliding clutch gear has six chamfered holes (three for the high range collar pins and three for the low range collar pins). Three springs in the high range collar press against the small-diameter portion of the low range collar's pins. This causes the high range collar

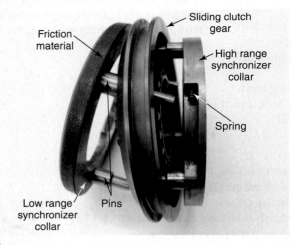
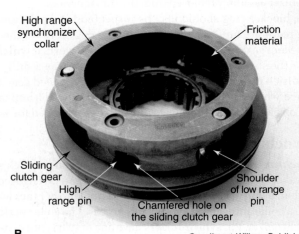

Figure 7-33. Fixed pin synchronizers perform the same function as block synchronizers. A—The high range synchronizer collar has three springs. The three pins from the low range synchronizer collar protrude through the three chamfered holes in the sliding clutch gear and compress the coiled springs into the high range synchronized collar. B—This synchronizer assembly has been shifted into the high range position. Notice that the low range pins are centered in the chamfered holes on the sliding clutch gear and the high range pins are off-center in the chamfered holes.

to rotate slightly in relation to the low range collar. As a result, none of the pins are centered in their holes in either of the collars or the sliding clutch disc. This prevents the large portions of the pins from being able to pass through the holes in the sliding clutch gear and keeps the high and low speed synchronizer collars in a neutral position.

As the auxiliary gearbox is shifted from neutral into gear (either high range or low range), a shift fork moves the sliding clutch gear toward one of the collars. Initially, the large-diameter portion of the pin is unable to pass through the hole in the sliding clutch gear. As a result, the synchronizer collar is pushed into contact with the range gear.

The synchronizer collars have a cone-shaped surface lined with friction material. When this friction material comes into contact with the cone on the high range gear or low range gear, the rotational speeds of the range gear and the collar equalize. Once the synchronizer collar is synchronized with the range gear, additional movement of the sliding clutch gear toward the range gear overcomes spring tension and causes the tapered shoulders of the collar pins to center themselves in the chamfered holes in the sliding clutch gear. When the pins are centered in their holes, the sliding clutch gear can advance past the shoulders of the pins. This allows the external splines on the sliding clutch gear to mesh with the internal splines on the range gear, coupling the range gear to the transmission shaft.

Split Pin Synchronizer

The *split pin synchronizer* is another style of pin synchronizer. See **Figure 7-35**. The synchronizer assembly has two cone-shaped discs at each end. These are called *synchronizer cones*, or *synchronizer cups*. Each synchronizer cone mates with a friction ring that is similar to the collar in the fixed pin synchronizer. The synchronizer cones spline to the speed gear.

The synchronizer sleeve, or sliding collar, is located in the center of the synchronizer. Its internal splines couple to a shift hub, also known as a clutch hub, which is splined to the transmission's shaft. The sleeve and hub rotate at the same speed as the transmission shaft.

Note

In some transmissions, the synchronizer sleeve is splined directly to the transmission shaft rather than a shift hub.

The synchronizer sleeve has chamfered holes through which friction ring pins pass. Notice in **Figure 7-36** that the split pin has an hourglass shape and is split down the center. Internal springs attempt to widen the split pin. When the sleeve is in the center position, the sleeve's chamfered holes are seated on the narrow portion of the split pins. As the sleeve is moved, it forces the chamfered holes to pass over the wider portion of the pins, squeezing the pin halves together and compressing their internal springs.

As the operator shifts the transmission, the sleeve first causes the split pins to push on the friction ring. As pressure is applied to the friction ring, it begins to grip the mating surface of the synchronizer cone. This creates a drag on the friction disc, which places a side load on the pins. Initially, the synchronizer pins are blocked

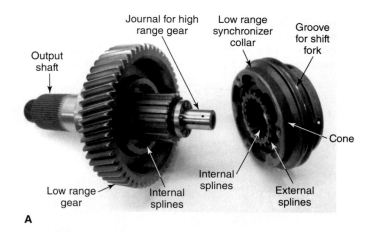

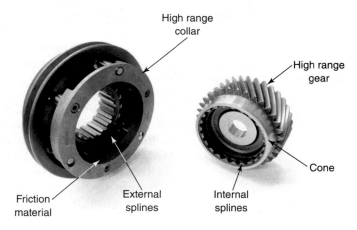

Goodheart-Willcox Publisher

Figure 7-34. The pin synchronizer is sandwiched between the low range gear and the high range gear. A—The low range side is shown here. The synchronizer is in the low range position. If it were slid onto the output shaft, the internal external splines on the synchronizer would mesh with the internal splines on the low range gear. Note the smooth journal where the high range gear would be installed on the other side of the synchronizer. B—The high range side of the assembly is shown here. The synchronizer is in the low range position. Notice the cone on the high range gear that mates with the friction material on the high range synchronizer collar.

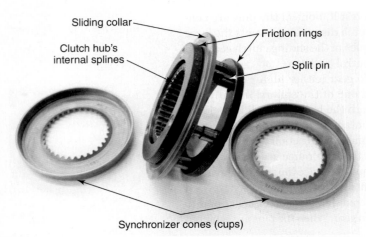

Figure 7-35. Split pin synchronizers have friction rings and cones. The pins are split apart. Springs keep the pin halves separated when the synchronizer is in neutral and must be compressed to engage the synchronizer with a gear.

Figure 7-36. The split pin synchronizer uses internal springs to resist movement of the sleeve out of the neutral position.

from engagement because the side load on the pins prevents the chamfered holes from squeezing over the wide part of the pins. Once the sleeve and the speed gear are rotating at the same speed, the side load on the pins is eliminated and the wide portion of the pins is able to pass through the chamfered holes and the friction rings can fully engage the cones.

The Caterpillar 424D backhoe is an example of a construction machine that uses a split pin synchronizer.

Friction Plate Synchronizers

Some old John Deere agricultural tractors were equipped with friction plate synchronizers, **Figure 7-37A**. A few examples of older John Deere tractors that could be equipped with friction plate synchronizers are 4010, 4320, 4650, and 4850. The synchronizers were used in the QUAD-RANGE transmission. The synchronizer assembly allowed the operator to select high range, low range, or neutral.

The synchronizer has some similarities to a multiple disc clutch, which is explained in Chapter 8, *Clutches and Planetary Controls*. Its friction discs spline to a steel synchronizer drum (**Figure 7-37B**). The synchronizer steel plates are placed alternatively between the friction discs. The steel plates spline to a single high-low speed blocker assembly. The blocker assembly's splines align with a hub that is splined to the transmission's drive shaft. A high-low range shifter surrounds the high-low range synchronizer assembly, **Figure 7-37A**.

The shifter is actuated by a shift rail. As the drum is moved by the shifter, the friction discs apply pressure to the steel plates, causing the drum to couple with the transmission shaft through the synchronizer blocker. Note that the drum engages its corresponding speed gear (high or low).

As the shifter is moved, the synchronized gear is engaged or released. The high-low range blocker assembly is shared by the high-low range synchronizer drums. When the shifter is in the center position, the synchronizer is in a neutral position. When the shifter is slid in one direction, the blocker assembly locks the transmission range drive shaft to the high range gear. When the shifter is moved in the opposite direction, it locks the transmission range drive shaft to the low range gear. The blocker assembly has four detent springs and detent balls that hold the synchronizer in its detented neutral position.

John Deere DirectDrive Dual Clutch Transmission

Although the vast majority of synchronized transmissions require the operator to press the clutch pedal and manually shift the transmission lever, John Deere manufactures an automatic synchronized transmission. It uses two automated clutches to automatically shift through eight synchronized speeds. John Deere labels the transmission a *DirectDrive (dual clutch) transmission*.

The dual clutch automated eight-speed transmission is an option offered in 6R series tractors, for example 6R175, 6R195, and 6R215 tractors. The eight-speed transmission is used in conjunction with a three-speed range transmission to provide a total of 24 forward and 24 reverse speeds.

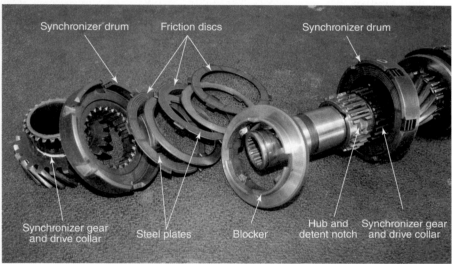

Figure 7-37. Plate synchronizers were used in older John Deere Quad-Range transmissions to provide low and high ranges. A—The shifter forces the two drums (high and low range) to actuate. B—As the synchronizer shifter is actuated, it forces the drum and friction discs to synchronize with the steel plates and blocker. The detent balls and springs are missing from the illustration.

Powertrain Power Flow

The tractor's transmission is comprised of three individual modules that are coupled to each other, forming the rear half of the tractor's integrated frame:

1. The PowrReverser module (the input to the transmission).
2. The dual path module (eight-speed gear transmission).
3. The range transmission (three speed ranges).

PowrReverser Module

The engine delivers power to the PowrReverser, which has a planetary gear set that is operated by a multiple disc clutch (forward clutch) and a multiple disc brake (reverse brake). A proportional solenoid is used to control the forward clutch and another solenoid is used to control the reverse brake. Note that multiple disc clutches are explained in

Chapter 8, *Clutches and Planetary Controls*, and planetary gear sets are explained in Chapter 9, *Planetary Gear Set Theory*. The PowrReverser will deliver forward or reverse power flow through its output shaft into the dual path module (eight-speed gear transmission).

Dual Path Module (Gear Transmission)

The PowrReverser module's output shaft is coupled to the dual path module's input shaft, which has two drive gears. Each drive gear delivers power to a driven gear on a countershaft. The two countershafts are known as side shaft 1 and side shaft 2. See **Figure 7-38**. Each side shaft has its own multiple-disc clutch. The clutch receives power by means of the drive gears. Each multiple-disc clutch is controlled by an individual proportional solenoid valve.

Side shaft 1 contains the odd speed gears: 1st, 3rd, 5th, and 7th. Side shaft 2 contains the even speed gears: 2nd, 4th, 6th, and 8th. Each side shaft uses two synchronizers to control its four speeds. Gears on the side shafts are in constant mesh with gears on the eight-speed gear transmission's output shaft.

When a gear is selected, one side shaft's multiple-disc clutch is engaged while the other side shaft's multiple-disc clutch is simultaneously disengaged. The tractor must have one of the side shaft multiple disc clutches engaged in order to move. The engaged clutch causes the synchronizer hubs and sleeves on that side shaft to rotate. A synchronizer sleeve will shift and lock the selected speed gear to its side shaft, which causes the selected speed gear

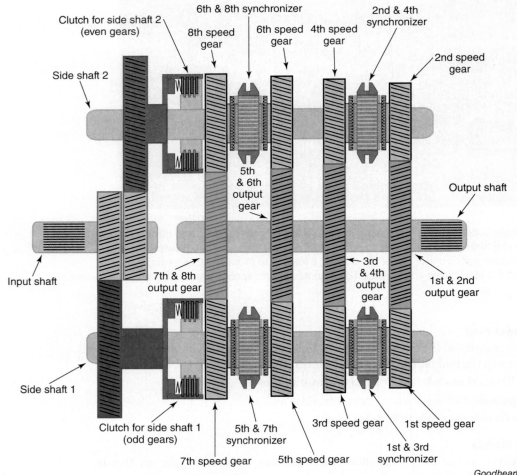

Figure 7-38. A top view of the John Deere Dual Path DirectDrive eight-speed gear transmission. The transmission has two multiple-disc clutches to control a pair of countershafts, known as side shafts. One side shaft has a clutch and even-numbered synchronizers and speed gears, and the other side shaft has a clutch and odd-numbered synchronizers and speed gears.

to drive the corresponding gear on the output shaft. The transmission will preselect the next synchronizer on the opposite side shaft in anticipation of the next shift. The next shift occurs by simultaneously releasing the engaged clutch (for example on the odd shaft) and engaging the opposite clutch (for example on the even shaft). While this is occurring, the transmission again preselects the synchronizer for the next anticipated shift and moves it into position. The application pressure of the side shaft multiple disc clutches is varied, known as modulated, to ensure power is smoothly transmitted. Modulation is further explained in Chapter 11, *Powershift and Automatic Transmission Controls, Service, and Repair*.

The output shaft gears are also known as idler gears. Each output shaft gear (idler gear) is shared between a speed gear on each side shaft. For example, the output gears can be labeled 1st-2nd, 3rd-4th, 5th-6th, and 7th-8th, based on which speed gears can transmit power through them.

Power will only flow through one side shaft at a time. As the transmission is upshifted from one gear to the next gear, one clutch will be engaged and the other disengaged to alternate power flow from one side shaft to the other side shaft.

Eight solenoids are used for operating the eight synchronized speeds. Two solenoids operate one shift pad (finger) which is linked to a shift fork. As the pad is operated, it causes the fork to actuate a synchronizer sleeve. The solenoids work in coordinated pairs. As the tractor is upshifted, one shift fork will disengage the lower gear speed, and another shift fork will engage the next higher gear speed. See **Figure 7-39**, **Figure 7-40**, and **Figure 7-41**

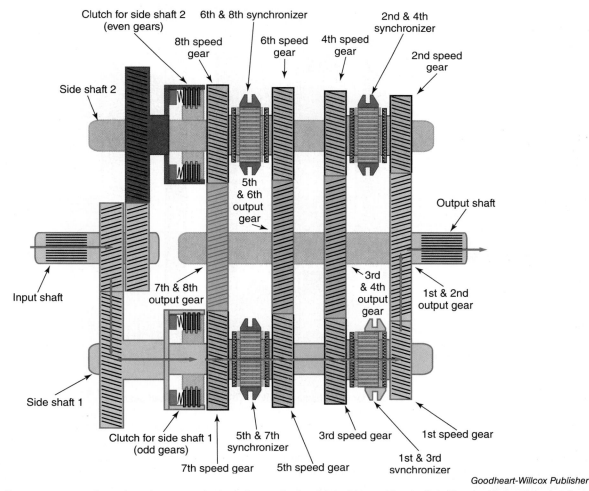

Goodheart-Willcox Publisher

Figure 7-39. In first gear, power flows into the transmission's input shaft, which drives side shaft 1. The multiple-disc clutch on side shaft 1 drives the odd-numbered synchronizer hubs. The 1st & 3rd synchronizer sleeve engages the 1st speed gear, which drives the 1st & 2nd output gear on the transmission's output shaft.

for power flow through the first three gears of the eight-speed gear transmission. The dual path eight-speed gear transmission module's output shaft is splined to the range transmission's input shaft, delivering power into the range transmission module.

Range Transmission

The range transmission produces three different gear ratios that can be used in combination with the eight-speed transmission. The result is three different ranges (A, B, and C) with eight gears in each range, for a total of 24 different gear ratios.

The range transmission's input shaft is one of two countershafts located inside the range transmission. Three drive gears, A, B, and C, spin freely on the input shaft. An A & B range synchronizer and a C range synchronizer are splined to the input shaft and used to couple and decouple the three range drive gears to the range input shaft. See **Figure 7-42**.

Once one of the range synchronizers is engaged, it connects the range drive gear to the range input shaft. The drive gear, in turn, powers the corresponding range driven gear on the differential drive shaft, which serves as the range output shaft. The differential drive

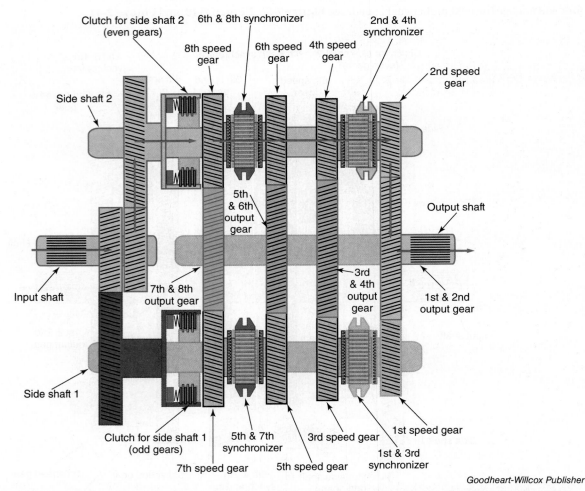

Goodheart-Willcox Publisher

Figure 7-40. In 2nd gear, power flows into the transmission's input shaft, which drives side shaft 2. The multiple-disc clutch on side shaft 2 drives the even-numbered synchronizer hubs. The 2nd & 4th synchronizer sleeve engages the 2nd speed gear, which drives the 1st & 2nd output gear on the transmission's output shaft.

shaft contains the pinion gear that drives the rear axle differential and also contains the drive gear for the front-wheel-drive (MFWD). The MFWD driven gear, clutch, and output shaft are located in the MFWD housing, which is attached to the bottom of the range module. Chapter 4, *Agricultural Equipment Identification* and Chapter 10, *Powershift and Automatic Transmission Theory* further explain MFWD.

The driver has the option of selecting one of three ranges, A, B, or B-C, by pressing one of the three range buttons on the right-hand armrest. In the B-C range, the tractor automatically shifts from the B to C range as needed. The range transmission can be shifted while the tractor is moving and under load without depressing the clutch pedal.

The three ranges provide the tractor the three ranges of travel speeds:
- Range A: .74–6.7 mph (2.7–10.8 km/h).
- Range B: 3.35–13.67 mph (5.4–22 km/h).
- Range C: 8.1–25 mph (13–40 km/h).

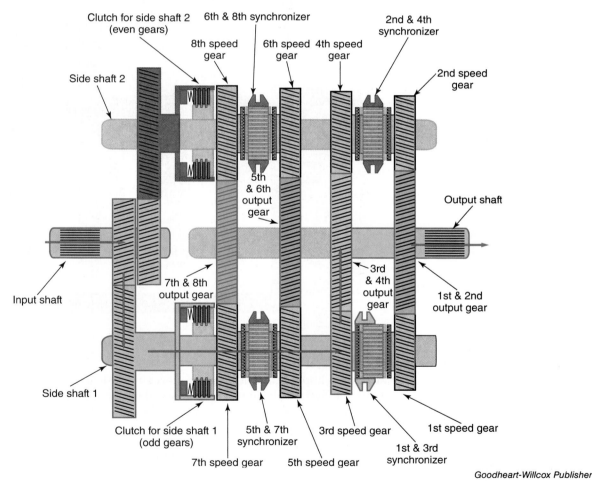

Figure 7-41. In third gear, power flows into the transmission's input shaft, which drives side shaft 1. The multiple-disc clutch on side shaft 1 drives the odd-numbered synchronizer hubs. The 1st & 3rd synchronizer sleeve engages the 3rd speed gear, which drives the 3rd & 4th output gear on the transmission's output shaft.

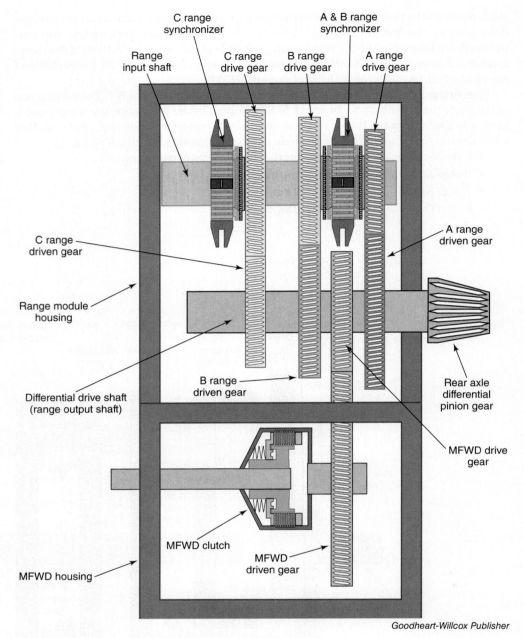

Figure 7-42. A simplified side view of the John Deere range transmission in a DirectDrive equipped tractor. The range transmission has three speeds, A, B, and C, controlled by two synchronizers. The MFWD housing is attached to the bottom of the range module.

Speed Control Lever

The tractor uses a speed control lever that moves in a T-shaped slot. The left side slot mode is used for manually controlling the eight-speed gear transmission, by pulsing the lever forward for upshifting and pulsing the lever rearward for downshifting. When the lever is in the right-hand position, the eight-speed gear transmission is in the automatic mode. The lever also has a thumb adjusted speed wheel that is used for increasing the tractor speed while operating in the automatic mode. See **Figure 7-43**.

Note

Chapter 15, *Continuously Variable Transmissions*, provides an example of a Caterpillar wheel loader that combines a hydrostatic drive, power shift, and synchronizers into one continuously variable transmission.

Shift Levers, Rails, and Cams

Manual transmissions use different mechanisms to move and hold the sliding gears, collars, and synchronizers. Examples include:

- Forks.
- Rails.
- Cams.
- Detents.

Shift forks are C-shaped and mate with a groove in a sliding gear, collar, or synchronizer sleeve. The fork is used to slide the sliding gear, shift collar, or a synchronizer sleeve to shift transmission gears. The shift forks are commonly actuated with steel rods called *shift rails*. See **Figure 7-44**.

The manual transmission's gearshift lever is held in place using a *detent mechanism*, consisting of a ball, a spring, and a notch in the shift rail. See **Figure 7-45**. The detent spring forces the detent ball to sit inside the shift rail's notch, which holds the transmission in gear. To shift gears, enough force must be applied to the rail to push the detent ball out of the shift detent groove, compressing the spring. As the rail moves, it forces the shift fork and sliding gear, collar, or synchronizer sleeve to move.

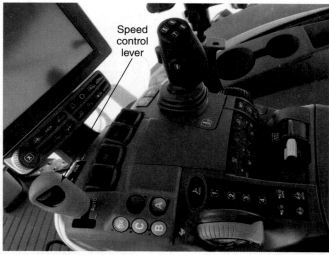

Philip McNew

Figure 7-43. When the DirectDrive speed control lever is placed in the left-hand slot it operates in the manual mode, and when placed in the right-hand slot it will operate in the automatic mode.

Note

The detent spring, like many other springs used in heavy equipment, often has specifications listed in the service literature, such as the overall uncompressed length or the compressed length with a given amount of compression force.

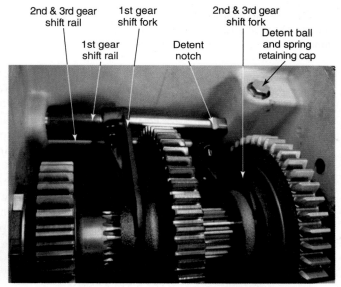

Goodheart-Willcox Publisher

Figure 7-44. These shift rails and forks are used in the Case IH combine sliding gear transmission discussed earlier in this chapter. Because the 2-3 shift rail is used for two different gears, it has three detents: second gear, third gear, and neutral. The first gear shift rail has only two detents: first gear and neutral. Only one detent notch is shown.

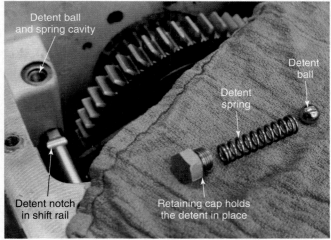

Goodheart-Willcox Publisher

Figure 7-45. A shift detent uses a spring to hold a detent ball inside a notch in the shift rail.

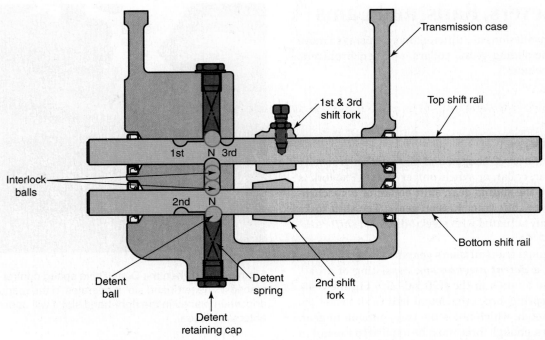

Case New Holland

Figure 7-46. An example of the two shift rails used in a Case IH 1620 and 1640 three-speed sliding gear transmission. The top shift rail is used to shift the transmission into first, neutral, and third gear. The bottom shift rail is used to shift the transmission into second gear and neutral.

Shift rails can also have an *interlock mechanism* that prevents two gears from being selected at one time. The interlock mechanism can use interlock balls or a pin. The image in **Figure 7-46** shows two interlock balls in a passageway between the two shift rails. At least one of the two shift rails must always be placed in a neutral position before the other shift rail can be shifted into gear. The rail that is shifted into neutral allows one of the interlocking balls to recess into the shift rail's notch. If the two shift rails were attempted to be simultaneously shifted into gear, the two interlock balls would prevent movement of the shift rails, because there would be insufficient room in the passageway for both balls.

Manual transmissions can also use a *cam shifter* to change gears. See **Figure 7-47**. The shifter consists of a cam plate with slots that cause the rollers of a shift fork to move along a set path as the cam plate is rotated. A shaft is often used to force the cam plate to pivot, which forces the shifter to actuate the shift rail and fork.

Bearings

Transmissions use different types of *bearings* to support rotating shafts and gears while minimizing drag, friction, resistance, heat, and wear. Bearings can be classified as friction or antifriction.

An *antifriction bearing* has two steel rings called *races* that surround rolling elements, such as balls, rollers, or needles. The primary purpose of the rolling elements is to reduce friction. Bearings can be used between a rotating member and a stationary member, or between two members that rotate at different speeds or in different directions. Some types of antifriction bearings are separable, meaning the races can be separated from the rolling elements. This makes the bearing easier to install and also allows adjustment of rolling element clearance. In one piece (non-separable) bearings, the rolling element clearance is set when the bearing is manufactured.

A *friction bearing*, also known as a plain bearing, does not have multiple rolling elements. Rather, it is a bushing, often made of brass or bronze, that supports a rotating shaft. A nylon or polymer friction bushing is sometimes used in light-load applications. Friction bearings are made of a soft material because they are designed to wear or absorb damage before the shafts that rotate in them. For example, if grit gets into the oil, the grit will embed in the soft bearing material rather than scoring the harder shaft material. Because a friction bearing is made of soft material, it gradually wears out. It must be continuously lubricated and rotated to spread the lubricant evenly.

A friction bearing can be a single solid bushing, known as a sleeve bearing, or it can be a split bearing, such as engine crankshaft main and connecting rod bearings. Friction bearings have a lot of surface area and are designed to support radial loads. They are less expensive than antifriction bearings.

Loads

The direction of the shaft, its load, and the type of gears installed (spur or helical) determine the type of load exerted on the bearings. There are three types of loads: radial, thrust, and angular. A *radial load* is applied perpendicular (at a right angle) to the shaft. Straight roller bearings are used in applications with little or no thrust load, such as supporting some axle shafts. Ball bearings are also used for supporting shafts with radial loads.

A *thrust load*, also known as an axial load, exerts force parallel with the shaft (axial to the shaft). See **Figure 7-48**. Some ball bearings can be used in applications with thrust loads, such as the thrust load created by the helical gear illustrated in **Figure 7-48**. However, tapered roller bearings are more commonly used in thrust load applications. Thrust bearings are also used in thrust load applications. An *angular load* is a combination of both radial and thrust loads.

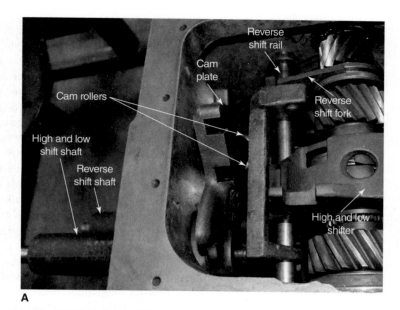

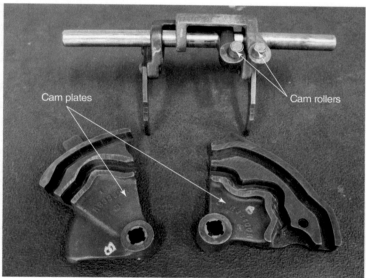

Goodheart-Willcox Publisher

Figure 7-47. Shift cams are used to actuate shifting collars. A—During a shift, the shift shaft rotates the cam plate. The cam rollers travel along the grooves in the cam plate, causing the shift fork to move along a set path. B—The shift rails and cam plates have been removed from the transmission.

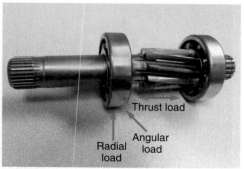

Goodheart-Willcox Publisher

Figure 7-48. Bearings can have radial loads (perpendicular to shaft), thrust loads (parallel to shaft), or angular loads (both perpendicular and parallel to shaft) exerted on the bearing.

Types of Bearings

Types of antifriction bearings include:
- Ball bearings.
- Roller bearings.
- Straight roller.
- Tapered roller.
- Needle bearings.
- Spherical roller bearing.
- Thrust bearing.

Ball Bearing

A *ball bearing* consists of several spherical rolling elements between inner and outer races. Both the inner and outer races have a groove for the balls. The outer race is often located in the seat of a housing. The inner race is often located on a shaft.

Some ball bearings have a *cage* (also known as a separator or retainer) that holds the balls. A cage does not carry any load, but is used to evenly space the balls. See **Figure 7-49**.

Ball bearings are the most popular type of bearing. They are used in radial load applications and are the best bearing for high-speed applications. Ball bearings do not support static loads as well as roller bearings.

Ball bearings designed for thrust load applications have deeply grooved races. The grooves support the rolling elements and help prevent the races from separating from the balls when perpendicular force is applied to the bearing. Ball bearings designed for angular loads have tall shoulders on the races. One shoulder is on the inner race and the other shoulder is on the outer race (diagonally opposed to the inner race's shoulder). Ball bearings do not perform well with shaft misalignment.

Ball bearings can be divided into two classifications based on how they are assembled: filling-notch ball bearings and non-filling-notch ball bearings. A *filling-notch bearing*, also known as a slot-filled bearing, has a notch or slot in the outside and inside races that is used to load the balls into the bearing assembly. See **Figure 7-50**. It is sometimes called a maximum capacity ball bearing because it can be designed to be completely filled, with no space between the balls. The more balls that are in a ball bearing, the greater the load-bearing surface area and heavier load the bearing can handle. Filling-notch bearings are good for angular loading applications and can handle a limited thrust load, but only in one direction. The bearing may or may not use a cage to separate the balls. Be sure to follow the manufacturer's service literature for installing the bearings.

A non-filling-notch ball bearing, also called a *Conrad bearing*, has no notches. The bearing is assembled by moving the inner and outer races close together at the top or bottom of the bearing to form a large gap through which the balls are loaded into the bearing. As the balls are added, the gap between the races narrows, limiting the number of balls that can be installed into this type of bearing. When the proper number of balls has been added, a two-piece cage is installed to keep the balls evenly spaced This also centers the inner race within the outer race. Conrad bearings are used in angular-load applications and can handle a limited thrust load in both directions.

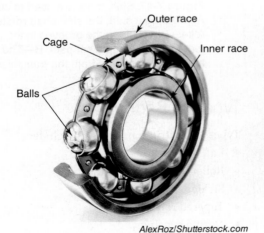

AlexRoz/Shutterstock.com

Figure 7-49. Ball bearings can be placed inside a cage to evenly space the bearings.

Opas Mitsom/Shutterstock.com

Figure 7-50. A filling-notch bearing has slots in the inside race and outside race used for loading the balls into the bearing. Notice that this bearing also uses a cage.

Roller Bearings

Roller bearings consist of rollers between inner and outer races. Four main types of roller bearings are available, based on the type of rollers used and their orientation to the shaft. The four types are straight roller, tapered roller, needle roller, and spherical roller.

Straight Roller Bearing

The **straight roller bearing**, also known as the **cylindrical roller bearing**, has multiple cylindrical rollers between inner and outer races. See **Figure 7-51**. The edges of a straight roller bearing race may have a rib that retains the rollers. The rollers may be contained in a separator, which is similar in form and function to the cage used in some ball bearings. Straight roller bearings handle radial loads much better than thrust loads. Some straight roller bearings (with short rollers) are used in high speed applications. Straight roller bearings do not perform well with shaft misalignment. The bearing's two races are often able to be separated, making the installation of the bearing easier.

Tapered Roller Bearing

A **tapered roller bearing** consists of multiple tapered (cone-shaped) rollers housed in a cage and positioned between two tapered races. The outside race is called the **cup**, and the inside race is called the **cone**. See **Figure 7-52**.

The cone rib is used to retain the rollers. Unlike other bearings, the races of tapered roller bearings can be separated. It is the only bearing that can have its clearances (both thrust and radial) adjusted by positioning the inner and outer races. Adjustments are explained later in the chapter.

Tapered roller bearings can handle heavy loads of all three types, thrust, radial, and angular. They perform well under a wide range of speeds. However, they are more expensive to manufacture than ball bearings and roller bearings.

Needle Roller Bearing

A **needle roller bearing** has thin cylindrical rollers, known as needles, between an inner and an outer race. The ratio of the rolling element's length to its diameter is normally greater than 4 to 1. The rolling elements in a straight rolling bearing have a length to diameter ratio that is normally less than 4 to 1. A needle bearing is thinner than other bearing types with the same diameter because its rollers are thinner.

The needle rollers may or may not be contained in a cage. Non-caged needle rollers have the rollers lining the inside diameter of the outer race. See **Figure 7-53**. Needle roller bearings are used for light-duty radial load applications. They are economical to manufacture. They are not used in bore diameters of 10 inches or larger.

Peter Ivanov Ishmiriev/Shutterstock.com

Figure 7-51. A straight roller bearing uses cylinder-shaped rollers that are parallel to the shaft.

SURAKIT SAWANGCHIT/Shutterstock.com

Figure 7-52. In tapered roller bearings, the rollers and both of the races are tapered.

A — *Stason4ik/Shutterstock.com*  B — *Goodheart-Willcox Publisher*

Figure 7-53. Needle roller bearings can be loose or placed inside a cage. A—Loose needle bearings in a universal joint trunnion cap. B—Caged needle bearings installed on a shaft.

Spherical Roller Bearing

Spherical roller bearing rollers have a slight barrel shape. The double-row spherical roller bearing is the most popular type of spherical roller bearing. It has two rows of rollers retained in a single cage, single inner race, and single outer race. See **Figure 7-54**.

The double spherical roller bearing is known as a self-aligning bearing because it can compensate for a small amount of misalignment between the shaft and the bearing. The rollers and cage can shift slightly in the races, allowing the bearing to compensate for a small amount of shaft wobble or deflection due to loads. It is the best choice for applications with a small amount of misalignment (up to .05 radians), but a poor choice when the shaft must be rigid. Spherical roller bearings can handle heavy radial loads. However, they are a poor choice for thrust load applications. They are the best choice for angular loading (combined radial and thrust loading). More energy is needed to rotate this type of bearing than the other types of antifriction bearings.

Thrust Bearing

A *thrust bearing* consists of rollers that are arranged so their axes are perpendicular to the shaft. See **Figure 7-55**. The bearings are often used in conjunction with helical gears that create side thrust loads. Thrust bearings are placed between gears, shafts, and housings to control shaft movement and end thrust.

Sealed Bearing

Some bearings have seals on the outside of the races. These bearings are known as *sealed bearings*. See **Figure 7-56**. The seals prevent contamination from entering the bearing and help retain the bearing's lubricant. A metal shield rather than a seal can be placed over one or both side openings on the bearing. The shield deters contaminants, but allows surplus lubricant to exit the bearing.

A *pillow-block bearing* is a bearing that has an integral housing that is designed to be bolted to a mounting surface. They are sometimes used in heavy equipment to support external shafts with high radial loads. See **Figure 7-57**. Spherical roller bearings are one type of bearing that are used in pillow-block applications.

Service and Repair

If an antifriction bearing becomes worn or damaged, it must be replaced. Manufacturers recommend replacing the cone, cup, and rollers as a set, even if the damage is apparent on only one element. This is true even for bearings that have separable races, like tapered roller bearings.

Monstar Studio/Shutterstock.com

Figure 7-54. A double set of spherical roller bearings are used in applications with a small amount of shaft misalignment.

Goodheart-Willcox Publisher

Figure 7-55. The rollers in thrust bearings are arranged so their axes are perpendicular to the shaft.

Goodheart-Willcox Publisher

Figure 7-56. A seal has been placed on each side of this bearing.

Note

A bearing race can be classified as a press fit or a slip fit. **Press fit bearing races** have an interference fit between the race and the mating shaft or housing. A press fit requires pressure to be applied to install or remove the race. A **slip fit race** has a small space between it and the shaft or housing. When installing or removing bearing races, do not apply pressure to a slip fit race.

Removal

Multiple types of tools are used for removing bearings: presses (arbor or hydraulic), drivers, slide hammers, and mechanical pullers (for example, a jaw-type puller). A press can be used to push a shaft out of a bearing or press a bearing off a shaft. Sometimes a component will not fit in a press and a puller must be used to remove the bearing. Special care must be used to ensure the puller is gripping the correct component and is gripping the full surface of the component. See **Figure 7-58**.

Some service literature recommends heating components (for example, a bearing race) to assist the installation process. However, open flame heating is not recommended due to the risk of fire with oil or fuel, and the concentrated heat can affect a part's heat treatment.

In some scenarios, a bearing's race is recessed in its cavity, which prevents the use of a press or a puller for removing the race. Some technicians will weld a bead around the inside circumference of the old bearing race cup, heating both the race and the surrounding housing. Then, the housing is turned over and tapped. Because the race cools faster than the surrounding housing, the tapping should cause the bearing race to fall out of the housing. However, if the housing has any combustible material on it, using this technique can cause a fire.

Some bearings will be removed to gain access to a failed component and later reused. Do not use a hammer and punch for removing bearings. The shaft or housing can be damaged. The bearing could also break, causing flying debris.

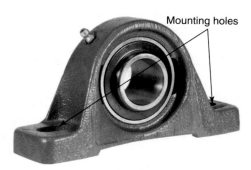

dcwcreations/Shutterstock.com

Figure 7-57. A pillow-block bearing is designed to be bolted to its mounting surface.

Installation

If a bearing is to be reused, it must be inspected and cleaned. Check for nicks, pits, cracks, and wear. To clean the bearings, suspend them in a container of solvent overnight. Rinse the bearing with clean solvent. Dry the bearing by letting it sit to dry. If compressed air is required, only use clean, dry compressed air and do not allow the compressed air to spin the bearing. Thoroughly lubricate the bearing and place it in a clean, sealed container.

Lubricate the race prior to installation. Inspect the shaft or housing bore for nicks, out-of-roundness, and to ensure it is smooth and clean. The side of the race with

Goodheart-Willcox Publisher

Figure 7-58. This puller is being used to remove a bearing. The puller's jaws are gripping under the gear, which acts as a driver to push the bearing off the shaft. The snap ring has already been removed.

the rounded corner should be slid onto the shaft first. If possible, avoid using a hammer to drive the bearing into position. If a press (hydraulic or arbor) cannot be used, do not use a hammer to directly drive a bearing race because it could fracture the bearing, dent it, bend it, or cause it to be installed unevenly.

Do not use a punch to drive a bearing because the race will not be installed evenly. Do not use a wooden driver because it can fracture or splinter and cause an injury. Use a driver that fits the bearing's race. Apply pressure only to press-fit bearing races, and never to slip fit races or bearing cages. Press the race onto the shaft or into the housing using steady pressure evenly applied across the race until it reaches the shoulder of the seat.

Service literature may specify to heat the bearing race to enable installation. Most shops have a dedicated oven for heating components. For example, the service literature may direct you to heat a bearing race in the oven until it reaches 250°F (121°C), and then slide the race into position. When the race cools, it forms an interference fit with the mating component. Always follow the service literature's procedures.

Warning

Always wear heavy leather gloves when handling any preheated parts.

Adjustment

Depending on the design, bearings, gears, and shafts may require some type of adjustment during the installation process. One adjustment is *endplay*, which is the distance a shaft or gear can move after bearing installation. In **Figure 7-59,** a dial indicator is being used to check the endplay of a gear after the tapered roller bearing has been installed.

Tapered roller bearings often require *preload*, which equals the amount of static force applied to the bearing rollers after they have been adjusted for zero endplay.

Backlash is the amount of clearance (play) between meshed gear teeth. Both backlash and preload will be further explained in Chapter 16, *Differentials*.

Contamination

During the installation process, the bearing must remain clean. Contamination is the number one cause of bearing failure. The bearing, lubricant, and oil sump must be clean to maximize the life of the bearing. Contaminants can bruise the races. This is caused when foreign particles or wear debris become wedged between the rollers and races and create fine indentations in the race.

Goodheart-Willcox Publisher

Figure 7-59. Endplay is often measured following the installation of a transmission shaft bearing. A—A dial indicator is placed so that its plunger is perpendicular to a gear on the shaft. The gear is pushed as far forward as possible, and the dial indicator is zeroed out. Next, a pry bar is used to push the gear as far rearward as possible as the endplay is read on the dial indicator's scale. B—The end of the shaft is supported with a tapered roller bearing. After the endplay has been measured, the bearing cap is removed. C—If the endplay is greater than specified, the bearing cap should be reinstalled with fewer or thinner shims between it and the transmission housing. If the measured endplay is less than the specification, additional shims or thicker shims should be installed. Shims are available in a variety of thicknesses.

Summary

- Gearing is used to increase or decrease speed, increase or decrease torque, provide direct drive, and reverse power flow.
- Gears can have straight cut teeth (spur gears), helical cut teeth (helical gears), or chevron-shaped teeth (herringbone gears). The teeth can be on the inside or outside of the gear.
- Planetary gear sets consist of a sun gear, planetary gears and carrier, and a ring gear.
- Rack-and-pinion sets convert rotary motion of the pinion gear into linear movement of the rack.
- Bevel gear sets consist of two gears with angled faces that are positioned perpendicular to each other. These gear sets change the direction of power flow by 90°.
- Worm gear sets consist of a screw-like worm gear and a spur gear, rack, or sector gear. They are used to change the direction of power flow.
- Gear ratios are determined by dividing the number of teeth on the driven gear by the number of teeth on the drive gear. Gear ratios larger than 1 result in an increase in torque and a decrease in speed. Gear ratios smaller than 1 are called overdrive ratios and they result in an increase in speed and decrease in torque. A gear ratio of 1 is referred to as direct drive and does not change speed or torque.
- Gear drives do not slip, do not require a tensioner, and can handle heavier loads than belts and chains. However, they are noisy and more expensive to manufacture.
- Sliding gear transmissions change gear ratios by using shift forks to move sliding gears into mesh with mating gears on a parallel shaft. These transmissions must be shifted only when stationary to prevent gear clash. They are used primarily in combination with hydrostatic drives.
- Collar shift transmissions consist of two sets of gears on parallel shafts. One set of gears is connected to its shaft; the other set of gears is not fixed to its shaft. The two sets of gears are in constant mesh. The gear ratio is set by using a shift fork to engage a collar into one of the free gears. The collar is splined to a shaft, and when it is coupled to a speed gear, locks that gear to the shaft. These transmissions are commonly used as a range transmission in conjunction with a synchronized transmission. To prevent gear clash, they must be shifted only when the machine is stationary.
- Synchronized transmissions use synchronizers to match transmission speed to engine speed before the input and output gears are meshed. This allows the transmission to be shifted while the machine is moving. Common types of synchronizers are block, fixed pin, split pin, and friction plate.
- Dual clutch transmissions have a single output shaft, two clutches, and two countershafts. Synchronizers engage a gear on each countershaft, but only one clutch is engaged at a time. Gear ratios are changed by disengaging a clutch while simultaneously engaging the other clutch. At the same time, a synchronizer on the disengaged shaft preselects the gear for the next anticipated shift.
- Shift forks move gears, synchronizers, or collars to engage and disengage gears in a transmission. In some transmissions, the forks are actuated by shift rails. In other transmissions, a cam plate is used to actuate the forks. Detent mechanisms are used to hold shift rails in the desired locations. Interlock mechanisms prevent two sets of gears from being engaged at the same time.
- Bearings support shafts and minimize friction resistance, heat, and wear. The two common categories of bearings are friction and antifriction. Common types of antifriction bearings include ball and roller. Roller bearings can be further classified as straight roller bearings, tapered roller bearings, needle bearings, spherical roller bearings, and thrust bearings.

Technical Terms

- amboid gear set
- angular load
- annulus gear
- antifriction bearing
- backlash
- ball bearing
- bearing
- bevel gears
- blocker ring
- block synchronizer
- bull-and-pinion
- bull gear
- cage
- cam shifter
- collar shift transmission
- cone
- cone synchronizer
- Conrad bearing
- cup
- cylindrical roller bearing
- detent mechanism
- DirectDrive (dual clutch) transmission
- endplay
- epicyclic gear set
- filling-notch bearing
- fixed pin synchronizer
- friction bearing
- helical gears
- herringbone gear
- hypoid gear set
- inserts
- interlock mechanism
- keys
- manual transmission
- needle roller bearing
- pillow-block bearing
- pinion gear
- pin synchronizer
- pitch circumference
- planetary gear set
- preload
- press fit bearing race
- race
- rack-and-pinion gear set
- radial load
- range transmission
- reverse idler
- ring-and-pinion gear set
- ring gear
- root
- sealed bearing
- shift collar
- shift fork
- shift gear
- shift rail
- sleeve
- sliding clutch
- sliding gear transmission
- slip fit race
- spherical roller bearing
- spiral bevel gear set
- split pin synchronizer
- spur gear
- straight roller bearing
- synchromesh transmission
- synchronized transmission
- synchronizer cone
- synchronizer cup
- synchronizer hub
- synchronizer rings
- tapered roller bearing
- thrust bearing
- thrust load
- worm drive gear set

Review Questions

Answer the following questions using the information provided in this chapter.

Know and Understand

1. Which type of gear is noisy and economical to manufacture?
   A. Helical.
   B. Hypoid.
   C. Spur.
   D. Worm.

2. A gear with 20 teeth is driving a gear with 20 teeth. What will be the result?
   A. Reduced speed.
   B. Increased speed.
   C. Increased torque.
   D. Direct drive.

3. A gear with 40 teeth is driving a gear with 20 teeth. What will be the result?
   A. Reduced speed.
   B. Increased speed.
   C. Increased torque.
   D. Direct drive.

4. The countershaft gears are constantly meshed with the output gears on all of the following transmission types, *EXCEPT*:
   A. collar shift.
   B. block synchronizer.
   C. plate synchronizer.
   D. sliding gear.

5. Which of the following is *not* a synchronized transmission?
   A. Collar shift.
   B. Sliding gear.
   C. Both A and B.
   D. Neither A nor B.

6. Which of the following is commonly used as a range transmission with hydrostatic transmissions?
   A. Block synchronizer transmission.
   B. Herringbone gear transmission.
   C. Friction plate synchronizer transmission.
   D. Sliding gear transmission.

7. Which of the following is an *incorrect* statement about agricultural collar shift transmissions?
   A. It has a shift fork.
   B. The output gears are constantly in mesh with the countershaft gears.
   C. It must be shifted while the machine is moving.
   D. It is used as a range transmission.

8. Which of the following is *not* a type of synchronizer?
   A. Block (cone).
   B. Friction plate.
   C. Pin.
   D. Sliding gear.

9. In a block synchronizer, the sleeve is designed to directly mesh with all of the following, EXCEPT:
   A. the blocker ring.
   B. the hub.
   C. the output shaft.
   D. a speed gear.

10. In a block synchronizer, three slots in the _____ hold the inserts (keys).
    A. hub
    B. output shaft
    C. sleeve
    D. speed gear

11. During a rebuild of a block synchronizer transmission, a feeler gauge is used to measure the _____ gap.
    A. blocker ring to speed gear
    B. blocker ring to sleeve
    C. blocker ring to hub
    D. blocker ring to output shaft

12. The John Deere DirectDrive transmission has what type of gear system inside its dual path eight-speed gear transmission module?
    A. Sliding gear.
    B. Planetary gear set.
    C. Shift collar.
    D. Synchronizers.

13. What is the purpose of a shift interlock?
    A. To hold two gears engaged at one time.
    B. To prevent two gears from engaging at one time.
    C. To hold a selected gear in position.
    D. To lock a synchronizer in gear.

14. Which type of bearing can have its clearances (both thrust and radial) adjusted during installation?
    A. Ball bearing.
    B. Spherical roller bearing.
    C. Straight roller bearing.
    D. Tapered roller bearing.

15. Which of the following is *not* used for bearing installation or removal?
    A. Jaw-type puller.
    B. Oven.
    C. Press.
    D. Torch.

16. Technician A states that gearing has the disadvantage of potential slippage during high loads. Technician B states that power transmitted through gears can be noisy. Who is correct?
    A. Technician A.
    B. Technician B.
    C. Both A and B.
    D. Neither A nor B.

Apply and Analyze

17. A large internal-toothed gear is known as a(n) _____ gear.
18. The bottom cavity between a gear's teeth is called a(n) _____.
19. A(n) _____ gear uses chevron-shaped teeth.
20. In a(n) _____ gear set, the pinion is positioned below the center axis of the ring gear.
21. In a(n) _____ gear set, the pinion is positioned above the center axis of the ring gear.
22. During a shift, the inserts in a block synchronizer directly press against the _____ ring.
23. _____ is the amount of clearance between meshed gear teeth.
24. The _____ bearing is best for high speeds.
25. The _____ roller bearing is used for heavy loads and all three types of loads (thrust, angular, and radial).
26. The double _____ roller bearing is known as a self-aligning bearing.
27. A(n) _____ load is exerted parallel to the shaft.

Critical Thinking

28. An input gear has 9 teeth and the output gear has 41 teeth. What is the gear ratio?
29. An input gear has 41 teeth and the output gear has 9 teeth. What is the gear ratio?

Chapter 8
Clutches and Planetary Controls

Objectives

After studying this chapter, you will be able to:
- ✓ Describe the different types of single- and dual-disc flywheel clutches.
- ✓ Explain flywheel clutch service and adjustment procedures.
- ✓ Explain the different types of multiple-disc clutches.
- ✓ Explain the operation of a one-way clutch.
- ✓ Explain the operation of electromagnetic clutches and brakes.
- ✓ Describe the operation of bands.
- ✓ Explain the purpose of a slip clutch.
- ✓ List applications that use centrifugal clutches.
- ✓ Identify mechanical schematic symbols.

Heavy equipment uses clutches of various types to engage and disengage power. Some examples of clutches include:

- Traction clutches—connect and disconnect power to the transmission.
- PTO clutches—connect and disconnect power to the PTO shaft.
- Bands or brakes—bring a rotating shaft/drum to a stop or hold it stationary.
- One-way clutches—allow a mechanical drive to deliver power in one direction and over-run in the opposite direction.

The most popular type of clutch is the *friction clutch*. In a friction clutch, friction material bound to a band or plate grips a mating drum or steel disc. Some friction clutches are designed to run dry. Other friction clutches, known as wet clutches, are cooled and lubricated with oil.

Traction Clutches

A *traction clutch,* sometimes called a *transmission clutch* or a *flywheel clutch*, is driven by the engine's flywheel and is used to disconnect engine power from the rest of the drive train, allowing the operator to shift the transmission. As explained in Chapter 7, *Manual Transmissions*, an operator must depress the clutch pedal to shift gears in a manual transmission. When the clutch pedal is depressed, the traction clutch releases. When the clutch is released, engine torque is disconnected from the transmission's input shaft, allowing the operator to shift gears. Older construction and agricultural equipment, as well as some newer, low-horsepower tractors, are equipped with a single-disc or dual-disc traction clutch.

Note

Traction clutches are often mounted directly to the engine's flywheel, which explains why they are sometimes called flywheel clutches. Some machines have both a PTO clutch and a traction clutch mounted directly to the engine's flywheel, even though the two clutches serve different purposes. As explained in Chapter 4, *Agricultural Equipment Identification*, the term *flywheel* can also be used to refer to the large PTO flywheel found on an agricultural hay baler. A slip clutch is used to protect the baler's PTO drive system, and this clutch can be called a flywheel slip clutch.

Single-Disc Traction Clutch

In a single-disc traction clutch application, the flywheel is bolted to the engine's crankshaft. A pressure plate is normally bolted to the flywheel and rotates with the flywheel. The ***clutch disc*** is a steel disc that is lined with friction material. It is located between the pressure plate and the flywheel. See **Figure 8-1**. When the clutch is engaged (clutch pedal released), the pressure plate pushes the clutch disc firmly against the flywheel. This causes the clutch disc to rotate with the pressure plate and flywheel. Because the clutch disc is splined to the clutch shaft, the clutch shaft turns when the clutch is engaged. The clutch shaft often also serves as the transmission input shaft. The clutch disc may or may not have ***dampening springs***, which help prevent vibrations produced by engine pulsations from being transferred to the rest of the driveline.

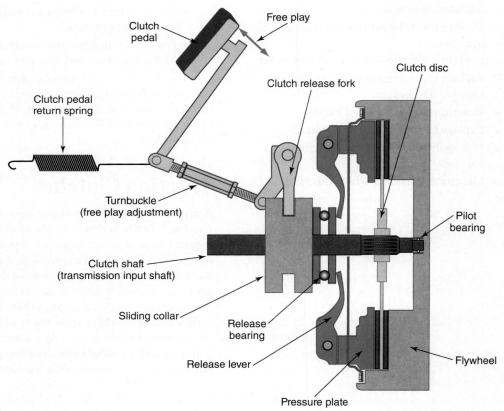

Goodheart-Willcox Publisher

Figure 8-1. Low-horsepower equipment and older heavy equipment use manually operated flywheel clutches consisting of a clutch pedal, linkage, release fork, release bearing, pressure plate, clutch disc, and flywheel. This drawing depicts a simple single-disc flywheel clutch.

Years ago, the clutch discs used friction materials that contained **asbestos**, which fared well against heat but is no longer used because it is a carcinogen, or cancer-causing material.

Caution
When working on older machines that contain asbestos-lined clutches and brakes, special care must be taken to avoid breathing in the clutch and brake dust.

Today, the friction material is made of safer substances, such as synthetic (man-made) fibers, bronze, or ceramic. The friction material can be woven, sintered, or molded. Although a single-disc clutch can be a wet clutch, it is more likely to be a dry clutch.

A **clutch release mechanism** is used to actuate the pressure plate, causing the clutch to release. The release mechanism often consists of a release fork and a sliding collar, which can be actuated mechanically or hydraulically by means of a clutch pedal. The sliding collar can be designed to press into the pressure plate to release the clutch, or it can be designed to pull away from the pressure plate to release the clutch.

Clutches in which the sliding collar pulls away from the pressure plate to release the clutch are called pull-type clutches. Pull-type clutches are used in medium- and heavy-duty on-highway trucks. As the clutch pedal is depressed, the release mechanism and release bearing move rearward (toward the transmission). The release bearing is connected to an internal sleeve retainer assembly, which causes the clutch levers to pivot, compressing the pressure plate springs and allowing the two clutch discs to separate from the flywheel, pressure plate, and intermediate plate. An intermediate plate is a steel plate with a smooth machined surface on both sides. It is placed in between two friction discs, providing additional surface area for the friction discs to mate against.

A **release bearing**, also known as a **throw out bearing**, is incorporated into the release mechanism and is used to transmit the force from the sliding collar to the pressure plate's release levers (or fingers) to release the clutch. As the release fork actuates, the sliding collar moves linearly and applies pressure to the release bearing. Because the release levers (fingers) are part of the rotating pressure plate that is spinning at engine speed, the release bearing is needed to transmit the linear force of the release fork/sliding collar to the spinning release levers. This causes the pressure plate to release its grip on the clutch disc, disconnecting engine power from the transmission. With the clutch disengaged, the operator can shift gears and then release the clutch pedal, causing the traction clutch to re-engage.

A **pilot bearing** is located in the center of the flywheel or crankshaft, and is used to support the end of the clutch shaft or the transmission input shaft.

Dual-Disc Traction Clutch

A dual-disc traction clutch used in old Caterpillar D5B track-type tractors is shown in **Figure 8-2**. Note that the flywheel surrounds the clutch assembly. The flywheel's internal splines align with the drive plate's external splines, causing the drive plate to always rotate at engine speed. Two driven discs, which are lined on both sides with friction material, are splined to the clutch hub. When the clutch is engaged, a cam link/roller assembly forces the loading plate against the pressure plate. This causes the two driven discs to drive the hub, which in turn drives the clutch shaft. The cam link/roller holds force on the loading plate through an over-center design.

The clutch is engaged when the cam link/roller assembly snaps past a vertical centered position, causing the clutch assembly to hold the clutch discs engaged.

In the D5B example, moving the sliding collar away from the pressure plate causes the cam link/roller to release the loading plate. The sliding collar is actuated by the operator through a clutch control lever. As the operator pushes the clutch control lever forward, the lever pulls the sliding collar away from the clutch, causing the cam link/roller

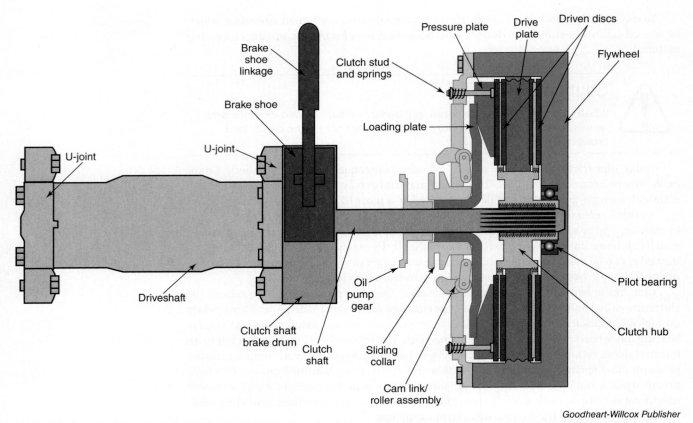

Figure 8-2. In this manually operated flywheel clutch, the flywheel drives the drive plate. When the clutch is engaged, the loading plate and pressure plate squeeze the driven discs. This causes the driven discs to drive the clutch hub and clutch shaft.

assembly to release the clutch. If the lever is pushed fully forward, a brake shoe will hold the clutch shaft's drum stationary. The clutch shaft drives the transmission's input shaft through means of a universal joint. When the clutch brake holds the clutch shaft, the operator can shift the transmission's sliding gears into mesh. As mentioned in Chapter 7, *Manual Transmissions*, sliding gear transmissions are shifted while the tractor is stationary. Without a clutch brake, the clutch shaft would continue to spin due to inertia, while the transmission's output shaft would be stationary. The clutch brake brings the clutch shaft to a stop, allowing the sliding gears to mesh without clashing.

 Note

Some machines use a slight variation of this application, in which the flywheel drives a single friction disc that is lined on both sides with friction material and is always rotating at engine speed. The friction drive disc mates with two steel driven plates. When the clutch is engaged, the steel driven plates will drive the clutch hub, which drives the clutch shaft.

PTO and Traction Clutch Assemblies

As explained in Chapter 4, *Agricultural Equipment Identification*, agricultural tractors with continuous PTO configurations have one clutch pedal to control two separate clutches, the PTO clutch and the traction clutch. Pressing the clutch pedal halfway down disengages the traction clutch. Fully depressing the clutch pedal disengages both the PTO clutch and the traction clutch.

In the simplified clutch shown in **Figure 8-3**, both the PTO clutch and the traction clutch are located in a flywheel clutch assembly. In this application, one clutch release collar is used to actuate both clutches.

As the clutch pedal is pressed down halfway, the release bearing presses on the levers, which pull the traction clutch's bolts. The bolts pull the traction clutch plate away from the clutch disc, releasing the traction clutch.

As the operator presses the clutch pedal all the way down, the levers push the PTO clutch pins. The pins push the PTO clutch plate, releasing the PTO clutch.

Note

Continuous PTO clutches are also known as dual-stage PTO clutches or two-stage PTO clutches.

An independent PTO clutch can be located in a flywheel clutch assembly. As mentioned in Chapter 4, *Agricultural Equipment Identification*, an independent PTO clutch is engaged and released independently (separately) from the traction clutch. When an independent PTO clutch and the traction clutch are both housed inside a flywheel clutch assembly, they will be controlled independently.

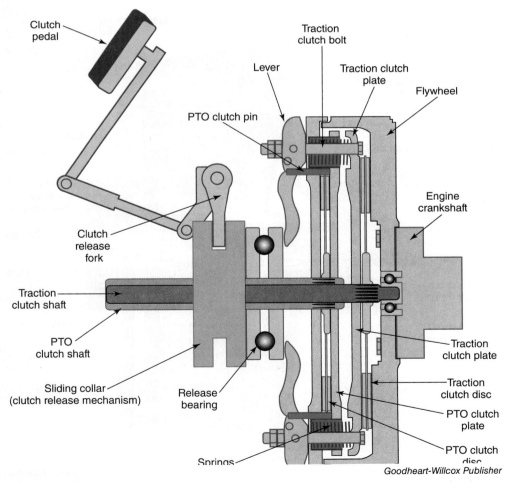

Goodheart-Willcox Publisher

Figure 8-3. This continuous PTO and flywheel clutch uses one clutch release mechanism to disengage both the traction clutch and the PTO clutch.

Figure 8-4. This John Deere PermaClutch assembly has two sets of operating levers. When the levers are pushed by a hydraulic piston, the clutch will engage.

PermaClutch

John Deere's *PermaClutch* is an oil cooled PTO and traction clutch assembly that uses two sets of operating levers actuated by two hydraulic pistons. See **Figure 8-4**. Each clutch is engaged when its hydraulic piston presses against the operating levers. The *operating levers* squeeze the clutch plates and discs. This causes the discs to drive the hub, which is splined to its corresponding shaft. Note that the levers in the PermaClutch are called operating levers rather than release levers because when actuated, they engage the clutch rather than release it.

As shown in **Figure 8-5**, the clutch plates are positioned alternately between the friction discs. The clutch plates are driven by the flywheel at engine speed. When the operating levers are hydraulically actuated, they squeeze the clutch, causing the clutch plates and friction discs to rotate at the same speed. The friction discs are mated to their corresponding hub, which is splined to the clutch shaft. Clutch lubrication oil flows between the traction clutch shaft and the PTO clutch shaft, and is routed through the PTO clutch hub. Centrifugal force directs oil from the hub to the clutches for cooling and lubrication.

A few examples of John Deere tractors that used a PermaClutch are old 30, 40, 50, 55, and 60 series tractors, such as the 4030, 4050, 4450, 4650, and 4850 agricultural tractors. See **Figure 8-6**. The PermaClutch was used in the tractors with a Quad Range transmission. The 50, 55, and 60 series tractors could

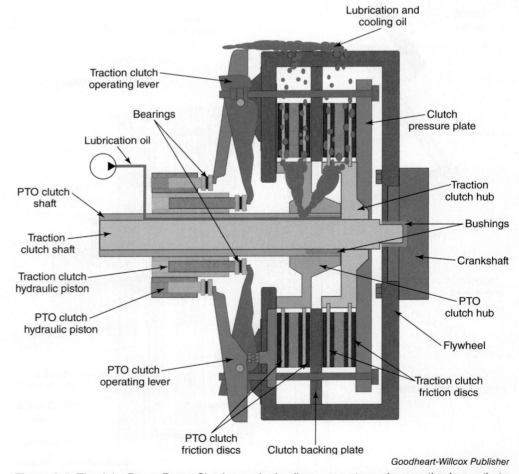

Figure 8-5. The John Deere PermaClutch uses hydraulic pressure to apply operating levers that engage the traction and PTO clutches independently. The clutch is oil cooled and lubricated.

Chapter 8 | Clutches and Planetary Controls

Figure 8-6. This figure shows a John Deere PermaClutch in various stages of disassembly. This particular clutch is used in the older 30 series John Deere tractors, such as 4430 and 4630 tractors. A—The cover plate has been removed, exposing the PTO pressure plate. B—The PTO pressure plate has been removed, exposing the PTO clutch disc and return springs. C—The PTO clutch disc has been removed, exposing the backing plate that separates the flywheel clutch and PTO clutch. D—The backing plate has been removed, exposing the traction clutch. Note that the PTO and traction clutch discs are different. Also note that the traction clutch disc has 22 holes to allow oil to escape, preventing traction clutch slippage.

be equipped with either the Quad Range or a Powershift transmission. The tractors with the Powershift transmissions used a different style of clutch that consisted of two separate traction clutches located in the flywheel clutch housing. These clutches were responsible for driving two different transmission input shafts.

Traction Clutch Service and Adjustment

Traction clutches require service and adjustment. The most common adjustments include clutch pedal free play adjustment and release lever or operating lever adjustment.

Clutch Pedal Free Play Adjustment

Mechanically operated clutches commonly require clutch pedal *free play adjustment*. This adjustment allows the clutch pedal to move a short distance (for example, 1″) before the release mechanism begins to actuate the pressure plate. If all free play is removed during adjustment, the clutch will slip and wear faster.

An example of a lever-operated clutch is illustrated in **Figure 8-7**. The clutch pedal free play is adjusted by loosening the jam nuts on each side of the turnbuckle. An eyelet with right-hand threads is inserted into one end of the turnbuckle, and an eyelet with left-hand threads is inserted in the opposite end of the turnbuckle. Turning the turnbuckle in one direction shortens the linkage, and turning the turnbuckle in the opposite direction lengthens the linkage. The turnbuckle should be turned until the pedal can be pressed to the free play specification before the release mechanism begins to actuate the pressure plate. The remaining pedal travel requires more force, as the remaining pedal travel overcomes the force required to disengage the clutch.

Some clutches use an adjustable cable in place of a mechanical linkage. An adjustable cable will have a jam nut that locks against an adjuster knob, wheel, or turnbuckle that is used to vary the cable's length and thereby adjust the clutch pedal's free play.

Release or Operating Lever Adjustment

Each clutch application has its own release or operating lever adjustment procedure. Be sure to follow the specific instructions found in the manufacturer's service information for the unit at hand.

The John Deere PermaClutch requires the use of a special tool when adjusting its operating levers. Note that the service information states that the clutch is designed for a long life and that it should not be disassembled for inspection. When the clutch discs are worn, the operating levers bottom out, resulting in the clutch slipping. Therefore, clutch inspection is unnecessary.

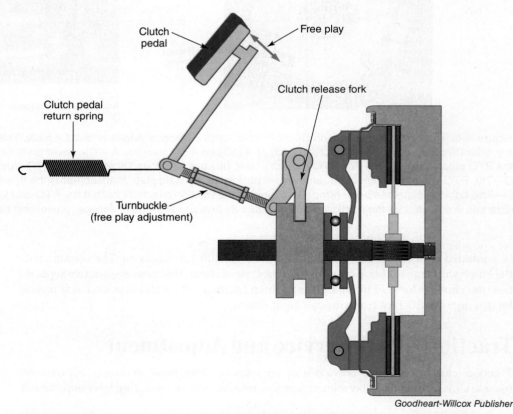

Goodheart-Willcox Publisher

Figure 8-7. Mechanically operated flywheel clutches often incorporate a turnbuckle that can be used to shorten or lengthen the clutch pedal linkage, depending on the direction the turnbuckle is rotated. The turnbuckle allows a technician to adjust free play, which is the distance the clutch pedal travels before pressure is applied to the clutch release mechanism.

If the clutch is disassembled, special tools must be used to adjust the height of the operating levers. The following sequence is an example of the steps required to adjust one style of John Deere PermaClutch on 4430 and 4630 tractors.

1. Label the three PTO clutch operating levers and the three traction clutch operating levers, indicating their position on the clutch.
2. Remove the PTO clutch operating levers by removing the retaining clips and pins. See **Figure 8-8**.
3. Loosen both the lock nuts and the adjusting nuts on all three traction clutch operating levers.
4. Loosen the jam nuts on all three PTO clutch operating lever adjusting screws and turn the screws inward, loosening the PTO adjustment. See **Figure 8-9**.
5. Remove the cover cap screw located next to the part number, as well as the cover cap screw located 180° from the part number.
6. Assemble the clutch adjustment tool. The adjustment tool includes selective rings of different thicknesses that must be installed on the tool. The rings used will depend on the model of the tractor being serviced. See **Figure 8-10**.
7. Install the two adjusting screw studs into the clutch cover holes that contained the cover cap screws removed in Step 5. Place the clutch adjustment tool onto the clutch and position the cross bar on top of the adjusting screw studs, allowing the long adjusting screw of the gauge tool to protrude through the cross bar's hole. See **Figure 8-11**.
8. Torque the cross bar adjustment screw in the center of the cross bar tool to 20 ft-lb.
9. Load the PTO pressure plate by applying 30 in-lb of torque to the three cap screws on the clutch adjustment tool.

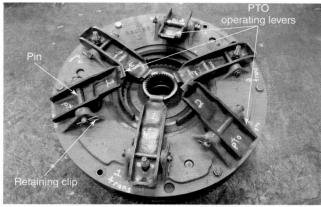

Figure 8-8. After labeling each operating lever and its position on the clutch, remove each PTO operating lever by first removing the lever's retaining clip and pin.

Figure 8-9. After removing the PTO operating levers, loosen the traction clutch operating lever adjusting screw lock nuts and adjusting nuts.

Figure 8-10. The selective rings must be fastened to the clutch adjustment tool before the tool is installed on the clutch. The service literature specifies the ring number based on the model of tractor.

Figure 8-11. Install the two adjusting screw studs, one in the hole next to the part number on the cover and the other diagonally opposed. The long adjusting screw of the clutch adjustment tool should extend through the cross bar's hole.

Figure 8-12. The PTO screw has been adjusted until the lever touches the adjusting tool.

Figure 8-13. The traction operating levers must be adjusted until they contact the gauge tool.

10. Place one PTO clutch operating lever on the clutch and insert the cross pin (pivot pin). Adjust the PTO adjusting screw by hand until the lever touches the gauge tool. See **Figure 8-12**. Make sure the lever is still clearly marked and then remove it so the lock nut can be tightened while ensuring that the adjusting screw remains stationary. The PTO clutch operating lever must be reinstalled and the clearance rechecked because the torque applied to the lock nut will cause the screw to stretch. The clearance between the lever and the gauge tool must be less than 0.010″ after the lock nut is tightened. If clearance is too large, the lever must be readjusted. Repeat this process for the remaining two PTO operating levers.
11. Unload the PTO clutch by backing off the three cap screws on top of the clutch tool.
12. Ensure the three PTO clutch operating levers are properly marked (labeled) and then remove them.
13. The traction clutch adjustment bolt heads fit inside a slot in the pressure plate to keep the bolts from spinning. Ensure the bolts remain in position and then adjust each lever by turning its adjusting nut by hand until the lever contacts the gauge tool. See **Figure 8-13**.
14. Torque the adjusting nuts to 70 in-lb and then re-torque them to 90 in-lb.
15. Torque the lock nuts.
16. Remove the gauge tool and reinstall the two cover cap screws. Tighten the screws to 35 ft-lb of torque.
17. Reinstall the PTO clutch operating levers.

Removal and Installation of a PermaClutch

Any time a PermaClutch assembly is installed or removed from a tractor, wedges must be placed between the operating levers and the clutch cover. The wedges place pressure on the levers, which hold the clutch in place. Without the wedges, the traction disc would fall off its hub.

Removal and Installation of a Conventional Single-Disc Traction Clutch

Removal and installation of a conventional single-disc traction clutch requires the use of a clutch alignment tool, which is a replica of the transmission input shaft. The alignment tool can be made of plastic or steel. It sometimes comes in a new clutch parts kit, which also includes the pressure plate, clutch disc, pilot bearing, and release bearing.

When the clutch pressure plate is installed onto the flywheel, the clutch disc is squeezed between the pressure plate and the engine's flywheel. If the pressure plate is being removed, the clutch disc will slide out of position as the pressure plate bolts are loosened, unless it is supported. The alignment tool is inserted through the center of the clutch disc and into the center of the pilot bearing. The alignment tool supports the clutch disc, allowing the technician to remove the pressure plate while keeping the clutch disc in position. The tool is also used to center the clutch disc during installation, which allows the transmission's input shaft to be fully inserted into the clutch.

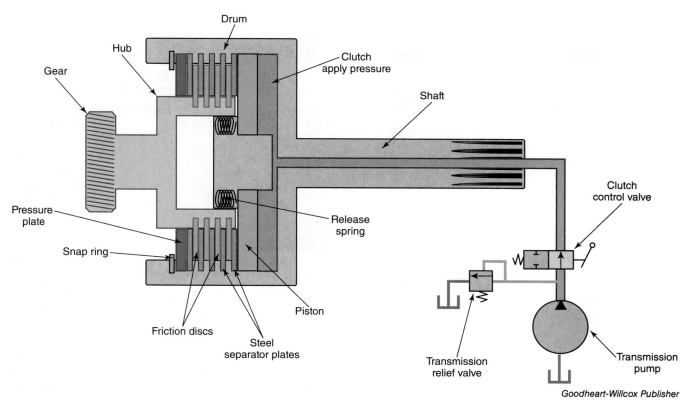

Figure 8-14. In this multiple-disc clutch, the transmission shaft is part of the clutch pack drum, which is splined to the steel separator plates. The hub is part of the gear, and the friction discs are splined to the hub.

Multiple-Disc Clutches

Many PTO and traction clutches use a different clutch design known as a multiple-disc clutch, unlike the clutches that are mounted to the flywheel and contain one or two large discs. The *multiple-disc clutch* is the most popular style of clutch used in heavy equipment power trains. It contains a set of friction discs positioned alternately between steel separator plates. A simple multiple-disc clutch is shown in **Figure 8-14**.

The *friction discs* can have internal or external splines. See **Figure 8-15**. Most multiple-disc clutches use friction discs with internal splines that engage matching splines in a *hub*, **Figure 8-16**. As a result, the hub and friction discs rotate as a unit. The hub is normally attached to a shaft or gear.

Manufacturers use different types of friction material bonded to the friction disc. Common friction materials include graphite, ceramic, bronze, elastomers, and paper fibers. The friction material can have a variety of groove patterns cut into it, but it is rarely smooth. See **Figure 8-17**. The grooves help cool the clutch and provide a passage for oil in wet clutch applications.

Some separator plates have internal splines and mate with the hub. However, most steel *separator plates* have external splines that are secured in a drum as shown in **Figure 8-18** or in the transmission case if the clutch is a holding clutch. The *drum*, also known as a *cylinder*, *basket*, or *carrier*, surrounds the multiple-disc clutch and rotates at the same speed as the separator plates, **Figure 8-16**.

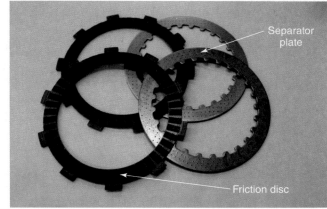

Figure 8-15. Some multiple-disc clutches use friction discs with external splines and steel separator plates with internal splines.

Figure 8-16. This is a multiple-disc input clutch from a Case IH Steiger Synchroshift transmission. It is spring released and oil applied. The friction discs align with the hub, which is part of the input gear. The separator plates align with the clutch drum, which is driven at a speed proportional to engine speed. Note one of the clutch piston return springs was intentionally removed.

Multiple-disc clutches are used as holding clutches, input clutches, or coupling clutches. ***Holding clutches***, also known as ***brakes*** or ***stationary clutches***, hold a planetary gear or shaft stationary. When the clutch is used to hold a component, a ***housing***, such as a transmission housing or a brake housing, is used in place of a drum. The multiple-disc clutch shown in **Figure 8-18** is an example of a Caterpillar holding clutch. ***Input clutches*** provide an input to a transmission. The input clutch causes a transmission component, such as a shaft, gear, or hub, to rotate. See **Figure 8-16**. Allison and Caterpillar call the input clutches used in their planetary transmissions ***rotating clutches***. Caterpillar rotating clutches can look similar to a holding clutch when the clutch housing is designed to rotate. See **Figure 8-19**. Not all rotating clutches are input clutches. For example, a ***coupling clutch*** couples two components. The first speed clutch in a Caterpillar D6R dozer attaches a planetary ring gear to the transmission's output shaft, which is explained in Chapter 9, *Planetary Gear Set Theory*.

Figure 8-17. These friction discs are both used in a Caterpillar planetary transmission. The clutch disc on the left has fiber clutch material with a crosshatch pattern. The clutch disc on the right has bronze friction material with a spiradial pattern.

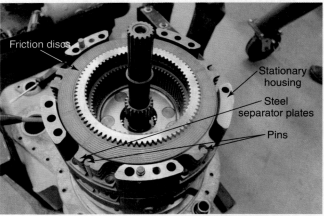

Figure 8-18. This clutch is from a Caterpillar planetary transmission. When the transmission case (not pictured) is lowered on top of the clutch pack, long bolts hold the housing stationary to the transmission case, resulting in a stationary clutch.

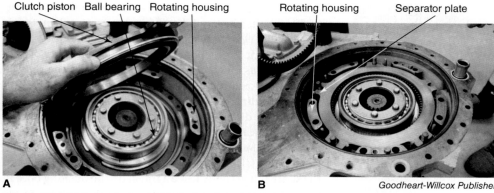

Figure 8-19. This Caterpillar planetary transmission has one rotating clutch at the bottom of the transmission. A—Notice the housing is designed to rotate on a ball bearing. B—The separator plate and housing look similar to those found in a stationary clutch, but the housing is not bolted to the transmission case. Instead, it rotates on the bearing.

Multiple-Disc Clutch Operation

The multiple-disc clutch can be engaged using a mechanical lever, hydraulic fluid pressure, or spring pressure. As the clutch is engaged, the plates and discs (known collectively as the clutch pack) are compressed against one another, coupling the drum and hub together so that they are both rotating at the same speed (clutch application) or both are held stationary (brake application).

Notice that the hub of the multiple-disc clutch shown in **Figure 8-14** is attached to a gear and the drum is part of the shaft. In terms of input, either the drum or the hub can be the input to the clutch, with the other member being the output member.

Multiple disc clutches offer the advantage of lots of potential surface area. The clutch engagement can be varied, and offers engineers future flexibility of adding or removing clutch discs for a given clutch size by resizing the other clutch components. One negative is that they require an outside mechanism like a lever or oil pressure to either apply or release the clutch.

Methods of Clutch Engagement

The two most common methods for compressing a multiple-disc clutch are oil pressure and spring pressure. Both are frequently used in heavy equipment power trains. Multiple-disc clutches can also be engaged manually through a lever mechanism.

 Note
Regardless of whether oil pressure is used to engage or release the clutch, the multiple-disc clutch normally uses oil for cooling and lubrication.

Oil-Applied Clutch

An ***oil-applied clutch*** relies on oil pressure to engage the clutch. It is the most common type of multiple-disc clutch used in heavy equipment power trains. Refer back to **Figure 8-14**. Oil is sent from either a spool valve or a solenoid through shaft passageways to apply the clutch piston.

The shaft normally has two types of passages, lube passages and pressure passages. The ***lube passages*** supply oil to lubricate the clutch discs. The ***pressure passages*** supply fluid to apply the pistons. The shaft in **Figure 8-20** has four drilled passages. The two pressure passages have steel check balls pressed into their bores to prevent pressure from leaking. The passages are precisely machined so the check balls can be properly pressed into the shaft with an interference fit. If the check balls leak or fall out of their bores, clutch pressure will drop and the clutch will slip, burn, and fail. If a check ball is pressed too

Figure 8-20. This shaft has four drilled passages. The two pressure passages have check balls pressed into their bores to allow pressure to build in the passages. This enables the pressure to apply the clutch piston. The two lube pressure passages do not have check balls.

Figure 8-21. This shaft has two pressure passageways and one lube pressure passage. Three square-cut plastic seals are located on the shaft. The seals are not a solid ring. Instead, they are cut in one spot allowing them to be spread apart to aid installation. The top two rings have been removed, leaving behind empty grooves.

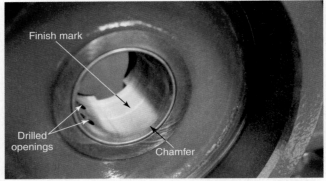

Figure 8-22. The plastic seal has left behind a finish mark in the bore of the bushing. The finish mark illustrates that the seal is precisely located between the bushing's two drilled openings. Note that the bushing has a slight chamfer, which helps during the installation of the countershaft into the bore to prevent damage to the seals.

Figure 8-23. Pistons are often made of cast aluminum. This clutch failed and was disassembled to analyze the damage.

far into the passage, it can prevent the fluid from entering the shaft. The two lube pressure passages do not have check balls. This shaft supplies lube and applies pressure to two modulated directional clutches, and therefore, it provides twice as much oil flow for lubrication as the shaft in **Figure 8-21**, which only has one lubrication passageway. A modulated directional clutch has the pressure varied during machine directional shifts, which eases the engagement of the clutch and prevents a sudden lurch in power flow. Too much modulation will cause the clutch to slip and overheat.

The plastic seal rings used to prevent leaking around the shaft can actually leave a finish on the bushing bore in which they ride, leaving evidence of their positions in the bore. A smooth finish causes no harm, but a grooved or scored finish can cause a fluid leak. See **Figure 8-22**.

The piston in **Figure 8-23** is made of aluminum. As fluid is fed into the clutch passage, **Figure 8-24**, it enters the clutch drum and forces the piston against the separator plates and friction discs, squeezing the plates and discs together and causing the drum to couple with the hub.

Rubber seals are used to prevent the fluid from leaking between the piston and the drum. Piston seals can be fitted to the piston, to the drum, or to both the drum and the piston. Piston seals vary in design. A few examples of seal designs are D-shaped, square-cut, O-ring type, and lip seal. See **Figure 8-25**.

Oil-applied clutches use one or more *return springs* to hold the piston in a released position when the clutch is disengaged (oil drained). See **Figure 8-26**. The return spring can consist of several small coil springs, one large coil spring, or Belleville washer(s). The *Belleville washers* illustrated in **Figure 8-27** have a cone shape. They are placed on top of each other with the large diameter of one washer placed against the large diameter of the next washer. The small diameter portion of the washer is placed against the small diameter of the next washer.

Figure 8-24. The screwdriver is pointing to the fluid passage in the bottom of the clutch drum. When the clutch is engaged, oil is sent through this passage to the bottom side of the piston, causing the piston to apply the clutch. The piston has been removed from the drum in this photo.

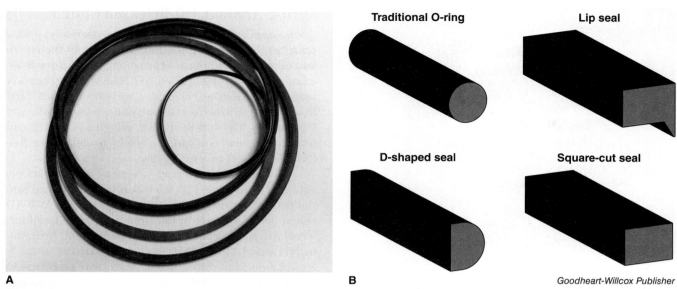

Figure 8-25. Piston seals vary in design. A—A variety of piston seal shapes and sizes are shown here. B—This drawing includes four common piston seal cross-sectional shapes.

This arrangement allows the washers to compress (flatten out) when the piston is hydraulically applied during clutch engagement.

The oil-applied clutch has a ***disengaged clutch clearance***, which ensures the clutch is fully released. The disengaged clutch clearance, also known as *piston free travel*, equals the distance between the pressure plate and the snap ring. It also equals the distance the piston must travel before it reaches the clutch pack. The disengaged clutch clearance can be checked using a variety of tools, including a depth micrometer, feeler gauge, or dial indicator. Adjusting disengaged clutch clearance will be explained in Chapter 11, *Powershift and Automatic Transmission Controls, Service, and Repair*.

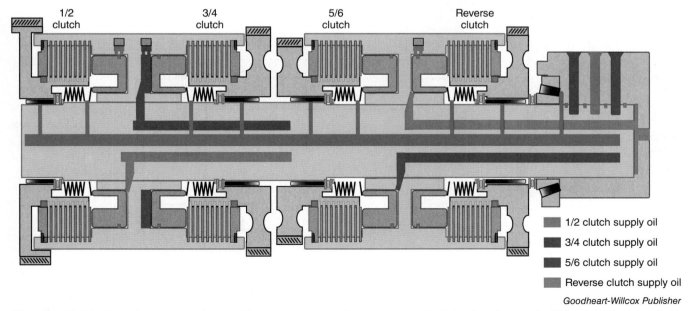

Figure 8-26. Multiple-disc clutches have drilled passageways for applying and lubricating the clutch. This simplified drawing shows a Case IH Magnum transmission, which will be explained in Chapter 10, *Powershift and Automatic Transmission Theory*. Notice the passageways for applying the clutches.

Figure 8-27. Belleville washers are cone-shaped springs used in heavy equipment clutches. They are commonly used to hold the piston in a retracted position. Eight Belleville washers are shown here. A snap ring retainer is illustrated on top of the Belleville washers. When the Belleville washers are compressed, the retainer provides a flat ledge for the snap ring to butt up against.

Spring-Applied, Oil-Released Clutch

Spring-applied, oil-released clutches, or *normally applied clutches*, use a spring mechanism to apply the clutch and oil pressure to release the clutch. See **Figure 8-28**. Examples of spring-applied, oil-released clutches are mechanical front-wheel drive (MFWD) clutches used in agricultural tractor transmissions, park brakes, and PTO brakes. The spring-applied, oil-released clutch can use several small coil springs, one large coil spring, Belleville washer(s), or a spring disc.

Agricultural row crop tractors are commonly equipped with a MFWD clutch that is incorporated into the transmission housing. When engaged, the MFWD clutch delivers power to the front axle, providing four-wheel drive. The clutch is spring applied and hydraulically released. The MFWD clutch defaults to the normally engaged position. Note that it takes two conditions to release the clutch: voltage to energize the solenoid and oil pressure to release the clutch. With the loss of voltage or oil pressure, the MFWD will default to the engaged position and the tractor will remain in four-wheel drive. See **Figure 8-29**.

Spring-Applied, Manually Released Clutch

John Deere 4200, 4300, and 4400 compact utility tractors can have one of three types of transmissions: a collar

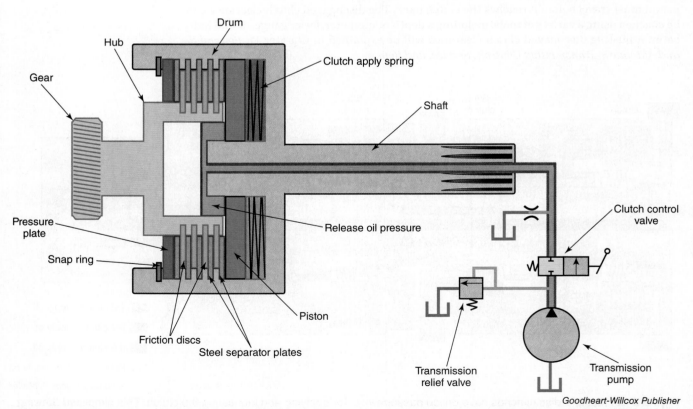

Figure 8-28. The oil-applied clutch shown in Figure 8-21 can be changed to a normally applied clutch by moving the spring to the opposite side of the clutch piston.

Chapter 8 | Clutches and Planetary Controls

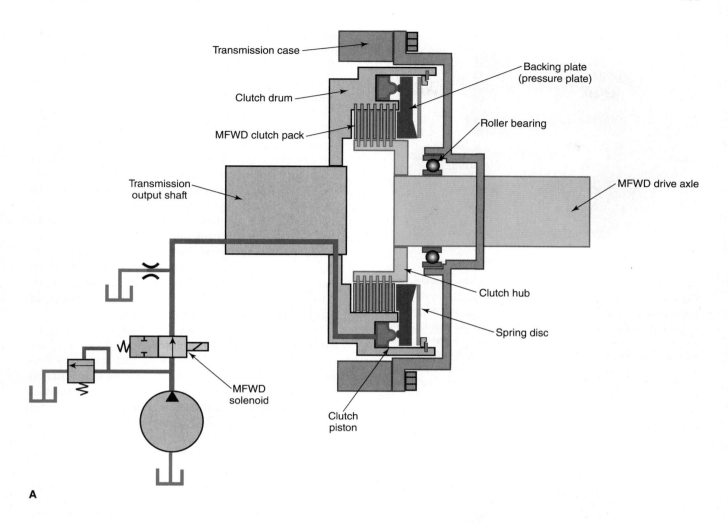

Figure 8-29. This is the MFWD clutch used in the 16-speed power shift transmission found in 8000 series John Deere tractors. A—A solenoid is used to deliver oil pressure to the clutch piston in order to release the MFWD clutch. B—The clutch is bolted to the front side of the Deere 16-speed power shift transmission.

shift, a SyncReverser, or a hydrostatic transmission. Two of the tractor types use one or two spring-applied, manually released clutches. The collar shift–equipped tractors have one spring-applied, manually released clutch. These tractors use a reverse gear range in the transmission to reverse the tractor's propulsion. See **Figure 8-30**.

A SyncReverser-equipped tractor has two spring-applied, manually released clutches—a forward clutch and a reverse clutch. The two clutches are responsible for rotating two clutch shafts, one in a clockwise direction and the other in a counterclockwise direction. When the tractor is moving, one clutch is engaged and the other is released. The clutches are oil lubricated through an internal passage. The back side of each input clutch has a drive gear that is driven by the transmission's input shaft. As a result, the traction clutch drum(s) rotate at speeds that are proportional to engine speed. See **Figure 8-31**. In this application, the clutch drum (basket) is the input member. The hub is the output member.

The clutch shown in **Figure 8-32** has been disassembled. To reassemble the clutch, place the clutch pack on top of the piston. Then, place the hub inside the center of the clutch pack, aligning it with the four posts on the piston. Finally, bolt the spring plate to the piston and place the assembled clutch pack inside the clutch drum (basket). The snubber plate is placed on top of the spring plate. The lifting pin is then inserted through the snubber plate and the bore of the bearing that is located inside the spring plate.

Note that in this application, the friction discs have external splines and are mated to the clutch drum. The separator plates have internal splines and are mated to the clutch hub. The clutch hub drives the clutch shaft, also known as the pinion shaft. Spring tension causes the piston to compress the clutch pack together. When the operator presses the clutch pedal down, a lift lever pushes a lifting pin, which pushes a snubber plate. The snubber plate then pushes the piston into the springs, causing the clutch to release.

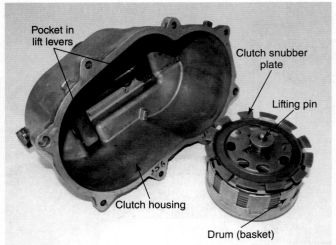

Goodheart-Willcox Publisher

Figure 8-30. John Deere 4200, 4300, and 4400 collar shift transmissions use a single multiple-disc traction clutch. SyncReverser™ transmissions have two multiple-disc clutches. The clutches are spring applied and manually released. Notice that this SyncReverser clutch housing has two lift levers that are used to release two input clutches. Only one clutch assembly is shown in the image. As the operator presses the clutch pedal, the lift lever actuates the lifting pin, which presses against the snubber plate. The snubber plate compresses the spring plate and springs, causing the clutch to release.

Goodheart-Willcox Publisher

Figure 8-31. This clutch receives its input from the input gear on the back of the drum (basket). The friction discs are splined to the drum. The steel plates are splined to the hub.

Note

Multiple-disc clutch service and repair, including piston free travel adjustment, is detailed in Chapter 11, *Powershift and Automatic Transmission Controls, Service, and Repair.*

One-Way Clutches

Mobile machinery can be equipped with a ***one-way clutch***, or ***overrunning clutch***. The one-way clutch is designed to lock when turned in one direction and unlock, or overrun, when turned in the opposite direction. Torque converter stator clutches, which will be explained in Chapter 12, *Hydrodynamic Drives*, are typically one-way clutches.

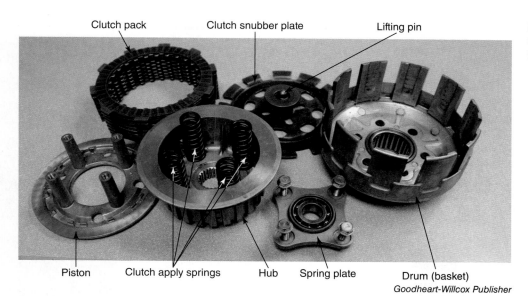

Figure 8-32. A disassembled spring-applied, multiple-disc traction clutch.

Goodheart-Willcox Publisher

Overrunning clutches have the advantages of compact size, quick engagement, and do not require electrical current or hydraulic pressure. A negative is that the actual clutch engagement cannot be slowed or varied. There are three common types of one-way clutches: one-way roller clutches, sprag clutches, and mechanical diodes.

One-Way Roller Clutch

One of the most popular types of one-way clutches is the one-way roller clutch. The *one-way roller clutch* has an outside race with cam-shaped pockets that contain rollers and springs, **Figure 8-33**. The inside race fits inside the outside race. Either the inside race or the outside race can be designed as a stationary member.

If the inside race is held stationary and the outside race is driven counterclockwise, the outside race will overrun and rotate. See **Figure 8-34**. Note that the rollers compress the springs and rotate into the larger portion of the cam-shaped pockets in the outside race. This provides room for the rollers to spin, allowing the outside race to rotate. If the inside race is held and a force attempts to rotate the outside race in a clockwise direction, the rollers are wedged into the narrow portion of the cam-shaped pocket. This causes the one-way clutch to lock, which holds the outside race stationary.

Sprag Clutch

A *sprag clutch* contains several dog bone–shaped pieces between inside and outside races. See **Figure 8-35**. Notice that the sprags have two diagonal distances, one short (blue line and arrowheads) and one long (red line and arrowheads).

When the rotation of the races causes the sprags to lean forward, or stand up, the sprags will lock the races together, **Figure 8-36A**. When the rotation of the races causes the sprags to lean back, or fall down, the sprags will overrun and the races will turn freely. See **Figure 8-36B**. **Figure 8-37** shows a torque converter stator equipped with a sprag clutch.

Mechanical Diode

A mechanical diode is an overrunning clutch commonly used in automotive applications. It has two plates, a pocket plate and a notch plate, that function as races. See **Figure 8-38**. The mechanical diode is designed to lock when driven in the direction that allows the struts in the pocket plate to lock into the cavities in the notch plate. It will overrun when driven in the opposite direction, which prevents the struts from locking in the notch plate's cavities. When

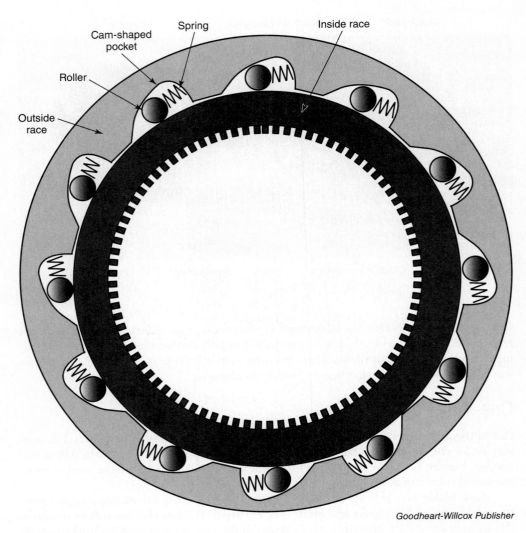

Figure 8-33. One-way roller clutches are the most popular type of overrunning clutch. A one-way roller clutch contains rollers and springs located between an outside race and an inside race. The outside race has cam-shaped pockets. Each pocket houses a roller and a spring.

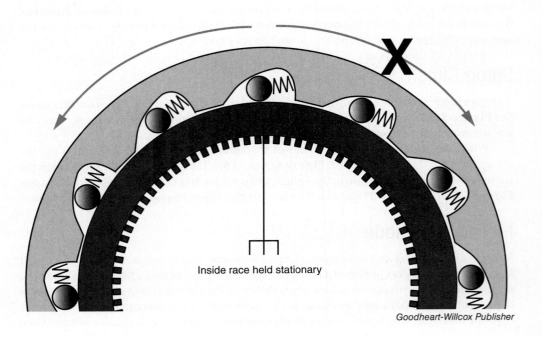

Figure 8-34. In this example, if the inside race of an overrunning clutch is held stationary, the outside race will overrun when turned counterclockwise and lock when turned clockwise.

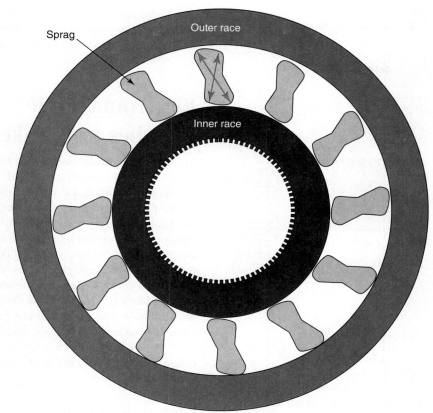

Figure 8-35. A sprag clutch is an overrunning clutch that contains several dog bone–shaped sprags.

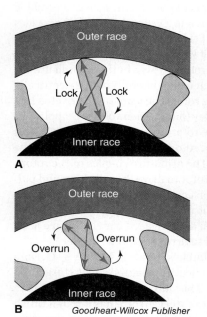

Figure 8-36. The sprags in a sprag clutch separate the inner and outer races. A—When the sprags stand up, the clutch locks. B—When the sprags roll over, the clutch overruns.

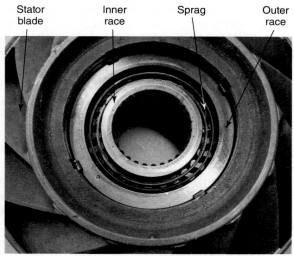

Figure 8-37. This torque converter stator has a one-way sprag clutch.

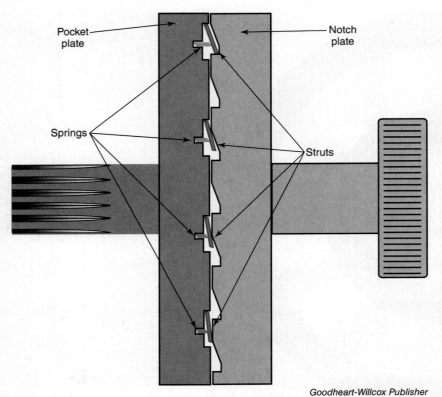

Figure 8-38. A mechanical diode uses spring-loaded struts that are forced into a notch plate, causing the struts to lock against the plate's shoulder. The mechanical diode will overrun when the plate is driven in the opposite direction.

this occurs, the struts are forced into the notches in the pocket plate, overcoming their spring tension. Either plate can be the driven member.

Electromagnetic Clutches and Brakes

An *electromagnetic clutch*, also known as an *electric clutch* or an *electromechanical clutch*, uses electricity to create a magnetic force that couples a rotating rotor to a clutch plate. The electricity causes the clutch to engage, allowing the clutch to mechanically deliver torque. Electromagnetic clutches are used on air conditioning compressor pulleys, on lawn tractor PTOs to drive the mower blade, and to engage the header drive belt on some combines. Two common electromagnetic clutches are the coiled wire clutch and the electromagnetic particle clutch.

Coiled Wire Clutch

Coiled wire clutches vary in size and application. In some applications, such as an air conditioning clutch, the input to the clutch is a pulley. In other applications, such as a combine belt drive, the clutch's output is a pulley. In the clutch shown in **Figure 8-39**, a shaft provides the input to the clutch assembly. A coil of wire, known as a field coil, serves as an electromagnet. The field coil can be stationary, or it can rotate. If it rotates, it will be equipped with slip rings and brushes. The field coil in **Figure 8-39** is stationary, and the rotor is driven by the input shaft.

The armature is the component within the clutch that moves laterally (when electrically engaged) to couple the output hub/pulley to the rotating rotor. The armature is riveted to a return spring, and the return spring is riveted to the pulley (hub). When voltage is applied to the field coil, the magnetic field created magnetizes the rotating rotor. The magnetic field extends across the air gap between the armature and the rotor's clutch plate and causes the armature to be drawn into engagement with the rotor. The armature couples to the rotor's clutch plate located on the face of the rotor, causing the rotor to drive the armature and pulley. The electric control (actuation) results in a mechanical coupling of the rotor, clutch plate, armature, and pulley (hub). When the electrical current is shut off, the flat return spring returns the armature to a retracted position, resulting in a disengaged clutch. **Figure 8-40** shows an electromagnetic coiled wire clutch that is used to drive a combine header.

A disassembled air conditioning electromagnetic clutch is shown in **Figure 8-41**. Note that there are two wire leads extending from the coil, indicating that the coil is stationary. The rotor (pulley) is rotated by the accessory drive belt. The clutch plate and armature are coupled to the compressor's input shaft. The clutch plate's steel surface has a rough finish designed to couple with the rotating rotor. When the field coil is energized, the armature moves laterally. This causes the clutch plate to engage the rotating rotor. The rotor drives the armature and clutch plate, which drives the compressor's input shaft.

A small *air gap* exists between the armature's clutch plate and the rotating rotor assembly. The service literature often lists an air gap specification that can sometimes be checked with a feeler gauge. See **Figure 8-42**.

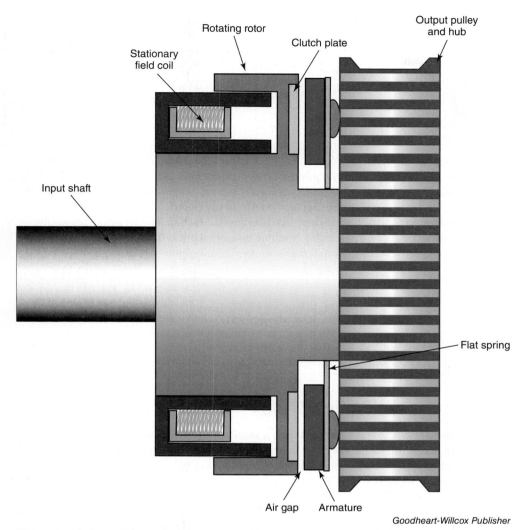

Figure 8-39. A simplified view of an electromagnetic coiled wire clutch.

Figure 8-40. This electromagnetic clutch drives a combine's header drive belt, which is used to power the combine's header PTO shaft located at the bottom of the combine's feeder house.

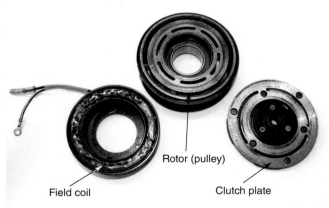

Figure 8-41. A conventional air conditioning compressor is driven by an electromagnetic clutch. When the coil is energized, it draws the clutch plate to engage with the rotating pulley. The clutch plate is coupled to the compressor's input shaft.

Goodheart-Willcox Publisher

Figure 8-42. Electromagnetic clutches have an air gap that must be checked to ensure proper operation. A shim is located under the clutch plate, between the clutch plate and the compressor's input shaft.

Warning

If a lip prevents the insertion of a feeler gauge, a dial indicator can be used to measure the travel of the clutch disc as it is energized. Never make any measurements to a drive clutch while the machine is running.

If the clutch is worn or if the air gap is misadjusted, the clutch may slip or it may not engage. If there is no air gap between the clutch plate and the rotor, the clutch cannot be released and will always be driving. Lawn mower PTO electric clutches can sometimes be adjusted by tightening or loosening nuts or cap screws to increase or decrease the clutch's spring tension. On many electric clutches, shims of different thicknesses are used to adjust the air gap. The air gap specification for one particular combine header clutch is 0.060″. The specification for one manufacturer's air conditioning clutch is 0.020″.

Note

The coiled wire clutch normally receives system voltage, either 12 volts or 24 volts, for engaging the clutch. The current is fixed based on the resistance of the field coil. However, the electromagnetic particle clutch can use a variable current to engage the clutch.

Electromagnetic Particle Clutch

The *electromagnetic particle clutch* has a small cavity between the input and output members that contains a magnetic powder. As electric current is applied to the field coil, the resulting magnetic field causes the particles to bind together. This causes the input and output members to engage. The electronic control module can vary the current based on operating conditions to increase or decrease clutch slip. When the ECU applies minimal current flow, the electromagnetic particle clutch is partially engaged and will slip if it encounters a large load. During high-torque applications, the ECU applies high current flow to fully engage the clutch. The ECM removes current to disengage the clutch. The clutch is unique in that it can be designed to slip without being damaged.

Electromagnetic Brakes

Electromagnetic controls can also couple a rotating component to a stationary component. This application is known as an electromagnetic brake. The brake is used to stop or hold a component. It can be designed to be held when energized and released by spring pressure, or can be spring-applied and electrically released, which is used in some aerial platforms to hold the electric drive motors stationary. In heavier-duty applications, like overhead bridge cranes, an electromagnetic coil can be used to apply friction-lined shoes to stop a rotating drum.

Bands

A flexible *band* can be used to hold a rotating drum. The band is lined with friction material. One end of the band rests on a stationary anchor. The other end of the band is actuated through a mechanical lever or a hydraulic piston, causing the band to squeeze against the drum. The band is also known as a brake. Examples of band applications in mobile equipment are dozer steering brake systems, traction-clutch shaft brakes, agricultural PTO brakes, and bands for holding a planetary gear. See **Figure 8-43**. Many planetary drives require a held member to produce an output.

Chapter 8 | Clutches and Planetary Controls

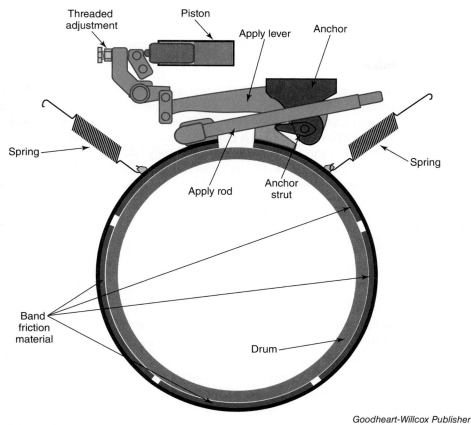

Goodheart-Willcox Publisher

Figure 8-43. A simplified dozer steering brake is shown here. Fluid pressure applies the piston, causing the apply lever to actuate the apply rod.

Old International 460 and 560 tractors use two bands in the PTO clutch. One band is used to hold the planetary ring gear, which is necessary to drive the PTO shaft. When the PTO is disengaged, the ring gear band is released and the PTO brake band is used to hold the PTO output shaft stationary.

To operate effectively, bands must have three qualities:
- High coefficient of friction—the band must have a high coefficient of friction so it can securely hold the drum.
- Wear resistance—the band must resist wear if it is to provide a long life.
- Resiliency—the band must be resilient so it releases fully while in the disengaged position. If the band does not release fully, it will drag on the drum.

Engineers design thin flexible bands that wrap around the majority of the drum's circumference so they have a ***self-energizing*** effect. This means that the band is drawn into the rotating drum to increase stopping force. The further the band tightens around the drum, the greater the self-energizing effect and the higher the stopping force applied to the drum. For example, if the drum pushes the band away at the 3 o'clock position, it will also pull the band against the drum at the 9 o'clock position.

Bands offer the advantage of taking little space and are effective for holding planetary gears or drums.

Slip Clutch

A ***slip clutch*** is a device designed to slip when it encounters an excessive load, such as when a PTO-driven implement encounters an immovable object. Slip clutches are used on

Goodheart-Willcox Publisher

Figure 8-44. This slip clutch is used on 540 rpm PTO-driven equipment, which is one of the three types of agricultural tractor PTOs that were explained in Chapter 4, *Agricultural Equipment Identification*.

Multiple-disc Clutch

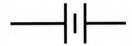

Stationary (held to case)

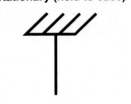

Band

One-way Clutch

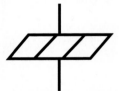

Goodheart-Willcox Publisher

Figure 8-45. Mechanical schematics can be used to explain power flow and troubleshoot transmissions. The four symbols are used to depict components inside transmissions and other powertrain components.

agricultural implements such as rotary cutters and balers. In its normal state, the clutch is compressed into its engaged position by spring tension. If the PTO-driven implement hits an obstruction, the clutch is designed to slip, allowing the tractor's PTO shaft to keep spinning but keeping the implement stationary until the drive is no longer obstructed. See **Figure 8-44.**

To prevent damage to driveline components, some implements use a shear bolt rather than a slip clutch. A *shear bolt* is a low-grade bolt that is designed to break (shear) before the mechanical drive fails. If a slip clutch fails to slip or a shear bolt fails to break when a PTO-driven implement hits an obstruction, the PTO shaft, gearbox, or other driveline components will likely be damaged.

Before each season, a slip clutch must be intentionally "slipped," which is also known as burnishing the clutch discs. If this is not done, the clutch discs can rust together, preventing slip protection. The following sequence is an example of steps to perform when "slipping" a clutch.

 Warning
Be sure to follow the implement manufacturer's recommended procedures.

1. Shut off the machine's engine.
2. Remove the PTO driveshaft to gain access to the slip clutch.
3. Mark the input and output of the slip clutch to provide a visual indicator that the clutch has successfully slipped.
4. Loosen the spring tension compressing the slip clutch.
5. Reinstall the PTO driveshaft.
6. Install all safety shields and covers.
7. Start the engine.
8. Engage and disengage the PTO multiple times.
9. Shut off the engine and inspect the slip clutch to determine if it has successfully slipped (you might have to remove the PTO shaft to visually inspect the indicator marks).
10. If the clutch has successfully slipped, tighten the spring tension on the clutch according to the service literature instructions.
11. If the clutch did not slip, either the clutch needs to be rebuilt or the spring tension is still too high.

Centrifugal Clutches

Centrifugal clutches engage based on an increase in shaft speed. Eaton Ultra-Shift first generation on-highway truck transmissions use a centrifugal clutch. When the engine speed is below 800 rpm, the centrifugal clutch is disengaged. At 800 rpm, the centrifugal clutch causes four arms to slide outward to begin forcing the clutch disc into engagement with the flywheel. The clutch slips at speeds between 800 to 1100 rpm and is fully engaged above 1100 rpm. Other styles of centrifugal clutches include non-variable centrifugal clutches, which are used in outdoor power equipment, and variable belt centrifugal clutches, which are used in utility vehicles (UTVs). A steel-belt–style variable centrifugal clutch is used in a compact tractor and is explained in Chapter 15, *Continuously Variable Transmissions*

Mechanical Schematics

Chapter 9, *Planetary Gear Set Theory*, will explain that plane-tary gear sets use holding members (brakes), input members (clutches), and coupling clutches to control the transmission. *Mechanical schematic symbols* are sometimes used to explain power flow through a planetary transmission. The schematic symbols shown in **Figure 8-45** represent clutches and bands.

Summary

- In a friction clutch, friction material bound to a band or plate grips a mating drum or steel disc.
- In a single-disc traction clutch application, the flywheel is bolted to the engine's crankshaft. A pressure plate is normally bolted to the flywheel and rotates with the flywheel. When the clutch is engaged (clutch pedal released), the pressure plate pushes the clutch disc firmly against the flywheel.
- In a dual-disc traction clutch, the flywheel surrounds the clutch assembly. The flywheel's internal splines align with the drive plate's external splines, causing the drive plate to always rotate at engine speed. Two driven discs, which are lined on both sides with friction material, are splined to the clutch hub. When the clutch is engaged, a cam link/roller assembly forces the loading plate against the pressure plate.
- Agricultural tractors with continuous PTO configurations have one clutch pedal to control two separate clutches, the PTO clutch and the traction clutch.
- Traction clutches require service and adjustment. The most common adjustments include clutch pedal free play adjustment and release lever or operating lever adjustment.
- Multiple-disc clutches contain a set of friction discs positioned alternately between steel separator plates. They can be engaged using a mechanical lever, hydraulic fluid pressure, or spring pressure.
- One-way clutches are designed to lock when turned in one direction and unlock, or overrun, when turned in the opposite direction.
- An electromagnetic clutch or brake uses electricity to create a magnetic force that couples a rotating rotor to a clutch plate.
- A flexible band lined with friction material can be used to hold a rotating drum. One end of the band rests on a stationary anchor. The other end of the band is actuated through a mechanical lever or a hydraulic piston, causing the band to squeeze against the drum.
- A slip clutch is a device designed to slip when it encounters an excessive load, such as when a PTO-driven implement encounters an immovable object.
- Centrifugal clutches engage when a shaft's speed increases.

Technical Terms

- air gap
- asbestos
- band
- basket
- Belleville washers
- brake
- carrier
- centrifugal clutch
- clutch disc
- clutch release mechanism
- coiled wire clutch
- coupling clutch
- cylinder
- dampening springs
- disengaged clutch clearance
- drum
- electric clutch
- electromagnetic clutch
- electromagnetic particle clutch
- electro-mechanical clutch
- flywheel clutch
- free play adjustment
- friction clutch
- friction discs
- holding clutch
- housing
- hub
- input clutch
- lube passage
- mechanical schematic symbols
- multiple-disc clutch
- normally applied clutch
- oil-applied clutch
- one-way clutch
- one-way roller clutch
- operating levers
- overrunning clutch
- PermaClutch
- pilot bearing
- pressure passages
- release bearing
- return springs
- rotating clutch
- self-energizing
- separator plates
- shear bolt
- slip clutch
- sprag clutch
- spring-applied, oil-released clutches
- stationary clutch
- throw out bearing
- traction clutch
- transmission clutch

Review Questions

Answer the following questions using the information provided in this chapter.

Know and Understand

1. Which one of the following is used to disengage the engine to allow the transmission to be shifted?
   A. One-way clutch.
   B. PTO clutch.
   C. Slip clutch.
   D. Traction clutch.

2. All of the following are examples of flywheel clutches, *EXCEPT*:
   A. band clutch.
   B. PTO clutch and traction clutch.
   C. single-disc clutch.
   D. dual-disc clutch.

3. Continuous PTO clutches are also known as _____ PTO clutches.
   A. single-stage
   B. dual-stage
   C. independent
   D. None of the above.

4. How often should a John Deere PermaClutch be disassembled for inspection?
   A. 1000 hours
   B. 2000 hours
   C. 4000 hours
   D. Never

5. A multiple-disc clutch can be engaged using all of the following methods, *EXCEPT*:
   A. centrifugal force.
   B. manual lever.
   C. hydraulic pressure.
   D. spring pressure.

6. Technician A states that the hub can provide the input into a multiple-disc clutch. Technician B states that the drum can provide the input into a multiple-disc clutch. Who is correct?
   A. Technician A.
   B. Technician B.
   C. Both A and B.
   D. Neither A nor B.

7. All the following are examples of one-way clutches, *EXCEPT*:
   A. mechanical diode.
   B. roller clutch.
   C. slip clutch.
   D. sprag clutch.

8. Electromagnetic clutches couple a rotating rotor to a _____.
   A. clutch plate
   B. roller
   C. sprag
   D. band

9. Electromagnetic clutches are used in which of the following applications?
   A. Combine header drive.
   B. Air conditioning compressor pulleys.
   C. Lawn tractor PTOs.
   D. All of the above.

10. Technician A states that the electromagnetic clutch air gap can be checked with a feeler gauge. Technician B states that the electromagnetic clutch air gap can be checked with a dial indicator. Who is correct?
    A. Technician A.
    B. Technician B.
    C. Both A and B.
    D. Neither A nor B.

11. All of the following can be used to hold a component stationary, *EXCEPT*:
    A. band.
    B. input clutch.
    C. brake.
    D. stationary clutch.

12. Which of the following can have a self-energizing effect?
    A. Band.
    B. Input clutch.
    C. Multiple-disc clutch.
    D. Stationary clutch.

13. All of the following are examples of bands used in machinery, *EXCEPT*:
    A. steering brake in a dozer.
    B. steering clutch in a dozer.
    C. ring gear in a tractor's planetary PTO.
    D. agricultural PTO brake.

14. Bands have all of the following qualities, *EXCEPT*:
    A. high coefficient friction.
    B. wear resistance.
    C. resiliency.
    D. ability to slip under heavy loads.

Apply and Analyze

For Questions 15–18, match the term with the correct mechanical symbol.

15. Band _____
16. Multiple-disc clutch _____
17. One-way clutch _____
18. Stationary _____

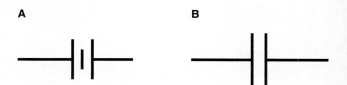

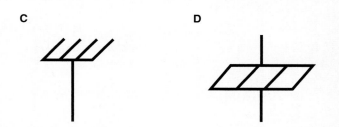

19. Any time a PermaClutch assembly is installed or removed from a tractor, _____ must be placed between the operating levers and the clutch cover.
20. Multiple-disc PTO brakes, park brakes, and MFWD clutches normally use _____ pressure to engage the clutch pack.
21. One-way clutches are designed to _____ in one direction.
22. A(n) _____ clutch uses several pieces resembling the shape of a dog bone.
23. A(n) _____ clutch uses cam-shaped pockets and springs.
24. A low-grade bolt designed to break before a mechanical drive fails is known as a(n) _____ bolt.
25. Electromagnetic clutches are actuated by _____.

26. A(n) _____ clutch is often used on agricultural implements to protect PTO shafts and implement drive components.
27. A(n) _____ clutch must be checked each season to ensure that it is not rusted together.
28. A(n) _____ clutch engages as shaft speed increases.

Critical Thinking

29. A technician checks a manual clutch free pedal adjustment and finds that it has no free play. Explain why this can cause the clutch to slip and wear faster.
30. List all of the mechanism found in this chapter that can be used to drive another component.

Chapter 9
Planetary Gear Set Theory

Objectives

After studying this chapter, you will be able to:

- ✓ Explain the different power flows through a simple planetary gear set.
- ✓ Calculate simple planetary gear set ratios.
- ✓ Trace power flow through a transmission containing multiple planetary gear sets.
- ✓ Describe Caterpillar planetary transmissions.
- ✓ List the common components used in Allison 1000, 2000, 3000, and 4000 series transmissions.

Planetary gear sets are used in most off-highway machines. They are used in transmissions, final drives, power take-offs (PTOs), transfer cases, and pump drives. It would be difficult to find a mobile machine that does not contain at least one simple planetary gear set.

Simple Planetary Gear Set

A *simple planetary gear set* consists of a sun gear, planetary pinion gears, a planetary carrier, and a ring gear. See **Figure 9-1**. The term *planetary gear set* is derived from the fact that this type of gear set resembles the orientation of the planets in our solar system. The centrally located *sun gear* is surrounded by the planetary pinion gears. The *planetary pinion gears* mesh with and revolve around the sun gear. They are secured in the *planetary carrier* by pins or shafts. The ring gear, also known as an annulus gear, surrounds the entire gear set and meshes with the planetary pinion gears. Planetary gears can be spur, helical, or herringbone.

Planetary gear sets are often chosen because of their compact design. Compared to the traditional external toothed gear design, the loads placed on planetary gear sets are offset by the multiple planetary pinion gears, which allows the use of smaller bearings. Planetary gear sets offer engineers multiple configurations to provide a variety of speeds and torques.

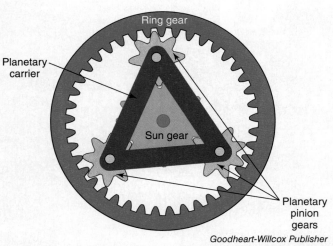

Figure 9-1. A simple planetary gear set contains a sun gear, a planetary carrier with planetary pinions, and a ring gear.

Simple Planetary Gear Set Principles

To understand power flow through a simple planetary gear set, you must understand a few important principles. These principles apply only to simple planetary gear sets, not to the compound planetary gear sets discussed later in this chapter.

State of the Planetary Members

The sun gear, planetary carrier, and ring gear can be driving members, driven members, held members, or coupled members. A driving member is a planetary component that provides the input power into the planetary gear set. A driven member is the output member of the planetary gear set. A held member is the planetary member that is held stationary, sometimes called a reaction member. Two members that are attached to one another are known as coupled. To get an output (driven member) from a planetary gear set, an input (driving member) must be received into the gear set.

A change of speed and/or direction in a planetary gear set requires that one member be held or that two input members be driving at different speeds and/or in different directions.

One of the most important factors for determining planetary gear set configuration is the state of the planetary carrier.

- If the planetary carrier is held, the gear set reverses the direction of power flow.
- If the planetary carrier is the input member, the gear set is in an overdrive mode.
- If the planetary carrier is the output member, the gear set is multiplying torque.

Any time two planetary members are driving (input) at the same speed and in the same direction, direct drive occurs. Direct drive can also be achieved by receiving one input member and coupling one planetary member to another. Note that this is not the same as holding a planetary member.

If a planetary gear set receives only one input member (by itself) and nothing is held or coupled together, the gear set is in neutral. If the planetary gear set has only one member held and no input, the gear set is in neutral

Planetary Member Size Relationships

The size relationships of the sun gear, ring gear, and planetary carrier determine the output speed of the gear set. The sun gear is the smallest diameter member, the planetary carrier is the middle-sized diameter member, and the ring gear is the largest diameter member. A small input gear driving a large output gear produces a slow output. A large input gear driving a small output gear will produce a fast output.

>
> **Note**
> The number of planetary pinions in a planetary gear set does not have an effect on the direction of power flow or the gear set's ratio. However, for a given size planetary gear set, more planetary pinions will make a stronger gear set because the torque is distributed across more teeth.

Chapter 9 | Planetary Gear Set Theory

Pro Tip

The term *epicyclical* is sometimes used to describe a planetary gear set. The word *epicycle* is derived from the second century astronomer Ptolemy, who hypothesized that planets spin in small cycles that he called epicycles. The center of the small epicycle follows a path inside a larger circle. The larger circle is a circumference that surrounds what was thought to be the earth. In a planetary gear set, the planetary pinions spin in small epicyclical motions. As the planetary carrier rotates, the center of the planetary pinion gears revolve in a set circumference that is centered on the sun gear's axis.

Planetary Gear Set Configurations

A simple planetary gear set can produce the following configurations:
- Reverse (slow or fast).
- Forward torque multiplication (slow or fast).
- Forward overdrive (slow or fast).
- Direct drive.
- Neutral.

Reverse

Reverse is the easiest simple planetary configuration to understand. The gear set can produce a slow reverse (torque multiplication) or a fast reverse (overdrive reverse). Any time the simple planetary gear set is in a reverse mode, the planetary carrier is the held member, or stationary member. See **Figure 9-2**. If the held member is known, then the next question posed must be, what are the input and output members? If a slow output is desired, the sun gear (small gear) should be the input member and the ring gear should be the output member. If a fast output is required, the ring gear (large gear) should be used as the input member and the sun gear should be the output member. See **Figure 9-3**.

Forward Torque Multiplication (Speed Reduction)

Any time the simple planetary gear set is in a forward torque-multiplication mode (speed reduction), the planetary carrier is the output member. See **Figure 9-4**. A tractor axle's inboard planetary final drive is shown in **Figure 9-5**. The inboard planetary is used for multiplying torque; therefore, the planetary carrier is the output member and is attached to the axle's output shaft.

If the output member is known, then the next question posed must be, what is the input member? If a slow output is desired, the sun gear (small gear) should be the input and the ring gear should be held stationary. If a fast output is required, the ring gear (large gear) should be the input and the sun gear should be held stationary. See **Figure 9-6**.

Forward Overdrive (Torque Decrease)

Any time the simple planetary gear set is in a forward overdrive mode, the planetary carrier is the input member, or driving member. See **Figure 9-7**. If the input

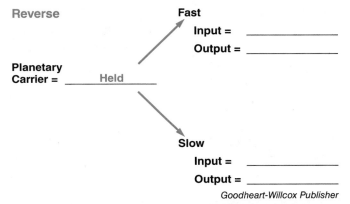

Figure 9-2. When learning simple planetary power flows, always start by answering the question, "What is the planetary carrier doing in the configuration?" The planetary carrier is held in reverse.

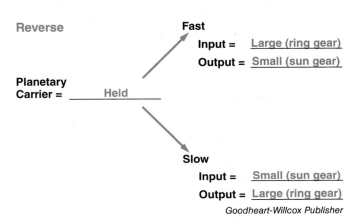

Figure 9-3. After establishing that the planetary carrier is held in reverse, you can choose the size of gears required for slow and fast configurations.

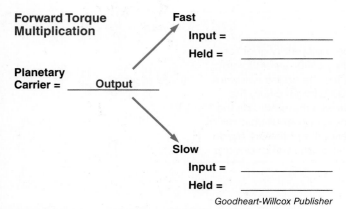

Figure 9-4. The planetary carrier is always the output member when a simple planetary gear set is producing a forward torque-multiplication power flow.

Figure 9-5. A tractor axle's inboard planetary gear set contains a stationary ring gear that is fixed to the axle housing, a sun gear that is driven by the differential's side gear, and a planetary carrier that is splined to the axle's output shaft.

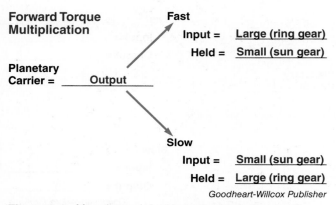

Figure 9-6. After determining the output members and input members, the remaining planetary member is held.

member is known, then the next question posed must be, what is the output member? If a slow output is required, the ring gear (large gear) should be used as the output and the sun gear should be held stationary. If a fast output is desired, the sun gear (small gear) should be the output and the ring gear should be held stationary. See **Figure 9-8**.

Direct Drive

A simple planetary gear set can achieve direct drive by two different methods:

- Having two input members driving at the same speed in the same direction.
- Having one input member and coupling two planetary members to each other.

In direct drive, the entire gear set rotates as a unit. The individual planetary pinions do not spin because the planetary carrier rotates at the same speed as the sun gear and ring gear. Direct drive delivers the same output torque, speed, and direction as received from the input member.

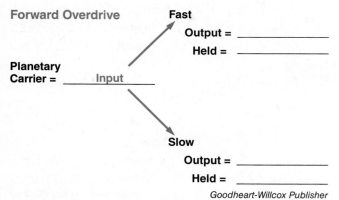

Figure 9-7. The planetary carrier is the input in forward overdrive. With the input known, the next step is to determine the output member.

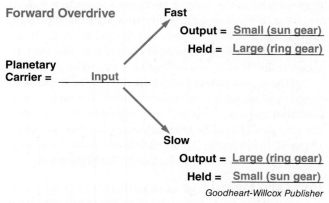

Figure 9-8. After determining the input members and output members for forward overdrive, the remaining planetary member is held.

Pro Tip

If a simple planetary gear set has a noise in all gears except for direct drive, the gear set most likely has a chipped tooth or a bad bearing. The pinions do not spin in direct drive, resulting in no noise.

Neutral

If the planetary gear set has only one input member and nothing is held or coupled, the gear set will remain in neutral. The planetary gear set will not be able to transfer power flow until another planetary gear is held or coupled.

Note

Some planetary transmissions have a planetary gear driving or held while the transmission is in a neutral position. Sometimes the engagement of the driving gear or the held gear is chosen because it is likely to be engaged when leaving neutral. For example, in **Figure 9-44**, Allison 1000 through 4000 series transmissions have Clutch #5 engaged in neutral, holding the #2 planetary carrier and #3 ring gear. The transmission will most likely shift from neutral to first forward or reverse, which would require only engaging clutch #1 (for first forward) or clutch #3 (for first reverse).

Planetary Gear Ratios

Planetary gear ratios are calculated based on the number of teeth on the sun gear and the ring gear. As explained in Chapter 7, *Manual Transmissions*, a gear ratio determines if the gear set is multiplying or reducing torque, multiplying or reducing speed, or producing direct drive. The first number of the gear ratio indicates how many revolutions the input gear must spin to achieve one output revolution. The last number (1) equals how many revolutions the output gear will spin as compared to the number of revolutions of the input gear. See the matrix in **Figure 9-9**.

How to Address the Planetary Carrier in Gear Ratio Calculations

As mentioned earlier, the number of planetary pinions in a planetary carrier has no effect on the gear ratio. Therefore, a carrier with four pinions will have more teeth than carrier with three pinions in an otherwise identical gear set. However, for the purpose of calculating planetary gear ratios, the carrier is treated as if it has the same number of teeth as the ring gear and sun gear combined.

Gear Range	Sun and Ring Gears	Example if Sun = 20 Teeth and Ring = 60 Teeth
Slow reverse	ring ÷ sun = ratio	60 ÷ 20 = −3:1
Fast reverse	sun ÷ ring = ratio	20 ÷ 60 = −0.33:1
Slow torque multiplication	(sun + ring) ÷ sun = ratio	(20 + 60) ÷ 20 = 4:1
Fast torque multiplication	(sun + ring) ÷ ring = ratio	(20 + 60) ÷ 60 = 1.33
Slow overdrive	ring ÷ (sun + ring) = ratio	60 ÷ (20 + 60) = 0.75:1
Fast overdrive	sun ÷ (sun + ring) = ratio	20 ÷ (20 + 60) = 0.25:1

Goodheart-Willcox Publisher

Figure 9-9. Gear ratios equals the output divided by the input. Calculating simple planetary gear ratios requires knowing the number of teeth on the sun gear and the number of teeth on the ring gear. The planetary carrier is treated as if it has the same number of teeth as the sun gear plus the ring gear. In this figure "(sun+ring)" has been used in place of "carrier."

For example, in a planetary gear set in which the ring gear has 42 teeth and the sun gear has 18 teeth, the planetary carrier would actually have 36 teeth if it has three pinions or 48 teeth if it has four pinions. However, in both cases, the carrier must be calculated as the sun gear teeth plus the ring gear teeth, equaling 60 teeth.

Planetary Schematics

Engineers sometimes use a mechanical schematic that consists of a line drawing composed of graphic symbols to show what components are connected to each other and what components are connected to clutches, brakes, bands, and/or a stationary case. Manufacturers' instructors and product support personnel can use mechanical schematics to explain power flow through a planetary transmission.

Understanding power flow in a simple planetary gear set will help you diagnose transmissions, especially complex transmissions with multiple planetary gear sets. Studying a faulty transmission's power flow and identifying the component used in each of the malfunctioning gears often enables a technician to pinpoint a failed component before disassembling the transmission. Consider a technician who is diagnosing a faulty agricultural PTO brake. The technician is unsure about the operation of the clutch and band because the service literature is lacking clear information. However, the technician found the mechanical schematic shown in **Figure 9-10**. After studying the schematic, the technician determines the following:

- The band is the PTO brake, which holds the shaft stationary while the PTO is disengaged.
- The clutch operates (drives) the PTO.
- The ring is held stationary to the case.

Therefore, if the symptom is that the PTO shaft rotates while the PTO is disengaged, the PTO band should be diagnosed. If the symptom is that the PTO will not engage, the clutch should be diagnosed to determine if it is engaging.

Note

An agricultural PTO brake is not designed to bring a fully operational PTO shaft to a stop, but rather to hold the shaft stationary while the PTO is disengaged.

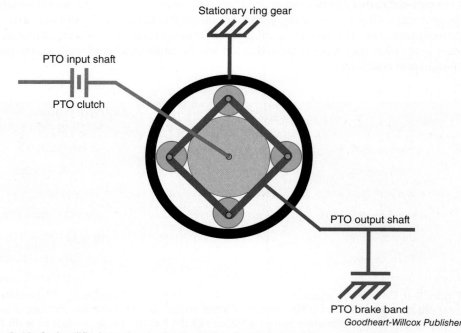

Figure 9-10. A simplified mechanical schematic of a planetary gear set in a PTO.

Multiple Planetary Gear Sets

Heavy equipment systems also combine multiple planetary gear sets within a single component. Examples of multiple planetary gear sets include double reduction final drives, powershift transmissions, and differential steering systems.

Double Reduction Final Drives

Some Caterpillar dozers use a double reduction final drive assembly that contains two simple planetary gear sets connected in series. Each gear set is configured to provide a slow forward torque multiplication gear ratio. The final drive receives input from the tractor's axle shaft, which drives the first planetary gear set's sun gear (#1 sun gear). The ring gear is shared by both gear sets and is held stationary. The first planetary gear set's carrier (#1 planetary carrier) drives the second planetary gear set's sun gear (#2 sun gear). The second planetary gear set's carrier (#2 planetary carrier) is the final drive output, which drives the track's drive sprocket. See **Figure 9-11**.

Note

When two simple planetary gear sets are combined into one assembly and when one of the planetary members is shared, such as the shared ring gear in the double reduction drive in **Figure 9-11**, the gear set is known as a *compound planetary gear set*.

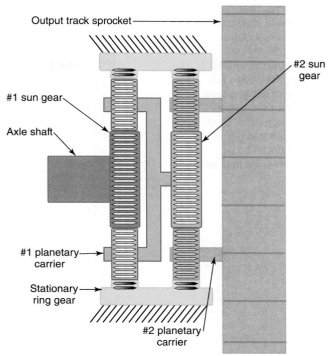

Goodheart-Willcox Publisher

Figure 9-11. This drawing shows a simplified double reduction final drive used in large Caterpillar elevated sprocket tractors. Notice the axle shaft drives the #1 sun gear as the input to the first gear set. The ring gear is shared and held in both gear sets. The #1 planetary carrier is the output of the first gear set and drives the #2 sun gear as the input of the second gear set. The combination planetary gear sets provide an overall final drive 20:1 gear reduction ratio.

Simplified Transmissions

Mobile equipment can use transmissions with multiple gear sets. A simplified transmission is shown in **Figure 9-12**. Notice that it uses three simple planetary gear sets in series. The combination of the three gear sets can provide multiple forward and reverse speeds.

Gear Set A

The sun gear is the input member of gear set A. It has 28 gear teeth. Clutch A is used to couple the 80-toothed ring gear to the planetary carrier. Brake A is used to hold the planetary carrier stationary. The ring gear is the output member of gear set A and it is coupled to the sun gear of gear set B.

When brake A is applied, it holds the planetary carrier stationary. This causes gear set A to reverse power flow with a –2.86:1 gear ratio. When clutch A is applied, gear set A is driving in direct drive.

Gear Set B

The sun gear is the input member of gear set B. It has 36 teeth. Clutch B is used to couple the ring gear to the planetary carrier to provide direct drive. Brake B is used to hold the 80-toothed ring gear stationary, which provides a forward 3.22:1 gear ratio.

Figure 9-12. This transmission illustration provides a foundation for learning power flows in more complex transmissions. Studying the power flow of one gear set at a time will result in the overall transmission ratio. Each gear set requires that either the clutch or brake be applied if the transmission is to deliver power flow.

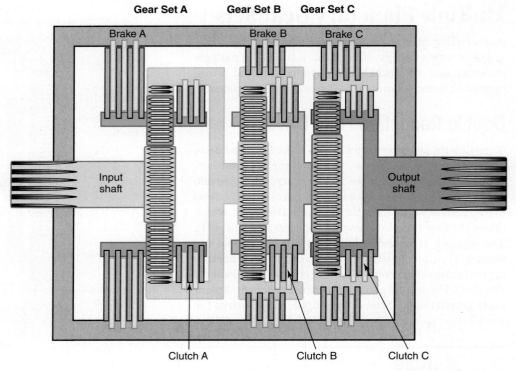

Goodheart-Willcox Publisher

Gear Set C

The sun gear is the input member of gear set C. It has 46 teeth. Clutch C is used to couple the ring gear to the planetary carrier to provide direct drive. Brake C is used to hold the 60-toothed ring gear stationary, providing a forward 2.3:1 gear ratio.

After determining the individual gear ratio for each simple planetary gear set, the ratios must be multiplied to calculate the overall transmission gear ratio. See the matrix in **Figure 9-13**.

Clutch Apply Charts

Manufacturers often place *clutch apply charts* similar to the chart shown in **Figure 9-13** in their service literature. These charts help technicians determine if the transmission has a common planetary control that is malfunctioning. For example, if a customer asked a technician to diagnose the transmission listed in **Figure 9-13** because it has no first gear, the technician would be able to see that it is possible to have problems with clutch A, brake B, and brake C. The technician would then determine if the transmission is having problems in any of the other gears that also use these three planetary controls. If the technician finds that the transmission operates properly in second gear forward but does not operate in third gear forward, then brake C or the controls used to operate brake C, such as the solenoid valve used to engage brake C, could be at fault. Clutch apply charts will be discussed further in Chapter 11, *Powershift and Automatic Transmission Controls, Service, and Repair*.

First Forward Gear

In first forward gear, clutch A is applied. This causes gear set A to produce a 1:1 direct drive power flow. Brake B is applied, causing gear set B to produce a 3.22:1 slow forward torque multiplication ratio. Brake C is also applied, causing gear set C to produce a 2.3:1 slow forward torque multiplication ratio. The overall gear ratio for first forward equals 7.4:1. See **Figure 9-14**.

Chapter 9 | Planetary Gear Set Theory

Gear Range	Clutch A	Clutch B	Clutch C	Brake A	Brake B	Brake C	Transmission Ratio
1st forward	X	—	—	—	X	X	1 × 3.22 × 2.3 = 7.4:1 forward ratio
2nd forward	X	—	X	—	X	—	1 × 3.22 × 1 = 3.22:1 forward ratio
3rd forward	X	X	—	—	—	X	1 × 1 × 2.3 = 2.3:1 forward ratio
4th forward	X	X	X	—	—	—	1 × 1 × 1 = 1:1 forward ratio
1st reverse	—	—	X	X	X	—	2.86 × 3.22 × 1 = −9.21:1 reverse ratio
2nd reverse	—	X	—	X	—	X	2.86 × 1 × 2.3 = −6.58:1 reverse ratio
3rd reverse	—	X	X	X	—	—	2.86 × 1 × 1 = −2.86:1 reverse ratio

Goodheart-Willcox Publisher

Figure 9-13. In the clutch apply chart, notice that clutch A is applied any time the transmission is in a forward gear, and could be more appropriately named the forward clutch. Brake A is applied any time the transmission is moving in reverse, and could be more appropriately named the reverse brake. Note that three planetary controls must be applied for the transmission to deliver power flow.

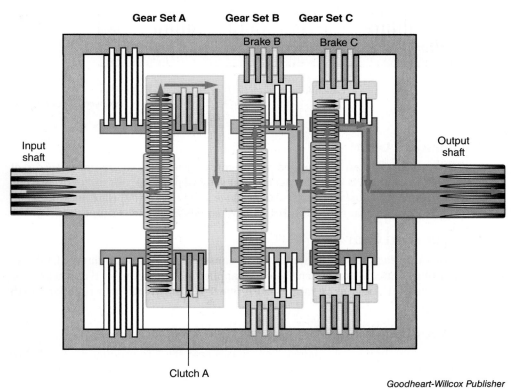

Figure 9-14. In first forward, clutch A causes gear set A to produce direct drive, brake B causes gear set B to produce slow forward torque multiplication, and brake C causes gear set C to produce slow forward torque multiplication.

Goodheart-Willcox Publisher

Note

The power flow arrows were intentionally left off the bottom half of the gear sets in **Figures 9-14** through **9-21** to simplify the drawings, making them easier to interpret.

Second Forward Gear

In second forward gear, clutch A is applied. It causes gear set A to produce a 1:1 direct drive power flow. Brake B is applied, causing gear set B to produce a 3.22:1 slow forward torque multiplication ratio. Clutch C is also applied, causing gear set C to produce a 1:1 direct drive ratio. The overall gear ratio for second forward equals 3.22:1. See **Figure 9-15**.

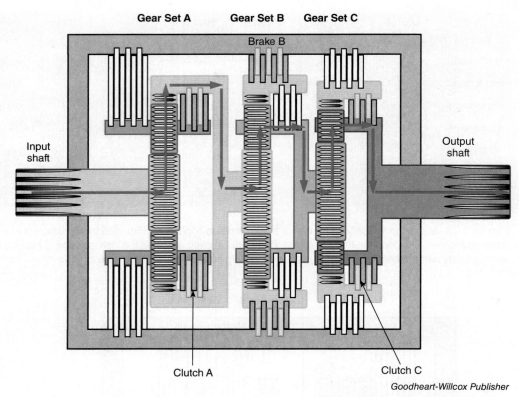

Figure 9-15. In second forward, clutch A causes gear set A to produce direct drive, brake B causes gear set B to produce slow forward torque multiplication, and clutch C causes gear set C to produce direct drive.

Third Forward Gear

In third forward gear, clutch A is applied, which causes gear set A to produce a 1:1 direct drive power flow. Clutch B is applied, causing gear set B to produce a 1:1 direct drive ratio. Brake C is also applied, causing gear set C to produce a slow forward torque multiplication 2.33:1 ratio. The overall gear ratio for third forward equals 2.33:1. See **Figure 9-16**.

Fourth Forward Gear

In fourth forward gear, clutch A is applied. This causes gear set A to produce a 1:1 direct drive power flow. Clutch B and clutch C are also applied, causing gear set B and gear set C each to produce a 1:1 direct drive ratio. See **Figure 9-17**. The overall gear ratio for fourth forward equals 1:1.

First Reverse Gear

In first reverse gear, brake A is applied. This causes gear set A to produce a −2.86:1 slow reverse ratio. Brake B is applied, causing gear set B to produce a 3.22:1 slow forward torque multiplication ratio. Clutch C is also applied, which causes gear set C to produce a 1:1 direct drive ratio. The overall gear ratio for first reverse equals −9.21:1. See **Figure 9-18**.

Second Reverse Gear

In second reverse gear, brake A is applied. It causes gear set A to produce a slow reverse ratio of −2.86:1. Clutch B is applied, causing gear set B to produce a 1:1 direct drive ratio. Brake C is also applied, causing gear set C to produce a 2.3:1 slow forward torque multiplication ratio. The overall gear ratio for second reverse equals −6.58:1. See **Figure 9-19**.

Third Reverse Gear

In third reverse gear, brake A is applied, causing gear set A to produce a slow reverse ratio of −2.86:1. Clutches B and C are also applied. This causes gear sets B and C to each produce a 1:1 direct drive ratio. See **Figure 9-20**. The overall gear ratio for third reverse equals −2.86:1.

Chapter 9 | Planetary Gear Set Theory

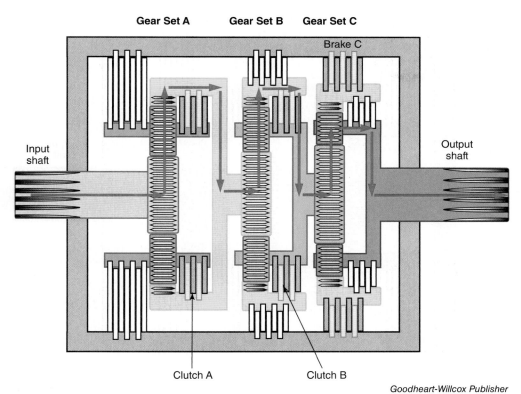

Figure 9-16. In third forward, clutch A causes gear set A to produce direct drive, clutch B causes gear set B to produce direct drive, and brake C causes gear set C to produce slow forward torque multiplication.

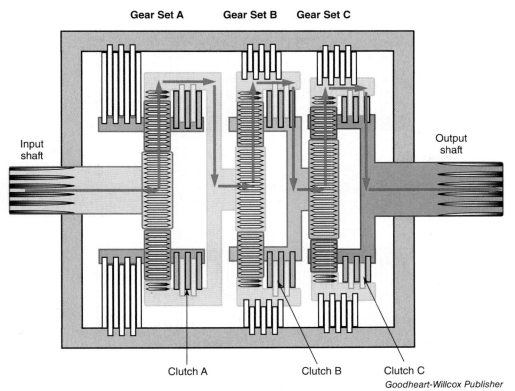

Figure 9-17. In fourth forward, clutch A causes gear set A to produce direct drive, clutch B causes gear set B to produce direct drive, and clutch C causes gear set C to produce direct drive.

Goodheart-Willcox Publisher

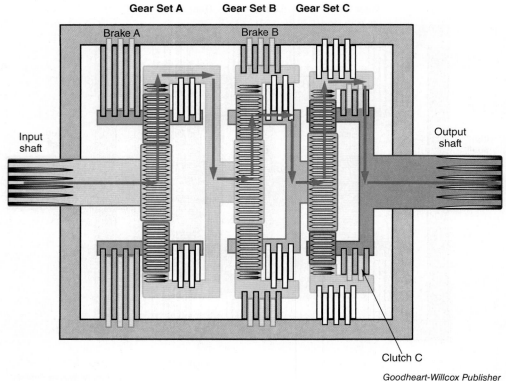

Figure 9-18. In first reverse, brake A causes gear set A to produce slow reverse, brake B causes gear set B to produce slow forward torque multiplication, and clutch C causes gear set C to produce direct drive.

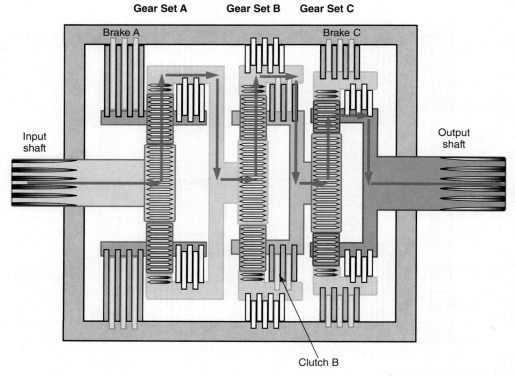

Figure 9-19. In second reverse, brake A causes gear set A to produce slow reverse, clutch B causes gear set B to produce direct drive, and brake C causes gear set C to produce a slow forward torque multiplication ratio.

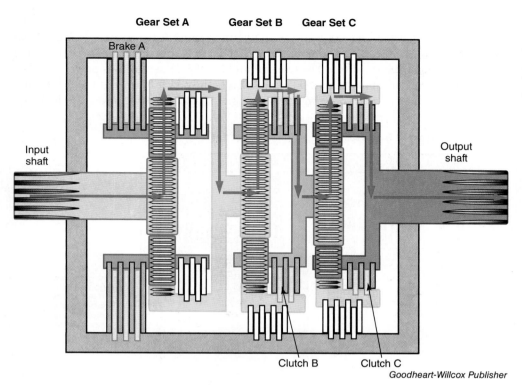

Figure 9-20. In third reverse, brake A causes gear set A to produce slow reverse, clutch B causes gear set B to produce direct drive, and clutch C causes gear set C to produce direct drive.

Four Simple Planetary Gear Sets

In **Figure 9-21**, a fourth gear set has been added to the simplified transmission. Notice that gear set D is a duplicate of gear set A. The matrix in **Figure 9-22** shows how gear set A and gear set D can be used together to provide a forward reduction. In first forward gear, brake A is engaged. This causes the power flow to be reversed through the first planetary gear set. With clutches B and C engaged, gear sets B and C each produce a 1:1 direct drive ratio. With brake D engaged, power flow is reversed back to a forward direction. The gear ratio is a result of the reversed torque multiplication of gear sets A and D and the direct drive of gear sets B and C, equaling a forward torque multiplication ratio of 8.18:1.

Caterpillar Elevated-Track Dozer, Three-Speed Transmissions

In 1977, Caterpillar introduced its first elevated track dozer, the D10, with a three-speed planetary transmission. Caterpillar has used this style of transmission in numerous models of elevated-track dozers and continues to use it in the largest dozers: D9, D10, and D11. These transmissions contain four planetary gear sets and five clutches. Although some manufacturers use the term *brake* to describe a planetary control that holds a planetary member stationary, Caterpillar calls all of their planetary controls *clutches*. **Figure 9-23** is a simplified drawing of a D6R transmission, which provides three forward speeds and three reverse speeds. The transmission is also known as a powershift transmission, which is explained in more detail in Chapter 10, *Powershift and Automatic Transmission Theory*. The transmission is unique in that the input shaft travels through the center of the output shaft and both the input shaft and output shaft protrude through the front of the transmission. See **Figure 9-23**.

The first two planetary gear sets are the inputs to the transmission (reverse and forward). The last two planetary gear sets and #5 clutch hub are the outputs of the transmission (first, second, and third speed). The #2/#3 planetary carrier (center planetary carrier) links the input planetary gear sets to the output planetary gear sets. The direction of rotation for the center planetary carrier depends on whether the reverse clutch (clutch #1) or the forward clutch (clutch #2) is applied.

The transmission shown in **Figure 9-24** is an actual three-speed Caterpillar track-type dozer D6R transmission. Depending on the gear selection, the D6R can provide travel speeds up to 7.3 miles per hour (mph) in forward and 9.1 mph in reverse. See **Figure 9-25**.

The transmission housing is located at the rear of the machine and contains both the planetary transmission and the transfer-and-bevel gear set. Chapter 11, *Powershift and Automatic Transmission Controls, Service, and Repair,* explains the process for removing the transmission from the dozer. The transfer-and-bevel gear set receives power from the transmission's output shaft, turns it 90°, and delivers the power to the right and left inner axle shafts. Chapter 24, *Track Steering Systems*, explains track steering power flow in detail.

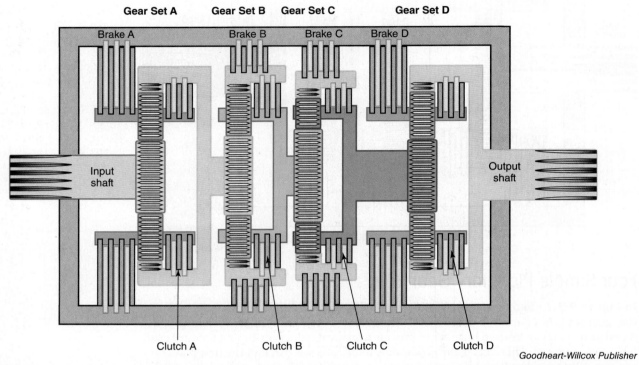

Goodheart-Willcox Publisher

Figure 9-21. A duplicate of gear set A has been added to this simplified transmission and labeled gear set D. If brake A and brake D are both applied, brake A causes a reverse power flow, and brake D reverses the power flow back to a forward direction.

Gear Range	Clutch A	Clutch B	Clutch C	Clutch D	Brake A	Brake B	Brake C	Brake D	Transmission Ratio
1st forward	—	X	X	—	X	—	—	X	2.86 × 1 × 1 × 2.86 = 8.18:1 forward ratio
2nd forward	X	—	—	X	—	X	X	—	1 × 3.22 × 2.3 × 1 = 7.4:1 forward ratio
3rd forward	X	—	X	X	—	X	—	—	1 × 3.22 × 1 × 1 = 3.22:1 forward ratio
4th forward	X	X	—	X	—	—	X	—	1 × 1 × 2.3 × 1 = 2.3:1 forward ratio
5th forward	X	X	X	X	—	—	—	—	1 × 1 × 1 × 1 = 1:1 forward ratio
1st reverse	—	—	X	X	X	X	—	—	2.86 × 3.22 × 1 × 1 = −9.21:1 reverse ratio
2nd reverse	—	X	—	X	X	—	X	—	2.86 × 1 × 2.3 × 1 = −6.58:1 reverse ratio
3rd reverse	—	X	X	X	X	—	—	—	2.86 × 1 × 1 × 1 = −2.86:1 reverse ratio

Goodheart-Willcox Publisher

Figure 9-22. This matrix shows the ratios that can be achieved when a fourth gear set is added to the simplified transmission. Four planetary controls must be applied to achieve power flow through the transmission. The planetary carriers in gear set A and gear set D can both be held in order to obtain a forward reduction. Gear set A causes a reverse power flow and gear set D reverses the power flow back to a forward direction. Note that this transmission does not contain a dedicated reverse brake or a dedicated forward clutch.

The transmission receives input from the torque divider. A torque divider is a torque converter with an internal simple planetary gear set. Torque dividers are explained in Chapter 12, *Hydrodynamic Drives*. After the operator has chosen the speed and direction, the transmission will deliver power to the output shaft.

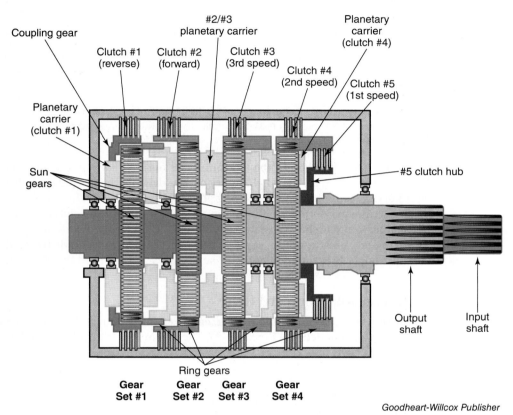

Figure 9-23. This simplified Caterpillar D6R transmission uses five clutches to control four planetary gear sets to provide three forward and three reverse speeds. Gear sets #1 and #2 receive input from the input shaft through the sun gear. It also shows that the output members are the #3 and #4 sun gears and the #5 clutch hub.

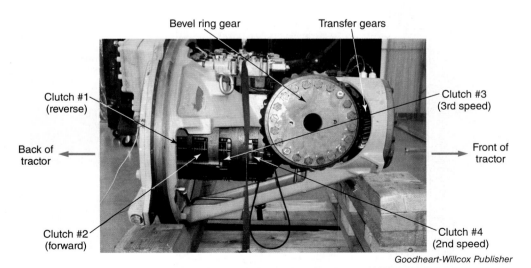

Figure 9-24. Clutches #1 through #4 are visible in this Caterpillar D6R three-speed track-type dozer transmission. The large bevel gear is used to turn the direction of power and will be explained in Chapter 24, *Track Steering Systems*.

Gear	Speed (mph)
1st forward	2.4
2nd forward	4.2
3rd forward	7.3
1st reverse	3
2nd reverse	5.3
3rd reverse	9.1

Goodheart-Willcox Publisher

Figure 9-25. A Caterpillar D6R track-type tractor has a maximum travel speed of 7.3 mph in forward and 9.1 mph in reverse.

Caterpillar D6R First Forward Gear

First forward gear power flow in a D6R transmission is shown in **Figure 9-26**. When the operator selects the first forward gear, clutches #2 (forward clutch) and #5 (first speed clutch) are applied. The #2 sun gear is the input member to the transmission and is driven in a clockwise direction as viewed from the front of the transmission. With the #2 ring gear held by its clutch, the #2/#3 planetary carrier rotates in a forward slow torque multiplication mode. The #3 sun gear is attached to the transmission's output shaft and acts like a held planetary member. With the #2/#3 planetary carrier rotating at a reduced speed, the #3 ring gear rotates. The #3 ring gear is attached to the #4 planetary carrier, which causes the #4 planetary carrier to rotate.

Note

The power flow arrows were intentionally left off the bottom half of the gear sets in **Figures 9-26** through **9-31** to simplify the drawings, making them easier to interpret.

The #4 sun gear is also attached to the transmission's output shaft, and it too acts like a held planetary member. With the #4 planetary carrier driving, the #4 ring gear also rotates. Clutch #5 is engaged. This causes the ring gear to drive the clutch hub, which is attached to the transmission's output shaft. Clutch #5 is the only rotating clutch in this transmission. The other four clutches are stationary clutches. Clutch #5 provides first gear by coupling the #4 ring gear and the output shaft. In first forward gear, power is distributed to the output shaft through three different paths, #3 sun gear, #4 sun gear, and the #5 clutch hub.

Caterpillar D6R Second Forward Gear

When the operator selects the second forward gear, clutch #2 (forward clutch) and clutch #4 (second speed clutch) are applied. See **Figure 9-27**. The #2 sun gear is the input member

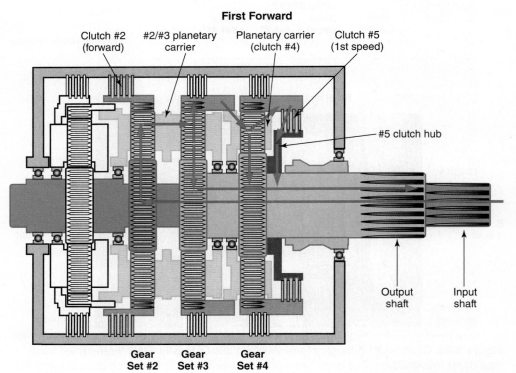

Figure 9-26. The D6R transmission in first forward receives an input through the #2 sun gear. Power is sent to the output shaft through #3 sun gear, #4 sun gear and #5 clutch hub.

Goodheart-Willcox Publisher

to the transmission. With the #2 ring gear held by its clutch, the planetary pinions begin to rotate. This causes the #2/#3 planetary carrier to rotate in a forward slow torque multiplication mode.

The #3 sun gear is attached to the output shaft and acts like a held planetary member, which causes the #3 ring gear to rotate. The #3 ring gear is attached to the #4 planetary carrier, causing it to rotate. Clutch #4 holds the #4 ring gear stationary, which causes the #4 sun gear to drive the output shaft.

In second forward gear, power is distributed to the output shaft through two different paths, the #3 sun gear and the #4 sun gear.

Caterpillar D6R Third Forward Gear

When the operator selects the forward third gear, clutches #2 and #3 are applied. The #2 sun gear is the input member to the transmission. With the #2 ring gear held by its clutch, the #2/#3 planetary carrier rotates in a forward slow torque multiplication mode. Clutch #3 is applied, which holds the #3 ring gear stationary. With the #2/#3 planetary carrier rotating, the #3 sun gear drives the output shaft. The third gear set is operating in a fast forward overdrive mode. The combination of slow forward torque multiplication (gear set #2) and fast forward overdrive (gear set #3) results in an overall transmission ratio of 1:1 direct drive. See **Figure 9-28**.

- Planetary gear set #2 produces a slow forward torque multiplication.

$$(\text{sun} + \text{ring}) \div \text{sun} = \text{ratio}$$

$$(42 + 90) \div 42 = 3.143:1$$

- Planetary gear set #3 produces a fast forward overdrive operation.

$$\text{ring} \div (\text{sun} + \text{ring}) = \text{ratio}$$

$$42 \div (42 + 90) = 0.318$$

- Forward third gear ratio.

$$3.143 \times 0.318 = 1:1 \text{ direct drive}$$

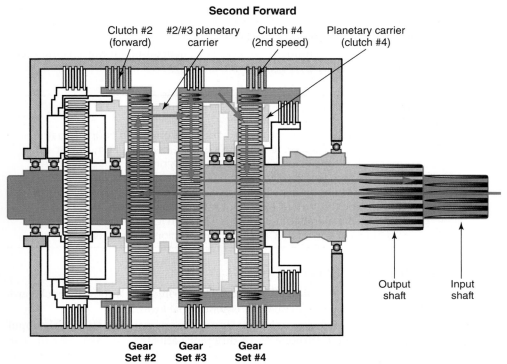

Figure 9-27. The D6R transmission in second forward receives an input through the #2 sun gear. Power is sent to the output shaft through the #3 sun gear and the #4 sun gear.

Goodheart-Willcox Publisher

Figure 9-28. The D6R transmission in third forward receives an input through the #2 sun gear. Power is sent to the output shaft through #3 sun gear.

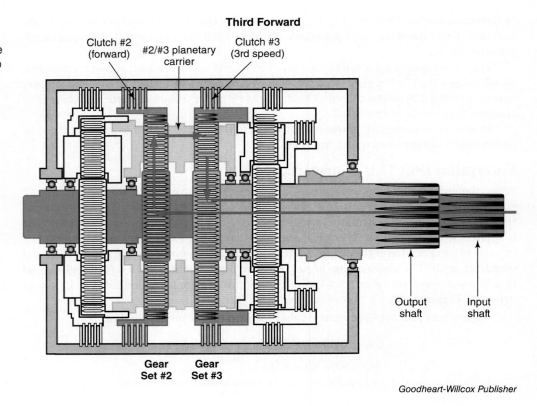

Goodheart-Willcox Publisher

Caterpillar D6R Reverse Gears

Reverse power flow through the D6R transmission is very similar to the forward power flow. The only difference is that the input shaft delivers power into the #1 sun gear and clutch #1 holds the coupling gear, which is attached to the #1 planetary carrier. The *coupling gear* couples the #1 clutch to the #1 planetary carrier. Because the #1 planetary carrier is held in all reverse gears, it causes the #1 ring gear to rotate backward. The #1 ring gear is connected to the #2/#3 planetary carrier. The #1 ring and the #2/#3 planetary carrier become the input to the remaining three gear sets, depending on the specific speed selected for reverse. See **Figure 9-29**, **Figure 9-30**, and **Figure 9-31**.

Note

Most Caterpillar 3-speed and 4-speed transmissions use a coupling gear in the #1 planetary gear set to hold the #1 planetary carrier any time the transmission is driving in reverse. In some rare instances, the Caterpillar service literature for a few specific machines has mislabeled and incorrectly explained the #1 ring gear as the coupling gear and the coupling gear as the #1 ring gear. This text will explain all reverse power flows with the coupling gear as the member that links the #1 clutch to the #1 planetary carrier, which is the most consistent Caterpillar explanation of power flow.

Caterpillar D6R First Reverse Gear

In first reverse gear, clutches #1 and #5 are applied. The #1 ring gear is rotating backward, which drives the center #2/#3 planetary carrier assembly backward.

With the center #2/#3 planetary carrier rotating in the opposite direction of the input shaft, power flow is distributed through three different paths (#3 sun gear, #4 sun gear, and #5 clutch hub) to the output shaft. First reverse gear is similar to first forward, but the output shaft is rotating backward. See **Figure 9-29**.

Chapter 9 | Planetary Gear Set Theory

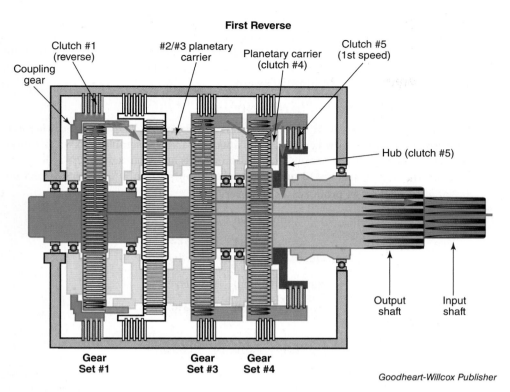

Figure 9-29. The D6R transmission in first reverse receives an input through the #1 sun gear. The #1 clutch holds the coupling gear, which holds the #1 planetary carrier. The #1 ring gear rotates backward while delivering power to the #2/#3 planetary carrier assembly. Power is sent to the output shaft through the #3 sun gear, #4 sun gear, and #5 clutch hub.

Caterpillar D6R Second Reverse Gear

In second reverse gear, clutches #1 and #4 are applied. The #1 ring gear is rotating backward, which drives the center #2/#3 planetary carrier assembly backward.

The #3 sun gear is attached to the output shaft and acts like a held planetary member. This causes the #3 ring gear to rotate backward. See **Figure 9-30**. The #3 ring gear is

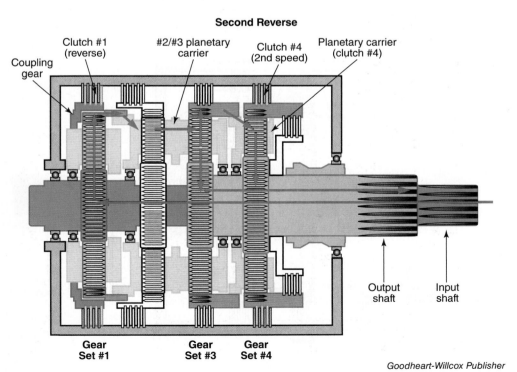

Figure 9-30. The D6R transmission in second reverse receives an input through the #1 sun gear. The #1 planetary carrier is held, which causes the #1 ring gear to rotate backward and deliver power to the #2/#3 planetary carrier assembly. Power is sent to the output shaft through the #3 sun gear and the #4 sun gear.

attached to the #4 planetary carrier, causing it to rotate backward. Clutch #4 holds the #4 ring gear stationary, which causes the #4 sun gear to drive the output shaft backward.

In second reverse gear, power is distributed to the output shaft through two different paths, the #3 sun gear and the #4 sun gear. Second reverse gear is similar to second forward, but the output shaft is rotating backward.

Caterpillar D6R Reverse Third Gear

In reverse third gear, **Figure 9-31**, clutches #1 and #3 are applied. The #1 ring gear is rotating backward, which drives the center #2/#3 planetary carrier assembly backward. Clutch #3 is applied, which holds the #3 ring gear stationary. Because the center #2/#3 planetary carrier is rotating backward, the #3 sun gear drives the output shaft in a reverse direction.

- Planetary gear set #1 produces a slow reverse.

$$\text{ring} \div \text{sun} = \text{ratio}$$
$$83 \div 34 = -2.44:1$$

- Planetary gear set #3 produces a fast forward overdrive.

$$\text{sun} \div (\text{sun} + \text{ring}) = \text{ratio}$$
$$42 \div (42 + 90) = 0.318:1$$

- Reverse third gear ratio.

$$-2.44 \times 0.318 = -0.776:1$$

Note that the #1 sun gear and #1 ring gear are smaller than the #3 sun gear and #3 ring gear.

Power flow through the Caterpillar dozer transmission can be traced by using the schematic in **Figure 9-32**.

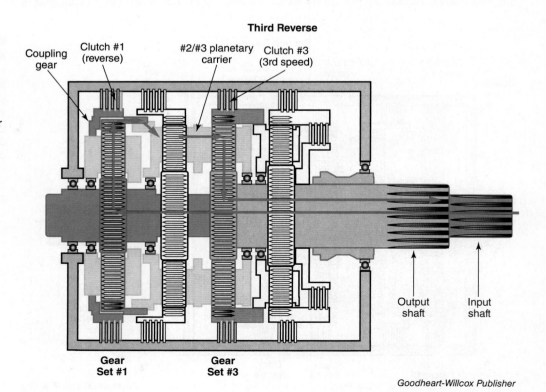

Figure 9-31. The D6R transmission in third reverse receives an input through the #1 sun gear. The #1 clutch is held, which causes the #1 ring gear to rotate backward and deliver power to the #2/#3 planetary carrier assembly. Power is sent to the output shaft through the #3 sun gear.

Goodheart-Willcox Publisher

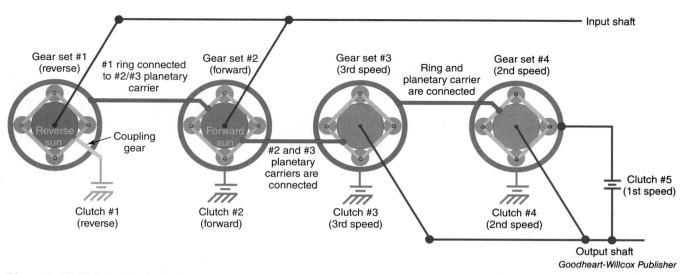

Figure 9-32. The mechanical schematic shows that the first two gear sets provide the input to the transmission. The coupling gear couples clutch #1 and #1 planetary carrier to each other. The #1 ring gear is connected to the #2 /#3 planetary carrier. The #3 ring gear is splined to the #4 planetary carrier. Output is achieved through the last two gear sets through the #3 sun gear, #4 sun gear, and #5 clutch hub.

Caterpillar D6R Clutch Apply Chart

Figure 9-33 is a clutch apply chart for the D6R dozer. Notice that clutch #1 is used only in reverse; therefore, Caterpillar calls this clutch the reverse directional clutch. Clutch #2 is used only in forward gears and is called the forward directional clutch. The last three clutches are called speed clutches because they are used only for specific transmission gear speeds:

- Clutch #3 is used only in third gear.
- Clutch #4 is used only in second gear.
- Clutch #5 is used only in first gear.

Note

Late-model three-speed Caterpillar dozers can function as six-speed or nine-speed dozers by using different engine speeds within each of the transmission's three speeds. This is discussed further in Chapter 11, *Powershift and Automatic Transmission Controls, Service, and Repair.*

—	Directional Clutch #1 (Reverse)	Directional Clutch #2 (Forward)	Speed Clutch #3 (3rd Speed)	Speed Clutch #4 (2nd Speed)	Speed Clutch #5 (1st Speed)
1st Forward	—	X	—	—	X
2nd Forward	—	X	—	X	—
3rd Forward	—	X	X	—	—
Neutral	—	—	X	—	—
1st Reverse	X	—	—	—	X
2nd Reverse	X	—	—	X	—
3rd Reverse	X	—	X	—	—

Goodheart-Willcox Publisher

Figure 9-33. The Caterpillar D6R clutch apply chart shows that clutches #1 and #2 are directional clutches, and clutches #3, #4, and #5 are speed clutches. Notice that two clutches must always be applied to achieve power flow through the transmission.

Caterpillar Wheel-Type Machine, Three-Speed Planetary Transmissions

Several other Caterpillar machines use the same style transmission as the three-speed track-type dozer. For example, the Caterpillar 528B wheel skidder, the 844K and 854K wheel dozers, and the 990K, 992K, and 993K mining wheel loaders also use four planetary gear sets to achieve three forward and three reverse speeds. The differences in these applications are that the input shaft is located at the front of the planetary gear sets, the output shaft exits at the rear of the planetary gear sets, and clutch #5 couples the hub with the #4 planetary carrier instead of the #4 ring gear. See **Figure 9-34**. However, these transmissions use the same clutch apply chart as the dozer, **Figure 9-33**. Some of the planetary components found in a Caterpillar 528B wheel skidder are shown in **Figure 9-35**, **Figure 9-36**, and **Figure 9-37**.

Note that a 994H mining wheel loader uses the same style transmission as the track-type dozers. Both the input shaft and the output shaft enter the front of the transmission, but the #5 clutch couples the #4 planetary carrier to the output shaft instead of the #4 ring gear.

Caterpillar Wheel-Type Machine, Four-Speed Planetary Transmissions

Caterpillar produces transmissions with an additional clutch and an additional planetary gear set located to the rear of the three-speed gear set. These transmissions are used on wheel-type machines and provide four forward speeds and three or four reverse speeds, depending on the model. They are used in the 966H–988K wheel loaders, 824K and 834K wheel dozers, and 825K and 826K landfill compactors. The apply chart in **Figure 9-38** shows that the additional clutch (clutch #6) becomes the first speed clutch. **Figure 9-39** shows a simplified four-speed planetary Caterpillar transmission with the additional planetary gear set and clutch.

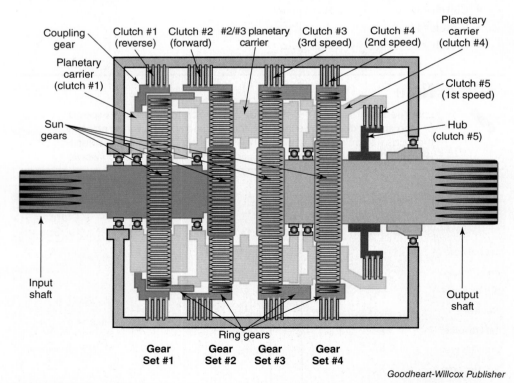

Goodheart-Willcox Publisher

Figure 9-34. Several Caterpillar wheel-type machines use a variation of the track-type dozer three-speed planetary transmission. The two differences are that the input and output shafts enter and exit at opposite ends of the transmission, and clutch #5 connects the hub to the #4 planetary carrier.

Chapter 9 | Planetary Gear Set Theory

Figure 9-35. This coupling gear is held in all reverse gears. It is splined to the #1 planetary carrier, which holds the carrier to reverse the direction of the power flow.

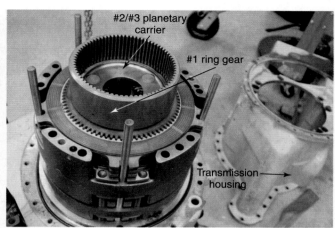

Figure 9-36. The #1 ring gear mates with the #2/#3 planetary carrier. The transmission housing holds the planetary drums stationary.

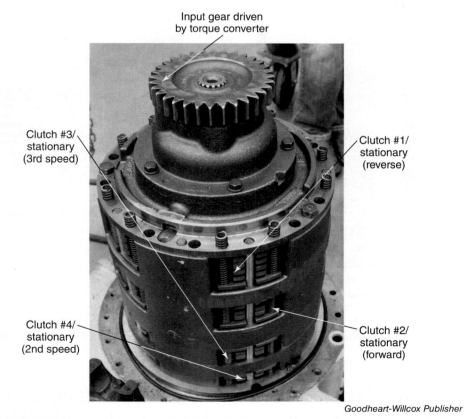

Figure 9-37. The input gear provides input into the transmission via the torque converter's turbine. The outside case has been removed to expose the four stationary clutches.

Gear	Directional Clutch #1 (Reverse)	Directional Clutch #2 (Forward)	Speed Clutch #3 (4th Speed)	Speed Clutch #4 (3rd Speed)	Speed Clutch #5 (2nd Speed)	Speed Clutch #6 (1st Speed)
F1	—	X	—	—	—	X
F2	—	X	—	—	X	—
F3	—	X	—	X	—	—
F4	—	X	X	—	—	—
N	—	—	X	—	—	—
R1	X	—	—	—	—	X
R2	X	—	—	—	X	—
R3	X	—	—	X	—	—

Goodheart-Willcox Publisher

Figure 9-38. The clutch apply chart for large Caterpillar wheel loaders is similar to that for the dozers, with the exception of clutch #6, which is used for first speed.

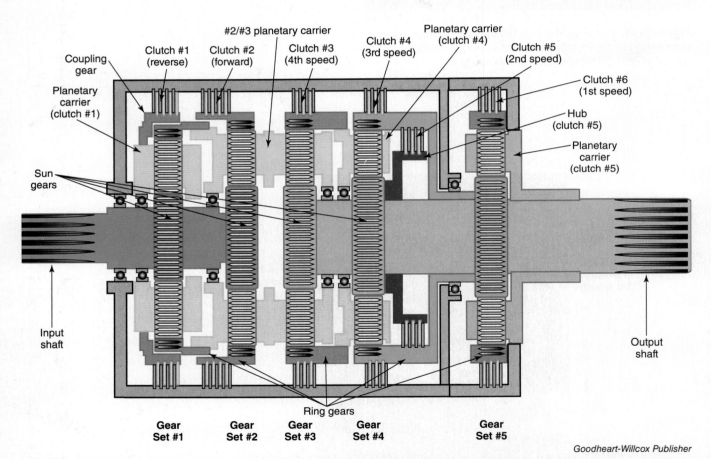

Goodheart-Willcox Publisher

Figure 9-39. Large Caterpillar wheel loaders use planetary transmissions that provide four forward speeds and three or four reverse speeds.

Caterpillar Elevated-Track Dozer, Four-Speed Transmissions

Caterpillar later-model D6T, D8T, and D6, D7, D8 dozers are configured with a four-speed transmission. The transmission's four-speeds, coupled with electronic controls, limit engine speeds to cause the machine to function as if it had additional speeds between the four different transmission speeds. Although the dozer's transmission is capable of having five forward speeds and five reverse speeds, the clutch apply chart in **Figure 9-40** shows the transmission is only allowed to have four forward speeds (2F, 3F, 4F, and 5F) and four reverse speeds (1R, 2R, 3R, and 4R). This is because the reverse gear set has a faster gear ratio, for example 108 ÷ 60 = -1.8:1, than the forward gear set, (93+42) ÷ 42 = 3.2:1. Depending on the dozer model and build date, the apply chart can also label the first four forward gears as F1–F4, rather than F2–F5.

The transmission uses five planetary gear sets, four holding clutches, and two rotating clutches. See **Figure 9-41**. The transmission shares several similarities with the three-speed transmissions used on track-type machines:

- Input and output shafts protrude through the front of the transmission.
- #1 and #2 sun gears provide the input to the transmission.
- Clutch #1 holds the coupling gear to hold the first planetary carrier for reverse gears.
- #2 and #3 planetary carrier are joined.
- #1 ring gear is splined to the #2/#3 planetary carrier
- Clutch #2 holds the second ring gear for forward gears.
- Clutch #4 holds the last ring gear.
- The last planetary carrier is splined to the previous ring gear.
- High-speed forward power flow is similar in both transmissions, entering through the second sun gear and exiting through the third sun gear.

See the mechanical schematic in **Figure 9-42**. Some differences of the four-speed as compared to the three-speed are:

- Three planetary controls must be applied to move forward or rearward (one directional clutch and two speed clutches).
- A sun gear shaft couples the #4 sun gear and the #5 sun gear to the #5 and #6 clutch hub.
- #4 planetary carrier is splined to the output shaft.
- #3 ring gear delivers power through a rotating assembly to the #3 clutch drum.

Gear Range	Directional Clutch #1 (Reverse)	Directional Clutch #2 (Forward)	Speed Clutch #3	Speed Clutch #4	Speed Clutch #5	Speed Clutch #6
1st Reverse	X	—	X	—	X	—
2nd Reverse	X	—	X	—	—	X
3rd Reverse	X	—	X	X	—	—
4th Reverse	X	—	—	X	—	X
Neutral	—	—	—	X	X	—
2nd Forward	—	X	—	—	—	X
3rd Forward	—	X	—	—	—	—
4th Forward	—	X	—	—	—	X
5th Forward	—	X	—	—	X	—

Goodheart-Willcox Publisher

Figure 9-40. Caterpillar elevated-track dozer, four-speed transmission clutch apply chart. Notice that the transmission's forward speeds are limited to 2F, 3F, 4F and 5F, and the reverse speeds are limited to 1R, 2R, 3R, and 4R. Three clutches must be applied to move forward or reverse.

24M Series Mining Motor Grader Six-Speed Transmission

The Caterpillar 24H, 24M, and 24 mining motor graders use a planetary-style transmission. Unlike the smaller motor graders, which do not have torque converters and use a countershaft design, the 24 uses a torque converter to deliver power into the transmission. Chapter 10, *Powershift and Automatic Transmission Theory*, details the smaller motor grader countershaft transmissions. The 24 planetary transmission uses three directional clutches: clutch #1—reverse, clutch #2—low forward, and clutch #3—high forward. The remaining clutches (#4, #5, and #6) are speed clutches. The transmission provides six forward and three reverse speeds. The clutch apply chart is shown in **Figure 9-43**.

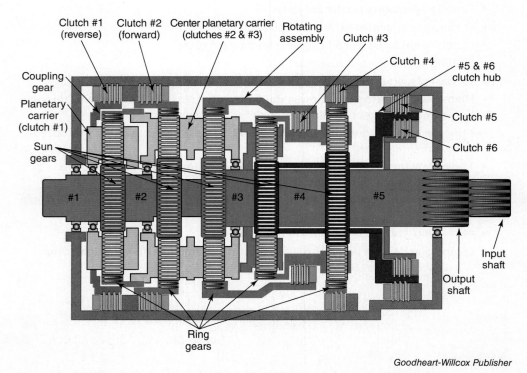

Figure 9-41. Simplified drawing of a Caterpillar elevated-track dozer, four-speed transmission. It uses five simple planetary gear sets and six clutches.

Figure 9-42. A mechanical schematic of a Caterpillar elevated-track dozer, four-speed transmission.

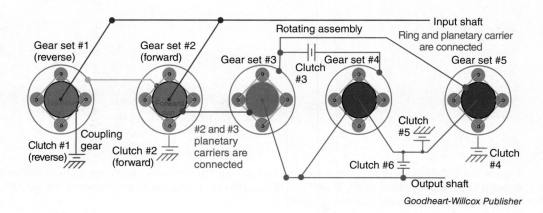

Gear Range	Directional Clutch #1 (Reverse)	Directional Clutch #2 (Low Forward)	Directional Clutch #3 (High Forward)	Speed Clutch #4	Speed Clutch #5	Speed Clutch #6
1st Forward	—	X	—	—	—	X
2nd Forward	—	—	X	—	—	X
3rd Forward	—	X	—	—	X	—
4th Forward	—	—	X	—	X	—
5th Forward	—	X	—	X	—	—
6th Forward	—	—	X	X	—	—
Neutral	—	—	—	X	—	—
1st Reverse	X	—	—	—	—	X
2nd Reverse	X	—	—	—	X	—
3rd Reverse	X	—	—	X	—	—

Goodheart-Willcox Publisher

Figure 9-43. Caterpillar 24H, 24M, and 24 series motor grader clutch apply chart.

Caterpillar Scraper and Haul Truck Planetary Transmissions

Caterpillar scrapers and haul trucks also use planetary transmissions. A tandem engine scraper can be equipped with a four-speed scraper transmission (4F/1F) and an eight-speed tractor transmission (8F/1F). Rigid-frame haul truck transmissions can have seven forward speeds and one reverse speed. Unlike the dozer and loader transmissions with one rotating clutch, scraper and haul truck transmissions can have two rotating clutches.

Allison Automatic Transmissions

Allison Transmission is well known for producing planetary transmissions with torque converters. However, Allison also manufactured a countershaft-style transmission, such as their TC10 series transmission, which is explained in Chapter 10, *Powershift and Automatic Transmission Theory*.

Allison transmissions can be used in a wide range of applications, ranging from 300 horsepower to more than 3300 horsepower and 10,000 foot-pounds of torque. Some examples of Allison transmission applications include:

- General Motors three-quarter ton pickup trucks.
- Buses.
- Cranes.
- Articulated dump trucks.
- Fire trucks.
- Trash trucks.
- Rigid frame, off-road quarry and mining trucks.
- Oil field industry (drilling, fracturing, and pumping trucks).
- Vocational trucks, such as 10-wheeled dump trucks and concrete trucks.
- Airport vehicles including fuel trucks and aircraft tow trucks.
- Military off-road transport vehicles.
- Recreational vehicles (RVs).
- On-highway class 8 trucks.

Allison 1000 through 4000 Series Automatic Transmissions

Allison automatic transmissions are offered in a range of models, from 1000 series up to a 9000 series transmission. This text will discuss the most popular Allison transmissions, ranging from 1000 to 4000 series transmissions. The 1000 and 2000 series are considered the light- and medium-duty transmissions. These transmissions use both electronic controls and a manually operated shift cable. The shift cable operates a manual spool valve and a neutral safety backup (NSBU) switch. The NSBU switch prevents the vehicle from starting in gear and controls the reverse lights and a backup alarm. The manual spool valve provides a limp home mode, which is the capability of the transmission providing one forward gear, reverse, and neutral any time the electronic controls fail. The 3000 and 4000 series transmissions are considered heavy-duty, and are labeled WT for World Transmissions family. The WT transmissions are electronically controlled but do not use a shift cable or a NSBU switch.

The 1000 through 4000 Allison transmissions use the following:

- Three helical-cut planetary gear sets (gear set #1, gear set #2, and gear set #3).
- Two rotating clutches (clutch #1 and clutch #2).
- Three stationary clutches (clutch #3, clutch #4, and clutch #5).

The transmissions offer six forward speeds with three forward reductions (one direct drive and two overdrives) and is available with an optional PTO.

The clutch apply chart for the 1000, 2000, 3000, and 4000 series Allison transmissions is shown in **Figure 9-44**. Notice that the two input clutches are applied in fourth forward gear, which is direct drive.

A simplified mechanical schematic of Allison 1000–4000 series automatic transmissions is shown in **Figure 9-45**. The clutches perform the following:

- Rotating clutch #2 provides an input to the planetary carrier in gear set #2.
- Rotating clutch #1 provides an input to the sun gears in planetary gear set #2 and planetary gear set #3.
- Stationary clutch #3 holds the ring gear of planetary gear set #1.
- Stationary clutch #4 holds the planetary carrier of gear set #1 and the ring gear of planetary gear set #2.
- Stationary clutch #5 holds the planetary carrier of gear set #2 and the ring gear of planetary gear set #3.

Gear Range	Rotating (Input) Clutch #1	Rotating (Input) Clutch #2	Stationary (Holding) Clutch #3	Stationary (Holding) Clutch #4	Stationary (Holding) Clutch #5
Neutral	—	—	—	—	X
1st Forward	X	—	—	—	X
2nd Forward	X	—	—	X	—
3rd Forward	X	—	X	—	—
4th Forward	X	X	—	—	—
5th Forward	—	X	X	—	—
6th Forward	—	X	—	X	—
1st Reverse	—	—	X	—	X

Goodheart-Willcox Publisher

Figure 9-44. Allison 1000 through 4000 series transmissions use two rotating clutches and three stationary clutches.

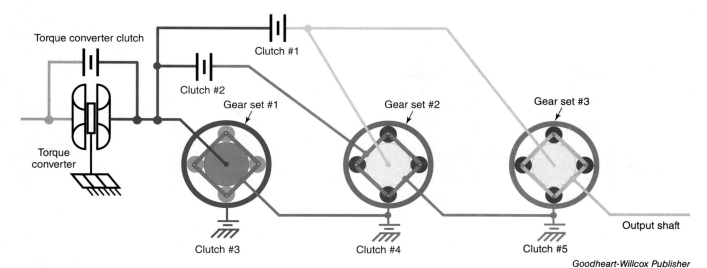

Figure 9-45. Allison 1000 through 4000 series transmissions use three planetary gear sets, two rotating clutches (clutch #1 and clutch #2), and three stationary clutches (clutch #3, clutch #4, and clutch #5).

Allison groups the transmissions into modules that are easily removable and serviceable. The input modules include the torque converter, converter housing, front support, and charging pump. The gearbox modules include the rotating clutch, converter housing, planetary gear set #1, planetary gear set #2, and the main shaft. The output modules include the rear cover, planetary gear set #3, and a retarder or transfer gear. A retarder is a finned rotor that spins inside a finned housing, and when flooded with oil will slow the machine. Chapter 12, *Hydrodynamic Drives*, further explains retarders.

1000 and 2000 Series

The 1000 and 2000 series transmission is the smallest Allison automatic transmission. It is used in three-quarter ton and one-ton trucks, buses, RVs, and other on-highway applications with up to 340 horsepower. See **Figure 9-46**. The torque converter used in Allison 1000 and 2000 series transmissions is a welded assembly, and it is not repairable. Some 1000 series Allison transmissions are limited by the electronic control module to five forward speeds. In certain applications, the sixth speed (the highest overdrive speed) is not needed and the transmission control module (TCM) is programmed to allow speeds one through five only. The transmission

Figure 9-46. The 1000 and 2000 series Allison transmission uses three planetary gear sets, two rotating clutches, and three stationary clutches.

Figure 9-47. This 1000 series Allison transmission is equipped with a PTO drive gear on the C1 and C2 clutch drum housing.

hardware exists and is capable of providing a sixth speed, but the software has limited the transmission to five speeds. At the customer's discretion, the TCM can be reprogrammed to provide the sixth speed. An example of a truck that might be electronically limited to five speeds is a small garbage truck.

Some 1000 and 2000 series transmissions have an optional torque converter–driven PTO. In these transmissions, the PTO is driven at the turbine speed. The turbine shaft drives the rotating clutch module, which includes the clutch drum for clutch #1 and clutch #2. A tone wheel (to read turbine speed) or a PTO drive gear (to deliver power to a PTO and to read turbine speed) is located on the outside of the drum. See **Figure 9-47**.

3000 Series

The 3000 series transmission is used in applications up to 450 horsepower. The transmission is available with a hydraulic retarder and an engine-driven PTO, which is driven by the torque converter's impeller (not the turbine). The 3000 series transmissions are used in medium-duty on-highway truck applications, such as RVs, buses, and vocational trucks. A vocational truck is used for a specific occupation (something other than the traditional hauling of cargo), such as a dump truck, garbage truck, or a concrete mixing truck. See **Figure 9-48**. Torque converters used in 3000 series and larger transmissions are assembled with bolts, allowing the assembly to be serviced.

4000 Series

The 4000 series Allison transmission is a large automatic transmission used in heavy-duty applications up to 800 horsepower. It can be equipped with a retarder and an engine-driven PTO. See **Figure 9-49**.

Allison 1000 through 4000 Series Power Flow

The physical size and design of the Allison 1000 through 4000 series automatic transmissions vary. However, they share the following commonalities. They all have:

- A torque converter.
- Three planetary gear sets (planetary gear set #1, planetary gear set #2, and planetary gear set #3).

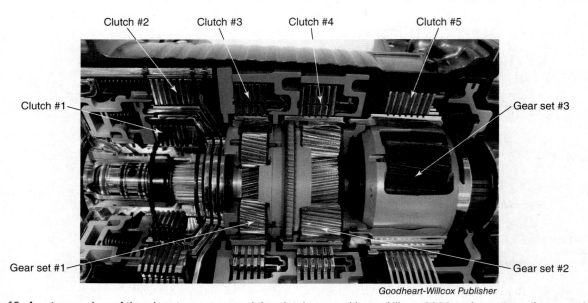

Figure 9-48. A cutaway view of the planetary gears and the clutches used in an Allison 3000 series automatic transmission.

Chapter 9 | Planetary Gear Set Theory

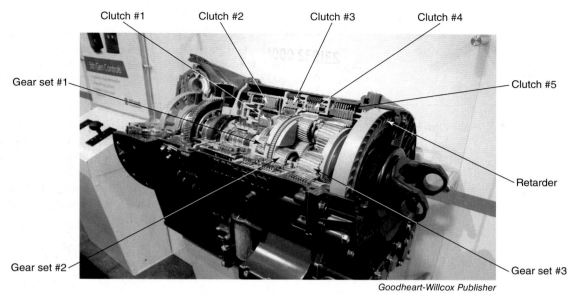

Figure 9-49. A cutaway view of an Allison 4000 series transmission with a hydraulic retarder mounted at the rear of the transmission.

- Two rotating clutches (clutch #1 and clutch #2).
  - Clutch #1 provides an input to the sun gears in planetary gear sets #2 and #3.
  - Clutch #2 provides an input to the planetary carrier in gear set #2, and the ring gear in gear set #3.
- Three stationary clutches (clutch #3, clutch #4, and clutch #5).
  - Clutch #3 holds the ring gear of gear set #1.
  - Clutch #4 holds the planetary carrier of gear set #1 and the ring gear of planetary gear set #2.
  - Clutch #5 holds the planetary carrier of gear set #2 and the ring gear of planetary gear set #3.

A simplified Allison transmission is shown in **Figure 9-50**.

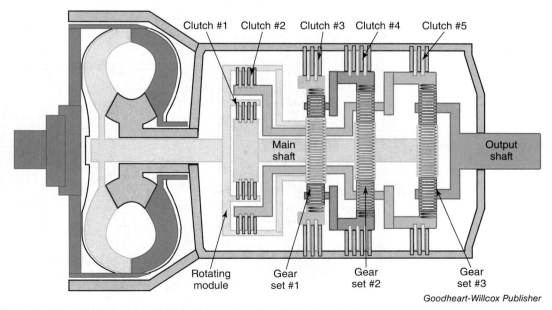

Figure 9-50. A simplified drawing of an Allison 6-speed automatic transmission.

Note

3700 and 4700 transmissions can be equipped with an additional planetary gear set to achieve a deeper first and reverse gear ratio. Deeper gear ratios provide more torque multiplication. This will be explained in more detail later in this chapter.

First Forward

In first forward, power is transmitted through planetary gear set #3. See **Figure 9-51**. Rotating clutch #1 and stationary clutch #5 are applied. Clutch #1 delivers power from the torque converter's turbine to the #3 sun gear via the main shaft. Stationary clutch #5 holds the #3 planetary ring gear. With the #3 sun gear rotating at a speed proportional to turbine speed and the #3 ring gear held, the #3 planetary carrier drives the output shaft in a first forward torque multiplication mode. The gear ratios are:

- 1000 series (close ratio) = 3.10:1.
- 2500 series (wide ratio) = 3.51:1.
- 3000 series (close ratio) = 3.49:1.
- 3500 series (wide ratio) = 4.59:1.
- 4000 series (close ratio) = 3.51:1.
- 4500 series (wide ratio) = 4.70:1.

Second Forward

In second forward gear, **Figure 9-52**, the power flow is transmitted through planetary gear sets #2 and #3. Rotating clutch #1 and stationary clutch #4 are applied. Clutch #1 delivers power from the torque converter's turbine to the sun gears located in planetary gear set #2 and planetary gear set #3 via the main shaft. Stationary clutch #4 holds the #2 planetary ring gear.

The #2 sun gear is rotating at a speed proportional to turbine speed, the #2 ring gear is held, and the #2 planetary carrier is driven in a forward torque multiplication mode.

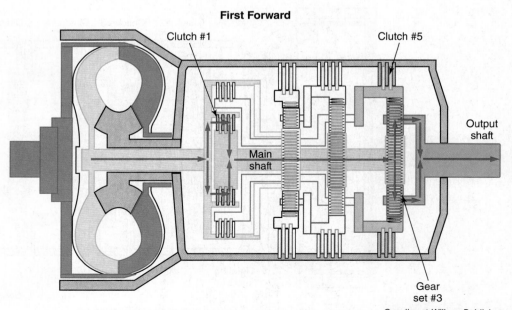

Figure 9-51. In first forward, clutches #1 and #5 are applied. Power flow is delivered through planetary gear set #3 in a forward torque multiplication.

Goodheart-Willcox Publisher

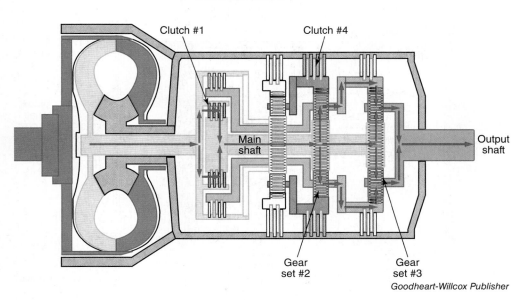

Figure 9-52. In second forward, clutch #1 and clutch #4 are applied. Power flow is delivered through #2 and #3 planetary gear sets.

In gear set #3, the #2 planetary carrier is coupled to the #3 ring gear, causing it to rotate at a slower speed than the #3 sun gear. The #3 ring gear acts like a held member, causing the #3 planetary carrier to rotate in a second gear torque multiplication mode. The gear ratios are:

- 1000 series (close ratio) = 1.81:1.
- 2500 series (wide ratio) = 1.90:1.
- 3000 series (close ratio) = 1.86:1.
- 3500 series (wide ratio) = 2.25:1.
- 4000 series (close ratio) = 1.91:1.
- 4500 series (wide ratio) = 2.21:1.

Third Forward

In third forward, the power flow is transmitted through all three planetary sets. See **Figure 9-53**. Rotating clutch #1 and stationary clutch #3 are applied.

In gear set #1, the #1 planetary sun gear receives an input proportional to turbine speed from the rotating drum assembly. The #1 ring gear is held stationary when the #3 stationary clutch is applied. This causes the #1 carrier to be driven in a forward torque multiplication mode.

In gear set #2, the #1 carrier drives the #2 ring gear at a reduced speed, causing the ring to act like a held member. With clutch #1 applied, the #2 sun gear rotates at a speed proportional to turbine speed, which results in the #2 carrier being driven at a reduced speed in a forward direction.

In gear set #3, applying clutch #1 causes the #3 sun gear to rotate at turbine speed. The #2 planetary carrier drives the #3 ring gear at a reduced speed, causing the #3 ring gear to act like a held member. As a result, the output shaft rotates at a reduced forward torque multiplication. The gear ratios are:

- 1000 series (close ratio) = 1.41:1.
- 2500 series (wide ratio) = 1.44:1.
- 3000 series (close ratio) = 1.41:1.
- 3500 series (wide ratio) = 1.54:1.
- 4000 series (close ratio) = 1.43:1.
- 4500 series (wide ratio) = 1.53:1.

Figure 9-53. In third forward gear, clutch #1 and clutch #3 are applied and power is transmitted through all three planetary gear sets.

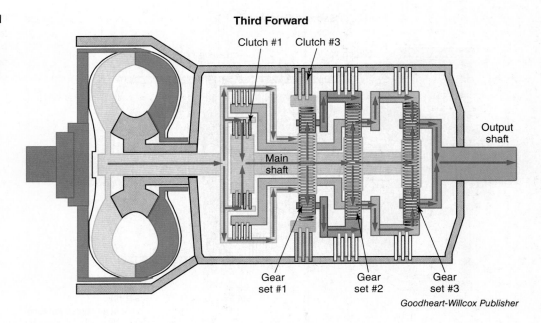

Fourth Forward (Direct Drive)

In fourth forward, both rotating clutches (clutches #1 and #2) are applied and no stationary clutches are applied.

In gear set #3, clutch #2 causes the #2 planetary carrier and the #3 ring gear to rotate at a speed proportional to turbine speed. Clutch #1 causes the #3 sun gear to rotate at a speed proportional to turbine speed. This causes the #3 gear set to lock in a direct drive mode, resulting in the #3 planetary carrier being driven proportional to turbine speed.

The fourth forward gear ratio for all of the 1000 through 4000 series transmissions is a 1:1 direct drive. In fourth forward gear, all the sun gears, planetary carriers, and ring gears revolve as a complete assembly, rotating at the same speed and direction as the turbine shaft. The planetary pinions are not spinning on their shafts due to the direct drive ratio. See **Figure 9-54**.

Fifth Forward (Overdrive 1)

In fifth forward, rotating clutch #2 and stationary clutch #3 are applied. Power is transmitted through all three planetary gear sets. See **Figure 9-55**.

Figure 9-54. In fourth forward gear, the transmission is in a 1:1 direct ratio. Both rotating clutches (clutch #1 and clutch #2) are applied. All three sets of gears are locked in a direct drive ratio.

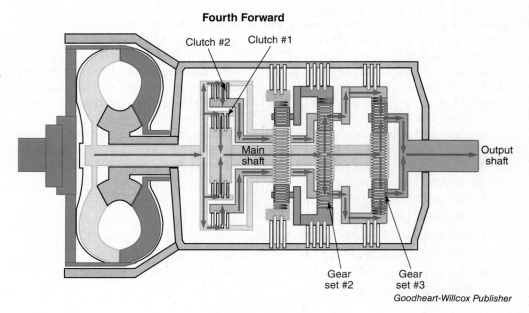

In gear set #1, the rotating clutch drum drives the #1 planetary sun gear at a speed proportional to turbine speed. With clutch #3 applied, the #1 ring gear is held, causing the #1 planetary carrier to drive at a reduced forward speed.

In gear set #2, the #2 ring gear is driven at a reduced speed (by the #1 carrier), which causes the #2 ring to act like a held member. The rotating clutch #2 is applied, causing the #2 planetary carrier to be an input into planetary gear set #2. The #2 sun gear is the output member and is driven at a faster speed.

In gear set #3, the #2 sun gear is coupled to the main shaft and the #3 sun gear, which is rotating faster than turbine speed. Because clutch #2 is applied, the #3 ring gear is rotating at turbine speed, acting as a held member. As a result, the #3 carrier rotates the output shaft at a faster speed than the turbine. The gear ratios are:
- 1000 series (close ratio) = 0.71:1.
- 2500 series (wide ratio) = 0.74:1.
- 3000 series (close ratio) = 0.75:1.
- 3500 series (wide ratio) = 0.75:1.
- 4000 series (close ratio) = 0.74:1.
- 4500 series (wide ratio) = 0.76:1.

Sixth Forward (Overdrive 2)

In sixth forward, rotating clutch #2 and stationary clutch #4 are applied. Power is transmitted through planetary gear sets #2 and #3. See **Figure 9-56**.

In gear set #2, clutch #2 is applied. This causes the #2 planetary carrier to be the input, rotating at turbine speed. Because stationary clutch #3 holds the #2 ring gear stationary, the #2 sun gear is the output, rotating in a forward fast overdrive speed.

The #2 sun gear is coupled to the main shaft and the #3 sun gear, causing the #3 sun gear to be the input to the planetary gear set #3. The #3 ring gear is coupled to the #2 planetary carrier, which is rotating at turbine speed, causing the #3 ring gear to act like a held member. With the #3 sun gear rotating faster than the ring gear, the #3 planetary carrier becomes the output member, driving the output shaft at the transmission's faster overdrive ratio. The gear ratios are:
- 1000 series (close ratio) = 0.61:1.
- 2500 series (wide ratio) = 0.64:1.
- 3000 series (close ratio) = 0.65:1.
- 3500 series (wide ratio) = 0.65:1.

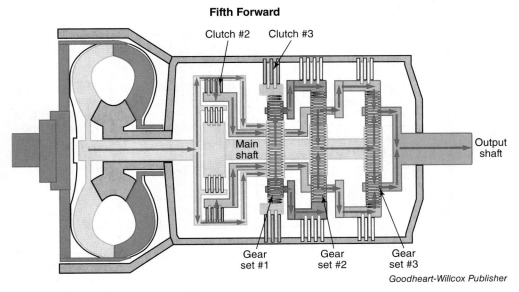

Figure 9-55. In fifth forward gear, clutch #2 and clutch #3 are applied. Power is transmitted through all three planetary gear sets.

Goodheart-Willcox Publisher

Figure 9-56. In the sixth forward gear, clutch #2 and clutch #4 are applied and power flow is transmitted through planetary gear sets #2 and #3.

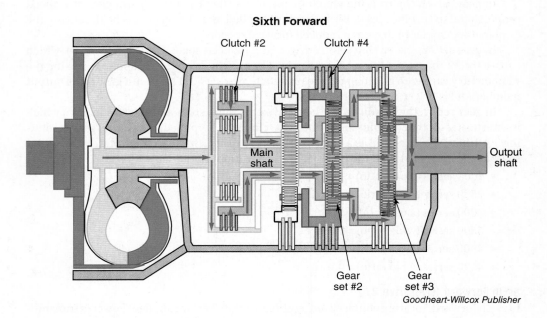

- 4000 series (close ratio) = 0.64:1.
- 4500 series (wide ratio) = 0.67:1.

First Reverse

In first reverse, two stationary clutches (clutch #3 and clutch #5) are applied but no rotating clutches are applied. Power is transmitted through all three planetary gear sets. See **Figure 9-57**.

In gear set #1, the rotating clutch drum rotates the #1 sun gear in the same direction and speed as the turbine. With the #1 ring gear held by clutch #3, the #1 planetary carrier is driven in a forward torque multiplication mode.

In gear set #2, the #1 planetary carrier is coupled to the #2 ring gear, which causes the #2 ring gear to be the input member in a forward torque multiplication mode. Clutch #5

Figure 9-57. In first reverse, two stationary clutches (clutch #3 and clutch #5) are applied and no rotating clutches are applied. Input power is received via the rotating clutch drum, which drives the #1 sun gear. Power is transmitted through all three planetary gear sets.

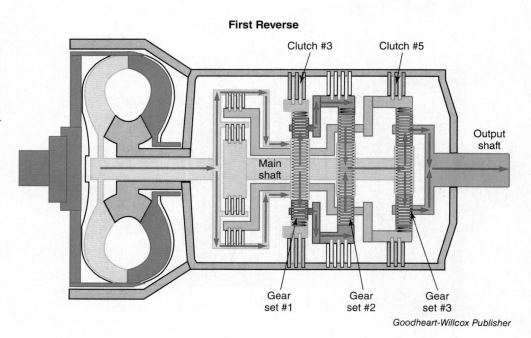

is applied, which holds the #2 planetary carrier stationary. With the #2 planetary carrier stationary, the #2 sun gear is driven in the opposite direction of the turbine.

In gear set #3, the #2 sun gear is coupled to the main shaft and the #3 sun gear. This causes the #3 sun gear to be the input to the last planetary gear set rotating backward. Clutch #5 holds the #3 ring gear stationary. The backward rotating #3 sun gear (input) causes the #3 planetary carrier to multiply torque. The #3 planetary carrier rotates in the same direction as the #3 sun gear. The gear ratios are:

- 1000 series (close ratio) = −4.49:1.
- 2500 series (wide ratio) = −5.09:1.
- 3000 series (close ratio) = −5.03:1.
- 3500 series (wide ratio) = −5.00:1.
- 4000 series (close ratio) = −4.80:1.
- 4500 series (wide ratio) = −5.55:1.

3700 and 4700 Allison Transmissions

The 3000 and 4000 series Allison transmissions can be designed with six or seven forward speeds. The six-speed transmissions covered earlier in this chapter have only one reverse speed. As explained, the traditional six-speed 3000 and 4000 series transmissions use the same clutch apply chart as the 1000 and 2000 series transmissions.

The 3700 specialty series (SPS) and 4700 rugged-duty series (RDS) transmissions have a seventh forward speed, which is a deeper first gear (6.93:1 or 7.63:1, for example). The first gear ratio used in traditional Allison six-speed transmissions range from 3.49:1 to 4.7:1, depending on the specific model. The seven-speed 4700 transmission has two reverse speeds, a traditional reverse gear ratio, such as a −4.80:1 ratio, and a deep reduction reverse gear ratio, such as −17.12:1.

The 4700 seven-speed transmission has an additional planetary gear set (gear set #4) located at the rear of the transmission. This gear set is controlled by a sixth clutch (clutch #6) that is stationary. A cutaway of a 4700 series transmission is shown in **Figure 9-58**. The seven-speed clutch apply chart is shown in **Figure 9-59**.

In low-speed forward, the 4700 uses clutch #1 to provide turbine input speed to the main shaft, which causes the #3 sun gear to be driving at turbine speed. The #3 planetary carrier is attached to the output shaft, which causes it to act like a held member, resulting

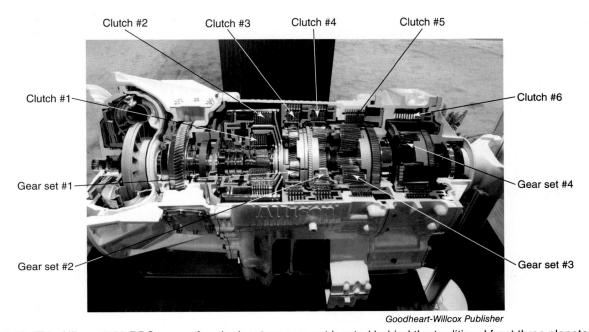

Goodheart-Willcox Publisher

Figure 9-58. This Allison 4700 RDS uses a fourth planetary gear set located behind the traditional front three planetary gear sets. Clutch #6 is used to hold the #4 planetary carrier.

in the #3 ring rotating backwards in a slow reduction. The #3 planetary ring is attached to the #4 sun gear, which provides input into the planetary gear set #4. Clutch #6 holds the #4 planetary carrier, causing the #4 ring gear to reverse the power flow back to a forward direction in a slow reduction. Because the #4 ring gear is attached to the #3 planetary carrier, the output shaft is driven by the #3 planetary carrier via the #4 ring gear. A mechanical schematic is shown in **Figure 9-60**.

The 3700 SPS transmission does not utilize a fourth planetary gear set, but instead uses a transfer gear (drop box) that drops the power down via a helical gear set and delivers power to the front and rear axles, enabling four-wheel drive. Its clutch #6 is responsible for holding the main shaft, which effectively holds the #2 and #3 sun gears in low-speed.

Note

The 3700 SPS low-speed forward gear uses clutch #3 and clutch #6, while the 4700 uses those two clutches to achieve low-speed reverse due to the transmission mechanical design differences. The 3700 uses clutch #6 to hold the main shaft. The 4700 uses clutch #6 to hold the #4 planetary carrier. The 4700 transmission applies clutch #1 and clutch #6 to achieve low-speed forward. The 3700 SPS and 4700 clutch apply chart remains the same for first through sixth forward gears. First reverse also remains the same. A mechanical schematic of the 3700 SPS is shown in **Figure 9-61**.

Allison 6625 ORS

Allison offers larger series transmissions, such as the 6625 transmission shown in **Figure 9-62**. It is designated an Off-Road Series (ORS) transmission. Transmissions of this type are used in rigid-frame haul trucks, articulated dump trucks, coal haulers, and stationary oil field equipment, as well as in ARFF trucks. ARFF stands for aircraft rescue and firefighting. The 6625 transmission can be used in applications with 1025 hp and 3300 lb-ft of torque. This transmission has six forward and two reverse speeds. Notice the retarder is located directly after the torque converter rather than at the rear of the transmission.

Gear Range	Rotating (Input) Clutch #1	Rotating (Input) Clutch #2	Stationary (holding) Clutch #3	Stationary (holding) Clutch #4	Stationary (holding) Clutch #5	Stationary (holding) Clutch #6
Neutral	—	—	—	—	X	—
3700 Low-Speed	—	—	X	—	—	X
4700 Low-Speed	X	—	—	—	—	X
1st Forward	X	—	—	—	X	—
2nd Forward	X	—	—	X	—	—
3rd Forward	X	—	X	—	—	—
4th Forward	X	X	—	—	—	—
5th Forward	—	X	X	—	—	—
6th Forward	—	X	—	X	—	—
1st Reverse	—	—	X	—	X	—
4700 Low Reverse	—	—	X	—	—	X

Goodheart-Willcox Publisher

Figure 9-59. The Allison 3700 SPS and 4700 RDS transmissions are offered with a deep low-speed gear ratio. Forward speeds first through sixth and first reverse remain the same as the other 1000 through 4000 series Allison transmissions. The 3700 and 4700 low-speed clutch applications are different due to the mechanical design differences in the transmissions.

Chapter 9 | Planetary Gear Set Theory

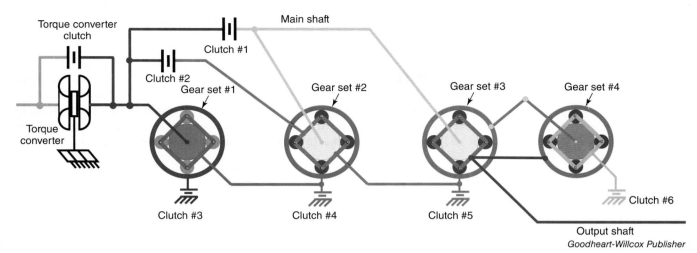

Figure 9-60. The 4700 Allison mechanical schematic shows the addition of planetary gear set #4. It uses clutch #6 to hold the #4 planetary carrier.

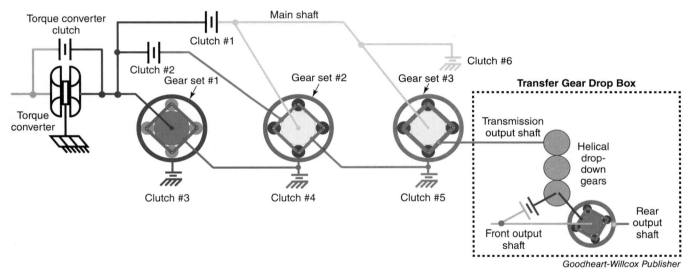

Figure 9-61. This 3700 SPS Allison mechanical schematic shows that clutch #6 holds the main shaft. Notice the absence of planetary gear set #4 and the addition of the transfer gear drop box.

Fracking Transmissions

The oil and gas mining industry use heavy duty transmissions for fracking, which is the process of pumping fracking fluid consisting of water, sand and chemicals, under high pressure to hydraulically fracture underground layers of rock so gas and oil can be recovered. A large diesel engine drives a fracking transmission, which in turn drives a fracking pump. Some fracking transmissions are designed to handle 3500 hp and 10,000 ft-lb of torque. They must be able to endure harsh vibrations and repetitive cycles. Fracking transmissions can have up to nine forward speeds and do not require reverse speeds. Caterpillar, Allison and Twin Disc are examples of manufacturers who produce large fracking transmissions. Caterpillar and Allison fracking transmissions use planetary gear sets.

Figure 9-62. The Allison 6625 Off-Road Series (ORS) transmission is used in rigid-frame haul trucks, articulated dump trucks, coal haulers, stationary oil field equipment, and in ARFF trucks.

John Deere 19-Speed Planetary Transmissions

Although it may appear challenging to trace power flow through the Caterpillar and Allison planetary transmissions described in this chapter, other mobile equipment can be equipped with transmissions that have a much larger range of planetary speeds. John Deere produced an older 19-speed planetary transmission for the 7610, 7710, and 7810 tractors. The transmission had 19 forward speeds and seven reverse speeds. It used five planetary gear sets and no countershaft gears. It was controlled by six brakes and four clutches. It required four planetary controls to be applied to move forward or reverse.

Summary

- A simple planetary gear set can produce eight configurations.
- A simple planetary gear set will produce a change of speed or a change of direction if it receives one input and has one member held, or if it has two inputs driving at two different speeds or driven in two different directions.
- When the planetary carrier is held, the simple planetary gear set is reversing power flow.
- When the planetary carrier is the output, the simple planetary gear set produces a forward torque multiplication.
- When the planetary carrier is the input, the simple planetary gear set produces a forward overdrive.
- When two inputs are driving at the same speed and in the same direction the gear set is in direct drive.
- In order to compute any gear ratio in a simple planetary gear set it requires knowing the number of teeth on the sun and ring gear.
- Tracing power flow through a planetary gear set requires knowing the input members, coupling members, and or held members.
- A Caterpillar D6R dozer transmission has three forward and three reverse speeds.
- The transmission on a Caterpillar elevated-track dozer is located at the rear of the tractor.
- The input shaft and output shaft on Caterpillar elevated-track dozer transmissions enter and exit the front of the transmission.
- A Caterpillar D6R dozer transmission has two directional clutches and three speed clutches. One of each must be applied for the dozer to move.
- Caterpillar uses four planetary gear sets to obtain three forward and reverse speeds in track-type dozers, wheel dozers, wheel skidders, and mining wheel loaders.
- Caterpillar uses five planetary gear sets to obtain four forward speeds and three or four reverse speeds in large wheel loaders, large compactors, and mid-sized dozers.
- Caterpillar and Allison build fracking transmissions that can handle 3500 hp and 10,000 ft-lb of torque and have up to nine speeds and no reverse speeds.
- Allison transmissions use three stationary clutches, two rotating clutches, and three planetary gear sets to achieve six forward speeds.
- John Deere 7610, 7710, and 7810 agricultural tractors used five planetary gear sets controlled by six brakes, four clutches to produce 19 forward and 7 reverse speeds.

Technical Terms

clutch apply charts
compound planetary gear set
coupling gear
epicyclical
planetary carrier
planetary pinion gears
simple planetary gear set
sun gear

Review Questions

Answer the following questions using the information provided in this chapter.

Know and Understand

1. A simple planetary gear set with two inputs rotating at different speeds will be in the _____ mode.
   A. direct drive
   B. overdrive
   C. torque multiplication
   D. Either B or C.

2. What is the advantage of a simple planetary gear set that has more than three planetary pinions?
   A. Torque is distributed across more teeth.
   B. It is quieter.
   C. It has a faster ratio.
   D. It has a slower ratio.

3. Based on the drawing provided, which set of the planetary controls must be applied in order to achieve fourth forward gear?
   A. Clutch A, Brake B, Clutch C.
   B. Clutch A, Brake B, Brake C.
   C. Clutch A, Clutch B, Brake C.
   D. Clutch A, Clutch B, Clutch C.

4. Caterpillar uses _____ planetary gear sets in its D6R transmission.
   A. three
   B. four
   C. five
   D. six

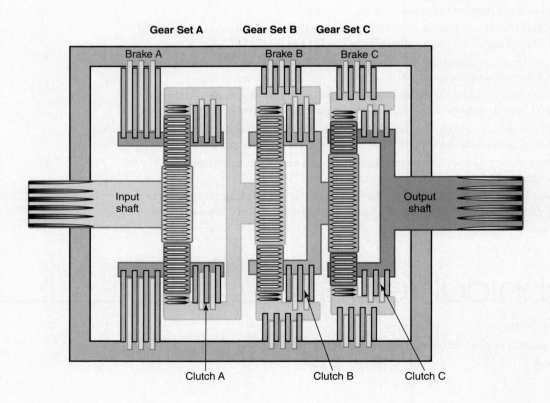

324 Heavy Equipment Power Trains and Systems

5. Which of the following are the input members to the Caterpillar D6R transmission?
   A. #1 and #2 ring gears.
   B. #1 and #2 sun gears.
   C. #3 ring gear, #4 ring gear, and #5 clutch hub.
   D. #3 ring gear, #4 sun gear, and #5 clutch hub.

6. Which of the following are the output members to the Caterpillar D6R transmission?
   A. #1 and #2 ring gears.
   B. #1 and #2 sun gears.
   C. #3 ring gear, #4 ring gear, and #5 clutch hub.
   D. #3 sun gear, #4 sun gear, and #5 clutch hub.

7. In the D6R transmission, the _____ planetary gear set is used only in forward.
   A. #1
   B. #2
   C. #3
   D. #4

8. In the D6R transmission, the _____ planetary gear set is used only for reverse.
   A. #1
   B. #2
   C. #3
   D. #4

9. When the coupling gear is held in a D6R transmission, the _____ is also held.
   A. #1 sun gear
   B. #1 planetary carrier
   C. #2 sun gear
   D. #2 planetary carrier

10. In a D6R transmission, the coupling gear couples _____ to the #1 planetary carrier.
    A. clutch #1
    B. clutch #2
    C. clutch #3
    D. clutch #4

11. How many directional clutches are in a D6R transmission?
    A. Zero.
    B. One.
    C. Two.
    D. Three.

12. Caterpillar 966H–988K wheel loaders and 825K and 826K compactors use _____ planetary gear sets in their transmissions.
    A. three
    B. four
    C. five
    D. six

13. How many stationary clutches are used in Allison 1000 through 4000 series transmissions
    A. One.
    B. Two.
    C. Three.
    D. Four.

14. Which speed is direct drive in Allison 1000 through 4000 series transmissions?
    A. Third.
    B. Fourth.
    C. Fifth.
    D. Sixth.

15. How many rotating clutches are applied in Allison 1000 through 4000 series transmissions to achieve reverse?
    A. Zero.
    B. One.
    C. Two.
    D. Three.

16. How many planetary gear sets are in the John Deere 7010 series planetary transmission?
    A. Three.
    B. Four.
    C. Five.
    D. Six.

17. How many planetary controls are engaged at one time to achieve power flow in the John Deere 7000 series planetary transmission?
    A. Two.
    B. Three.
    C. Four.
    D. Five.
18. How many countershafts are in the John Deere 7000 series planetary transmission?
    A. Zero.
    B. One.
    C. Two.
    D. Three.

Apply and Analyze

19. A simple planetary gear set in the _____ mode has two inputs rotating at the same speed and in the same direction.
20. Any time a simple planetary gear set has one input and the remaining two gears are locked together, the resulting power flow will be _____.
21. Any time a simple planetary gear set only has one input and nothing else (applied), the resulting configuration will be _____.
22. If a simple planetary gear set's planetary carrier is the output, the gear set's configuration will be _____.
23. The D6R transmission provides _____ forward speeds.
24. In a Caterpillar three-speed transmission, the planetary _____ is shared between the second and third planetary gear sets.
25. In a D6R transmission, the number of clutches that must be engaged at one time in order to propel the machine is _____.
26. When the D6R transmission is in first gear (forward or reverse), power flow is distributed (sent to the output shaft) through _____ different path(s).
27. An Allison 1000 through 4000 series transmission has a total of _____ planetary gear sets.
28. An Allison 1000 through 4000 series transmission has a total of _____ rotating clutches.

Critical Thinking

29. A simple planetary gear set has a 30-tooth sun gear as the input. The planetary carrier is the output. The 70-tooth ring gear is held. What is the gear speed, direction, and gear ratio?
30. A simple planetary gear set has a 70-tooth ring gear as the output. The planetary carrier is the input. The 30-tooth sun gear is held. What is the gear speed, direction, and gear ratio?

Chapter 10
Powershift and Automatic Transmission Theory

Objectives

After studying this chapter, you will be able to:

✓ Describe on-highway countershaft automatic transmissions.
✓ Describe the different types of off-highway transmissions.
✓ Describe countershaft powershift transmissions.
✓ Describe powershift transmissions that contains both planetary and countershaft gearing.

Automatic and powershift transmissions are found in a variety of off-highway machines, including agricultural, construction, and mining equipment.

On-Highway Countershaft Multiple Disc Clutch Transmissions

Allison manufactured the TC10 series of transmission to compete with the automated manual transmissions used in the class 8 on-highway truck market. The TC prefix is an abbreviation for *twin countershaft*. A cutaway of a TC10 is shown in **Figure 10-1**. The TC10 transmission has two clutches on each countershaft in the main gearbox, along with a forward-reverse synchronizer. Countershaft #1 contains clutches #3 and #1. Countershaft #2 contains clutches #5 and #2. A simplified drawing is shown in **Figure 10-2**. Each of the countershafts' clutches receive input from a hub that is driven by one of four gears on the main shaft. When a countershaft clutch is engaged, its clutch drum is splined to the countershaft, causing the countershaft to spin. A gear is located at the rear of both countershafts and delivers power to the front gearbox's output gear. The output gear provides input to the planetary gear set located in the rear gearbox.

Clutch #4 is located in the rear gearbox and is used to provide direct drive through the front gearbox to the rear gearbox, bypassing the countershafts. The rear gearbox also contains two additional clutches (clutch #6 and clutch #7) that operate a simple planetary gear set, either low speed (clutch #6) or direct drive (clutch #7).

The transmission has 10 forward speeds and two reverse speeds. A clutch apply chart is shown in **Figure 10-3**. Two clutches must be engaged for the output shaft to move, one of the first five clutches in combination with either clutch #6 or clutch #7. Even though the forward

Figure 10-1. A cutaway of the Allison TC10 transmission. It uses both a twin countershaft in the main gearbox and a planetary gear set in the rear gearbox.

synchronizer is engaged for all forward gears, it is only used in conjunction with clutch #2, similar to the reverse synchronizer that is also only delivering power flow in conjunction with clutch #2. The forward and reverse synchronizer is linked to the engagement of clutch #2.

Similar to an automated manual transmission, the TC10 shifts shortly after the engine load has been reduced. However, the engine load does not have to be completely reduced as it does with an automated manual transmission. The biggest difference between an automated manual transmission and the TC10 is that the TC10 has a torque converter that allows slippage to take place between the engine's crankshaft and the transmission's input shaft.

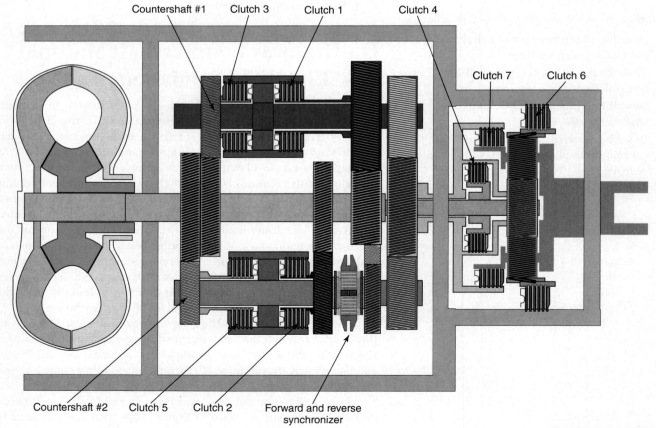

Figure 10-2. A simplified view of an Allison TC10 transmission.

Gear Range	Clutch #1	Clutch #2	Clutch #3	Clutch #4	Clutch #5	Clutch #6 (Low)	Clutch #7 (Direct)	Forward Synchronizer	Reverse Synchronizer
1st Reverse	—	X	—	—	—	X	—	—	X
2nd Reverse	—	X	—	—	—	—	X	—	X
Neutral	—	—	—	—	—	X	—	—	X
1st Forward	X	—	—	—	—	X	—	X	—
2nd Forward	—	X	—	—	—	X	—	X	—
3rd Forward	—	—	X	—	—	X	—	X	—
4th Forward	—	—	—	X	—	X	—	X	—
5th Forward	—	—	—	—	X	X	—	X	—
6th Forward	X	—	—	—	—	—	X	X	—
7th Forward	—	X	—	—	—	—	X	X	—
8th Forward	—	—	X	—	—	—	X	X	—
9th Forward	—	—	—	X	—	—	X	X	—
10th Forward	—	—	—	—	X	—	X	X	—

Goodheart-Willcox Publisher

Figure 10-3. The Allison TC10 transmission requires that either the forward or reverse synchronizer and two clutches (the clutch #1, clutch #2, clutch #3, clutch #4 or clutch #5) and clutch #6 or clutch #7 be engaged in order to move.

Off-Highway Transmission Applications

An *automatic transmission* automatically upshifts or downshifts based on the machine's travel speed and load. Automatic transmissions are common in applications with higher travel speeds, up to 45 mph. Two examples of machines that use automatic transmissions are haul trucks and self-propelled scrapers. Manufacturers also use electronic controls to allow a powershift transmission to function as an automatic, which is explained further in Chapter 11, *Powershift and Automatic Transmission Controls, Service, and Repair*. Some examples of machines using electronically controlled powershift transmissions include dozers, agricultural MFWD and articulated tractors, and motor graders.

A *powershift transmission* is a semi-automatic transmission that relies on the command of the operator to upshift and downshift. It is different from a manual transmission in that it does not require the operator to depress the clutch pedal to interrupt engine power for completing a shift. A *clutch pedal* is a foot-operated pedal used to disengage engine power from the drive train. In a powershift transmission, the operator simply commands the tractor to upshift or downshift (by pressing the upshift or downshift buttons) and the powershift transmission makes the shift without the need to depress a clutch pedal.

Machines equipped with a direct drive powershift transmission are often equipped with a clutch pedal. A direct drive powershift transmission does not have a torque converter, and the engine directly drives the input shaft of the powershift transmission. The clutch pedal on direct drive powershift transmissions must be depressed when bringing the machine to a stop. Otherwise, if the service brake pedal is depressed and the powershift remains in gear, the engine will stall and die. The clutch pedal is also sometimes called an *inching pedal* because when the machine is stopped and the transmission is in gear, as the operator slowly releases the clutch pedal, the machine slowly inches forward or rearward. Examples of machines that can use direct drive powershift transmissions with a clutch pedal are agricultural tractors and motor graders.

Any time the clutch pedal is depressed, engine power is decoupled from the drivetrain, using one of several different clutches within the powershift transmission. Decoupling the engine in a powershift transmission is a little different than decoupling the engine from a traditional manual transmission. Machines with traditional manual transmissions typically use a single dedicated traction clutch to decouple the engine from the transmission. Depending on the powershift design, when the clutch pedal is depressed, a control module disengages one clutch or one of multiple different clutches. For example, later in this chapter the John Deere 16-speed powershift is explained. It has three output clutches (BC, CC, DC) and brake A that can serve as the transmission's master clutch.

The *master clutch* is a clutch that is responsible for decoupling the engine from the powertrain when the clutch pedal is depressed. Master clutches are modulated, meaning that the clutch apply pressure is varied to ease the clutch engagement to make a smoother transition from neutral to forward, or neutral to reverse.

The master clutch is not always an input clutch. An *input clutch* is a clutch that, when engaged, provides input into the transmission. It transfers power from the engine to the transmission's countershaft. In some applications, an input clutch can be called a *directional clutch* because (depending on which input clutch is engaged) it causes the machine to move forward or in reverse. Input clutches can be located on the input shaft or on a countershaft. The master clutch can be a single clutch, such as the one in a Case IH Magnum powershift transmission, which is explained later in this chapter. The master clutch can also be an output clutch that decouples the transmission from the axles. *Output clutches* are often located on the output shaft of a countershaft transmission and transfer power from the transmission's countershaft to the transmission's output shaft. The John Deere 8R 16-speed powershift transmission (which is explained later in this chapter) uses one of three output clutches or brake A as the master clutch. Later in this chapter, it is explained that the Caterpillar H, M, and non-suffix series motor graders use the clutch pedal (inching pedal) to modulate the oil to the powershift's directional clutches.

Some powershift-equipped machines are equipped with torque converters and no clutch pedal. Examples include large dozers (track type and wheel type), large wheel loaders, and large articulated compactors. When the powershift is in gear and if the service brake pedal is depressed, the engine will not die because the torque converter allows for slippage. Chapter 12, *Hydrodynamic Drives*, explains torque converter operation.

Note
Some large torque-converter-equipped Caterpillar wheel loaders have three foot pedals even though they do not have a master clutch. The left pedal controls an impeller clutch within the torque converter.

Automatic and powershift transmissions share many characteristics:
- They have fixed gear ratios for a given transmission gear selection.
- They can upshift or downshift without the operator pressing a clutch pedal and without interrupting the machine's power.
- Their design may include planetary gear sets.
- Their design may include countershaft gears.
- Their design may include both planetary and countershaft gears.

Note
Because of the many similarities between automatic and powershift transmissions, some people may use the terms automatic and powershift interchangeably to describe a transmission.

Transmissions can be further classified as hauling or cycling based on machine operation. A *hauling transmission* is found in machines that haul payloads long distances, such as a mile or more, at higher travel speeds (up to 45 mph). Two examples of machines that require hauling transmissions are scrapers and haul trucks. See **Figure 10-4**.

Machines that operate for short distances and frequently cycle between forward and reverse propulsion are equipped with *cycling transmissions*. Examples of machines that use cycling transmissions are wheel loaders and dozers. Cycling transmissions often have the same number of gears for forward propulsion and for reverse propulsion. However, some cycling transmissions have more forward gears than reverse gears.

Powershift transmissions and automatic transmissions can be either hydraulically controlled or electrohydraulically controlled by an electronic control module (ECM). Older machines use a mechanical lever to control the transmission. See **Figure 10-5**. Late-model electronically controlled machines normally use push button and joystick controls. See **Figure 10-6**.

Some powershift transmissions have special control features, including speed matching, preselecting a gear, and skip shifting. The *speed matching* feature, also known as ground speed matching, matches the transmission gear to the tractor's travel speed. This occurs when the clutch pedal is depressed, the tractor slows, and then the clutch pedal is released. Speed matching normally occurs when the tractor is decreasing speed (downshifting). The *preselect gear* feature allows the operator to preselect a specific forward or reverse gear. The machine initiates its propulsion from a stop in the preselected gear.

Goodheart-Willcox Publisher

Figure 10-4. This transmission gear shift lever is in a Caterpillar mining haul truck. Notice the shift lever has the same shift positions found in an automotive automatic transmission. The lever on the right is the hoist lever. It has four positions for controlling the dump bed. From front to rear, the positions are lower, float, hold, and raise.

The *skip shifting* feature allows the operator to quickly shift into higher gears by intentionally skipping some of the transmission gears while upshifting. The old Case IH 9100-9300 series Steiger 12-speed powershift transmissions had a skip shift button and a pulser lever. The *pulser lever* is a three position control lever used to upshift and downshift the transmission. It is held in a neutral centered position. When the operator presses the lever forward (pulsed forward), it signals the ECM to upshift the transmission. When the lever is pulsed rearward, the ECM downshifts the transmission. When the skip shift button was depressed and the pulser lever was pulsed forward, the transmission could be commanded to skip shift from first gear to fourth gear, fourth to sixth gear, and sixth to eighth gear. In reverse, the tractor could skip to third reverse.

Goodheart-Willcox Publisher

Figure 10-5. This Caterpillar D5H dozer uses one mechanical lever to control the speed and direction of the dozer's three-speed planetary powershift transmission. The lever slides inside the C-shaped slot. The top of the slot is the transmission's neutral position. The left side of the slot is for first, second, and third gear reverse propulsion. The right side of the slot is for first, second, and third gear forward propulsion.

A B *Goodheart-Willcox Publisher*

Figure 10-6. Caterpillar M series motor grader transmission controls. A—The right-hand joystick controls the powershift transmission. The two yellow buttons in the center are used for upshifting (top yellow button) and downshifting (bottom yellow button). B—The yellow toggle switch on the front of the joystick allows the operator to toggle between forward, neutral, and reverse. The transmission has eight forward speeds and six reverse speeds.

Caution

Care should be taken when using the preselect gear on older agricultural tractors. An inexperienced operator who starts in too high of a gear may be thrown out of the seat if the seat belt is not being worn. The sudden lunge of the machine may not be as severe in some late-model tractors with modulated multiple-disc clutches and electronic controls.

Countershaft Powershift Transmissions

Many off-highway machines have a *countershaft powershift transmission*. This type of transmission resembles a traditional manual transmission often containing three or more parallel shafts, such as an input shaft, countershaft, and an output shaft. However, the countershaft powershift transmission uses multiple-disc clutches rather than shift collars or synchronizers to lock the gears to the shafts and distribute power flow. A countershaft powershift transmission takes up more space than a planetary powershift transmission.

Case IH Magnum Countershaft Powershift Transmission

In 1987, Case IH produced the 7100 series Magnum tractors with a countershaft powershift transmission. The Magnum tractor continued with the 7200 series, 8900 series, MX Magnum series, and Magnum 180-340 tractors. Depending on the model, Magnum tractors can be purchased with a countershaft powershift transmission or with a continuously variable transmission (CVT). Late-model, high-horsepower Magnum tractors are offered only with CVT. CVTs will be explained in Chapter 15, *Continuously Variable Transmissions*.

The early Magnum countershaft powershift transmissions had 18 forward speeds and four reverse speeds. The transmission also offered an optional creeper feature that provided six forward creeper speeds and two reverse creeper speeds. A *creeper transmission* is required when a tractor needs to travel at very slow speeds, such as at a rate of feet per minute instead of miles per hour. The slow speeds enable the engine to operate at high speeds so the tractor can deliver maximum PTO speed and hydraulic flow. Creeper transmissions are used for planting and harvesting vegetables. See **Figure 10-7**.

Note

If a creeper transmission is needed, the tractor should be ordered with this option. Adding the option later typically requires splitting the tractor, which is labor intensive and cost prohibitive.

Late-model Magnum countershaft powershift transmissions without the creeper option can be equipped with an overdrive 19th speed. The 19-speed powershift transmission provides a maximum road speed of 31 mph (50 kph), as compared to the 18-speed with 24.8 mph (40 kph) maximum travel speed. Customers needing a creeper function and a fast travel speed can purchase the CVT-equipped Magnum tractor, which offers both.

Note

For years, European agricultural tractors have been designed with a road speed up to 31 mph (50 kph). Today, North American CVT-equipped tractors offer higher travel speeds, 31 mph (50 kph), than tractors equipped with traditional powershift transmissions, 24.8 mph (40 kph). See Chapter 15, *Continuously Variable Transmissions*, for more information on CVT-equipped tractors.

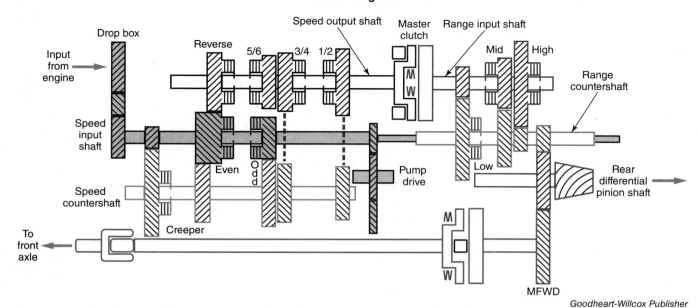

Figure 10-7. The Case IH Magnum countershaft powershift transmission requires a minimum of four clutches be applied to achieve power flow. A creeper transmission is an option for customers who need high engine rpm and slow travel speeds. The dashed lines between the speed countershaft and the speed output shaft indicate that the gears on the two shafts are in mesh with each other.

The Magnum powershift transmission is electronically controlled with pulse-width modulating (PWM) solenoids. PWM solenoids are variable solenoids controlled by an ECM to vary the clutch's apply pressure. PWM is explained in greater detail in Chapter 11, *Powershift and Automatic Transmission Controls, Service, and Repair*. The transmission has shuttle shift capability, allowing the operator to switch between forward and reverse by actuating the shuttle shift controls without shifting the transmission or depressing a clutch pedal. The transmission can be programmed to shuttle shift while in forward gears (one through twelve) to reverse gears (one through four). Early Magnums had a shuttle shift lever on the steering column. Late-model Magnums have forward and reverse electric push buttons on the shift lever.

 Note

At least one off-highway transmission manufacturer inhibits shuttle shifting based on the transmission output-shaft speed. The maximum output shaft speed allowable for shuttle shifting varies depending on the transmission and application. If the maximum output shaft speed is set as 350 rpm by the manufacturer, the machine will be prevented from shuttle shifting any time the transmission output -haft speed is greater than 350 rpm.

The Magnum transmission also has **auto shift control**, which automatically shifts the transmission based on engine load, tractor speed, and the current gear selected.

The transmission has two sub-modes of operation: auto road operation and auto field operation.

- **Auto road operation** enables the operator to control the speed of the tractor with the engine's throttle. This feature is used for reaching maximum road speed.
- **Auto field operation** maintains engine speed while operating in the field. The operator chooses the maximum transmission gear for field operation. If the engine speed drops, the auto field operation automatically downshifts to maintain engine speed. This ensures there is not loss of hydraulic flow or PTO speed. When the tractor's load lightens, the auto field operation automatically upshifts until the selected maximum gear is reached.

Power Flow

Because the Magnum engine's crankshaft is not in alignment with the speed transmission's input shaft, the engine delivers power to a drop box. The **drop box** connects the engine to the speed transmission. The order of power train components is:

1. Engine.
2. Drop box.
3. Speed transmission.
4. Range transmission.
5. Differential.
6. Final drives.

Power flows through the transmission in the following order:

1. Speed input shaft.
2. Speed countershaft.
3. Speed output shaft.
4. Master clutch.
5. Range input shaft.
6. Range countershaft.
7. Differential pinion shaft and mechanical front-wheel-drive (MFWD) clutch.

The clutch apply chart in **Figure 10-8** shows that the Magnum countershaft powershift transmission requires three separate clutches and the master clutch to be applied in order to achieve power flow:

- The odd, even, creeper, or overdrive clutch must be applied.
- The 1-2, 3-4, 5-6, or reverse clutch must be applied.
- The low, mid, or high clutch must be applied.
- The master clutch must be applied.

Mechanical Front-Wheel Drive

The Magnum is an agricultural mechanical front-wheel-drive (MFWD) tractor. The **MFWD clutch** is the clutch pack used to engage four-wheel drive. When the MFWD clutch is released, the tractor is in two-wheel drive and propelled by the rear axle.

Like most MFWD tractors, the Magnum MFWD clutch is a spring-applied clutch pack inside the transmission. The clutch is applied when there is no oil pressure and released when there is oil pressure. When the clutch is applied, power flow is sent to the front axle through the MFWD clutch driveshaft.

Note

MFWD clutches are normally located after the transmission gears. The MFWD clutch is driven by the same gear or shaft that drives the rear differential pinion gear. As the transmission speed and direction change, the MFWD driveshaft turns at the same speed as the rear differential pinion.

Warning

It is dangerous to operate an MFWD tractor when the rear axle is on jack stands and the front wheels are on the ground. Even when the clutch is released, the MFWD clutch can engage the front axle if the service brake is applied. This could cause the front axle to pull the tractor off the rear jack stands.

Case IH Magnum Countershaft Powershift Clutch Apply Chart

Gear Range	Master	Odd	Even	Creeper	Overdrive	1-2	3-4	5-6	Reverse	Low	Mid	High
Neutral	—	—	—	—	—	—	X	X	—	—	—	—
Park	—	—	—	—	—	—	—	—	—	—	—	—
1F	X	X	—	—	—	X	—	—	—	X	—	—
2F	X	—	X	—	—	X	—	—	—	X	—	—
3F	X	X	—	—	—	—	X	—	—	X	—	—
4F	X	—	X	—	—	—	X	—	—	X	—	—
5F	X	X	—	—	—	—	—	X	—	X	—	—
6F	X	—	X	—	—	—	—	X	—	X	—	—
7F	X	X	—	—	—	X	—	—	—	—	X	—
8F	X	—	X	—	—	X	—	—	—	—	X	—
9F	X	X	—	—	—	—	X	—	—	—	X	—
10F	X	—	X	—	—	—	X	—	—	—	X	—
11F	X	X	—	—	—	—	—	X	—	—	X	—
12F	X	—	X	—	—	—	—	X	—	—	X	—
13F	X	X	—	—	—	X	—	—	—	—	—	X
14F	X	—	X	—	—	X	—	—	—	—	—	X
15F	X	X	—	—	—	—	X	—	—	—	—	X
16F	X	—	X	—	—	—	X	—	—	—	—	X
17F	X	X	—	—	—	—	—	X	—	—	—	X
18F	X	—	X	—	—	—	—	X	—	—	—	X
19F	X	—	—	—	X	—	—	X	—	—	—	X
1R	X	X	—	—	—	—	—	—	X	X	—	—
2R	X	—	X	—	—	—	—	—	X	X	—	—
3R	X	X	—	—	—	—	—	—	X	—	X	—
4R	X	—	X	—	—	—	—	—	X	—	X	—
Creep-1F	X	—	—	X	—	X	—	—	—	X	—	—
Creep-2F	X	—	—	X	—	—	X	—	—	X	—	—
Creep-3F	X	—	—	X	—	—	—	X	—	X	—	—
Creep-4F	X	—	—	X	—	X	—	—	—	—	X	—
Creep-5F	X	—	—	X	—	—	X	—	—	—	X	—
Creep-6F	X	—	—	X	—	—	—	X	—	—	X	—
Creep-1R	X	—	—	X	—	—	—	—	X	X	—	—
Creep-2R	X	—	—	X	—	—	—	—	X	—	X	—

Goodheart-Willcox Publisher

Figure 10-8. This clutch apply chart is for a Case IH Magnum countershaft powershift transmission that is equipped with the optional creeper transmission or the optional 19th speed overdrive gear.

Magnum Countershaft Powershift 1F Power Flow

When the first forward gear is selected, the odd clutch, 1-2 clutch, and low range clutch are applied. See **Figure 10-9**. Power flow is delivered to the transmission's input shaft by the drop box. With the odd clutch engaged, the speed input shaft sends power to the speed countershaft. The speed countershaft sends power to the speed output shaft, which has the 1-2 clutch engaged.

The master clutch is applied as the operator releases the clutch pedal, delivering power from the speed output shaft to the range input shaft. The range input shaft delivers power to the range countershaft, which has the low clutch applied. The range countershaft sends power to the rear differential pinion shaft. The rear differential pinion shaft drives the rear axle's differential assembly. The MFWD clutch is engaged and sends power to the front axle.

In the next three sections, examples of the power flow through a Magnum Countershaft Powershift transmission in forward gears is explained. Each of the examples is based on the engagement of a different range clutch. The remaining gears within each range are achieved by engaging the odd or even clutch and a speed clutch in different combinations.

Magnum Countershaft Powershift 10F Power Flow

When 10th forward gear is selected, the even clutch, 3-4 clutch, and midrange clutch are applied. See **Figure 10-10**. Power flow is delivered to the transmission's speed input shaft by the drop box. With the even clutch engaged, the input shaft sends power to the speed countershaft. The speed countershaft sends power to the speed output shaft, which has the 3-4 clutch engaged.

The master clutch is applied as the operator releases the clutch pedal, delivering power from the speed output shaft to the range input shaft. With the mid clutch engaged, the range input shaft delivers power to the range countershaft. The range countershaft sends power to the rear differential pinion shaft. The rear differential pinion shaft drives the rear axle's differential assembly. The MFWD clutch is engaged and sends power to the front axle.

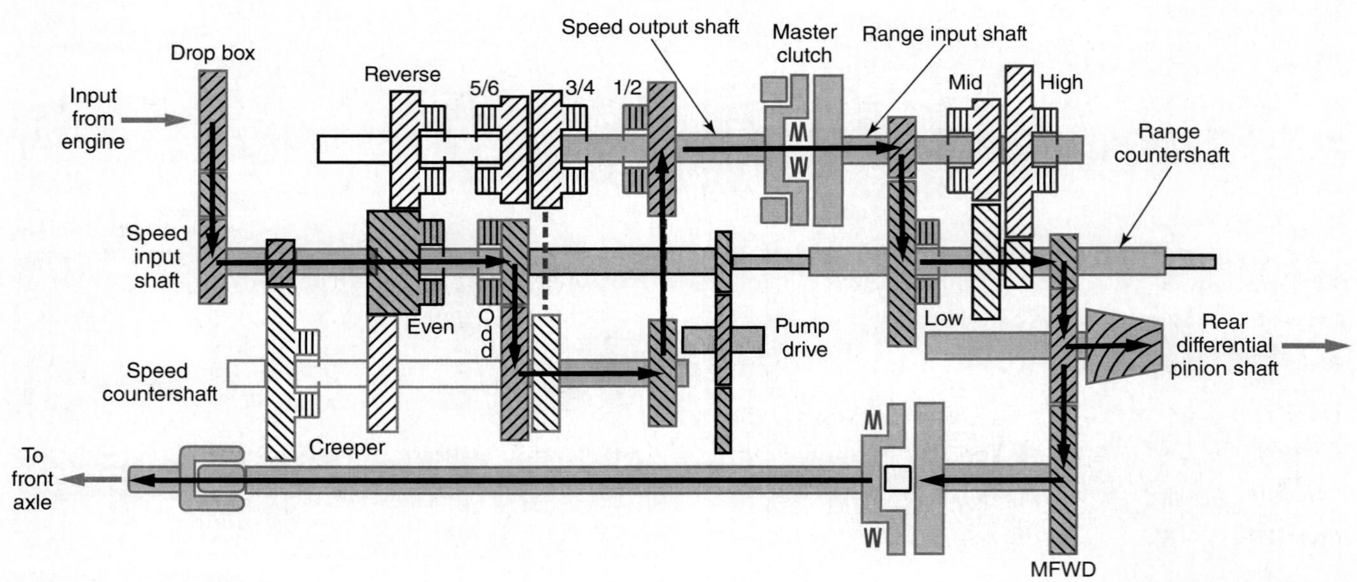

Goodheart-Willcox Publisher

Figure 10-9. In first forward gear, power flows into the speed input shaft. With the odd clutch engaged, power travels to the speed countershaft, then to the speed output shaft. With the 1-2 clutch engaged, power is delivered to the master clutch. When the clutch pedal is released, power is sent to the range input shaft, which delivers power to the range countershaft. With the low clutch applied, power is sent to the rear differential pinion shaft. The MFWD clutch is engaged, and power is sent to the front axle.

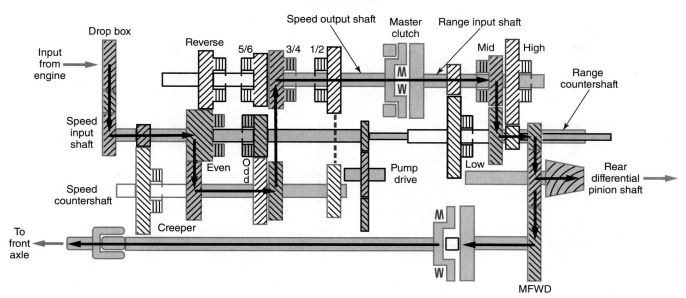

Figure 10-10. In 10th forward gear, power flows through the speed input shaft. With the even clutch engaged, power travels to the speed countershaft, then to the speed output shaft. With the 3-4 clutch engaged, power is delivered to the master clutch. With the clutch pedal released, power is sent to the range input shaft. The mid clutch is applied, sending power to the range countershaft then to the rear differential pinion shaft. The MFWD clutch is engaged and power is sent to the front axle.

Magnum Countershaft Powershift 17F Power Flow

When 17th forward gear is selected, the odd clutch, 5-6 clutch, and high clutch are applied. See **Figure 10-11**. Power flow is delivered to the transmission's speed input shaft by the drop box. With the odd clutch engaged, the speed input shaft sends power to the speed countershaft. The speed countershaft sends power to the speed output shaft, which has the 5-6 clutch engaged.

The master clutch is applied as the operator releases the clutch pedal, delivering power from the speed output shaft to the range input shaft. With the high clutch engaged, the range input shaft delivers power to the range countershaft. The range countershaft sends power to the rear differential pinion shaft. The rear differential pinion shaft drives the rear axle's differential assembly. The MFWD clutch is engaged and sends power to the front axle.

Magnum Countershaft Powershift 18F and MFWD Clutch Disengaged

When 18th forward gear is selected, the even clutch, 5-6 clutch, and high clutch are applied. See **Figure 10-12**. Also, notice that in this example the MFWD has been disengaged by the operator. Fluid pressure has overcome the spring-applied MFWD clutch, resulting in the front axle being disengaged.

Power flow is delivered to the transmission's speed input shaft by the drop box. With the even clutch engaged, the speed input shaft sends power to the speed countershaft. The speed countershaft sends power to the speed output shaft, which has the 5-6 clutch engaged.

The master clutch is applied as the operator releases the clutch pedal, delivering power from the speed output shaft to the range input shaft. With the high clutch engaged, the range input shaft delivers power to the range countershaft. The range countershaft sends power to the rear differential pinion shaft. The rear differential pinion shaft drives the rear axle's differential assembly.

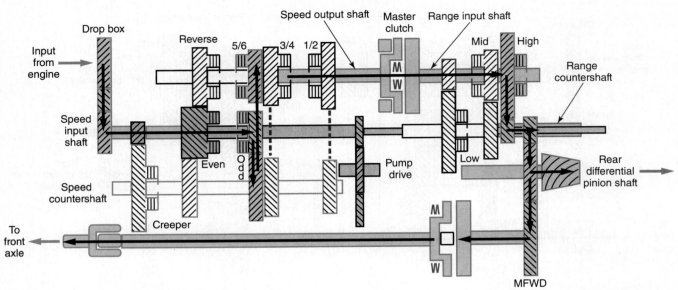

Figure 10-11. In 17th forward gear, power flows through the speed input shaft. With the odd clutch engaged, power travels to the speed countershaft, then to the speed output shaft. With the 5-6 clutch engaged, power is delivered to the master clutch. With the clutch pedal released, power is sent to the range input shaft. With the high clutch engaged, power flows to the range countershaft, which delivers power flow to the rear differential pinion shaft. The MFWD clutch is engaged and power is sent to the front axle.

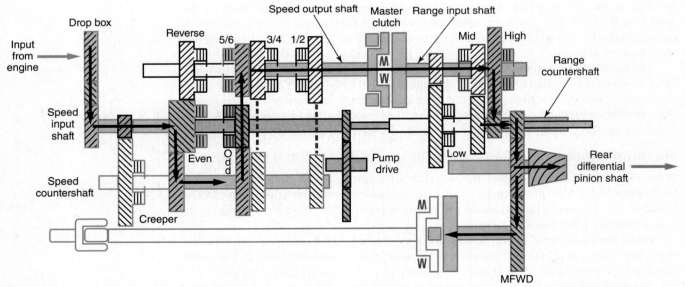

Figure 10-12. In 18th forward gear, power flows through the speed input shaft. With the even clutch engaged, power travels to the speed countershaft, then to the speed output shaft. With the 5-6 clutch engaged, power is delivered to the master clutch. With the clutch pedal released, power is sent to the range input shaft. The high clutch is engaged, sending power to the range countershaft then to the rear differential pinion shaft. The MFWD clutch is disengaged, so no power is sent to the front axle.

Magnum Countershaft Powershift 1R Power Flow

When first reverse gear is selected, the odd clutch, reverse clutch, and low clutch are applied. See **Figure 10-13**. Power flow is delivered to the transmission's speed input shaft by the drop box. With the odd clutch applied, power flows to the speed countershaft. Power then flows to the even-speed driven gear on the speed input shaft. The even clutch is disengaged, so the even-speed driven gear rotates freely on the speed input shaft and acts like a reverse idler. The even-speed driven gear sends power to the speed output shaft where the reverse clutch is engaged. Power is then sent to the master clutch.

The master clutch is applied as the operator releases the clutch pedal, delivering power from the speed output shaft to the range input shaft. The range input shaft delivers power to the range countershaft, which has the low clutch applied. The range countershaft sends power to the rear differential pinion shaft. The rear differential pinion shaft drives the rear axle's differential assembly. When the MFWD clutch is applied, power is sent to the front axle.

Magnum Countershaft Powershift 2R Power Flow

When second reverse gear is selected, the even clutch, reverse clutch, and low clutch are applied. See **Figure 10-14**. Power flow is delivered to the transmission's speed input shaft by the drop box. With the even clutch applied, power flows directly to the speed output shaft. Because power flow was not sent to the speed countershaft, the power flow is already driving in the reverse (opposite) direction.

The master clutch is applied as the operator releases the clutch pedal, delivering power from the speed output shaft to the range input shaft. The range input shaft delivers power to the range countershaft, which has the low clutch applied. The range countershaft sends power to the rear differential pinion shaft. The rear differential pinion shaft drives the rear axle's differential assembly. When the MFWD clutch is engaged, it drives the front axle.

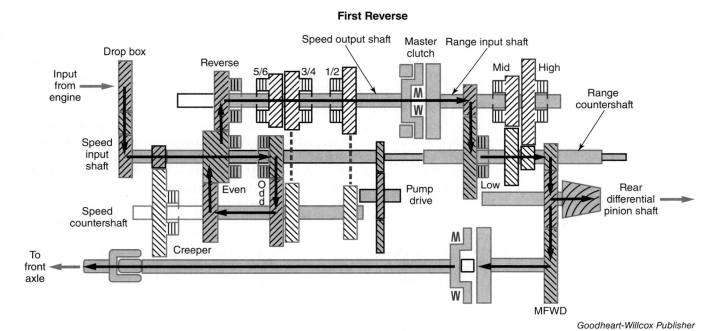

Goodheart-Willcox Publisher

Figure 10-13. In first reverse gear, power flows through the speed input shaft to the speed countershaft via the odd clutch. Power then flows to the even-speed driven gear on the speed input shaft acting like a reverse idler. Power is sent to the speed output shaft, the master clutch, the range input shaft, the range countershaft, and the rear differential pinion shaft. When the MFWD clutch is engaged, power flow is also sent to the front axle.

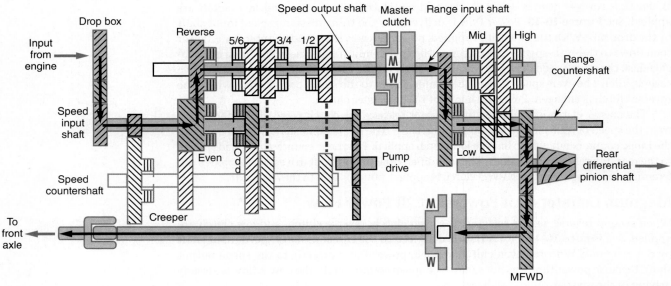

Figure 10-14. In second reverse gear, power flows through the speed input shaft and directly to the speed output shaft via the even clutch. Power then flows to the master clutch, the range input shaft, the range countershaft, and the rear differential pinion shaft. When the MFWD clutch is engaged, power flow is also sent to the front axle.

Magnum Countershaft Powershift 1F Creeper Power Flow

When creeper first forward gear is selected, the creeper clutch, 1-2 clutch, and low clutch are applied. See **Figure 10-15**. Power flow is delivered to the transmission's speed input shaft by the drop box. Power flows to the speed countershaft with the creeper clutch applied. The speed countershaft delivers power to the speed output shaft where the 1-2 clutch is applied.

The master clutch is applied as the operator releases the clutch pedal, delivering power from the speed output shaft to the range input shaft. The range input shaft delivers power to the range countershaft, which has the low clutch applied. The range countershaft sends power to the rear differential pinion shaft. The rear differential pinion shaft drives the rear axle's differential assembly. When the MFWD clutch is applied, power is sent to the front axle.

Magnum 19-Speed Powershift

As mentioned earlier, late-model Magnum powershift transmissions may include an optional 19th speed instead of the creeper gear. See **Figure 10-16**. The overdrive clutch, 5-6 clutch, and high clutch are engaged. Power flow is delivered to the transmission's input shaft by the drop box. With the overdrive clutch engaged, the speed input shaft sends power to the speed countershaft. The speed countershaft sends power to the speed output shaft, which has the 5-6 clutch engaged.

The master clutch is applied as the operator releases the clutch pedal, delivering power from the speed output shaft to the range input shaft. With the high clutch engaged, the range input shaft delivers power to the range countershaft. The range countershaft sends power to the differential pinion shaft. The differential pinion shaft drives the rear axle's differential assembly. This figure also shows a spring-applied parking brake that is hydraulically released. The MFWD clutch is engaged and sends power to the front axle.

Caterpillar H, M, and Non-Suffix Series Motor Grader Countershaft Powershift Transmission

Construction equipment can also have countershaft powershift transmissions. Examples include Caterpillar H, M, K, and non-suffixed series motor graders. A direct-drive

Chapter 10 | Powershift and Automatic Transmission Theory

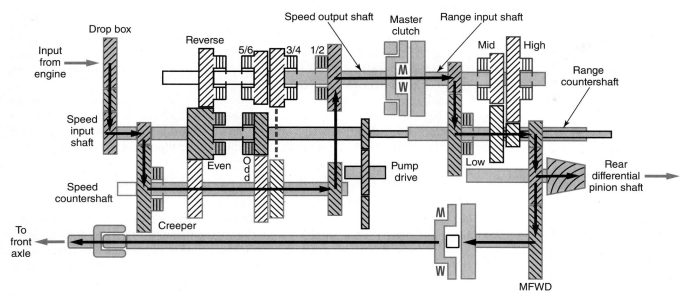

Figure 10-15. In creeper first forward gear, power flows into the speed input shaft. With the creeper clutch engaged, power travels to the speed countershaft, then flows to the speed output shaft. With the 1-2 clutch engaged, power is delivered to the master clutch. With the clutch pedal released, power is sent to the range input shaft, which delivers power to the range countershaft. The low clutch is applied, sending power to the rear differential pinion shaft. The MFWD clutch is engaged and power is sent to the front axle as well.

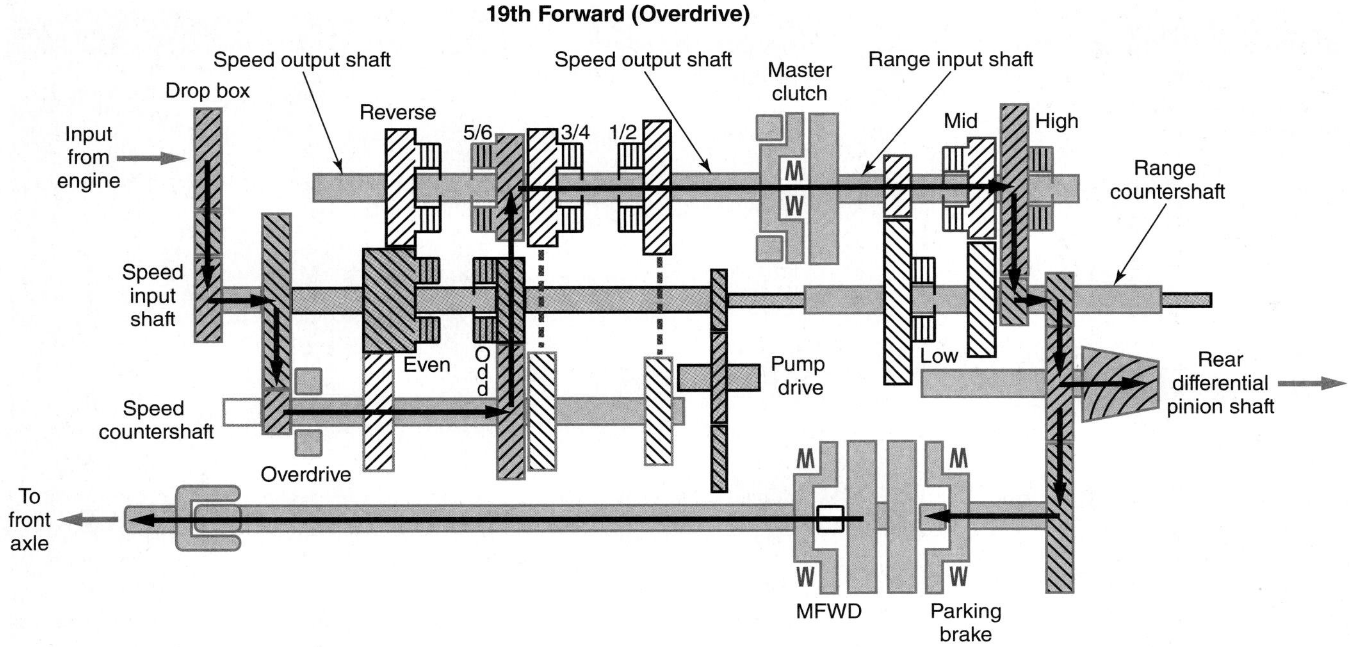

Figure 10-16. Late-model Magnum powershift transmissions can be equipped with a 19th speed overdrive gear, which fits in place of the creeper gearing. This illustration also shows a spring-applied parking brake.

countershaft powershift transmission is found on Caterpillar 120-, 12-, 140-, 14-, 160-, and 16-sized motor graders. Not all Caterpillar motor graders use a countershaft transmission. As explained in Chapter 9, *Planetary Gear Set Theory*, the Caterpillar 24-sized mining motor graders have a planetary transmission and a torque converter to provide input to the transmission. Caterpillar also produces a GC series motor grader with a countershaft powershift transmission that has six forward and three reverse speeds and uses a torque converter to eliminate the clutch pedal.

H, M, and non-suffix series' countershaft powershift transmissions use eight clutches to provide eight forward speeds and six reverse speeds. The clutch apply chart for this transmission is shown in **Figure 10-17**. Three clutches must be engaged to drive forward or reverse. One of the clutches from each of the following groups must be engaged:

- Forward high, forward low, or reverse.
- First speed, second speed, or third speed.
- High range or low range.

The transmission contains five countershafts. See **Figure 10-18**. The input shaft has no clutches and receives input directly from the engine through a flywheel-mounted torsional coupling with dampening springs that dampen engine pulsations. The front (right) gear on the input shaft meshes with gears on both shaft 2 and shaft 3. The rear (left) gear on the input shaft meshes with a gear on shaft 2. The clutch 1 (forward high) and clutch 2 (forward low) driven gear meshes with the clutch 4 (second speed) and clutch 3 (reverse) drive gear on shaft 3. Gears on shaft 3 are in mesh with gears on shaft 4. Gears on shaft 4 are in mesh with gears on shaft 5, which is the transmission's output shaft. The output shaft has the parking brake and delivers power to the motor grader's differential.

Caterpillar H and M Series Countershaft Powershift Clutch Apply Chart								
Gear Range	Clutch 1 (forward high)	Clutch 2 (forward low)	Clutch 3 (reverse)	Clutch 4 (second speed)	Clutch 5 (third speed)	Clutch 6 (first speed)	Clutch 7 (low range)	Clutch 8 (high range)
N	—	—	—	—	X	—	—	X
F1	X	—	—	—	—	X	X	—
F2	—	X	—	X	—	—	X	—
F3	—	X	—	—	X	—	X	—
F4	X	—	—	—	X	—	X	—
F5	X	—	—	—	—	X	—	X
F6	—	X	—	X	—	—	—	X
F7	X	—	—	X	—	—	—	X
F8	X	—	—	—	X	—	—	X
R1	—	—	X	—	—	X	X	—
R2	—	—	X	X	—	—	X	—
R3	—	—	X	—	X	—	X	—
R4	—	—	X	—	—	X	—	X
R5	—	—	X	X	—	—	—	X
R6	—	—	X	—	X	—	—	X

Goodheart-Willcox Publisher

Figure 10-17. The clutch apply chart for the transmission on Caterpillar H and M series motor graders.

Chapter 10 | Powershift and Automatic Transmission Theory

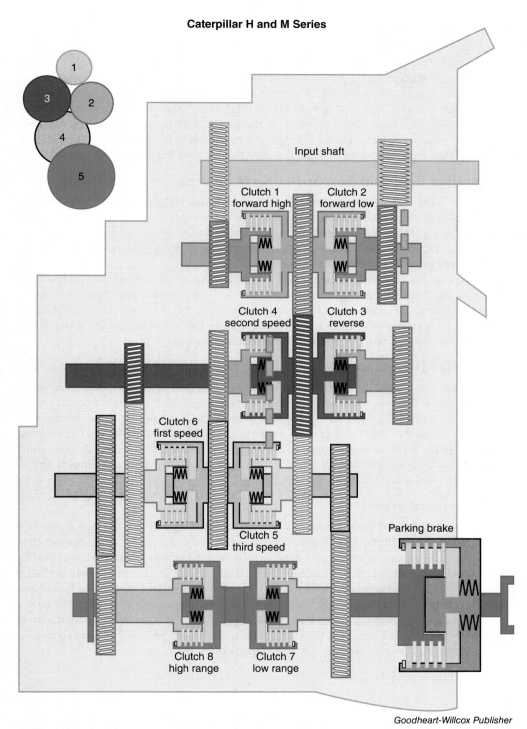

Figure 10-18. Countershaft transmission on Caterpillar H, M, and non-suffix series motor graders. The circles in the upper-left-hand corner depict the ends of the countershafts viewed as if standing in front of the motor grader.

Caterpillar H, M, and Non-Suffix Series Clutch Pedal

Caterpillar H, M, and non-suffix series motor graders use a clutch pedal to actuate a pulse-width modulating (PWM) sensor. The clutch pedal, or inching pedal, controls the fluid pressure to the directional clutches. This is known as modulating the directional clutch. The PWM sensor determines the position of the clutch pedal, and an ECM uses the PWM input to vary the application of the directional clutches. If the PWM sensor fails, a limit switch controls the application pressure of the directional clutches while the machine operates at reduced power. Clutch modulation is explained in further detail in Chapter 11, *Powershift and Automatic Transmission Controls, Service, and Repair*.

Caterpillar H, M and Non-Suffix Series Valves

Caterpillar H, M, and non-suffix series motor graders use eight electronic pressure control solenoids to vary clutch modulation pressures to the eight transmission clutches. The overall function of the solenoid control valves is the same in both series, but the valves are constructed differently for H series machines and for M series machines. This means the valves are not interchangeable between series.

The transmission has 14 different gear combinations, 8 forward speeds and 6 reverse speeds. To avoid repetition, only four forward speeds (F1–F3 and F5) and one reverse speed (R1) are explained. This selection of gears addresses the application of all the transmission clutches. The speeds omitted from this chapter do not use any other unexplained clutch and function similarly to the speeds that are explained in this chapter.

Caterpillar Direction of Shafts in Forward

The transmission's input shaft rotates in the same direction as the engine's crankshaft, clockwise (CW) when viewed from the front of the transmission. Any time the motor grader is moving forward, either the forward high clutch or the forward low clutch is engaged. This causes shaft 2 to rotate counterclockwise (CCW). Shaft 2 drives the shaft 3, causing shaft 3 to rotate in the same direction as the engine and shaft 1 (CW). Shaft 3 drives shaft 4, causing it to rotate CCW. Shaft 4 drives shaft 5, causing it to rotate in the same direction as the engine and shaft 1 (CW).

First Forward Speed

In first forward, clutch 1 (forward high), clutch 6 (first speed), and clutch 7 (low range) are applied. As shown in **Figure 10-19**, power is delivered to the input shaft. The input shaft is splined to its two gears. With clutch 1 applied, power is transferred from the input gear to the forward high gear, which is attached by the clutch to shaft 2. The center gear of shaft 2 drives shaft 3. The rear gear on shaft 3 is in mesh with the gear that drives the hub of clutch 6 on shaft 4. With clutch 6 applied, power is transferred from the hub to shaft 4. Shaft 4 delivers power to the drive gear and hub of clutch 7 on shaft 5. With clutch 7 applied, power is transferred to shaft 5, which is the output shaft. With fluid pressure applied to release the parking brake, power is transferred out of the transmission.

Second Forward Speed

In second forward, clutch 2 (forward low), clutch 4 (second speed), and clutch 7 (low range) are applied. Second forward is shown in **Figure 10-20**. Power is transferred to the input shaft, and the front input-shaft gear drives the drive gear and hub of clutch 2 on shaft 2. With clutch 2 applied, its drum delivers power to the middle gear of shaft 2, which delivers power to the drive gear and drum of clutch 4 on shaft 3. With clutch 4 applied, its hub and driven gear delivers power to shaft 4. The front gear on shaft 4 delivers power to the drive gear and hub of clutch 7. With clutch 7 applied and the parking brake released, power is delivered through shaft 5 to the differential.

Third Forward Speed

In third forward, clutch 2 (forward low), clutch 5 (third speed), and clutch 7 (low range) are applied. Third forward is shown in **Figure 10-21**. Power is transferred to the input shaft, and the front input-shaft gear drives the drive gear and hub of clutch 2 on shaft 2. With clutch 2 applied, the

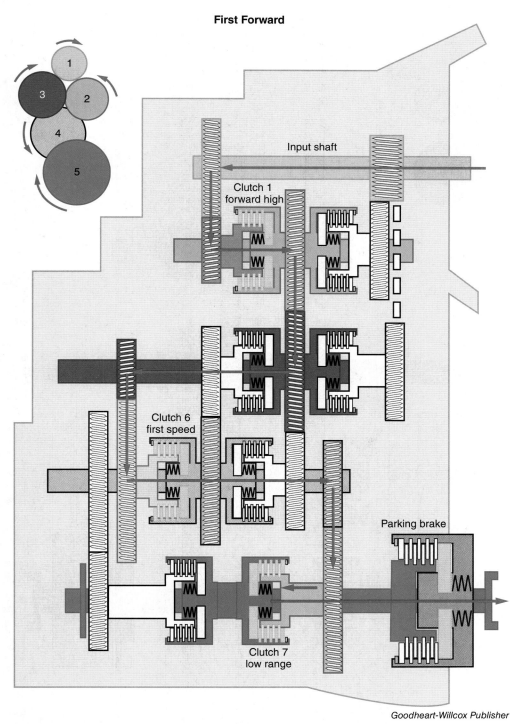

Figure 10-19. First forward power flow in Caterpillar H, M, and non-suffix series motor grader transmissions.

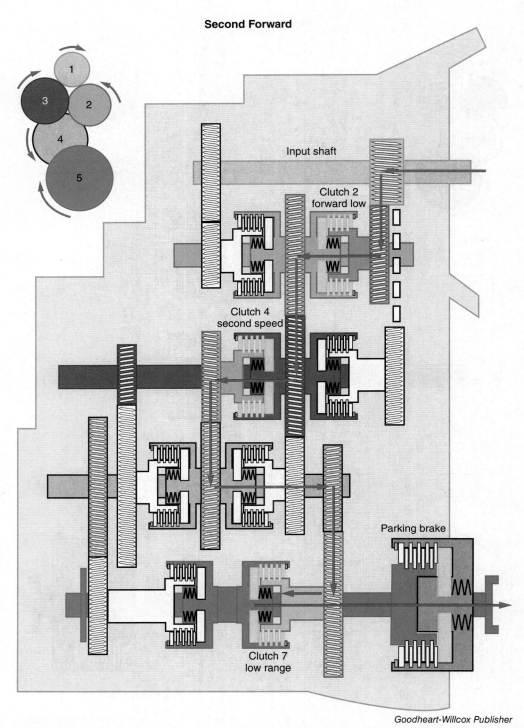

Figure 10-20. Second forward power flow in Caterpillar H, M, and non-suffix series motor grader transmissions.

Chapter 10 | Powershift and Automatic Transmission Theory

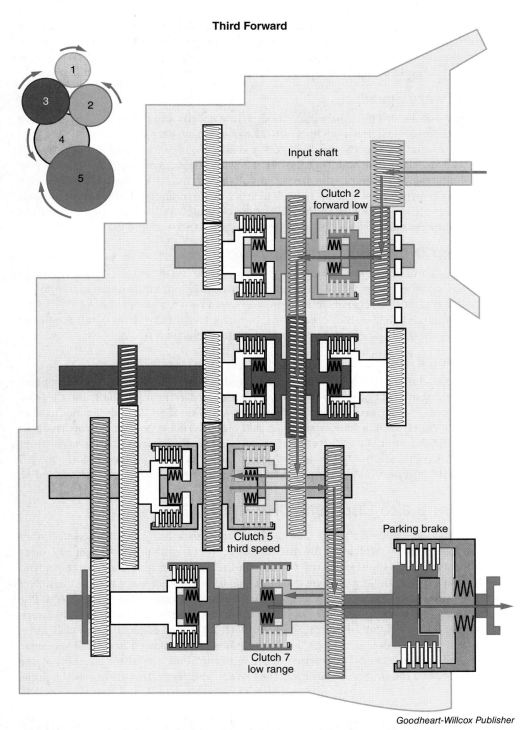

Figure 10-21. Third forward power flow in Caterpillar H and M series motor grader transmissions.

middle gear of shaft 2 delivers power to a drive gear on shaft 3, which is in mesh with the drive gear and hub of clutch 5 on shaft 4. With clutch 5 applied, its hub delivers power to its drum, which is splined to shaft 4. Shaft 4 delivers power to the drive gear and hub of clutch 7. With clutch 7 applied and the parking brake released, power is delivered through shaft 5 to the differential.

Fifth Forward Speed

In fifth forward, clutch 1 (forward high), clutch 6 (first speed), and clutch 8 (high range) are applied. Fifth forward is shown in **Figure 10-22**. Power is transferred to the input shaft, and the rear input-shaft gear drives the drive gear and hub of clutch 1 on shaft 2. With clutch 1 applied, its drum drives the middle gear of shaft 2, which delivers power to shaft 3. The rear gear on shaft 3 delivers power to the drive gear and hub of clutch 6 on shaft 4. With clutch 6 applied, its hub delivers power to its drum, which is splined to shaft 4. The rear gear on shaft 4 delivers power to the drive gear and hub of clutch 8 on shaft 5. With clutch 8 applied and the parking brake released, power is delivered through shaft 5 to the differential.

Caterpillar Direction of Shafts in Reverse

The transmission's input shaft rotates in the same direction as the engine's crankshaft, clockwise (CW) when viewed from the front of the transmission. Any time the motor grader is moving in reverse, the reverse clutch is engaged. As a result, shaft 1 rotates CW, driving shaft 3 CCW. Shaft 3 drives shaft 4, causing it to rotate CW. Shaft 4 drives shaft 5, causing it to rotate in the opposite direction of the engine and input shaft (CCW).

First Reverse Speed

In first reverse, clutch 3 (reverse), clutch 6 (first speed), and clutch 7 (low range) are applied. First reverse is shown in **Figure 10-23**. Power is transferred to the input shaft, and the front input-shaft gear drives the drive gear and hub of clutch 3 on shaft 3. Shaft 2 is not used in reverse. With clutch 3 applied, the rear driven gear of shaft 3 delivers power to the drive gear and hub of clutch 6 on shaft 4. With clutch 6 applied, power is delivered to shaft 4. The front gear of shaft 4 delivers power to the drive gear and hub of clutch 7 on shaft 5. With clutch 7 applied and the parking brake released, power is delivered through shaft 5 to the differential.

John Deere e23 Countershaft Transmission

In 2014, John Deere offered the e23 countershaft transmission in the 7R tractors. Today, it is available in 7R, 8R, 8RT, and 8RX tractors. It receives engine power through a torque dampener and has one input shaft, two countershafts, and one output shaft. For the output shaft to rotate, the high, low, or reverse clutch, one of the four speed clutches, and one of the three range clutches must be engaged. The transmission has 12 reverse speeds. The 12th reverse speed is only possible in the auto shift mode. The transmission has 24 forward speeds: F1–F7, F8a, F8b, and F9–F23. See **Figure 10-24**. F8a is engaged when upshifting and F8b is used when downshifting. The transmission is not allowed to shift in sequence from F8a to F8b, because it would require disengaging and engaging three elements at one time resulting in a rough shift. In addition, the F8a and F8b gear ratios are nearly the same.

Combination Planetary Countershaft Powershift Transmission

Mobile equipment can have powershift transmissions that contain both planetary gears and countershaft gears. Some examples are John Deere 8R series 16-speed tractors, Challenger MT 700 series rubber track tractors (through 2017), MT800 series rubber track tractors (through 2020), and MT900 articulated tractors (through 2021). **Figure 10-25** shows the John Deere 8R clutch apply chart. See the Appendix for the *Challenger MT800 Clutch Apply Chart*.

Chapter 10 | Powershift and Automatic Transmission Theory

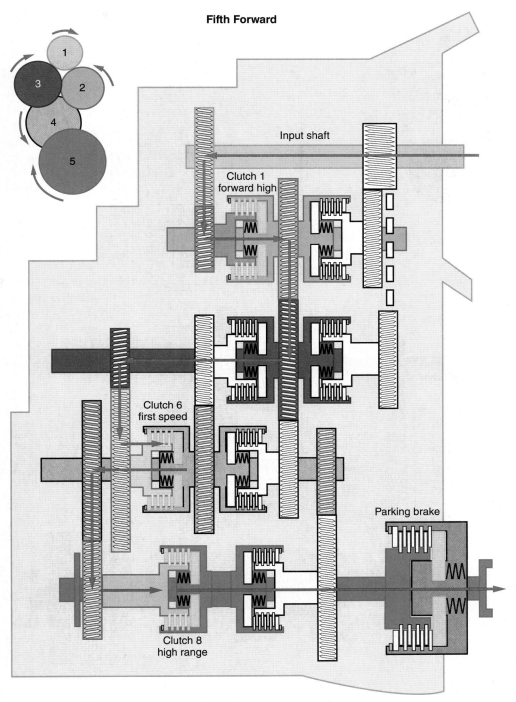

Goodheart-Willcox Publisher

Figure 10-22. Fifth forward power flow in Caterpillar H and M series motor grader transmissions.

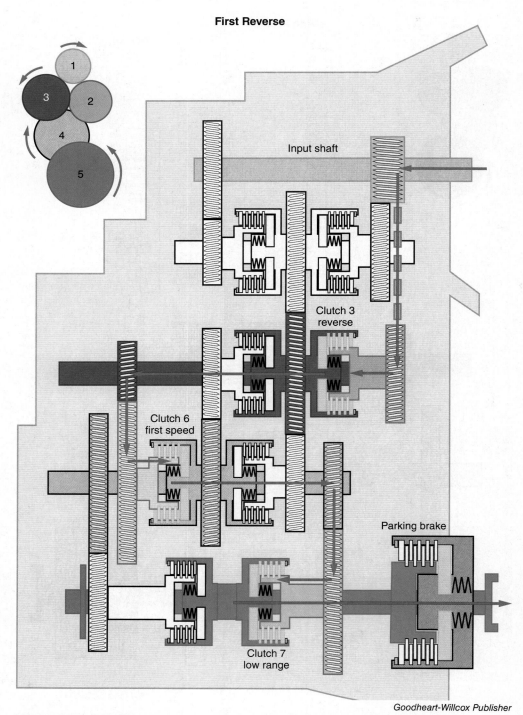

Figure 10-23. First reverse power flow in Caterpillar H and M series motor grader transmissions.

Gear Range	Reverse	Fwd Low	Fwd High	Speed 1	Speed 2	Speed 3	Speed 4	Range 1	Range 2	Range 3
F1	—	X	—	X	—	—	—	X	—	—
F2	—	—	X	X	—	—	—	X	—	—
F3	—	X	—	—	X	—	—	X	—	—
F4	—	—	X	—	X	—	—	X	—	—
F5	—	X	—	—	—	X	—	X	—	—
F6	—	—	X	—	—	X	—	X	—	—
F7	—	X	—	—	—	—	X	X	—	—
F8a	—	—	X	—	—	—	X	X	—	—
F8b	—	X	—	X	—	—	—	—	X	—
F9	—	—	X	X	—	—	—	—	X	—
F10	—	X	—	—	X	—	—	—	X	—
F11	—	—	X	—	X	—	—	—	X	—
F12	—	X	—	—	—	X	—	—	X	—
F13	—	—	X	—	—	X	—	—	X	—
F14	—	X	—	—	—	—	X	—	X	—
F15	—	—	X	—	—	—	X	—	X	—
F16	—	X	—	X	—	—	—	—	—	X
F17	—	—	X	X	—	—	—	—	—	X
F18	—	X	—	—	X	—	—	—	—	X
F19	—	—	X	—	X	—	—	—	—	X
F20	—	X	—	—	—	X	—	—	—	X
F21	—	—	X	—	—	X	—	—	—	X
F22	—	X	—	—	—	—	X	—	—	X
F23	—	—	X	—	—	—	X	—	—	X
R1	X	—	—	X	—	—	—	X	—	—
R2	X	—	—	—	X	—	—	X	—	—
R3	X	—	—	—	—	X	—	X	—	—
R4	X	—	—	—	—	—	X	X	—	—
R5	X	—	—	X	—	—	—	—	X	—
R6	X	—	—	—	X	—	—	—	X	—
R7	X	—	—	—	—	X	—	—	X	—
R8	X	—	—	—	—	—	X	—	X	—
R9	X	—	—	X	—	—	—	—	—	X
R10	X	—	—	—	X	—	—	—	—	X
R11	X	—	—	—	—	X	—	—	—	X
R12	X	—	—	—	—	—	X	—	—	X

Goodheart-Willcox Publisher

Figure 10-24. The clutch apply chart for John Deere's e23 countershaft powershift transmission.

John Deere 8R Series Tractor Clutch Apply Chart									
Gear Range	Clutch 1	Clutch 2	Clutch 3	Clutch 4	Reverse clutch	Brake A	Clutch B	Clutch C	Clutch D
1F	X	—	—	—	—	X	—	—	—
2F	—	X	—	—	—	X	—	—	—
3F	—	—	X	—	—	X	—	—	—
4F	—	—	—	X	—	X	—	—	—
5F	X	—	—	—	—	—	X	—	—
6F	X	—	—	—	—	—	—	X	—
7F	—	X	—	—	—	—	X	—	—
8F	—	X	—	—	—	—	—	X	—
9F	—	—	X	—	—	—	X	—	—
10F	—	—	X	—	—	—	—	X	—
11F	—	—	—	X	—	—	X	—	—
12F	—	—	—	X	—	—	—	X	—
13F	X	—	—	—	—	—	—	—	X
14F	—	X	—	—	—	—	—	—	X
15F	—	—	X	—	—	—	—	—	X
16F	—	—	—	X	—	—	—	—	X
1R	—	—	—	—	X	X	—	—	—
2R	—	—	—	—	X	—	X	—	—
3R	—	—	—	—	X	—	—	X	—
4R	—	—	—	—	X	—	—	—	X

Goodheart-Willcox Publisher

Figure 10-25. The clutch apply chart for a John Deere 8R series tractor. The powershift transmission has 16 forward speeds and four reverse speeds. Brake A is located on the output shaft and holds the ring gear of a planetary gear set, causing the carrier to be the output member. The remaining clutches are countershaft clutches.

John Deere 8R 16-Speed Powershift Transmission

John Deere produced the first 8000 series tractors in 1994. The 8000 series continued with the 8010 series, 8020 series, 8030 series, and 8R series, which was released in 2009. All of the 8000 series and some 8R tractors can be equipped with a 16-speed powershift transmission. The transmission's input shaft is directly driven by the engine through a heavy duty rubber torsional damper. Like nearly all agricultural tractors, it does not use a torque converter.

The transmission contains an input shaft with three clutches, a countershaft with two clutches, and an output shaft with three clutches and two brakes. See **Figure 10-26**. One brake is the parking brake. The other is brake A, which holds the one planetary ring gear in the first four forward speeds and first reverse. See **Figure 10-27**.

One of the five input clutches (clutch 1, clutch 2, clutch 3, clutch 4, or the reverse clutch) and one of the four output elements (the brake A, clutch B, clutch C, or clutch D) must be engaged for the tractor to move forward or rearward. The five input clutches consist of four forward clutches and a reverse clutch:

- Clutch 1.
- Clutch 2.

Chapter 10 | Powershift and Automatic Transmission Theory

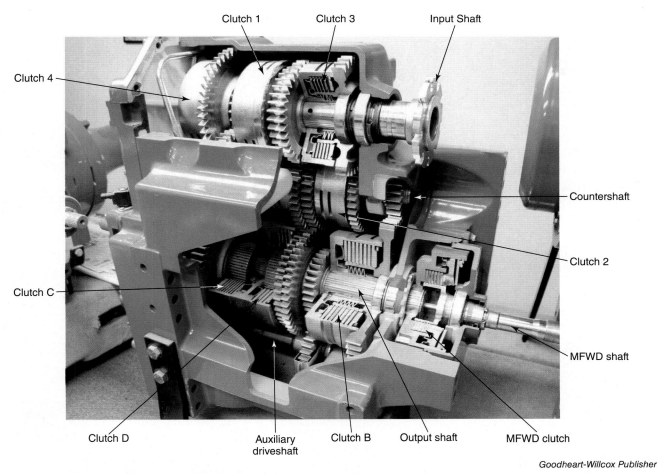

Figure 10-26. This cutaway transmission shows the three shafts: input shaft, countershaft, and output shaft. The parking brake and brake A are not seen in this photograph.

- Clutch 3.
- Clutch 4.
- Reverse clutch.

The input shaft is driven at engine speed. The drums to clutch 1, clutch 3, and clutch 4 are splined to the input shaft and rotate at engine speed. When one of these clutches is engaged, the clutch's hub drives a gear that is meshed with a gear on the countershaft.

The other two input clutches, clutch 2 and the reverse clutch, are on the countershaft. Clutch 2 receives power from the input shaft through a drive gear. The reverse clutch receives power from the input shaft through a reverse idler.

The remaining clutches and brakes are located on the output shaft:

- Clutch B.
- Clutch C.
- Clutch D.
- Brake A.
- Parking brake.

Clutches B, C, D, and brake A are the four output elements. One of these must be engaged for the tractor to move. The MFWD clutch is at the front of the transmission, splined to the transmission's output shaft.

When the engine is running, the auxiliary driveshaft is driven to provide power to the PTO clutch and to drive the hydraulic pumps and the scavenge pump. The scavenge pump

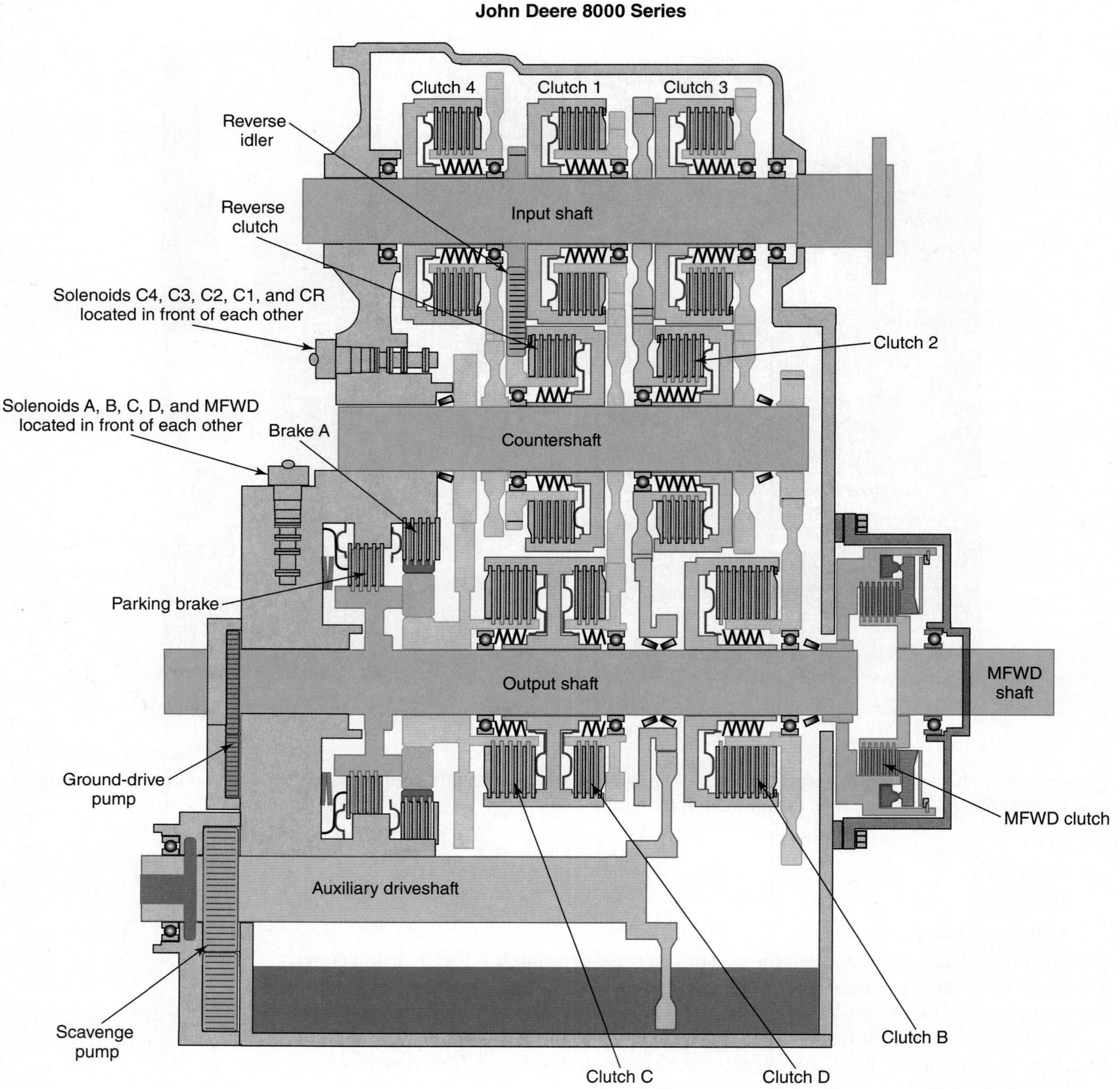

Goodheart-Willcox Publisher

Figure 10-27. This simplified drawing of a powershift transmission shows the three shafts used in a 16-speed John Deere 8000 series tractor: input shaft, countershaft, and output shaft. The input shaft and countershaft have the five input clutches. The output shaft has the output clutches, brake A, and the parking brake.

draws oil out of the transmission case and directs it through the center of the auxiliary driveshaft to the pump drive case.

Powershift transmissions in the John Deere 8010 series and 8020 series tractors had a ***ground-drive pump (GDP)*** splined to the output shaft. This meant that any time the tractor was moving, the GDP provided oil for tractor steering and brakes in case of a loss of hydraulic flow. See **Figure 10-28**. Later models use the differential shaft in the rear axle to power the ground-drive pump only when needed, which reduces power loss.

In the examples presented in the following sections, the parking brake is released and the MFWD clutch is applied. The parking brake must receive oil pressure to release its spring. The oil pressure has been removed from the MFWD control circuit, allowing the spring disk to apply the MFWD clutch, which drives the front axle driveshaft.

The transmission has 20 different gear combinations, 16 forward speeds and 4 reverse speeds. To avoid repetition, only seven forward speeds (F1–F6 and F13) and one reverse speed (R1) are explained. This selection of gears addresses the application of all the transmission clutches (1, 2, 3, 4, B, C, D) and brake A. The speeds omitted from this chapter do not use any other unexplained clutch element and function similarly to the speeds that are explained in this chapter.

First Forward Speed

The John Deere 16-speed powershift transmission achieves first forward speed by applying clutch 1 and brake A. Clutch 1 is on the input shaft and its drum is driven at engine speed. When clutch 1 is engaged, its hub drives the clutch 1 drive gear, which drives the countershaft-driven gear. See **Figure 10-29**.

The countershaft has a drive gear that drives the sun gear of the planetary gear set. With brake A holding the ring gear, the planetary carrier is forced to drive the output shaft. See **Figure 10-30**.

Second Forward Speed

Second forward speed is achieved by applying clutch 2 and brake A. Clutch 2 is on the countershaft. The hub of clutch 2 is driven by the gear on the input shaft. When clutch 2 is engaged, the hub of clutch 2 drives the drum, which drives the countershaft. See **Figure 10-31**.

The countershaft has a drive gear that drives the sun gear of the planetary gear set. With brake A holding the ring gear, the planetary carrier is forced to drive the output shaft. See **Figure 10-32**.

Third Forward Speed

Clutch 3 and brake A are applied for third forward speed. Clutch 3 is on the input shaft. The drum of clutch 3 is driven by the input shaft. When clutch 3 is engaged, it causes the hub of clutch 3 to drive the drive gear, which drives the countershaft. See **Figure 10-33**.

The countershaft has a drive gear that drives the sun gear of the planetary gear set. With brake A holding the ring gear, the planetary carrier is forced to drive the output shaft.

Fourth Forward Speed

To achieve fourth forward speed, clutch 4 and brake A are applied. Clutch 4 is on the input shaft. The drum of clutch 4 is driven by the input shaft. When clutch 4 is

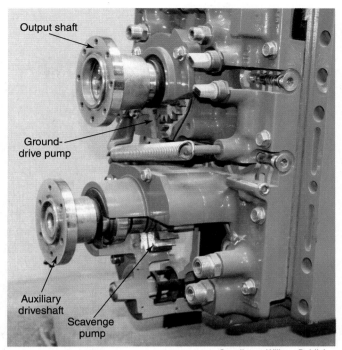

Figure 10-28. The John Deere 16-speed powershift transmission contains a scavenge pump on the end of the auxiliary driveshaft, which rotates at engine speed. The scavenge pump draws oil out of the transmission and delivers it to the pump drive case. The 8010 series and 8020 series tractors have a ground-drive pump on the rear of the transmission output shaft that provides oil to steering and brakes in case there is a loss of main hydraulic flow.

Figure 10-29. In first forward, clutch 1 on the input shaft is engaged. Clutch 1 rotates the drive gear, which is in mesh with the driven gear on the countershaft.

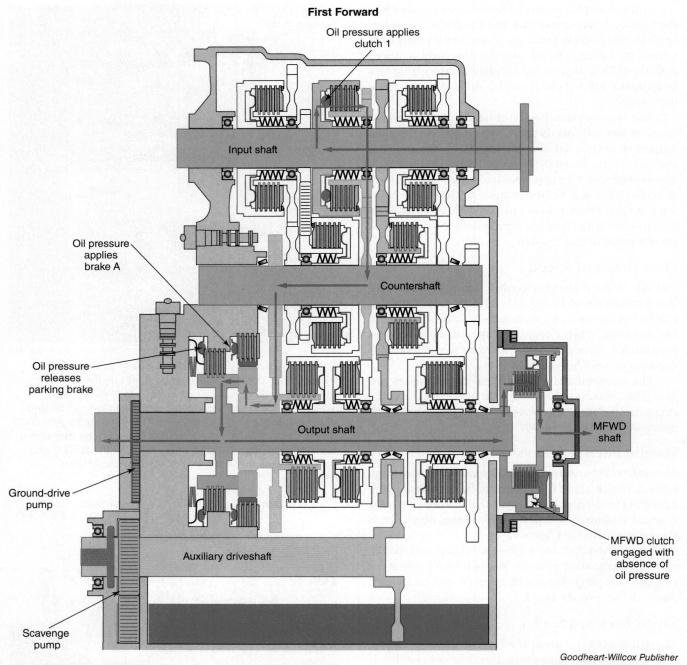

Figure 10-30. In first forward, clutch 1 and brake A are applied. The parking brake is hydraulically released and the MFWD clutch is spring applied.

engaged, it causes the hub of clutch 4 to drive the drive gear, which drives the countershaft. See **Figure 10-34**.

The countershaft has a drive gear that drives the sun gear of the planetary gear set. With brake A holding the ring gear, the planetary carrier is forced to drive the output shaft.

Fifth Forward Speed

The tractor achieves fifth forward speed when both clutch 1 and clutch B are applied. Clutch 1 is on the input shaft. The drum of clutch 1 is driven by the input shaft. When clutch 1 is engaged, it causes the hub of clutch 1 to drive the drive gear, which drives the countershaft. See **Figure 10-35**.

The countershaft has a drive gear at the front of the shaft that drives the hub of clutch B. With clutch B applied, the drum drives the output shaft. See **Figure 10-36**.

Sixth Forward Speed

Sixth forward speed is achieved by applying clutch 1 and clutch C. Clutch 1 is on the input shaft. The drum of clutch 1 is driven by the input shaft. When clutch 1 is engaged, it causes the hub of clutch 1 to drive the drive gear, which drives the countershaft.

The countershaft has a drive gear located at the rear of the shaft that drives the drive gear and hub to clutch C. See **Figure 10-37**. With clutch C applied, it causes the drum to drive the output shaft. See **Figure 10-38**.

Goodheart-Willcox Publisher

Figure 10-31. In second forward, power is delivered to clutch 2 on the countershaft via the input shaft drive gear. Once clutch 2 is engaged, it causes the drum to drive the countershaft, which delivers power to the sun gear in the planetary gear set on the output shaft.

Thirteenth Forward Speed

Thirteenth forward speed is achieved by applying clutch 1 and clutch D. Clutch 1 is located on the input shaft. The drum of clutch 1 is driven by the input shaft. When clutch 1 is engaged, it causes the hub of clutch 1 to drive the drive gear, which drives the countershaft.

The drive gear in the middle of the countershaft drives the drive gear and hub to clutch D located on the output shaft. See **Figure 10-39**. With clutch D applied, the drum drives the output shaft. See **Figure 10-40**.

First Reverse Speed

First reverse speed is achieved by applying the reverse clutch and brake A. The input shaft drives a drive gear that is meshed with the reverse idler gear. The reverse idler gear drives the driven gear and hub to the reverse clutch at the rear of the countershaft. See **Figure 10-41**.

With the reverse clutch engaged, the reverse clutch drum drives the countershaft in a reverse direction. The drive gear at the rear of the countershaft drives the sun gear to the planetary gear set on the output shaft. With brake A holding the ring gear, the planetary carrier is forced to drive the output shaft in an opposite direction. See **Figure 10-42**.

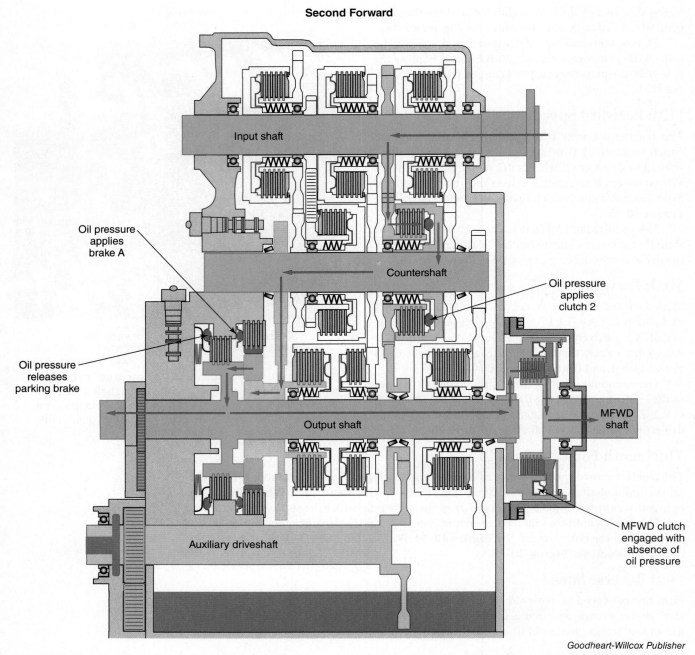

Figure 10-32. Second forward is achieved by applying clutch 2 on the countershaft and brake A on the output shaft. The parking brake is hydraulically released and the MFWD clutch is spring applied.

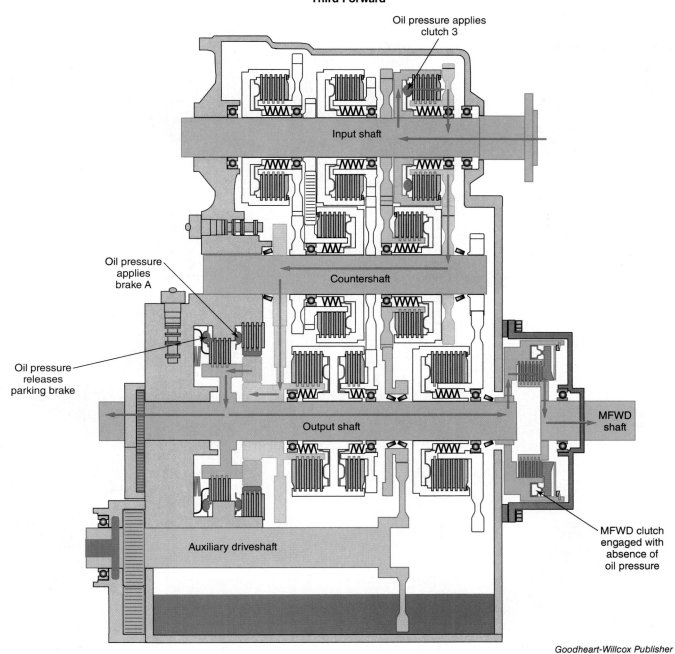

Figure 10-33. Third forward is achieved by applying clutch 3 on the input shaft and brake A on the output shaft. The parking brake is hydraulically released and the MFWD clutch is spring applied.

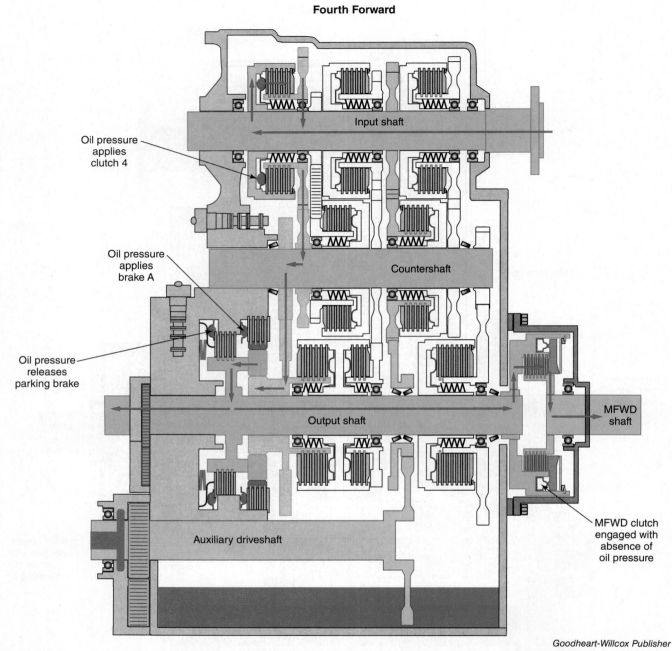

Figure 10-34. Fourth forward is achieved by applying clutch 4 on the input shaft and brake A on the output shaft. The parking brake is hydraulically released and the MFWD clutch is spring applied.

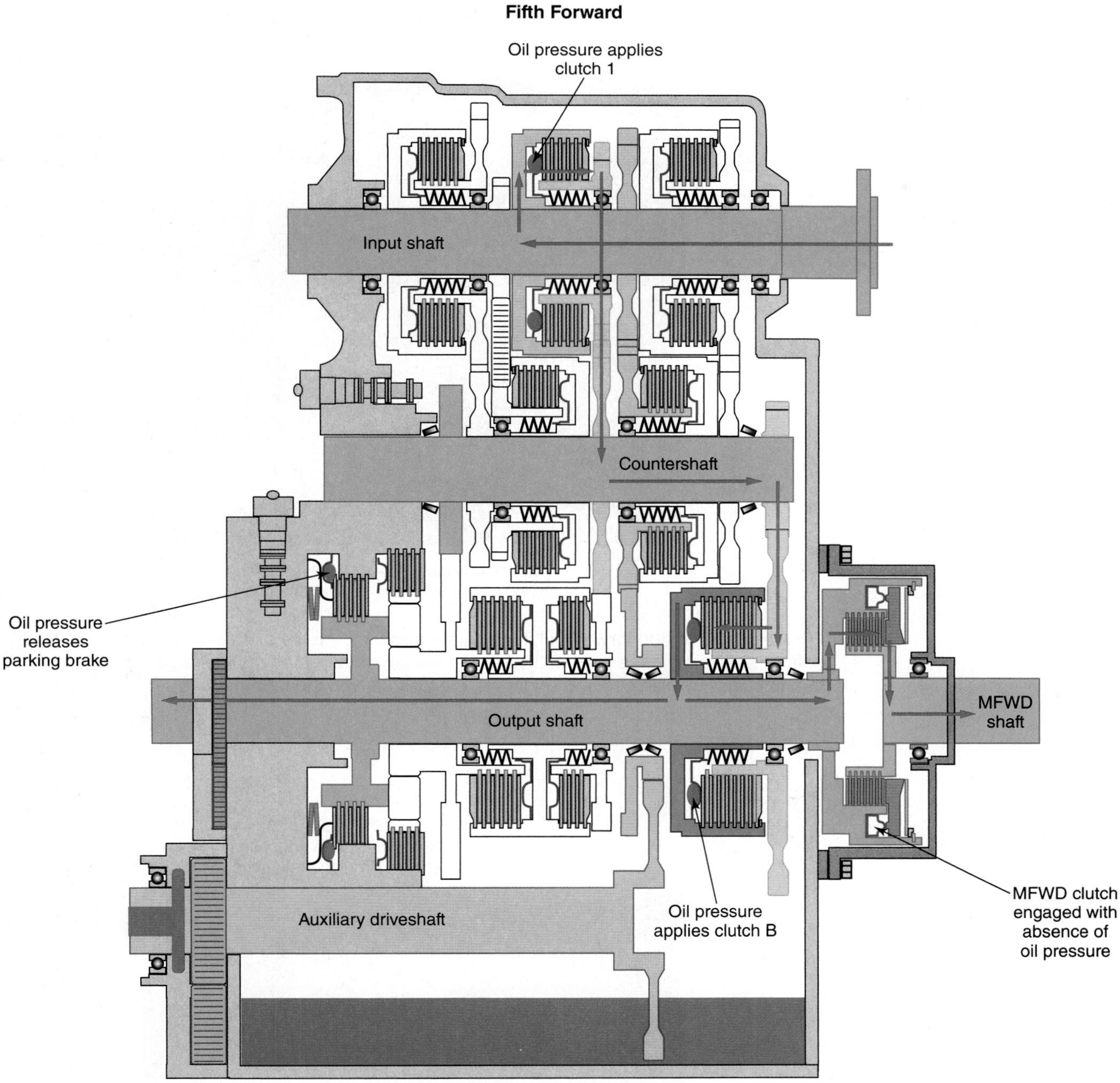

Chapter 10 | Powershift and Automatic Transmission Theory

Fifth Forward

Figure 10-35. Fifth forward is achieved by applying clutch 1 on the input shaft and clutch B on the output shaft. The parking brake is hydraulically released and the MFWD clutch is spring applied.

Figure 10-36. In fifth forward, clutch 1 is engaged and drives a drive gear, which drives the countershaft. The drive gear at the front of the countershaft drives the large driven gear and hub to clutch B. With clutch B engaged, the drum drives the output shaft.

Figure 10-37. Clutch C is on the output shaft. Its hub and gear are driven by the countershaft. When clutch C is engaged, the clutch drum is driven, which drives the output shaft.

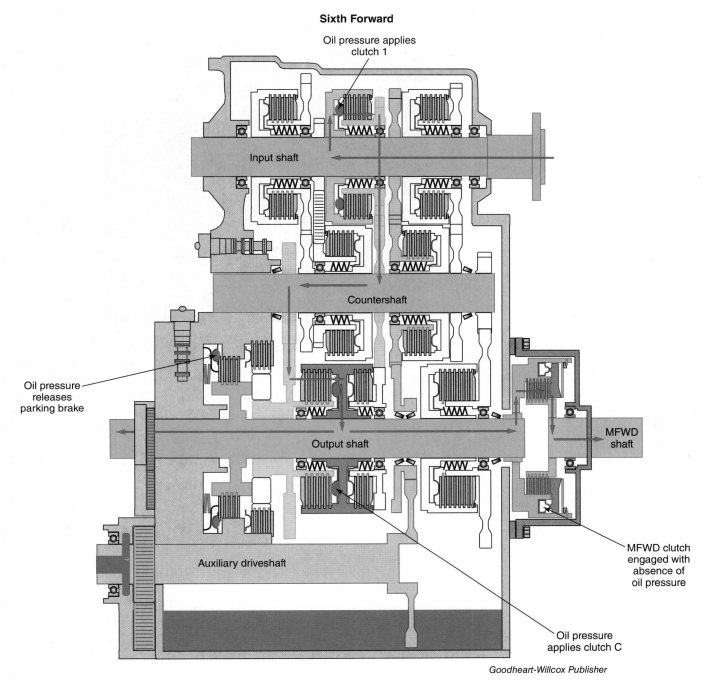

Figure 10-38. Sixth forward is achieved by engaging clutch 1 on the input shaft and clutch C on the output shaft.

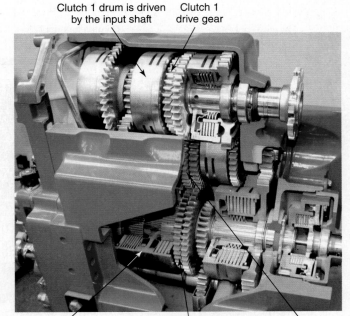

Figure 10-39. Clutch D is on the output shaft. Its hub and gear are driven by the countershaft. When the clutch is engaged, the clutch drum is driven, which drives the output shaft.

Goodheart-Willcox Publisher

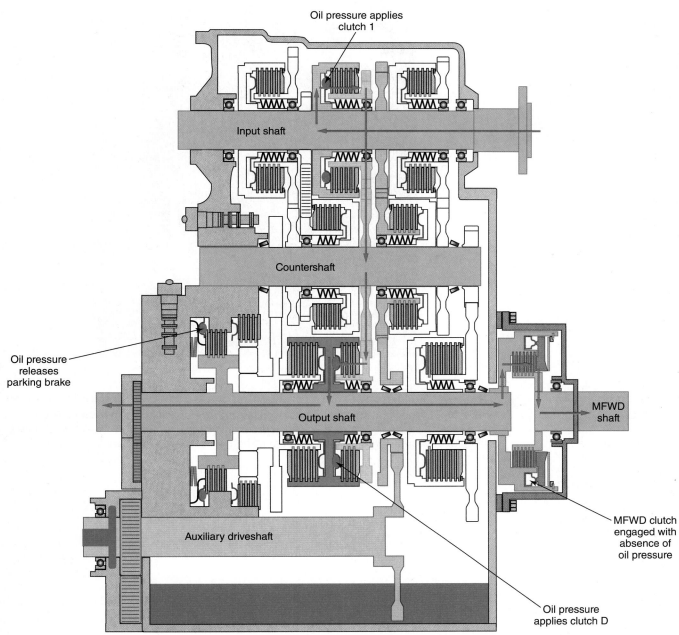

Figure 10-40. Clutch 1 on the input shaft and clutch D on the output shaft are engaged to provide thirteenth forward.

Figure 10-41. The input shaft's drive gear drives the reverse idler. The reverse idler drives the reverse clutch gear. The reverse clutch drum drives the countershaft. The countershaft's drive gear drives the large gear on the output shaft, which drives the planetary sun gear.

First Reverse

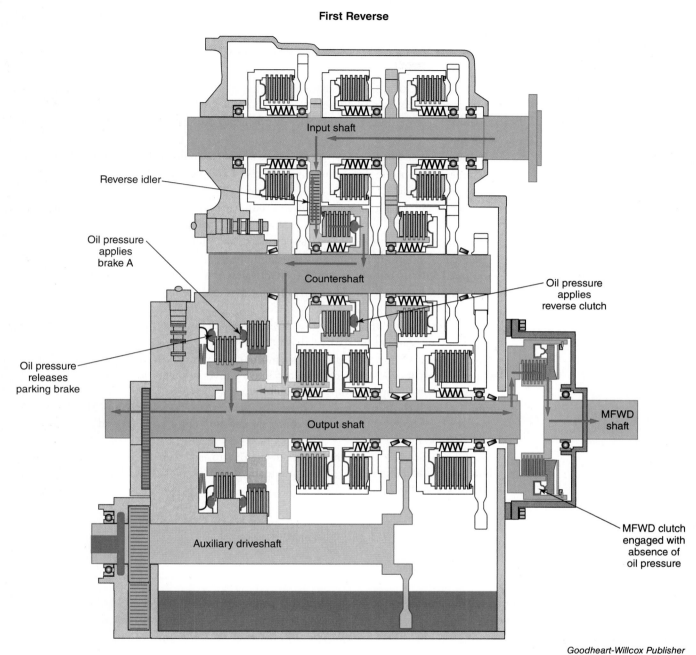

Goodheart-Willcox Publisher

Figure 10-42. In first reverse, the reverse clutch and brake A are applied. The input shaft has a drive gear that is in mesh with the reverse idler gear. The reverse idler drives the drive gear and hub to the reverse clutch on the countershaft. With the reverse clutch engaged, the reverse clutch drum drives the countershaft, which drives a gear located at the rear of the countershaft. This gear drives the sun gear of the planetary gear set. With brake A applied, the ring gear is held and the planetary carrier drives the output shaft.

Summary

- The Allison TC10 is an automatic countershaft transmission that uses a torque converter and contains a planetary gear set at the back of the transmission.
- Automatic transmissions automatically upshift or downshift based on the travel speed and load.
- Operators must command powershift transmissions to upshift and downshift.
- Automatic and powershift transmissions have fixed gear ratios for a given transmission gear selection.
- Automatic and powershift transmissions can use planetary gear sets, countershafts, or both.
- Countershaft transmissions are less compact than planetary transmissions.
- The Case IH Magnum tractors use a countershaft powershift transmission. It can be equipped with an optional creeper transmission or an optional 19th speed.
- Case IH Magnum tractors require three clutches and the master clutch to be applied to achieve power flow.
- MFWD agricultural tractors have a spring-applied clutch in the transmission for engaging the tractor's four-wheel drive.
- Caterpillar H, M, and non-suffix series motor graders use a countershaft powershift transmission. It produces eight forward and six reverse speeds and requires three clutches to be applied to propel the machine.
- John Deere 8R series tractors and many Challenger MT700, MT800, and MT900 tractors have powershift transmissions that use both countershafts and planetary gear sets.
- The John Deere 8000 series 16-speed powershift transmission requires that either one input clutch and one output clutch or one input clutch and brake A be applied to move forward or reverse.

Technical Terms

auto field operation	creeper transmission	hauling transmission	powershift transmission
automatic transmission	cycling transmission	inching pedal	preselect gear
auto road operation	directional clutch	master clutch	pulser lever
auto shift control	drop box	MFWD clutch	skip shifting
clutch pedal	ground-drive pump (GDP)	output clutch	speed matching
countershaft powershift transmission			

Review Questions

Answer the following questions using the information provided in this chapter.

Know and Understand

1. The TC10 is an on-highway automatic transmission that uses _____.
   A. countershafts
   B. a planetary gear set
   C. Both A and B.
   D. Neither A nor B.

2. The major advantage of using a powershift or automatic transmission rather than a manual transmission is _____.
   A. increased fuel efficiency
   B. high torque capacity
   C. high horsepower capacity
   D. uninterrupted power flow during shifts

3. A creeper transmission is primarily used to enable the tractor to achieve _____ while traveling at slow speeds.
   A. maximum PTO speed and maximum hydraulic flow
   B. maximum fuel efficiency
   C. low engine rpm
   D. low PTO speed and low hydraulic flow

4. The auto field operation upshifts or downshifts in order to maintain which of the following?
   A. Engine speed.
   B. Fuel efficiency.
   C. Travel speed.
   D. Transmission life.

5. Where is the MFWD clutch normally located in relation to the transmission gears?
   A. Transmission input shaft.
   B. Transmission output shaft.
   C. Transmission countershaft.
   D. Both A and B.

6. What is the meaning of the dashed lines between the speed countershaft and the speed output shaft in the drawings of the Magnum powershift transmission?
   A. They show optional power flow.
   B. They show missing gears.
   C. They show that two gears that are not adjacent in the drawing are actually meshing gears.
   D. They indicate that multi-disc clutch packs have been intentionally left out of the drawing.

7. If 200 psi of oil pressure is supplied to an MFWD clutch, what is the status of the tractor?
   A. Two-wheel-drive mode.
   B. Four-wheel-drive mode.
   C. Tractor is power hopping.
   D. Front tires are slipping.

8. In addition to the master clutch, how many other clutches must be engaged to obtain power flow in the Magnum powershift transmission?
   A. One.
   B. Two.
   C. Three.
   D. Four.

9. On the Magnum powershift transmission, power flow passes through which clutches first?
   A. 1-2, 3-4, and 5-6.
   B. Creeper, even, and odd.
   C. Low, mid, and high.

10. How many clutches must be engaged to deliver power through the Caterpillar H, M, and non-suffix series transmission?
    A. One.
    B. Two.
    C. Three.
    D. Four.

11. How many planetary gear sets are used in the Caterpillar H, M, and non-suffix series transmission?
    A. Zero.
    B. One.
    C. Two.
    D. Three.

12. How many shafts are used in the Caterpillar H, M, and non-suffix series transmission?
    A. Two.
    B. Three.
    C. Four.
    D. Five.

13. Which shaft is not used in reverse in the Caterpillar H, M, and non-suffix series transmission?
    A. Shaft 1.
    B. Shaft 2.
    C. Shaft 3.
    D. Shaft 4.
14. How many total clutches are in Caterpillar H, M, and non-suffix series motor grader transmissions?
    A. Seven.
    B. Eight.
    C. Nine.
    D. Ten.
15. The parking brake on Caterpillar H, M, and non-suffix series motor grader transmissions is on which shaft?
    A. Shaft 2.
    B. Shaft 3.
    C. Shaft 4.
    D. Shaft 5.
16. A John Deere 8000 series powershift transmission's input clutches are on the _____.
    A. input shaft
    B. countershaft
    C. Both A and B.
    D. Neither A nor B.
17. A John Deere 8000 series powershift transmission's output clutches and brakes are on the _____.
    A. countershaft
    B. output shaft
    C. Both A and B.
    D. Neither A nor B.
18. A John Deere 8000 series powershift transmission's auxiliary shaft provides power to all of the following, EXCEPT:
    A. the transmission scavenge pump.
    B. the PTO.
    C. the hydraulic pumps.
    D. the ground drive pump.

Apply and Analyze

19. MFWD clutches are typically applied with _____ pressure?
20. Most MFWD clutches are located in the _____ housing.
21. The _____ clutch is eliminated in a Magnum powershift transmission equipped with an overdrive clutch.
22. There are _____ number of planetary gear sets used in the Magnum powershift transmission.
23. The maximum number of forward speeds for the Case IH Magnum powershift transmission is _____.
24. A John Deere 8000 series powershift transmission has a total of _____ brakes.
25. Brake A in a John Deere 8000 series powershift transmission holds the _____.
26. A John Deere 8000 series 16-speed powershift transmission uses a total of _____ planetary gearset(s).
27. A John Deere 8000 series powershift transmission contains a total of _____ input clutches.
28. A total of _____ clutches and brakes must be applied to obtain a forward or reverse gear in a John Deere 8000 series 16-speed powershift transmission.

Critical Thinking

29. Describe the three different types of transmissions that can be used in late-model Caterpillar motor graders.
30. Why does the John Deere e23 powershift transmission use two different gears for eighth forward?

Chapter 11
Powershift and Automatic Transmission Controls, Service, and Repair

Objectives

After studying this chapter, you will be able to:

- ✓ Explain hydraulic flow through a powershift transmission.
- ✓ List the different methods for controlling powershift transmissions.
- ✓ List tools used for diagnosing powershift transmissions.
- ✓ Describe the repair processes for servicing powershift transmissions.
- ✓ List steps taken to remove a powershift transmission from a Caterpillar elevated-sprocket track-type tractor.

Transmission Fluids, Pumps, Reservoirs, and Circuitry

Transmissions require fluid for engaging clutches, maintaining lubrication, operating the torque converter, and cooling. Without the proper fluid, the transmission will not operate. Using the appropriate fluid for the application is critical. Tractors can have sudden transmission failures if equipped with the wrong fluid. Manufacturers will specify the type of transmission fluid for the transmission.

Some transmissions have their own oil reservoir; other transmissions share the reservoir used for the machine's hydraulic system.

Transmission Fluids

As technicians fill or replace the transmission fluid, they need to ensure that the transmission fluid meets the manufacturer's specification. The manufacturer's fluid specification lists multiple properties. Some examples of transmission fluid properties are:

- Pour point—the lowest temperature at which the fluid will flow.
- Viscosity—the thickness of the oil, also known as the resistance to fluid flow, in units of Saybolt Universal Seconds (SUS), or centistokes (cSt).
- Viscosity grade—the International Standards Organization (ISO) viscosity range for a fluid based on a 104°F (40°C) temperature; the ISO VG number is an average viscosity value for that ISO range.

- SAE viscosity ranges—a grade of oil specified by the Society of Automotive Engineers (SAE) based on a fluid temperature of 212°F. The grade of oil is specific to summer (non W) or winter (with a W). If the grade has both, for example 5W20, it is a multigrade oil.
- Viscosity index—a rating used to indicate how much or how little the oil's viscosity changes across a wide temperature range.
- Stability—the resistance to chemical reaction with other substances.
- Anti-wear—a fluid property that ensures a full film of lubricant between components during boundary lubrication, when the component surfaces are in close contact with each other.
- Seal compatibility—a fluid's compatibility with the materials used in transmission seals, such as nitrile rubber.

Manufacturers often specify lower *viscosity*, or thinner, transmission oil that flows more freely when the machine is operating in colder climates. Examples of Caterpillar transmission fluids include Transmission/Drive Train Oil (TDTO) in SAE 10W, 30 and 50 viscosities; Transmission Multi-Season oil (TDTO-TMS); TDTO-Cold Weather; Cat Special Application Transmission Oil (SATO); and Cat ATF-HD2.

TDTO-TMS is used if the machine operates in a broad range of temperatures. TDTO-Cold Weather is recommended for operating in temperatures as cold as -40F (-40C). Cat SATO is recommended for medium wheel loaders and can extend the life of the fluid from 1000 hours to 2000 hours, but cannot be used in final drives, axles, and differentials. Cat ATF-HD2 is used in heavy duty automatic transmissions.

Allison has specified automatic transmission fluids over the years, including TES 295, TES 389, TES 468, TES 295, and the latest fluid, TES 668.

Case Study
Determining the Cause of Fluid Loss

A Chevrolet 5500 truck with an Allison 2000 series rolled upside down. The owner called the dealer and body shops, because the truck would not move. He asked the dealer and body shop if it was possible that the transmission became electronically disabled because the side airbags deployed. Even if the electronics failed, the truck should still have reverse and third gear as a limp-home mode. This truck had no forward gears or reverse. There was no evidence of transmission leaks. However, the transmission was low on fluid. It was determined that the fluid leaked out of the transmission's breather, which is mounted on top of the transmission, when the truck was upside down.

Transmission Pumps

Most transmissions receive oil from a *fixed-displacement pump*, which pumps a fixed volume of oil during each revolution. However, some transmissions have a *variable-displacement pump*, which reduces flow when the oil demands have been met. The pump can be dedicated to the transmission, or it can provide oil for other functions, such as steering or charging the implement piston pump.

If the pump provides oil to only the transmission, it is often integrated with the transmission in such a way that it is driven proportionally to engine speed. For example, even if the transmission oil pump is located at the back of the transmission, it is not driven by the output shaft. Instead, it is linked to an input shaft or the torque converter. The pump must spin proportionally to engine speed to provide the torque converter live, continuous flow of hydraulic oil. The pump's volume must be large enough to meet the requirements of the transmission and the torque converter, even when the engine runs at low speeds. At higher speeds, excess oil is directed to the reservoir. The transmission housing may serve as the reservoir for the transmission oil, or the machine may be equipped with a separate reservoir.

Scavenge Pumps

Some transmissions use a *scavenge pump*, which transfers fluid from the transmission's case to a different oil reservoir. Scavenge pumps will pump a small percentage of air, around 20% air and 80% oil. Traditional hydraulic pumps will quickly fail if subjected to aeration, but scavenge pumps will not. This is because they transfer oil at a very low pressure, while traditional positive-displacement pumps operate at higher pressures. Refer to Chapter 10, *Powershift and Automatic Transmission Theory*, which discusses a John Deere 16-speed powershift transmission that uses a scavenge pump.

Transmission Filtration

Transmissions use screens and filters to remove fluid contaminants. *Suction screens*, also called *suction strainers*, are wire mesh filtration devices that prevent large particles from entering the inlet of the transmission pump. See **Figure 11-1**. Filters remove smaller particles. Powershift transmissions may filter the pump's outlet oil, which is under charge pressure. This filter must have a burst pressure rating two to three times the system pressure. A filter's burst pressure is the pressure at which the filter will deform to the point of failure. The filter in **Figure 11-2** has a burst pressure of 700 psi (48 bar).

A

B *Goodheart-Willcox Publisher*

Figure 11-1. Suction screens are often used to prevent large contaminants from entering the inlet of the transmission pump. A—Filter screen installed in its housing. B—The filter screen has been removed.

Case E and F Series Wheel Loader Hydraulic Circuitry

The transmission hydraulic circuit in Case E and F series wheel loaders is shown in **Figure 11-3**. The main relief valve regulates the transmission's main system pressure of 230 psi to 265 psi (16 bar to 18 bar). The main system oil is directed to a pressure-reducing valve and to the six transmission clutch controls.

The remainder of the oil dumped across the main relief valve is sent to charge the torque converter. After the torque converter, the fluid is cooled and used to lubricate the transmission. Torque converters will be discussed further in Chapter 12, *Hydrodynamic Drives*.

The *pressure-reducing valve* is a normally open pressure control valve that senses pressure downstream. The valve reduces the pressure downstream and maintains that pressure. In the example shown in **Figure 11-3**, the valve reduces pressure downstream to between 130 psi and 160 psi (9 bar to 11 bar). The reduced downstream pressure is used for pilot system pressure. Manufacturers use pilot pressure as a low-pressure medium for actuating control valves. The lower pressure range enables engineers to better control the response of the control valves. A pilot pressure that is too high would result in a hydraulic control system that is too responsive. As shown in

Goodheart-Willcox Publisher

Figure 11-2. This powershift transmission is a laboratory trainer. The filter must have a burst pressure rating two to three times the 230 psi operating pressure.

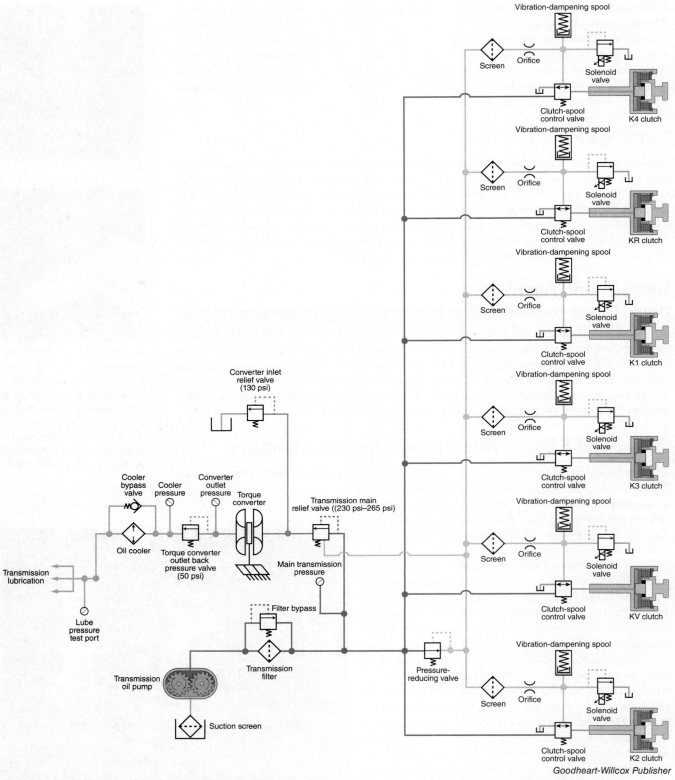

Figure 11-3. This schematic shows the powershift transmission in Case 521–921E and F series wheel loaders. The transmission can be configured with four or five speeds.

Figure 11-4, for each clutch, the pilot system oil flows through a screen and orifice and then acts on the following:

- Vibration-dampening spool.
- Clutch-spool control valve.
- Solenoid valve.

The vibration-dampening spool works in conjunction with the clutch-spool control valve to modulate the shift. This dampens or softens the engagement of the clutch as the clutch piston reaches its end of travel and the clutch fully engages. When the solenoid valve is de-energized, the pilot oil pressure overcomes the solenoid valve's spring pressure and the pilot oil is dumped to the reservoir. The orifice maintains the pilot oil pressure for the other clutch controls. Without it, pilot oil pressure would drop to nearly zero any time the solenoid valve is de-energized.

Notice that the clutch-spool control valve is a normally closed spool valve that blocks the main transmission pressure from engaging the clutch and drains the oil in the clutch to the reservoir. When the solenoid valve is energized, the pilot oil is no longer dumped to the reservoir. After the orifice, the pilot oil pressure builds and opens the clutch-spool control valve so the main transmission pressure engages the clutch. See **Figure 11-5**.

Caterpillar D6R Oil Flow

Figure 11-6 shows transmission oil flow for a Caterpillar D6R track-type dozer. The transmission provides the oil sump for the transmission, transfer gears, bevel gears, steering

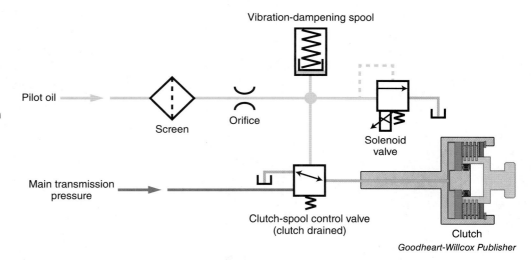

Figure 11-4. Pilot system pressure acts on the vibration-dampening spool, solenoid valve, and clutch-spool control valve. When the solenoid valve is de-energized, the clutch-spool control valve blocks the main transmission pressure from engaging the clutch and drains the oil in the clutch.

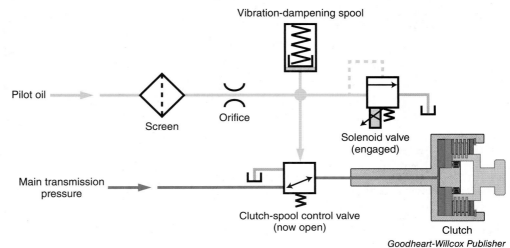

Figure 11-5. When the solenoid valve is energized, pilot oil pressure is no longer dumped to the reservoir. Pressure builds and opens the clutch-spool control valve, allowing main transmission pressure to engage the clutch.

clutches, steering brakes, and torque converter. Chapter 24, *Track Steering Systems*, explains the power flow of the bevel gears, steering clutches, and steering brakes. Chapter 12, *Hydrodynamic Drives*, discusses torque converters.

The transmission uses a triple oil pump assembly driven by an engine auxiliary shaft. The three oil pumps are labeled A, B, and C.

Pump A provides pressurized oil to the steering clutches, steering brakes, and transmission clutches. It draws oil from the bevel gear case and a suction screen. The oil is filtered and sent to the steering and brake control valve and to the transmission modulating valves and main relief valve. A portion of the oil flowing after the relief valve provides transmission lubrication.

Pump B provides oil to the torque converter and lubrication for the steering clutches and steering brakes. It draws oil from the bevel gear case and delivers it to the priority valve. The priority valve distributes supply oil to the torque converter and lubrication oil to the steering clutches and brakes. When the electronic control module (ECM) energizes the priority valve's solenoid, pump B's flow is blocked at the priority valve and flows through the check valve into the same passage used by pump A. The oil is used for steering clutches, steering brakes, and transmission operation, with priority given to steering and brakes.

Pump C is the scavenge pump. It pulls oil from the transmission housing and torque converter housing and delivers it to the bevel gear housing.

Transmission Operating Pressure

As mentioned, most transmissions use a fixed-displacement pump. A relief valve is used to maintain *transmission operating pressure*, also referred to as *main pressure*, *regulated pressure*, or *transmission charge pressure*. Transmission operating pressure is used to

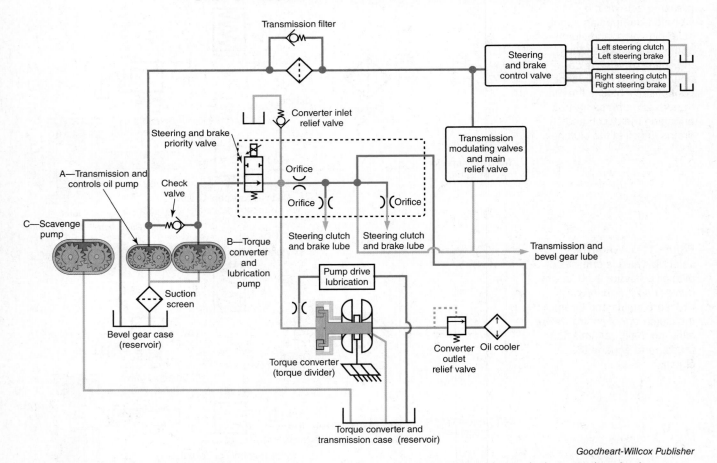

Figure 11-6. A Caterpillar D6R track-type dozer uses a triple oil pump assembly for the transmission, steering clutches, steering brakes, and torque converter.

apply the transmission's clutches. Some machines operate with a pressure as low as 200 psi (14 bar) and some with a pressure as high as 500 psi (35 bar). The operating pressure for many off-highway transmissions is around 300 psi (21 bar). The service literature will specify how to check and adjust the pressure. This may involve warming the hydraulic fluid, transmission fluid, or engine coolant to a certain temperature, placing the transmission in park, and running the engine at a specific speed. The fluid temperature and fluid pressure might be displayed on a monitor inside the cab or on an electronic service tool, such as the Caterpillar electronic technician (ET). If a problem is suspected, it is best to measure the pressure with a pressure gauge. See **Figure 11-7**.

An adjusting screw can be used to adjust the pressure on some transmissions, as shown in **Figure 11-7**. Other transmissions require that the technician replace springs or add or remove shims to adjust the pressure. Always ensure the system has been depressurized before adjusting or removing hydraulic valves and components.

Warning

As mentioned in Chapter 1, *Shop Safety and Practices*, as little as 100 psi of pressure can cause a fluid injection injury. Do not adjust hydraulic valves while the system is pressurized.

Transmission Controls

There are three types of controls that can be used in powershift transmissions:
- Mechanical controls.
- Electrical controls.
- Electronic controls.

Goodheart-Willcox Publisher

Figure 11-7. This Challenger MT845 rubber track tractor's transmission is operating at a pressure of 320 psi (22 bar). Notice that the cap has been removed from the end of the transmission relief valve. Shut off the machine and ensure the pressure gauge shows zero pressure before making adjustments. A wrench is placed on the adjusting screw to make adjustments to the transmission operating pressure. After making adjustments, restart the machine and determine if the pressure is within specification.

Mechanical Controls

The first off-highway powershift transmissions were manually operated with mechanical controls. A mechanical lever, through linkage or cables, shifted a control valve. Mechanical controls have been used in scrapers, dozers, motor graders, agricultural tractors, and several other off-highway machines. With the rise of electrical and electronic controls in the 1990s, the use of mechanical controls has steadily declined. It is rare to find late-model off-highway machines with mechanical controls.

A cross-section of a manually operated transmission control valve used in older Caterpillar track-type dozers, such as the D5H, is shown in **Figure 11-8**. The valve is shown in the neutral position. Chapter 9, *Planetary Gear Set Theory*, provides an explanation of power flow through this powershift transmission. Notice that two manually operated

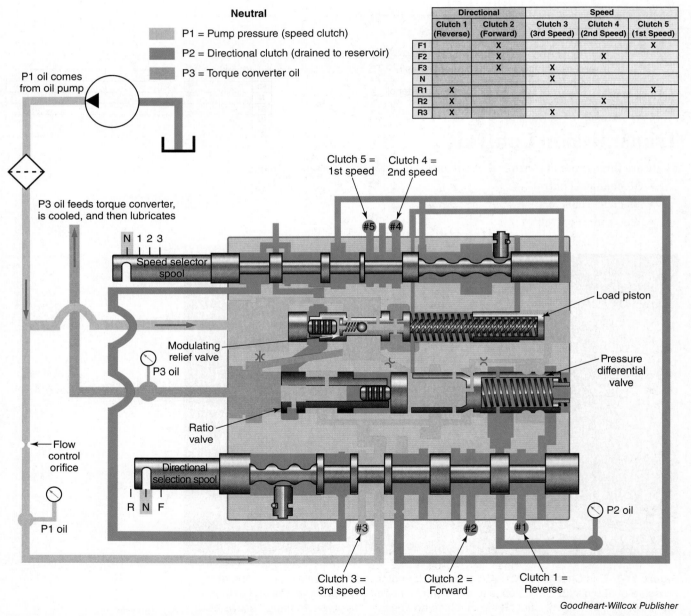

	Directional		Speed		
	Clutch 1 (Reverse)	Clutch 2 (Forward)	Clutch 3 (3rd Speed)	Clutch 4 (2nd Speed)	Clutch 5 (1st Speed)
F1		X			X
F2		X		X	
F3		X	X		
N			X		
R1	X				X
R2	X			X	
R3	X		X		

Goodheart-Willcox Publisher

Figure 11-8. This is a cross-section of the manually operated powershift transmission control valve of a Caterpillar elevated track-type dozer in neutral. Clutch 3 is engaged in neutral, and all other clutches are drained to the reservoir. The priority valve is not shown to simplify the drawing.

valves are used: a four-position speed selector spool for neutral, first, second, and third speeds, and a three-position directional spool for reverse, neutral, and forward.

> **Note**
>
> Earlier versions of this transmission did not have a neutral position on the directional spool. The spool was designed to be in either forward or reverse, while the speed selector spool provided the only neutral position.

The transmission control valve distributes oil to the appropriate directional clutch, speed clutch, and torque converter based on the gear selected by the operator. As depicted in the legend in **Figure 11-8** and **Figure 11-9**, P1 oil supplies oil to the speed clutches, P2 oil supplies oil to the directional clutches, and P3 supplies oil to the torque converter.

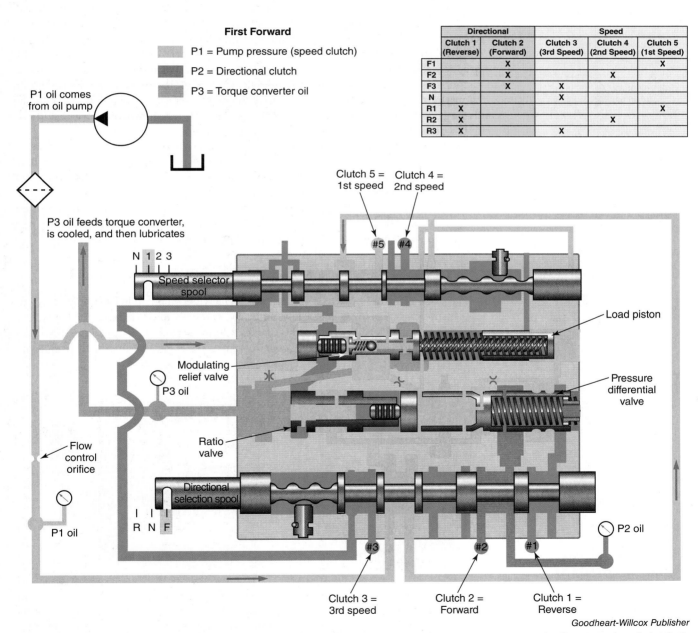

Figure 11-9. In first forward, the two spool valves shift one position to the right to engage clutch 5 for first speed and clutch 2 for forward.

The transmission pump delivers oil through a filter and a priority valve. After steering and brake oil demands are met, the oil supply, labeled P1, is directed to the transmission control valve to engage the speed clutches. The P1 oil is sent to the modulating relief valve and through the flow control orifice to the directional spool valve. The oil then flows to the ratio valve and the pressure differential valve. The *ratio valve* limits the maximum pressure to the torque converter, which is necessary for starting the engine in cold climates. When the torque converter's flow demands are met, pressure rises and the ratio valve directs the remaining oil to the reservoir. The *pressure differential valve* maintains the pressure in the speed clutch circuit so that it is 55 psi (4 bar) above the pressure in the directional clutch circuit. This ensures oil is sent to the speed clutch before the directional clutch.

When the transmission control valve is in a neutral position, the speed selector spool drains oil pressure from clutch 4 and clutch 5. The directional spool drains oil pressure from clutch 1 and clutch 2 and directs P1 oil to apply clutch 3. Because P2 oil is routed to the reservoir, it is represented in green in **Figure 11-8**. Clutch 3 is the only clutch applied in neutral (represented in pink). The directional spool valve also sends oil to the pressure differential valve, causing it to shift and open the right-hand passage to the reservoir.

With the transmission control valve in the neutral position and clutch 3 engaged, P1 pressure increases. This is sensed at the modulating relief valve and at the right side of the load piston. The load piston and modulating relief valve work in unison to modulate the P1 oil pressure by slowly allowing the P1 pressure to increase as the speed clutch piston reaches its end of travel. When the modulating relief valve reaches its maximum pressure, it shifts to the right to divert the remaining pump oil to the torque converter circuit (P3 oil).

When the operator shifts the transmission, the speed selector spool and the directional selection spool are actuated. **Figure 11-9** shows the transmission being shifted into first forward. The speed selector spool shifts to first speed and the directional selection spool shifts to forward. When the speed selector spool shifts one position to the right, it drains oil from clutch 3 and clutch 4, and applies P1 oil to clutch 5, the first speed clutch (represented in pink). When the directional spool valve shifts to the right, it drains clutch 1, the reverse clutch, and applies P2 oil to clutch 2, the forward clutch (represented in dark orange).

Electrical Controls

In the 1990s, machine manufacturers began using electrically controlled solenoid valves to control transmissions. The first electrically controlled transmissions used electrical switches. The powershift control lever in **Figure 11-10** uses a rotary switch to energize the transmission's solenoids (**Figure 11-11**). Electrical controls can be placed away from the transmission,

A B *Goodheart-Willcox Publisher*

Figure 11-10. This control lever is used to operate a powershift transmission with six forward speeds and three reverse speeds. A—As the lever is moved, the rotary switch engages the appropriate transmission solenoid. B—The lever can be moved into three reverse positions, six forward positions, and the neutral position. The lever is spring loaded so that it is held in the different positions.

since they are connected to the transmission with a wire harness instead of bulky mechanical linkage.

Electronic Controls

Electronic controls began being used around the 1990s, depending on the specific off-highway industry (agriculture, construction, forestry) and specific manufacturer. ECMs incorporate a wide range of electrical inputs, such as engine speed, engine load, vehicle speed, ground speed, clutch pedal position, pressure switches, pressure sensors, and temperature sensors. The ECM uses the input data along with its software to control the transmission. Today, nearly all manufacturers choose electronic computer controls to operate their transmissions. See **Figure 11-12**.

Clutch Pedal Controls

Many motor graders and most agricultural tractors equipped with a powershift transmission do not use a torque converter. Chapter 12, *Hydrodynamic Drives*, explains the functions of a torque converter. As mentioned in Chapter 10, *Powershift and Automatic Transmission Theory*, a transmission without a torque converter is a ***direct-drive powershift transmission***, meaning that the engine directly supplies the input to the transmission through a flywheel-driven dampener assembly.

Goodheart-Willcox Publisher

Figure 11-11. This Funk powershift transmission, used as a trainer for students, has six solenoids. Because there were no individual clutch pressure ports, a machinist made a plate using the control valve's gasket so that the passages could be tapped and the individual clutch pressures measured.

Direct-drive powershift transmissions require a master clutch pedal, or inching pedal. Prior to electronically controlled transmissions, the clutch pedal actuated a mechanical cable that operated a hydraulic clutch control valve that controlled a transmission's clutch, such as a master clutch. Today, most transmissions use an ***inching pedal potentiometer***, which is a variable resistor that provides a variable voltage indicating the position of the inching pedal as an input to the ECM.

A

B

Goodheart-Willcox Publisher

Figure 11-12. Caterpillar manufactures a CX automatic transmission that was used in their on-highway trucks and recreational vehicles. The CX31-P600 is still used in the oil and gas industry. A—This Caterpillar CX28 transmission has external solenoids that can be changed without removing covers or oil pans. B—The transmission has external pressure taps on the bottom (left side of the oil pan) to quickly connect and measure transmission pressures.

In addition to the potentiometer, many inching pedal assemblies also include switches, such as a top-of-travel switch and a bottom-of-travel switch. The potentiometer is commonly used to modulate clutch engagement, such as for the master clutch. Some manufacturers use the ***top-of-travel switch*** for certain transmission features, such as to initiate speed matching, as described in Chapter 10, *Powershift and Automatic Transmission Theory*. The ECM monitors the change of the inching pedal position based on the input from the potentiometer. When the pedal is fully released, the top-of-travel switch closes, allowing the ECM to determine that the pedal has been fully released.

The ***bottom-of-travel switch*** is a normally closed switch that acts as a neutralizer. When the ECM recognizes that the inching pedal has been fully depressed, the switch opens and neutralizes the transmission by draining clutch fluid. Applications vary. Depending on the design, the bottom-of-travel switch can remove the common power supply or the common ground to the shift solenoids. The switch can cause the ECM to drain both the directional clutches and speed clutches. In some machines, it drains only the directional clutches. **Figure 11-13** is a screenshot of an electronic service tool being used to check the travel of the master clutch pedal and the position of the bottom-of-travel switch.

Electronic Control Advantages

Electronic controls offer several advantages over mechanical controls, including:
- Calibration.
- Modulation.
- Adaptive learning.
- Reflashing.
- Virtual Operating Speeds.

Calibration

Most electronically controlled transmissions have a ***calibration mode*** that allows the ECM to learn transmission clutch values, such as clutch fill and clutch hold. Variables like piston free travel, oil passage restriction, clutch wear, and piston seal leakage can affect the time it takes for a clutch to engage. The ECM uses data from speed sensors, pressure sensors, and commanded gear shifts to calibrate itself and establish the clutch fill and clutch hold values. **Figure 11-14** shows an electronic

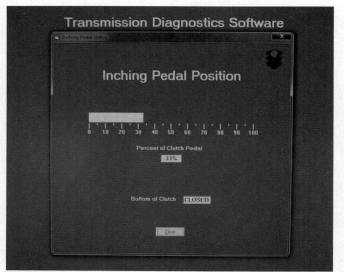

Goodheart-Willcox Publisher

Figure 11-13. This Funk transmission electronic service tool checks the travel of the inching pedal. Notice that the inching pedal has been depressed 33%, and the bottom-of-travel switch is closed. When the pedal reaches the bottom of travel and the switch opens, the ECM will neutralize the transmission.

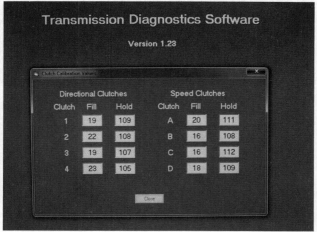

Goodheart-Willcox Publisher

Figure 11-14. The electronic service tool's screen lists transmission calibration values. Each directional and speed clutch has a fill value and a hold value.

service tool listing the clutch values of a Funk transmission. ***Electronic service tools*** are programs installed on computers or other digital devices and are explained later in this chapter.

The ***clutch fill*** value is the amount of time it takes to fill a clutch. The clutch fill numbers listed in **Figure 11-14** represent the number of counts it takes to fill the clutch. For example, if the controller is operating at a 5 millisecond count, a clutch fill value of 18 would equal 90 milliseconds (18 × 5 milliseconds). A clutch is considered full when the cavity behind the clutch piston is filled with oil but the clutch piston has not applied enough force on the friction elements to transfer significant torque.

Note

The clutch fill values for Case IH Magnum transmissions are specified in amperage rates, such as 190 milliamps. The values represent the amount of current that must be applied to the solenoid to achieve the correct fill rate.

Once the clutch is filled, the ECM keeps the clutch from draining without fully applying the clutch. A shift problem will occur if the clutch is engaged too quickly. During the clutch fill period, the ECM energizes the solenoid using the maximum current, in this example 750 milliamps for a 12-volt system. When the clutch is filled, the solenoid's amperage drops and the modulation period starts. See **Figure 11-15**. During the modulation period, the ECM increases the current to the solenoid to fully engage the clutch.

The clutch hold value is directly tied to torque. After the fill period, the ECM controls the solenoid to slow the incoming oil. This holds the clutch while the clutch's transfer of torque begins. The ***clutch hold values*** in **Figure 11-14** represent the current that the ECM driver uses to hold the clutch.

Note

Keep in mind that the clutch hold value represents torque, and torque is achieved when fluid pressure is applied to the clutch. Pressure is achieved by applying electrical current to the solenoid valve.

It may be helpful to record clutch values for future use. If a customer has experienced a rough shift, old values can be compared with values from a new calibration. An unusually large clutch value may indicate a potential problem, such as a leak in the clutch piston seal, solenoid valve, or drum/cylinder seal.

Some manufacturers specify intervals for calibrating the transmission. For example, the Case E and F series wheel loaders should be calibrated after the first 250 hours of operation and then every 1000 hours.

Calibration may be recommended when a component has been replaced or when a machine has experienced a rough shift. Two examples of improper shifts are

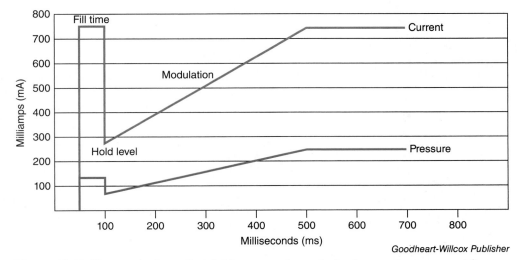

Figure 11-15. The graph shows that fluid pressure is applied using maximum current flow to fill the clutch. The fill pressure is half of the final engagement pressure because of the distance the piston moves during the fill period. The clutch is held to prevent it from draining, and then the ECM slowly applies the clutch for its final engagement.

a slipping condition and a stall. A slipping condition occurs when an off-going, or releasing, clutch releases too soon or when an on-coming, or engaging, clutch takes too long to fill. Both cause the engine speed to increase. A stall occurs when the off-going clutch releases too slowly or when the on-coming clutch fills too early. Both cause the machine to noticeably stall.

Manufacturers have different calibration procedures. The vehicle may need to remain stationary and specific fluid temperatures and engine speeds may need to be maintained. If controls, such as engine throttle, clutch switch, brake switch or forward and reverse propulsion switch, are adjusted during calibration, the machine may abort the calibration or the transmission may calibrate incorrectly. The calibration may also stop if the program requirements fail. Examples that can cause a program to fail are a clutch fill pressure did not occur or an output shaft speed sensor that indicates rotation when the shaft should be stationary.

Case Study
Transmission Input Shaft Speed

A Caterpillar 160M motor grader was unable to complete a transmission fill calibration due to an "abnormal incorrect transmission input speed" fault. The powershift transmission was a direct drive transmission, meaning it did not have a torque converter. The input shaft was directly driven from the engine through a vibration dampener. The transmission input shaft speed should have matched engine speed. However, when the engine was running at a low idle of 800 rpm, the transmission input shaft speed sensor was only reading 354 rpm. The transmission input shaft speed sensor was a two-wire, permanent magnet sensor. The technician measured the frequency, resistance, and AC voltage of the transmission input shaft speed sensor and had the luxury of comparing the readings taken from another 160M motor grader that was the same vintage, with the machine serial number being one number higher than the other machine. At 800 rpm, the sensor read 266 hertz (0.93 AC volts). At 1340 rpm, the sensor read 455 hertz (1.53 AC volts). The two machines had comparable input shaft speed readings. The two ECMs were swapped between the two motor graders, and it was determined that the fault stayed with the ECM. Based on this information, the technician determined the motor grader required a new ECM.

Caution
Always follow the service literature and make sure the machine is clear of any obstruction when calibrating the transmission. Tractors may lunge forward or backward during the calibration process.

Note
If an error, such as a speed sensor failure, occurs during calibration, many transmissions will operate in a default mode. The ECM might limit horsepower and show an error code on the machine's monitor.

Modulation

Transmission clutches are often engaged with **modulation**, meaning that the pressure is varied (slowly increased) to optimize the quality of the shift, or shift feel, and clutch life. Early transmissions used modulating control valves consisting of spool valves and springs to modulate the shift. Some early transmission control valve banks have only one modulating valve, such as the Caterpillar 1985 D5H dozer control valve bank shown in **Figure 11-9**. In 1998, Case IH MX Magnum agricultural tractors also used one modulating valve within a transmission control valve bank, **Figure 11-16**. Other transmissions can have multiple modulating valves. Caterpillar calls their multiple valve system *Individual Clutch Modulation (ICM)*.

An ICM transmission control valve bank used on an older Caterpillar scraper (for example, 657E scraper) or haul truck is shown in **Figure 11-17A**. It operates in conjunction with a transmission control valve with three solenoids, an upshift solenoid, a downshift solenoid, and a low-speed solenoid. See **Figure 11-17B**. The solenoids direct oil pressure to rotate a rotary actuator. The rotary actuator is coupled to a rotary spool valve inside the transmission control valve, **Figure 11-17C**. As the rotary spool is rotated, it directs oil to the individual ICM valves, which modulate the transmission shifts.

Figure 11-18 shows the complete transmission being tested on a transmission stand. While the transmission runs on the stand, a technician records the clutch pressures. After the pressures are measured, the ICM valve block is removed. The technician counts the shims and measures their thicknesses. Shims are added or removed to calibrate the transmission shift quality to meet the service literature's pressure specifications. When the transmission is being rebuilt, the valves are carefully placed in a tray to keep the ICM valve spools and springs together. See **Figure 11-19**.

Today, most machines use solenoid control valves for shifting transmissions. A *solenoid control valve* consists of an electromagnetic coil and a movable iron plunger rod. When electrical current is applied to the coil, a magnetic field is created, causing a plunger rod to move a small distance to open or close a fluid passage. *On/off solenoids* are not variable. They either block oil or allow oil to flow through the valve when system voltage is applied to the coil (either 12 or 24 volts). See the 1998 Case IH MX Magnum tractor example in **Figure 11-16**.

Figure 11-16. This early Case IH MX Magnum transmission's control valve bank has four on/off solenoids (top vertical positioned valves) and one modulating valve (lower horizontal positioned valve). The tractor has three of these valve banks. One such control valve bank controls the odd clutch, even clutch, creeper clutch, and parking brake. An identical valve bank controls the MWFD clutch, the low clutch, mid clutch, and high clutch. Another valve bank controls speed clutches 1, 3, 5, and reverse. All four solenoids are shown in a de-energized position.

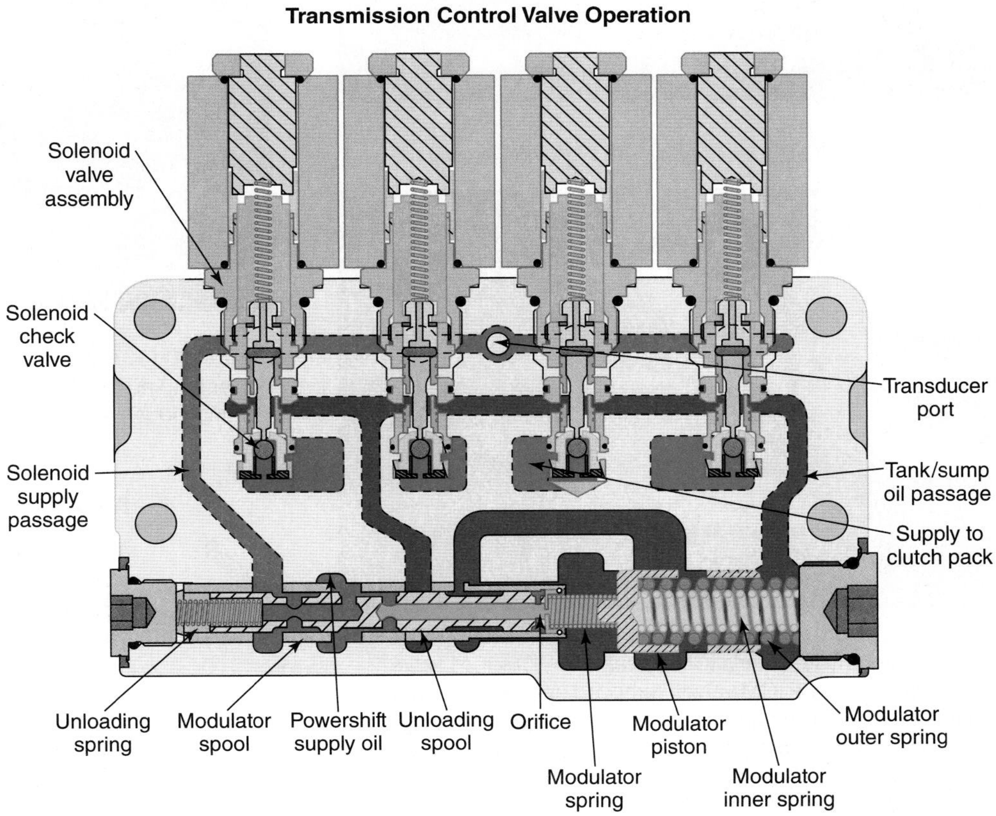

Case New Holland

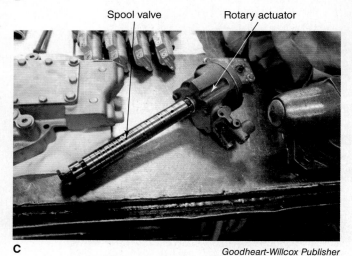

Figure 11-17. Caterpillar's ICM system. A—This ICM valve bank has one of the ICM valves removed and disassembled on the left. B—The solenoids on the transmission control valve direct oil to a rotary actuator. C—The rotary actuator is hydraulically rotated by the solenoids. The rotary spool rotates and directs oil to the ICM valves to shift the transmission. The gear at the end of the shaft is a detent to hold the spool in gear.

Figure 11-18. This Caterpillar scraper transmission is being tested on a transmission testing stand. The solenoids are electronically shifted and the ICM valves are pressure tested with the attached gauges.

Figure 11-19. A group of ICM valves have been disassembled to be cleaned during a transmission rebuild. The individual valves are labeled on the far left of the tray.

Note

A powershift transmission may use several on/off solenoid valves in conjunction with a hydraulic modulating spool valve. The early Case IH MX Magnum powershift transmission is an example. It has three control valve assemblies. Each assembly contains four on/off solenoid control valves and one modulating valve. The four on/off solenoids in the odd/even valve assembly control the even clutch, odd clutch, creeper clutch (if equipped), and the parking brake. The four on/off solenoids in the range valve assembly control the MWFD clutch, low clutch, mid clutch, and high clutch. The four on/off solenoids in the speed valve assembly control clutches 1, 3, 5 and reverse. The transmission has one proportional solenoid control valve to control the master clutch. Late-model Magnum powershift transmissions use individual proportional solenoid valves.

Many late-model machines have a dedicated proportional solenoid control valve for each clutch to modulate clutch engagement. A *proportional solenoid control valve*, also known as a *variable solenoid control valve*, varies the degree to which it opens or closes based on the current sent by an ECM. **Figure 11-20** shows an MT 845 Challenger rubber track tractor's transmission with nine proportional solenoids used to control nine clutches.

Today, manufacturers use electronic inputs and computer controls to modulate clutch engagement to optimize clutch life and shift feel. Caterpillar refers to this as *electronic clutch pressure control (ECPC)*. The transmission ECM monitors transmission inputs, such as clutch pressures, shaft speeds, and fluid temperature, and communicates with the engine ECM. Additional inputs may be monitored in some applications. These may include ground speed and wheel slip from a radar and terrain grade from a gyroscope. Based on the inputs, the transmission ECM monitors machine load, vehicle speed, grade of the terrain, and altitude to vary clutch apply pressures. As the ECM energizes the proportional solenoid, the rod pushes the ball toward the drain orifice, increasing the pressure between the drain orifice and the spool valve. The spool valve overcomes the spring tension on its right side and shifts to the right. This opens a passage for fluid pressure to apply the clutch. See **Figure 11-21**.

Goodheart-Willcox Publisher

Figure 11-20. On this MT 845 Challenger tractor transmission, pressure gauges are mounted directly to each solenoid to measure clutch apply pressures.

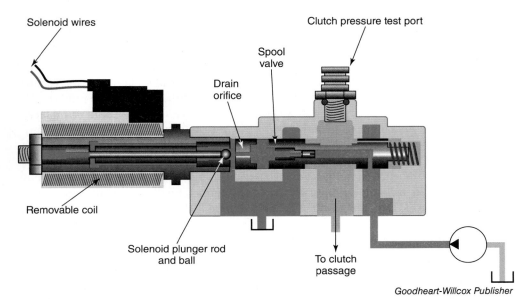

Goodheart-Willcox Publisher

Figure 11-21. This Caterpillar ECPC transmission control valve is electronically controlled with a proportional solenoid. The ECM energizes the coil, causing the valve to shift and send oil to the clutch.

When the solenoid is de-energized, spring pressure shifts the valve to the left, blocking pressure from applying the clutch. The spool vents the fluid inside the clutch to the reservoir.

ECMs do not directly vary the amperage to solenoids in an analog form. Rather, they often use *pulse width modulation (PWM)* to control proportional solenoids. PWM varies the off time and on time of the full system voltage for a given cycle. The ratio of on time to off time during a given period (called a *cycle*) is referred to as the duty cycle and is expressed as a percentage. If the solenoid is energized at a low duty cycle, it results in lower average DC voltage being applied to the solenoid. Based on Ohm's law, which explains a fixed relationship among voltage, amperage, and resistance, it takes one volt of electromotive force to push one amp of current through one ohm of resistance. If the solenoid resistance is fixed, and if the average DC voltage drops (as a result of the lower PWM duty cycle), the current applied to the solenoid drops. See **Figure 11-22**. As the ECM increases the duty cycle, the circuit's amperage increases. A higher percentage duty cycle means that the solenoid valve has a larger opening.

Pro Tip

Many digital volt ohm meters (DVOMs) have a duty-cycle scale to measure the variable PWM. The unit of measurement for duty cycle is a percentage, equaling the percentage of solenoid on time. Technicians using a meter without a duty cycle scale may use the traditional DC voltage scale, which measures *average DC voltage* applied to the solenoid. For example, a low-duty-cycle 12-volt solenoid could read two or three DC volts, and a higher-duty-cycle 12-volt solenoid could read nine or ten DC volts. Note that measuring average DC voltage is not the preferred method. It is much less precise than using a duty-cycle scale, and the controller technically is *not* varying the voltage. The controller varies the on and off time using a fixed voltage of either 12 or 24 volts, which would be seen if output were measured on an oscilloscope.

Caution

Agricultural tractors often operate at 12 volts, and many construction machines operate at 24 volts. Regardless of the machine, never apply full system voltage to the coil of a proportional solenoid, as it can damage the valve. Only on/off solenoids are designed to receive full system voltage.

Solenoid Testing

There are several ways to test the electrical operation of a solenoid. The simplest is to measure the coil's resistance. The manufacturer's service literature will specify a range for the coil's resistance, and if the coil's resistance is higher or lower, the coil should be replaced. An example range is 2 ohms to 10 ohms.

Perhaps one of the best tests for determining solenoid condition is to measure the solenoid's current. One machine can have different ranges of solenoid current flow for the various solenoids used in it. Generally, solenoid current flow ranges are low, often below 1 amp, such as 350 to 850 milliamps. If the coil is open, the ammeter will read no amperage. If the circuit has a ground fault ahead of the coil, it will read little or no amperage. If the coil is internally shorted, its resistance will be too low, causing the amperage to read higher than the specification.

Measuring the duty cycle allows the technician to determine whether the ECM is properly generating the output to the solenoid. A meter with a duty cycle scale or an oscilloscope will display the percentage of on time during the duty cycle. The least accurate test is to measure the average DC voltage across the solenoid because, as mentioned earlier, the ECM does not actually vary the voltage.

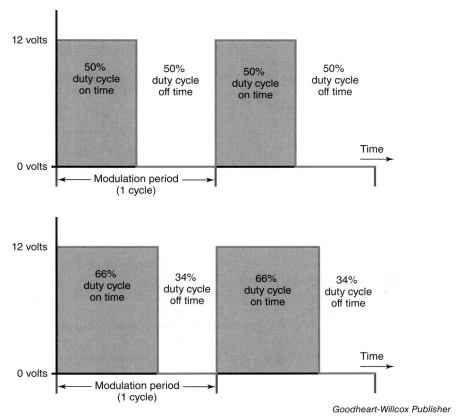

Figure 11-22. ECM pulse width modulation (PWM) of a solenoid varies the amount of on time per cycle.

The coil in many solenoids can be removed by removing a jam nut and sliding the coil off the valve's armature. See **Figure 11-23**. If the coil fails, it can be replaced without removing the valve.

Adaptive Learning

A transmission ECM senses when an element (such as a transmission clutch, brake, or torque converter clutch) slips, is slow to engage, or engages too fast. When a clutch has been commanded to engage, the ECM monitors the speed sensors, including the engine speed sensor, transmission input shaft speed sensor, countershaft speed sensor, intermediate shaft sensor, and output shaft speed sensor. The ECM can determine if a clutch is slipping and add more oil to the slipping element or command a solenoid to respond more quickly. The ECM can also determine if the clutch is engaging too fast and can respond by slowing the clutch engagement. This is referred to as ***adaptive learning***. A ***slow adapt*** takes place during normal operation. The ECM determines the volume of oil needed to engage the clutch and the speed of clutch engagement, and updates those values, making minor adjustments to the ECM's software.

Figure 11-23. The coil on many solenoid valves can be replaced by removing the jam nut. This design allows the valve to remain installed while the coil is exchanged.

Note

An intermediate shaft is sometimes used in a countershaft transmission, between the input shaft and output shaft. It spins at a different speed than the input shaft and output shaft. Funk transmissions often have an intermediate cylinder speed sensor instead of an intermediate shaft speed sensor. The intermediate cylinder is the equivalent of a clutch drum assembly located after the input shaft and before the output shaft. It has a toothed reluctor ring to provide the transmission's intermediate speed. Teeth are spaced around the reluctor ring. As the reluctor ring spins past a magnetic speed sensor, each tooth that passes by generates a pulse in the sensor. The sensor measures the shaft speed based on the speed at which the reluctor's teeth pass by.

A *fast adapt* is performed after the transmission has been rebuilt or a major component has been replaced, or if the transmission is malfunctioning. The technician uses an electronic service tool to trigger the ECM's calibration mode and forces the ECM to learn the volume of oil and time it takes for the clutches to engage. A fast adapt is equivalent to the forced calibration explained earlier in the chapter.

Reflashing

The electronics can be easily modified by *reflashing*, or reprogramming, the computer. Reflashing consists of installing new software to fix, modify, or improve the machine's performance. Reflashing the ECM with improved software can provide transmission updates and solutions.

Caution

Be sure to follow the service literature closely. Reflashing may require a steady system supply voltage. If this is not provided, a failure may result that could require replacing the ECM.

Virtual Operating Speeds

Some machines, for example Caterpillar dozers, have electronic controls to allow a transmission to function as if it has additional speeds by subdividing the engine speed ranges within each transmission gear range to create virtual gear ranges between the traditional transmission's three speeds. Each of the virtual gears has an electronically controlled engine-speed range. This allows the operator to select one of the virtual gear ranges to keep the engine operating in the most efficient range for the type of work being performed.

Early models like the Caterpillar D5N and D6N limited engine speeds to low and high ranges within first and second gears, allowing the three-speed transmission to function as a five-speed. Early D6T and D8T dozers, have a three-speed transmission that can function as a nine-speed, six-speed, or the standard three-speed. To function as a nine-speed, the engine speed ranges for each of transmission's three speeds is subdivided into three different engine speed ranges. Examples include .5, .7 and 1 speed, 1.5, 1.7 and 2 speed, 2.5, 2.7, and 3 speed. To function as a six-speed, the engine speed range for each of transmission's three speeds is subdivided into two separate engine speed ranges. The three-speed functions can also function as a traditional three-speed powershift transmission. Three-speed transmissions in D9, D10, and D11s can also function as if the dozer has additional speeds.

Late-model Caterpillar dozers with three or four speed transmissions allow the operator to request faster machine speed using a scroll wheel. The electronic controls manage engine speeds and the transmission gear speeds to provide incremental increases in machine speed (such as 0.1 mph increases), giving the feel of partial gear speed changes or virtual gear speed changes. The importance of the incremental virtual gears is to improve the machine's fuel efficiency and convenience by allowing the operator to match the engine and transmission settings to the field application.

Depending on the model year and the size of the dozer, it can have numerous different transmission electronic control descriptors. The D5N, D6N, and D6T dozers can have the option of *Multi Velocity Program (MVP)*, which allows an operator to choose one of five different operating ranges for forward and reverse. While operating in the specific range, the engine automatically varies to provide the best production results and fuel efficiency. MVP helps when working in light loads. Large dozers also have the option of selecting one of five different operating ranges, but it is not labeled. It is simply described in the portion of the Enhanced AutoShift controls.

Early machines have *Autoshift,* which is used for directional shifts, allowing the operator to choose the specific speeds for forward and reverse any time the dozer is making a directional shift. Examples include first forward and second reverse, second forward and second reverse, and second forward and first reverse.

Operators can also select the *auto-kickdown* mode, also known as auto downshift. This mode automatically downshifts the transmission when the load is too heavy.

Enhanced AutoShift (EAS) is a mode that can be chosen by the operators of late-model machines that makes the powershift transmission behave like an automatic transmission. Once the operator has selected one of the five operating ranges, the operator can adjust a scroll wheel to increase machine speed or decrease machine speed. The machine automatically maintains machine speed by automatically shifting the transmission (up or down) and adjusting engine speed to match the load to improve fuel efficiency and improve productivity.

EAS on late-model mid-sized dozers like the D6, D7 and D8 use four-speed transmissions, variable engine speeds, and lockup torque converters/torque dividers to optimize performance and improve efficiency. EAS on late-model D9, D10, and D11 use three-speed transmissions, variable engine speeds, and a lockable stator clutch. Chapter 12, *Hydrodynamic Drives*, explains lock up clutches and lockable stator clutches.

Transmission Diagnostics

Technicians use a variety of tools for diagnosing transmissions, including:
- Service literature.
- Pressure gauges.
- Electronic service tools.
- Manufacturer databases.

Service Literature

Service literature varies from manufacturer to manufacturer and may be provided in the form of manuals, bulletins, letters, and online software. Each manufacturer has a different name for their online system:
- Caterpillar—*Service Information System (SIS)*.
- John Deere—*Service Advisor*.
- Komatsu—*Customer Support System (CSS)*.
- CNH—*Electronic Technical Information Manuals (eTim)*.
- Liebherr—*Liebherr Information and Documentation Services (LIDOS)*.

Service literature often includes:
- Clutch apply charts.
- Solenoid apply charts.
- Diagnostic troubleshooting flowcharts.
- Clutch pressure specifications.
- Error codes.

Using Clutch Apply Charts to Diagnose Automatic and Powershift Transmissions

Clutch apply charts may be one of the most useful transmission diagnostic tools. Chapter 9, *Planetary Gear Set Theory*, and Chapter 10, *Powershift and Automatic Transmission Theory*, provide examples of clutch apply charts for different off-highway machines. The clutch apply chart is a matrix that lists transmission elements in the top row and transmission speeds in the left column. The cells within the matrix indicate when the individual elements are applied. See **Figure 11-24**.

Case 721F, 821F, and 921F wheel loaders can be equipped with a four- or five-speed, German-built Friedrichshafen (ZF) countershaft powershift transmission. **Figure 11-24** shows the clutch apply chart for the five-speed transmission. The gear ratios are also included. Notice that two clutches must be applied for the loader to move forward or backward. Also, notice that the KR clutch is used only in reverse.

The clutch apply chart is used to determine whether a brake or clutch is causing a problem in multiple speeds. Refer to **Figure 11-24**. If the transmission slips in second forward, the first step is to determine the clutches applied in that gear: K2 and KV. The next step is to check for problems in the other gears that use those clutches. The K2 clutch is used in second forward, third forward, and second reverse. The KV clutch is used in first forward, second forward, and fourth forward. If the K2 clutch is slipping, the operator should notice problems in second forward, third forward, and second reverse. If the KV clutch is slipping, the operator should notice problems in first forward, second forward, and fourth forward.

Note

A few off-highway transmissions must disengage three clutches and simultaneously apply three different clutches to shift to the next gear. Reference the 6th to 7th forward gear shift in **Figure 11-25**. This causes a noticeably harsh shift. Because this is part of the transmission's design, many manufacturers have no solution for the technician. Chapter 10 explained how the John Deere e23 transmission is actually a 24 speed that uses 23 specific forward speeds when upshifting (1F–8aF, 9F–23F) and 23 specific forward speeds when downshifting (23F–F8b, F7–F1) to avoid a three-element shift.

Friedrichshafen (ZF) 5-Speed Powershift Clutch Apply Chart							
Gear	K1	K2	K3	K4	KV	KR	Ratio
1st Forward	Applied	—	—	—	Applied	—	3.921
2nd Forward	—	Applied	—	—	Applied	—	2.255
3rd Forward	—	Applied	—	Applied	—	—	1.466
4th Forward	—	—	Applied	—	Applied	—	0.942
5th Forward	—	—	Applied	Applied	—	—	0.613
1st Reverse	Applied	—	—	—	—	Applied	–3.718
2nd Reverse	—	Applied	—	—	—	Applied	–2.138
3rd Reverse	—	—	Applied	—	—	Applied	–0.894

Goodheart-Willcox Publisher

Figure 11-24. This chart shows the clutch application for the optional five-speed powershift transmission used in Case 721F, 821F, and 921F wheel loaders.

Case Study: Identifying a Dragging or Warped Clutch

If an older transmission without electronic controls has an overheated clutch that warps or welds the clutch plates together, the clutch may be engaged all the time, even without oil pressure. Normally clutch pressures are measured to determine if a clutch is engaged. However, in the case of a warped clutch pack, the clutch is engaged even in the absence of oil pressure. Prior to electronic controls, the technician had no easy way to identify a warped clutch pack. An extra clutch that is warped and engaged all the time can cause the engine to lug or die.

If the tractor is capable of being started in each gear, a technician can start the machine in each gear and then release the clutch pedal to determine, which gear does not lug the engine or cause the engine to die. To eliminate loads, the technician jacks the machine up off the ground or removes the drive shafts. Next, the technician starts the machine in each of the different gears. For instance, referencing **Figure 11-25**, if the engine does not die when operating in 1st reverse, 2nd reverse, 3rd reverse, it is most likely the 1F clutch is warped and always engaged. This can be deduced because that clutch is normally engaged for 1st reverse, 2nd reverse, and 3rd reverse, so there is no extra load on the engine in those gears. If the transmission is electronically controlled, the ECM might have the ability to identify a clutch that is dragging when it should be released and provide a useful error code.

Solenoid Apply Charts

Solenoid apply charts are used the same way as clutch apply charts. Transmissions often have solenoids dedicated to specific clutches. The solenoid apply chart shows the solenoids that are engaged for each gear. The solenoid and clutch apply charts may be combined along with the transmission's gear ratios. See **Figure 11-25**.

When there is a problem in a certain gear, the solenoid apply chart is used to determine the solenoids used in that gear and the other gears that use the same solenoids. The technician can test those gears for problems and find the solenoid responsible. Some manufacturers recommend swapping solenoids to isolate and identify the problem solenoid.

The chart in **Figure 11-25** is for a 12-speed countershaft powershift transmission (model number is EW-16) used in Case IH Steiger four-wheel-drive articulated tractors from 1986 to 1999. The clutches on this chart are indicated by a number and a letter. The number indicates the clutch shaft, and the letter indicates whether the clutch is at the front or rear of the shaft. The 1R clutch, for example, is on the first countershaft at the rear of the shaft. The clutch letter does *not* indicate the direction of tractor travel. Three clutches must be engaged for the transmission to propel the tractor.

Case IH Steiger 9100, 9200, 9300 Clutch and Solenoid Apply Chart			
Gear	Clutches	Solenoids	Ratio
1st Reverse	1F 4F 5F	YVS	−5.874
2nd Reverse	1F 4R 5F	YTS	−3.248
3rd Reverse	1F 4F 5R	YVQ	−1.848
1st Forward	1R 4F 5F	ZVS	7.681
2nd Forward	2R 4F 5F	XVS	6.403
3rd Forward	2F 4F 5F	WVS	5.203
4th Forward	1R 4R 5F	ZTS	4.248
5th Forward	2R 4R 5F	XTS	3.541
6th Forward	2F 4R 5F	WTS	2.877
7th Forward	1R 4F 5R	ZVQ	2.416
8th Forward	2R 4F 5R	XVQ	2.014
9th Forward	2F 4F 5R	WVQ	1.636
10th Forward	1R 4R 5R	ZTQ	1.336
11th Forward	2R 4R 5R	XTQ	1.113
12th Forward	2F 4R 5R	WTQ	0.905

Goodheart-Willcox Publisher

Figure 11-25. This chart combines clutch-apply, solenoid-apply, and gear-ratio information for a 12-speed powershift transmission used in Case IH Steiger 9100, 9200, and 9300 series four-wheel-drive, articulated agricultural tractors.

The chart also identifies the specific solenoids (Q, S, T, V, W, X, Y, Z) that must be engaged for each gear. Looking closely at the chart reveals that each solenoid is linked to a specific clutch (Q:5R, S:5F, T:4R, V:4F, W:2F, X:2R, Y:1F, Z:1R). The transmission's five countershafts and eight clutches are shown in **Figure 11-26**.

Diagnostic Troubleshooting Flowcharts

Service literature often contains troubleshooting flowcharts, also known as troubleshooting trees. The charts provide technicians with questions that have yes or no answers. Based on the answers, the chart shows the next step in troubleshooting. An example chart is shown in **Figure 11-27**.

Clutch Pressure Specifications

Nearly all manufacturers provide clutch pressure specifications in their service literature. The specifications can be used to determine if clutch pressure is too low. It is also possible for a clutch pressure to be too high, such as due to a restricted return or a faulty regulating valve. If a control valve has a fracture or an internal leak, it can allow pressure to apply a clutch when it should be released. However, it is much more likely for an operator to notice a faulty clutch, such as a clutch slipping due to the pressure being too low, resulting in a slipping gear. Clutch pressure is often measured by installing a pressure gauge to a diagnostic pressure tap. A diagnostic pressure tap is a quick coupler designed to allow technicians to quickly connect a pressure gauge to a hydraulic system for testing.

In some machines, a panel must be removed to access the diagnostic pressure tap. Older machines require removing a plug and installing a diagnostic pressure tap. Pressure taps are shown in **Figure 11-12B**.

If the transmission uses a pressure sensor, the machine might display the transmission pressure on the machine's monitor or on the electronic service tool. A pressure sensor is an electronic sensor in a component that sends an electrical input to the ECM indicating the system's pressure or circuit's fluid pressure. Pressure sensors measure a wide range of pressures, while a pressure switch only opens or closes at a specific pressure setting. Pressure switches can provide the ECM only very limited data, such as whether the pressure is high enough to close the switch or not.

Error Codes

One of the advantages of electronically controlled transmissions is their ability to be diagnosed electronically. Today's ECMs provide various error codes that help service personnel find solutions. *Error codes* are diagnostic reference numbers tied to specific faults. Error code descriptions and solutions are provided in service manuals and electronic service tools.

Power Flows and Mechanical Schematics

It is a good practice to try to pinpoint a transmission failure prior to disassembling the transmission. Otherwise, if a problem cannot be found at the time of disassembly, additional transmission tests cannot be performed until the transmission is re-assembled and reinstalled in the machine. Clutch apply charts are a great tool for identifying soft part failures such as a bad clutch or piston seal. However, learning transmission power flows and learning mechanical schematics are useful tools for pinpointing hard part failures, such as worn splines; broken spot

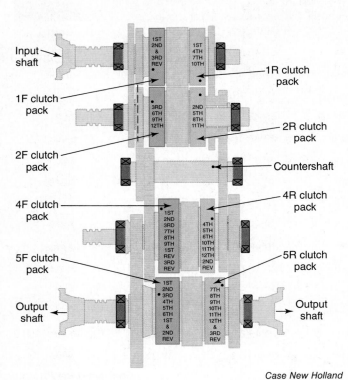

Case New Holland

Figure 11-26. The Case IH Steiger 12-speed powershift transmission has five shafts. The clutch number indicates the shaft and the letter indicates whether the clutch is in the front or rear.

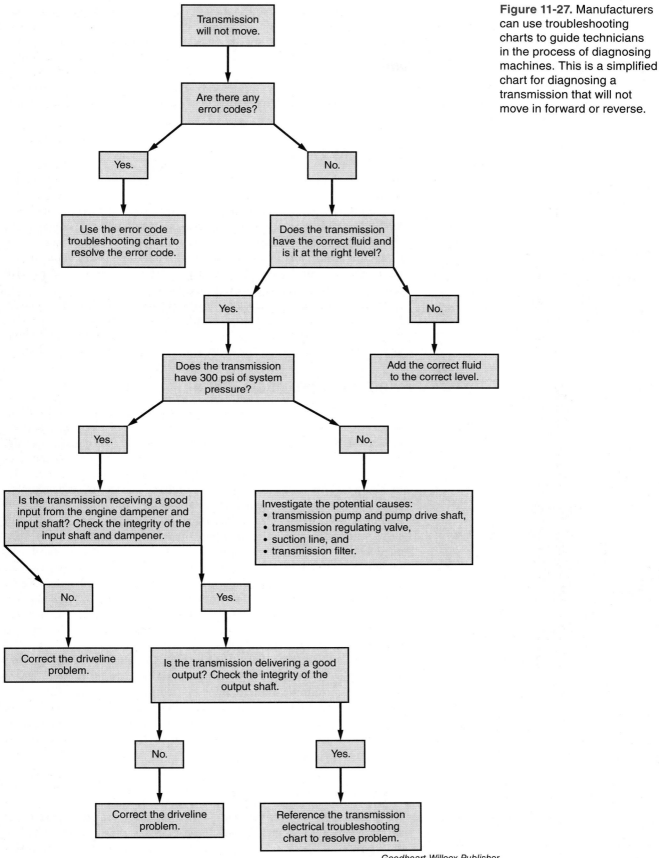

Figure 11-27. Manufacturers can use troubleshooting charts to guide technicians in the process of diagnosing machines. This is a simplified chart for diagnosing a transmission that will not move in forward or reverse.

welds; and broken gears, hubs, or shafts. For example, if the spot welds fail on a drum or a hub, making the one-piece component now a two-piece component, and a technician was quickly rebuilding the transmission, it would be possible for the technician to overlook the failure. As a result, the transmission would need to be rebuilt again. Another example is the Caterpillar three-speed dozer transmission. If the transmission incurred a failure that caused a loss of all forward gears and all reverse gears, a study of the power flows and mechanical schematics would show that three suspect areas: the input shaft, output shaft, and the #2/#3 planetary carrier.

Pressure Gauges

As explained in Chapter 2, *Tools and Fasteners*, technicians can choose from a variety of instruments for measuring pressures. The most common type is the Bourdon tube pressure gauge. Technicians can also use a MICO Quadrigage, which is similar to the Caterpillar Tetra Gauge. These combination gauges can be used to measure a wide pressure range, from a vacuum all the way up to 6000 psi (414 bar). Some technicians use an electronic pressure meter. Although these meters are more expensive than traditional gauges, they can provide additional features, as detailed in Chapter 2, *Tools and Fasteners*. These additional features include MIN/MAX, differential pressure, and *Delta Zero*. An electronic pressure meter's number one advantage is safety, allowing the technician to hold a low voltage meter rather than a live pressure gauge in his or her hand.

Electronic Service Tools

Nearly all of today's manufacturers design their machines to be diagnosed by some type of electronic service tool. An electronic service tool consists of a computer or other digital device that connects to the machine through cables and a communication adapter, or via a wireless technology. This tool allows the technician to communicate with the machine's ECMs over a data bus. Caterpillar's electronic service tool is called Electronic Technician (ET). A Caterpillar communication adaptor is shown in **Figure 11-28A**. A data bus is an electrical communication system normally consisting of a pair of twisted wires that meets standard Society of Automotive Engineers (SAE) communication standards, known as protocols. The data bus provides the electrical path (wires) that the ECMs use to share data. The data is distributed as small pulses of current, transmitted across the bus from one ECM to another. The data cannot be interpreted with a digital volt ohmmeter. The electronic service tool or machine monitor must be used to communicate or interpret the data passing through the data bus.

John Deere's electronic service tool is called Service Advisor, which is the same name as their online service literature. John Deere calls their communication adaptor an Electronic Data Link (EDL). See **Figure 11-28B**. CNH calls their service tool the Electronic Service Tool (EST). Komatsu machines use their onboard monitors, the same monitors the operator uses to control and monitor the machine's systems, as an electronic service tool.

Manufacturer Solutions

Many machine manufacturers have a ***manufacturer solutions group*** consisting of personnel assigned to assist licensed dealerships with difficult technical problems. CNH calls their group the ***Technical Service Group (TSG)***,

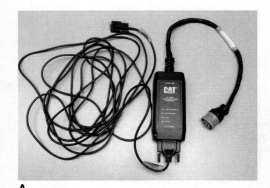
A

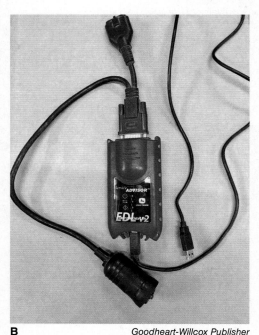
B
Goodheart-Willcox Publisher

Figure 11-28. Manufacturers design their machines to be diagnosed with electronic service tools, consisting of software installed on a computer or other digital device. A communication adaptor is required to connect the service tool to the machine. A—One example of a Caterpillar communication adaptor used for the CAT ET service tool. B—The John Deere EDL electronic data link is used to connect Service Advisor to a machine.

and John Deere calls their group the *Diagnostic Technical Assistance Center (DTAC)*. While questions were originally handled over the phone, most are now answered through web-based software. Technicians can search the manufacturer's database for solutions to similar questions or submit a new inquiry.

Multiple-Disc Clutch Service and Repair

Multiple-disc clutches have a long life when properly operated and maintained, but they will wear over time. Oil used for clutches must be free of contaminants and moisture, and the oil must be cooled. Heat will cause clutch piston seals to fail and the clutch to slip. This leads to worn discs and burnt separator plates.

Disassembly

When a countershaft powershift transmission is disassembled, the case is split apart. The countershafts are often intermingled with one another so they cannot be removed individually. A lifting plate is attached to each of the countershafts so that all of the shafts can be removed from the transmission case at once. The top plate shown in **Figure 11-29** is the lifting plate used to lift the countershafts out of an 18-speed John Deere powershift transmission after the transmission case has been removed. It also contains a stand on the bottom that holds the countershafts. Once the shafts are removed and placed on a table, the top plate is removed, and the shafts can be handled individually.

Goodheart-Willcox Publisher

Figure 11-29. The top lifting plate is attached to the countershafts, allowing the shafts to be removed from a John Deere 18-speed powershift transmission used in a 9RT tractor. The bottom plate is a stand that holds the shafts after they are removed from the transmission case.

 Warning
Always follow the manufacturer's service literature when disassembling transmissions. They often have components loaded with spring tension that can cause injury if the clutch is improperly disassembled.

Depending on the clutch design, the clutch may be spring applied or hydraulically applied. If the clutch is spring applied, a special tool, such as a press or a threaded compression tool, may be needed to disassemble the clutch pack. The clutch in **Figure 11-30** is an MFWD clutch in a John Deere 8000 series 16-speed powershift transmission. The clutch pack is spring applied, which means that with no oil pressure, the clutch plates and discs will not freely spin. A tool must be used to apply pressure to the clutch in order to remove the clutch pack. See **Figure 11-31**. With the clutch compressed, the retaining ring or snap ring can be removed. The pressure is then slowly released, decompressing the spring and providing access to the clutch pack, piston, and apply spring.

Goodheart-Willcox Publisher

Figure 11-30. Most agricultural MFWD clutch packs are spring applied and oil released. The clutch must be compressed in order to remove the retaining ring, clutch plates, and discs.

Figure 11-31. Clutch compression tools are used in conjunction with a press. The tool often has an opening so snap-ring pliers can be used to remove the retaining ring or snap ring.

If the clutch is hydraulically applied, the clutch pack is held in a released position by a piston-return-spring mechanism so the clutch plates and discs can be removed. However, the piston is under tension from the return spring and a special tool might be needed to depress the piston. See **Figure 11-32**.

 Warning

Any time a compression tool, such as a hydraulic press, arbor press, or clutch compression tool, is used to remove a retaining ring, the spring can easily get jammed or bound, meaning that it will be held momentarily in a compressed position, even with the retaining ring removed. After the retaining ring is removed, always ensure that the return spring is evenly releasing while simultaneously releasing the pressure on the press. Otherwise, the spring can launch out of the clutch unexpectedly when it becomes unbound.

Inspection

After the clutch has been disassembled, it must be inspected. Check the pressure plates, separator plates, and friction discs for heat discoloration, warpage, and wear. If the clutch has failed with low hours, it is important to identify the cause of failure.

A number of things can cause a burnt clutch, including incorrect machine use, wrong fluid, torn or damaged piston seal, porous casting, broken seal ring, missing check ball, and low supply pressure caused by a faulty pump, faulty valve, or fluid leak. If the clutch is rebuilt without remedying the cause of the failure, a repeat failure is likely.

A

B

Figure 11-32. Clutch 4 of a John Deere 16-speed powershift transmission. A—A press and special tool are used to compress the piston-return mechanism so the retaining ring can be removed. B—After the retaining ring is removed, the press should be allowed to slowly retract. Be sure the piston does not become bound and is fully released before attempting to remove the spring and piston. This clutch uses seven Belleville washers as the piston-return mechanism.

Piston Travel

Hydraulically applied clutches have a disengaged-clutch clearance, also known as ***piston free travel***. Procedures for measuring this distance vary. A Case IH 9350 Steiger 12-speed synchroshift transmission uses a multiple-disc clutch as its master clutch. The service manual directs technicians to use a 1/8″ pipe tap to cut threads into the input shaft bearing retainer housing so that an 1/8″ pipe plug can be installed to block the port. The passage on the flat surface of the bearing retainer that mounts directly to the transmission case is normally a transmission fluid passage. The threaded plug allows pressure to build inside the retainer to apply the master clutch. See **Figure 11-33**. Next, the input-shaft bearing retainer housing is placed on top of the clutch and an air pressure regulator is threaded into a service port on the bearing retainer. See **Figure 11-34**.

The literature states that at 40 psi the clutch piston should begin to move, and at 80 psi the clutch piston should be fully extended. At 80 psi, the clutch's hub should no longer be able to rotate. With 80 psi applied to the clutch through the bearing retainer, use multiple leafs of a feeler gauge stacked together, and place them between the piston and the clutch drum at all eight openings to measure the piston travel. See **Figure 11-35**. The piston travel for this application should be between 0.276″ to 0.315″ (7 mm to 8 mm). If the piston travel is out of specification, replace the original steel separator plates with new selective steel separator plates as needed to ensure the clearance falls within specifications. Two different thicknesses of replacement steel separator plates are available, 0.268″ (6.8 mm) and 0.283″ (7.2 mm). A clearance that is too large can cause the clutch to slip. A clearance that is too small can cause the clutch to drag or be engaged when it is supposed to be in a released position.

A feeler gauge is often the preferred tool for measuring a disengaged clutch's clearance, also known as piston travel. To measure clearance in smaller hydraulically applied clutches, with the clutch assembly removed from the transmission, a feeler gauge is placed between the clutch's retaining ring and the pressure plate. If necessary, the feeler gauge blades of appropriate thickness can be stacked to measure the clearance.

Goodheart-Willcox Publisher

Figure 11-33. A tap was used to cut threads into this bearing retainer so that a pipe plug could be inserted into the cavity to block regulated shop air from exiting the retainer.

Goodheart-Willcox Publisher

Figure 11-34. A pressure gauge and regulator are installed on the bearing retainer and used to actuate the clutch's piston so that a feeler gauge can be used to check the clutch piston's free travel.

Feeler gauge blades

Goodheart-Willcox Publisher

Figure 11-35. This clutch has bad seals and cannot build more than 60 psi of pressure. The clutch piston begins to move at 45 psi, which is too high.

Some oil-applied clutches, due to the design of the pressure plate, prevent a feeler gauge from being inserted between the pressure plate and the snap ring. Some service literature might recommend the use of a dial indicator to measure the clutch pack clearance. With the clutch drum resting on a bench and the clutch pressure plate facing upward, the dial indicator is placed perpendicular to the face of the pressure plate with the tip of the plunger contacting the pressure plate. The dial indicator is then zeroed out. Next, a small screwdriver is used to lift up on one of the clutch plates so the dial indicator can measure the pressure plate's travel, also known as the clutch pack's disengaged clearance.

A depth micrometer is required in some automotive applications for measuring a clutch pack disengaged clearance. However, it is more likely that a depth micrometer will be used in heavy-equipment transmissions for choosing the correct selective shim to use between two components, such as two different planetary gear sets. This is explained later in this section.

In some Allison transmissions (for example, the 3000/4000 series), instead of measuring the disengaged-clutch clearance, technicians must measure the thickness of the friction discs, the steel separator plates, and the pressure plate. For example, the C4 clutch has five friction discs and five separator plates that would need to be measured.

> **Pro Tip**
>
> Allison refers to the pressure plate as a piston return plate. They also refer to separator plates as steel reaction plates. Be aware that component terminology may differ among manufacturers.

Each friction disc must have a minimum of 0.087″ (2.21 mm) thickness. The friction discs have oil grooves that must have a minimum groove depth of 0.008″ (0.20 mm). The friction discs cannot be coned more than 0.016″ (0.40 mm). The cone measurement is a measurement of how much the plate has warped due to heat, causing the disc to take a shape similar to a Belleville washer. If the friction discs are too thin, have oil grooves that are too shallow, or have too much cone, the disc must be replaced.

The disc's cone measurement is used to determine the flatness of the disc. The disc must be placed on a flat surface. A depth micrometer can be placed across the disc to measure its cone. Another technique is to place a straightedge across the disc and insert a feeler gauge blade under the center portion of the straightedge.

The steel separator plates must be at least 0.095″ (2.41 mm) thick and have coning of no more than 0.016″ (0.40 mm). The piston return plate must be at least 0.135″ (3.41 mm) thick and have cone of no more than 0.016″ (0.40 mm). If the plate is too thin or too much coned, it must be replaced. After assembly, the transmission is calibrated so the ECM can learn the new clutch piston free travel and determine the clutch fill rate.

Case Study
Failed Forward Directional (High and Low) Clutches

The clutch drum and friction material of a failed powershift transmission are shown **Figure 11-36**. Notice the drum housing and clutch pack have been overheated and a discolored. This transmission is a direct-drive countershaft powershift transmission, mea ing that it does not use a torque converter.

After the transmission was disassembled, the technician inspected the component The transmission oil pump, drum, and piston were fine. The countershaft lube passagewa were checked by applying air pressure and were found to be unrestricted. The countersh pressure passageway check balls were in place. If they were missing, it would cause a dr in clutch pressure. Because the components were not defective and only the forward dire tional clutches failed, the technician determined that the operator caused this low-ho failure (600 hours).

In this transmission, the speed clutches are quickly engaged, but the direction clutches are modulated, meaning they are engaged more slowly than the speed clutche The operator was operating the machine with his or her foot on the inching pedal, causir the forward low and high clutches to constantly slip. In addition, a review of the transmi sion control module history revealed machine overspeeding and transmission overheatin

Selective Shims

Selective shims can be used to set the correct clearance between planetary gear sets or to set a shaft's endplay. Shaft endplay is the distance that the shaft can move axially. The proper endplay specified by manufacturers is typically a very short distance, such as 0.025″. Too much or too little endplay will cause excessive wear on the shaft, its bearings, seals, or its gears.

In an Allison 3000 series transmission example, to determine the proper shims to be used in the transmission, multiple measurements must be taken at the rear of the main transmission module and at the front of the rear cover module. A simplified drawing is shown in **Figure 11-37**. With the rear of the main transmission module upward (#2 planetary gear set facing upward), a straightedge is placed across the case where the rear cover module or retarder mounts. A depth micrometer is used to measure the distance from the straightedge to the bearing spacer surface on which the selective shim mounts. Because the measurement is taken from the bottom of the straightedge, the thickness of the straightedge must be subtracted from the overall depth that was measured.

Next, the rear cover module or retarder module is positioned with the #3 planetary carrier facing upward. The gasket must be installed on the mating surface of the rear cover module housing, where it mounts to the main case module. A straightedge is placed across the #3 planetary carrier, and the distance between the straightedge and the gasket is measured. Again, the thickness of the straightedge is subtracted from the measurement. The selective shim is chosen by subtracting the second depth measurement (made at the rear cover module) from the first depth measurement (made on the main module).

Goodheart-Willcox Publisher

Figure 11-36. This directional clutch (low forward and high forward) failed. Notice the clutch drum is discolored. The clutch pressure plate and clutch pack are also burnt. Sometimes plates will warp when overheated, but these are still straight. A new friction disc from another clutch pack is shown for comparison.

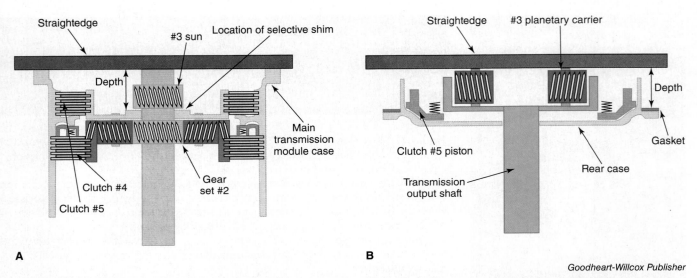

Figure 11-37. A simplified drawing of an Allison 3000 series transmission with the rear module removed from the main module case. A—A depth is measured from the end of the main case. B—A second depth measurement is made from the front of the rear case. The difference in the measurements is used to select the correct shim thickness.

One of seven different thicknesses of selective shims is selected and installed to correctly set the clearance between the two planetary gear sets. The shims are available in a range of thicknesses, from 0.180″–0.190″ through 0.245″–0.254″. The thicknesses of the shims are identified by the number of notches in the shim, with a one-notch shim having a thickness of 0.180″–0.190″ and a seven-notch shim having a thickness of 0.245″–0.254″.

Assembly

Clutch pistons must have a seal on the inside circumference and the outside circumference. Sometimes the seal is inside the drum instead of on the piston. The piston may have to align with a check valve inside the drum, like the one in **Figure 11-38**. The piston might have individual seals or an integral seal. If the piston has an integral seal, the entire piston is replaced if the seal is bad.

If the piston has a lip seal, the seal may roll outward, making it difficult to install in the drum. Clutch assembly tools may be used to guide the piston's lip into the drum, **Figure 11-39**. Some technicians use a feeler gauge to guide the seal into place. See **Figure 11-40**.

Figure 11-38. The small opening in the piston must align with the check valve inside the clutch drum. This is clutch #4 from a John Deere 16-speed powershift transmission.

Soaking the Clutch Pack

During a rebuild, the friction discs and separator plates must be soaked in clean oil for at least 15 minutes prior to assembly. See **Figure 11-41**. If the discs and plates are not soaked, the lack of oil can cause the clutch to slip and burn at startup. The piston seals must also be lubricated with clean oil prior to installation.

Installing and Aligning the Clutch Pack

Clutch discs and plates are installed in an alternating order so that a friction disc is always next to a steel plate surface. After the clutch pack is installed, the discs may be difficult to align with the corresponding clutch hub. Some manufacturers recommend special tools and specific procedures for aligning the clutch pack.

Clutch B from a John Deere 8000 series 16-speed powershift transmission is shown in **Figure 11-42**. After the clutch pack is installed in the clutch drum, the technician must do the following:

1. Install the black alignment tube inside the clutch pack. The alignment tube has a single external spline that aligns with the friction discs.
2. Position an air supply manifold under the clutch drum. Bolt the mandrel through the center of the clutch drum and into the air supply manifold located below the clutch drum. Connect an air line, equipped with a shutoff valve, to the air supply manifold. Open the air-supply shutoff valve to apply air pressure to the clutch.
3. After air pressure has applied the clutch pack, remove the alignment tube.
4. Remove air pressure by shutting the valve off and then remove the alignment tube from the clutch.
5. Remove the mandrel and air supply manifold.
6. The clutch must remain in a vertical position. Install the clutch onto the transmission's output shaft, which is in a vertical position, to ensure that the clutch is not tipped and the friction discs do not become misaligned. The clutch drum splines to the transmission's output shaft.
7. The clutch hub is part of the drive gear, and the clutch-hub drive-gear assembly is installed last onto the clutch assembly.

Transmission Removal and Installation

Depending on the machine, removing and installing a transmission can be quite challenging. As explained in Chapter 4, *Agricultural Equipment Identification*, some agricultural tractors require the operator's cab to be removed and the tractor to be split to gain access to the transmission. Tractors that do not have to be split, including some four-wheel-drive articulated tractors, might require using a service bay with a deep pit. The transmission is lowered into the pit and the tractor is towed out of the bay. The transmission can then be lifted out of the pit for service.

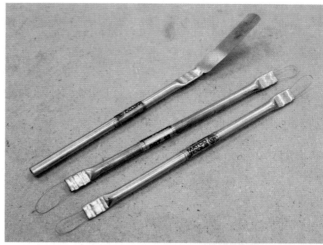

Goodheart-Willcox Publisher

Figure 11-39. These clutch tools are used to guide the piston's lip seal during installation.

Goodheart-Willcox Publisher

Figure 11-40. If the piston has a lip seal, some technicians use a feeler gauge to guide the piston's lip seal as the piston is installed in the drum.

Goodheart-Willcox Publisher

Figure 11-41. This clutch pack is being soaked prior to installation. This is the traction clutch used in a comp utility tractor.

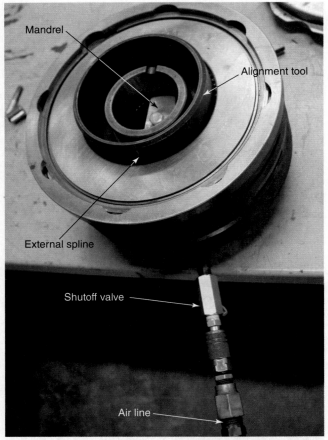

Figure 11-42. The black tube is used to align the clutch friction discs. The mandrel is used to attach the clutch to an air supply and to block off the fluid passageway. After the discs are aligned, air pressure is applied to the clutch through the mandrel. The alignment tool is removed. Air pressure is removed. The mandrel is removed. The clutch remains in the vertical position and is installed on the clutch shaft.

Caterpillar Elevated Track-Type Tractor Transmission Removal

The design of the Caterpillar elevated track-type tractors, like the D6R dozers, makes transmission removal easier. Transmissions in steering clutch and brake tractors can be removed more quickly than those in differential steering tractors. Chapter 24, *Track Steering Systems*, explains both steering clutch and brake machines and differential steering machines.

The service literature provides steps for removing the transmission, including draining oil and disconnecting certain components, like drive shafts, plumbing, linkages, and electrical connectors. Some differential steering dozers require removing the differential steering planetary gear set from the axle. See **Figure 11-43**. The drive axles are easily removed by removing the center plates from the final drives and using a slide hammer to pull the axle shafts. See **Figure 11-44**.

> **Warning**
> The tractor can move when the axles have been removed. If the tractor is on a hill, it can roll down the hill if it is not properly chocked.

To remove the transmission and bevel gear set as an assembly, first remove the six-point hex head cap screws securing the transmission to the axle housing. These cap screws have already been removed in **Figure 11-45**. The 12-point cap screws (two visible and two not visible in **Figure 11-45**) should remain installed in the recessed holes of the transmission case. Next, install two cap screws into specific holes in the transmission case, one on each side of the case. The specified holes do not extend into axle housing. As the cap screws are threaded into the holes, they press against the axle housing and push the transmission case away from the axle housing. See **Figure 11-45**. Notice the lifting fixture attached to the end of the transmission.

The transmission housing has two rollers at the bottom front that ride on rails. See **Figure 11-46**. The sole purpose of the rollers and rails is serviceability. The rollers slide on two rails inside the dozer's frame, allowing the transmission to be rolled out of the tractor and reinstalled in the tractor.

Chapter 11 | Powershift and Automatic Transmission Controls, Service, and Repair 405

Figure 11-43. This steering differential planetary gear set has been removed from a Caterpillar D6R dozer. This required separating the track and removing the left side final drive. The axle shaft is also shown.

Figure 11-45. This transmission is being removed from a Caterpillar D6R track-type dozer. The two visible 12-point cap screws must remain installed. Two "pusher" cap screws (one visible on the left) are threaded into special holes to separate the transmission from the rear frame of the dozer. A lifting fixture has been attached to the rear of the transmission.

Figure 11-44. The axles can be removed from this Caterpillar D6R dozer by removing the plate from the final drive and using a slide hammer to pull the axles. Be sure to follow the service literature and chock the tracks to prevent the machine from moving.

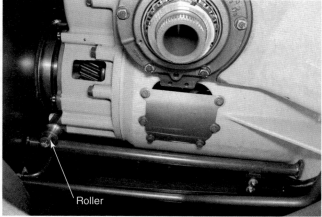

Figure 11-46. A Caterpillar D6R dozer transmission has two rollers mounted on the front of the transmission, one on the front left side (visible) and one on the front right side (not visible). The transmission rolls out of the tractor using two side rails that are fixed to the tractor's axle housing.

Summary

- Powershift transmissions commonly use fixed-displacement pumps to supply oil to the transmission and torque converter.
- Scavenge pumps transfer oil from one oil reservoir to another oil reservoir.
- Operating pressure for many off-highway powershift transmissions is around 300 psi. Some machines operate with a pressure as low as 200 psi and others with a pressure as high as 500 psi.
- Powershift transmissions can be controlled manually, electrically, or electronically.
- Electronically controlled transmissions can be calibrated to allow the ECM to learn the transmission clutch values, such as clutch fill and clutch hold. Calibration improves the shift feel.
- Transmission clutch pressure is modulated to optimize the shift feel and clutch life.
- Adaptive learning is the process in which a transmission ECM determines if a clutch is engaging too slowly or too quickly and commands a solenoid to respond more quickly or more slowly. Technicians perform fast adapts when a major transmission component is replaced.
- Reflashing, or reprogramming, a transmission ECM involves installing new software to fix, modify, or improve the machine's performance.
- Technicians have a variety of tools available to diagnose transmissions. These tools include service literature, pressure gauges, electronic service tools, and manufacturer databases.
- Clutch apply charts are matrices that list each clutch used in each of the speeds.
- Some multiple disc clutches require special tools to remove the clutch pack if the clutch is spring applied. If the clutch is hydraulically applied and spring released, a special tool might be required to remove the clutch piston.
- Clutches should be inspected before rebuild to determine the cause of the failure.
- Clutch packs must be soaked in clean oil before installation.
- Some transmissions are difficult to remove. Caterpillar elevated track-type tractors have been designed so that the transmission is easier to remove and service.

Technical Terms

- adaptive learning
- auto-kickdown
- Autoshift
- bottom-of-travel switch
- calibration mode
- clutch fill
- clutch hold value
- Customer Support System (CSS)
- Diagnostic Technical Assistance Center (DTAC)
- direct-drive powershift transmission
- electronic clutch pressure control (ECPC)
- electronic service tool
- Electronic Technical Information Manuals (eTim)
- Enhanced AutoShift (EAS)
- error code
- fast adapt
- fixed-displacement pump
- inching pedal potentiometer
- Individual Clutch Modulation (ICM)
- Liebherr Information and Documentation Services (LIDOS)
- main pressure
- manufacturer solutions group
- modulation
- Multi Velocity Program (MVP)
- on/off solenoid
- piston free travel
- pressure differential valve
- pressure-reducing valve
- proportional solenoid control valve
- pulse width modulation (PWM)
- ratio valve
- reflashing
- regulated pressure
- scavenge pump
- Service Advisor
- Service Information System (SIS)
- slow adapt
- solenoid control valve
- suction screen
- suction strainer
- Technical Service Group (TSG)
- top-of-travel switch
- transmission charge pressure
- transmission operating pressure
- variable-displacement pump
- variable solenoid control valve
- viscosity

Review Questions

Answer the following questions using the information provided in this chapter.

Know and Understand

1. Technician A states that scavenge pumps are designed to transfer oil from one case to another. Technician B states that scavenge pumps pump a small percentage of air. Who is correct?
   A. Technician A.
   B. Technician B.
   C. Both A and B.
   D. Neither A nor B.

2. Which of the following is a common operating pressure for off-highway powershift transmissions?
   A. 30 psi.
   B. 300 psi.
   C. 1000 psi.
   D. 3000 psi.

3. In a Caterpillar dozer with a manually controlled powershift transmission, _____ oil is used for directional clutches.
   A. P1
   B. P2
   C. P3
   D. Both A and B.

4. In a Caterpillar dozer with a manually controlled powershift transmission, _____ oil is used for speed clutches.
   A. P1
   B. P2
   C. P3
   D. Both A and B.

5. In a Caterpillar dozer with a manually controlled powershift transmission, _____ oil is used for the torque converter circuit.
   A. P1
   B. P2
   C. P3
   D. Both A and B.

6. Clutch fill values are most related to the _____.
   A. transfer of torque
   B. time it takes for oil to enter the clutch
   C. speed of the input shaft
   D. speed of the output shaft

7. Clutch hold values are most related to _____.
   A. transfer of torque
   B. time it takes for oil to enter the clutch
   C. the speed of the input shaft
   D. the speed of the output shaft

8. What does Caterpillar call the transmission control system that uses multiple modulating valves?
   A. ECPC.
   B. ET.
   C. ICM.
   D. SIS.

9. A technician is testing a proportional solenoid. Technician A states to measure the coil's resistance and the circuit's amperage. Technician B states to apply system voltage to the coil to see whether the clutch receives full system pressure. Who is correct?
   A. Technician A.
   B. Technician B.
   C. Both A and B.
   D. Neither A nor B.

10. A common range of solenoid coil resistances is _____.
    A. 2 ohms to 10 ohms
    B. 20 ohms to 60 ohms
    C. 200 ohms to 600 ohms
    D. 2 K ohms to 8 K ohms

11. Which of the following can safely receive full system voltage?
    A. On/off solenoid.
    B. Proportional solenoid.
    C. Both A and B.
    D. Neither A nor B.

12. When a solenoid's current flow is being measured and the coil is internally shorted, the measured current flow will be _____.
    A. higher than specified current flow
    B. the same as specified current flow
    C. lower than specified current flow
    D. zero amps

13. A transmission solenoid coil has failed. Technician A states that most solenoid valves must be replaced as a complete unit. Technician B states that many solenoid valves are designed so the coil can be replaced without removing the valve. Who is correct?
    A. Technician A.
    B. Technician B.
    C. Both A and B.
    D. Neither A nor B.

14. Which manufacturer uses an electronic service tool called *ET*?
    A. Caterpillar.
    B. CNH.
    C. John Deere.
    D. Komatsu.

15. Which manufacturer uses an online software called *Service Advisor*?
    A. Caterpillar.
    B. CNH.
    C. John Deere.
    D. Komatsu.

16. Which manufacturer uses an online software called *eTim*?
    A. Caterpillar.
    B. CNH.
    C. John Deere.
    D. Komatsu.

17. What is the minimum amount of time a clutch pack should be soaked in clean oil before installation?
    A. 1 minute.
    B. 15 minutes.
    C. 15 hours.
    D. 24 hours.

18. Which manufacturer designed their powershift transmission to roll out of its housing on two rollers?
    A. Caterpillar.
    B. CNH.
    C. John Deere.
    D. Komatsu.

Analyze and Apply

19. A transmission main oil pump rotates proportionally to _____ speed.
20. A pressure-reducing valve establishes a(n) _____ pressure.
21. The _____-of-travel switch on a clutch pedal control can be used to neutralize the transmission.
22. _____ is defined as optimizing shift feel and clutch life by varying clutch pressure.
23. _____ is defined as the process of installing new software on an ECM.
24. _____ occurs when the transmission ECM, which monitors speed sensors, senses that a clutch is slipping and commands a solenoid to respond more quickly.
25. A(n) _____ adapt occurs when an ECM determines the amount of oil needed to engage a clutch and updates the ECM's software.
26. _____ varies the off and on times a solenoid is energized for a given cycle.
27. A hydraulically applied clutch has a disengaged clutch clearance known as _____ travel.
28. A(n) _____ chart is a matrix used to determine if a clutch is causing problems in multiple speeds.

Critical Thinking

29. Explain how to measure the disengaged clutch pack clearance on an oil-applied clutch pack using a feeler gauge. How would you adjust the clearance if it were out of specification?
30. A disengaged oil-applied clutch pack clearance specification is .050″ to .075″. Answer whether the low value or the high value is the best clearance to set the clutch to after rebuilding it, and explain why.

Pecold/Shutterstock.com

A waterwheel is a very basic example of a hydrodynamic drive system.

Chapter 12
Hydrodynamic Drives

Objectives

After studying this chapter, you will be able to:

✓ List the different types of fluid drives.
✓ Explain torque converter operation.
✓ Describe the operation of dynamic retarders.
✓ Explain the process of contamination control.

Two styles of fluid drives are used in off-highway equipment, hydrostatic and hydrodynamic. The two drives are sometimes confused with one another. Both use oil as the means of transmitting power. However, the two fluid drives have significant operating differences.

Hydrostatic and Hydrodynamic Drives

A hydrostatic drive uses a hydraulic pump to drive a hydraulic motor. Drive pressures can exceed 7000 psi. The pump and motor have tightly sealing surfaces and are designed not to slip. When the fluid drive is stalled, high-pressure relief valves protect the system. Chapter 13, *Hydrostatic Drives*, and Chapter 14, *Hydrostatic Drive Service and Diagnostics*, detail hydrostatic transmissions.

A **hydrodynamic drive** is a fluid drive that contains an impeller and a turbine and is used to transmit power through a rotating shaft. The engine drives the input member known as the *impeller*. The impeller rotates at a speed proportional to engine speed. The impeller is sometimes called the pump because it sends oil to the turbine to drive it. See **Figure 12-1**.

The *turbine* is the output member of the hydrodynamic drive. It has the responsibility of driving the transmission's input shaft. The turbine can be directly splined to the transmission's input shaft or the turbine can spline to a driveshaft that drives the transmission's input shaft. When the engine-driven impeller develops enough fluid energy, it forces the turbine to rotate. See **Figure 12-2**.

The impeller and turbine are sealed in a single donut-shaped shell. The donut-shaped shell is directly connected to the engine (usually bolted to the flywheel) and rotates at engine speed. The impeller is affixed to the shell and therefore also rotates at engine speed. The turbine, however, is contained within the shell, but not attached to it. As a result, the turbine is free to rotate at a different speed than the shell.

Hydrodynamic drives can be classified as fluid couplings or torque converters. A **fluid coupling** contains only an impeller and a turbine, both of which have straight finned blades. See **Figure 12-3**. Fluid couplings are not commonly used in heavy equipment because they do not multiply torque. However, fluid couplings can be found in mining conveyor systems, crushers, mixers, industrial drives, and some limited locomotive applications.

Goodheart-Willcox Publisher

Figure 12-1. The impeller is driven by the engine. Its fluid energy drives the turbine.

Goodheart-Willcox Publisher

Figure 12-2. The turbine is the output member of the hydrodynamic drive. The turbine splines to the transmission's input shaft. This turbine has been balanced. Notice the 17 divots that were machined on the left side of the turbine to balance the assembly.

The advantage of a fluid coupling is the same as for all hydrodynamic drives; it can serve as an automatic clutch, allowing the transmission to remain in gear when the machine is brought to a stop. The fluid coupling allows slippage between the impeller and turbine. If a tractor has a manual clutch and the transmission stays in gear, the engine dies if the tractor is brought to a stop (if the clutch is engaged). With a fluid coupling, the turbine would slow and stop as the tractor stops, but the impeller would be able to continue rotating. This slippage would allow the engine to continue running.

Torque Converters

The most common type of hydrodynamic drive is the *torque converter*. The torque converter serves multiple purposes: acts like an automatic clutch, multiplies engine torque, and when not operating in the lockup mode, reduces vibrations caused by engine pulsations. Lockup mode is explained later in the chapter. Some torque converters are also responsible for driving the transmission's oil pump.

Like a fluid coupling, the torque converter contains an impeller and turbine. However, a torque converter also contains a *stator* (also known as a *reactor*) that acts like a recycler and redirects the oil exiting the turbine back to the impeller. See **Figure 12-4**.

A stator is a recycling device that receives oil expelled from the turbine and sends the oil back into the impeller. The force of the oil striking the impeller helps to turn it, multiplying the input torque. Without the stator, the oil exiting the turbine would be directed back into the impeller, but against its direction of rotation. The stator prevents this counteracting force by redirecting the turbine's oil back into the impeller in the same direction as impeller rotation. This recycled turbine oil enables the torque converter to re-harness the fluid energy, multiplying torque.

Torque converters are used in several types of construction machines, industrial machines, on-highway trucks and buses, and automobiles. They are rarely used in agricultural machines. One rare example is self-propelled Versatile SX 240, 275, and 280 agricultural sprayers that use an Allison transmission and torque converter. Some old agricultural articulated tractors also used Allison transmissions with torque converters.

Torque Converter Mountings

Torque converters can have a few different mounting designs. The torque converter can be directly mounted to the engine and enclosed in a bell housing. The transmission is bolted

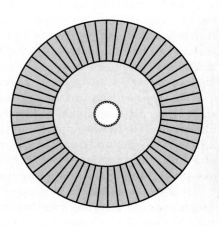

A B *Goodheart-Willcox Publisher*

Figure 12-3. Fluid couplings have straight fins on the impeller and turbine and do not have a stator. A—This simplified drawing shows an example of a fluid coupling turbine. B—This fluid coupling is used in a belt drive system in a mining conveyor.

to the end of the bell housing opposite the engine. See **Figure 12-5**. Wheel loaders, and on-highway trucks typically have a direct-mounted torque converter. Generally, with this arrangement, the torque converter shell is bolted to the engine flywheel, and the turbine inside the shell is splined to the transmission's input shaft.

In other applications, such as large Caterpillar track-type dozers and mechanical drive haul trucks (non-electric drive trucks), the torque converter is attached directly to the engine, but sends its output through a driveshaft to a transmission. The image in **Figure 12-6** shows a Caterpillar dozer with the operator floor plate removed. Note that a driveshaft couples the torque converter to the transmission.

A torque converter can be remote mounted, away from the engine, and receive its input from the engine through a driven shaft. A remote mounted converter can have its own housing, mounted separately from the transmission, or sit inside a traditional bell housing attached directly to the transmission case. The two images in **Figure 12-7** show that a Caterpillar scraper has a remote mounted torque converter at the back of the tractor. The engine delivers power through a drive shaft to the torque

Goodheart-Willcox Publisher

Figure 12-4. A stator acts like an oil recycler, redirecting oil expelled from the turbine back to the impeller to help rotate it.

Goodheart-Willcox Publisher

Figure 12-5. A separate bell housing containing a torque converter is directly mounted to the engine. The Funk powershift transmission bolts to the bell housing.

Goodheart-Willcox Publisher

Figure 12-6. This top view of a Caterpillar dozer (with the operator floor plate removed) shows the driveshaft located between the direct mounted torque converter and the powershift transmission.

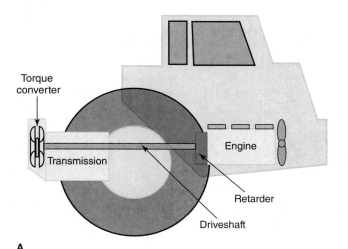

A

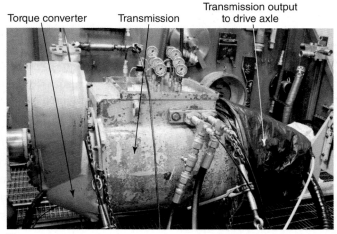

B

Goodheart-Willcox Publisher

Figure 12-7. The transmission and torque converter on a Caterpillar scraper are mounted at the back of the tractor. A—The torque converter is located furthest from the engine. B—This scraper torque converter and transmission are being tested on a transmission test stand. Notice the test stand is driving the torque converter from the rear portion of the housing. The front portion of the transmission (is normally located in the axle housing) has been covered in plastic to prevent oil from slinging.

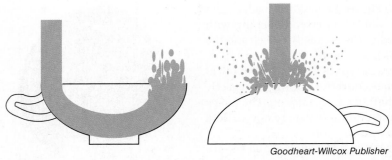

Figure 12-8. A teacup turned right side up resembles the curved vanes in a turbine, which absorb the maximum fluid energy coming from the impeller. When the cup is inverted, the fluid energy is not harnessed as well.

Figure 12-9. A split guide ring is located in the center of the turbine and in the center of the impeller.

converter at the rear of the tractor. The torque converter delivers power to the transmission, which sends power to the axle.

A torque converter differs from a fluid coupling in that the torque converter has a stator and has curved vane members. The turbine has the most aggressively curved vanes. Consider the analogy of using a water faucet to fill a teacup. When the cup is placed upside down, the cup cannot harness the water's full fluid energy because the water splashes in multiple directions. See **Figure 12-8**. However, when the faucet directs the water into the cup that is positioned right side up, the cup can absorb the fluid energy as well as redirect the water in the opposite direction, providing maximum force to the cup. This is the reason the turbine has aggressively curved vanes.

Split Guide Rings

Most torque converters and some fluid couplings have *split guide rings* on both the impeller and the turbine. The split guide rings form a donut-shaped void between the two members in an assembled torque converter, which helps direct the oil as it cycles through the torque converter. See **Figure 12-9**.

External Vanes

Some large torque converters used in heavy machinery have a different internal design. A Caterpillar 938G wheel loader torque converter is shown in **Figure 12-10**. The turbine vanes are on the outside of the turbine shell rather than the inside. The impeller also has external vanes and is bolted to the impeller housing. The impeller and housing assembly surrounds the turbine. A gear attached to the transmission side of the housing assembly drives the oil pump. A ring gear cast into the engine side of the impeller housing provides the input to the torque converter. An output shaft is splined to the turbine and passes through, but is not connected to, the stator and oil pump drive gear. Because it passes through the center of the torque converter, the output shaft is not visible in **Figure 12-10**.

The stator vanes are straight, arranged radially around its center, and oil flow enters the stator radially at its outer circumference rather than axially from the center as in a conventional torque converter. This stator is also unique in that it is always fixed. Fixed stators are explained later in this chapter.

Torque Converter Operation

The order of power flow in a torque converter is:
1. Engine drives the impeller.
2. Oil from the impeller drives the turbine.
3. Turbine drives the transmission input shaft.
4. Stator redirects the oil exiting the turbine back to the impeller.

Within the torque converter, the turbine is located closest to the engine, the stator is in the middle and the impeller is located the furthest from the engine, which is closest to the transmission. See **Figure 12-11**. The impeller rotates clockwise when viewed from the engine, which is the same direction as crankshaft rotation.

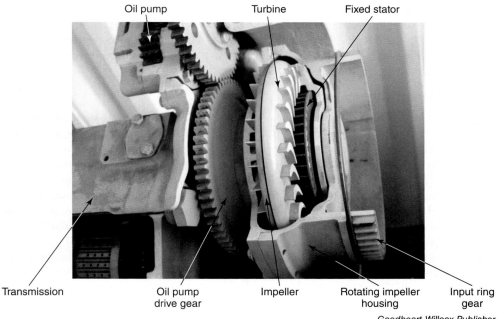

Figure 12-10. A torque converter equipped with external vanes. Notice the drive gear attached to the rear of the impeller housing. This gear is responsible for driving the transmission's oil pump. The ring gear at the front of the impeller housing provides the input to the torque converter.

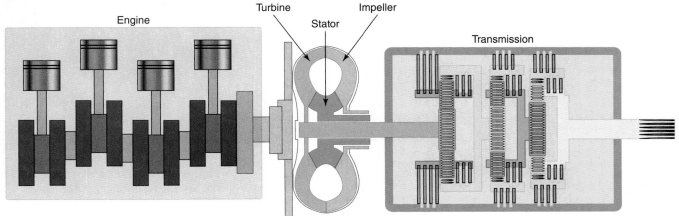

Figure 12-11. The location of the components in a torque converter is backward from the order of power flow. The turbine is located closest to the engine, but it delivers the output of the torque converter. The impeller is located closest to the transmission, but it is the input of the torque converter.

Torque converters have different phases of operation, which are determined by the speed relationships between the impeller and turbine. A conventional torque converter has a torque-multiplication phase and a coupling phase. A lockup torque converter has a third phase, known as the lockup phase. See **Figure 12-12**. Lockup torque converters and the lockup phase will be discussed later in the chapter.

Torque-Multiplication Phase

If there is a large speed difference between the impeller and turbine, such as when the equipment encounters a heavy load like a dozer's blade digging into a pile of dirt, the torque converter is operating in the *torque-multiplication phase*. Because of the difference in relative speeds of the impeller and turbine, there is a *vortex oil flow* in the torque converter. The term *vortex oil flow* describes a short cyclical flow of oil, with the oil rapidly recirculating from the impeller, to the turbine, to the stator, and back to the impeller. See **Figure 12-13**. Vortex oil flow causes the fluid temperature to increase. Because of the heat generated during this phase of torque converter operation, an oil cooler is needed to lower the operating temperature of the oil.

Stators are splined to a stationary shaft, that connects the stator's inside hub to the transmission housing. The shaft can be called a reactionary shaft, stationary shaft, or stator support shaft. The stationary shaft is hollow, and the transmission's input shaft (also known as the turbine shaft) passes through it and can rotate inside it.

Fixed Stator

Conventional torque converters can use a fixed stator, an overrunning stator, or a multiple clutch stator. A *fixed stator* is locked to the stator shaft and cannot rotate in either direction. See **Figure 12-14**. As a result, it is effective only during the torque-multiplication

Torque Converter Phase	Impeller and Turbine Speed Relationship
Torque multiplication phase	Significant speed differences
Coupling phase	Turbine speed 90% of impeller speed
Lockup phase	Turbine rotating at the same speed as impeller

Goodheart-Willcox Publisher

Figure 12-12. The speed relationship between the impeller and turbine determines the torque converter's phase of operation.

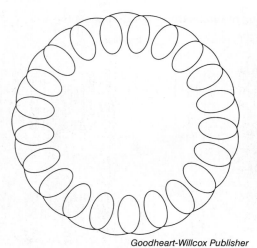

Goodheart-Willcox Publisher

Figure 12-13. Vortex oil flow is the short cyclical flow of oil that makes multiple passes in a coiled fashion from the impeller to the turbine to the stator and back to the impeller.

Goodheart-Willcox Publisher

Figure 12-14. A fixed stator design does not use an overrunning clutch.

phase and produces drag as the impeller and turbine speeds equalize. Fixed stators are not as common as overrunning stators.

Overrunning Stator

An ***overrunning stator*** has an overrunning clutch that holds the stator stationary during the torque-multiplication phase and allows it to overrun during the coupling phase. As a result, as the impeller and turbine rotational speeds equalize, the stator begins to rotate with them, reducing drag. The overrunning clutch can be either a roller clutch or a sprag clutch, which were explained in Chapter 8, *Clutches and Planetary Controls*. See **Figure 12-15**.

When the turbine and impeller have significant speed differences, the oil leaving the turbine hits the concave side of the stator causing the stator's overrunning clutch to lock the stator. An example of a significant speed difference occurs when the tractor's transmission is in gear, the engine is at a high speed, and the drive tires or tracks are locked stationary, causing the turbine to be held stationary. Any time the turbine is stationary and the impeller is spinning at a high speed, the stator locks and the torque converter stalls. The ***stall mode*** of operation causes the oil to be sheared between the fast-spinning impeller and the stationary turbine, which generates high fluid temperatures.

The stator is stationary any time the impeller and turbine are at significant speed differences. For example, if a tractor has a substantial increase in load, the engine will drive the impeller at a speed faster than the turbine speed, and the stator will lock, causing the torque converter to increase torque. Typically, a conventional torque converter has a maximum of a 2.5:1 torque multiplication factor. Torque dividers have more torque multiplication and are explained later in the chapter.

Goodheart-Willcox Publisher

Figure 12-15. The sprag clutch in this overrunning stator allows the stator to overrun in the coupling phase.

 Note

Any time a stator with an overrunning clutch is stationary, it is multiplying torque.

Multiple Disc Stator Clutch

Late-model Caterpillar D9, D10, and D11 dozers use a multiple-disc stator clutch. The D9 has a torque converter and lockable stator. The D10 and D11 use a torque converter with a torque divider and a lockable stator. See **Figure 12-16**. When the transmission is operating under light loads, the ECM applies full electrical current to an ECPC solenoid, which drains the oil from the stator clutch piston. This increases the machine's fuel efficiency. The stator clutch defaults to being engaged, with maximum oil pressure, when the ECPC solenoid is de-energized. The stator clutch is engaged when the transmission is operating under heavy and normal loads, during engine braking, and when the transmission is being shifted.

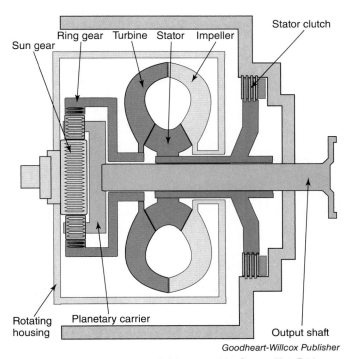

Goodheart-Willcox Publisher

Figure 12-16. This torque divider, used in Caterpillar D10 and a D11 dozers, has a multiple-disc stator clutch.

Coupling Phase

When the tractor's load decreases, the speed of the impeller and turbine begin to equalize. During this phase, also known as *coupling phase*, vortex oil flow diminishes. Instead, the converter's oil is slung to the outer circumference, in the same fashion that fluid is trapped in a spinning bucket due to centrifugal force. See **Figure 12-17**. This oil flow is known as *rotary oil flow*. Rotary oil flow acts on the convex side of the stator vanes, which causes the stator to overrun. During the coupling phase, oil temperature decreases.

Torque Converter Design Variations

Torque converter designs vary based on the specific needs of the equipment. For example, for on-highway haul trucks, it is important to have a torque converter design that maximizes efficiency at high speeds. For a wheel loader, the ability to limit torque during directional shifts is important. Various torque converter designs have been used over the years to improve the performance of equipment.

Lockup Clutches

During the coupling phase, a torque converter cannot achieve a true 1:1 direct drive but drives the turbine up to 90% of the impeller speed. To address this inefficiency, some converters have a *lockup clutch*. The purpose of the lockup clutch is to mechanically couple the turbine with the impeller or torque converter's shell so that the turbine will spin at exactly the same speed as the engine. This direct drive through the torque converter eliminates the 10% inefficiency of the coupling phase and provides an increase in machine speed. This *lockup phase* occurs only when the machine has a light load. Lockup clutches are used in wheel loaders, wheel dozers, track-type dozers, landfill compactors, scrapers, on-highway trucks, and off-road haul trucks. A simplified drawing of a torque converter with a lockup clutch and impeller clutch is shown in **Figure 12-18**. Impeller clutches will be explained in the next section of the chapter.

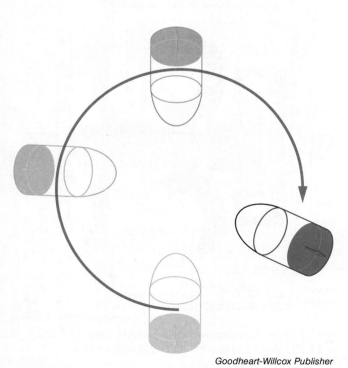

Goodheart-Willcox Publisher

Figure 12-17. A bucket spun in a 360° motion traps fluid due to centrifugal force.

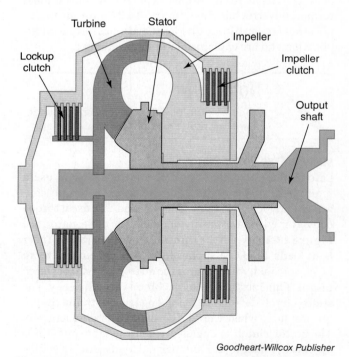

Goodheart-Willcox Publisher

Figure 12-18. A lockup clutch locks the turbine to the impeller, causing the turbine to rotate at the same speed as the engine and eliminating the slippage that takes place during the coupling phase.

The lockup clutches on large machines have multiple friction discs and separator plates. See **Figure 12-19**. Smaller on-highway trucks use a single friction disc for a lockup clutch.

Note
Caterpillar D6–D8 dozers use lockup torque converter clutches. However, later in the chapter, it is explained that large D9–D11 dozers instead use multiple-disc stator clutches.

Impeller Clutch

Caterpillar manufactures a torque converter with an impeller clutch that is used in large wheel loader applications. The ***impeller clutch*** consists of a multiple disc clutch that can modulate impeller engagement to reduce the converter's output torque. See **Figure 12-18** for a simplified image. A reduction in output torque at the torque converter is translated through the transmission and ultimately reduces rimpull. ***Rimpull*** is a term used to describe the amount of torque delivered at the ground, through the tire or track shoe. The clutch allows the operator to slow the propulsion of the machine while maintaining higher engine speeds to provide higher hydraulic flow. Through modulation of the impeller clutch, wheel slip and tire wear are reduced.

Older wheel loaders used a mechanical hand lever to control a valve that engaged and released the impeller clutch. Today's Caterpillar wheel loaders use an electronically controlled solenoid valve to engage and disengage the impeller clutch. The ECM controls the impeller clutch solenoid based on several inputs, which vary among the different generations of wheel loaders:

- Position of the left foot pedal, known as the neutralizer pedal, or the left brake pedal.
- Change of the loader's direction.
- Change of the loader's speed.
- Engine speed.
- Position of the reduced rimpull selection switch.
- Position of the on/off rimpull switch.
- Condition of the torque converter's output shaft.

Note
A reduced rimpull switch has four settings, **Figure 12-20**:
- 85% rimpull.
- 70% rimpull.
- 55% rimpull.
- 45% rimpull.

When the clutch's fluid pressure is decreased, the torque converter reduces its output torque. The clutch is controlled by a solenoid. When the solenoid is deactivated, fluid pressure applies the impeller clutch. When the electronic control module (ECM) applies current to the solenoid, the fluid pressure applying the impeller clutch is reduced. This allows the impeller to slip, reducing the amount of torque out of the converter. If the solenoid circuit were to have an open circuit fault, the torque converter impeller would default to maximum torque due to maximum oil pressure engaging the impeller clutch.

Multiple Impellers

Caterpillar offered a ***variable-capacity torque converter*** in some older machines, such as a 988B wheel loader and 623B elevating scraper. The

Goodheart-Willcox Publisher

Figure 12-19. Lockup converters on large machines use large multiple-disc clutch packs. A—This torque converter's lockup clutch's discs were replaced due to the failed and missing friction material on the discs. B—This torque converter uses a lockup clutch. The hub and turbine are visible, but the clutch pack is absent.

Goodheart-Willcox Publisher

Figure 12-20. This Caterpillar wheel loader's rimpull switch is located on the right side of the steering column. Other rimpull switches can be a toggle switch located on the dash console.

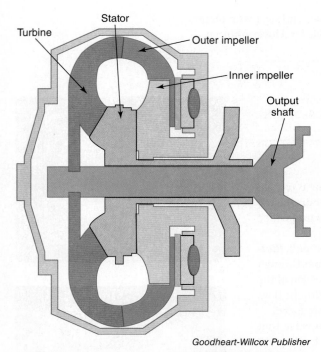

Figure 12-21. A Caterpillar variable-capacity torque converter can increase or decrease torque by controlling oil pressure to engage or release a clutch that engages the outer impeller.

variable-capacity torque converter's purpose is to reduce wheel slippage by limiting the torque converter's torque multiplication. By reducing wheel slip, tire life is extended and the machine's engine can use the extra power to produce more hydraulic power.

The converter contains a standard turbine and stator, but also contains a two-member impeller assembly, consisting of an inner impeller and an outer impeller. The inner impeller is connected to the torque converter shell and always rotates at engine speed. The outer impeller is hydraulically controlled with a clutch. See **Figure 12-21**. When the clutch receives maximum clutch oil apply pressure, the outer impeller engages and rotates along with the inner impeller. When engaged, the outer impeller delivers additional oil to the turbine, which increases torque. When the outer impeller is slowed, the converter's torque is reduced. The operator controls the second impeller with a manually operated torque lever that provides variable torque control. The wheel loader also contains an electric on/off switch that allows the operator to vary the torque from maximum to minimum with the flip of the switch.

Multiple Turbines

A torque converter can also be equipped with two turbines. In **Figure 12-22**, each of the turbines is splined to its own turbine shaft. The turbine shafts are concentric. The second turbine's shaft is larger and hollow, and the first turbine's shaft passes through it.

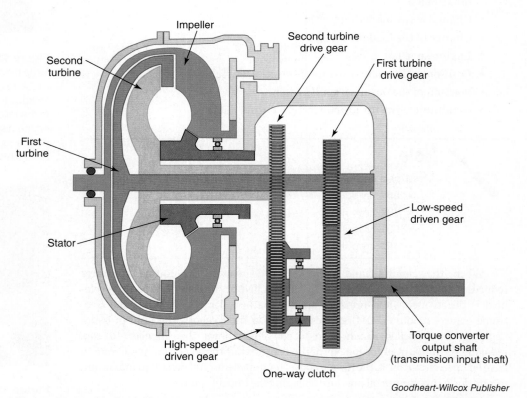

Figure 12-22. A twin-turbine torque converter contains two concentric turbine shafts that drive their respective low-speed and high-speed driven gear. A one-way clutch locks at low speeds and overruns at high speeds. The torque converter has only one output shaft, also known as the transmission input shaft.

Each of the turbine shafts is mated with a different-sized drive gear. Notice that the torque converter has only one output shaft, also known as the transmission input shaft. The torque converter acts like a two-speed transmission depending on the load (light load or heavy load).

As the machine begins to move from a stop, the torque converter operates under a heavy load. The oil from the impeller drives the first turbine. The first turbine's drive gear is small, and it drives the large driven gear on the torque converter output shaft, multiplying torque. The low-speed driven gear drives the high-speed driven gear by locking the one-way clutch, which powers the torque converter's output shaft (transmission's input shaft). The high-speed driven gear is also in mesh with the second turbine drive gear, which causes the second turbine to react and spin at a slower speed than the first turbine.

As the machine builds speed and the demand for torque decreases, the oil from the impeller begins applying more force to the second turbine, which drives the high-speed driven gear. As the second turbine receives more of the fluid force, it speeds up. The one-way clutch overruns, allowing the first turbine and low-speed driven gear to freewheel.

Note
Years ago John Deere used the Allison twin-turbine torque converter in the A through D series 444, 544, and 644 wheel loaders. The model numbers of Allison twin turbine transmissions are TT2001, TRT2001, TTB2001, TT2421, TT3420, TT4700, and TRT4800.

Multiple Stage Torque Converters

A torque converter can be defined by its number of stages, which equals the number of times the oil cycles through the turbine before it is sent back to the impeller. A traditional torque converter is a single-stage torque converter because after the oil leaves the turbine, it flows through the stator, and back to the impeller. Heavy equipment can be equipped with a three-stage torque converter that can multiply engine torque up to five times. The converter is often described as a variable-speed gear reducer. See **Figure 12-23**.

A *three-stage torque converter* contains a turbine with three layers of turbine blades and a stator with two layers, which allows the torque converter's oil to be recycled through the following sequence:

1. Impeller.
2. Turbine (stage 1).
3. Stator layer 1 (stage 1).
4. Turbine layer 2 (stage 2).
5. Stator layer 2 (stage 2).
6. Turbine layer 3 (stage 3).
7. Back to the impeller.

The impeller is driven by the engine. The turbine is the output member and is comprised of a turbine wheel with three rings of blades. The two layers of stator vanes are integrated into the stationary housing, called a turbine housing. The reason it is called a turbine housing rather than a stator housing is because the turbine wheel spins inside of it.

Figure 12-24 shows an impeller with one ring of blades, a turbine wheel with three rings of blades, and a stationary turbine housing that contains two rings of stator blades. The turbine wheel spins inside the stationary turbine housing. The 2nd stage stator blades are set inside the circumference of the 2nd stage turbine ring. The outer flat portion (circumference) of the turbine

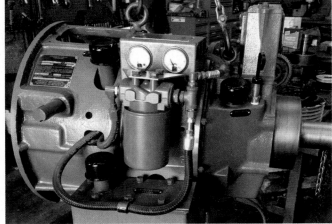

J & M Clutch and Converter

Figure 12-23. This three-stage torque converter has three sumps. The breather caps on the upper-right and upper-left sides are sumps used for lubricating bearings inside of the torque converter. The lower-left breather cap is used for the torque converter sump. Three-stage torque converters use a very thin viscosity fluid close to the consistency of diesel fuel.

Figure 12-24. Disassembled three-stage clutch. A—The primary components of a three-stage torque converter include the impeller, a turbine wheel with three stages of turbine blades, and a turbine housing with two stages of stator vanes. B—The turbine wheel and impeller rotate inside the turbine housing. The turbine housing is bolted to a bell housing to create a sealed unit.

wheel rides within a clearance of 0.030″ inside of the 1st stage stator blades. The impeller rides inside the 1st stage of the turbine ring.

Three-Stage Applications

Three-stage converters are an older technology used in three types of applications: duty cycle, non-duty cycle, and continuous. *Continuous applications* include oil-well drilling applications, where the converters are operated continuously. *Duty cycle applications* include draglines and clam shell crane applications for driving the winch and boom drums. The operator momentarily places the machine under load while loading the bucket, then swings to dump and swings back unloaded, completing one cycle. *Non-duty cycle applications* include applications that are subjected to overrunning loads, such as lifting cranes lifting and lowering the load placed on the crane's hook.

The Twin-Disc three-stage torque converter in **Figure 12-25A** has a large external toothed drive gear (sometimes called a spider gear) on its input shaft. The torque converter is used in an oil-drilling rig application. The spider gear meshes with an internal toothed ring gear on the engine's flywheel. As a result, the spider gear always rotates at a speed proportional to engine speed. In this application, a clutch on the output shaft of the torque converter is used to disconnect the engine's torque from the oil rig drill's transmission. The benefit of this design is that the torque converter's charge pump is always driven at engine speed, providing constant oil to the torque converter. The disadvantage of decoupling the torque converter's output shaft rather than its input shaft, is that the output shaft's torque is multiplied by the torque converter, whereas an input shaft clutch decouples an engine's torque that has not been multiplied. As a result, a heavier-duty clutch must be used to decouple the output shaft.

Figure 12-25B shows the input shaft and a clutch fork on a Twin-Disc three-stage torque converter. This converter is used in a crawler crane application. A large 17-inch over-center flywheel clutch (as described in Chapter 8) is used to connect and disconnect engine power from the input to the torque converter.

Three-stage torque converters have a fixed stator that is riveted to the turbine housing. The turbine housing is stationary, but the turbine wheel rotates. The harder a three-stage converter works (builds torque), the cooler it runs. This occurs because a higher percentage of horsepower is used for multiplying torque rather than generating heat. When a three-stage converter is not multiplying torque, the horsepower is transformed into heat energy, which can overwhelm the oil cooler. Under high-speed light-load applications, the converter sends the wasted horsepower into the oil cooler, and often it is more heat than the cooler can handle.

 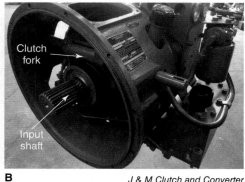

A   **B**   *J & M Clutch and Converter*

Figure 12-25. Three-stage torque converters are used in oil drill rigs and older crawler cranes. A—A three-stage torque converter used in an oil drill rig application. The large external-toothed gear meshes with an internal toothed ring gear located on the engine's flywheel. B—A three-stage torque converter used in a crawler crane application. The converter's splined input shaft mates with the splined hub of a friction disc on a flywheel-mounted clutch. The flywheel clutch would be manually engaged and disengaged using the clutch fork.

During light load applications, the operator must reduce the engine's throttle to reduce the horsepower. Over speeding a three-stage under light loads causes the oil to overheat. If the load is increased, the turbine will slow down, increasing torque converter efficiency.

Crawler three-stage converters used in lifting crane applications contain an internal one-way clutch that acts as a load-lowering brake. This one-way clutch is different from the stator one-way clutch in a conventional torque converter. The three-stage torque converter's stator's fins (inside the turbine housing) are always stationary and do not use a one-way clutch. The sprag one-way clutch assembly inside a three-stage torque converter couples the turbine to the impeller any time the crane operator attempts to lower a heavy load. This prevents the turbine from spinning faster than the impeller, which is rotating at engine speed. Without a one-way clutch, lowering a heavy load would cause the turbine to over speed resulting in a tremendous amount of oil turbulence. The clutch locks based on three variables: engine speed, horsepower, and the crane's load. It prevents an overrunning load (a heavy load that is lowering on the crane's winch) from over speeding the engine.

During normal operation when the crane is lifting, the sprag one-way clutch is overrunning, allowing the impeller to rotate faster than the turbine. Although the one-way clutch is only locked during overrunning load conditions, it is doing the same thing as a late-model torque converter's lockup clutch, which is coupling the turbine to the impeller. But it is coupling the impeller and turbine when lowering the overrunning load rather than lifting the load.

The one-way clutch's outside race is a press fit. Because it has an interference fit, the hub is heated to 500°F and the race is chilled in a freezer immediately before installation. If the race is overloaded and cracks in service, it can only be removed by welding a bead around the race and allowing it to cool, similar to the method of bearing removal detailed in Chapter 7, *Manual Transmissions*.

The torque converter is used as a load-lowering brake so the machine's brake bands are not solely responsible for lowering the load. The three-stage torque converters in oil rigs and duty cycle applications, such as draglines and clam shell cranes, are not equipped with a one-way clutch.

Three-stage converters used in duty-cycle applications (clam shell and draglines) have a governor on their tail shafts. The governor drives a speed type cable proportional to the torque converters output shaft speed. The governor speed is inputted into the engine's throttle control and is used by the engine throttle control to either increase or decrease engine speed. If the governor speeds up, it will reduce the engine's throttle to reduce overheating.

Variable-Pitch Stator

Some torque converters have a two-position variable-pitch stator. During normal operation, the stator's vanes are in a fully opened, high-capacity, position. When the vanes are in this position, the torque converter captures all of the engine's power, multiplies it, and transfers it to the transmission for propulsion. When the stator vanes are in the partially closed (low capacity) position, the torque converter cannot capture all of the engine's power. The balance of the engine power (as much as 40%) becomes available to power auxiliary equipment on the machine, such as an elevator or pump.

One end of a crank arm (thin rod with offset ends) is attached to each stator vane. The other end of each crank arm is located inside a groove of a hydraulically actuated piston. A constant low oil pressure is directed to one side of a piston causing the stator vanes to be held in the fully opened position.

A manual lever-actuated or pneumatic-pilot control valve can be used to control the stator vane position. As the manual lever or air cylinder is actuated, a hydraulic check valve opens, causing the stator control valve to shift. When the control valve is actuated, it directs a higher oil pressure to the other side of the actuating piston, causing it to move against the low oil pressure. As the piston is actuated by the higher oil pressure, the crank arms force the stator vanes into the partially closed position. Variable stators were available in older Allison CLBT 750 and VCLT/VCLBT 4000, 5000, and 6000 series transmissions for off-road haul truck, airport crash truck, and scraper applications. See **Figure 12-26**.

Torque Divider

Caterpillar has used torque dividers for decades in medium to large track-type tractors. A *torque divider* is a torque converter that contains an internal planetary gear set. The torque converter delivers 75% of the torque and the planetary gear set provides 25% of the torque. See **Figure 12-27**.

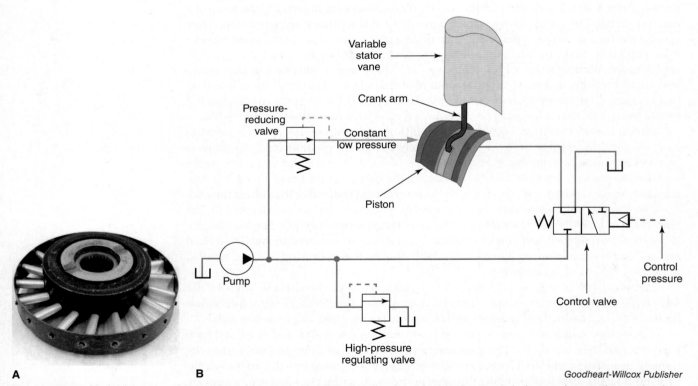

Goodheart-Willcox Publisher

Figure 12-26. Variable stators have adjustable vanes. A—A hydraulic piston inside the stator hub can change the angle of the stator vanes, varying the torque converter's torque capability. B—An example of a stator vane mounted on a crank arm. As the hydraulically operated piston moves forward or back, it causes the crank arm and vane to pivot. Constant low oil pressure acting on the piston holds the vanes in an open position. The control valve is shown in the high capacity position, draining the high-pressure side of the piston to the reservoir. As the control valve is actuated, it directs high-pressure oil to the high-pressure side of the piston, causing the stator vanes to partially close, reducing the converter's capacity.

Chapter 12 | Hydrodynamic Drives

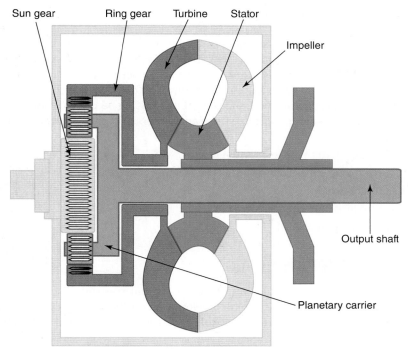

Goodheart-Willcox Publisher

Figure 12-27. A torque divider is a torque converter that contains a simple planetary gear set.

The engine directly drives two components, the impeller and the planetary sun gear. Both rotate at engine speed. The fluid-driven turbine is splined to the ring gear. The planetary carrier is the torque divider's output member.

During high torque multiplication, the engine spins faster than the torque divider's output shaft, the stator locks, and the turbine drives the ring gear slowly while the engine drives the sun gear at engine speed. As a result of the two inputs, the planetary carrier is driven (as the output member) at a slow speed while multiplying the engine's torque.

Under light load conditions, the impeller (and sun gear) and the turbine (and ring gear) approach close to the same speed, while the stator overruns. As a result, the planetary carrier (the output member) achieves nearly a 1:1 ratio. The planetary gear set will produce direct drive any time the turbine and the impeller are rotating at the same speed. Keep in mind that the speed of the sun gear and the speed and direction of the ring gear dictate the torque and speed of the planetary carrier and converter output shaft.

 Note

The amount of load on the torque divider's output affects the speed and direction of the turbine. For example, if the torque divider is being stalled, the output to the transmission is stationary. Because the planetary carrier is splined to the stalled output shaft, the held planetary carrier causes the ring gear and the turbine to rotate slowly in a reverse direction.

A torque divider serves the same purpose as a conventional converter: it multiplies torque, dampens engine pulsations, and acts like an automatic clutch. In addition, a torque divider's planetary gear set provides an increase in torque multiplication over a conventional torque converter. Since dozers tend to operate under heavier loads for longer durations than hauling machines, like trucks and scrapers, the increased torque multiplication provided by torque dividers improves their performance.

Torque Converter Oil Flow

Torque converters often receive their supply of oil from the same pump used by the transmission. See **Figure 12-28**. Typically, the transmission's relief valve sends return oil to the torque converter instead of the reservoir. Torque converter circuits can also include a torque converter inlet relief valve and a torque converter outlet relief valve.

Torque converters generate heat and require the exiting oil to be cooled. After the oil is cooled, it is normally sent to lubricate the transmission. A common torque converter oil flow path is:

1. Pump delivers oil to the transmission circuit.
2. Relief valve routes excess oil to charge the torque converter.
3. Oil is sent to a cooler.
4. Oil is sent to lubricate the transmission.
5. Oil returns to the reservoir.

Torque Converter Installation

When installing a direct-mounted torque converter and transmission into a machine, install the converter onto the transmission before attempting to couple the transmission to the engine. Most torque converters typically have at least three sets of splines that must be aligned during assembly:

- Stator onto the stator support shaft.
- Turbine onto the transmission input shaft.
- Impeller into the transmission oil pump.

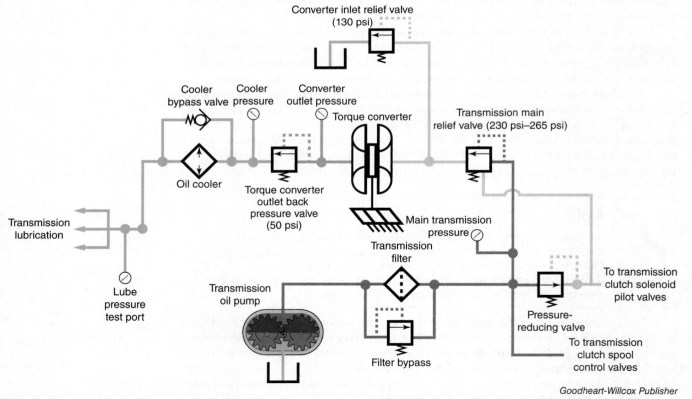

Figure 12-28. This transmission oil flow schematic is an example of the Case 521-921E and F series wheel loaders. It illustrates that the transmission relief valve directs the excess oil to the torque converter, then the oil is cooled and used for lubrication.

If a technician attempts to bolt the torque converter to the engine and then couple the transmission to the engine, the three sets of splines will not easily align. For this reason, the torque converter is typically splined onto the transmission first, the transmission is then bolted to the engine, and finally the torque converter is bolted to the engine's flywheel. For this technique to work, the transmission's bell housing or engine must have an access cover that provides an opening for bolting the torque converter to the engine's flywheel.

Early Case and Deere loader backhoes, rough terrain forklifts, and skip loaders, do not have a torque converter access point. These machines first require bolting the torque converter to the engine's flywheel, installing the bell housing to the engine, then slowly attach the ***power shuttle transmission***, also known as a ***reverser***, into the bell housing. Some technicians attempt to turn the transmission's output shaft hoping for a little bit of clutch drag to cause the input shaft to rotate. Sometimes it requires another technician slowly rotating the engine's crankshaft manually.

A common backhoe power shuttle transmission uses a single forward and reverse clutch to control a single planetary gear set, allowing the machine to quickly shuttle between forward and reverse, essentially a one-speed transmission. The power shuttle transmission is often combined with a mechanical four-speed range transmission. A ***barring tool,*** a small pinion gear resembling the pinion gear in a starter motor, can be inserted into mesh with the engine's flywheel, allowing a ratchet to slowly rotate the engine's flywheel manually. If the transmission's shafts do not fully insert into the torque converter splines, a gap will remain between the transmission and the bell housing. Do not use an air ratchet to pull the transmission into mesh with the bell housing, or you will damage the splines.

Torque Converter Testing

If a torque converter problem is suspected, the first step is to check the fluid. Follow the manufacturer's service literature to correctly check the level, check for aeration, determine if it is burnt or oxidized, and ensure the torque converter has the correct fluid.

Pro Tip

Some torque converters share the same reservoir as the transmission, while other torque converters have their own reservoir. Manufacturers specify steps for checking the fluid level. Most light-duty, light-truck automatic transmissions specify the fluid level be checked with the engine running and the fluid at full operating temperature. However, heavy equipment manufacturers often specify that the fluid level be checked with the engine stopped and the oil cold. Some machines have a dipstick while other machines have a sight gauge for checking the fluid level. A Caterpillar 777G haul truck's torque converter reservoir has a cold oil sight gauge and a hot oil sight gauge. Some older Allison transmissions have a fill tube that can contain one or two plugs or petcocks that are used for checking the level and adding fluid. Be sure to follow the machine's service literature when checking fluid levels.

Depending on the torque converter's design, it can exhibit a variety of symptoms. See the matrix in **Figure 12-29**.

Because the impeller, turbine, and stator are made of aluminum, another beneficial test is to inspect the fluid

Symptom	Fault
No transmission output in any speed or in either direction	Turbine splines sheared, or torque converter output shaft sheared
Low power or low torque	Stator will not lock
Torque converter overheating	Stator locked all the time
Reduced fuel economy	Lockup clutch will not lock
Torque converter and transmission overheating	Oil cooler restricted
	Locked stator that will not overrun
	Slipping lockup clutch
	Operator abuse: Operating in too high of a gear under too heavy of a load

Goodheart-Willcox Publisher

Figure 12-29. Torque converters can exhibit several different faults that might lead a new technician to condemn the transmission instead of the torque converter. Be sure to follow OEM service literature to diagnose the symptom.

and filter for evidence of aluminum debris. Be sure to trend fluid samples to look for an uptick in aluminum contaminants within the fluid.

Stall Testing

A torque converter stall test consists of chocking the tires, applying the machine's parking brakes and service brakes, placing the transmission in the highest gear, and increasing the throttle to maximum engine speed. The highest engine speed achieved during this test is the stall speed.

Warning

Stalling a torque converter not only causes damaging heat in a transmission but can endanger neighboring equipment if it powers through its applied parking brake. A technician must follow the manufacturer's service literature. Do not trust that a machine's parking brakes will hold for a stall test. Because the engine's torque is multiplied by the torque converter/torque divider, transmission, and final drives, the machine's total torque output could overcome the brakes, and propel the machine over anything in its path. For this reason, if an off-highway manufacturer provides specifications for stall testing a torque converter, the test should be performed only using the transmission's highest gear.

Be sure to closely follow the manufacturer's service procedures. Some manufacturers recommend against a stall test of the torque converter due to the risks imposed on the machine, neighboring equipment, and personnel.

Before beginning a stall test, obtain the stall test specification. An example might be 1350 rpm. The actual stall procedure should be performed for no longer than five seconds due to the heat generated during the test. After performing the test, be sure to follow the cool down procedure. **Figure 12-30** shows a matrix for evaluating a stall test.

Inspecting Torque Converters

Heavy equipment torque converters are often disassembled and reconditioned. The bearings and bushings are inspected and replaced if needed. Seals are replaced. The torque converter's members also need to be inspected. Depending on the torque converter design, it can include any of the following components:

- Impeller.
- Turbine.
- Stator and overrunning clutch or multiple disc clutch.
- Impeller clutch.
- Lockup clutch.
- Planetary gear set.

Most heavy-duty torque converters have cast aluminum turbines, impellers, and stators. If a lighter-duty torque converter is used, it is less likely that it can easily be disassembled for service. Check the vanes for wear. Also check the outer circumference of the turbine for wear, **Figure 12-31**. If the torque converter is equipped with a clutch, inspect the discs for wear. Refer back to **Figure 12-19**. After a torque converter has been reconditioned, it is often tested on a test stand.

Sources of Torque Converter Failure

Two common causes of torque converter failure are fluid contamination and operator misuse, such as making directional shifts at full engine speed or operating in too

Test Results	Possible Cause
rpm within specification	Indicates only that the stator is locked for the test. It does not indicate if the stator will unlock to achieve the coupling phase.
Stall rpm too low	Stator not locking
	Engine low power
Stall rpm too high	Stripped turbine splines
	Weak transmission clutch or brake (causing transmission to slip)
	Aerated transmission fluid (causing the transmission to slip)
	Low transmission regulated pressure (causing transmission to slip)

Goodheart-Willcox Publisher

Figure 12-30. Stall testing can achieve different results: rpm within specifications, rpm too low, or rpm too high.

high of a gear under too much load. Oil contamination can result from normal everyday machine use and lack of good service and maintenance practices.

Torque converters wear during every day normal use; therefore, it is a good maintenance practice to perform oil analysis every 250 to 500 operating hours. This interval allows the technician to chart trends to determine if fluid contamination, such as aluminum, is rising over time. Another good maintenance practice is to inspect and replace the oil filters before they become completely plugged and begin bypassing contaminated oil.

Many manufacturers recommend a filter change interval. However, the recommended schedule is based on typical use of the machinery. It does not take into consideration operating conditions or severity of use. The general nature of the schedule results in many filters being replaced either too early or too late. A filter that is replaced too early has service life left when it is replaced, wasting money. A filter that is replaced too late is already plugged and no longer functioning, resulting in excessive wear. The best scenario is to check the filter's indicator daily, if the machine has one. The indicator is a bypass valve that allows personnel to visually determine if the filter has service life left or if it must be replaced. See **Figure 12-32**. However, many large contractors prefer to change the filters early. They do this because, for them, ensuring machine uptime and availability outweighs the cost of changing a filter too early.

Many machines have a bypass switch that is monitored by electronic control modules (ECMs). When the bypass switch closes, the operator receives a warning on the machine's monitor that the filter needs to be replaced. Cold thick oil can cause the fluid to bypass the filter.

A *kidney-loop filtration cart* is an external hydraulic pump and filtration system that pulls oil out of a reservoir, filters the oil using a high-efficiency filter, and returns the oil back to the machine's reservoir. See **Figure 12-33**. It is a good practice to use kidney-loop filtration cart whenever a good opportunity presents itself. For example, if a machine is brought in for air conditioning system service, the filtration cart could be cleaning the transmission and torque converter fluid as the A/C service is being performed.

Goodheart-Willcox Publisher

Figure 12-31. This turbine had to be replaced due to the amount of wear around its outer circumference. The heavy wear caused excessive oil to leak past the turbine, dropping the converter's efficiency.

Dynamic Retarders

Some heavy equipment such as haul trucks and scrapers are equipped with a dynamic retarder, also known as a hydraulic retarder. The purpose of the retarder is to assist in slowing the machine's travel speed. A *dynamic retarder* consists of a finned rotor that spins inside a finned stationary housing called the stator. See **Figure 12-34**. When oil floods the stator housing, it acts like a brake, slowing the rotating rotor and causing the machine to slow its travel speed.

Retarder Applications

The retarder can be installed before the transmission or after the transmission. For example, Caterpillar scrapers have the retarder attached to the engine. Allison 1000-4000 series transmissions have their retarders at the rear of the transmission. The 6625 Allison transmission in **Figure 12-35** has the retarder between the torque converter and the transmission. Voith manufactures a retarder that is designed to be bolted to the back of the transmission's housing.

Retarder Controls

Retarders are often actuated with a lever on the steering column or with a foot-operated pedal. The steering column retarder lever on some machines actuates a pneumatic valve that directs air pressure to a hydraulic control valve that directs oil into the retarder housing. Other retarders use an

Goodheart-Willcox Publisher

Figure 12-32. Filter indicators allow personnel to maximize the life of the filter. When the filter becomes plugged the valve moves into the red portion on the indicator demonstrating that the filter is bypassing the dirty oil and the filter needs to be replaced.

Figure 12-33. A kidney-loop filtration cart is designed to clean a fluid reservoir by using its own electrically driven pump.

electric switch that is actuated with the steering lever or brake pedal. The switch often provides an input to an ECM which then directly controls a solenoid valve that directs oil into the retarder housing.

The braking ability of a retarder is determined by the amount of oil or oil pressure inside the retarder housing. Some designs direct more oil into a retarder to increase its braking ability. Once the retarder is full of oil, increasing the pressure inside the retarder provides more braking ability.

Retarder Advantages

Retarders allow machines to travel at higher speeds due to the retarder's ability to slow the machine's travel. The use of a retarder allows the operator to slow a machine faster. Retarders also lower operating costs due to the increased life of the machine's service brakes.

Contamination Control

Rebuilding of transmissions, valves, and torque converters should be completed with good contamination control practices. The following list is an example of good practices to follow to ensure components are rebuilt in a clean environment.

Education: All dealership personnel, including parts, service, sales, and managers, must be properly educated about contamination control, including the causes of contamination, the consequences of contamination, and the correct methods of preventing contamination.

Cleaning: Prior to bringing the machine into the shop, the entire machine, including all of the components, must be meticulously cleaned to ensure that all mud, grime, and dirt are removed. Mud removal requires a high volume of water. Grease and challenging dirt require pressurized hot soapy water.

Shop Environment: Contamination control requires a high standard of shop cleanliness. Poor shop practices, such as cluttered, dirty, and greasy working areas with poor lighting must be eliminated. Today's shops must maintain impeccable shop cleanliness.

Figure 12-34. A dynamic retarder can be seen just ahead of the output yoke on this cutaway of an Allison 4000 series transmission.

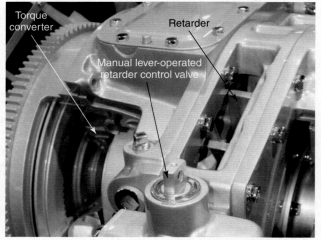

Figure 12-35. This Allison 6625 transmission has the hydraulic retarder located between the torque converter and the transmission. Notice the manually operated retarder control valve.

- The shop's wash bay, the apron leading into the wash bay, and the path into the shop must be thoroughly cleaned after every machine or component has been cleaned.
- The shop doors should be fully operational.
- The shop floors should be clean and sealed.
- Machine stalls and work areas should be clearly defined with stripes that are not flaking or peeling.
- Clean component stands and cribbing blocks should be used effectively to keep the shop floor clean.
- Workbenches should have clean protective tops to provide a clean working environment.
- Shop practices that generate constant debris, such as grinding, welding, cutting, and deburring, should be performed in a different shop location than transmission or other hydraulic system work. Activities that produce intermittent debris should be performed in a screened enclosure to contain contaminants.

Component Removal: Prior to removing components, technicians must have access to a clean stock of plugs, caps, and plastic wrap.

- Hose ends and tubing must be capped with the proper plug or cap.
- Components with machined surfaces must be wrapped in plastic that contains a rust inhibitor, especially if components are stored long term or in a high-humidity climate. Flash rust can appear on a machined surface relatively quickly.
- During breaks such as lunches and shift changes, components must be covered with clean towels or cloths.
- New components should be kept in their original, clean packaging until they are ready to be installed on the machine.

Clean Rooms: Individual components should be disassembled, inspected, and rebuilt in clean rooms.

- Clean rooms are specially designed rooms that use a high-efficiency particulate air (HEPA) filter to remove contaminants from the air.
- They use specially designed heating, air conditioning, and ventilation systems to ensure that the room has a clean, dry atmosphere with a positive air pressure. The positive air pressure prevents dirt and dust from entering when the door is opened.

Cleaning Solution: Components must be cleaned only with approved cleaners, such as mineral spirits or clean solvent. Brake cleaner and other non-approved cleaners are harmful to seals, O-rings, and hoses.

- Solvent tanks must be filtered to meet stringent ISO cleanliness levels.
- Air guns must use air that has been filtered for moisture and contaminants. Power tools should use clearly marked air lines that contain oilers.

Summary

- Heavy equipment use two types of fluid drives: hydrodynamic and hydrostatic.
- Hydrodynamic drives operate at lower pressures than hydrostatic drives.
- Hydrodynamic drives use fluid energy to drive a turbine.
- Fluid couplings contain an impeller and turbine.
- Torque converters contain an impeller, turbine, and stator.
- The stator redirects the oil exiting the turbine and routes it back to assist in rotating the impeller.
- The torque-multiplication phase occurs any time the stator is locked and the impeller and turbine are rotating at significantly different speeds. Fluid temperature increases during the torque-multiplication phase.
- The coupling phase occurs when the turbine and impeller are rotating at nearly the same speed. The stator overruns, and the fluid temperature decreases.
- During the coupling phase, vortex oil flow diminishes, replaced with rotary oil flow.
- Impeller clutches are used to slow a machine's propulsion by slipping the clutch.
- Lockup clutches are used to eliminate the 10% slip that occurs during the coupling phase.
- A three-stage torque converter routes the oil back to the turbine three times before the oil starts over at the impeller.
- Torque dividers are torque converters with a simple planetary gear set integrated. They are often used in Caterpillar mid-sized and large dozers.
- Torque converters often receive their oil flow from the transmission's relief valve. The oil exiting the torque converter is cooled and then used for lubrication.
- During torque converter installation, the torque converter is first mounted into the transmission. Next, the transmission and torque converter are mounted to the engine. Lastly, the torque converter is bolted to the flywheel.
- Some manufacturers provide torque converter stall testing procedures, which require applying the parking and service brakes, chocking the tires, placing the transmission in a high gear, and increasing the engine speed to its highest attainable level. Always follow the manufacturer's specifications.
- Many large off-highway machines have torque converters that can be disassembled and inspected. Smaller on-highway torque converters might be welded together and not repairable.
- Two common sources for torque converter failure are misuse and fluid contamination.
- A dynamic retarder consists of a rotor that rotates inside a fixed stator housing. The volume and pressure of oil in the housing determines the amount of retarding.
- Contamination control is a systematic process to ensure the torque converter reaches its maximum usable life.

Technical Terms

- barring tool
- coupling phase
- dynamic retarder
- fixed stator
- fluid coupling
- hydrodynamic drive
- impeller
- impeller clutch
- kidney-loop filtration cart
- lockup clutch
- lockup phase
- overrunning stator
- power shuttle transmission
- reactor
- reverser
- rimpull
- rotary oil flow
- split guide rings
- stall mode
- stator
- three-stage torque converter
- torque converter
- torque divider
- torque-multiplication phase
- turbine
- variable-capacity torque converter
- vortex oil flow

Review Questions

Answer the following questions using the information provided in this chapter.

Know and Understand

1. All of the following are considered hydrodynamic drives, *EXCEPT*:
   A. fluid coupling.
   B. hydrostatic transmission.
   C. torque converter.
   D. torque divider.

2. Which of the following will generate the most heat?
   A. Coupling phase.
   B. Lockup phase.
   C. High rotary oil flows.
   D. High vortex oil flows.

3. A conventional torque converter typically has a maximum torque multiplication ratio of _____.
   A. 1.25:1
   B. 2.5:1
   C. 4.0:1
   D. 5.5:1

4. When is the converter in the coupling phase?
   A. During conditions of high vortex oil flow.
   B. When the impeller and turbine are at significant speed differences.
   C. When the tractor's load decreases.
   D. When the stator is locked.

5. High rotary oil flow causes _____.
   A. high oil temperatures
   B. the stator to lock
   C. the impeller to lock
   D. the turbine to approach the same speed as the impeller

6. Which of the following is used to describe the relationship between the impeller speed and the turbine speed?
   A. Stage.
   B. Phase.
   C. Efficiency.
   D. Energy.

7. When does a lockup torque converter apply its lockup clutch?
   A. Under heavy loads.
   B. During torque multiplication.
   C. When the impeller and turbine have significantly different speeds.
   D. When the impeller and turbine have close to the same speed.

8. In a torque converter, where does the oil flow after it leaves the impeller?
   A. Stator.
   B. Turbine.
   C. Both A and B.
   D. Neither A nor B.

9. In a torque converter, where does the oil flow after it leaves the turbine?
   A. Stator.
   B. Impeller.
   C. Both A and B.
   D. Neither A nor B.

10. The number of times the oil cycles through the turbine before it returns to the impeller is known as _____.
    A. phases
    B. stages
    C. lockups
    D. releases

11. Which of the following statements is *not* true?
    A. Stators can be mounted to an overrunning clutch.
    B. Stators can be designed to be stationary all the time.
    C. Stators can have variable vane operation.
    D. Stators multiply torque when they are overrunning.

12. In a torque divider, what two components rotate at engine speed?
    A. Impeller and ring gear.
    B. Stator and sun gear.
    C. Turbine and planetary carrier.
    D. Impeller and sun gear.

13. When the torque divider is being stalled, the _____ will rotate backward.
    A. planetary carrier
    B. turbine
    C. impeller
    D. sun gear

14. Which of the following is considered the output member of a torque divider?
    A. Sun gear.
    B. Planetary carrier.
    C. Ring gear.

15. When installing a torque converter, what is the most common practice?
    A. Install the converter onto the transmission before installing the transmission to the engine.
    B. Install the converter onto the engine flywheel before installing the transmission.
    C. Install the torque converter last.
    D. Fill the transmission with oil before installing the torque converter.

16. Which transmission gear is used for stall testing an off-highway machine's torque converter?
    A. Low-speed gear.
    B. Mid-range gear.
    C. High-speed gear.
    D. Only reverse.

17. If a stall test results in a low stall rpm, what could be the cause?
    A. Aerated fluid.
    B. Low transmission pressure.
    C. Weak planetary transmission control.
    D. Slipping stator.

18. Which of the following dynamic retarder statements is *not* true?
    A. The volume of oil inside the retarder determines its rate of retarding.
    B. The rotor rotates.
    C. The stator is stationary.
    D. Some dynamic retarders do not require a cooler.

Analyze and Apply

19. The _____ is the torque converter member known as a pump.
20. The _____ is torque converter member known as the recycler.
21. The _____ is the output member of the torque converter.
22. The _____ is the input member of the torque converter.
23. The torque converter's _____ rotates at engine speed.
24. The _____ is the torque converter member that is splined to the transmission input shaft.
25. During stall mode, the turbine is _____.
26. During the torque-multiplication phase, the impeller spins _____ than the turbine.
27. During the torque-multiplication phase, the stator is _____.
28. The torque converter member closest to the engine is the _____.

Critical Thinking

29. Describe all of the different types of clutches that can be integrated into a torque converter and the purpose of the clutch.
30. Explain the fail-safe condition of a Caterpillar dozer's stator multiple disc clutch.

Chapter 13
Hydrostatic Drives

Objectives

After studying this chapter, you will be able to:

- ✓ Explain hydrodynamic drives.
- ✓ Describe attributes of hydrostatic drives.
- ✓ List examples of off-road HST applications.
- ✓ Explain HST charge pumps and main piston pumps.
- ✓ Explain the flow of oil in an HST circuit.
- ✓ Explain the operation of variable-speed HST motors.
- ✓ Describe the operation of inching valves, manual bypass valves, pressure-release solenoids, IPOR valves, pressure limiter valves, and anti-stall control valves.

As discussed in Chapter 12, *Hydrodynamic Drives*, off-highway machines use two styles of fluid drives: hydrostatic and hydrodynamic. Both drives require the use of fluid as a means of transmitting power, but the two fluid drives have significant operating differences.

Hydrodynamic Drives

A hydrodynamic drive system is a fluid drive system that operates at a relatively low fluid pressure and relies on the fluid's mass and velocity for transmitting power. The most common type of hydrodynamic drive in the off-highway industry is the torque converter. The torque converter serves two purposes: it functions as an automatic clutch and it multiplies engine torque.

Attributes of Hydrostatic Drives

A *hydrostatic drive*, also known as a *hydrostatic transmission (HST)*, uses fluids under pressure to drive a machine or a component and change its speed, torque, and direction of travel. (The term *hydrostatic* indicates the ability to transfer power via fluids at rest or under pressure.) A hydrostatic drive uses a hydraulic pump and a hydraulic motor. See **Figure 13-1**. The HST is used in place of other styles of propulsion systems, such as the powershift, automatic, or manual-clutch synchronized transmissions.

Most HSTs use a variable-displacement reversible engine-driven hydraulic pump. Variable displacement pumps can vary the effective volume of their chambers to adjust the pump's flow. Reduced flow causes the machine to travel at slow speeds and increased flow causes the machine to travel at fast speeds. *Reversible pumps* can internally reverse the direction of the pump's flow. When the pump's flow is reversed, the machine's direction of travel is reversed.

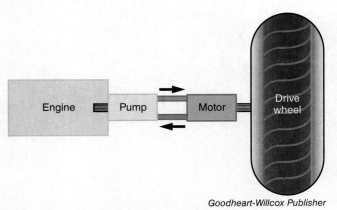

Figure 13-1. A hydrostatic transmission is simply a hydraulic pump and a hydraulic motor that provides forward and reverse propulsion and variable speed.

Caution

The term reversible pump also describes a pump that can be rotated in either direction, clockwise or counterclockwise. However, hydrostatic pumps are normally designed to be spun in only one direction. When replacing a hydrostatic pump, it is critical to ensure that the replacement pump has the correct direction of rotation or it will fail.

The hydraulic motor can be fixed-displacement or variable-displacement. A *fixed-displacement motor* will operate at a constant speed for a given amount of input flow. *Variable-displacement motors* can vary the effective volume of their chambers to adjust the motor's output speed and torque. The displacement of the motor has an inverse relationship with the motor's output speed. As the motor's displacement is decreased, the output of the speed of the motor will be increased. Motor displacement is discussed in further detail later in this chapter.

HST systems are used on agricultural equipment, such as combines, cotton harvesters, tractors, swathers, sprayers, and lawn tractors. Construction equipment that uses HSTs includes skid steers, dozers, track loaders, wheel loaders, trenchers, excavators, and concrete mixing trucks. Note that differential steer tractors, such as Caterpillar dozers, Challenger rubber track tractors, and John Deere agricultural twin-track tractors, use a hydraulic pump and motor that is very similar to a traditional HST. However, the pump and motor in these machines are used only for hydraulic steering input and not for propulsion. Differential steering is explained in Chapter 24, *Track Steering Systems*.

Open-Loop and Closed-Loop HSTs

HSTs are classified as open-loop or closed-loop. An *open-loop HST* routes the motor's return oil back to the reservoir. In these applications, the pump must draw all of the inlet oil from the reservoir. Excavators commonly use this style of hydraulic propulsion. See **Figure 13-2**.

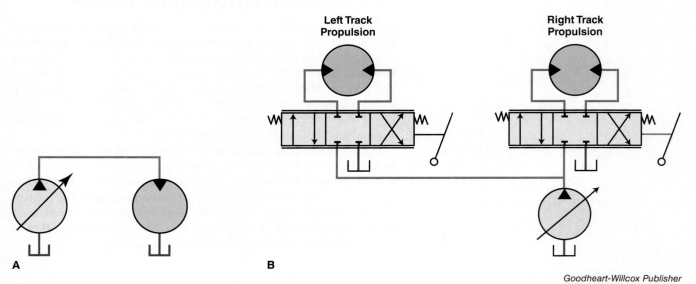

Figure 13-2. A—An open-loop HST draws new pump oil directly from the reservoir and is not assisted by the hydraulic motor's return oil. B—Excavators typically use unidirectional, open-loop pumps to supply oil to DCVs to propel the machine.

Unlike the pump in an open-loop system, the closed-loop pump does not have to draw all of the inlet oil from the reservoir. The *closed-loop HST* pump is designed to drive a closed-loop motor. The motor's return oil is sent directly back to the pump's inlet instead of the reservoir. This arrangement helps charge the piston pump's inlet. Closed-loop HSTs also normally have a charge pump that helps supercharge the piston pump's inlet. **Figure 13-3**. Charge pumps provide makeup oil to compensate for losses due to pump and motor inefficiencies. Charge pump circuits are discussed later in this chapter.

When a closed-loop pump or motor fails or becomes contaminated, both the pump and the motor must be rebuilt. Many customers or technicians want to condemn the pump or the motor and attempt to simply replace the one part. However, in a closed-loop HST, the pump feeds the motor and the motor feeds the pump. Therefore if the pump or motor becomes contaminated, it will inject contaminants into the other component. One exception is if a closed-loop HST uses closed-loop filters. See **Figure 13-4**. These filters are rarely used because the filters must be able to withstand very high drive pressures, for example up to 7000 psi (483 bar), and allow oil flow in both forward and reverse directions.

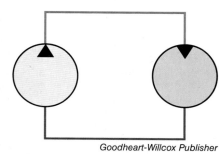

Goodheart-Willcox Publisher

Figure 13-3. A closed-loop HST sends the motor's return oil directly to the pump's inlet.

Applications of Hydrostatic Drives

A variety of hydrostatic drive designs are available to meet the demands of different applications. Several common designs are discussed in the following sections.

Single-Path HST

The simplest HST design is a *single-path HST*, which uses one variable-displacement reversible hydraulic pump and either a fixed-displacement or variable-displacement hydraulic motor. See **Figure 13-5**. The single-path design is used in compact utility tractors, combine harvesters, cotton harvesters, concrete mixing trucks, and lawn tractors, **Figure 13-6**.

Dual-Path HST

A *dual-path HST* drive uses two separate hydrostatic transmissions to propel and steer the machine. One pump and one motor are used to drive the left side of the machine and one pump

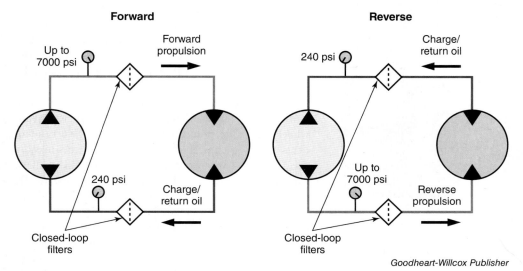

Goodheart-Willcox Publisher

Figure 13-4. Closed-loop HST filters must be able to handle high pressure and oil flow in both forward and reverse directions.

Figure 13-5. This single-path HST consists of a variable-displacement pump that drives a fixed-displacement motor. This trainer has the traditional components found in an older combine. The motor provides input into a three-speed mechanical transmission that contains a differential. The differential side gears deliver power out of the transmission's left and right output shafts, which can be held by the service brake and parking brake.

Figure 13-6. A common HST-propelled lawn tractor uses a belt-driven HST pump that drives an HST motor. The pump and motor are often part of a single transaxle assembly that also includes a differential. The HST motor delivers power to the differential, which drives the rear left and right wheels. The HST pump normally has a fan mounted to the input shaft to cool the transmission.

and one motor are used to drive the right side. Machines that use a dual-path system include skid steers, track loaders, dozers, swathers, and zero turning radius (ZTR) lawn mowers. See **Figure 13-7** and **Figure 13-8**.

Dual-path HSTs will propel the machine straight forward or reverse when both motors are driven in the same direction at the same speed. When the motors are driven at different speeds or directions the machine will steer to the left or to the right.

One system that is similar to a dual-path HST is found on an all-wheel-drive motor grader, such as an all-wheel-drive Caterpillar motor grader. The rear tandem drive wheels are powered by a power shift transmission. The front drive wheels are powered by two HSTs. Like a traditional dual-path HST, the two HSTs independently control each front wheel, one HST controls the left front wheel and one controls the right front wheel. The difference is that these two HSTs provide only propulsion and do not steer the machine. However, when the machine is turning, the ECM that controls the two HSTs compensates, similar to a differential, and drives the inside wheel at a slower speed than the outside wheel.

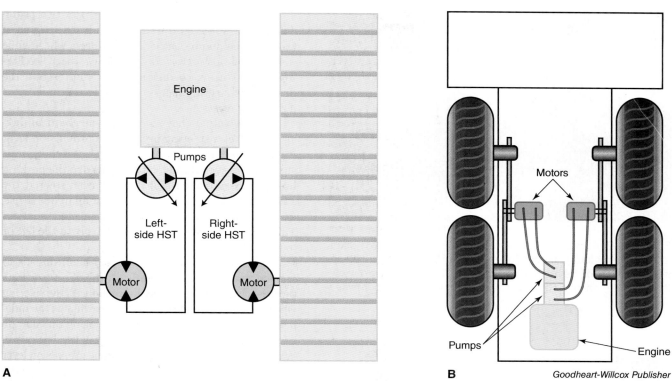

Figure 13-7. Dual-path systems. A—HST-propelled dozers and track loaders use a pump and motor for each track. B—On a skid steer, each side is driven by a separate pump and motor.

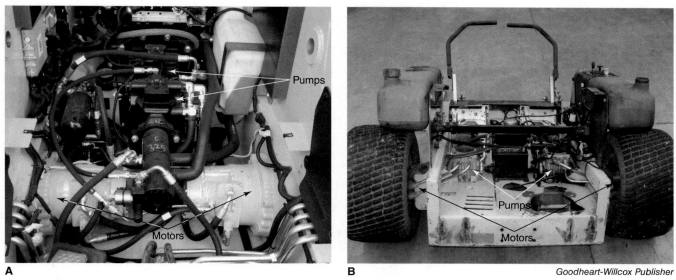

Figure 13-8. Dual-path system pump and motor arrangements. A—Skid steer pumps normally are driven in tandem. The two HST piston pumps are also coupled to the two implement pumps. Notice this dual path HST uses two cam lobe motors. B—A zero turning radius (ZTR) mower is equipped with a dual-path pump and motor used to drive each side of the mower. The left propulsion lever controls the left HST and the right propulsion lever controls the right HST.

Swathers

Self-propelled swathers, or windrowers, use dual-path HST drives. The dual-path HST propels the front wheels, while the back wheels are mounted on casters and swivel as needed. See **Figure 13-9**.

Figure 13-9. A swather is a self-propelled windrower that is used to cut hay and forage crops. The swather will cut the crop, gather it to the center of the header, and deposit it in a single narrow row that can later be baled by a hay baler or harvested by a combine.

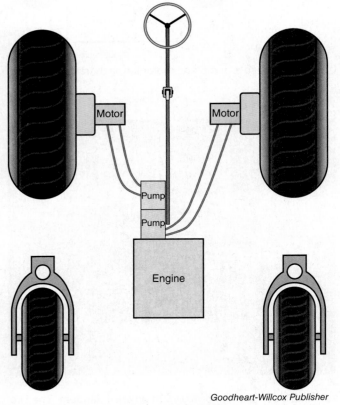

Figure 13-10. A swather commonly has two pumps driven in tandem. One pump controls the left drive wheel, and the other pump controls the right drive wheel.

The dual-path HST on a swather uses two piston pumps that are often directly driven by the engine in tandem. One pump controls the left drive wheel while the other pump controls the right drive wheel. See **Figure 13-10**. This type of dual-path HST controls both the direction of steering (left or right) and machine travel (forward or reverse).

Older swathers have a complex mechanical linkage that uses a steering input and a propulsion lever input. The two input combination operates the pumps simultaneously but independently. When the parking brake is released and the steering wheel is steered, the steering shaft linkage strokes one pump in one direction and the other pump in the opposite direction. This causes the swather to turn sharply to the left or to the right.

The steering shaft must have left-hand threads to operate one pump and right-hand threads to operate the other pump in the opposite direction. As the steering wheel is turned, the steering shaft coordinates the independent control of the two pumps, moving one pump's swashplate forward and the other swashplate backward. When the operator is actuating only the propulsion lever and the steering wheel remains in a neutral position, the two pump swashplates stroke in the same direction at the same swashplate angle.

The swather in **Figure 13-11** uses mechanical linkage for controlling the HST pumps. The steering wheel turns the steering shaft. If the steering shaft is rotated when the machine is sitting still, the HST pump control arms pivot inward (toward each other) or outward (away from each other), depending on the direction of steering wheel rotation.

When the propulsion lever is actuated, it operates the DCV (direct control valve), which causes the actuation cylinder to operate the pivot plate. The pivot plate strokes the control strap, which causes the HST pump control arms to operate in the same direction, either both forward or both rearward, depending on the direction the propulsion lever is moved. When the propulsion lever and the steering wheel are operated at the same time, the control arms operate independently of one another.

Note that MacDon manufactures an M series dual-direction swather that allows the operator station to be rotated. This action reverses the operator station and heading of the machine. The traditional front-drive wheels are located at the rear of the machine and the rear caster wheels are located at the front of the machine. MacDon also manufactures the John Deere W155 and W170 swather with a reversible operator station. John Deere calls this feature Engine Forward Transport because the operator station is rotated 180°, causing the machine to have the engine in a forward heading.

Figure 13-11. This older swather HST has controls consisting of a steering shaft, a propulsion lever, DCV and actuation cylinder, pivot plate, centering spring, and a control strap.

Multiple Pump and Motor Applications

The single-path and dual-path HST configurations described earlier dedicate one pump to drive one motor. Some HSTs use one or more hydraulic pump(s) to drive two or more hydraulic motors. Machines that use this type of HST configuration include excavators, agricultural sprayers, and Caterpillar wheel loaders.

Excavators

The HSTs used in excavators are similar to dual-path systems. Most excavators use two implement pumps rather than a dedicated hydrostatic pump for propulsion. In addition to track propulsion, the implement pumps often control other hydraulic functions, such as boom, stick, bucket, and swing. An excavator design may also use one pump drive to drive one or both track motors. The hydraulic motor drives a planetary final drive, which is responsible for driving the track's drive sprocket. The implement pump is a unidirectional pump (non-reversible pump), which requires the use of a DCV for controlling the direction and speed of oil flowing to the motor. Refer to **Figure 13-2**. Although most excavators use open-loop pumps to provide oil flow for both propulsion and implement functions, it is possible to find an excavator, like the Bobcat 430, that uses a dual-path HST for track-propulsion.

Figure 13-12. This image shows the hydraulic hoses that are connected to the sprayer's drive wheels. The sprayer has a drive motor at each of the four wheels.

Agricultural Sprayer

Four-wheel sprayers commonly have one hydrostatic drive motor located at each wheel. The John Deere 30 series and 4 series sprayers use two hydrostatic pumps to supply oil to four hydrostatic motors. The pumps drive two motors that are situated diagonally from each other.

- The 30-series sprayers use the front pump to drive the left front motor and the right rear motor, and the rear pump to drive the right front motor and the left rear motor.
- The 4-series sprayers use the front pump to drive the right front motor and the left rear motor, and the rear pump to drive the left front motor and the right rear motor.

The sprayers also have a traction control feature. When the traction control is engaged it will sense when one of the drive wheels has less resistance based on the increase in wheel speed. As a result, the control will electronically slow the speed of the faster spinning motor by increasing its displacement. The elevated agricultural sprayer in **Figure 13-12** has narrow tires that allow it to travel through a field without causing any damage to the crops.

Fixed- and Variable-Displacement Hydrostatic Motors

Most off-highway HSTs use a variable-displacement pump. However, the motor can be a fixed- or variable-displacement motor. The simplest application is a fixed-displacement motor, sometimes called a ***single-speed motor***. This application has a fixed swashplate angle and can be found in compact utility tractors, older combines, cotton harvesters, and concrete mixer trucks. See **Figure 13-13**.

Other off-highway machines use some type of variable-displacement motor to provide a wider range of machine speeds and torque. A larger degree of swashplate angle will result in slower speed and increased torque, because it takes more oil to complete one motor revolution. A smaller swashplate angle will equal a faster travel speed and reduced output torque. Some combines have the option of a ***two-speed motor***, which provides the customer two specific swashplate angles, such as a high-speed (15°) swashplate angle and a low-speed (18°) swashplate angle.

Figure 13-13. A single-speed motor uses a fixed swashplate. The motor's valve block assembly has been removed from the back of the motor.

John Deere X9 combines and Caterpillar 924K, 926M, 930K, 930M, 938K, and 938M wheel loaders use a unique variable-displacement motor application. The machines use one variable-displacement pump and two variable-displacement motors. Both motors provide power into a single gearbox. When the machine requires a slower travel speed or higher torque, both hydraulic motors provide an input to the gearbox. When the machine requires a higher travel speed or lower torque, only one motor provides an input to the gearbox. The system operation is explained later in this chapter.

Hydrostatic Transmission Advantages

HSTs provide numerous advantages. The output shaft speed can be maintained even as the engine speed varies. Output shaft speed and direction can be controlled remotely and accurately and can be infinitely variable. The machine speed can be varied without having to change other functional speeds. The machine direction can be quickly reversed without having to mechanically shift gears and without a shock load to the machine. The drive offers overload protection in the form of relief valves. The transmissions use relatively small components for the amount of power they are capable of transferring.

The transmissions do not coast or freewheel. Instead, they provide hydrostatic engine braking. The braking occurs when the operator reverses the propulsion lever, which causes the hydraulic motor to act like a pump and the pump to act like a motor, resulting in engine braking. This form of braking virtually eliminates wear on the machine's service brakes.

All of these advantages lead to the most common reason why engineers and customers choose HSTs over traditional mechanical drives, which is increased overall machine productivity. With no need to disengage a clutch or shift gears or brakes, the operator can make the most efficient use of the machine.

Hydrostatic Transmission Disadvantages

The HST however does have disadvantages. It is not as energy efficient as a mechanical transmission. A mechanical transmission can have an overall efficiency of 92% or higher and an HST will have an overall efficiency of 85% or lower. As a result, a machine equipped with an HST will use more fuel than a machine equipped with a traditional mechanical transmission. HSTs are also noisy and sensitive to contamination and heat.

Configurations

HSTs can be categorized into two configurations: split and integral. A *split HST* is one of the most common designs found in the off-highway industry. It allows the engine-driven pump to be located a considerable distance away from the HST motor(s). Some examples of machines that use split HSTs are combines, cotton harvesters, and dozers. Notice in **Figure 13-14** that the split HST configuration requires hydraulic hoses or tubing to route oil from the pump to the motor.

Integral HSTs eliminate the need for external hoses or tubing. The pump and the motor are directly connected to each other. Integral configurations may be inline, U-shaped, or S-shaped. See **Figure 13-14**.

Note

John Deere produced the 6600, 6620, 7700, and 7720 combines in the 1970s and 1980s. These combines use an integral inline HST. The pump is driven by a double V-belt, and the integral motor is directly coupled to the mechanical transmission gearbox. The charge pump is located separate from the HST. Some late-model machines have integral HSTs inside the CVT. The Case IH Magnum and New Holland T8 variator uses an inline integral HST and is discussed further in Chapter 15, *Continuously Variable Transmissions*.

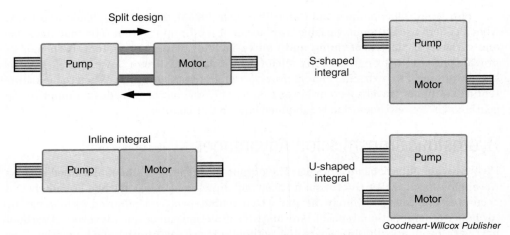

Figure 13-14. HSTs can be split or integral. Split HSTs require the use of hydraulic hoses or tubing. Integral HSTs can be S-shaped, inline, or U-shaped.

Hydrostatic Charge Pumps and Main Piston Pumps

An HST commonly uses a charge pump. A fixed-displacement gear pump is normally used as the transmission's charge pump. It is typically driven in tandem off the back of the transmission's piston pump, but it can be located separate from the HST. See **Figure 13-15**.

 Pro Tip

Always follow the manufacturer's service literature to ensure the equipment has been properly identified. The Case IH 10, 20, 30, 240, and 250 series flagship combines use Sauer Danfoss (early 10 series) and Bosch Rexroth (for the other models) hydrostatic pumps that have an integral charge pump. However, this integral charge pump has its flow directed for a different purpose other than the HST. It is used to lubricate the PTO drive gear case. The HST uses a separate (externally located) pump that serves as the HST charge pump.

The charge pump serves the following four purposes:
- Supercharges the transmission's piston pump.
- Provides make-up oil to the closed loop to prevent cavitation.
- Supplies oil to the transmission's directional control valve (DCV).
- Cools the transmission by replenishing case lubrication.

Refer to the schematic in **Figure 13-16**. The charge pump flows oil into the HST's closed loop through two check valves that are also known as make-up valves. When the HST is operating, one check valve prevents high-pressure oil from flowing into the charge pump circuit, while the other check valve allows charge pump flow into the charge leg of the closed loop.

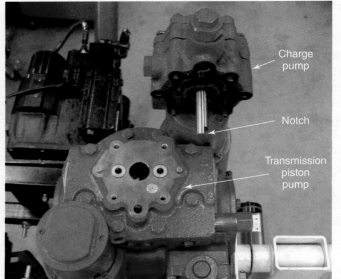

Figure 13-15. The charge pump has been removed from the Eaton HST piston pump. Notice the charge pump's drive shaft has a notch that fits into a matching notch on the piston pump's shaft.

 Pro Tip

Bosch Rexroth calls the HST charge pump a boost pump.

Sizing a Charge Pump

Charge pumps are sized to prevent piston pump cavitation. Pumps *cavitate* when they lack a good supply of inlet oil. The fluid vaporizes to form bubbles, and as the fluid vaporizes, the bubbles implode, causing shock waves to erode and pit the metal in the pump. Charge pump flow is most critical when the transmission is operating at high drive pressures. As a rule of thumb, charge pumps are sized to provide 19% of the piston pump's displacement. This value enables the charge pump to provide cavitation-free operation if the pump and motor each fall to a 90% volumetric efficiency. For example:

0.90 (pump efficiency) × 0.90 (motor efficiency) = 0.81 (transmission volumetric efficiency).

1 − 0.81 = 19% charge pump displacement

If a piston pump was flowing 50 gpm at the rated engine speed, the charge pump would need to provide the following flow:

50 gpm × .19 = 9.5 charge gpm

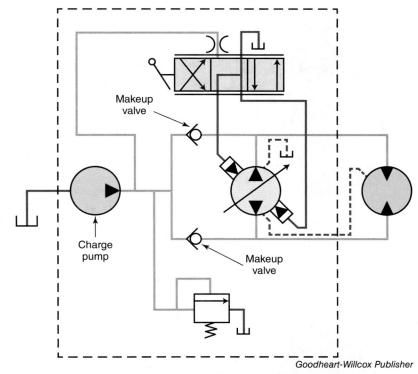

Goodheart-Willcox Publisher

Figure 13-16. An HST charge pump will charge the piston pump and closed loop, supply oil to the HST control valve, and cool the HST by replenishing make-up oil.

The Eaton *Heavy Duty Hydrostatic Transmission Pump and Motor Sizing Guide* (No. 3-409) recommends assuming a maximum volumetric efficiency of 96% for a pump and 97% for a fixed-displacement motor. Based on the previous piston pump example of 50 gpm and a charge pump flow of 9.5 gpm, how much extra flow would the charge pump be flowing in a new hydrostatic transmission application?

0.96 (pump) × 0.97 (motor) = 0.9312 overall transmission volumetric efficiency

1− 0.9312 = 0.0688 (necessary charge pump displacement for a new transmission)

50 gpm × 0.0688 = 3.44 gpm (necessary charge flow for a new transmission)

9.5 gpm (charge pump flow)− 3.44 gpm (necessary charge flow) = 6.06 gpm extra flow

Styles of Piston Pump Frames

Hydrostatic pumps are normally inline axial piston pumps, but they can be designed as bent-axis. Inline axial piston pumps are either a single-servo piston design or a pump that uses two servo pistons.

Two-Servo-Piston Inline Axial Piston HST Pump

The older housing used two servo pistons for stroking a swashplate that pivoted on two trunnion-tapered roller bearings. See **Figure 13-17**. This style of pump has been around for decades. Old Sundstrand Sauer Danfoss Series-20 pumps utilize the two-servo-piston design having a swashplate with tapered roller trunnion bearings. This design is still available from Eaton, for use in "heavy-duty" hydrostatic applications and is labeled a Series-1 pump. One servo is used to stroke the pump for forward propulsion and the other servo is used to stroke the pump for reverse operation.

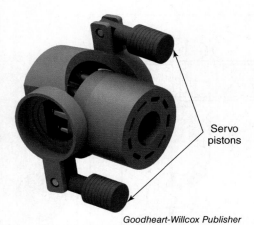

Figure 13-17. A dual-servo HST pump uses a reverse servo piston and a forward servo piston. Examples include Eaton Series-1 pumps and Danfoss Series-20 pumps.

Figure 13-18. A single-servo cradle bearing hydrostatic piston pump achieves forward and reverse propulsion by moving the single piston. The piston moves up for one direction and moves down for the reverse direction. An example is a Rexroth A4VG pump.

Single-Servo-Piston Inline Axial Piston HST Pump

The newer style of hydrostatic pump frame uses a single servo piston to rotate a cradle bearing swashplate. See **Figure 13-18**. The single servo piston is stroked in one direction for propelling the machine forward and stroked in the opposite direction for propelling the machine in reverse. Eaton labels their single-servo-piston pumps as Series-2. For a given amount of flow, the single-servo-piston pump housing is smaller than the dual-servo-piston pump housing.

Danfoss Series-40 and Series-90 HST piston pumps also utilize a single servo piston and a cradle-bearing swashplate. The Danfoss Series-90 HST pump's servo, linkage, and swashplate are shown in **Figure 13-19**. The servo piston inside the Danfoss Series-90 HST pump is enclosed by two end caps. Each cap has a threaded adjustment, as shown in **Figure 13-20**. The adjustment acts like a displacement limiter that allows a technician to limit the maximum swashplate angle. The figure also shows the pump's single side cover with a lever bolted to the cover. Two swashplate spring seats are on one side of the lever. Each spring seat holds one end of a large swashplate spring. The opposite end of each swashplate spring rests against the pump's end cover. The other side of the lever has two arched contact points that apply pressure to the swashplate. When the pump is in a neutral position, the springs hold the swashplate in the centered position. As the swashplate pivots, it must overcome the spring tension.

> **Note**
>
> Sauer purchased Sundstrand in 1990. In 2000, Danfoss merged with Sauer, becoming Sauer-Danfoss. In 2021 Danfoss purchased Eaton.

Types of Hydrostatic Pump Controls

Hydrostatic pumps are commonly actuated by one or two servo pistons that are controlled with a manual servo control valve, electronic proportional solenoids, or an electronic servo valve. It is also possible to find HST pumps that use a mechanical lever to directly control the pump's swashplate. The mechanically controlled HST pumps are normally used in lower power applications or require some type of mechanical advantage, such as a long lever, to actuate the swashplate.

Manually Controlled Hydrostatic Pumps

The manually controlled servo valve, also called an HST DCV, has been around the longest. The actuation of the servo control valve will control the speed and the direction of the HST. A propulsion lever is located in the machine's cab or operator station, and a cable connects the propulsion lever to the pump's control valve. The propulsion lever is sometimes called a *FNR lever* for *forward and reverse lever*.

Chapter 13 | Hydrostatic Drives

Figure 13-19. A photo of a Danfoss Series-90 pump with the servo piston and linkage held connected to the pump's swashplate.

Figure 13-20. A photo of a Danfoss Series-90 pump. One of the servo caps and its displacement limiter adjustment is illustrated and the side cover and lever are shown in the photo.

As the propulsion lever is stroked, a cable or manual linkage moves a spool valve located inside the HST control valve. The charge pump supplies oil to the servo control valve through an orifice. See **Figure 13-21**.

In a dual-servo pump, the spool valve receives oil through the orifice and sends control oil to the appropriate servo piston to stroke the pump forward or reverse. See **Figure 13-22**.

Manually operated servo control valves can include a *neutral safety switch*, which will block electrical current flow preventing the engine from starting any time the lever is outside of neutral. See **Figure 13-23**.

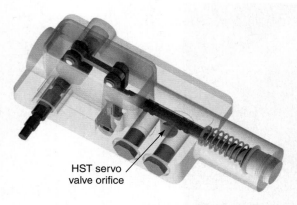

Goodheart-Willcox Publisher

Figure 13-21. An example of an Eaton Series-1 manual HST servo valve contains a spool valve that is spring centered. An orifice meters the oil into the spool valve.

 Warning

If the switch fails or the servo valve switch assembly becomes misadjusted, the machine can start even when the propulsion lever is in a forward or reverse position. Always be prepared for a component to fail or become misadjusted.

Electronically Controlled Hydrostatic Pumps

Many late-model HSTs are electronically controlled, with either proportional solenoids or a servo-motor. Manufacturers use three types of electronically controlled motors: force motor, torque motor, or stepper motor. Electric servo-motors are more expensive and less common than proportional solenoids.

Solenoid-controlled HSTs require one solenoid for each direction of propulsion, for example, one reverse solenoid and one forward solenoid. See **Figure 13-24**. Each solenoid has a power wire and a ground. The electronic control module can control either the power or the ground. An example range of solenoid current flow is 350 milliamperes to 850 milliamperes.

The hydraulic schematic of a solenoid-controlled hydrostatic transmission looks similar to that of a manually controlled transmission. In **Figure 13-25**, the dual-servo hydrostatic pump schematic illustrates a pump that uses two proportional solenoids.

Feedback Link

Many hydrostatic pumps use a *feedback link*. The link provides a direct mechanical feedback of the position of the swashplate to the pump's control valve. When the operator actuates the control valve, the spool directs servo oil to the servo piston, which causes the swashplate to pivot, resulting in the pump flowing oil and the machine moving forward or reverse. As the control valve directs oil to one servo piston, the other servo piston is ported to the tank. If the spool valve was allowed to continue to direct oil to the servo piston, the pump's swashplate would continue to pivot, resulting in increased machine travel. To limit an uncontrolled acceleration of machine

Goodheart-Willcox Publisher

Figure 13-22. A manually operated servo valve receives oil through an orifice for the purpose of actuating the forward servo piston or the reverse servo piston.

Chapter 13 | Hydrostatic Drives

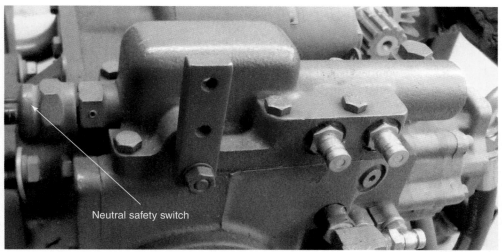

Goodheart-Willcox Publisher

Figure 13-23. An example of an Eaton Series-1 HST control valve's neutral safety switch, which is designed to prevent the engine from starting when the lever is in a forward or reverse position.

travel, as the swashplate moves, the feedback link repositions the spool so that it can hold the pump swashplate at the exact angle that the operator requested. See **Figure 13-26**.

 Pro Tip

Some late-model dual-path hydrostatic pumps use swashplate angle sensors instead of mechanical feedback levers. The ECM monitors the feedback of the swashplate angle sensor to determine when the pump has reached its commanded flow rate. It maintains the swashplate angle as commanded by input from the operator.

Goodheart-Willcox Publisher

Figure 13-24. This Bosch Rexroth A4VG HST pump uses two solenoids to control the pump.

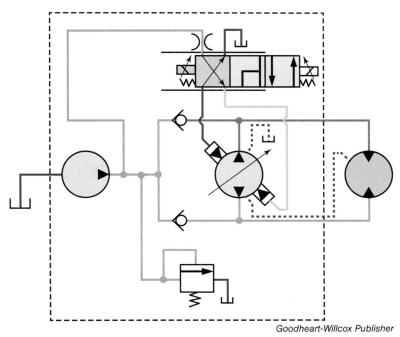

Goodheart-Willcox Publisher

Figure 13-25. An example of an Eaton Series-1 electronically controlled HST pump with two solenoids to control the pump's two servo pistons. The schematic shows the pump propelling the machine in a forward direction.

Speed-Sensing Valve

As mentioned earlier in the chapter, a manually actuated HST uses a single DCV that controls both the speed and the direction of the HST. A compact Caterpillar wheel loader, for example a 902C, uses two separate controls to perform the same two tasks:

- A forward neutral reverse (FNR) valve that is operated with dual solenoids that only controls the machine's direction.
- A speed-sensing valve that controls the speed of the machine.

The three-position *FNR valve* is not infinitely variable; it simply has three fixed positions: forward, neutral, and reverse. The wheel loader uses an FNR electric switch, which controls the FNR valve, resulting in one of the following:

- Both solenoids are de-energized in neutral.
- The forward solenoid is energized in forward gear.
- The reverse solenoid is energized when in reverse.

Unlike the manually actuated HST's DCV, which receives its supply oil directly from the charge pump and orifice, this FNR valve receives its supply oil from a *speed-sensing valve*. The speed-sensing valve is manually actuated by a lever from the engine's governor. As the accelerator pedal is depressed further, the engine speed increases and the governor's lever actuates the speed-sensing valve further, which causes an increase in the *speed-sensing signal pressure*. The speed-sensing signal pressure is directed to the HST pump's FNR valve, which directs the oil (also known as *signal oil*, *signal pressure*, or *servo oil pressure*) to actuate the HST pump servo piston either forward or reverse. The amount of signal pressure coming from the speed-sensing valve determines how far the servo piston is actuated and how much oil flow is produced by the HST pump. The speed-sensing signal pressure is also used to control the variable-displacement motor. The pump's FNR valve determines only the machine's direction. See **Figure 13-27**. This hydrostatic drive operates similarly to an automotive automatic transmission. As the wheel loader's accelerator pedal is pressed, the engine speed increases and the HST travel speed increases.

The schematic also shows that the signal oil pressure can be connected to a brake master cylinder (inching valve) or to a creeper inching valve. Chapter 19, *Hydraulic Brake Systems*, details a unique combination brake master cylinder and inching valve that applies the brakes and slows the HST by draining the signal oil. This slows the HST propulsion (inching or creeping) while allowing the engine speed to remain high so that the machine will have maximum hydraulic flow for implement hydraulic functions. Some machines may not have the unique combination brake master cylinder HST creeper valve, but might be equipped with the single creeper (inching valve). Inching valves are explained later in this chapter.

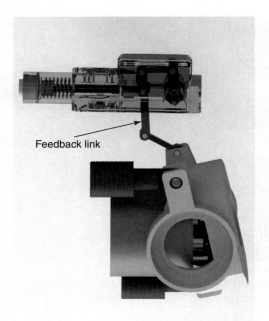

Goodheart-Willcox Publisher

Figure 13-26. An example of an Eaton Series-1 HST pump's feedback link, which moves based on the position of the swashplate. The link will hold the control valve's spool in the last commanded position.

Bosch Rexroth DA Control

A late-model Bosch Rexroth speed-sensing HST is labeled DA control, which is sometimes described as "speed dependent." The pump has a cartridge valve that varies pilot pressure based on engine speed. The pilot pressure can also be called control pressure or servo supply pressure. The DA cartridge valve receives oil flow from the charge pump, which Bosch Rexroth calls the boost pump. This oil flow is proportional based on engine speed. As engine speed increases, so will the charge pump flow increase.

The DA cartridge valve receives the charge pump flow and uses variable orifices to generate a pilot pressure that is proportional to engine speed. This oil pressure is sent to the FNR solenoid valve. The FNR solenoid valve is a non-variable directional control valve. When actuated, the FNR solenoid valve simply chooses forward or reverse. The variable pilot pressure determines how far the servo piston moves, which shifts the pump's swashplate proportional to the engine speed. See **Figure 13-28**.

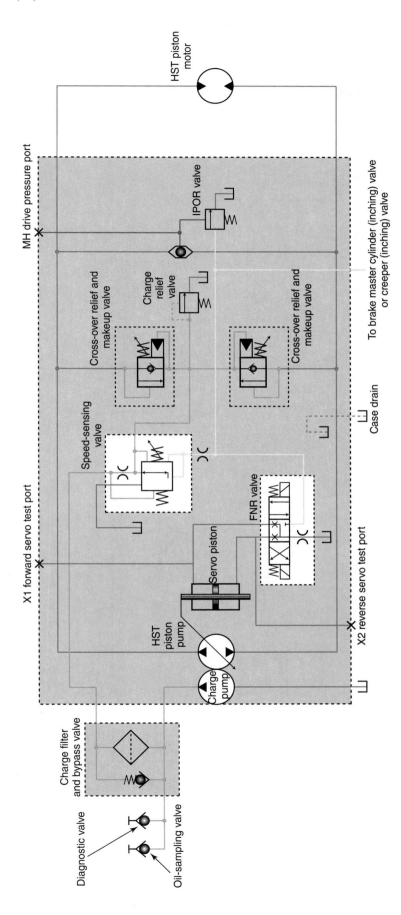

Figure 13-27. This schematic illustrates a speed-sensing valve and an FNR valve used in an HST pump. The speed-sensing valve develops a signal pressure proportional to engine speed. The FNR valve controls the direction of the HST. The individual components normally used in an HST motor, such as the flushing spool valve and relief valve, have been intentionally left out of the schematic to focus on the pump's controls.

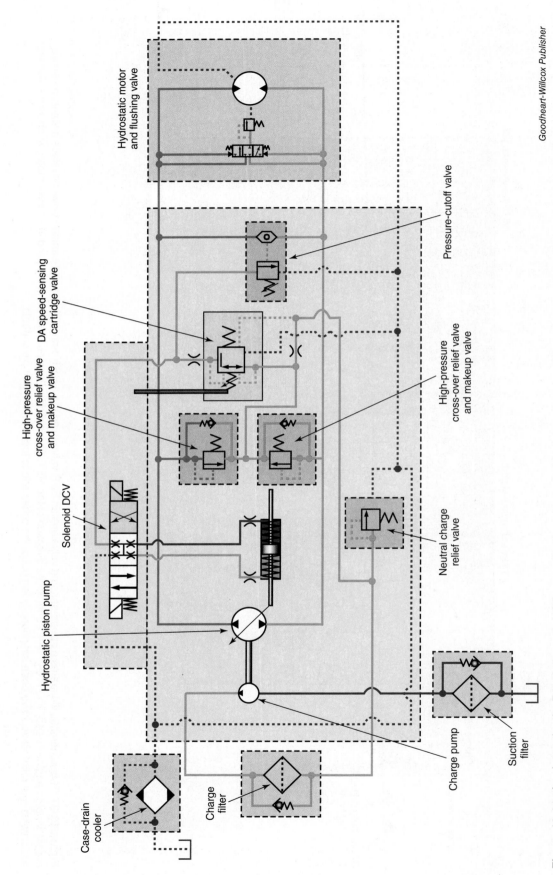

Figure 13-28. A schematic of a Bosch Rexroth HST pump that is equipped with electronic solenoids and a speed sensing DA cartridge valve.

The table in **Figure 13-29** shows how engine speed affects pilot control pressure, the pump's swashplate angle, which is also called pump swivel angle, especially if the pump is a bent-axis pump. At idle, the pilot pressure is approximately 72 psi (5 bar) and the pump swivel angle is approximately 2%. At wide-open throttle, the pilot control pressure is approximately 406 psi (28 bar) and the pump's swivel angle is 100%.

The pump with DA control also has a unique small adjustment that can be made to the rotating group's valve plate. This adjustment is explained in the next chapter.

Hydrostatic Drive Operation and Oil Flow

A sequence of hydraulic events takes place to propel an HST. The charge pump draws oil from the reservoir into the charge pump's inlet. The charge pump delivers charge oil to the following three areas:

- The main piston pump's inlet.
- The control valve orifice.
- The transmission's closed loop.

Servo Oil

The control valve's orifice supplies oil to the spool valve. If the propulsion lever is in the neutral position, the spool valve blocks oil flow. When the operator strokes the propulsion lever, the spool valve is actuated and routes **servo oil**, sometimes called **control oil**, to actuate a servo piston. As the servo piston moves, it causes the piston pump's swashplate to stroke the pump. See **Figure 13-30**.

Note that the piston pump is a reversible pump. The engine drives the pump in a specific direction, either clockwise or counterclockwise. The operator chooses which direction the oil flows by moving the propulsion lever either forward or reverse. HST pumps are the most common application of reversible pumps.

As the piston pump swashplate is actuated, the piston pump draws oil into its inlet and pushes oil out its outlet. The pump's inlet and outlet swap functions when the swashplate is reversed.

The drive oil leaving the pump is sent to the motor to propel the machine. As oil enters the hydrostatic motor, it exerts pressure against the pistons inside the motor's rotating group. Once the drive pressure reaches the threshold necessary to propel the machine, the pistons are pushed against the swashplate. This causes the rotating group to spin, which, in turn, rotates the motor's output shaft.

Effect of Engine Speed on Pump			
Engine Speed	**Pilot Control Pressure**	**Swivel Angle**	**Flow**
1050 rpm	72 psi (5 bar)	2%	0 gpm (0 lpm)
1200 rpm	116 psi (8 bar)	7%	0 gpm (0 lpm)
1500 rpm	174 psi (12 bar)	17%	4 gpm (15 lpm)
1900 rpm	290 psi (20 bar)	71%	20 gpm (76 lpm)
2250 rpm	406 psi (28 bar)	100%	33 gpm (126 lpm)

Goodheart-Willcox Publisher

Figure 13-29. A graph showing how a Bosch Rexroth DA-controlled HST pump's pilot control pressure, pump flow and pump swivel angle all increase as the engine speed increases.

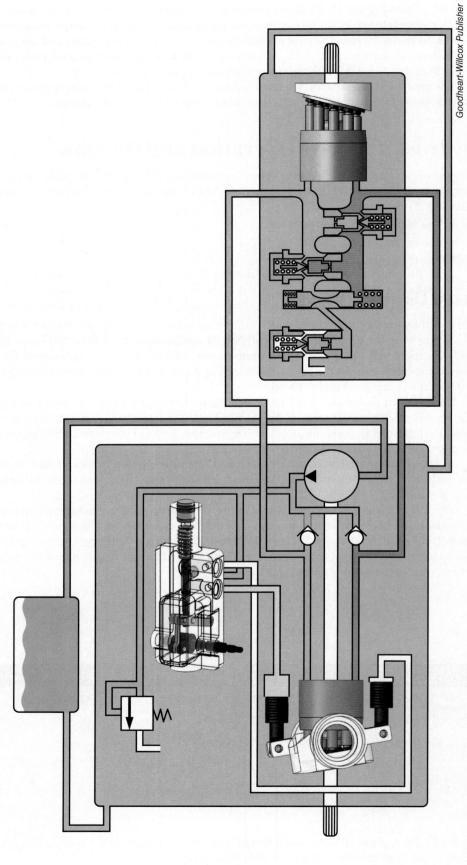

Figure 13-30. An example of an Eaton Series-1 dual-servo HST pump operated by a control valve. The pump directs oil to the motor causing it to rotate.

Flushing Valve

Most closed-loop HSTs are equipped with a shuttle valve that is used to flush the oil from the motor and pump, which aids in cooling the transmission. The shuttle valve may be referred to as a *flushing valve*, *replenishing valve*, or *hot-oil purge valve*.

The flushing valve is normally located inside the motor because the HST motor can have more bearings than the HST pump, especially if it is a bent-axis HST motor. Bearings generate heat and raise oil temperatures. However, in rare applications, the manufacturer places the flushing valve inside the pump. It is critical to know the location of the flushing valve for diagnostic purposes. This is explained in Chapter 14, *Hydrostatic Drive Service and Diagnostics*.

The flushing valve senses both legs of the closed-loop transmission. When the transmission is propelling the machine, one leg becomes drive pressure and the other leg remains as charge pressure. Because drive pressure is higher than charge pressure, the shuttle valve will shift. As soon as the shuttle valve shifts, the shuttle valve opens, allowing the charge oil to act on a lower-pressure charge relief valve. This lower-pressure charge relief valve is similar to the charge relief valve in the pump, except that it is set approximately 30 psi (2 bar) lower.

The charge relief valve in the pump can be called the *neutral relief valve* because it is in command whenever the transmission is in neutral. The charge pressure is highest when the machine is in neutral and the engine is at high idle. The lower-pressure charge relief valve inside the motor acts as the charge relief for forward and reverse propulsion. Although some manufacturers call this relief a *charge relief*, the valve may also be called the *flushing relief valve*. See **Figure 13-31**. The lower-pressure flushing relief is located in the same housing as the flushing shuttle valve, both normally in the motor or in a rarer scenario, both in the pump.

If the transmission is in neutral, the shuttle valve remains in a balanced state with charge pressure acting on both sides of the shuttle valve. In this state, the shuttle valve blocks charge oil from acting on the flushing relief valve. As a result, the neutral charge relief valve keeps HST charge pressure at the 30 psi (2 bar) higher value.

Any time the propulsion lever is stroked, drive pressure builds, causing the shuttle valve to shift, resulting in a 30 psi (2 bar) pressure drop in charge pressure. In **Figure 13-30**, the excess charge oil, which is not needed by the transmission, is dumped into the case of the motor to purge the transmission's hot oil. The case drain oil in the motor is then routed to the case of the pump, and the case drain of the pump is then routed to the reservoir. As a result, oil is purged from both the motor and the pump any time the transmission is in forward or reverse. See **Figure 13-32**.

The schematic in **Figure 13-33** shows the flushing shuttle valve and the lower-pressure flushing relief valve in the motor.

Deceleration

When the machine is propelled down a slope or if the propulsion lever is returned to the neutral position, the hydrostatic transmission

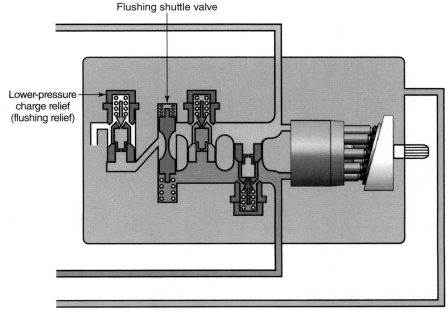

Goodheart-Willcox Publisher

Figure 13-31. The flushing shuttle valve senses both closed-loop pressures. As soon as the HST moves forward or reverse, the shuttle valve shifts, causing the charge pressure to be controlled by the lower-pressure charge relief valve.

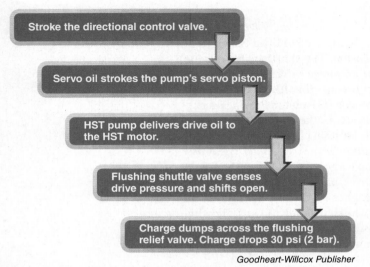

Figure 13-32. A sequence of actions is required in order for the HST charge pressure to drop 30 psi (2 bar).

decelerates, providing engine braking. During this condition, the motor is driven by the machine's momentum. The shuttle valve shifts in the opposite direction. After deceleration, once the pump begins driving the motor again, the shuttle valve shifts back to the original position.

High-Pressure Relief Valves

If the machine's load becomes excessive or stalls, two high-pressure relief valves provide circuit protection by relieving the high pressure. One relief valve protects the system from forward drive pressure, and the other relief valve protects the system from excessive reverse drive pressure. The relief valves are sometimes called cross-over reliefs because they dump directly into the opposite leg of the closed-loop transmission, which is the charge circuit. See **Figure 13-34**. Note that some machine manufacturers design the wheels or tracks to lose traction before the HST drive pressure stalls. Although the slipping traction greatly increases wear on tires and tracks and should be avoided, it does reduce the number of HST stalls. This reduces excessive heat and wear, helping to prevent premature transmission failure.

In Eaton heavy-duty Series-1 HST motors and Sauer Sundstrand (Danfoss) Series-20 motors, the high-pressure relief valves are in the motor block along with the shuttle valve and

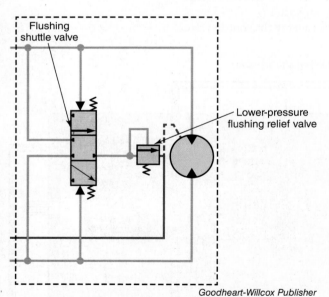

Figure 13-33. The flushing shuttle valve senses both legs of a closed-loop HST. When the HST is in forward or reverse, the shuttle valve will shift up or down, which will allow charge oil to be controlled by the lower-pressure flushing relief valve.

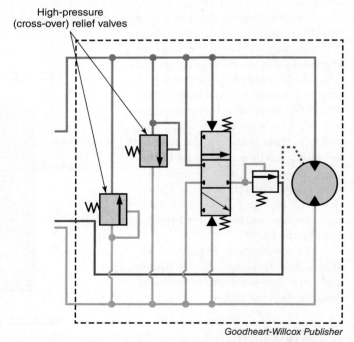

Figure 13-34. The yellow highlighted cross-over relief valves will dump the high-pressure drive oil into the charge loop if the drive pressure reaches the high-pressure relief valve setting. These cross-over relief valves are illustrated inside the motor. Examples include Eaton Series-1 motors and Danfoss Series-20 motors.

the flushing relief valve. See **Figure 13-35**. High-pressure relief valves can also be in the pump, depending on the manufacturer. Late-model transmissions use electronic controls to prevent stalling of the motor. These transmissions rely on only the high-pressure relief valves for spike or surge protection. Late-model electronic controls are discussed later in this chapter.

Figure 13-36 provides flow charts for the path of oil flowing through an HST.

Multi-Function Valves

Cross-over reliefs are commonly designed as cartridge valves. The Danfoss Series-90 HST pumps integrate multiple valves within a single cartridge valve, including a make-up check valve, a bypass valve, a pressure limiter valve, and a cross-over relief valve. Danfoss calls the

Goodheart-Willcox Publisher

Figure 13-35. An example of an Eaton Series-1 fixed-displacement HST motor that is commonly used on older combines, cotton harvesters, and concrete mixer trucks. The rectangle block contains two high-pressure relief valves, a lower-pressure flushing relief valve, and a shuttle valve.

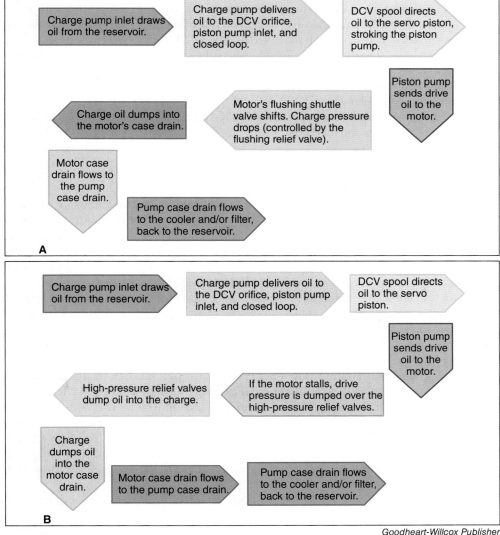

Goodheart-Willcox Publisher

Figure 13-36. HST oil flow diagrams. A—This flowchart depicts the low-pressure oil path through an HST. B—This flowchart depicts oil flowing through an HST, including the high-pressure oil path.

valve assembly a *multi-function valve*. Multi-function valves are used in other applications, such as hydraulic cylinders. The Danfoss Series-90 multi-function valve, as it relates to its pressure limiter valve, is explained later in this chapter.

Reverse Propulsion

When the operator reverses the propulsion lever, the servo valve's spool sends oil to the opposite servo piston causing it to reverse the pump's swashplate angle. The pump reverses its oil flow, causing the motor to rotate in the opposite direction. See **Figure 13-37**.

Hydrostatic Transmission Filtration and Cooling

Earlier in the chapter, it was explained that most closed-loop HSTs do not use closed-loop filtration because the filters would have to filter oil in both directions and be able to withstand high pressures. On older machines, transmissions commonly had suction strainers placed in the reservoir and a suction filter prior to the charge pump. With those old designs, the pump inlet would initially be protected from contamination. However, it is difficult to determine if a suction strainer is plugged. Suction strainers are also difficult to replace because they require the reservoir to be drained. Plugged suction strainers and suction filters and poor maintenance practices can cause pump cavitation, resulting in catastrophic harm. If the transmission has suction filtration, a bypass valve should be used to reduce the risk of catastrophic failure. See **Figure 13-38**.

Many machines that incorporate filter bypass valves will use a warning system to alert the operator when the filter is bypassing the oil. Some suction filters that have a bypass valve will use a filter indicator. The indicator needs to be regularly checked to determine if the filter needs to be replaced.

If the transmission does not use closed-loop filtration or suction filtration, it can employ three other filtration methods:

- Case drain filtration.
- Charge pressure filtration.
- Off-line kidney-loop filtration.

Some manufacturers choose to filter case drain. There are two challenges with filtering case drain. The first is that case drain is also commonly used for cooling. The backpressure of filtering case drain plus the backpressure caused by the case drain oil cooler can cause the case pressure to rise. Most case drain pressures for pumps and motors are quite low, for example 15 psi (1 bar) or less. However, cooling and filtering case drain flow can cause case pressures to rise higher than 45 psi (3 bar). The rise in case pressure poses two problems. One concern is the potential for a motor or pump shaft seal to start leaking. The second concern is that high case pressure is often the result of too much internal leakage. This can cause a technician to incorrectly diagnose a transmission as faulty when, in reality, the filter or the cooler is plugged, causing too much backpressure. See **Figure 13-39**.

Filtering the charge pump flow also creates two potential challenges. The first is that the filter must be able to withstand a pressure as high as 300 psi (20 bar). The second is that most charge pumps are driven in tandem directly off the back side of the piston pump, and the pump flow is often routed internally through the housing. If the charge pump flow needs to be filtered prior to the pump's inlet, the charge pump flow must be routed outside of the tandem pump housing and then back into the pump housing. See **Figure 13-40**. Late-model machines are commonly configured with charge-pump filters.

Many of today's machines use *off-line kidney-loop filtration*, which is a bypass filtration system that is separate from the main hydraulic system and hydrostatic drive system. It

Chapter 13 | Hydrostatic Drives

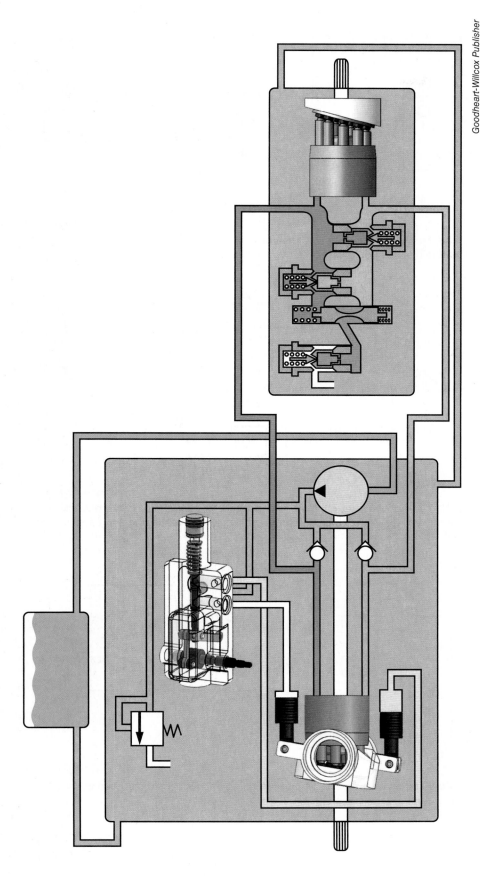

Figure 13-37. An example of an Eaton Series-1 HST. When the HST lever is stroked in a rearward direction, the reverse servo piston causes the pump to deliver reverse drive pressure to the motor. As a result, the machine moves in a reverse direction.

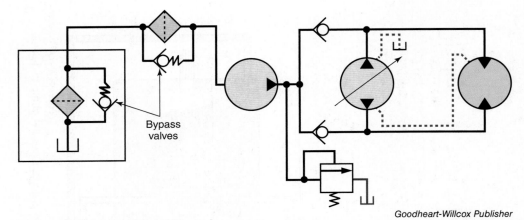

Figure 13-38. Many HSTs use suction strainers and/or suction screens. At a minimum, any suction filtration device should contain a bypass valve.

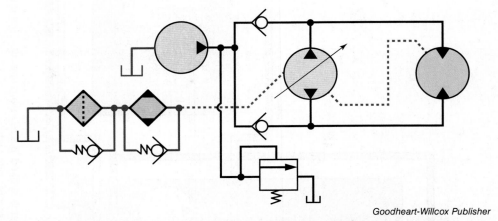

Figure 13-39. If a manufacturer chooses to cool and filter case drain oil, the restriction of the cooler and filter will elevate the case drain pressure. As a result, the shaft seal is more likely to leak.

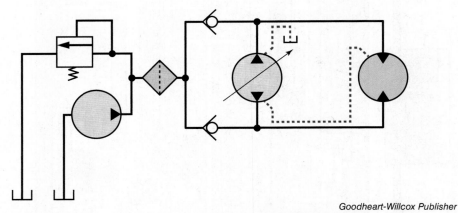

Figure 13-40. Filtering an HST charge pump requires the filter to be plumbed in series between the pump and the closed loop.

is powered independently. It uses a very efficient filter to filter the fluid inside the reservoir. A filter cart can be connected to the hydraulic reservoir to filter the reservoir's oil when the machine is shut down for inspection, maintenance, or repair. See **Figure 13-41**.

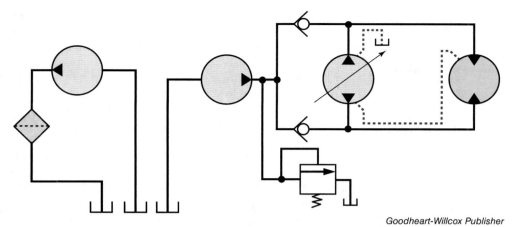

Figure 13-41. Off-line kidney-loop filtration can filter particles to a very small micron rating as compared to the other filtration options.

Variable Hydrostatic Drive Motor Applications

Practically all hydrostatic drives use a variable-displacement pump that allows the operator to change the machine's travel speed. However, sometimes the pump does not provide enough variation of travel speed. For that reason, some manufacturers offer a variable-displacement hydraulic motor.

Two-Speed Motor Applications

Many machines are equipped with a two-speed HST that uses a variable-displacement pump to drive a variable-displacement motor that contains two specific displacements. The motor is configured with a large displacement for low speed and a smaller displacement for high speed. The two specific displacements give the HST two operating ranges, low speed and high speed. Technically, the motor has variable displacement. However, the motor is not infinitely variable. It has only a low-speed position and a high-speed position. Sometimes the low-speed position is called *field speed*, as it is used during the normal operation of the machine while operating in the field. The high-speed position can be called *road speed*, because it is often used when the machine is driving from one field to another field.

Some examples of machines and manufacturers that currently use or have used two-speed motors are skid steers, track loaders, John Deere dozers (450J, 550J, and 650J), Case IH combines (1600, 2100, and 2300 series), and excavators. Two-speed motors can have one of three types of motor frames. Inline axial piston motors are commonly used on combines and dozers. Bent-axis piston motors are used in dozers. Radial piston (cam-lobe) motors are used in skid steers and combine rear power drive axles.

Two-speed HST transmissions can also achieve two separate speeds by having an ECM vary the pump to provide two specific pump displacements. However, most two-speed HSTs use a two-speed HST motor to achieve the different speeds.

Eaton Two-Speed Axial Piston Motor

Eaton has offered axial two-speed piston motors for decades. These motors are commonly used in agricultural combines. On the Case IH combine, the low-speed swashplate angle is 18° and the high-speed swashplate angle is 15°. A solenoid is used to control the motor's two displacements. The Case IH 2300 series combine uses a separate oil source, with regulated pressure, to supply the oil to the motor's control valve. See **Figure 13-42**.

When the solenoid is de-energized, a regulated-gear pump supplies 300 psi (21 bar) of pressure to act on the spool, causing the spool to shift against its spring. As the spool shifts, it directs servo oil to the low-speed servo (S1), which places the motor in the low-speed position.

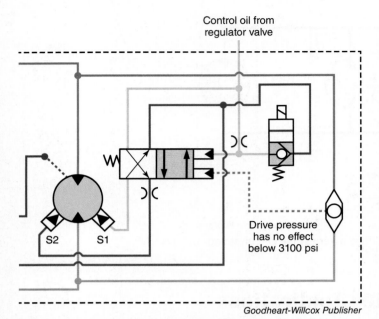

Figure 13-42. An Eaton Series-1 two-speed motor. When the two-speed solenoid is de-energized, the control pressure causes the spool valve to shift. This results in servo oil actuating the low-speed servo (S1), which increases swashplate angle.

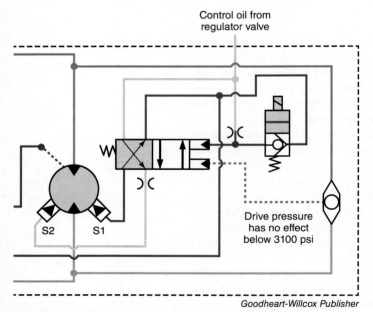

Figure 13-43. An Eaton Series-1 two-speed motor. When the two-speed solenoid is energized, control oil is drained from the spool valve, causing the spring to shift the spool to the right. Control oil is then directed to the high-speed servo (S2), causing the motor's swashplate to shift to the decreased angle.

When the operator selects the motor's high position, the motor's solenoid is energized. See **Figure 13-43**. The solenoid drains the control oil that was previously acting on the spool. The spool's spring shifts the valve, which causes the valve to direct servo oil pressure to the motor's high-speed servo (S2). The solenoid shifts the swashplate to the high-speed position, which is a decreased swashplate angle.

While in the high-speed position, a shuttle valve is used to sense reverse or forward drive pressure. When drive pressure reaches a threshold of 3100 psi (214 bar), it causes the valve to downshift the motor to the low-speed position. To summarize:

- 3100 psi (214 bar) of drive pressure will force the motor to shift to low.
- If the solenoid is de-energized, 300 psi (21 bar) of regulated pressure will cause the motor to shift to low.

Based on that summarization, a person might ask how it is possible that a pressure as low as 300 psi (21 bar) has the same effect as a pressure ten times that amount, 3100 psi (214 bar). The answer is the pressures are being applied to different surface areas. See the cross-sectional drawing in **Figure 13-44**.

When the motor is in the low-speed position, the solenoid is de-energized. As a result, 300 psi (21 bar) of control oil pressure acts on the control spool, which shifts the pressure response spool. The pressure response spool then directs servo oil to the S1 low-speed servo piston.

The solenoid must be energized to shift to the high-speed position. See **Figure 13-45**. When the solenoid is energized, it connects the two passageways, allowing the control oil to drain to the reservoir. The pressure response spool spring shifts the spool to the right, which opens the passageway for servo oil to be sent to the high-speed servo (S2).

The cross-sectional drawing in **Figure 13-45** shows that the control spool is shifted to the right when the motor is in the high-speed position. Notice a shuttle valve and needle roller are located on the right side of the control valve. The shuttle valve senses if the vehicle is moving forward or reverse. When drive pressure reaches 3100 psi (214 bar), the drive pressure forces the needle roller to the left, shifting the control spool and the pressure response spool, resulting in the motor downshifting to the low-speed position.

Cross-sectional drawings provide service personnel a view that is helpful for diagnostics. For example, a combine equipped with an Eaton two-speed motor repeatedly blew the gasket that was located between the control spool block and the needle roller block. Every time a new gasket was installed, the gasket would fail each time the propulsion handle was stroked. Using a cross-sectional drawing, it was determined that if the needle roller was missing, the high pressure could no longer be isolated from the lower, regulated pressure. Upon disassembly of the valve, the technician found that the needle roller was indeed missing. See **Figure 13-46**.

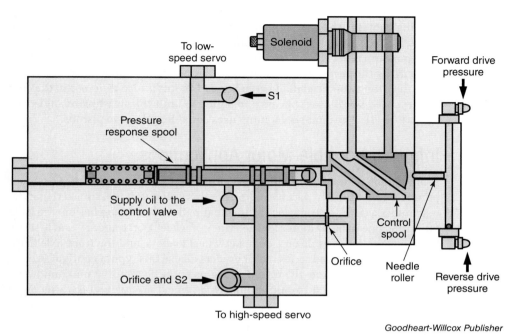

Figure 13-44. This cross-sectional drawing of an Eaton Series-1 two-speed motor control valve illustrates that the solenoid is de-energized, which allows the solenoid to direct oil to act on the control spool. Notice that drive pressure acts on a needle roller which has an area ten times smaller than the control spool.

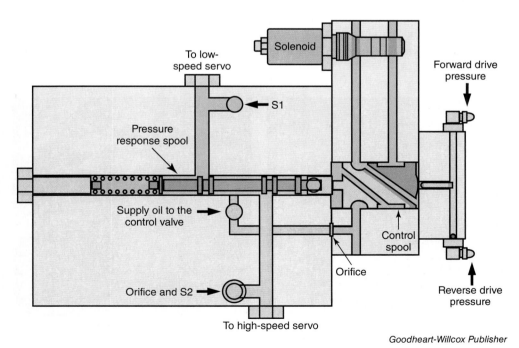

Figure 13-45. An Eaton Series-1 two-speed motor control valve. When the solenoid is energized, the control oil is drained to the reservoir. The spring shifts the spool, causing servo oil to be sent to the high-speed servo.

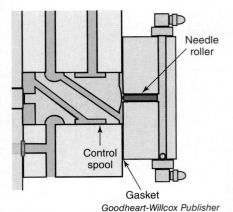

Figure 13-46. A cross-sectional drawing can help diagnose problems. In this case, it allowed a technician to determine that if the needle roller was missing, high-pressure oil would be directed into a low-pressure cavity, causing the gasket to rupture.

Identifying an Eaton Dual-Servo Pump and Eaton Dual-Servo Motor

The Eaton dual-servo two-speed Series-1 motor housing looks similar to the Eaton dual-servo Series-1 piston pump housing. At first glance, the average person will not be able to distinguish between the two housings. However, the pump housing contains the charge pump, and the motor housing contains the valve block. The valve block includes the flushing shuttle valve, flushing relief valve, high-pressure relief valves, and two pressure jumper hoses used to sense drive pressure for the purpose of downshifting the motor. See **Figure 13-47**.

An Eaton two-speed dual-servo Series-1 motor is also different in that the S1 low-speed servo uses spacers and shims to limit the swashplate angle. See **Figure 13-48**. The dual-servo pump uses equal-length servo pistons.

Infinitely Variable Motor Applications

Numerous machines are configured with an infinitely variable-displacement pump and an infinitely variable motor. For this example, it is assumed that if the machine is a dozer or track loader, the dual-path transmission has one variable pump and one variable motor for each track or wheel. Caterpillar D- and K-series track loaders, K-series dozers, compact wheel loaders, and the front axle drive on M- and non-suffixed series motor graders all use this type of configuration.

When accelerating from a stop, the machine begins with the motor at maximum displacement and the pump stroked to minimum displacement. While trying to increase travel speed, the ECM will first fully upstroke the pump before destroking the motor. If the pump has achieved maximum displacement and the operator has requested more travel speed, the ECM will begin to reduce the motor's displacement. See **Figure 13-49**.

Variability by Using One or Two Variable-Displacement Motors

Earlier in the chapter, it was explained that the John Deere X9 combines and Caterpillar 924K, 926M, 930K, 930M, 938K, and 938M wheel loaders use one variable-displacement pump and two variable-displacement motors. In the Caterpillar wheel loader example,

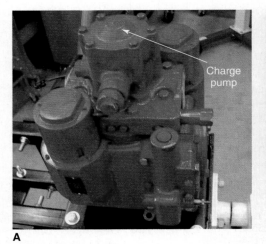

Figure 13-47. A—An Eaton dual-servo Series-1 HST pump has a charge pump coupled to the back of the pump housing. B—An Eaton Series-1 two-speed HST motor uses the same motor housing as the pump housing, except the motor contains a rectangular block with the flushing valves, high-pressure relief valves, and the two high-pressure hoses.

Figure 13-48. An Eaton Series-1 two-speed HST motor has spacers on one servo and no spacers on the opposite servo.

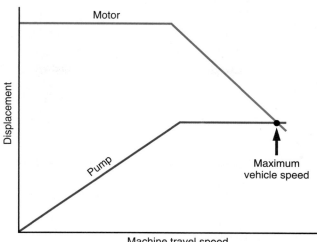

Figure 13-49. This graph illustrates that an ECM will first fully upstroke the pump to gain travel speed. If more travel speed is required after the pump has been fully upstroked, the motor displacement will be decreased.

both motors provide power into a single gearbox that splits power flow to the front and rear axles. One motor has a small displacement, and the other motor has a larger displacement.

Both motors are plumbed in parallel and always receive oil from the same hydrostatic pump. However, at high travel speeds the ECM nullifies the large-displacement motor, making it ineffective. The ECM must perform two tasks to nullify the large-displacement motor. The large-displacement motor is coupled to the gearbox through a clutch mechanism. The ECM disengages the clutch to prevent the larger motor from providing input into the gearbox. If the motor were allowed to simply freewheel, the pump's oil flow would take the path of least resistance. Therefore, the second task the ECM must perform is to destroke the motor's swashplate to a neutral angle. When an axial piston motor swashplate is placed in an exact neutral angle, it can no longer rotate even if it is receiving pump flow.

The small motor is always in use and will be rotating any time the wheel loader is moving, regardless of the travel speed. When the wheel loader requires more torque, the pump drives both hydrostatic motors, which increases the total motor displacement. As a result of using both motors, torque is increased and travel speed is reduced. When the machine is traveling at high speed, only the small motor is used. The large motor does not provide power into the gearbox. See the graph of the displacements of the pump and motors in **Figure 13-50**.

The wheel loaders also offer a ***creeper control***, which provides very slow travel speeds while allowing a large amount of implement-hydraulic flow to operate attachments such as brooms, brush cutters, or snow blowers. In range 1, the customer can limit the maximum travel speed to as slow as 0.6 mph or as fast as 8 mph. The default maximum travel speed in range 1 is 4.4 mph. The operator can choose one of four travel speed ranges listed in **Figure 13-51**.

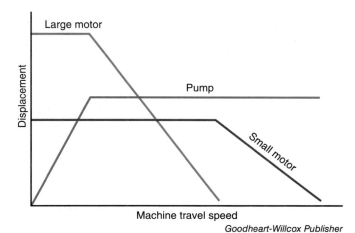

Figure 13-50. On the Caterpillar 924K, 926M, 930K, 930M, 938K, and 938M wheel loaders, the ECM will first fully upstroke the pump to increase travel speed. If the loader needs more travel speed, the ECM will begin reducing the large motor's displacement. If more travel speed is required, the ECM will finally reduce the small motor's displacement.

Mode	Maximum Travel Speed (mph)	Maximum Travel Speed (km/h)	Application
Range 1	0–8.0 mph	0–13 km/h	Creeper–operations requiring high hydraulic flows, like brooms and snowblowers
Range 2	0–17 mph	0–27 km/h	Truck loading
Range 3	0–25 mph	0–40 km/h	Load and carry
Range 4	0–25+ mph	0–40+ km/h	Roading (transport)

Goodheart-Willcox Publisher

Figure 13-51. Caterpillar 924K, 926M, 930K, 930M, 938K, and 938M wheel loaders offer four ranges of travel speeds. Each range allows an operator to vary the machine speed from a stop to the maximum travel speed for that range.

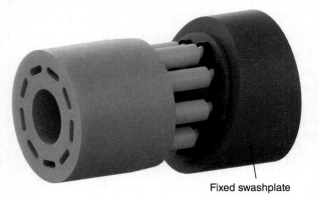

Fixed swashplate

Goodheart-Willcox Publisher

Figure 13-52. A fixed-displacement, single-speed HST motor. Examples include Eaton Series-1 fixed-displacement motors and Danfoss Series-20 fixed-displacement motors.

Note
The loader has two foot pedals that enable the operator to control the engine speed and travel speed independently. The left foot pedal controls the machine's travel speed. The right foot pedal controls the engine speed.

Chapter 15, *Continuously Variable Transmissions*, explains the power flow through mechanical gearboxes on John Deere X9 combines and Caterpillar 924K through 938M series wheel loaders.

Single-Speed, Fixed-Displacement Hydrostatic Motor

Many hydrostatic transmissions use a simple fixed-displacement motor, which is also known as a single-speed motor. If the motor is an inline axial motor, **Figure 13-52**, it contains a fixed swashplate. With this type of motor, travel speed is changed by adjusting the variable pump.

A fixed-displacement motor is the simplest design used in off-highway equipment. It does not require any controls such as electrical current or control oil, because the pump is responsible for the change of speed. The motor simply rotates at the speed and direction commanded by the pump. A single-speed motor poses one potential problem when it is being replaced. Due to the symmetrical shape of the motor, it is easy for the technician to accidentally install the new motor upside down, which can cause a dangerous situation if not noticed.

Warning
If a motor is installed upside down, the swashplate will also be upside down. This will cause the propulsion lever to work opposite of the expected way. This problem can also occur if the technician disassembled the motor to replace the output shaft and accidentally reinstalled the swashplate upside down. It is possible, but less likely, for the same symptom to occur if a technician installs the closed-loop drive hoses to the wrong ports. To prevent this problem from occurring, the inlets and outlets of the motors are labeled, for example port "A" and port "B". Be sure to install the motor and drive lines in the original configuration to prevent the HST from operating backwards. When disassembling a motor, be sure to reinstall the swashplate back in its original position.

Additional Hydrostatic Valving

HSTs can be equipped with additional types of control valves that offer different features. The following controls are commonly used in mobile HST applications.

Inching Valve

Some HSTs are equipped with an *inching valve*. The valve is commonly operated by a foot pedal and acts similarly to a transmission clutch. The valve can perform the following functions:

- Provide a method to control the steady acceleration from a stop, by slowly releasing the foot pedal after the propulsion lever has been actuated.
- Enable the operator to slowly inch up to an implement to ease the installation of the implement.
- Allow an operator to coast to a stop.
- Enable the operator to disengage the transmission and use the service brakes for stopping.

Some machines use the hydrostatic motor to drive a two-, three-, or four-speed gearbox that provides the machine additional operating ranges. On these machines, the gearing typically does not use synchronizers, and the operator's manual specifies to change the ranges only when the machine is stopped. If the machine is equipped with a two-, three-, or four-speed gearbox, depressing the inching valve can help the operator change ranges.

Inching Valve Located inside the Pump

The inching valve can be designed to work in conjunction with the pump or the motor. If the inching valve is incorporated with the pump, it is located inside the pump's servo control valve. See the schematic in **Figure 13-53**. As the inching valve is operated, it connects the forward servo port to the reverse servo port, which causes the servo pistons to return to the balanced neutral position.

Inching Valve Used in Conjunction with the Motor

The foot-and-inch valve on older Case IH combines, for example 1600, 2100, and 2300 series, was used in conjunction with the hydrostatic motor. For this style foot-and-inch valve system, the pedal valve worked in combination with the motor's two high-pressure relief valves. Together, all three valves formed a pilot relief system. See **Figure 13-54**.

The foot-and-inch valve sets the pilot pressure that is held against both high-pressure relief valves. When the pedal is pressed, the pilot oil is dumped, as a result, the high-pressure relief valves drop to a very low value and the motor is no longer able to develop enough torque to propel the machine.

On the style of foot-and-inch valve system shown in **Figure 13-54**, technicians can adjust the transmission's high-pressure setting for both forward and reverse by adding or removing shims inside the foot-and-inch pedal valve. As a word of caution, HSTs have been damaged as a result of a technician adjusting the pressure too high. In one scenario, a customer capped off the line going to the foot-and-inch valve, which resulted in the customer destroying the transmission.

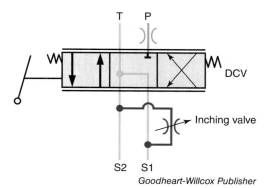

Figure 13-53. If the inching valve is placed inside the pump, it will connect the forward and reverse servo ports when actuated, neutralizing the pump's flow.

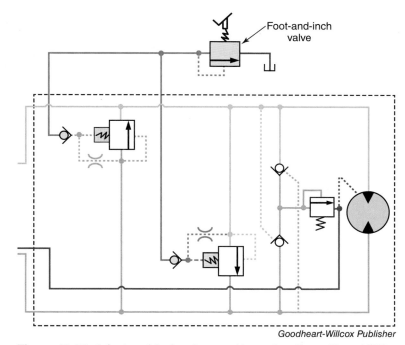

Figure 13-54. A foot-and-inch valve used in conjunction with an HST motor will hold pilot pressure against the high-pressure relief valves. When the foot-and-inch pedal is depressed, the high-pressure relief valve values fall to practically no pressure, which neutralizes the HST.

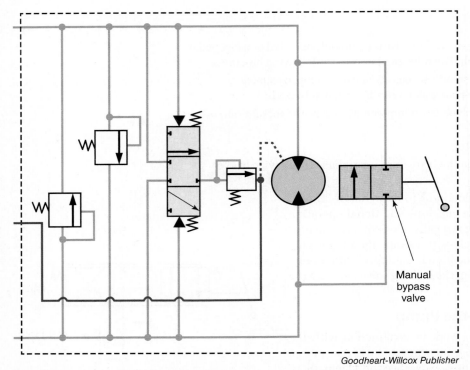

Figure 13-55. A manual bypass valve will connect the two drive loops of a closed loop HST, neutralizing the HST.

Manual Bypass Valve

A motor can also be equipped with a *manual bypass valve*. See **Figure 13-55**. The valve allows the fluid from one leg of the drive loop to be bypassed into the other leg of the loop. The valve is used in several situations:

- To ease shifting of a two-, three-, or four-speed multi-gear transmission.
- To allow an inoperative machine to be pulled into a shop.
- To deactivate the hydrostatic transmission for safety purposes.

The manual bypass valve is a hand-operated rotary valve. It is designed to be either fully open or fully closed. The valve is to be used only when the pump is in neutral and the machine is stationary. The valve will unlock the motor's output shaft, enabling it to be rotated. During normal hydrostatic transmission operation, the valve will be closed.

The Danfoss Series-90 HSTs are equipped with two multi-function valves, and each contains a manual bypass valve. The multi-function valve is explained later in this chapter. The next chapter explains how to properly operate the manual bypass valve. Manual bypass valves are not designed for long-distance high-speed towing, but simply to move a broken machine a short distance at a low speed.

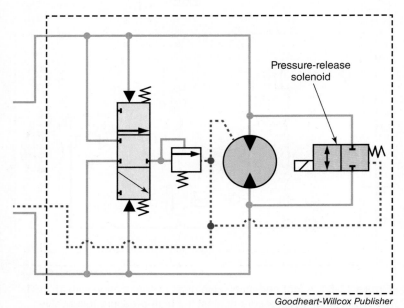

Figure 13-56. A pressure-release solenoid will connect the two drive loops together, neutralizing the HST.

Electronic Pressure-Release Solenoid

HSTs can use a solenoid similar to the manual bypass valve. On a Case IH 7120-9120 series combine, the solenoid is called a *pressure-release solenoid*. When energized, the solenoid connects the two legs of the closed loop, allowing the motor to freewheel, which makes it easier to shift the transmission gearbox. When the propulsion lever is in neutral and the operator requests a new speed range in the gearbox, the ECM will automatically energize the pressure-release solenoid to make it easier to shift the gearbox. See **Figure 13-56**.

Eaton Internal Pressure Override

Eaton Series-1 HST pumps can be equipped with an *internal pressure override (IPOR) valve*. The valve is located inside the pump housing and placed in series between the charge pump and the inlet to the drive pump's servo control valve. The IPOR senses forward and reverse drive pressures. When the pressure reaches the IPOR valve's setting, the valve limits the supply oil to the servo control valve, which causes the pump to return to a neutral state.

A cross-sectional drawing of the valve is shown within a schematic in **Figure 13-57**. Drive-loop leg A and drive-loop leg B are acting on small pins. When one of the drive pressures is high enough to overcome the IPOR valve spring's value, the drive pressure forces the IPOR valve to move to the left. When the IPOR valve shifts it depletes the supply control oil that was previously being delivered to the HST DCV and neutralizes the pump. The IPOR valve acts like a high-pressure limiter. **Figure 13-58** is a simplified schematic of an IPOR valve. Notice that the orifice is located before the HST DCV. It is necessary, otherwise if the IPOR valve were to shift, it would deplete charge oil. The orifice allows the IPOR valve to deplete only the supply oil to the DCV, while maintaining charge pressure.

The IPOR valve is used to protect the machine from high pressure overloads for extended periods of time. Combines and cotton harvesters are a couple of examples that have used IPOR valves. An operator unfamiliar with this style of control might state that the machine is malfunctioning because the transmission loses power at high pressure settings. In this scenario, if the machine is equipped with a multispeed gearbox, the transmission

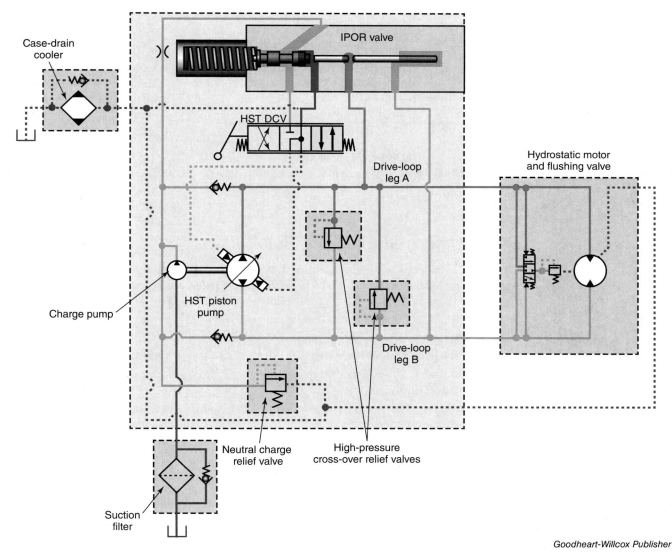

Goodheart-Willcox Publisher

Figure 13-57. A cross sectional drawing of an Eaton internal pressure override valve inside an Eaton Series-1 HST pump.

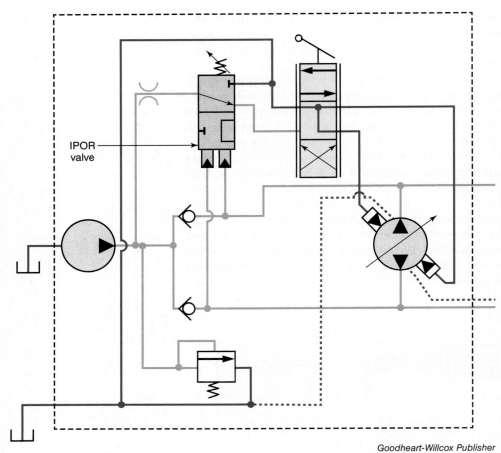

Figure 13-58. The Eaton Series-1 HST pump's IPOR valve senses high pressure and limits the supply oil to the servo control valve when the high pressure reaches the IPOR setting.

should be shifted to a lower gear ratio.

Older Case IH rice combines, such as 1600-, 2100-, and 2300-series, utilized IPOR valves. However, the corn and grain combines did not utilize IPOR. Older Case IH cotton harvesters also were equipped with an IPOR valve. Some older John Deere combines, for example 50-series STS, also utilized an IPOR valve. However, the description is simply listed as pressure override, not IPOR, and the machine is not distinguished by a specific crop designation, like corn, grain, or rice.

Danfoss Series-90 180cc pumps offer, as an option, a pressure override (POR) valve, which acts similarly to the Eaton IPOR valve. The pump uses a shuttle valve that senses both closed-loop drive legs. The shuttle valve sends the drive pressure to the POR valve. When drive pressure exceeds the POR valve setting, it will deplete the control oil to the servo DCV.

Bosch Rexroth Pressure Cutoff Valve

Case IH flagship combines, for example 7120-9120, 7230-9230, and 7240-9240 series combines, use a similar valve called a *pressure cutoff valve*, manufactured by Bosch Rexroth. It is similar to the Eaton IPOR in the following ways:

- The valve is located inside the pump housing.
- It senses forward and reverse drive pressure.
- It is placed in series between the charge pump and the supply to the servo control valve.
- It limits the supply oil to the servo control valve if drive pressure reaches the cutoff limit.

In this application, the pressure cutoff valve controls the transmission's high pressure during gradual pressure buildup. The pump also contains high-pressure relief valves, known as cross-over relief valves, which protect the system from sudden or rapid pressure increases. See **Figure 13-59**.

The high-pressure relief valves are set 10% higher than the high-pressure cutoff valve setting. Once a high-pressure relief valve opens from a sudden pressure spike, it is possible to see the high pressure drop 10% to the high-pressure cutoff valve setting.

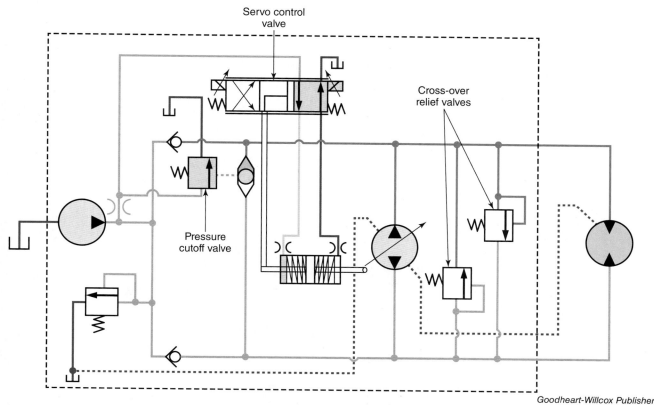

Figure 13-59. A Bosch Rexroth's pressure cutoff valve is similar to an Eaton IPOR. It senses high oil pressure and limits the supply to the servo control valve when pressure reaches the relief setting.

One benefit of the system in **Figure 13-60** is that it contains two separate high-pressure controls in the pump and not the motor. Therefore, if a blockage like a pinched high-pressure line were to occur between the pump and motor, the pump would still have high-pressure relief protection. If the system contained high-pressure relief valves only in the motor, any type of blockage between the pump and motor would result in over pressurization, damaging the pump and/or high-pressure drive plumbing.

The 7250–9250 Case IH flagship combines do not have pump pressure cutoff valves. The same control of limiting pump high pressure by neutralizing the pump is performed electronically instead of using the pressure cutoff valve inside the pump.

Danfoss Series-90 Pressure-Limiter Valve

Earlier in the chapter the Danfoss Series-90 multi-function valves were introduced. These Series-90 pumps contain two multi-function valves. Each includes a make-up valve, a manual bypass valve, and a cross-over relief valve. Some multi-function valves also include a pressure-limiter valve.

The pressure-limiter valve is similar to the Eaton IPOR valve and the Bosch Rexroth pressure-cutoff valve. The pressure-limiter valve is inside the multi-function valve, which is in the piston pump. The pressure-limiter valve senses drive pressure and acts like a high-pressure relief. However, instead of draining the supply oil to the DCV, it performs two other actions to neutralize the pump. When the pressure-limiter valve initially opens, it first acts like a pilot relief valve to the main cross-over relief valve. When the pressure limiter opens, the pilot oil acting on the back side of the cross over relief is depleted, causing the setting of the cross-over relief valve to drop in pressure, similar to the foot-and-inch pedal valves explained earlier in the chapter. The other action that takes place is that the oil dumping across the pressure limiter is directed to the opposite end of the servo piston,

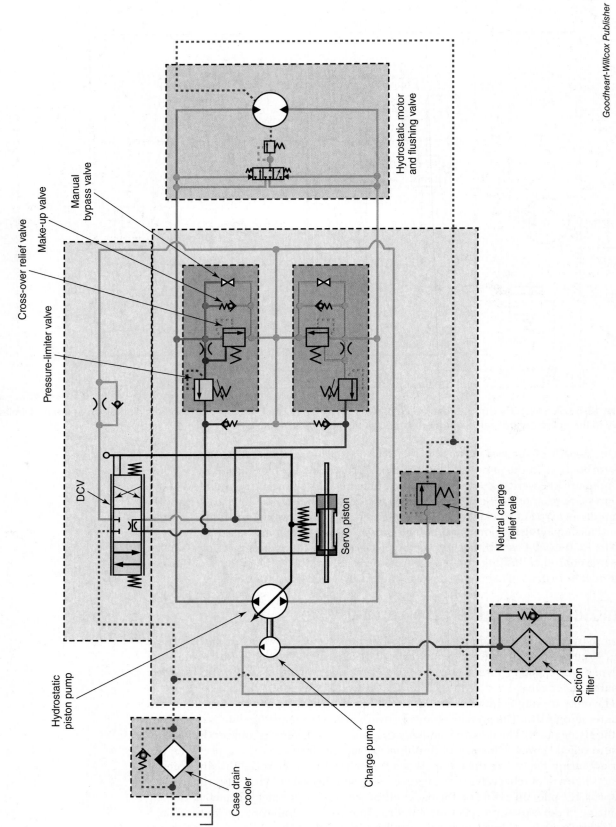

Figure 13-60. A schematic of the Danfoss Series-90 HST pump with multi-function valves that include a pressure limiter valve, cross-over relief valve, manual bypass valve and makeup valve.

which neutralizes the pump when drive pressure reaches the pressure-limiter setting. The cross-over relief valves are also designed to be fast-acting relief valves for high-pressure spikes. See **Figure 13-61**.

Each multi-function valve also contains a manual bypass valve. Its adjustment is explained in the next chapter.

Electronic Anti-Stall Control

Transmissions can be equipped with an *electronic anti-stall control*. The two-position destroke solenoid is normally closed. An ECM monitors the machine's engine speed. When engine speed drops, the ECM modulates the solenoid. As the solenoid is energized, it hydraulically connects the pump's forward and reverse servo circuit, which reduces the pump's displacement. The control allows an ECM to electronically limit stalling of the pump. The larger the difference between the engine speed setting and the actual engine speed, the more the ECM modulates the solenoid. The control enables the operator to operate the machine at full power and reduces the possibility of lugging the engine. See **Figure 13-61**.

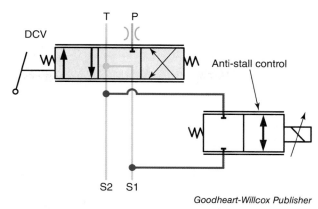

Figure 13-61. An electronic anti-stall control solenoid will prevent the HST from lugging the engine by combining the two servo pressures.

Electronically Controlled Displacement

Figure 13-61 shows an electronic anti-stall solenoid used in conjunction with a manual directional control valve. If the HST pump is electronically actuated, the ECM can simply reduce the current to the HST pump's solenoids, eliminating the need for an anti-stall control solenoid.

Note

John Deere compact utility tractors that are equipped with electronic-controlled HSTs, known as eHydro, are equipped with an engine anti-stall feature called **LoadMatch**. When LoadMatch is turned on, the computer prevents stalling of the engine when the tractor is working under heavy loads. This is helpful when working the loader in heavy applications or pulling an exceptionally heavy load.

Over-Speed Limit Valve

Earlier in this chapter it was mentioned that John Deere X9 combines use one HST pump to drive two HST motors in parallel. This system, like all hydrostatic drives, can have the motor(s) drive the pump during two conditions: when the machine is driving forward and the propulsion is quickly reversed, or when the machine is driving forward down steep hill and the momentum of the machine causes the motors to begin driving the pump. During these conditions, the forward charge pressure becomes reverse drive pressure due to the motors driving the pump. This momentum and high pressure cause pump over speeding and is harmful to the pump, engine, and potentially other drivetrain components. To prevent damage, the X9's ECM actuates the pump to maximum displacement. An integrated speed limit valve is also included in series between the charge closed loop leg and the pump inlet. A pair of orifices route charge pressure to act on both sides of the speed limit valve, balancing it in a normally open position. This pressure, also known as pilot pressure, holds the speed limit valve normally open, allowing the closed loop charge pressure leg to supercharge the pump. A simplified example is shown in **Figure 13-62**.

As the charge pressure oil travels through the normally open speed limit valve, it branches through one orifice that sends oil to deadhead on one end of the speed limit valve.

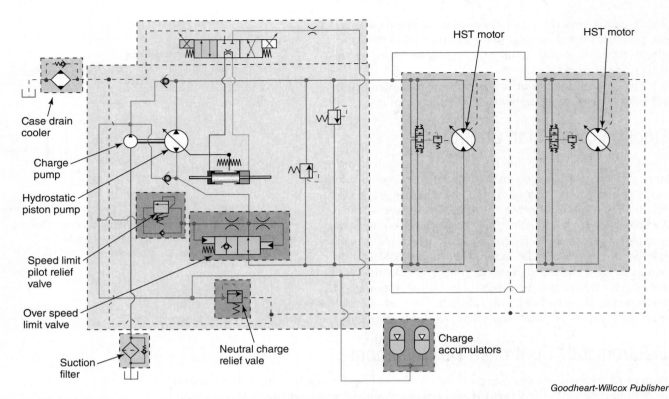

Figure 13-62. An HST schematic of the John Deere X9 combine with an over-speed limit valve.

The other branch has another orifice route the charge oil to the opposite end of the speed limit valve. This end of the spool has a spring, and its charge oil is exposed to an integrated speed limit pilot valve, set at 5080 psi (350 bar). The integrated speed limit valve and the integrated pilot valve work in unison. If the motors start driving the pump, causing the charge pressure to become reverse drive pressure exceeding 1600 psi (110 bar), the speed limit pilot valve opens and the speed limit valve closes. The speed limit valve then blocks the reverse drive pressure from driving the pump and limits the over speeding condition. The high-pressure pilot oil is dumped into the charge circuit. The HST has two charge pressure accumulators to prevent cavitation.

Accumulator Effect

This chapter has detailed numerous mechanisms used to control or limit high drive pressure, such as IPOR, pressure cutoff valves, pressure-limiter valves, cross-over relief valves, and anti-stall destroke solenoids. If a traditional hydrostatic transmission is allowed to stall, and if the closed loop uses hoses instead of steel tubing, it is possible for the pump to cavitate. The condition occurs when extremely high drive pressures are achieved, causing the hydraulic hoses to swell. This effect is called *volumetric expansion*.

During normal operation, the motor is rotating and exhausting the return oil into the closed-loop low-pressure leg. The return loop helps supercharge the piston pump. However, if the motor stalls while building high pressure, the high pressure can cause the hydrostatic drive hose to swell, which can cause a momentary point of pump cavitation. During the brief moment of cavitation, the pump is not receiving any of the return oil while the motor is stalled. In some unique applications, accumulators are added to the charge circuit to prevent pump cavitation any time a motor is stalled.

Summary

- Hydrodynamic and hydrostatic drives use fluid energy to propel a machine.
- Torque converters are hydrodynamic drives.
- Open-loop HSTs draw intake oil from only the reservoir.
- The intake of a closed-loop HST pump draws oil from a charge pump and the HST motor's return.
- A single-path HST uses one pump and one motor.
- Dual-path HSTs use one pump and one motor to drive each side of a machine.
- HSTs increase the machine's productivity.
- HSTs are noisy, sensitive to heat, and mechanically inefficient.
- HST charge pumps supply oil to the servo control valve, supercharge the piston pump, provide make-up oil to the closed loop, and help cool the circuit by replenishing the closed loop's oil.
- The feedback link is actuated by the swashplate and holds the servo control spool in the last commanded position.
- A flushing valve senses the two legs of a closed-loop HST.
- When the HST is operated, the drive pressure causes the flushing shuttle valve to shift, which allows charge oil to be controlled by the lower-pressure flushing relief valve.
- The flushing relief valve dumps excess charge oil into the motor's case. The oil is then routed to the pump's case for the purpose of purging and replacing the hot oil in the HST pump and motor.
- An HSTs charge pressure is highest when the transmission is in neutral and the engine is at high idle.
- HST high-pressure relief valves dump the oil into the opposite leg of the closed loop.
- Closed-loop HST filtration is rare because the filter must be able to filter oil in two directions and filter high-pressure oil.
- HST case drain is often cooled and sometimes filtered.
- Variable-speed motors can be inline axial, bent axis, or radial piston style.
- Variable-speed motors can have two specific speeds or can be infinitely variable.
- An HST with a variable-displacement pump and a variable-speed motor will vary the pump from minimum displacement to maximum displacement, and then vary the motor from maximum displacement to minimum displacement to increase the machine speed from slow to fast.
- A single-speed motor is sometimes accidentally installed 180° upside down, causing the HST to operate backwards.
- Inching valves and manual bypass valves can be used to help shift a two-, three-, or four-speed manual gearbox while the machine is sitting still.
- Inching valves can be used in conjunction with the HST pump or the HST motor.
- Eaton IPOR valves sense high oil (drive) pressure and limit the supply oil to the servo control valve, neutralizing the HST.
- Bosch Rexroth pressure-cutoff valves sense high drive oil pressure and limit the supply oil to the servo control valve, neutralizing the HST pump.
- Danfoss Series-90 pumps with pressure-limiter valves sense drive pressure. When pressure is high, they dump the pilot oil held behind the cross-over relief valve and direct oil to the opposite end of the servo piston to neutralize the pump.

- Danfoss Series-90 pumps use a side cover, lever, and two springs to hold the pump centered while it is in the neutral position.
- Many late-model HSTs use electronic controls to prevent the HST from stalling or lugging the engine.
- Hydraulic drive hoses can swell due to excessive drive pressure, causing the HST pump to cavitate.

Technical Terms

cavitate
charge relief
closed-loop HST
control oil
creeper control
dual-path HST
electronic anti-stall control
fixed-displacement motor
flushing relief valve
flushing valve
FNR lever
FNR valve
forward and reverse lever
hot-oil purge valve
hydrostatic
hydrostatic drive
hydrostatic transmission (HST)
inching valve
integral HST
internal pressure override (IPOR) valve
LoadMatch
manual bypass valve
neutral relief valve
neutral safety switch
off-line kidney-loop filtration
open-loop HST
pressure cutoff valve
pressure-release solenoid
replenishing valve
reversible pump
servo oil
single-path HST
single-speed motor
speed-sensing signal pressure
speed-sensing valve
split HST
two-speed motor
variable-displacement motor

Review Questions

Answer the following questions using the information provided in this chapter.

Know and Understand

1. Which of the following machines does not have a dual-path hydrostatic drive?
   A. Skid steer.
   B. Dozer.
   C. Track loader.
   D. Combine.

2. Which type of machine uses a hydrostatic motor at each of the drive wheels, a complex steering mechanism, and rear caster tires?
   A. Combine.
   B. Swather.
   C. Sprayer.
   D. Dozer.

3. Which of the following is an advantage of a hydrostatic drive?
   A. Quiet.
   B. Component cost.
   C. Mechanical efficiency.
   D. Overall productivity.

4. A hydrostatic charge pump generally should provide at least how much flow in relationship to the pump's flow?
   A. 19%.
   B. 42%.
   C. 85%.
   D. 100%.

5. Actuating the HST control valve results in what two actions?
   A. Relieve case drain and relieve charge pressure.
   B. Relieve drive pressure and relieve charge pressure.
   C. Increase/decrease speed and change forward or reverse directions.
   D. Isolate high pressure/charge pressure circuits and cool the HST.

6. What purpose does the linkage that connects the control valve to the swashplate serve?
   A. To reposition the spool so that it can hold the pump swashplate at the exact angle requested by the operator.
   B. To tell the pump to generate flow.
   C. To relieve charge pressure.
   D. All of the above.

7. All of the following are used to describe the hot-oil purge valve, *EXCEPT*:
   A. flushing valve.
   B. replenishing valve.
   C. shuttle valve.
   D. case drain valve.

8. Which of the following pressures will *not* act on the hot-oil purge valve?
   A. Drive pressure.
   B. Charge pressure.
   C. Servo pressure.
   D. All of the above.

9. Whenever the HST is driven forward, which charge pressure relief valve is controlling charge pressure?
   A. Pump charge relief valve.
   B. Flushing relief valve.
   C. High-pressure relief valve.
   D. IPOR valve.

10. What conditions result in the highest charge pressure?
    A. Low idle, neutral.
    B. Low idle, forward or reverse.
    C. High idle, neutral.
    D. High idle, forward or reverse.

11. What has to happen in order to get charge pressure to drop 30 psi (2 bar)?
    A. The high-pressure shuttle valve must shift.
    B. The HST pump control valve must be actuated.
    C. High-pressure relief oil must act directly on top of the charge relief.
    D. Engine speed must be raised to high idle.

12. What is the reason the charge pressure drops 30 psi (2 bar)?
    A. To aid in cooling the HST.
    B. To save horsepower.
    C. To lower pump inlet vacuum.
    D. To seat the motor's high-pressure relief valves.

13. The Case 2100–2300 series combine's two-speed HST motor will downshift from high speed to low speed when a specific drive pressure is reached. This several thousand psi of drive pressure will act on the control valve assembly and cause the control spool to shift back to low. When the control valve is in the neutral position, why is only 300 psi required to shift the control valve assembly to the low position?
    A. Pressure intensification.
    B. Length differences of the control spool and needle roller.
    C. Difference in areas of the needle roller and control spool.
    D. Because of the sequencing valve.

14. When an inline hydrostatic piston motor's swashplate is in a neutral angle, what will be the result?
    A. The pump will drive the motor at a higher speed.
    B. The pump will drive the motor at a lower speed.
    C. The pump will not drive the motor.
    D. Motor direction will reverse.

15. A technician just replaced a single-speed hydrostatic motor. After installing the motor, he or she found that the propulsion lever works backward. What is wrong?
    A. The motor was installed 180° upside down.
    B. The wrong motor was installed.
    C. The linkage was connected backward.
    D. The wrong servo control valve was installed.

16. A combine's foot-and-inch pedal is similar to what type of pedal found on tractors?
    A. Service brake.
    B. Transmission clutch.
    C. Engine decelerator.
    D. None of the above.
17. Once the pressure reaches the IPOR spring setting, which of the following pressures will the IPOR valve dump to the tank?
    A. Case pressure.
    B. Servo control valve supply oil.
    C. Drive pressure.
    D. None of the above.
18. How does the anti-stall solenoid hydraulically reduce drive pressure?
    A. It dumps charge pressure.
    B. It dumps high-pressure relief.
    C. It connects both legs of the closed loop.
    D. It connects both servo pressures.

Analyze and Apply

19. A(n) _____ drive uses high velocities and low pressures for transmitting power.
20. Most excavators use _____-loop hydrostatic drives.
21. In a closed-loop hydrostatic transmission, after the oil leaves the motor, it is sent to the _____.
22. A trunnion bearing HST uses a total of _____ servo piston(s).
23. A cradle-bearing-style HST uses a total of _____ servo piston(s).
24. The _____ link connects the servo control valve to the swashplate.
25. A single-path closed-loop HST uses a total of _____ charge relief valves.
26. When high-pressure oil opens the high-pressure relief valve, the oil is dumped into the _____ circuit.
27. The IPOR valve senses _____ pressures.
28. Caterpillar 924K, 930K and 938K wheel loaders use one pump and a total of _____ HST motors.

Critical Thinking

29. An HST pump has an efficiency of 85%, and the HST motor has an efficiency of 85%. If the HST pump flow at high idle is 50 gpm, what would be the required charge pump flow rate?
30. You have been asked to assist a technician who just made a repair to a hydrostatic drive. The propulsion lever now works backward. What are the questions you are going to ask the technician and why?

Chapter 14
Hydrostatic Drive Service and Diagnostics

Objectives

After studying this chapter, you will be able to:
- ✓ List the steps required to adjust the neutral position on two styles of hydrostatic pumps.
- ✓ Explain how to adjust the Danfoss Series-90 manual bypass valve.
- ✓ Describe the Bosch Rexroth DA valve plate adjustment.
- ✓ Explain how to measure HST shaft runout.
- ✓ List the steps for starting a hydrostatic transmission after replacing a pump or motor.
- ✓ Describe the process for troubleshooting the following hydrostatic transmission symptoms:
  - Overheating.
  - Transmission creeps forward or backward while in neutral.
  - Low power or sluggish in forward and reverse.
  - Low power or sluggish in just one direction.
  - Coasting or freewheeling transmission.
  - Machine will not move in either direction.
  - Charge pressure is too high.
- ✓ Explain why you should not attempt to isolate a hydrostatic pump.
- ✓ Explain a hydrostatic drive test stand.

As discussed in Chapter 13, *Hydrostatic Drives*, hydrostatic transmissions (HSTs) do not coast or freewheel by design. If a hydrostatic pump is misadjusted, the pump will not have a true neutral, but instead will creep either forward or rearward. Regardless of whether the pump is a single-servo cradle bearing style or a dual-servo trunnion bearing style, there are two types of adjustments for setting the pump to a neutral position. One adjustment is to set the single-servo piston or dual-servo pistons to neutral, and the other adjustment is to set the control valve to neutral.

Pro Tip

Always adjust the position of the servo piston(s) *before* adjusting the control valve. If the servo piston(s) are out of adjustment, the control valve will be incorrectly adjusted to compensate for the lack of neutral due to the misadjusted servo piston(s). This misstep will require another adjustment to center the control valve after the servo piston(s) are centered correctly.

Centering Adjustments

Safety precautions must be taken prior to making any centering adjustments, which will cause the machine to move. Follow the manufacturer's safety procedures when making pump adjustments. Many manufacturers specify to place the machine on jack stands to prevent personnel or equipment from being run over. If the pump is electronically controlled, the manufacturer may specify to unplug one or more speed sensors to prevent error codes for uncommanded machine propulsion.

In addition to a misadjusted pump, a misadjusted pump linkage or a shorted solenoid circuit can cause a pump to creep. Both of these conditions must be considered as a possible cause for a creeping transmission. In some manually controlled, dual-path hydrostatic drives, such as those on older skid steers and older swathers, the process for adjusting the dual-pump linkage can be complex. The manufacturer's service literature describes detailed procedures for centering both pumps.

On several electronically controlled HST pumps, the solenoid coil can be removed by removing the solenoid's retaining nut, which allows the solenoid's coil to be slid off the armature without requiring the solenoid's wires to be disconnected. If possible, remove the pump control valve solenoid coils to see if the machine still creeps in neutral. If the machine still creeps, a shorted solenoid circuit or a failed ECM is not at fault.

Single-Servo Cradle Bearing Pump

Newer HST pumps use a single-servo piston to control a cradle bearing swashplate. The servo piston will actuate the pump's swashplate to provide forward and reverse propulsion. Examples include Eaton Series-2 pumps, Bosch Rexroth A4VG pumps, and Danfoss Series-90 pumps. The pump can have two adjustments for centering the pump to neutral:

- Centering the servo piston.
- DCV null adjustment.

Centering the Servo Piston

A Bosch Rexroth single-servo pump, for example the A4VG pump, includes a threaded rod that is used for placing the servo piston in an exact neutral position. See **Figure 14-1**. Note that a jam nut holds the threaded rod in position. As the threaded rod is screwed in or out, the position of the servo piston shifts, changing the angle of the pump's swashplate. See **Figure 14-2**.

Prior to adjusting the servo piston, a technician must first hydraulically loop the two servo piston control pressures together. A hydraulic jumper hose is attached to the forward servo test port and the reverse test port, S1 and S2. This step hydraulically balances the servo control pressures (connecting forward servo pressure to reverse servo pressure).

Warning

Note that as this adjustment is made, the machine will move. *Always* follow the manufacturer's instructions for lifting the machine and placing it on jack stands to prevent it from moving or shifting.

Figure 14-3 shows a pump on a stand, not a running transmission. The photo also shows the S1 and S2 pressure ports looped together. This setup equalizes the two servo control pressures so the servo piston's position can be adjusted. Gauges are installed to measure pressure

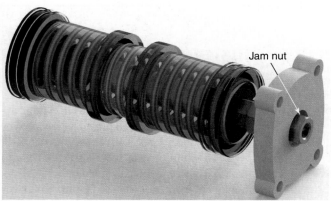

Goodheart-Willcox Publisher

Figure 14-1. A Bosch Rexroth single-servo cradle bearing hydrostatic pump contains an end plate with a threaded adjustment for centering the servo piston.

Goodheart-Willcox Publisher

Figure 14-2. A Bosch Rexroth HST pump. As the threaded adjustment is turned, it causes the swashplate to actuate. The threaded rod is adjusted until the swashplate is centered in a neutral position. A keyway links the servo piston to the swashplate. It is pinned to the swashplate and rides in the center grooves of the servo piston.

at the two drive pressure ports. The service literature will specify the appropriate size of pressure gauge for measuring the two drive pressures. Because most manufacturers recommend having the machine placed on jack stands, it normally takes only a small amount of drive pressure to propel the unloaded tires or tracks. If the wheels or tracks are off the machine, it will take even less pressure to propel the axle forward or in reverse, due to the lack of load on the axle. While adjusting the position of the servo piston and looking for the slight increase in pressure on the closed-loop legs, you can also observe the wheels (or axle or hubs if the wheels are removed) to see when they begin turning. However, the adjustment is often made at a low engine speed, and if the wheels are removed, the axle movement is very slow and sometimes difficult to recognize.

Figure 14-3. A Bosch Rexroth HST pump. The servo piston is adjusted by first connecting the forward and reverse servo control pressures together with a looped hydraulic hose. While monitoring the drive pressures, adjust the servo piston until the pressures are equal.

 Caution
If a pressure gauge with a low pressure range is being used, be sure to avoid applying the brakes. Applying the brakes would cause the drive pressure to reach stall pressure, which would damage low-pressure gauges.

While adjusting the servo piston's threaded adjustment, technicians monitor pressures on both legs of the closed-loop circuit. With the servo control pressures hydraulically looped together, the technicians adjust the servo piston until the pressures of both legs of the closed loop are equal and the drive wheels are stationary. The pump in **Figure 14-3** has the benefit of being able to simultaneously monitor both forward and reverse drive pressures.

Single-Servo Pumps with a Single Drive-Pressure Test Port

Some single-servo pumps use a shuttle valve and a single drive-pressure test port. See **Figure 14-4**. A technician cannot measure both forward and reverse drive pressures

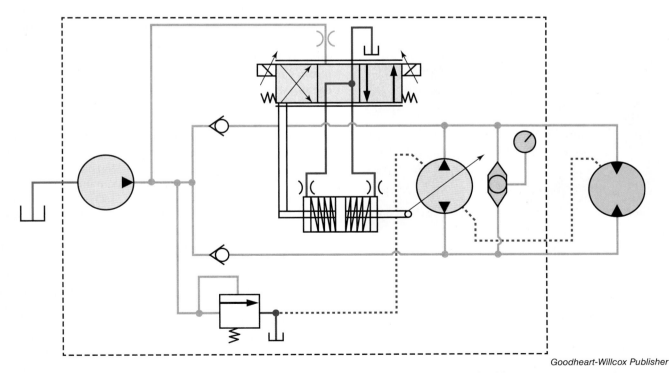

Figure 14-4. This single-servo hydrostatic pump uses a shuttle valve to direct oil to a single drive-pressure test port.

simultaneously with separate pressure gauges because the shuttle valve directs drive pressure to a single test port, limiting the technician to measuring only drive pressure rather than individual forward drive pressure or individual reverse drive pressure. To center a pump with this configuration, a couple of extra steps are required. As previously mentioned, any time a technician is making pump neutral adjustments, the machine will literally drive forward or rearward. Precautions must be taken to prevent the machine from running over personnel or equipment. Place the machine on jack stands as directed by the manufacturer's service procedure.

Case Study
Centering the Servo Piston

A technician was centering the single servo piston on a Bosch Rexroth pump used as the steering pump on an MT 800 series Challenger tractor. When centering a pump, technicians can watch the tires or tracks to easily see when they begin to spin. The tracks were removed from the tractor so the technician only had a 5″ diameter axle to view rather than a large track or 5′ diameter rotating drive. While adjusting the position of the servo piston, the technician did not identify the slight increase in pressure on the legs of the closed loop. The technician also did not observe the axles turning and further adjusted the position of the servo piston. Unfortunately, the technician adjusted the servo piston too far, which caused the servo piston to detach from the swashplate as well as bend the mechanical feedback link. The technician then had to drain 60 gallons of hydraulic oil in order to remove the pump. The pump was disassembled to allow the servo piston to be reattached to the swashplate and for a new mechanical feedback link to be installed. When centering the servo piston, it does not take much movement to force the pump to flow oil. Be careful and only make slight adjustments on the pump.

Goodheart-Willcox Publisher

Figure 14-5. If centering the servo piston on a Bosch Rexroth single-servo hydrostatic pump with just one drive-pressure test port, the position of the adjustment rod must be marked on the pump at the point where the pump initially begins to drive forward. It must also be marked where the pump initially begins to drive rearward. The adjustment rod is then centered between the two marked lines.

Although this type of pump has only one drive-pressure test port, it is still equipped with two servo control pressure test ports. After connecting the hydraulic jumper hose to the two servo control pressure test ports, loosen the jam nut to the servo piston's adjustment rod. On this Bosch Rexroth A4VG pump, the adjustment rod is turned with an Allen wrench. A flathead screwdriver is used on some smaller Bosch Rexroth pumps. Gradually turn the adjustment rod in one direction until the pump reaches maximum drive pressure. As mentioned earlier, drive pressure will be lower than normal because the machine is off the ground and has very little load. Stop moving the adjustment rod when the machine reaches maximum drive pressure. This is the point when the tires or tracks first begin to rotate. Use a marker to mark the exact location of the Allen wrench. Next, move the adjustment rod in the opposite direction while monitoring the drive pressure gauge. The goal is to gradually sweep the adjustment in the opposite direction until maximum drive pressure is reached. This is the point when the tires or tracks first begin rotating in the opposite direction. Mark the exact position of the Allen wrench. See **Figure 14-5**. Next, place the servo piston in the exact neutral position by moving the Allen wrench exactly between the two marked locations. This procedure is used for centering the differential steering hydraulic pump on an MT 800 series Challenger tractor.

DCV Null Adjustment

The second neutral adjustment on a single-servo piston is called the null adjustment. The null adjustment centers the spool inside the solenoid valve assembly. The null adjustment on a Bosch Rexroth pump, such as that on the A4VG pump shown in **Figure 14-6,** uses two Allen-head screws. One screw acts like a jam nut and locks the null adjustment in a neutral position. A solenoid spool valve and two Allen-head screws are in the center of the housing.

The null adjustment screw is technically not an adjustment screw, but an eccentric shaft. As the eccentric is rotated, the solenoid's spool valve is adjusted in and out of neutral. See **Figure 14-7.**

Two low-pressure gauges must be installed in the servo control pressure test ports (S1 and S2) before the adjustment is made. See **Figure 14-8.** The goal is to adjust the null adjuster until the two servo control pressures equalize. After the outside locking screw is loosened, an Allen wrench is used to adjust the center null eccentric. The null eccentric is very sensitive and requires slow and careful minute adjustments.

Goodheart-Willcox Publisher

Figure 14-6. A Bosch Rexroth single-servo hydrostatic piston pump contains an eccentric shaft that is centered with an Allen wrench or screwdriver. The adjustment process is called the null adjustment.

 Caution

The pressure gauges in **Figure 14-8** do not have an adequate pressure rating for measuring maximum servo pressure, such as full forward or full reverse. In those cases, use gauges that can safely handle the equivalent of charge pressure, for example 600-psi gauges. However, when measuring the null adjustment pressures, the adjustment is very sensitive, and pressure readings are less precise when using 600-psi gauges. If you are not careful, you will damage a low-pressure gauge by over pressurizing it.

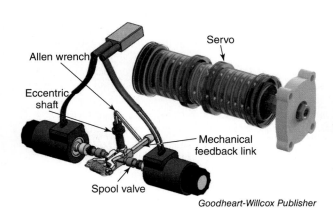

Goodheart-Willcox Publisher

Figure 14-7. The null adjustment on a Bosch Rexroth pump. An Allen wrench or screwdriver rotates an eccentric shaft to center the solenoid's spool valve. The adjustor does not have threads and can be rotated freely through 360° of rotation. The adjustment can be very sensitive, causing the servo pressures to build with very little movement to the eccentric shaft.

Goodheart-Willcox Publisher

Figure 14-8. Adjusting a Bosch Rexroth pump's null adjustment requires monitoring the forward servo pressure and reverse servo pressure. This photo was taken with the pump off the tractor. When the adjustment is made on a live machine, the eccentric is rotated slightly until the pressures equalize; for example, both gauges read 30 psi (2 bar). Any time the pressures are unequal, the pump is stroked in either a forward or reverse position, and the drive tires move.

Warning

If a lock screw is backed out too far, the eccentric rod can shoot out of the housing. If this happens, oil will spew out of the housing and the machine will default to maximum forward or reverse propulsion. It will also require finding all of the parts and reassembling them. The eccentric shaft, spool valve, a washer, the feedback link, two lever arms, and the spring must be assembled inside of the housing. *Always* follow the manufacturer's service and safety specifications.

Note

When making the null adjustment, if the servo pressures are unbalanced but the tires or tracks are stationary, the servo piston is not centered and the control valve has been adjusted to hold the HST in a neutral position. The servo piston will need to be centered first. The null adjustment will need to be adjusted last. It is likely that the servo piston was slightly adjusted out of neutral when the jam nut was tightened. After centering the servo piston, always be sure that the closed loop leg pressures are equal, and the tires or tracks are truly stationary at high engine speeds. Then proceed to adjust the null adjustment.

Case Study
Uncommanded Actuation of a Hydraulic Pump

An MT 845 Challenger uses a hydraulic steering pump and motor (essentially a hydrostatic transmission) as the hydraulic steering input for the differential steering system (covered in Chapter 24, *Track Steering Systems*). The machine uses a permanent-magnet speed sensor to detect rotation in the hydraulic steering motor. A customer complained that her MT 845 Challenger would steer without command during engine startup. After the fault code was cleared and the machine was restarted, the machine performed as follows:

- Steered in one direction for a brief moment.
- The ECM de-energized the steering pump solenoids.
- The machine stopped steering.
- A fault code was recorded on the monitor, preventing the tractor from steering or moving.

The steering pump's servo piston and null adjustment were both checked and found to be centered. The steering pump solenoid wires were also checked for shorts. When the steering pump solenoid coils were removed and the machine's engine was started, the machine would not actuate the steering pump without command. The repair required a new ECM. Although rare, it is possible for an ECM to cause an uncommanded actuation of a hydraulic pump.

Dual-Servo Trunnion Bearing Pump Adjustments

The Eaton Series-1 dual-servo trunnion bearing hydrostatic pumps also have two different adjustments for centering the pump to a neutral position. The pump's control valve has a neutral adjustment and the position of the two servo pistons affect the pump's swashplate angle. If the servo pistons are not equally positioned to the same depth, the pump's swashplate will not be at the 0° position.

Adjusting the Dual Servo Pistons

The Eaton Series-1 dual-servo piston pumps have threaded servo caps installed over both spring-loaded servo pistons. As the servo caps are installed, a depth micrometer or a dial indicator is used to measure the position of the swashplate. Both sides of the swashplate are measured. Eaton states that the two measurements should be within 0.0005″ (0.0127 mm) of each other. It is critical to mark the caps before removal to aid the assembly process. See the Eaton Series-1 dual-servo piston pump in **Figure 14-9**. When rebuilding this pump on a workbench, most technicians can correctly adjust the servo pistons and will not have to perform a fine tune adjustment of the servos when the pump is being test run on the pump test stand.

Adjusting the Pump DCV Neutral Position

The Eaton Series-1 pump's DCV uses a centering spring to hold the control spool in a neutral position. Similar to the Bosch Rexroth null adjustment, the control spool needs to be adjusted so that the servo control oil pressures are balanced.

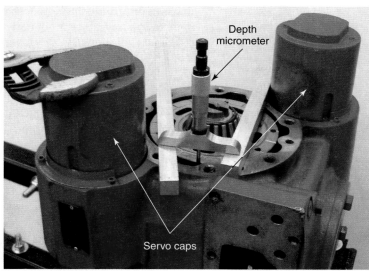

Figure 14-9. The installation of the servo caps adjusts the Eaton Series-1 pump's swashplate angle. Prior to removing the caps, remove the servo retainers (which are still installed on this pump), mark the servo cap's location with a permanent marker. A depth micrometer or a dial indicator is used to center the swashplate while installing the servo caps.

Some transmissions, such as those with Eaton Series-1 dual-servo pumps, are designed to have zero forward and reverse servo pressures while in neutral. Other transmissions, such as those with Bosch Rexroth A4VG pumps, have positive but balanced forward and reverse servo pressures, such as 40 psi forward and 40 psi reverse, while in neutral. A cap commonly covers the Eaton Series-1 DCV centering spring. A set screw may be used to hold the cap in place and must be removed before the cap can be removed. The centering spring has an Allen-head adjustment screw for placing the spool in a neutral position. See **Figure 14-10**.

Opening a Danfoss Series-90 Manual Bypass Valve

The Danfoss Series-90 pump is equipped with two multi-function valves. Each valve contains a manual bypass valve. A technician can use a combination wrench or socket to turn the external hex head on the manual bypass valve. To open the bypass valve, each valve must be turned counterclockwise three revolutions, which effectively connects drive leg A and drive leg B. Backing the valves out three and a half revolutions or more can cause a fluid leak. If the pump uses the manual displacement control (MDC), the manual control lever must also be held in the fully forward position while moving the

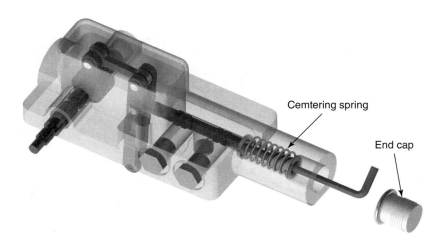

Figure 14-10. An Eaton Series-1 dual-servo pump can have a manual control valve that is centered with a spring. First, remove the set screw. The end cap is then removed to gain access to the centering spring adjustment. The spring is adjusted with an Allen wrench.

machine or turning the motor's shaft in the manual bypass mode. Danfoss cautions that the manual bypass feature is not designed for towing a machine long distances at high speeds, but should only be used for moving a dead machine slowly for a short distance.

Bosch Rexroth DA Valve Plate Adjustment

A Bosch Rexroth DA-controlled pump has a cartridge valve that develops pilot control pressure proportional to engine speed. As the engine speed increases, the increase of pilot control pressure causes the pump swashplate to increase its swivel angle.

The Bosch Rexroth DA pump can incorporate a fine-tune adjustment that rotates the pump's valve plate, or control lens, by a very small degree of rotation. In a traditional closed-loop HST, like the Bosch Rexroth EP pump control, the valve plate is pinned stationary. However, the DA-style pump has a small eccentric screw pin that is held in place by a jam nut. When the eccentric screw pin is rotated a fraction of a turn, it causes the valve plate to slightly twist, or turn, in one direction or the other, depending on the direction of rotation of the eccentric pin. See **Figure 14-11**.

The end of the pump's rotating barrel is positioned against the valve plate. In a pump containing nine pistons, there would be five pistons on the high-pressure side and four pistons on the low-pressure side for a portion of a rotation. As the barrel rotated a few more degrees, there would be four pistons on the high-pressure side and five pistons on the low-pressure side. After a few more degrees of rotation, there would once again be five pistons on the low-pressure side. After a few more degrees of rotation, there would once again be five pistons on the high-pressure side. This continuous shift in the number of pistons on the high-pressure and low-pressure sides would cause a pulsation. To prevent this, pump manufacturer commonly positions the valve plate so that it divides the pistons so there are always four and a half pistons on each side, high-pressure and low-pressure.

The eccentric pin directly acts on the valve plate so that as the eccentric pin is turned it adjusts the angle of the valve plate. The eccentric screw pin must not be turned more than 1/4 turn clockwise or 1/4 turn counterclockwise. If the eccentric screw pin is turned more than this, the valve plate can be damaged.

Goodheart-Willcox Publisher

Figure 14-11. Eccentric pin and valve plate used on a Bosch Rexroth DA-controlled HST pump.

The standard position of the valve plate is turned slightly, for example four degrees, from zero-degrees center. When the pump is manufactured, it is tested by the Rexroth pump technician on a test stand. The technician makes the fine adjustment to the eccentric screw pin to tune the position of the valve plate to optimize the pump based on the machine manufacturer's requirements. It should not be adjusted in the field nor by a traditional technician, but only by a properly trained pump specialist with a test stand. If the pump's eccentric screw pin or valve plate were damaged during adjustment, the pump would need to be disassembled for repair. A properly trained and experienced pump technician can also make this adjustment on the machine in the field.

When the HST is operating at the maximum efficiency and the operator begins to request the HST pump to maximum torque, the pressure rises in the drive pressure circuit. Because the valve plate is twisted in a four-degree position, it causes the load of the high-pressure pistons to have a larger area on one side of the swashplate compared to the pistons on the other side of the swashplate. Due to this difference in areas, as the drive pressure increases, that high pressure (larger area) places a force on the swashplate causing it to rotate a fraction of an angle back towards the neutral position.

The pump's servo control pressure moves the servo piston, which, in turn, moves the swashplate. As servo pressure increases, hydraulic horsepower output increases. When the larger area of high pressure increases on the swashplate, causing it to move a fraction of an angle back toward a neutral angle, servo pressure attempts to hold the swashplate in the last commanded position. Making a fine adjustment to the eccentric screw pin on the pump allows the technician to slightly rotate the valve plate, which increases or decreases the area where the high-pressure pistons are acting upon each side of the swashplate. Too much of an adjustment past four degrees will cause the pump to destroke too soon, resulting in a lack of torque because the pump is swiveling too early. Adjusting the angle less than four degrees causes the pump to destroke too late.

Before adjusting the eccentric screw pin, the technician should perform what Bosch Rexroth calls a "block graph." It consists of stalling the HST pump and not allowing the HST motor to spin. An experienced pump technician working on a wheel loader might drive a wheel loader's bucket into a large pile of material so that the machine cannot move, and apply the park brake and service brake so that the machine is stalled. Next the technician will graph the hydraulic high drive pressure at various engine speeds. For example, the technician may load the HST with the machine stalled at an engine speed of 1200 rpm and record the drive pressure, increase engine speed to 1400 rpm and record the pressure, and then increase engine speed to 1600 rpm and record the pressure:

- HST reaches 670 psi (46 bar) at 1200 rpm engine speed.
- HST reaches 1885 psi (130 bar) at 1400 rpm engine speed.
- HST reached 3625 psi (250 bar) at 1600 rpm engine speed.

When the technician makes an adjustment, it is critical that they mark the eccentric screw pin prior to loosening the jam nut. This ensures they know the initial position the eccentric screw pin, because it will be damaged if it is turned more than 90° in either direction. The technician would then adjust the eccentric screw pin, such as turning the eccentric shaft counterclockwise 45°, and then complete another block graph.

An example of the next results might be:
- HST reaches 940 psi (65 bar) at an engine speed of 1200 rpm.
- HST reaches 2540 psi (175 bar) at an engine speed of 1400 rpm.
- HST reaches 4640 psi (320 bar) at an engine speed of 1600 rpm.

When comparing the results after making the adjustment, the wheel loader reaches a higher drive pressure at lower engine speeds. As a result, the drive pressure will be reached sooner for a given engine speed and the engine could potentially stall too soon. If the adjustment caused the pressures to move in the opposite direction, the pump might have difficulty achieving torque or response at higher engine speeds. The technician would have to compare the test results to the requirements of the machine manufacturer to determine if the pump has been adjusted correctly.

Shaft Run-Out

Hydrostatic pumps and motors are mounted in different configurations. If the mount, adaptor, flywheel, or drive coupler is incorrectly machined or not flat, the shaft splines can wear prematurely. Some manufacturers provide detailed instructions for checking the pump mount surface and the shaft drive for run-out. The measuring tool of choice is a dial indicator. If the shaft's splines fail prematurely, the pump or motor can be disassembled and the shaft replaced. The problem that caused the shaft to fail prematurely must be fixed. Installing a new shaft is typically cheaper than purchasing a new pump or motor. If the failure is prematurely worn shaft splines, most technicians will simply replace the shaft, even if the hydrostatic transmission is a closed-loop drive, because the rotating group is usually unrelated to the worn shaft splines.

Figure 14-12 provides numerous drawings for checking alignments with a dial indicator. Place the dial indicator on the housing and flywheel as illustrated and set the dial to zero. For each measurement slowly turn the engine's crankshaft by hand using a breaker bar and socket. The best results have the dial indicator's needle moving very little. The total runout equals the total needle movement. For example, if the dial moves below zero 0.003″ and moves above the zero 0.003″, the total runout equals 0.006″. Be sure to follow the manufacturer's service literature for the correct specification.

The following describe where to position the dial indicator for various measurements:

- To check for flywheel housing face squareness, place the dial indicator's base mount on the flywheel, and place the probe of the dial indicator on the face of the flywheel housing.
- To check the flywheel face for squareness, place the dial indicator's base mount against the flywheel housing face, and place the dial indicator's probe against the face of the flywheel.
- To check the pump mounting plate flatness, place the dial indicator's base mount in the center of the flywheel, and place the dial indicator's probe against the face of the pump mounting plate.
- To check the flywheel for eccentricity, place the dial indicator's base mount on the face of the flywheel housing, and place the dial indicator's probe at on the axial edge of the flywheel.
- To measure the coupling plate flatness, place the dial indicator's base mount on the flywheel housing. Place the dial indicator's probe against the face of the pump drive coupler.
- To measure the pump drive coupler's flatness, place the dial indicator's base mount on the flywheel housing. Place the dial indicator's probe against the face of the pump drive coupler.

Pump and motor shaft splines that are lubricated in a gearbox with bath lubrication or splash lubrication suffer less wear than splines operating in a dry environment. If the splines operate outside of a lubricated gearbox, manufacturers often specify that a special grease, consisting of high-temperature grease and an additive of molybdenum disulphide powder, be applied to the splines. Some brand examples include Molycote, Metaflux, Never Seeze, and Optimol. The special grease limits fretting wear and increases shaft-spline life.

Note

With proper maintenance, service, and contamination control, a pump can often last well over 10,000 hours.

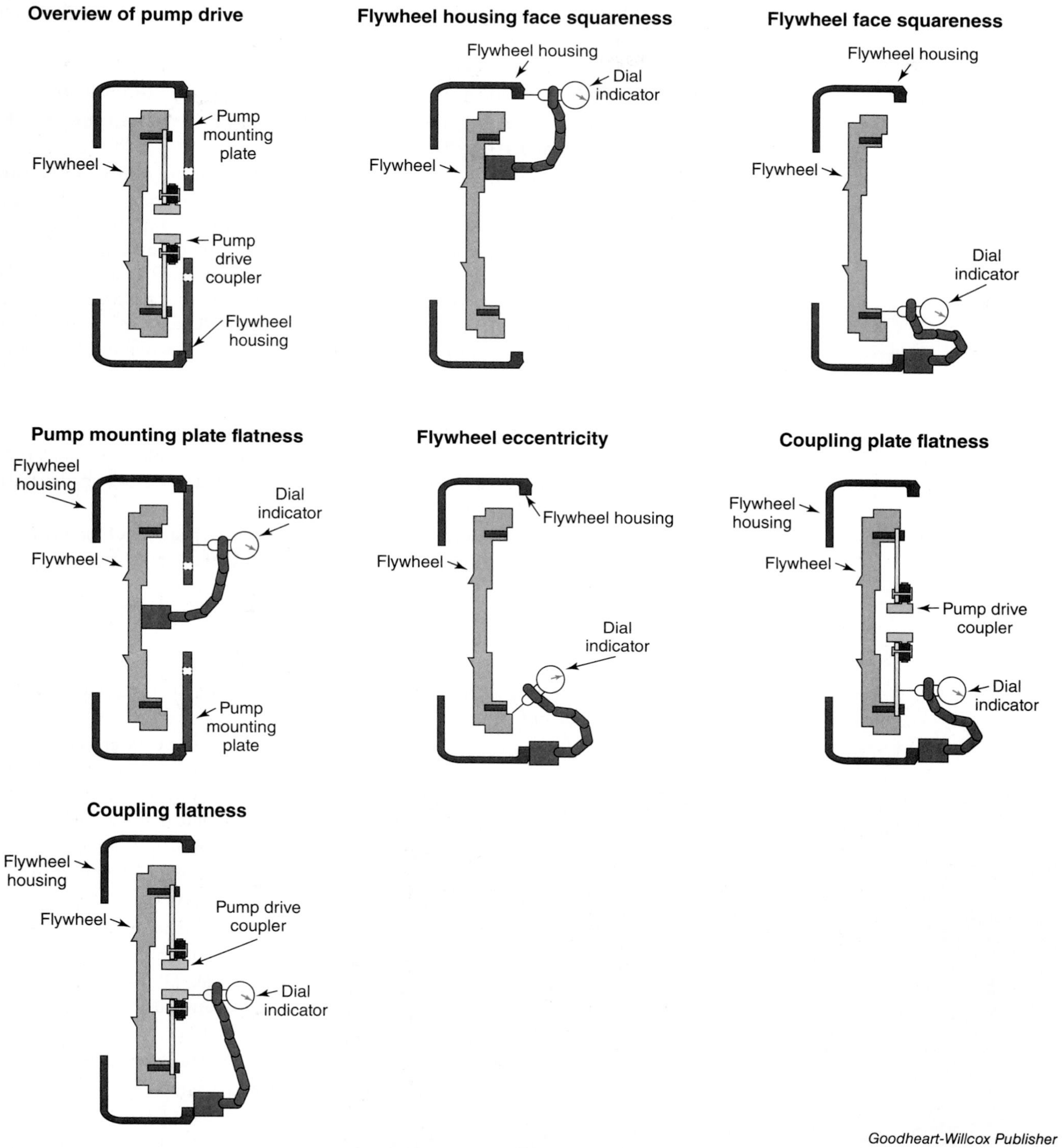

Goodheart-Willcox Publisher

Figure 14-12. Setups for checking for causes of premature spline wear using a dial indicator.

Startup (Commissioning) of a Hydrostatic Drive

Many mistakes can be made during the installation of a new pump or motor. The International Fluid Power Society (IFPS) labels the process of returning a system to service following pump or motor replacement as ***commissioning*** the hydraulic system. If the component is not installed correctly, the life of the component can be drastically reduced. Be sure to follow the manufacturer's startup procedures when installing a new hydrostatic pump or motor, even if the component was removed only to install a new shaft.

The following is an example of one manufacturer's startup procedure:

1. Prevent the machine's engine from running by removing a key or pulling a fuse.
2. Fill the case (of the pump and/or motor) with oil through the case drain port, which might have a hose connected to it.

   If the pump and/or motor case is below the fluid level inside the reservoir, the case-drain lines from the reservoir can fill the pump and motor case, presuming that other components like case-drain filters or case-drain coolers do not limit the ability of oil to flow freely backward. Even so, air must be bled out of the pump and motor case. Pumps and motors can have a vent plug. After the pump or motor has been filled with oil, the technician removes the vent plug to bleed the air out of the case and ensure that the pump and motor case housing is full of oil. The port is labeled "R" on this Bosch Rexroth AA10VG pump in **Figure 14-13**. Bleeding is complete when oil flows out of the vent hole.

3. Install a pressure gauge in the charge-pressure test port.
4. Place the multi-speed range gearbox in the neutral position. (Manufacturer's instructions differ regarding whether or not to also release the brake and chock the machine's wheels.) The startup procedure should be designed so that the hydrostatic transmission has no load on the motor's shaft.
5. Start and run the engine at low idle while monitoring charge pressure. If pressure fails to build to specification in less than 10 seconds, shut off the machine and diagnose the cause.
6. If charge pressure meets specification, increase the engine speed to 1500 rpm to purge air through the system.
7. Slowly move the propulsion lever halfway forward and allow it to operate for 4 to 5 minutes. Charge pressure should drop a little due to the flushing relief when the hydrostat is rotating.
8. Slowly move the propulsion lever back through neutral and halfway through reverse and allow it to operate 4 to 5 minutes.
9. Return the propulsion lever back to neutral and set the parking brake.
10. Repeat the startup procedure three times.
11. Operate all the control pressure circuits to purge air from the system.
12. Refill the reservoir as needed.
13. Check all pressures.
14. Replace the hydraulic filters after 10 to 15 hours of operation.

Goodheart-Willcox Publisher

Figure 14-13. The top bleed (vent) port in a Bosch Rexroth AA10VG HST pump. A ratchet is being used to remove the bleed port plug.

Hydrostatic Drive Diagnostics by Symptom

Hydrostatic transmissions can exhibit numerous types of symptoms, making it a challenge to accurately diagnose a problem. The last half of this chapter provides instruction for diagnosing HSTs based on the following symptoms and presents important tips relating to troubleshooting HSTs:

- Overheating.
- Transmission creeps forward or backward while in neutral.
- Low power or sluggish in forward and reverse.
- Low power or sluggish in just one direction.
- Coasting or freewheeling transmission.
- Transmission will not move in either direction.
- Charge pressure is too high.

Overheating

Many off-highway machines have a transmission overheating indicator light, monitor, and/or buzzer to alert the operator that the machine is overheating. When the transmission's overheating light is illuminated, a technician needs to first determine if the problem is with the indicator or an actual overheating transmission. The manufacturer service literature should provide the temperature threshold for illuminating the overheating light, as well as the location of the temperature sensor. One off-highway machine will illuminate the light once the oil reaches 200°F (93.3°C).

If manufacturer service literature lacks this information, a rule of thumb is that the oil should not be more than 100°F (55.6°C) warmer than the ambient temperature. For example, if a combine is harvesting in July and it is 97°F (36.11°C) outside, the rule of thumb states the oil should not exceed 197°F (92°C). Use an infrared thermometer to measure the temperature near the oil temperature sending unit. If the temperature is well below the manufacturer's specified threshold, diagnose it as an indication problem rather than true overheating. In **Figure 14-14**, the sending unit could be shorted and a short to ground could exist between the sending unit and the instrument panel.

If the transmission oil temperature is indeed too hot, it could be caused by a variety of factors, including a worn pump or motor, a restricted oil cooler, a lazy oil cooler bypass, or a purge relief valve swapped with the neutral charge relief valve. Many transmissions cool the case drain oil as explained in Chapter 13, *Hydrostatic Drives*. If the case drain oil

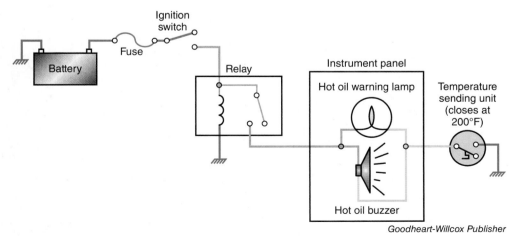

Goodheart-Willcox Publisher

Figure 14-14. HSTs commonly employ instrumentation to warn operators when the oil temperature has exceeded the normal operating range.

cooler becomes internally restricted or externally plugged, the transmission will overheat. The transmission case drain pressure should be measured. If the case pressure is the same value as the oil cooler bypass valve, for example 60 psi (4 bar), the next step would be to diagnose the oil cooler. If possible, measure the case drain oil flow after the bypass and after the oil cooler. See **Figure 14-15**.

If the oil cooler is not plugged, inspect the oil cooler bypass valve. More than one technician has made the costly mistake of presuming the excessive leakage in the pump and motor are causing the system to overheat. They needlessly replaced the pump and motor, only to later find a malfunctioning oil cooler bypass valve. The oil will take the path of least resistance, and if it is easier for the oil to bypass the cooler and go straight to the reservoir, then it will.

Measuring Case Drain Flow

If the oil cooler and oil cooler bypass valve are good, the next focus is to determine if the transmission has excessive internal leakage. Flow rating HST case drain requires first fully understanding the system's design. If the HST is a traditional closed-loop design with a flushing relief valve in the motor and the motor's flushing relief set 30 psi (2 bar) lower than the pump, then the motor case drain oil flow equals the internal leakage of the motor (case drain flow) plus the balance of the excess charge flow. The sum of these two oil flows exiting the motor cannot be separated. For example, when the machine is moving, the motor's flushing relief valve dumps the excess balance of charge oil into the motor's case. Therefore, if the flushing relief valve is located in the motor, it is only possible to isolate the pump's case drain flow and determine if the pump has excessive case drain oil flow. Or, if the flushing relief valve is in the pump, the case drain of the pump will equal the excess balance of charge pump flow plus the pump's case drain oil flow, and it will only be possible to isolate the motor's case drain to determine the motor's case drain flow.

The following are some design variations that affect measurement of case drain flow:

- Some HST circuits do not route the motor case drain oil flow in series through the pump but route the motor's case drain oil flow and pump case drain oil flows separately back to a cooler or reservoir.
- Some charge pumps are used for other machine functions, such as pilot controls.
- Some hydrostatic drives use a separate pump (shared with other systems) for supercharging the hydrostatic closed loop.
- Some hydrostatic drives direct the flushing relief valve's oil to a pressure compensator valve or an orifice to regulate the flushing oil flow.
- Some HSTs do not direct the balance of excess charge pump flow to a single component, like the motor, but distribute the balance of excess charge pump oil flow equally across multiple components: the HST pump and the HST motor(s).

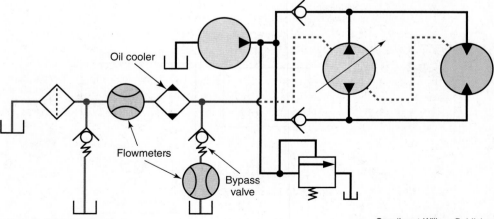

Goodheart-Willcox Publisher

Figure 14-15. Measure case drain after the bypass valve and after the oil cooler to determine if the oil cooler is plugged.

Case Study
Overheating Hydraulic Oil When Driving Under Light Loads

A skid steer manufacturer had experienced overheating hydraulic systems in skid steers any time the machine was propelled under light loads. The machine manufacturer found that the HST motor's flushing valves were not opening because drive pressures were relatively low. To remedy the situation, the machine manufacturer recommended driving the machine in a fast figure-eight pattern to cool the oil.

The machines were equipped with dual-path HSTs (one pump and one motor for each side). If such a machine is propelled fast and steered in a tight figure-eight for 15 minutes, it causes both shuttle valves in the HST motors to open alternately, cooling the dual-path HST. Keep in mind that the symptom only occurred under light loads. Also, note that the HST supplier was hesitant to use flushing shuttle valve springs with a weaker pressure setting because the weaker springs might prevent the HST from reaching proper operating temperature in certain climates. In addition, this was the only machine application that was experiencing the overheating problem under light loads.

Due to the potential complexity of some charge circuits, it is critical to use solid OEM data prior to measuring any transmission's case drain flow. At least one manufacturer has confirmed that measuring the hydrostatic motor's case drain flow (that contains a flushing relief valve) is a valid test and is indeed able to determine if the transmission has excess case drain flow.

- To fully understand the complexity of measuring case drain in a traditional HST with a motor flushing valve set at a lower pressure setting than the pump's neutral charge relief valve, consider the example of the single-path HST shown in **Figure 14-16**. In this example, the charge pump delivers 10 gpm (37.85 lpm) when the engine is running at wide-open throttle. The internal clearances within the HST pump results in 1 gpm (3.785 lpm) of case drain oil leaking from the HST pump, and the HST motor is leaking 1 gpm (3.785 lpm) of case drain oil. The charge pump flow that is left after making up the pump and motor deficiencies is 8 gpm (30 lpm). In **Figure 14-16A**, the motor's case drain is fed to the pump and the flowmeter is measuring the total case drain flow, which is equal to the charge pump flow of 10 gpm (37.85 lpm) as it is exiting the HST pump housing.

- In **Figure 14-16B**, two separate flowmeters have been installed, one for the oil exiting the pump and one for the case drain oil exiting the motor. The bottom schematic shows that if the flushing valve (with a lower pressure setting) is located in the motor, it is easy to see the net case-drain oil flow from the pump. However, the oil flow exiting the motor is 9 gpm (34 lpm), and a technician cannot determine how much of that oil is motor case drain leakage and how much of it is the balance of the charge pump oil is flowing across the flushing relief valve. In the example, it was given that both the pump and the motor had 1 gpm (3.785 lpm) of leakage, but **Figure 14-16B** depicts 9 gpm (34 lpm) leaking out of the motor. However, if the motor case drain leakage was 5 gpm (19 lpm) and the pump was still leaking 1 gpm (3.785 lpm), the test shown in Figure 26-16B would still depict a total case drain flow from the motor of 9 gpm (34 lpm). The 1 gpm (3.785 lpm) leaking out of the pump would register on the first flowmeter, and the 5 gpm (19 lpm) motor leakage plus the remaining 4 gpm (15 lpm) of charge pump flow would be register as 9 gpm (34 lpm) on the second flowmeter at the motor.

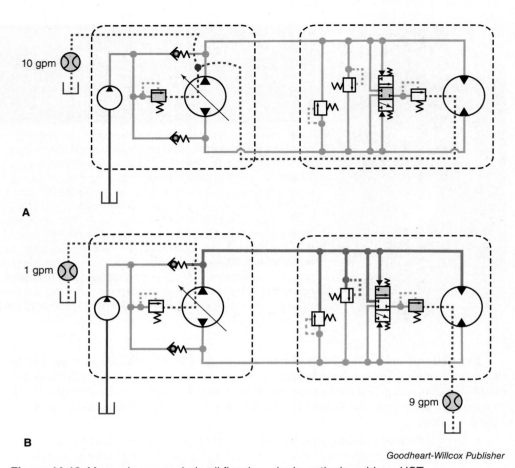

Goodheart-Willcox Publisher

Figure 14-16. Measuring case-drain oil flow in a single-path closed-loop HST.

- However, there is one scenario where it is more helpful to measure HST motor case-drain oil flow even if the flushing valve is in the motor. The scenario is an HST with one piston pump, one charge pump, and multiple HST motors. One such example is a combine harvester with a rear power guide axle, (four-wheel drive). In this example, one HST motor is responsible for driving the front drive axle, one HST motor is responsible for driving the right rear wheel, and one HST motor is responsible for driving the left rear wheel. See **Figure 14-17**.
- In this example, each of the three HST motors has a flushing valve set at a pressure value lower than the neutral charge relief valve and flowing a portion of the total charge pump flow into the motors' case drain. Four different flowmeters provide more useful information regarding case drain oil than in the single-path HST example in **Figure 14-16**.
- In **Figure 14-17**, the HST pump is leaking 1 gpm (3.785 lpm), the front axle HST motor is leaking 2 gpm (7.57 lpm), and the rear left HST motor is leaking 2 gpm (7.57 lpm). However, the most noticeable and useful information is that the right rear HST motor is leaking a 5 gpm (19 lpm). It is very useful to know that one of the three motors has a significantly larger amount of case drain oil flow than the other motors and pump. However, due to the flushing valve inside the motor, it is still difficult to determine how much of the 5 gpm (19 lpm) flowing out of the right rear HST motor is from case drain leakage and how much is from the flushing valve. It would also be helpful to know the setting of the cracking pressure for all three motor flushing relief valves and the pump's neutral charge relief valve, as this can influence the case drain oil flow from the three HST motors and pump.

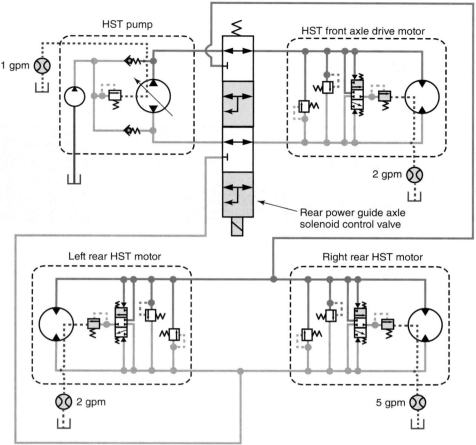

Figure 14-17. Measuring case-drain oil flow in a combine harvester that has an HST pump, a front-axle HST motor, and two rear HST motors.

Due to these complexities, perhaps one of the best tests for diagnosing a worn-out transmission is to focus on the charge pressure, which is by far the most important pressure to measure when diagnosing hydrostatic transmissions. Too much internal leakage in the pump and motors causes charge pressure to drop and case pressure to rise any time the HST is at operating temperature and placed under a load. This occurs because the charge pump does not have enough oil flow to overcome the excessive internal leakages.

However, it is possible that an abnormally high case pressure is due to a restriction in the case drain oil cooler or case drain filter, misleading a technician into thinking the pump and motor are excessively worn. Charge pressure is the key to diagnosing a suspect transmission. The role of charge pressure in diagnosis will be discussed in more depth later in this chapter.

Hot-Oil Purge Valve Mislocated

If the hot-oil purge valve is swapped in place of the neutral charge relief in a traditional closed loop HST, the transmission will not cool. When the transmission is operating properly and the propulsion lever is moved past neutral, the charge pressure should drop by 30 psi (2 bar). If the charge pressure is 30 psi (2 bar) below normal when the transmission is in neutral and does not change significantly as the propulsion lever is moved, it is possible that the hot-oil purge relief valve and the neutral charge relief valve were swapped. Be sure to study the manufacturer's literature to determine the correct placement and cracking pressures of the hot-oil purge relief valve and neutral relief valve. If the valves are misplaced or incorrectly set, the transmission will overheat.

> **Note**
>
> The relief valves in an Eaton Series-1 HST have three digits stamped on the ends of the valve assemblies. This value multiplied by ten equals the valve's pressure setting, for example 024 equals 240 psi.

Case Study
Charge Pressure Will Not Drop 30 PSI in Forward or Reverse

An instructor was demonstrating a HST simulator to a group of students. When the pump was actuated to forward or reverse, charge pressure did not change. The instructor discovered that the neutral charge relief valve and the motor flushing valve located in the Eaton Series 1 HST were stamped with the same number and set to same pressure setting.

Creeping Hydrostatic Transmission

A misadjusted transmission may continually creep forward or in reverse, indicating that the transmission will not remain in neutral. With the propulsion lever in the neutral position, both legs of the closed loop should have equal pressure, which equals charge pressure. If the propulsion lever is in neutral and one of the legs of the closed loop has a higher pressure than the other, the transmission is not truly in neutral. Details were provided earlier in this chapter for centering the hydrostatic pump. The focus should be on the control linkage or wiring, pump, the servo piston(s), and the directional control valve. It is very unlikely that a motor would cause a transmission to creep forward or rearward.

Low Power or Sluggish in Forward and Reverse

One of the most common and difficult HST symptoms to diagnose is an HST with low power in both forward and reverse. The operator may describe the transmission as sluggish or unable to pull under a load. The first step should be to determine if the low power occurs in both forward and reverse. The following seven areas must be investigated to determine the root cause of the lack of power.

Engine

Before suspecting that the transmission is at fault, a technician needs to be sure that the engine is not at fault. A poor-performing engine can be mistaken for a weak hydrostatic transmission. Therefore, the engine must be investigated before the transmission is diagnosed.

One manufacturer's service literature states to place the machine's gearbox in the high gear range, with the brakes applied, and then stroke the propulsion lever. If the engine speed drops due to the heavy load, the transmission is performing properly, and the engine should be the focus of diagnosis.

Transmission Controls and Other Features and Options

Depending on the transmission's configuration, the transmission can respond differently to the test listed above. For example, Chapter 13, *Hydrostatic Drives*, explains that if the machine has an IPOR valve, pressure cutoff valve, or a pressure limiter valve, the pump will be destroked once drive pressure reaches the IPOR, pressure cutoff, or pressure limiter valve's setting. If the pump is configured with an ECM anti-stall mode, the transmission would also appear to have low power when the ECM destrokes the pump.

If the transmission did not respond properly by drawing the engine speed down, the next step would be to eliminate other hydrostatic drive functions. For example, if the machine is a combine equipped with a power guide axle (PGA), the PGA should be shut off and the transmission should be retested. If the transmission can then draw down engine speed, the PGA must be diagnosed. However, keep in mind that when the PGA is engaged, the pump is sending oil to the main hydrostatic motor, plus the two rear PGA motors. The additional two rear motors should provide much more torque with the PGA engaged, rather than when the PGA is disengaged. Engaging the PGA will also lower drive pressure due to the increased motor displacement.

If the engine produces adequate power and if no other control, such as a power guide axle, is affecting the lower power symptom, then the following areas need to be investigated:

1. Oil (check type, level, and look for fluid aeration/bubbles).
2. Pump inlet (check for cavitation and/or aeration).
3. Charge (check for proper pressure).
4. Filtration (take an oil sample from case drain and inspect the oil filter).
5. High-pressure relief valves.

Reservoir

Before proceeding too far into diagnosing a poor performing hydrostatic transmission, be sure to check the reservoir. Look to see if the oil level is low or the oil is aerated. Determine if the correct type of oil is being used in the machine. For example, some manufacturers require an oil with a higher viscosity index number if the machine is equipped with a hydrostatic transmission. If standard viscosity oil is being used, it could be affecting the pump's performance.

Charge Pump Inlet

After ensuring the machine has the correct oil at the right level, check to see if the charge pump is receiving a good supply of oil. Some technicians will listen for cavitation before installing a vacuum gauge. Some manufacturers might not list a pump inlet specification. A rule of thumb is less is best, but preferably the pump has a positive head pressure and no pump inlet vacuum. One pump manufacturer states the charge pump inlet vacuum should not exceed 7″ (178 mm) of mercury. See **Figure 14-18**.

Some manufacturers today are less likely to use suction strainers or suction filters due to their tendency to starve the pump of oil when clogged. However, some machines still have a lengthy suction hose, for example, a concrete mixing truck's drum drive. If charge pump inlet vacuum is a concern, be sure to measure the vacuum directly at the charge pump inlet. This might require fabricating a special fitting to tap into the pump's inlet.

Charge Pressure

The next step is to investigate charge pressure. When the transmission's piston pump and motor become excessively worn, the charge pump volume will not be able to overcome the internal leakage losses. As a result of the worn hydrostatic transmission, charge pressure will drop. When charge pressure drops, the piston pump will cavitate, causing further damage to the hydrostatic transmission.

Low charge pressure can be a telltale sign that the transmission has excessive leakage, resulting in a low power complaint. When a low power complaint is being diagnosed on a hydrostatic transmission, charge pressure is arguably the most important pressure to monitor. See Chapter 13, *Hydrostatic Drives*, for a refresher on the charge system and flushing valve.

With the oil at operating temperature, measure charge pressure when the transmission is under load and compare it to the manufacturer's specification. In a traditional closed loop HST, the charge pressure should drop approximately 30 psi (2 bar) when the propulsion lever is shifted from neutral to either forward or reverse. The manufacturer will normally specify an engine speed for the test, 1900 rpm for instance.

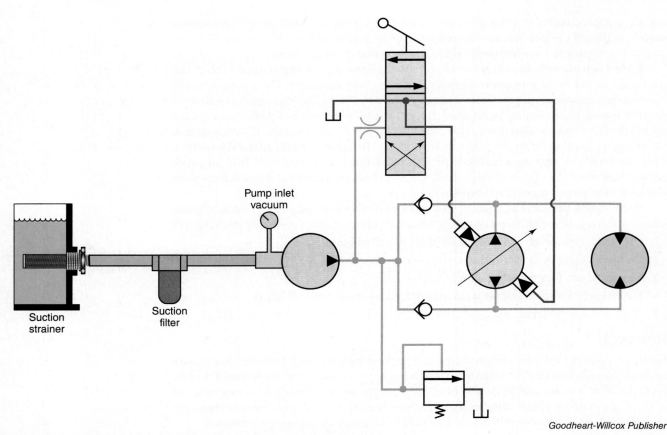

Figure 14-18. HST pump vacuum is measured at the inlet to the pump.

If the charge pump and the rest of the charge circuit are okay, the charge pressure measurement in neutral should be close to the manufacturer's specification. Explained another way, if the hydrostatic pump and motor are weak, it is possible for the neutral charge pressure to be within specification and the forward/reverse charge pressure to drop below specification. See **Figure 14-19** for an example of two transmissions, one good and one bad.

Inspect Oil and Filter for Contamination

If the charge pressure drops below specification when the system is under load, inspect the oil and filter for contamination. One manufacturer recommends removing the motor's case drain plug to obtain a sample of oil and cutting open the transmission oil filter. If contamination is found in the filter, the pump case, or the motor case, and if the transmission is a closed loop-drive, both the pump and the motor must be replaced. In addition,

Good Transmission		
Measurements	Specification	Measured Pressure
Neutral Charge Pressure	300 psi (20 bar)	300 psi (20 bar)
Forward/Reverse Charge Pressure (under load)	270 psi (18 bar)	270 psi (18 bar)
Good Charge Pump, Bad Piston Pump and Motor		
Measurements	Specification	Measured Pressure
Neutral Charge Pressure	300 psi (20 bar)	300 psi (20 bar)
Forward/Reverse Charge Pressure (under load)	270 psi (18 bar)	75 psi (5 bar)

Goodheart-Willcox Publisher

Figure 14-19. Examples of two charge pressure measurements, one good and one bad. Make sure the machine is at operating temperature when measuring the pressures.

the entire system, including the hoses, reservoir, and cooler, must be flushed. Install new oil and a filter. Some manufacturers recommend also replacing the cooler due to the difficulty in properly flushing a contaminated oil cooler.

Drive Pressure

If the motor's oil sample and filter have no contamination, the next step is to measure the system's drive pressure. Notice that this book lists measuring drive pressure last. Many technicians, especially those unfamiliar with hydrostatic transmissions, will want to measure drive pressure first. If drive pressure is low, study the schematic and inspect any valve that can limit drive pressure. Examples include the internal pressure override (IPOR) valve, pressure cutoff valve, pressure limiter valve, high-pressure relief (cross-over relief valves), foot-and-inch valve, anti-stall solenoid, manual bypass valve, and electronic pressure-release solenoid. Note that some shops have the capacity to individually test the high-pressure relief valves. See **Figure 14-20**.

After everything listed above (engine, transmission controls, reservoir, pump inlet, charge pressure, oil and filter, drive pressure, and other valves that can limit drive pressure) have been found to be operating properly, the solution is for the customer to shift the gearbox to a lower operating range. It is possible that the customer will desire the machine speed of second gear and the torque of first gear. The customer might have purchased a machine that is too small to meet his or her needs. For a fixed amount of horsepower, the customer can get an increase in torque or an increase in speed, but not both. The only way to increase both speed and torque is to increase engine horsepower. See **Figure 14-21**.

Goodheart-Willcox Publisher

Figure 14-20. Relief valves can be bench tested to determine their relief pressure setting.

Goodheart-Willcox Publisher

Figure 14-21. For a fixed amount of horsepower, speed is inversely proportional to torque

Case Study: Weak HST

A 1999 Case IH 2388 combine with more than 3900 engine hours would not pull itself up a hill, even with the header removed from the machine. The HST was weak in both forward and reverse. The customer had owned the machine for many years and knew the machine was previously capable of pulling itself up the hill. The machine lacked so much torque that, when the brakes were applied, the HST could not pull down (lug) the engine.

The machine had a foot-and-inch valve. Because the foot-and-inch valve sets the drive circuit's pressure in both forward and reverse, it was suspected as the problem. Also, keep in mind the relatively low hours, less than 4,000 engine hours. The foot-and-inch valve was disassembled and inspected and new components were ordered and installed due to the findings. The technician drained the hoses, pump, and motor, letting the hoses drain overnight. When he looked at all the drained oil, he could see only a small amount of glitter. After replacing the poppet inside the foot-and-inch valve, the HST was still weak.

This machine was equipped with a two-speed HST motor and a rear power-guide axle. This made it more challenging to measure charge pressure. The Eaton Series-1 two-speed HST motor did not have individual drive pressure ports, forward and reverse. It had only one drive pressure port. If the machine had test ports on both closed-loop legs, one port could be used to measure charge pressure while the other port could be used to measure drive pressure. The Eaton Series-1 HST pump also did not have a charge-pressure test port for measuring charge pressure. The port that is normally used for measuring charge pressure was used to supply pilot oil (via an orifice elbow) to the rear power-guide-axle control valve. This was another reason that drive pressure was measured first, instead of charge pressure. The machine was also at the customer's farm and not at the dealership, where the technician could find the appropriate fittings needed to tap into and measure charge pressure. When the charge pressure gauge was installed in place of the orifice elbow, even with the power guide axle off, oil would leak backwards through the open power-guide axle pilot line.

The technician went and got the necessary tee fitting to tap into the charge pressure. The technician found that the charge pressure would drop to 75 psi (5 bar) when the machine was hot and loaded. Before buying a new HST pump, motor, and oil cooler, the customer wanted the technician to ensure that a malfunctioning relief valve could not be the cause of the poorly performing HST.

Technically the flushing valve inside the HST motor is in command of charge pressure when the machine is outside of neutral. Although it would be incredibly rare that the flushing relief would be at fault, if it had a lazy spring, it could cause the charge pressure to drop when propelling in forward or reverse. The technician swapped the flushing relief valve with the pump's neutral charge relief to prove to the customer that the flushing relief was not at fault and that the pump and motor needed to be replaced. Keep in mind, this was an extreme measure taken to justify to a machine's owner that the charge relief valves were okay. It is not a recommended procedure. If the charge relief valves remained swapped, the transmission would likely overheat due to charge oil no longer flushing the hot oil out of the motor, but only flushing the pump's case.

The technician explained to the customer that, during the test, the charge pressure would be approximately 20 to 30 psi lower than normal in neutral, but, due to the worn-out pump and motor, would once again drop off significantly when the system was hot and loaded. The test results were as expected. As a result, the pump and motor were condemned. The pump, motor, and oil cooler were replaced. The oil was drained, and the filters were replaced. The new hydrostatic transmission components fixed the problem. Charge pressure no longer fell off, and the machine could pull itself up the hill and draw down the engine (causing it to lug) when the brakes were applied and the propulsion lever was actuated.

>
> **Note**
>
> The Eaton Series-1 pump has multiple test ports that can easily be tapped to measure pressure, such as servo pressures and charge pressure. See **Figure 14-22**. Because the manufacturer used charge pressure via an elbow orifice to supply pilot oil to the power-guide-axle control valve, the arrangement can cause problems during diagnostics and when adding an aftermarket power-guide axle. If a technician placed a charge pressure gauge in the reverse servo port, thinking it was the charge port, he or she would be confused by the bizarre results. In addition, some technicians have added aftermarket power-guide-axle kits to combines and accidentally placed the elbow orifice in the reverse servo port. This caused the machines to stumble in reverse any time the power-guide axle was engaged. This occurred because reverse pilot oil was taken away from the reverse servo piston and used for the power-guide-axle control valve.

Low Power or Sluggish in Just One Direction

If a hydrostatic transmission is noticeably underpowered in only one direction, one of the high-pressure relief valves may be at fault. Most transmissions use the same type of high-pressure relief valve for both forward and reverse propulsion. If the manufacturer uses the same type of valve for forward and reverse, the forward and reverse high-pressure relief valves can be swapped to see if the low power symptom changes to the opposite direction. If the transmission is then underpowered in the opposite direction, the high-pressure relief valve is at fault. As previously mentioned, the neutral charge relief should never be swapped with the flushing relief valve, otherwise the transmission will overheat, greatly reducing the life of the transmission.

It is also possible for a make-up valve to cause an HST to exhibit low power in only one direction. A leaky make-up valve will also cause charge pressure to be too high in one direction.

Goodheart-Willcox Publisher

Figure 14-22. The side view of an Eaton series 1 HST pump. The control valve has two pressure ports, forward and reverse servo pressure. Another reverse servo pressure port is located on the case of the pump and the other pump port is charge pressure.

Coasting or Freewheeling Transmission

Hydrostatic transmissions provide the benefit of hydrostatic braking. If an operator notices the transmission freewheeling or coasting, the machine should also lack power. A machine with these symptoms should be diagnosed. See the recommendations previously listed for diagnosing low power in both directions.

Transmission Will Not Move in Either Direction

If a transmission will not move in either direction and the pump is electronically controlled, check to see if the solenoids are being energized. If the problem is electrical, use the manufacturer's service information to diagnose the problem. If the solenoids are receiving electrical power or if the pump is manually controlled, the next step is to check for proper charge pressure because the charge is responsible for supplying oil to the directional control valve.

If charge pressure is good, see if the pump can build servo pressure. Manufacturers rarely publish any specifications for servo pressure. The process of balancing servo pressures to center a pump's null adjustment was explained earlier in this chapter.

When the pump's control valve is actuated, the DCV receives supply oil from the charge pump through the orifice. If the pump has good charge pressure but is unable to build any servo pressure, the supply orifice is likely plugged.

Case Study: Plugged DCV Orifice

Figure 14-23 shows a plugged orifice on a hydrostatic transmission. Multiple technicians had attempted to diagnose why this machine would not propel itself. Once it was determined that the machine could not develop servo pressure, a technician checked to see if the DCV orifice was plugged. It was a pleasant surprise to find that the machine could not develop servo pressure because it is almost a given that this is an indication that the DCV orifice is plugged. In this case, a portion of an O-ring had plugged the orifice. The plugged orifice would not allow the servo to be actuated, which caused the drive pressure to build. This should have led to charge pressure dropping 30 psi (2 bar) but both legs of the closed loop had the same pressures. Once the orifice was fixed, servo pressure could be developed, causing the pump's swashplate to pivot. This resulted in the development of drive pressure, which caused the flushing shuttle valve to shift and the charge pressure to drop.

However, because untrained technicians had previously attempted to fix the machine, two more problems had to be fixed after the plugged orifice was fixed. The HST DCV was misadjusted and attempting to propel the machine while in neutral. The IPOR valve was also adjusted too low to the point that the machine would not pull itself. Both the DCV and the IPOR valve were properly adjusted and the HST was restored back to proper operation.

Dual-Path HST Skid Steer

Commonly, the charge pump delivers oil to the HST control valve via an orifice to provide servo control oil. Some machines can provide servo control oil through different valving. For example, CNH has produced skid steers with a Bosch Rexroth A22VG twin HST pump. See **Figure 14-24**. The skid steers have been equipped with a separate solenoid block containing, one, two, or three solenoids. In the example containing three solenoids, one of the solenoids (parking brake) is used for two separate purposes, to release the parking brake and to direct servo control supply oil into the pump inlet YST port, which is the supply to the HST pump's electronically controlled solenoids. This means that if the parking brake release solenoid or electrical circuit fails, the parking brake cannot release, nor can the dual-path HST piston pump be upstroked, due to the lack of servo supply oil. Another solenoid within the triple solenoid block controls the high- and low-speed displacements of the two-speed motor. The third solenoid is the loader solenoid. It controls the pilot oil to the implement DCVs for enabling the loader controls. The actual solenoid-control valve block is shown in **Figure 14-25** being used on a skid-steer trainer. When the operator presses the operate button on the skid steer's instrument panel the electronics energize the parking brake release solenoid and the loader enable solenoid to allow the machine to operate both the HST and the loader.

Goodheart-Willcox Publisher

Figure 14-23. The control valve's orifice was plugged with material from an O-ring.

A = Spring-applied parking brake
B = Two-speed motor
C = Two-speed control spool
D = Flushing relief valve
E = Flushing spool valve
F = Brake pressure switch
G = Brake solenoid valve
H = Two-speed solenoid valve
I = Charge filter
J = High-pressure relief valves
K = Pump DCV solenoid valves
L = Pump servo piston
M = HST variable displacement pump
N = Charge pump
O = Neutral charge relief valve

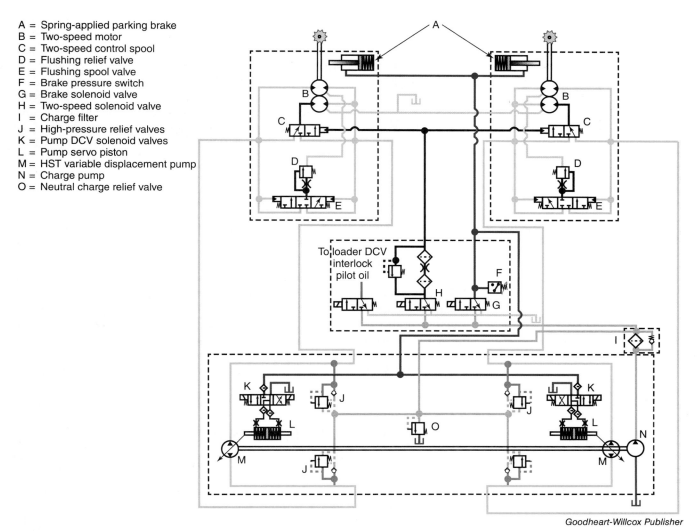

Goodheart-Willcox Publisher

Figure 14-24. A schematic of a CNH skid-steer dual-path HST that is configured with a two-speed motor, parking brake, and loader-enable valve.

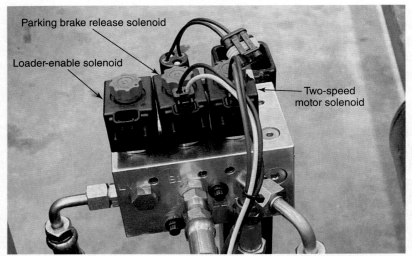

Figure 14-25. A triple-solenoid valve block used on a school trainer's dual-path HST.

Effects of the Directional Control Valve Orifice

The DCV orifice controls the responsiveness of the transmission. The larger the orifice, the more responsive a transmission will be. A smaller orifice causes the transmission to be less responsive. A machine's service literature may provide part numbers for different orifice options.

The Eaton Series-1 HST pump typically has an orifice on the bottom of the DCV. However, if it is equipped with an IPOR valve, the orifice can be on the pump's end cover, which houses the IPOR valve. The orifices in an Eaton Series-1 HST pump are staked to retain their position within its housing. If the orifice is removed, a new orifice will need to be staked into the housing.

Charge Pressure Is Too High

If a technician encounters a machine with charge pressure that is too high when the machine is operated in one direction, one of the closed-loop makeup valves could be allowing high-pressure oil to leak back into the charge circuit. In this scenario, the charge pressure would be considerably higher than specification. See **Figure 14-26**. If one of the make-up valves is leaking, it can also cause low power in one direction.

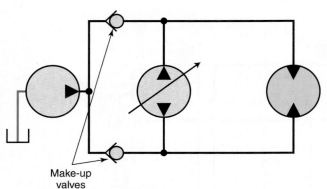

Figure 14-26. Make-up valves are check valves that allow the charge pump to supercharge the closed loop.

Do Not Attempt to Isolate a Hydrostatic Pump

If a technician is diagnosing a weak closed-loop hydrostatic transmission, he or she should never attempt to isolate the pump. There are two reasons for this. First, if the motor contains the high-pressure relief valves (such as Eaton Series-1 motors and Danfoss Series-20 motors), the technician can rupture a pump or drive hose while trying to isolate the pump. This is likely to happen if the technician caps off the closed-loop drive lines. Second, it should be unnecessary to isolate a pump. If the transmission is weak and the pump is bad, the motor should be replaced as well, because the HST is a closed-loop transmission. This concept was explained in Chapter 13, *Hydrostatic Drives*.

Using a Hydrostatic Transmission Test Stand

Some hydraulic shops have the means for testing a hydrostatic transmission's pump and motor. The test stand in **Figure 14-27** is powered by a diesel engine. Notice that the test stand uses one pump and motor to drive the rebuilt pump and another pump to load the transmission's rebuilt motor. In this example, the test stand uses three separate fluid reservoirs, one for each HST: the driving HST, the HST being tested, and the HST placing the load on the motor.

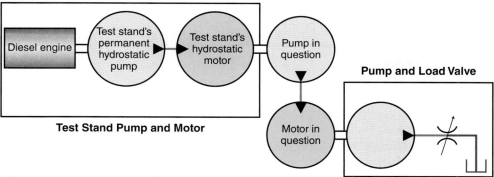

A

B

Goodheart-Willcox Publisher

Figure 14-27. A—A hydrostatic drive test stand consists of an engine that drives a test stand pump and motor assembly. The test stand's hydrostatic motor drives the pump being tested. Notice the test stand also uses a pump and a load valve to load the transmission. B—A hydrostatic transmission test stand at B&B Hydraulics in Hutchinson, Kansas.

Summary

- Servo-operated hydrostatic pumps can have two separate adjustments for setting the pump to neutral: DCV null and servo piston(s). The machine must be safely placed on jack stands prior to adjusting the pump because the adjustments will cause the machine to move. Adjust the servo piston(s) before adjusting the control valve.
- The Danfoss Series-90 manual bypass valve is designed only to allow a machine to move a short distance at slow speeds.
- The Bosch Rexroth DA valve plate adjustment is designed to rotate the lens plate a very small degree of rotation.
- HST pump and motor shaft runout requires using a dial indicator to make six different measurements on the HST pump/motor frame and drive.
- Two of the most important steps for recommissioning a HST is to be sure to fill the pump and motor case with oil and to start the system with no load.
- Charge pressure is the most important pressure to measure when diagnosing a weak HST. It will drop approximately 30 psi (2 bar) when the transmission is shifted from neutral to forward or reverse. A worn hydrostatic transmission typically results in a drop in charge pressure and will contaminate the oil and the filter.
- If the charge pump is believed to be cavitating, measure the pump inlet's vacuum. The pump inlet vacuum should be no more than 7″ (178 mm) of mercury.
- If the transmission is overheating, perform the following checks:
  - Ensure the problem is not an indication problem.
  - Check case drain and oil cooler flow.
  - Measure charge pressure to see if the transmission is excessively worn.
  - Determine if someone mistakenly swapped the neutral charge relief valve with the hot-oil purge relief valve.
- HST pumps should not be isolated because the high-pressure relief valves might be in the motor.
- Hydrostatic drive test stands use either a large electrical motor or a diesel engine to drive a HST that propels the HST being tested. A reversible pump is connected to the HST being tested so that it can be loaded.

Technical Term

commissioning

Review Questions

Answer the following questions using the information provided in this chapter.

Know and Understand

1. All of the following can cause a hydrostatic transmission to creep, *EXCEPT*:
   A. a misadjusted pump DCV.
   B. a misadjusted pump servo piston.
   C. a shorted pump solenoid.
   D. a shorted two-speed motor solenoid.

2. Opening the Danfoss Series-90 manual bypass valve _____.
   A. allows for towing at high speeds
   B. allows for towing long distances
   C. requires the motor to be dismantled
   D. is done by rotating the adjuster three revolutions CCW

3. A(n) _____ is used to measure HST pump and motor shaft runout.
   A. feeler gauge
   B. dial caliper
   C. dial indicator
   D. DVOM

4. A technician has installed a new hydrostatic motor. All of the following are necessary steps for starting (commissioning) the transmission, *EXCEPT*:
   A. filling the motor's case with oil.
   B. monitoring charge pressure.
   C. starting the machine with the gearbox in the neutral position.
   D. monitoring case pressure.

5. A hydrostatic transmission is overheating. Which of the following hydraulic pressures should be investigated first?
   A. Case drain.
   B. Charge.
   C. Drive.
   D. Servo.

6. A hydrostatic transmission has good charge pressure but will not move forward or backward. A technician strokes the propulsion lever but cannot obtain any servo pressure. What is most likely at fault?
   A. A bad high-pressure relief valve.
   B. A plugged DCV orifice.
   C. A bad flushing shuttle valve.
   D. A stuck-open make-up valve.

7. Which of the following has a large effect on the responsiveness of the hydrostatic transmission?
   A. Make-up valve.
   B. DCV orifice.
   C. Flushing relief valve.
   D. Pressure rating of the drive loop hoses.

8. A transmission is sluggish in forward but operates strongly in reverse. Which of the following steps is the most common procedure performed by veteran technicians to diagnose this problem?
   A. Swap the high-pressure relief valves.
   B. Swap the charge relief valves.
   C. Swap the make-up valves.
   D. Swap the drive hoses.

9. When is it okay to swap the charge relief valves?
   A. When charge pressure is low.
   B. When charge pressure is high.
   C. When the transmission lacks performance in one direction.
   D. Never.

10. A technician has been investigating a hydrostatic transmission with a low power complaint in both directions. All of the following are diagnostic steps for investigating a low power symptom, *EXCEPT*:
    A. cutting open a filter.
    B. pulling a sample from the motor's case drain.
    C. measuring charge pressure.
    D. measuring servo pressure.

11. A technician has been investigating a hydrostatic transmission with a low power complaint in both directions. What is the recommended procedure if all of the test results are good?
    A. Increase the setting for high-pressure relief valves.
    B. Change the fluid and filters.
    C. Shift to a lower gear.
    D. Swap the charge relief valves.

12. All of the following are normal hydrostatic transmission characteristics, *EXCEPT*:
    A. it coasts or freewheels.
    B. it is loud during high drive pressures.
    C. the charge pressure drops 30 psi (2 bar) when moving.
    D. it provides hydrostatic braking.

13. A hydrostatic transmission is overheating. All of the following can cause overheating, *EXCEPT*:
    A. a lazy oil cooler bypass.
    B. a restricted oil cooler.
    C. a worn hydrostatic pump and motor.
    D. a faulty indicator lamp.

14. Charge pressure is too high when driving in one direction. Which of the following is most likely at fault?
    A. A leaking make-up valve.
    B. A leaking charge relief valve.
    C. A leaking drive relief valve.
    D. A leaking DCV spool.

15. A hydrostatic transmission uses a flushing shuttle valve located in the hydrostatic motor. It is possible to accurately measure which case drain?
    A. Pump case drain.
    B. Motor case drain.
    C. Both A and B.
    D. Neither A nor B.

16. A technician is testing a hydrostatic transmission's charge pressure. At wide-open throttle in neutral, the charge pressure measures 300 psi (20 bar). Which of the following charge pressure readings measured under load in forward or reverse would indicate a worn transmission?
    A. 330 psi.
    B. 300 psi.
    C. 270 psi.
    D. 75 psi.

17. When is it okay to isolate a hydrostatic pump and cap off its drive lines?
    A. When charge pressure is low.
    B. When charge pressure is high.
    C. When the transmission lacks performance in one direction.
    D. Never.

Apply and Analyze

18. When centering the servo piston on a Bosch Rexroth HST pump that is equipped with only one servo piston, the _____ pressures must be measured.

19. When making the DCV null adjustment on a Bosch Rexroth HST equipped with a single-servo piston pump, the _____ pressures must be measured.

20. When centering the servo piston on a Bosch Rexroth HST pump that is equipped with only one servo piston, the _____ pressures must be hydraulically looped together.

21. When making the DCV null adjustment on a Bosch Rexroth HST equipped with a single-servo piston pump and electronic solenoids, the Allen wrench is turning/adjusting the _____ shaft.

22. A technician is installing the servo piston caps on a dual-servo Eaton Series-1 piston hydrostatic pump. The technician is measuring the swashplate to ensure the pump is neutrally centered. The maximum difference allowed from one side to the other is _____.
23. The maximum allowable adjustment on the Bosch Rexroth DA valve plate adjustment in one direction is _____ turn.
24. A technician has installed a new hydrostatic pump. During the startup procedure, the transmission should be operated under _____ load.
25. If a hydrostatic transmission sounds as if it is cavitating, _____ pressure should be measured.
26. A hydrostatic transmission lacks power. The most important pressure to measure during the diagnostic process is _____ pressure.
27. Total HST case drain flow essentially equals _____.
28. A hydrostatic drive test stand can have a total of _____ hydraulic reservoirs.

Critical Thinking

29. A transmission will not drop charge pressure when the DCV is stroked. Describe all of the things that could cause this symptom.
30. A single-path closed-loop HST has a flushing relief valve in the motor. Explain what can be tested for case drain and explain why.

meunierd/Shutterstock.com

Case IH offers continuously variable transmissions (CVTs) in their line of high-horsepower agricultural MFWD tractors, the Case IH Magnum (235–380 HP). The Magnum 380 model with rear rubber tracks is shown here.

Chapter 15
Continuously Variable Transmissions

Objectives

After studying this chapter, you will be able to:

✓ Describe the different types of mechanical variators used in CVTs.

✓ Describe the different types of CVTs that use both a mechanical input and a hydraulic variator.

✓ Explain transmissions that use a single hydrostatic pump to drive two hydrostatic motors.

Mobile equipment has been using *continuously variable transmissions (CVT)* since 1996. A CVT, also known as a *stepless transmission* or *infinitely variable transmission (IVT)*, is a transmission that provides an infinite range of speed control rather than a limited number of discrete gear ratios. The CVT normally uses a *variator*, which is a variable drive that produces an infinite number of speeds. The variator can be a mechanical drive, in the form of a variable belt drive or a variable roller drive. CVTs can also use a hydraulic variator consisting of a hydraulic pump and motor.

The CVTs can use a combination of a mechanical input plus a variator, or the CVT might use the variator by itself with no other mechanical input. The CVT might deliver a single output, such as driving the rear axle, or it might deliver two outputs, for example driving the front and rear axles. CVTs can be used for propelling the machine, or for other drives like rotors, headers, and feeders. The types of CVTs used in mobile equipment include variable-diameter pulley (VDP) system, toroidal, combination mechanical drive with a hydraulic variator, and hydrostatic drive transmission (HST).

Mechanical Variators

CVTs can have one of two types of mechanical variators, the variable diameter pulley (VDP) and a toroidal CVT. Both are used in mobile equipment.

Variable Diameter Pulley (VDP) CVT Drive

Chapter 6, *Belts and Chain Drives* and Chapter 8, *Clutches and Planetary Controls* explain that rubber belt drives can use a variable diameter pulley (VDP) drive. The system uses two variable sheaves that widen and narrow resulting variable pulley diameters and a fixed diameter drive belt. Two examples of rubber belt VDPs include a combine rotor drive and the transmission in a utility terrain vehicle (UTV), such as a John Deere Gator.

The VDP CVT, also known as a Reeves drive, uses a flexible steel belt with either a push or a pull design. *Push belts* drive by pushing the

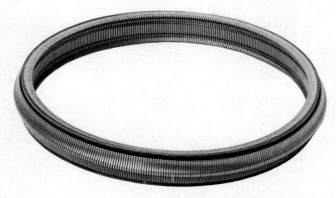

Figure 15-1. A flexible steel push-type belt used in a mechanical CVT variator. The belt is comprised of hundreds of steel segments with serrated edges held together with two steel ring packs.

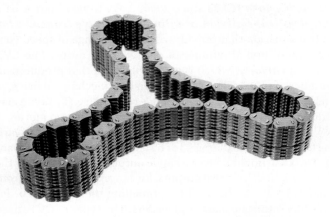

Figure 15-2. An example of a silent chain (pull-type belt) that can be used in a mechanical CVT variator.

belt to the output pulley. *Pull belts* drive by pulling the sheave. Both are used in automobiles and the pull belt is used in compact tractor applications.

A push belt is a directional belt that has an arrow on the belt pointing in the direction of belt rotation. The push belt is comprised hundreds of steel segments, also known as elements. See **Figure 15-1**. The elements have ridged serrations on the sides that drive the variable steel sheaves. The elements are also directional and are centered between two sets of steel ring packs. A ring pack has multiple thin steel rings, for example 9 to 12 thin rings, on each side of the segments.

The pull belt, also known as a chain drive or Luk-brand chain, is comprised of hundreds of chain links and attaching pins, similar to a silent chain. The variable sheaves have a steeper taper than push belt sheaves. See **Figure 15-2**. Power is transmitted by the pins contacting the pulley surfaces. The chain drive CVT is used in a New Holland compact utility Boomer series tractor called the EasyDrive CVT.

Toroidal CVT Drive

A *toroidal drive* uses one or two sets of cone shaped rollers that pivot between two toroidal-shaped discs. A simplified drawing is shown in **Figure 15-3**. The input disc drives a set of power rollers. As the rollers pivot, the drive ratio varies based on where the power rollers contact the input disc and output disc. Manual shift controls can be used to move mechanical pivot the power rollers. Electronic controls can also be used to enable the transmission to function like a stepless automatic transmission.

If one set of rollers are used as shown in **Figure 15-3**, the output disc rotates in the opposite direction of the input disc. When one set of power rollers drive a second set of power rollers, the second set of rollers reverse the direction of rotation back to the same direction as the input roller. The CVT CORP produces and licenses the toroidal CVTs marketed for mobile equipment. Skyjack uses the toroidal CVT in telehandlers.

Mechanical Input Plus Hydraulic Variator CVT

The most popular style of CVT used in mobile equipment combines a mechanical input and a hydraulic variator input. The mechanical input can vary among manufacturers, and can be coupled to the transmission by synchronizers, multiple disc clutch packs, planetary gear sets, countershaft gearing, or a direct input from the engine via a drive coupler with a dampener assembly.

The most common hydraulic variator is a hydrostatic transmission; however, the variator can use a traditional hydraulic motor that receives flow from an implement pump. A CVT normally does not use any type of a hydrodynamic drive like a torque converter, due to its inefficiencies.

The combined mechanical and hydraulic variator CVT can be held in neutral, driven forward or reverse, and change speed or torque without the traditional feel of upshifting or downshifting. The hydraulic variator is the input that allows the transmission to be infinitely variable.

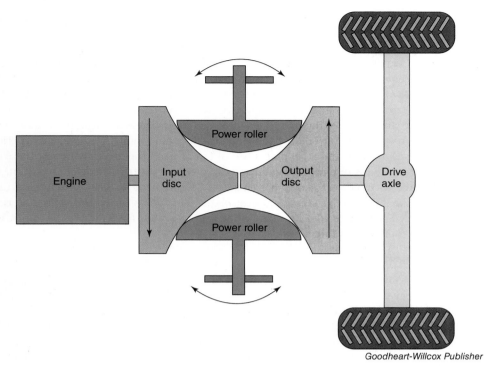

Figure 15-3. An example of a toroidal type mechanical variator with one set of power rollers.

Unlike traditional mechanical transmissions, CVTs allow a continuous range of travel speeds. They can provide speeds ranging from creeper speeds to fast road speeds without requiring a shift. As a result, operators do not need to use a clutch pedal or shift gears.

An airport walkway conveyor can be used to explain the operation of the mechanical, hydraulic-variator-type CVT. A person that steps onto a moving conveyor moves forward at a set speed. The direction and speed of the person standing still on the conveyor is similar to the fixed direction and speed of the engine's mechanical input to the CVT. The variable speed and direction of the hydraulic input is similar to the person choosing to move forward or rearward on the conveyor while it is moving. This variable input can change the overall speed and direction that the person moves relative to the surroundings. For example, if the person decides to walk backward on the conveyor at a slow speed, the person would remain in the same position relative to the surroundings. This is analogous to the power neutral in a CVT. If the person moves backward quickly, the person can overcome the motion of the conveyor and begin moving backward relative to the surroundings. If the person chooses to walk or run forward on the moving conveyor, the person moves forward faster than a person standing still on the conveyor.

Case IH Combine Feeder and Rotor CVT

In 2003, Case IH introduced their 8010 flagship combine. The models have evolved over the years from 10 series, 20 series, 30 series, and 240 series to 250 series combines (7250, 8250, 9250). The flagship combine offers a Power Plus CVT rotor drive and a Power Plus CVT feeder drive. See **Figure 15-4**.

Figure 15-4. The PTO gearbox on a Case IH combine is used to drive both the rotor CVT and the feeder CVT.

The feeder CVT contains a bevel gear set, a simple planetary gear set, two clutches, and a bidirectional fixed-displacement hydraulic motor. The ring gear is either held stationary or driven by the engine. The sun gear is either held stationary or driven by the hydraulic motor. The planetary carrier is the output of the CVT. The bevel gear set provides the mechanical input from the engine-driven PTO gearbox. One of the two clutches is the engine-to-ring (ETR) clutch, which is an input clutch for driving the ring gear. See **Figure 15-5**. The ETR friction plates have external splines and the steel plates have internal splines.

The other clutch is the ring-to-frame (RTF) clutch, which is a stationary clutch. The RTF clutch is responsible for holding the ring gear to frame (housing). The construction of the RTF friction discs and steel plates are different from the construction of the ETR clutch. The steel separator plates are externally splined and the friction discs are internally splined, **Figure 15-6**. When the clutch is engaged, CVT housing acts like the clutch drum, mating with the RTF clutch's steel separator plates. This holds the clutch drum and ring gear stationary. See **Figure 15-7** for a simplified drawing of the feeder CVT. The rotor CVT is similar, but does not contain a bevel gear set.

Clutch Functions and Planetary Operation

Any time the ETR clutch is applied, the engine provides a mechanical input into the CVT through the bevel gear set and drives the ring gear. When the ETR clutch is applied, the hydraulic motor can either hold the sun gear stationary or can rotate it forward or backward.

If the RTF stationary clutch is applied, holding the ring gear stationary, the hydraulic motor is the sole input into the CVT. It drives the sun gear in forward or reverse, or holds the sun gear stationary to hold the feeder drive stationary. Note that the ETR (input clutch)

Goodheart-Willcox Publisher

Figure 15-5. The feeder CVT on a Case IH combine uses a clutch that couples the engine to the ring gear to provide the mechanical input to the CVT. This clutch is called an ETR clutch. A—The ETR clutch's friction discs have external splines. The steel separator plates have internal splines. B—The clutch drum mates with the friction discs. C—The ring gear is part of the ETR drum.

Goodheart-Willcox Publisher

Figure 15-6. The RTF clutch is responsible for holding the ring gear to frame (housing).
A—The steel separator plates use external splines. The friction discs use internal splines.
B—The CVT housing acts like the clutch drum, mating with the rtf clutch's steel separator plates.

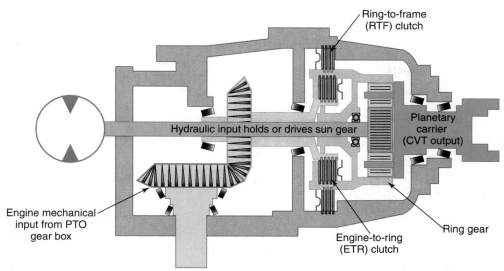

Figure 15-7. Case IH's flagship combine uses one CVT to power the feeder and another CVT to power the rotor. A bevel gear set provides the mechanical input into the feeder CVT. A hydraulic motor holds or drives the sun gear, and two clutches either hold the ring gear or allow the mechanical input to drive the ring gear.

and the RTF (holding clutch) can never be applied at the same time because it would stall the engine. Only one clutch can be engaged when the engine is running.

Feeder Drive Line

The feeder CVT output drives a drive shaft on the left-hand side of the combine. See **Figure 15-8**. The drive shaft delivers power to a feeder gearbox located at the top of the feeder house. The feeder gearbox drives the feeder house and delivers power to the bottom of the feeder to the header gearbox.

Engine Input

If the hydraulic motor is held stationary and the ETR input clutch is applied, it causes the ring gear to be an input and the planetary carrier to be the output in a fast torque-multiplication mode. This mode causes the CVT output speed to be proportional to the engine's speed. It is not used during field operation due to the speed being fixed in proportion to the engine speed.

Engine and Hydraulic Input

If the ETR input clutch is applied and the hydraulic motor is driven, the planetary carrier output speed can be increased if the hydraulic motor is driven forward. The planetary carrier output speed can be decreased if the hydraulic motor is driven in reverse.

Hydraulic Input

If the RTF clutch is engaged, which holds the ring gear, the hydraulic motor is the sole input to the CVT. If the hydraulic motor is spun forward, the feeder will drive forward. If the hydraulic motor is held stationary, the feeder will remain stationary. If the hydraulic motor is driven backward, the feeder drive will rotate backward.

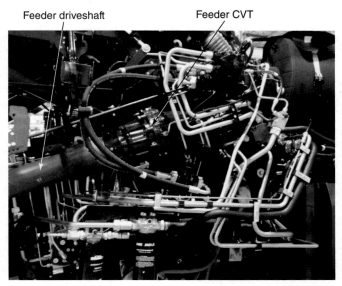

Figure 15-8. The feeder CVT delivers power to the feeder house through a drive shaft that is positioned on the left side of the combine.

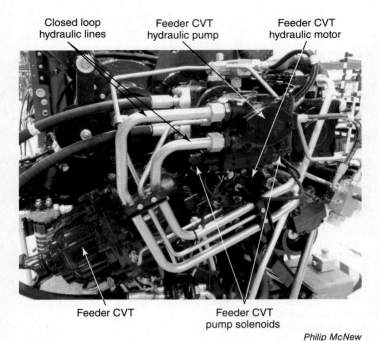

Figure 15-9. The Case IH combine feeder CVT uses a variable-displacement reversible hydraulic pump to vary the hydraulic input into the CVT drive. Two solenoids control the direction and quantity of pump flow. The dedicated feeder CVT pump and motor are connected through the two closed loop hydraulic lines.

Any time the RTF clutch is engaged, the feeder will rotate in the same direction as and at a speed proportional to hydraulic motor rotation. Note that the speed variability and the direction of the motor is controlled directly by the feeder drive pump. The pump has a forward solenoid and a reverse solenoid. The combine control module #1 (ccm1) varies the pulse width modulation (PWM) to the solenoids to increase pump flow, which varies the CVT hydraulic motor rotation. See **Figure 15-9**.

Feeder Modes of Operation

The Case IH flagship combine's feeder has six modes of operation: off, calibration, feeder engagement manual mode, feeder engagement auto mode, reverse, and passive deceleration.

Feeder Off

In the off mode, the feeder drive is held stationary. The ETR clutch is disengaged. The RTF clutch is engaged and the hydraulic motor is hydrostatically held in neutral.

Calibration Mode

The calibration mode is performed so the controller known as the combine control module #1 (ccm1) can learn the ETR and RTF clutch fill times. The calibration mode also allows the controller to learn how much current is required to actuate the feeder's hydrostatic pump.

The CVT has three controllers that work together for operating the feeder: right hand module (rhm), ccm1, and ccm2. The rhm receives inputs from the feeder reverse switch, feeder speed potentiometer, reel speed potentiometer, and auto/manual switch. The ccm2 receives inputs from the ground speed sensor and the seat switch. The ccm1 communicates with the rhm and ccm2 across the controller area network (CAN). The ccm1 receives input from the feeder engage switch through the k-16 relay. The ccm1 controls CVT outputs by engaging and disengaging the ETR clutch solenoid, RTF clutch solenoid, and the feeder pump solenoids. See **Figure 15-10**.

Feeder On (Manual)

When the feeder has been engaged, the ccm1 controls the solenoid to modulate oil pressure to apply the ETR clutch, which causes the feeder to begin to rotate. The ccm1 energizes the feeder hydrostatic pump solenoids, causing the pump to drive the hydraulic motor to achieve the feeder speed requested by the operator. The feeder will achieve its desired speed within the first five seconds that the feeder is engaged. The speed is set based on the adjustment of the feeder-speed potentiometer.

Feeder On (Automatic)

In the automatic mode, the controller drives the feeder similarly to the way it does in the manual mode, except it varies the speed of the feeder's CVT based on the ground speed of the combine and the ratio established by the operator. The ratio is adjusted by varying the same feeder-speed potentiometer that was used in the manual mode. The difference is that the feeder switch is in the automatic mode, and not the manual mode.

Reverse

If the feeder house or the header becomes plugged, the operator can direct the CVT to reverse the feeder drive. The reverse position on the feeder switch is a momentary position, meaning that the operator must hold the switch in this position to keep it selected. As the

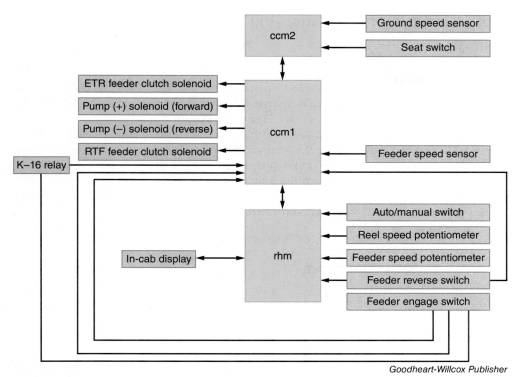

Figure 15-10. The central ccm1 is responsible for controlling the outputs of the feeder CVT circuit: the ETR and RTF clutch solenoids and the feeder pump solenoids. The ccm1 communicates with the rhm and ccm2 through the controller area network (CAN).

operator holds the feeder switch in the reverse position, the ETR clutch is disengaged, the RTF clutch is engaged, and the feeder drive pump rotates the CVT's hydraulic motor in reverse. The maximum header reverse speed is 130 rpm with the engine at high idle.

Passive Deceleration

As the operator shuts off the feeder, if the feeder speed is higher than 50 rpm, the controller disengages the ETR input clutch and monitors the feeder speed. When the feeder speed drops below 50 rpm, the controller engages the RTF clutch.

Note

If the hydraulic motor is stationary and the RTF clutch is engaged, the feeder is held in a locked position.

Agricultural Tractor CVTs

For years, farmers have wished that their tractors could shift to a slightly different gear ratio to achieve the optimum field speed and tractor efficiency. CVTs have become popular in agricultural tractors because they offer farmers a transmission that is infinitely variable. This allows the farmer to maximize the tractor's efficiency by operating the tractor at the precise speed needed for any situation.

An agricultural tractor CVT also often allows a customer to set the CVT controls based on the preferred usage of the tractor. For example, a CVT normally allows the operator to set the controls to either maximize fuel economy, to improve fuel economy while ensuring the tractor has enough pulling power to power through heavy loads, or to maintain an engine speed to ensure the PTO and/or hydraulic flow needs are always met.

Case IH Magnum and New Holland T8 CVT Inputs and Outputs

In 2011, Case New Holland (CNH) began offering a CVT in their high-horsepower MFWD tractors, the Case IH Magnum (235–380 hp), and the New Holland T8. The CVT contains one compound planetary gear set that receives two inputs, one mechanical input from the engine and a variable hydraulic input. See **Figure 15-11**.

The engine's crankshaft provides the mechanical input to the CVT. The tractor's direction is selected by engaging either the forward clutch or the reverse clutch. Both the forward and reverse clutch drive the input sun gear (large sun gear). However, the reverse clutch receives its input via a reverse idler gear, causing the tractor to reverse its direction. A variable, reversible, hydraulic pump and a fixed hydraulic motor provide the variability of the CVT by changing the speed and the direction of the planetary gear set's ring gear. See **Figure 15-12**.

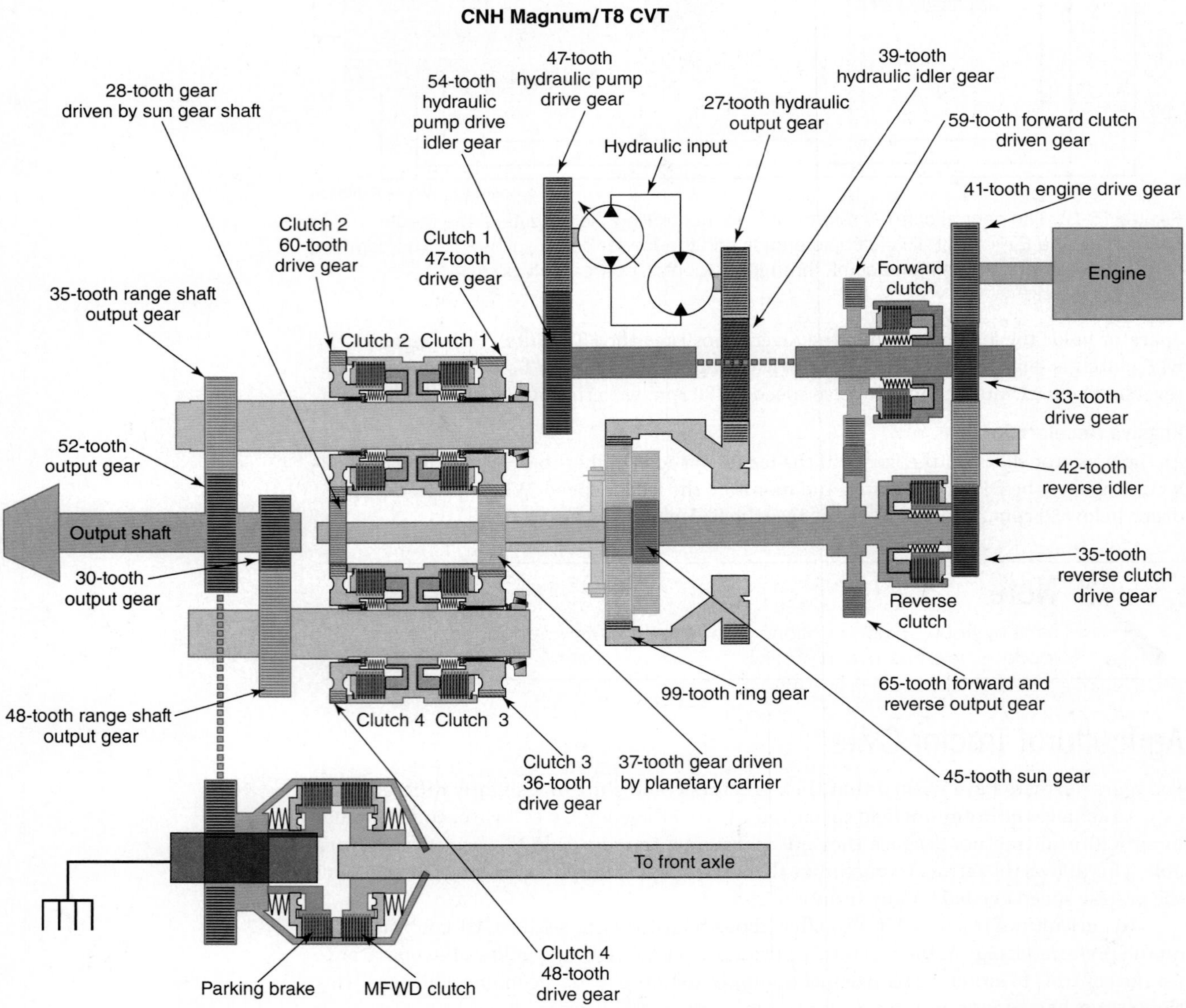

Goodheart-Willcox Publisher

Figure 15-11. The Case IH Magnum CVT and the New Holland T8 CVT use six clutches, a compound planetary gear set, and a variable hydraulic input.

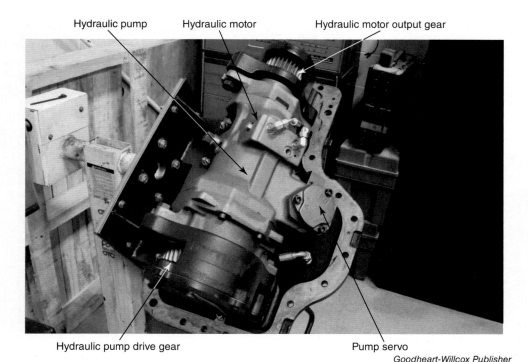

Figure 15-12. The Case IH Magnum and New Holland T8 CVT uses an integral hydrostatic transmission that is responsible for varying the speed and direction of the compound planetary ring gear. The pump and motor are located inside a single housing, making it difficult to see where the pump ends and the motor begins. This hydraulic pump and motor assembly bolts into the top of the transmission's case. In this image, the hydraulic pump and motor have been mounted on a revolving engine stand. The hydraulic pump drive gear rotates proportional to engine speed. The pump servo piston (behind the plate) varies the quantity and direction of pump flow, and the hydraulic motor output gear drives the ring gear through the 39-tooth idler gear.

The universal control module (UCM) energizes a single solenoid for controlling the CVT's variable-displacement reversible hydrostatic pump. The pump's servo piston is positioned by a spring to a negative swashplate angle (SPA) of approximately 100%. As the UCM varies the amperage to the solenoid (400 milliampere to 1220 milliampere), the swashplate angle changes from a wide negative angle to a wide positive angle. A current of approximately 850 milliampere, causes the pump swashplate to be in a neutral position. See **Figure 15-13**.

The compound planetary gear set drives two output members, a planetary carrier shaft (also called the outboard output shaft) and a sun gear shaft (also called high-speed inboard output shaft). The gear set's two sizes of planetary pinions form a ***cluster gear*** assembly. Each cluster gear (pair of large and small pinions) rotates at the same speed and in the same direction. The planetary carrier output shaft delivers power to the number one and number three range clutch hubs, which will drive the respective clutch drum when the clutch is engaged. The sun gear output shaft delivers power to the number two and number four range clutch hubs, which will drive the respective clutch drum when the clutch is engaged. The four drums are splined to their respective range countershaft. See **Figure 15-14**.

Clutches one and two are on one range shaft and clutches three and four are on the other range shaft. During CVT operation, as the tractor increases in speed, the universal control module (UCM) alternates the clutch pack engagements in consecutive order from clutch one through clutch four by pulse-width modulating the solenoids. Each range shaft contains a gear at the end of the shaft that drives a gear on the CVT's output shaft. The transmission's output shaft delivers power to the rear axle. The transmission's output shaft

Signal Current to Solenoid	Swashplate Angle (percentage of maximum angle)
400 mA	−99.7 %
450 mA	−88.3 %
500 mA	−77.2 %
550 mA	−66.5 %
600 mA	−56.0 %
650 mA	−45.6 %
700 mA	−32.2 %
750 mA	−24.7 %
800 mA	−13.9 %
850 mA	−2.9 %
900 mA	8.6 %
950 mA	20.7 %
1000 mA	33.4 %
1050 mA	46.9 %
1100 mA	61.3 %
1150 mA	76.6 %
1200 mA	93.0 %
1220 mA	99.9 %

Goodheart-Willcox Publisher

Figure 15-13. The single solenoid is used to control the pump swashplate from a negative 100% angle to a positive 100% angle. A—Location of the swashplate angle solenoid. B—At the lowest amperage, 400 milliampere, the swashplate is at a full 100% negative angle (equaling maximum reverse hydraulic flow). At approximately 850 milliampere, the swashplate angle is brought to a neutral angle. At approximately 1220 milliampere, the swashplate is positioned at a 100% positive angle producing maximum forward hydraulic flow.

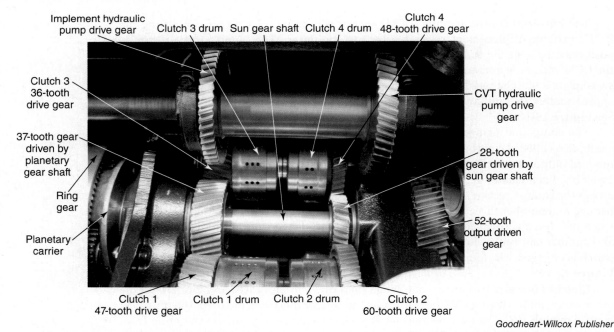

Goodheart-Willcox Publisher

Figure 15-14. The Case IH Magnum and New Holland T8 CVT uses a compound planetary gear set that provides two outputs, the planetary carrier shaft and a sun gear shaft. The planetary carrier delivers power to clutch 1 and clutch 3. The sun gear shaft delivers power to clutch 2 and clutch 4.

also contains a gear that is in mesh with the parking brake and MFWD clutch assembly (**Figure 15-11**). Both are spring applied and hydraulically released.

The CVT hydraulic pump/motor assembly is installed in the top of the transmission housing. See **Figure 15-15**.

Case IH Magnum and New Holland T8 CVT Operator Controls

The Case IH Magnum CVT and New Holland T8 CVT function the same, but the operator controls are a little different. Both tractor brands use a *multifunction handle (MFH)* that functions as a propulsion lever. The MFH also contains several buttons and a rotary encoder. See **Figure 15-16**.

Most agricultural CVT tractors have *preselect modes*. A preselect mode is an adjustable preprogrammed setting that determines the maximum travel speed for a particular situation. Multiple preselect modes allow the operator to switch between different transmission settings for different situations with the push of a button. A preselect mode is adjusted by pressing the desired preselect mode on the tractor's monitor (or switches on the MFH) and rotating the *rotary encoder*, which is a rotary speed dial on the MFH. The tractor attempts to achieve the maximum travel when the operator moves the MFH forward or rearward and the clutch pedal is released. If the implement load is light enough and the engine throttle is high enough, the tractor will achieve the preselect mode's maximum speed. The preselect mode maximum travel speed can normally be adjusted while moving by rotating the rotary encoder. Preselect modes are discussed later in this chapter. Most agricultural CVT tractors have multiple forward preselects. Some only have one reverse while others have multiple reverse preselects.

MFH Controls

The T8 MFH has springs that return the MFH to its centered (middle) position when the operator releases it. The Magnum MFH does not have these return springs and remains in its last commanded position when it is released. This means that if a Magnum MFH is

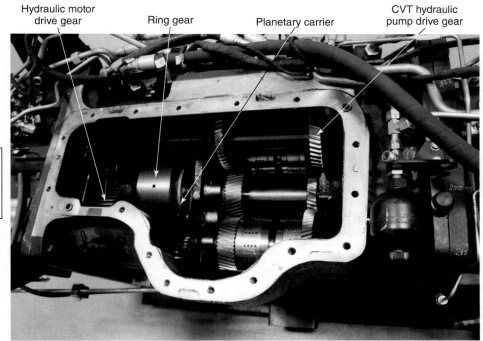

Goodheart-Willcox Publisher

Figure 15-15. The CVT hydraulic pump and motor have been removed from the transmission case, exposing the CVT gearing. Notice how the hydraulic pump and motor assembly shown in **Figure 15-12** would align and bolt up to this transmission.

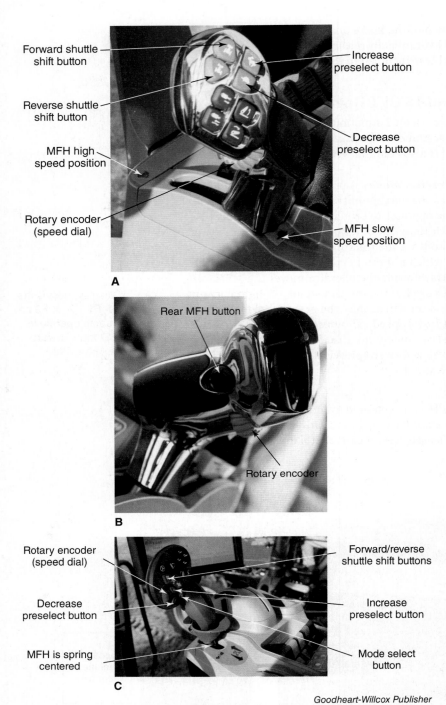

Figure 15-16. The Case IH Magnum MFH is actuated within a single slot. A—The Magnum MFH uses the following controls for operating the CVT: two shuttle shift buttons, two speed preselect buttons, and a rotary encoder (speed dial). B—On the backside of the Magnum MFH is a button used in conjunction with the shuttle shift buttons. C—The New Holland T8 CVT MFH is spring centered in its own slot. Like the Magnum MFH, it has forward/reverse shuttle buttons, a rotary encoder (speed dial), and speed preselect buttons.

moved forward, it remains in that position until the operator repositions it.

For example, suppose a Magnum operator chooses forward preselect 1 and preselect 1 has been set to a maximum speed setting of 5 mph. If the operator moves the MFH fully forward, the tractor will attempt to increase speed to 5 mph, depending on the tractor load and the engine throttle setting. To slow the tractor, the operator must pull the Magnum MFH rearward.

The T8 MFH operates similarly, except the operator only momentarily pushes the MFH forward. For example, the operator presses the MFH all the way forward when wanting to achieve maximum speed, then releases the MFH allowing it to return to the middle position. To reduce the tractor speed, the MFH is momentarily pulled rearward. The degree of rearward movement varies the tractor's travel speed within its preselected speed range. After the operator releases the MFH from the rearward position, the MFH moves forward to its spring-centered position.

CNH Magnum and T8 Preselects

Both the Magnum and the T8 have three forward and three reverse preselect modes of operation. **Figure 15-17** shows the transmission display in the front corner display monitor (known as an *A-post display* or *corner post display*) in a New Holland T8 tractor. The monitor is displaying the maximum travel speed for three forward preselects: 19 mph, 17 mph, and 1.6 mph. Also notice in this example the tractor is operating in the F2 preselect mode.

The operator can adjust the preselect mode's maximum speeds. However, the preselect modes must be set in a consecutive order, meaning that F2 must be higher than F1, and F3 must be higher than F2. Reverse preselects must also be set in the same way, with R2 higher than R1, and R3 higher than R2.

Both the Magnum MFH (**Figure 15-16A**) and the T8 MFH (**Figure 15-16C**) have two pre-select buttons. When the operator presses the increase preselect button, it causes the tractor's preselect mode to shift upward, for example from F1 to F2. The A-post display highlights the selected preselect mode. Pushing the decrease preselect button causes the tractor to shift downward, for example from F2 to F1. When a preselect mode is highlighted on the A-post display, the operator can use the rotary encoder to increase or decrease the maximum speed for the selected preselect mode. For example, in **Figure 15-17**, the operator could adjust the rotary encoder to either increase or decrease the F2 setting of 17.0 mph.

Pro Tip
The maximum preselect speed can be adjusted whether the tractor is moving or sitting still.

Transmission Forward, Neutral, Reverse, Park Lever

The Magnum and T8 CVT tractors use the same forward, neutral, reverse, park (FNRP) lever. See **Figure 15-18**. The lever is in front of the steering wheel. The lever has two spring-detented positions, neutral and park.

The tractor is in the park position when the lever is pushed forward and to the right. When the lever is shifted to the left and pulled toward the steering wheel, the tractor's transmission is placed in the neutral position. From the neutral position, the lever can be momentarily pulsed to the left (for reverse) or pulsed to the right (for forward).

The operator can use the MFH lever to shuttle shift by pressing the button on the back of the MFH and pressing the MFH forward/reverse shuttle shift buttons. A shuttle shift occurs when the tractor is moving in one direction (for example, forward) and the operator commands the tractor to move in the opposite direction (reverse). Many modern tractors, including CVTs, do not require pressing the clutch pedal to shuttle shift.

Split Throttle

The Magnum CVT uses two throttle position levers, known as a dual throttle or split throttle. See **Figure 15-19**. The left throttle lever adjustment sets the minimum engine operating speed. The right throttle lever sets the maximum engine speed. The operator chooses the throttle conditions, then the engine and the transmission work together (in automatic mode) to obtain the best efficiency, performing more work with less fuel.

Loader Application

If the tractor is being used in a loader application, the left-hand throttle can be adjusted to the minimum desired engine speed. The right lever can be pushed further ahead so that the engine can achieve a higher speed. See **Figure 15-20A**. Using the foot accelerator pedal, the operator can maximize the tractor's productivity.

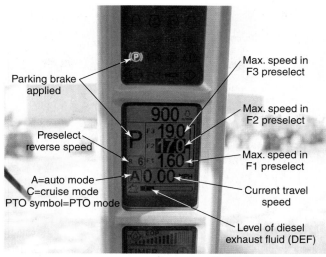

Goodheart-Willcox Publisher

Figure 15-17. The tractor's monitor (A-post display) shows the current preselect settings, which will be either F1, F2, or F3 or R1, R2, or R3. In this example, the tractor is set to operate in preselect F2 and is limited to a maximum of 17 mph forward travel speed. The display also lists the maximum reverse speed if the tractor is shuttle shifted into reverse, in this example reverse maximum speed is set to 6 mph (R6).

Goodheart-Willcox Publisher

Figure 15-18. The forward, neutral, reverse, park lever is located on the steering column. The lever is currently in the park position.

Figure 15-19. The split throttles set the range of allowable engine speeds. The left lever sets the minimum engine speed. The right lever sets maximum engine speed.

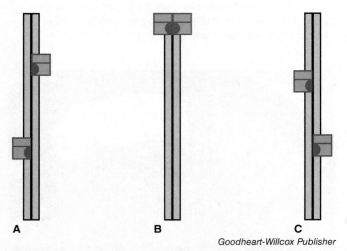

Figure 15-20. The Magnum CVT uses dual throttle levers. The positions of the throttle levers determine how the engine reacts to varying loads. A—Operators working in a loader application may choose to move the right throttle lever ahead of the left lever and use the accelerator pedal. B—If the operator wants the engine to run at a constant speed, both levers must be set to that constant speed. C—If the tractor is operating with varying engine loads and the operator wants the engine to be able to lug (drop with the load), the right lever must be positioned behind the left lever.

Fixed Engine Speed

If a specific engine speed is desired at all times, both throttle levers must be moved to that desired engine speed position. If both levers were moved fully forward, then the tractor would always be able to deliver maximum hydraulic flow and good PTO speed, but the tractor would be operating at its least efficient setting. See **Figure 15-20B**.

Ability to Lug the Engine

The throttles can be set so that the engine has the ability to be pulled to a lower speed due to an increasing engine load. In this example, the left throttle would be set further ahead of the right throttle, allowing the engine to lug down when the tractor's load becomes difficult, such as tilling a field that changes from light soil to heavy soil, or baling hay and the crop becomes denser. The engine could lug to the speed set by the right throttle setting. See **Figure 15-20C**.

New Holland T8 Throttle and Droop Control

The New Holland T8 tractor uses a single throttle lever in conjunction with an adjustable engine droop control. See **Figure 15-21**. The lever sets the desired engine speed. Depending on the transmission's mode of operation, *droop control* is the electronic engine control that allows the operator to set either the maximum or the minimum engine speed. In the normal mode, the droop sets the maximum engine speed setting. In the PTO mode, the droop control sets the minimum engine speed threshold.

Acceleration Response Control

The Magnum and T8 CVT has an acceleration response control (ARC) that allows the operator to adjust how responsively the tractor accelerates and decelerates. The operator chooses one of three modes: 1, 2, or 3.

- Mode 1 provides minimum acceleration response.
- Mode 2 provides medium acceleration response.
- Mode 3 provides the maximum acceleration response.

The ARC button is identified in **Figure 15-19**. The button contains three small green lights (LEDs) that illuminate to indicate the specific acceleration mode that has been selected. By pressing the button, the operator cycles through the three settings and the number of lights illuminated indicates which of the three settings has been selected.

Manual Mode

Operation of a tractor with the Magnum and T8 CVT in manual mode is similar to operation of a tractor with a powershift transmission. The operator uses the throttle lever or foot pedal to control the engine speed and the MFH to control the transmission ratio.

Note

In manual mode, both Magnum dual throttles must be moved simultaneously because the dual throttle function is not available.

Auto Mode

If the CVT is in the auto mode and the load on the tractor increases, the engine speed increases to automatically compensate. If the increase in engine speed is insufficient to overcome the tractor's load, the transmission changes gear ratios to increase torque.

Cruise Mode

The cruise mode is available when the tractor is operating in the auto mode. It is activated by pressing the mode select button on the MFH. Cruise mode allows the operator to set the tractor to operate at a constant travel speed.

Figure 15-21. The New Holland T8 CVT tractor uses a hand throttle lever and an engine droop control adjustment.

PTO Mode

In the PTO mode, the transmission and engine work together to operate at a set engine-speed range. This allows the tractor to maintain a specific PTO speed or hydraulic system flow rate. The engine droop control can be adjusted to set the minimum engine speed. When operating a large square baler, it is important for a tractor to maintain a PTO speed of 1000 rpm. During this application, the engine's hand throttle is adjusted to approximately 1750 rpm, which provides a PTO speed of approximately 1000 rpm. The engine's droop control knob is adjusted to approximately 85%, which equals roughly 1800 engine rpm. As the tractor experiences heavier loads, the tractor's ground speed (not engine speed) is reduced, allowing the PTO to maintain the 1000 rpm.

If the PTO is being used to inject manure, it is not as critical to maintain a PTO speed of 1000 rpm. The operator can set the engine hand throttle to 1750 rpm, but set the engine droop to 65%, which allows the engine to fall to 1400 rpm. As the tractor encounters a heavier load, the engine can drop to 1400 rpm (known as lugging the engine) with a proportional drop-off in PTO speed as the tractor tries to achieve the target ground speed.

Normal Mode

In a T8 tractor in normal mode, the engine droop control sets the maximum engine speed. This mode is commonly used for tillage applications. The T8's engine produces maximum horsepower and torque at an engine speed near 1800 rpm. Limiting the engine's maximum speed to 1800 rpm ensures the engine will be able to produce maximum torque and horsepower and not waste fuel by running at a higher speed. In a tillage application, the operator sets the minimum engine throttle by moving the engine throttle lever to 1200 rpm. The droop control knob is adjusted to 85%, which sets the engine's maximum speed to roughly 1800 rpm.

If the tractor encounters a hard soil as it moves through the field, making it difficult to pull the tillage tool, the engine droop control does not allow the engine to increase above 1800 rpm. Instead, the tractor's travel speed is reduced to keep the engine operating within the power band of 1200 to 1800 rpm set by the operator. As the tractor travels into softer soils, the tractor increases speed until it reaches its target ground speed, while the engine speed is varied to maintain maximum efficiency.

Case IH Magnum and New Holland T8 CVT Power Flow

Depending on the speed requested by the operator, the CVT will be in one of the following modes: park, power zero, range 1 forward, range 2 forward, range 3 forward, range 4 forward, range 1 reverse, range 2 reverse, or range 3 reverse.

Park

When the tractor's controls have been placed in the park position, the clutches are released and the parking brake is applied. If the MFWD clutch is disengaged, the rear wheels or tracks are held stationary, but the front wheels are free to roll. If the MFWD clutch is applied, all four wheels (and/or tracks) are held stationary.

Power Zero (pz)

If the parking brake is released and the MFH is positioned at 0%, the CVT holds the tractor stationary. The transmission's clutches are engaged (for example forward clutch and clutch 1), but the CVT adjusts the hydraulic input so the ring gear rotates backward. As a result, the tractor is powered but remains stationary. This is known as *power neutral*, *power zero (pz)*, or active stop. The tractor will remain stationary without requiring the operator to press the clutch pedal or brake pedals. The CVT holds the tractor stationary regardless of the ground slope. The tractor will *not* stay in the pz mode indefinitely. After 90 seconds, it will automatically apply the parking brake and shift to neutral.

Range 1 Forward

When the tractor begins to move forward from a stop, the forward clutch provides mechanical input into the compound planetary gear set through the large sun gear. The hydrostatic pump begins at a wide negative swashplate angle (–90%), which initially holds the tractor in a power neutral position. When the operator commands the tractor to move forward, the hydrostatic pump's negative swashplate angle moves from the –90% angle toward neutral, based on the target ground speed chosen by the operator. This slows the backward spinning ring gear, resulting in the planetary carrier's output shaft delivering power to clutch one.

When clutch 1 is engaged, the clutch drum drives the range shaft, which drives the output shaft. In "range 1 forward" mode, the hydrostatic pump starts at –90% swashplate angle and varies through a neutral angle up to +84% swashplate angle. When the pump is at an exact neutral angle, all of the input power being transmitted into the planetary gear set is driven mechanically by the engine through the sun gear shaft (large sun gear).

Range 2 Forward

If the operator requests a faster travel speed than "range 1 forward" can produce, the UCM disengages clutch 1 and engages clutch 2. This causes the range transmission to receive power from the sun gear output shaft (small sun gear) instead of the planetary carrier output shaft.

In "range 1 forward" mode, the last hydrostatic pump swashplate angle was +84%. At the start of "range 2 forward", the UCM begins to reduce the +84% pump swashplate angle, crossing through neutral, until the pump reaches a –10% swashplate angle. The –10% swashplate angle maximizes the tractor speed for "range 2 forward".

Range 3 Forward

If the operator requests a faster travel speed than "range 2 forward" can produce, the UCM disengages clutch 2 and engages clutch 3. This causes the range transmission to receive power from the planetary carrier output shaft instead of the sun gear output shaft.

In "range 2 forward" mode, the last hydraulic pump swashplate angle was –10%. At the start of "range 3 forward", the UCM begins to reduce the –10% swashplate angle, crossing through neutral, until the pump reaches a +80% swashplate angle, which maximizes the tractor speed for "range 3 forward".

Range 4 Forward

If the operator requests a faster travel speed than "range 3 forward" can produce, the UCM disengages clutch 3 and engages clutch 4. This causes the range transmission to receive power from the sun gear output shaft instead of the planetary carrier output shaft.

In "range 3 forward" mode, the last hydraulic pump swashplate angle was +80%. At the start of "range 4 forward", the UCM begins to reduce the +80% swashplate angle, crossing through neutral until the pump reaches a –100% swashplate angle. This maximizes the tractor speed for "range 4 forward."

The graph in **Figure 15-22** shows the changing swashplate angle of the hydrostatic pump as the tractor increases forward speed and progresses through the four forward range modes.

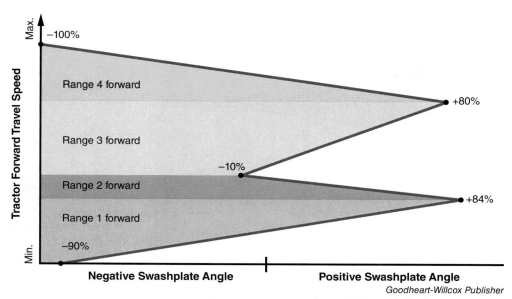

Figure 15-22. The Magnum and T8 CVT forward propulsion from first through four is depicted with the corresponding hydrostatic pump swashplate angle.

Range 1 Reverse

When the tractor is shifted into reverse, the reverse clutch provides mechanical input into the compound planetary gear set through the large sun gear. The hydrostatic pump begins at a wide positive swashplate angle (+90%), which initially holds the tractor in a power neutral position. When the operator commands the tractor to move in reverse, the hydrostatic pump's positive swashplate angle begins to decrease, based on the target ground speed chosen by the operator. This slows the ring gear, resulting in the planetary carrier's output shaft delivering power to clutch 1.

With clutch 1 engaged, the clutch drum drives the range shaft, which drives the output shaft. In "range 1 reverse" mode, the hydrostatic pump's swashplate angle starts at +90% and varies through a neutral angle, to a −84% swashplate angle.

Range 2 Reverse

If the operator requests a faster travel speed than "range 1 reverse" can produce, the UCM disengages clutch 1 and engages clutch 2. This causes the range transmission to receive power from the sun gear output shaft instead of the planetary carrier output shaft.

In "range 1 reverse" mode, the last hydrostatic pump swashplate angle was −84%. At the start of "range 2 reverse," the UCM begins to reduce the −84% pump swashplate angle, crossing through neutral until the pump reaches a +10% swashplate angle. This maximizes the tractor speed for "range 2 reverse."

Range 3 Reverse

If the operator requests a faster travel speed than "range 2 reverse" can produce, the UCM disengages clutch 2 and engages clutch 3. This causes the range transmission to receive power from the planetary carrier output shaft instead of the sun gear output shaft.

In the "range 2 reverse" mode, the last hydraulic pump swashplate angle was +10%. At the start of "range 3 reverse," the UCM begins to reduce the +10% swashplate angle, crossing through neutral until the pump reaches a −100% swashplate angle. This maximizes the tractor speed for "range 3 reverse."

The graph in **Figure 15-23** shows the changing swashplate angle of the hydrostatic pump as the tractor increases reverse speed and progresses from range 1 reverse through range 3 reverse modes.

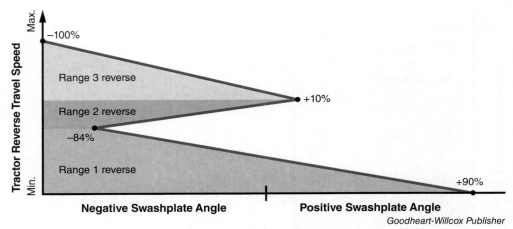

Figure 15-23. The Magnum and T8 CVT reverse propulsion from first through third is depicted with the corresponding hydrostatic pump swashplate angle.

Figure 15-24. AGCO's Terra Gator is used to spread fertilizer or can be used to spray chemicals, such as herbicides and insecticides.

Figure 15-25. The JCB Fastrac agricultural tractor uses a Fendt Vario CVT. It is one of the fastest tractors in the agricultural industry, with travel speeds up to 40 mph (65 kph). It uses four-wheel anti-lock disc brakes.

Fendt CVT

Fendt is a German tractor manufacturer owned by AGCO. In 1996, Fendt began producing the Vario CVT-equipped tractor for the European market. Today, the Vario is available worldwide, with models including the 200, 300, 400, 500, 600, 700, 800, 900, 1000, and 1100 series, and horsepower ranging from 81 to 670. AGCO also offers the Vario transmission in their Challenger tractors, Massey Ferguson tractors, and in their Terra Gator applicators that are used for spreading manure or spraying chemicals. See **Figure 15-24**. JCB is a non-AGCO manufacturer that uses the Fendt Vario CVT in their Fastrac agricultural tractors. See **Figure 15-25**.

Fendt's Vario CVT uses a simple planetary gear set instead of compound planetary gear set. It also does not use any clutch packs. **Figure 15-26** shows that the engine directly drives the planetary gear set's planetary carrier. One of the most unique features of this transmission is that the CVT's ring gear is used to drive the hydrostatic pump. Note that the speed of the ring gear is affected by the tractor speed, which will be explained later.

The CVT's mechanical output is the sun gear. The sun gear shaft contains a gear that drives the collector shaft, also known as the collection shaft. The collector shaft is directly splined to the one or two hydrostatic motors (depending on the model of tractor). If the CVT has two motors, the motors are plumbed in parallel.

Like other agricultural tractor CVTs, the Vario uses a variable reversible hydrostatic pump, which controls the direction and quantity of oil flow to the hydrostatic motor(s). The Vario's one or two hydrostatic motors are a bidirectional variable design. Both the pump and motor(s) are a bent axis design.

Neutral

When the transmission is in neutral, the engine is driving the planetary carrier, and the hydrostatic pump is at a neutral angle. Because the pump is not delivering any

Fendt CVT

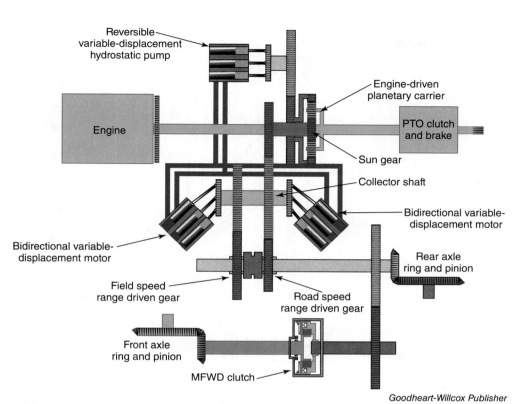

Figure 15-26. The engine in a Fendt Vario CVT-equipped tractor directly drives the planetary carrier in a single planetary gear set. The CVT has a reversible variable-displacement hydrostatic pump that is driven by the planetary gear set's ring gear. It also has one or two bidirectional variable-displacement hydrostatic motor(s) that are indirectly linked through a collector shaft to the planetary gear set's sun gear. In this drawing, the CVT is in neutral, resulting in the hydrostatic pump at a neutral angle and the motors at maximum displacement.

flow to the hydrostatic motors, the hydrostatic system has practically no drive pressure. The lack of resistance on the hydrostatic drive allows the planetary gear set's ring gear to drive the pump, which is waiting to be commanded to pivot. Neither range gear is coupled to the range shaft.

Forward Slow Speed

The Vario CVT has a pair of range gears on the range shaft. The operator must select the field gear or the road gear. Some Vario CVTs have a field speed range of 0–20 mph and a road speed range of 0–30 mph, though the field and road speed ranges vary depending on the Vario model.

After selecting the range gear, the operator chooses forward or reverse. If the operator selects the forward position, the pump pivots to a positive angle, causing oil to flow to the dual motors (or single motor). Because the motors begin at a maximum displacement, the hydrostatic pressure is at its lowest pressure setting, resulting in a faster ring gear speed.

As pump flow is directed to the hydrostatic motor(s), the collector shaft is hydraulically driven. As the collector shaft rotates, it drives the pair of range drive gears. The range shaft delivers power to the rear axle and to the MFWD clutch through whichever range shaft gear the operator has selected. See **Figure 15-27**. As the tractor speed increases, the planetary ring gear begins to experience resistance due to the increased hydrostatic pressure. This resistance causes the planetary sun gear to increase speed. As the tractor speed is commanded to increase, the hydrostatic pump angle is increased to increase flow to the hydrostatic drive motor(s).

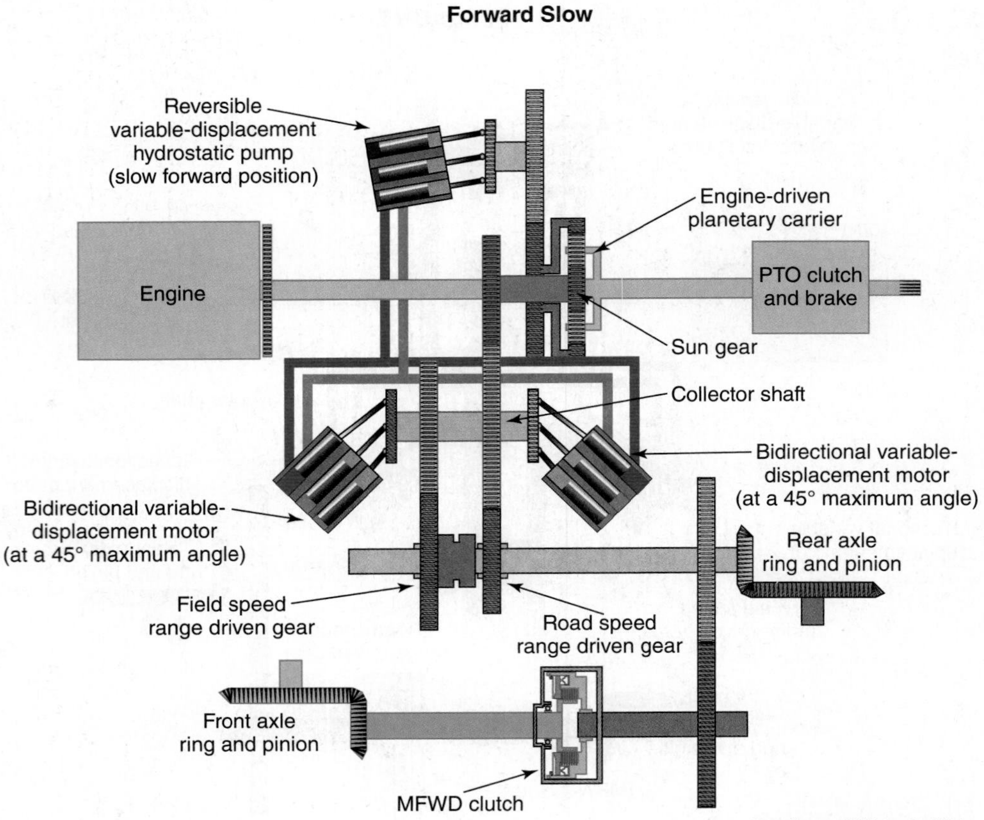

Forward Slow

Figure 15-27. When the tractor begins to move forward, the pump flows oil to the hydrostatic motors.

Forward Fast Speed

As the tractor continues to increase its forward speed, the pump pivots to its maximum positive (forward) angle. If additional forward speed is commanded, the hydrostatic motors pivot to decrease their displacement. As pump flow increases and motor displacement decreases, the hydrostatic pressure increases. This causes further resistance on the planetary ring gear. As the ring gear encounters increased resistance, the sun gear experiences less resistance and its speed increases. Because the sun gear speed increases, it contributes a larger percentage of mechanical power through the CVT. The decreased hydrostatic drive motor angle causes the motor's speed to increase resulting in an increase in speed of the collector shaft. See **Figure 15-28**.

At the fastest travel speed, the hydrostatic motors are fully destroked to a neutral angle. The pump cannot drive the neutralized motors, which causes the drive pressure to increase to its maximum. This high hydrostatic pressure causes the pump and ring gear to stop rotating, resulting in 100% mechanical power flow through the CVT.

As the tractor is commanded to slow its forward travel speed, the hydrostatic motors increase their angle, which reduces the hydrostatic pressure. This drop in pressure allows the ring gear to increase its speed. After the motors reach their maximum angle, the pump angle is decreased. Once the pump reaches a neutral angle, the tractor stops moving. When the pump is in the neutral position, the carrier drives the ring gear at full speed, but the sun gear is stationary.

Reverse

When the operator commands the tractor to move in reverse, the hydrostatic pump pivots to a negative (reverse) angle. This causes oil flow to drive the hydrostatic motor(s) in reverse. The reverse rotation is transferred to the collector shaft, then through the range

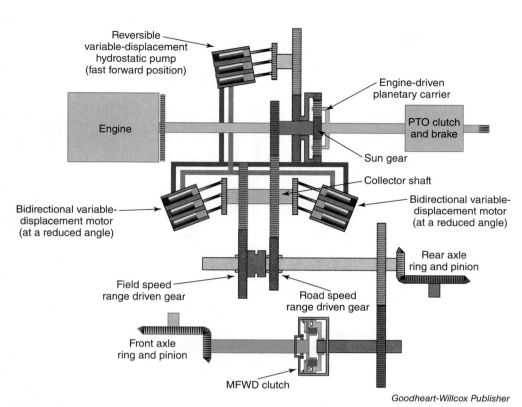

Forward Fast

Figure 15-28. During a forward fast travel speed, the pump is actuated to its maximum positive angle. If the operator requests a faster travel speed, the motors' angles are decreased. At the fastest travel speed, the motors are at a neutral angle, causing high hydrostatic pressure to hold the pump and ring gear stationary.

shaft to the axles. The rate of reverse speed is determined solely by the angle of the reversed pump because the motors remain fixed at the maximum angle. When the CVT is in reverse, all power to the axles is produced hydraulically. See **Figure 15-29**.

Deere IVT

In the early 2000s, John Deere released 6000 series and small frame 7000 series tractors with a ZF-built CVT that is labeled as an infinitely variable transmission (IVT). John Deere later designed their own IVT for a large frame 7000 series tractor and also designed their own IVT for the 8000 series tractors. The 8030 IVT was first produced in 2004 and continues to be used in the 8R series tractors.

John Deere 8R AutoPower/IVT

Like most heavy-equipment CVTs, the John Deere 8R IVT, also known as AutoPowr, contains a mechanical input from the engine and a hydraulic input. The IVT is designed to receive the majority of the power flow through the mechanical input to achieve maximum mechanical efficiency. The IVT contains two planetary gear sets, a hydrostatic module, two clutches, one reverse brake, and a synchronizer assembly. See **Figure 15-30**.

Compound Planetary Gear Set

The compound planetary gear set contains two sun gears, one ring gear, and one planetary carrier. The large sun gear is known as the input sun gear and is driven by the input driven gear, which is driven by the input shaft drive gear. The large sun gear rotates whenever the

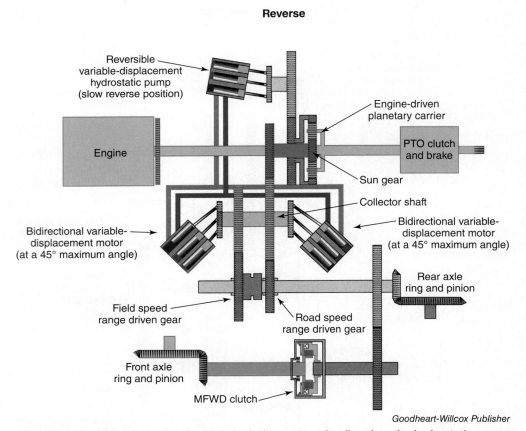

Figure 15-29. When the tractor is commanded to reverse its direction, the hydrostatic pump angle is reversed. The CVT power is 100% hydraulic. The tractor's speed is determined by the angle of the pump.

engine is running. The small sun gear is driven or held by the C2 driven gear. The small sun gear is also known as the high-speed output sun gear because the sun gear shaft is coupled to the high-speed clutch hub. The forward ring gear is held or driven by the ring unit (RU) hydrostatic unit through an idler gear. Each small-diameter planetary pinion is attached to a large-diameter planetary pinion forming a cluster gear assembly, and rotates on a planetary carrier pinion shaft. See **Figure 15-30**.

Reverse Planetary Gear Set

The compound planetary gear set's planetary carrier is shared with another planetary gear set, the reverse planetary gear set. The reverse planetary gear set's portion of the carrier houses two sets of planetary pinions. The reverse idler pinions mesh with the reverse sun gear. The reverse pinions mesh with the reverse idler pinions and the reverse ring gear. See **Figure 15-31**. The reverse brake (RB) is used to hold the ring gear when the tractor is in reverse.

Hydrostatic Module

The IVT has a hydrostatic module consisting of two members. Both can act like a pump or a motor, depending on the mode of operation. One is labeled the ring unit (RU) and is responsible for driving or holding the compound planetary gear set's forward ring gear through the RU idler. The other hydrostatic module member is the clutch unit (CU). It is responsible for holding or driving the C1/C2 shaft by means of the CU drive gear. When the tractor is initially started, the RU hydrostatic member acts like a pump and is at a neutral angle, and the CU member acts like a hydraulic motor and is at a $-44.5°$ angle. IVT operation will be discussed later.

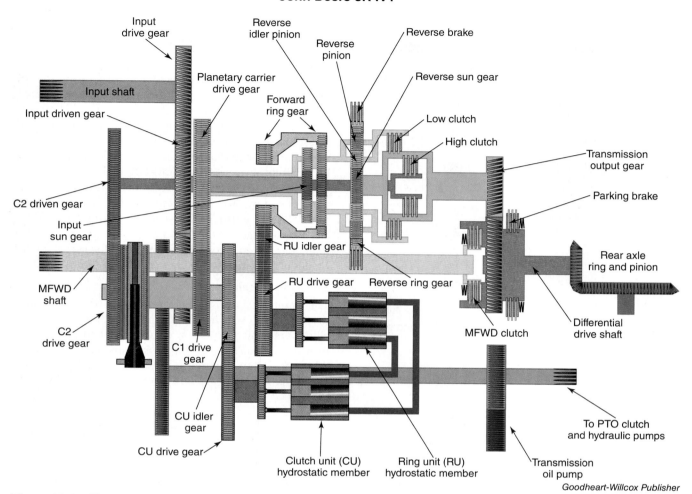

Figure 15-30. The John Deere 8R IVT contains two planetary gear sets, a dual synchronizer assembly, a low clutch, a high clutch, a reverse brake, and a hydrostatic module.

Synchronizer Assembly

The synchronizer shaft has three gears: C2, C1, and the clutch unit (CU) idler gear. The shaft is splined directly to the CU idler gear and the synchronizer hub. The synchronizer can engage the C1 drive gear, the C2 drive gear, or neither. The C1 and C2 drive gears rotate freely on the synchronizer shaft when the synchronizer is in a neutral position. When the synchronizer engages the C1 drive gear, it links the CU hydrostatic member to the planetary carrier. When the synchronizer engages the C2 drive gear, it links the CU hydrostatic member to the high-speed (small-diameter) sun gear in the compound planetary gear set through the C2 driven gear.

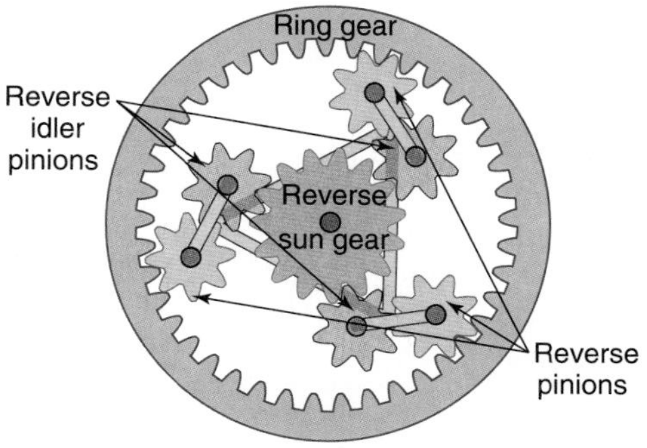

For the tractor to move, the synchronizer must be engaged with one of the drive gears. The synchronizer's shift fork is attached to a shaft that is shifted by one hydraulic piston. Three different oil pressures can act on the piston. When C1 oil is applied, it forces the shift fork to move the synchronizer so it engages the C1 drive gear. When C2 oil is applied, it forces the shift fork to move the synchronizer so it engages the C2 drive gear. When oil is cold, it is resistant to flow. This can slow down the shifting process. In situations like this, C3 oil works in conjunction with C1 oil to increase the speed of the shift. The

Figure 15-31. This simplified drawing shows the two sets of pinions used in the reverse planetary gear set.

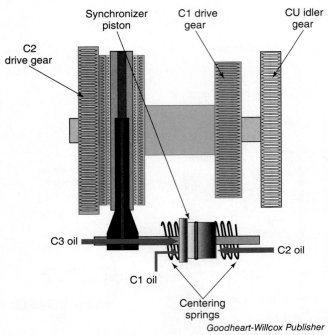

Figure 15-32. The synchronizer shift fork is operated by applying C1/C3 and C2 oil pressures.

C3 oil passes through the center of the shaft, working in unison with C1 oil to help engage the C1 drive gear. Late-model machines always use C1 and C3 oil in conjunction with each other. See **Figure 15-32**.

Clutches

Located after the reverse planetary gear set are the low clutch and the high clutch. The low clutch couples the sole planetary carrier to the transmission's output shaft. The high clutch couples the compound planetary gear set's small sun gear and the C2 driven gear to the transmission's output shaft. For the tractor to move, the low clutch, high clutch, or reverse brake must be applied. The IVT can operate in a PowerZero mode, four forward modes, and two reverse modes. The reverse clutch is engaged when the IVT is in either of the two reverse modes. The low clutch is engaged in forward 1 and forward 2 modes. The high clutch is engaged in forward 3 and forward 4 modes. Modes of operation are explained later.

Transmission Output

The transmission's output gear drives the differential drive shaft, to which the parking brake and the MFWD clutch are attached. Both are spring applied and hydraulically released. The parking brake must be hydraulically released to allow the tractor to move and the MFWD must be spring engaged to drive the front axle. The track-type tractor (8RT) is rear wheel drive only and does not have a MFWD clutch.

PTO Pump Shaft

The shaft at the bottom of the transmission is used to drive the transmission's scavenge pump, implement pump, and PTO clutch. This PTO shaft passes through the hydrostatic module but is not splined or connected to it. The scavenge pump transfers oil from the IVT housing to the differential housing through the center of PTO output shaft. See **Figure 15-33**.

Modes of Operation

As mentioned, the 8R IVT can operate in PowerZero, forward 1, 2, 3, 4, or reverse 1 or 2. **Figure 15-34** specifies the action of different IVT elements for each mode of operation.

PowerZero

In the PowerZero mode, the synchronizer is engaged with the C1 drive gear and the low clutch is engaged. When the tractor is located on a level surface and is in the PowerZero mode, the RU member is acting like a pump and the CU member is acting like a motor. The RU hydrostatic member is destroked to a 0° yoke angle and the CU hydrostatic member is positioned at a –44.5° angle. The RU hydrostatic member is not pumping oil because it is destroked to a neutral angle. The CU hydrostatic member is coupled to the tractor's drive wheels through the CU idler gear, C1 drive gear, planetary carrier, low-speed clutch, transmission output shaft, and differential drive shaft.

The input sun gear is rotating clockwise (as viewed from the rear of the tractor), proportional to engine speed. The planetary carrier is locked stationary because the RU hydrostatic member is in a neutral angle and cannot pump oil to the CU hydrostatic unit. As a result, the forward ring gear rotates counterclockwise (as viewed from the rear of the tractor.)

In the PowerZero mode, any time the tractor is on an incline, the torque exerted by the wheels is transmitted to the CU hydrostatic member, causing the CU member to act like a pump. However, because the RU hydrostatic member (now acting like a motor) remains at a neutral angle, the CU hydrostatic member's oil flow cannot rotate the neutralized RU unit. As a result, the hydrostatic system is held stationary, keeping the tractor stationary.

Chapter 15 | Continuously Variable Transmissions

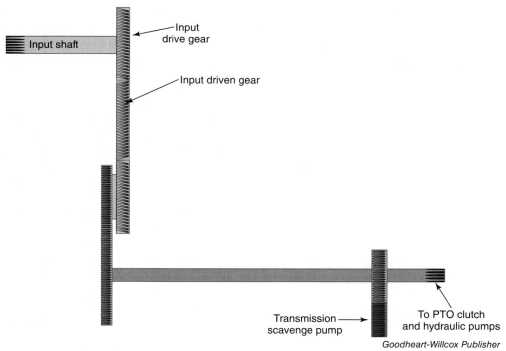

Figure 15-33. The PTO and pump driveshaft is at the bottom of the IVT transmission and rotates at a speed proportional to engine speed.

Forward 1

When the forward 1 mode is selected, the C1 drive gear is engaged by the synchronizer and the low clutch is engaged. The power train IVT control unit (PTI) commands the RU hydrostatic member to move its yoke from 0° to 1.2°, causing the RU hydrostatic member to begin pumping oil to the CU hydrostatic member. The yoke of the CU begins at a maximum angle of −44.5°. Because the RU yoke is at a minimal angle and the CU yoke is at a maximum angle, the CU hydrostatic member has more torque capacity than the RU hydrostatic member. The forward ring gear (that is spinning counterclockwise) must exert torque on the RU drive gear to get it to pump oil to the CU hydrostatic member. This creates drag on the ring gear, which slows it down.

Mode	RU Yoke Angle Range	CU Yoke Angle Range	Drive Gear Engaged	Clutch Applied	Direction of Forward Ring Gear
PowerZero	0°	−44.5°	C1	Low	CCW
Forward 1	1.20° to 44.50°	−44.5° to 0°	C1	Low	CCW-stationary
Forward 2	44.50° to 14.70°	0° to 44.5°	C2	Low	Stationary-CW
Forward 3	14.70° to 44.50°	44.5° to 0°	C2	High	CW-stationary
Forward 4	44.50° to 10.30°	0° to −30.4°	C1	High	Stationary-CCW
Reverse 1	1.20° to 44.50°	−44.5° to 0°	C1	Reverse	CCW-stationary
Reverse 2	44.50° to 14.70°	0° to 44.5°	C2	Reverse	Stationary-CW

Figure 15-34. The John Deere 8R IVT can operate in several modes. The chart lists the yoke angles of the hydrostatic members for each range, the position of the synchronizer, which drive gear is engaged, and the direction of the forward ring gear rotation.

As the forward ring gear speed slows, the rotating planetary pinions force the planetary carrier to begin rotating in a clockwise direction. Because the low clutch is applied, power is transferred to the output shaft. The planetary carrier's speed is directly influenced by the forward ring gear's speed. As the forward ring gear reduces speed, the carrier increases speed, and vice versa. Power from the CU hydrostatic member is added to the compound planetary gear set through the engaged C1 drive gear.

If the operator requests the tractor to continue increasing speed in the forward 1 mode, the PTI increases the RU yoke angle. The increased volume of RU oil increases the resistance on the forward ring gear, further slowing the ring gear and increasing the speed of the planetary carrier. When the RU yoke angle reaches 13.8°, the CU yoke begins to pivot. Both the RU yoke and CU yoke pivot in unison, with the RU yoke angle increasing and the CU yoke angle decreasing. When the RU yoke angle reaches 44.5° and the CU yoke angle reaches 0.0°, the ring gear stops because the CU's yoke has been neutralized, preventing it from receiving oil from the RU. With the sun gear as the input, the ring held, and the planetary carrier as the output, the IVT power flow is 100% mechanical.

Because the CU yoke is in the neutral position, the planetary carrier is no longer contributing power to the system (through the CU hydrostatic member). The carrier is simply being driven (with minimal resistance) by the sun gear. The spinning planetary carrier drives the engaged C1 drive gear, which rotates the CU hydrostatic member freely because the 0.0° yoke angle has eliminated oil flow. The planetary pinions also drive the small sun gear. Since the C2 driven gear shares a shaft with the small sun gear, it drives the C2 drive gear, causing it to spin freely on its shaft at a speed close to that of the C1 drive gear, allowing the PTI to shift the synchronizer. When the PTI shifts the synchronizer clutch from C1 to C2, the IVT enters the forward 2 mode of operation.

Note

When the Deere 8R IVT is operating in the forward 1 mode, the tractor's travel speed can range from 0 to 6 mph.

Forward 2

As the operator requests a faster forward travel speed, the IVT enters the forward 2 mode. The low clutch remains engaged and the synchronizer disengages the C1 drive gear and engages the C2 drive gear. As a result, the small sun gear indirectly drives the CU hydrostatic member through the C2 driven and drive gears and the CU idler and drive gears.

As the operator requests the tractor to continue increasing speed, the PTI gradually increases the CU yoke angle from 0° to a positive angle (up to 44.5°) and decreases the RU's wide yoke angle (down to 14.7°). With the CU hydraulic member generating flow, the RU hydraulic member causes the forward ring gear to rotate in a clockwise direction, which is opposite of the direction it spun in forward 1 mode. The input sun gear, planetary carrier, small sun gear, and forward ring gear all rotate clockwise. The engine-driven input sun gear causes the planetary pinions to revolve inside the spinning ring gear, causing the planetary carrier to increase speed. The low-speed clutch continues to deliver power to the output shaft.

As the operator continues to request more speed, the PTI continues to gradually increase the CU yoke angle up to 44.5° and decrease the RU yoke angle down to 14.7°. When the CU and RU reach those angles, the compound planetary gear set is operating in direct drive. The PTI engages the high-speed clutch, and when the high-speed clutch is completely applied, the PTI releases the low-speed clutch. This places the IVT in the forward 3 mode.

Forward 3

In the forward 3 mode, the C2 synchronizer clutch and the high-speed clutch are engaged. The RU yoke begins at the positive 14.7° angle and the CU begins at the 44.5° angle. As in the forward 1 mode, the RU hydrostatic member is pumping oil to the CU hydrostatic member. However, the flow is in the reverse direction because the forward ring gear is rotating clockwise rather than counterclockwise. Also, because the high clutch is engaged, the small sun gear, rather than the planetary carrier, is the output member.

As the operator requests an increase in tractor speed, the RU yoke angle is pivoted from 14.7° to 19.7° and then the PTI begins to reduce the CU yoke angle. Both hydrostatic members continue to pivot in unison, just as in forward 1 mode, causing the forward ring gear to slow its speed.

The engine's mechanical input is delivered through the input sun gear. The decrease in ring gear speed results in a slower planetary carrier speed, causing the small sun gear to increase its speed in an overdrive mode. The CU member also delivers power to the planetary gear set's small sun gear through the C2 driven gear.

At the top speed of the forward 3 mode, the RU's yoke arrives at a 44.5° angle while the CU yoke ends at 0°, locking the ring gear and RU hydrostatic member stationary. Although the CU hydrostatic member is driven through the C2 gear, its 0° yoke angle prevents it from pumping oil to the RU hydrostatic member. The C1 and C2 drive gears are approaching the same speed. When they reach the same speed, the shift fork moves the synchronizer to disengage the C2 drive gear and engage the C1 drive gear, placing the IVT into forward 4 mode.

Forward 4

In forward 4 mode, the high clutch remains engaged, and the C1 drive gear is engaged. The RU hydrostatic member begins the mode with a 44.5° yoke angle, and the CU hydrostatic member begins the mode with a 0° yoke angle. The forward ring gear is initially held stationary. As additional speed is requested, the PTI increases the negative yoke angle of the CU hydrostatic member to deliver oil flow in the opposite direction to the RU hydrostatic member. In turn, the RU hydrostatic member drives the forward ring gear counterclockwise. This is similar to forward 2 mode, except in forward 4 mode, the CU hydrostatic member's yoke pivots in the opposite direction, to –30.4°. Because the CU yoke pivots to a negative angle, it drives the RU hydrostatic member counterclockwise, which causes the forward ring gear to rotate counterclockwise as well. As the PTI increases the CU hydrostatic member's yoke angle, the CU and RU work in unison, causing the forward ring gear to increase its counterclockwise speed. As the ring gear speed increases, the planetary pinions spin faster, causing the small sun gear to increase its overdrive speed. The CU's yoke quits pivoting when it is at –30.4°, and the RU yoke quits pivoting when it is at 10.3°. Under these conditions, the tractor achieves its top forward speed.

Reverse 1

When the operator requests the tractor to reverse its propulsion, the IVT is placed into reverse 1 mode. The reverse brake is engaged, and the C1 drive gear is engaged by the synchronizer. The RU yoke begins at a 1.2° angle, and the CU yoke begins at a –44.5° angle.

Note

If the tractor was previously in the PowerZero mode, the low clutch will be disengaged as the reverse brake is engaged.

The reverse brake holds the reverse ring gear. As in the forward 1 mode, the CU and RU hydrostatic members work in unison to control the speed of the planetary carrier. With the reverse ring held stationary, the planetary carrier spins clockwise, and the dual set of pinions reverse the direction of the reverse sun gear, which drives the transmission output gear in reverse.

As the tractor is commanded to increase its speed in reverse, the PTI decreases the CU's negative yoke angle and increases the RU yoke angle. When the CU yoke reaches 0° and the RU yoke reaches 44.5°, the forward planetary ring gear stops rotating. The C1 and C2 drive gears are rotating at nearly the same speed, and the PTI is ready to shift the synchronizer from the C1 drive gear to the C2 drive gear.

Reverse 2

When the synchronizer engages the C2 drive gear, the IVT enters the reverse 2 mode. The reverse brake remains engaged. The RU yoke begins at the 44.5° angle, and the CU yoke begins at the 0° angle. As in the forward 2 mode, the PTI gradually decreases the RU yoke angle and increases the CU yoke angle. This reverses the rotation of the forward planetary ring gear. Note that this action affects only the speed and direction of the forward planetary ring gear rotation, which increases the tractor's speed. As in reverse 1 mode, the IVT still reverses the tractor's direction through the held reverse ring gear.

Note

The planetary carrier rotates in the same direction for both forward and reverse propulsion. The tractor's propulsion is reversed by holding the reverse ring gear.

Operator Controls

There are two different types of controls used in IVT tractors. The first is a single right-hand transmission control lever that has an integrated reverser. See **Figure 15-35**. The second is a right-hand transmission control lever with a left-hand reverser lever. See **Figure 15-36**.

Note

An older Deere 7000 IVT tractor has a selector switch with four positions (0, 1, 2, and 3).

- "0" or "off" position—maximizes PTO or hydraulic pump operation.
- "1"—maximizes tractor's pulling power, for tasks such as pulling a heavy disc harrow (less fuel efficient).
- "2"—improves fuel economy when pulling lighter loads.
- "3"—used under lightest load conditions to provide maximum fuel economy.

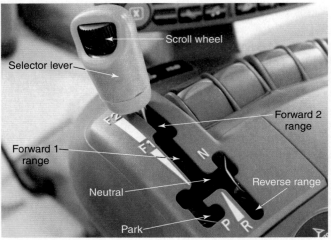

Goodheart-Willcox Publisher

Figure 15-35. When the John Deere 8R IVT is equipped with only the right-hand transmission control lever, the transmission reverser is integrated with the right-hand lever. The control lever can be positioned in one of five slots in its housing: park, neutral (PowerZero), reverse, forward 1, and forward 2.

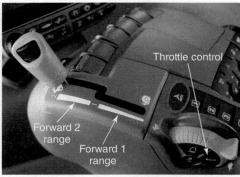

A   **B**   *Goodheart-Willcox Publisher*

Figure 15-36. Many John Deere 8R IVTs have an optional left-hand reverser lever along with a right-hand transmission control lever. A—The left-hand reverser lever can be positioned in park, neutral (PowerZero), reverse, or forward. B—When the machine is equipped with the left-hand reverser, the right-hand selector lever has only two range slots, forward 1 and forward 2.

The lever in **Figure 15-35** has five positions: park, PowerZero, reverse, forward 1, and forward 2. The operator sets the maximum speed for each of the three preselect modes: forward 1, forward 2, and reverse. The maximum speed for each of the three modes can be set while the IVT is in the neutral position by clicking on the different modes (F1, F2, and R) on the monitor and adjusting the encoder wheel (scroll wheel) to the desired maximum speed. See **Figure 15-37**. The maximum speed for each mode can also be adjusted while operating in that specific mode. For example, while the tractor is traveling in forward 1, the operator can adjust the maximum speed of forward 1 by adjusting the encoder wheel.

Goodheart-Willcox Publisher

Figure 15-37. This Deere IVT monitor shows that the tractor has been commanded to operate in the preselect F2 mode and its maximum speed is set to 12.4 mph. The maximum speed for preselect F1 has been set to 2.5 mph and the maximum speed in reverse has been set to 2.1 mph.

Note

The IVT modes of operation are different from the operator control preselect choices. When moving forward, the IVT transmission can be operating in any one of the four forward modes of operation to achieve the speed commanded by the operator. The operator does not know in which of the four modes the tractor is operating. The operator can choose one of two forward preselect controls. The machine uses the four different forward modes of operation based on how the operator sets the forward preselect controls. When moving in reverse, the IVT transmission can operate in either one of the two reverse modes of operation to achieve the speed commanded by the operator. The operator does not know in which of the two reverse modes of operation the tractor is operating. The IVT's operator controls only allow for the operator to vary one preselect reverse control.

The operator normally sets the engine throttle to maximum speed, and once the selector lever is moved into one of the three modes (F1, F2, or R), the tractor immediately begins moving. The operator can hold the service brake pedals to prevent the tractor from moving and then slowly release the brake pedals to slowly begin propelling the tractor.

Three methods can be used to slow the tractor:

- Moving the selector lever rearward within its range.
- Rotating the encoder wheel counterclockwise.
- Pressing the service brake pedals.

Caution

As with most CVT tractors, if the tractor is commanded to start at its maximum speed, for example 19 mph in forward 2, the machine will very quickly reach that speed. If the operator does not understand the controls, he or she might panic and forget how to stop the tractor. Before operating a CVT tractor, be sure to receive proper operator training to avoid injury or damage to other machines.

The tractor has a hand throttle control on the right-hand console, as shown in **Figure 15-36A**, but can also be equipped with a foot-operated throttle control.

Caterpillar Wheel Loader

In late 2012, Caterpillar began offering the 966K XE wheel loader with a CVT transmission. In 2014, Caterpillar released a second model, 972M XE wheel loader, with the CVT transmission. The XE classification often designates that the loader is equipped with the CVT but can also designate a loader with an electric drive.

The Caterpillar 966K XE, 972M, XE, 966 XE, and 972 XE wheel loaders use the CVT explained in this chapter and shown in **Figure 15-38**. The Caterpillar 980XE and 982XE wheel loaders use a CVT, but the transmission's design is a little different from the 966XE and 972XE loaders. The Caterpillar 988K XE and 988XE do not have a mechanical and hydraulic CVT but use an electric drive powertrain.

Like other CVTs, this transmission increases fuel efficiency up to 35%. The CVT enables the loader to maintain a constant speed even when climbing an incline. When the machine is used for loading trucks, the transmission eliminates the hesitations that occur in powershift transmissions, resulting in smoother operation.

The CVT uses a compound planetary gear set, three electro-hydraulically controlled synchronizers (low forward/high reverse, low reverse/high forward, and auxiliary), two multiple disc clutches (A and B), a reversible variable-displacement hydraulic pump, and a fixed-displacement hydraulic motor, which Caterpillar labels as the variator. The variator provides the infinite variability in the transmission by varying the speed and direction of the ring gear in the secondary planetary gear set. See **Figure 15-38**. The variator's pump can act like a motor, and the variator's motor can act like a pump, depending on the CVT operation. The variable-displacement reversible pump starts at a maximum negative displacement to start the tractor moving from neutral.

CVT Planetary Inputs

The compound planetary gear set consists of two planetary gear sets. The first gear set receives its input through the planetary carrier that is driven by the engine through an idler gear. The idler drives the carrier at a speed proportional to engine speed. The first planetary gear set's ring gear is integrated to the carrier of the second gear set and the low range output gear.

The second planetary gear set has a ring gear assembly consisting of an external-toothed ring gear joined to an internal-toothed ring gear. The external-toothed ring gear meshes with a gear on the variator, which provides the system's hydraulic input. The internal-toothed ring gear serves as the ring gear for the second planetary gear set. This combined ring gear assembly rotates as a single unit and is driven by the variator. As in other CVTs, the variator is used to decrease, hold, or increase the transmission's speed.

Planetary Outputs

The gear set has three outputs. Both sun gears are connected to a common shaft that drives the auxiliary gear drive shaft, which drives the *high* range output gear. The high range output gear drives the high forward gear. The high forward gear drives the high reverse gear.

Note

In **Figures 15-38** through **15-43**, note that a small separation exists between the high range output and high reverse gears. The dashed lines signify that the high range output gear is actually in mesh with the high forward gear. The high forward gear is in mesh with the high reverse gear. Power flow is explained later in the chapter.

Another output is the *low* range output gear, which is driven by the first ring gear and the second planetary carrier. The low range output gear is in mesh with the low forward gear. The low forward gear is in mesh with the low reverse gear.

Chapter 15 | Continuously Variable Transmissions

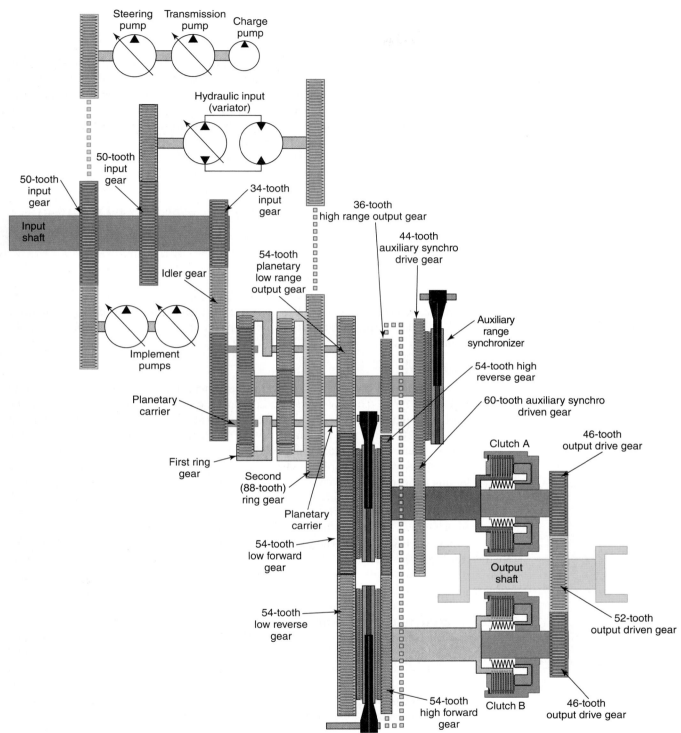

Figure 15-38. Caterpillar's 966XE and 972XE wheel loaders have a CVT drive. The variator consists of a reversible variable-displacement pump and a fixed-displacement motor. The CVT contains a compound planetary gear set, two multiple-disc clutches (A and B), and three synchronizers (low forward/high reverse, low reverse/high forward, and auxiliary range) to achieve five ranges.

The last output is the auxiliary drive gear which is in mesh with the auxiliary driven gear. When the auxiliary synchronizer is engaged, the auxiliary drive gear drives the auxiliary driven gear.

Intermediate Power Flow

Depending on which gear is engaged by the low forward/high reverse, low reverse/high forward synchronizer, or auxiliary, the power is distributed from the low range gears, the high range gears, or the auxiliary gear. After the synchronizers, power is sent to a multiple-disc clutch, either A or B. The multiple-disc clutch (A or B) then drives the CVT's output shaft, which sends power to the front and rear axles. At a minimum, the CVT requires one synchronizer and one multiple-disc clutch to be engaged to propel the loader.

The order of power flow through the CVT is as follows:

1. Engine (mechanical) and hydraulic (variator) inputs.
2. Compound planetary gear set.
3. Synchronizers (low forward/high reverse, low reverse/high forward, or auxiliary)
4. Clutch A or B.
5. Output shaft.

CVT Component Operation

The CVT uses three synchronizers, two multiple-disc clutches, and one variator to achieve the desired travel speed. See the Appendix for *Status of Members during Caterpillar CVT Operation*. This table lists the status of key parts of the CVT during different modes of operation.

Low Forward

When the wheel loader starts moving forward from a neutral position, the CVT is operating in a low forward mode. The low forward synchronizer gear and Clutch A are engaged. The engine provides power into the CVT's input shaft through a torsional coupler. The input shaft drives the variator and the planetary carrier through an idler gear. The variator pump initially begins at the maximum negative yoke angle, and decreases its displacement (toward a neutral angle) to increase machine travel speed.

The low forward/high reverse synchronizer engages the low forward gear, which is driven by the planetary gear set's low range output gear. The low forward shaft transmits power to the Clutch A hub. With Clutch A engaged, the output drive gear drives the output shaft, which delivers power to the front and rear axles. See **Figure 15-39**.

To increase travel speed, the ECM varies the current flow to the variator solenoids, which decrease the yoke angle. As the yoke angle moves from a maximum negative angle toward the neutral angle, it causes the second ring gear (which is spinning backward) to decrease its speed, which causes the planetary carrier to increase its speed. The increase in carrier speed is translated to the first planetary gear set's ring gear, which is integrated into the carrier, resulting in an increase of the low range output gear's speed.

When the variator pump yoke reaches neutral, the second planetary ring gear stops rotating. As the operator requests more travel speed, the variator actuates the pump's yoke in a positive angle, causing the second planetary ring gear to begin spinning in a forward direction. As a result, the speed of the carrier and output gear increase. As more machine speed is requested in the Low Forward mode, the variator continues increasing the pump's yoke angle to a maximum positive angle.

High Forward

As the operator requests faster forward travel speed, the low reverse/high forward synchronizer engages the high forward gear and Clutch B is engaged. As Clutch B is engaged, Clutch A is released. Notice in **Figure 15-40** that the arrows are depicting power flow between the high range output gear and the high forward gear. With Clutch B engaged, the output drive gear drives the output shaft, which delivers power to the front and rear axles.

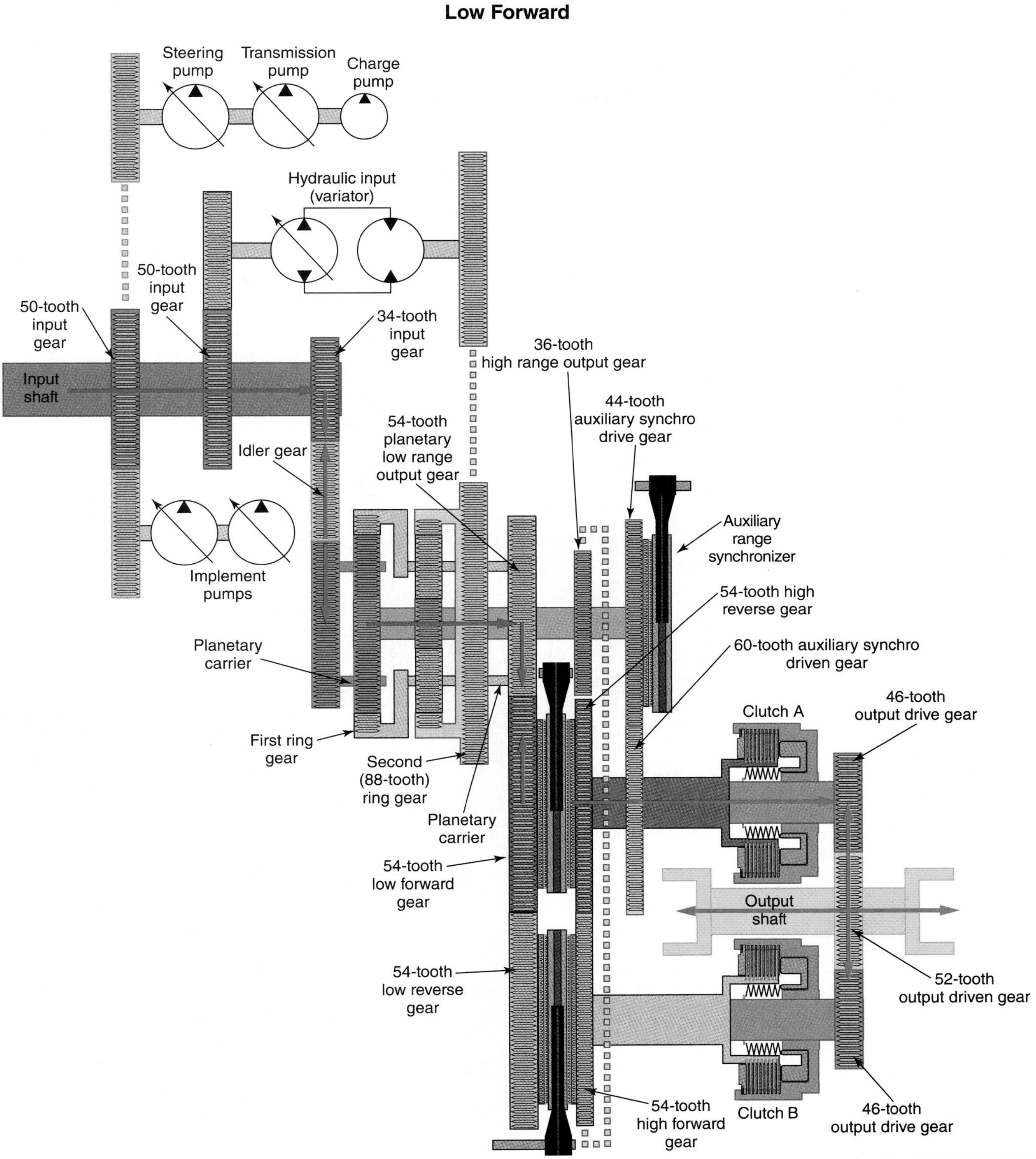

Chapter 15 | Continuously Variable Transmissions

Low Forward

Goodheart-Willcox Publisher

Figure 15-39. In low forward mode, the low forward/high reverse synchronizer is engaged to the low forward gear. The planetary low range output gear drives the low forward gear, which transmits through a shaft to Clutch A. Because Clutch A is engaged, it transfers power to the output shaft.

High Forward

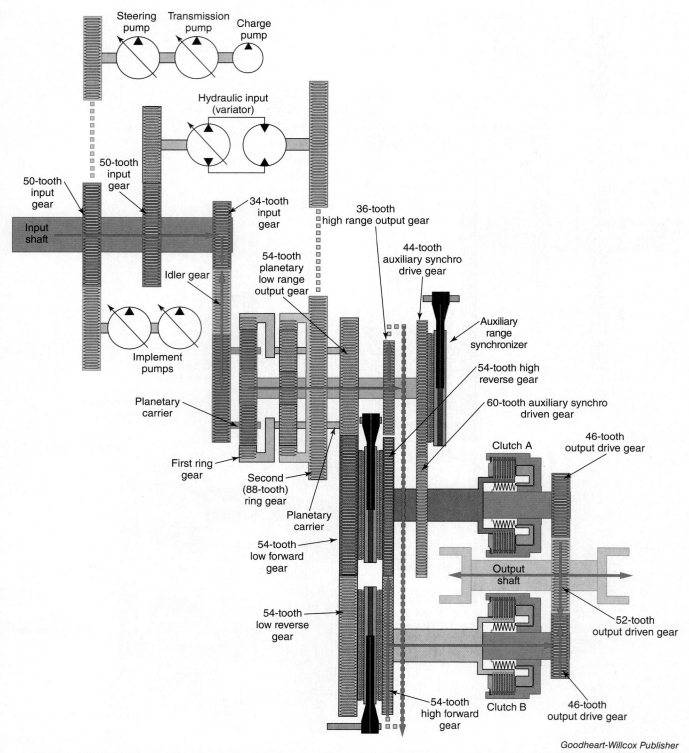

Figure 15-40. In the high forward mode, the sun gear/auxiliary shaft drives the high range output gear. The arrows show that the high range output gear drives the high forward gear. With Clutch B engaged, power is transferred to the output shaft.

To increase travel speed in the high forward mode, the ECM varies the current flow to the variator solenoids. In turn, the solenoids decrease the yoke angle from a maximum positive angle, through neutral, to the maximum negative angle. As the yoke angle moves, it causes the sun gears (auxiliary drive shaft) to increase speed, which causes the high range output gear to increase the high forward output gear's speed.

Auxiliary Mode

If the operator continues to request more forward travel speed, the ECM places the CVT in the fastest mode for forward travel speed, the auxiliary mode. The CVT enters the auxiliary mode by engaging the auxiliary synchro drive gear, releasing Clutch B, and engaging Clutch A. See **Figure 15-41**. The auxiliary shaft drives the auxiliary driven gear, and power is delivered through Clutch A to the output shaft.

During the shift to the auxiliary mode, the yoke of the variator has to be quickly repositioned. The yoke is at the maximum negative angle and, when the transmission is shifted into the auxiliary mode, the ECM actuates the solenoids to rapidly pivot the variator's yoke to the maximum positive angle. As more travel speed is requested, the variator actuates the pump yoke from a maximum positive angle, through neutral, to a maximum negative angle to achieve the fastest forward travel speed. The auxiliary mode is only available for forward travel.

Low Reverse Mode

When the loader is commanded to move in reverse, the low reverse/high forward synchronizer engages the low reverse gear, and Clutch B is engaged. The planetary carrier drives the low range output gear, which drives the low forward gear. The low forward gear acts as a reverse idler as it drives the low reverse gear. Power is sent through Clutch B to the output shaft. See **Figure 15-42**.

The variator's yoke begins at a maximum negative angle. As more speed is commanded, the ECM energizes the solenoids to force the variator yoke to move from maximum negative angle, through neutral, to a maximum positive angle. This causes the ring gear to slow, reverse direction, and gain speed, which increases the CVT's output speed.

High Reverse Mode

As the operator continues to request a faster reverse travel speed, the ECM engages the high reverse gear and Clutch A. As when the CVT is operating in the high forward range, power is sent from the auxiliary shaft to the high forward reverse gear, which is shown with arrows in **Figure 15-43**. During the high reverse mode, the high forward gear acts like a reverse idler as it drives the high reverse gear. Clutch A is engaged and sends power to the output shaft.

When entering the high reverse mode, the variator's yoke is initially at a maximum positive angle. As the operator requests additional speed, the ECM energizes the solenoids to gradually decrease the yoke angle, causing it to move through neutral to a maximum negative angle.

John Deere Two-Speed Automatic ProDrive Transmission

John Deere manufactures a two-speed automatic ProDrive transmission that was initially released with self-propelled forage harvesters around 2004. It is also used in later-model STS combines, S-Series combines, cotton pickers, and cotton strippers. Although the transmission is not considered a CVT, it provides a wide range of travel speeds and integrates common components and technology that are often used in off-highway CVTs:

- Variable hydrostatic transmission.
- Planetary gear set.
- A multiple disc rotating clutch.
- A multiple disc holding brake.

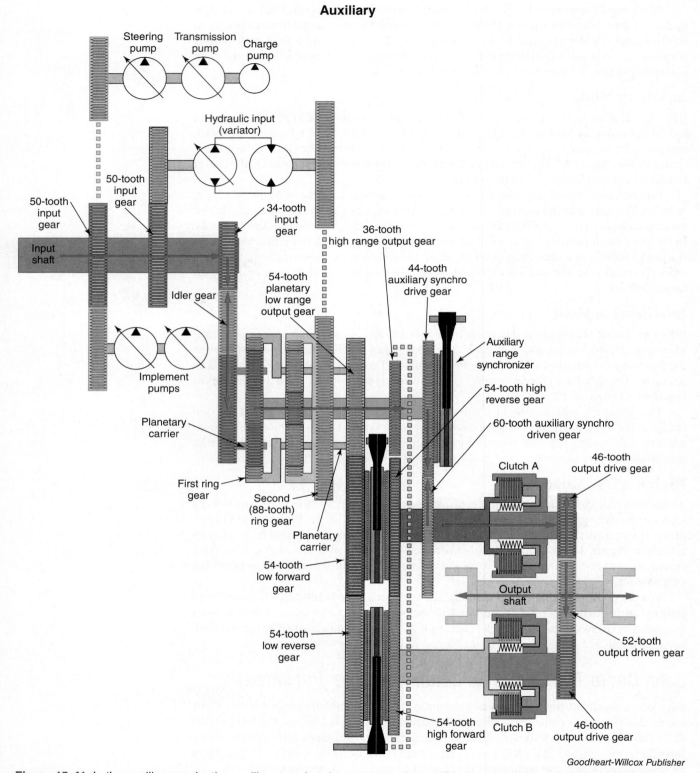

Figure 15-41. In the auxiliary mode, the auxiliary synchronizer engages the auxiliary synchro drive gear, causing it to drive the auxiliary driven gear. Clutch A is engaged, and transfers power output drive gear, which drives the output shaft. Maximum machine forward travel speed is achieved in this mode.

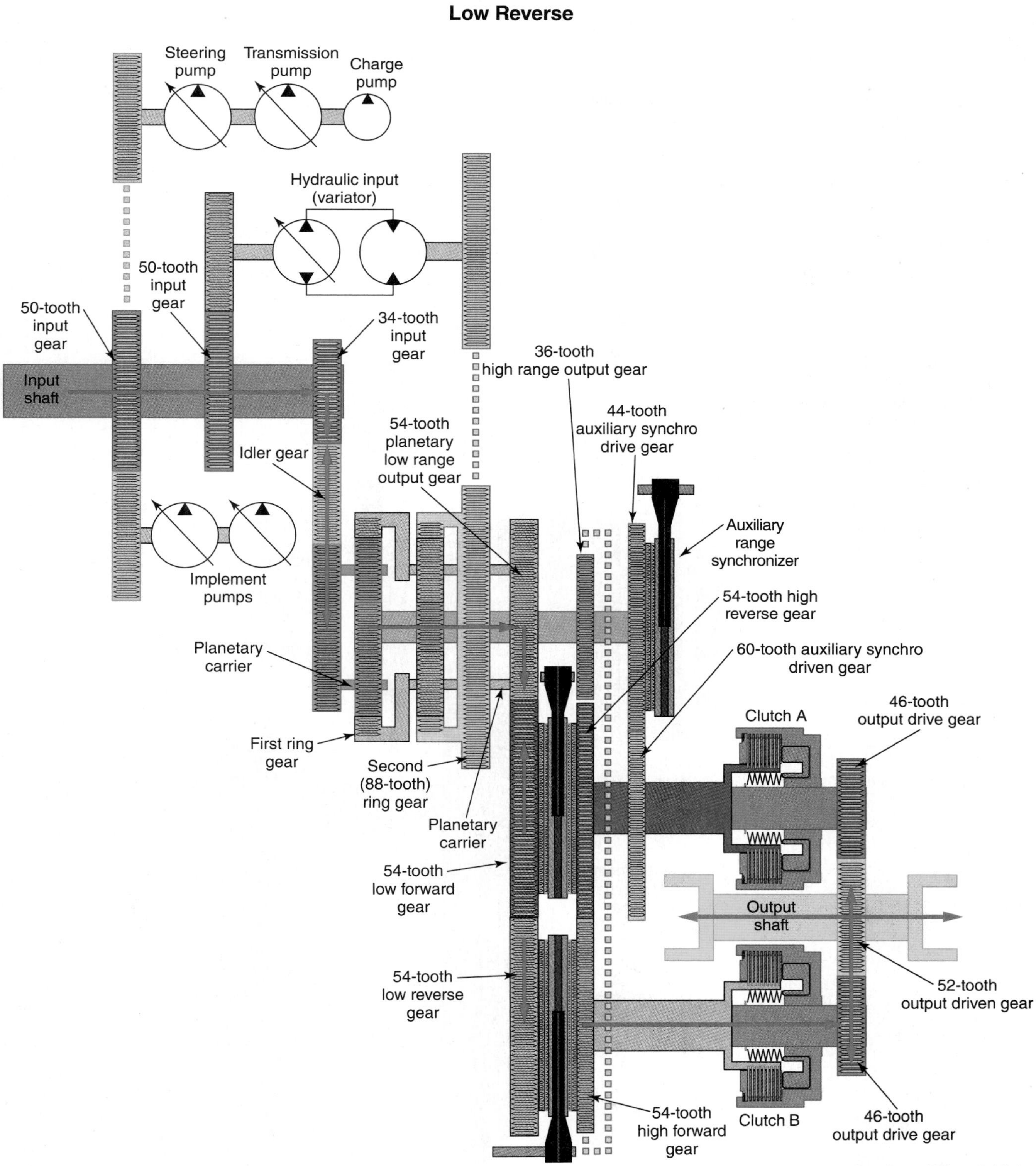

Chapter 15　Continuously Variable Transmissions

Low Reverse

Figure 15-42. In the low reverse mode, the low forward gear acts like a reverse idler as it drives the low reverse gear. With the low reverse synchronizer engaged, power is sent to Clutch B to the output shaft.

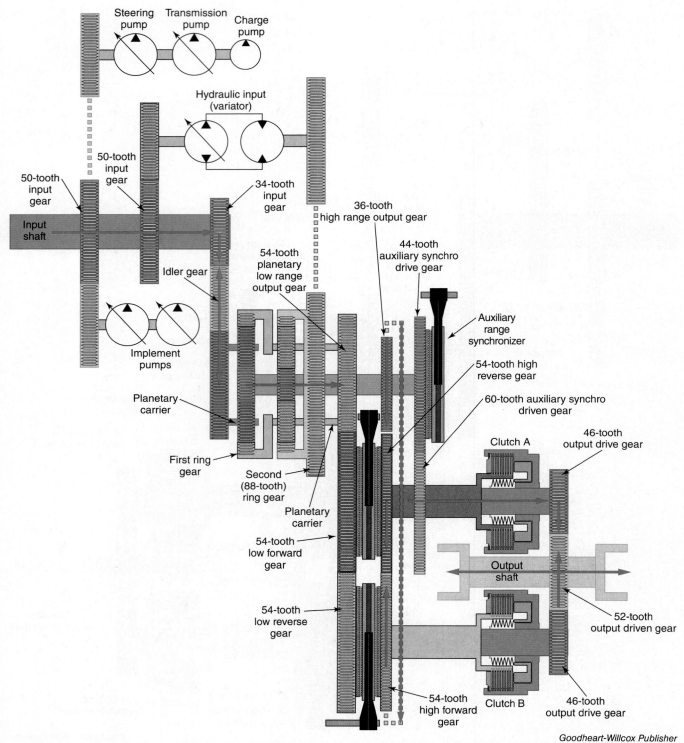

Figure 15-43. In the high reverse mode, power is sent from the auxiliary shaft/auxiliary drive gear to the high forward gear, which acts like a reverse idler. The high forward gear drives the high reverse gear. Clutch A is engaged sending power to the output shaft.

As in other hydrostatic power trains, the two-speed ProDrive transmission has a mechanical gearbox that provides an additional speed range. The ProDrive is unique in that it provides an automatic two-speed range; therefore, it is sometimes called a two-range automatic transmission or an automatic powershift transmission. It is different from traditional powershift transmissions because it uses a variable hydrostatic drive to perform three functions:

- Provide the input into the transmission.
- Vary machine speed.
- Reverse machine propulsion.

The ProDrive is different from other hydrostatically propelled harvesters because the machine can automatically upshift or downshift the speed range. Prior to ProDrive, a three-speed manual transmission required the customer to stop the machine to change the speed range.

In addition to needing to stop the machine to change gears, another drawback of the manual three-speed gearbox was that any time the four-wheel drive was engaged, the hydrostatic drive's displacement would dramatically increase. This resulted in the machine speed drastically slowing down. Likewise, any time the four-wheel drive was disengaged, the machine would speed up considerably.

As mentioned in Chapter 4, *Agricultural Equipment Identification*, a self-propelled forage harvester does not store the crop it harvests. It must be able to blow the crop into a wagon or truck that is attempting to match the harvester's ground speed. If the forage harvester is blowing the crop into a truck at a rate of 222 pounds per second and a sudden change in the harvester's speed causes the truck to lose pace, a substantial amount of crop can be lost quickly.

ProDrive provides customers the following benefits:

- Computer-controlled acceleration and deceleration, including antilock brake application during rapid deceleration.
- Maximum productivity in multiple modes of operation:
  - Front-wheel drive.
  - Optional four-wheel drive.
  - Field speed.
  - Road speed.
- Ability to maintain a steady speed as four-wheel drive is disengaged or engaged.
- Ability to maintain a steady travel speed while traveling downhill.
- Maximum travel speed in four-wheel drive.
- Anti-slip regulation (ASR)/traction control (the ability to send drive oil to the front or rear axle based on wheel slip).
- Automatic parking brake control based on the position of the propulsion lever.
- Engine overspeed prevention during panic stops.

ProDrive Controls

To select a speed range, the operator presses button #1 or button #2 on the ProDrive controls. Each button has an associated maximum travel speed setting. These preselected speed ranges should not be associated with the actual 2-speed automatic shift that takes place inside a ProDrive.

For example, depending on the machine, crop, and field conditions, the operator might set button #1 for a maximum speed of 4 mph, and set the maximum speed for button #2 at 8 mph. The multi-function handle (propulsion lever) works the same as that on a traditional hydrostatic drive. If the operator moves the lever from neutral to maximum forward travel speed, the machine attempts to increase the speed to the preselected maximum speed. In this example, the maximum speed would be 4 mph (button #1) or 8 mph (button #2). The maximum speed for button #1 must be set lower than the maximum speed for button #2. Also, the button #1 maximum travel speed cannot be set lower than 3 mph (5 km/h).

The transmission is 100% electronically controlled, including the operation of these key components:
- Planetary clutches.
- Displacement of the hydrostatic pump, front motor, and rear four-wheel drive motors.
- Parking brakes.

ProDrive Components

The ProDrive transmission has the following key components:
- Variable-displacement hydrostatic pump.
- Variable-displacement hydrostatic front motor.
- Optional variable-displacement rear hydrostatic motors (four-wheel drive).
- Compound planetary gear set.
- A multiple-disc rotating clutch.
- A multiple-disc holding brake.
- Electronically controlled solenoid valves.
- An accumulator.
- Lockable differential.
- Dual combination parking brake and service brake assemblies, or single wet parking brake and separate service brake assemblies.

Some machines, like a combine, also have other features, called a ***proportionator pump/motor***. These two items work similar to a scavenge pump. Lubrication oil starts in the engine separator gear case. If the machine has the five-speed feeder gearbox, lubrication oil is fed from the separator gearbox to the five-speed feeder gearbox, and then the oil is finally sent to the ProDrive transmission gearbox to lube the components. The proportionator pump and motor draw the lube oil out of the ProDrive transmission and return it to the engine separator gear case.

Variable-Displacement Hydrostatic Transmission

The hydrostatic transmission uses an electronically controlled engine-driven variable-displacement reversible hydrostatic pump that is responsible for changing the machine's torque, speed, and direction of travel. An electronically controlled variable-displacement hydrostatic front motor receives oil from the pump and provides the rotational input into the ProDrive transmission.

Four-Wheel Drive

Many harvesters are equipped with optional four-wheel drive. Forage harvesters use infinitely variable bent-axis hydrostatic motors on their rear wheels (four-wheel drive). Combine harvesters use a two-speed cam lobe design hydrostatic motors on their rear wheels (four-wheel drive).

Compound Planetary Gear Set

The ProDrive transmission uses a compound planetary gear set with two gear ratios:
- Low range (torque multiplication).
- High range (direct drive).

The planetary gear set contains three large planetary pinions and three small planetary pinions. Each large planetary pinion is attached to a small planetary pinion forming a cluster gear assembly. Each cluster gear (pair of large and small pinions) rotates at the same speed and in the same direction. See **Figure 15-44**.

The three planetary pinion cluster gears are pinned to a single planetary carrier. The different-sized pinion gears mate with different size sun gears.

A small sun gear meshes with the large planetary pinions and is splined to the planetary gear set's input shaft. It rotates at the same speed and in the same direction as the hydrostatic motor.

The large sun gear meshes with the small planetary pinions and is attached to the planetary output gear. The planetary gear set does not contain an internal-toothed ring gear. See **Figure 15-45**.

The planetary output gear meshes with a countershaft gear. The countershaft has a pinion gear that drives the differential's ring gear. See **Figure 15-46**.

Planetary Clutch and Brake

The ProDrive transmission contains two clutches: a rotating clutch that is used for direct drive and a holding clutch, also known as a brake, used for low speed.

Direct Drive (Rotating Clutch)

The rotating clutch drum is bolted to the planetary carrier and rotates at the same speed and direction of the planetary carrier. The rotating clutch's hub is splined directly to the planetary gear set's input shaft and rotates at the same speed and the same direction as the hydrostatic motor. See **Figure 15-47**.

When the rotating clutch is applied, it couples the planetary input shaft to the planetary carrier. Due to the large sun gear being in mesh with the smaller planetary pinion (cluster gears), it causes a locking action that results in direct drive (1:1 ratio). The large sun gear then rotates at the same speed and in the same direction as the hydrostatic motor. Note that the rotating clutch drum also serves as the holding brake's hub. See **Figure 15-48**.

Figure 15-44. This photo shows two planetary cluster gear assemblies.

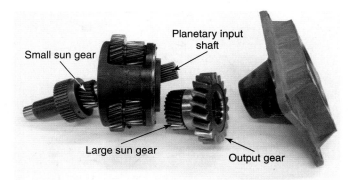

Figure 15-45. The small sun gear is splined to the planetary gear set's input shaft. The large sun gear is the output member of the gear set, driving a gear that meshes with the countershaft.

Figure 15-46. The planetary output gear drives the countershaft. The countershaft drives the differential's ring gear. Note that the large sun gear assembly has been slid out of mesh with the planetary cluster gears so the output gear is visually aligned with the countershaft's gear for illustration purposes.

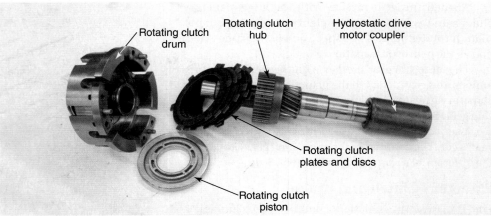

Figure 15-47. The rotating clutch's hub is splined to the planetary input shaft. The hydrostatic drive motor coupler is splined to the end of the input shaft. The rotating clutch drum mates with the rotating clutch's steel plates. The rotating clutch drum also serves as the holding brake's hub. The rotating clutch drum is bolted to the planetary carrier. High speed is used for travel speeds ranging from 12 mph–25 mph (19 km/h–40 km/h).

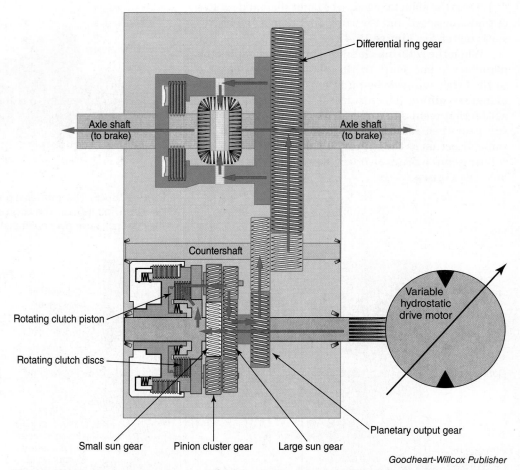

Figure 15-48. An example of a simplified illustration of a John Deere two-speed automatic ProDrive transmission when the rotating clutch is engaged causing the planetary to be in direct drive.

Low-Speed (Holding Brake)

The holding brake drum is coupled to a side cover that bolts to the transmission's case. See **Figure 15-49**. As mentioned, the holding brake's hub also serves as the rotating clutch's drum and is bolted to the planetary carrier. When the holding brake is applied, it holds the planetary carrier stationary. As the hydrostatic motor drives the planetary input shaft, it drives the small sun gear. When the planetary carrier is held stationary, the cluster gears spin as the sun gear rotates. The small sun gear drives the cluster gear and the cluster gear drives the large sun/planetary output gear, providing low-speed torque multiplication at a ratio of 2.38:1. Low-speed mode provides travel speeds ranging from 0–12 mph (0–19 km/h). See **Figure 15-50**.

Figure 15-51 shows the operating range of the transmission based on the application of the ProDrive low range brake and high range clutch.

Goodheart-Willcox Publisher

Figure 15-49. The holding brake's drum is bolted to a side cover, which is bolted to the transmission's housing. The clutch plates, friction discs, and piston have been removed.

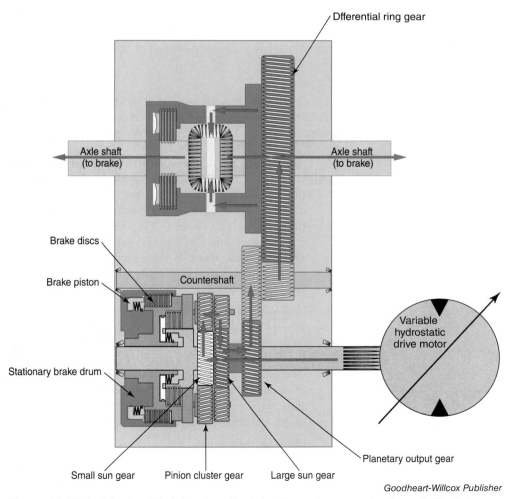

Goodheart-Willcox Publisher

Figure 15-50. In this simplified drawing of a John Deere two-speed automatic ProDrive transmission, the planetary brake is applied, causing the planetary gear set to be in the low-speed position.

Gear Range	Low Brake (holding brake)	High Clutch (stationary clutch)
Neutral	Released	Released
Low	Applied	Released
High	Released	Applied
Transmission locked	Applied	Applied

Goodheart-Willcox Publisher

Figure 15-51. The two clutches offer four functions: neutral, low speed, high speed, or locked transmission.

ProDrive Operation

Depending on the maximum preselected travel speed, the ProDrive is designed to operate in either a low range (low-speed torque multiplication) or the high range (direct drive). The machine's travel speed is affected by the following:

- Tire diameters.
- ProDrive planetary gear range (low or high).
- Displacement of the hydrostatic pump.
- Displacement of the hydrostatic front motor.
- Displacement of the optional rear hydrostatic motors (four-wheel drive).

Planetary Gear Set—Low Speed

When starting from a stop, the ProDrive planetary gear set operates in the low-speed position. The hydrostatic pump begins at a neutral yoke angle (zero displacement) and the front hydrostatic motor starts at maximum displacement (high torque and low speed). When the operator moves the propulsion lever forward, the hydrostatic pump displacement increases. If the machine's speed does not match the operator's speed request and the hydrostatic pump displacement has reached its maximum displacement, the electronic controls begin to decrease the front hydrostatic motor's displacement. This occurs around 6 mph, and as the motor displacement continues to decrease, the machine's travel speed continues to increase. The motor displacement continues to decrease until the machine reaches a maximum speed of 12 mph (19 km/h) with the planetary gear set in the low-speed range.

Planetary Gear Set—High Speed

If the machine still has not achieved the travel speed requested by the operator but has reached 12 mph, the ProDrive shifts from low-speed to high-speed configuration. Unlike CVTs that have a stepless range shift that is not felt, the ProDrive range shift can be felt by the operator.

During the low-speed to high-speed automatic shift, the pump displacement is momentarily reduced to smooth the shift and the hydrostatic front motor displacement is increased to maximum. After the shift, the pump's displacement reverts back to its maximum displacement and remains at maximum displacement. The electronic controls then increase the machine's travel speed by decreasing the motor's displacement. See **Figure 15-52**.

Four-Wheel Drive

When four-wheel drive is engaged, the total displacement of the hydrostatic drive motors is increased. This slows the machine's travel speed. To compensate, the electronic controls reduce the ProDrive front hydrostatic motor's displacement and increase the hydrostatic pump's displacement so the operator does not notice a change in the machine's travel speed.

Lockable Differential

The ProDrive has a lockable front axle differential. When the differential control is in the manual mode, the differential is allowed to lock as long as the machine is traveling less than 6 mph (10 km/h) and the steering is operating at angles less than 9°. If operating in the

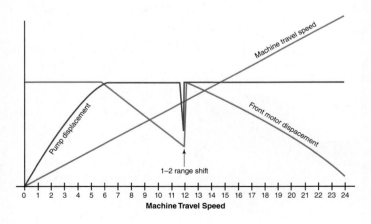

Goodheart-Willcox Publisher

Figure 15-52. An example of the hydrostatic pump and motor displacements used to increase machine travel speed. The graph does not depict the use of optional four-wheel drive.

Chapter 15 Continuously Variable Transmissions

manual differential lock mode, the differential releases if the machine speed exceeds 7.5 mph (12 km/h) or the steering angle exceeds 9°.

When the differential is placed in the automatic mode, the differential automatically locks when it senses a difference of 10% in speeds between the front-right and front-left drive tires. The four-wheel-drive machines also have an *anti-slip regulation (ASR)* traction control feature. It monitors wheel slip and varies front and rear hydraulic motor displacements as needed to provide maximum traction.

Wet Parking Brake and Service Brake

Some two-speed ProDrive transmissions, like those in STS series combines, contain a pair of combination brakes. Chapter 19, *Hydraulic Brake Systems*, explains that hydraulic combination brakes commonly use a single clutch pack that serves as both the parking and service brakes. The brakes are bolted directly to the ProDrive transmission housing. See **Figure 15-53**.

The parking brakes are spring applied with Belleville washers and hydraulically released. The service brakes are hydraulically applied and spring released. If the machine cannot run and needs to be towed, the service literature provides instructions for manually releasing the brakes. This procedure typically consists of tightening cap screws to compress the parking brake spring.

Goodheart-Willcox Publisher

Figure 15-53. The two brakes have been removed from an STS series combine two-speed automatic ProDrive transmission. Each brake assembly contains a single clutch pack that serves as a parking brake and a service brake.

Note

Machines equipped with four-wheel drive will have the anti-lock brake system (ABS) in operation whenever the four-wheel drive is in operation.

Early STS series combines use a traditional three-speed transmission with drum-type service brakes. The service brakes on the S-Series ProDrive transmissions do not contain the parking brakes. The park brake is integrated within the ProDrive transmission.

Optional Rear Drive (Four-Wheel Drive)

As mentioned earlier, the machine can be equipped with optional rear wheel motors that provide four-wheel drive. Although the two rear hydraulic drive motors are not located on the ProDrive transmission housing, their operation is integrated with the ProDrive transmission. As explained earlier, when the rear hydraulic drive motors are engaged, the ProDrive hydrostatic pump displacement is increased and the front hydrostatic motor's displacement is decreased to compensate for the total change of hydrostatic motor displacement.

Hydrostatic Drives

Hydrostatic drive transmissions (HST) have similarities to CVTs and are sometimes labeled CVTs. HSTs provide stepless increase and decrease of speeds. Two unique HST designs that provide a wide range of travel speeds similar to CVTs are explained in this chapter. They are the John Deere X9 ProDrive combine transmission and the Caterpillar K-Series wheel loader transmission. Both the combine and the wheel loader use a single HST pump that drives two infinitely variable displacement HST motors in parallel. The pair of HST motors drive a single external-toothed ring gear inside a mechanical transmission gearbox.

John Deere Dual Motor ProDrive Transmission

This chapter explained the two-speed automatic ProDrive transmission found in some STS and S-Series combines, forage harvesters, and cotton harvesters. The X9 series combine has a different style of ProDrive transmission. It uses a single HST pump to drive two infinitely variable displacement motors in parallel. Both HST motors drive the differential ring gear, which is attached to the differential case. See **Figure 15-54**.

The sequence for increasing travel speed requires first increasing the pump displacement, then decreasing motor 1. After motor #1 reaches minimum displacement, a clutch disengages the motor from delivering power into the transmission, and the motor is set to a neutral angle. The displacement of motor 2 is next reduced to continue increasing machine travel speed.

Caterpillar K Series Wheel Loader

The Caterpillar 924K, 926M, 930K, 930M, 938K, and 938M wheel loaders also uses one infinitely variable displacement HST pump to drive two infinitely variable displacement motors plumbed in parallel. Both HST motors deliver power into a single gearbox that splits power flow to the front and rear axles. One motor has a small displacement and the other motor has a larger displacement. As shown in **Figure 13-55,** motor 1 delivers power into the external toothed output gear at a $(62 \div 37) = 1.676$ ratio. When motor 2 has its clutch engaged it delivers power into the output gear at a ratio of: $(41 \div 29) \times (62 \div 22) = 1.414 \times 2.818 = 3.985:1$.

Like the John Deere example, the Caterpillar wheel loader at high travel speeds will have the ECM nullify the large-displacement motor, making it ineffective by disengaging its clutch and placing the motor at a neutral swivel angle. The small motor is always in use and will be rotating any time the wheel loader is moving, regardless of the travel speed. When the wheel loader is travelling at slower speeds, the pump drives both hydrostatic motors. When the machine is traveling at high speed, only the small motor is used. The large motor does not provide power into the gearbox.

When initially looking at **Figure 15-54** and **Figure 15-55**, it might seem odd that the two different motors' output gears are driving the same gear at two different gear ratios, with two different motor displacements, and no overrunning clutches are being used. The question might be asked, "What reconciles the differences in motor displacements and gear ratios to avoid breaking gear teeth?" Since both motors are always hydraulically tied together, receiving pump oil flow in parallel, one HST motor will have the most leverage and will be driving the transmission's external-toothed ring gear and the other motor will be driven by the ring gear causing the motor to act like a pump.

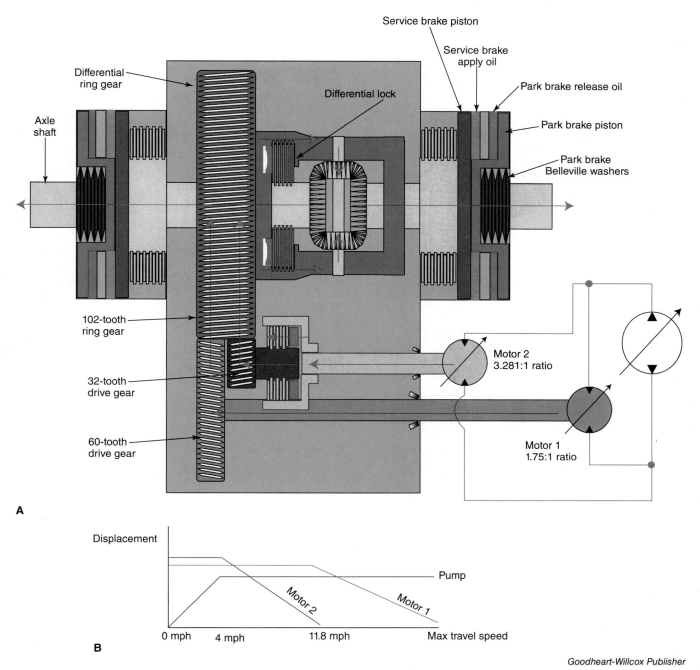

Figure 15-54. A—Simplified drawing of a John Deere X9 ProDrive dual-motor combine transmission. B—Graph of ProDrive transmission pump and motor displacements at varying travel speeds.

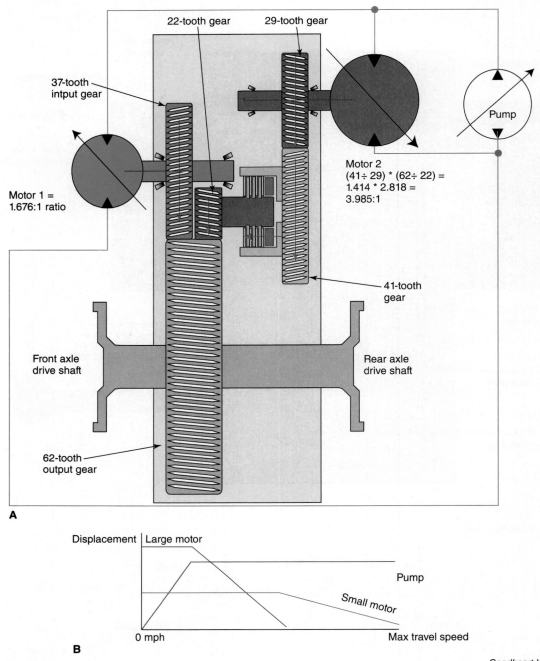

Figure 15-55. A—Simplified drawing of a Caterpillar 924K, 926M, 930K, 930M, 938K, and 938M wheel loader transmission. B—Graph of transmission pump and motor displacements at varying travel speeds.

Summary

- Mechanical variators can use a variable diameter pulley (VDP) or a toroidal variable drive.
- The most popular mobile equipment CVT uses two inputs: a mechanical engine input and a hydrostatic input.
- CVTs commonly use planetary gear sets, clutches, and a hydrostatic drive. Some also use synchronizers.
- The Case IH flagship combine has two CVTs: one to drive the rotor, one to drive the feeder/header.
- The CNH Magnum and T8 CVTs have a compound planetary gear set, six multiple-disc clutches, and an integral hydrostatic pump and motor.
- Tractor CVTs can be held stationary by a hydraulic input even while receiving mechanical input from the engine. This operating mode is known as power zero or power neutral. It prevents the tractor from moving while providing uninterrupted power to the implement hydraulic systems.
- Fendt's Vario CVT has one simple planetary gear set and no clutches. The operation of the CVT's planetary gear set is controlled by a reversible variable-displacement pump and one or two bidirectional variable-displacement motor(s).
- John Deere's 8R IVT has a hydrostatic unit, a compound planetary gear set, a reverse planetary gear set, a reverse brake, a low clutch, a high clutch, and a C1/C2 synchronizer assembly.
- The Caterpillar XE CVT wheel loader has a compound planetary gear set, a reversible variable-displacement hydrostatic pump and fixed-displacement hydrostatic motor, three synchronizers (providing five ranges), and two multiple-disc clutches (A and B).
- John Deere's STS and S-Series ProDrive transmission is used in their self-propelled forage harvesters, combines, and cotton harvesters. It provides an automatic two-speed range transmission that is driven by a variable-displacement hydrostatic drive.
- The John Deere X9 ProDrive combine transmission uses one HST pump to drive two HST motors in parallel that drive a differential ring gear. One motor can be disengaged from the transmission using a clutch.
- The Caterpillar 924K, 926M, 930K, 930M, 938K, and 938M wheel loaders uses one HST pump to drive two HST motors in parallel that drive a transmission output ring gear. One motor can be disengaged from the transmission using a clutch.

Technical Terms

- anti-slip regulation (ASR)
- A-post display
- cluster gear
- continuously variable transmission (CVT)
- corner post display
- droop control
- infinitely variable transmissions (IVT)
- multifunction handle (MFH)
- power neutral
- power zero (pz)
- preselect mode
- proportionator pump/motor
- pull belt
- push belt
- rotary encoder
- stepless transmission
- toroidal drive
- variator

Review Questions

Answer the following questions using the information provided in this chapter.

Know and Understand

1. What must pivot inside a toroidal mechanical variator to cause the ratio to change?
   A. Power rollers.
   B. Discs.
   C. Belt.
   D. Pump/motor.

2. Off-highway CVTs can contain all of the following, *EXCEPT*:
   A. hydrostatic transmission.
   B. planetary gear set.
   C. torque converter.
   D. multiple disc clutches.

3. The Case IH combine CVT is used to drive all of the following, *EXCEPT*:
   A. feeder.
   B. header.
   C. rotor.
   D. ground drive transmission.

4. The ETR clutch in the Case IH combine CVT is a(n) _____.
   A. holding clutch
   B. input clutch
   C. Both A and B.
   D. Neither A nor B.

5. The CNH Magnum and T8 CVT delivers power via the compound planetary output shaft(s) to _____.
   A. one range shaft
   B. two range shafts
   C. one planetary gear set
   D. two planetary gear sets

6. The planetary carrier in the CNH Magnum and T8 CVT delivers power to _____.
   A. clutch 1 and clutch 3
   B. clutch 2 and clutch 4
   C. clutch 5 and clutch 7
   D. clutch 6 and clutch 8

7. Which of the following controls the direction of a tractor equipped with a CNH Magnum or T8 CVT?
   A. Forward and reverse clutches.
   B. Hydrostatic pump yoke angle.
   C. Forward and reverse synchronizers.
   D. Clutch A and clutch B.

8. The Fendt Vario CVT uses _____.
   A. no planetary gear sets
   B. one simple planetary gear set
   C. one compound planetary gear set
   D. dual simple planetary gear sets

9. The Fendt Vario CVT has _____.
   A. zero multiple-disc clutches
   B. one multiple-disc clutch
   C. two multiple-disc clutches
   D. three multiple-disc clutches

10. What drives the hydrostatic pump in a Fendt Vario CVT?
    A. Engine.
    B. Forward clutch.
    C. Planetary carrier.
    D. Ring gear.

11. How does the Vario CVT control the direction of the tractor?
    A. Forward and reverse clutch.
    B. Hydrostatic pump yoke angle.
    C. Forward and reverse synchronizers.
    D. Clutch A and clutch B.

12. Which of the following consists of two sets of pinions coupled together to spin together as a single assembly?
    A. Cluster gear.
    B. Club gear.
    C. Clump gear.
    D. Clutch gear.

13. On a Caterpillar XE CVT, the loader's direction of travel is reversed by actuating the _____.
    A. forward and reverse clutch
    B. hydrostatic pump yoke angle
    C. forward and reverse synchronizers
    D. clutch A and clutch B

14. For the Caterpillar XE CVT loader to move, at least _____ must be engaged.
    A. one synchronizer and one multiple disc clutch
    B. two synchronizers and one multiple disc clutch
    C. one synchronizer and two multiple disc clutches
    D. two synchronizers and two multiple disc clutches

15. In a Caterpillar XE CVT in low reverse, the _____ acts like a reverse idler.
    A. low forward gear
    B. high forward gear
    C. Both A and B.
    D. Neither A nor B.

16. The John Deere ProDrive transmission is used in all of the following, *EXCEPT*:
    A. combines.
    B. self-propelled forage harvesters.
    C. cotton harvesters.
    D. 8R tractors.

17. The John Deere STS and S-Series Combine ProDrive transmission has all of the following components, *EXCEPT*:
    A. a planetary gear set.
    B. a synchronizer.
    C. a variable-displacement hydrostatic pump.
    D. a variable-displacement hydrostatic motor.

18. When the John Deere X9 combine and the Caterpillar 924K, 926M, 930K, 930M, 938K, and 938M loaders are increasing travel speed, all of the will occur to one of the HST motors, *EXCEPT*:
    A. clutch disengaged
    B. motor swivel angle pivoted to a neutral angle
    C. drive oil continues to be sent to the motor
    D. motor is adjusted to maximum displacement

Apply and Analyze

19. The New Holland Boomer-series compact utility tractor uses a CVT with a(n) _____-type belt.
20. The Case IH combine CVT uses a hydraulic motor to drive or hold the planetary _____.
21. A total of _____ multiple-disc clutch packs are used in the CNH Magnum and T8 CVT.
22. The CNH Magnum and T8 CVT uses a compound planetary gear set that delivers power through _____ outputs.
23. The John Deere 8R IVT has a CU hydrostatic module member, and it is linked to the _____ shaft.
24. The John Deere 8R IVT has a RU hydrostatic module member, and it is linked to the forward _____ gear.
25. The Caterpillar XE CVT's variator pump yoke begins at a(n) _____ angle when the loader begins to move from a neutral position.
26. The Caterpillar XE CVT wheel loader has a total of _____ synchronizers.
27. The Caterpillar XE wheel loader CVT has a total of _____ multiple-disc clutches.
28. The Caterpillar 924K, 926M, 930K, 930M, 938K, and 938M wheel loaders' transmission uses a total of _____ infinitely variable HST motors.

Critical Thinking

29. The HST motors in Caterpillar 924K, 926M, 930K, 930M, 938K, and 938M wheel loaders drive the same output ring gear, but do it using different-sized gears with different gear ratios, and different motor displacements, without any overrunning clutches. How is it possible that the motors' different ratios and displacements do not fight each other and break gear teeth?
30. How is it possible that a tractor with CVT, containing a mechanical input and a hydraulic variator, can be held stationary on a hill in a power neutral position with the engine running?

Nikonaft/Shutterstock.com

The purpose of a differential is to drive the axles while allowing them to rotate at different speeds if one axle is subjected to greater resistance than the other.

Chapter 16
Differentials

Objectives

After studying this chapter, you will be able to:
- ✓ Explain the functions of a differential.
- ✓ Explain the operation of various types of differentials.
- ✓ Diagnose differential symptoms.
- ✓ Describe the procedure for adjusting a differential.

A *differential* is a gear system that allows two drive wheels to rotate at different speeds. This is especially useful when the machine is turning, as the outside drive wheel must travel a greater distance in the turn than the inside drive wheel.

Introduction to Differentials

Differentials are found in most heavy equipment machines. The differential allows two axles to do the following:
- Rotate at different speeds when a machine is turning a corner.
- Rotate at different speeds when the two drive wheels have different amounts of load or resistance.
- Rotate at different speeds when the tires have different diameters due to tread wear or low air pressure.
- Drive both wheels at the same speed when the wheels have the same resistance, have the same traction, and are the same size.

A differential is designed to drive two wheels at different speeds while turning a corner and drive both wheels at the same speed (with the same traction) when the machine is moving in a straight line.

Parts of a Differential

A differential assembly contains the following components, shown in **Figure 16-1**.
- Differential case.
- Ring gear.
- Two, three, or four spider gears.
- Two side gears.

The ***differential case*** contains or supports the other components of the differential assembly. The case can be installed and removed from the axle's housing as an assembly. It contains differential components, such as the side gears, ring gear, spider gears, cross shaft, and bearings. The ***spider gears*** and ***side gears*** are straight cut bevel gears. The differential ***ring gear*** is attached to the differential case and is normally driven by a spiral bevel pinion gear.

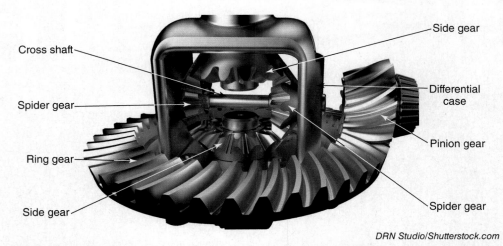

Figure 16-1. Differentials consist of a ring gear, spider gears, and two side gears. The ring gear is typically driven by a spiral bevel pinion gear but can be driven by an external-toothed spur or helical gear.

Spiral bevel drives are classified as *amboid* if the pinion is above the center axis of the ring gear and *hypoid* if the pinion is below the center axis of the ring gear. Either design can be used in heavy equipment differentials. The spiral bevel pinion and ring gear provide torque multiplication and change the power flow at a 90° angle.

A **differential carrier** is the structure that holds the differential case in proper alignment with the pinion gear. In some differential designs, the pinion gear and differential case are contained together in a *removable differential carrier*. This type of differential carrier keeps the pinion in alignment with the ring gear during assembly and disassembly. The differential assembly and differential carrier are removed or installed from the front of the axle housing as a single unit. See **Figure 16-2**.

In other designs, the differential carrier is an integral part of the axle housing. This type of differential carrier is referred to as an *integral* or *unitized differential carrier*. In these designs, the differential assembly must be removed through the back of the axle housing before the pinion gear can be removed, **Figure 16-3**. In both types of differential carriers, the differential case is attached to the carrier by side bearing caps.

Note

Differentials can be located in something other than an axle housing, such as a harvester transmission housing. Differentials that are integrated into a harvester transmission often have a ring gear that is driven by an external-toothed spur gear or external-toothed helical gear rather than a spiral bevel pinion gear. See **Figure 16-4**.

As the ring gear forces the differential case to rotate, the spider gear shafts revolve forward or rearward with the case. As noted earlier, differentials have two, three, or four spider gears. In a differential with four spider gears, the spider gears are often mounted on a single shaft called a **cross shaft**.

The side gears attach to the axle shafts and drive the wheels. The two side gears are the outputs of the differential. In light-duty trucks, the side gears directly drive the axles that turn the wheels. In most heavy equipment, final drives multiply the torque from the differential before it is applied to the wheels. Final drives are covered in Chapter 17, *Final Drives*.

Although the pinion-to-ring gear ratio multiplies torque, this torque multiplication function is separate from the main function of a differential, which is to allow drive wheels to spin at different speeds when turning corners. The resistance of the drive wheels determines the speed of the side gears.

Figure 16-2. This removable differential carrier has been removed from a Caterpillar 771D quarry truck. The pinion gear is enclosed in the differential carrier and the differential case is attached to the carrier by side bearing caps.

Figure 16-3. This differential is installed in an integral carrier. The back cover of the axle housing has been removed to access the differential.

Overview of Differential Operation

To understand how a differential works, you must be able to visualize how the various components work together. As each component is introduced, visualize how it behaves and how its behavior affects the components already introduced.

As mentioned earlier, the ring gear is attached to the differential case. As the ring gear rotates, the differential case rotates with it. The spider gear cross shaft is pinned to the differential case. As the differential case rotates, the cross shaft and its spider gears revolve with the rotating differential case. Keep in mind that as the spider gears revolve, they may or may not be spinning (rotating) on the cross depending on the resistance of the drive tires, **Figure 16-5A**. The purpose of the spider gears is to transfer the rotational energy of the differential case to drive the side gears.

Figure 16-4. This differential assembly in a Case IH 2100 series combine three-speed transmission has a ring gear that is driven by an external-toothed spur gear, not by a spiral bevel gear.

Next, add the side gears and axles. The axles pass through the center axis of the differential case, but are not connected to it. The side gears at the end of the axles are in mesh with the spider gears. If the machine is being driven in a straight forward direction with equal resistance on the drive tires, as the ring gear and differential case rotate, the spider gears revolve around the central axis of the differential *without* rotating on their own axes. Because the side gears are meshed with the spider gears, the spider gears drive the two side gears at equal speeds. The side gears and axles revolve around the center axis of the differential. See **Figure 16-5B**.

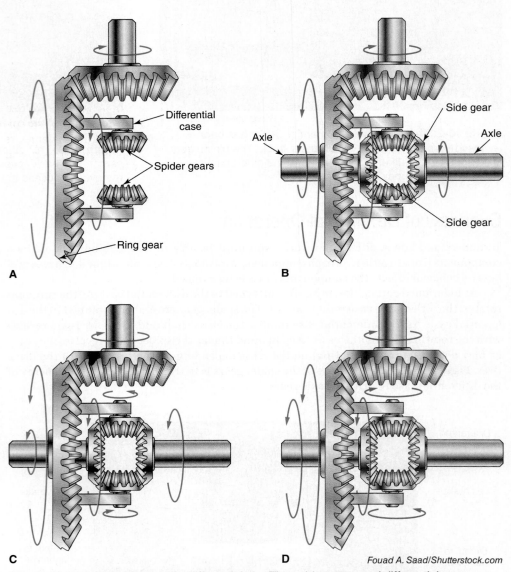

Fouad A. Saad/Shutterstock.com

Figure 16-5. Basic operation of a differential. A—The spider gears and differential case revolve around the ring gear axis. B—If the axles have the same resistance, the spider gears revolve with the differential case, but do not rotate on their own axes. They drive the side gears. C—If one axle has more resistance than the other, they are driven at different speeds. The left axle has greater resistance and slows down. The spider gears rotate on their axes, driving the right axle at a higher speed. D—If there is a large difference in resistances between the axles, the axle with the high resistance (the right axle in this example) is not driven and the spider gears walk around the stationary side gear. The spiders rotate on their axes very quickly, driving the low resistance axle at twice the normal speed.

Now imagine you apply a load to one of the axles. The revolving differential case and spider gears continue to drive the side gears. However, the side gear on the axle with the load requires more torque to rotate, and therefore provides more resistance to the spider gear than the other side gear does. Because of the difference in resistance, the spider gears begin to rotate on their own axes as they revolve around the central axis of the differential. Now, because the side gears are not only driven by the revolution of the spider gears around the differential axis but also by the spider gears' rotation around their own axes, the side gear on the low-load axle speeds up by the same amount that the high-load side gear slows down. This state is the equivalent of a machine going around a corner. The inside drive wheel is subject to greater resistance (load) than the outside drive wheel. See **Figure 16-5C**.

Next, imagine an equal load is applied to both axles. The side gears offer the same resistance to the spider gear because they require the same torque to rotate. Because the spider gears have the same resistance on both sides, they quit rotating on their own axes, but continue to revolve around the center axis of the differential, driving both side gears equally. This state is the equivalent of the machine driving in a straight line. See **Figure 16-5B**.

Now imagine that the load is suddenly released from one of the axles. The drop in resistance causes the spider gears to begin rotating toward that side gear as they revolve around the differential's central axis. As the spider gear rotation accelerates, it drives the low-load side gear at higher speed. However, on the other side gear, the backward rotation of the spider gears around their own axes counteracts the push of the revolution of the spider gears around the differential's central axis. As a result, the axle with the load is not driven, and the spider gears walk around the stationary side gear and drive the other side gear (axle) rapidly. See **Figure 16-5D**. This is the equivalent of a machine driving in a straight line when one wheel hits a slick patch and begins spinning.

Note

While the explanation provided here should give you some insight into the operation of a differential, the surest way to understand how they work is to observe one in action. There are a number of useful educational videos online. Also, if you have an opportunity to work with a demonstration model, experiment with varying the load on the drive axles while observing the behavior of the spider gears.

Types of Differentials

Heavy equipment can be equipped without a differential or with up to seven differentials on one machine, such as the Liebheer LTM 1750-9.1 mobile crane. Examples of machines that do not use a differential include four-wheel-drive hydrostatically propelled sprayers, steering clutch and brake dozers, and John Deere 944K electric drive wheel loaders. Most machines have at least one differential to allow the drive axle to power the drive wheels at different speeds while turning a corner.

Some off-highway machines have three drive axles (6×6), such as an articulated dump truck, or four drive axles (8×8), such as an all-terrain crane. See **Figure 16-6**. A 6×6 articulated dump truck can be equipped with up to five differentials—three drive axle differentials (cross-axle differentials), and two inter-axle differentials (IADs). IADs are discussed later in this chapter.

Goodheart-Willcox Publisher

Figure 16-6. All-terrain cranes can be equipped with four drive axles, each containing a differential, as well as longitudinal differentials, and a transmission differential.

Goodheart-Willcox Publisher

Figure 16-7. A rear drive axle from a Deere 8000 series tractor. The left final drive and axle assembly has been removed from the housing. Like most traditional drive axles, the differential delivers power to the left and right final drives. A brake is located between the differential and each of the final drives.

The most common location of a differential is inside the rear axle housing, as in a two-wheel-drive truck or tractor. See **Figure 16-7**. The differential's spiral bevel pinion gear is the input to the differential and is normally driven by a drive shaft via the transmission's output shaft. The spiral bevel pinion gear drives the spiral bevel ring gear.

On many machines, a second differential is located in the machine's front axle housing. This differential is also driven by a spiral bevel pinion gear. Second differentials are used in four-wheel-drive tractors and trucks, wheel loaders, loader backhoes, and telehandlers.

In harvesters, such as combines, self-propelled forage harvesters, and cotton harvesters, the differential is not placed inside the traditional rear axle housing, nor does the harvester use a spiral bevel gear set. Harvesters often have a hydrostatic motor that delivers power into a two-speed, three-speed, or four-speed transmission. The multispeed transmission delivers power to the differential's ring gear, which is a spur gear design. The differential side gears deliver power to two shafts, one inside each brake housing. Each shaft can be held by the brake or released to deliver power to its respective final drive. Refer to **Figure 16-4**.

Non-Locking Differential

The differential shown in **Figure 16-1** is a non-locking differential, also called an ***open differential*** or *standard differential*. An open differential increases delivery of power to the drive wheel with the lower resistance and decreases power delivery to the drive wheel with greater resistance. If the difference in resistances between the two wheels is great, as when one wheel is in mud and the other is on packed gravel, the machine will lose traction and possibly be unable to move. For this reason, many heavy equipment manufacturers offer locking differentials and limited-slip differentials to improve machine traction while operating in wet or slick conditions.

Locking Differential

Locking differentials lock one of the side gears to the differential case, which causes the side gear to rotate at the same speed as the differential case. When one side gear is locked to the differential case, the spider gears can no longer rotate (spin on their shaft). The non-rotating spider gears drive the other side gear at the same speed as the differential case. The result is a locked differential that drives both axles at the exact same speed, regardless of the resistances exerted on the two drive wheels.

A locking differential used to lock the right and left drive wheels is called a ***cross-axle differential lock***. A ***locking inter-axle differential*** is designed to lock front and rear or tandem drive axles so they drive at the same speed.

Mechanical Locking Differential

The most economical locking differential is the mechanical locking differential, or *jaw clutch differential lock*. It commonly has a foot-operated pedal that is pressed with the operator's toe or heel. See **Figure 16-8A**. The lever actuates a mechanical linkage that forces a collar to couple either a side gear or an axle shaft (attached to a side gear) to the differential carrier. The collar's splines are sometimes called *dog teeth*. When the pedal is released, a spring returns and holds the differential lock collar in a released position. See **Figure 16-8B**.

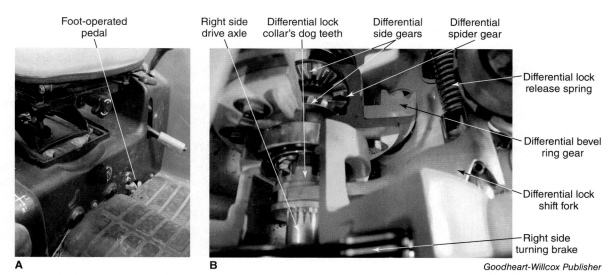

Figure 16-8. In mechanical locking differentials, a manually operated linkage slides a collar to couple one axle or side gear to the differential housing. A—This foot-operated pedal, located directly below the front edge of the operator's seat, is actuated by the operator's heel. B—This cutaway image reveals the components in a mechanical differential lock, including the shift fork, release spring, shift collar, side gear, and drive axle.

The manufacturer's service literature provides specific operating instructions for safety and to avoid harming the differential. The following is an example set of instructions:

- Do not apply the mechanical differential lock under these conditions:
  - At high speeds.
  - When turning.
  - When driving on side slopes.
  - When driving on dry, hard surfaces, such as concrete or asphalt.
  - When the two drive wheels are spinning at significant speed differences.
- Apply and release the differential lock under these conditions:
  - When driving at slow speeds.
  - When the drive wheels are spinning at the same speed.
  - Only when necessary.
- After the differential lock has been engaged, it is kept engaged by drive wheel slippage. When the drive wheels experience the same resistance, the differential lock releases (if the mechanical foot pedal has been released). If the operator prefers to keep the differential lock engaged (and if the resistance on the drive wheels equalize), the operator must keep pressing the foot pedal.
- If a mechanical differential lock does not release, press one of the turning brakes and then the other brake.

Controls for Differential Locks

In addition to mechanical linkage, hydraulic, pneumatic, electrical, or electronic controls can actuate the differential lock's shift collar. If electrical controls are used, an electrical switch can directly energize a solenoid, energize a relay that operates a solenoid, or provide an input to an electronic control module (ECM) that energizes a differential lock solenoid. The differential lock switch is on the floor (foot operated) or on the console. When energized, the differential lock solenoid opens a passageway that allows air pressure (electro-pneumatic) or hydraulic pressure (electro-hydraulic) to actuate the differential lock.

Some early machines used a mechanical linkage that directly controlled a pneumatic or hydraulic valve. The valve directed pressure to actuate the differential lock shift collar.

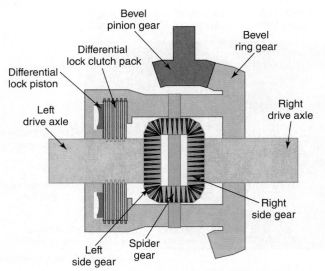

Figure 16-9. In a multiple disc clutch locking differential, the friction discs are splined to the drive axle and the separator plates are splined to the differential case. When the clutch is applied, the axle and case become locked and rotate together.

Multiple Disc Clutch Differential Lock

Off-highway machines can also have a multiple disc clutch that locks the differential. See **Figure 16-9.** Examples of machines that may have multiple disc clutch differential locks are agricultural mechanical front-wheel-drive (MFWD) tractors, harvesters, articulated dump trucks, and motor graders.

Electronic Controls

On late-model off-highway equipment, an electronic control module energizes a differential lock solenoid. In addition to the input received from the differential lock toggle switch, other variables affect the engagement and release of the differential lock.

> **Note**
> On some agricultural tractors, an end-of-row option called a *headland management system* allows the operator to program functions that are automated in a specific sequence by the tractor when it reaches the end of a row. These functions include changing the engine speed, lifting the planter and shutting off the seed metering units, and shifting gears. The differential lock can be included in the headland management system function so the differential unlocks to make turning easier at the end of the row.

Automatic Differential Lock

Agricultural tractors can be equipped with an automatic rear differential lock feature. See **Figure 16-10.** John Deere 6R series tractors, in the automatic differential lock mode, use input from a steering angle sensor to engage and release the rear differential lock. A tractor equipped with this feature has a limited-slip differential in the front axle. The automatic rear differential lock can be programmed to be triggered by one of four steering angles: 6°, 9°, 12°, and 15°. When the automatic differential lock switch has been selected, the rear differential lock remains engaged as long as the tractor is driving straight. When the steering angle sensor provides input indicating that the tractor is steering, the differential lock is released.

In the manual differential lock mode, the ECM disengages the differential lock under these conditions:

- The tractor exceeds 14.5 mph (23.3 km/h).
- The manual differential lock switch (on the floor or on the right-hand console) is pressed.
- The brake pedal is pressed.

Automatic Traction Control

Late-model machines, such as articulated dump trucks, can be equipped with *automatic traction control (ATC).* An ECM monitors wheel slip and modulates solenoid valves to vary the apply pressures to the cross-axle differential locks and inter-axle differential locks, reducing wheel slip. The ECM also automatically disengages the differential locks based on the traction of the drive wheels.

Figure 16-10. This John Deere 6120R utility tractor has two differential lock switches. The switch on the left is for automatic differential lock mode. The switch on the right is for manually engaging the differential lock. The differential lock is located in the rear axle.

ATC enables the operator to concentrate on operating the truck without having to manually engage and disengage the differential locks. ATC also reduces wear on the machine.

Limited-Slip Differential

A ***limited-slip differential***, also called a *positive traction differential*, has two multiple-disc clutch packs, one for each side gear. The clutch friction discs and plates have a small amount of clearance. When the machine is being driven straight ahead, the limited-slip differential operates as a standard differential and drives both side gears at equal speeds.

When one drive wheel loses traction and begins to spin, that side gear moves away (inward) from its clutch pack, allowing the wheel to spin. The other side gear moves away from the spider gears due to the driving force of the side gear separating from the spider bevel gear set. As this side gear moves outward, it squeezes the clutch pack, locking the side gear to the differential case or ring gear. As a result, the drive wheel with the most resistance/traction receives all the power from the limited-slip differential. See **Figure 16-11**. Limited-slip differentials can be found in wheel loaders, loader backhoes, garden tractors, compact utility tractors, utility tractors, agricultural MFWD tractors and articulated dump trucks.

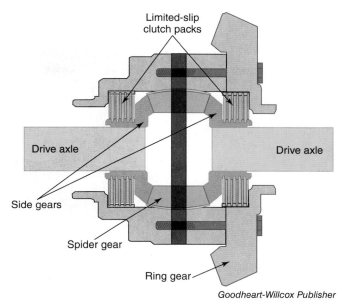

Goodheart-Willcox Publisher

Figure 16-11. Limited-slip differentials have a clutch pack at each side gear. When one drive wheel loses traction, the torque exerted through the other side gear forces it outward, causing it to squeeze its clutch. As a result, power is distributed through the drive wheel with traction.

Note

During a turn, the side gear thrust load is low. This enables the drive wheels to overcome the friction between the clutch pack's plates and friction discs. As a result, the differential functions as a standard differential. The inside wheel slows and the outside wheel increases speed.

No-Spin Differential

Heavy equipment can be equipped with a ***no-spin differential***, which delivers power to both drive wheels when the machine is driven in a straight line. When the machine turns, the outside wheel is allowed to free-wheel and spin faster, while the inside drive wheel is positively driven by the differential.

If a machine with an open differential had one drive wheel suspended freely in the air, all of the power would be sent to the wheel off the ground. However, with a no-spin differential, the drive wheel on the ground would not slip but would be driven by the no-spin differential. A no-spin differential can be used in place of a standard differential inside the same differential case. No-spin differentials are offered as an option in wheel loaders, wheel dozers, landfill compactors, articulated dump trucks, and on-highway dump trucks.

Note

Other terms for a no-spin differential include automatic locking differential, Detroit locker, ratcheting differential, and full-locking differential.

No-Spin Spider

At the heart of a no-spin differential is the ***no-spin spider***, a circular ring with four steel dowels protruding from it. See **Figure 16-12A**. These dowels fit in place of the standard

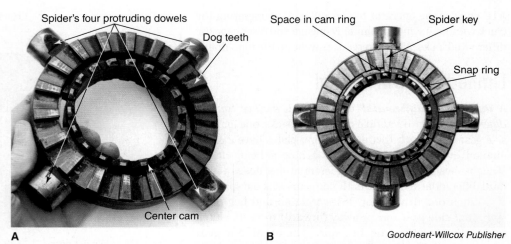

Figure 16-12. A spider ring is located in the center of a no-spin differential. A—A no-spin spider contains four protruding dowels that are inserted into the differential case. B—The spider contains a long tooth known as the *spider key*. It fits in the space located on the cam ring.

differential's cross shaft. The no-spin spider replaces the spider pinions found in a standard differential. The spider is driven at the same speed as the differential case. Dog teeth on both sides of the spider mate with right and left side driven clutches.

A *center cam* in the center of the spider has a groove for a snap ring that retains the cam inside the spider. The cam can rotate a few degrees inside the spider, but its rotation is limited by the space in between the cam and the *spider key* (the long tooth on the spider). See **Figure 16-12B**.

Driven Clutches

The spider has slightly rounded dog teeth that mesh with slightly rounded dog teeth on left and right *driven clutches*, also known as *jaw clutches*. The driven clutches are the differential components that couple and uncouple the drive axle side gears from the rest of the differential. The assembly of the spider and driven clutches is known as a *dog clutch*. The driven clutch on the right side of the differential splines to the side gear on the right, and the driven clutch on the left side of the differential splines to the side gear on the left. See **Figure 16-13**. A spring between each side gear and driven clutch holds the driven clutch in mesh with the spider under normal conditions.

Center Cam and Holdout Ring

A *holdout ring* is a spacer ring between the driven clutch and the spider. Raised portions of the holdout ring fit into spaces in the cam ring. When one of the drive wheels is forced to turn faster than the spider gear, as when going around a corner, the rounded dog teeth on the driven clutch ride up and over the rounded teeth on the cam ring, pushing the driven clutch away from and out of engagement with the spider. As a result, the side gear is free to spin at the necessary speed. As long as the wheel continues to turn faster than the spider, the holdout ring is unable to reseat in the cam ring, which keeps the driven clutch and spider separated. When the affected drive wheel slows to the appropriate speed, the clutch spring pushes the clutch and holdout ring back into engagement with the cam ring. As the raised portions of the holdout ring drop back into the slots in the cam ring, the dog teeth on the driven clutch reengage with the dog teeth on the spider. See **Figure 16-14**.

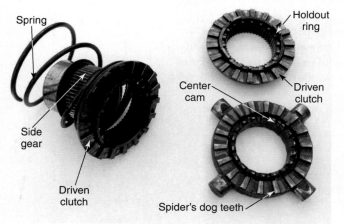

Figure 16-13. In a no-spin differential, a spider with dog teeth meshes with two driven clutches with dog teeth.

Inter-Axle Differential

Off-highway machines can be equipped with an *inter-axle differential (IAD)*, also called a *power divider* or *longitudinal differential*. This type of differential is designed to allow speed differences between different drive axle assemblies, such as between the front and rear drive axles or between the tandem rear drive axles on an on-highway semi-truck, **Figure 16-15**, or on an articulated dump truck (ADT). Most IADs are equipped with a locking mechanism.

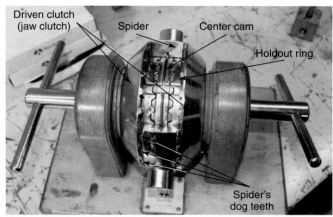

Figure 16-14. In this no-spin differential cutaway training aid, the driven clutches, spider, center cam, and holdout rings are visible. Notice the spider is cutaway, but is in mesh with the driven clutches and holdout rings.

 Note

One of the IADs in articulated dump trucks can be located in a gear housing. The gear housing may be called a transfer gearbox, OTG (output transfer gear), output transfer drive, or drop box. However, not all output transfer gear assemblies contain an IAD, but rather have a single shaft that drives both the front and rear axles. Examples of machines with this arrangement include articulated wheel loaders and articulated compactors.

Figure 16-15. A cutaway of an IAD that allows speed differences between two rear tandem drive axles in an on-highway truck application. The IAD sends power to the lower helical gear to drive the front drive axles and sends power out the yellow output shaft to the rear drive axles.

The output transfer gears receive input from the transmission. The transfer gearbox directs power to the front axle and to the rear axles. See **Figure 16-16**. The center axle contains a *center axle through drive*, which consists of a gear-shaft assembly. The shaft receives its input power from the OTG and directs power out of the center axle to the rear axle. A gear on the shaft also provides input to the pinion gear shaft, which drives the center axle's cross differential.

A Case articulated dump truck previously manufactured by Astra has five differentials. The center differential is an IAD inside the transmission. The center differential sends power to the front drive axle and to the rear inter-axle differential. The center differential has a differential lock. The rear inter-axle differential is a locking differential that sends power to the two rear drive axles. All three drive axles contain a limited-slip cross-axle differential. See **Figure 16-17**.

A Doosan 6×6 articulated dump truck has only three differentials—two cross-axle differentials and one inter-axle differential. The rear cross-axle differential sends power to a pair of tandem driven tires, known as a free-swinging bogie axle that allows the left and right tandems to pivot. See **Figure 16-18**. Chapter 21, *Suspension Systems*, further explains the free-swinging bogie axle.

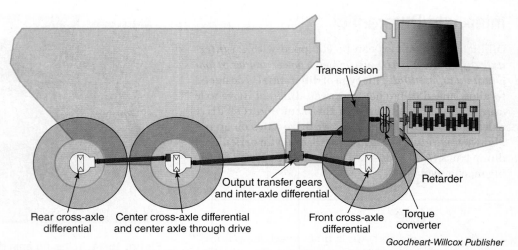

Figure 16-16. A simplified side view of a Caterpillar 6×6 articulated dump truck (such as B and C series ADTs). The truck has four differentials—one inter-axle differential and three cross-axle differentials.

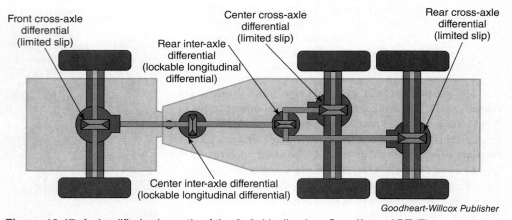

Figure 16-17. A simplified schematic of the 6×6 driveline in a Case/Astra ADT. There are two inter-axle (longitudinal lockable) differentials and three cross-axle limited-slip differentials.

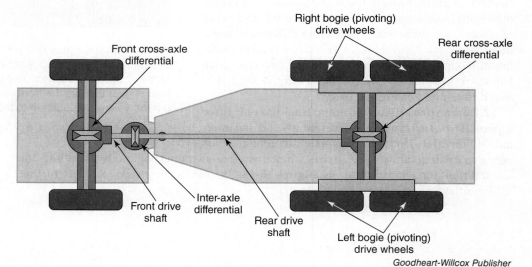

Figure 16-18. A top view of a Doosan 6×6 ADT with three differentials. There are two cross-axle differentials and one inter-axle differential. The rear wheels are a pivoting bogie design.

Planetary Differential

Most inter-axle differentials in ADTs do not have side gears and spider bevel gears. The inter-axle differential is normally constructed using a compound planetary gear set. There are several planetary differential designs. The gear set in **Figure 16-19** consists of the following components:

- An external-tooth ring gear. The ring gear has six planetary pinion shafts and serves as the planetary carrier. The ring gear provides input into the planetary differential.
- Two sets of planetary pinion gears. There are three pinion gears in each set.
- An output sun gear that drives the rear axle output shaft.
- An output sun gear that drives the front output shaft.
- A differential locking mechanism that couples the planetary carrier to the front output drive shaft.

Note

Inter-axle planetary differential locking mechanisms can use a pneumatically or hydraulically actuated multiple-disc clutch or a piston-actuated lever that engages and releases a mechanical sliding collar.

Manufacturers design the planetary differential to distribute a specific power ratio. The following are three examples:

- 40% of the power to the front drive axle and 60% to the two rear drive axles (6×6 ADT).
- One-third power to the front drive axle and two-thirds power to the two rear drive axles (6×6 ADT).
- 50% power to the front drive axle and 50% to the rear drive axle (4×4 ADT).

The differences in ratios is due to the size of the output sun gears, resulting in a different number of sun gear teeth.

During normal unlocked operation, when the front and rear drive axles are not slipping (axle resistances are equal), the planetary pinions are not rotating on their shafts. However, when one of the drive axles encounters resistance, its set of planetary pinion gears begin to spin on their shafts. These revolving pinions are in mesh with the other set of pinions that are driving the output gear/axle shaft with less resistance. As a result, the pinions driving the axle with minimal resistance increase speed to the axle with the least resistance and the axle with more the resistance will not receive power. When the planetary differential lock is applied, equal torque is delivered to both axles, 50% to the front axle and 50% to the rear axle.

The Case and Astra articulated dump trucks use ZF powershift transmissions. The ZF 260 transmission has a planetary inter-axle differential. The ring gear is the input, the sun gear is the front axle output, and the planetary carrier is the rear axle output. The differential has two sets of planetary pinions known as *satellite gears*. See **Figure 16-20**.

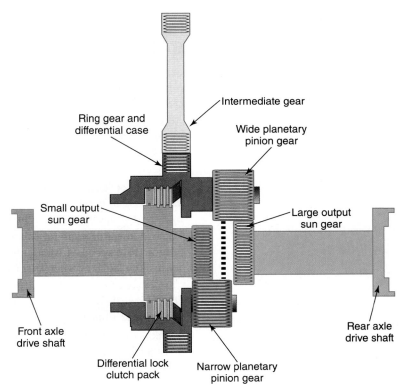

Goodheart-Willcox Publisher

Figure 16-19. A simplified drawing of a planetary gear set used as an inter-axle differential in a Caterpillar D25D, D30D and D350D articulated dump truck. The D25D and D30D are 4×4 and the D350D is a 6×6.

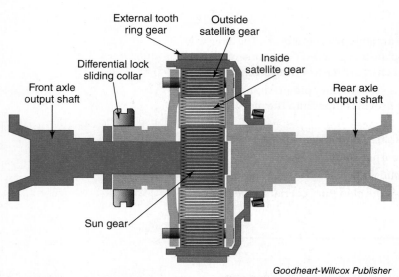

Figure 16-20. The ZF 260 powershift transmission uses a planetary inter-axle differential.

Goodheart-Willcox Publisher

Pro Tip

Many twin-track tractors are equipped with differential steering, in which planetary gear sets propel and steer the tractor. Do not confuse this design with a planetary differential in an articulated dump truck. Differential steering is explained in Chapter 24, *Track Steering Systems*.

As stated earlier, a transmission can contain an inter-axle differential that supplies power to an articulated dump truck's front and rear axles. Powershift transmissions also have other designs for powering front and rear axles. The following are examples:

- A no-slip inter-axle differential.
- No differential, but geared to a single shaft that drives both the front and rear axles in four-wheel-drive articulated agricultural tractors, articulated wheel loaders, and articulated compactors).
- A mechanism to disengage and engage the front drive axle (agricultural tractor MFWD).

Diagnosis

Differentials can suffer a variety of failures. The following list is not intended to be a replacement of the manufacturer's service literature, but to provide general guidelines for diagnosing differentials:

- **Noise on deceleration.** If the differential exhibits noise during deceleration, such as whirring or howling, investigate the pinion gear bearings for wear and the pinion gear's preload to see if it is too loose. The pinion gear and the ring gear are not suspect.
- **Noise on acceleration.** If the differential exhibits noise during acceleration, such as whining or howling, check the ring and pinion gear for wear. Also check for incorrect ring and pinion adjustment (setup).
- **Noise at lower or moderate speeds.** If a howling or rumbling noise occurs at lower or moderate speeds, investigate the differential case bearings. This symptomatic noise can also vary when turning a corner.
- **Clunk every few feet.** If the differential exhibits a consistent clunking, banging, or clicking while barely creeping forward or rearward, inspect for a failed ring gear or pinion gear. Look for missing teeth.
- **Clunk while turning.** If the differential exhibits a clunking, popping, or banging noise during a sharp turn, inspect for failed spider gears, low fluid in a limited-slip differential, or worn clutches in a limited-slip differential. Also inspect the spider gears.
- **Rumble on turns.** If a rumbling noise is exhibited while the machine is turning, inspect the axle housing for bad wheel bearings.
- **Vibration increasing with speed.** If a vibration occurs that is proportional to travel speed, investigate the driveline balance and check universal joints for wear. (Chapter 18, *Axles and Drive Shafts*, provides further instruction on drivelines.)

- Clunks during initial movement. If a clunk occurs during the initial movement of the machine or during an aggressive acceleration from a coast, check the yokes and universal joints for wear. Investigate the potential for internal transmission component wear or failure.

Repairing Differentials

Differential gears, bearings, and shafts constantly wear and will eventually fail, requiring service. Prior to removing and disassembling a differential, perform a thorough inspection. The inspection can direct the repair process to avoid a repeat failure. Check the gears, shafts, and bearings for wear, cracks, pitting, and broken or chipped teeth.

Prior to disassembly, use a dial indicator to measure and record the backlash between the pinion gear and the ring gear. Backlash is the amount of play between gear teeth. It equals the amount of clearance between meshed gear teeth. Checking backlash is detailed later in this chapter.

If the differential has side bearing caps, mark the caps prior to removal to ensure that the left cap is used on the left side of the differential housing and the right bearing cap is used on the right side. See **Figure 16-21**.

Pro Tip
Follow the manufacturer's service literature for replacing the differential's gears. Some require specific procedures, such as correct placement of thrust washers. The ring gear may need to be heated before installing it onto the differential housing.

Some technicians also check the spiral bevel gear set's tooth contact pattern. After cleaning the differential's teeth, the technician applies gear marking compound to the teeth. The process of checking tooth contact pattern is discussed later in this chapter.

Adjusting (Setting) a Differential

Rebuilding is also referred to as *adjusting* or *setting* a differential. The differential's design dictates the manufacturer's specific process for adjustment. Generally, the sequence for rebuilding a differential includes the following steps:

1. Set the pinion bearings' preload.
2. Set the pinion depth (may require readjustment later in the sequence).
3. Install the differential case assembly.
4. Set the differential case side bearings' preload.
5. Set pinion and ring gear backlash.
6. Check tooth pattern.
7. Move the ring or pinion to achieve the correct tooth pattern.

Note
The preceding steps are not intended to supersede the manufacturer's service literature. They provide only a general sequence of steps for adjusting differentials.

Goodheart-Willcox Publisher

Figure 16-21. This Carraro differential is used in the front axle of construction equipment, such as a Case loader backhoe. One bearing cap is hidden behind the ring gear.

Setting the Pinion Bearings' Preload

The spiral bevel pinion gear shaft is supported with two tapered roller bearings. Preload equals the amount of static force applied to the bearing rollers above zero end-play. The correct preload prevents the pinion gear from moving. With no preload, the pinion attempts to move forward during acceleration. During deceleration, the pinion attempts to move rearward.

Preload can be checked in various ways. The pinion shaft on some differentials can be rotated with a dial torque wrench or a flexible beam torque wrench and a socket. The torque wrench is turned steadily and measures the amount of turning torque (rolling torque) it takes to turn the pinion gear shaft. See **Figure 16-22**.

Goodheart-Willcox Publisher

Figure 16-22. Using a dial torque wrench to measure pinion preload.

Note
A clicker torque wrench cannot be used to check preload.

Some manufacturers specify checking preload with a spring scale. A string or thin soft wire with a spring scale is attached to the pinion bearing cage (pinion carrier). The technician pulls the scale to determine the amount of force required to turn the cage. See **Figure 16-23**.

The pinion shaft installation can include a spacer and shims, which are used to set the bearings' preload. A technician will use shims or spacers of different thicknesses to adjust the preload. The manufacturer's literature specifies the thickness of the spacer or shim used to adjust the pinion bearings' preload. The service literature might include a matrix that converts the spring scale measurement into a rolling torque value, accounting for the diameter or radius of the bearing cage. For example, 3–7 spring pounds (2–3 kilograms) equal 10–20 inch-pounds of rolling torque.

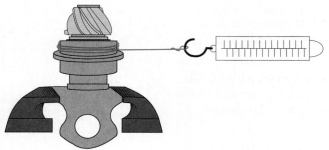

Goodheart-Willcox Publisher

Figure 16-23. A spring scale can be used to measure the amount of rolling torque required to spin a pinion bearing cage. The pinion shaft is held in a vise.

If the preload measurement is too high, such as 30 inch-pounds when the specification is only 10–20 inch-pounds, consult the manufacturer's literature. Follow instructions for reducing the rolling torque, such as installing a thicker spacer shim. The manufacturer might require checking the initial preload during the assembly process while the pinion shaft is in the press without the seal installed, and then rechecking the preload after final assembly of the pinion.

Some pinions use a *crush sleeve* instead of a spacer and shims. The sleeve is designed to crush (collapse) as the pinion nut is tightened to set the correct pinion bearing preload. Be careful to avoid setting the preload too high while tightening the nut. If the preload is accidentally set too high, a new crush sleeve will need to be installed and the preload will need to be reset.

Setting the Pinion Depth

The next step in adjusting a differential is setting the pinion gear depth. Pinion gear depth determines where, between the root and top land, the pinion teeth contact the ring gear's teeth. A proper pinion gear depth centers the contact point between the root and top land. A pinion depth that is too deep has a contact pattern that is too close to the root. A pinion gear depth that is too shallow has a contact pattern too close to the top land.

The pinion gear normally has a ***pinion cone variation number*** stamped or scribed on the end of the gear, **Figure 16-24**. Replacing the pinion gear with another pinion gear with a higher or lower variation number changes the pinion gear depth. The number is preceded by a positive (+) or negative (–) symbol. The number equals the variance in thousandths of an inch that the face of the pinion gear protrudes too far toward (+) or away from (–) the ring gear's center axis.

If the old housing, ring gear, and pinion gear will be reused, the pinion's selective shims will be reused with the pinion gear to initially set the pinion gear's depth. The pinion depth will be checked later when the tooth contact pattern step is performed. If the pinion depth is incorrectly set, a different selective shim pack will be needed to correct the tooth contact pattern.

If a new pinion gear and ring are being installed, it is helpful to know the thickness of the old shim pack and the pinion depth variation of the old pinion gear and the new pinion gear. Note that not all pinion gears have depth variation numbers and will require measurement tooling to determine the new shim pack thickness starting point.

Use the following formula as a starting point for the new pinion-depth shim pack: *old variation number minus the new variation number* and be sure to pay attention to the positive and negative signs. For example, if the old number was –4 and the new number is +2, the formula equals (–4) – (+2) = –6. This means the new shim pack should be 0.030″ (old shim pack) minus .006″ (difference between pinion variations) equaling 0.024″ (new shim pack). Another example is the old pinion was +2 and the new pinion is -2, and the original shim pack was 0.021″ thickness. The new shim pack would be +2 – (–2) = +4. The new shim thickness is 0.021″ + 0.004″ equaling 0.025″.

Goodheart-Willcox Publisher

Figure 16-24. Spiral bevel pinion gears have a pinion cone variation number on the pinion (for example, –4). Refer to this number to adjust the shim pack thickness in order to set the correct pinion depth.

Note

Ring and pinion gears are a matched set. If either the ring gear or the pinion gear must be replaced, the other gear must be replaced as well. Failure to replace both gears in the set will cause the new gear to wear excessively and fail quickly.

Installing the Differential Case in the Carrier

After the pinion preload and initial pinion depth have been set, the differential case assembly is installed into the differential's carrier. The side bearings should be lubricated prior to installation. If the differential carrier has side bearing caps, ensure that the left and right caps are installed on the correct side of the differential carrier. Avoid cross-threading the adjuster rings when installing them into the carrier.

Note

On some mobile equipment, the carrier must be inserted through the side rather than lowered from the top. A special lifting fixture is required to load the differential carrier into the axle housing.

Install the adjuster rings by hand until they are snug. Loosen the adjuster on the ring side of the carrier until one thread is visible (protruding from the carrier housing) to provide a small amount of temporary endplay. Next, rotate the opposite bearing adjuster ring (on the same side of the ring gear) until the differential ring gear and the pinion gear have no backlash.

Setting the Differential Case Side Bearing Preload

The differential case is mounted on two side bearings that are normally tapered roller bearings. These bearings must be properly preloaded to ensure long life and reduce gear noise. Proper preload also keeps the differential case from moving side to side during operation.

Side bearing preload on small differentials may require selective washers to be inserted between the bearing cups and the bearings. Most large differentials have adjustable retaining rings that are rotated with spanner wrenches. See **Figure 16-25**.

Each manufacturer has individual guidelines for setting preload. In general, the procedure involves the following steps:

1. Install the bearing retainers hand tight.
2. Loosen a specified retainer, such as the non-ring gear side bearing retainer, to introduce a temporary small amount of play. (The literature might specify using a dial indicator to measure the play).
3. Adjust the opposite side bearing retainer until there is zero backlash (play). Rock the ring gear back and forth while making the adjustment until there is no play.
4. Tighten the adjuster rings the additional amount specified in the service literature, such as two or three more notches.

Setting Pinion and Ring Gear Backlash

The adjustment of the differential side bearings' preload also sets the pinion and ring gear backlash. After the side bearings have been preloaded, check the pinion gear and ring gear backlash and readjust if needed.

To check backlash, place the plunger of a dial indicator perpendicular to the outside edge of a ring gear tooth's heel, **Figure 16-26**. Load the plunger against the ring gear with the dial's indicator set to zero. While viewing the dial indicator, slightly rotate the ring gear back and forth and note how far the dial moves.

The service literature specifies the pinion and ring gear's backlash, such as 0.006″ to 0.016″ (0.15 mm–0.40 mm). Both Meritor and DanaSpicer recommend adjusting a used pinion and ring gear backlash to the setting it had prior to disassembly.

Although the dial indicator's plunger is physically contacting the outer edge of the ring gear, it is technically measuring the gap between a pinion gear's tooth and a ring gear's

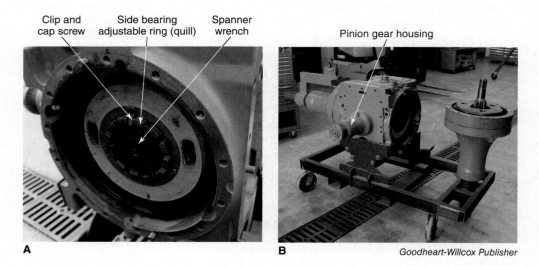

Figure 16-25. An axle from an 8000 series John Deere agricultural MFWD tractor. A—The left side bearing in the rear tractor axle has a retaining nut with twelve lugs separated by twelve notches, allowing the triangular spanner wrench to be inserted into three of the notches. Bearing preload is adjusted with the spanner wrench. A clip and cap screw prevent the ring from rotating. B—The pinion gear housing.

tooth. An example pinion and ring gear tooth count is 11 and 54, respectively. Because of possible ring gear tooth variation, the pinion gear and ring gear backlash should be checked at three or more equally spaced locations on the ring gear.

Moving Ring or Pinion to Achieve the Correct Tooth Pattern

If the backlash measurement is not within specification, an adjustment is required. Moving the ring gear away from the pinion gear increases the backlash. Moving the ring gear towards the pinion gear decreases the backlash. Since the differential case bearing preload has already been set correctly, the pinion and ring gear backlash adjustment must keep the case bearing preload the same. For example, if one side is loosened one notch, the other side must be tightened one notch to maintain the case bearing preload.

Checking Tooth Pattern

After the pinion and ring gear backlash have been adjusted, the tooth contact pattern must be checked by painting gear marking compound or Prussian blue dye onto the ring gear's teeth. See **Figure 16-27**. Paint 12 teeth of the ring gear. Roll the gear to obtain a contact pattern.

Goodheart-Willcox Publisher

Figure 16-26. Backlash is measured with a dial indicator. The plunger is placed at the edge of the ring gear's tooth.

Pro Tip

Although Prussian blue is commonly used, some technicians prefer a yellow gear marking compound, which can be more easily seen on the ring gear's teeth.

The painted compound leaves an impression where the pinion teeth mesh with the ring's teeth. The vertical location of the impression on the tooth reveals the accuracy of the pinion gear's depth. The horizontal position of impression on the tooth reveals whether

A B *Goodheart-Willcox Publisher*

Figure 16-27. Apply gear marking compound on the ring gear's teeth to check the pinion-to–ring gear pattern. A—Three ring gear teeth painted with yellow gear marking compound. B—Prussian blue can also be used to check the tooth patterns, but the darker blue is more difficult to see than yellow tooth contact pattern.

the ring gear is correctly located next to the pinion gear or needs to be moved toward or away from the pinion gear.

Prior to studying the tooth contact pattern, the technician should be familiar with ring gear tooth terminology. See **Figure 16-28**. The following terms refer to different areas of a gear tooth:

- *Heel*—the portion of the ring gear's tooth closest to the outer edge of the gear.
- *Toe*—the portion of the ring gear's tooth closest to the inside diameter of the ring gear.
- *Root*—the cavity or space between the base of a ring gear tooth and the base of the adjacent tooth.
- *Top land*—the flat surface at the peak of the ring gear's tooth.
- Hypoid coast side—the concave side of the gear tooth.
- Hypoid drive side—the convex side of the gear tooth.

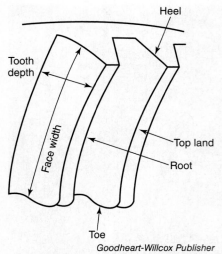

Figure 16-28. Gear tooth terminology.

> **Note**
> An even, rectangular tooth contact pattern is found on new gears. However, a used gear shows a "pocket" at the toe end of the teeth. See **Figure 16-29**.

A properly adjusted differential leaves a tooth pattern (impression) centered in the ring gear's tooth on both the drive side and the coast side of the ring gear. The tooth pattern will not touch the toe edge of the tooth. If the teeth patterns were obtained by rolling the ring gear without a load, the size of the impression will be one third to two thirds the length of the tooth. See **Figure 16-30**. Note that a hypoid gear set drives forward on the convex side of the ring gear. An amboid gear set drives forward on the concave side of the ring gear. Rotating a differential to reveal its contact pattern requires holding resistance on one of the gears (either ring or pinion) while the other gear (ring or pinion) is being rotated. If the tooth contact pattern is too close to the inside edge, touching the toe, backlash must be increased. See **Figure 16-31**. If the contact pattern is too far to the outer edge (heel), backlash must be decreased. See **Figure 16-32**.

If the tooth contact pattern is located towards the root of the tooth, the pinion is too close to the ring gear and must be moved away from the ring gear. See **Figure 16-33**. If the contact pattern is located too close to the top land, the pinion gear is too far from the ring gear and must be moved closer to the ring gear. See **Figure 16-34**.

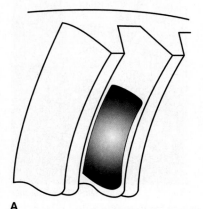
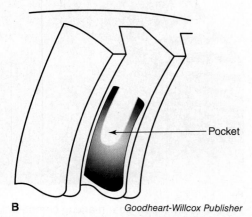

Figure 16-29. A ring and pinion gear set's contact pattern changes as the gears wear. A—A new gear set has a rectangular contact pattern. B—As the gears wear, a pocket develops in the contact pattern.

Chapter 16 | Differentials

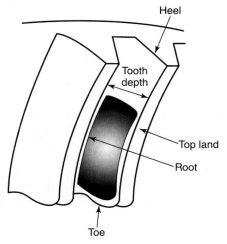

Figure 16-30. A good tooth contact pattern does not touch the toe edge of the tooth. It extends from the top land to the root and is approximately one third to two thirds the length of the tooth.

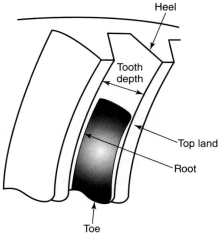

Figure 16-31. This contact pattern is touching the inner edge (toe) of the tooth. Backlash needs to be increased.

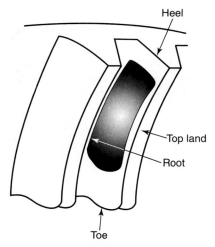

Figure 16-32. The tooth contact pattern is too close to the outer edge (heel). Backlash needs to be decreased.

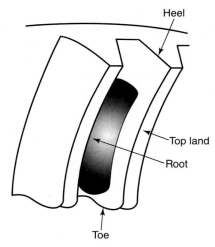

Figure 16-33. The tooth contact pattern is too close to the root. The pinion gear needs to be moved away from the ring gear.

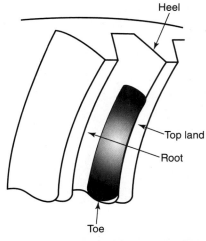

Figure 16-34. The tooth contact pattern is too close to the top land. The pinion needs to be moved toward the ring gear.

Summary

- Differentials allow drive wheels to rotate at different speeds when a machine is turning a corner.
- A standard differential loses traction when one wheel encounters a slick surface.
- Locking differentials can be engaged with a sliding collar or a multiple disc clutch.
- In locking differentials, mechanical levers, electro-hydraulics, or electro-pneumatics engage the differential lock.
- Automatic traction control systems apply cross-axle and inter-axle differential locks to prevent drive wheel slippage. A cross-axle differential lock couples the left and right drive wheels. A locking inter-axle differential lock couples two separate drive axles, such as front and rear drive axles or tandem rear drive axles.
- A limited-slip differential has two multiple disc clutches, one for each side gear. Power is delivered to the non-slipping drive wheel when the machine experiences slick conditions.
- A no-spin differential uses dog clutches to drive or release the axles as needed. Power is always delivered to the wheel with traction.
- Inter-axle differentials use a planetary differential design in articulated dump trucks to differentiate power between two drive axles.
- Differentials and axles can exhibit noises, such as howling, clunking, whining, rumbling, or popping. To diagnose the symptom, determine if the noise occurs during acceleration, deceleration, or when turning.
- The sequential steps for adjusting a differential are to set the pinion bearings' preload, set the pinion depth, install the differential case assembly, set the differential housing side bearings' preload, set pinion and ring gear backlash, check tooth pattern, and move the ring or pinion to achieve the correct tooth pattern.
- A properly adjusted differential leaves a tooth pattern (impression) centered in the ring gear's tooth on both the drive side and the coast side of the ring gear.

Technical Terms

- automatic traction control (ATC)
- center axle through drive
- center cam
- cross-axle differential lock
- cross shaft
- crush sleeve
- differential
- differential carrier
- differential case
- dog clutch
- driven clutch
- heel
- holdout ring
- inter-axle differential (IAD)
- limited-slip differential
- locking differential
- locking inter-axle differential
- no-spin differential
- no-spin spider
- open differential
- pinion cone variation number
- ring gear
- root
- side gears
- spider gears
- spider key
- toe
- top land

Review Questions

Answer the following questions using the information provided in this chapter.

Know and Understand

1. A differential allows two axles to rotate at different speeds in each of the following conditions, *EXCEPT*:
   A. when the two drive wheels have the same resistance.
   B. when the tires have different diameters.
   C. when the two drive wheels have different loads.
   D. when the machine is turning a corner.

2. All differentials have at least _____.
   A. two side gears
   B. two spider gears
   C. one cross shaft
   D. one differential lock

3. Open differentials must have all of the following components, *EXCEPT*:
   A. spider gears.
   B. side gears.
   C. a spiral bevel pinion gear.
   D. a ring gear.

4. All of the following are used in heavy equipment systems to engage a differential lock, *EXCEPT*:
   A. centrifugal clutch.
   B. electro-pneumatic controls.
   C. electro-hydraulic controls.
   D. mechanical linkage.

5. In a limited-slip differential, _____ causes the clutch pack to engage.
   A. the slowing of the pinion gear
   B. hydraulic pressure
   C. the force of the side gear moving away from the spider gears
   D. the centrifugal force of the ring gear

6. What takes place in a no-spin differential when the machine turns a corner?
   A. The clutch pack slips.
   B. The spider gears spin on their axes.
   C. A spring releases the side gear.
   D. The outside drive wheel causes the driven clutch to overrun.

7. If noise on deceleration is heard, the technician should _____.
   A. inspect the spider gears
   B. look for missing teeth on the pinion gear
   C. check the pinion gear bearings for wear
   D. check the ring gear

8. If a rumbling noise occurs at low or moderate speeds, the technician should _____.
   A. investigate the driveline balance
   B. inspect for a failed ring gear
   C. investigate the differential case bearings
   D. check for incorrect ring and pinion adjustment (setup)

9. Each of the following should be performed before disassembling a differential, *EXCEPT*:
   A. mark the differential side bearing caps (right and left).
   B. measure back pinion gear and ring gear backlash.
   C. paint the spider gears with gear marking compound.
   D. inspect the gears for damage.

10. The first step when adjusting a differential is to set the _____.
    A. pinion preload
    B. pinion depth
    C. differential housing side bearings' preload
    D. pinion and ring gear backlash

11. When setting a differential, what is performed after setting the pinion depth?
    A. Setting the pinion preload.
    B. Installing the differential case assembly.
    C. Setting the differential housing side bearings' preload.
    D. Setting pinion and ring gear backlash.

12. When setting a differential, what is performed after setting the differential housing side bearings' preload?
    A. Setting the differential housing side bearings' preload.
    B. Setting pinion and ring gear backlash.
    C. Checking tooth pattern.
    D. Moving the ring or pinion to achieve the correct tooth pattern.

13. When setting a differential, what is performed after checking the tooth pattern?
    A. Installing the differential case assembly.
    B. Setting the differential case side bearings' preload.
    C. Setting pinion and ring gear backlash.
    D. Moving the ring or pinion to achieve the correct tooth pattern.

14. All of the following are used to measure a pinion gear's rolling torque, *EXCEPT*:
    A. dial torque wrench.
    B. flexible beam torque wrench.
    C. clicker torque wrench.
    D. string and spring scale.

15. A differential's tooth contact pattern is too close to the toe. What should be performed?
    A. Increase backlash.
    B. Move the pinion gear away from the ring gear.
    C. Decrease backlash.
    D. Move the pinion gear toward the ring gear.

16. A differential's tooth contact pattern is too close to the heel. What should be performed?
    A. Increase backlash.
    B. Move the pinion gear toward the ring gear.
    C. Move the pinion gear away from the ring gear.
    D. Decrease backlash.

17. A differential's tooth contact pattern is too close to the root. What should be performed?
    A. Increase backlash.
    B. Decrease backlash.
    C. Move the pinion gear away from the ring gear.
    D. Move the pinion gear toward the ring gear.

18. A differential's tooth contact pattern is too close to the top land. What should be performed?
    A. Increase backlash.
    B. Decrease backlash.
    C. Move the pinion gear away from the ring gear.
    D. Move the pinion gear toward the ring gear.

Apply and Analyze

19. The _____ gear drive(s) the differential's case.
20. The _____ directly drives the cross shaft.
21. The _____ directly drive(s) the side gears in an open differential.
22. Differentials commonly contain a total of _____ side gears.
23. Detroit locker is another name for a(n) _____ differential.
24. A limited-slip differential contains a total of _____ clutch packs.
25. The _____ is located in the center of a spider (and retained by a snap ring) in a no-spin differential.
26. In a no-spin differential, when one tire spins, the _____ separates from the cam ring.
27. The _____ is the most common type of inter-axle differential design in ADTs.
28. The term _____ is another name for an inter-axle differential.

Critical Thinking

29. In a locking differential, what occurs when the left side gear is locked to the differential case?
30. When a machine with a limited-slip differential experiences a drive wheel slip, what occurs?

Chapter 17
Final Drives

Objectives

After studying this chapter, you will be able to:

✓ Explain the purposes of a final drive.
✓ Describe final drives integrated with axle housings.
✓ Describe final drives that are not integrated in axle housings.
✓ Explain the processes for servicing and repairing final drives.
✓ List examples of component failures.

Heavy equipment systems multiply torque in several locations, including the torque converter, transmission, differential pinion and ring gear, and the final drive.

Introduction to Final Drives

A *final drive* is a gearing system that multiplies the machine's torque and reduces the speed one final time prior to the drive wheel or drive sprocket being driven. Heavy-duty bearings in a final drive support the machine's load and absorb torque and shock loads.

Final drives also serve the following purposes:
- Drop the drive's power flow to a lower axis (traditional dozer).
- Elevate the drive's power flow to a higher axis (elevated rubber-track tractor).
- Drive tandem wheels through drive chains (skid steers and motor graders).
- Change the direction of power flow by 90° (spiral bevel gear set).

In a traditional automotive light-duty truck application, the differential's spiral bevel ring and pinion gear set is the powertrain's only final drive. However, most off-highway machines have separate final drives. Off-highway machines may or may not use a spiral bevel gear set.

If torque multiplication were provided by the transmission only, the machine would require a much heavier transmission. In addition, stronger shafts would be needed to deliver the torque from the transmission to the drive wheels or tracks. The use of final drives allows for smaller transmission components (bearings, gears, and shafts) to deliver torque to the final drives, where the torque is multiplied one more time prior to driving the wheel or sprocket.

Several types of final drive gearing are used on off-highway machines. They include the following:
- Spiral bevel.
- Bull-and-pinion.
- Planetary gears.
- Chain drive.

> **Note**
>
> The gearing systems found in final drives were detailed earlier in this textbook.
> - Chapter 6, *Belt and Chain Drives*—chain drives
> - Chapter 7, *Manual Transmissions*—gearing
> - Chapter 9, *Planetary Gear Set Theory*—planetary gear sets
> - Chapter 16, *Differentials*—spiral bevel gear sets

Number of Gear Reductions

Machines can have single-reduction, double-reduction, or triple-reduction final drives. The number of reductions equals the number of different gear pairings that work together to provide the final gear ratio. For example, a ***double-reduction final drive*** might have two gear sets that work together to produce a final gear ratio of 20:1. The torque entering the final drive is multiplied by 20 and the speed is divided (reduced) by a factor of 20. Refer to Chapter 7, *Manual Transmissions*, for an in-depth explanation of gear ratios.

Final drives may be used in conjunction with a spiral bevel ring and pinion gear set in the differential. For example, a mining truck has a spiral bevel ring and pinion gear set in the differential that provides a torque multiplication and turns the power flow at a 90° angle. The mining truck also has either a single-reduction or double-reduction final drive at the drive wheels. An excavator, on the other hand, uses a pair of hydraulic motors that each drive either a double-reduction or triple-reduction final drive. The excavator does not use a spiral bevel ring and pinion gear set.

Location of the Final Drives

A final drive can be a stand-alone component or integrated into the axle housing. Final drives that are integrated in the axle housing are either inboard or outboard. Inboard final drives are located right next to the differential. Outboard final drives are located near the drive wheels or drive sprockets. Tandem chain final drives are located near the drive wheels on skid steers and motor graders.

Final Drives Integrated into the Axle Housing

The most common location for integrated final drives is within the drive axle housing. Inboard final drives have a planetary design or a bull-and-pinion gear design.

Inboard Single-Reduction Planetary Final Drive

An ***inboard single-reduction planetary final drive*** attaches directly to the axle's differential housing. At each side, one of the differential's side gears drives the planetary sun gear. The ring gear is bolted stationary to the axle housing, and the planetary carrier is the output member and attaches directly to the axle's driveshaft. See **Figure 17-1**. The stationary outer axle housing is sometimes referred to as the *trumpet housing* because of its shape. Machines with inboard planetary final drives include agricultural tractors, wheel loaders, and loader backhoes.

Inboard Double-Reduction Planetary Final Drives

John Deere manufactures 9R series four-wheel-drive agricultural wheeled tractors with inboard double-reduction planetary final drives. The differential's side gear delivers power to the first planetary gear set's sun gear. The ring gear is bolted stationary to the axle housing. The first planetary carrier drives the secondary planetary sun gear. The second planetary ring gear is also stationary, and the second planetary carrier drives the bar axle. See **Figure 17-2**.

Inboard Bull-and-Pinion Gear Final Drives

In an *inboard bull-and-pinion gear final drive*, the pinion gear receives its power directly from the differential side gear. Usually, the service brakes are integrated on the pinion gear shaft. The pinion gear drives the large ring gear (*bull gear*). The center of the bull gear is directly splined to the tractor's axle shaft, which propels the drive wheel. This type of drive is sometimes used in compact utility tractors. See **Figure 17-3**.

A

B

Goodheart-Willcox Publisher

Figure 17-1. Inboard single-reduction final drives attach to the differential housing. A—A wheel loader axle with inboard planetary gear sets. The wheels mount to the drive hub flanges. B—A John Deere 8000 series rear axle with inboard planetary final drives. The left side inboard planetary final drive has been removed and inverted in a stored position. The ring gear is bolted to the differential housing and held stationary. The sun gear is machined into the inboard axle shaft, which is driven by the differential's side gear. The planetary carrier is the output and is attached to an outboard axle shaft.

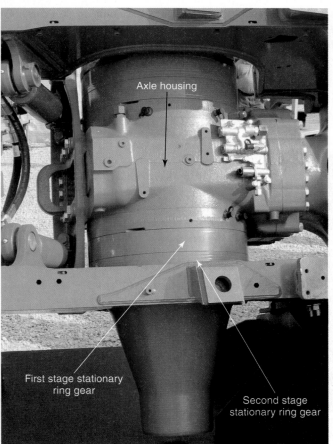

Goodheart-Willcox Publisher

Figure 17-2. John Deere 9R series four-wheel-drive agricultural wheeled tractors have inboard double-reduction planetary final drives. This front axle assembly is on display at the US National Farm Progress show, providing a clear view to the underside of the front axle assembly. The bar axles are missing from the axle assembly.

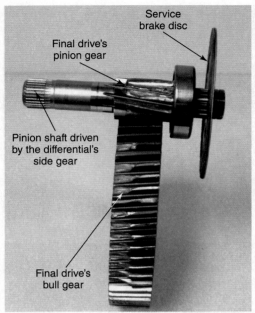

Figure 17-3. This bull-and-pinion final drive was removed from a compact utility tractor after the gears' teeth failed. The splines on the left side of the pinion shaft mate with the differential's side gear. The splines on the right side mesh with the service brake disc.

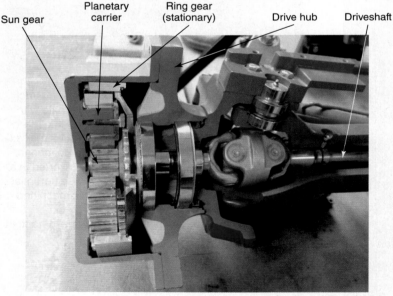

Figure 17-4. A cutaway of the front axle in an 8000 series John Deere tractor. The planetary carrier is coupled to the drive wheel's hub. The right side driveshaft is driven by the MFWD differential's side gear and splined to the planetary gear set's sun gear. The sun gear is the input member to the outboard planetary final drive. The axle housing holds the ring gear stationary, and the planetary carrier is the output member and is coupled to the drive hub.

Outboard Single-Reduction Planetary Final Drive

On many off-highway machines, outboard planetary final drives are integrated into the axle housing. The final drive is still attached to the axle housing but is located outboard, near the drive wheel. *Outboard single-reduction planetary final drives* use a single planetary gear set. They are found in both steering drive axles and rigid drive axles. Outboard single-reduction final drives are used in the front drive axle in MFWD agricultural tractors, as well as in drive axles in large construction equipment such as four-wheel-drive wheel loaders, articulated dump trucks, and rigid frame haul trucks.

In a John Deere 8000 series MFWD tractor front steer axle, the axle shaft provides input to the outboard planetary gear set's sun gear. The ring gear is held stationary to the axle's housing. The planetary carrier is the output member and is coupled to the wheel's drive hub, which drives the wheel. See **Figure 17-4**.

Many rigid-frame quarry trucks use outboard single-reduction planetary final drives in the drive axle (rear axle). A rigid-frame quarry truck final drive normally propels dual tires. See **Figure 17-5**.

Large wheel loaders can have outboard single-reduction planetary final drives in both the front and rear drive axles. These axles are nonsteer axles because the loader has articulation steering. See **Figure 17-6**. Articulated dump trucks and mining wheel loaders can also be equipped with outboard single-reduction final drives. See **Figure 17-7**.

Outboard Double-Reduction Planetary Final Drives

Mining wheel loaders, mining trucks, and articulated agricultural tractors can be equipped with *outboard double-reduction planetary final drives* that are incorporated into the axle housing. The two planetary gear sets are connected together in series. See **Figure 17-8**.

Caterpillar Outboard Double-Reduction Planetary

In a Caterpillar outboard double-reduction planetary final drive, the differential's side gear delivers power to the first reduction sun gear. Both ring gears are coupled to the spindle and are held stationary. The first reduction planetary carrier delivers power to the second reduction gear set's sun gear. The second reduction planetary carrier propels the drive wheels.

Chapter 17　Final Drives

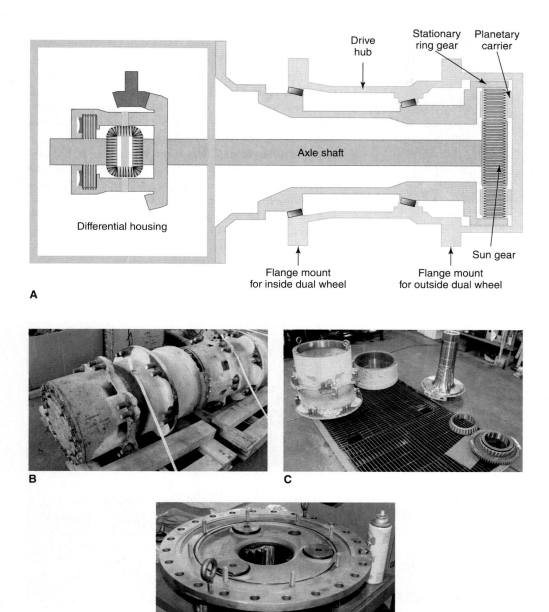

Goodheart-Willcox Publisher

Figure 17-5. A quarry truck can be equipped with outboard single-reduction planetary final drives. A—The differential side gear delivers power to the planetary sun gear. The ring gear is held stationary, and the planetary carrier drives the wheel. B—These final drives were salvaged from a Caterpillar 771D quarry truck to be used in another truck. C—This final drive is being rebuilt to be reinstalled in a 769C quarry truck. D—The single planetary gear set from a 769C quarry truck's final drive.

Figure 17-6. This front axle was removed from a Caterpillar 988F wheel loader. The left side outboard final drive is still attached to the axle. The left-side final drive housing has a barrel-shaped design.

Raba 694 Axle Outboard Double-Reduction Planetary

Case IH Steiger tractors have used different axle brands, including Allis Chalmers, Clark, Dana, Massey, Raba, Rockwell, Steiger, and ZF. The Raba 694 axle is found in the 9350, 9370, and 9380 Steiger tractors. The axle uses double-reduction outboard planetary final drives.

The differential side gear delivers power to the first stage planetary sun gear. The ring gear is attached to the wheel drive hub and acts like a held member. The first stage planetary carrier sends power to the second stage sun gear. The second stage planetary carrier is attached to the spindle and is held stationary, which forces the second stage ring gear to rotate in the opposite direction of the differential's side gear. See **Figure 17-9**.

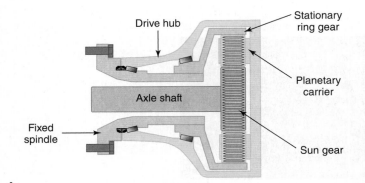

Figure 17-7. Articulated dump trucks and mining wheel loaders can use outboard single-reduction planetary final drives. A—A diagram of an outboard single-reduction planetary final drive with a trumpet-shaped design. B—The final drives are attached to the axle housing. C—Outboard single-reduction final drives that were removed from the front axle of a Caterpillar 992G mining wheel loader to be salvaged for use in another loader.

Pro Tip

The axle manufacturer can control the direction of differential rotation by the placement of the ring gear in relation to the pinion gear—for example, placing the ring gear on the right or left side of the pinion gear. The direction of differential rotation is controlled so the second stage planetary can force the power to rotate back in the correct direction.

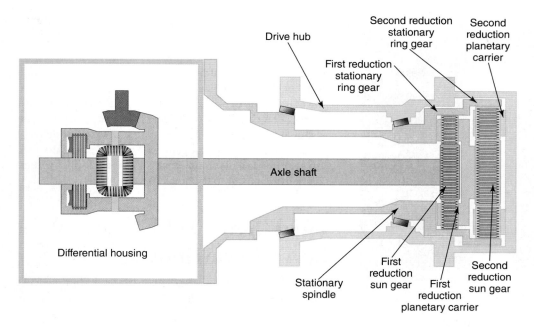

Figure 17-8. In outboard double-reduction planetary final drives, the axle drives the first reduction sun gear, and the first reduction planetary carrier drives the second reduction sun gear. The second reduction planetary carrier drives the wheel. The two ring gears are held stationary via the spindle.

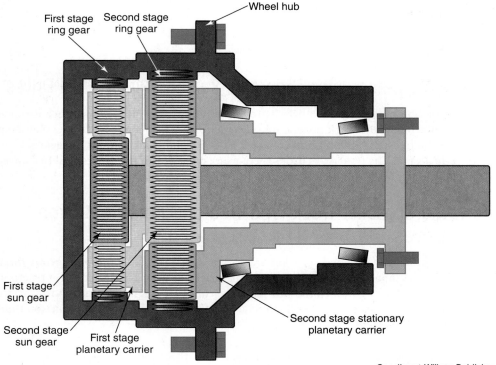

Figure 17-9. In this double-reduction outboard planetary final drive, the inner second stage planetary carrier is held, causing the power flow to rotate in the opposite direction of the axle shaft.

Elevated Drives in Articulated Rubber-Track Tractors

Some articulated rubber-track tractors have a unique drive mechanism consisting of two spur gears that elevate power up to the drive sprocket instead of lowering it like a drop axle. A Case IH STX Steiger Quadtrac undercarriage is shown in **Figure 17-10**. The axle's differential delivers power to the bottom spur gear. The top spur gear drives a sun gear inside a planetary gear set that multiplies torque and delivers power to the track drive sprocket.

Goodheart-Willcox Publisher

Figure 17-10. This Case IH STX Steiger Quadtrac undercarriage uses two external-toothed spur gears to elevate the power to the track's drive sprocket.

Final Drives Not Integrated into an Axle Housing

Unlike traditional inboard or outboard final drives, *nonintegrated final drives* are not part of the front or rear axle housing. Examples are:
- Dozer final drives.
- Single-reduction drop-axle bull-and-pinion final drives.
- Single-reduction planetary final drives.
- Double-reduction planetary final drives.
- Harvester final drives.
- Single-reduction bull-and-pinion final drives.
- Tandem chain drives.
- Excavator final drives.
- Compact rubber-track loader final drives.

Dozer Single-Reduction Drop-Axle Bull-and-Pinion Final Drives

Single-reduction bull-and-pinion drives are found in low-track dozers and track loaders, such as Caterpillar low-track dozers, John Deere dozers, Leibherr dozers, and Komatsu dozers. When bull-and-pinion gears are used in dozers and track loaders, the gear set is a *drop-axle* design. The final drive drops the power coming from the differential to a lower axis on the machine, in addition to multiplying torque. See **Figure 17-11**.

Elevated-Sprocket Track-Type Tractor Single-Reduction Final Drives

Midsized Caterpillar elevated-sprocket track-type tractors use **single-reduction final drives**. Like the inboard single-reduction planetary final drives in agricultural tractors, the axle shaft delivers power to the sun gear. The ring gear is held. The planetary carrier propels the drive sprocket. A simplified image of a Caterpillar D6R single-reduction final drive is shown in **Figure 17-12**.

Elevated-Sprocket Track-Type Tractor Double-Reduction Final Drives

As explained in Chapter 9, *Planetary Gear Set Theory*, large Caterpillar elevated-sprocket track-type tractors use double-reduction final drives. Like the outboard double-reduction planetary final drives in mining loaders and trucks, the axle shaft delivers power into the

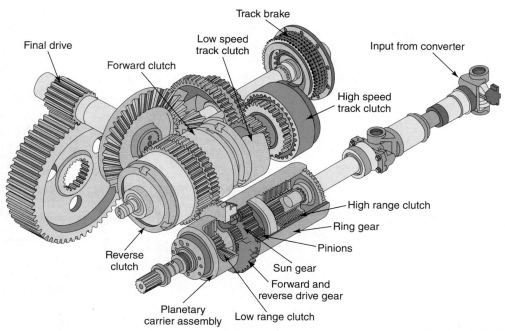

Figure 17-11. The Case 1150B track loader used bull-and-pinion final drives. The left side final drive is shown here.

first reduction sun gear. Both ring gears are held. The first reduction planetary carrier drives the second reduction sun gear. The second reduction planetary carrier propels the drive sprocket. See **Figure 17-13**.

Harvester Final Drives

Cotton pickers, cotton strippers, self-propelled forage harvesters, and combines have unique powertrains. A *harvester final drive* is mounted on the rigid axle housing. The harvester's hydrostatic transmission consists of a hydraulic pump and motor that provide the input into a two-speed, three-speed, or four-speed gearbox, which includes a differential. Chapter 7, *Manual Transmissions*, details a three-speed gearbox, and Chapter 15, *Continuously Variable Transmissions*, details a John Deere two-speed ProDrive gearbox.

The transmission's right and left differential side gears deliver power to the right and left transmission output shaft respectively. The machine's right and left brakes are often mounted to the transmission output shafts. A coupler is used to couple the left transmission's output shaft to the left driveshaft and another coupler is attached to the end of the left driveshaft to couple the driveshaft to the left final drive. Two couplers and a driveshaft are also used to connect the right transmission output shaft to the right side final drive. Each final drive is mounted to the front rigid axle housing.

A Case IH 10, 20, 30, and 240 series flagship combine front axle is shown in **Figure 17-14**. The transmission is

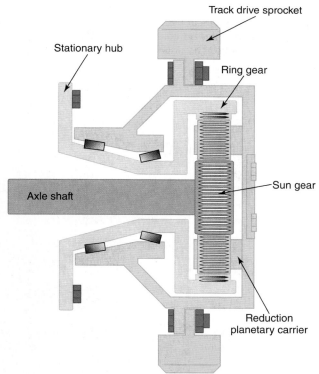

Figure 17-12. A simplified drawing of a Caterpillar dozer single-reduction final drive.

a four-speed countershaft gearbox. Each transmission output shaft is splined to its own service brake rotor, which can be held by applying its two hydraulically actuated service brake calipers. The parking brake is not directly attached to the transmission's output shaft, but is attached to the transmission's countershaft. The transmission output shafts deliver power to the final drive through the couplers and driveshafts.

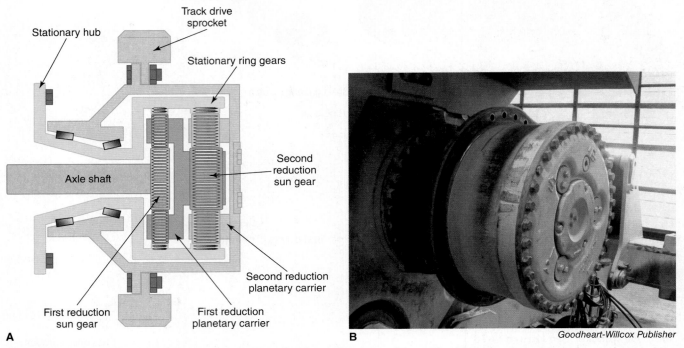

Figure 17-13. Double-reduction final drives are used in Caterpillar large dozers. A—Diagram of an elevated-sprocket undercarriage double-reduction final drive. B—A Caterpillar D9 dozer with the undercarriage removed, exposing the double-reduction final drive. The drive sprocket segments have been removed from the final drive.

Figure 17-14. On Case IH flagship 10, 20, 30, and 240 series combines, the final drives are mounted to the rigid axle housing. The transmission sends power through the output shafts to the final drive assemblies.

Harvester Bull-and-Pinion Final Drives

Bull-and-pinion final drives and planetary final drives are found in combines, cotton pickers, cotton strippers, and self-propelled forage harvesters. The combine bull-and-pinion final drive shown in **Figure 17-15** consists of a ten-tooth pinion that drives a 95-tooth bull gear for a 9.5:1 speed reduction, torque multiplication ratio. It is used in an S670 John Deere combine.

Some CNH combines also have bull-and-pinion final drives. They are the only final drives offered on midrange 2016 Case IH combines, such as the 5140, 6140, and 7140. Bull-and-pinion final drives are also used in different generations of flagship combines, including Case IH 7240, 8240, and 9240, and New Holland's CR 9.90 and CR 10.90.

Harvester Planetary Final Drives

Other harvesters use planetary final drives. A combine planetary final drive can integrate a bull-and-pinion, which serves as the first reduction, and a planetary gear set as the second reduction. In the final drive shown in **Figure 17-16**, a 16-tooth pinion drives a 73-tooth bull gear for a 4.56:1 ratio. Case IH labels the large bull gear a *reducer gear*. This final drive is the heavy-duty final drive used in class 8, 9, and 10 combines. The planetary ring has 75 teeth, and its sun gear has 33 teeth, for a 3.27:1 ratio. Case IH combines with factory-equipped tracks use a bull-and-pinion final drive with a different ratio because the track frame carries the structural load.

John Deere 9000 series and 9010 series combines also use a final drive that has a planetary gear set. A bull-and-pinion produces the first reduction and a planetary gear set produces the second reduction. See **Figure 17-17**.

Tandem Chain Drives

In some off-highway machines, **tandem chain drives** propel tandem drive wheels. A tandem chain drive consists of a pair of drive sprockets that drive a pair of chains and driven sprockets that are splined to a drive wheel. Each side of the machine has two chains used to propel a set of tandem wheels. Examples include skid steers and motor graders.

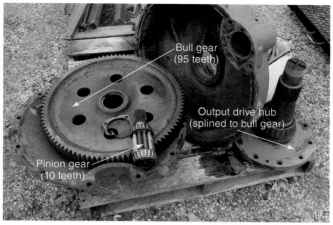

Figure 17-15. This bull-and-pinion final drive is used in a John Deere S670 combine.

Figure 17-16. This Case IH combine final drive uses a bull-and-pinion gear set for the first reduction and a planetary gear set for the second reduction. A—The pinion gear has been set beside the bull gear to show how the pinion drives the bull gear. B—On the back side of the bull gear is the 33-tooth planetary sun gear, which meshes with the three planetary pinion gears.

Figure 17-17. This John Deere 9600 combine uses a combination bull-and-pinion and planetary gear set final drive.

Skid steer chain drives have a hydrostatic motor on each side of the machine that provide the input to the chain drive. In many skid steers, a single-speed or two-speed hydraulic motor directly drives the two small drive sprockets. See **Figure 17-18.**

The skid steer's drive sprockets have their own separate chain and propel a large wheel sprocket (driven sprocket). A driven sprocket is splined to each drive wheel's axle. See **Figure 17-19.**

In motor grader tandem chain drives, a traditional power shift transmission sends power to a differential. The differential delivers power to the right and left tandem drives. See **Figure 17-20.** The tandem housing is attached to the differential housing. Each tandem's housing pivots forward and rearward. Each tandem drive contains a planetary gear set, two drive sprockets, two chains, two driven sprockets, and two output wheel spindles. The planetary sun gear receives power from the differential's side gear. The ring gear is stationary. The planetary carrier is splined to the two drive sprockets. Each drive sprocket has its own separate chain and propels a large wheel sprocket, which is splined to the output wheel spindle. The service brakes are located on the spindles. See **Figure 17-21.**

Figure 17-18. A skid steer hydrostatic drive motor. This Rexroth MCR hydraulic drive motor has integrated sprockets. It is used on Case SR200, SR210, SR220, SR250, SV250, SV280, SV300, New Holland L221, L225, L228, and L230 skid steers. The rear of the hydraulic motor has one Belleville washer spring that applies a multiple disc parking brake that must be hydraulically released for the skid steer to move.

Figure 17-19. A tandem drive chain on each side of a skid steer propels the drive wheels. A hydrostatic motor drives two small sprockets that propel the two drive chains. The drive chains propel the driven sprockets, and the driven sprockets drive the skid steer's axles.

Chapter 17 Final Drives

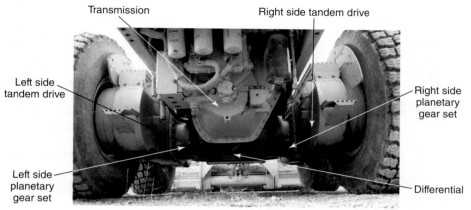

Figure 17-20. A motor grader's differential sends power to the right and left planetary drives. The planetary drives send power to the right and left tandem chain drives to propel the drive wheels.

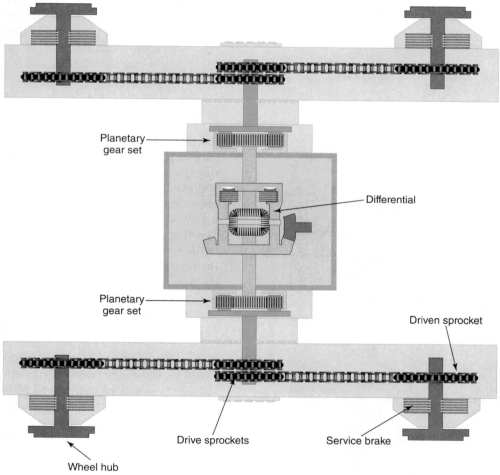

Figure 17-21. A top view of the powertrain used in a motor grader. A differential delivers power to the right and left side tandem drives. Each tandem drive uses a planetary gear set to multiply torque. The planetary carrier powers two drive sprockets. Each drive sprocket powers a drive chain. The drive chain propels the driven sprocket, which drives the wheel.

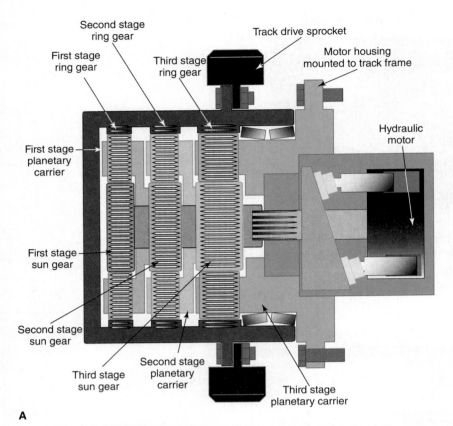

Figure 17-22. Excavators can use triple-reduction final drives. A—A side view of an excavator final drive. The hydraulic motor drives the first stage sun gear. The three ring gears are attached to the track drive sprocket. The third planetary carrier is held stationary. B—A cutaway of a final drive.

Note

The Doosan Da30 and Da40 articulated dump trucks have a tandem drive similar to that found on motor graders. However, the trucks use gears in place of the chain drives.

Excavator Final Drives

Excavator final drives receive their input from a hydraulic motor. The final drive often has two or three planetary gear sets. Each final drive on Caterpillar 330D, 336D, 374F, and 385C excavators has three planetary gear sets. The three planetary gear sets are incorporated into a single final drive. See **Figure 17-22**. The machine has one final drive for the left track and one final drive for the right track.

All three ring gears are part of the track drive sprocket housing. The hydraulic motor provides the input into the final drive by propelling the first stage sun gear (outboard sun gear). The ring gear is part of the sprocket housing and acts like a held member. The first reduction planetary carrier is driven in a slow forward torque multiplication. The first planetary carrier is coupled to the second stage (center) sun gear in the second reduction gear set, causing the sun gear to be the input member to the second gear set. The second ring gear is also part of the sprocket housing and acts like a held member, causing the second planetary carrier to drive in a slow forward torque multiplication. The second planetary carrier drives the last sun gear located in the third planetary gear set, causing the sun gear to be the input to the last gear set. The third planetary gear set uses a stationary planetary carrier that is attached to the housing, which is bolted to the track frame. The stationary planetary carrier causes the ring gear to propel slowly in a direction opposite of the hydraulic motor shaft rotation. Therefore, to move forward, the hydraulic motor is driven counterclockwise. To move backward, the hydraulic motor is driven clockwise.

Excavators can also use two planetary gear sets in the final drive, **Figure 17-23**. The final drive operates similarly to the final drive in **Figure 17-22**, but it does not contain the third-stage planetary gear set. The last planetary carrier is also held stationary, causing the ring gear to propel slowly in a direction opposite of the hydraulic motor shaft rotation. Examples of excavators that use double planetary gear sets in the final drives are Caterpillar 312E and 319D.

Compact Rubber-Track Loader Final Drives

In many compact rubber-track loaders, a hydraulic motor drives the planetary final drive, which propels the drive sprocket. See **Figure 17-24**. The hydraulic motor commonly is either an inline axial piston motor or a cam lobe motor. The final drive assembly can also have an integrated parking brake mechanism.

Case New Holland rubber-track loaders have a single-reduction planetary final drive. See **Figure 17-25**. The hydraulic cam lobe motor spins the shaft with an integrated parking brake, which drives the planetary sun gear. The planetary carrier is held stationary, and the ring gear is the output member, which is propelled in a slow reduction in the opposite direction of the hydraulic motor.

Note

Compact rubber-track loaders are sometimes incorrectly called *skid steers*. Compact rubber-track loaders have a hydraulic motor and a single-reduction final drive that propels a track, but they do not use chain drives and have only one drive sprocket on each side of the loader.

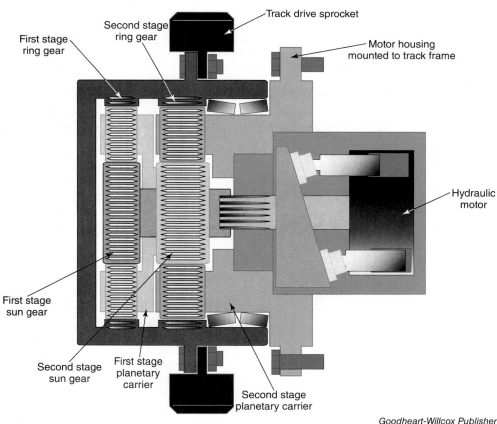

Figure 17-23. This simplified excavator final drive uses two planetary gear sets. The hydraulic motor drives the first sun gear. The first planetary carrier drives the second sun gear. The second planetary carrier is held, which causes the ring gear to reverse the direction of rotation.

Goodheart-Willcox Publisher

Figure 17-24. These two final drive assemblies were rebuilt to be reinstalled in a Caterpillar 259B rubber-track loader.

Servicing and Repairing Final Drives

Final drives require maintenance, service and repair. Daily and scheduled preventative maintenance inspections are used to inspect the conditions of final drives and check the fluid levels. Some service procedures are also made during the scheduled intervals such as cleaning breathers and adjusting chain tension. If a leak is severe or if other problems are encountered, the machine must be scheduled for repair.

Checking Final Drive Oil Level

Many final drives in heavy equipment have a stand-alone fluid housing. The final drive reservoir must be checked separately from other components, such as the transmission and differential. Examples of machines with stand-alone final drives are dozers with elevated-sprocket undercarriages, and harvesters. On machines with integrated outboard final drives, such as large wheel loaders and haul trucks, each final drive fluid level should be checked separately from the differential housing because the final drives are sealed separately from the differential housing.

Final drives should be inspected for leaks during daily visual walk-around inspections. Some final drives are equipped with a sight glass for checking the fluid level without the need for special tools. See **Figure 17-26**.

Pro Tip
Be sure to view the sight glass closely. The glass can become discolored over time, causing a misreading of the final drive's fluid level.

Checking Outboard Planetary Fill Levels

On Caterpillar dozers with elevated-sprocket undercarriages, several wheel loaders, and haul trucks, the final drive must be oriented in a specific position

A

B

Figure 17-25. A single-reduction planetary final drive is used in a Case rubber-track loader. A—The sun gear is driven by the hydraulic motor's output shaft. The planetary carrier is held and the ring gear propels the sprocket housing. B—The backside of the final drive housing. The parking brake and hydraulic cam lobe motor are missing.

Figure 17-26. This Case IH combine final drive has a sight glass. The final drive has been removed, drained, and inverted. The sight glass and breather are normally located at the top of the final drive.

before the fill level is checked. The final drive has a *horizontal oil level mark* across the face of the housing that must be rotated until it is horizontal to the ground, with the wording right side up. When the final drive is correctly positioned, the fill plug can be removed. The fluid should be level with the bottom of the fill plug opening. See **Figure 17-27**. When the final drive oil level line is horizontal to the ground, the drain plug at the bottom of the final drive can be used to drain the oil if needed.

Some machines do not have a horizontal oil level mark on the final drive. They have only two plugs, a fill plug and a drain plug. To check the final drive level, orient the final drive so that the drain plug is positioned at the bottom of the final drive. The fill plug can then be removed to check the fluid level.

Caterpillar recommends checking the final drive's fluid level and taking a fluid sample every 250 hours or monthly. The fluid change interval is once every 2000 hours or annually.

Final Drive Inspection Cover

Some late-model Caterpillar dozers (for example a D9T WH) have a final drive inspection cover in the final drive housing. Removing the cover's four bolts allows a technician to inspect inside the housing to see if any debris has become wrapped around the shaft, potentially causing an oil seal failure. See **Figure 17-28**. The recommended interval for removing the cover and inspecting the final drive is every 500 hours, or sooner if the dozer is used in severe applications (such as landfills).

Tandem Drive Fill Levels

Chain tandem drives have their own oil compartment. Periodic fluid level checks are necessary.

Motor Grader Chain Drive Fill Levels

Caterpillar recommends the following service intervals for motor grader tandem chain drives:

- Check the tandem oil level once every 100 hours.
- Clean or replace the tandem breather every 500 hours or three months.
- Take an oil sample from the tandem housing every 500 hours or three months.
- Change the tandem oil every 2000 hours or two years.

Skid Steer Chain Drive Fill Levels

Skid steer chains have their own oil sump. The side of the chain drive housing contains a fluid level plug. Methods for draining, checking, and filling the oil reservoir vary among manufacturers. Consult the manufacturer's service literature for specific instructions.

Most commonly, fluid level is checked by removing the fill plug on the side of the chain drive housing. Be sure the skid steer is on level ground. The oil should be level with the bottom of the fill plug hole. See **Figure 17-29**. Caterpillar recommends checking the chain drive oil level every 500 hours or three months and replacing the oil every 1000 hours or six months.

Skid Steer Chain Drive Housing Breaker

Skid steer chain drive housings have a breather that must be checked along with the chain drive fluid level. When the machine operates in wet climates, water accumulates in the chain drive housing if the breather becomes plugged. In severe cases, the water will freeze in cold weather, preventing the skid steer from driving.

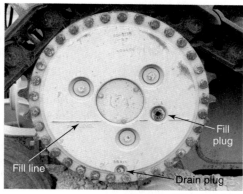

Goodheart-Willcox Publisher

Figure 17-27. On several Caterpillar machines, a line on the housing must be horizontal, the writing right side up, and the drain plug at the bottom before the fill plug is removed to check the final drive fluid level.

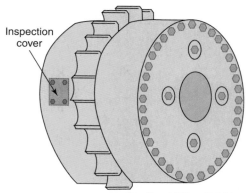

Goodheart-Willcox Publisher

Figure 17-28. A drawing of a Caterpillar D9T waste handler dozer's final drive with an inspection cover that can be removed to check for debris wrapping around the final drive components.

Goodheart-Willcox Publisher

Figure 17-29. The chain drive fill plug is near the bottom of the side of the skid steer. The drain plug is not visible because it is located underneath the frame.

Tensioning Drive Chains

Skid steer chain drives wear over time, causing the chain drive to loosen, make noise, and potentially fall off the sprockets. Severely loose chains can cause the sprockets to slip on the chain.

Most skid steers have a provision for adjusting the drive chain tension. Case, New Holland, and Caterpillar skid steer axle housings have elongated holes. A technician can loosen the axle housing mounting bolts and move the axle housing forward or rearward to adjust the chain drive tension.

Skid steer chain tension should be checked every 500 hours or every three months. If the machine is operated on hard surfaces, the tension should be checked more often. Steel tracks installed around the skid steer rubber tires can cause more stress and wear and tear on the chain drives and require the chain tension to be checked more often.

Adjusting Chain Tension on Caterpillar Skid Steers

Be sure to follow the manufacturer's service procedures for adjusting skid steer chain tension. The process for adjusting a Caterpillar skid steer chain drive is as follows:

1. Place the skid steer on jack stands.
2. Remove the drive chain case cover. Note that the cover is located above the oil level. If the chain case is properly filled, the oil does not need to be drained from the chain case.
3. Loosen the eight nuts that secure the axle housing to the skid steer.
4. Rotate the drive wheel to eliminate the slack at the bottom of the chain.
5. Place a straightedge across the top of the sprockets and measure the amount of chain slack (chain deflection). The chain should have 0.6″ (15 mm) of deflection.
6. Place an adjusting tool between the two axle housings to spread the housings in order to achieve proper tension on the chain.
7. When the chain tension is within specification, tighten and torque the axle housing's eight nuts to specification.

Note

Late-model Bobcat skid steers have fixed axles and maintenance-free chain cases. Technicians do not need to adjust the drive-chain tension on these machines.

Repair

Final drives, like other heavy equipment power train components, are very heavy. Overhead cranes are needed for removal and installation. Some manufacturers provide service procedures for disassembling outboard planetary final drives while the final drive housing remains attached to the machine. If a final drive needs repair, read the service literature before starting the repair. It may not be necessary to split a track or remove a wheel or final drive housing if the goal is to remove and replace the final drive gears. However, special lifting tools are needed to pull and remove heavy components.

Replacing Skid Steer Drive Chains

Drive chains must eventually be replaced. Always follow the manufacturer service literature when making repairs. Some skid steer drive chains do not have a master link, requiring the chains to be removed and installed without being split.

Gaining access to the drive chain requires removing the tires, possibly draining the oil, and removing the chain drive access cover and drive axles. To remove the axle housing, a bolt may be needed to push the axle shaft out of the driven sprocket. For example, in a

Case 90XT skid steer, a cover on the inside frame needs to be removed in order to access the pusher bolt. The pusher bolt is threaded into the housing to push the axle off the skid steer's frame. If the chain has a tensioning mechanism, the tensioner must be loosened. The axle housing, the drive sprocket, driven sprocket, and chain (from the chain cover housing) are removed next. The chain is reinstalled in the reverse order.

Replacing Motor Grader Drive Chains

Always refer to the manufacturer's service literature for the specific steps required to replace tandem chain drives on a motor grader. The following are general steps for replacing the chains:

To remove the chains, do the following:
1. Chock the front wheels.
2. Remove driveshaft.
3. Lift the rear frame and place it on jack stands.
4. Drain the tandem drive oil reservoir.
5. Remove the walkway from the top of the tandem drive and the access covers.
6. Place the plate (tooling) under the chain.
7. The master link must be positioned at the top. Depending on the size of the motor grader, the chain can by heavy, ranging from 70 pounds on an average sized motor grader to 300 pounds on a Caterpillar 24M mining motor grader. A special tool is sometimes required to hold pressure on the two ends of the chain. The tool is used to add or remove tension from the two ends of the chain. After removing the master link cotter pin, remove the master link and separate the chain.
8. Using a lift, remove the chain.

To install the chains, do the following:
1. Using a lift, install the chain.
2. Install the master link using a chain tool to add or remove enough tension so that the two ends of the chain can be connected with a master link. The master link will have to be blocked into position.
3. Install cotter pins.
4. Install access covers.
5. Lower the machine to the ground.
6. Fill the fluid reservoir.
7. Install the driveshaft.
8. Remove wheel chocks.

Preload and Endplay

As in most transmissions, shims are used to set the preload or endplay in final drives. A dial indicator is used to check the pinion shaft's endplay, **Figure 17-30**. If the endplay needs adjustment, the shims need to be changed to a different thickness. See **Figure 17-31**.

Failure Analysis

Components fail for a variety of reasons. Identifying the root cause of the failure is crucial in order to avoid repeat failures. Failures can be the result of improper use or maintenance, contamination, or manufacturing mistakes. Wear, fracture, distortion, and corrosion can occur.

Goodheart-Willcox Publisher

Figure 17-30. A dial indicator is used to check the pinion shaft's endplay.

Goodheart-Willcox Publisher

Figure 17-31. If final drive pinion shaft endplay needs adjustment, the shims are changed to a different thickness.

Wear

Components can wear, causing the outer surface of the component to erode. There are three common types of wear failures:

- Adhesive wear.
- Abrasive wear.
- Corrosion wear.

Adhesive wear occurs when two adjacent component surfaces contact each other. Friction between two heated components results in the two surfaces welding to each other. As the components attempt to move, material lifts off both components, tearing off or transferring material. Adhesive wear causes a component to fail quickly.

Adhesive wear can occur due to lack of proper heat treatment. When a component is heat treated, the outside surface hardens and becomes wear-resistant. A component that lacks proper heat treatment experiences adhesive wear but will not be discolored if it was properly lubricated. See **Figure 17-32**.

 Pro Tip

A file can be used as a hardness gauge. If the file digs into the component and scratches the metal when only light force is applied, the component is too soft and lacks heat treatment. If the file skates across the surface without digging in, the component has a hardened surface.

When adhesive wear occurs due to a lack of lubrication, the component is darkened and discolored. A colored oxide layer known as a heat stain will form on carbon steel due to overheating. See **Figure 17-33**. A color scale can be used as a reference to gauge the temperature reached by the component. Examples of color and temperature are as follows:

- Temperatures above 700°F/371°C form a dark gray oxide layer and cannot be precisely determined by the color of the oxide layer.
- Gray (700°F/371°C).
- Light blue (650°F/343°C).
- Dark blue (590°F/310°C).
- Purple (540°F/282°C).

Figure 17-32. A component that exhibits adhesive wear and has not been properly heat treated will not be discolored if it has proper lubrication. A—This bevel gear lacked proper heat treatment and shows adhesive wear. B—The outside race to the tapered roller bearing lacked proper heat treatment, causing it to exhibit adhesive wear. The inside race and tapered roller bearings show little wear because they were properly heat treated.

A

B

Goodheart-Willcox Publisher

- Brown (500°F/260°C).
- Straw color (400°F/204°C).

> **Note**
>
> A different color temperature chart is used for metal that is being actively heated. However, that chart is not used for failure analysis because a technician is not able to observe the component in the process of being overheated.

Foreign particles, or contaminants, in the lubricant cause ***abrasive wear*** to components. Abrasive wear normally takes longer to cause component failure than adhesive wear. To detect abrasive wear, follow proper oil sampling techniques at the recommended manufacturer's service interval. By comparing the sample results over a period of time, it can be determined whether the final drive fluid meets cleanliness specifications or whether it contains contaminants. Analysis of the sample can identify any contaminants present. The equipment personnel can then determine whether the fluid needs to be changed to extend the life of the component and take appropriate action to prevent further contamination.

Corrosion wear is the damage to a component caused by wear that occurs in a corrosive environment due to the effects of chemical reaction and mechanical action. Water is the substance that most commonly causes corrosion wear. See **Figure 17-34**.

Goodheart-Willcox Publisher

Figure 17-33. These gears have been overheated due to a lack of lubrication. The gear on the left shows adhesive wear. Heat caused a softening of the metal and a darkening of the gear.

Fractures

A component fracture is a failure that occurs when a single part breaks into two or more pieces. Fractures propagate from an area of concentrated stress called a ***stress riser***. Stress risers can be caused by part design, material imperfections, poor material handling processes, and poor manufacturing processes.

Overloads can cause a fracture. Overloads are generally the result of a single stress load that exceeds the component's tensile strength or yield strength. ***Tensile strength*** is the amount of stress that can be placed on a component (such as a bolt or shaft) before it stretches and breaks. ***Yield strength*** is the point at which a component deforms plastically (does not return to its original condition after deformation).

A cyclical ***overload failure*** can occur when an operator repeatedly exceeds the machine's operating capacity. For example, a ring gear that has been overloaded can fracture into multiple pieces. A spiral bevel gear set may have multiple fractures on the concave or convex side of the teeth.

Fractures vary in appearance based on their type. Fracture types include the following:
- Fatigue fracture.
- Impact fracture.
- Ductile fracture.
- Brittle fracture.
- Combination fracture.

A ***fatigue fracture*** occurs due to repeated stress cycles. A common example of a fatigue fracture is a piece of metal that is bent back and forth until it breaks. Curved lines called ***beach marks*** radiate from one or more origin points of a slowly occurring fatigue fracture. See **Figure 17-35**. A forged component

Goodheart-Willcox Publisher

Figure 17-34. When water entered this transmission case through a faulty seal, the oil turned to a thick sludge, resulting in gear noise caused by corrosive wear.

Figure 17-35. This spiral bevel pinion gear has a fatigue fracture. The wavy (curved-shaped) beach marks radiate from the origin of the initial fracture.

may have both beach marks and chevron marks. A forged component is formed using heat and force. See **Figure 17-36**.

Ratchet marks are sometimes located at the origin of the failure. These short abrupt lines indicate a very high stress load. The final fracture of a component is sometimes a brittle fracture, as evidenced by a dark woody appearance. See **Figure 17-37** and **Figure 17-38**.

An *impact fracture* is caused by an abrupt, extreme force (shock load). The fracture can produce an immediate failure or a delayed failure. Different forces, such as a bending stress or torsional stress, can cause an impact fracture.

A *torsional stress fracture* is a fracture that occurs due to applied torque or twisting action. This type of fracture leaves a 45° break that indicates the torsional twisting that occurred during the failure. See **Figure 17-39**.

A *ductile fracture* is a slow type of fracture that results in plastic deformation, meaning the component stretches and elongates before it fails. This type of fracture may cause bending or narrowing of the component. See **Figure 17-40**.

A *brittle fracture* is a fast type of fracture. The material does not elongate, but breaks apart quickly. A brittle fracture in wrought metals often has chevron-shaped marks that point to the origin of the fracture. A wrought metal is a metal that is worked during its formation, such as a forged component. A cast component (formed by pouring molten metal into a casting) will not have chevron-shaped marks. A brittle fracture that follows the grain boundaries leaves a darker, rough surface. A brittle fracture that occurs through the grains has a sparkled appearance when a light shines on the surface. See **Figure 17-41**.

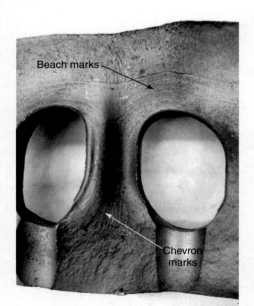

Figure 17-36. A failure started at the center of this forged track link. The beach marks radiate away from the center toward the top of the link. The final brittle fracture has chevron-shaped marks that point to the origin of the fracture.

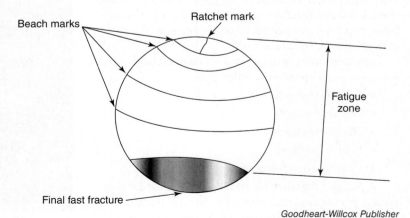

Figure 17-37. The fatigue failure shown in this drawing originated at the single ratchet mark and radiated outward, as shown by the beach marks. The fatigue zone is an area of slow crack growth. The final fast fracture is at the bottom

Figure 17-38. A shaft failure exhibiting multiple ratchet marks.

Figure 17-40. A ductile fracture causes the component to deform. Shown here is one half of a piston connector-rod bolt that was over-heated, resulting in a ductile fracture.

Figure 17-39. This shaft has a torsional stress fracture, as evidenced by the 45° line at the initial point of fracture.

Figure 17-41. The threaded shaft of this clevis is fractured. The iron-based fracture leaves behind a rough, "woody" surface texture.

A *combination fracture*, as its name implies, occurs to a component as the result of a combination of fatigue fracture, impact fracture, ductile fracture, or brittle fracture.

Distortion

Shafts, bolts, rods, and other ductile materials can distort. Examples of distortion are torsional twisting or bending. A distortion failure differs from a fracture because the component distorts but does not fracture into two or more pieces. See **Figure 17-42**.

Corrosion

Water and other contaminants can cause components to corrode. The corrosion can result in wear, distortion, or fractures. Corrective actions include preventing the intrusion of moisture and the use of proper lubricants.

Spalling

Spalling is a surface contact stress fatigue in which the mating surfaces of a component or components (such as gear teeth) chip, break, or flake off in pieces. See **Figure 17-43**. Common causes of spalling are lack of proper lubrication, extended periods of excessive loading, or load misalignment. Spalling can also occur as a result of initial surface pitting or bearing brinelling. To prevent future occurrences, ensure that the component receives proper lubrication, is properly aligned, and is not excessively loaded.

Figure 17-42. This bolt exhibits distortion due to being tightened beyond its yield strength. As a result, the bolt is stretched and narrowed.

Brinelling

Brinelling is the result of a hardened metal surface exhibiting one or more indentations. Bearings are the most common component with brinelling type failures. The needle, roller, or ball bearing create a valley in the component, such as an indentation in the bearing's race or in the yoke of a universal joint.

The precision grinding of bearing races leaves fine lines in the race that are difficult to see. If the fine lines (precision grinding marks) are visible in the valley of the brinelling indentations, the condition is known as ***true brinelling***. True brinelling results in permanent indentations in the bearing race without the loss of metal. It occurs due to an impact or shock overload that causes the bearing to indent the race. If brinelling is due to a load, investigate whether it is a static overload or a shock load. When the overload source has been determined, take steps to correct the cause of the overload.

Pro Tip

Hammering a bearing into position can cause true brinelling. This cause of brinelling can be prevented by using a hydraulic press to properly push the bearing into place.

If the brinelling indentations show no evidence of precision grinding marks, the condition is known as ***false brinelling***. False brinelling occurs over time and is often caused by excessive vibration that occurs when the bearing is stationary. The indentations are worn into the race and the material has been removed. False brinelling indentations can also be caused by electrical current flow through the bearing, large particle contamination, and corrosive etching. Preventative action includes eliminating bearing contamination and ensuring that components have good electrical grounds to prevent current flow through the bearing. See **Figure 17-44**.

Fretting Corrosion

Fretting corrosion is a surface wear failure that occurs on the surfaces of two close-tolerance stationary components that have moved. Examples are the outside race of a bearing pressed into a housing, or the inside race of a bearing pressed onto a shaft. Vibration or movement causes the two normally stationary components to make minor oscillations in the absence of lubrication. The slight back-and-forth movement of the two components causes wear on microscopic ***asperities*** (tiny high spots on the component's surface). Fretting corrosion often results in brownish-red oxide on the component's surface, **Figure 17-45**. The color depends on the type of lubricant and the metal oxide formed by the adhesive wear. Corrective action is needed to restore proper fit so the components remain stationary.

A

B

C

Goodheart-Willcox Publisher

Figure 17-43. Spalling is the flaking or breaking away of the outside surface of a component. A—Spalling on the outside edge of planetary pinion gear teeth. The gear was misaligned, causing the outside tooth contact surface to overload and spall. B—Spalling on tapered roller bearings. C—This bearing's inside race was installed on the rear axle of an agricultural tractor. The race is fixed stationary in the axle housing. Spalling occurred on one side of this bearing and on one side of the bearing on the opposite side of the drive axle. The identical spalling failures indicate that the tractor's axle was overloaded.

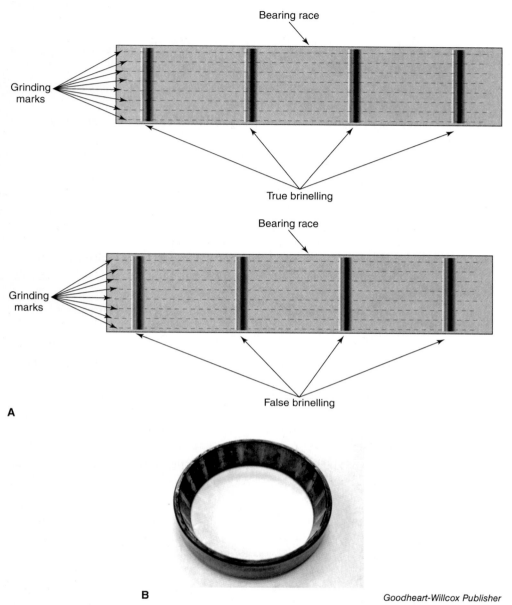

Figure 17-44. Examples of true and false brinelling. A—Machining marks are visible in the indentations in true brinelling but not in the indentations of false brinelling. B—This bearing race has false brinelling

Figure 17-45. The areas of brownish-red oxide on this bearing race indicate fretting corrosion.

Summary

- Final drives provide one last speed reduction and torque multiplication before propelling the drive wheel or track sprocket.
- Inboard final drives can have a planetary gear set design or a bull-and-pinion gear design.
- Outboard planetary final drives can be single reduction or double reduction.
- Low-track dozers can use a bull-and-pinion final drive, known as a *drop axle*.
- Caterpillar elevated-sprocket undercarriages use single- or double-reduction planetary final drives.
- Combines, cotton pickers, cotton strippers, and self-propelled forage harvesters use either bull-and-pinion final drives or planetary final drives.
- Skid steers and motor graders use two drive chains to propel tandem drive wheels.
- Excavator final drives receive their input from a hydraulic motor and have two or three planetary gear sets.
- Skid steer chain drive tension should be checked every 500 hours or once every three months.
- Tandem chain drives have their own oil reservoir.
- Some outboard final drives can be replaced without the need to split a track or remove a drive wheel.
- Some skid steer chains do not have a master link and are removed and installed as a complete chain assembly.
- Shims are often used to adjust bearing preload or shaft endplay.
- Component failures include wear, fracture, corrosion, and distortion.
- Component fracture types include fatigue, brittle, ductile, and combination fractures.

Technical Terms

- abrasive wear
- adhesive wear
- asperities
- beach marks
- brinelling
- brittle fracture
- bull gear
- combination fracture
- corrosion wear
- double-reduction final drive
- drop axle
- ductile fracture
- excavator final drive
- false brinelling
- fatigue fracture
- final drive
- fretting corrosion
- harvester final drive
- horizontal oil level mark
- impact fracture
- inboard bull-and-pinion gear final drive
- inboard final drive
- inboard single-reduction planetary final drive
- nonintegrated final drive
- outboard double-reduction planetary final drive
- outboard final drive
- outboard single-reduction planetary final drive
- overload failure
- ratchet marks
- single-reduction final drive
- spalling
- stress riser
- tandem chain drive
- tensile strength
- torsional stress fracture
- true brinelling
- yield strength

Review Questions

Answer the following questions using the information provided in this chapter.

1. Final drives do all of the following, *EXCEPT*:
   A. Provide one last torque multiplication.
   B. Provide one last speed reduction.
   C. Allow the two drive wheels to drive at two different speeds.
   D. Use planetary gears or bull-and-pinion gears.

2. Which type of final drive is *not* commonly used in off-highway equipment?
   A. Inboard planetary single reduction.
   B. Inboard planetary triple reduction.
   C. Outboard planetary single reduction.
   D. Outboard planetary double reduction.

3. _____ final drives are located next to the differential.
   A. Midboard
   B. Outboard
   C. Tandem chain
   D. Inboard

4. What planetary member is connected to the axle shaft in a single-reduction planetary gear set?
   A. Sun gear.
   B. Planetary carrier.
   C. Planetary pinion gear.
   D. Ring gear.

5. The inboard planetary final drive housing is sometimes called a _____ housing.
   A. drum
   B. trombone
   C. trumpet
   D. barrel

6. In what type of equipment are inboard bull-and-pinion gear final drives used?
   A. Compact utility tractors.
   B. Haul trucks.
   C. Elevated-sprocket undercarriages.
   D. Wheel loaders.

7. Outboard single-reduction planetary final drives are used in all of the following applications, *EXCEPT*:
   A. articulated dump trucks.
   B. skid steers.
   C. rigid frame haul trucks.
   D. four-wheel-drive wheel loaders.

8. All of the following use nonintegrated final drives, *EXCEPT*:
   A. articulated dump truck.
   B. combine.
   C. cotton picker.
   D. self-propelled forage harvester.

9. All of the following are *true* statements about harvester final drives, *EXCEPT*:
   A. they can have planetary gears.
   B. they can have bull-and-pinion gears.
   C. they can have bevel gears.
   D. they are nonintegrated final drives.

10. All of the following are *true* statements about skid steer final drives, *EXCEPT*:
    A. they are propelled by a hydraulic motor.
    B. they have a pair of drive chains on each side.
    C. they have a chain oil reservoir.
    D. they have planetary gear sets.

11. Motor grader tandem drives include all of the following, *EXCEPT*:
    A. bull-and-pinion gears.
    B. chains and sprockets.
    C. differentials.
    D. planetary gear sets.

12. All of the following are *true* statements about excavator final drives, *EXCEPT*:
    A. they can contain two planetary gear sets.
    B. they can contain three planetary gear sets.
    C. they rotate in the opposite direction of the hydraulic motor.
    D. they are driven by a differential side gear.

13. In a Caterpillar excavator final drive with three planetary gear sets, _____.
    A. all three ring gears are part of the track drive sprocket housing
    B. the third planetary gear set has a stationary planetary carrier
    C. the hydraulic motor propels the first stage sun gear
    D. All of the above.

14. All of the following are used for checking final drive fluid levels, *EXCEPT*:
    A. horizontal oil level mark.
    B. fill plug.
    C. pressure sensor.
    D. sight glass.
15. How are Caterpillar skid steer chain drives adjusted?
    A. Adjustable idler sprocket.
    B. Adjustable nylon gauge blocks.
    C. Movable axle housing with slotted holes.
    D. They are self-adjusting.
16. If skid steers are operating on hard surfaces, how often should the chain tension be checked as compared to the normal interval?
    A. Less frequent.
    B. Same interval.
    C. More frequent.
    D. None of the above.
17. How are late-model Bobcat skid steer chain drives adjusted?
    A. Adjustable idler sprocket.
    B. Adjustable nylon gauge blocks.
    C. Movable axle housing with slotted holes.
    D. They are nonadjustable.
18. How are motor grader drive chains removed?
    A. Through the tandem chain access covers.
    B. By removing four master links.
    C. By removing the tandem chain housings from the motor grader.
    D. Through the differential housing.

Apply and Analyze

19. The _____ is bolted to the axle housing in a single-reduction planetary final drive.
20. Drop-axle final drives are used in low-track _____.
21. Elevated drives in some articulated rubber-track tractors use _____ gears to elevate the power up to the drive sprocket.
22. Final drive fluid levels should be checked every _____ hours.
23. Final drive fluids should be changed every _____ hours.
24. Caterpillar _____ have a final drive inspection cover that is used to look for debris wrapped around the final driveshaft.
25. _____ failure occurs when two adjacent component surfaces contact each other and friction welds the surfaces together, causing material to lift off when the components attempt to move.
26. _____ failure occurs due to foreign particles or contaminants in the lubricant.
27. _____ is the amount of stress that can be placed on a component before it stretches and breaks.
28. _____ are fracture patterns that radiate from one or more origin points of a slow-occurring fatigue fracture.

Critical Thinking

29. Regarding final drives, what are two different methods employed on machines that reverse the direction of a final drive's output?
30. What are some mistakes that can be made when checking final drive fluid levels?

Chapter 18
Axles and Driveshafts

Objectives

After studying this chapter, you will be able to:
- ✓ Identify the different types of axles used in heavy equipment.
- ✓ Explain the function and installation of Duo-Cone seals.
- ✓ Describe heavy equipment driveline.
- ✓ List the different types of PTOs used in heavy equipment powertrains.

An axle assembly is a housing that connects the machine's frame to the wheels. It serves the purpose of attaching the wheels, transferring the load of the machine to the wheels, and providing the means by which the wheels are driven. The axle assembly contains the axle shafts and wheel bearings used to drive and support the machine's wheels. Some axles also contain a differential assembly and final drives.

Axles

Off-highway equipment uses different types of axle assemblies that are described using a host of different terminologies. Axles can be categorized in various ways, including the following:
- How the wheels are mounted to the axle.
- Whether or not the axle shaft and its bearings directly support the weight of the machine.
- Whether or not the axle propels the drive wheels.
- Axles can also be defined based on whether or not they are used for steering.

The following sections describe various types of axles.

Bar Axle and Flange-Mount Axle

Many agricultural tractors ranging from 60 hp to 670 hp are equipped with a **bar axle**, which is a solid bar that serves as the drive axle. The solid bar protrudes from the inboard planetary final drive, driven by the planetary carrier. Examples of bar diameters are 100 mm, 110 mm, 120 mm, and 145 mm.

Bar axles allow the drive wheels to be repositioned by sliding the wheels inward or outward as needed based on crop row spacing. For that reason, the axle can be called an *adjustable bar axle*. The axle is either smooth or notched. The wheel is secured to the axle with a hub and one or two tapered wedges, which serve as clamps. The notches form a rack, and the hub contains a pinion gear, forming a rack and pinion that aids the technician in repositioning the wheel.

A

B

Goodheart-Willcox Publisher

Figure 18-1. Agricultural row crop tractors often have rear bar axles. A—These tapered wedges have been removed from the tractor's axle. They are used to attach the drive hub to the drive axle on a MT 845 Challenger rubber track tractor bar axle. B—A hub and a pair of tapered wedges serve as a clamp that allows the wheel to be fastened to the bar axle. The wheels can be adjusted to different row widths.

Goodheart-Willcox Publisher

Figure 18-2. This John Deere twin rubber-track tractor uses a rear bar axle. The tracks on this tractor have been slid outward to provide a wide stance.

Goodheart-Willcox Publisher

Figure 18-3. This wheel loader axle has a flange for mounting the drive wheel.

See **Figure 18-1**. The process of adjusting wheel spacing on a bar axle is explained in Chapter 22, *Tires, Rims, and Ballasting*.

Bar axles are commonly used on the rear axle of MFWD and row crop tractors. However, small articulated tractors and twin rubber track tractors also use bar axles. See **Figure 18-2**.

Caution

Operators must be cautious when operating a bar axle tractor when the wheels are slid inward. The bar axles still protrude and can easily be overlooked. Damage can result when the tractor is operated near other objects, such as door openings and fence posts.

The other axle style used by agricultural manufacturers is the flange mount axle. The limitation of this design is that the wheels must be mounted at a fixed location on the axle. See **Figure 18-3**. Although the wheels cannot be slid inward or outward, the wheels' inside flanges are offset. Flipping the rim and mounting the wheel using the other side of the flange offers a limited amount of variability. The center of the rim that mounts to the axle's flange is known as the disc. The distance from the center line of the rim to the outside face of the wheel's disc is known as the wheel's offset. Single-piece rims are designed with a fixed amount of offset and, when the wheel is flipped, will place the tire either a few inches closer to the tractor frame or a few inches farther away. If the wheel's disc is bolted to the rim, it can be removed, flipped, and reinstalled on the rim to change the wheel's offset. Changing the wheel's offset provides wheel spacing adjustment, placing the tire either closer or farther away from the machine. Chapter 22, *Tires, Rims, and Ballasting*, provides details on tires and wheel positioning.

Semi-Floating Axle and Full-Floating Axle

Axles can also be classified based on whether or not the axle shaft and its bearings directly support the weight of the machine. In a *semi-floating axle*, **Figure 18-4**, the machine's weight is exerted on the axle bearings. The weight is then transferred through the axle shaft and onto the drive wheel. Semi-floating axles are commonly used in conjunction with inboard final drives. See **Figure 18-5**.

In a *full-floating axle*, the axle shaft is responsible only for driving the wheel. The weight of the machine is exerted on the axle housing instead of the axle shaft, allowing the axle shaft to float. See **Figure 18-6**.

Many heavy equipment powertrains use outboard final drives in combination with full-floating axle designs. See **Figure 18-7**. Examples of outboard planetary final drives are explained in Chapter 17, *Final Drives*.

Goodheart-Willcox Publisher

Figure 18-4. This bar axle is a semi-floating axle.

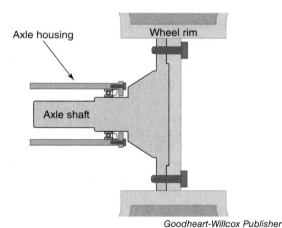

Goodheart-Willcox Publisher

Figure 18-5. This simplified drawing depicts a semi-floating axle that transfers the machine's weight to the axle bearings and to the axle shaft.

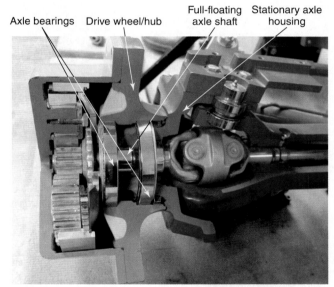

Goodheart-Willcox Publisher

Figure 18-7. This front axle cutaway is of an axle on a John Deere 8000 series tractor. The axle is a full-floating design. The drive wheel/hub transfers the load onto the bearings, which transfers the load onto the stationary axle housing.

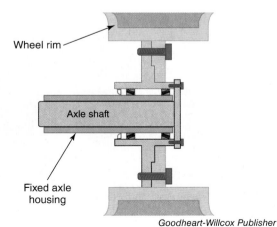

Goodheart-Willcox Publisher

Figure 18-6. In a full-floating axle, the machine's weight is transferred from the wheel bearings to the fixed axle housing.

Dead Axle and Live Axle

Axles can also be classified based on whether or not they propel the drive wheels. A *live axle* is a drive axle that receives power from the transmission and transfers the power to the drive wheels to propel the machine. Off-highway machines generally are equipped with one to four live axles. Examples of machines that use one drive axle are rigid frame trucks, two-wheel-drive agricultural tractors, traditional forklifts, and two-wheel-drive loader backhoes. Examples of machines that use two live axles are wheel loaders, four-wheel-drive agricultural tractors, and some articulated dump trucks. Examples of machines that use three live axles are articulated dump trucks. All-terrain cranes can have three or four live axles.

A *dead axle*, also known as a *lazy axle*, supports the weight of the machine but does not propel the drive wheels. All-terrain cranes often use multiple dead axles. A dead axle located ahead of a drive axle is called a *pusher axle*. A dead axle located behind a drive axle is called a *tag axle*.

Steer Axle and Rigid Axle

Axles can also be classified based on whether or not they are used for steering. A *steer axle* allows the wheels to pivot. For example, the steering knuckle can pivot to allow the machine to turn a corner. Steer axles can be located in different areas, including the front, middle, or rear of the machine. The most common location is the front axle. Steer axles are located in the front axle in traditional agricultural tractors. See **Figure 18-8**.

Goodheart-Willcox Publisher

Figure 18-8. This steer axle is the front axle of a John Deere 5325 agricultural tractor.

The steer axle can be a live axle or a dead axle. A single steer axle can be used on the rear axle, such as those found on combines, cotton pickers, forage harvesters, and traditional off-highway forklifts. Some all-terrain cranes use steer axles located in the middle of the machine, in addition to the front steer axle.

Years ago, manufacturers produced rigid frame four-wheel-drive machines that used both front and rear steer axles known as *four-wheel steer*, **Figure 18-9**. Today, JCB manufactures the Fastrac agricultural tractor and loader backhoes with a rigid frame and four-wheel steer. See **Figure 18-10**. Telehandlers are also equipped with four-wheel steer, **Figure 18-11**. Vermeer manufactures an RTX1250 trencher with four-wheel steer, equipped with either four tracks or four wheels.

Goodheart-Willcox Publisher

Figure 18-9. Case produced four-wheel-drive agricultural tractors with rigid frames and four-wheel steer, like this 4890 tractor.

Goodheart-Willcox Publisher

Figure 18-10. This JCB 3CX loader backhoe uses four-wheel steer. All four tires are the same size, which is less common in loader backhoes.

Chapter 18 | Axles and Driveshafts

Most articulated four-wheel-drive machines do not use steer axles. They have **rigid axles** in which the wheels are fixed to a wheel hub or axle shaft, and there is no steering knuckle. These tractors have a front frame hinged to a rear frame by means of an articulation joint. Hydraulic cylinders pivot the two frames to perform the steering. Examples of articulated machines with rigid axles are articulated dump trucks, agricultural four-wheel-drive tractors, four-wheel-drive wheel loaders, and scrapers. Articulation steering is explained in Chapter 25, *Wheeled Steering Systems*.

Twin track tractors do not use steer axles. They use a variety of designs to steer the machine. Track steering systems are covered in Chapter 24, *Track Steering Systems*.

Duo-Cone Seal

Off-highway axles, final drives, and other powertrain components, such as track rollers, require a heavy-duty seal. The seal retains fluids and prevents contamination between a rotating component and a static component. See **Figure 18-12**. The **Duo-Cone seal** is commonly used in these applications. The seal consists of two metal face seal rings that are held compressed against each other by means of two rubber **torics**. A toric is a large rubber ring that applies pressure to one of the Duo-Cone metal seals. The term toric is derived from the geometric shape of the ring, which is a torus. See **Figure 18-13**. One metal ring and one rubber toric are placed in one side of the component, such as the stationary axle housing, and the other ring and toric are placed inside the other component, such as the rotating final drive.

Taina Sohlman/Shutterstock.com

Figure 18-11. Most telehandlers use four-wheel steer.

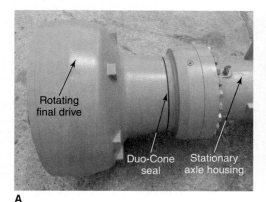

Goodheart-Willcox Publisher

Figure 18-12. Duo-Cone seals use two metal flat faces to seal a fixed housing against a rotating housing. A—The Duo-Cone seal is located on an axle assembly. B—Cutaway of a planetary winch drive. The two metal flat face seals hold against each other with the force of the two rubber torics.

Figure 18-13. Two metal seal rings and one rubber toric. The other toric is not shown.

Both the seal ring and the seal housing have a ramp. The housing also has a retaining lip. See **Figure 18-14**. The toric rides between the ramp of the seal ring and the housing and is retained by the retaining lip.

The face of one metal seal rides against the face of the opposite metal seal. The two rubber torics place pressure on the two metal seal rings, holding the metal faces against each other and providing a metal-to-metal seal. See **Figure 18-15**.

As shown in **Figure 18-16**, the Duo-Cone seals at three areas:

- Between the two flat faces.
- Between the toric and the housing ramp.
- Between the toric and the seal ring ramp.

Installation

As shown in **Figure 18-14**, a special seal installation tool is used to install the Duo-Cone seal. The tool is sized appropriately to the specific seal. Rebuild shops have a variety of Duo-Cone seal installation tools based on size and ramp angle. For a given seal diameter, the ramp can have different angles, for example, 8°, 15°, and 20°.

During installation, the seal rings and housings must be exceptionally clean and free from dirt, oil, grease, and burrs. Clean surfaces with a nonpetroleum-based solvent and a lint-free towel. Thoroughly inspect ramps to ensure that they do not have any burrs and have a good surface finish. Inspect the rubber torics to see if they are cut, torn, or worn. During installation, make sure that the toric is not twisted as it is positioned on the metal seal ring. See **Figure 18-17**.

After placing the seal assembly onto the tool, lubricate the bottom half of the rubber toric with the specified lubricant. Examples of specified lubricants are QUAKERCLEAN 68 RAH, Houghto-Grind 60 CT, and isopropyl alcohol. An example of lubricant that should not be used is Stanisol because it leaves an oily film and does not evaporate quickly.

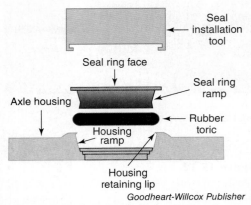

Figure 18-14. An exploded side view of one half of a Duo-Cone seal. The seal housing in this example is the axle housing.

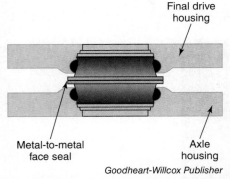

Figure 18-15. The Duo-Cone pushes the flat face of one seal against the flat face of the other seal.

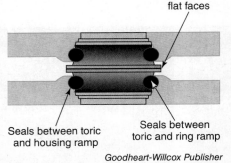

Figure 18-16. The Duo-Cone seals at three different areas.

Figure 18-17. The Duo-Cone seal ring is placed inside the installation tool. The rubber toric must not be twisted when it is installed on the seal ring.

Place the tool and seal assembly on top of the axle's housing. Tap on the tool with a rubber mallet to force the seal assembly into the housing so the rubber toric passes by the retaining lip. Tap evenly on the tool to install the seal into the housing. Do not strike the installation tool where its two ends meet.

After the seal is installed, check the seal height with a caliper at four equally spaced points around the circumference of the seal. See **Figure 18-18**. If the four measurements have more than one millimeter of difference, the seal is not installed evenly. The tool can be reinstalled on the seal to adjust the seal by pushing and lifting. Removal and reinstallation of the seal may be required.

Goodheart-Willcox Publisher

Figure 18-18. Check the height of the seal to ensure that it is installed evenly in the housing.

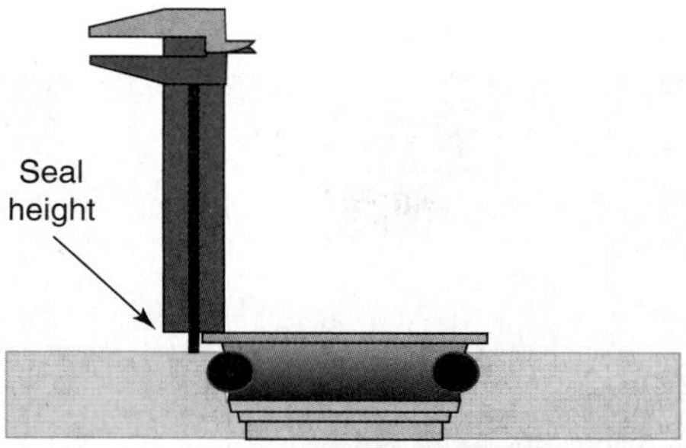

Place a light film of oil on the face of the seal ring before the rotating housing is installed onto the fixed housing. Be sure the oil does not touch any surface except for the seal ring face. Ensure that housings are installed evenly and do not tip.

Another metal-to-metal face seal is the heavy-duty dual face seal. The seal functions similarly to the Duo-Cone seal, except that Belleville washers keep the metal faces compressed against each other. A Belleville washer is a cone-shaped washer made of spring steel.

Drivelines

Heavy equipment systems require driveshaft assemblies to connect powertrain components, such as the engine, hydrostatic pumps, torque converter, transmission, and drive axles. A driveshaft assembly, **Figure 18-19**, can incorporate the following components:

- Driveshafts.
- Slip splines.
- Yokes.
- Universal joints (U-joints).
- Center support bearings.

Goodheart-Willcox Publisher

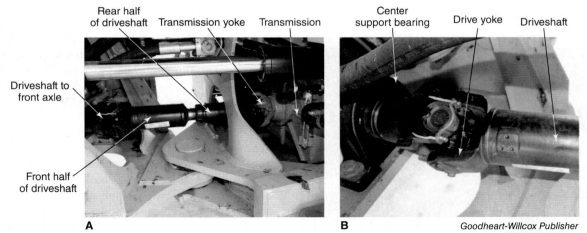

Figure 18-19. In this Caterpillar 950M wheel loader, a driveshaft connects the transmission located on the rear frame to the drive axle attached to the front frame. A—The transmission output shaft yoke drives the rear half of the driveshaft. B—The driveshaft to the front axle is supported by a center support bearing.

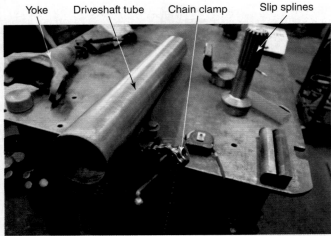

Figure 18-20. This driveshaft is being fabricated. The shaft's tube is resting in the saddle of a chain-type clamp. The slip splines and yoke will be inserted into the tube, aligned, and welded into the tube.

Figure 18-21. This two-piece driveshaft has external splines on the longer shaft and internal splines on the short shaft (slip yoke).

Figure 18-22. This driveshaft assembly is used in an on-highway truck. The corrugated boot keeps dirt and debris away from the slip splines.

Driveshaft Assembly

A *driveshaft*, also known as a *propeller shaft*, is a hardened steel tube used to couple two drivetrain components. Driveshafts vary in diameter, tube thickness, and length. See **Figure 18-20**. A single-piece driveshaft has a yoke located on each end.

Slip Splines

A *slip spline* is a sliding joint located on a two-piece driveshaft. It consists of external splines on one end of a driveshaft and internal splines on the mating driveshaft. See **Figure 18-21**. The slip spline joint allows one driveshaft to drive the other driveshaft. When the two driveshafts are coupled, one driveshaft slides within the other. This allows the overall length of the driveshaft assembly to change as the axle assembly raises and lowers as the machine travels over bumps and ditches. Each piece of the driveshaft has a yoke on one end and splines on the opposite end.

The splines are lubricated with grease. A corrugated boot is sometimes placed over the splines to keep out contaminants. See **Figure 18-22**.

Yokes

A *yoke* is a Y-shaped fixture attached to one end of the driveshaft or powertrain component that allows it to be connected through a universal joint (U-joint) to another driveline component. For example, yokes and a universal joint are used to connect a transmission to a driveshaft or a driveshaft to an axle. Yokes can have one of three designs:

- Full-round yoke.
- Half-round yoke.
- Wing-type yoke.

The *full-round yoke* consists of two round lobes that attach to two of the U-joint bearing cups. See **Figure 18-23A**. The cups are held in the yoke by retaining rings or bearing plates. The *half-round yoke* has two partial lobes that act as cradles for the U-joint bearing cups. The U-joint is secured in the yoke by two straps or two U-bolts. See **Figure 18-23B**. The *wing-type yoke* has four threaded bolts used to bolt the wing type U-joint to the yoke. See **Figure 18-23C**.

Conventional U-joints

A *universal joint (U-joint)* is a component that connects together two driveline components, such as a driveshaft to a transmission, or a driveshaft to an axle. In addition to connecting the two components, it is capable of transferring torque from one component to the other driveline component through a range of angles. The main part of the U-joint is the *cross* (also called a *spider*), which has four trunnions. The trunnions are smooth, machined journals on each end of the cross. A bearing cup fits over each of the trunnions, so there are four bearing cups per U-joint. The inner diameter of each cup is lined with needle bearings. The trunnions serve as the inner

Figure 18-23. Yokes are used in conjunction with U-joints to attach driveshafts to drivetrain components such as a transmission or axle. A—This full-round yoke is used in conjunction with a bearing plate U-joint. The round balancing weight has been welded to the yoke with three spot welds. B—This U-joint is used in conjunction with a half-round yoke, requiring a strap to retain the yoke to the U-joint. C—This wing-type yoke is designed to attach to a wing-type U-joint. The notches on each side of the yoke mate with two notches on the wing-type U-joint.

races for the needle bearings, and the cups serve as the outer races. See **Figure 18-24** and **Figure 18-25**. The bearings allow the cross to pivot inside the yoke as the U-joint rotates. Each cup has a seal that prevents dirt from entering the U-joint and retains the lubricant inside the U-joint's cup. A conventional U-joint is also called a Cardan U-joint or Hooke U-joint.

A variety of methods can be used to attach U-joints to driveshafts:
- Bearing plate.
- Bearing strap.
- Snap ring.
- U-bolt.
- Wing-type.

Bearing plate U-joints have bearing cups that are fitted with plates on the outside surface and are bolted to the yoke. ***Bearing strap U-joints*** have two metal straps that are bolted around the circumference of the bearing cup to retain the U-joint in the yoke. See **Figure 18-26**.

Figure 18-24. U-joints can have retaining rings located towards the center of the cross. This small U-joint has all four bearing cups and seals installed on the cross. It is used in small, light-duty applications.

Figure 18-25. This bearing cup has been removed from a high wing U-joint, revealing the needle rollers.

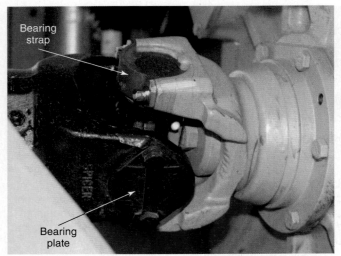

Figure 18-26. On this Caterpillar 950M wheel loader, the transmission yoke uses the bearing strap method of retaining the bearing cups. On the driveshaft side, the bearing cups are retained with a bearing plate.

Snap-ring U-joints have a retaining ring that holds the bearing cup inside the yoke. The retaining ring can be located either inboard or outboard of the bearing cup. See **Figure 18-27**. Depending on the design of the retaining ring, it may be able to be removed with needle-nose pliers and a standard screwdriver, or they may require the use of snap ring pliers.

Older machines may be equipped with U-bolt U-joints. These joints rely on U-bolts to clamp the bearing cups to the yoke. They are no longer commonly used in off-highway machines.

Wing-type U-joints have bearing cups that are bolted directly to a yoke, **Figure 18-28**. These U-joints generally do not require special tools to replace. Different wing-type U-joints include the following:

- Low block (low wing).
- Delta wing.
- High block (high wing).

Wing U-joints can use drilled holes or threaded holes. Low wing U-joints are found in smaller applications, such

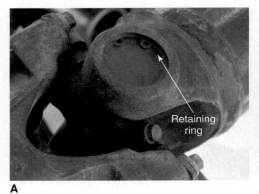

Figure 18-27. Snap rings are sometimes used to hold bearing cups within driveshaft yokes. A—This driveshaft is attached to the front axle of a New Holland telehandler. The snap ring can be removed with snap ring pliers. B—A U-joint on an on-highway truck application. The snap ring can be removed with a pair of needle-nose pliers and a flat-tip screwdriver.

Figure 18-28. Wing-type U-joints are commonly used in construction equipment. A—A high wing U-joint that has been removed from the driveshaft. B—This high wing U-joint is bolted to a yoke.

as smaller wheel loader applications. Delta wing U-joints are used in midsize equipment. High wing U-joints are used in high-horsepower applications. See **Figure 18-29**.

 Caution
Whenever a wing-type driveline or a half-round yoke is removed, always install new hardware, including retaining bolts and straps. Even if the old U-joint will be reused, install new hardware. The bolts have Loctite threadlocker on the threads. Failure to install new hardware can cause a driveline failure.

A U-joint's cross has one or two grease zerks (fittings). Drilled passageways in the cross allow technicians to grease all four cups simultaneously. Remove grease zerks prior to disassembly to prevent damage. The U-joint in **Figure 18-30** is used in low-horsepower applications, such as small telehandlers and light-duty on-highway trucks. During operation, the hot grease turns to liquid. Some U-joints contain *stand pipes*, which are check valves that prevent the lubricant from draining out of the trunnion that is pointing upward. This helps ensure that all four cups are full of grease during the next machine operation. See **Figure 18-31**.

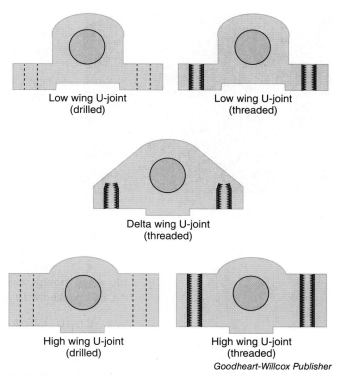

Figure 18-29. Wing-type U-joints include low wing, delta wing, and high wing types. The holes can be drilled or threaded.

Constant Velocity U-joints

Some manufacturers use a ***constant velocity U-joint (CV joint)***, which, as its name implies, is designed to rotate at a constant velocity. CV joint designs include the double Cardan U-joint, Bendix-Weiss U-joint, and Rzeppa U-joint.

A ***double Cardan U-joint*** consists of two conventional U-joints inside a center yoke, **Figure 18-32**. This joint is used in agricultural PTO applications. PTOs are discussed later in this chapter.

Sharomka/Shutterstock.com

Figure 18-30. The grease zerk on this small U-joint must be further threaded into the U-joint's cross. The snap rings shown hold the U-joint into the yokes.

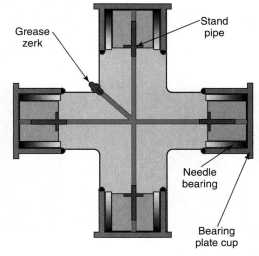

Goodheart-Willcox Publisher

Figure 18-31. Stand pipes are check valves that prevent lubricant from draining out of the vertically positioned trunnion.

Goodheart-Willcox Publisher

Figure 18-32. This CV joint consists of two conventional U-joints inside a center yoke.

A **Bendix-Weiss U-joint** has two ball yokes rather than conventional U-joint yokes. Four balls transmit power from one yoke to the other. A smaller fifth ball is a spacer between the two shafts.

A **Rzeppa U-joint** (*ball-and-cage joint*), also does not use the conventional U-joint yokes. This U-joint contains an inner race with six channels in which six balls ride. A cage retains six ball bearings. As the inner race is driven, it causes the balls to drive, which propel the outer bearing race. Rzeppa joints are used for drive axles in utility terrain vehicles (UTVs), such as John Deere Gators. Agricultural technicians need to periodically replace these joints. See **Figure 18-33**.

Bendix-Weiss U-joints and Rzeppa U-joints are not commonly found in heavy equipment. Although they provide the advantage of a constant velocity, they cannot handle heavy loads like a conventional U-joint can, nor are they as economical.

Rubber boots are used to retain the grease in CV joints. It is important to periodically check the boots for cuts, cracks, and tears. If the boot tears, the grease will be lost and the U-joint will fail. If the tear is found soon enough, the U-joint will need to be completely disassembled, thoroughly cleaned, and packed with grease. A new boot and boot clamps will need to be installed.

Center Support Bearings

Center support bearings are used to support and limit the vertical movement of driveshafts across long distances. The bearing is often mounted in rubber to reduce driveline vibrations. They are commonly found in articulated vehicles to support the driveshafts between the two articulating frames. They are also called *carrier, hanger,* or *midship support bearings*. Refer to **Figure 18-19B**.

When replacing a center support bearing, be sure to reuse the shims. Ensure that the same thickness of shims is used with the new bearing. The shims were installed by the manufacturer to adjust the driveline angle. If the proper shims are not used, driveline vibration can occur.

Driveline Angles

Drivelines normally spin with operating angles, consisting of the angle created by the U-joint connecting two driveline components, such as a transmission and a driveshaft. The driveline angles on the machine can be in a vertical plane, a horizontal plane as in **Figure 18-34**, or a combination of both. These angles often change when a component, such as the axle, is allowed to move up and down as the machine travels through the field. Improper driveline angles, magnified with increased shaft speeds, can cause severe vibrations and significantly reduce the life of U-joints. Drivelines are designed so that opposing

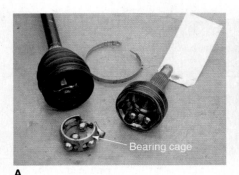
A

B

C

Goodheart-Willcox Publisher

Figure 18-33. This CV axle shaft was removed from a John Deere Gator. It has a Rzeppa U-joint on both ends. A—The bearing cage is broken. B—The rubber boot has been removed from the opposite U-joint, which has not failed. The cage still retains the six ball bearings. C—Rubber boots are used to retain the grease. Two boot clamps secure each boot.

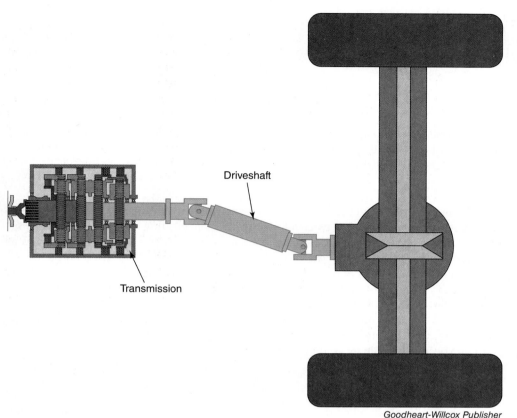

Figure 18-34. Some drivelines can be positioned off center in the horizontal plane (offset).

pairs of driveline angles are equal in magnitude but opposite in orientation. The vibrations that would be caused by speed changes in one U-joint are canceled by the speed changes in the other U-joint, minimizing vibration.

Consider the driveline in **Figure 18-35**. The angle of yoke A to the driveshaft is the same as the angle of yoke B to the driveshaft, except that the angles are in the opposite direction. The two yokes are in parallel with each other. In addition, the two yokes are operating in the same axis, meaning that the driveline angles are all in the vertical plane rather than a combination of vertical and horizontal planes. Yoke A rotates at a steady speed, known as a *constant velocity*.

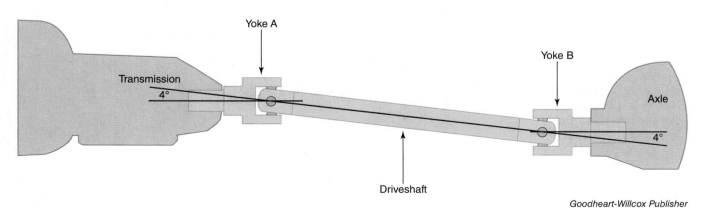

Figure 18-35. A driveline with parallel yokes with equal angles in the opposite direction allows the second yoke to offset the angularity of the first yoke, resulting in a constant output speed.

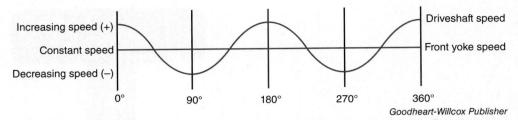

Figure 18-36. The yoke that is propelling a driveshaft rotates at a constant speed. However, a U-joint that connects the driveshaft assembly to the front or rear yoke has an operating angle that causes the driveshaft to increase and decrease speed twice per revolution.

The angle of the U-joint between the transmission and the driveshaft causes the driveshaft to increase and decrease speed twice for every revolution, creating a pulsing effect. See **Figure 18-36**. The driveshaft is said to rotate at a nonuniform velocity. A U-joint that rotates at nonuniform velocities is called a *conventional U-joint*, *Hooke U-joint*, or **Cardan U-joint**.

Since yoke B operates in parallel with yoke A and is in the same axis plane as yoke A, the angle of yoke B counteracts the angularity of yoke A and causes yoke B to rotate consistently at the same constant speed as yoke A. See **Figure 18-37**.

The two U-joints must be in-phase in order to drive the output yoke at the same speed as the input yoke. A one-piece driveshaft will always be in-phase, but a two-piece driveshaft can be accidentally reassembled out-of-phase.

Driveline Phases

A driveshaft is ***in-phase*** when the front and rear yokes are in straight alignment with each other. An ***out-of-phase*** driveshaft has yokes that are not in alignment. See **Figure 18-38**.

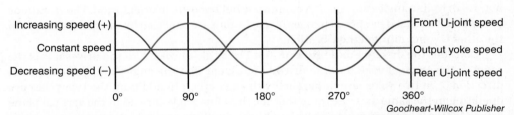

Figure 18-37. The front U-joint increases and decreases speed twice per revolution. The rear joint also increases and decreases speed twice per revolution. However, the rear U-joint speed fluctuation counteracts the front U-joint fluctuation, resulting in a constant yoke output speed.

Figure 18-38. With the exception of some applications with compound operating angles, a two-piece driveshaft must be positioned in-phase, which means that the front yoke and rear yokes are aligned with each other.

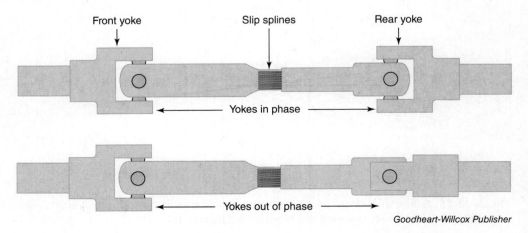

If the two driveshafts are separated and accidentally installed out-of-phase and the yokes are rotating at an operating angle, the output yoke will spin at a nonuniform speed, even if the yokes are in the same axis and parallel to each other. Therefore, prior to removing a two-piece driveshaft, mark the driveshafts to ensure that the shafts will be in-phase during installation. If an out-of-phase driveline is operating in a straight line with no U-joint operating angles, no vibrations will occur.

Note

A driveshaft can have a compound angle due to a horizontal plane angle (offset) and a vertical plane angle. The driveline might be intentionally assembled out-of-phase to compensate for the driveline's compound angle.

U-joints have a maximum operating angle specification. The maximum angle depends on the size of the U-joint and the operating speed. See **Figure 18-39**.

Driveline Repairs and Maintenance

U-joints must be replaced before they fail, otherwise a broken driveshaft will cause significant damage and potential injury. The driveline should be periodically inspected as part of the preventative maintenance inspection.

Warning

Prior to inspecting any driveline, disable the machine so no one else can start the machine while you are inspecting the driveline. Also post a *do not operate* tag on top of the ignition switch. A rotating driveline can cause a severe injury or death.

During the inspection, check for the following:
- Broken, loose, bent or missing mounts, hardware, or components.
- Loose yokes.
- Worn center hanger bearings or missing hanger bearing shims.
- Loose U-joints.
- Weights that have fallen off the driveshaft, evidenced by a discolored circle shape and remnants of spot welds.

Driveline Vibrations

Driveline vibration can be caused by many problems, including the following:
- Worn U-joints.
- Worn driveline bearings.
- Improper U-joint operating angles.
- Out-of-phase yokes.
- Worn or bent suspension components.
- Out-of-balance shafts.
- Bent driveshafts.

Vibration Analysis

Vibrations are classified in units of *order*. A ***first order vibration*** is a vibration that occurs once during every revolution of the shaft. Out-of-balance driveshafts and out-of-round driveshafts exhibit first order vibrations. For example, if a truck is run on a dynamometer, and a

Driveshaft Speed	Maximum Operating Angle
5000 rpm	3.250°
4500 rpm	3.670°
4000 rpm	4.250°
3500 rpm	5.000°
3000 rpm	5.830°
2500 rpm	7.000°
2000 rpm	8.670°
1500 rpm	11.50°

Goodheart-Willcox Publisher

Figure 18-39. U-joints have a maximum operating angle that is reduced as the drive speed increases.

photo tachometer measures the driveshaft spinning at 1500 rpm, the driveshaft first order vibration frequency would be 25 hertz. **Second order vibrations** occur twice during every revolution of the component. A second order vibration can be caused by driveshaft angularity problems.

Technicians diagnose vibrations with a vibration analyzer that displays the vibration in frequency or rpm. The frequency indicates the number of times the shake occurs for a given time period, such as per second. The rpm scale indicates the number of times the shake occurs for a given minute. The analyzer also displays a numerical value of the magnitude of the vibration in gravitational force (Gs) and sometimes analyzers also display a graph which provides a visual indication of the amplitude of the vibration.

Analyzers come in a variety of designs, from inexpensive tachometers such as a mechanical sirometer (approximately $35) or a vibrating reed tachometer that costs a few hundred dollars, to a portable electronic analyzer that can cost several thousands of dollars.

The sirometer is a mechanical tachometer consisting of an extendable wire with a loop at the end, **Figure 18-40**. When the sirometer is held on a hard surface of the machine, vibrations in the machine are transferred through the sirometer, causing the wire to vibrate. A scale on the sirometer indicates the frequency of the vibration. A dial is turned to extend the wire loop out of the sirometer. The wire is extended out or retracted back in as needed to achieve the widest possible fan pattern with the vibrating wire. The vibration frequency is then read from the scale. The sirometer displays the rate of vibration on two scales, rpm and hertz.

The vibrating reed tachometer is a mechanical tachometer with a series of reeds that each vibrate when exposed to a different frequency vibration. The reed tachometer is placed on a hard surface of the machine. As the machine is operated, the technician determines which reed on the tachometer's scale is vibrating. Depending on its design it will either display rpm or frequency. If tachometer uses a frequency scale, such as 10 to 800 hertz, you can convert the hertz to rpm by multiplying the hertz value by 60.

A portable electronic analyzer is an electronic device that uses a vibration sensor consisting of an accelerometer that is often installed on a machine using a magnet. The analyzer measures vibrations and displays the number of vibrations (in rpm or hertz) and the amplitude of the vibration (in Gs). It is often powered by the machine's 12-volt DC outlet power socket. Analyzers come with one or more vibration sensors.

A vibration can occur in three axes—horizontal, lateral, and vertical. The sensor is designed to measure a vibration in one axis depending on its position relative to the vibration. However, a large vibration can often be measured regardless of the sensor's placement. It is also very quick and easy to simply move the sensor from one hard surface to another while investigating machine vibrations.

It is important to place the sensor on a solid point, for example, an axle housing rather than a seat cushion. The cushion will dampen the vibration reducing its amplitude, although the frequency will remain the same.

While measuring vibrations, it is sometimes helpful to move the sensor to a different location on the machine to see if the amplitude changes. For example, if the sensor is placed at four different corners of the machine, the location on the machine near the origin of the vibration will exhibit a larger G force. Technicians often place the sensor on the metal mount on the operator's seat, because it is the operator that normally has sensed a vibration and the seat is where the operator

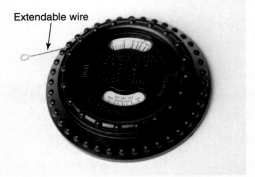

Goodheart-Willcox Publisher

Figure 18-40. A sirometer is a simple device for measuring vibration frequency. The meter is placed against the machinery and the wire is extended out until it creates the widest possible fan. The vibration's frequency can then be read on the instrument's scales.

sensed the vibration. When moving the sensor to check other locations be sure to give the analyzer enough time to register a consistent reading.

When making measurements, keep track of the frequency that is causing the vibration, the amplitude of the vibration, and the engine's speed. For example, the engine, flywheel, and sometimes a driveshaft (on a direct drive powershift machine), all rotate at crankshaft speed, allowing a technician to investigate potential causes of a first order vibration. If the vibration is slower than a first order vibration, the technician must consider driveline angles, wheels, hubs, and other areas on the machine as a possible cause.

Years ago Kent-Moore (OTC) tools sold a common electronic vibration analyzer known as an EVA. Many automotive manufacturers, as well as Caterpillar, sold the tool to dealers as a special tool for diagnosing noise, vibration, and harshness (NVH). See **Figure 18-41.** These instruments are normally equipped with one sensor but could be equipped with two sensors. Additional attachments include an inductive timing light and inductive clamp. The inductive clamp attaches to a wire loop on the EVA. The looped wire is a trigger for the timing light and the two attachments work together for balancing drivelines. The EVA analyzer can also record ten different snapshots that can later be viewed.

The PicoScope, which is a specific brand of oscilloscope, is becoming a more popular tool in automotive, light truck, and heavy-duty on-highway truck shops. The PicoScope can be configured as a vibration analyzer. It requires purchasing the PicoScope, a NVH kit that includes an interface noise sensor, a vibration sensor (accelerometer), and a laptop to operate the PicoScope. Analyzing light-duty trucks requires using a data link connector to communicate with the truck's onboard computers.

Note
An app is available that enables a smartphone to be used for measuring vibrations.

Out-of-Balance Shafts
Driveshafts can become bent or imbalanced, causing a driveline vibration. Although manufacturers provide information on the process of installing hose clamps on driveshafts and adjusting the clamps to balance a driveshaft, the labor rate of most repair shops makes it impractical for technicians to spend time balancing driveshafts. It is more practical to send an out-of-balance driveshaft to a driveline repair facility that specializes in balancing driveshafts.

A

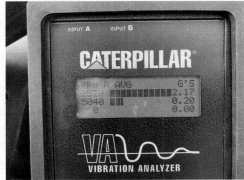
B

Goodheart-Willcox Publisher

Figure 18-41. An electronic analyzer measures machine vibration. A—This analyzer receives its power from the machine's 12-volt power outlet. One of two magnetic sensors is shown plugged into the analyzer. B—This analyzer displays rpm or hertz and the gravitational force (Gs) of the vibration. Notice it measured a vibration of 2.17 Gs that occurred at 2520 rpm.

Case Study
Diagnosing a Caterpillar D5H Dozer with Vibration Problems

A Caterpillar D5H dozer began experiencing a significant vibration. The machine had approximately 12000 hours on it. During its life, the machine had set for 10 years and the engine had been rebuilt. The vibration caused the HVAC compressor brackets to break multiple times, fracturing the refrigerant lines. It also caused the engine fan's shaft to fracture into two pieces, and the engine's starter to vibrate loose twice. The torque converter had been replaced and the three-speed transmission had been rebuilt. The engine was a four-cylinder Caterpillar 3304 engine with two balancing shafts.

A vibration analyzer was used to measure the engine's vibration. The largest vibration measured was a second-order vibration, with a frequency twice the engine rpm. At an engine idle of 1250 rpm, the frequency of the vibration was 2500 hertz. The vibration analyzer probe was placed on multiple locations on the machine, including the frame in front of the engine and on top of the torque converter housing. Each of those yielded a second order vibration around 0.4 to 0.7 Gs.

The probe was then placed at multiple locations on the front of the engine, including directly on top of the engine, the front of the engine and the side of the engine. The readings on the front and the side of the engine ranged from 0.4 to 0.7 Gs. The reading on the top of the engine measured 2.17 Gs. See **Figure 18-41B**.

Based on these results it was concluded that the vibration was an engine vibration. During each revolution of the crankshaft, two cylinders fire individually on a power stroke, and causing a 2.17 G vibration during each cylinder's power stroke. The engine was disassembled, and the technician found the balance shafts were out of time. The engine was rebuilt, the balance shafts were correctly timed, and the vibration was eliminated.

Calculating Operating Angles

Technicians measure and check driveline slopes in order to compute U-joint operating angles. Devices used to measure the slopes include spirit level protractors, digital protractors, inclinometers, and smartphone angle finders.

A protractor that is used to measure driveline slopes must be accurate to within one quarter of a degree. Attachments allow the protractor to be placed directly on the U-joint's cup, which provides a slope of the yoke.

Be sure that the machine is on level ground. Then, place the protractor on the yoke that is propelling the driveshaft, for example, the transmission's output yoke. Measure the slope with the protractor. Next, place the protractor on the driveshaft and measure the slope. If the driveline contains a second driveshaft, measure the slope of that driveshaft. Finally, measure the slope of the axle's pinion shaft.

An example is shown in **Figure 18-42**:
- Transmission output yoke A has a 1° downward slope.
- Driveshaft A has a 3° downward slope.
- Yoke and driveshaft B have a 6° downward slope.
- Axle's pinion yoke C has a 10° downward slope.

To compute the U-joint angles, identify the two slopes that intersect at the U-joint. If both slopes are inclined in the same direction, subtract the lesser slope from the greater slope. If one slope is downward and the other is upward, add the two slopes.

In the previous example, the operating angles are as follows:
- The first U-joint operating angle is 3°–1° = 2°.
- The second U-joint operating angle is 6°–3° = 3°.
- The third U-joint operating angle is 10°–6° = 4°.

A second example is shown in **Figure 18-43**.

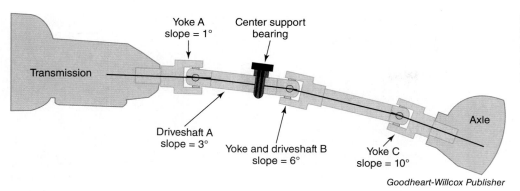

Figure 18-42. This driveline has three U-joints, two driveshafts, and one center support bearing. All slopes point downward.

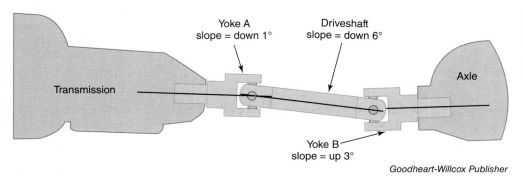

Figure 18-43. This driveline uses two U-joints and one driveshaft. Two slopes point downward and one points upward.

- Transmission output yoke A has a 1° downward slope.
- The driveshaft has a 6° downward slope.
- Yoke B has a 3° upward slope.

In this example, the operating angles are as follows:

- The first U-joint operating angle is 6°–1° = 5°.
- The second U-joint operating angle is 6°+3° = 9°.

The calculated operating angles can now be compared to the manufacturer's specifications. Maximum U-joint operating angles are listed in **Figure 18-39**. As a rule of thumb, the difference between the front and rear U-joint angles should not exceed 1° to 2° to achieve 100% life expectancy. If the difference is more than 3°, the life of the U-joints will be reduced.

If operating angles are excessive, check the chassis for loose, broken, missing, or bent mounts. These conditions would cause the operating angles to change.

Replacement

A heavy-duty driveline can weigh in excess of 100 pounds. Be sure to use a lifting device to remove the driveline.

 Note

When replacing a U-joint, always install a complete new cross and four bearing cups as an assembly. Do not mix and match used bearing cups and crosses.

Many heavy-duty wing-type U-joints are used in heavy equipment systems. Wing-type U-joints can normally be replaced without requiring any special tools. However, bearing plate and snap-ring type U-joints normally require a U-joint removal tool.

Tiger Tool manufactures a heavy-duty U-joint removal tool that is slid over one end of the cross. This tool is used with an air impact wrench to push the bearing cup on the opposite end of the cross out of the yoke. If the U-joint is a bearing plate design, remove the two cap screws that retain each bearing plate cup in the end of the yoke. Place the tool on the U-joint so the tool presses against the cross. See **Figure 18-44**. An air impact is used

Figure 18-44. The Tiger U-joint puller is used in conjunction with an air impact wrench to drive a U-joint out of the yoke. A—After the two cap screws are removed from the bearing cup, the removal tool is positioned so the bottom edge of the puller is under the top part of the yoke and bearing cap it holds. The screw is adjusted until the two arms of the puller press against the cross. B—An air impact wrench is used to drive the U-joint out of the yoke. As the tool pulls the top part of the yoke upward, the tool arms push down on the cross, driving the bottom bearing cup out of the yoke. C—The bearing cup has been fully pressed out of the yoke and the removal tool screw has been retracted.

to drive the cross and one bearing cup out of the yoke. The tool can then be placed on the opposite side of the cross and used to press the opposite bearing cup out of the yoke.

Fabricating a Driveshaft

If a driveline fails, the old driveshaft can be modified or a new one can be fabricated. A driveline repair shop will order parts to repair or fabricate the driveshaft as needed, including the driveshaft tube with a specific diameter, thickness, and overall length, yoke(s), and a slip splines attachment.

Prior to inserting the new components into the driveshaft tube, a flapper grinding wheel is used to prepare the tube's outside circumference to provide a clean welding surface. The inside circumference is also prepped to remove any obstructing burrs or surface rust that would limit the installation of the yoke and slip spline components. See **Figure 18-45**.

The yoke and slip spline components have a tight fit into the tube. A hammer or press is used to drive the components into the tube. See **Figure 18-46**.

The fabricator is responsible for ensuring that the driveshaft is aligned so the yokes are in-phase. The fixed yoke is driven into the tube first. Then, before the slip spline fitting is driven into the tube, it must be aligned and marked. The slip yoke is slid onto the spline fitting and aligned so the two yokes are in-phase. After the slip spline fitting has

Figure 18-45. A grinding wheel is used to prep the driveshaft tube prior to inserting the yoke.

Figure 18-46. A hammer is used to drive a fixed yoke into the tube.

been marked, the slip yoke is removed and the slip spline fitting is driven into the tube, **Figure 18-47**. Because the fitting will move slightly out of alignment, it is rechecked and adjusted prior to being welded to the tube.

After the fitting is driven into the tube, its alignment is checked. The slip yoke is slid into mesh with the fitting. Two cross bars are placed on top of the yokes to determine if the yokes are out-of-phase. See **Figure 18-48**. If necessary, a hammer can be used to tap the fittings back into phase.

After the yokes are aligned, the driveshaft's runout is measured. A shaft's *runout* indicates how much the shaft moves radially from its center axis. Runout is checked by placing the driveshaft in a lathe and measuring the high and low spots of the shaft with a dial indicator. See **Figure 18-49**. A punch is used to reposition the fittings to ensure that the driveshaft has little runout, which is described as *truing a shaft*. See **Figure 18-50**. The least amount of runout is best. An example of maximum allowable driveshaft runout is 0.005″.

After the shaft runout has been minimized, the fitting is spot welded to the tube. The next fitting is checked and adjusted for runout and then spot welded in place. See **Figure 18-51**. Lastly, the driveshaft remains in the lathe, and the fabricator slowly spins the shaft and welds the fitting to the tube in a single, smooth pass. See **Figure 18-52**. Some driveshafts are welded with automated lathes and welders.

Straightening a Driveshaft

If a fabricated driveshaft has excessive runout, driveshaft repair shops can use heat to straighten a manageable amount of runout, such as 0.040″. The driveshaft's runout is measured and the shaft's high side is marked. The

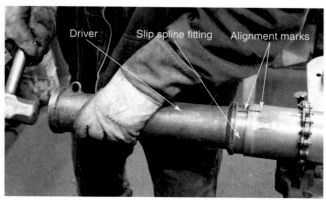

Goodheart-Willcox Publisher

Figure 18-47. After the yokes are aligned and the slip spline fitting and driveshaft are marked, the slip yoke is removed. A driver is placed over the slip spline fitting. With the fitting and tube markings aligned, the fitting is driven into the tube.

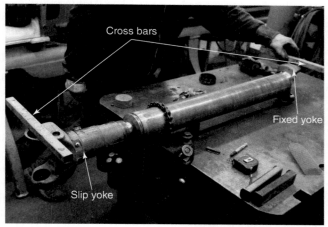

Goodheart-Willcox Publisher

Figure 18-48. Before the slip spline fitting is welded to the driveshaft, cross bars are placed on top of the yokes to check that the yokes are still in-phase.

Goodheart-Willcox Publisher

Figure 18-49. The shaft's runout is checked with a dial indicator before the shaft is welded. The seam that runs laterally across the entire length of the tube causes the indicator to dip as the seam passes by the indicator. However, the runout measurement should focus on the entire shaft.

Goodheart-Willcox Publisher

Figure 18-50. A hammer and punch are used to adjust the fitting so the fitting and shaft are running in a true alignment.

Figure 18-51. Four evenly spaced spot welds are used to hold the slip spline fitting in the driveshaft's tube.

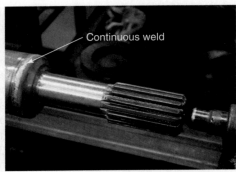

Figure 18-52. After both fittings are spot welded in place, the lathe is slowly turned while the fitting is welded in a single smooth pass.

specialist uses a torch to heat the high side in a small concentrated area, such as the area of a small coin. The heat causes the tube to increase in height. After it is heated, the tube is immediately cooled with coolant, which causes the tube to shrink. The specialist then rechecks the runout to determine if the shaft is within an acceptable limit, such as 0.005″.

Balancing a Driveshaft

Driveshaft specialists use a specially designed lathe for spin balancing a driveshaft. Large driveshafts are normally balanced at a speed no greater than 1500 rpm. A marker in a stationary fixture is used to mark the lathe spindle on the end of the shaft that is the most out-of-balance.

The driveshaft is brought to a stop. The specialist uses a rubber strap to secure a weight to the side of the driveshaft so it aligns to the largest portion of the mark made on the spindle. The balancer again spins the shaft to determine if it is still out-of-balance. The specialist adjusts the position of the weight and changes the size of the weight until the out-of-balance is minimized.

Once one end of the driveshaft is properly balanced, the procedure is repeated for the other end of the driveshaft. When the amount and locations of corrective weights have been determined, small welds are used to fix the weights to the shaft.

Lubrication

U-joints require periodic greasing. Drivelines may need to be greased every 250 hours, monthly, or at the manufacturer's recommended interval. Use only OEM-approved grease, which contains extreme-pressure (EP) agents and is rated as grade 1 or 2 by the National Lubricating Grease Institute.

Most U-joints are fitted with one or two grease zerks. To apply grease to the U-joint, slowly pump grease into one of the zerks. Continue applying grease until a small amount of grease pushes out from each of the four bearing cup trunnion seals. If grease does not push out from all of the seals, one of the bearing cups is not getting greased and will fail.

If a U-joint trunnion has trouble accepting grease, do the following:

- Use a pry bar to move the driveline back and forth while applying grease.
- If the U-joint has two zerks, grease the other zerk.
- If the U-joint has bearing-plate type cups, loosen the retaining bolts, lift the bearing cup out of the cross trunnion approximately 1/8″, and apply grease to the U-joint.

Power Take-Off (PTO) Driveshaft

Many off-highway machines incorporate a power take-off drive (PTO). The PTO drive allows the machine to mechanically drive another component. A dozer can use a PTO to drive a winch. As explained in Chapter 4, *Agricultural Equipment Identification*, agricultural

Chapter 18 | Axles and Driveshafts

tractors use PTOs to drive balers, mowers, augers, tillers, posthole diggers, and grain grinders.

PTOs require some type of clutch mechanism that allows the operator to engage the PTO drive. On late-model PTOs, multiple-disc clutches engage the drive. The input to the PTO clutch receives its power either directly or indirectly from the engine. This input rotates proportional to engine speed.

Agricultural tractor PTOs are designed to rotate at either 540 rpm or 1000 rpm. Small tractors and smaller horsepower applications use the 540 rpm. Many midsize tractors can be equipped with both a 540 rpm and 1000 rpm PTO. Large agricultural tractors equipped with a PTO are designed to rotate at 1000 rpm. The tractor's tachometer often has an indicator that shows the specific engine rpm that will achieve the specified PTO speed. See **Figure 18-53**.

Agricultural PTOs can be equipped with one of three shafts. These shafts vary in size and the number of splines:

- 540 rpm PTO, 1 3/8″ diameter shaft with 6 splines.
- 1000 rpm PTO, 1 3/8″ diameter shaft with 21 splines.
- 1000 rpm PTO, 1 3/4″ diameter shaft with 20 splines.

Goodheart-Willcox Publisher

Figure 18-53. The tachometer on this Case IH Farmall 75c tractor identifies two usable engine speeds (1100 rpm and 2300 rpm) and one unusable engine speed (3600 rpm) for the PTO. An operator can achieve the first two engine speeds by using range one or range two. Range three is used for road speed and not for operating a 540 rpm PTO.

 Warning

Always shut off the tractor before working on a PTO. Never attempt to couple a PTO when the tractor is running. A failed component, such as an electric switch or solenoid, can cause the PTO to suddenly engage. As a result, you could get caught in the shaft and be severely injured or killed.

Agricultural PTO shafts can be coupled to an implement using a bolt, spring-loaded pin, or sliding collar. A single bolt can be used to couple the PTO shaft to a gearbox. The bolt is a low grade bolt (also called a ***shear pin***) that is designed to fail when the implement encounters an obstruction. See **Figure 18-54**.

Spring-loaded pins can be used to couple a PTO shaft to a tractor. The spring holds the pin engaged. To couple the shafts, the spring must be pressed inward. The two shafts are then coupled, the pin is released, and the shaft is pulled rearward to ensure that the pin is locked in place. See **Figure 18-55**. If the spring-loaded pin is rusted or inoperative, replace the coupling.

Goodheart-Willcox Publisher

Figure 18-54. This PTO shaft's coupling is secured to the implement's gearbox with a low-grade bolt referred to as a shear pin. The shear pin is designed to fail if the implement encounters an obstruction.

Goodheart-Willcox Publisher

Figure 18-55. This PTO shaft has two spring-loaded pins. Either can be pressed, and then the coupler can be slid onto the stub shaft.

A sliding collar can also be used to couple a PTO shaft to a tractor. To couple the shafts, slide the collar rearward. Slide the splined coupling over the shaft and release the collar. After the collar is released, attempt to slide the PTO shaft rearward to ensure that the coupling is locked in place. See **Figure 18-56**.

Goodheart-Willcox Publisher

Figure 18-56. The PTO coupler contains an outside collar that must be slid inward before attempting to couple the driveshaft to the tractor's stub shaft.

Summary

- Axle housings use either a bar axle design or a flange mount design for mounting wheels to the axle.
- In a semi-floating axle, the weight of the machine is exerted onto the axle bearings, which is transferred to the axle shaft and to the drive wheel.
- The axle shaft of a full-floating axle is only responsible for driving the wheel. The weight of the machine is exerted on the axle housing and not the axle shaft.
- A live axle is a drive axle that propels the machine. A dead axle is a nondrive axle used only to support the weight of, and in some cases, steer the machine.
- Steer axles use a steering knuckle that pivots, allowing the machine to steer.
- Duo-Cone seals use two metal flat face seal rings and two rubber torics to form a seal between a fixed housing and a rotating housing.
- Drivelines consist of driveshafts, slip splines, yokes, U-joints, and center support bearings.
- U-joint designs include bearing plate, bearing strap, snap ring, U-bolt, and wing-type designs.
- A driveshaft (with conventional U-joints) increases and decreases its speed twice for every revolution due to the operating angle of the U-joint.
- A rear U-joint operates at a constant velocity if it has the same angle (in the opposite direction) as the front U-joint and if the yokes are operating in the same axis.
- A two-piece driveshaft must be in-phase (both yokes are matched in the same position). If out-of-phase, the output yoke will rotate at a nonuniform speed.
- U-joints have a maximum operating angle that depends on the size and speed of the U-joint.
- To calculate operating angles, if the angle contains two downward slopes, subtract the smaller slope from the larger slope. If the angles contain a downward and upward slope, add the slopes.
- Driveline vibrations can be caused by worn U-joints, worn driveline bearings, improper U-joint operating angles, suspension or driveline components and mounts that are worn or bent, and out-of-balance or bent driveshafts.
- PTO drive allows a machine to mechanically drive another component. Agricultural tractors use PTOs to drive balers, mowers, augers, tillers, posthole diggers, and grain grinders.

Technical Terms

- bar axle
- bearing plate U-joint
- bearing strap U-joint
- Bendix-Weiss U-joint
- Cardan U-joint
- center support bearing
- constant velocity U-joint (CV joint)
- cross
- dead axle
- double Cardan U-joint
- driveshaft
- Duo-Cone seal
- first order vibration
- four-wheel steer
- full-floating axle
- full-round yoke
- half-round yoke
- in-phase
- live axle
- out-of-phase
- pusher axle
- rigid axle
- runout
- Rzeppa U-joint
- second order vibration
- semi-floating axle
- shear pin
- slip splines
- snap-ring U-joint
- spider
- stand pipe
- steer axle
- tag axle
- toric
- truing a shaft
- universal joint (U-joint)
- wing-type U-joint
- wing-type yoke
- yoke

Review Questions

Answer the following questions using the information provided in this chapter.

Know and Understand

1. Technician A states that bar axles allow drive wheels to be repositioned by sliding the wheels inward or outward. Technician B states that wheels can be flipped to make tire spacing adjustments on flange-mounted axles. Who is correct?
   A. Technician A.
   B. Technician B.
   C. Both A and B.
   D. Neither A nor B.

2. Which machine often has one live axle?
   A. Rigid frame haul truck.
   B. Wheel loader.
   C. Articulated dump truck.
   D. Swather.

3. A _____ often has two live axles.
   A. rigid frame haul truck
   B. wheel loader
   C. motor grader
   D. swather

4. A(n) _____ often has three live axles.
   A. rigid frame haul truck
   B. wheel loader
   C. articulated dump truck
   D. swather

5. Technician A states that all off-highway machines are equipped with at least one steer axle. Technician B states that some off-highway machines are equipped with two live axles. Who is correct?
   A. Technician A.
   B. Technician B.
   C. Both A and B.
   D. Neither A nor B.

6. All of the following machines use rigid axles, *EXCEPT*:
   A. wheel loaders.
   B. articulated agricultural tractors.
   C. scrapers.
   D. telehandlers.

7. Technician A states that Duo-Cone installation tools vary in diameter. Technician B states that the special installation tool varies based on the seal ramp's angle. Who is correct?
   A. Technician A.
   B. Technician B.
   C. Both A and B.
   D. Neither A nor B.

8. All of the following are types of U-joints, *EXCEPT*:
   A. bearing strap.
   B. bearing plate.
   C. welded cup.
   D. wing-type.

9. Which type of U-joint is rarely used in off-highway equipment?
   A. Bearing strap.
   B. Bearing plate.
   C. U-bolt.
   D. Wing-type.

10. Which type of U-joint is commonly used in high-horsepower applications?
    A. Bearing strap.
    B. Bearing plate.
    C. U-bolt.
    D. High wing.

11. A double Cardan U-joint consists of _____.
    A. two ball yokes rather than conventional U-joint yokes
    B. a ball and cage
    C. two metal straps that are bolted around the circumference of the bearing cup
    D. two conventional U-joints inside a center yoke

12. When replacing a center support bearing, the old shims should be _____.
    A. replaced with 0.010″ thinner shims
    B. reused with the new bearing
    C. replaced with 0.010″ thicker shims
    D. double the thickness and install with new bearing

13. A two-piece driveshaft must be marked prior to removal to _____.
    A. ensure it is installed in-phase
    B. ensure proper operating angles
    C. Both A and B.
    D. Neither A nor B.

14. All of the following cause driveline vibrations, *EXCEPT*:
    A. bent mounts.
    B. in-phase yokes.
    C. out-of-balance driveshafts.
    D. worn U-joints.

15. A two-piece driveshaft has been installed out-of-phase. The driveline has no operating angles, meaning that it is rotating in a straight line. Technician A states that the driveline will have a vibration. Technician B states that if the driveshaft spins above 1500 rpm it will have a vibration. Who is correct?
    A. Technician A.
    B. Technician B.
    C. Both A and B.
    D. Neither A nor B.

16. A driveshaft _____ vibration is a first order vibration.
    A. out-of-balance
    B. out-of-round
    C. angularity problem
    D. Both A and B.

17. All of the following are PTO shafts used in agricultural tractors, *EXCEPT*:
    A. 540 rpm PTO, 1 3/8″ diameter shaft with 6 splines.
    B. 540 rpm PTO, 1 3/8″ diameter shaft with 20 splines.
    C. 1000 rpm PTO, 1 3/8″ diameter shaft with 21 splines.
    D. 1000 rpm PTO, 1 3/4″ diameter shaft with 20 splines.

18. All of the following are used to couple a PTO shaft to an implement, *EXCEPT*:
    A. a bolt.
    B. a cotter pin.
    C. a spring-loaded pin.
    D. a sliding collar.

Apply and Analyze

19. A(n) _____ axle has the weight of the machine exerted on the axle's shaft.

20. A(n) _____ axle is an axle shaft that does not support the weight of the machine, but simply drives the wheel.

21. The term _____ best describes a nonpowered axle (a nondrive axle).

22. A drive axle that receives power from the transmission and transfers the power to the drive wheels is called a(n) _____ axle.
23. As driveshaft speed increases, the maximum allowable operating angle _____.
24. Two rubber _____ applies pressure to the two metal face seals in a Duo-Cone seal.
25. A Duo-Cone seal functions by sealing at a total of _____ area(s).
26. The flat face of a Duo-Cone metal seal is placed flush against the flat face of the opposing _____.
27. _____ is lightly brushed onto the flat face of a Duo-Cone metal seal prior to installation.
28. A second order vibration occurs _____ during every revolution of the component.

Critical Thinking

29. Yoke A has a downward slope of 4°. The mating yoke has an upward slope of 6°. What is the operating angle of the U-joint?
30. Yoke A has a downward slope of 4°. The mating yoke has a downward slope of 6°. What is the operating angel of the U-joint?

Chapter 19
Hydraulic Brake Systems

Objectives

After studying this chapter, you will be able to:

- ✓ List the different types of brakes used in heavy equipment.
- ✓ Explain Pascal's law.
- ✓ Describe the operation of a master cylinders and hydraulic brake control valves.
- ✓ Explain the operation of hydraulic slack adjusters.
- ✓ List the types of mechanical and hydraulic brakes used in heavy equipment.
- ✓ Describe the operation of a gas accumulator as it is used in a brake circuit.
- ✓ List the order of fluid flow through a brake cooling circuit.
- ✓ Describe methods for towing a vehicle with a spring-applied parking brake.
- ✓ Describe how to adjust an internal wet disc brake compensator.

Brakes are applied when the operator presses the service brake pedal. They use friction-lined pads, shoes, bands, or plates to slow a rotating drum or disc, which converts moving energy into heat energy. The friction lining is often on the stationary component, such as the shoe, pad, or band. However, in some brake designs, it is on the rotating component, such as the rotating brake disc.

Brake Types

Heavy equipment is equipped with both service brakes and parking brakes. *Service brakes* are considered dynamic brakes because they are used to bring a moving machine to a stop. If the machine has two brake pedals that can be locked together, the individual pedals are also known as *turning brakes*. They get this name because applying just one brake (left or right) can help the machine make a tight turn. See **Figure 19-1**.

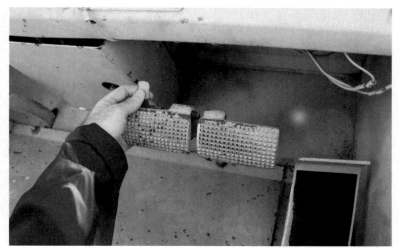

Goodheart-Willcox Publisher

Figure 19-1. This image shows the pin that joins the left and right service brake pedals. When the pin joins the two pedals, the operator applies both left and right service brakes by pressing only one pedal.

Figure 19-2. This Case forklift uses a manual lever to apply the parking brake. The parking brake is shown in the applied position.

A *parking brake* is considered a static brake because it is used to hold a stationary machine in a parked position, not to slow a moving machine. It is applied when the transmission is in a neutral position. Parking brakes are applied when the operator engages the parking brake switch, lever, or pedal. See **Figure 19-2**.

A parking brake inside a skid steer's hydraulic motor is shown in **Figure 19-3**. The brake discs and steel plates both lack friction material, as the brake is used only for static holding and not dynamic braking. A Belleville washer normally provides the spring pressure to apply the brake. Hydraulic fluid pressure is required to actuate a piston to overcome the Belleville washer spring tension allowing the skid steer to move.

The parking brake and service brakes can be in separate locations on a machine. In **Figure 19-4**, the forklift's parking brake is located at the rear of the transaxle housing, and is used to hold the transaxle's output shaft stationary. The right side service brake (turning brake) is located on the right side of the transaxle. The left side service brake is located on the left side of the transaxle.

Some machines use a combination brake. In a combination brake, the friction plates are used for both the service brakes and parking brakes, but different pistons are used to apply the service brake and the parking brake. An example is the brake on Case IH midrange 2100, 2300, 2500, 88 series, 130 series, and 140 series combines. The brake is also used in Case IH and New Holland flagship combines with the common ground drive two-speed transmission. The brake contains wet discs that can be applied with a parking brake piston by a spring, or with a service brake piston by means of fluid pressure. See **Figure 19-5**. Wet disc brakes and towing machines with spring-applied parking brakes will be explained later in this chapter.

Different braking systems use different methods to apply a machine's brakes, including:
- Mechanical leverage.
- Air pressure.
- Hydraulic pressure.

Mechanically actuated brakes use mechanical linkage that is actuated by the operator to apply the brakes. An example of this type of brake is the mechanically operated ball

Figure 19-3. This Rexroth MCR hydrostatic motor is used to propel a skid steer. It contains a spring-applied parking brake. Notice neither the discs (internal splines) nor the steel plates (external splines) have friction material. The parking brake is used only as a static holding brake and is not designed to be a dynamic service brake.

Figure 19-4. This Case forklift uses a manually applied parking brake that squeezes two brake disc pads to hold the parking brake disc (rotor) stationary. The parking brake is used to hold the transaxle's output shaft. The individual service brakes are located on each side of the transaxle.

ramp disc brake. Ball ramp disc brakes are explained later in this chapter. Mechanical brakes are used for service brakes in low-horsepower applications and for parking brakes. They are not commonly used for service brakes in large heavy equipment because the leverage is too small to stop heavy machinery.

Hydraulic brakes use fluid pressure to apply a piston, which actuates the brake mechanism. Fluid pressure can be developed by a mechanically operated master cylinder or an engine-driven positive-displacement pump. Hydraulic brakes are the most common type used in heavy equipment. Air brakes are used on some large heavy equipment, such as wheel loaders, haul trucks, and scrapers. Air brake systems are explained in Chapter 20, *Air Brake Systems*. Electric drive machines can use dynamic braking. It occurs when electrical energy from coasting electric wheel motors is sent to braking resistors that convert it to heat energy. This causes the electric motors to act like generators under load, providing braking action. Electric drive systems are explained in Chapter 26, *Electric and Hybrid Drive Systems*.

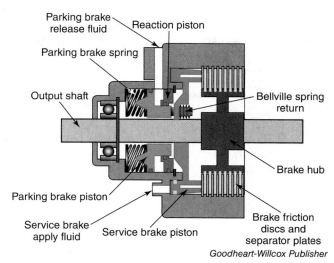

Figure 19-5. Case IH midrange 2100, 2300, 2500, 88, 130 and 140 series combines, and Case IH and New Holland flagship combines with the common ground drive two-speed transmissions have one brake assembly on each side of the transmission for both the parking brake and the service brake.

Hydraulic Principles

Hydraulic brakes work on the principle of ***Pascal's law***, which states that when a force is placed on confined liquid in a container, pressure develops, and that pressure acts equally in all directions. See **Figure 19-6**. It is this pressure that provides the foundation for hydraulically applied brakes.

A simple hydraulic system contains a small-diameter input cylinder and a large-diameter output cylinder. See **Figure 19-7**. When a small force is applied to the input cylinder's piston, a pressure develops in the system. When this pressure acts on an output cylinder with an area larger than the input cylinder, it causes the piston in the output cylinder to move with a larger output force. The pressure developed in the system is applied to the output cylinder's larger surface area, equaling a larger force.

To compute the output force, you must first find the cross-sectional areas of the pistons. Two formulas can be used to calculate the cross-sectional area of a piston:

$$\text{Area} = D^2 \times 0.7854$$

D = the diameter of the piston (See **Figure 19-8**).

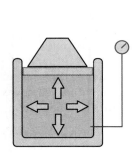
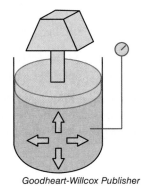

Figure 19-6. Pascal's law states that fluid in a container will exert pressure equally in all directions when a force is applied to that confined body of fluid.

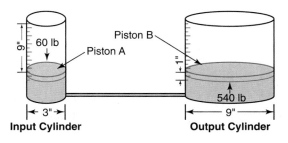

Figure 19-7. A simplified hydraulic system contains a small-diameter input cylinder and a large-diameter output cylinder. The cross-sectional area of the output cylinder is nine times larger than the cross-sectional area of the input cylinder, so the input force is multiplied nine times.

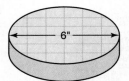

Figure 19-8. To find the area of a circle using the diameter, square the diameter (multiply the diameter times itself) and multiply the result by 0.7854.

 Pro Tip

Some prefer to use the πr^2 formula for computing the area of a circle. However, a piston's diameter is often given instead of its radius. To use the πr^2 formula, the radius will need to be found from the diameter. For example, if a piston has a diameter of 2.75″, then the radius can be found by calculating 2.75″ × 0.5 = 1.375″ or 2.75″ ÷ 2 = 1.375″.

$$\text{Area} = \pi r^2$$

r = the radius of the piston

π = 3.14

Determine the cross-sectional area of the input piston (piston A) in **Figure 19-7**.

$$\text{Area} = D^2 \times 0.7854$$
$$\text{Area} = 3^2 \times 0.7854$$
$$\text{Area} = 9 \times 0.7854$$
$$\text{Area of piston A} = 7.07 \text{ in}^2$$

The output force cannot be determined unless the input force is known. For this example, 60 pounds of force is applied to the input cylinder's piston (piston A).

To determine the potential pressure, divide the input force by the cross-sectional area of the input cylinder's piston. The pressure will only reach its full potential (maximum pressure) when the output cylinder is loaded at its maximum weight. Therefore, if the output cylinder has a smaller load (weight), the pressure will not reach its maximum potential.

$$\text{Pressure} = \text{Input force}/\text{Area of piston A}$$
$$\text{Pressure} = 60 \text{ pounds}/7.07 \text{ in}^2$$
$$\text{Potential pressure} = 8.49 \text{ psi}$$

Next, determine the cross-sectional area of piston B.

$$\text{Area} = D^2 \times 0.7854$$
$$\text{Area} = 9^2 \times 0.7854$$
$$\text{Area} = 81 \times 0.7854$$
$$\text{Area of piston B} = 63.62 \text{ in}^2$$

The potential output force can now be determined by multiplying the potential pressure by the output piston's cross-sectional area. Do not be confused regarding which piston area should be used. When computing the potential output force of the output piston, use the cross-sectional area of the output piston.

$$\text{Force} = \text{Pressure} \times \text{Area}$$
$$\text{Force} = 8.49 \text{ psi} \times 63.62 \text{ in}^2$$
$$\text{Potential output force} = 540 \text{ lb}$$

For this example, notice that the input force of 60 lb was multiplied nine times to get an output force of 540 lb. This multiplication effect is the result of the output cylinder's cross-sectional area being nine times the size of the input cylinder's cross-sectional area.

 Note

Considering the example in **Figure 19-7**, a higher output force could be achieved by one or more of the following means:

- Using a stronger input force (more than 60 lb).
- Using an input cylinder with a smaller diameter (smaller than 3″).
- Using an output cylinder with a larger diameter (larger than 9″).

The multiplication of force comes at the cost of a reduced output stroke, or shortened travel distance. In **Figure 19-7**, the input cylinder is pushed a distance of 9″, but the distance the output cylinder travels is reduced nine times, resulting in a stroke of 1″.

Hydraulic Brake Controls

Manufacturers use one of two methods to apply hydraulic brakes, brake control valves and brake master cylinders. Both have been used for decades and both are still being used in late-model products.

Master Cylinder Brakes

A *master cylinder* consists of an actuator assembly with a small hydraulic piston that acts as the input cylinder to the brake's hydraulic system. See **Figure 19-9**.

As the operator presses the service brake pedal, the brake pedal linkage actuates the master cylinder piston, which forces fluid to act on the actuating piston in the service brakes. The actuating cylinder is known as the output cylinder, *wheel cylinder*, or *slave cylinder*. In drum brakes, the wheel cylinder has two pistons that apply pressure to the brake shoes. See **Figure 19-10**. A master cylinder can be used to actuate one or more wheel cylinders.

The brake pressure is low when the piston in the wheel cylinder is initially extended. As the brake shoes press against the rotating drum, resistance increases, causing the pressure to increase and the drum to slow. Pressure continues to increase as the brake pedal is pressed farther. When the shoes can move no farther, the drum slows to a stop. See **Figure 19-11**. Master cylinders can be used to actuate different types of brakes, such as drum, ball-ramp disc, and rotor disc brakes.

A single master cylinder has three fluid ports connected to the cylinder's bore, **Figure 19-9**:

- A tapered port between the master cylinder and the reservoir, known as the compensating port, or vent port.
- A straight drilled port between the master cylinder and reservoir, known as the replenishing port, inlet port, bypass port, filler port, or breather port.
- An outlet port that connects the master cylinder to the brake's wheel cylinder.

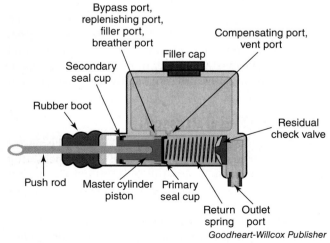

Figure 19-9. This single master cylinder has a mechanically operated piston that directs fluid to a wheel cylinder.

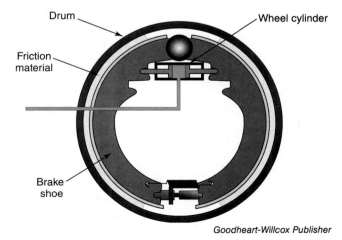

Figure 19-10. This drum brake uses a hydraulically actuated wheel cylinder to push two brake shoes against a rotating drum.

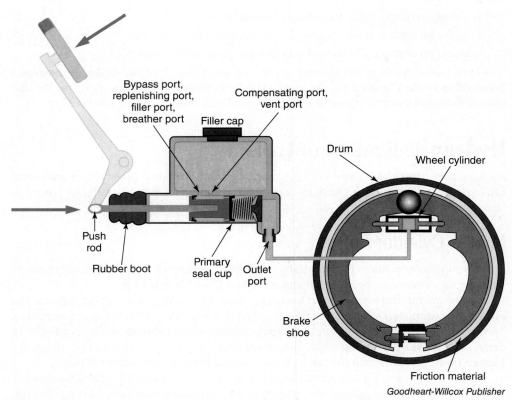

Figure 19-11. As the operator depresses the service brake pedal, the master cylinder piston pushes fluid to extend the wheel cylinder's pistons.

The master cylinder piston has two cup seals, a primary and a secondary seal. The primary cup seal is located on the front of the piston and is the sealing mechanism that enables the piston to develop pressure as it is extended. The secondary cup seal prevents the fluid from leaking around the master cylinder piston and falling behind the piston, which could then leak out around the push rod.

A flexible rubber boot is attached to the rear of the master cylinder and the push rod. The boot keeps dirt and other contaminants from getting into the master cylinder through the small gap between the push rod and master cylinder body.

Master Cylinder Operation (Brake Pedal at Rest)

When the master cylinder is at rest, the flexible primary seal cup is positioned between the compensating port and the replenishing port, **Figure 19-9**. The *compensating port* enables fluid in the outlet port to compensate and flow back into the reservoir as the fluid temperature rises and the fluid expands. If the master cylinder did not have the compensating port, an increase in fluid temperature and pressure could apply the brakes, even if the operator's foot was off the pedal.

Master Cylinder Operation (Brake Pedal Pressed)

As the operator presses the brake pedal, the linkage causes the push rod to push the master cylinder piston into its bore. The piston and primary seal cup move past the compensating port blocking off the reservoir. The primary seal cup pushes the fluid through the master cylinder's bore. The fluid moves through the residual check valve and out of the outlet port to extend the wheel cylinder's pistons and apply the brakes.

As the master cylinder piston is pressed into the bore, fluid is free to flow between the reservoir and the area behind the head of master cylinder piston through the *replenishing port*. This maintains equalized pressure between the reservoir and the area behind the piston.

Master Cylinder Operation (Brake Pedal Released)

Once the brake pedal is released, a brake pedal spring returns the brake pedal to the fully released position. The spring in the bore of the master cylinder pushes the master cylinder piston to its retracted position. As the piston is retracted, the fluid between the primary and secondary cup seals can pass to the reservoir or pass across the primary cup seal to the front side of the master cylinder piston.

If the brake pedal is released slowly, the fluid behind the piston head flows to the reservoir through both the replenishing port and compensating port until the compensating port is blocked off. Then, the majority of the fluid moves to the reservoir through the replenishing port. When the piston continues to retract (opening the compensating port in front of the primary seal), a low pressure is developed in the master cylinder, allowing fluid from the wheel cylinder to open the residual check valve and flow through the check valve and compensating port into the reservoir.

If the brake pedal is released quickly, the master cylinder spring forces the piston to retract quickly. The fluid behind the primary cup seal passes across the primary cup seal, causing the lip of the seal to fold inward, which allows the fluid to pass by the front of the piston. Some master cylinder pistons have small drilled passageways that help route the fluid past the primary piston cup seal.

With the brake pedal in a released position, the master cylinder bore remains full of fluid and ready for the next brake application. When the brake is in the released position, the residual check valve holds a small amount of pressure (6 to 18 psi) in the brake line leading to the wheel cylinder. This residual pressure keeps the wheel cylinder seals sealed, preventing air intrusion and fluid leaks at the wheel cylinder. Late-model master cylinders may not have a residual check valve due to the wheel cylinder cup seal design, which is able to hold its seal at a lower pressure.

Types of Master Cylinders

A variety of master cylinder designs are used in heavy equipment:
- Single master cylinder.
- Tandem master cylinder.
- Dual master cylinder.

Single master cylinders have just one piston, **Figure 19-12**. They can be found in some older, smaller machines, such as forklifts and loader backhoes.

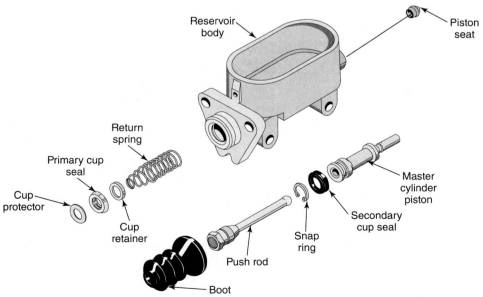

Case New Holland Industrial

Figure 19-12. This single master cylinder is one of two used on an older Case 586E forklift.

A *fluid reservoir* is used to supply fluid to the master cylinder piston. The fluid reservoir can be attached directly to the master cylinder housing located on top of the master cylinder, or it can be located remotely from the master cylinder. The Case IH 1600 series, 7230–9230 series, and 7240–9240 series combines use a single brake reservoir that supplies fluid to two single master cylinders.

Remote master cylinder reservoirs can also receive fluid from a sealed main hydraulic reservoir or the main reservoir's return line. The sealed reservoir builds a small amount of pressure due to fluid circulation and heat that causes the fluid to expand. This small amount of pressure is sufficient to keep the master cylinder reservoir charged with fluid. See **Figure 19-13**.

A *tandem master cylinder* contains two pistons that are located one in front of the other. As the service brake pedal is pressed, the push rod acts on both pistons. This master cylinder design contains two separate fluid reservoirs, each dedicated to its own master cylinder piston.

Since 1967, tandem master cylinders have been required in light- and medium-duty on-highway trucks. The dual circuit enables the operator to apply service brakes with one of the circuits even if the other circuit has a problem, such as a leak. See **Figure 19-14**.

A machine equipped with *dual master cylinders* has two separate single master cylinders. The right master cylinder operates the right-side brake (turning brake) and the left master cylinder operates the left-side brake (turning brake). Forklifts and loader backhoes are two examples that use this application, with the brakes located in the drive axle. See **Figure 19-13**.

If the backhoe is four-wheel drive, then four-wheel braking is achieved any time the front-wheel drive clutch in the transmission is engaged, which is the same effect explained

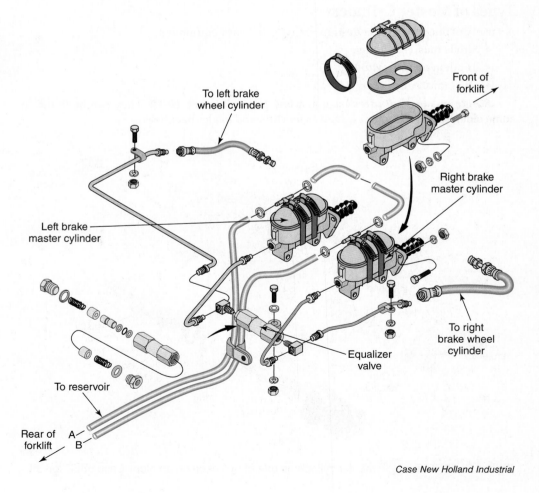

Figure 19-13. This Case 586E forklift's brake master cylinders receive a supply of fluid from the main hydraulic system's hydraulic reservoir.

Case New Holland Industrial

in Chapter 10, *Powershift and Automatic Transmission Theory*, regarding agricultural MFWD tractors.

Compensation Valve

A machine equipped with dual master cylinders can use one or more compensation valves. The *compensation valve*, also known as an equalizer valve, connects the two separate pressures from the master cylinders in order to equalize them. This allows the same amount of pressure to be applied to each brake when the master cylinders are engaged simultaneously.

When only one brake pedal is pressed, the compensation valve isolates the brake pressure so that the single master cylinder can apply its brake, either the right turning brake or the left turning brake, depending on the pedal pressed. **Figure 19-15** provides an example of dual master cylinders with compensation valves used in a Caterpillar 416C loader backhoe.

Combination Master Cylinder and Hydrostatic Transmission Inching Valve

A master cylinder can be combined with a hydrostatic transmission inching valve. This allows the service brake pedal to operate both controls. The hydrostatic transmission inching valve, when actuated, drains the hydrostatic system signal pressure. This prevents the hydrostatic transmission servo pistons from upstroking to propel the machine. The valve acts like a hydrostatic transmission inching pedal, enabling the machine travel speed to be reduced without affecting engine speed (refer to Chapter 13, *Hydrostatic Drives*). The ability to maintain high engine speed provides high hydraulic flow to maximize the functions of the loader. See the example of the master cylinder in a Caterpillar 902 compact wheel loader shown in **Figure 19-16**. The first 1″ to 2″ of brake pedal travel causes the hydrostatic transmission inching spool valve to drain hydrostatic drive system signal pressure to the reservoir.

The rest of the brake pedal travel causes the brake piston to act as in a normal master cylinder. As the pedal continues to be depressed, the piston directs fluid to apply the front axle's service brakes. Pressure is maximized as the brake pedal is fully depressed. When the operator releases the brake pedal, the springs return the piston and spool to the retracted position.

Boost-Assist Master Cylinders

Some machines are equipped with a master cylinder that uses a boost piston to increase the brake pressure. The boost piston receives fluid pressure from another source, such as a hydraulic pump. This type of master cylinder is a *boost-assist master cylinder*. **Figure 19-17** shows an example used in Challenger 35, 45, and 55 series agricultural rubber-track tractors. In this application, only one brake pedal and one master cylinder are used.

When the brake pedal is in the at-rest position, the two return springs keep the master cylinder plunger and the boost piston retracted. The *center valve* allows fluid from the

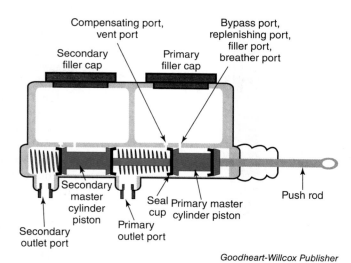

Goodheart-Willcox Publisher

Figure 19-14. Tandem master cylinders are used in on-highway applications to provide separate dual circuits, for example a primary circuit for front disc brakes and secondary circuit for rear drum brakes.

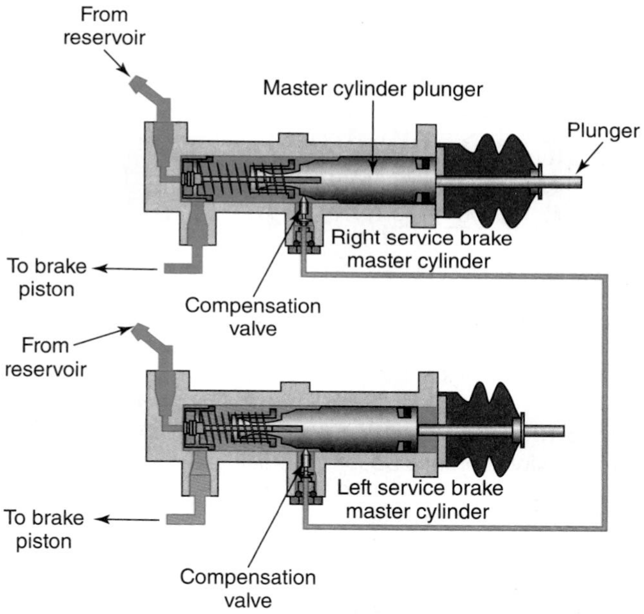

Goodheart-Willcox Publisher

Figure 19-15. In this example, the left brake master cylinder has been actuated, causing the left brake to apply. Notice the compensation valve in the left brake master cylinder has opened to direct fluid to the right brake master cylinder. However, the right brake master cylinder's compensation valve is closed, blocking brake pressure from the left brake master cylinder.

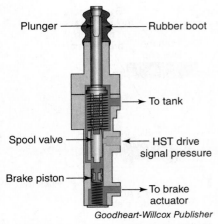

Figure 19-16. A Caterpillar 902 compact wheel loader's master cylinder drains the hydrostatic drive signal pressure during the initial movement of the brake pedal and applies the brakes as the brake pedal is pressed farther.

outlet port to charge the master cylinder's reservoir. As the input plunger is actuated, it presses against the boost piston. As the boost piston moves, inlet fluid from the transmission control circuit acts on the boost piston. This force is combined with the force applied by the input plunger to increase the brake pressure. The boost pressure varies based on the leakage between the tip of the input plunger and the boost piston's cross-drilled passage. When the plunger is placed directly against the piston, little leakage occurs, resulting in the highest boost-assist pressure.

The fluid to the left of the boost piston is low-pressure fluid and is directed into the master cylinder's reservoir. After the reservoir is charged full of fluid, the remaining fluid is drained to the tank. The *flapper valve* keeps the boost piston chamber full of fluid. During fast brake pedal applications, the flapper valve remains open to allow fluid from the low-pressure left side of the piston to charge the boost fluid cavity on the right side. When there is sufficient pressure on the right side of the boost piston, the flapper valve closes.

As the master cylinder plunger is pushed to the left, the center valve closes off the passageway to the reservoir, building pressure in the fluid being directed out of the master cylinder to apply the brakes. See **Figure 19-18**.

The brakes can still be applied when the engine is not running. However, the master cylinder will only produce the pressure applied by the brake pedal since the transmission pump cannot supply fluid to the boost piston.

Master Cylinder Fluid Types

Master cylinders normally use one of two types of fluid, either traditional machine hydraulic fluid or DOT brake fluid. The two fluids cannot be mixed. It is important to use the fluid specified in the owner's manual.

If the machine uses traditional hydraulic fluid, the owner's manual will specify the fluid type, such as John Deere Hy-Gard, Case IH Hy-Tran, or Caterpillar HYDO. If the master cylinder is designed for brake fluid, the owner's manual will specify the type, such as DOT 3, DOT 4, DOT 5, or DOT 5.1.

DOT 3, DOT 4, and DOT 5.1 are glycol-based brake fluids and are *hygroscopic*, which means they will absorb water. If a small amount of water enters the brake system, the fluid will absorb it. For example, if the brake fluid is left unsealed, it will absorb water at a rate of 2% by volume from the atmosphere, which reduces the fluid's boiling point. When brake temperatures rise, the fluid's temperature increases and moisture in the system vaporizes, causing the brakes to become dangerously spongy. Therefore, it is important to avoid leaving DOT 3, DOT 4, and DOT 5.1 brake fluid containers unsealed to the atmosphere.

Examples of brake fluid boiling points are:

- Dry DOT 3 (new, unopened fluid, with no water) = 400°F
- Wet DOT 3 (with 3.7% absorbed water) = 284°F
- Dry DOT 4 (with no water) = 446°F
- Wet DOT 4 (with 3.7% absorbed water) = 311°F

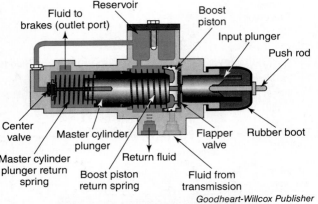

Figure 19-17. This master cylinder uses fluid from the transmission circuit to boost the brake pressure.

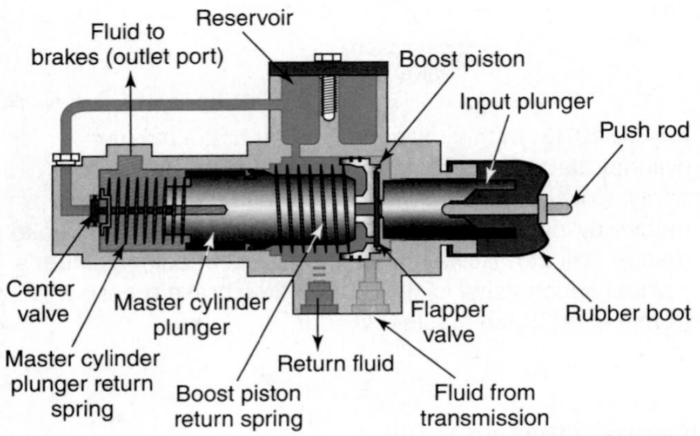

Figure 19-18. As the brake pedal is pressed, the boost piston uses fluid from the transmission circuit to boost the fluid pressure used to engage the brakes.

Note

Some agricultural dealerships service ATVs, and some ATVs require DOT 5 brake fluid. DOT 5 is a silicone-based brake fluid that is **hydrophobic**, or non-hygroscopic, meaning that it repels water instead of absorbing it. Special care must be taken to prevent water from entering the system, or corrosion will occur. DOT 5 fluids are prone to absorbing a small amount of air, making the fluid more compressible. DOT 5 fluids should not be used in anti-lock braking systems.

Caution

Handle DOT 3, DOT 4, and DOT 5.1 brake fluids with care because they will damage paint. A technician who is not careful could harm a machine's paint finish, causing customer dissatisfaction.

Hydraulic Service Brake Control Valves

Many late-model hydraulically actuated brakes use a hydraulic pump as the source for the brake system pressure along with a hydraulic brake control valve. The pump may be dedicated solely to the brake system, or it may be used for other circuits, such as steering, transmission, pilot controls, or implement hydraulics. The pump must be a positive-displacement pump so it can build sufficient pressure to apply the brakes. As mentioned in Chapter 4, *Agricultural Equipment Identification*, a positive-displacement pump uses tight sealing surfaces to produce a constant flow of fluid for a given revolution. The pump loses little fluid flow when pressure increases, as compared to a non-positive-displacement pump, which uses an impeller and works on the principle of centrifugal force. Non-positive-displacement pumps lose flow when they encounter pressure, and for that reason, are not used for applying brakes.

The hydraulic pump can be a fixed-displacement pump or a variable-displacement pump. A fixed-displacement pump produces a specific volume of fluid for each revolution. Fixed-displacement pump circuits use a relief valve to prevent excess pressure from building in the circuit. A variable-displacement pump varies the flow. For example, it decreases the system flow when flow is not needed.

A *service brake control valve* directs pump flow to apply the brakes when the operator depresses the brake pedal. The service brake control valve can be located away from the brake pedal and mechanically actuated, or it can be included with the brake pedal. Some wheel loaders use two separate brake pedals (right and left) that are connected through a linkage to actuate a single brake valve. Variations of the simple brake service valve shown in **Figure 19-19** are commonly used in Caterpillar machines, including medium and large wheel loaders, wheel dozers, and motor graders.

In this application, the valve has two separate supply lines, providing the advantage of two separate circuits. If a problem occurs in one of the circuits, the other circuit will continue to provide braking. In wheel loaders, wheel dozers, and trucks, one circuit is used for the front brakes and the other circuit is used for the rear brakes. Both circuits are controlled by one brake pedal. In motor graders,

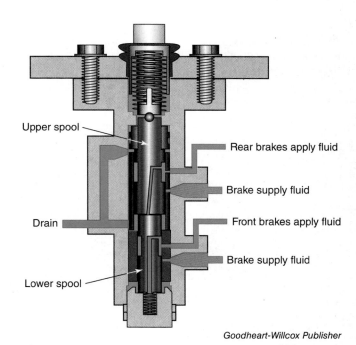

Goodheart-Willcox Publisher

Figure 19-19. An example of a foot-operated brake control valve with two separate supply lines. The valve controls both the front and rear brakes.

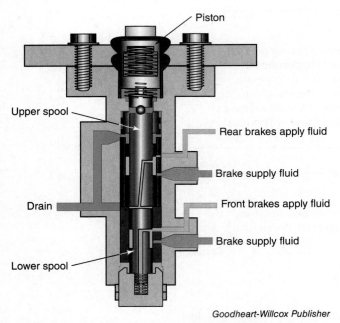

Figure 19-20. The brake pedal has been depressed causing the two spools to shift downward, which allows fluid to apply the front and rear brakes.

one circuit is used to apply the right tandem service brakes and the other is used to apply the left tandem service brakes. The term *tandem*, as applied to motor graders, describes the pair of tires driven on the right rear side of the machine by two chains, and another pair of tires driven by chains on the left rear of the machine.

In **Figure 19-19**, the brake pedal has been released. The valve contains two spools, an upper spool and a lower spool. The upper spool controls the fluid to apply the rear brakes, and the lower spool controls the fluid to apply front brakes. Springs hold the two spools in a centered position to block the supply fluid from applying the brakes.

When the operator depresses the brake pedal, the piston pushes the springs, which causes the upper and lower spools to shift downward. As the operator depresses the pedal farther, the spools shift farther, creating a larger opening and applying more fluid pressure to the front and rear brakes. See **Figure 19-20**. Notice that both the upper and lower spools have cross-drilled passageways that allow the brake apply pressure to be exerted on the bottom of the spools. This pressure is felt as resistance in the brake pedal allowing the operator to sense the application of the brakes.

A hydraulic schematic depicting a dual circuit brake control valve is shown in **Figure 19-21**. Notice that two separate supplies of fluid are used to feed the single brake control valve. Each supply has its own *accumulator*, which is a hydraulic energy storage device that provides a supply of hydraulic fluid to apply the brakes in the event of an engine or hydraulic pump failure. The two check valves prevent the accumulators from driving the hydraulic pump like a hydraulic motor when the engine is off. Accumulators are discussed later in this chapter.

Figure 19-21. Two separate supplies of fluid feed the single brake control valve.

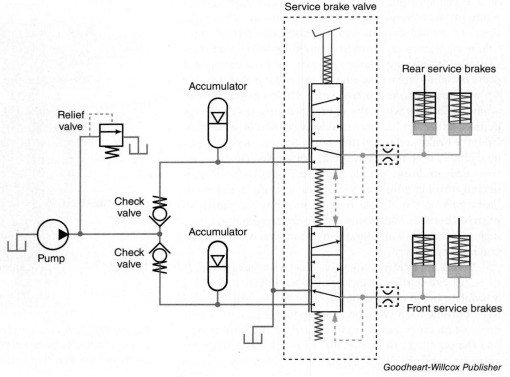

Hydraulic Slack Adjusters

Brakes will wear over time. A *hydraulic slack adjuster* is used, as its name implies, to compensate for the brake slack that occurs due to brake wear. Brake slack is the additional distance the brake wheel cylinder (brake actuator) must move before the brake begins to slow the wheel. The hydraulic slack adjuster is located between the brake control valve or master cylinder and the brake assembly. See **Figure 19-22**. Note that if the slack adjuster is used in conjunction with a master cylinder, the master cylinder is an air and hydraulic actuator, which will be explained in Chapter 20, *Air Brake Systems*.

Two hydraulic slack adjusters can be located in one slack adjuster housing and placed on an axle to supply fluid to both the left and right brakes. A slack adjuster can also be placed at each wheel assembly and used to supply fluid directly to each brake. Large rigid frame haul trucks are an example of machines that uses hydraulic slack adjusters.

Each slack adjuster is filled with fluid and contains two pistons, a large piston and a small piston. As the operator presses the service brake pedal, fluid from the brake valve enters the slack adjuster and pushes the large piston, causing it to move in conjunction with its large spring. As the large piston extends, it pushes fluid out of the slack adjuster to apply the brake. See **Figure 19-23**.

When brakes are new, the large piston can apply the brake without reaching the end of the slack adjuster housing (**Figure 19-23A**). As the brake discs wear, the large piston will move farther inside the bore of the slack adjuster. While the operator presses the brake pedal, pressure builds inside the slack adjuster. If the large piston reaches the end of its travel and the pressure inside the slack adjuster rises higher than the brake-application pressure at the brake wheel cylinder (brake actuator), the small piston will actuate. The small piston will move inward (toward the center of the slack adjuster), opening a passageway that allows fluid flow from the brake valve to pass through the slack adjuster and apply the brake. See **Figure 19-23B**.

When the brake pedal is released, the brake piston (located at the wheel) is retracted by its return spring, which forces fluid to flow back into the slack adjuster. Brake piston operation in a multiple disc wet brake is explained later in this chapter. The fluid entering the slack adjuster causes the large piston to retract, compressing the large spring. As the large piston is retracted, it forces fluid to flow back to the brake valve or master cylinder. See **Figure 19-24**.

The slack adjuster's large spring keeps a small amount of fluid pressure held against the brake's piston or wheel cylinder. However, the brake cooling fluid pressure overcomes this pressure to prevent brake wear. The result is that, as the operator presses the service brake pedal, the brake piston is already extended and ready to make a quick braking application without having to make up slack due to brake wear.

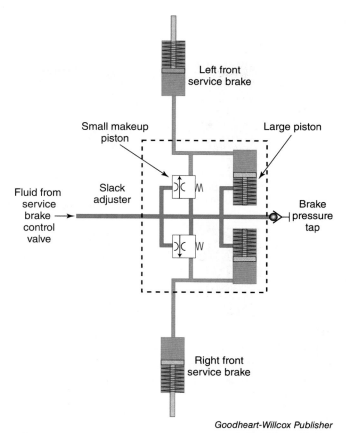

Goodheart-Willcox Publisher

Figure 19-22. This schematic shows a single housing that contains two slack adjusters for the right and left front brakes.

Types of Mechanical and Hydraulic Brakes

Off-highway machines use different types of mechanical brakes and hydraulic brakes, including:
- Drum brakes.
- Ball ramp brakes.
- External caliper brakes.
- Internal wet disc brakes.

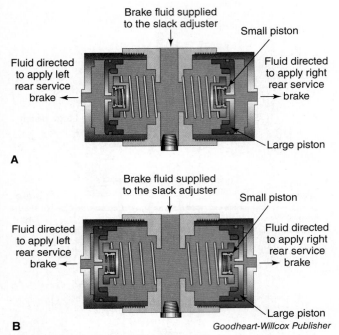

A

B

Figure 19-23. This slack adjuster is used to control both the left and right brakes on one axle. A—Slack adjuster shown with the brakes fully applied without the large piston reaching the end of its travel. B—Slack adjuster shown with the brakes applied and the large piston at the end of its travel. The small piston has moved inward toward the center of the slack adjuster.

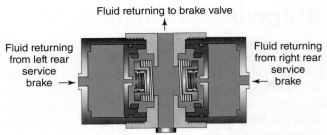

Figure 19-24. This slack adjuster is shown with the brakes released.

Off-highway machines use a variety of different parking brake designs. Some examples include multiple wet disc brake, drum brake, and external caliper parking brake. Many off-highway machines have a dedicated parking brake either inside the transmission or directly outside the transmission on the output shaft. When the parking brake is engaged, the output shaft is held. Since the output shaft is connected to the axles through the drive shaft(s), the engaged parking brake holds the machine stationary. Examples of machines with this type of brake include motor graders, agricultural tractors, and wheel loaders.

Other off-highway machines have a brake assembly that serves as both the service brake and the parking brake. The dual brake can be placed between the differential and the drive axle. Examples include the ball ramp brake used on some compact agricultural tractors. Multiple disc combination brakes are used on some combines and are located on the transmission's output shafts. Other machines, like rigid frame haul trucks, have a combination park and service brake at each wheel.

 Note

A combination service brake rotor/parking brake drum is attached to transmission output shaft on a Caterpillar 528B wheel skidder. See **Figure 19-25**. The drum brake is used as a parking brake and secondary brake assembly. The parking brake should only be applied as a secondary brake if the service brakes fail. The parking drum brake assembly is spring applied and released with air pressure.

The rotor brake assembly is called a driveline service brake. It is engaged after the wheel service brakes are fully applied, if additional service brake stopping power is required. The driveline service brake is engaged by the same foot-operated service brake control valve as the wheel service brakes. After the operator presses the pedal to fully engage the wheel service brakes, he or she can press the pedal farther to engage the driveline service brake. The driveline service brake's hydraulic caliper is actuated by a hydraulic master cylinder with an air chamber. Air brakes are explained in Chapter 20, *Air Brake Systems*.

Drum Brakes

Drum brakes contain two brake shoes that are wedged against a rotating drum. Refer to **Figure 19-11**. A hydraulic wheel cylinder contains two pistons inside a bore. As the master cylinder is actuated, fluid extends the pistons inside the wheel cylinder, which forces the

Figure 19-25. A Caterpillar 528B wheel skidder has a combination brake drum/rotor assembly attached to the transmission's output shaft. A—The spring-applied parking brake applies the drum brake's shoes. Air pressure must be used to compress the spring to allow the machine to move. B—The rotor is part of the driveline pneumatic-hydraulic actuated service brake that is applied after the wheel service brakes are fully actuated.

brake shoes to apply pressure to the rotating cast drum. Friction material attached to each brake shoe slows the drum. The hydraulically actuated drum brake is not as common as the internal wet disc brake. However, John Deere combines (specifically 100 series, 10 series, 50 series, 60 series, 70 series, and small S series, such as S550 and S660) and light-duty trucks use hydraulically actuated drum brakes. See **Figure 19-26**.

Drum brakes have different designs, including nonservo and duo-servo. The **nonservo drum brake** is also known as the **leading-trailing shoe drum brake**. The bottoms of the shoes are anchored to the backing plate by a pin or an anchor strap. Although both shoes are simultaneously extended by the wheel cylinder's pistons, only one shoe slows the drum, based on the direction of machine travel.

When the brake is applied while the machine is moving forward, the leading, or front, shoe slows the drum. As the wheel cylinder moves the shoes, the top of the leading shoe (front shoe) pushes forward against the drum, creating friction and providing the braking action. As the drum rotates, it attempts to rotate leading shoe (front shoe). However, the shoe is anchored at the bottom and cannot rotate. Therefore, as the drum rotates the leading shoe, it further wedges the shoe against the drum, providing more braking. This effect is called *self-energizing braking*. See **Figure 19-27**.

Figure 19-26. This John Deere 9600 combine has the right drive wheel removed, making it easier to see the three-speed mechanical transmission with the right drum brake. Notice the drive axle protruding from the drum brake. It delivers power to the outboard final drive, as explained in Chapter 17, *Final Drives*.

The trailing, or rear, shoe is pressed against the drum with only the force applied by the wheel cylinder. As a result, it generates significantly less braking action than the self-energizing leading shoe.

When moving in reverse, the trailing shoe provides the braking action while the leading shoe de-energizes. The John Deere 8570–8970 four-wheel drive agricultural tractors with a power shift transmission had a parking brake that used a mechanically actuated leading-trailing shoe drum brake.

In a **duo-servo drum brake**, both shoes are attached to the backing plate by pins or hold-down clips. The tops of the shoes are held against an anchor pin with a return spring. A shoe spring and an adjuster wheel are attached at the bottom of the shoes. See **Figure 19-28**. The shoe spring, adjuster wheel, and adjuster linkage work together to remove slack from the brake as the brake shoes wear. Any time the machine moves in

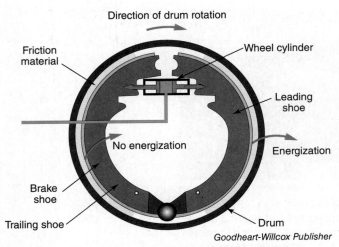

Figure 19-27. Nonservo drum brakes have one shoe applying the majority of the braking force to the drum, depending on the direction of drum rotation. As the wheel cylinder piston is hydraulically actuated, it causes the shoes to move outward. As the machine is moving forward, the drum attempts to rotate the leading shoe, but the bottom of the shoe is anchored at the bottom. As a result, the leading shoe rotates into the drum, providing a self-energizing action.

reverse and the brakes are applied, the secondary shoe moves downward, away from the anchor pin at the top of the backing plate. As a result, the adjuster lever turns the adjuster wheel to remove brake slack. The adjuster wheel can also be manually adjusted through an opening in the backing plate, after the rubber grommet is removed.

When the machine is moving forward, as the brake pedal is pressed, the wheel cylinder's pistons apply pressure to both shoes. As the *primary shoe* is forced against the drum, it attempts to rotate due to the spinning drum. The bottom of the primary shoe transfers a force through the adjuster wheel to the bottom of the secondary shoe. This pushes the bottom of the *secondary shoe* into contact with the drum. As a result, both shoes are forced into the rotating drum (duo-servo) in a self-energizing braking action. It is the self-energizing action of both shoes that gives duo-servo brakes an advantage over nonservo brakes.

Figure 19-28. John Deere combines have used drum brakes for several decades. A—This drum brake is used for both the service brake and the parking brake. It is located on the right side of a John Deere 9660 STS combine. B—Notice the drum has internal splines that spline to the axle shaft.

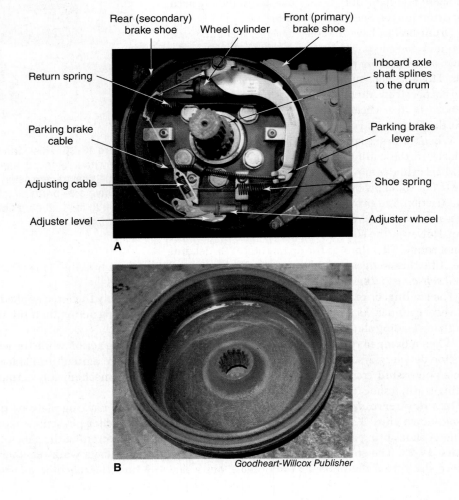

> **Note**
>
> Both nonservo and duo-servo drum brakes can have a parking brake mechanism that actuates one or both shoes. The John Deere combines mentioned earlier in this chapter use the duo-servo service brake and parking brake drum design. See **Figure 19-28**.
>
> The combine has three brake pedals: a right service brake pedal, a left service brake pedal, and a parking brake pedal. The right service brake pedal operates a master cylinder that hydraulically actuates the right service brake. The left service brake pedal operates another master cylinder that hydraulically actuates the left service brake. When the parking brake pedal is depressed, it mechanically actuates two cables that apply both the right and left brakes.

Ball Ramp Disc Brakes

Ball ramp disc brakes use an actuating assembly composed of two actuating discs, steel balls, and springs. See **Figure 19-29**. The steel *actuating discs* have a flat surface on one side, which mates with a brake friction disc. The other side of the actuating disc has pockets, one for each of the steel balls in the assembly. Each pocket has a ramp in the shape of a teardrop. See **Figure 19-30**.

The brake assembly shown in **Figure 19-29** through **Figure 19-33** is used in a Case 586E forklift. It is a *dry disc brake assembly*, meaning that the friction material is dry and does not use fluid for cooling or lubrication. Similar to drum brakes with dry brake shoes, the heat is dissipated into the atmosphere through the dry brake friction disc and actuating discs.

The two actuating discs sandwich the steel balls in the disc pockets. The actuating discs are held together with springs. As the operator presses the service brake pedal, the master cylinder directs fluid to the actuator, which causes the pull rod to lift the brake links. See **Figure 19-31**. As the brake links lift, they force the two actuating discs to rotate. On the bottom of each actuating disc is a notch that rides on an anchor pin. See **Figure 19-32**.

As the actuating discs rotate, the steel balls roll up the ramps in their pockets. This causes a wedging action that forces the actuating disc to apply pressure to each brake disc. Each brake assembly (right and left service brakes) contains two friction discs:

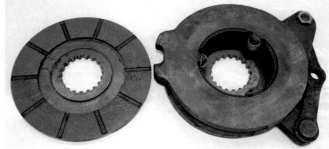

Goodheart-Willcox Publisher

Figure 19-29. One of the two friction discs on the ball ramp brake actuating assembly is shown. The actuating assembly has three springs that clamp together the two actuating discs and five steel balls.

A B

Goodheart-Willcox Publisher

Figure 19-30. An actuating assembly contains two actuating discs. These two discs were removed from a Case forklift. A—One side of each actuating disc has a flat surface that mates with a brake friction disc. B—The other side of the disc has five pockets that each retain a steel ball. Each pocket has a ramp. Only one of the five steel balls is shown.

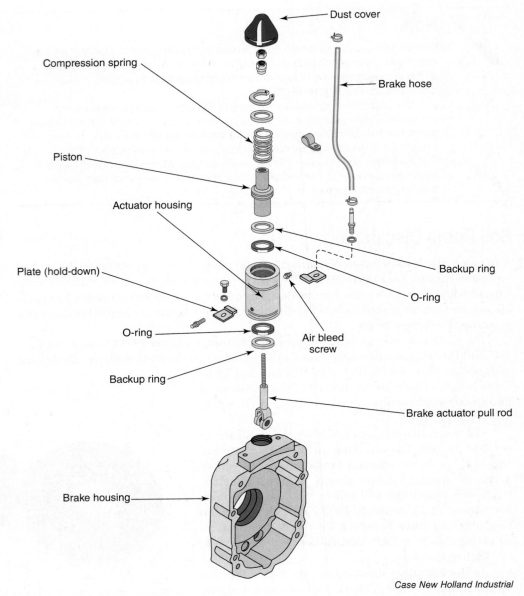

Figure 19-31. As the master cylinder directs fluid to the actuator, the pull rod lifts the links, forcing the actuating discs to rotate and apply the service brakes. Each side of the transaxle has an actuator.

Case New Holland Industrial

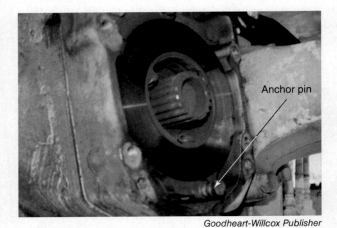

Goodheart-Willcox Publisher

Figure 19-32. The actuating discs are contained in the brake housing and have notches that ride on an anchor pin.

an inboard friction disc and an outboard friction disc. See **Figure 19-33**. Both brake discs are splined to the differential's side gear shaft assembly.

As the brake is applied, the actuating disc presses the friction discs outward against a flat steel surface. On the Case 586E, the two flat surfaces are the brake housing (inboard flat surface) and the brake cover (outboard flat surface). See **Figure 19-34**.

Examples of machines that use hydraulically actuated ball ramp disc brakes are:

- Case 586E forklift.
- Case 580D, 580F, and 580G loader backhoes.
- Case IH 1600 series combines.
- Case IH 1844 cotton harvester.

Some ball ramp disc brakes contain additional friction discs and plates. For example, the brakes in the Case IH 1600 series combines contain three friction discs, two traditional flat actuating discs (flat on one side and ball ramps on the opposite side), a flat brake housing cover plate, and a flat intermediate plate. See **Figure 19-35**.

A

B

Goodheart-Willcox Publisher

Figure 19-33. Ball ramp disc brake components. A—A disc brake actuator and two friction discs. B—The actuator has been separated, showing the five steel balls inside their pockets. When assembled, the three springs hold the halves of the actuator together but stretch as the balls move up their ramps.

 Note

Ball ramp disc brakes in lower-horsepower applications do not require the use of hydraulics, but instead are mechanically actuated. One example is John Deere 4000 series compact utility tractors. The brakes are wet brakes and serve as both parking brakes and service brakes.

External Caliper Disc Brakes

Off-highway machines may use external caliper disc brakes, which use a caliper to squeeze two brake pads against a single rotating disc known as a *rotor*. The *caliper* is a cast housing that contains one or more pistons that apply pressure to the brake pads. The pads are lined with friction material. The rotor is normally a dry rotor. Dry rotors do not use fluid for lubrication or cooling because they are air cooled. Many caliper disc brakes have two calipers and one brake rotor. See **Figure 19-36**. Caliper disc brakes can be used as parking or service brakes.

Caliper disc parking brakes are commonly located one of three places:
- The transmission's output shaft, **Figure 19-4.**
- On the transmission's countershaft, **Figure 19-37**.
- The drive axle's input shaft, **Figure 19-38**.

The brake caliper can be manually applied, spring applied or hydraulically applied.

Hydraulic External Caliper Disc Brakes

Off-highway machines may use ***hydraulically actuated caliper disc brakes***, which use one or more pistons used to squeeze the brake pads against the rotor. A hydraulically applied caliper brake is commonly used for service brakes. Typically, hydraulic calipers, like the one shown in **Figure 19-36**, are located at each wheel. However, they can also be located next to the transmission, as in **Figure 19-37**. Examples of machines that use caliper brakes for service brakes are combines and articulated trucks.

Goodheart-Willcox Publisher

Figure 19-34. The right-hand brake cover has been removed. Notice the flat surface, which is meshed against the right outboard brake friction disc.

The two types of calipers are fixed and sliding. A *fixed caliper* uses two hydraulic pistons that are opposed to one another. Hydraulic fluid is directed to both pistons through the caliper's passageways so that fluid acts on the pistons in parallel. The pistons squeeze

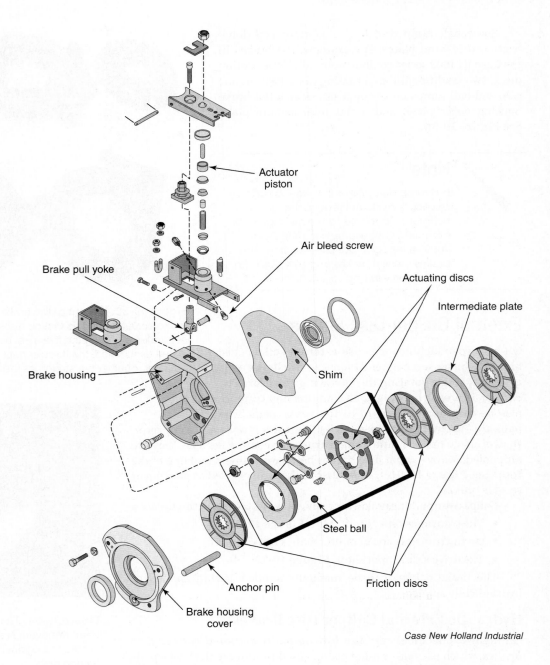

Figure 19-35. A Case IH 1600 series combine uses three friction discs, two actuating discs, and an intermediate plate.

Case New Holland Industrial

Figure 19-36. This is an Auburn Gear Power Wheel final drive and brake assembly that is used on agricultural sprayers and street sweepers. A—The assembly uses a dual caliper service brake. It also contains an internal static parking brake that is not visible. The final drive is a double reduction. B—The back side of the wheel drive is bolted to the machine. A hydraulic motor is used as an input to the final wheel assembly.

Goodheart-Willcox Publisher

Figure 19-37. This Case IH combine uses three caliper brakes. Each side of the transmission has a service brake rotor splined to the transmission's output shaft. Both of the service brake rotors use twin calipers. The machine also has a spring-applied caliper parking brake, which is the smaller rotor splined to the transmission's countershaft. The brakes are used on Case IH 10, 20, 30, and 240 series flagship combines.

Goodheart-Willcox Publisher

the friction material against the single disc brake, **Figure 19-39**. Some Caterpillar articulated dump trucks use fixed calipers at each wheel for service brakes. A caliper can use as many as six pistons, three on each side of the caliper.

A *sliding caliper*, also known as a *floating caliper*, has one or more pistons on only the inboard side of the caliper assembly. As the pistons are applied, they force the caliper to slide inboard while pushing their own pads outboard. This squeezes both the inboard and outboard brake pads against the rotating rotor. An example of a sliding caliper is shown in **Figure 19-36**.

A piston seal on a caliper serves multiple purposes:
- To retain fluid pressure.
- To prevent leaks.
- To retract the piston when the brake pedal is released.

The seal is a square-cut seal. As the fluid pressure forces the piston to extend, the square-cut seal deforms. When fluid pressure is drained away from the piston, the seal goes back to its original form and the piston retracts. A boot is often used to prevent dust or other contamination from entering between the piston and the seal.

Goodheart-Willcox Publisher

Figure 19-38. This Caterpillar 926M wheel loader has a caliper parking brake on the front axle. When the spring-applied caliper holds the rotor, the machine is held stationary.

Mechanically Applied External Caliper Disc Brakes

Some external caliper parking brakes are mechanically engaged and do not require hydraulic fluid. The caliper brake in **Figure 19-40** is a mechanically actuated parking brake that is placed on the output shaft of a powershift transmission.

Case IH 9300 Steiger four-wheel drive agricultural tractors also use a mechanical lever in the operator's cab to apply an external caliper parking brake. The mechanically actuated parking brake can be located on the rear of the transmission (9330) or can be located on the front axle (9350–9380). The mechanically applied parking brake is integrated with the hydraulically applied service brake, sharing the same rotor and caliper.

John Deere 9100, 9200, 9300, and 9400 four-wheel drive agricultural tractors with powershift transmissions had a parking brake rotor on the back of the transmission. The caliper was mechanically actuated by a lever in the operator's cab.

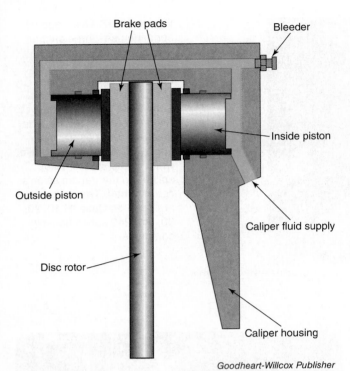

Figure 19-39. A fixed caliper has pistons on both sides of the brake rotor. The pistons are plumbed in parallel, allowing them to press against the brake pads simultaneously.

Spring-Applied External Caliper Disc Brakes

The Case IH 10, 20, 30, and 240 series flagship combines use a caliper parking brake, as shown in **Figure 19-37**. It is spring applied and hydraulically released. As described in Chapter 17, *Final Drives*, it is attached to the transmission's countershaft, which is in mesh with the differential's ring gear. The parking brake is engaged when the parking brake switch is engaged, the engine is off, or the four-speed transmission is shifting gears (which requires the machine to be stationary). In order to propel the machine, the parking brake switch is moved to the "off" position. This causes a controller to energize a solenoid that sends fluid to the parking brake piston, which compresses the spring and releases the rotor. See **Figure 19-41**. A procedure is presented later in this chapter for manually releasing the parking brake.

Wheel loaders and articulated dump trucks sometimes can also use an external caliper, spring-applied, and hydraulically released, disc parking brake. A Caterpillar 926M wheel loader with the parking brake rotor and caliper on the input to a drive axle is shown in **Figure 19-38**.

Internal Wet Disc Brakes

One of the most popular types of brakes used in heavy equipment is the ***internal wet disc brake***. Like a multiple disc clutch, internal wet disc brakes contain friction discs, steel separator plates, a piston, and a spring mechanism. The design determines whether the piston is spring applied or hydraulically applied. The design also determines whether the friction discs hold the shaft exiting the differential, the final drive sun gear, or the drive axle. Examples of internal wet disc brake locations include:

- Brake located between the rear axle's differential and inboard final drive (agricultural MFWD tractor).
- Brake located outboard, after differential, and integrated with the outboard final drive (wheel loader and articulated dump truck).
- Brake located between the transmission and outboard final drive (combine).
- Brake located outboard, beyond the drive sprocket, on the drive axle (motor grader).

Figure 19-40. This caliper parking brake is located at the rear of a powershift transmission. A—The mechanical lever actuates a cam. The lever is shown in the released position. B—When the lever is actuated, the cam forces the caliper to apply the brake pads.

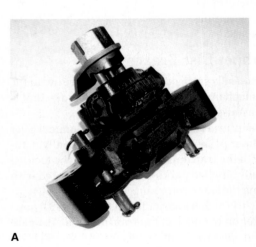

A

B

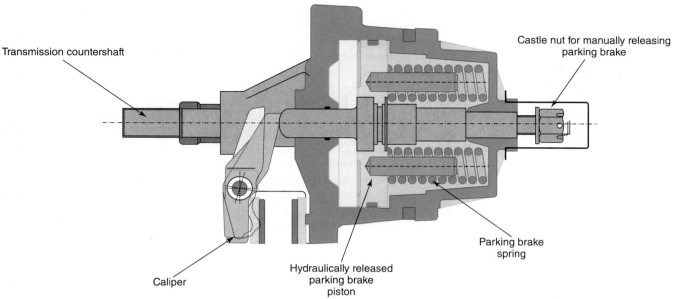

Figure 19-41. Case IH 10, 20, 30, and 240 series flagship combines use a spring-applied caliper parking brake that is used to hold the transmission's countershaft. The piston must be hydraulically actuated for the brake to be hydraulically released. Instructions for using a wrench to manually release the parking brake are given later in this chapter.

Many agricultural MFWD tractors, loader backhoes, and some wheel loaders use *inboard hydraulically actuated wet disc brakes*, which are located between the rear differential and the inboard planetary final drive. The brake consists of a hydraulically actuated piston, a friction disc, and a backing plate. Manufacturers often use a spring to hold the service brake piston in a released position. When the brakes are applied, fluid pressure causes the piston to overcome the spring's pressure. **Figure 19-42** shows the right-hand service brake used in a John Deere 8000 series MFWD tractor's rear axle. Both the piston and the backing plate are pinned so that they will not rotate. The friction disc is splined to the final drive sun gear. As the operator presses the brake pedal, the piston applies pressure to the friction disc, which is squeezed against the backing plate. The left brake pedal controls the left brake, and the right brake pedal controls the right brake. The brake disc is cooled by the fluid in the axle housing.

Note

The Caterpillar 950G wheel loader also uses a single inboard brake disc that is wedged between a plate and a piston to hold the planetary sun gear.

As explained in Chapter 10, *Powershift and Automatic Transmission Theory*, many agricultural MFWD tractors and loader backhoes have service brakes located in the rear axle only. The tractor achieves four-wheel braking by engaging the MFWD clutch, which provides drive power to all four drive wheels. Applying the rear service brakes causes all four drive wheels to brake.

The power flow through the John Deere 8000 series countershaft powershift transmission is also explained in Chapter 10, *Powershift and Automatic Transmission Theory*. This transmission contains the tractor's parking

Figure 19-42. A John Deere 8000 series tractor uses an internal brake disc that stops the final drive sun gear.

brake, which is spring applied and hydraulically released. See **Figure 19-43**. Notice the parking brake is designed to hold the transmission's output shaft. Many heavy equipment machines use spring-applied parking brakes, which are a fail-safe brake. The *spring-applied parking brake* requires both electricity to energize the parking brake solenoid and hydraulic pressure directed through the parking brake solenoid to release the parking brake. If either the electricity or the fluid pressure fails, the parking brake will apply.

As stated in Chapter 17, *Final Drives*, many machines use outboard planetary final drives. In these applications, the internal wet disc brakes are often located near the final drive. These brakes are known as *outboard hydraulically actuated internal wet disc service brakes*. Examples of machines that use outboard internal wet disc service brakes are wheel loaders and articulated dump trucks.

Figure 19-44 shows a simplified drawing of a Caterpillar 740B articulated dump truck service brake. Notice the service brake is outboard (away from the differential) and is integrated with the planetary final drive. The illustration depicts one of the guide pins and release springs that keep the brake piston retracted when the service brake is disengaged.

Another example of a machine that uses outboard internal wet disc service brakes is a motor grader. A Caterpillar M series motor grader service brake is shown in **Figure 19-45**. Notice that the service brake is located at the drive wheel. This internal wet disc brake contains two additional features: a compensator adjustment and a wear indicator. Both will be explained later in the chapter.

Wet disc brakes can also be located between the transmission and the outboard final drive, as shown in **Figure 19-46**. This brake is used on Case IH midrange 2100, 2300, 2500,

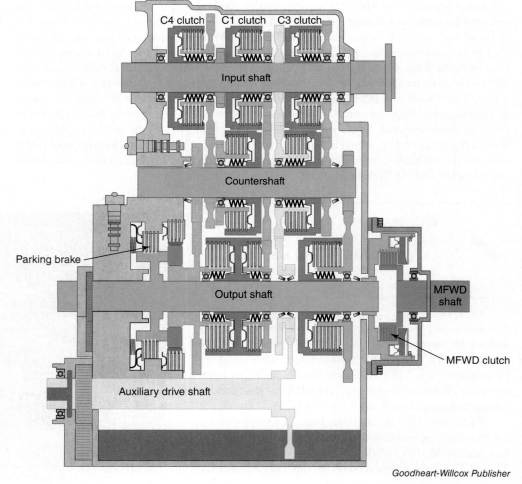

Figure 19-43. The John Deere 8000 series powershift transmission contains the tractor's spring-applied parking brake, which is designed to hold the transmission's output shaft.

Goodheart-Willcox Publisher

88, 130, and 140 series combines with three-speed transmissions and Case IH and New Holland flagship combines with the two-speed common drive transmissions. This brake is a *combination wet disc brake*, meaning it is used for both the parking brake and service brake.

Notice in **Figure 19-47** that the orange portion shows that parking brake fluid pressure has been directed into the brake, which causes the reaction piston to seat against its snap ring on the right side of the reaction piston. The parking brake piston moves to the left, which compresses the parking brake spring. If the service brake has no fluid pressure, the parking brake piston pulls the service brake piston to the left in a disengaged position via the Belleville return springs. The combine is then free to move forward or in reverse.

When the service brake pedal is depressed, fluid pressure is directed to the service brake piston, causing it to compress the Belleville return springs and apply the brake friction discs. See **Figure 19-48**. The amount of service brake fluid pressure determines the brake force. Note that the combine has a hydrostatic drive transmission. Therefore, the most common means of stopping the machine is through the hydrostatic braking effect of the hydrostatic transmission.

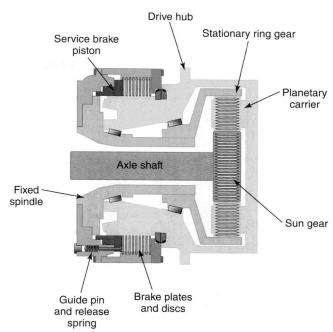

Figure 19-44. A simplified drawing of a Caterpillar 740B articulated dump truck service brake.

 Warning

As mentioned in Chapter 11, *Powershift and Automatic Transmission Controls, Service, and Repair*, disassembling a spring-applied brake requires using a press to compress the spring. See **Figure 19-49**. Special caution must be taken. After the spring has been compressed and the retaining ring has been removed, make sure the piston extends out of its bore evenly as the pressure is released on the press. If the piston becomes wedged inside its bore with the spring still under pressure, the piston and springs can shoot out of the brake housing when it becomes dislodged.

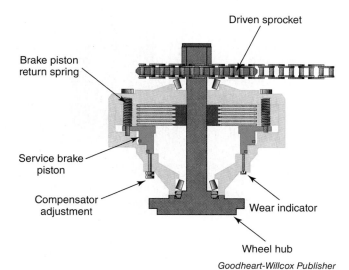

Figure 19-45. A Caterpillar M series motor grader contains internal wet disc brakes at the drive wheel.

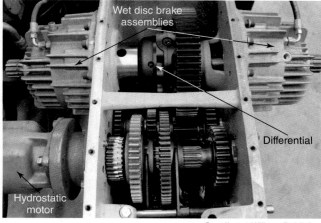

Figure 19-46. The hydrostatic motor provides an input into the three-speed mechanical transmission. The differential delivers output to the right and left combination brakes. The drive shafts deliver power to the right and left final drives.

Caution

Many machines equipped with spring-applied parking brakes have enough power to propel the machine even if the parking brake is applied. When this occurs, it normally burns the parking brake plates. See **Figure 19-50**.

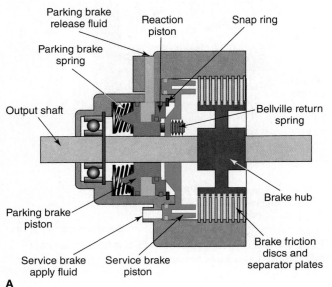
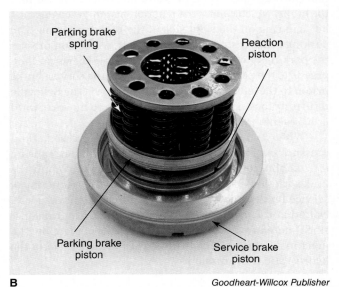

Figure 19-47. Case IH midrange 2100, 2300, 2500, 88 series, 130 series, and 140 series combines, and Case IH and New Holland flagship combines with the common ground drive two-speed transmission use a combination wet disc brake on each side of the transmission. A—When parking brake fluid pressure is directed into the brake assembly, it causes the parking brake piston to compress the parking brake spring. If the service brake has no fluid pressure, the parking brake piston will hold the service brake piston in a disengaged position. B—The parking brake spring, parking brake piston, reaction piston, and service brake piston have been removed from the brake housing.

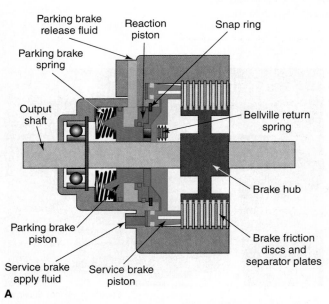
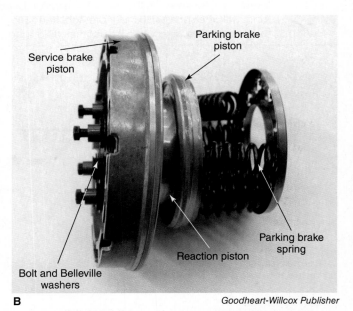

Figure 19-48. When fluid pressure is used to release the parking brake, the parking brake piston uses Belleville washers to hold the service brake piston retracted until the service brakes are applied. A—The combination dual brake is drawn with the parking brake released (orange fluid) and the service brake applied (red fluid). B—The service brake piston is retracted by Belleville washers that are attached to the parking brake piston.

Goodheart-Willcox Publisher

Figure 19-49. This Case IH combine brake housing has been placed in a press. Pressure has been applied to the parking brake so the retaining ring can be removed. After the snap ring has been removed, slowly and carefully release the pressure on the press. Stop releasing pressure on the press if the piston quits moving. If the piston quits moving, it has become wedged and is still under load from the compressed parking brake spring.

Goodheart-Willcox Publisher

Figure 19-50. If an operator accidentally propels a machine while the parking brake is applied or if the release fluid is leaking internally, the brake plates will likely burn. Notice the middle plates of this brake assembly are burnt. This brake had low hours, but was replaced after the operator accidentally drove the machine with the parking brake applied.

Combination wet disc brakes are also used in rigid frame haul trucks. They are used on both the non-drive steer wheels and the rear drive wheels. In these applications, the brakes are located outboard and are integrated with the wheel assembly. This type of combination brake is shown in **Figure 19-51**. Notice the parking brake is spring applied and hydraulically released. The service brake is hydraulically applied. These trucks often use a left-hand ***secondary brake pedal*** that allows the operator to apply the parking brake in the event the service brake is inoperative.

Band Brakes

A ***band brake*** consists of a flexible band that is located around the circumference of a drum. See **Figure 19-52**. The band is often hydraulically actuated. The band brake can be used as a parking brake, service brake, or PTO brake. Some older track-type tractors use bands as a steering brake. Track-type steering is explained in Chapter 24, *Track Steering Systems*.

Brake Swept Area

The amount of brake drum or rotor area that sweeps past the friction material during a single revolution of the wheel is known as the ***swept area***. A rotor has more swept area than a drum because both sides of the rotor sweep past the friction material. A drum has only one side of swept area. More swept area provides better heat transfer, resulting in improved cooling.

Accumulators

Brake accumulators store hydraulic energy, just as a battery or a capacitor stores electrical energy. The stored hydraulic energy ensures the machine has fluid available to apply the brakes if the brake hydraulic pump fails to deliver hydraulic flow. In addition, the stored energy means the brake pump does not have to maintain a constant high pressure when the brakes are not in use.

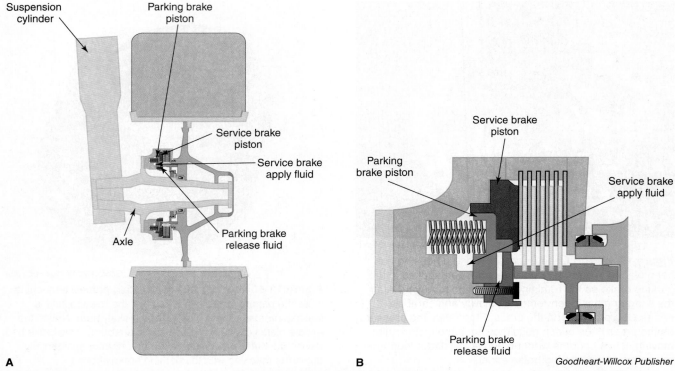

Figure 19-51. These drawings depict a combination brake used on the non-drive steer wheel on a Caterpillar rigid frame mining truck. A—View of the brake in relation to the suspension cylinder and wheel. B—View of the combination brake.

Gas accumulators, also called pneumatic accumulators or hydro-pneumatic accumulators, are used within the hydraulic brake circuit in many heavy equipment machines. They are more common than spring accumulators or weighted accumulators in heavy equipment. Gas accumulators are filled with a gas to a *precharge pressure*, which is the level of gas pressure in the accumulator when no brake fluid is present. The precharge pressure affects the volume of liquid that is admitted into the accumulator.

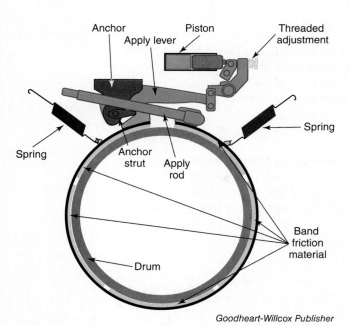

Figure 19-52. Example of a band used as a steering brake on a track-type tractor

 Warning

Oxygen will react with brake and hydraulic fluids, which are fuel sources. Oxygen should *never* be used as a gas for precharging accumulators. Atmospheric air, which contains oxygen, must also *never* be used for charging accumulators.

Dry nitrogen gas is the preferred gas for filling accumulators for multiple reasons:

- Nitrogen is an inert gas and will not react with brake or hydraulic fluid.
- Nitrogen is not a fuel source for a potential fire.
- Nitrogen is economical. The atmosphere (by volume) is made up of approximately 78% nitrogen, 21% oxygen, and 1% argon and miscellaneous gases. Nitrogen, which is much more plentiful, economical, and compatible than other suitable gases, has proven to be the best solution. Therefore, use only nitrogen gas for charging accumulators.

The accumulator housing has a gas port for checking and filling the accumulator with nitrogen. Accumulators may use a Schrader-type valve with a valve core that is similar to those used on tires. This makes it tempting for the untrained technician to use shop air to charge an accumulator, but because it contains oxygen, shop air should never be used. Some accumulators use a gas port that requires an adapter fitting and a wrench to check or fill the nitrogen pressure.

Machine manufacturers use different accumulator arrangements. Depending on the available space, a large accumulator may be used for a single brake circuit or several smaller accumulators may be plumbed in parallel and used for a single brake circuit.

The three types of gas accumulators are bladder accumulator, piston accumulator, and diaphragm accumulator. Bladder accumulators are the most popular style of gas accumulator. They are used in both mobile machinery and industrial applications.

Bladder Accumulator

A *bladder accumulator* consists of a synthetic rubber bladder, also called a bag, inside a metal housing. The bladder is charged with nitrogen. See **Figure 19-53**. The bladder accumulator can be equipped with one of two types of protection features, a safety poppet valve or a safety button. A safety poppet valve, also known as an anti-extrusion valve, may be located at the bottom of the housing. This spring-loaded poppet valve closes when the fluid is exhausted out of the accumulator to prevent the rubber bladder from tearing. A safety button is sometimes placed on the actual bladder.

Whether the bladder accumulator has a poppet valve or a safety button, both are used to protect the bladder as the bladder forces fluid out of the accumulator's housing. They also close off the accumulator, preventing fluid from entering until the hydraulic pressure can overcome the bladder's precharge nitrogen pressure.

The bladder can be damaged if the precharge pressure is set too high or too low. If precharge pressure is set too high, the bladder and poppet valve can be damaged. If precharge pressure is set too low, the bladder can be damaged by the gas charging valve when high-pressure fluid fills the accumulator. The process of charging a bladder with nitrogen should be completed slowly; otherwise the synthetic rubber bladder can become brittle due to the sudden cooling effect of the fast nitrogen charge.

A bladder accumulator is a quicker-responding gas accumulator than a piston gas accumulator and can handle contamination better than piston accumulators. However, if a bladder accumulator is in a horizontal position, it is possible for a little fluid to become trapped between the bladder and the housing when pressure drops and fluid is expelled

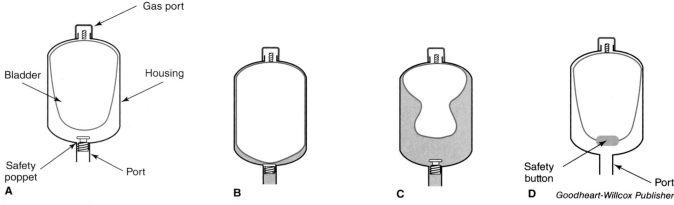

Figure 19-53. Bladder accumulators contain a synthetic rubber bladder inside a metal housing. A—Parts of the accumulator. B—Bladder is fully charged with nitrogen, and hydraulic fluid pressure is lower than nitrogen pressure. The bladder presses on the top of the safety poppet, closing it. C—Hydraulic fluid pressure is high enough to reopen the safety poppet, allowing fluid to fill the accumulator and compress the bladder. D—This bladder includes a safety button. The safety button is a reinforced area, larger than the port, that prevents the bladder from being ripped by the edges of the port.

from the accumulator. The bladder can wear unevenly when in the horizontal position, especially if the fluid is contaminated. The best orientation for bladder accumulators is the vertical position, with the gas port on top.

Bladder accumulators commonly range in sizes from 1/4 gallon to 15 gallons. However, one manufacturer offers a 120-gallon, low-pressure bladder. Some bladder accumulators are designed with a 10,000 psi (690 bar) pressure capacity.

Piston Accumulator

A *piston accumulator* consists of a cylinder, with an internal piston and seals, and a gas fill valve. See **Figure 19-54**. The piston accumulator resembles a hydraulic cylinder without a rod. Piston seals are used to separate the nitrogen gas from the fluid.

Piston accumulators offer the largest volumes, up to 160 gallons. They can also be designed as small as 15 in^3. Piston accumulators have high-pressure capabilities; some are capable of 15,000 psi (1034 bar). A piston accumulator is most often mounted in a vertical position with the gas charging valve on top. However, the accumulator can be mounted in a horizontal position. Contamination control is the key to piston accumulator longevity.

Piston accumulators can be twice as expensive as bladder accumulators with the same capacity. Some designers prefer piston accumulators over bladder accumulators because they wear gradually over a period of time, whereas the bladder accumulator might have a sudden catastrophic failure. Both styles will wear due to contamination, but bladder accumulators are less sensitive to contamination.

Diaphragm Accumulator

The *diaphragm accumulator* is a small vessel that contains a flexible synthetic rubber diaphragm that separates the nitrogen gas and brake fluid. See **Figure 19-55**. The diaphragm flexes as nitrogen pressure fills the vessel or as fluid pressure compresses the nitrogen.

Diaphragm accumulators are more common in the aircraft industry because they are compact and lightweight. They range in volume from 6 in^3 to 150 in^3. Some diaphragm accumulators have a pressure capability up to 5000 psi (345 bar).

Accumulator Safety

Accumulators are considered one of the most dangerous hydraulic components on a machine. Any type of device that is capable of storing energy, such as a compressed spring, a charged capacitor, or a full accumulator, can cause serious injury or death. Accumulators, whether large or small, can contain a tremendous amount of stored energy.

Warning

An untrained technician who attempts to remove a charged accumulator can cause the accumulator to become a projectile, rocketing through the air. Only properly trained technicians should work on accumulators.

If a machine is shut off but the brake's hydraulic system contains an accumulator that is charged full of fluid, the system must be treated as a live operating hydraulic system. The charged accumulator can supply fluid pressure, causing fluid to be injected into the skin of a technician. Hydraulic systems must have the fluid drained from

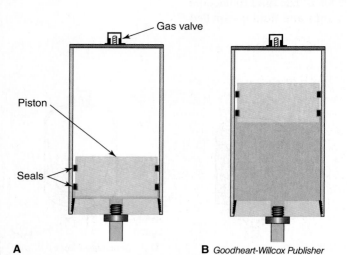

Figure 19-54. Piston accumulators resemble a hydraulic cylinder without a rod. Piston seals separate the nitrogen gas and hydraulic fluid. A—Hydraulic pressure is not high enough to overcome nitrogen pressure. B—Hydraulic pressure is high enough to compress the nitrogen.

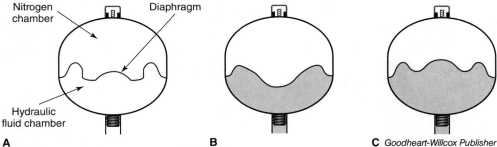

Figure 19-55. A diaphragm gas accumulator uses a synthetic rubber diaphragm to separate the nitrogen and hydraulic fluid. A—When there is no hydraulic fluid or nitrogen pressure, the diaphragm is relaxed. B—When nitrogen pressure exceeds hydraulic pressure, the diaphragm expands into the hydraulic fluid chamber. C—As hydraulic pressure overcomes nitrogen pressure, the diaphragm pushes into the nitrogen chamber.

the accumulator before performing any work on the system. Follow the manufacturer's literature to properly discharge the accumulator.

After the oil has been drained, the precharge of nitrogen still needs to be discharged before the accumulator can be removed, serviced, or replaced. Always follow the manufacturer's service literature for depleting the precharge nitrogen pressure.

Precharge an accumulator only when it is installed on the machine. Just as capacitors should not be handled or transported while charged with electricity, accumulators should not be handled or transported when charged with nitrogen. The high-pressure gas can cause an injury if the bladder or seals burst. Many freight companies will not ship a charged accumulator due to the inherent risk of the pressurized vessel.

Accumulators should never be exposed to heat. A technician who disregards this critical precaution can cause the pressure vessel to explode, easily causing a fatality. Never use a welder, torch, or any other heat-generating device near accumulators. As previously mentioned, never use oxygen or compressed air for charging an accumulator.

If discharging nitrogen from a large-volume accumulator, be sure to perform the procedure in a ventilated area. Do not discharge the nitrogen in a small, confined area. The nitrogen released could displace air in the space, and the absence of oxygen could cause difficulty breathing.

Never use an automotive valve core in an accumulator. The automotive valve core is a lower-pressure valve and is not designed to withstand the high pressures used in gas accumulators.

Servicing and Repairing Accumulators

Gas accumulators will lose their nitrogen precharge over time. Precharge pressure must be checked periodically. In the weeks following installation, or if a leak is suspected, the precharge pressure should be checked daily. If the accumulator is operating correctly, the precharge pressure should be checked once every three to six months or during a 500-hour preventative maintenance (PM) inspection.

If the precharge pressure is low, the accumulator is likely leaking. Soapy water is often used to help locate nitrogen leaks. The source of the leak should be identified, and the accumulator should be repaired or replaced before it is precharged again. If the accumulator is empty, there is a chance that the bladder has burst and cannot hold any nitrogen pressure. Always follow the manufacturer's service literature for checking precharge pressure and precharging the accumulator.

Checking Precharge Pressure with a Nitrogen Service Tool

A dedicated service tool can be used to check the precharge pressure and to charge the accumulator with nitrogen gas. This tool physically taps into the accumulator's valve core to accurately measure the nitrogen pressure. See **Figure 19-56**.

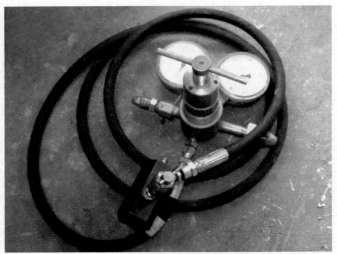

Goodheart-Willcox Publisher

Figure 19-56. This accumulator service tool contains a high-pressure regulator valve with two gauges, a fitting for attaching it to a nitrogen supply tank, and a fitting that threads into the accumulator.

Although this is the most precise way to measure precharge pressure, there is a slight risk that the accumulator's valve core will stick in an open position, causing a loss of nitrogen. **Figure 19-57** shows a service tool connected to a nitrogen supply tank.

Warning
Nitrogen supply tanks contain high volumes of high-pressure nitrogen gas, with pressures exceeding 3000 psi.

When the tool in **Figure 19-57** is used to measure an accumulator's precharge pressure, it must be connected to a nitrogen supply tank. The tool does not have a check valve, which means if it were only connected to only an accumulator, the tool would vent the accumulator's nitrogen gas to the atmosphere.

The following steps can be followed for checking the precharge pressure of an accumulator using the service tool in **Figure 19-57**.

Caution
Always follow the manufacturer's service instructions. Accumulator service tools and their proper operation vary from manufacturer to manufacturer.

1. Be sure that all of the fluid pressure has been depleted from the accumulator.
2. Be sure the cylinder valve on the nitrogen supply tank is closed. You will know it is closed if the cylinder pressure gauge reads zero pressure. See **Figure 19-57**.
3. As a good habit, be sure the regulator handle is backed out (turned counterclockwise). The reason for this step is explained in more detail later in the chapter.
4. Remove the dust cover from the accumulator's gas port.
5. Before attaching the service tool's valve chuck to the accumulator, back out the tool's T-handle fully by turning it counterclockwise. See **Figure 19-58**.
6. Thread the service tool's valve chuck fitting onto the accumulator's gas port. Many of these ports resemble a tire's valve stem core. See **Figure 19-59**.
7. An extension may need to be added between the gas port and valve chuck if the valve core is situated inside a deep valley, preventing the direct attachment of the service tool. See **Figure 19-60**.
8. Special fittings may need to be installed to the chuck to adapt to the gas port on some accumulators. See **Figure 19-61**.
9. With the service tool attached to the nitrogen supply tank and the accumulator, both gauges should still read zero pressure.
10. Slowly turn the T-handle on the valve chuck clockwise. When the valve is fully open, the working pressure gauge (on the left in **Figure 19-57**) will read the accumulator's precharge pressure.

Charging an Accumulator with Nitrogen

Continuing from step 10, the following steps can be followed to charge the accumulator with nitrogen gas.

Figure 19-57. The accumulator service tool is connected to a nitrogen supply tank. The tool contains two pressure gauges, a high-pressure nitrogen regulator valve, a hose, and a valve chuck with a fitting to thread into the accumulator's Schrader valve.

 Caution

As mentioned, precharge an accumulator *only* when it is installed on a machine. Some customers may bring in an empty accumulator, asking for it to be recharged. Not only is it a hazard to charge the accumulator with nitrogen while it is off the machine, but if the bladder has previously failed, the precharge will spew residual fluid out of the port. See **Figure 19-62**.

Figure 19-58. The service tool contains a threaded fitting on the valve chuck that attaches to the accumulator. Prior to attaching the fitting to the accumulator, turn the T-handle fully counterclockwise.

Figure 19-59. Some accumulators have a gas port that resembles the valve core of a car tire.

Goodheart-Willcox Publisher

Figure 19-60. An extension has been threaded onto this accumulator. If the accumulator's valve core is located inside a valley, it might require the use of an extension. This accumulator would not require the extension, but it has been added for illustration purposes.

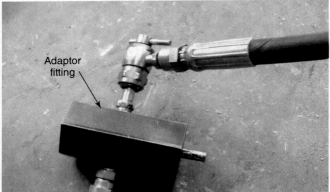

Goodheart-Willcox Publisher

Figure 19-61. Some accumulators require the use of an adaptor fitting to connect the valve chuck to the accumulator.

Goodheart-Willcox Publisher

Figure 19-62. Do not attempt to precharge an accumulator off a machine. Leave the accumulator installed on the machine. If the bladder bursts, it can cause an injury. If the bladder has already failed, it can cause fluid to spew out of the accumulator as shown.

1. Prior to charging the accumulator, ensure the nitrogen tank's cylinder valve is closed (turned fully clockwise), the regulator valve is backed out (turned fully counterclockwise), and the cylinder pressure gauge (on the right) reads zero pressure. The working pressure gauge, on the left, will read accumulator precharge pressure. If the accumulator is completely discharged, it will read zero.
2. Double-check the precharge pressure specification.

 Warning

If the regulator valve is mistakenly left fully open (turned clockwise until seated), high-pressure nitrogen will be sent to the accumulator immediately when the cylinder valve is opened. For example, if the tank supply pressure is 3000 psi and the accumulator precharge setting is 90 psi, the accumulator will receive 3000 psi of nitrogen, which could rupture the bladder and accumulator housing, and potentially injure the technician.

3. Open the nitrogen supply tank's cylinder valve. The cylinder pressure gauge (on the right) will read the nitrogen tank pressure. If the tank pressure is lower than the required accumulator precharge pressure, the nitrogen supply tank must be replaced with one that is fully charged.
4. Slowly turn the service tool's regulator valve clockwise while closely monitoring the working pressure gauge (on the left).

 Caution

Charge the accumulator at a slow rate to avoid damaging it.

5. When the precharge pressure has been met, immediately close the valve on the valve chuck by turning it counterclockwise. Then, close the cylinder valve.
6. Remove the valve chuck from the gas port on the accumulator. Keep in mind that a small amount of nitrogen pressure will be lost while removing the service tool.
7. If the charging tool has a bleeder valve, open it to deplete the nitrogen in the service tool.
8. Back out the regulator valve by turning it counterclockwise until the valve handle is loose.

Accumulator Repair

The life expectancy of a gas accumulator is approximately 12 years. If an accumulator develops a leak, it must be repaired or replaced. Some bladder, piston, and diaphragm accumulators are designed to be repairable, others are not.

Repairable bladder accumulators can be configured as bottom-repairable or top-repairable. The bottom-repairable design is an older style and requires removing the accumulator in order to replace the bladder. The top-repairable design allows a technician to replace the bladder while the accumulator is attached to the system. As previously mentioned, the oil and the nitrogen pressure must be drained from the accumulator before it can be repaired.

Hydraulic Brake Circuit Operation

Hydraulic brake circuits vary among different machines and manufacturers. **Figure 19-63** shows a brake circuit that is similar to the circuit used in Case E and F series wheel loaders. A fixed-displacement hydraulic pump delivers fluid to the brake system. The large dashed box indicates that a single brake assembly valve contains numerous individual brake valves, including the following:

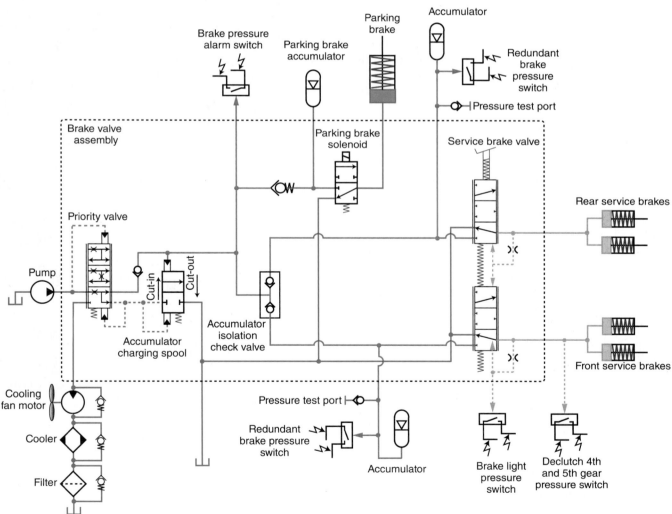

Figure 19-63. This is an illustration of a brake circuit similar to those used in Case E and F series wheel loaders. One fixed-displacement hydraulic pump is used to operate the brakes and the cooling fan's hydraulic motor.

- Priority valve.
- Accumulator charging spool valve.
- Accumulator isolation check valve.
- Parking brake solenoid valve.
- Service brake valve.

The priority valve directs the fluid to charge the brake accumulators first. The spring at the bottom of the priority valve ensures that the spool is initially shifted upward. This causes the pump's flow to be directed to charge all three brake accumulators.

The accumulator charging spool valve is responsible for filling the accumulators. The spring at the bottom of the accumulator charging spool valve ensures that the charging spool is also initially shifted upward. This prevents the pilot fluid that works in conjunction with the priority valve's spring from draining to the reservoir. When the charging spool valve is in this position, the *cut-in* position, the accumulators are being charged with fluid. The accumulator *cut-in pressure* is the pressure at which the accumulator is not fully charged and is capable of receiving pump flow. This also means the brake supply pressure (made available to the inlet of the service brake control valve) is lower than the accumulator's nitrogen pressure.

The accumulator pressure is sensed at the top of the accumulator charging spool. Once the accumulators are charged, the pressure rises and forces the charging spool to shift downward to the *cut-out* position. The accumulator *cut-out pressure* is the pressure at which the accumulator is fully charged and fluid is prevented from further charging the accumulators.

When the brake circuit's fluid flow demand has been met, the priority valve directs the balance of the fluid to the cooling fan motor. The fluid is then routed to the cooler, the filter, and then to the reservoir. Notice that this brake circuit in **Figure 19-63** does not use a pump relief valve or an unloading valve, as the priority valve is regulating the excess flow to the tank through the cooling fan motor. Some brake circuits that use a fixed-displacement pump can have both a relief valve and an unloading valve, **Figure 19-64**.

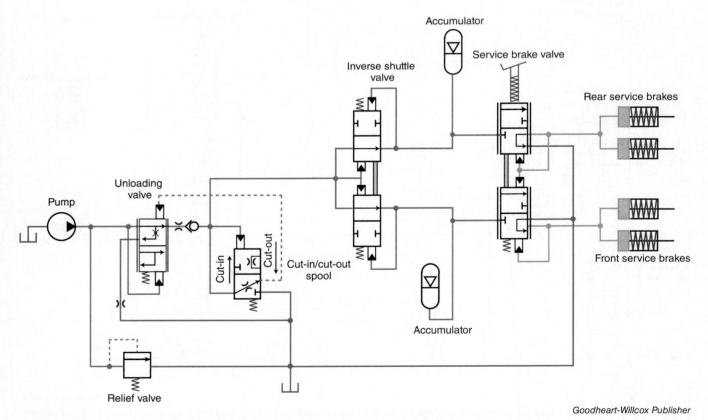

Figure 19-64. This drawing shows an inverse shuttle valve used in an accumulator brake circuit.

The ***unloading valve*** is designed to dump excess pump flow to the tank at a pressure value set by the unloading valve's spring pressure value.

The accumulator isolation check valve assembly in **Figure 19-63** contains two separate check valves. One check valve isolates the rear service brake circuit and the other isolates the front service brake circuit. The circuit with higher pressure will cause its isolation check valve to be seated. As a result, the lower pressure circuit is sensed on the top of the accumulator charging spool, which directly controls whether the accumulators receive fluid (cut-in or cut-out). If a leak occurred in either the rear brake circuit or the front brake circuit, the accumulator isolation check valve in the good circuit would seat, which would enable the fluid charged in the good brake circuit to apply the brakes. Some Caterpillar machines use a similar valve, called an ***inverse shuttle valve***, in place of the isolation check valve. It too, senses both the front and rear service brake accumulator pressures, but when both accumulators are fully charged, the inverse shuttle valve blocks flow to both front and rear service brake circuits. See **Figure 19-64**.

The schematic in **Figure 19-63** also shows that the parking brake is spring applied. A two-position solenoid valve controls the parking brake. When the solenoid is de-energized, the spring holds the valve shifted in the upward position, which drains the fluid from the parking brake and allows the parking brake spring to apply the parking brake. When the parking brake solenoid is energized, it forces the solenoid valve to shift downward. Fluid stored in the parking brake accumulator compresses the parking brake spring and allows the machine to travel.

Cooling Internal Wet Disc Brakes

Some heavy equipment equipped with internal wet disc brakes uses one or more pumps dedicated to cooling the wet disc brakes. An example of a brake cooling circuit is shown in **Figure 19-65**. It is similar to the system used on a Caterpillar 777G rigid frame haul truck. In this example, multiple pumps are integrated into the brake cooling circuit. One pump is dedicated to cooling the brakes. Three other pumps work in combination for truck bed hoist operation, service brake operation, and brake cooling.

Fluid is pulled from the reservoir, pumped through two coolers, and then the cooled fluid is directed through the machine's wet disc brakes. The brake coolers are water-type coolers that use the engine's coolant to cool the hydraulic fluid.

The brake cooling circuit uses a low-pressure relief valve that is bench tested at 85 psi (6 bar). Under normal conditions, the brake cooling relief valve is closed. The valve commonly opens when the fluid is cold. The brake temperature warning light comes on when the fluid reaches 255°F (124°C).

The truck's hoist hydraulic control valve is controlled with a four position lever, but has five operating positions.
- Raise the bed.
- Lower the bed.
- Hold the bed.
- Float.
- Snub.

Note

The operator can only select raise, lower, hold, or float position. The control valve's snub position is solely actuated (electronically) by the ECM any time the operator has selected float or lower and if the truck body (bed) switch is in a down position.

When the truck's hoist (bed) control valve is in the float, hold, or snub position, the hoist pump's flow is directed to the brake cooling circuit, where it is combined with the

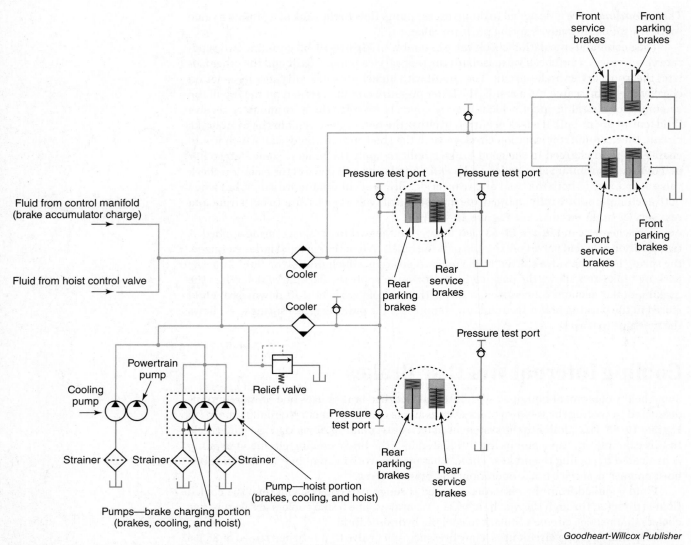

Figure 19-65. Off-highway trucks may use several fixed-displacement pumps for cooling internal wet disc brakes. This example is similar to a Caterpillar 777G truck. It has one pump dedicated to brake cooling and three other pumps that are used for brakes and hoist, as well as brake cooling.

flow from the dedicated brake cooling pump. The hoist pump's flow is not cooled when the hoist control valve is in the lower or raise position.

During normal truck operation, the hoist control valve is left in the float mode. This allows the bed to sit fully on the truck frame pads rather than resting on the hoist cylinder, which would cause high pressure in the hoist cylinder. When the bed control valve is placed in the lower position, the truck body (bed) is hydraulically lowered. As the bed approaches the fully lowered position, the ECM will actuate the hoist control valve to the snub position. This slows the final retraction of the truck's dump bed. Technically, when the truck is traveling in normal operation, the operator places the control valve in the float position. And, as mentioned earlier, any time the lever is in the float position (or lower position) and when the body switch is in the down position, the ECM electronically actuates the hoist control valve to the snub position (even though the operator can only recognize that the control valve is in the lower position). While in the snub position, the pump's flow is directed to the brake cooling system.

Towing

Many heavy equipment machines use spring-applied parking brakes, which require fluid pressure to compress the parking brake spring, releasing the parking brake. If a machine must be towed and the engine will not start, the parking brake spring must be compressed in some other way.

Warning
Always follow the instructions in the manufacturer's service literature for towing a machine.

Some machines use accumulators to compress the parking brake spring. If the parking brake accumulator is charged and does not leak, the accumulator can supply the necessary fluid pressure to compress the spring, releasing the parking brake and allowing the machine to be towed.

Pro Tip
Towing a dead machine with the parking brake disengaged will cause the wheels to spin. Spinning drive wheels force the axles and drive shafts to spin. With the engine not running, the transmission will not be lubricated. Therefore, be sure to follow the manufacturer's service literature regarding recommended towing procedures.

Instructions for towing a dead machine may include removing the drive shafts or releasing the parking brake. In these scenarios, the parking brake will be completely ineffective. Special care must be taken to ensure that the towed machine is positively attached (rigidly attached) to the tow vehicle, which is responsible for stopping the towed machine.

Case IH Combine

Case IH 2300 series combines are equipped with a ***tow valve***. The tow valve enables the operator to release the parking brake on a machine with a dead engine by pressing the left service brake pedal. The batteries must be fully charged.

Be sure to follow the manufacturer's service literature to avoid damaging the machine or causing injury. The following steps are an example of how to manually release the parking brake:

1. Turn on the ignition.
2. Shut off the parking brake switch.
3. Place the hazard/tow switch in the "tow" position.
4. Rapidly press the left service brake pedal until the brake lamp turns off.

Note
The amount of internal leakage within the brake assembly will determine how long the parking brake remains released. If the brake has moderate internal leakage, the parking brake will soon reapply. If the parking brake solenoid loses power, the parking brake will apply, which could cause injury or machine damage. Therefore, special caution must be taken when towing a machine.

Another style of Case IH combines—flagship combines from 2008 to 2016—use the spring-applied piston to hold the parking brake rotor, as shown in **Figure 19-41**. This parking brake cannot be released using the service brake pedal, but it can be released manually using a wrench to compress the parking brake spring. Be sure to follow the manufacturer's service literature. The following steps are an example of how to release the parking brake:

1. Raise the header and place the header cylinder lock (safety latch) in place.
2. Chock the wheels.
3. Remove the rubber boot.
4. Remove the cotter pin from the castle nut.
5. Use a 30 mm wrench to tighten the castle nut until the parking brake piston reaches the end of its travel, effectively compressing the parking brake spring.

Caterpillar 784C and 785C Trucks

The 784C and 785C Caterpillar mining trucks have air-over-hydraulic brakes, which use air to actuate hydraulic brakes. The minimum amount of air pressure is 80 psi. If the truck is incapable of building the 80 psi of air pressure, an external source can be used. Care should be taken, as the system can be damaged by pressures that exceed 120 psi.

Caterpillar 777G Truck

The Caterpillar 777G rigid frame haul truck has an electric-powered hydraulic pump that is used for secondary steering and parking brake release. The pump operates in two modes: auto and manual. The manual mode is used for towing the truck. The pump will provide fluid to release the parking brake and to steer the truck.

In the auto mode, the pump will provide fluid for steering and parking brake release when the fluid pressure in the primary steering and parking brake release system drops. Like on most machines, primary fluid for steering and parking brake release is powered by an engine-driven hydraulic pump. Any time the secondary pump is operating, an indicator light is on in the operator's cab.

Warning

The secondary steering and brake pump is only for emergency use. The system is designed to be used to move the machine to a safe area and should not be used for a long period. If the system is used to tow the machine a long distance, the electric motor will fail.

Case Study
Parking Brake Adjustment

A New Holland telehandler's parking brake was inoperable. It was a spring-applied, hydraulically released parking brake, actuated with a lever. When the lever was actuated, the parking brake light on the dash would correspond with the movement of the lever. However, the parking brake would not engage, allowing the machine to roll on a slope. The technician studied the service literature and found that the parking brake was inside the front axle, inboard. The literature stated that the brake could be mechanically released by loosening two jam nuts on each side (right side parking brake and left side parking brake) and turning the cap screws inward. See **Figure 19-66**. In order to set the parking brake back to the operable state, the literature stated to set the adjustment cap screws to a 30 mm length. The technician found that the jam nuts for the adjusting screws were all loose. The technician backed out the parking brake release cap screws a little past the 30 mm length setting and retightened the jam nuts and the parking brake worked perfectly.

Adjusting Internal Wet Disc Brakes

As mentioned earlier, some heavy equipment incorporates wear indicators and adjustments with the internal wet disc brakes. **Figure 19-45** shows a motor grader brake. The wear indicator is covered with a cap. After the cap is removed, a technician can compare how far the indicator protrudes from the housing. Service literature lists three types of brake conditions based on the position of wear indicator:

- Indicator in the middle depth = brakes are new.
- Indicator sunken into the bore = brakes are worn.
- Indicator protruding from the bore = brakes are warped.

The brake compensator consists of a threaded bolt with a jam nut. The service literature provides the sequence for inspecting and adjusting the brake. The following is the procedure recommended by one manufacturer:

1. Move the machine away from personnel and other equipment.
2. Place the transmission in neutral.
3. Apply the parking brake.
4. Place the machine's implements on the ground.
5. Shut off the engine.
6. Chock the wheels.
7. Remove the cap from the brake wear indicator.
8. Apply the service brakes while inspecting the wear indicator.
9. With the service brakes fully applied, loosen the jam nut.
10. Unthread the compensator bolt, and then rethread the bolt into the brake housing until the bolt touches the brake piston.
11. Unthread the compensator bolt the number of revolutions and degrees indicated in the service literature, such as two revolutions and 60°.
12. While holding the compensator bolt in the set position, torque the jam nut to the specification.

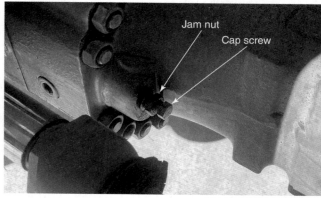

Goodheart-Willcox Publisher

Figure 19-66. This New Holland telehandler has inboard spring-applied parking brake that can be disengaged by loosening the jam nut and tightening the cap screw. When the cap screws are tightened, the parking brake is mechanically released and the telehandler will roll.

Summary

- Service brakes are used for dynamic braking, bringing a moving machine to a stop.
- Individual service brakes on the left or right can be used to make a tight turn.
- Parking brakes are used for static braking, holding a stationary machine in a parked position.
- A combination brake is used for both a parking brake and a service brake.
- Pascal's law states that when a force is placed on liquid enclosed in a container, pressure develops that acts equally in all directions.
- A master cylinder is an actuator assembly that contains a mechanically operated piston used to direct fluid to a brake's wheel cylinder.
- A single master cylinder has just one piston.
- A tandem master cylinder contains two pistons, lined up with one in front of the other.
- Dual master cylinders use two single master cylinders, one to control the right-side brakes and one to control the left-side brakes.
- Boost-assist master cylinders use another hydraulic source to increase the brake pressure.
- A service brake control valve is used to direct pump flow to apply service brakes.
- A hydraulic slack adjuster is used to adjust the brake slack that occurs due to brake wear.
- Drum brakes use two brake shoes to stop a rotating drum.
- In a nonservo drum brake, one of the two shoes provides the majority of the braking action, based on the direction of machine travel.
- When the machine is moving forward, the leading shoe provides the braking action in a nonservo drum brake.
- When the machine is moving in reverse, the trailing shoe provides the braking action in a nonservo drum brake.
- In a duo-servo drum brake both shoes apply braking force when the brakes are applied.
- Ball ramp disc brakes contain two actuator discs with steel balls that roll up ramps to apply pressure to the brake discs.
- External caliper brakes squeeze brake pads against a single rotating disc known as a rotor.
- A fixed caliper uses two hydraulic pistons that are opposed to one another.
- A sliding caliper uses one or more pistons located on only the inboard side of the caliper.
- Caliper brakes can be used for parking brakes or service brakes.
- Some caliper parking brakes are mechanically applied and some are spring applied.
- Internal wet disc brakes are commonly used in heavy equipment, normally placed inboard or outboard.
- A secondary brake pedal is a left-hand brake pedal used to apply the parking brake in the event the service brakes are inoperative
- A band brake can be used as a parking brake, service brake, PTO brake, or a track steering brake.
- Swept area is the amount of brake drum or rotor area that sweeps past the friction material during a single revolution.
- An accumulator is a hydraulic storage device used to store hydraulic energy that can be used to apply service brakes.
- Gas accumulators are charged with nitrogen gas.
- Bladder accumulators have a synthetic rubber bladder.
- A piston accumulator looks similar to a hydraulic cylinder without a rod.
- A diaphragm accumulator has a flexible rubber diaphragm.
- Accumulators should only be charged while they are mounted to the machine.

- An accumulator must have its hydraulic fluid drained before the technician can safely measure, check, or add nitrogen.
- Cut-in pressure is the spring value of an accumulator charging spool that allows the fluid to charge an accumulator.
- Cut-out pressure is the spring value of an accumulator charging spool that shuts off fluid from charging an accumulator.
- Heavy equipment can use multiple hydraulic pumps to cool brakes.
- Accumulators can be used to compress the parking brake spring to allow the machine to be towed.
- Some parking brakes are mechanically released using a wrench.
- Internal brake compensators are set while the service brakes are fully applied.

Technical Terms

accumulator
actuating discs
ball ramp disc brake
band brake
bladder accumulator
boost-assist master cylinder
caliper
combination wet disc brake
compensating port
compensation valve
cut-in pressure
cut-out pressure
diaphragm accumulator
dry disc brake assembly
drum brakes
dual master cylinder
duo-servo drum brake
fixed caliper
floating caliper
fluid reservoir
gas accumulators
hydraulically actuated caliper disc brakes
hydraulic brakes
hydraulic slack adjuster
hydrophobic
hygroscopic
inboard hydraulically actuated wet disc brakes
internal wet disc brake
inverse shuttle valve
leading-trailing shoe drum brake
master cylinder
mechanically actuated brakes
nonservo drum brake
outboard hydraulically actuated internal wet disc service brakes
parking brake
Pascal's law
piston accumulator
precharge pressure
primary shoe
replenishing port
rotor
secondary brake pedal
secondary shoe
self-energizing braking
service brake control valve
service brakes
single master cylinder
slave cylinder
sliding caliper
spring-applied parking brake
swept area
tandem master cylinder
tow valve
turning brakes
unloading valve
wheel cylinder

Review Questions

Answer the following questions using the information provided in this chapter.

Know and Understand

1. What is the least common method used for applying the service brakes in large heavy equipment?
   A. Mechanically applied.
   B. Hydraulically applied.
   C. Air applied.

2. Which of the following is used to equalize brake pressure from two different master cylinders for the purpose of equalizing brake pressure to both wheel cylinders?
   A. Compensation valve.
   B. Replenishing port.
   C. Vent port.
   D. None of the above.

3. A *hygroscopic* fluid _____.
   A. repels water
   B. has no effect on water
   C. absorbs water
   D. is approximately a 50% mixture of water

4. Brake systems can use all of the following types of hydraulic pumps, *EXCEPT*:
   A. positive displacement.
   B. non-positive displacement.
   C. fixed displacement.
   D. variable displacement.

5. Nonservo brakes use _____.
   A. primary and secondary shoes
   B. rotors and pads
   C. leading and trailing shoes
   D. five steel balls

6. Duo-servo brakes use _____.
   A. primary and secondary shoes
   B. rotors and pads
   C. leading and trailing shoes
   D. five steel balls

7. Caliper disc brakes are used for _____ brakes.
   A. service
   B. parking
   C. Both A and B.
   D. Neither A nor B.

8. Which style of caliper uses pistons that oppose one another?
   A. Fixed.
   B. Sliding.
   C. Both A and B.
   D. Neither A nor B.

9. Which style of caliper uses one or more pistons on only the inboard side of the caliper?
   A. Fixed.
   B. Sliding.
   C. Both A and B.
   D. Neither A nor B.

10. Combination internal wet disc brakes are used for _____ brakes.
    A. parking
    B. service
    C. Both A and B.
    D. Neither A nor B.

11. The purpose of a left-hand secondary brake pedal is to _____.
    A. increase the service brake apply force
    B. apply the parking brake in the event the service brakes are inoperative
    C. provide primary service braking action
    D. apply trailer brakes

12. The _____ area is the amount of brake drum or rotor surface that passes the friction material during a single revolution.
    A. swell
    B. slept
    C. slew
    D. swept

13. A _____ accumulator contains a synthetic rubber bag.
    A. piston
    B. bladder
    C. diaphragm
    D. spring

14. A _____ accumulator resembles a hydraulic cylinder without a rod.
    A. piston
    B. bladder
    C. diaphragm
    D. spring

15. All of the following actions must be followed in order to perform safe accumulator service, *EXCEPT*:
    A. discharge fluid pressure before service.
    B. discharge gas pressure before repair.
    C. apply heat to dislodge an accumulator with seized threads.
    D. never weld on an accumulator.

16. If a gas accumulator is suspected of having a nitrogen leak, what can be used to detect the leak?
    A. Electronic leak detector.
    B. Butane torch leak detector.
    C. Soap and water.
    D. Dye and a black light.

17. A(n) _____ directs fluid to fill an accumulator.
    A. charging spool valve in the cut-in position
    B. charging spool valve in the cut-out position
    C. unloading valve
    D. inverse shuttle valve

18. A(n) _____ senses both the front and rear service brake accumulator pressures and blocks flow to both circuits when the accumulators are fully charged.
    A. charging spool valve in the cut-in position
    B. charging spool valve in the cut-out position
    C. unloading valve
    D. inverse shuttle valve

Apply and Analyze

19. The _____ brakes are dynamic brakes used to bring a machine to a stop.
20. The term _____ brake is used to describe a right brake pedal controlling a right-side brake.
21. The _____ brakes are static brakes used to hold a machine stationary.
22. Pascal's law stipulates that when a force is placed on liquid enclosed in a container, it develops _____ that acts equally in all directions.
23. The _____ cylinder is used to deliver fluid to the hydraulic actuating piston (actuating cylinder).
24. A(n) _____ master cylinder contains two pistons (one in front of the other).
25. Hydraulic slack adjusters are used to adjust for brake _____.
26. _____ brakes use steel balls and pockets in the shape of teardrops.
27. _____ brakes have a rotor.
28. Rotors are normally _____ cooled.

Critical Thinking

29. Describe how to properly release a parking brake spring on a Case IH flagship 240 series combine.
30. Describe how to adjust the internal compensator on a Caterpillar wheel motor grader's service brake.

Ake Apichai Chumsri/Shutterstock.com

Air brake systems are commonly used on on-highway trucks and some types of off-highway equipment.

Chapter 20
Air Brake Systems

Objectives

After studying this chapter, you will be able to:

✓ Explain the purpose and operation of the components used in an air brake supply system.

✓ Explain the advantage of a dual circuit air brake system.

✓ List examples of machines that use a single circuit air brake system.

✓ Explain the operation of common valves used in air brake systems.

✓ Identify the types of foundation brakes used in air brake systems.

✓ Explain the operation of an air-over-hydraulic brake actuator.

✓ Compare the differences between air brakes and hydraulic brakes.

While hydraulic brakes are more common than air brakes in off-highway machines, air brake systems are used in some machines, including loader backhoes, wheel loaders, large compactors, scrapers, and haul trucks. While hydraulic brakes are more common than air brakes in off-highway machines, air brake systems are still used in some older machines, including loader backhoes, wheel loaders, large compactors, scrapers, and haul trucks. Dry fertilizer applicators like AGCO's TerraGator, Case IH's Titan Floater, and John Deere 800R floater are examples of late-model products that use air brakes. On-highway class 8 trucks primarily use air brake systems.

In an air brake system, a compressor supplies compressed air that is used to apply the brakes. The compressed air is stored in one or more air tanks, and is directed by brake valves to brake actuators, which apply the foundation brakes. The foundation brakes are either drum brakes or caliper disc brakes. Air brake systems consist of the following:

- Air brake supply system.
- Primary circuit.
- Secondary circuit.
- Foundation brakes.

Air Brake Supply System Components

An *air brake supply system* consists of a compressor, governor, air dryer, and supply tank. The engine-driven air compressor compresses air and is regulated by a governor. The governor is discussed later in the chapter. The compressor directs the compressed air to an air dryer, which removes moisture and contaminants from the air. The dryer sends the clean, dry air to an *air supply tank*, also known as a *supply reservoir* or wet tank. The air supply tank stores the compressed air so that it can be used to apply the machine's brakes. A simplified air brake supply system is illustrated in **Figure 20-1**.

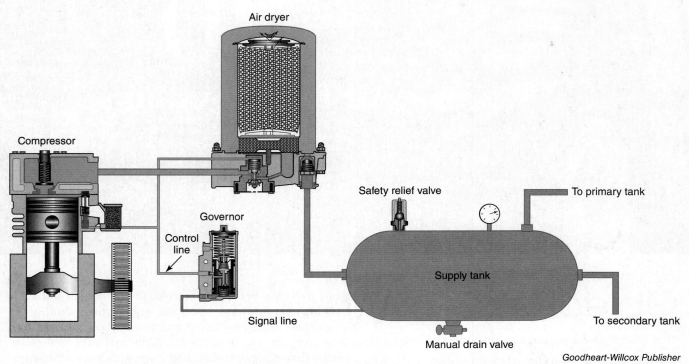

Figure 20-1. A simplified air brake supply system consists of an engine-driven air compressor, a governor, an air dryer, and a supply tank.

Air Compressor

The *air compressor* generates the compressed air used to apply the air brakes. The compressor is normally driven by the engine's accessory gear drive and rotates at a speed proportional to engine speed. In some rare applications, the compressor is belt driven.

Compressors vary in size and displacement. They are typically equipped with one or two cylinders, but can have up to four cylinders. Like the cylinders in a traditional internal combustion engine, each of the compressor's cylinders contains a piston. Each piston is attached to the crankshaft's offset throw by a connecting rod. As the engine runs, it forces the compressor's crankshaft to spin, and the crankshaft's offset throw causes the piston to slide up and down inside its cylinder. Piston rings seal the small gap between the piston and the cylinder wall while allowing the piston to slide in the cylinder.

The compressor is cooled with the engine's coolant and lubricated with the engine's crankcase oil. Coolant is routed through the compressor's cylinder head. The compressor receives pressurized engine oil at the crankshaft rear cover. The oil is directed to the rear main bearing and to the connecting rod bearings. Gravity causes oil to drain out of the bottom of the compressor and back to the engine's sump. If the oil drain passage becomes restricted and the compressor fills with oil, oil will bypass the piston rings and enter the air brake supply system.

Each cylinder has three valves: an inlet valve, a discharge valve, and an unloader valve. These valves allow the piston to draw in air, compress it, and deliver it out of the compressor's discharge port. The inlet and discharge valves are one-way check valves that are either open or closed. See **Figure 20-2**.

Intake Cycle

The *inlet valve* is a one-way check valve that allows air into the cylinder and blocks air from exiting the cylinder. As the piston moves downward in its bore during the intake cycle, the cylinder's volume expands. This generates a pressure that is lower than atmospheric pressure, also known as a vacuum. The low pressure causes the inlet valve to open, which allows air to be drawn into the cylinder. **Figure 20-2** shows the intake stroke of a Bendix air compressor. Machine manufacturers choose to draw inlet air from one of three locations: a dedicated compressor air filter, the engine's air filter, or the boost line of a turbocharger.

Compression Cycle

As the piston begins its upward stroke during the compression cycle, the inlet valve closes to prevent air from exiting the cylinder. **Figure 20-3** shows a simplified Bendix air compressor with the piston on the compression stroke. The piston's upward movement compresses the air in the cylinder. When the pressure inside the cylinder is higher than the pressure in the discharge line, the discharge valve opens and the compressed air exits the compressor. The ***discharge valve*** is a one-way check valve that allows the compressed air to flow out of the compressor. The inlet and discharge valves on a two-cylinder Bendix compressor are shown in **Figure 20-4**. From the compressor, the air moves to the air dryer where it is dried and cleaned before moving to the air tanks.

Note

In an air compressor with two cylinders, when one piston is on the intake cycle, the other piston is on the compression cycle.

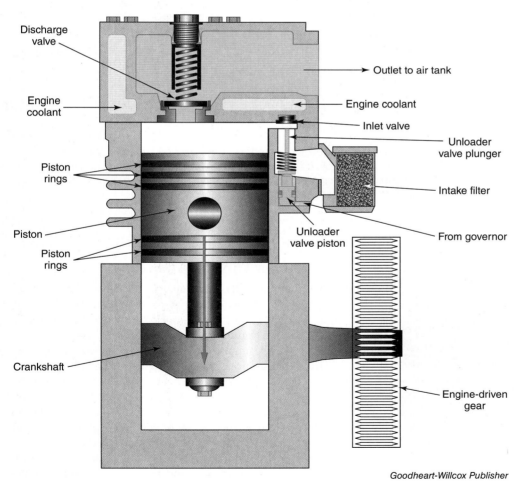

Goodheart-Willcox Publisher

Figure 20-2. This simplified drawing of a Bendix air compressor shows the piston on the intake stroke. Air is drawn into the chamber from the intake filter. The inlet valve opens, and the discharge valve remains seated.

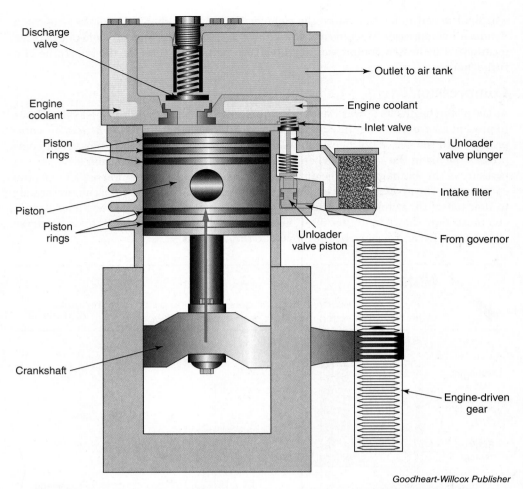

Figure 20-3. In this simplified Bendix air compressor, the piston is on the compression stroke. As the piston begins its upward movement, the inlet valve seats. As the piston compresses the air, the pressure increases. The discharge valve opens once the pressure in the cylinder becomes higher than the pressure in the discharge port.

Unloader Valve and Governor

Each cylinder has an unloader valve consisting of an unloader piston and plunger. The unloader piston is air operated by the compressor's governor. The *governor* senses and regulates the air system's pressure. The governor can be mounted directly to the compressor or away from the compressor. See **Figure 20-5** and **Figure 20-6**.

The *operating pressure*, or *system pressure*, ranges from the minimum, or cut-in, pressure to the maximum, or cut-out, pressure. If the cut-in pressure value is set too close to the cut-out pressure value, the compressor will cycle too often. The cut-in pressure value is not adjustable, and if it is incorrect, the governor must be replaced. The cut-out pressure is adjustable. Be sure to follow the manufacturer's service literature.

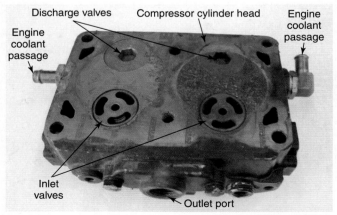

Figure 20-4. The cylinder head of a two-cylinder Bendix air compressor has been removed and turned upside down, exposing the two inlet valves and two discharge valves.

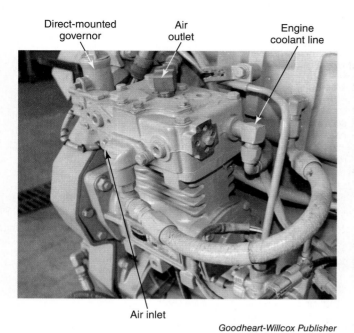

Figure 20-5. This Bendix compressor is mounted to a Caterpillar engine. The governor is mounted directly to the compressor. It is a two-cylinder compressor.

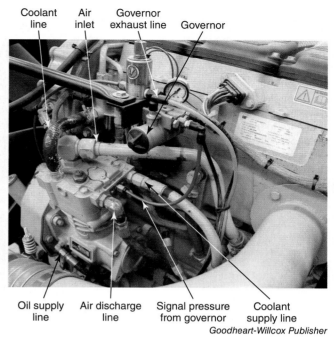

Figure 20-6. Notice this governor is mounted away from the air compressor. The compressor is a single-cylinder Bendix.

Cut-Out Pressure

When the system reaches the ***cut-out pressure***, approximately 125 psi, the governor sends a ***signal pressure*** to the compressor's unloader valves. This pressure signals the compressor to cut out, or unload. Different machines can have different governor cut-out and cut-in pressure specifications. For example, a Caterpillar 657G scraper's governor cut-out pressure specification is 120 psi and the cut-in pressure is 95 psi. A Caterpillar 926E wheel loader can be equipped with two different governors: one with a 105 psi cut-out pressure and a cut-in pressure of 85 psi, and the other with a 95 psi cut-out pressure and a cut-in pressure of 75 psi.

In **Figure 20-7**, system pressure is acting on the bottom of the governor's piston. Due to the elevated pressure, the piston moves upward. At the cut-out pressure setting, the governor's exhaust stem moves downward, opening the governor's two-way ***inlet/exhaust valve***. The inlet/exhaust valve is a two-position valve. When the governor reaches cut-out pressure, the inlet/exhaust valve allows system pressure to pass through the governor's piston through its cross-drilled passage to the unloader valve. When the governor reaches cut-in pressure, the inlet/exhaust valve blocks off system pressure while allowing the unloader valve to exhaust the air to the atmosphere. During cut out, air pressure from the supply tank moves around the inlet/exhaust valve and flows through the center channel and the cross-drilled passage of the governor's piston. It is then sent out the center port to the compressor's unloader valves.

When the unloader valves receive the governor's signal pressure, the unloader pistons move upward and their plungers open the compressor's inlet valves. See **Figure 20-8**. With the inlet valves open, the compressor cannot build pressure. Air is drawn into the cylinder during the intake stroke but, during the compression stroke, it exits the same passage it entered (the inlet) without being compressed. The compressor is operating in an unloaded state.

Cut-In Pressure

The governor has a ***cut-in pressure*** value that is approximately 25 psi less than the cut-out pressure. If the cut-out pressure is 125 psi, the cut-in pressure is 100 psi. When the operating pressure drops to the cut-in pressure, the governor blocks the system pressure and drains the air pressure in the unloader port.

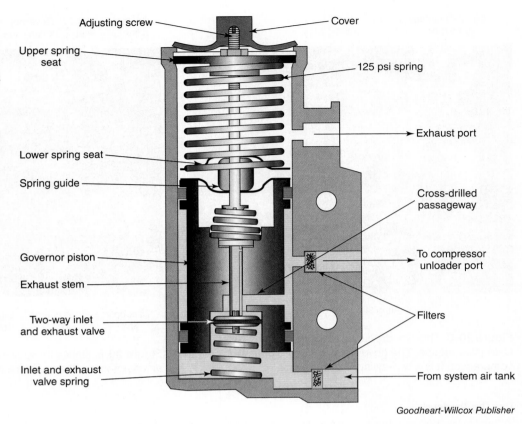

Figure 20-7. The governor regulates the compressor's air pressure. The governor senses the air tank pressure. When the pressure increases to the cut-out pressure, for example 125 psi, the governor directs a signal pressure to the compressor's unloader valves, which unloads the compressor.

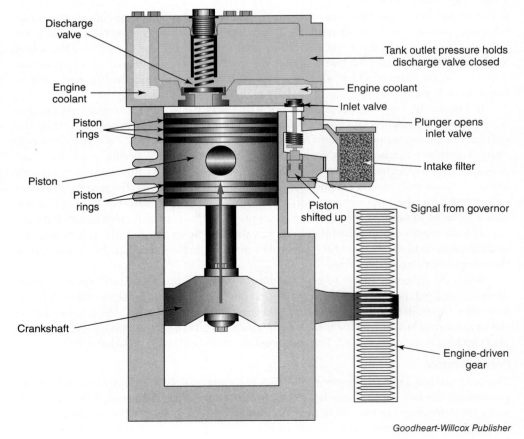

Figure 20-8. When the governor senses 125 psi system pressure, it sends a signal to the piston's unloader valves, causing the unloader plungers to shift upward. The plungers lift the inlet check valves off their seats. When the inlet valves are held off their seats, the compressor's pistons cannot build pressure because the air is pushed out through the inlet passage.

When system pressure drops below cut-in pressure, the governor's upper spring pushes the governor's piston downward in its bore, **Figure 20-9**. The inlet/exhaust valve blocks system pressure. The air from the compressor's unloader port travels through the center of the governor's exhaust stem and out of a cross-drilled passageway near the top of the exhaust stem. The air is then sent out of the governor's exhaust port.

The air brake system often contains one air pressure gauge with two needles or two separate air pressure gauges that display the primary air tank pressure and the secondary air tank pressure. As the governor is cut in, it allows the compressor to send compressed air to the supply tank. The supply tank directs compressed air to the primary and secondary air tanks. The increase in pressure in the primary and secondary air tanks is indicated on the pressure gauge(s). When the pressure reaches the governor's cut-out pressure, the gauges will read the pressure level of the governor's cut-out pressure, for example (125 psi). Primary and secondary air tanks are explained later in the chapter.

When the air pressure drops below 100 psi, the governor vents air pressure instead of sending it to the unloader. The machine's pressure gauge will rise during the cut-in mode. When the air pressure increases to approximately 125 psi, the governor directs signal pressure to the compressor's unloader valves to unload the compressor. If the machine's pressure gauge continues to rise above the manufacturer's specified cut-out pressure, there is a problem with the compressor or the governor.

Air Dryer

During the air compression process, the air is heated. Heated air can hold more water vapor than cooler air. That water vapor would condense out and form liquid water as the compressed air cools. To prevent this, from the compressor, the air is routed to an *air dryer*.

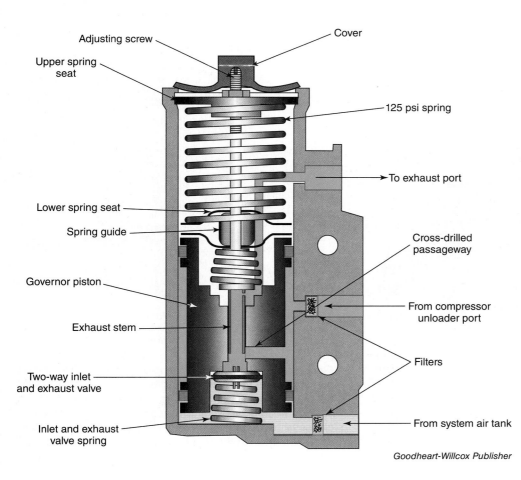

Figure 20-9. When the system is operating at or below the cut-in pressure, the governor drains air pressure from the compressor's unloader valves.

Goodheart-Willcox Publisher

The air dryer removes three types of contaminants from the air: solids, liquids, and vapor. Without an air dryer, moisture would cause corrosion within the air brake system. In cold environments, the moisture would freeze and cause inoperative air brakes.

Air dryers remove moisture by two means: condensing water vapor and draining the resulting liquid, and trapping remaining water vapor with a desiccant material. Some of the compressed air cools prior to entering the dryer, which causes moisture to condense. This water and contaminants settle in the base of the dryer, where they can later be purged. The compressed air will still contain some water vapor. This leftover water vapor is adsorbed and trapped by the desiccant material in the dryer.

A traditional air dryer has three ports that connect to air lines:

- Supply port—connects to air line coming into the dryer from the air compressor.
- Delivery, or discharge, port—delivers air to the supply tank (wet tank) through an air line.
- Control port— receives a signal pressure from the governor to purge the heavy contaminants from the dryer.

Charge Cycle

While the compressor builds air pressure, it sends air into the dryer's end cover at the base of the dryer. Some end covers have a spiral cavity that circulates and cools the air. As the air is cooled, moisture condenses and liquid water forms. The liquid water and contaminants fall to the bottom of the end cover, where it will be purged during the purge cycle. A disassembled Bendix AD-9 air dryer is shown in **Figure 20-10**.

From the end cover, the air passes through the dryer's oil separator, which removes liquid water, oil, and solid contaminants. The air next passes through a desiccant drying bed, which removes 95% of water vapor in the air. See **Figure 20-11**.

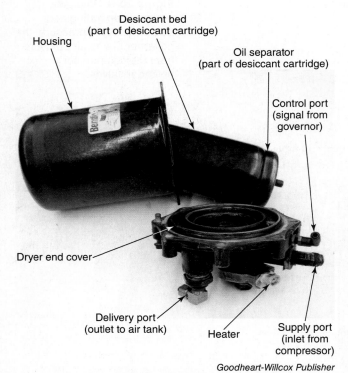

Goodheart-Willcox Publisher

Figure 20-10. This Bendix AD-9 air dryer was replaced because the desiccant cartridge came loose from the end cover due to worn threads. The housing has been removed from the end cover.

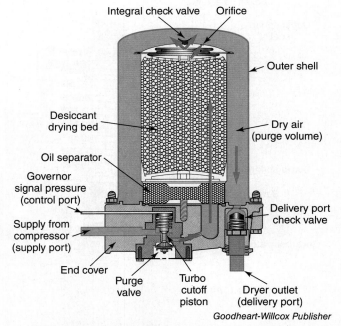

Goodheart-Willcox Publisher

Figure 20-11. A simplified drawing of a Bendix AD-9 integral purge air dryer in the charge cycle. The compressor has not reached the cut-out pressure and a signal has not been sent to the dryer. The compressed air travels through the end cover, oil separator, desiccant drying bed, purge volume area, and then out the dryer outlet check valve, where it is routed to the supply air tank.

Most of the air exits the desiccant cartridge through an integral check valve at the top of the desiccant drying bed. A small portion of air exits through an orifice at the top of the desiccant drying bed. The dry air from the desiccant enters the *purge volume*, which is the area between the desiccant cartridge and the dryer's outer shell.

The air then moves through the ***delivery port check valve***, a one-way check valve that allows dry air to exit the dryer. The air exiting the delivery port check valve is sent to the supply tank. The delivery port check valve also prevents air from the supply tank from being dumped during the dryer's purge cycle. See **Figure 20-11** and **Figure 20-12**.

Purge Cycle

When the system pressure reaches governor cut-out pressure, the governor sends a signal pressure to the compressor's unloader valves and to the air dryer's control port, which operates the purge valve. The dryer's ***purge valve*** is designed to expel the heavy contaminants, including solids and liquids, from the dryer. See **Figure 20-12** and **Figure 20-13**.

When the governor's signal pressure actuates the purge valve, the tapered portion of the valve, called the turbo cutoff piston, blocks the compressor's supply air. If this port is not blocked on air compressors that receive inlet air from the turbocharger, actuating the purge valve would cause a drop in turbo boost pressure.

The purge valve opens a passage to the dryer's exhaust port, and the air in the dryer reverses its direction. The dry pressurized air inside the dryer's purge volume is allowed to pass from the purge volume (high pressure) across the orifice, resulting in a pressure drop. The drop in pressure causes the air to expand. As mentioned earlier, the air in the purge volume between the desiccant cartridge and the outer shell is 95% free of moisture. This dry air dries even further as it expands and is then used in a regenerative process to pull moisture from the desiccant. It travels through the top of the desiccant cartridge, drawing moisture from the desiccant drying bed. The air continues through the oil separator and

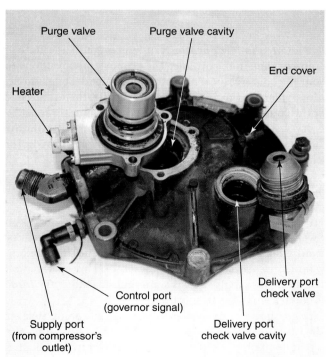

Goodheart-Willcox Publisher

Figure 20-12. The end cover has been removed from this Bendix AD-9 air dryer. The delivery port check valve and the purge valve have both been removed from their cavities.

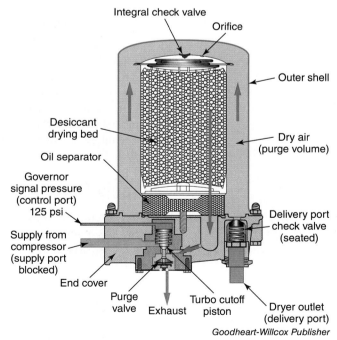

Goodheart-Willcox Publisher

Figure 20-13. When the air pressure reaches approximately 125 psi, the governor sends a signal pressure to the compressor's unloader valves and the dryer's purge valve. The purge valve shifts and closes off the supply from the compressor. The dryer then purges the air out the exhaust port.

expels the contaminants out of the exhaust port in the bottom of the dryer's end cover. During the purge cycle, the delivery port check valve seats, preventing the supply tank air from flowing out through the dryer.

Pro Tip

A normally closed electric solenoid valve can be used to direct signal air pressure to the governor to unload the compressor and purge the dryer. Caterpillar G series scrapers, such as the Caterpillar 637G, use three pressure sensors. When the three pressure sensors measure a pressure equal to the cut-out pressure value, the transmission control module energizes the solenoid valve, which directs air pressure to the governor, causing it to unload the compressor and purge the dryer. If one of the pressure sensors measures a pressure below the cut-out value, the solenoid returns to a closed position and the compressor builds air pressure. A schematic of the Caterpillar 637G system is shown later in this chapter.

Traditional dryers have one desiccant cartridge, but some have two cartridges. The dual cartridge dryers can operate continuously in both modes: one cartridge dries the air while the other regenerates. The cartridges alternate modes, switching between drying and regenerating.

The Bendix AD-9 is an *internal purge air dryer*, which means that the air inside the dryer is used to regenerate the desiccant material. Some single-cartridge air dryers use a purge tank or a system purge design, but these are less popular. Some dryers, such as Cummins dryers, are equipped with a fourth air line. When the purge valve is opened on three-line dryers, the pressure in the dryer's discharge line is depleted. This lowers the compressor's cylinder head pressure and causes compressor oil loss. A fourth line connects the dryer's supply line and the discharge line so the compressor is always running under a load and oil consumption is reduced.

Alcohol Evaporators

Some older air brake systems have an *alcohol evaporator* that injects small amounts of alcohol into the air system to minimize icing in the air line. The evaporators inject one to two ounces of alcohol for every hour the compressor is operating under load. For example, if the machine averages four hours of compressor loading during a shift, the evaporator injects 4–8 ounces of alcohol over the course of that shift. The alcohol evaporator must be mounted so it operates after the dryer. Alcohol injected before an air dryer will ruin the desiccant by causing it to gel. If the air system does not have a dryer, the alcohol evaporator is placed in the compressor's discharge line.

Supply Tank

After the compressed air has been cleaned and dried, it is sent to the air supply tank, also known as a supply reservoir or wet tank. The term *wet tank* was originally coined for machines without air dryers, because moisture would collect in the tank as the hot compressed air cooled. The air supply tank may also contain oil vapor if the compressor has oil control issues. An air tank located after the supply tank is sometimes called a *dry tank*, because most of the moisture has condensed in the supply tank.

Depending on the machine's design, the air brake system might have one or multiple air tanks. The Case 680H loader backhoe uses a single air tank to control its air brake system. See **Figure 20-14**. Dual brake circuits that use multiple tanks are explained later in this chapter.

Drain Valves

Air brake systems require a *drain valve* to remove moisture (vapor) from the air brake system. Drain valves can be manually operated or automatic.

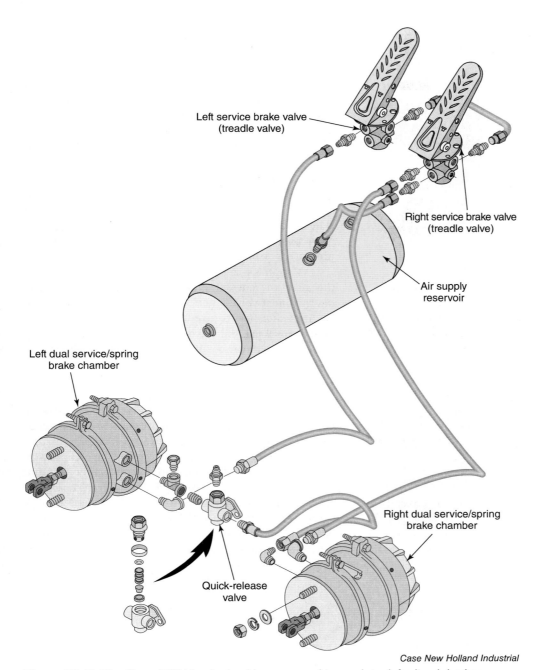

Figure 20-14. The Case 680H loader backhoe uses only one air tank for its air brakes.

Manual Drain Valves

Some manual drain valves are spring closed and require pulling a lanyard to expel moisture from the tank. See **Figure 20-15**. Moisture should be removed from the tanks at the end of each shift.

Other manual drain valves have a small knob or handle that must be rotated counterclockwise to drain the tank. This type of valve is called a drain cock. While the pull valve described in the previous paragraph should be opened after each shift to drain the reservoir, a threaded drain cock is especially helpful for draining the reservoir during repairs. The drain cock can be fully opened and left fully opened during repairs, such as fixing a leaky air line or replacing an air dryer, air valves, or other components. Manual drain valves can be located on the tank or away from the tank. See **Figure 20-16**.

Figure 20-15. Notice the metal lanyard used to open the drain valve on the air tank.

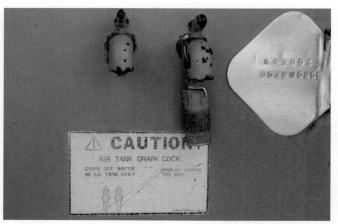

Figure 20-16. This Kawasaki wheel loader has two manual drain valves on the left rear side of the loader.

Automatic Drain Valves

Automatic drain valves contain an inlet/exhaust valve, a top reservoir port, a side reservoir port, and an exhaust port on the bottom. The drain valve is normally mounted directly to the bottom or end of the reservoir. If the machine is used in a freezing environment, the valve will have a 12- or 24-volt electric heating element.

In **Figure 20-17**, the inlet/exhaust valve is seated because the system has no air pressure. When the system builds pressure, air enters the automatic drain valve and causes the inlet/exhaust valve to flex, allowing air and moisture to settle at the bottom of the valve. The valve is designed to allow air, moisture, and contaminants to move past the inlet/exhaust valve any time the system pressure is rising. See **Figure 20-18**.

As soon as the pressure stabilizes, such as when the governor cut-out pressure has been reached, the inlet/exhaust valve will close and block system pressure from passing. See **Figure 20-19**. When the system pressure drops 2 psi, the valve opens and contaminants are exhausted to the atmosphere. See **Figure 20-20**.

Safety Valves

Air brake systems have a *safety pressure relief valve*, or *pop-off valve*, that vents system pressure when it reaches 150 psi. If this valve is venting 150 psi of air pressure, there is a problem with the governor or the compressor. The safety valve can be located on the supply air tank, compressor, air dryer, or remotely, between the compressor and the air tank. See **Figure 20-21**. When manufacturers place safety valves in the compressor's cylinder head and in the supply tank, they offer dual protection if the discharge line between the compressor and the tank becomes clogged due to freezing moisture.

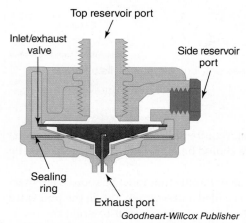

Figure 20-17. An automatic drain valve can be mounted to the bottom of a reservoir through the top port, or it can be mounted to the end of a reservoir through the side port.

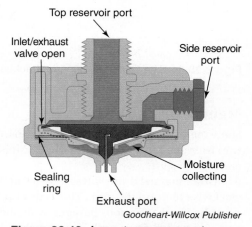

Figure 20-18. As system pressure rises, the inlet/exhaust valve opens, allowing air and moisture to settle at the bottom of the valve.

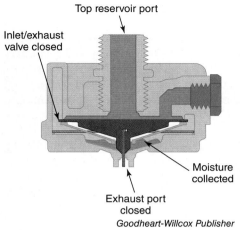

Figure 20-19. When the system pressure stops rising, the inlet/exhaust valve closes.

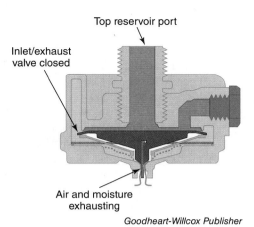

Figure 20-20. When the system pressure drops, the valve opens the exhaust path, allowing air and moisture to escape from the valve.

As part of periodic service, the safety valve should be tested by lifting the valve stem. The valve should vent system pressure when the stem is lifted and reseat when the stem is released.

Warning

As explained in Chapter 1, *Shop Safety and Practices*, compressed air poses risks to personnel. If compressed air enters your bloodstream, it can cause an embolism, leading to a stroke or a heart attack and resulting in death. Do not use your fingers to release compressed air from a safety valve. Instead, use a tool to open the safety valve.

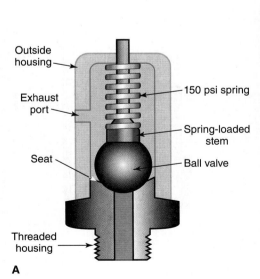

Figure 20-21. Safety pressure relief valves are used to protect air brake systems when the governor fails to unload the compressor. The safety valve can be located on the compressor, dryer, air tank, or on a line between the compressor and the air tank. A—The safety valve consists of a ball valve, a spring, and a stem. The spring and stem hold the ball valve closed against the seat. B—This shows the safety valve on top of a Bendix air compressor.

Low-Pressure Switch

Air brake systems are equipped with one or more *low-pressure switches*. This type of switch is commonly mounted in the supply tank or in both the primary and secondary tanks. The use of primary and secondary air tanks varies among machines and industries (on highway versus off highway). In general, two tanks or two sets of tanks (primary and secondary tanks) are isolated from one another through the use of check valves and/or a pressure-protection valve. This allows the primary tank to control one circuit and the secondary tank to supply air for the other circuit.

When the system pressure drops below the set pressure, such as 60 psi, the spring inside the low-pressure switch causes the electrical contacts to complete the electrical circuit, which illuminates a warning light in the cab. When the pressure increases to approximately 75 psi, the pressure forces the contacts to open, shutting off the warning light. See **Figure 20-22**.

Dual Brake Circuit

Since 1977, *on-highway* air brake systems have had to meet FMVSS 121, a law stipulating the use of a dual brake circuit. A *dual brake circuit* contains two separate circuits: a primary brake circuit and a secondary brake circuit. Each circuit has its own dedicated air tank. If a failure occurs in one circuit, the other circuit is isolated from the other by means of a check valve and/or a pressure-protection valve, allowing the good circuit to provide air pressure for bringing the machine to a stop.

In on-highway trucks, the *primary air brake circuit* applies the truck's rear tandem service brakes and the trailer service brakes via the service brake foot pedal control valve. The *secondary air brake circuit* operates the truck's front service brakes through the service brake valve and, if equipped, can apply the trailer service brakes via the trailer hand control valve. A simplified drawing depicting a supply air tank, primary air tank, and secondary air tank is shown in **Figure 20-23**.

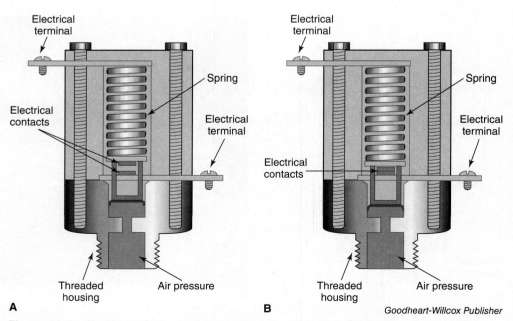

Figure 20-22. Example of a low-pressure switch. A—Pressure above 75 psi causes the piston to open the electrical contacts. B—Pressure below 60 psi causes the spring to close the electrical contacts.

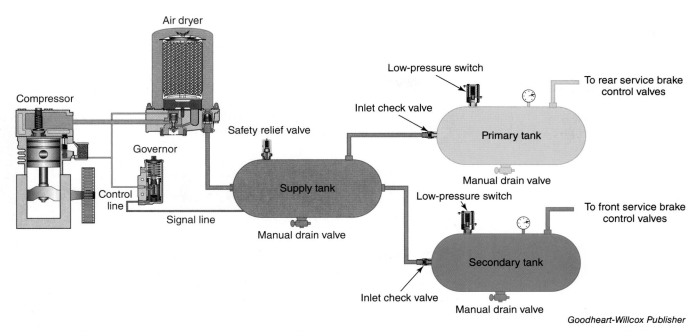

Figure 20-23. Dual air brake circuits use a primary air tank to supply the rear brakes and a secondary tank to supply the front brakes. Both the primary and secondary tanks are fed from the supply, or wet, tank.

Off-highway machines vary in how they use primary and secondary circuits. In a Caterpillar 777D haul truck, two primary air tanks, known as the service brake reservoirs, provide air for applying the machine's service brakes that are used for stopping a moving machine. The machine also has one secondary air tank, known as the parking brake reservoir. Secondary brakes are used as emergency brakes in the event that the service brakes fail. Parking brakes are used to hold the machine stopped. The service, secondary, and parking brake can all be used to apply the same friction element, such as a multiple disc brake, or they can operate separate braking elements. In a Caterpillar 637G scraper, the machine has two sets of air tanks separated by check valves. Two tanks are called the tractor air tanks and the other two are called the scraper air tanks.

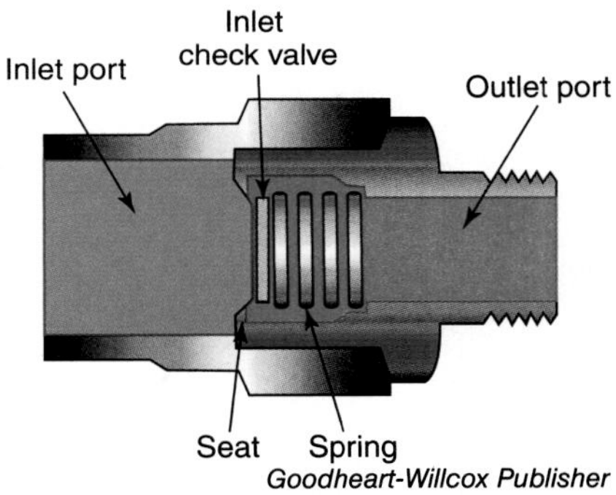

Inlet Check Valves

In the event of an air leak in either circuit of a dual brake circuit, both tanks have an *inlet check valve* that prevents the circuit with good air pressure from bleeding to the leaking air circuit. See **Figure 20-24**. Depending on the severity of the leak, the air pressure in the good circuit might only be enough to bring the truck or machine to a stop.

Figure 20-24. Inlet check valves are placed on the inlet side of air tanks. They isolate each tank in the event that a leak occurs in one of the circuits.

Dual Circuit Treadle Valve

Air brake systems use a foot-operated service brake control valve, also known as a *treadle valve*, to apply the service brakes. An example of a Bendix dual circuit treadle valve is shown in **Figure 20-25**.

A dual circuit treadle valve receives air from both the primary tank and the secondary tank. The primary and secondary supply ports are shown on the right side of the valve. The treadle valve is closed when the treadle, or brake pedal, is released. The closed valve enables the air tanks to maintain air pressure.

The treadle pivots on a pin, causing the roller to depress the plunger. As the plunger is forced downward, it moves the spring seat, which applies pressure to the graduating rubber spring. The *graduating rubber spring* provides brake pedal resistance that is felt

Figure 20-25. A treadle valve is a foot-actuated brake control valve. A—When the brake pedal is released, the primary and secondary supply ports are sealed off from the delivery ports. B—This treadle valve is displayed on an air brake trainer board.

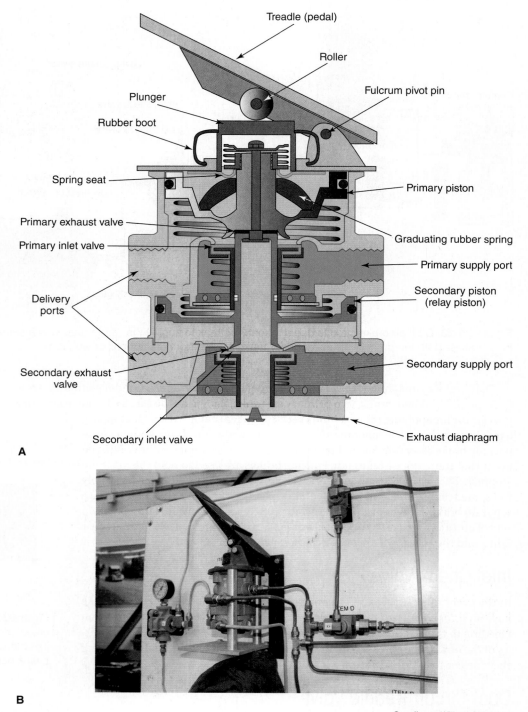

Goodheart-Willcox Publisher

by the operator. Without the graduating rubber spring, the operator would have difficulty feeling and controlling the amount of brake pressure being applied.

As the operator depresses the brake pedal, the graduating rubber spring compresses and forces the primary piston to move downward. The primary exhaust valve seat is located on the bottom of the primary piston. The initial downward movement of the primary piston closes the primary exhaust valve. As the primary piston continues moving downward, it opens the primary inlet valve. This allows air from the primary circuit to flow out of the primary delivery port to apply the brakes in the primary circuit. See **Figure 20-26**.

Chapter 20 | Air Brake Systems

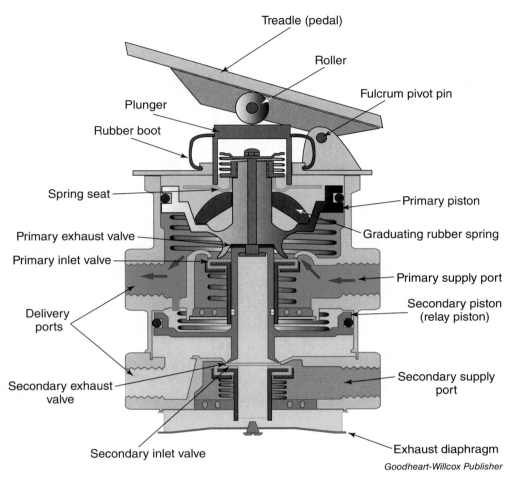

Figure 20-26. As the pedal is depressed, the primary piston closes the primary exhaust valve and opens the primary inlet valve. This allows the primary circuit air to be directed out of the primary delivery port.

Goodheart-Willcox Publisher

The air in the primary delivery port passes through a bleed passageway and shifts the secondary, or relay, piston downward. The secondary exhaust valve closes and the secondary inlet valve opens. Air pressure from the secondary supply port is then directed out of the secondary delivery port to apply the secondary brakes. See **Figure 20-27**.

During brake application, the primary piston achieves a balanced state, which means that the primary air pressure equals the pressure applied by the operator's foot on the pedal. The primary piston then moves upward, closing the inlet valve and blocking the primary supply port. The secondary piston achieves a balanced state when the air pressures on both sides of the piston stabilize to the same pressure. The secondary piston then closes the secondary inlet valve and blocks the secondary supply port. When the operator releases the brake pedal, the springs force the pistons upward and air is exhausted out the bottom of the treadle valve. See **Figure 20-28**. If a panic application is made and the pedal is fully depressed, both pistons are mechanically pushed down and maximum reservoir air pressure is applied to both circuits.

Notice that the secondary piston isolates the primary side circuit from the secondary side circuit. If a leak occurs in one circuit, the other circuit will continue to apply brake pressure. If the leak occurs in the primary circuit, the secondary piston will be mechanically actuated rather than air actuated.

One off-highway machine that uses a dual circuit air brake treadle valve is the Caterpillar G series scraper. In a Caterpillar 637G scraper, the primary circuit controls the tractor's service brakes and the secondary circuit controls the scraper's service brakes. See **Figure 20-29**.

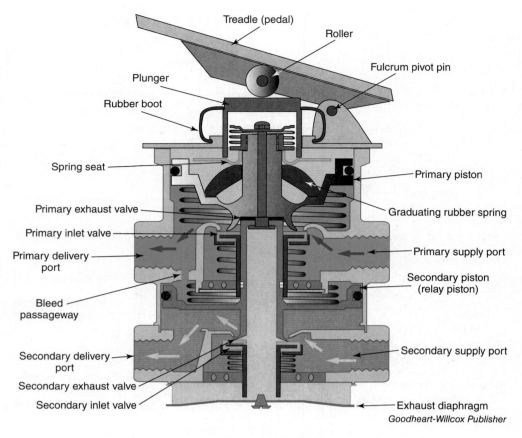

Figure 20-27. Air pressure in the primary delivery port passes through the bleed passageway and shifts the secondary piston downward. The secondary piston closes the secondary exhaust valve and opens the secondary inlet valve. Air pressure from the secondary supply port flows out the delivery port to apply the secondary brakes.

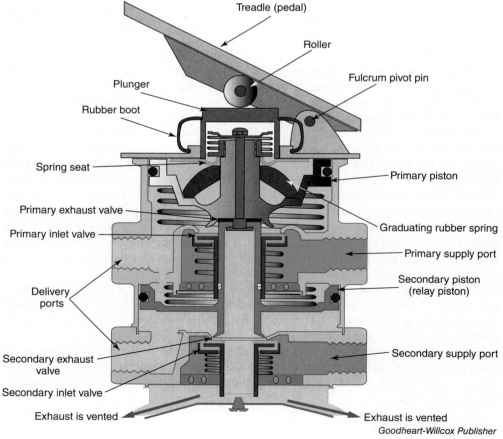

Figure 20-28. When the operator releases the treadle, the air in the delivery ports is exhausted out the bottom of the treadle valve.

Chapter 20 | Air Brake Systems

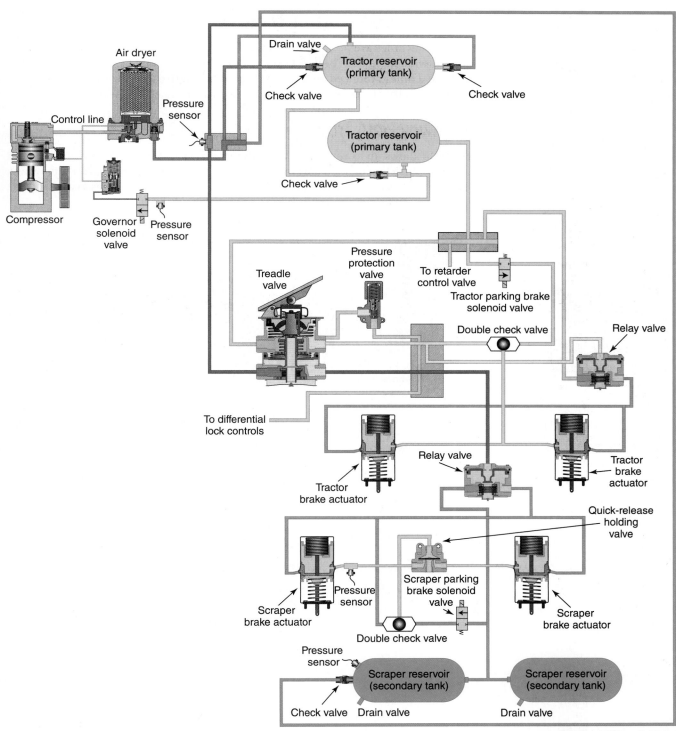

Figure 20-29. An example of a dual circuit air brake system used on a Caterpillar 637G scraper. Air brake actuators are explained later in this chapter. The brake actuators are drawn with the parking brake spring compressed (released).

Note

In off-highway applications, the treadle valve can be mounted to a firewall or to the floor.

Off-Highway Single Circuits

Off-highway machines are not required to conform to FMVSS 121, so not all off-highway air brake systems use a dual brake circuit. The systems vary among manufacturers and among the machines produced by each manufacturer. **Figure 20-14** shows a loader backhoe brake system that uses a single air tank to operate the machine's service brakes.

Off-Highway Haul Truck

Figure 20-30 shows an example of a 777D Caterpillar haul truck. It has a single circuit treadle valve, which is explained next in this chapter. The truck does not have a supply tank but instead uses three air tanks. Two tanks are used for the service brakes and a retarder,

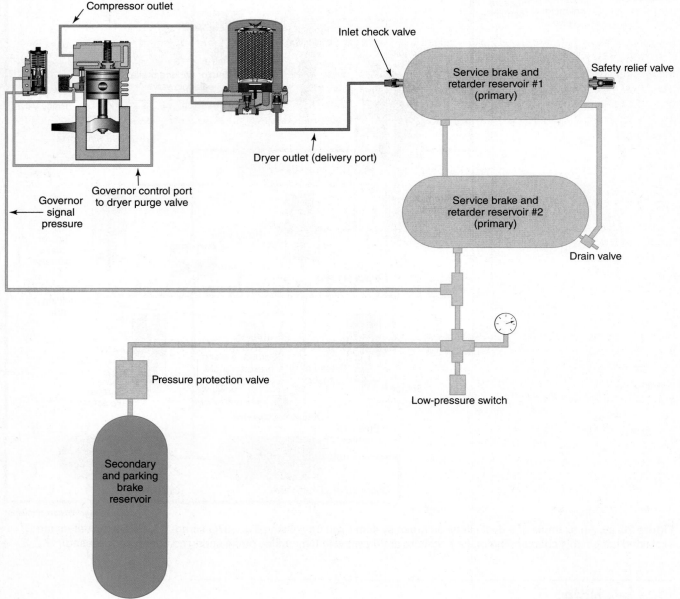

Figure 20-30. Some heavy equipment, such as Caterpillar D series haul trucks, use two tanks for service brakes and one tank for the parking brake. The truck does not have a supply tank.

and another tank is used for the secondary and parking brake. A retarder is used to slow the machine, but not bring it to a stop. Retarders help the operator to keep a safe speed when driving down slopes. The 777D truck's retarder consists of a hand-lever-operated air-control valve that, when actuated, applies the service brakes to retard (slow) the machine.

As previously mentioned, some machines might not have an air dryer. In these cases, water collects in the supply tank as the hot air cools. If a machine has an air dryer, like the 777D, the manufacturer might not use a supply tank. Although the truck uses a single circuit treadle valve to apply the service brakes, the truck still has two separate circuits, a service brake circuit and a secondary/parking brake circuit.

The truck uses rear multiple-disc combination brakes for service, secondary, and parking brakes. The parking brake portion of the brake has springs that are normally applied to engage the parking brake piston and an air-over-hydraulic control valve to release the parking brake.

The parking brake control valve is a two-position, hand-operated air-control valve on the right side of the operator, in the transmission control console. The secondary brake control valve is a hand-lever-operated air-control valve located on the left side of the steering column. Some haul trucks use a second foot-operated brake pedal to control the secondary brakes.

Either of the air-operated valves (secondary and parking) can be activated to block the air to the hydraulic secondary and parking brake control valve. See the schematic in **Figure 20-31**. When the parking brake control valve or secondary brake control valve is actuated, the air-operated valve will block air from entering the hydraulic control valve. When air is blocked from entering the hydraulic control valve, oil is vented from the parking brake piston, allowing the springs to apply the secondary/parking brake piston. The parking brake control valve is a two-position valve, on or off. The secondary brake control valve is variable, and varies the amount of pressure that is drained from the hydraulic control valve.

Single Circuit Treadle Valve

Treadle valves in single circuit air brake systems have a simpler design than those in a dual circuit system. Caterpillar 777D rigid frame haul trucks and Caterpillar 926F wheel loaders use single circuit treadle valves. **Figure 20-32** shows a simplified version of a single circuit treadle valve in the released position. Notice that a spring holds the piston in the lifted position and air from the outlet passage passes through the top of the valve. The air moves through the valve's center and out the bottom through the exhaust outlet. The air from the service brake reservoir is blocked and cannot flow through the service brake outlet passage.

As the operator depresses the treadle, the rubber spring pushes the piston downward, sealing the exhaust passageway and preventing air from exhausting out of the valve. When the piston reaches the inlet valve, it pushes the valve downward causing it to overcome spring pressure. The inlet valve passageway opens, and supply air is directed to the service brakes. Like the dual circuit treadle valve, the single circuit valve will balance. When air pressures on both sides of the inlet valve equalize, the inlet valve lifts and blocks air from the supply reservoir. See **Figure 20-33**.

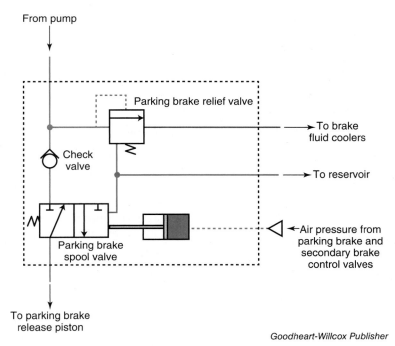

Goodheart-Willcox Publisher

Figure 20-31. A hydraulic schematic showing how air controls the release of the parking brake/secondary brake.

Figure 20-32. A simplified drawing of a single circuit treadle valve used in Caterpillar 777D rigid frame haul trucks and Caterpillar 926F wheel loaders.

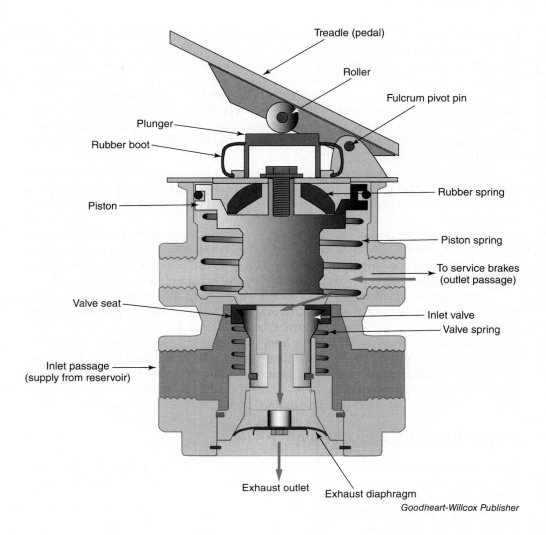

Air Brake Valves

Air brake systems often incorporate the use of different valves. These include relay valves, quick-release valves, and pressure-protection valves.

Relay Valve

A *relay valve* is used in applications that have the service brakes mounted away from the treadle valve. The relay valve works in conjunction with the treadle valve and is mounted in close proximity to the service brake actuators. The valve reduces the reaction time for applying and releasing the service brakes, so the brakes are applied and released more quickly.

As shown in **Figure 20-34**, a relay valve is connected to multiple air lines:

- Control pressure line coming from the treadle valve.
- Supply line coming from the air reservoir.
- Delivery lines going to the brake actuators.

The treadle valve directs control pressure from the treadle valve to the relay valve. Control pressure is the service brake application pressure that is developed by the brake treadle valve and is proportional to the position of the brake treadle valve. The further the treadle pedal is depressed, the higher the control pressure. The control pressure operates the relay valve, which sends air pressure from the reservoir to the brake actuators. The relay valve acts similarly to an electrical relay.

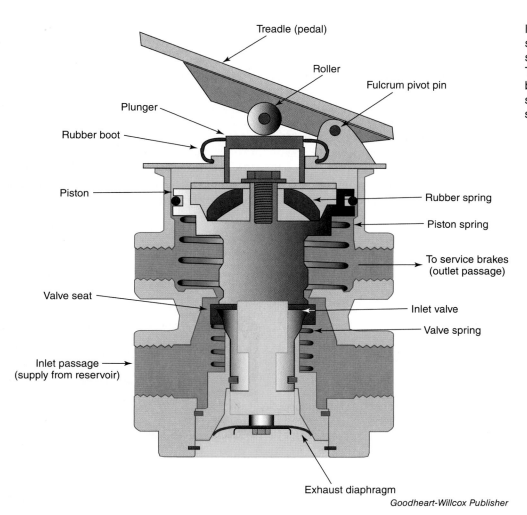

Figure 20-33. This simplified single circuit treadle valve shows the treadle depressed. The exhaust passage is blocked, and the inlet air supply is directed out to the service brakes.

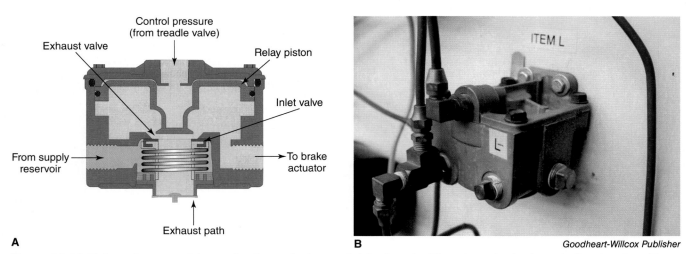

Figure 20-34. Relay valves speed the application and release of air brakes. A—The relay valve is shown in the at-rest position. B—A relay valve displayed on an air brake trainer board.

The control pressure from the treadle valve causes the relay piston to move downward. The piston's initial downward travel blocks the brake actuator air passage from the exhaust passage. See **Figure 20-35**. As the relay piston moves down farther, it pushes the inlet valve down, allowing the supply air to be directed to the brake actuator. See **Figure 20-36**.

As the brakes are applied, the air pressures on each side of the relay piston stabilize and the piston moves upward slightly. This closes the inlet valve while the exhaust valve remains closed. The relay valve is in a hold, or balanced, state when the control pressure equals the brake pressure. See **Figure 20-37**.

When the operator releases the treadle valve, the control pressure in the relay valve travels back through the treadle valve, where it is exhausted to atmosphere. Because the air pressure in the brake actuator is now higher than the vented control pressure, the relay piston moves upward and the air pressure in the brake actuator is exhausted out the bottom of the relay valve. See **Figure 20-38**.

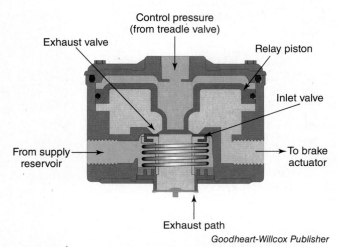

Goodheart-Willcox Publisher

Figure 20-35. When the treadle valve is depressed, it causes a control pressure to move the relay piston downward, which closes the exhaust passageway.

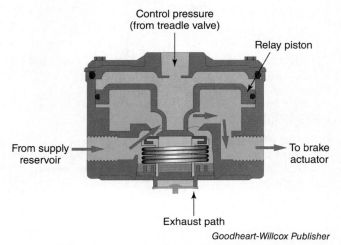

Goodheart-Willcox Publisher

Figure 20-36. As the relay piston moves down farther, it opens the inlet valve, which allows the air supply pressure to be directed to the brake actuator.

Goodheart-Willcox Publisher

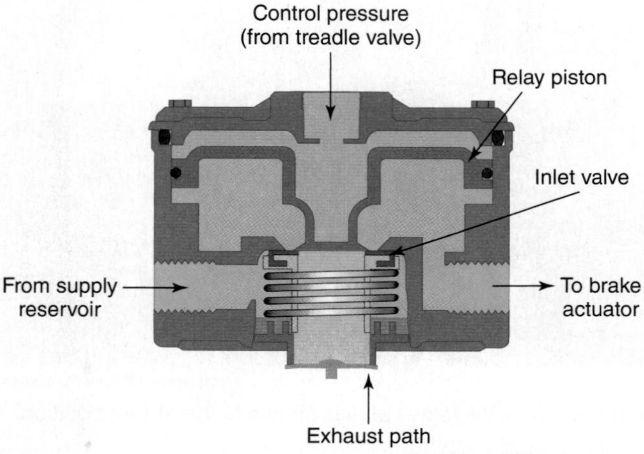

Figure 20-37. When the pressures on both sides of the relay piston equalize, the relay valve is in a hold state. The control pressure from the treadle valve equals the pressure in the brake actuator.

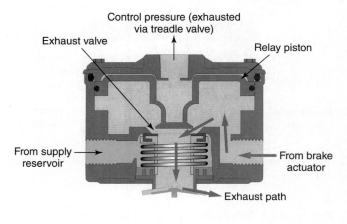

Goodheart-Willcox Publisher

Figure 20-38. When the operator releases the treadle valve, the control pressure is vented via the treadle valve and the brake actuator pressure is vented out the bottom of the relay valve.

Quick-Release Valve

A *quick-release valve* provides a method for quickly venting air pressure. While quick-release valves are more commonly used in parking brake circuits today, they have been used for service brake applications in the past and are still found on the steering axle of Meritor on-highway trucks. When used for service brakes, the quick-release valve vents the service brake pressure when the operator releases the treadle valve. The valve is placed in series between the treadle valve and the brake actuators. In the unfortunate event that the valve fails in a parking brake circuit and releases air pressure, the parking brake spring will apply the brakes. A failure in a service-brake circuit would cause the service brakes to lack the air pressure to apply.

A quick-release valve used for service brakes has four ports:
- Supply, or delivery, port coming from the treadle valve.
- Left brake actuator port.
- Right brake actuator port.
- Exhaust port.

The diaphragm is the only moving component in a quick-release valve. See **Figure 20-39**. When the treadle valve is depressed, air pressure pushes the diaphragm downward, blocking the exhaust passage. The ends of the diaphragm flex downward so air pressure flows around the diaphragm and out of the valve to the brake actuators. See **Figure 20-40**.

The air pressures above and below the diaphragm equalize. While the diaphragm continues to block the exhaust passage, it seals the inlet passage, putting the valve in a holding, or balanced, position. See **Figure 20-41**.

When the operator releases the treadle valve, the air pressure above the diaphragm is vented by the treadle valve. Air pressure in the brake actuators moves the diaphragm upward, which opens the exhaust passage and vents the brake actuator pressure. See **Figure 20-42**.

Parking Brake Control Valve

A *parking brake control valve* is used to apply and release the parking brake. Operators manually actuate a parking brake control valve by pushing or pulling the control knob. When the knob is pushed, the valve releases the parking brake. When the knob is pulled, the valve applies the parking brake. See **Figure 20-43**.

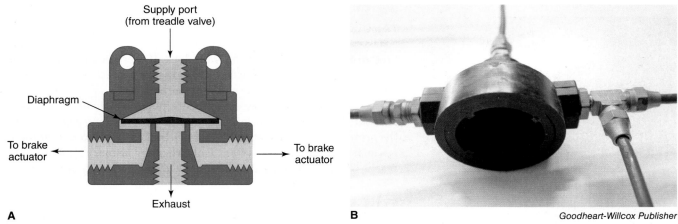

Goodheart-Willcox Publisher

Figure 20-39. A quick-release valve is used to quickly vent brake pressure. A—This valve is shown in an off state. There is no air pressure in the valve. B—This quick-release valve is displayed on an air brake trainer board.

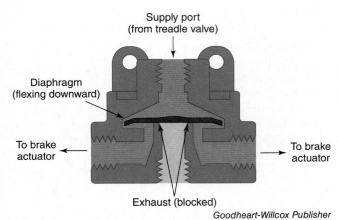

Figure 20-40. When the operator depresses the treadle valve, air pressure pushes the diaphragm downward, sealing the exhaust passage. The diaphragm flexes, allowing the pressure to be routed to the actuators to apply the service brakes.

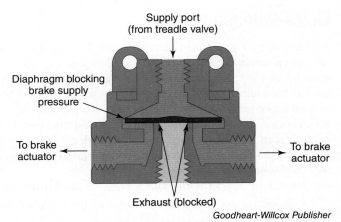

Figure 20-41. When the pressures above and below the diaphragm equalize, the diaphragm blocks the inlet and outlet. The valve is in a holding state.

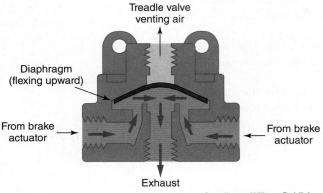

Figure 20-42. When the operator releases the treadle valve, the treadle valve vents the air that was previously acting on top of the diaphragm. The pressure from the brake actuators causes the diaphragm to flex upward, and the brake pressure is vented out the exhaust passageway.

When the operator pushes the knob, the valve directs system pressure to the brake chamber. The pressure compresses the heavy-duty spring in the chamber and releases the parking brake. Brake chambers are explained later in this chapter. If the system pressure drops below 40 psi, the parking brake control valve will automatically vent air pressure from the brake chamber, releasing the spring and applying the brake. **Figure 20-44** shows a Bendix parking brake control valve.

Pressure-Protection Valve

A *pressure-protection valve* isolates the main air brake system from accessory circuits. See **Figure 20-45**. Some examples of accessory circuits used in a haul truck are an air horn, air suspension seat, and an air starter circuit that uses its own air tank and control valve.

The pressure-protection valve on the Caterpillar 637G scraper system shown in **Figure 20-29** will not make air pressure available to the accessory circuits (differential lock control) unless it receives 75 psi of air pressure. At 75 psi of air pressure, the valve opens and air pressure passes to the accessory circuit. In this example, the accessory circuit provides air pressure for the tractor's differential lock. If there is a leak in the air supply and 75 psi is not reached, the pressure-protection valve closes and air is not supplied to the accessory circuit.

A pressure-protection valve on a Caterpillar 777D haul truck is used to isolate the primary and secondary air tanks, providing priority air to the primary air tanks to apply the service brakes in the event that there is a leak in the secondary system. Only when the primary tank pressure reaches 80 psi will air be allowed to flow to the secondary tank.

Foundation Brakes

This chapter has explained how the air compressor, reservoirs, treadle valve, relay valves, and quick-release valve supply air pressure to the brake actuators. Air brake systems use different brake actuators, which

Figure 20-43. The knob for the parking brake control valve can be located on the dash or near the steering column.

Chapter 20 | Air Brake Systems

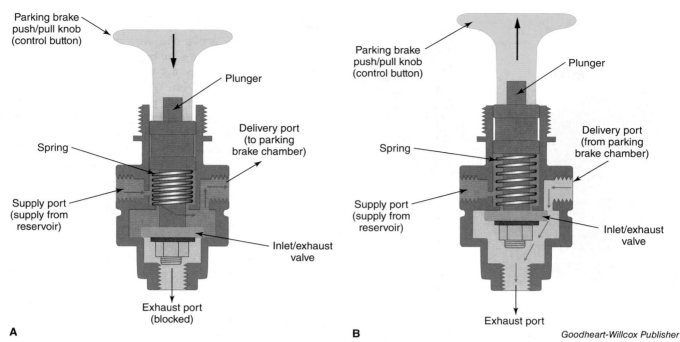

Figure 20-44. A manual push/pull parking brake control valve is actuated by the operator to apply and release the parking brake. A—In the disengaged position, the plunger is pressed downward and air from the reservoir is directed to the parking brake chamber. The air compresses the parking brake spring and releases the parking brake. B—In the engaged position, the plunger is lifted and the supply port is blocked from directing air to the parking brake chamber. The delivery port is vented, allowing the spring in the chamber to apply the brake.

are part of the foundation brakes. *Foundation brakes* are the mechanical components of the air brake system that apply the friction material to the rotating wheel assembly.

Foundation brakes contain an *air brake actuator*, also known as an *air chamber* or *brake chamber*, which converts the air pressure into linear force. The actuator has a clevis that attaches to a slack adjuster. The slack adjuster translates the linear force into a rotational torque that is applied to a cam mechanism. See **Figure 20-46**.

Brake Actuators

Brake chambers use air pressure and spring pressure to apply a linear force to the brakes. Examples of brake chambers include:

- Single service brake chamber.
- Single parking brake chamber.
- Rotochamber.
- Dual service/spring brake chamber (double chamber).

Brake chambers are sometimes identified with a number that indicates their effective area in square inches. For example, a type 20 brake chamber has 20 square inches of effective area.

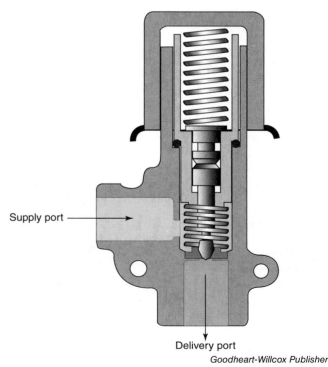

Figure 20-45. A pressure-protection valve allows air to flow out of the delivery port to an accessory circuit only if the pressure is high enough to overcome the piston's spring pressure. This simplified drawing depicts the pressure-protection valve used in a Caterpillar 637G scraper.

Single Service Brake Chamber

A *single service brake chamber* contains a pressure plate, non-pressure plate, rubber diaphragm, push plate, rod assembly, return spring, and a clamp ring that couples together the assembly. When the brake chamber is in an at-rest position, no air pressure is supplied to the brake chamber's inlet. The return spring holds the push plate and rod in a retracted position, and the brake is released. See **Figure 20-47**.

When air pressure is directed into the inlet port, the rubber diaphragm moves the push plate, which extends the rod. See **Figure 20-48**. The rod is attached to a slack adjuster with a clevis. The slack adjuster is splined to the S-cam, which forces the shoes against the drum, applying the brakes. The single service brake chamber is sometimes used to actuate the front axle brakes.

The amount of air pressure multiplied by the effective area of the brake chamber equals the amount of force exerted on the rod. For example, if the brake chamber's effective area is 15 square inches and 20 psi of air pressure is applied to the chamber, the rod will exert 300 pounds of force.

Air brakes are less responsive than hydraulic brakes because, unlike hydraulic fluid, air is compressible. A hydraulic brake application will appear to be instantaneous, whereas an air brake application will lag as the air builds sufficient pressure to actuate the brakes. The overall air brake delay is not great, typically less than a second, but it is still slower than a hydraulic application.

Goodheart-Willcox Publisher

Figure 20-46. Air brakes use brake chambers to convert the air pressure to linear force. The linear force is applied to the slack adjuster, which twists the camshaft and, in this case, causes the S-cam to apply the brake shoes against the rotating drum (not shown).

Single Parking Brake Chamber

Some machines are equipped with a *single parking brake chamber*. The chamber contains a heavy-duty spring that applies the parking brake when no air is supplied to the brake chamber. Air must be supplied to the chamber to compress the spring and release the parking brake. See **Figure 20-49**. When the operator presses the parking brake control

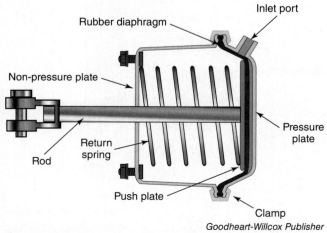

Goodheart-Willcox Publisher

Figure 20-47. This single service brake chamber is shown in the at-rest position. No air pressure is supplied to the inlet port. The return spring keeps the push plate and rod in a retracted position.

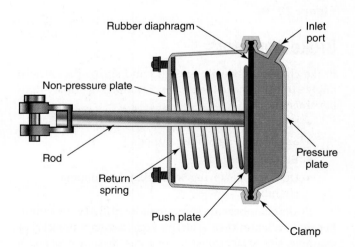

Goodheart-Willcox Publisher

Figure 20-48. When air pressure enters the inlet port, the push plate and rod extend to apply the brake mechanism.

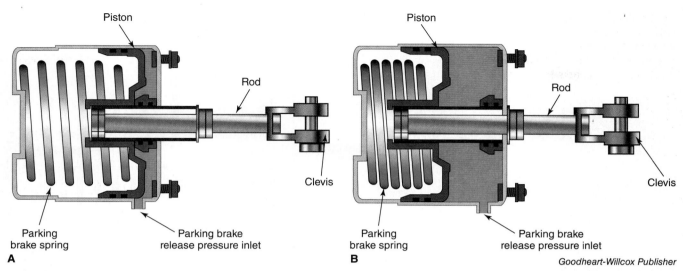

Figure 20-49. A single parking brake chamber is used as a parking brake actuator. A—When no air pressure is supplied to the chamber's inlet, the spring pushes the rod to apply the parking brake. B—When the operator releases the parking brake, air pressure is supplied to the chamber's inlet. This forces the piston to compress the parking brake spring and release the parking brake.

valve knob, the valve directs system pressure to the brake chamber, compressing the heavy-duty spring and releasing the parking brake.

Notice that this brake chamber uses a piston instead of a diaphragm. A single parking brake chamber is sometimes called a *hold-off brake chamber*. This brake is also used as a secondary brake to stop the vehicle if necessary. As mentioned earlier, if the system pressure drops below 40 psi, the parking brake control valve will automatically vent the air from the brake chamber, causing the brake to stop the machine. Caterpillar G936 and 926F wheel loaders use this brake chamber design to actuate the drum brake located on the front drive shaft.

Rotochamber

A *rotochamber* does not use a piston or a flat diaphragm to actuate the brakes. Instead it uses a lobed diaphragm that is designed to roll as it actuates. See **Figure 20-50**. The rotochamber provides a more consistent apply force than a non-lobed diaphragm chamber, such as the dual service/spring brake chamber. Caterpillar 768/769B and C rigid frame haul trucks use rotochambers to control service brakes. Some Caterpillar scrapers, such as 627B, 651B, and 657, use dual chamber rotochambers to control the parking brake and the service brake.

Dual Service/Spring Brake Chamber (Double Chamber)

Heavy equipment may use a *dual service/spring brake chamber* for both the parking brake and the service brake. The assembly contains two chambers. The front chamber is the service brake chamber and contains a service brake piston, a diaphragm, and a small return spring. The rear brake chamber is the parking brake chamber. It is sometimes called the piggyback chamber. It contains a parking brake piston, a diaphragm, and a heavy-duty spring that is normally applied (with no air pressure). Air pressure is required to compress the spring and release the parking brake. See **Figure 20-51** and **Figure 20-52**.

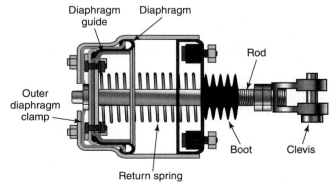

Figure 20-50. A rotochamber uses a rolling diaphragm that exerts a consistent force.

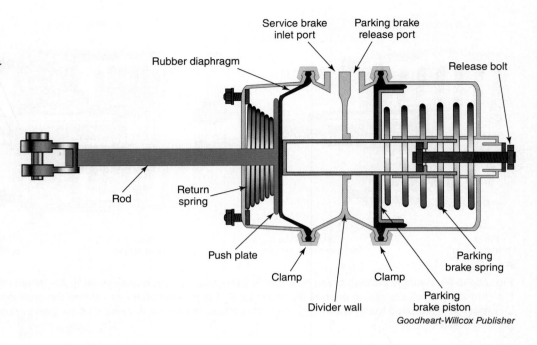

Figure 20-51. This dual service brake chamber is shown with the parking brake applied. There is no air pressure in the chamber.

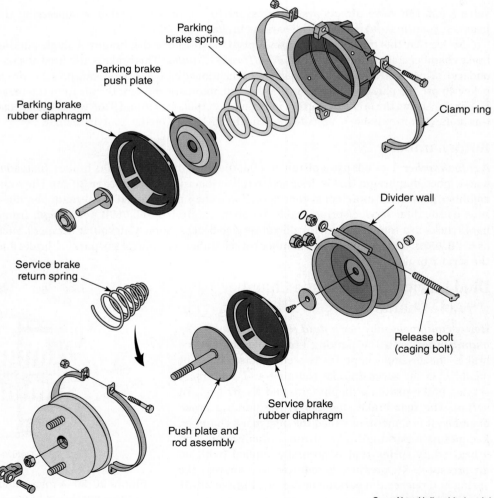

Figure 20-52. An exploded view of the dual service brake chamber used on Case 680H backhoes.

When the parking brake control valve is pressed, air pressure is routed to the rear chamber, causing the parking brake piston to compress the heavy-duty parking brake spring. **Figure 20-53** shows the parking brake released and the service brake not applied. There is no air pressure applied to the service brake piston. **Figure 20-54** shows the parking brake released and the service brake applied. There is maximum air pressure applied to the service brake piston.

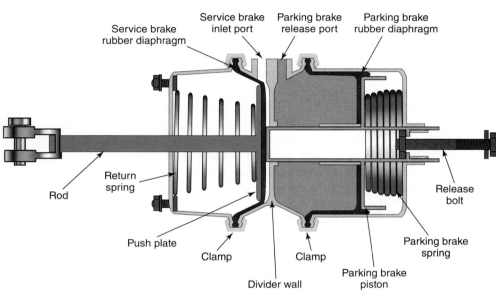

Figure 20-53. The parking brake is released and no air pressure is applied to the service brake piston.

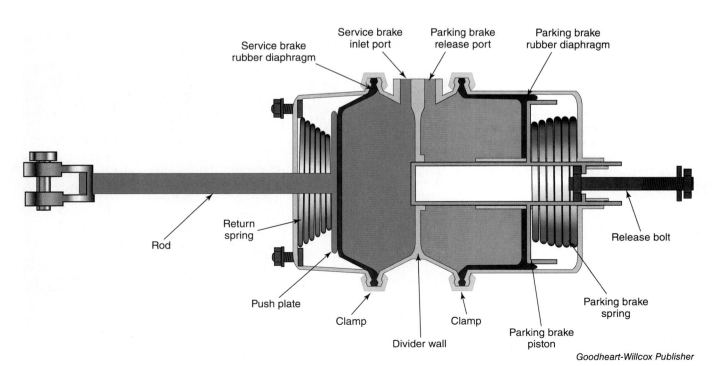

Figure 20-54. The parking brake is released and maximum air pressure is applied to the service brake piston.

Cam Brakes

Cam brakes are actuated by a camshaft, **Figure 20-55**. When the camshaft is rotated, the cam spreads and forces two rollers to push the brake shoes outward. This causes the shoes to apply friction to a rotating drum. The cam can be flat or an S-shape. See **Figure 20-56**. The S-cam is the most popular style of cam brakes used in heavy equipment.

Slack Adjusters

Cam brakes require *slack adjusters*, which have two functions:
- Transforming force—The slack adjuster transforms the linear force of the brake actuator rod into a rotating torque that is applied to the brake camshaft. The length of the slack adjuster acts like a lever, changing the brake chamber's linear force and applying it to the brake camshaft to develop rotating torque.
- Adjusting slack—As its name implies, the slack adjuster is used to eliminate the brake slack that occurs due to brake shoe wear.

Slack adjusters can be either manual or automatic. The manual slack adjuster shown in **Figure 20-57** is on a Caterpillar scraper.

A slack adjuster is attached to the brake chamber by a clevis threaded onto the brake chamber's push rod. The clevis is pinned to the slack adjuster. A worm gear drive in the slack adjuster is used to adjust the rotational position of the camshaft in relation to the brake chamber push rod clevis. As the worm shaft is rotated, it turns a spur gear that is splined to the camshaft. See **Figure 20-58**. This causes the cam to increase or decrease its clearance between the brake shoes and the drum, depending on which way the worm shaft is rotated.

In a *manual slack adjuster*, the worm shaft has a hex head that allows a technician to rotate the worm shaft to adjust the brake's slack. The worm shaft has a lock to hold it in place. The Caterpillar 637E scraper uses a lock

Goodheart-Willcox Publisher

Figure 20-55. An S-cam brake is actuated by a shaft with an S-cam on the end. The S-cam wedges the brake shoes against the drum.

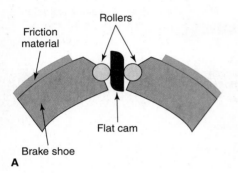

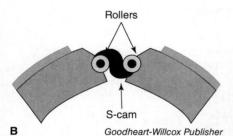

Goodheart-Willcox Publisher

Figure 20-56. Cam brakes have a cam between two brake shoes. As the brake camshaft is rotated clockwise, the cam forces the rollers outward, which forces the friction material against the rotating drum. A—Flat cam. B—S-cam.

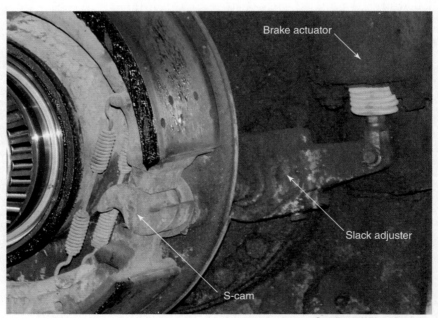

Goodheart-Willcox Publisher

Figure 20-57. This Caterpillar 623E scraper has the drum removed, exposing the S-cam and brake shoes.

screw to hold the worm shaft stationary. After the lock screw is loosened, the worm shaft can be rotated to meet the manufacturer's push rod extension specification. For example, one application specifies that the push rod should move 1.62″ (±0.12″) during brake application. Be sure to follow the manufacturer's service literature when replacing, servicing, or adjusting the brakes.

On-highway trucks must adhere to Federal Motor Vehicle Safety Standards (FMVSS) 121, which requires the use of automatic slack adjusters. An *automatic slack adjuster (ASA)* automatically adjusts the slack as the brake shoes wear. Technicians do not need to manually adjust the brakes for wear. Automatic slack adjusters should still be inspected and may need to be adjusted if automatic adjustment fails. An automatic slack adjuster that requires manual adjustment should be replaced. Two types of automatic slack adjusters are stroke-sensing adjusters and clearance-sensing adjusters.

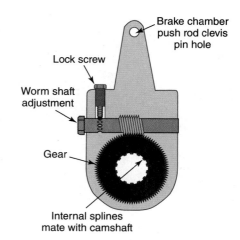

Figure 20-58. The manual slack adjuster has a lock screw that holds the worm shaft stationary. As the brake shoes wear, the slack adjuster must be adjusted to maintain the correct stroke on the brake push rod.

Stroke-Sensing ASA

An example of a Meritor stroke-sensing ASA is shown in **Figure 20-59**. The *stroke-sensing ASA* has a worm shaft, actuating sleeve, and ratcheting pawl that adjust for wear during brake applications. The actuating sleeve's internal splines mate with the worm shaft splines. The actuating sleeve's external helical splines work in conjunction with a spring-loaded pawl. When the brake is applied, the actuator rod is pulled out of its cavity and the pawl ratchets across the helical splines. As the brake is released, the actuator rod is pushed back into its cavity. If the actuator adjusting sleeve is able to move one more tooth in relation to the pawl, as the actuator rod moves back into the slack adjuster, the slack adjuster makes the physical slack adjustment. If the brake has wear, the pawl and helical splines on the actuator sleeve remove the slack.

The slack adjuster also contains a manual adjusting nut. To add tension, simply turn the manual adjusting nut. To remove tension, lift the pull-pawl mechanism outward. The manual adjusting nut can be used to do the following:

- Adjust the brakes during the installation of a slack adjuster.
- Adjust the brakes when installing new brakes shoes.
- Adjust for slack when the ASA is not adjusting properly.
- Back off the brakes, or remove tension, when removing a brake drum.

As mentioned earlier, if the ASA is not properly adjusting for slack and requires manual adjustment, it should be replaced.

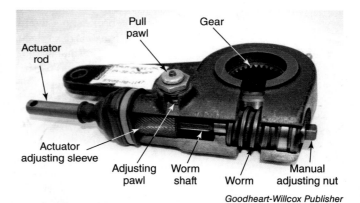

Figure 20-59. This stroke-sensing ASA has the actuator adjusting sleeve pulled out to reveal where it mates with the worm shaft.

Clearance-Sensing ASA

A *clearance-sensing ASA* uses a gear-clutch mechanism that automatically compensates for wear. The ASA can be designed to adjust during the brake rod extension (brake application) or during the brake rod retraction (brake release). When the shoe friction material presses against the drum, the clutch causes the adjuster to become stationary and prevents further adjustment. See **Figure 20-60**.

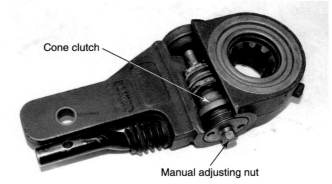

Figure 20-60. A clearance-sensing ASA uses a clutch to stop the adjuster's movement when the friction material is pressed against the drum.

Air Disc Brakes

External disc brake systems are used in construction equipment and agricultural equipment, but this type of equipment traditionally uses hydraulic disc brakes or air-over-hydraulic disc brakes. *Air disc brakes*, like hydraulic external disc brakes, use a caliper, often a sliding caliper, with pistons that press friction-lined brake pads into a rotating disc rotor. Air disc brake systems are commonly used in on-highway trucks.

External disc brakes do not have the self-energizing effect that drum brakes have; therefore, they require a larger input force to achieve the same braking effect. Air disc brake manufacturers use a variety of mechanical multipliers to increase force, including levers, cams, wedges, and power screws.

One example of an air disc brake is the Meritor power screw actuator. See **Figure 20-61**. The air chamber's push rod actuates the slack adjuster, which rotates the camshaft. As the camshaft rotates, it mechanically forces the powershaft nut to press against the caliper's brake piston, which applies the brake pads.

An air-over-hydraulic brake system that uses a caliper disc brake is discussed later in the chapter. The caliper portion in this type of brake circuit is hydraulic instead of air applied.

Wedge Brakes

An *air wedge brake* uses an air brake chamber that actuates a push rod that extends a wedge with two rollers. The rollers press against two angled plungers. See **Figure 20-62**. As the push rod extends, the rollers are wedged against the plungers. The plungers then force the brake shoes against the drum.

The brake is designed to automatically adjust. An adjusting bolt is located in the actuator. The actuator, or adjusting sleeve, moves inside the plunger. The actuator has helical grooves, or external teeth. The adjusting pawl plunger is spring loaded against the actuator's grooves. As the plunger is extended, the adjusting pawl slides over the adjuster's teeth and adjusts for brake wear. If the brake is properly adjusted, the adjusting pawl will not move to the next groove.

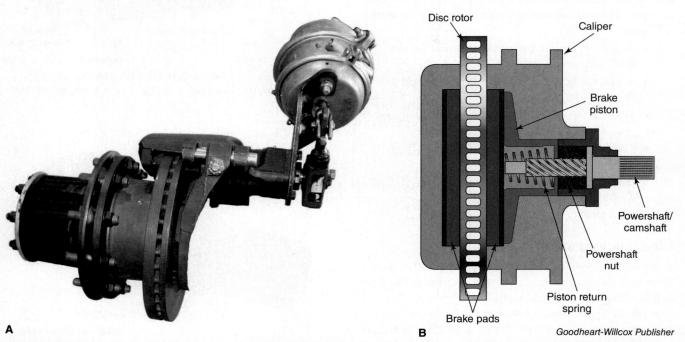

Figure 20-61. This air disc brake is an example of a Meritor power screw actuator. A—The brake is actuated by a dual brake chamber and a slack adjuster. B—A powershaft is rotated, forcing a nut to apply pressure to the brake pad. This causes the caliper to squeeze both brake pads into the rotor.

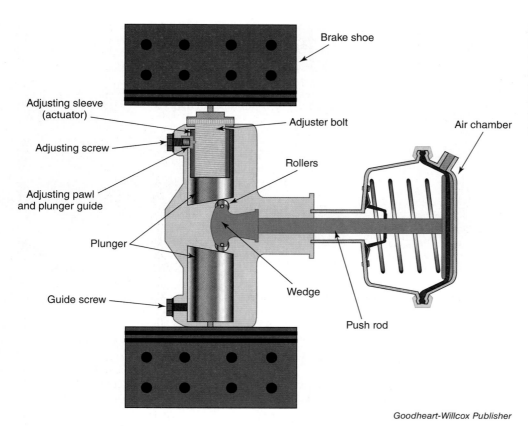

Figure 20-62. An air wedge brake uses a brake chamber to extend a push rod that is connected to a wedge with two rollers. As the push rod is extended, the rollers force the plungers to apply braking force to the brake shoes.

Single Wedge Actuator

A wedge brake assembly can use one or two air chambers to operate a pair of brake shoes. If one chamber is used, the shoes are pinned to the backing plate opposite the single wedge actuator. See **Figure 20-63**. As the single wedge is actuated with the single brake chamber, one end of each shoe is extended outward against the rotating drum. The single wedge actuator is a single servo actuated brake.

Twin Wedge Actuator

When a wedge brake uses two air chambers, both ends of each shoe are extended by wedge actuators. See **Figure 20-64**. The twin wedge actuator is a dual servo brake providing self-energization.

 Note

Some wedge brakes use a hydraulic wheel cylinder in place of an air chamber to extend the wedge into the plungers. Air wedge brakes were used in older on-highway trucks, but became unpopular and are no longer used in on-highway truck applications. They did not self-adjust well, nor did they actuate well. They made it difficult to achieve brake balance.

Figure 20-63. This wedge brake uses only one wedge actuator to apply the brake shoes. Notice that each shoe is pinned to the backing plate (spider) at the opposite end of the wedge actuator.

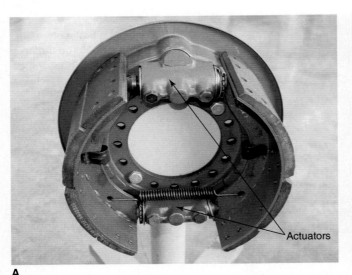

Figure 20-64. A drum brake can use two wedge actuators in a single brake assembly. A—Notice the two wedge actuators. B—Notice the two air chambers, one with a spring-applied parking brake.

Air-Over-Hydraulic Brake Actuator

Some heavy machinery systems use *air-over-hydraulic brakes*, which use both air and hydraulics to actuate the service brakes. As described earlier in the chapter, the air system consists of a compressor, governor, dryer, reservoir, treadle valve, relay valves, and quick-release valves. The air system actuates an air cylinder that, in turn, operates a hydraulic master cylinder. The air pressure pushes the air cylinder piston, which is connected to a rod that actuates a piston in the hydraulic master cylinder. The hydraulic oil is drawn from a stand-alone reservoir dedicated to the brakes, known as a makeup reservoir. See **Figure 20-65**.

Examples of machines that use air-over-hydraulic brakes are Caterpillar 785C rigid frame haul trucks and Caterpillar 926F wheel loaders. In the Caterpillar 926F wheel loader, twin single circuit treadle valves are used. The right treadle valve operates the service brakes. The initial pedal movement of the left treadle valve disengages the transmission, and the remainder of the treadle movement applies the service brakes. The front service

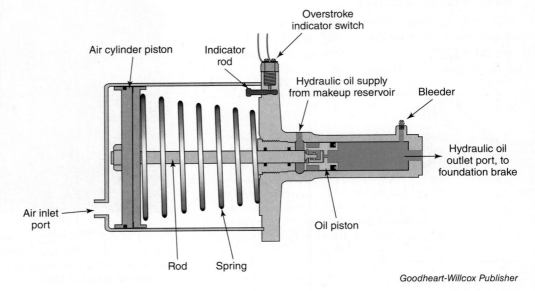

Figure 20-65. Caterpillar wheel loaders and compactors, such as the Caterpillar 926F wheel loader, use an air cylinder to actuate the brakes in the air-over-hydraulic brake system.

brakes and the rear service brakes each have an air cylinder, a hydraulic master cylinder, and a makeup reservoir that supplies oil to the master cylinder. The hydraulic master cylinders direct oil pressure to operate external caliper disc service brakes. See **Figure 20-66**.

The air cylinder has an area 16 times larger than the area of the hydraulic master piston (16:1 ratio). If the air cylinder is actuated with 20 psi of air pressure, it develops 320 psi of hydraulic pressure.

Air Brakes versus Hydraulic Brakes

As mentioned earlier in the chapter, hydraulic brakes are more common than air brakes in late-model heavy equipment. Air brake systems tend to have more disadvantages than advantages when compared to hydraulic braking systems.

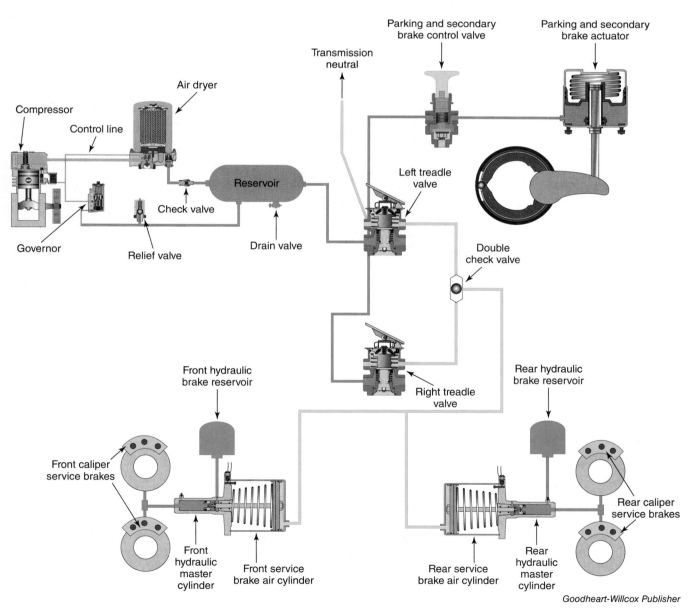

Goodheart-Willcox Publisher

Figure 20-66. An example of a Caterpillar 926F wheel loader that uses air-over-hydraulic brake controls.

Compressibility

Air compresses, whereas hydraulic fluids are practically incompressible, compressing only 0.5% for every 1000 psi. Due to the compressibility of air, a large volume of air is required to stop a machine. In addition, the compressibility results in longer braking distances compared to hydraulic braking systems.

Moisture

Compressing air creates heat, and hot air contains moisture. When hot air cools, the moisture in the air condenses to form liquid water. Liquid water in the system causes corrosion, but more importantly, it will freeze if a machine is operating in cold environments. Frozen brake lines and brake components render the brakes completely inoperative.

Machine Costs

Hydraulic brake systems often have fewer and smaller components resulting in a more cost-effective design. A machine that already has a hydraulic pump for steering or implements can use the same pump for brakes.

Reliability and Maintenance

Hydraulic brake systems are more reliable than air brake systems. They also require less maintenance. Air brake systems may have manual drain valves that need to be drained daily and alcohol evaporators that need to be maintained.

Advantages of Air Brakes

Air brakes do have a few advantages. For one, air is abundant and free. In addition, a leak in an air brake system will not harm the environment. The air compressor can overcome small air leaks, allowing the system to still stop a machine. In contrast, a leak in a hydraulic brake system results in the loss of hydraulic fluid, resulting in brake failure. Most hydraulic brake systems use fluid that is not environmentally friendly.

Summary

- An air brake supply system contains a compressor, governor, air dryer, and supply tank.
- A governor is used to send a signal pressure to the compressor's unloader valves, causing the compressor to unload when the system pressure reaches the cut-out pressure value.
- An air dryer is used to remove moisture and contaminants from the air system. The dryer is located between the compressor and the air reservoir.
- The supply tank, the first tank located after the compressor, is often called a wet tank.
- A primary air circuit is used to apply an on-highway truck's rear tandem service brakes and trailer service brakes via the treadle valve.
- A secondary air circuit is used to apply an on-highway truck's front service brake and might be used to supply air to the trailer service brakes via the trailer hand control valve.
- Some off-highway machines use a dual air brake circuit, while others use a single air brake circuit.
- Air brake systems use a foot-operated service brake control valve, also known as a treadle valve, to apply the service brakes.
- A relay valve speeds the application and release of service brakes so they apply and release with quicker reaction times.
- A quick-release valve provides a method for quickly venting the brake application pressure when the operator releases the treadle valve.
- A parking brake control valve is used to release the parking brake.
- A pressure-protection valve is used to isolate the main air brake system from accessory circuits.
- Foundation brakes are the mechanical components of the air brake system that apply the friction material to the rotating wheel assembly.
- A single service brake chamber is released when it is receiving no air pressure.
- A single parking brake chamber contains a heavy-duty spring that is responsible for applying the parking brake.
- A dual service/spring brake chamber is used for both the parking brake and the service brake.
- Slack adjusters serve two purposes: to act as a lever transforming force and to remove slack from the brake as the shoes wear.
- Air disc brakes use a caliper with pistons that press brake pads into a rotating disc rotor.
- Air-over-hydraulic brakes use both air and hydraulics to actuate service brakes.
- Air compresses, whereas hydraulic fluids are practically incompressible and require less volume to stop a machine.
- Hydraulic brake systems use smaller components, making them more economical.
- Hydraulic brake systems are more reliable.
- Air is abundant and environmentally friendly, unlike hydraulic fluids.

Technical Terms

air brake actuator
air brake supply system
air chamber
air compressor
air disc brakes
air dryer
air-over-hydraulic brakes
air supply tank
air wedge brake
alcohol evaporator
automatic slack adjuster (ASA)
brake chamber
clearance-sensing ASA
cut-in pressure
cut-out pressure
delivery port check valve
discharge valve
drain valve
dry tank
dual brake circuit
dual service/spring brake chamber
foundation brakes
governor
graduating rubber spring
hold-off brake chamber
inlet check valve
inlet/exhaust valve
inlet valve
internal purge air dryer
low-pressure switch
manual slack adjuster
operating pressure
parking brake control valve
pop-off valve
pressure-protection valve
primary air brake circuit
purge valve
purge volume
quick-release valve
relay valve
rotochamber
safety pressure relief valve
secondary air brake circuit
signal pressure
single parking brake chamber
single service brake chamber
slack adjuster
stroke-sensing ASA
supply reservoir
system pressure
treadle valve
wet tank

Review Questions

Answer the following questions using the information provided in this chapter.

Know and Understand

1. All of the following are part of an air brake supply system, *EXCEPT*:
   A. air compressor/governor.
   B. dryer.
   C. reservoir.
   D. relay valve.

2. An air dryer is located _____.
   A. before the compressor
   B. after the wet tank
   C. after the compressor
   D. None of the above.

3. All of the following are examples of drain valves, *EXCEPT*:
   A. manual pull-type drain valve.
   B. manual threaded drain cock.
   C. automatic drain.
   D. solenoid operated.

4. What is an example pressure that causes a low-pressure switch to close?
   A. 5 psi.
   B. 20 psi.
   C. 60 psi.
   D. 100 psi.

5. In a treadle valve, the graduating rubber spring _____.
   A. provides brake pedal resistance
   B. closes the primary piston
   C. actuates the secondary inlet valve
   D. keeps the secondary piston closed

6. Relay valves are mounted close to the _____.
   A. air dryer
   B. brake actuators
   C. relief valve
   D. treadle valve

7. Which valve is added to a circuit to speed the application and release of service brakes when the brake actuators are a long distance from the treadle?
   A. Automatic drain valve.
   B. Pressure-protection valve.
   C. Relief valve.
   D. Relay valve.

8. The _____ supplies the control pressure to a relay valve.
   A. compressor
   B. quick-release valve
   C. supply reservoir
   D. treadle valve

9. A relay valve has all of the following ports (air lines), *EXCEPT*:
   A. control pressure line.
   B. delivery line.
   C. purge line.
   D. supply line.

10. The _____ is used to isolate the main brake system from accessory circuits.
    A. pressure-protection valve
    B. quick-release valve
    C. relay valve
    D. safety-relief valve

11. What is the name for the mechanical components of the air brake system that apply the friction material to the rotating wheel assembly?
    A. Foundation brakes.
    B. Primary brakes.
    C. Secondary brakes.
    D. Parking brakes.

12. If a single service brake chamber has no air applied to it, what is its state?
    A. Brake released.
    B. Brake fully applied.
    C. Brake partially applied.
    D. None of the above.

13. If a single parking brake chamber has no air applied to it, what is its state?
    A. Brake released.
    B. Brake fully applied.
    C. Brake partially applied.
    D. None of the above.

14. The _____ is the most popular style of cam brake used in heavy equipment.
    A. flat cam
    B. O-cam
    C. S-cam
    D. W-cam

15. Technician A states that some slack adjusters are designed to be manually adjusted. Technician B states that some slack adjusters are designed to adjust automatically. Who is correct?
    A. Technician A.
    B. Technician B.
    C. Both A and B.
    D. Neither A nor B.

16. An air wedge brake uses all of the following, *EXCEPT*:
    A. camshaft.
    B. plungers.
    C. push rod.
    D. rollers.

17. All of the following are advantages of air brake systems, *EXCEPT*:
    A. air is abundant.
    B. air is free.
    C. air is environmentally friendly.
    D. air brake systems use smaller components.

18. All of the following are disadvantages of air brake systems, *EXCEPT*:
    A. air compresses.
    B. air is an economical medium.
    C. air brake systems use larger components.
    D. air brake systems require the removal of moisture.

Apply and Analyze

19. If a compressor contains one piston, it has _____ valves.
20. On a compressor, the _____ valve directs air to the air dryer or air reservoir.
21. The _____ is responsible for directly controlling the compressor's unloader valve.
22. Governor cut-out pressure is often set at _____ psi.
23. A governor's cut-in pressure value is typically _____ psi less than its cut-out pressure.
24. An air reservoir located directly after the compressor is often called a(n) _____ tank.
25. A safety relief valve is designed to open at approximately _____ psi.
26. A(n) _____ is placed in an air system to prevent a good air circuit from leaking into a bad air circuit.
27. A foot-operated service brake valve is known as a(n) _____.
28. The _____ valve is placed in between the treadle valve and the brake actuator and is used to quickly vent brake pressure.

Critical Thinking

29. Explain the advantage of a dual-circuit air brake system.
30. Explain the operation of an air-over-hydraulic brake actuator.

Chapter 21
Suspension Systems

Objectives

After studying this chapter, you will be able to:
- ✓ Describe the different types of construction suspension systems.
- ✓ Describe the different types of agricultural suspension systems.
- ✓ Describe the different types of cab suspension systems.

Construction and agricultural equipment are built with different styles of *suspension systems* to absorb shocks as the machine travels over rough terrain and as the machine handles loads while performing work. The tire deflection on rubber tire machines provides a slight degree of shock load cushioning. However, the tough demands placed on modern heavy equipment require manufacturers to design and equip their machines with more complex, rugged, and supportive suspension systems.

Depending on the type of equipment, the suspension system can include leaf springs, coil springs, hydraulic shock absorbers, hydropneumatic shock absorbers, rubber springs, or air springs. A suspension system design may be used solely for supporting the operator's seat, cab, or the machine's frame. Many machines use a combination of designs to support the three parts of the machine. A machine with no suspension, or a poorly operating suspension, will ride extremely rough, which can lead to machine damage and operator fatigue.

For decades, on-highway trucks have used leaf springs, shock absorbers, and/or air bags for absorbing shock loads. A *leaf spring suspension* is made with multiple flat strips of rectangular steel, called leaves, that are placed one on top of another to form an arch. The center of a leaf spring is usually attached to one side of the axle housing with U-bolts. The front and rear of the leaf spring are attached to a front and rear section of the frame with bolt-on brackets and shackles, **Figure 21-1**. A hydraulic suspension cylinder, also called a shock absorber, is often integrated into a leaf-spring suspension. Shock absorbers are explained later in this chapter. Many off-highway machines, however, operate in more demanding conditions and rougher terrains than on-highway trucks and, therefore, use more complex and sophisticated suspension systems than leaf springs.

Figure 21-1. The front suspension on this on-highway truck is equipped with leaf springs and a shock absorber.

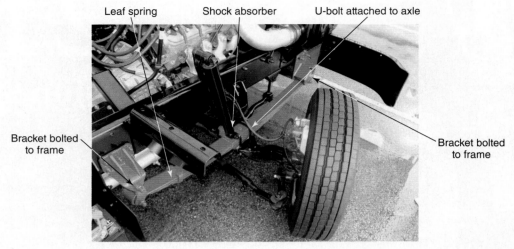

Goodheart-Willcox Publisher

Construction Suspension Systems

Construction and mining equipment operate in the most difficult off-highway terrain. The machines use different types of systems to absorb the shock loads. Based on the type of machine, suspension system designs can include the following:

- Hydro-pneumatic suspension cylinders and elastomer springs—used on rigid frame haul trucks.
- Hydro-pneumatic suspension cylinders and elastomer springs—used on articulated dump trucks.
- Cushioned hitch suspension—used on tractor scrapers.
- Leaf spring roller suspension—used on Bobcat compact track loaders.
- Implement ride control on several different types of machines.
- Air ride seats in several different types of machines.

Rigid Frame Haul Truck Hydro-Pneumatic Suspension

Large rigid frame haul trucks, such as those used in mining, have *hydro-pneumatic suspension cylinders* to serve as the spring suspension for the truck. The cylinders are also known as *nitrogen-over-oil cylinders*, *oil-pneumatic hydraulic cylinders*, or *shock absorbers*. Each cylinder has a self-contained volume of oil and nitrogen that works together to absorb chassis shock loads. As loads are placed on the wheels, the nitrogen compresses to provide shock absorption.

Most trucks contain four suspension cylinders, one at each front wheel and two on the rear axle. All four provide shock absorption independently from one another. Caterpillar, Liebherr, Komatsu, and Terex are among the manufacturers that use hydro-pneumatic suspension cylinders in their haul truck suspension systems.

Rigid Frame Haul Truck Front Suspension

The front suspension cylinders used on rigid frame haul trucks provide independent suspension at the left and right wheel, **Figure 21-2A**. On rigid frame haul trucks (Caterpillar models 769 through 797, for example), the front suspension cylinders are called *struts* because they form a structural part of the suspension and are incorporated into the steering system. One strut is located on each side and connects the frame to the wheel, with the cylinder also acting as the steering pivot shaft, known as the *kingpin*. See **Figure 21-2B**. The strut is angled to provide the steering's inclination angle, optimizing the truck's tight turning radius.

Chapter 21 Suspension Systems

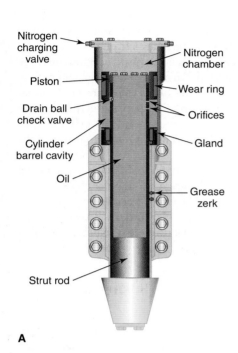
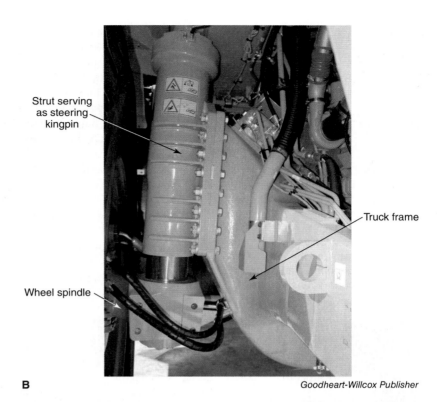

Figure 21-2. Mining trucks have hydro-pneumatic suspension cylinders as part of the front and rear suspension. A—A hydro-pneumatic suspension cylinder is filled with oil, and nitrogen is introduced at the top of the cylinder. B—A Caterpillar haul truck's front strut is bolted to the truck's frame and wheel spindle.

Each strut's housing is bolted to the frame of the truck. The strut's rod is attached to the front wheel. When the wheel encounters a load and the rod compresses into the cylinder, oil inside of the rod forces the piston to compress the nitrogen, which causes the nitrogen pressure to increase. This action causes oil in the rod to flow through the two orifices and a drain ball check valve to the barrel of the cylinder where it provides a shock-absorbing action. When the strut is compressing, it is in the ***jounce*** phase. Jounce occurs when a load is placed on the suspension. See **Figure 21-3**. The strut provides resistance (shock absorption) during compression and extension.

When the strut experiences a drop in load, the rod attempts to move down due to the nitrogen pressure above the piston. This extension is the ***rebound*** phase of the strut. As the rod moves down in the cylinder, the piston moves closer to the gland, decreasing the barrel volume and forcing oil out of the barrel. The flow of oil from the barrel back into the rod causes the drain ball check valve to seal the drain passage. At this point, fluid can travel from the barrel into the rod only through the two orifices. As the rod continues to extend, the bottom orifice is blocked, which causes oil to travel through only the top orifice. See **Figure 21-4**. As a result, the strut is allowed to compress (jounce) faster than it is allowed to extend (rebound). The rod is slowest to extend when it nears full extension.

Figure 21-3. When the strut encounters a load, the rod and piston move up in the cylinder (compress), which causes nitrogen pressure to increase. As the distance between the piston and gland increases, the barrel volume grows, causing oil to move into the barrel of the cylinder through the two orifices and a drain ball check valve.

Rigid Frame Haul Truck Rear Suspension

Two examples of rigid frame haul truck rear suspensions are the A-frame design and the four-link design. The rear axle on the A-frame suspension design has a triangular frame that extends from the front of the axle. A flexible joint (ball joint or a pivoting pin) is used at the front of the A-frame to attach the drive axle to the truck. At the rear of the machine,

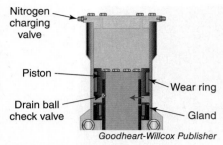

Figure 21-4. When the strut's load decreases, the strut rod attempts to move down in the cylinder (extend). During extension, the drain ball check valve prevents oil from entering the strut rod. Oil can only travel through the two orifices, and the lower orifice is blocked during the final extension. This slows, or dampens, rod travel.

two independent hydro-pneumatic suspension cylinders (attached to the back of the drive axle housing) and a tag link connect the axle to the frame. The tag link is a heavy-duty bar pinned to the top of the axle housing and to one side of the truck frame. The flexible joint, two suspension cylinders, and tag link allow the axle to oscillate and move separately from the truck's frame.

The twin suspension cylinders absorb the chassis shock loads as the truck is traveling or being loaded. See **Figure 21-5**. On some machines, the rods of the suspension cylinders are covered with rubber boots to reduce contamination of the cylinder's seals, **Figure 21-6**. Some rear suspension cylinders have the rods facing toward the ground (inverted), which keeps the cylinder rods cleaner by preventing contaminants from continuously resting on top of the rod seals.

A four-link rear suspension design has four dog-bone shaped links attached to the front of the axle. Upper and lower links are located on each side of the axle, connecting the axle to the truck frame. The two suspension cylinders are attached to the back of the axle. The left suspension cylinder is attached to the left side of the axle and to the left side of the truck frame. The right suspension cylinder is attached to the right side of the axle and to the right side of the truck frame.

Servicing Mining Truck Suspension Cylinders

Hydro-pneumatic suspension cylinders must be periodically inspected for leaks and extension height. Some machines have an onboard monitor, such as the Caterpillar Truck Payload Measurement System (TPMS), that alerts the operator when the cylinders have bottomed out. Many factors, including the following, affect the extension of a hydro-pneumatic suspension cylinder:

- Volume of oil.
- Volume of nitrogen.
- Temperature.
- Load on the machine (including fuel and debris stuck to the machine).
- Location of the cylinder on the machine and the proximity to the machine's center of gravity. (For example, a tractor with a cab on the left side can cause the front left cylinder to be more compressed than the right cylinder.)

Due to these factors, it is important to determine an accurate benchmark of the normal cylinder extension. Using the benchmark, technicians can clearly identify when the cylinder has a problem. Five factors—along with the benchmark—lead technicians to inspect the cylinders more closely to determine if a problem exists:

- Machine provides a rough ride during operation.
- Cylinders have or show signs of external leaks.
- TPMS fault codes are present.
- Cylinder's rod exhibits full travel as evidenced by the exposed rod having the majority of its chrome wiped clean by the wiper seal.
- Cylinders' rods are overcharged and fully extended when the truck has no payload.

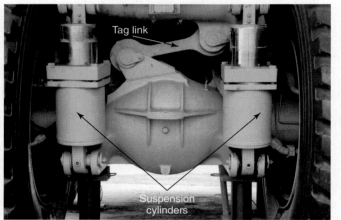

Figure 21-5. This mining truck's rear suspension is outfitted with twin hydro-pneumatic suspension cylinders and a tag link.

Figure 21-6. This Caterpillar 771D truck has boots around the rear suspension cylinders to prevent contaminants from entering the cylinders. The bed has been removed from the truck.

Chapter 21 | Suspension Systems

If a problem is identified after a closer inspection, a technician should strictly adhere to the following guidelines before servicing the hydro-pneumatic suspension cylinders:
- Follow all warning labels.
- Do not remove any plugs, caps, or components from the cylinder if the cylinder is extended.
- Check the oil in the cylinders only after the nitrogen has been depleted.
- Ensure that all technicians are completely clear of the machine while testing or removing or adding oil and nitrogen to the cylinders. The machine frame and components will move during these procedures, potentially causing injury.
- Ensure that the truck is empty and is placed on level ground before servicing the cylinders.
- Apply the parking brake.
- Charge or service the cylinders in pairs (either the front pair or rear pair), not one at a time.

Warning
Follow the manufacturer's service literature for servicing and maintaining hydro-pneumatic suspension cylinders. Failure to follow the manufacturer's service literature can cause machine damage or technician injury or death.

Charging Hydro-Pneumatic Suspension Cylinders

Before charging suspension cylinders, check the service literature for instructions on bleeding the oil and nitrogen currently located in the cylinders. Be sure to follow the specific manufacturer's service literature for bleeding and charging cylinders. The following procedure provides an example of one set of instructions and is listed here only for educational purposes:

1. Connect the tools (charging hoses and valves) to the two front or rear hydro-pneumatic suspension cylinders.
2. Connect the opposite end of the hoses to a vented storage tank.
3. Deplete the oil and nitrogen from the cylinders.
4. Leave the suspension in the drain state for five minutes after the oil has drained to allow the pressures to stabilize.
5. Close the charging valves.
6. Reference the manufacturer's service literature to determine how far the cylinders must extend while adding oil.
7. Attach a reference (such as banding from a pallet) to the cylinder's webbing with a reference mark as specified by the service literature (1″ of cylinder extension, for example).
8. Use the oil pump to charge both cylinders while watching the reference mark. If one cylinder reaches its reference mark first, close the valve to prevent further oil from entering and extending that cylinder. Continue to charge the remaining cylinder until it reaches the reference mark.
9. Shut off the oil pump and disconnect the oil pump from the circuit.
10. Locate the appropriate gauge blocks and shims for the truck's suspension based on the truck model number and ambient temperature.
11. Connect the nitrogen bottle to the charging lines.
12. Adjust the nitrogen regulator to 600 psi.
13. Open the charging valves and lift the cylinders to the height of the gauge blocks (8″ high, for example).
14. Deplete the nitrogen so the cylinders will rest on the gauge blocks.

15. Set the nitrogen regulator to the pressure specification.
16. Refill the nitrogen to the specified nitrogen pressure and leave for 5 minutes to allow pressures to equalize in the cylinders.

> **Note**
>
> If a nitrogen leak is suspected on a cylinder, spray soapy water around the nitrogen charging valve to identify leaks. Also, when charging a suspension cylinder, avoid adding too much nitrogen (especially to compensate for the added compression of the cylinder used to support the cabin side of the machine). An overfilled cylinder responds stiffly to suspension shocks, which results in a rough ride.

17. Close the nitrogen charging valves, shut off the nitrogen supply from the bottle, and remove the nitrogen charging tools.
18. In order to remove the gauge blocks, the cylinders must be extended. Raise the truck bed and turn the front wheels from side to side to extend the cylinders.
19. Operate the truck for a few load cycles. Then measure the cylinder extension according to the description in the following section to gain a benchmark for future reference.

Measuring Hydro-Pneumatic Suspension Cylinders

After a cylinder has been filled with oil and charged with nitrogen, its cylinder rod extension height must be checked and recorded as a benchmark for future service reference. As mentioned earlier, some variables influence the cylinders' rod height, including the weight of fuel and debris stuck to the truck.

To properly check the height, roll the truck to a gradual stop without braking and position it on flat ground. Use a tape measure to measure the exposed chrome on each rod of the four cylinders. Record this measurement as the benchmark for each cylinder.

Elastomer Suspension Cylinders

Miner Elastomer Products Corporation manufactures a TecsPak *elastomer suspension cylinder* that contains no oil or nitrogen, thus eliminating the potential for leaks and the need to recharge the cylinder. The suspension cylinder uses a stack of elastomer springs (rubber springs) that last ten to twenty times longer than traditional rubber and urethane springs respectively. See **Figure 21-7**.

Articulated Dump Truck Suspension

Articulated dump trucks (ADTs) are a type of construction haul truck that use a variety of suspension designs to absorb shocks and travel over rough terrains. Based on the machine's setup, articulated dump trucks have an *articulation joint* (hitch) that allows the front and rear frames to pivot in a horizontal plane, which provides steering. An *oscillation joint* between the front and rear frames allows each frame (front and rear) to rotate (twist/oscillate) clockwise or counterclockwise in a vertical plane, independent from one another. Refer to **Figure 21-8**.

An ADT front suspension often comprises a rigid axle housing attached to the front frame via some type of oscillating assembly. The oscillating assembly commonly includes an A-frame, stabilizer rod, rigid front axle, and two independent suspension cylinders. Like the rigid frame haul truck, the suspension cylinders are self-contained hydro-pneumatic cylinders. See **Figure 21-9**.

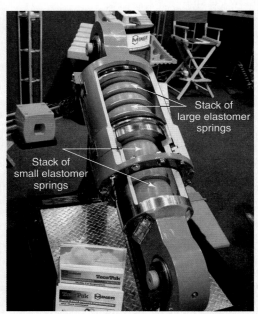

Goodheart-Willcox Publisher

Figure 21-7. Mining trucks may be equipped with an elastomer-style suspension cylinder that does not contain oil or nitrogen gas. The design features a stack of small elastomer springs to cushion the ride when the truck is empty and a stack of larger, high-capacity elastomer springs to handle the shock loads when the truck is full. A TecsPak elastomer suspension cylinder manufactured by Miner Elastomer Products is shown here on display.

Chapter 21 | Suspension Systems 737

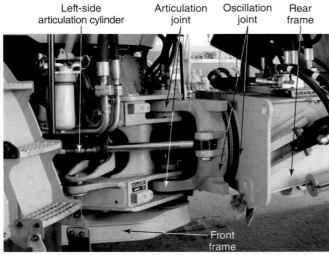

Figure 21-8. Articulated dump trucks have an articulation hitch that allows the two frames to steer the truck to the left and right. The truck's oscillation hitch allows the two frames to swivel to the left and right, independent from each other.

Figure 21-9. The left front hydro-pneumatic suspension cylinder on a Caterpillar 740B ADT.

Note

John Deere E and E-II series ADTs have hydraulic front struts that are connected to pneumatic accumulators.

The two rear axles are also rigid axles and have a suspension system that allows them to oscillate in relation to the rear frame. A combination of a trailing arm and walking beam suspension system keeps the machine operating efficiently. The Caterpillar ADT shown in **Figure 21-10** is built with a ***walking beam suspension*** that has two equalizer beams—one beam is located on each side of the rear frame of the truck. Each equalizer beam has a center pivot and attaches by elastomer springs to both the front tandem axle and rear tandem axle.

Each tandem axle has a trailing arm attached to the front of the axle and to the machine's frame. The trailing arm allows the axle to pivot up and down in response to terrain changes. If one side of the tandem axle is lifted, one end of the walking beam will pivot upward, resulting in the beam's opposite end exerting the reactionary force on the other tandem axle. The force is transferred from the axle to the equalizer beam through an elastomer spring, consisting of a stack of rubber springs that can be laminated or bolted together.

Comparison of ADT Suspensions

Articulated dump truck suspensions can vary depending on manufacturer. Some examples are hydro-pneumatic, hydraulic with electronic leveling, and rear walking beam. See **Figure 21-11** for a comparison of ADT suspensions.

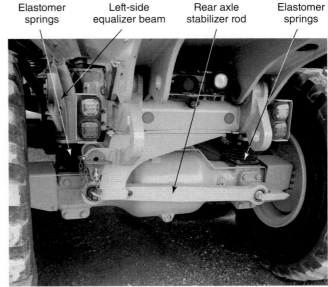

Figure 21-10. This Caterpillar ADT uses a walking beam suspension for the rear axles. The left-side equalizer beam pivots on the truck's rear frame and is attached via elastomer springs to the left side of the front tandem axle and rear tandem axle. The right-equalizer beam (not shown) also pivots on the truck's rear frame and is attached by elastomer springs to the right side of the front tandem axle and rear tandem axle.

Figure 21-11. Examples of ADT manufacturers and the types of suspensions used on their trucks.

Manufacturer	Front Suspension	Rear Suspension
Caterpillar (745C series)	Hydro-pneumatic cylinders	Walking beam/rubber suspension
Doosan (DA 40)	Hydro-pneumatic cylinders	Free-swinging bogie suspension
John Deere E series (370E–460E)	Hydraulic cylinders with pneumatic accumulators	Walking beam/laminated blocks
Komatsu (HM400-3, HM400-5)	Hydro-pneumatic cylinders	Hydro-pneumatic cylinders/rubber suspension
Volvo (A35GFS, A40GFS)	Hydraulic cylinders with electronic self-leveling	Hydraulic cylinders with electronic self-leveling

Goodheart-Willcox Publisher

The Bell, Caterpillar, John Deere, Volvo (standard), and Komatsu ADTs have rear tandems that use a walking beam suspension design with shock absorbers.

A Volvo FS ADT has a hydraulic suspension with a hydraulic *full suspension switch* in the tractor cab. The switch is turned to the *On* position (operating position) before the truck is operated. This causes the suspension cylinders to fully extend and lift the truck. The cylinders also provide *self-leveling*, a feature that uses an electronic control module to automatically adjust the height of the suspension cylinders based on different operating conditions.

When the full suspension switch is placed in the *Off* position, the suspension cylinders fully retract, which lowers the truck. The switch must be placed in the *Off* position before the truck is serviced, transported, parked, or the operator exits the truck.

A Doosan ADT uses a free-swinging bogie axle assembly. As explained in Chapter 17, *Final Drives*, swinging bogie drive wheels operate like the tandem drive wheels found on motor graders, except the Doosan ADT bogie drive wheels are powered by gears instead of chains. Each side of the bogie axle assembly is allowed to twist freely (pivot or swing). As the front or rear bogie drive wheel needs to lift or lower, the bogie axle swings to allow the drive wheels to remain in contact with the ground. An example of a free-swinging bogie axle assembly manufactured by NAF is shown in **Figure 21-12**.

Charging ADT Hydro-Pneumatic Suspension Cylinders

Be sure to follow the manufacturer's instructions for servicing, repairing, and charging hydro-pneumatic suspension cylinders. The front suspension cylinders on a Caterpillar ADT are charged in a similar fashion to the cylinders used on a rigid frame haul truck. However, the nitrogen pressure can be higher on an ADT, and the suspension height measurement is not measured at the suspension cylinders but is instead measured from the top of the wheel's rim to the bottom edge of the fender.

Scraper Cushioned Hitch

As explained in Chapter 3, *Construction Equipment Identification*, scrapers can have a cushioned hitch. A cushioned hitch consists of one double-acting hydraulic hitch-leveling cylinder, two nitrogen gas-filled accumulators, a hitch-leveling valve, and an electronic control module. After the inception of the cushioned hitch in the late 1960s, shock loads were reduced on tractor scraper hitches, which increased the life of the hitch frames and components. Machines could also be operated at faster speeds with reduced operator fatigue.

Goodheart-Willcox Publisher

Figure 21-12. Although manufacturer NAF's bogie axle is not the equipment installed on a Doosan ADT, its design is similar. A single limited-slip rear differential positioned in the center of the axle delivers power to the right- and left-side swinging tandem drives.

Chapter 21 | Suspension Systems

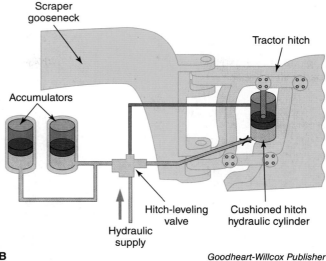

Figure 21-13. A cushioned hitch is used to smooth a scraper's ride during transit. A—Two accumulators dampen the cushioned hitch's hydraulic cylinder as the scraper travels. B—An electronic control module on the scraper controls the hitch-leveling valve. The valve can extend or retract the leveling cylinder, and either allow the accumulators to dampen the circuit or be isolated from the circuit. Isolation from the circuit causes the hitch to remain in a rigid position.

During transport, the two accumulators cushion the cylinder to dampen the hitch. See **Figure 21-13**. The operator locks out the cushioned hitch while the scraper is loading the bowl to ensure that the cutting edge remains in a fixed position.

Bobcat Compact Track Loader Suspensions

Machines equipped with undercarriage track rollers rely on the track rollers to transfer the weight of the machine to the ground. Bobcat T550 through T770 compact track loaders have the track rollers mounted on leaf springs. The leaf springs deflect and absorb shocks that are normally transmitted to the track roller frame, **Figure 21-14**. Bobcat T62 through T76 and T870 loaders use a five-link torsion suspension system. The left and right undercarriage roller frames are attached to front and rear torsion axles. Each axle has a square steel tube surrounded by four rubber torsion rods that allow torsional twisting of the axle. The ends of the axles attach to the track roller frame through a swinging fifth link that allows the torsional axle to function as an independent suspension.

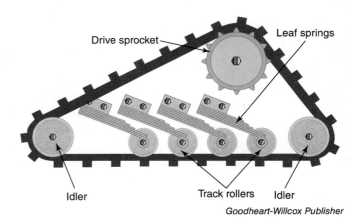

Figure 21-14. Bobcat offers a compact track loader with the track rollers mounted on leaf springs.

Implement Ride Control

Large construction machines have implements that can cause severe bouncing when the machine is traveling in the field or at high road speeds during transport. Some machines have one or more accumulators that are plumbed in parallel to the implement's lift cylinder to dampen the implement's movement. This system is known as ***ride control***. Manufacturers list "improved cycle times" as an additional advantage of ride control because the operator can operate the machine more productively with less bouncing. Wheel loaders, loader backhoes, wheeled excavators, and skid steers are among the machines that are equipped with ride control.

Refer to the simplified hydraulic schematic in **Figure 21-15**. A two-position solenoid valve turns the ride control on and off. When the machine is performing tasks that require

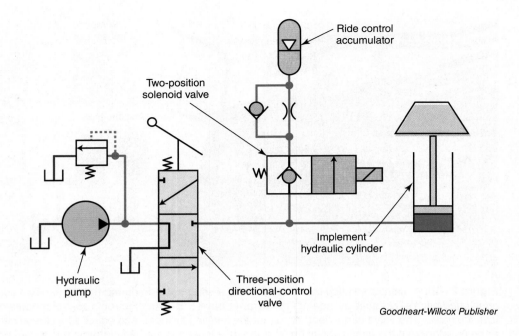

Figure 21-15. Ride control has a solenoid control valve that allows an accumulator to be plumbed in parallel to an implement cylinder, such as a loader lift cylinder.

precise implement control, such as grade work, the ride control accumulator is de-energized. Ride control can be included as a new machine option or sometimes ordered and installed as a kit after a machine has been sold.

Machines can also integrate *automatic ride control*, a feature that engages the ride control at a specific travel speed programmed into the electronic control module. The feature disengages when the machine is traveling at speeds below the preset speed. Automatic ride control is designed with a three-position toggle switch for the following operations:

- *Off*—the accumulator solenoid valve is de-energized, isolating the accumulator from the system.
- *On*—the accumulator solenoid valve is energized and the lift cylinder's operation is dampened, regardless of the machine travel speed.
- *Automatic*—the accumulator solenoid valve is energized when the machine is traveling above the preset travel speeds.

Seat Suspension Systems

Many off-highway machines are equipped with an *air suspension seat*, sometimes called an *air ride seat*, which contains an air bag for seat height adjustment and seat cushioning. The M series Caterpillar motor grader seat includes the following parts:

- An air bag, which serves as the seat's spring.
- An electric-powered compressor used to fill the air bag.
- A scissor-shaped frame connecting the seat to the base.
- A hydraulic shock absorber.

If the height of the seat must be raised, the operator activates a switch to turn on the electric compressor and fill the air bag. When the seat height needs to be lowered, air is released from the air bag through a valve. The pressure in the air bag is based on the weight of the operator. The hydraulic shock absorber dampens the movement of the seat. This seat design is known as a *passive air ride seat* since it does not feature any type of sensors, mechanical feedback lever, or electronic controls to vary the seat's suspension based on the terrain and machine operation.

Semi-active seat suspensions commonly use hydraulic cylinders filled with magnetorheological fluids. Due to microscopic iron particles suspended in the fluid, *magnetorheological fluids* can change their viscosity (fluid thickness) by means of magnetism. The seat does not require an actuator that exerts a mechanical force, but rather the system applies a magnetic field to the fluid to thicken it, thereby increasing resistance in the cylinders and stiffening the suspension.

Caterpillar Advanced Ride Management (CARM)

Caterpillar manufactures a seat suspension system called Caterpillar Advanced Ride Management (CARM), which is an *active air ride seat*. The CARM, and other active air ride seats, have multiple sensors (position, pressure, displacement, and temperature) to provide input to an electronic control module that controls a dampening assembly.

The CARM's dampening assembly consists of an air bag, electric-powered compressor, control valve, and voice coil. A voice coil functions like an electric actuator and moves in a linear motion as electrical current is applied to its assembly. The control valve is operated by the voice coil.

The electronic control module defaults the seat to a maximum dampening setting. This means that if the system has a problem, all electric power is cut, creating a stiff seat suspension that the operator will notice and report to a technician who can investigate the problem.

Agricultural Suspension Systems

Agricultural equipment, like construction equipment, operates in rough terrain. In addition, agricultural producers often struggle to find enough qualified personnel to operate the machines, leaving more work to be completed with fewer operators. As a result, agricultural machine manufacturers have invested tremendous amounts of resources in designing and manufacturing machines that optimize the machine's ride quality. Significantly improving a machine's ride quality allows the operator to work more comfortably and for longer hours each day.

Suspensions on agricultural tractors have evolved over the decades. Older tractors were equipped with steel wheels that lacked the deflection of rubber tires. They also had a metal operator seat supported only by a steel coil spring. Modern tractors have computer-controlled suspension systems for optimizing the machine's ride quality. Many late-model agricultural machines have a suspension system on at least one axle, a seat suspension setup, and a cab suspension system.

Oscillating Axle with Stops

Some agricultural machines are equipped with an oscillating axle (enabling one wheel to rise while simultaneously lowering the opposite wheel) and a rigid (stationary) axle. The rigid axle can be located on the front, as on an agricultural combine or cotton harvester, or it can be located at the rear of the machine, as on mechanical front-wheel-drive (MFWD) agricultural tractors.

The *oscillating axle* contains an oscillating tube in the center of the axle housing to allow the axle to oscillate as the machine travels over sloped or rough terrain. This design enables all four machine tires to continually transfer the load to the ground. See **Figure 21-16**. The oscillating axle is

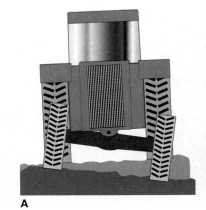 

Goodheart-Willcox Publisher

Figure 21-16. The oscillating front axle can be strictly a steering axle without the ability to drive the vehicle, or it can be a driving steer axle, such as those found on MFWD tractors. A—The rear axle can adjust to one terrain, and the front axle can oscillate to follow terrain sloped in the opposite direction. B—A John Deere 6115D utility tractor's front axle is attached to the front frame through two oscillating points (bushings), which allow the axle to oscillate to the left and right.

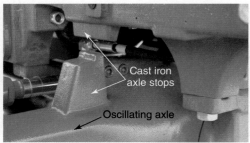

Figure 21-17. This John Deere 5085E tractor has cast iron axle stops.

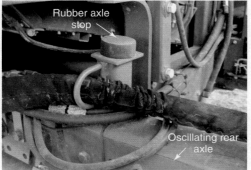

Figure 21-18. This Gleaner combine's rear axle is an oscillating axle fitted with rubber axle stops.

often the steering axle, such as on a combine or a utility tractor, and can also be powered to produce four-wheel drive.

Some oscillating axles have cast iron stops to limit maximum axle travel, **Figure 21-17**. Other designs have rubber stops. Combines have an oscillating axle (rear axle) and axle stops. A combine's front drive axle is mounted rigidly to the machine's frame, but the rear axle is allowed to oscillate to the left and right. The combine in **Figure 21-18** has rubber stops to limit maximum axle travel.

Massey Ferguson 9800 Series Swather GlideRider Suspension

AGCO offers self-propelled swathers, such as their Massey Ferguson 9800 and 9900 series for example, with a GlideRider rear axle suspension system. A spring and two gas-filled shock absorbers are located in the center of the axle housing. A pair of swinging parallel bars is located on each side of the axle. Refer to **Figure 21-19**. This suspension requires no adjustment and allows the axle to float as it travels through the field.

John Deere Swather Torsion Suspension

John Deere's self-propelled swather has rear wheels that feature a torsion-style suspension. Two collars clamp together the caster wheel arm and four cylindrical rubber isolators (rubber rods). The caster arm has an internal square tube with a flat surface on each of its four sides. The rubber rods are sandwiched between caster arm's flat spots and the fixed collar clamps that form a non-rotating tube. The caster wheel arm rotates inside of the non-rotating tube housing. The rubber deforms and deflects to absorb shock loads. See **Figure 21-20**.

The self-propelled swather also has a double-acting hydraulic cylinder connected to the spindle of each caster wheel. The cylinder has three modes of operation: no dampening, dampening, and hydraulic steering assist. John Deere calls this system *IntelliAxle*. The design assists steering and reduces caster shimmy at high speeds, allowing the machine to travel at speeds as fast as 22 mph (35 km/h).

Mechanical Front-Wheel-Drive (MFWD) Front Suspension Systems

Mechanical front-wheel-drive (MFWD) tractors are some of the most common workhorse tractors used in the agricultural industry. Most late-model MFWD tractors are equipped with some type of front suspension, a seat suspension setup, and some type of cab suspension.

Figure 21-19. A—An AGCO GlideRider rear axle suspension has a center-mounted spring with twin gas-filled shock absorbers that are accessible by lifting the rear engine cooler panel and swinging out a cooler. B—A pair of swinging parallel arms are located at each side of the axle. The caster arm has a friction brake that reduces high-speed caster wobble.

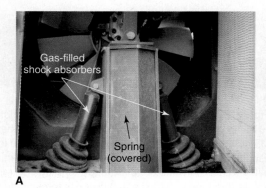

Note

The next three topics explain two types of John Deere MFWD suspension systems—Triple Link Suspension (TLS) and Independent Link Suspension (ILS)—and one John Deere four-wheel-drive suspension system—HydraCushion. The three systems all use double-acting hydraulic suspension cylinders that are dampened by nitrogen gas-filled accumulators and are electronically operated by the suspended front axle (SFA) control unit. However, the three types of suspension systems are different in the mechanical design. In John Deere's electronic service literature, all three systems are described under the heading "Suspended Front Axle". For the purpose of clarity, this textbook will not use the term *suspended front axle* for these suspensions, but instead use the John Deere trademark names, TLS Plus, ILS, and HydraCushion.

John Deere Triple Link Suspension (TLS)

John Deere first offered their optional *Triple Link Suspension (TLS)* on 6010 series tractors in the early 1990s. The option soon became available in the 7000 series tractors. Today, the TLS Plus is an option on 6R and 7R John Deere tractors.

TLS Plus Components

A tractor equipped with the TLS Plus system has a hydropneumatic suspension system to improve machine stability, improve traction, increase operator comfort, increase productivity, and reduce power hop. *Power hop* is a condition that causes a tractor to severely bounce (known as pitching fore and aft) while operating in the field with high draft loads. Dry soils, too little tractor ballast, improper weight distribution between the front and rear axles, and tires that are too small are additional contributing factors to power hop. Chapter 22, *Tires, Rims, and Ballasting*, discusses power hop, tires, and ballasting in further detail.

The TLS Plus system contains the following components:
- MFWD powered front axle.
- Torque arm (long draft arm).
- Pair of double-acting hydraulic suspension cylinders.
- Panhard rod.
- Three nitrogen gas-filled accumulators.
- Hydraulic control valves, including five electrohydraulic solenoids.
- Position sensor.
- Pressure sensor.
- Suspended front axle (SFA) control module.

The front axle is connected to the tractor frame in the TLS system at three points: the torque arm, the pair of double-acting hydraulic suspension cylinders, and the panhard rod.

The hollow cast *torque arm* is a long draft arm that allows the MFWD driveshaft to extend through the center of the housing, **Figure 21-21**. The front of the torque arm is bolted directly to the front axle housing, and the rear of the torque arm is attached near the tractor's center of gravity at the bottom center of the tractor's frame via an oscillation joint, **Figure 21-22**.

Goodheart-Willcox Publisher

Figure 21-20. A John Deere W235 self-propelled swather has a rear suspension with four rubber isolator rods that are clamped to the flat sides on the caster arm. As the arm rotates, the rubber rods deflect and deform, providing shock absorption. Note the double-acting hydraulic cylinder that is used for dampening and assisting with steering.

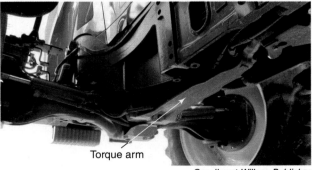

Goodheart-Willcox Publisher

Figure 21-21. The torque arm attaches the front axle to the center of the tractor.

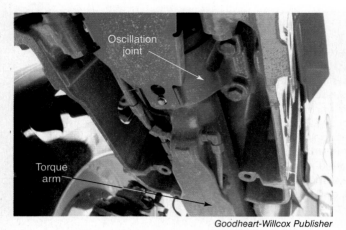

Figure 21-22. The torque arm attaches to the tractor's frame through an oscillation joint, which allows the front axle to oscillate to the left and right.

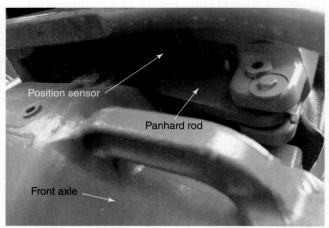

Figure 21-23. The TLS Plus panhard rod connects to the top of the front axle and the tractor's frame.

A *panhard rod* helps locate and stabilize the front axle, limiting axle lateral movement. It is attached to the top of the axle and to the tractor's frame, **Figure 21-23**. The TLS Plus position sensor (potentiometer) is connected to the panhard rod. The panhard rod, along with a pair of double-acting hydraulic suspension cylinders, **Figure 21-24**, suspends the front axle from the tractor's frame.

Three accumulators and two control valves control the suspension height and spring rate. The two accumulators shown in **Figure 21-25** are used to fully dampen the suspension when the dampening solenoid valve is energized. When energized, the valve fully ports the two accumulators to the cap end of both of the suspension cylinders. The other accumulator is located inside the engine compartment area and is plumbed into the rod side of both of the suspension cylinders.

Figure 21-24. Each side of the TLS Plus design has a double-acting hydraulic suspension cylinder.

Note

Many off-highway machine manufacturers label the head end of a hydraulic cylinder as the opposite end of the cylinder rod. The International Fluid Power Society and National Fluid Power Society label the rod side as the head end of the cylinder. This textbook intentionally avoids using the term *head end* due to the different uses of that term. Instead, this textbook will label the opposite end of the rod as the cap end of the cylinder.

TLS Plus Hydraulic Controls

The TLS Plus hydraulic control valve receives oil from the steering priority valve's excess flow port, which is the same circuit as the selective control valve (SCV) and three-point hitch. Oil is sent to the suspension manifold valve block, which contains several valves including the following:

- Suspension spring rate increase solenoid.
- Suspension spring rate decrease solenoid.
- Suspension raise solenoid.
- Suspension lower solenoid.
- Pressure-relief (regulating) valve. See **Figure 21-26**.

Chapter 21 | Suspension Systems

> **Note**
> John Deere's service manuals and literature use the term *selective control valve (SCV)* in place of directional-control valve.

The pressure inside the suspension cylinders can be manually relieved. The suspension lower solenoid valve has a manual knob that when turned will vent oil from the cap end of the suspension cylinders. The suspension spring rate decrease solenoid has a manual knob that when turned will vent oil from the rod end of the suspension cylinders.

The suspension dampening valve block contains the dampening solenoid valve. When the dampening solenoid valve is energized, it connects the two suspension cylinders' cap ends directly to the cap end accumulators. When the solenoid is de-energized, the cap ends of the cylinders are still connected to the two accumulators, but the fluid is restricted by a dampening orifice.

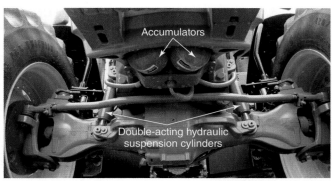

Goodheart-Willcox Publisher

Figure 21-25. Three accumulators are used in the TLS Plus. Two of the TLS Plus accumulators are visible underneath the frame in the front of the tractor. The third accumulator is not easily visible. It is located in the engine compartment.

TLS Plus Self-Leveling Operation

The TLS Plus has a self-leveling feature with a travel range of ±2″ (total of 4″). After the machine is started, the system will automatically self-level, regardless if the three-point hitch is raised or lowered. The TLS Plus will also self-level any time the tractor is traveling faster than 0.9 mph (1.5 km/h).

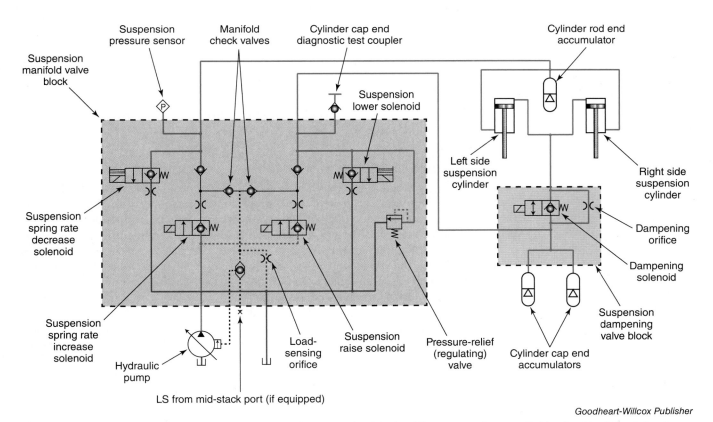

Goodheart-Willcox Publisher

Figure 21-26. The TLS Plus hydraulic system contains two valve blocks. The suspension manifold valve block contains four solenoid valves, and the suspension dampening valve block contains one solenoid valve.

TLS Plus Modes of Operation

When TLS Plus is engaged, the hydraulic suspension cylinders receive cushioning via the accumulators. The operator can set two different levels of firmness. In addition, the TLS Plus offers three different modes of operation: *Auto*, *MAX*, and *Manual*.

The TLS Plus becomes operable when the transmission lever is shifted out of the *Park* position. When the tractor speed exceeds 0.3 mph (0.5 km/h), the TLS Plus is activated, causing a delay in the control of the suspension.

In the *Auto* mode, the controller monitors multiple inputs (such as travel speed, brake pedals, implement weight, ground conditions, and the use of implements) in order to automatically adjust the suspension to provide the best ride.

In the *MAX* mode, the TLS Plus applies the maximum amount of spring rate (suspension stiffness) to the front axle. If the tractor exceeds 18 mph (30 km/h), the controller will force the TLS Plus back to the *Auto* mode, and if the tractor speed slows below 12 mph (20 km/h), the controller will revert back to the *MAX* mode of operation. The *MAX* mode is used when the tractor is lifting or carrying heavy front loads and a stiff front suspension is desirable.

The *Manual* mode enables the operator to manually raise or lower the tractor's front end suspension. When the tractor speed exceeds 3 mph (5 km/h), the controller places the TLS Plus back in the *Auto* mode.

John Deere Independent Link Suspension (ILS)

John Deere first offered the optional **Independent Link Suspension (ILS)** on its 8020 series MFWD tractors. The option continues to be available today on the 8R series tractors.

ILS Components

The ILS is a hydro-pneumatic independent front suspension system that increases ride comfort and machine productivity by improving traction and reducing power hop. The ILS system contains the following components:

- MFWD powered front axle.
- Constant velocity telescoping driveshafts.
- Pair of upper and lower control arms.
- Pair of double-acting hydraulic suspension cylinders.
- Four nitrogen gas-filled accumulators.
- Hydraulic control valves, including six electro-hydraulic solenoids.
- Two position sensors.
- Pressure sensor.
- Suspended front axle (SFA) control module.

At the heart of the ILS is a pair of upper and lower control arms, similar to the dual A-shaped arms found in automotive short-long arm suspension systems, **Figure 21-27**. On agricultural tractors equipped with dual front wheels, the dual wheels do not remain perpendicular to the ground as the suspension compresses and rebounds, causing a loss of traction. The ILS upper control arm is shorter than the lower control arm. As the ILS suspension moves, the upper control arm travels in a smaller curve than the lower control arm's curve, which allows the suspension's camber to keep the dual front wheels perpendicular to the ground to improve traction.

Camber is the tire angle formed between the vertical axis of the tire and the vertical axis of the tractor when viewed from the front or back of the axle. See **Figure 21-28**. When the top edge of the tire leans outward, the camber angle is positive. If the bottom edge of the tire leans outward, the camber angle is negative.

A double-acting hydraulic suspension cylinder is located on each side of the ILS. The SFA control module controls the suspension cylinders to provide the desired suspension height, as well as to keep the differential housing centered vertically between the wheel hubs. The ILS has a suspension travel range of ±5″ (total of 10″). Two accumulators provide shock absorption to the twin cylinders. See **Figure 21-29**.

Chapter 21 | Suspension Systems

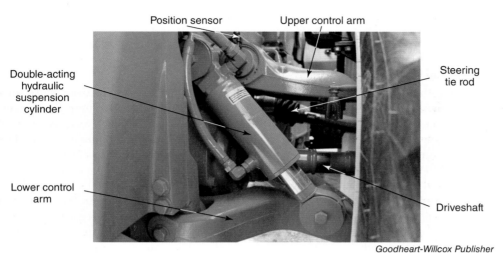

Figure 21-27. The ILS control arms, suspension cylinder, steering tie rod, driveshaft, and position sensor are shown on a John Deere 8320R tractor.

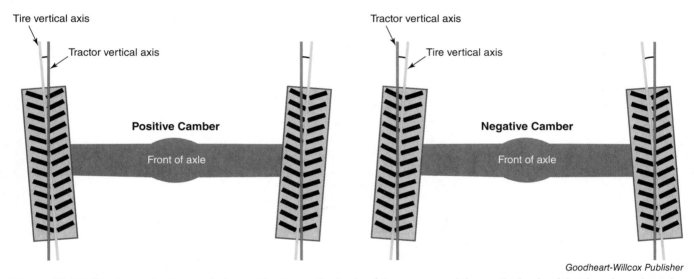

Figure 21-28. Camber is the tire angle formed by the vertical axis of the tractor and the vertical axis of the tire when viewed from the front or back of the axle. The top of the tire will angle slightly outward for positive camber, and the bottom of the tire will angle slightly outward for negative camber.

ILS Hydraulic Controls

The hydraulic pump flow is made available to the suspension manifold valve block from the tractor's hitch/SCV circuit. The oil supply enters the suspension manifold valve block, which contains the following ILS solenoid valves:

- Suspension spring rate increase solenoid.
- Suspension spring rate decrease solenoid.
- Suspension raise solenoid.
- Suspension lower solenoid.
- Pressure-relief (regulating) valve.

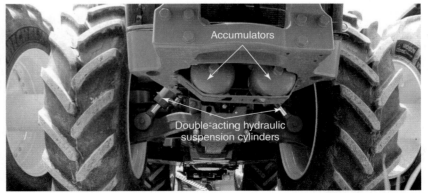

Figure 21-29. Two of the four ILS accumulators can be viewed directly underneath the front of an 8R John Deere tractor. These accumulators are used on the cap end of the suspension cylinders. The other two accumulators not shown are located above the ILS manifold valve.

The suspension manifold valve block also contains three other solenoid valves:
- Front differential lock solenoid.
- Front brake solenoid.
- Front brake enable solenoid. See **Figure 21-30**.

When the ILS is hydraulically lifted (cylinders extended), the suspension raise solenoid is energized by the SFA control module. Refer to **Figure 21-31**. Pump flow passes through the suspension raise solenoid valve and then through the check valve, causing oil to flow to the cap end of both suspension cylinders. The hydraulic flow causes the cylinders to extend and lift the front suspension. The load-sensing spring-loaded check valve also opens, allowing a signal pressure to develop in the signal pressure circuit used by the priority valve and the pump flow compensator control valve.

The suspension spring rate pressure sensor monitors the pressure in the rod end of the suspension cylinders. As the cylinders extend, the spring rate can increase to above the operator's commanded setting. If this occurs, the suspension spring rate decrease solenoid will energize to lower the spring rate pressure back to its commanded setting.

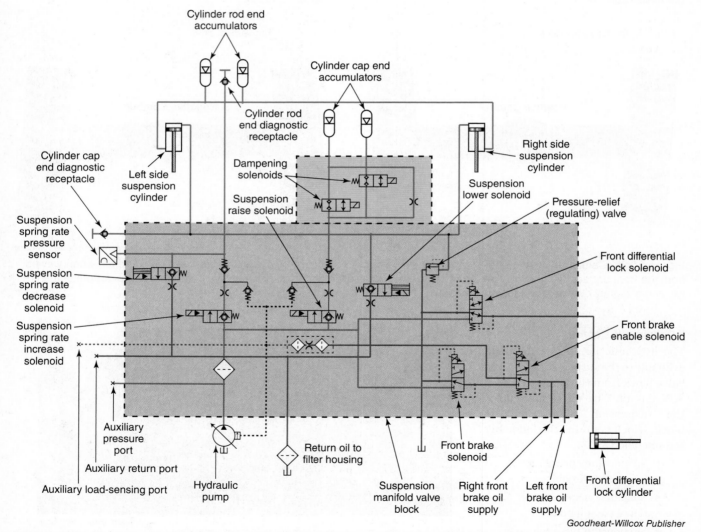

Goodheart-Willcox Publisher

Figure 21-30. The hydraulic schematic shows the valves used in a John Deere 8R ILS front axle. The hydraulic pump has been placed in the circuit to make it easier to understand the pump supply. However, the supply must first pass through a steering priority valve before it can be sent to the three-point hitch and SCV circuit (not shown), which is the circuit that supplies the ILS.

Chapter 21 | Suspension Systems

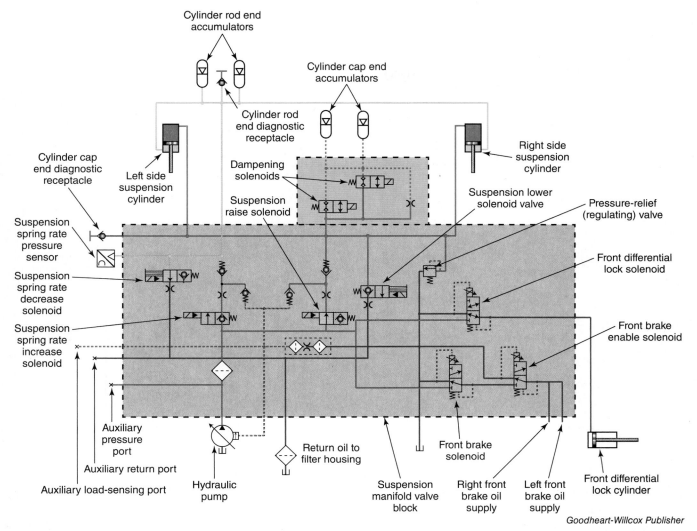

Figure 21-31. The hydraulic schematic shows the ILS suspension lifting and the suspension spring rate pressure decreasing.

When the ILS is hydraulically lowered (cylinders retracted), the suspension lower solenoid is energized by the SFA control module. See **Figure 21-32**. Oil from the cap end of the cylinder is allowed to pass through the suspension lower solenoid valve. As the oil is vented, the suspension cylinder retracts and lowers the front suspension. The two dampening solenoids are also energized to allow the equalization between the two cylinders and the cylinder rod end accumulator while the suspension is lowering.

The suspension spring rate pressure sensor monitors the pressure in the rod end of the suspension cylinders. As the cylinders retract, the spring rate can decrease to below the operator's commanded setting. If this occurs, the suspension spring rate increase solenoid is energized to raise the spring rate pressure back to its commanded setting. As the suspension spring rate increase solenoid is energized, hydraulic pump flow is used to refill the oil in the rod end of the suspension cylinders.

Fluid pressure is always maintained in the rod end of the suspension cylinders, which prevents the cap end fluid pressure from overcorrecting in rough terrain. As a result, the suspension spring rate increase solenoid and the spring rate decrease solenoid are energized as needed to maintain the rod end pressure that is set by the operator.

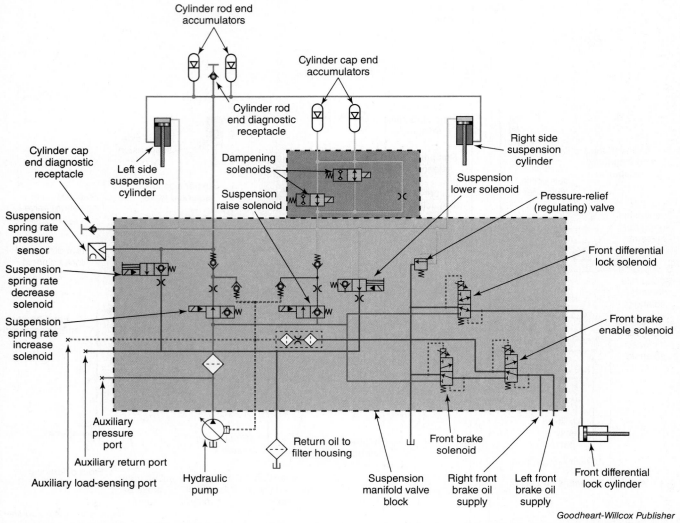

Figure 21-32. When the ILS front axle needs to lower, the SFA control module energizes the suspension lower solenoid, which vents oil from the cap end of the cylinder and causes the suspension to lower. The dampening solenoids are also energized to provide dampening while the suspension is lowering. If the spring rate drops too much, the controller energizes the suspension spring rate increase solenoid.

ILS Self-Leveling Operation

The ILS will automatically self-level and center itself vertically once tractor speed exceeds 0.3 mph (0.5 km/h). Any time the suspension settles, the SFA control module will raise the front end, as much as 1″, to achieve a centered position.

ILS Mode of Operation

The ILS system is integrated with the tractor's electronics and does not have an electric control switch. The ILS has three different dampening modes of operation: *Restricted*, *Unrestricted*, and *Auto*. The system will be either active (*Unrestricted* or *Auto*) or inactive (*Restricted*) based on the inputs to the tractor's electronic control system.

When the ILS is in the *Restricted* mode, the two dampening solenoids are de-energized, and the dampening orifice restricts the oil flow between the cap end of the suspension cylinders and the accumulators. As a result, the suspension is rigid and has less travel when traveling over rough terrain. The schematic in **Figure 21-31** shows the dampening solenoid valves de-energized and in the *Restricted* mode.

Chapter 21 | Suspension Systems 751

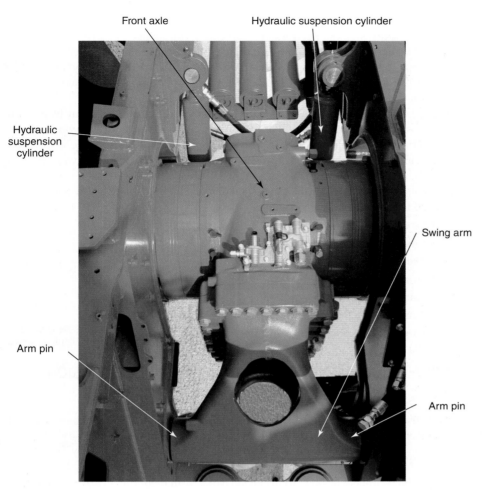

Figure 21-33. Note in the top view of this HydraCushion that the front axle is attached to the swing arm and the two suspension cylinders.

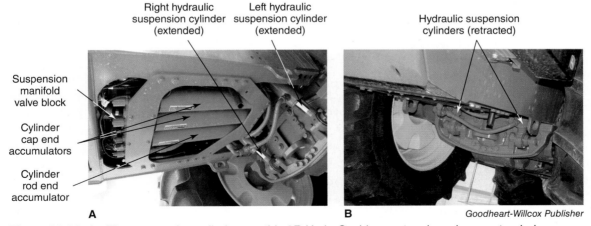

Figure 21-34. A—The suspension cylinders on this 9R HydraCushion system have been extended, causing the suspension to be lifted. B—The suspension cylinders have been retracted, causing the suspension to be lowered.

When the ILS is in the *Unrestricted* mode, the two dampening solenoids are energized, allowing oil to flow unrestricted between the cap end of the suspension cylinders and the accumulators. The accumulators can absorb oil when the suspension cylinders compress (jounce) due to rough terrain, which allows the suspension to move more freely throughout its operating range. The schematic in **Figure 21-32** shows the dampening solenoid valves energized and in the *Unrestricted* mode.

When the ILS is in the *Auto* mode, the SFA control module receives input from both control arm position sensors and will individually control the energization and de-energization of both dampening solenoids. The SFA control module allows the suspension to achieve maximum travel but limits the suspension jounce and rebound to provide a quality ride.

The ILS is placed back into the *Restricted* (suspension locked) mode by the SFA control module if any of the following actions occur:

- Tractor transmission shifted into *Park*.
- Both service brake pedals depressed.
- Rear power take-off (PTO) engaged.
- SCVs actuated.
- Three-point hitch actuated.
- Speed slows below 0.3 mph (0.5 km/h).
- Suspension corrects itself for out-of-level condition.

Large Four-Wheel-Drive Agricultural Tractor Suspension Systems

Large agricultural tractors can also suffer from power hop. Manufacturers have designed suspension systems to improve operator ride comfort and machine productivity.

John Deere 9R HydraCushion Suspension

John Deere introduced a suspension system called HydraCushion for the 9R series four-wheel-drive articulated tractors in 2014. *HydraCushion* is designed to reduce power hop, reduce road lope, improve customer ride, and improve machine productivity by increasing traction.

Road lope is similar to power hop in that it causes severe suspension bouncing. However, road lope does not occur in the field. Instead, it occurs when transporting the tractor (roading) on hard surfaces at higher speeds with no draft load. It normally occurs at a specific road speed and is often the result of an out of round tire or wheel, which can sometimes be corrected by repositioning the tire on the wheel. Chapter 22, *Tires, Rims, and Ballasting*, will discuss road lope and rim service in further detail.

Similar to ILS, HydraCushion is programmed to self-level the suspension and keep the front differential (axle housing) vertically centered to the tractor's chassis. The HydraCushion suspension has a travel range of ±2″ (total of 4″).

HydraCushion Components

Similar to TLS Plus and ILS, the HydraCushion system is a hydro-pneumatic front suspension that increases the operator's ride comfort and machine productivity. The system contains the following parts:

- Double-reduction axle assembly.
- Axle swing arm.
- Pair of double-acting hydraulic suspension cylinders.
- Three large, nitrogen gas-filled accumulators.
- Hydraulic control valves, including six electro-hydraulic solenoids.
- Position sensor.
- Pressure sensor.
- Suspended front axle (SFA) control module.

The double-reduction axle housing is attached to a *swing arm* that is pinned to the front frame allowing the front axle to swing up and down. Double-acting hydraulic suspension cylinders are attached to the front of the axle, one on the left and one on the right. See **Figure 21-33**.

The three accumulators are located in front of the two suspension cylinders, and the suspension manifold valve block is located in front of the accumulators. **Figure 21-34** shows both the cylinders in an extended position and in a retracted position.

HydraCushion Hydraulic Controls

The HydraCushion has six electronically actuated solenoid valves to actuate and dampen the two double-acting suspension cylinders. Two accumulators are plumbed to the cap end of the suspension cylinders, and one accumulator is plumbed to the rod end of the suspension cylinders, **Figure 21-35**.

The HydraCushion controls operate in a nearly identical method as the controls of the previously explained ILS system. When the HydraCushion is hydraulically lifted, the suspension raise solenoid is energized. As the suspension cylinders are extending, the SFA control module lowers the spring rate pressure as needed. See **Figure 21-36**.

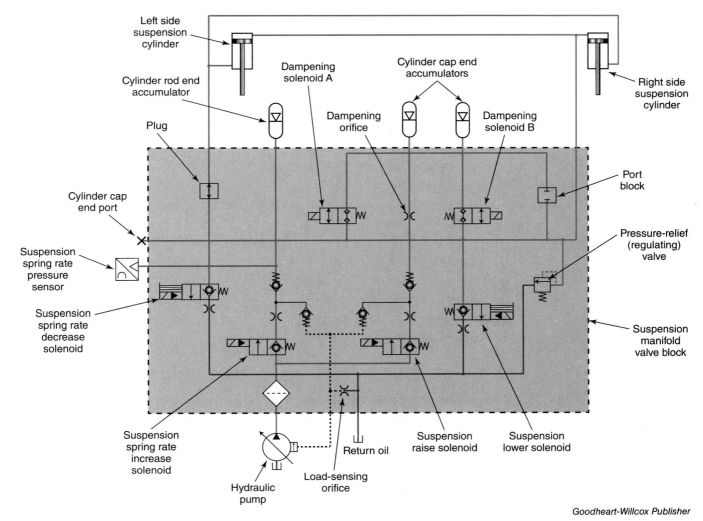

Goodheart-Willcox Publisher

Figure 21-35. The HydraCushion, like the TLS Plus and ILS, has four solenoids to raise, lower, and vary the suspension system's spring rate. Like the ILS, the HydraCushion has two dampening solenoids installed in the system. This hydraulic schematic shows the system in a neutral position.

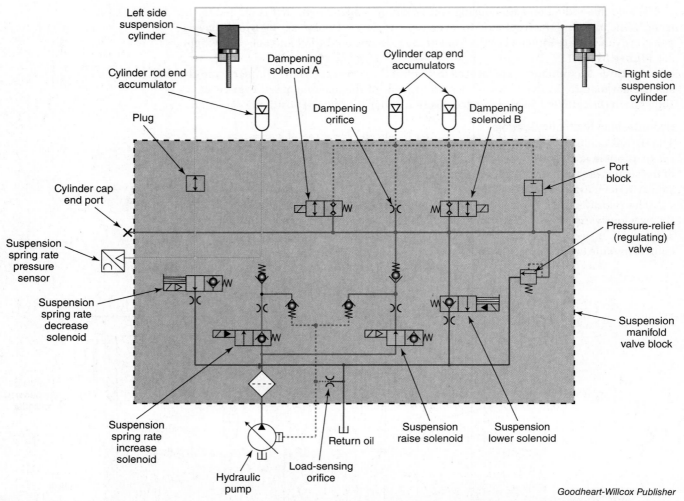

Figure 21-36. The hydraulic schematic shows the HydraCushion suspension lifting and the spring rate pressure decreasing.

When the HydraCushion is hydraulically lowered, the suspension lower solenoid is energized. As the suspension cylinders retract, the SFA control module increases the spring rate pressure as needed. See **Figure 21-37**.

The three different dampening modes of operation for the HydraCushion mirror the modes of the ILS system: *Restricted*, *Unrestricted*, and *Auto*. The dampening modes also perform the same functions as those in the ILS system.

Large Twin-Rubber-Track Tractor Suspension Systems

The agricultural industry has been producing large twin-rubber-track tractors since the late 1980s and early 1990s, beginning with Caterpillar's introduction of the Challenger Model 65 and 75 tractors in 1987 and 1991 respectively. Although the machines do not have four wheels or four tracks, they are often placed in the same category as four-wheel-drive agricultural tractors. This is because the twin-rubber-track tractors often use the same engine and transmission, deliver the same horsepower, and are used to perform the same work—such as pulling large implements with heavy draft loads—as four-wheel-drive tractors. The John Deere 9RT, Challenger MT 700 and 800 series, and Fendt 900 and 1100 Vario MT series are modern examples of these machines.

Chapter 23, *Undercarriages*, provides an example of an MT 845 Challenger equipped with a hard bar that attaches to the front of the left and right undercarriages. The hard bar

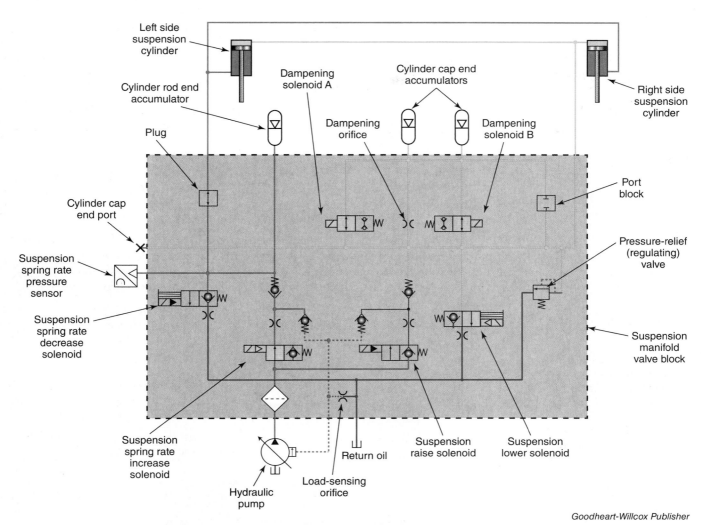

Figure 21-37. When the HydraCushion front axle needs to lower, the SFA control module energizes the suspension lower solenoid, which vents oil from the cap end of the cylinder and causes the suspension to lower. As the suspension lowers, the dampening solenoids are also energized to provide cushioning. The spring rate increase solenoid is energized if needed.

is suspended to the tractor frame through a large rubber spring, one on the left side of the hard bar and one on the right. See **Figure 21-38**.

ERA Manufacturing offers an aftermarket axle assembly that includes a pair of rubber cushions installed between the front axle and the tractor's frame to provide dampening to the front axle. The kit is used on the John Deere 9000 through 9020 twin-rubber-track tractors.

John Deere 8RT and 9RT Walking Beam AirCushion Suspension

John Deere 8030T, 8RT, 9030T and 9RT series tractors are equipped with a walking beam *AirCushion* suspension system that serves as the tractor's front axle assembly. The AirCushion suspension system is fully automatic. The operator cannot activate it or adjust it. The suspension improves the comfort of the ride in rough terrain

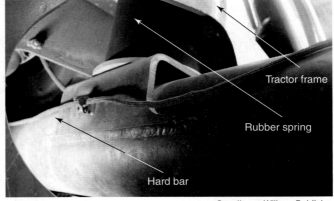

Figure 21-38. On this Challenger MT 845, a large rubber spring attaches the right side of the hard bar to the tractor's frame. The left side of the hard bar is equipped with the same type of spring.

and makes field speeds faster. See **Figure 21-39**. The AirCushion suspension contains the following parts:

- Walking beam.
- Upper front axle support.
- Lower front axle support (swing arm).
- Two spindles.
- Rubber isolator.
- Pivot pin.
- Electric-powered air compressor.
- Pneumatic control valve.
- Two air springs.
- Shock absorber.

The walking beam assembly mounts to the front of the left and right undercarriages via a spindle (pin), **Figure 21-40**. An electric-powered air compressor fills the air springs. The two air springs, shock absorber, rubber isolator, and walking beam assembly allow the suspension to independently cushion the left and right undercarriages, **Figure 21-41**.

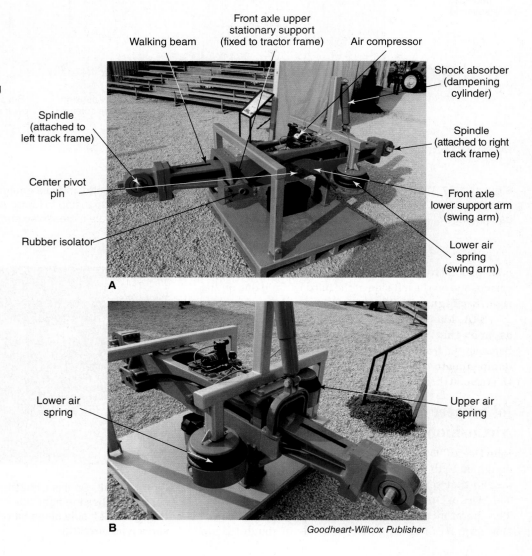

Figure 21-39. A—This John Deere RT series tractor AirCushion suspension system is a farm show training aid that allows customers to see the working components of the system. B—This view of the system shows both the upper and lower air springs.

Goodheart-Willcox Publisher

Chapter 21 Suspension Systems

Figure 21-40. This John Deere 8360RT tractor has the walking beam attached to the track frame via the spindle.

Twin-Coil and Hydraulic Shock Absorbers

Late-model Challenger MT700 and MT800 series tractors and Fendt 900 and 1100 Vario MT series tractors use a *SmartRide* suspension system. The system allows the track rollers to pivot. The front hard bar is suspended with two coil springs and heavy-duty hydraulic shock absorbers.

Air Suspension Systems

Agricultural machines can have air suspension systems to absorb shock loads to the chassis, such as the John Deere twin-rubber-track tractors (8RT and 9RT series), some sprayers, and a Case IH swather. An air suspension may have the following components:

- Self-contained air spring without its own compressor (Case IH swather or John Deere pull-type rotary mower).
- Multiple air springs with an electric-powered compressor (John Deere walking beam AirCushion) or an engine-driven compressor (agricultural sprayer).

Figure 21-41. A—The front axle suspension system on this John Deere 8370RT tractor connects to the right-side and left-side track frames. B—A closer look at the air spring on the right side of the tractor. An identical air spring is installed on the suspension on the opposite side of the tractor.

 Note
Air suspensions can also be used for cab suspensions and for seat suspensions.

Air springs have one of two types of designs: a rolling lobe or a double-convoluted air spring. An *air spring*, also known as an *air bag*, consists of a rubber element that is filled with compressed air. Either style of air spring can have a solid rubber bumper inside to act as a stop if the air spring loses pressure.

Case IH Swather Suspension

One of the simplest air suspension systems is used on Case IH WD 3, 4, and 5 series swathers. This system is shown in **Figure 21-42**. The rear independent suspension consists of a rolling lobe air spring and a dampening shock absorber on each of the two caster wheels. A rolling lobe air spring is a pliable barrel-shaped rubber air spring that is crimped (fastened) at the top to a bead plate and crimped at the bottom to a piston.

Figure 21-42. A Case IH WD 3 series swather has rear independent suspension. Each air spring incorporates the rolling lobe design and provides suspension to a caster wheel.

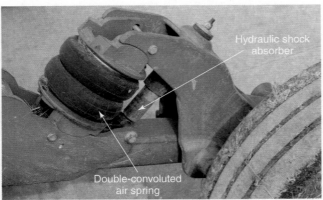

Figure 21-43. This rotary mower is equipped with a double-convoluted air spring and hydraulic shock absorber on its trailing wheel suspension.

As air is pumped into the air spring, air pressure forces the air spring to extend away from the piston. As air is vented from the air spring, the bottom of the spring (lobe) rolls over the piston as the spring deflates and collapses.

The Case IH swather does not have an onboard compressor. The operator simply uses a shop air compressor to fill the machine's two air springs.

John Deere Rotary Mower Suspension

Rotary mowers can also use simple self-contained air springs for suspension. The John Deere CX20 mower shown in **Figure 21-43** is equipped with a double-convoluted air spring and a shock absorber at each of its trailing wheels. A double-convoluted air spring has a top bellow and bottom bellow connected in the center with a girdle. The spring has a top bead plate and a bottom bead plate but does not have a piston. As air is pumped into the air spring, the bellows lengthen and the air spring extends. As air is vented from the air spring, the bellows retract and the air spring collapses. The double-convoluted air springs on a John Deere CX20 mower are individually filled with shop air.

Sprayer Suspension

Agricultural sprayers may have air suspension systems. Manufacturers who have produced sprayers with air suspensions include Hagie, John Deere, and RoGater. John Deere R series sprayers have a dual strut independent air suspension system that automatically self-levels. See **Figure 21-44**.

A belt-driven compressor is installed on the John Deere R4038 sprayer. Like the air systems described in Chapter 20, *Air Brake Systems*, the compressor contains a governor, dryer, reservoir, and control valves, **Figure 21-45**.

Each wheel contains its own mechanically operated air control valve that operates independently of the other valves. See **Figure 21-46**. When the suspension requires more air as it settles, the linkage pushes upward, causing the air control valve to direct air into the air spring. If the air spring is overfilled, the linkage moves downward, causing the air control valve to exhaust air from the air spring.

Figure 21-44. A John Deere R4038 sprayer is designed with a dual strut independent air suspension system.

Figure 21-45. This John Deere R4038 sprayer has the dryer, governor, and air reservoir mounted on the outboard right side of the sprayer.

Chapter 21 | Suspension Systems

Figure 21-46. A mechanically operated air control valve is located at each wheel to control the volume of air in the air spring.

Figure 21-47. This Case IH sprayer has a four-wheel independent suspension system with a coil spring and shock absorber at each wheel.

Diagnosing Air Suspensions

If a sprayer's suspension begins to sag due to a leak, the leak must be located before it can be repaired. Leaks can occur in supply lines, fittings, air springs, or control valves. Soapy water can be sprayed on air lines and fittings to look for air leaks.

If it is difficult to determine if the leak is in the air spring or located upstream, clamp the air line feeding the air spring. If the spring continues to leak, the air spring is at fault. If the spring stops leaking, the air control valve is the source of the problem.

Coil Spring and Shock Absorber Suspensions

Other sprayer manufacturers use coil springs and shock absorbers for suspensions. The Case IH sprayer shown in **Figure 21-47** features coil springs and shock absorbers as part of its four-wheel independent suspension.

Cab Suspension Systems

A few different suspension system designs are used for operator cabs, including rubber mounts, coil springs and shock absorbers, or an air suspension system. Most off-highway cab suspensions are passive.

Rubber Cab Mount Suspension

The most basic style of cab suspension system is the rubber cab mount. The design has rubber springs between the cab and the machine's frame. The rubber cab mount provides a very limited amount of suspension travel and cushioning. See **Figure 21-48**.

Coil Spring and Shock Absorber Cab Suspension

One of the more popular types of cab suspension systems is the coil spring and shock absorber system. The spring can be placed around the outside of the shock absorber as an assembly, **Figure 21-49**, or can be located separately from the shock absorber, **Figure 21-50**.

Figure 21-48. This older Ford 8670 agricultural tractor has rubber cab mounts.

Figure 21-49. This John Deere 6130M agricultural tractor is equipped with coil springs and shock absorbers for its cab suspension.

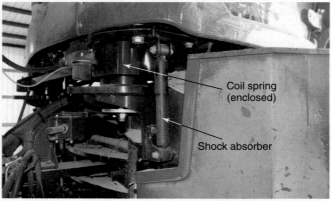

Figure 21-50. Four coil springs and four shock absorbers are installed as part of the Case IH Quadtrac's cab suspension system. Here a shock absorber attaches one corner of the Quadtrac's cab to the tractor frame.

Cab Air Suspension

Like on-highway trucks, off-highway cabs can be equipped with cab air suspension systems. However, cab air ride suspension systems are less common on machines delivered directly from the manufacturer's factory. Multiple aftermarket manufacturers offer kits for incorporating cab air ride systems on agricultural tractors.

John Deere Adaptive HCS Plus Cab Suspension

John Deere 7R and 8R series tractors can be equipped with a cab-suspension system called the Adaptive Hydro-pneumatic Cab Suspension (HCS) Plus. It contains two double-acting hydraulic cylinders, four accumulators, two solenoids, and a panhard rod. The tractor's control module receives inputs from the front axle suspension position sensor, engine throttle, clutch, brake, and transmission to control the Adaptive HCS Plus. As the tractor slows down and approaches the end of the field, the cab suspension stiffens. The cab suspension is also stiffened to prevent pitching any time the brake or clutch pedal is actuated or if the tractor has a change of speed commanded by the transmission control or engine throttle control. The tractor stiffens the cab based on a front axle suspension preset value.

The Adaptive HCS Plus has three settings for the stiffness of the cab suspension:

- *Auto soft*—features the most cab suspension movement but ideal for transport.
- *Auto medium*—ideal for working in the field.
- *Auto maximum*—features the stiffest cab suspension and ideal for loader work.

Summary

- A machine's suspension system absorbs shocks and handles loads as the machine travels over rough terrain and performs work.
- A suspension system can include leaf springs, coil springs, hydraulic shock absorbers, hydro-pneumatic shock absorbers, rubber springs, or air springs.
- Hydro-pneumatic suspension cylinders are used on rigid frame haul trucks.
- Elastomer suspension cylinders do not use oil or nitrogen and do not require recharging.
- An articulation joint allows the front and rear frames of an ADT to pivot and steer to the left or right.
- An oscillation joint allows the front and rear frames on an ADT to rotate (twist/oscillate) clockwise or counterclockwise in a vertical plane, freely independent from one another.
- A scraper's cushioned hitch contains one double-acting hydraulic hitch-leveling cylinder, two nitrogen gas-filled accumulators, a hitch leveling valve, and an electronic control module, which are used to reduce shock loads on the tractor scraper's hitch.
- Ride control consists of an accumulator that dampens the movement of large machine implements when the machine travels in the field or is transported at faster road speeds.
- Seat suspensions can be classified as passive, semi-active, or active.
- A magnetorheological fluid will change its fluid viscosity (thickness) by means of magnetism.
- Many agricultural machines use an oscillating axle suspension that is mounted to an oscillating tube, enabling all four tires to transfer the load to the ground.
- The TLS Plus suspension system equipped on John Deere 6R and 7R tractors uses a panhard rod, a pair of double-acting hydraulic suspension cylinders, a torque arm, and three gas-filled accumulators.
- The ILS suspension system offered on John Deere 8R tractors uses a pair of upper and lower control arms at each front drive wheel, a pair of double-acting hydraulic suspension cylinders, and four gas-filled accumulators.
- The AirCushion suspension system on John Deere 8RT and 9RT tractors has a walking beam assembly mounted to the front of the left and right undercarriages via a spindle. An electric-powered air compressor fills the two air springs. Along with a shock absorber and rubber isolator, the walking beam assembly and air springs allow the suspension to independently cushion the left and right undercarriages.
- Agricultural machines use air springs for machine suspensions, seat suspensions, and cab suspensions.
- Air springs can have a rolling lobe design or a double-convoluted design.
- Some air springs are self-contained and require shop air to be filled.
- Cab suspension system designs include rubber mounts, coil springs and shock absorbers, or an air suspension system.

Technical Terms

- active air ride seat
- air bag
- AirCushion
- air ride seat
- air spring
- air suspension seat
- articulation joint
- automatic ride control
- camber
- elastomer suspension cylinder
- full suspension switch
- HydraCushion
- hydro-pneumatic suspension cylinders
- Independent Link Suspension (ILS)
- IntelliAxle
- jounce
- leaf spring suspension
- kingpin
- magnetorheological fluids
- nitrogen-over-oil cylinders
- oil-pneumatic hydraulic cylinders
- oscillating axle
- oscillation joint
- panhard rod
- passive air ride seat
- power hop
- rebound
- ride control
- road lope
- self-leveling
- semi-active seat suspensions
- shock absorbers
- SmartRide
- struts
- suspension systems
- swing arm
- torque arm
- Triple Link Suspension (TLS)
- walking beam suspension

Review Questions

Answer the following questions using the information provided in this chapter.

Know and Understand

1. Hydro-pneumatic suspension cylinders can be referred to by all of the following names, *EXCEPT*:
    A. shock absorbers.
    B. oil-pneumatic hydraulic cylinders.
    C. rebound hitches.
    D. Nitrogen-over-oil cylinders.

2. Technician A states that hydro-pneumatic suspension cylinders should be charged individually, one at a time. Technician B states that hydro-pneumatic suspension cylinders should be charged two at a time, either the front pair of cylinders or the rear pair of cylinders. Who is correct?
    A. Technician A.
    B. Technician B.
    C. Both A and B.
    D. Neither A nor B.

3. All of the following factors affect the extension of a hydro-pneumatic suspension cylinder, *EXCEPT*:
    A. temperature.
    B. load on the machine.
    C. volume of oil.
    D. material used to construct the machine's frame.

4. Technician A states that ADTs have an articulation joint that allows the front and rear frames to pivot. Technician B states that ADTs have an oscillation joint that allows the front and rear frames to oscillate independently from one another. Who is correct?
    A. Technician A.
    B. Technician B.
    C. Both A and B.
    D. Neither A nor B.

5. Technician A states that agricultural combines often have an oscillating steering axle. Technician B states that utility tractors often have an oscillating steering axle. Who is correct?
    A. Technician A.
    B. Technician B.
    C. Both A and B.
    D. Neither A nor B.

6. Technician A states oscillating axles have only rubber stops installed to limit maximum axle travel. Technician B states oscillating axles can have cast iron stops to limit maximum axle travel. Who is correct?
    A. Technician A.
    B. Technician B.
    C. Both A and B.
    D. Neither A nor B.

7. What company manufactures a swather with a GlideRider suspension system?
   A. AGCO.
   B. Case IH.
   C. New Holland.
   D. John Deere.

8. Which John Deere suspension system is equipped with an electric-powered air compressor, two air springs, and a walking beam to improve the ride quality of 8RT and 9RT twin-rubber-track tractors?
   A. TLS Plus.
   B. ILS.
   C. HydraCushion.
   D. AirCushion.

9. Which John Deere suspension system is equipped with an axle swing arm, two double-acting hydraulic suspension cylinders, and three large, gas-filled accumulators to improve ride quality and reduce power hop on a 9R articulated tractor?
   A. TLS Plus.
   B. ILS.
   C. HydraCushion.
   D. AirCushion.

10. Which John Deere suspension system is equipped with dual control arms, two double-acting hydraulic suspension cylinders, and four gas-filled accumulators to improve ride quality and reduce power hop on an 8R MFWD tractor?
    A. TLS Plus.
    B. ILS.
    C. HydraCushion.
    D. AirCushion.

11. Which John Deere suspension system is equipped with a panhard rod, two double-acting hydraulic suspension cylinders, a torque arm, and three gas-filled accumulators to improve ride quality and reduce power hop on a 6R and 7R MFWD tractor?
    A. TLS Plus.
    B. ILS.
    C. HydraCushion.
    D. AirCushion.

12. On a MFWD tractor equipped with a TLS Plus system, which part helps locate and stabilize the front axle, limiting axle lateral movement?
    A. Articulation joint.
    B. Panhard rod.
    C. Torque arm.
    D. Walking beam.

13. All of the following are TLS Plus system modes of operation, *EXCEPT*:
    A. auto.
    B. MAX.
    C. manual.
    D. MINIMUM.

14. When the TLS Plus, ILS, or HydraCushion system lifts the front suspension, what other action might the suspended front axle control module perform?
    A. Energize the suspension lower solenoid.
    B. Energize the suspension spring rate increase solenoid.
    C. Energize the suspension spring rate decrease solenoid.
    D. Apply the parking brake.

15. When the TLS Plus, ILS, or HydraCushion system lowers the front suspension, what other action might the suspended front axle control module perform?
    A. Energize the suspension raise solenoid.
    B. Energize the suspension spring rate increase solenoid.
    C. Energize the suspension spring rate decrease solenoid.
    D. Apply the parking brake.

16. All of the following are ILS system dampening modes of operation, *EXCEPT*:
    A. manual.
    B. restricted.
    C. unrestricted.
    D. auto.

17. Air suspension systems can be used to absorb shocks for each of the following machine components, *EXCEPT*:
    A. seat.
    B. chassis.
    C. cab.
    D. scraper hitch.

18. Technician A states that hydro-pneumatic suspension cylinders can have a rolling lobe design. Technician B states that coil springs installed on a machine's suspension can have a double-convoluted design. Who is correct?
    A. Technician A.
    B. Technician B.
    C. Both A and B.
    D. Neither A nor B.

Apply and Analyze

19. A(n) _____ can also serve as a steering system kingpin.
20. The term _____ describes the compression of a suspension spring or a hydraulic suspension cylinder.
21. The term _____ describes the extension of a suspension spring or a hydraulic suspension cylinder.
22. Some Bobcat compact track loaders have _____ mounted on the rubber track rollers.
23. Ride control is designed to absorb shocks and dampen the movement of the _____ hydraulic circuit.
24. A traditional air suspension seat that does not use any magnetorheological fluids, electronic sensors, or feedback is referred to as a(n) _____ system.
25. On a MFWD tractor equipped with a TLS Plus system, the MFWD driveshaft passes through the center of the _____.
26. Severe tractor suspension bouncing and swaying that occurs when transporting a tractor on pavement at increased speeds is referred to as _____.
27. The ILS suspension has a total travel range of _____ inches.
28. The HydraCushion suspension has a total travel range of _____.

Critical Thinking

29. A scraper uses a cushioned hitch to achieve what outcomes?
30. What is the difference between power hop and road lope?

Chapter 22
Tires, Rims, and Ballasting

Objectives

After studying this chapter, you will be able to:

- ✓ Describe the safety practices and warnings for servicing and working with heavy equipment tires.
- ✓ Describe the different types of tires and their applications on off-highway equipment.
- ✓ Identify and interpret tire size nomenclature.
- ✓ Explain how to manually calculate wheel slip.
- ✓ List the actions to perform when inspecting tires and identify ways to maximize tire life.
- ✓ Recognize how the ton-mile per hour (TMPH) tire value is calculated and its importance to heavy equipment tire selection.
- ✓ Describe the different types of rims used in off-highway systems and their terminologies.
- ✓ Explain the process for changing a tire on a multi-piece rim.
- ✓ Explain the different methods used for adjusting agricultural tractor wheel spacing.
- ✓ Name the different ballasting methods and the factors necessary to calculate the correct ballast on an agricultural tractor.

The second-highest operating cost for rubber tire heavy equipment machines, next to fuel expenditures, is tire replacement. The abuse, misuse, neglect, and overall mismanagement of tires leads to substantially higher operating costs for the machine owner. Therefore, it is important to understand the various types of tires used in heavy equipment systems, how to properly use them, and the inspection and maintenance procedures necessary to ensure safe and efficient operation.

Tire construction and design vary widely based on the type of machine and its application. Tires can be classified as pneumatic, non-pneumatic, or solid tires. Heavy equipment tires and wheels are sometimes labeled as *off-the-road (OTR) tires*.

 Note

In the United States and Canada, *tire* is spelled with an *i* (tire), and in England, Australia, and most of Europe, it is spelled with a *y* (tyre). The spelling used in service literature and manuals depends on the location of the manufacturer's headquarters and where the machine/tire is sold.

Tire Safety

Tires pose some of the greatest safety risks to heavy equipment personnel and technicians. A pneumatic tire on a large agricultural or construction machine holds enough potential energy to launch a person hundreds of feet if the tire ruptures or explodes. Failure to follow service warnings and manufacturer guidelines can lead to a dangerous situation or violent accident. Review and practice the following safety tips when working around heavy equipment tires:

- Follow the manufacturer's service literature.
- Wear the appropriate personal protective equipment (PPE).
- Perform inspection and service tasks only after receiving proper training from a tire and wheel expert.

- Never expose the tire or wheel to excessive heat.
- Do not perform work on tire assemblies if the brakes are hot or overheated.
- Do not take breaks around a machine's inflated tires.
- Deflate tires before removing the wheel assembly from the machine or before servicing a wheel assembly installed on a machine.
- Before removing wheel lug nuts, be sure the wheel assembly is secured with the appropriate lifting equipment.
- Do not attempt to heat, weld, or braze a rim or wheel to make a repair. Replace all worn, cracked, or broken rim and wheel components.
- Do not mix or match components from different rims on a multi-piece rim.
- Do not attempt to install a tire on the wrong size rim. Seek expert advice on matching the correct tire size to the correct rim size.
- Never use a steel hammer to hammer on wheels or rims. If the components of a multi-piece rim require adjustment with a hammer after the tire has been deflated, use a rubber mallet, plastic mallet, or brass mallet. The preferred tool, however, is a lock ring removal tool that uses a threaded adjustment to apply pressure evenly to both sides of a lock ring.
- Do not use fire or combustibles to inflate tires.
- Carefully inspect a tire and rim assembly before inflation if the tire has been run flat.
- Never hammer on multi-piece rim components during tire inflation.
  - When replacing tires or performing tire repair work, always place the tire inside a tire safety cage before inflating the tire. See **Figure 22-1**.
  - Use a long enough hose with a clip-on air chuck and remote pressure gauge to allow technicians to stand away from the trajectory of the tire while it is inflating.

Types of Tires

The off-highway industry uses a wide variety of tires. The two main categories are pneumatic tires and non-pneumatic tires.

Pneumatic Tires

A ***pneumatic tire*** requires compressed air or high-purity nitrogen to inflate the rubber tire so that it can support the machine's load. Pneumatic tires can be equipped with an inner tube or be tubeless. An ***inner tube tire*** has an inflatable rubber tube that is placed inside the body of the tire. To protect the tube, tube tires typically have a ***flap*** installed between the inner tube and the rim around the tube's inside circumference. In heavy equipment systems, tubeless tires are much more common than tube tires.

Pneumatic tires can also be classified as bias-ply and radial-ply. These two classifications will be discussed later in this chapter.

Goodheart-Willcox Publisher

Figure 22-1. This tire cage is used for inflating on-highway truck tires. Larger cages are available to accommodate larger tires used on off-highway machines.

Pneumatic Tire Construction

Figure 22-2 shows that the makeup of a tire consists of the following parts:
- Carcass (plies and bead).
- Inner liner (tubeless tires).
- Sidewalls.
- Tread.
- Crown.
- Shoulder.

Carcass

A tire's *carcass*, or rubber body, contains the tire's plies and beads. The *plies* are layers of fabric-type or steel cords that are bonded into the rubber of the carcass to give strength to the tire. Decades ago, the cords were made of cotton. Today's tires have ply cords made of nylon, polyester, rayon, Kevlar, steel, or a variety of other materials.

The tire's plies wrap around one or more beads at the edges of the tire. A *bead* is comprised of a group of high-tensile-strength steel wires encased in rubber. The tire's plies attach to the bead or beads to provide a firm circular base for attaching the tire to the rim of the wheel. See **Figure 22-3**. The carcass is formed with a rubber compound that is poured into a mold containing the plies and bead and then heated.

Inner Liner

Tubeless tires have a rubberized layer known as the *inner liner* that surrounds the inside of the tire. The inner liner is used in place of the inner tube found in inner tube tires. The liner ensures the tire will hold its pressure when inflated.

Sidewalls

A tire's *sidewall* is the flexible rubber exterior that covers the sides of a tire from the bead to the tire's shoulder. Markings on the tire sidewall provide different information (characteristics, ratings, etc.) about the tire. When the machine is operating, the tire sidewalls are designed to perform the following jobs:
- Protect the sides of the tire from cuts and abrasions.
- Resist cracking.
- Cushion the plies.
- Flex without cracking when a load is placed on the machine.

Tread

The *tread* is the outer surface of a tire that contacts the soil or pavement. Tire treads vary in design, pattern, and depth. The tread is designed to resist wear and protect the tire from cuts, punctures, abrasions, heat, water, and oil. A tire's tread wears at a variable rate based on operating factors that include tire slip, machine load, haul-road conditions, operator habits, and usage (distance traveled).

Tread wear is measured in 32nds of an inch. A light-duty truck tire can have 12/32" to 14/32" of tread when new. A new mining tire with an 11′ diameter can have 172/32" of tread, which is more than 5" of tread. In severe applications, large off-road tire treads can wear as much as 12/32" of an inch per day.

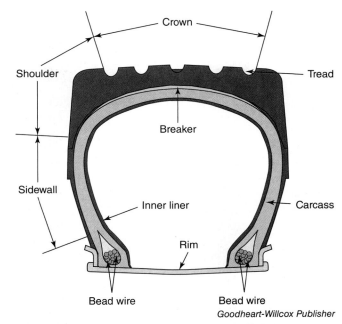

Figure 22-2. The fundamental parts of an OTR tire.

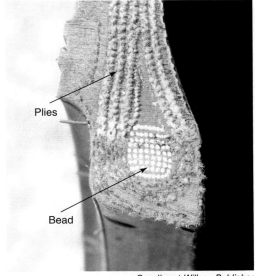

Figure 22-3. This tire cutaway exposes its single bead, which retains the cords of the tire's plies.

Breaker

A *breaker* is a rubberized layer of cords between a bias-ply tire's carcass and tread for increased protection. In applications where tires are at a high risk for punctures and penetration, a steel breaker improves a bias-ply tire's resistance to tread penetration. A steel breaker can also extend into the sidewall to decrease the risk of sidewall penetration.

Crown and Shoulder

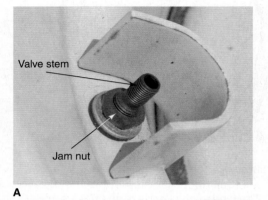

A tire's *crown* is the center portion of the tire's tread that contacts the ground surface. On a tire with a standard aspect ratio, the crown's width is 50% of the tread width. In wide-base tires, the crown is 60% of the tread width. Aspect ratio is explained later in this chapter.

A tire's *shoulder* is the portion of the tire where the tread joins the sidewall. The outer edge of the tread is part of the tire's shoulder.

On passenger car tires, only punctures in the crown area can be safely repaired. However, on OTR tires, some tire repair companies offer kits and repair processes for repairing the shoulder and portions of the sidewalls.

Valve Stems

Tubeless tires require a valve stem to be mounted in the wheel's rim. A *valve stem* is a pneumatic valve that contains a removable core often called a *Schrader valve*. When an air chuck is connected to the valve stem, the air chuck compresses the core and allows air or nitrogen to inflate the tire.

Large machines have valve stems inserted through the rim. A rubber seal ensures a leak-free stem. A jam nut secures the valve stem onto the wheel, **Figure 22-4A**. Small wheels installed on garden tractors or light-duty trucks have valve stems that do not require a jam nut. Instead, the valve stem has a recessed groove on the rubber stem that seals the stem from leaking. If this style of stem leaks, the whole valve stem must be replaced. Larger valve stems, however, can sometimes have the seal replaced.

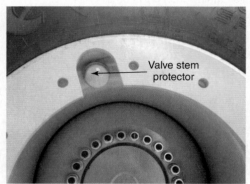

Large machines, such as log skidders, that work in areas with hazardous debris may be equipped with a valve stem protector. Although designs can vary, a common valve stem protector consists of a two-inch pipe nipple that is attached to the wheel and surrounds the valve stem. A two-inch pipe cap threads onto the nipple. The protector fully encloses the valve stem and shields it from debris like tree limbs and branches, **Figure 22-4B**.

Figure 22-4. A—A tubeless tire's valve stem mounted in the wheel's rim. Note the jam nut. B—A valve stem protector installed around and over the valve stem on a log skidder's rear tire.

Bias-Ply Tires

A *bias-ply tire* has the cords of each fabric-type ply layer angled at 45° to the tire's centerline. The individual layers cross in opposite directions, with the bottom layer angled in one direction (to the right) and the next layer angled in the opposite direction (to the left). When viewed from the top of the tire, the cords form a diagonal crisscross pattern, **Figure 22-5**. No steel cord plies are used in a standard bias-ply tire.

Bias-ply tires have multiple plies, which provide a strong, stiff, and thick sidewall that flexes very little. However, the overlapping crisscrossing layers in a bias-ply tire create internal friction that leads to higher tire temperatures than those produced by a radial-ply tire under the same conditions.

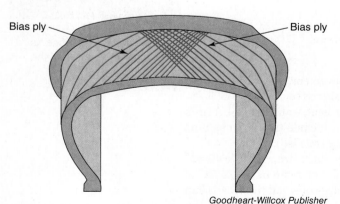

Figure 22-5. Bias-ply tires have the plies laid angularly in a diagonal direction. The next ply crosses over the previous one in the opposite diagonal direction.

Radial-Ply Tires

A *radial-ply tire* has each layer of steel cord plies arranged in an orbital pattern around the axis of the tire. The steel cord plies are routed from bead to bead at a perpendicular angle (90°) to the tire tread. Compared to the diagonal, fabric-type ply construction of a bias-ply tire, a radial-ply tire's perpendicular ply construction and use of steel cord plies are the primary differences in design. A radial-ply tire also has steel belts between the tread and the top of the carcass. See **Figure 22-6**.

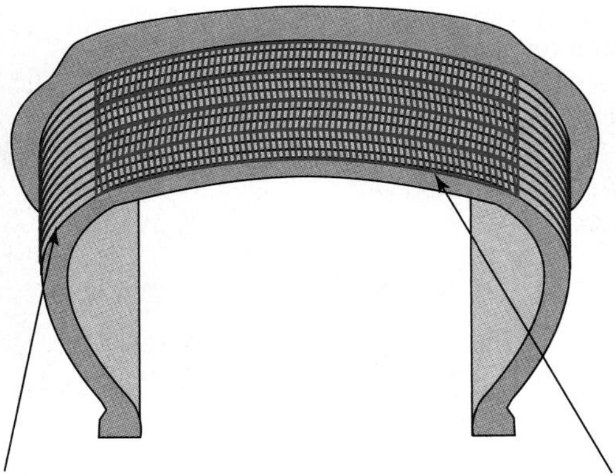

A radial-ply tire has fewer sidewall plies than a bias-ply tire, which creates more sidewall flex (including bulging on agricultural low-pressure applications) and less rolling resistance. Radial tires increase machine fuel economy and reduce tread wear. The radial tread produces a wide footprint where the tire contacts the ground surface.

If a construction machine will be frequently traveling on a paved road surface at high speeds, radial-ply tires are often preferred over bias-ply tires because they produce less road lope. As explained in Chapter 21, *Suspension Systems*, road lope is a condition that causes the tractor suspension to bounce severely. It commonly occurs when the tractor is traveling on hard surfaces at higher speeds with no draft load (roading). Road lope normally occurs at a specific road speed and is often the result of an out-of-round tire or wheel.

Figure 22-6. Radial-ply tires have steel cords running perpendicular to the bead and tire tread. Steel belts are also located between the carcass and tread.

Ply Rating

Large heavy equipment tires can have 20 or more plies. A small agricultural implement tire might only have three or four plies. Many bias-ply tires have a *ply rating* that indicates the tire's strength and load-carrying capacity.

Today, ply-rated tires often have fewer plies than their rated number, such as a 6-ply tire having a 10-ply rating. For a given tire size and application, a tire with a higher ply rating can support more machine load and hold a higher air pressure than a tire with a lower ply rating. **Figure 22-7** shows a large tire used on a mining wheel loader that has a 58-ply rating.

In an agricultural application requiring tires to be inflated to a relatively low pressure, the stiff-walled bias-ply tire has little sidewall bulge compared to a radial-ply tire. Some operators use bias-ply tires in applications where premature tire failure due to sidewall cuts is common, working in shot rock for example. The thicker sidewall provides more protection and helps extend the life of the tire. However, radial-ply tires have better resistance to tread penetration. As mentioned earlier, a bias-ply tire can be made with a steel-ply breaker in the sidewall to reduce sidewall penetration. Steel breaker tires are more sensitive to heat due to adhesion between the steel breaker and the fabric cords. For this reason, they should not be used in applications that generate a lot of heat.

Star Rating

Similar to a bias-ply tire's ply rating, off-road radial-ply tires can have a *star rating* that indicates the tire's strength, **Figure 22-8**. The star rating provides the minimum amount of tire pressure that is required for a given tire load. The table in **Figure 22-9** provides the minimum

Figure 22-7. The 58-ply rating of this mining wheel loader tire is listed on the tire's sidewall.

Figure 22-8. Tire manufacturers can use stars to rate a radial-ply tire's strength. A—A one-star-rated tire on an ADT. B—On the same ADT, a tire rated with two stars.

Goodheart-Willcox Publisher

Figure 22-9. Examples of minimum tire inflation pressures based on star ratings for earthmoving construction tires.

Tire Base	Number of Stars	Minimum Inflation Pressure (30 mph)	Minimum Inflation Pressure (5 mph)
Narrow-base tires	1 star	69 psi	80 psi
Narrow-base tires	2 stars	102 psi	120 psi
Narrow-base tires	3 stars	116 psi	131 psi
Wide-base tires	1 star	54 psi	73 psi
Wide-base tires	2 stars	76 psi	94 psi
Wide-base tires	3 stars	N/A	145 psi

Goodheart-Willcox Publisher

air pressure required for earthmoving (E-type) construction tires based on the tire's star rating. The star rating on an agricultural R-type tire indicates the maximum allowable load the tire can handle based on a specific tire pressure. Refer to **Figure 22-10**. R-type and earthmoving (E-type) construction tires are explained later in this chapter.

Note

On OTR tires, ply rating is normally used to describe the strength of a bias-ply tire, and star rating is normally used to describe the strength of a radial-ply tire. However, light-duty truck radial-ply tires often use ply ratings.

Load Index

Manufacturers assign *load index values* to a tire to indicate the amount of load a tire can handle at the tire's specified speed rating. The manufacturers provide the values in a load index table. The table indicates the load weight the tire can handle in pounds or kilograms with a specified load index value. See the *Appendix* for *Tire Load Index Values*.

Figure 22-10. Agricultural drive tires are star rated to specify the tire's maximum load rating. The rating is based on a specific tire pressure listed in the table.

Number of Stars	Specific Tire Pressure for a Tire-Rated Load
1 star	Maximum tire load rating at 18 psi
2 stars	Maximum tire load rating at 24 psi
3 stars	Maximum tire load rating at 30 psi

Goodheart-Willcox Publisher

Speed Rating

Tires have a ***speed rating*** that is specified by one or two digits marked on the tire's sidewall. The speed rating assigned by the manufacturer designates the top speed that the tire is designed to travel. See **Figure 22-11A**.

The tire in **Figure 22-11B** has two speed ratings. A tire's speed rating is also tied to its load index value. A dual speed rating allows manufacturers to simultaneously list two "speed and load index ratings," which highlights the tire's ability to be used in two different applications: a slower speed/heavier load or a faster speed/lighter load.

Nitrogen-Filled Tires

Nitrogen is an inert (nonreactive) gas and will not burn or cause corrosion. Although traditional compressed air-type tires have a large portion of nitrogen (78% by volume), the remaining gas is 21% oxygen and 1% argon and miscellaneous gases. The nearly one-fifth of remaining oxygen gas is destructive to the rubber and other tire materials. Under the right conditions, the oxygen inside a tire can support combustion within the tire, potentially causing damage or physical injury. As a result, the tires on heavy equipment machines, especially mining equipment, are filled with high-purity nitrogen.

A nitrogen-filled tire holds its pressure longer—up to three to four times longer—than a tire filled with compressed air. The gradual loss of gas through a tire's carcass over time is reduced due to the size of the nitrogen molecule, which is larger than an oxygen molecule. Additional benefits of using high-purity nitrogen gas to inflate tires include the following:

- Improved machine fuel economy (accurate tire pressure maintained during operation).
- Safer due to the absence of oxygen (risk of combustion reduced), less operating risk due to stable steering tire pressures, and less chance of unstable braking due to low tire pressure.
- Improved tire life (reduced tread wear).
- Improved handling.
- Reduced wheel corrosion (no moisture buildup inside the tire).

Compared to using compressed air, the disadvantages of using nitrogen to fill tires are the high cost of the gas and cylinder, the special equipment required, and the dangers associated with storing and handling high-pressure gas. Tires can be filled with nitrogen using a nitrogen gas bottle equipped with a regulator and a pneumatic hose. Nitrogen generator machines that draw atmospheric air into the machine and remove the non-nitrogen gases can also be used to fill tires with pure nitrogen.

Speed Rating	Maximum Tire Speed	Speed Rating	Maximum Tire Speed
A1	2.5 mph (5 km/h)	A8	25 mph (40 km/h)
A2	5 mph (10 km/h)	B	30 mph (50 km/h)
A3	10 mph (15 km/h)	C	35 mph (60 km/h)
A4	12.5 mph (20 km/h)	D	40 mph (65 km/h)
A5	15 mph (25 km/h)	E	43 mph (70 km/h)
A6	20 mph (30 km/h)	F	50 mph (80 km/h)
A7	22.5 mph (35 km/h)	G	55 mph (90 km/h)

A

B

Goodheart-Willcox Publisher

Figure 22-11. A—Maximum tire speed rating examples are listed in this table. B—This implement tire has two speed ratings (A8 = 25 mph/B = 30 mph) and a 159 load index (9650 lb/4375 kg).

Caution

Only properly trained technicians should be allowed to handle and use a nitrogen gas bottle. The nitrogen bottle must include a relief valve and a regulator valve that is set no higher than 20 psi above the tire inflation pressure.

Foam-Filled Tires

Foam-filled tires have a chemical solution, instead of air or nitrogen, pumped into the tire. That chemical solution expands to fill the interior of the tire with a solid foam material. These tires eliminate the possibility of machine downtime caused by one of a machine's pneumatic tires going flat. Common applications for foam-filled tires are skid steers, commercial lawn mowers, and three-point-mounted and trailing finish mowers.

The desire to eliminate flat tires and the problems they cause is a main reason some machine owners choose to use foam-filled tires. The advantages of eliminating flat tires, however, are offset with some disadvantages:

- Increased expense to foam fill a tire.
- Increased strain on a machine's chassis and final drives due to the additional weight of foam-filled tires.
- Decreased tire cushioning due to the lack of tire flex.
- Increased difficulty and machine downtime to remove and replace a worn foam-filled tire from the rim. A heavy-duty lathe-type machine is needed to cut the tire off the rim. The machine first cuts down the center of the tire to split the tire in half. Next, the machine removes each tire half before cutting the foam off the rim.

Filling a Tire with Foam

The process to turn a pneumatic tire into a foam-filled tire includes the following steps:

1. Remove the tire from the rim and inspect the tire while thoroughly cleaning it.
2. Remount the tire on the rim.
3. Place the tire securely in a safety cage.
4. Attach the hose from the foam-filling machine onto the tire's valve stem. The machine draws two separate liquids from supply tanks, mixes the foam-filling liquid, and pumps the foam-filling liquid into the tire through the valve stem.
5. Drill a vent in the top of the tire and place a vent needle and hose in the top of the tire.
6. Monitor the foam-filling machine as it fills the tire completely full of the foaming solution.
7. Remove the vent needle and hose from the top of the tire and insert a screw to close the vent hole.
8. Allow the tire solution to sit for the specified drying time at the specified temperature. One tire foam manufacturer recommends 12 hours at 77°F.
9. Store the tire in a horizontal position to ensure the foam is equally distributed and the tire does not become lopsided.

Tire Chains

In some applications, a chain netting is wrapped around a machine's tires to protect the tires on a rugged worksite and improve traction in slick conditions, **Figure 22-12**. The chain apparatus consists of multiple rings that are interconnected with one another. The tire chain system is sometimes called chain mail. Conditions that would require tire chains include working in a shot rock quarry, in volcanic rock, or in a steel mill around molten metal.

Figure 22-12. A—This wheel loader's tires are covered with chains to protect the tires from punctures. B—This log skidder is equipped with tire chains to improve traction and to protect the tires.

Non-Pneumatic Tire Types

Non-pneumatic tires are still made of rubber but are not filled with compressed air, high-purity nitrogen, or foam. Solid tires and airless tires are the two categories of non-pneumatic tires.

Airless Tires

Airless tires have pockets (holes), arranged in honeycomb or spoke-shaped patterns within the tire sidewall and body, that allow the rubber webbing to cushion and support the load placed on the tire. See **Figure 22-13**. The tires are pressed onto the rim with a tire press machine. Telehandlers, skid steers, wheel loaders, and commercial lawn mowers are examples of machines that use airless tires.

Airless tires have several different trade names coined by the various tire manufacturers. The tire in **Figure 22-14** is a SolidAir-branded tire manufactured by Camso. Trelleborg calls their airless tires Pneu-Trac, **Figure 22-15**. Caterpillar named their airless tires Flexport.

Figure 22-13. Airless tires often have pockets (holes) in the tire sidewall to provide some flexing. Camso manufactures airless tires including this tire that is used on skid steers.

Figure 22-14. This telehandler is equipped with airless SolidAir tires manufactured by Camso.

Figure 22-15. A—An agricultural MFWD tractor with a Trelleborg PneuTrac tire installed on the front axle. Note how the tire flexes against the ground to increase the tire's footprint. B—A closeup of the PneuTrac tire's sidewall.

Michelin manufactures a combination wheel and airless tire assembly that they call a *Tweel*. The tire never has to be installed on a rim assembly because the Tweel is a one-piece tire and wheel design. Refer to **Figure 22-16**.

Solid Tires

The entire tire body and interior of *solid tires* are made completely of solid rubber. The solid rubber provides very little flex or cushion during operation. Like airless tires, solid tires are replaced by using a tire press machine to press the tires off and onto the rim. Warehouse forklifts, personnel scissor lifts, track rollers on rubber-track tractors, and some agricultural implements (gauge wheel on a round baler's pickup) are a few examples of machines that are equipped with solid tires. See **Figure 22-17**.

Skid steers, wheel loaders, and loader backhoes can be equipped with solid tires or airless tires. In severe applications like demolition work or metal recycling, solid and airless tires offer the advantages of a longer tire life (up to three times more life) and an elimination of the downtime caused by flat tires. However, the price per tire is considerably higher than for pneumatic tires. A smooth-tread, airless tire on a machine that operates in a refuse worksite is shown in **Figure 22-18**.

Rotary cutters, also known as brush hogs, have a type of solid tire known as a *laminated tire*, **Figure 22-19**. The tire is made by attaching similar-sized rectangular rubber scraps to a steel rim and fusing the rubber together with extreme heat and pressure. The final product is a wheel with a long service life that will not run flat. The disadvantage of laminated tires, however, is a limited operating speed. Operating the machine at too fast of road speeds (speeds above 15 mph, for example) can generate too much heat and cause the laminated tire to come apart. Always refer to the machine's owner's manual to determine the maximum allowable operating speed.

A
Goodheart-Willcox Publisher

B
Phillip McNew

Figure 22-16. Michelin manufactures a wheel and airless tire combination assembly called a Tweel. Different applications require slight changes to the overall design and construction. A—A skid steer equipped with Tweels. B—This Tweel is installed on a zero-turn mower.

Goodheart-Willcox Publisher
Figure 22-17. A solid tire installed on a forklift.

Goodheart-Willcox Publisher
Figure 22-18. This Caterpillar 938H wheel loader is operated in a refuse job site where the use of pneumatic tires would create problems. The tires have approximately 10,000 hours of use. If the machine were operated in a metal recycling yard, the tire life would dramatically decrease.

Tire Nomenclature

Tires can be sized in inches or millimeters. The dimensions for basic tire measurements are shown in **Figure 22-20A**. An example of a MFWD agricultural rear tractor tire with US inch nomenclature is shown in **Figure 22-20B**. The first set of digits "18.4" indicates the approximate width of the tire's cross section in units of inches. The second set of digits "38" indicates the nominal diameter of the rim in units of inches. The dash between the numbers indicates that the tire is a bias-ply tire. If an "R" is included between the two numbers instead of the dash, the tire is a radial-ply tire.

A metric tire used on the rear of a MFWD agricultural tractor is shown in **Figure 22-21**. The first set of numbers "480" indicates the approximate cross sectional width of the tire in units of millimeters. Dividing the millimeters by 25.4 converts the width to inches.

480 mm ÷ 25.4 = 18.9″ (tire width)

The next set of numbers "70" is the tire's ***aspect ratio***, which determines the tire's section height. The section height of the tire is computed by multiplying the tire's section width by the aspect ratio number written as a percentage. This result equals the tire's section height in millimeters. Dividing the section height in millimeters by 25.4 converts

Goodheart-Willcox Publisher

Figure 22-19. Laminated tires are often installed as a trailing wheel on rotary cutters.

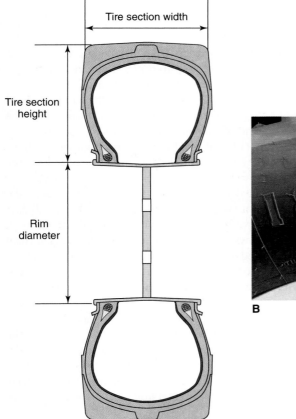

Goodheart-Willcox Publisher

Figure 22-20. A—Basic tire dimensions. B—The "18.4–38" on the tire sidewall indicates that the tire has a section width of 18.4″ and a rim diameter of 38″.

the section height to inches. The "R" listed after the "70" indicates that the tire is a radial tire. The "38" indicates the nominal diameter of the rim in units of inches.

$$480 \text{ mm} \times 0.70 = 336 \text{ mm}$$
$$336 \div 25.4 = 13.22'' \text{ (tire section height)}$$

Some tires list the tire size in both inches and metric. See **Figure 22-22**. In this example, the tire is a radial-ply and is mounted on a 30″ diameter rim. The metric size provides the tire's section width, aspect ratio, and rim diameter. The inch size provides only the tire section's width and the rim's diameter.

$$380 \div 25.4 = 14.9'' \text{ (tire section width)}$$
$$380 \text{ mm} \times .85 = 323 \text{ mm}$$
$$323 \div 25.4 = 12.72'' \text{ (tire section height)}$$

Agricultural Tire Service Classifications

Manufacturers and owners can equip agricultural machines with a wide variety of different tires based on the type of work and the working conditions. Tires are categorized by the purpose and application they best serve. For both agricultural and construction equipment, tire classifications based on function and type of machine are known as the tire's *service code*. The service code is designated with a letter and a numeral (R-4 or E-3, for example). Some codes also feature a second letter (F-2T or L-5S, for example). Classifications for agricultural equipment include R, F, and I.

R Tires

Main agricultural tractor tires are classified as *R tires*. R tires are most commonly used as drive tires and are often manufactured with lugs, also known as cleats, to provide traction. They are also used on construction machines. The tire pictured in **Figure 22-23** is an R-1 9.5–24 tire used on the front axle of a utility tractor. As mentioned earlier, the dash in the middle of 9.5 and 24 indicates the tire is a bias-ply tire.

Four-wheel-drive agricultural tractors have R-type tires on both front and rear axles. Some four-wheel-drive machines have all the same size R tires while others have R tires of different sizes on each axle.

Although many R-type tires are used for drive tires, some are used in non-drive applications. R-type tires can also be installed on all four wheels of a two-wheel-drive machine. Despite the fact that many of today's combines are four-wheel drive, the most common agricultural example of R-type tires used on a two-wheel-drive machine is on a two-wheel-drive combine. Large R-type tires are located on the front wheels, while smaller R-type tires are located on the rear wheels (non-drive tires).

The lugs on R-type tires form a chevron pattern. When R-type tires are installed on the wheels of a drive axle, the chevrons formed by the lugs must be pointing forward. The tire also has an arrow on the sidewall that indicates the direction of rotation, **Figure 22-24**. If the tire is installed on a non-drive axle, such as the rear axle of a front-wheel-drive combine, the chevrons formed by the lugs and arrow on the sidewall should point backward.

Goodheart-Willcox Publisher

Figure 22-21. A metric tire lists the tire's section width in millimeters (480), the aspect ratio (70), and the rim's diameter in inches (38).

Goodheart-Willcox Publisher

Figure 22-22. A tire can have a dual tire size, listing both the inch size and the metric size. This tire is a radial tire with a 30″ diameter rim, a section width of 14.9″ (380 mm), and an aspect ratio of 85.

Goodheart-Willcox Publisher

Figure 22-23. This utility tractor has an R-1 tire size 9.5–24 (bias-ply) on the front axle. The lugs on each half of the tire form a chevron pattern.

R-type tires have several classifications ranging from R-1 to R-4, each with different tread depths. The depths of the tire lugs, however, do not increase as the R number increases. For example, R-2 has the deepest tread, and R-3 has the shallowest tread for R tires.

The R-1 tire is the most common R-type agricultural tractor tire in North America. It is used in moderate, dry soil climates. It has an ordinary appearance for a drive tire but has deep-angled drive lugs.

An R-1W tire has approximately 25% deeper tread in the center of the tire than an R-1 tire. Although it is used in North America, the tire is designed for wetter soils and is more commonly used in Europe. See **Figure 22-25**.

An R-2 tire has the deepest tread of the available agricultural tires. When used on combines, it is sometimes called a rice tire, due to how it operates in wet rice fields.

As a non-directional tread tire that has 50% less tread depth than an R-1 tire, the R-3 tire is sometimes used in turf applications where it is important to avoid leaving tire ruts. The tire provides the least amount of traction of R-type tires because it has the shallowest lug depth, **Figure 22-26**.

An R-4 tire is often described as a heavy duty, industrial-type tire. Its tread depth is approximately 70% of the R-1 tire, but it normally has a higher load-carrying capacity than an R-1 tire. The drive lugs are not as angled as those on an R-1 tire. See **Figure 22-27**. R-4 tires are commonly used on compact utility tractors, loader backhoes, and skid steers, where deep tread is not as important as a tire with stiffer sidewalls that can handle heavy loads.

Goodheart-Willcox Publisher

Figure 22-24. R-type tires have an arrow on the tire sidewall that points the direction the tire rotates in forward gear on a drive wheel. If it is installed on a non-drive wheel, the tire should be mounted so that the arrow is reversed.

Goodheart-Willcox Publisher

Figure 22-25. This R-1W tire is on the front axle of a MFWD tractor.

Goodheart-Willcox Publisher

Figure 22-26. This Caterpillar compactor has R-3 tires on the rear axle.

Figure 22-27. A compact utility tractor with R-4 tires equipped on the front and the rear of the tractor is used to handle heavier jobs.

Goodheart-Willcox Publisher

Figure 22-28. This two-wheel-drive tractor has F-2 type front tires.

Figure 22-29. This I-1 tire is used on a round baler.

Figure 22-30. A Goodyear LSW tire requires a larger-diameter wheel to maintain the same tire radius.

F Tires

Agricultural two-wheel-drive tractors often use a tire with an F classification, which signifies the tire as a front tread pattern tire. *F tires* do not have lugs but are manufactured with one or more ribs that surround the circumference of the tire, making the tire easier to steer. See **Figure 22-28**. F-type tires range from F-1 to F-3. A letter can be used to signify the number of ribs, such as a "D" for double and a "T" for triple. An example would be an F-2T, which signifies the F-2 tire is configured with a triple rib. An F tire can also have the letter "M" designation to signify the tire has multiple ribs.

F-1 tires are used in very wet fields and often in conjunction with rear R-2 drive tires. In average soil conditions, F-2 tires are used with R-1 drive tires. F-3 tires are an industrial front tire normally used with rear R-4 industrial drive tires.

I Tires

Agricultural implements use *I tires*, which stands for implement tires. The most common are I-1 through I-3. The I-1 has rib treads and is commonly used for trailing implement tires. The I-2 tire has a button-type tread for traction. An I-3 tire provides the most traction and uses bar-type tread. Refer to **Figure 22-29**.

Low Sidewall (LSW) Tires

Chapter 21, *Suspension Systems*, mentioned that agricultural tractors can experience severe bouncing in the field (known as power hop) or on the road (known as road lope). One factor that increases the chances of power hop and road lope is the height of the tire's sidewall. Due to their ability to flex and recoil, tall sidewalls are more prone to power hop and road lope.

Goodyear manufactures tires with a short sidewall that they call *low sidewall (LSW) tires*. See **Figure 22-30**. The shorter sidewall tire requires a larger-diameter rim to maintain the same tire radius, but changing these two elements improves the tire's stability. In addition to reducing power hop and road lope, the tire reduces soil compaction.

Construction Tire Service Classifications

As mentioned earlier in this chapter, service codes are used for both agricultural and construction equipment tires. Construction equipment tires have several different service code classifications. Examples of these classifications are E, L, C, LS, and G.

> **Note**
> In addition to agricultural machines, R-type classified tires are also sometimes installed on construction equipment.

E Tires

Construction equipment can be equipped with *E tires*, which stands for earthmoving equipment tires. Examples of machines equipped with E-type tires are rigid frame haul trucks, articulated dump trucks, and scrapers. The table in **Figure 22-31** shows common E-type tires and also includes information on L tires, G tires, and C tires.

L, G, LS, and C Tires

Figure 22-32 shows a bias-ply tire on a rough-terrain forklift. The tire has a dual classification—an E-3 and an L-3 tire classification. Along with the earthmoving tire classification, the L-3 classification designates it as a loader tire. Additional construction tire service codes are identified in the following list:

- *L tires* are classified as loader and wheel dozer tires.
- *G tires* are classified as motor grader-type tires. **Figure 22-33** shows a G-2 tire installed on a telehandler.
- *LS tires* are classified as log skidder tires. An LS-2 tire is an intermediate tread tire.
- *C tires* are classified as compactor tires.

Retread

New tires are a major expense for a machine owner. To reduce the cost, an alternative option for some heavy equipment tires is retreading. A *retread tire*, also known as a *recap*, is created with a process that removes the worn tire tread from the used tire casing and adheres a new rubber tire tread onto the old tire casing. The tire *casing* is the leftover rubber shell of a worn tire. The markings indicate the tire in **Figure 22-34** has been retreaded.

Earthmoving Equipment		
Industry Code	Tread Type	Tread Depth
E-1	Rib	100%
E-2	Traction	100%
E-3	Rock	100%
E-4	Rock deep tread	150%
E-7	Floatation	80%
Loader and Wheel Dozers		
Industry Code	Tread Type	Tread Depth
L-2	Traction	100%
L-3	Rock	100%
L-3S	Smooth	100%
L-4	Rock deep tread	150%
L-4S	Smooth deep tread	150%
L-5	Rock extra deep tread	250%
L-5S	Smooth extra deep tread	250%
Motor Graders		
Industry Code	Tread Type	Tread Depth
G-1	Rib	100%
G-2	Traction	100%
G-3	Rock	100%
G-4	Rock deep tread	150%
Compactors		
Industry Code	Tread Type	Tread Depth
C-1	Smooth	100%
C-2	Grooved	100%

Figure 22-31. The type of tread used for the different construction tire categories, including earthmoving, loader and wheel dozer, motor grader, and compactor tire classifications.

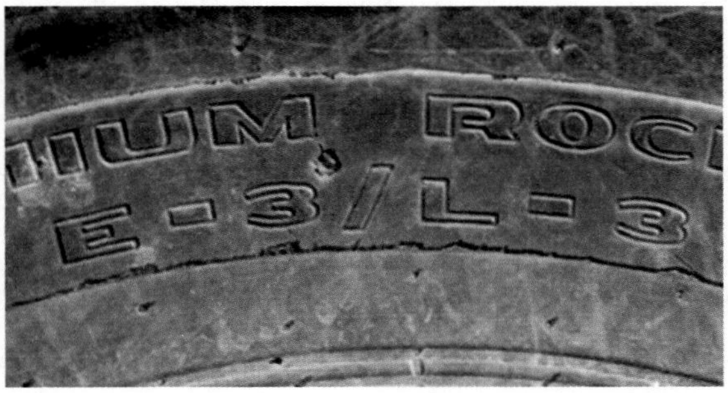

Figure 22-32. E-type tires are classified as earthmoving tires. This tire has a dual classification, E for earthmoving and L for loader and dozer.

Figure 22-33. G-type tires are classified as motor grader tires. This G-2 tire is installed on a telehandler.

Goodheart-Willcox Publisher

Figure 22-34. Retreaded tires will often be labeled on the tire's sidewall.

Wheel Slip

Agricultural tractors optimize their performance by allowing limited wheel slip, or traction loss. However, construction machinery should operate with no wheel slip.

Construction Machine Wheel Slip

Construction machines often operate with the maximum tire air pressure, which produces the following characteristics:

- Maximum load-carrying capability.
- Reduced contact surface area (reduced footprint).
- Increased tire-to-ground pressure.
- Increased soil compaction.

The construction industry intentionally uses soil compactors to create solid soil foundations. Therefore, maximum tire air pressure in construction equipment provides the benefits of increased maximum payload and increased soil compaction.

A highly inflated tire reduces the tire's contact surface area, increases tire-to-ground pressure, and increases the chances of disadvantageous wheel slip. Wheel slip causes increased tire wear as well as more strain and wear on the machine and drive train. On construction work sites under normal conditions, little to no wheel slip is necessary to maximize overall machine operating efficiency. Superior drive tire traction is desired when digging, grading, loading trucks, or towing payload up and down elevations.

In an effort to recognize excessive wheel slip, some construction firms use spray paint to paint a line from the center of a machine's tire to the drive tire's sidewall edge. The line allows personnel to observe the machine from a distance, making it easier to visualize the speed of the tire in relation to the machine's travel speed. To reduce wheel slip, the operator should decrease the load and/or decrease the machine speed. As explained in Chapter 12, *Hydrodynamic Drives*, wheel loaders can be equipped with rimpull electronic controls that are used to reduce the machine's torque and thereby reduce wheel slip.

Agricultural Tractor Wheel Slip

Agricultural tractor operators desire some degree of wheel slip to provide peak tractive efficiency. As little as 60% to 70% of expended tractor engine horsepower is effectively delivered to the ground due to power train losses, accessory drive losses, and wheel slip. Although it would seem that no wheel slip would provide maximum horsepower efficiency, an agricultural tractor reaches peak tractive efficiency by means of a specific percentage of wheel slip. Agricultural tractors are rated in engine horsepower (power delivered by the engine's crankshaft), PTO horsepower (power delivered by the PTO shaft) and drawbar horsepower (power delivered by the tractor's drawbar). Peak tractive efficiency ensures the highest amount of engine horsepower is transferred to drawbar horsepower. The desirable amount of wheel slip is dependent on the type of tractor and the type of ground surface on which the machine is operating. See **Figure 22-35**.

Late-model agricultural tractors are equipped with ***ground speed radar***, which is an electronic device used to measure the tractor's true ground speed, **Figure 22-36**. The tractor's control module subtracts the machine's true ground speed from the actual wheel speed to determine the percentage of wheel slip. Many late-model agricultural tractors can display the tractor's percentage of wheel slip on the cab monitor, allowing the operator to make adjustments to achieve the optimum wheel slip.

 Caution

To avoid possible eye injury, do not look directly into the radar face when working near the ground speed radar installed on a tractor.

Type of Agricultural Tractor	Desirable Amount of Wheel Slip on Normal Soil Surfaces
Two-wheel drive	10%–15%
Four-wheel-drive MFWD tractor	8%–12%
Four-wheel-drive articulated tractor	8%–12%
Rubber track tractor	2%–4%

Goodheart-Willcox Publisher

Figure 22-35. Depending on the tractor type, agricultural tractors maximize their performance with a specific amount of wheel slip. Note that operating in very firm soils and soft, sandy soils will alter the desired amount of wheel slip listed here.

Goodheart-Willcox Publisher

Figure 22-36. A ground speed radar that has been removed from a Challenger tractor.

On older agricultural tractors without a radar system, wheel slip can be determined by completing the following steps:

1. Move the machine to a field with normal operating conditions. Attach the implement to the machine.
2. Mark one tire with a reference line.
3. Measure the distance it takes to make 20 tire revolutions with the implement in the ground.
4. Move the machine back to the same starting point. Take the implement out of the ground.
5. Count how many tire revolutions it takes to travel the same distance (under no implement load).
6. Compute the wheel slip using the following equation. As an example, assume the number of loaded tire revolutions is 20 and the number of unloaded revolutions is 19.

$$[(\text{Loaded revolutions} - \text{unloaded revolutions}) \div \text{loaded revolutions}] \times 100 = \% \text{ wheel slip}$$
$$[(20 - 19) \div 20] \times 100 = \% \text{ wheel slip}$$
$$(1 \div 20) \times 100 = 5\% \text{ wheel slip}$$

As noted from the example, each tire revolution under implement load loses 5% of its rotation to wheel slip.

Inspecting Tires

Tires should be inspected daily before usage while the tires are cool. Tire inspection should not take place after a shift when the tires have retained heat. At a minimum, the tires should be stationary for at least three hours before being inspected. The following items should be checked for during the inspection:

- Air pressure.
- Tread depth.
- Uneven wear.
- Cracks.
- Punctures.
- Tire separation.
- Bent or cracked wheels.
- Material wedged between dual wheels and caked onto wheel lugs.

Tires need to meet the recommended air pressure for their application. Underinflation and overinflation both create problems. A tire that is underinflated by as little as 10% can decrease the life of the tire by 10% to 15%. A tire that is overinflated by 30% can decrease the tire life by as much as 50%.

Tire treads should be checked for wear and depth in multiple locations. The wear should be consistently even. ***Tread wear rate*** is the number of 32nds of an inch of rubber tread worn off the tire per the number of hours of use. The measurement should be benchmarked to maximize tire life. If tires are wearing prematurely, further investigation is required to find the problem.

While inspecting tread depth during a preventative maintenance (PM) inspection, evaluate the past wear rate based on the application and determine if the tire has enough tread to safely make it to the next PM inspection. Doing so prevents the machine from incurring unforeseen downtime due to tire repair. If the tire tread is expected to wear below the minimum before the next PM, the tire should be replaced during the current PM.

Maximizing Tire Life

The life of a tire in a construction application can vary. In severe applications, the tire might last only 500 hours. The correct tire for an application that is operating in ideal conditions can last more than 8000 hours. Multiple steps can be taken to maximize the life of a tire, including the following actions:

- Operate the tires with the correct tire pressure based on the application and conditions.
- Avoid steering the tires when the machine is stationary.
- Avoid all wheel slip in construction applications, including quick starts and stops.
- Operate the machine with the correct amount of wheel slip in agricultural traction applications.
- Do not overload the machine. For example, owners often overload their machines by adding spill boards (lumber) to the sides of 10-wheeler dump truck trailers and tractor scrapers to increase the volume of dirt the machine is capable of hauling.
- Avoid driving over the uneven surfaces found on road shoulders.
- Reduce travel speed during turns.
- Remove dirt, rocks, and debris that have become lodged in the tires.
- Do not operate the machine on rocks when loading or dumping, if possible.
- Do not drive through grease or oil spills.
- Keep load areas clean and free of rock and spilled dirt.
- Load the center of truck beds.
- Do not use bent or damaged wheels.
- Avoid working in waterways when possible. A wet tire is 10 times easier to cut than a dry tire, and the lack of ground visibility in water increases the possibility of tire damage.
- Study each worksite to find the optimum tire for that specific worksite. For example, several loaders of all the same size are operated daily for a construction company. Each loader, however, is equipped with a different type of tire based on its application:
  - Turf tires are installed on the loaders working at a local golf course.
  - Traditional loader tires are installed on the loader that loads and fills at the company's stockpile operation.
  - R5 mining tires are chosen for the loaders operating in a landfill worksite.

Ton-Mile per Hour Value

During the tire manufacturing process, the tire is vulcanized. The vulcanization process consists of heating the rubber and chemicals to a temperature of approximately 270°F, which causes the materials to form into their final compound. Caterpillar developed a tire performance measurement, known as ***ton-mile per hour (TMPH)*** or ***ton-kilometer per hour (TKPH)***, which helps maintain appropriate tire temperatures. The TMPH (or TKPH) value equals the average tire load multiplied by the average machine speed per shift. The measurement is now used by most off-road tire manufacturers.

The value applies to tires used on large earthmovers, such as haul trucks and scrapers. The value is used to ensure the tires are not overheated. When a tire is overworked, the tire's temperature rises, which causes the tire's rubber to revert back to its original state, before the vulcanization process was performed. This can cause tire separation and ply separation.

For example, if a tire supported the loads listed below, the average tire load would equal the following:

Average tire load

$$\text{Machine empty} = 12 \text{ tons} \quad \text{Machine loaded} = 14 \text{ tons}$$
$$(12 \text{ tons} + 14 \text{ tons}) \div 2 = 13 \text{ tons average tire load}$$

The next step requires computing the machine's average speed per shift by multiplying the average distance traveled (in miles) times the number of trips traveled per shift and divide that product by the number of hours per shift.

Average speed per shift

$$(10 \text{ miles per trip} \times 14 \text{ trips per shift}) \div 7.5 \text{ hours per shift} = 18.67 \text{ mph}$$

The last step is to multiply the average tire load by the machine's average speed per shift.

Ton-mile per hour

$$13 \text{ tons} \times 18.67 \text{ mph} = 242 \text{ TMPH}$$

To convert the TMPH to a TKPH value, multiply the TMPH by 0.685 or divide the TMPH by 1.459.

Ton-kilometer per hour

$$242 \text{ TMPH} \times 0.685 = 165.77 \text{ TKPH}$$

After computing the worksite ton-miles per hour, compare that number with the tire manufacturer's TMPH to determine if the tire is rated for that capacity. If the calculated TMPH exceeds the manufacturer's recommended TMPH, the tire is at a greater risk to fail prematurely under the current work conditions.

Caterpillar offers TMPH tire monitoring. The electronic software monitors ambient temperature, machine speed, and original equipment tire manufacturer TMPH. The software continuously calculates the variables and sets an alarm to alert the operator when tire temperatures rise too high.

Wheels and Rims

Heavy equipment tires can be mounted on a variety of different types of wheels and rims. A *wheel* is a rotating assembly that couples the drive hub (or non-driving hub) to the tire. On a drive hub, it transfers the torque from the hub to the rim and supports the machine's load. The wheel assembly can be designed in two ways. The first type consists of a single, solid rim. The second type is designed with a rim that is welded or bolted to a disc that, in turn, is bolted to the axle's hub. Refer to **Figure 22-37**. The *rim* is the flanged portion of the wheel against which the tire mounts. Rims can be a one-piece design or a multi-piece design.

Goodheart-Willcox Publisher

Figure 22-37. This tractor's wheel consists of a rim and a disc. The disc bolts to the tractor's hub and to the rim of the wheel.

Single Piece Rim

A *single piece rim* consists of one complete wheel. Single piece rims are most common with small tire applications but can be found in some large agricultural applications. They are not commonly used in large construction applications. The rim can have different types of profiles (contours), which can create a slight variation in basic wheel dimension measurements, **Figure 22-38**. One example is the *drop-center rim*. A drop-center rim is a single piece rim with a recessed (dropped) center to aid in removal and installation of the tire.

Multi-Piece Rims

Large construction and mining equipment tires are often mounted on a *multi-piece rim*, also known as a *split rim*. Multi-piece rim assemblies are made up of two to seven separate pieces. A locking ring secures the tire and the rim assembly parts onto the wheel by interlocking the parts together after the tire is inflated. Refer to **Figure 22-39** and **Figure 22-40**. On a multi-piece rim for a tubeless tire, an O-ring seal is necessary to maintain the tire's air pressure. The components used to make a three-piece rim include the following:

- Rim base.
- Locking ring.
- O-ring (seal).
- Removable outer flange (with a bead seat band as part of the flange).

A five-piece rim includes all of the three-piece rim components but also includes a removable inner flange and a separate bead seat band (detached from the removable outer flange). When installed, the inner flange holds one side of the tire's bead. The separate bead seat band works in conjunction with the removable outer flange, rim base, and locking ring to hold the tire's other bead, and to allow the wheel assembly to be disassembled.

Multi-piece rims contain a *driver key* that fits in the pocket of the bead seat band (or outer flange on three-piece rims) and the pocket of the rim base to retain the locking ring.

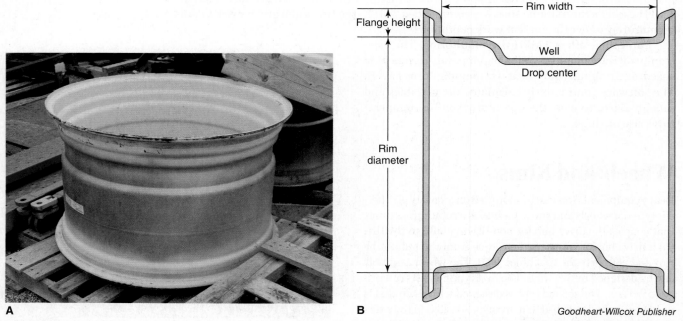

Goodheart-Willcox Publisher

Figure 22-38. A—This single piece rim is used on large combines. B—This drawing shows a drop-center rim with dimension lines drawn to measure flange height, rim diameter, and rim width. The drop-center design aids in removal and installation of the tire.

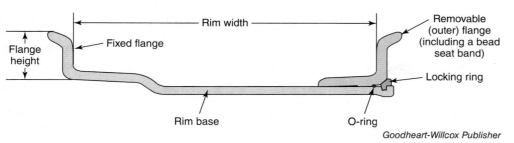

Figure 22-39. The parts of a three-piece rim.

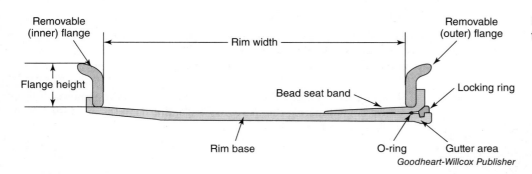

Figure 22-40. The parts of a five-piece rim.

The example in **Figure 22-41** is a loose-style driver, meaning that the key is not attached to the wheel. Rims can also have crimped-on driver keys, which are attached to the wheel.

Changing a Tire on a Multi-Piece Rim

Heavy equipment tires impose a great risk of injury or possible death to the technician servicing them. It is recommended that only a properly trained tire technician replace a tire on a multi-piece rim. Manufacturer service procedures must always be followed. The safety precautions mentioned earlier in this chapter should also be strictly followed.

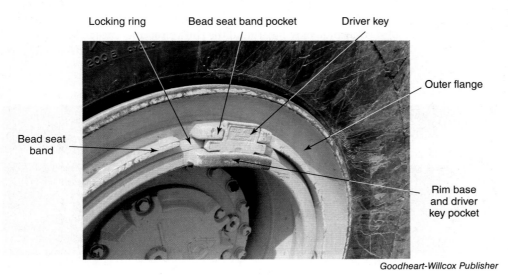

Figure 22-41. A driver key retains the multi-piece locking ring.

Tire Removal and Demounting from the Rim

An example procedure for replacing a tire on a multi-piece rim includes the following steps:

1. Use appropriate personal protective equipment (PPE) including eye protection, gloves, and steel toe boots.
2. Chock the opposite wheel.
3. Raise the axle with a jack and block the machine so that the tire is safely suspended in the air.

> ⚠️ **Warning**
> Never remove an inflated tire from a machine. Always fully deflate tires before working on tires and wheels. An inflated tire has the potential to explode with tremendous force, causing injury and possible death.

4. Completely deflate the tire by removing the tire's valve core. Do not stand near the tire while it is deflating. If the machine is equipped with an inner dual wheel that will remain on the machine, remove its valve core to release its air as well. Use a wire to probe through the valve stem's hole to ensure it is completely free of debris or ice so the tire will not have any issues deflating.

> **Pro Tip**
> Replace large tires with the wheel remaining mounted on the machine. Technicians use a truck-mounted boom, forklift, or loader for removing and installing large tires.

5. After the tire has been deflated, some tire technicians use a truck crane's boom to push inward on the deflated tire to unseat the outside tire bead from the outside rim, **Figure 22-42**. If a boom is not available to unseat the tire, a hydraulic tire demounting tool is used instead.
6. Place the lip of the hydraulic tire demounting tool into the wheel assembly's pry bar pocket and adjust the threaded adjustment so the tool is perpendicular to the tire. If the tool is improperly seated, it can fly off the wheel and cause injury or death. While standing away from the trajectory, actuate the hydraulic tool. If required, release the tool from the pocket and readjust the threaded adjustment. Lift the flange approximately 3/4″. Then, use a pry bar to pry between the flange and the bead seat band. Continue moving around the rim's circumference in large 24″ increments, avoiding the butt weld and keeping hands and fingers clear, until the tire's bead is unseated from the band. See **Figure 22-43**.

Goodheart-Willcox Publisher

Figure 22-42. After the tire has been deflated, a truck boom crane is often used to push the outside tire bead off the outside rim.

Figure 22-43. A hydraulic tire demounting tool is used on multi-piece rims. A—The lip of the tool and the pocket where the tool is inserted are shown. B—This tire has been removed from the machine and flipped upside down to remove the inner rim components.

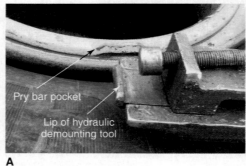
A

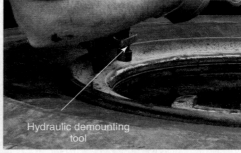
B

Goodheart-Willcox Publisher

7. If the rim is fitted with a driver key, remove the driver key.
8. Remove the locking ring by prying near the split portion of the ring.
9. Remove and discard the O-ring.
10. Remove the bead seat band.
11. Remove the flange.
12. If the wheel has a multi-piece rim design on the other side of the tire, complete the steps for removing the inner-side rim components.

Tire Remounting onto the Rim

During assembly, be sure to first check that the multi-piece rim components are properly installed and aligned. Use a clip-on air chuck to avoid standing next to the tire or near its trajectory while it is inflated. Do not overinflate the tire.

1. Thoroughly clean the rim components. Replace any component if it is cracked, worn, or damaged (heavily rusted, for example).
2. After cleaning the components, coat them with an anti-rust paint or inhibitor.
3. If the rim's rear flange is a single piece, the tire will be prepped and slid onto the rim that remains mounted on the machine. If the rim has a multi-piece flange on both sides, the rim must be placed on blocks on the ground and the inside flange components installed first. The instructions provided here are for rims with the multi-piece components located only on the outside of the tire. See the manufacturer's procedure for rims with multi-piece flanges on both sides.
4. Slide the tire over the rim.
5. Thoroughly lubricate the flange's bead seating area and tire beads with approved vegetable-based lubricant.
6. Install the bead seat band and flange and push them out of the way to expose the O-ring groove/seat.
7. Lubricate the new O-ring and the rim's O-ring groove with the vegetable-based lubricant. Install the O-ring. Do not roll the O-ring. Push back the flange if needed to allow enough space to install the O-ring.
8. Install the locking ring and align the components. It is critical that components are properly seated or injury or death can result. If the rim is fitted with a driver key, install the driver key.
9. Use a clamp-on air chuck so you can stand away from the tire and rim's trajectory (outlined in the manufacturer's warnings) during inflation. If the tire is not being inflated on the machine, install the tire and wheel in a safety cage before inflation.
10. Inflate tire to about 3–5 psi and check the tire and rim assembly. If there is a problem with the assembly, fully deflate the tire again and correct the issue. If the assembly is right, continue to inflate the tire to specification.

Adjusting Agricultural Tractor Wheel Spacing

As mentioned in Chapter 18, *Axles and Driveshafts*, wheels can be mounted on a tractor's adjustable bar axle or a flange-mounted axle. Regardless of the type of axle, always be sure to adjust the wheels on an agricultural tractor at equal spacing so that the tires are located in the center between two crop rows.

Offset

Wheel assemblies have a dimension known as *offset*, which is the distance of the rim's centerline to the outside disc surface. See **Figure 22-44**. The offset can be negative, positive, or zero. Dual wheels often have a large offset, which is either positive (inside dual) or negative (outside dual).

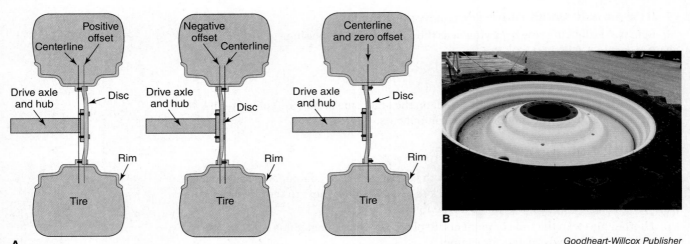

Figure 22-44. Wheels have an offset dimension. A—The offset equals the distance between the wheel's centerline and the face of the hub. B—Machines equipped with dual wheels require wheels with large offsets.

The wheel offset determines the position of the tire relative to the hub, which establishes wheel spacing. Agricultural tractors have provisions for adjusting the wheel spacing. Row crop tractors have the wheel spacing adjusted to ensure the tires travel in between the rows of crops. Municipality-owned tractors mow along roads, highways, and steep ditches. The wheel spacing of these tractors is adjusted as wide as possible to provide the tractor with the most stable operating stance possible. Small utility tractors have the wheel spacing narrowed to ensure the tractor can be driven onto a flatbed trailer and fit inside the trailer's wheel wells for transport.

Adjusting Flange-Mounted Wheel Spacing

On many tractors, the wheels are fixed, which does not allow a variable amount of adjustment for tire spacing. Many compact utility tractors and agricultural utility tractor owner's manuals list wheel spacing specifications that are based on reversing the wheels. For example, one compact utility owner's manual lists eight different rim and disc offsets that provide rear tire width dimensions from a narrow 51″ dimension to a wide 75″ dimension.

Note

If the wheel's disc is welded to the rim, the wheel only has two different spacing options. As mentioned earlier in this chapter, many drive tires are directional. To avoid having to remove and reverse the tires on the rim, the tire rim assemblies can be swapped to the opposite side of the tractor.

Adjusting Bar Axle Wheel Spacing

Agricultural MFWD tractors, some agricultural articulated tractors, and many twin-rubber-track tractors have a bar axle, which was described in Chapter 18, *Axles and Driveshafts*. Within its minimum width to its maximum width, the bar axle provides an infinitely variable amount of wheel or track spacing.

The wheel is secured to the axle with a hub and a single wedge or pair of tapered cast iron *wedges*, also called a bushing or a sleeve. See **Figure 22-45**. The wedges are tapered and are designed to fit snuggly inside the hub's tapered hole. Manufacturers use a variety of different wedge designs. A group of bolts clamp the wedge into the hub. **Figure 22-46** shows all of the bolts inserted through the face of the wedge. In **Figure 22-47A**, the right

Chapter 22 | Tires, Rims, and Ballasting

Figure 22-45. These tapered wedges have been removed from the tractor's axle. They are used to attach the drive hub to the drive axle on a MT 845 Challenger rubber-track tractor bar axle.

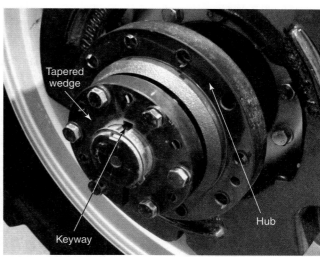

Figure 22-46. This Case IH MFWD tractor has a one-piece tapered wedge and hub to secure the wheel onto the bar axle. The notch on the bar axle is where a piece of keystock is to be fitted. Note that all of the bolts securing the wedge to the hub are inserted through the face of the tapered wedge.

rear wheel of the tractor has three clamping bolts and two jack bolts inserted through the face of the hub. When the clamping bolts are loosened, the jack bolts can be turned to push the wedge away from the hub, separating them. In **Figure 22-47A,** the three outboard clamping bolts secure one wedge to the hub. **Figure 22-47B** shows two of the three inboard clamping bolts that secure the other wedge to the hub. A piece of keystock is sometimes placed inside the wedge, into matching keyways in the wedge and the bar axle.

Normally only one wedge needs the be removed to adjust the wheel spacing. To adjust the wheel spacing on a tractor like the one shown in **Figure 22-47A**, lift (jack up) the axle so that the wheel is not supporting any weight of the tractor. Loosen the three clamping bolts on the outboard side of the wheel and then turn the two jack bolts to push the wedge out of the hub.

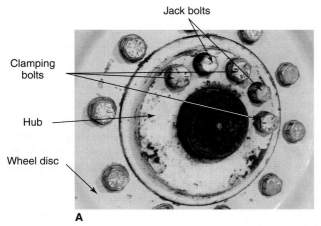

Figure 22-47. This John Deere 4640 (right rear wheel) is equipped with a pair of wedges. A—Three outboard clamping bolts secure one of the wedges to the hub. Two jack bolts are used to help push the wedge out of the hub during disassembly. B—Two of three inboard clamping bolts are visible and are used to secure the second wedge to the hub.

> **Note**
>
> If the tractor is lifted in the air, the weight of the tire and wheel (or the track of rubber-track machines) can hold the wedge tight against the hub. It might be easier to remove a wedge that has been positioned at the bottom of the hub. Some wedges do not have jack bolts, but have threaded holes that are used with pusher bolts. After the clamping bolts are removed, two of the clamping bolts are placed into the threaded pusher holes and tightened to push the wedges out of their bore.

The bar axle might be smooth or it might have notches. On the notch-style axle, the notches form a rack, and the hub contains a pinion gear. This rack and pinion arrangement aids the technician in repositioning the wheel. First, a wedge is loosened as described in the previous paragraph. Then, as the pinion bolt is turned, it causes the pinion to rotate, which slides the wheel inward or outward. See **Figure 22-48**.

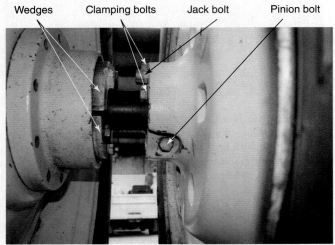

Goodheart-Willcox Publisher

Figure 22-48. This John Deere 4640 (left rear wheel) has a rack and pinion design to help move the wheels inward and outward. After the tractor has been lifted, one wedge must first be loosened from the hub before the wheels can be slid inward or outward on the axle. First, the clamping bolts are loosened. Then, the jack bolts are turned to separate the wedge from the hub. Then, the pinion bolt is turned to move the wheel on the axle.

> **Note**
>
> Wedges that have not been adjusted or removed in a long period of time will most likely be extremely difficult to remove (seized). It is common when trying to remove seized wedges to accidentally break a wedge, wheel hub, pinion bolt, or pinion gear. Penetrating oil can be used to help free the seized wedges. Spray and let the penetrating oil soak before turning the wedge 180°. Next, spray the wedges and the pinion bolt (if equipped) again with penetrating oil. Some veteran technicians can use a hammer to loosen the wedges without breaking components.

Weighting and Ballasting Agricultural Tractors

As explained earlier in this chapter, agricultural tractors achieve peak tractive efficiency based on specific percentages of wheel slip. The wheel slip is achieved by adjusting the weight of the tractor, which is known as *ballasting* the tractor.

Types of Ballasting

Tractors can have a variety of different types of weight installed on the machine. The most common type of weight is cast iron weight, often in the *suitcase weight* design, which resembles a suitcase with a handle. This type of weight is usually placed on the front or rear of the tractor, but can also be placed on the side, **Figure 22-49A**. An average suitcase weight weighs approximately 100 lb. The other popular type of cast iron weight is *wheel weights*, which are the heaviest and most difficult to remove or install. They are bolted to the tractor's wheels. Wheel weights vary widely in weight, ranging from 100 lb to 2000 lb, **Figure 22-49B**.

The least favorable type of tractor weight is liquid added to tires, normally in the form of calcium chloride. Although liquid calcium is economical when compared to cast iron weight, it causes corrosion of wheels and valve stems and prevents the tire from being able to provide any type of air suspension. If liquid ballast is used, only 75% of the tire's volume should be filled with the liquid ballast.

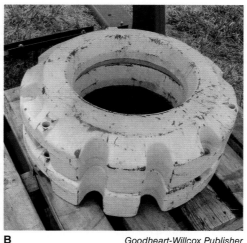

Figure 22-49. Agricultural tractors are equipped with cast iron weights to ballast the tractor. A—Suitcase weights are often placed on the front or rear of the tractor. This Challenger tractor has 30 suitcase weights at the front of the tractor. B—Wheel weights can weigh up to 2000 pounds.

Rimguard sells a byproduct of sugar-beet juice for liquid ballast. This solution is popular in northern regions where sugar beets are grown. It is biodegradable, nontoxic, animal safe, noncorrosive, can be used in tubeless tires, and has a low freezing point of –35°F. Some equipment owners have used an antifreeze solution for ballast, which is harmful to animals.

Ballasting Calculators and Weight Ratios

Tractor manufacturers provide tractor weighting and ballasting tools to help technicians optimize machine performance. The tools can range from an old paper circular sliding calculator to a computer software program, such as Microsoft Excel. Some manufacturers offer free online electronic calculators on their website. The operator enters information such as the following into the calculator:

- Total load (weight).
- Type of draft (low drawbar load, semi-mounted, or high three-point-hitch load).
- Type of soil condition (firm soil, soft soil, sandy soil).
- Tractor speed.
- Type (singles, duals, triples) and size of tires.

The calculator or software provides a ballasting starting point based on the information provided. This starting point includes the total machine weight, percentage of weight on the front axle, and percentage of weight on the rear axle. If service literature is unclear, **Figure 22-50** provides average starting points for tractor weight ratios.

Weight Ratios (Front/Rear) for Different Draft Types			
Tractor Type	Low (Towed) Drawbar	Semi-Mounted	Three-Point (Fully Mounted)
Two-wheel drive	25%/75%	30%/70%	35%/65%
Mechanical front-wheel drive	35%/65%	35%/65%	40%/60%
4WD articulated rubber tire tractor	55%/45%	55%/45%	60%/40%

Goodheart-Willcox Publisher

Figure 22-50. This table provides a general starting point for ballasting agricultural tractors.

>
> **Note**
> When used in construction applications, four-wheel-drive articulated scraper tractors have more weight on the front than when they are used in agricultural applications. Articulated scraper tractors used in construction application have a front axle weight ratio of 65% to 70% and a rear axle weight ratio of 30% to 35%.

Total Weight for a Given Horsepower

If the service literature does not provide a total recommended tractor weight, **Figure 22-51** provides tractor weights based on tractor horsepower. The weights are listed in pounds per one horsepower. For example, if an articulated four-wheel-drive tractor was operating at less than 4.5 mph and had 500 horsepower, the total tractor weight should not exceed 55,000 lb.

$$110 \text{ lb} \times 500 \text{ hp} = 55,000 \text{ lb}$$

Figure 22-51. A tractor's maximum weight can be determined by using this table, which lists the amount of pounds the tractor should weigh for each engine horsepower.

Weight-to-Horsepower Ratios Based on Travel Speed			
Tractor Type	**Less than 4.5 mph**	**5 mph**	**Greater than 5.5 mph**
Two-wheel drive	130 lb/1 hp	120 lb/1 hp	110 lb/1 hp
Mechanical front-wheel drive	130 lb/1 hp	120 lb/1 hp	110 lb/1 hp
Four-wheel drive	110 lb/1 hp	100 lb/1 hp	90 lb/1 hp

Goodheart-Willcox Publisher

Summary

- Always follow all manufacturer's procedures and safety warnings when servicing heavy equipment tires. Request the services or help of a tire professional if possible.
- Pneumatic tires are inflated with compressed air or nitrogen.
- Bias-ply tires have multiple layers of plies made of nylon, polyester, rayon, Kevlar, or a variety of other materials angled 45° to the tire's centerline.
- Radial-ply tires have steel cord plies and steel belts perpendicular to the tire tread to increase the strength of the tire.
- Tire tread is measured in 32nds of an inch.
- A ply rating indicates the strength and load-carrying capacity of a bias-ply tire.
- A star rating indicates the minimum amount of tire pressure required for a given tire load on a construction tire, or the maximum allowable load an agricultural tire can handle based on a specific air pressure.
- Nitrogen-filled tires improve fuel economy, tire life, handling, and safety while reducing wheel corrosion.
- Foam-filled tires reduce machine downtime, but are expensive and difficult to replace.
- To cushion and support the machine's load, airless tires have pockets formed in a honeycomb or spoke-shaped rubber pattern within the tire sidewall and body.
- The entire body and interior of solid tires is solid rubber and provides very little flex or cushion.
- Common off-the-road tires are classified with the designations R, F, I, E, L, G, LS, C, and LSW.
- Tires can be specified in millimeters or inches.
- A certain amount of wheel slip is needed to achieve peak tractive efficiency in agricultural traction applications.
- Construction equipment tires should operate with no wheel slip.
- Inspect tires only when they are cool, such as before a work shift (after the tires have been stationary for a minimum of three hours).
- Caterpillar developed the ton-mile per hour measure to prevent overheating tires.
- A wheel couples the drive hub to the tire.
- The rim is the flange portion of the wheel against which the tire mounts.
- Rims can be a single-piece design or multi-piece design.
- Multi-piece rims pose great dangers for the technician servicing them. Only properly trained technicians should work on them.
- Multi-piece rims have a driver key that is placed in the pocket of the bead seat band (or outer flange on three-piece rims) and the pocket of the rim base to retain the locking ring.
- Wheels have an offset dimension (positive, negative, or zero) that is the distance of the rim's centerline to the outside disc surface.
- A flange-mounted wheel can have up to eight different wheel spacing positions if the wheel's disc is bolted to the rim. The wheel spacing is adjusted by moving the wheel's disc position on the rim.
- Bar axles provide an infinitely variable amount of wheel spacing between the minimum and maximum limits.
- Tapered wedges are used to lock a wheel's hub to a bar axle.
- Tractors can be ballasted with suitcase weights, wheel weights, or liquid ballast in the tires.

Technical Terms

airless tires	foam-filled tires	offset	solid tires
aspect ratio	F tires	off-the-road (OTR) tires	speed rating
ballasting	ground speed radar	plies	split rim
bead	G tires	ply rating	star rating
bias-ply tire	inner liner	pneumatic tire	suitcase weight
breaker	inner tube tire	radial-ply tire	ton-kilometer per hour (TKPH)
carcass	I tires	recap	ton-mile per hour (TMPH)
casing	laminated tire	retread tire	tread
crown	load index values	rim	tread wear rate
C tires	low sidewall (LSW) tires	R tires	Tweel
driver key	LS tires	service code	valve stem
drop-center rim	L tires	shoulder	wedges
E tires	multi-piece rim	sidewall	wheel
flap	non-pneumatic tires	single piece rim	wheel weights

Review Questions

Answer the following questions using the information provided in this chapter.

Know and Understand

1. All of the following are safe practices for working around heavy equipment with tires, EXCEPT:
   A. never expose the tire or wheel to heat.
   B. do not perform work on tires if the brakes are overheated.
   C. do not take breaks around inflated tires.
   D. inspect tires at the end of a shift.

2. Which of the following parts is designed to protect a tire's sides from cuts and abrasions and flex without cracking when a load is placed on the machine?
   A. Bead.
   B. Inner liner.
   C. Sidewall.
   D. Tread.

3. All of the following are disadvantages of foam-filled tires, EXCEPT:
   A. increased number of flat tire incidents.
   B. decreased tire cushioning due to lack of tire flex.
   C. increased strain on machine's chassis and final drives due to the additional weight of foam-filled tires.
   D. increased expense to foam-fill a tire.

4. Technician A states that a tire with a 58-ply rating must have 58 plies. Technician B states that a tire with a 58-ply rating can hold a higher air pressure than a tire with a 30-ply rating. Who is correct?
   A. Technician A.
   B. Technician B.
   C. Both A and B.
   D. Neither A nor B.

5. What is the highest star rating for off-road radial-ply tires?
   A. One star.
   B. Two stars.
   C. Three stars.
   D. Four stars.

6. All of the following are *advantages* of using nitrogen to fill tires, *EXCEPT*:
   A. more economical than using compressed air.
   B. tires hold pressure longer than tires filled with compressed air.
   C. improved tire life (reduced tread wear).
   D. reduced wheel corrosion.

7. Which tire design has pockets arranged in honeycomb or spoke-shaped patterns within the tire sidewall and body?
   A. Airless tire.
   B. Foam-filled tire.
   C. Pneumatic tire.
   D. Laminated tire.

8. Which type of tire has one or more ribs around the circumference of the tire?
   A. F tires.
   B. G tires.
   C. R tires.
   D. None of the above.

9. In reference to an 18.4–38 tire, which of the following statements is true?
   A. It is a radial-ply tire.
   B. It requires an 18.4″ diameter rim.
   C. It has a 38 mm cross sectional width.
   D. It requires a 38″ diameter rim.

10. I tires are used on what type of heavy equipment?
    A. Forklifts.
    B. Rigid frame haul trucks.
    C. Agricultural implements.
    D. Articulated dump trucks.

11. Which of the following R tires has the deepest tread?
    A. R-1.
    B. R1-W.
    C. R-2.
    D. R-3.

12. If a low-sidewall tire and a standard tire have the same overall diameter, the rim diameter used for the low sidewall (LSW) tire is _____ the rim diameter of the standard tire.
    A. smaller than
    B. equal to
    C. larger than
    D. None of the above.

13. The desirable amount of wheel slip for four-wheel-drive agricultural tractors on normal soil is _____.
    A. 0%
    B. 2% to 4%
    C. 8% to 12%
    D. 20% to 25%

14. If a heavy equipment tire is overinflated by 30%, the tire's life can be reduced by as much as _____.
    A. 10%
    B. 20%
    C. 30%
    D. 50%

15. All of the following guidelines for inspecting tires are correct, *EXCEPT*:
    A. inspect the tire treads for consistent wear and proper depth.
    B. inspect the tires for cracks, punctures, or tire separation.
    C. check the air pressure of the tires.
    D. inspect the tires after they are hot from operation.

16. Technician A states that a wheel is the rotating assembly that couples the hub to the tire. Technician B states that a rim is the flange portion of a wheel to which the tire mounts against. Who is correct?
    A. Technician A.
    B. Technician B.
    C. Both A and B.
    D. Neither A nor B.

17. All of the following items are used for ballasting an agricultural tractor, EXCEPT:
    A. liquid ballast.
    B. high-purity nitrogen.
    C. suitcase weights.
    D. wheel weights.

18. All of the following factors are needed to calculate optimal tractor ballasting, EXCEPT:
    A. tractor speed.
    B. type and size of tires.
    C. ambient temperature.
    D. total load (weight).

Apply and Analyze

19. A(n) _____ tire has multiple layers of fabric-type plies (nylon, polyester, rayon, Kevlar, etc.) in a 45° diagonal, crisscrossing pattern.

20. The portion of a tire that is comprised of the plies and bead is called the _____.

21. For protection on rugged worksites and improved traction in wet conditions, a(n) _____ may be wrapped around a machine's tires.

22. A(n) _____ is manufactured into a bias-ply tire to reduce the risk of punctures to the tire's tread.

23. A(n) _____ tire has steel cord plies perpendicular to the tire's tread and additional steel belts.

24. The _____ is a group of steel wires used to attach and anchor the tire's plies.

25. The 70 marking on a 480/70R38 tire is the tire's _____.

26. Adhering to the ton-miles per hour value helps prevent _____ overheating.

27. Multi-piece rims require a(n) _____ sealing element.

28. A tapered wedge is used to attach the wheels on a(n) _____-axle design.

Critical Thinking

29. What is the section height in inches of a 460/85R38 tire?

30. Calculate the wheel slip of an agricultural tractor that made 20 tire revolutions with the implement in the ground and had 18 revolutions unloaded.

Chapter 23
Undercarriages

Objectives

After studying this chapter, you will be able to:

✓ Describe different types of undercarriage classifications.

✓ List the common components used on undercarriages.

✓ Explain how to inspect and adjust track tension.

✓ List the two methods used for splitting tracks.

✓ Describe the track inspection process.

✓ Describe how and why undercarriages wear.

✓ Recall undercarriage operating tips.

✓ Describe the unique features of the Caterpillar SystemOne Undercarriage.

✓ List the different types of rubber track designs.

Excavators, loaders, dozers, cranes, log skidders, trenchers, pavers, and pipe layers are among the wide variety of heavy equipment that can be equipped with track undercarriages. *Undercarriages* function to support and propel the machine and are constructed as side frames that house and support the drive sprockets and idlers, the carrier rollers and track rollers, the track tensioner and recoil mechanism, and the track chain and shoes. Each of the components of an undercarriage assembly is described later in this chapter.

Manufacturers choose to install track undercarriages instead of tires on their machines for different reasons. Tracks provide increased traction and enable operators to work the tractors more efficiently. Tire-equipped machines are less stable in soft soil conditions and do not offer the high flotation ability of tracks. Undercarriages lower a machine's center of gravity and allow machines to operate on steeper grades and slopes. Track-type machines also offer improved maneuverability, including the ability to make sharp turns. In applications where pneumatic tires can fail prematurely due to heat, abrasions, or rocky conditions, the use of track-type machines is preferred.

Track-type machines are better suited than wheeled machines for operation in snow, loose soil, extremely rocky, wet, or muddy conditions. To prevent soil compaction (especially of the subsoil) in agricultural applications and to increase flotation, a farmer can equip his or her machine with tracks. Tracks decrease the machine's *ground pressure*, which is the amount of pressure the machine's weight exerts on the soil.

Increasing the length of the undercarriages and widening the track shoes lowers the machine's ground pressure. **Figure 23-1** shows the inverse relationship between track width or undercarriage length

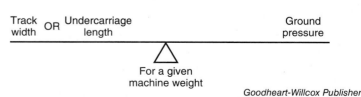

Goodheart-Willcox Publisher

Figure 23-1. As the track width or undercarriage length is decreased, the machine's ground pressure increases.

and machine ground pressure. Although the ground pressure can be as high as 40 psi for some track-type machines, others can exert ground pressures as low as 3 psi. Wheel-type tractors can exert ground pressures as high as 80 psi.

Cost is one of the primary disadvantages of track undercarriages. Up to 20% of the machine's initial purchase price can be attributed to tracks and up to 50% of the machine's maintenance costs after purchase can be attributed to the undercarriages. Track machines also have slower travel speeds. Although most off-highway machines are transported on flatbed trailers from worksite to worksite, if two sites are close to each other, it is possible that a machine equipped with rubber tires could be driven from one site to the other. The track machines, however, would normally be hauled to the worksite to reduce wear to the undercarriages and prevent potential road damage.

Note

Tracks are designed to lose traction when the tractor approaches maximum drawbar pull, within 3% of maximum drawbar pull, for example. Slipping tracks dramatically increases operating costs due to increased wear and should be avoided. Note that tire slippage was covered in Chapter 22, *Tires, Rims, and Ballasting*.

Undercarriage Classifications

Caterpillar uses the following undercarriage classifications, which may have slight design variations within each category:

- Standard—Traditional crawler undercarriage that operates best on firm soil.
- XR—Extended-to-rear undercarriage is preferred for high drawbar pulling applications, like pulling implements (dozer ripper shank) or winching applications.
- XL—Extra-long track undercarriage that features improved floatation.
- XW—Extra-wide track that enables the tractor to operate in muddy conditions.
- LGP—*Low-ground-pressure undercarriage* has both a wider track and a longer undercarriage that enables the tractor to operate in marshy wet conditions.
- TSK—Track skidder undercarriage that features narrow tracks and is used for drawbar applications and log skidding. Refer to **Figure 23-2**.

Note

Manufacturers publish specifications in performance handbooks. The performance handbook contains different types of data, and some data is available in an electronic format on the manufacturer's website. A few examples related to undercarriages are:

- Machine weight (based on model, including variations of undercarriages).
- Ground pressures (based on model, including variations of undercarriages).
- Maximum drawbar pull (based on transmission gear and machine speed).

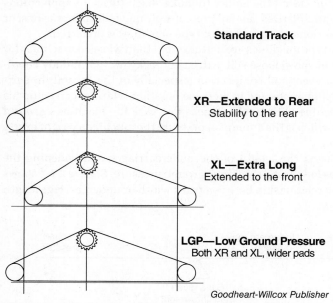

Goodheart-Willcox Publisher

Figure 23-2. Caterpillar uses multiple track classifications to describe the length and direction of the undercarriage extension, as well as the track width.

Track-type machines can also be more broadly classified based on the type of work the undercarriage performs: carrier or tractor. Cranes, excavators, and hydraulic shovels have a type of undercarriage called a *carrier undercarriage*, which is used to propel the machine and to support the machine with a stable footing.

A carrier-type undercarriage on an excavator provides the machine with a firm base to ensure it has a stable footing to perform its job. Traction or drawbar pull is much less important than stability because the majority of the work is completed by the excavator's boom, stick, and bucket. The undercarriage is only used for moving or repositioning the machine. If the excavator needs to drive down a steep slope, the boom is often used to assist in moving down the incline, **Figure 23-3**. There are a variety of other heavy equipment machines that use this type of undercarriage for moving or repositioning the machine.

A *tractor undercarriage* is installed on dozers, track loaders, and log skidders. This type of undercarriage, like the carrier undercarriage, propels the machine and provides a stable footing. They are also designed for strong pulling power and provide excellent traction. They exhibit great maneuverability, which includes being able to travel up or down steep inclines or push dirt on steep-sided slopes. A machine equipped with a tractor undercarriage is often referred to as a *crawler tractor*. See **Figure 23-4**.

Goodheart-Willcox Publisher

Figure 23-3. Excavators commonly work on top of a stockpile to make it easy to load trucks. The excavator's boom and stick are used to safely climb down the stockpile.

Goodheart-Willcox Publisher

Figure 23-4. Tractor undercarriages feature superb maneuverability and provide excellent traction to give the machine high drawbar pulling power.

Undercarriage Components

Undercarriages, also known as track assemblies, are comprised of multiple components. Most crawler tractor undercarriages contain a track frame, pivot shaft, equalizer bar, track rollers, one or more carrier rollers, a track chain assembly, a drive sprocket, one or two idlers, a tensioner, right and left side track frames, a recoil mechanism, and optional roller guards.

Track Frame

The *track frame*, also called the roller frame, is a large housing that provides a mounting place for the undercarriage components, such as the track rollers, carrier rollers, a track tensioner/recoil mechanism, a front idler, and a rear idler (on elevated-sprocket undercarriages). See **Figure 23-5**. Elevated-sprocket undercarriages will be discussed later in this chapter. The track frame attaches to the tractor frame through the pivot shaft and hard bar or equalizer bar.

Hard Bar

The track frame can be rigid or it can be designed to pivot. Examples of rigid frame undercarriages are older track loaders, excavators, pipe-laying tractors, and twin-rubber-track agricultural tractors. Some rigid frame

Goodheart-Willcox Publisher

Figure 23-5. This partially disassembled Caterpillar track frame shows the carrier roller mounted on the top and the track rollers mounted to the bottom of the frame.

Figure 23-6. This Challenger tractor has its undercarriage removed to expose the rear drive axle and the front hard bar. The rubber spring is not clearly visible in this view.

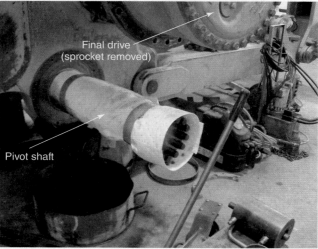

Figure 23-7. The track frame has been removed from this Caterpillar dozer with an elevated-sprocket undercarriage. Note that the pivot shaft has been covered for protection while the machine is being rebuilt. The pivot shaft is below the final drive. In this photo, the sprocket has been removed from the final drive.

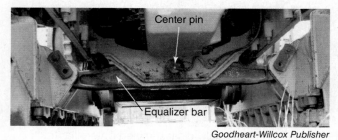

Figure 23-8. The equalizer bar attaches to both the right and left track frames, as well as the dozer frame.

undercarriages use a *hard bar*, which is a metal tube that connects to the front of both the left and right track frames and attaches to the tractor's frame. As explained in Chapter 21, *Suspension Systems*, the hard bar in **Figure 23-6** is attached by a large rubber spring on each side of the tractor.

Pivot Shaft

Elevated-sprocket undercarriages are designed to pivot on a separate *pivot shaft* at the rear of the tractor. The pivot shaft extends across the tractor's entire width. The design of these undercarriages enables the tractor to travel over large objects (boulders, small gravel piles, etc.) with agility. The right- and left-side track frames are attached to the machine's pivot shaft near the rear of the roller track frames. See **Figure 23-7**. The track roller frame can serve as a pivot shaft reservoir. A fill plug must be removed to check the oil level of the reservoir. Periodically checking this oil level ensures the seals are holding the oil in the pivot shaft cavity within the track frame.

Traditional, non-elevated (oval-shaped) undercarriages often incorporate the pivot shaft into the drive sprocket assembly. These systems also have two diagonal arms or braces, also known as swing arms, with one positioned on each side of the machine. One arm end attaches at the rear of the machine inboard from the track roller frame and drive sprocket, and the other arm end attaches to the front of the track roller frame. The arms keep the track frames aligned with the machine as the track frames pivot up and down.

Equalizer Bar

Pivoting undercarriages, either elevated or traditional in design, have an equalizer bar that connects the left- and right-side track frames to the tractor's frame. An *equalizer bar* attaches to the front of both roller frames. It is similar to the hard bar on rigid frame undercarriages and pivots on a center pin that is attached to the tractor's frame, **Figure 23-8**. The equalizer bar allows the tracks to oscillate, enabling the right or left track frame to pivot upward or downward.

Links, Pins, and Bushings

The heart of an undercarriage is the *track chain*, which is comprised of track links, pins, and bushings. **Figure 23-9** shows the top view of two sections of track chain: an old flush-mounted design and a late-model interlocking design. The old track links were known as flush mounted because the bushing mounted flush and did not pass through both sets of links. Late-model undercarriages use an interlocking design in which the bushing protrudes into the outer link to improve the sealing of the links.

On a single track link, a track bushing is pressed into a bore on one end and a track pin is pressed into a bore on the other end. When the chain is assembled, each track pin is inserted into a bushing and can rotate freely within the bushing. Track links are specified as left hand or right hand and cannot be interchanged.

The tractor's drive sprocket propels the machine by driving the track's bushings. The bushings and drive sprocket will wear over time and are one of the many undercarriage areas that require close inspection during maintenance. Undercarriage component inspection is discussed later in this chapter. The empty holes on the top of the links are used to bolt the track shoes to the links.

The link has flat surfaces on the top and bottom, known as rails or flanges. The distance between the two rails is the **link height**. One of the link's rails has two mounting holes for attaching the track shoe. The link's other rail rides on the idler(s) and rollers, which help keep the chain aligned, **Figure 23-10**. This rail, along with the idler(s) and rollers, require inspection as part of a maintenance routine. The links must be hardened to minimize wear. As the rails wear, a cupped surface is worn into the rail. Depending on track alignment, machine use, and field conditions, the idlers and rollers will wear also.

Some undercarriages are **sealed and lubricated track (SALT) undercarriages**. SALT pins have a passageway in the center of the pin that provides a path for lubricant. The SALT design prevents internal wear between the pins and bushings. Seals keep the lubricant contained between the pin and bushing and prevent contamination from entering. The bushings on SALT tracks are inserted into a recess in the link (refer back to the interlocking design in **Figure 23-9**), whereas the bushings are flush with the adjacent track link on older non-SALT tracks. The seals are placed inside the links along with thrust washers, and the links, pins, and bushings are pressed together to form the track chain. A plug is inserted in one end of the pin to retain the lubricant.

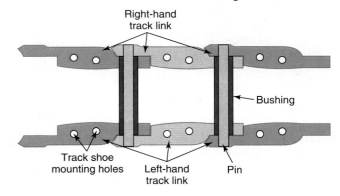

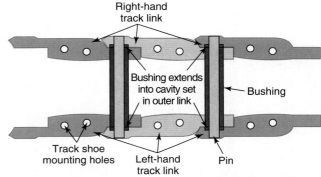

Figure 23-9. On both track chain designs, one pin passes through four links, with two links overlapping each other on the right and on the left. The pin is located inside the center of the bushing. The pins are pressed into the outer track links and the bushings are pressed into the inner track links. On the interlocking design, the bushing is set in a cavity inside of the outer links to prevent debris from getting in between the links and wearing the pin.

Note

Sealed undercarriages offered by manufacturers contain lubricant between the pin and bushing. However, a solid metal pin is used in these types of undercarriages instead of an internally bored pin. The passageway for lubricant within the pin is eliminated in a sealed undercarriage design.

Caterpillar manufactures other types of track chains. A **positive-pin-retention track** uses a snap ring at the end of the track pins to positively retain the pin within the chain link, reducing pin endplay. A **heavy-duty track** is a sealed and lubricated track with links, bushings, and pins that are enhanced for higher stress and impact loading environments. The retention and sealing of the track are also improved. The **heavy-duty extended life (HDXL) with Duralink track** uses taller, crowned links to achieve a longer life. The bushings are also larger in diameter, extending their service life. Tri-link and quad-link tracks have one or two additional track chains bolted to the end of the track shoes in heavy-duty applications to obtain longer track life in logging or rocky terrain. The System-One Undercarriage is another style that is discussed later in this chapter.

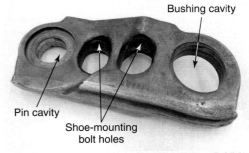

Figure 23-10. Each track link has four holes for housing a pin, bushing, and shoe-mounting bolts. Note that the two large center holes are part of the link's casting.

Drive Sprocket

Undercarriages are driven by *drive sprockets*. A drive sprocket is mounted to the tractor's final drive on both sides of the machine. Older machines had a one-piece drive sprocket that was welded to the final drive. Today, however, most manufacturers use bolt-on sprockets due to the wear the drive sprocket experiences during use. The bolted sprocket can be a one-piece design or have five separate pieces (*drive segments*). A segmented drive sprocket makes it possible for a technician to install or replace one piece at a time without splitting the track. See **Figure 23-11**.

The drive sprocket drives the track chain's bushings. The sprocket is a hunting-tooth design, which means that the bushings are driven by every other sprocket tooth. It takes two revolutions for the sprocket to drive each of the chain's bushings. The design helps deter wear because several revolutions are required for the same sprocket tooth to drive the same bushing.

Goodheart-Willcox Publisher

Figure 23-11. A single drive segment is one of five matching pieces that comprise the drive sprocket on an undercarriage. When the drive sprocket is divided into five segments in this manner, a technician can replace the segments without having to split the track.

Note

The drive sprocket on a carrier undercarriage can be called a **drive tumbler**. On cranes, it may be driven by a chain that is powered by a hydraulic motor or gearing.

Idler

The *idler* helps to keep the track aligned when the tractor reverses its direction. The idler spins on two bearings on the idler's shaft. The idler shaft is mounted to a yoke and the yoke assembly is attached to the track tensioner. Refer to **Figure 23-12** and **Figure 23-13**.

Low-track undercarriages (refer back to **Figure 23-4**) have one front idler that is level with the drive sprocket located

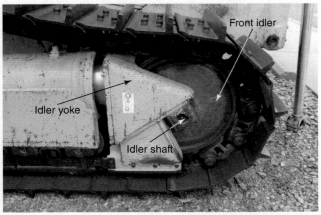

Goodheart-Willcox Publisher

Figure 23-12. The front-mounted idler is installed on a shaft that is attached to a yoke.

Goodheart-Willcox Publisher

Figure 23-13. This yoke assembly, also known as the front portion of the track frame, has been removed from the track frame. The idler has been removed from the yoke.

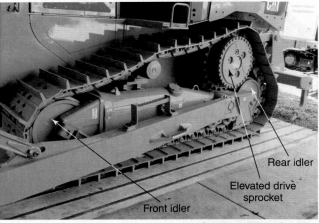

Goodheart-Willcox Publisher

Figure 23-14. Elevated-sprocket undercarriages have a front and rear idler. Note the position of the drive sprocket in this undercarriage design.

at the rear of the track frame. ***High-track undercarriages***, also known as ***elevated-sprocket undercarriages***, have the drive sprocket raised and positioned slightly forward from a second idler located at the rear of the track, **Figure 23-14**. Unlike the front idler, which is attached to a movable track tensioner, the rear idler shaft is fixed to the rear of the track frame.

The track links ride along the idler's two stepped surfaces on the idler's outer edges. Both the idler and the links are surfaces susceptible to wear and require inspection. The raised center circumference of the idler is known as the ***idler flange***, or ***idler rib*** area. It rides between the track links. See **Figure 23-15**. The idler's outside circumference has a slight taper, which helps keep the track running in alignment.

Track Tensioner

The track's chain tension is maintained with a ***track tensioner***. Many undercarriages have a hydraulic piston to hold tension on the track chain. The tension mechanism contains a fill valve (adjustment valve) and a relief valve, **Figure 23-16**. The ***fill valve*** serves as a hydraulic inlet port for filling the tensioner's piston. The ***relief valve*** releases track tension. When the relief valve is opened, hydraulic pressure is released, causing the track chain to sag. The hydraulic pressure is often manually applied with a grease gun. The procedure for adjusting track tension is addressed later in the chapter.

On rubber-track tractors, the machine's hydraulic system is sometimes used to tension the track. Once the track has been tensioned, a check valve holds the fluid pressure, which maintains the track's tension.

Note

Older track-type tractors were equipped with a threaded mechanical adjustment for tensioning the tracks. The idler's yoke would be extended or retracted using a turnbuckle-style sleeve. A lock collar prevented the adjuster from loosening.

Recoil Mechanism

Undercarriages have a ***recoil mechanism*** installed and connected to the front idler. It allows the front idler to retract within the track frame when debris becomes wedged between the track chain and the idlers or sprocket, or when a force outside the undercarriage tries to move the idler rearward.

Most undercarriages have either a very large spring or a nitrogen-filled accumulator installed as a recoil mechanism, **Figure 23-17**. Some older tracks feature a series of cone plates, similar to large Belleville washers, as a recoil spring. The recoil mechanism, track tensioner, and front idler yoke assembly all work together to maintain track tension.

Goodheart-Willcox Publisher

Figure 23-15. Idlers have a stepped surface on both sides of the idler flange. The links ride on the stepped portions of the idler.

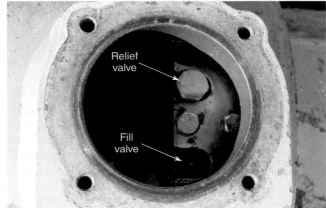

Goodheart-Willcox Publisher

Figure 23-16. The track tensioner has two valves: a relief valve and a fill valve. Grease is pumped into the fill valve to tighten the track, and the relief valve is turned counterclockwise to decrease track tension.

Goodheart-Willcox Publisher

Figure 23-17. A nitrogen-filled accumulator serves as the recoil mechanism on this Challenger rubber-track tractor.

Track Rollers

Track rollers are installed below the track frame, between the bottom of the track frame and the track chain links. They evenly transfer the load of the machine to the chain's links. The rollers guide the track links.

Each track roller consists of a roller shell, shaft, a pair of bearings or bushings, seals, and contain lubricant. The rollers are sealed and filled with lubricant when they are manufactured or rebuilt. Each shell has outside flanges that are machined with a slight taper (4 degrees) to guide the chain's links onto the rollers. Rollers are labeled *single flange* if the roller has one flange for guiding the left link and one flange for guiding the right link. Other track rollers have additional inside flanges and are labeled *double-flange* rollers. Refer to **Figure 23-18**.

An undercarriage can be equipped with both single- and double-flange rollers. Single-flange rollers are installed next to the drive sprocket due to their design. Double-flange rollers are placed on the frame based on allowable space to provide the best track alignment. Both types of rollers are made of hardened steel for durability.

Figure 23-18. These track rollers (single-flange and double-flange) are ready to be installed on a roller frame. The double-flange rollers have inside flanges. The two parallel track links ride in between the roller's inside and outside flange on both ends of the roller.

Bogies

Instead of the traditional track rollers, some undercarriages are equipped with *bogies*, which consist of four small rollers on a pivoting frame. The bogie frame has the ability to oscillate.

Agricultural rubber-track tractors commonly use bogies, **Figure 23-19**. On these machines, each bogie roller rides on one side of the lugs that line the center of the rubber track's interior surface. Two rollers ride on the right side of the rubber track, and the other two rollers ride on the left side of the rubber track.

Carrier Rollers

The *carrier roller* is located on top of the track frame. It helps support and align the top half of the track. The carrier roller's flanges are in the center of the roller, and not on the exterior edges. The roller is often mounted to a carrier roller pedestal, **Figure 23-20**. Not

Figure 23-19. Bogies consist of four small rollers (two sets of paired rollers) on a pivoting frame. This tractor uses two sets of bogies. Four of the rollers are not visible because they are on the inside of the track.

Figure 23-20. Carrier rollers are located on the top of the track frame and are used to guide and support the top half of the track.

all undercarriages are equipped with carrier rollers, but those that are typically have one or two carrier rollers per track.

Carrier rollers share the same basic construction as track rollers (shell, shaft, a pair of bearings or bushings, seals, and lubricant). Both types of rollers often have Duo-Cone seals. As explained in Chapter 18, *Axles and Driveshafts*, the heavy-duty seals are used to retain fluids and prevent contamination between a rotating component and a static component. The seal consists of two metal face seal rings that are held compressed against each other by means of two rubber torics, **Figure 23-21**.

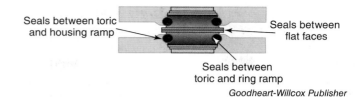

Figure 23-21. A Duo-Cone seal has two rubber torics that apply pressure on the two metal flat face seals. The seals are used in track rollers and carrier rollers.

The shaft is bolted to the track frame, and the roller shell spins on the shaft's bearing. The Duo-Cone seals retain the fluid between the rotating shell and stationary shaft.

Guides and Guards

Undercarriages can be equipped with *guides and guards* to help direct the track chain, prevent debris from packing between the rollers and the links, and protect the rollers, idlers, and sprockets from impact damage. The guards are located at the bottom of the track frame, between the frame and the shoes, **Figure 23-22**. A rear idler guide is shown in **Figure 23-23**.

Track Shoes

Track shoes, also called *pads*, are bolted to the track links. The shoes support the tractor, protect the chain, and provide traction to propel the tractor. **Figure 23-24** shows shoes that have a single *grouser*, which is the vertical steel lug protruding from the shoe. Track shoes are offered in a variety of different designs:

- Single standard grouser.
- Multi-grouser.
- Open-center grouser.
- Self-cleaning grouser.
- Rubber track pads.

Figure 23-22. Roller guards protect the track rollers from rocks and debris that can damage the assembly.

Figure 23-23. This rear idler guide is welded to the track roller frame. It directs the links alignment and protects the rear idler.

Single Standard Grouser

A shoe with a single grouser is called a ***standard grouser*** and is the most common track shoe used on crawler tractors (refer to **Figure 23-24**). It provides excellent traction because its single tall grouser digs deep into the soil. As the grouser wears over time, machine traction will decrease.

The leading edge of the shoe is curved downward and the trailing edge of the shoe is curved upward so that one shoe can ride below the next shoe. The front edge of a single grouser shoe has two notches removed, known as link reliefs, so that as the shoe rotates around the idler, the front of the track shoe does not bind with the track's links, **Figure 23-25**.

Caution

Operating single grouser shoes on asphalt and concrete surfaces will damage the surface and increase the rate of wear of the shoes. Many operators place mats, tires, or lumber down on the road surface and drive over that material to prevent damaging the road's surface.

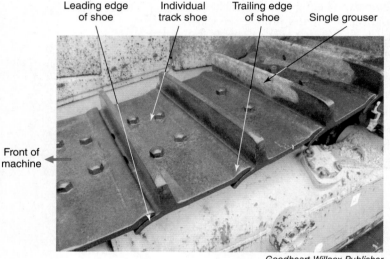

Goodheart-Willcox Publisher

Figure 23-24. In order to operate without any obstruction, the track shoes overlap each other with a lip in the front that points downward and a lip in the rear of the shoe that angles upward.

Multi-Grouser

A ***multi-grouser*** shoe has two or more short grouser bars on each track shoe, **Figure 23-26**. They are used on excavators and track loaders. Although they provide less traction than standard grouser shoes, multi-grouser shoes offer increased maneuverability and turning ability.

Open-Center Grouser

When a machine is operated in snow or muddy conditions, equipment owners must account for the mud or snow becoming packed between the chain and rollers, idlers, and sprockets and causing the track to become too tight. An ***open-center grouser*** shoe has an opening in the center of the shoe that allows the drive sprocket to push material out of the shoe, which prevents increased pin and bushing wear. See **Figure 23-27**.

Goodheart-Willcox Publisher

Figure 23-25. Notches, known as link reliefs, are removed from the front portion of the shoe. As the track rotates around the idlers and sprockets in the undercarriage, the notches prevent the shoe's leading edge from binding with the links.

Goodheart-Willcox Publisher

Figure 23-26. The undercarriages of track loaders and excavators have multi-grouser shoes installed on the links. The two or more grousers on each shoe are shorter than the ones used on single grouser track shoes.

Self-Cleaning Grouser

A *self-cleaning grouser* shoe has a triangular profile when viewed from the side of the machine, **Figure 23-28**. The shoe design enables dirt and mud to fall freely out of the tracks when the shoes separate slightly during the track's rotation around the drive sprocket and idler. They are used on loose, marshy soil where high machine flotation is necessary. This design is also sometimes referred to as a low-ground-pressure (LGP) track. However, a single grouser shoe can also be installed on a LGP track.

Rubber Track Pads

Some undercarriages are configured with rubber track pads. **Rubber track pads** consist of steel pads with rubber bonded to the pad. They are used to prevent scuffing, scoring, or damage to the hard surface the machine is operating on. See **Figure 23-29**.

Pro Tip

Although wide track shoes provide excellent floatation, extra stress and strain is placed on the undercarriage when wider than necessary track shoes are installed. Machines should be equipped with the narrowest shoes capable of providing stable footing and propelling the machine in the worksite conditions.

Track Tension

Proper track tension eliminates the additional track wear caused by an overtightened track and prevents the chain from falling off or whipping if the track is too loose.

Measuring Track Tension

Before measuring track tension, the operator must bring the machine to a stop by coasting forward. The brakes or the blade cannot be used to stop, and the machine cannot

LuckyPhoto/Shutterstock.com

Figure 23-27. Open-center grouser shoes allow the drive sprocket to force mud through the grouser's center hole. If material is left packed around the inside of the track during operation, track tension can become too tight.

Goodheart-Willcox Publisher

Figure 23-28. Self-cleaning shoes allow dirt and mud to drop between them while the shoes rotate around the drive sprocket and idler. Note the shoes have triangular profiles.

Figure 23-29. This excavator's undercarriage is configured with rubber track pads, which prevent scoring the pavement.

Figure 23-30. A straightedge is placed across the top of the track as part of the method to measure track tension. Note how the steel ruler is positioned on the top of the lowest grouser to take the measurement.

be backed up in reverse. The machine must also be in the field conditions in which it will be operating. Track tension *cannot* be properly measured on a concrete floor in a repair shop. Be sure to follow the manufacturer's service literature that can provide additional steps, such as whether or not a track pin should be located over the carrier roller or centered on an idler.

Use a straightedge or draw a tight string across the top of the track. Measure the track sag by locating the lowest grouser and measuring from the bottom of the straightedge to the top of the grouser, **Figure 23-30**. If the machine has a carrier roller, measure the two lowest grouser points on each side of the roller and take an average of those two measurements. Compare the measured track sag measurement to the service literature specification. If necessary, adjust the track tension according to the manufacturer's procedures.

Warning

Never perform any type of undercarriage service or repair without receiving the proper training beforehand and following all manufacturer service procedures and warnings. The complexity and weight of undercarriage assemblies require a trained technician familiar with the manufacturer's equipment to prevent personal injury and damage to the machine.

Adjusting Track Tension

Most modern undercarriages require a grease gun or the machine's hydraulic system to adjust the track's tension. The most common method for adjusting track tension requires a grease gun. Many technicians use a battery-powered or pneumatic grease gun. Opening an access cover on the track frame allows a technician to place the grease gun's hose on the fill valve's grease zerk. Pumping the grease gun forces grease into the track tensioning cylinder, extending the cylinder and adding tension to the track. If the track tension is too tight, the release valve is opened and grease escapes the cylinder.

Some rubber track machines use the machine's hydraulic system along with the track tensioning tool hose and pressure gauge kit to add or remove hydraulic oil from the track tensioning cylinder. With the machine shut off, the hose/pressure gauge kit is installed to the track tension port and to a hydraulic coupler, often located at the back of the tractor. The machine is started and the machine's hydraulic control valve is activated. This causes oil to flow from the machine's control valve, through the hose/pressure gauge tool kit, and into the track tensioning cylinder to fully extend the cylinder. Some procedures require actuating the machine's control valve for a short period of time to ensure that the track tensioning cylinder was charged to the hydraulic system's high-pressure setting. The tension cylinder has a one-way check valve that prevents the oil from escaping the cylinder once it is properly tensioned. A release valve is used in conjunction with the hose kit to drain the cylinder and release the track tension.

Warning

High-pressure grease or oil escaping a track tensioning cylinder can penetrate a human's skin and cause a serious or life-threatening fluid injection injury. Always follow manufacturer's safety standards for adjusting track tension.

Splitting the Track

Many, but not all, undercarriage repairs require the track chain to be split in order to perform service on the track and undercarriage components. One section of the track chain features either master links or a master pin as the means to split the track. Different manufacturers have different designs for each style.

A *master link* consists of a two-piece mating link with notched teeth that is bolted securely together. See **Figure 23-31**. Compared to the rest of the pins used in tracks, a *master pin* can be longer, have a slightly smaller diameter, and its end can have a different design such as a step, mark, or machined hole. The master bushing can be shorter.

Although the master pin has a slightly smaller diameter that is supposed to make it easier to remove than a traditional pin, the master pin is an interference fit and can be very difficult to remove. A hydraulic press is normally used to remove the pin. Sometimes heat or a sledgehammer and punch are used to remove the pin. The master bushing is removed after the master pin is removed.

Undercarriage Inspection

Prior to performing an undercarriage inspection, be sure the machine's attachments have been lowered to the ground, the machine has been disabled per manufacturer's instructions, and a "Do not operate" tag has been placed on the controls in the operator's cab. The undercarriage must be thoroughly cleaned prior to inspection. While taking measurements, carefully check the accuracy of the measurement. Even a seemingly insignificant error in measurement can cause a significant error in estimating the remaining service life of a part.

As part of scheduled maintenance, a thorough inspection is performed in addition to making multiple measurements. The visual inspection includes checking for cracked or excessively worn components, broken welds, leaks, and loose or missing nuts and bolts.

Calipers, rulers, tape measures, and adjustable depth gauges are the tools used to manually measure undercarriage wear, **Figure 23-32A**. Be sure to print or copy the undercarriage specification sheet from the machine's service literature so that it can be referenced during the inspection. After taking a manual measurement with calipers, gauge bars, and rulers, record the results on the measurement sheet so that it can be compared to the specifications.

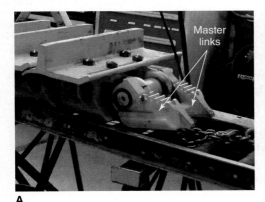
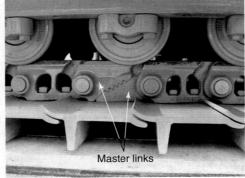

Figure 23-31. Most track chains have either master links or a master pin to aid the process of splitting the track chain. The master link design is more common in late-model machines. A—A master link design has two mating links with notched teeth bolted to each other on both the right and left sides of the chain. B—Master links assembled on a machine.

Goodheart-Willcox Publisher

Figure 23-32. Measuring tools. A—Calipers and an adjustable depth gauge are used to measure different undercarriage components. B—An ultrasonic measurement tool is used to measure track wear.

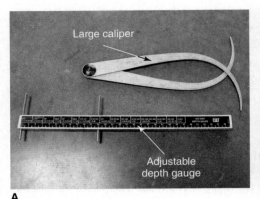
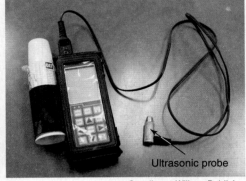

Goodheart-Willcox Publisher

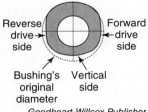

Figure 23-33. The ultrasonic tool is used to calculate track link wear. Note the position of the ultrasonic probe.

Many technicians use an *ultrasonic measurement tool*, which is an electronic tool used for measuring undercarriage wear. The tool measures the thickness of metal track components, such as track links and rollers. Gel must be applied to the ultrasonic probe before each measurement is taken. The tool must also be calibrated before measuring the track wear. See **Figure 23-32B**.

Track Links Inspection

The ultrasonic measurement tool can be used to measure the track link wear. The probe must be placed on the bushing end of the link, against the link's bottom rail, and aligned with the center of the bushing hole. See **Figure 23-33**. The tool measures the distance from the link's bottom rail to the bushing's bore. The probe must be slid along the length of the link's bottom rail under the bushing in order to obtain the smallest measurement. The smallest measurement indicates the distance between the bottom of the rail to the outer most edge of the bushing's bore in the rail.

Bushing Inspection

Bushings will wear on three sides: the forward drive side, vertical side, and reverse drive side. See **Figure 23-34**. Using an ultrasonic probe, a technician must measure all three sides and compare the readings to specifications to determine the bushing's wear. Choosing the smallest measurement of all three measurements should indicate the bushing's percentage of wear on all three sides.

Machine owners want to maximize the life of their machine's undercarriage but do not want to waste money on unnecessary service. Determining when the bushings should be turned (not too early or too late) is an important undercarriage service decision. A *bushing turn* is the process of removing the track from the tractor, putting the track on a track link press, pressing the bushings out of the links, and turning the bushings 180° before reinserting them into the links. See **Figure 23-35**. For the machine owner to maximize the life of the bushings, the bushings should be turned when they are approximately 80% worn.

Note

Track link presses can be stationary models in heavy equipment repair shops or trailer-mounted or portable models that can be transported to worksites.

Figure 23-34. A track bushing will wear as it is driven by the drive sprocket. This simplified end view of a bushing illustrates the large amount of wear that will occur across the three sides. The outside dashed line represents the bushing's original outside diameter.

If the machine owner has the bushings turned too soon, money is wasted on the service because the maximum life of one side of the bushing is not used. If the machine owner waits too long to turn the bushing, the lubricant will leak out of the bushing and damage will result, **Figure 23-36**. New pins and bushings will have to be purchased and installed.

Figure 23-35. Track link presses are used to disassemble and assemble track chains. A—The press has an air wrench that is used to remove the bolts that secure the shoes to the links. B—A 200-ton hydraulic press is used to remove and install the pins and bushings. A conveyor belt supports the track chain during the process.

Track service specialists measure the temperature of the bushings with an infrared temperature gun after the undercarriage has been run for more than an hour. A warm or hot bushing indicates the bushing has lost its lubricant.

Drive Sprocket Inspection

Drive sprockets can wear and cause the end of the teeth to form a thin sharp edge. A ruler is used to measure across three teeth on each drive segment. The results are compared to the specification to determine the amount of wear. See **Figure 23-37**.

Some drive sprockets are checked for wear using a hexagonal "no-go" gauge that is placed between two drive teeth. The shape of the gauge is similar to that of a large machinery nut. If three points of the gauge touch the drive sprocket segment, the segment needs to be replaced. If only two points of the gauge touch, the drive sprocket has more usable life. Refer to **Figure 23-38**.

Figure 23-36. The drive side of a bushing wears as it is driven by the sprocket's teeth. The bushings need to be turned at approximately 80% wear. This bushing was not turned in time, which caused severe wear to the bushing and damage to the pin.

Idler Inspection

A *depth gauge* is a precision tool equipped with sliding probes that measure the linear distance below a specific point. It is used to measure the wear of an idler wheel. See **Figure 23-39**. Place the flat edge of the depth gauge against the idler's flange. Push the

Figure 23-37. Sprocket wear is measured with a ruler across three teeth on the drive segment. Compare the results with the specifications. The measurement is normally taken when the drive segment is mounted on the final drive.

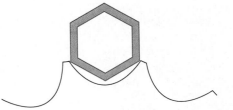

Figure 23-38. If the three points of the hexagonal gauge touch the drive segment, the segment should be replaced. Only two points of this gauge are touching the drive segment, so the drive segment has additional wear life.

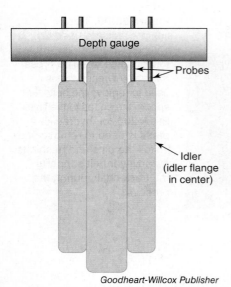

Figure 23-39. Depth gauges with probes are used to measure the amount of wear on idler wheels. After the probes are extended to press against the idler wheel, a ruler is used to measure the protrusion length of the probes and an average is calculated of the measurements.

probes through the depth gauge until they rest against the idler's wear surface on each side of the flange. Measure the distance that the probes protrude through the depth gauge. Calculate an average of the multiple probe measurements before recording the results.

The ultrasonic probe can be used to measure the wear on the idler's flange if the idler is a manufactured idler. If the idler is a cast idler, the track has to be removed to determine the amount of flange wear.

As the idler wears, the material on the idler's edges decreases, which causes the flange height to increase. Worn idlers can be resurfaced using a submerged arc welder as long as they are not too far worn, **Figure 23-40**. After the idler has been resurfaced, it must be allowed to cool slowly to avoid any cracking, **Figure 23-41**. Once the idler has sufficiently cooled, the weld is ground down to a smooth surface.

 Warning

If the idler has a hollow center, heat will build inside the idler during the welding process. The increased air pressure from the heat will cause the idler to explode and cause potential injury or death. Before welding an idler with a hollow center, a hole must be drilled in the side of the idler to allow the hot air to escape during the welding process. The hole can be closed later.

Roller Inspection

Inspect the carrier and track rollers for wear, leaks, and loose or missing hardware. Check the position of the carrier roller and the wear to determine if the roller is out of alignment. Although the ultrasonic probe can be used for measuring wear on the carrier roller, a large caliper is normally the preferred measuring tool. The caliper is opened until it slides across the opening of the roller. After obtaining the diameter with the caliper, use a ruler to measure the caliper's opening.

The ultrasonic probe is used to measure wear on the track rollers. Be sure to measure the wear on both the inside (interior flange) and the outside portions of the roller (outer

Figure 23-40. An idler can be resurfaced by using a submerged arc welder that is especially designed for resurfacing idlers. A submerged arc welding machine is shown here.

Figure 23-41. A dealership covered this idler after it was resurfaced to allow it to cool slowly. It was temporarily uncovered to take a photo. Note the shiny new metal added on each side of the idler flange.

flange), **Figure 23-42**. Move the probe across the wear surface area and note the smallest measurement.

Track Shoes Inspection

Track shoes need to be inspected for grouser wear. A crawler with worn grousers will spin tracks and push less dirt. Check the shoes to see if they are bent, warped, cracked, and if the trailing edges are scalloped from wear. Scalloped trailing edges allow rocks to become wedged between the shoes, resulting in bent shoes. One bent shoe can place additional stress on the entire undercarriage.

When measuring a grouser's height, the measurement must be taken away from the edges of the grouser because the edges will be rounded from wear. A good place to measure is on the top of the grouser, about one-third of the distance from the edge of the shoe (close to the track shoe mounting bolts). When using an ultrasonic probe, the probe can be placed on the top edge of the grouser.

Goodheart-Willcox Publisher

Figure 23-42. This track roller's inside flange has severe wear.

Pitch Measurement

Undercarriages have three different pitches: link pitch, track pitch, and sprocket pitch. ***Link pitch*** is the distance from the center of the bushing hole to the center of the pin hole on a single track link. The link pitch is always fixed and will not change. See **Figure 23-43**.

Track pitch accounts for wear between a pair of track sections. The track's pitch will lengthen as the track wears. Track pitch is measured from the center of one pin to the center of the next pin in a track chain. Sometimes technicians measure track pitch from the leading edge of one pin to the leading edge of the next pin. Refer to **Figure 23-44**.

Manufacturers can provide specifications for measuring track pitch across several pins. For example, on a SALT undercarriage, a technician might need to measure the distance from the center of one pin to the center of the fifth pin, across a distance of four track links.

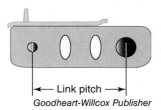

Goodheart-Willcox Publisher

Figure 23-43. Link pitch is the distance from the center of the pin to the center of the bushing on a single track link.

Note

Before making the track pitch measurement, the undercarriage must be tight. A separate track pin or something similar is normally placed between the drive sprocket and the link, and the tractor is moved forward in order to place tension on the track.

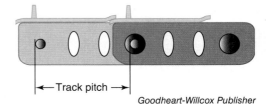

Goodheart-Willcox Publisher

Figure 23-44. Track pitch is the distance from the center of one pin to the center of the next pin. Some track specialists find that it is easier to measure the distance between the front edge of both pins, as illustrated.

Sprocket pitch is measured from the center of one bushing located in the sprocket to the center of the adjacent bushing on the sprocket. See **Figure 23-45**. An empty sprocket space will be located between the two bushings. Sprocket pitch does not change as the track wears.

Track Wear Gauge

Some manufacturers offer a special track wear gauge for a technician to use during inspection, **Figure 23-46A**. The gauge is used to measure grouser height, track pitch, rail height, sprocket wear, and idler wear. The shape of the gauge is designed for specific undercarriages and fits like a puzzle piece between different undercarriage components to visually check their wear, **Figure 23-46B**.

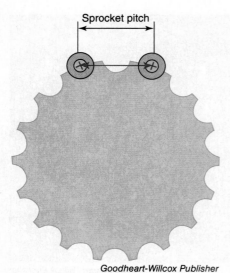

Figure 23-45. Sprocket pitch is measured from the center of one bushing on the sprocket to the center of the adjacent bushing on the sprocket.

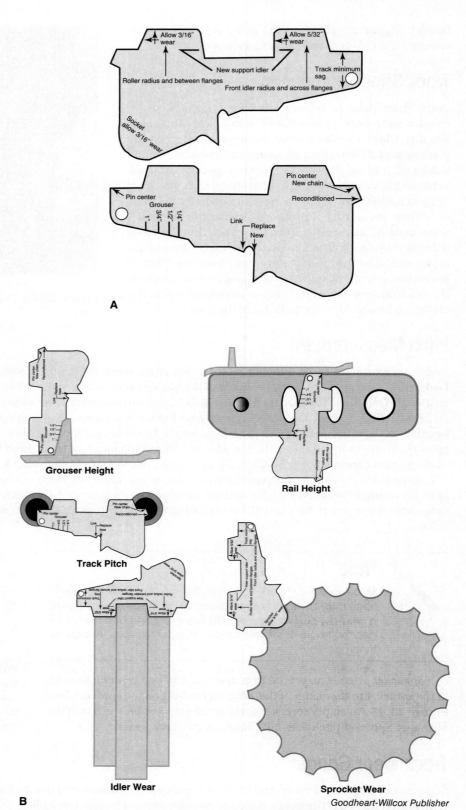

Figure 23-46. A track wear gauge is a special measuring tool that is designed to measure different wear points on a specific make and model of undercarriage. A—One manufacturer's track wear gauge tool. B—By orientating the gauge in the correct position, a technician can visually measure grouser height, track pitch, rail height, sprocket wear, and idler wear.

Undercarriage Wear

The three main factors that affect undercarriage wear are contact, load, and relative movement. When a properly tensioned track is propelling a machine forward, the bushing enters the sprocket, and contact between the bushing and the drive sprocket is made. In addition, load is present, with approximately 80%–85% of the load being exerted between the sprocket and first bushing. Refer to **Figure 23-47**. However, there is no relative movement between the bushing and the sprocket, which minimizes wear. All three factors (contact, load, and relative movement) must be present and substantial for wear to occur. As the bushing rotates out of the sprocket, relative movement occurs. However, because there is very little load on the bushing at this point of forward rotation, substantial wear does not occur.

If the track tension is too tight, load is present throughout the entire track. Therefore, as the bushing rotates out of the sprocket, substantial wear occurs. If dirt and material are packed tight into the undercarriage, they increase track tension. Track shoe selection (open-center shoes) can help by pushing the material out of the track shoes. In addition, track tension should be adjusted and continuously monitored and maintained in the field, due to the tendency of tracked machines to have some degree of debris packing in the tracks.

The pins rotate inside the bushings. However, even though there is contact, load, and relative movement, there is minimal wear because of the internal lubrication between the pin and bushing.

When the track is propelling the machine in reverse, as the bushing enters the sprocket, there is contact, load, and relative movement, causing substantial wear to occur. See **Figure 23-48**. Therefore, operating at high speeds in reverse should be avoided, due to the accelerated undercarriage wear this causes. An allowable exception to this rule is if a tracked machine is using high-speed reverse to get into position to assist another machine, such as a scraper waiting on a push dozer, for example. In this situation, the wear is justifiable. If the track is overtightened and the machine is operating in reverse, the undercarriage components wear even more quickly than they would during reverse operations with a properly tensioned track.

As the undercarriage components wear, the track pitch increases. When the track pitch no longer matches the sprocket pitch, the bushings will slide and move when being propelled forward and in reverse. The movement of the bushings results in a faster wear rate than a new track would have.

Other factors have an influence on track wear. Operation in sand or abrasive soils increases track wear. Tracks operating on larger, heavier machines with greater horsepower also experience increased wear.

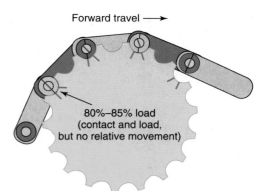

Goodheart-Willcox Publisher

Figure 23-47. When a properly tensioned track is propelling a machine forward, the first bushing entering the sprocket will have contact and load, but minimal wear results because there is no relative movement. The green bushing alignment marks on the sprocket depict the position of the bushing (black marks) as it enters the sprocket and the movement of the bushing as it rotates out of the sprocket. The red pin alignment marks depict the rotation of the pins inside of the bushing.

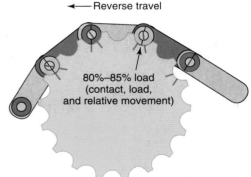

Goodheart-Willcox Publisher

Figure 23-48. In reverse, the track experiences all three factors related to undercarriage component wear: load, contact, and relative movement. The green bushing alignment marks on the sprocket depict the rotation of the bushing (black marks) as the sprocket is reversed. The red pin alignment marks depict the rotation of the pins inside of the bushing.

Undercarriage Operating Tips

As previously mentioned, undercarriage tracks account for 50% of a tractor's maintenance costs. Owners and operators can follow several practices for reducing undercarriage wear:

- Choose the narrowest track shoe that provides adequate traction and a stable footing. Wider shoes increase undercarriage wear.
- Decrease the load on the machine to avoid slipping the tracks.
- Operate a tracked machine at high speed only when it is critically necessary. Avoid reversing at high speed, unless justified by the style of work or type of worksite. High speed travel increases the rate of track wear.
- Avoid always turning to the left or always turning to the right in order to prevent one track from wearing more than the other.
- Alternate directions when working on a side slope to wear both sides of the undercarriage evenly.
- Work downhill rather than uphill, if possible. Not only will track wear be reduced but production speed will increase.
- Choose the right undercarriage features for the worksite, such as using open grousers on snow or mud, installing wear-resistant shoes in tough conditions, removing roller guards when in the mud, and using optional carrier rollers.
- Do not operate a machine if the tracks are too tight or too loose. Correct the track tension before operating.
- Do not dig off the side or the rear of an excavator. The machine is designed to dig with the idlers facing forward because it places less wear on the final drive and drive sprocket.
- Drive a track-type machine once every two months to prevent the pins from seizing, which could happen if the machine sits stationary for a long period of time. This process is referred to as *exercising the tracks*.

Caterpillar SystemOne Undercarriage

In 2005, Caterpillar introduced a new undercarriage system for D3 through D8 tractors called *SystemOne Undercarriage*. It is now offered on some track loaders as well. The undercarriage system is designed to provide longer life, while reducing maintenance and downtime related to servicing undercarriages. It also reduces operating costs.

The track links are manufactured in a one-piece cartridge that is sealed for the life of the component. The cartridges contain a synthetic oil that cuts seal friction in half. The links have a unique stronger straight-link design, as opposed to the older curved design. The track shoe bolt pattern is straight, as compared to the old offset bolt pattern.

The undercarriage is equipped with a wide *center tread idler* with twice the material as the old design. The new design allows the idlers to be used through two track-chain life cycles before replacement. In this design, the link rails do not contact the idler. The idler rides inside the link rails, making contact with only the bushings. This eliminates link rail scalloping and vibration and improves rail guidance. The majority of the machine weight is carried on the track rollers, which further reduces wear on the idler(s). The rear idler contains a shim that is removed once the links and rollers are half worn.

The drive sprockets are also designed to be used through two track-chain life cycles. The carrier rollers have more material, adding 50% more life to the carrier rollers. The track rollers have a larger flange, which improves track guidance.

The bushings are a rotating design that makes it possible for the center tread idlers and drive sprockets to be reused through a second undercarriage life. Because the bushings rotate, wear is reduced due to the lack of relative motion. This eliminates the need to turn the bushings.

The master link, which Caterpillar calls a *clamp master link*, has a center strut added to the link. The center strut increases the strength of the track's master link.

Rubber Track Systems

Numerous types of off-highway machines, including the following, use rubber track systems:
- Skid steers.
- Excavators.
- High-speed dozers.
- Agricultural tractors (4WD, twin-track tractors, and MFWD).
- Combines.
- Grain carts.

In the agricultural industry, rubber tracks provide advantages over traditional rubber tire machines, such as a larger footprint, which reduces the ground pressure and soil compaction. Although rubber tracks cost more than rubber tires, rubber tracks improve machine traction and will never suffer downtime from a flat tire.

Rubber Track Drives

Rubber track systems can have three different drive designs:
- Non-drive rubber tracks.
- Driven by track tension and friction.
- Driven with a drive sprocket, similar to a steel track system.

Non-drive rubber track systems are used in non-drive axle applications, such as the grain cart in **Figure 23-49**. The inside portion of the rubber track might have lugs, but in these applications, the lugs are used for belt alignment and not for driving the track.

Twin-rubber-track machines such as MT Challenger tractors, Fendt Vario MT tractors, John Deere RT tractors, and Caterpillar AT1055F pavers, use tension and friction to drive the tracks. The drive wheels have rubber bonded around them, and when the track is fully tensioned, the tension and friction of the drive wheel against the rubber track propel the track. The original Challenger tractors had small-diameter drive wheels, **Figure 23-50**. Late-model twin-rubber-track tractors have large drive wheels, **Figure 23-51**.

Goodheart-Willcox Publisher

Figure 23-49. Rubber track systems on grain carts are not driven by sprockets but require a tractor to pull the cart through the field.

Goodheart-Willcox Publisher

Figure 23-50. The original Challenger tractors have a small-diameter idler (front) and drive wheels (rear) lined with rubber. The friction between the rubber track and drive wheels provide the positive drive of the track. The left drive wheel assembly is removed from this tractor's final drive.

Goodheart-Willcox Publisher

Figure 23-51. This twin-track Challenger tractor has narrow tracks that are spaced far apart to run between rows of crops. Note the large drive wheels equipped in the rubber track undercarriage on the tractor.

These tracks also have lugs on the inside of the tracks, but they are only used for belt alignment purposes.

Many compact track loaders and agricultural tractors equipped with four-rubber-track undercarriages are driven with the belt's drive lugs. The drive sprocket is designed to mesh with the lugs on the interior of the rubber belt to provide a positive drive. See **Figure 23-52**.

Figure 23-52. Some rubber tracks are driven by the belt's drive lugs. A—This John Deere compact track loader has a drive wheel that looks similar to the drive sprocket on steel track undercarriages. The track also has a separate set of lugs that are used for alignment. B—This Case IH Quadtrac undercarriage shows how the drive wheel meshes with the belt's drive lugs.

Goodheart-Willcox Publisher

Summary

- Caterpillar track classifications include standard, XR, XL, XW, LGP, and TSK.
- Undercarriages can be designed as a carrier undercarriage or a tractor undercarriage.
- Track frames can be designed to pivot or be rigid.
- Track chains are constructed of track links, pins, and bushings.
- The drive sprocket is manufactured in a one-piece design or split up into five segments that bolt onto the machine's final drive.
- Recoil mechanisms work in conjunction with the front idler yoke assembly and the track tensioner mechanism to maintain track tension.
- Track rollers are installed below the track frame to evenly transfer the load of the machine onto the track chain's links.
- Carrier rollers are positioned on top of the track frame to help support and align the top half of the track.
- Track shoes are available in standard grouser, multi-grouser, open-center grouser, self-cleaning grouser, and rubber track pads.
- Tracks are often tensioned with a grease gun.
- Manufacturers normally use a master pin or master link to help split a track link.
- The common tools used for track inspection are a tape measure, straightedge or string, calipers, adjustable gauge bars, track wear gauge, and an ultrasonic measurement probe.
- Track bushings and pins should be turned when they are 80% worn.
- Measure across three drive teeth when measuring sprocket wear.
- Track pitch is the distance from the center of one pin to the center of the next pin in the track chain and will lengthen as the track chain wears.
- Tracks must be tight when measuring track pitch, which is often achieved by placing a track pin in the chain and propelling the track to wedge the pin between the sprocket and chain.
- When measuring track wear, record the smallest measurement.
- The majority of bushing wear occurs on three sides: vertical, reverse drive side, and forward drive side.
- Undercarriage components wear when there is contact, load, and relative movement. Increased track tension, machine weight, horsepower, abrasive soil conditions, and poor operating habits increase track wear.
- To promote long undercarriage life, owners should choose the narrowest track shoe possible and avoid slipping tracks.
- Caterpillar SystemOne Undercarriages are designed to provide longer life, while reducing maintenance and downtime related to servicing undercarriages, as well as a reduction in operating costs.
- Rubber-track tractors are driven using a friction drive design or a toothed drive wheel design.

Technical Terms

- bogie
- bushing turn
- carrier roller
- carrier undercarriage
- center tread idler
- clamp master link
- crawler tractor
- depth gauge
- drive segments
- drive sprocket
- drive tumbler
- elevated-sprocket undercarriage
- equalizer bar
- exercising the tracks
- fill valve
- ground pressure
- grouser
- guides and guards
- hard bar
- heavy-duty extended life (HDXL) with Duralink track
- heavy-duty track
- high-track undercarriages
- idler
- idler flange
- idler rib
- link height
- link pitch
- low-ground-pressure undercarriage
- master link
- master pin
- multi-grouser
- open-center grouser
- pads
- pivot shaft
- positive-pin-retention track
- recoil mechanism
- relief valve
- rubber track pads
- sealed and lubricated track (SALT) undercarriage
- self-cleaning grouser
- sprocket pitch
- standard grouser
- SystemOne Undercarriage
- track chain
- track frame
- track pitch
- track rollers
- track shoes
- track tensioner
- tractor undercarriage
- ultrasonic measurement tool
- undercarriage

Review Questions

Answer the following questions using the information provided in this chapter.

Know and Understand

1. Cranes, excavators, and hydraulic shovels are equipped with undercarriages designed for all of the following purposes, *EXCEPT*:
   A. support.
   B. stability.
   C. propulsion.
   D. drawbar pull.

2. Technician A states that excavators are equipped with carrier undercarriages. Technician B states that dozers are equipped with tractor undercarriages. Who is correct?
   A. Technician A.
   B. Technician B.
   C. Both A and B.
   D. Neither A nor B.

3. Technician A states that track links are designed as left-hand and right-hand links. Technician B states that track shoes are bolted to the bushings. Who is correct?
   A. Technician A.
   B. Technician B.
   C. Both A and B.
   D. Neither A nor B.

4. Technician A states that pins are pressed into track links. Technician B states that bushings are pressed into track links. Who is correct?
   A. Technician A.
   B. Technician B.
   C. Both A and B.
   D. Neither A nor B.

5. Which undercarriage component is mounted to a yoke and attached to the track tensioner?
   A. Drive sprocket.
   B. Idler.
   C. Track roller.
   D. Carrier roller.

6. Which of the following components transfers the machine's weight evenly onto the track links?
   A. Track roller.
   B. Carrier roller.
   C. Drive sprocket.
   D. Idler.

7. Technician A states that a track roller with both outer and inner flanges is labeled as a *single-flange roller*. Technician B states that track rollers are mounted below the track frame, between the bottom of the track frame and the track shoes. Who is correct?
   A. Technician A.
   B. Technician B.
   C. Both A and B.
   D. Neither A nor B.

8. Which of the following components supports the upper portion of the track and helps align the track's chain?
   A. Track roller.
   B. Carrier roller.
   C. Drive sprocket.
   D. Equalizer bar.

9. Technician A states that multi-grouser track shoes offer increased maneuverability and turning capability compared to standard grouser shoes. Technician B states that undercarriage guards and guides are used to direct and protect rollers, idlers, and sprockets. Who is correct?
   A. Technician A.
   B. Technician B.
   C. Both A and B.
   D. Neither A nor B.

10. Which track shoe design enables material to fall out of the tracks when the shoes separate slightly during the tracks' rotation?
    A. Single grouser.
    B. Multi-grouser.
    C. Open-center grouser.
    D. Self-cleaning grouser.

11. All of the following components work together to maintain track tension, *EXCEPT*:
    A. track tensioner.
    B. track shoes.
    C. front idler yoke assembly.
    D. recoil mechanism.

12. Technician A states that a drive sprocket with five bolt-on drive segments will require splitting the track chain for all undercarriage service. Technician B states that drive sprocket teeth drive every bushing on each revolution of the track chain. Who is correct?
    A. Technician A.
    B. Technician B.
    C. Both A and B.
    D. Neither A nor B.

13. Technician A states that a master pin is permanently pressed into the master bushing and cannot be removed. Technician B states that the master link design for uncoupling a track chain is no longer used by manufacturers in late-model machines. Who is correct?
    A. Technician A.
    B. Technician B.
    C. Both A and B.
    D. Neither A nor B.

14. Technician A states that a ruler should be placed across three teeth when measuring a drive sprocket segment for wear. Technician B states that an ultrasonic measurement tool probe should be positioned on top of a track shoe grouser to measure bushing wear. Who is correct?
    A. Technician A.
    B. Technician B.
    C. Both A and B.
    D. Neither A nor B.

15. Track pitch is measured from the _____.
    A. top rail to the bottom rail of one track link
    B. center of one pin to the end of the next track link
    C. center of one pin to the center of the next pin
    D. leading edge of one pin to the middle of the same track link

16. All of the following must be present to cause undercarriage wear, *EXCEPT*:
    A. contact.
    B. load.
    C. relative motion.
    D. noise.

17. A tractor can be equipped with four different track shoe widths. Which one will exert the highest ground pressure?
    A. 24″.
    B. 28″.
    C. 32″.
    D. 36″.
18. All of the following are rubber track designs, *EXCEPT*:
    A. driven with a quintuple bogie system.
    B. driven with a drive sprocket.
    C. driven by track tension and friction.
    D. non-drive rubber tracks.

Apply and Analyze

19. Rigid frame undercarriages have a(n) _____ that attaches to the tractor's frame and to the front portion of both the left and right track frames.
20. Traditional, non-elevated undercarriages often incorporate a(n) _____ into the drive sprocket assembly and have two diagonal arms or braces positioned on each side of the machine.
21. On pivoting undercarriages, the _____ attaches to the front of both track roller frames and pivots on a center pin that is attached to the tractor's frame.
22. Drive sprockets propel the track by driving the _____ in forward or reverse.
23. A(n) _____ is inserted through the center of a track bushing and can rotate freely in the bushing.
24. Track shoes are bolted to the _____.
25. The _____ ride on the two stepped surfaces on the idler's outer edges.
26. The most common method or tool used for adjusting track tension on the undercarriage is a(n) _____.
27. Ideally, bushing should be turned when they are approximately _____% worn.
28. SystemOne Undercarriages use a(n) _____-shaped link design.

Critical Thinking

29. What is a bushing turn?
30. Explain the proper process used for checking track tension.

Chapter 24
Track Steering Systems

Objectives

After studying this chapter, you will be able to:

- ✓ Describe the types of twin-track turning radiuses.
- ✓ Explain the operation of the three most popular twin-track steering designs.
- ✓ Describe the power flow through a two-speed planetary track steering system.
- ✓ Describe the power flow through an independent-geared twin-track tractor.
- ✓ Describe steering systems for tractors equipped with four rubber tracks.

As explained in Chapter 23, *Undercarriages*, many off-highway machines are equipped with track undercarriage systems. Based on the type of machine, either two- or four-track undercarriages are used on the machine to provide mobility and steering. Machines with four tracks use articulation steering or front-axle Ackerman steering. Chapter 25, *Wheeled Steering Systems*, explains both steering systems in greater detail. Machines equipped with two tracks (twin-track machines) require steering systems that are unique to off-highway machines.

Twin-Track Turning Radiuses

Tractors equipped with twin tracks require the operator to rotate the track undercarriages at different speeds and/or in opposite directions in order to steer the tractor. The method the operator uses to steer the tractor determines how sharp of a turn the tractor performs. Each of the following track steering methods creates a different turning radius:

- Counter rotation.
- Brake steer.
- Power turn.
- Neutral steer.

The operator determines the machine's turning radius by using one of the methods to steer the tractor in a circle. The circle made by the widest point of the machine's undercarriage forms the turning circumference. The distance straight from the edge of the circumference to the center of the circumference is the tractor's *turning radius*. Not all track-type tractors are capable of performing all of the steering methods.

Counter Rotation

Applying the *counter rotation* method, the operator propels one track forward and propels the other track in reverse. See **Figure 24-1**. Not all twin-track tractors are capable of counter rotating. Counter rotation provides the sharpest turn of the four steering methods, with a turning radius that is equal to half the machine's overall length. A machine's maneuverability is maximized with counter rotation steering, but the method accelerates track component wear.

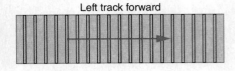

Counter Rotation to the Right

Goodheart-Willcox Publisher

Figure 24-1. When the tracks are propelled in opposite directions, the machine operator is performing a counter rotation turn.

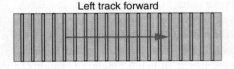

Brake Steer to the Right

Goodheart-Willcox Publisher

Figure 24-2. The brake steer method involves braking either the right or left track while propelling the other track forward or in reverse. The tractor steers to the side with the stationary track.

Figure 24-3. A power turn is completed by propelling both tracks in the same direction, but with one track moving faster than the other track. The tractor steers to the side with the slower track speed.

Brake Steer

The *brake steer*, or *pivot turn*, method has the operator propel one track in either forward or reverse, while holding the other track stationary by applying a steering brake, **Figure 24-2**. The tractor steers toward the side with the stationary track. Brake steer provides the second sharpest turning radius of the four steering methods. The turning radius is slightly wider than the overall length of the tractor. Track-type tractors equipped with steering-clutch and brake systems use brake steering. Steering-clutch-and-brake systems are discussed later in this chapter.

Power Turn

An operator performs a *power turn* by driving both tracks in the same direction, but one track is propelled faster than the other track, **Figure 24-3**. The tractor steers toward the track that is rotating slower. The turning radius is determined by the difference in track speeds and soil conditions. For example, a two-speed planetary track steering system (explained later in this chapter), can drive one track in a fixed high-speed gear ratio and drive the other in a fixed low-speed gear ratio, resulting in a power turn. Also, if one track is operating in a firm soil and the other track is moving in a slicker soil, the machine steers more toward the slicker soil condition.

Neutral Steer

The *neutral steer* method requires the operator to power one track and shift the other track into neutral, **Figure 24-4**. The tractor steers toward the side that has the track shifted into neutral. The resistance of the neutral track determines the length of the turning radius.

Common Track Steering System Designs

The three most common steering systems equipped on tractors with twin tracks are the following:
- Dual-path hydrostatic drive.
- Steering clutch and brake.
- Differential steer.

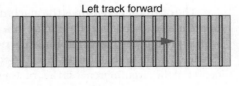

Power Turn to the Right

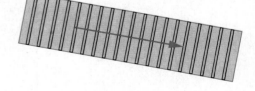

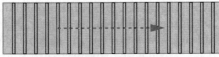

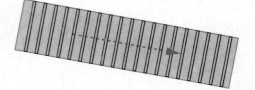

Goodheart-Willcox Publisher

Dual-Path Hydrostatic Drive Systems

Hydrostatic-propelled track-type tractors are equipped with dual-path hydrostatic transmissions. As mentioned in Chapter 13, *Hydrostatic Drives*, dual-path hydrostatic transmissions use two separate hydrostatic transmissions. One pump and one motor are responsible for driving the left track and one pump and one motor are responsible for driving the right track. See **Figure 24-5**. Most late-model hydrostatic-propelled track-type tractors use joystick controls, **Figure 24-6**.

Dual-path transmissions provide forward and reverse propulsion as well as steering. When the left and right hydraulic motors are propelled in the same direction and at the same speed, the tractor moves either forward or in reverse. When the motors are propelled in the same direction but at different speeds, the tractor performs a power turn.

Dual-path hydrostatic transmissions are capable of counter rotating. The operator can perform a counter rotation turn at variable speeds, ranging between a slow counter rotation to a fast counter rotation. Some machines without dual-path hydrostatic transmissions, such as an older Case track loader that is described later in this chapter, are capable of counter rotating, but only at fixed speeds.

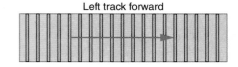

Figure 24-4. The neutral steer method involves shifting one track into neutral and propelling the other track in either forward or reverse.

Dual-Path Hydrostatic Drive Advantages

Hydrostatic-propelled track-type tractors offer the following advantages:

- A simpler power train (two pumps and two motors) compared to a steering-clutch-and-brake tractor power train that normally contains a torque converter, driveshaft, powershift transmission, a bevel gear axle assembly, axle shafts, steering clutches, and brakes.

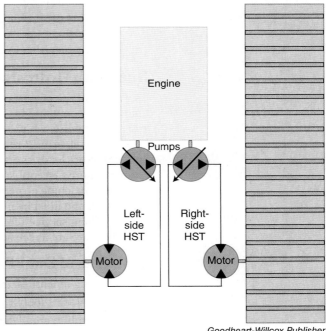

Figure 24-5. Dual-path hydrostatic transmissions use a pump and motor to drive the left track and a separate pump and motor to drive the right track.

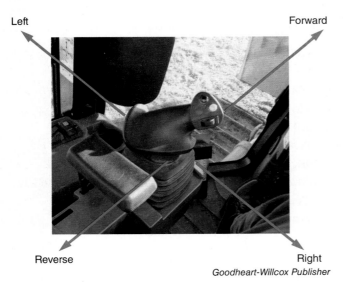

Figure 24-6. This Caterpillar D5K dozer is propelled using the left-hand joystick. The tractor moves in the same direction as the operator directs the joystick: forward, reverse, left, or right.

- Design simplicity for engineers to locate the pump and motor in separate locations. The pumps can be directly coupled to the engine, and the motors can be individually placed at each track, typically splining to a final drive.
- Independent control of each track.
- Infinitely variable track speed controls for the operator to precisely manage propulsion and steering.
- Capable of counter rotation.
- Capable of maintaining a straight forward motion when the machine is digging with just one side of the dozer blade.

Note

When a dozer blade is cutting on just one side, the dozer normally pulls to the same side because of a loss of traction on that side. With hydrostatic steering, the operator can easily steer the tractor to offset the pulling motion and maintain a straight forward motion. A dozer equipped with a steering-clutch-and-brake system does not have the same ability.

- Hydraulic pumps and motors take less space than other designs.
- No gear shifting.
- Hydrostatic braking. As explained in Chapter 13, *Hydrostatic Drives*, when the propulsion levers are reversed, the motor acts like a pump and the pump acts like a motor, resulting in engine braking slowing the machine.

Dual-Path Hydrostatic Drive Disadvantages

As with any type of system, dual-path hydrostatic transmissions also have some disadvantages. First, compared to the other track-type steering systems, hydrostatic propulsion generates more noise, which may cause problems for operators. Second, any time the machine is moving or steering it requires drive pressure, which can be very high pressure depending on the tractor's load. As a machine's service hours accumulate, the hydraulic drive lines and hoses are more susceptible to leaks due to the constant high pressure. Lastly, and perhaps the costliest to machine owners, is that hydrostatic transmissions are very sensitive to fluid contamination and must be maintained with extreme cleanliness in order for the transmission to have a long service life. Although steering-clutch-and-brake systems must also be kept clean, they are not as sensitive to contamination as hydrostatic transmissions.

Examples of Dual-Path Hydrostatic Track-Type Tractors

The following machines are examples of hydrostatic-propelled track-type tractors:
- Case dozers (650M, 750M, 850M, 1150M, 1650M, 2050M).
- Caterpillar small dozers (D3K2, D4K2, D5K2, D6K2, D1-D4).
- Caterpillar track loaders (953, 963, 973K).
- John Deere dozers (450K, 550K, 650K, 700L, 750L, 850L, 950K, 1050K).
- John Deere track loaders (655K, 755K).
- Komatsu dozers (D37, D39, D51, D61 and D71 includes EX, PX, EXi, PXi,-24 models).
- Liebherr dozers (PR 716, PR 726, PR 736, PR 746, PR 756, PR 766, PR 776).
- Liebherr track loaders (LR 626 and LR 636).

Note

The Liebherr PR776 is one of the highest-horsepower HST dozers. It has an elevated-sprocket track and produces 757 hp.

Steering-Clutch-and-Brake Systems

A tractor with a ***steering-clutch-and-brake system*** is normally equipped with a torque converter and powershift transmission. **Figure 24-7** shows the basic power flow through a large Caterpillar steering-clutch-and-brake dozer as the power moves along the following path:

1. Engine.
2. Torque divider (torque converter).
3. Driveshaft.
4. Transmission (operator selects speed and direction).
5. Transfer gears.
6. Bevel gears.
7. Inner axle shafts.
8. Steering clutches and brakes.
9. Outer axle shafts.
10. Double reduction final drives.

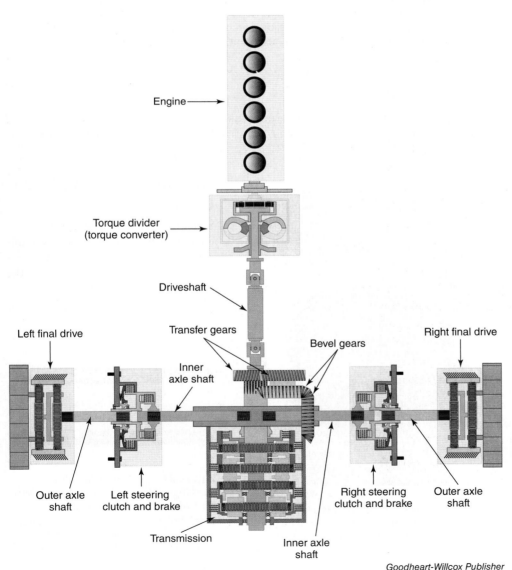

Figure 24-7. The torque divider sends power through a driveshaft to the rear of the machine where the transmission is located. The design of the transmission input and output shafts allows power to enter through the input shaft and exit via the output shaft to the transfer gear set, which sends power to the bevel gear set.

Goodheart-Willcox Publisher

As mentioned, a hydrostatic-propelled track-type tractor has a much simpler system design and can eliminate items 2 through 9.

In Chapter 9, *Planetary Gear Set Theory*, the power flow through a Caterpillar three-speed dozer transmission is explained. Once the operator selects the transmission speed and direction, the power exits the transmission and is delivered to two helical transfer gears, which deliver power to a bevel gear set. See **Figure 24-8**. The bevel ring gear drives the hollow shaft to which the inner axle shafts are internally splined.

Both inner axle shafts transfer power to the accompanying left and right steering clutches and brakes. Based on how the operator moves the dozer's drive controls, power moves through and then exits the steering clutches and brakes via the outer axle shafts. The outer axle shafts deliver the power to the right and left final drives, which function as the mechanisms driving the track undercarriages on both sides.

Pro Tip

There are instances when a technician must remove the axle shafts so the tractor can be towed. For example, if the engine or transmission is inoperative and the machine needs to be moved into the shop, such a procedure would be necessary. On a Caterpillar steering-clutch-and-brake dozer with an elevated sprocket, the axles (**Figure 24-9**) are removed through the center of the final drives. A slide hammer can be used to remove the axle shafts. Be sure to follow the manufacturer's service procedures.

Warning

The tractor's parking brake is no longer effective once the axle shafts are removed. If the dozer is positioned on a slope, the tractor will roll down the slope if the axle shafts are removed.

Steering-Clutch-and-Brake Controls

Steering-clutch-and-brake tractors commonly have left and right mechanical levers (or electronic fingertip control (FTC) sensors) for steering the dozer. See **Figure 24-10**.

Figure 24-8. The transfer gear and bevel gear sets are incorporated into the transmission housing. After the transmission speed and direction are selected by the operator, power is sent to the transfer gear set, which sends power to the bevel gear set. The bevel ring gear drives the shaft that mates with the two inner axles. Only one of the two transfer gears is visible in the photo.

Figure 24-9. A Caterpillar D6R steering-clutch-and-brake tractor with an elevated sprocket has two inner axle shafts and two outer axle shafts that are both removed through the center of the final drives. A—Inner axle shaft. B—Outer axle shaft.

Chapter 24 | Track Steering Systems

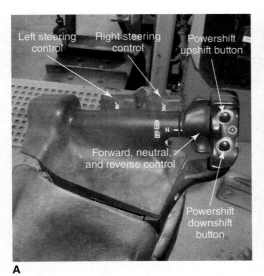
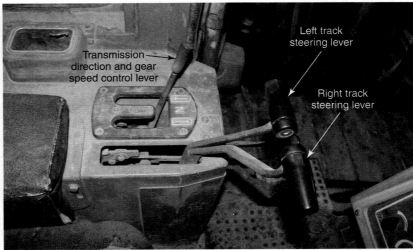

Figure 24-10. A—This D6R steering-clutch-and-brake control consists of an electronic fingertip control. B—This D5H steering-clutch-and-brake dozer has mechanical control levers the operator must use to move the machine.

The operator selects the transmission controls for the direction and speed, such as forward first gear. The positions of the steering-clutch-and-brake controls dictate if the tractor will be propelled in a straight line or a steered direction. For the example in **Figure 24-11**, the steering-clutch-and-brake controls have just three positions.

If both steering levers remain all the way forward, the tractor travels straight in one direction, either forward or in reverse depending on the transmission control lever. If one steering lever is pulled halfway or fully backward, the tractor steers. See the table in **Figure 24-12** and the mechanical control levers in **Figure 24-13**.

Steering Lever Position	Steering Clutch Status	Steering Brake Status
Completely forward	Engaged	Released
Halfway back	Released	Released
Fully backward	Released	Engaged

Figure 24-11. The status of the steering clutches and brakes on the left or right side of the machine is determined by the position of the corresponding steering control lever.

Transmission in Forward 1st Gear		
Type of Steer	Left Steering Lever Position	Right Steering Lever Position
Left gradual steer	Halfway back	Completely forward
Left sharp steer	Fully backward	Completely forward
Right gradual steer	Completely forward	Halfway back
Right sharp steer	Completely forward	Fully backward
Forward (straight path)	Completely forward	Completely forward

Figure 24-12. The operator's movement and positioning of the steering levers determine the directional steering of the tractor.

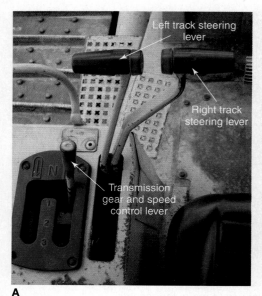 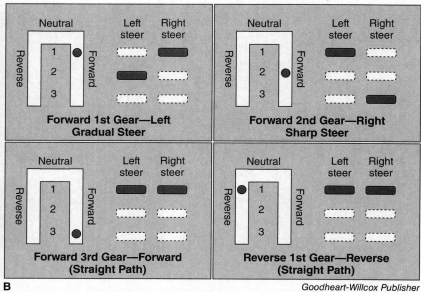

Figure 24-13. An example of an older Caterpillar dozer with a mechanical transmission and steering control levers. Note the position of the transmission lever inside the C-shaped slot and the position of the two steering levers. A—The mechanical levers used to control the transmission and the steering clutches and brakes. B—A diagram of different transmission and steering control positions and their effects.

Note

On late-model steering-clutch-and-brake tractors, the proportion the steering control levers are moved determines the proportion of fluid pressure applied to or released from the steering clutches and brakes. For example, if the steering lever is gradually pulled rearward, the steering clutch pressure drops slightly and causes the steering clutch to slip.

Steering-Clutch-and-Brake Operation

When a steering-clutch-and-brake tractor is moving forward or in reverse in a straight path, both steering clutches will be engaged and both steering brakes will be released. The steering-clutch-and-brake designs vary by manufacturer but typically have the following features:

- The steering clutches are normally a multiple-disc design. The steering clutch can be spring applied and hydraulically released, or hydraulically applied and spring released.
- The steering brake can consist of a large band and rotating drum or can be a multiple-disc design.
- The multiple-disc brake can be spring applied and hydraulically released, or can be hydraulically applied and spring released.
- The steering brake can also be used for both the parking brake and the service brake.

An example of a Caterpillar D9R steering clutch and brake is shown in **Figure 24-14**. The steering clutch is hydraulically applied and the brake is spring applied and hydraulically released. The inner axle shaft drives the steering clutch's input hub. When fluid pressure is applied to the steering clutch pressure cavity, the steering clutch engages, which causes the steering clutch drum to deliver power to the steering brake (output hub). When fluid pressure is applied to the steering brake pressure cavity, the brake releases, which allows the brake's output hub to drive the outer axle shaft.

In order for the tractor to move, oil pressure must be supplied to both the steering clutch to engage it and the steering brake to release it. Oil pressure measurements to the steering clutch and brake based on the steering control levers and foot brake pedal are shown in **Figure 24-15**.

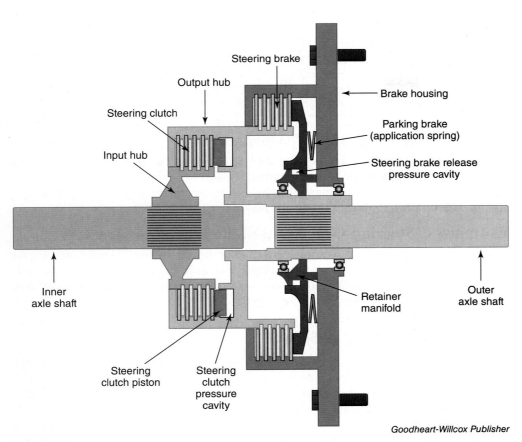

Figure 24-14. A simplified illustration of a Caterpillar D9R steering-clutch-and-brake mechanism. The clutch is hydraulically applied and the brake is spring applied.

On a traditional steering-clutch-and-brake tractor, the machine can be held stationary by three different means:
- Parking brake applied.
- Service brake applied.
- Both steering levers simultaneously pulled fully backward.

Tractor Operation	Steering Clutch	Steering Brake
Forward or reverse (straight path)	Full pressure	Full pressure
Neutral steer	0 psi	Full pressure
Brake steer	0 psi	Decreases to 50 psi Brake fully applied
Foot brake applied Steering levers forward	Full pressure	Decreases to 0 psi Brake fully applied Torque converter in stall mode

Figure 24-15. Fluid pressures applied to or released from the steering clutch and steering brake are listed based on the position of the steering levers and foot brake pedal. If the tractor is brought to a stop by pulling both steering levers fully backward, the torque converter will not enter stall mode because the steering clutches will be released. However, if the steering levers are moved completely forward and the service brake pedal is pressed, the torque converter will enter stall mode because the steering clutches are engaged and the brakes are applied.

Caution

The steering brakes can be used for both a parking brake and service brakes. As described in Chapter 12, *Hydrodynamic Drives*, a dozer torque converter should not be stall tested in a low transmission gear because the tractor often has enough torque to drive through the parking brake.

Steering-Clutch-and-Brake Disadvantages

Steering-clutch-and-brake tractors must slow or completely stop one track in order to steer. They are not capable of counter rotating. Hydrostatic-propelled tractors and differential-steered tractors have power applied to both tracks during a turn, which allows the tractor to maintain a constant speed while making a turn.

Examples of Steering-Clutch-and-Brake Track-Type Tractors

Some manufacturers still use steering clutch and brakes for controlling track steering on their highest horsepower tractors. Caterpillar's two largest dozers, the D10T2 and D11, have steering clutches and brakes. Komatsu's two largest dozers, the D375A-8 and D475A-8, also have steering clutches and brakes equipped.

Differential Steering Systems

Multiple manufacturers offer a twin-track steering system that is often referred to as *differential steering*. A differential steering system does not contain a traditional differential. It instead uses planetary gear sets, bevel gears, spur or helical gears, and a reversible hydraulic motor to provide variable power to twin-track tractors. The tractor's operator can infinitely vary the speed between the right and left tracks. As the operator steers the tractor, the steering differential increases the speed of one track and proportionately reduces the speed of the other track. Tractors equipped with differential steering are capable of counter rotation. See **Figure 24-16** for examples of tractors with differential steering systems. Although differential steering system designs and nomenclature vary, some similarities do exist between the different systems.

Differential Steering Operation

In a differential steering system, the steering differential receives either one or both of the following inputs. The first is a hydraulic steering motor input, which determines the direction of the turn (left or right) and how sharp of a turn is made. The second is a CVT, powershift, or electric drive motor, which determines the speed and the direction of the tractor's propulsion (forward or reverse). The steering differential delivers uninterrupted power to the tracks through two output shafts, a right and left axle shaft, which drive a right and left planetary final drive respectively.

Manufacturer	Tractor	Steering Nomenclature
AGCO	MT Challenger and Fendt twin-rubber-track tractors	Differential steering
Caterpillar	D6, D7, D8GC, and D9T dozers	Differential steering
Case New Holland Industrial (CNHi)	Case 1650K and 1850K dozers Fiat Kobelco D180 New Holland DC150 and DC180	Power steer
John Deere	8RT and 9RT twin-rubber-track tractors	Differential steering
Komatsu	D65EX-18, D85EX-18, and D155AX-8 dozers	Hydrostatic steering system (HSS)

Goodheart-Willcox Publisher

Figure 24-16. Examples of differential-steered twin-track tractors.

When the transmission delivers the only input into the steering differential, the tractor drives forward or in reverse in a straight path. When the hydraulic steering motor delivers the only input into the steering differential, the tractor counter rotates to the left or to the right. If the steering differential receives input from both the transmission and the hydraulic steering motor, the tractor moves forward or in reverse and steers either to the left or right.

AGCO and Caterpillar Differential Steering

The differential steering systems found in AGCO Challenger tractors, Fendt tractors, and Caterpillar dozers are very similar. The steering differential contains three planetary gear sets that are connected with a sun gear shaft, also known as the inner axle shaft. See **Figure 24-17**.

One outer planetary gear set, known as the equalizing planetary gear set, uses a fixed ring gear. The other outer planetary ring gear, known as the steering planetary gear set, is either held or driven by the hydraulic steering motor. The middle planetary gear set, also called the drive planetary gear set, has the planetary carrier driven by the powershift transmission. The middle planetary ring gear is connected to the steering gear set's planetary carrier. All three of the planetary gear sets are connected through the sun gear shaft. Late-model AGCO twin-rubber-track tractors use CVTs instead of powershift transmissions.

When the machine travels in a straight path, the hydraulic steering motor holds the steering ring gear stationary. As the transmission drives the middle gear set, power is distributed equally to the left and right axle shafts and on to both drive sprockets.

If the operator chooses to steer the tractor when it is driving forward or in reverse, the hydraulic steering motor rotates, either clockwise or counterclockwise. The rotation of the steering motor causes one axle shaft to increase speed and the other axle shaft to proportionately decrease speed. The tractor begins to turn in the direction of the track that is moving slower.

If the operator chooses to steer the tractor while the transmission is in neutral, one axle shaft spins in one direction and the other axle shaft spins in the opposite direction, which causes the tractor to counter rotate.

Large Caterpillar dozers equipped with differential steering have the double-reduction final drives shown in **Figure 24-17**. The Challenger and Fendt twin-rubber-track tractors, however, have a single-reduction inboard planetary final drive.

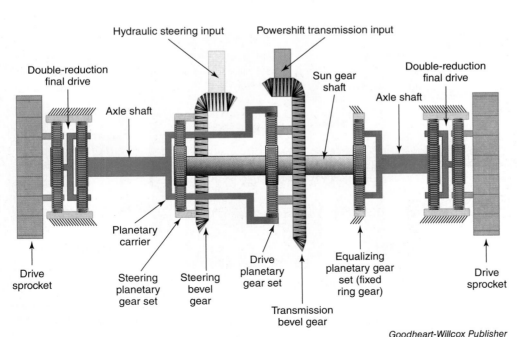

Figure 24-17. A differential steering system used on large Caterpillar dozers. The D7, D8GC, and D9T use double-reduction final drives as explained in Chapter 17, *Final Drives*.

Figure 24-18. A Caterpillar differential steering control joystick is called a steering tiller. The top push button is for upshifting the powershift transmission, and the lower push button is for downshifting the powershift transmission.

Caterpillar Differential Steering Joystick Controls

Caterpillar differential steering controls consist of a single joystick called a *steering tiller* for controlling the machine's direction, speed, and steering. Refer to **Figure 24-18**.

When the operator presses the tiller forward, the tractor steers to the left. The tractor steers to the right when the operator pulls the tiller backward. When the operator rotates the steering tiller grip forward, the tractor is propelled forward. The tractor is propelled in reverse when the operator rotates the tiller grip backward.

Caterpillar Electric-Drive Dozers

Caterpillar D7E and D6XE dozers use electric drives to propel the machine, which is further explained in Chapter 26. Both dozers use differential steering with three planetary gear sets tied together with the common sun gear shaft. The D7E uses two electric motors to drive a single bull gear that drives the planetary carrier to provide the propulsion input into the differential steering gear set. The D6XE uses one electric motor to directly drive the planetary carrier to provide the propulsion into the differential steering gear set. See **Figure 24-19**. A hydraulic motor provides the steering input to the differential steering system.

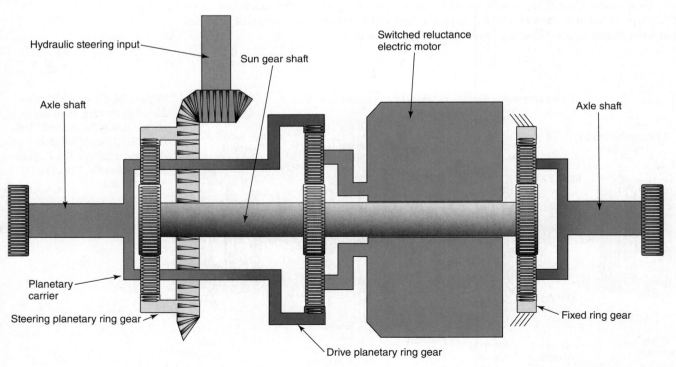

Figure 24-19. A Caterpillar D6XE dozer uses a single switched-reluctance electric motor to drive the planetary carrier for propelling the tractor forward and rearward.

Case, New Holland, and Fiat Power Steering

The Case 1650K and 1850K (Series I, II, III), New Holland DC150, DC150.B, DC180, and DC180.B, and Fiat Kobelco D150 and D180 dozers are equipped with ***power steering***, which is the term Case New Holland industrial (CNHi) uses to describe differential steering. Note that later CNHi model dozers, such as Case 750M–2050M dozers, and the New Holland D125C, D150B, D150C, D180B, and D180C dozers are equipped with dual-path hydrostatic drives rather than differential steering.

The steering differential receives input from the powershift transmission through the bevel gear set. The bevel gear set delivers input to the left and right ring gears. The hydraulic steering motor is linked to both sun gears through a series of spur gears and external-toothed shafts. Output to the left and right axles is provided by the planetary carriers. See **Figure 24-20.**

The hydraulic steering motor drives the sun gear short shaft that in turn drives the right sun gear. The sun gear short shaft also drives the sun gear long shaft, which acts like a reverse idler causing the sun gear long shaft to reverse the direction of the left sun gear.

The design of the sun gear short shaft meshing with the sun gear long shaft results in the hydraulic steering motor driving the right sun gear in one direction and the left sun gear in the opposite direction, causing the tractor to steer. When the hydraulic steering motor reverses its direction of rotation, the tractor steers in the opposite direction.

When the tractor is driving forward or in reverse along a straight path, the hydraulic steering motor holds the two sun gears stationary, which causes the planetary carriers to drive both tracks forward or backward at the same speed. When the hydraulic steering motor is the only input into the steering differential, one sun gear is driven clockwise and the other sun gear is driven counterclockwise, causing the tractor to counter rotate. The direction of counter rotation is determined by the direction of the hydraulic steering motor's rotation. When both the transmission and the hydraulic steering motor are delivering power into the steering differential, the tractor moves and steers simultaneously.

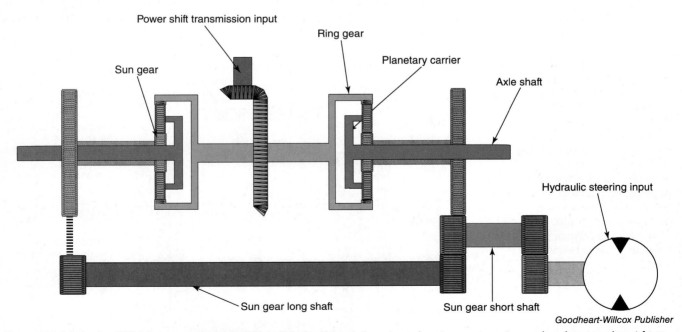

Goodheart-Willcox Publisher

Figure 24-20. In the CNHi "power steering" differential steering system, two planetary gear sets receive ring gear input from the powershift transmission. The two sun gears receive input from the hydraulic steering motor. Output to the left and right axle shafts is delivered by the planetary carriers. The axle shafts drive a bull-and-pinion final drive that also drops down the power flow to the drive sprocket commonly used on low-sprocket crawlers (non-elevated dozers) as explained in Chapter 17, *Final Drives*.

John Deere Twin-Rubber-Track Differential Steering

John Deere manufactures twin-rubber-track tractors in their two largest agricultural horsepower applications, the 8RT and 9RT tractors. The twin-rubber-track tractors are equipped with differential steering, **Figure 24-21**.

The steering differential receives input from the transmission via the transmission output shaft and a drive gear and driven gear. The driven gear propels the bevel pinion gear, which propels the bevel ring gear. The bevel ring gear drives the bevel gear shaft, which drives the two planetary carriers. If the machine is propelled in forward or reverse along a straight path, the planetary carriers drive the two ring gears in the same direction (either forward or reverse). The hydraulic steering motor is linked to the two planetary sun gears, which are held stationary by the hydraulic steering motor while the tractor is moving on a straight path in forward or reverse.

When the operator steers the tractor, the hydraulic steering motor drives one planetary sun gear clockwise and the other sun gear counterclockwise. For example, if the hydraulic steering motor rotates clockwise, it causes the steering cross shaft to rotate counterclockwise. The steering cross shaft drives the reverse idler clockwise, which then drives the left sun gear counterclockwise. The steering motor also drives a pair of external-toothed gears, similar to a cluster gear, counterclockwise. The cluster gear drives the right sun gear clockwise. If the hydraulic steering motor is the only input into the steering differential, the tractor will counter rotate.

The planetary ring gears are the right and left output members of the steering differential. The ring gears directly drive the final drive sun gears, which are the inputs to the right and left inboard single-reduction final drives. The original 8000T tractors through the 8030T series tractors, however, have outboard single-reduction final drives. See **Figure 24-22**.

Komatsu Hydrostatic Steering System

Komatsu manufactures a dozer that uses a differential steering system that is labeled as a *hydrostatic steering system (HSS)*. The Komatsu hydrostatic steering system and the differential steering system in John Deere twin-rubber-track tractors have a similar design, in that the pinion and ring bevel gear set drives a bevel gear shaft, which drives both planetary carriers. The hydraulic steering motor also controls the steering planetary sun gears.

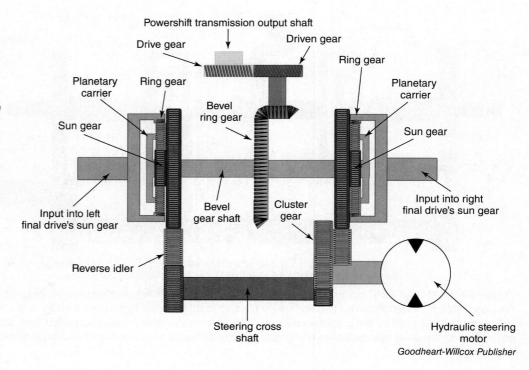

Figure 24-21. Deere twin-rubber-track tractors use differential steering. The transmission is responsible for driving the planetary carriers. The hydraulic steering motor is responsible for holding or driving the planetary sun gears via the large external-toothed ring gears. The planetary ring gears (internal-toothed ring gears) are the outputs that drive the input to the final drives.

Goodheart-Willcox Publisher

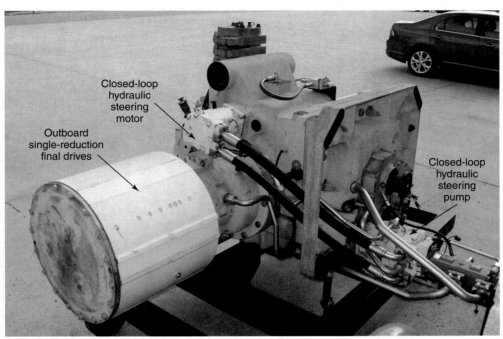

Figure 24-22. This differential steering axle is used on John Deere 8000T–8030T series tractors. Note the outboard final drives and the hydraulic steering motor. The dedicated closed-loop hydraulic pump and motor provide input to the axle any time the tractor is steering. Later-model John Deere 8RT tractors use inboard planetary final drives, which provide a more accessible drive axle for easier repositioning of the tracks when their spacing needs to be changed.

On the D275AX-5 Komatsu dozer, the largest Komatsu differential-steered dozer, the left and right service brakes can be used as steering brakes if the operator actuates a *turn mode switch*. The right or left brake is applied when the differential steering is unable to make the dozer complete a tight turn. See **Figure 24-23**.

The hydraulic steering motor drives a pinion gear that meshes with gear A, which is in mesh with gear B and gear C. Gear C is connected to the same shaft that has gear D on its opposite end. Gear D is in mesh with gear E. Gears B and E are coupled to the left and right sun gears respectively.

If, for example, the operator steers the dozer to the right, the hydraulic steering motor and its pinion gear rotate clockwise. The pinion gear causes gear A to rotate counterclockwise. Gear A drives gear B, gear C, the transfer shaft, and gear D clockwise. Gear D rotates gear E counterclockwise. This power flow through the gears ends with the left sun gear rotating clockwise and the right sun gear rotating counterclockwise to power their respective planetary carriers and final drives to turn the machine to the right.

Hydraulic Steering Motor

At the heart of a differential steering system is a reversible hydraulic motor that provides an input into the steering differential any time the operator directs the tractor to steer. Many differential-steered tractors use a stand-alone hydraulic pump and hydraulic motor that are dedicated solely for providing a hydraulic input to the steering differential.

If the hydraulic schematics for a differential steering system and a traditional closed-loop hydrostatic transmission (HST) are compared, the two will look identical. However, the hydraulic steering pump and motor in the differential are not used for propulsion, but are used only when the tractor is being steered. The HST closed-loop reverse and forward drive lines mirror the steer left and steer right pressure lines in the differential steering system.

The first Caterpillar differential-steered dozer, the D8N, did not use a closed-loop hydraulic steering pump and motor. The system obtained steering oil from the implement pump and used a directional-control valve for controlling the direction of oil flow to the reversible hydraulic motor. Case New Holland differential-steered tractors also do not use a stand-alone closed-loop steering pump and motor, but use oil flow from the implement pump.

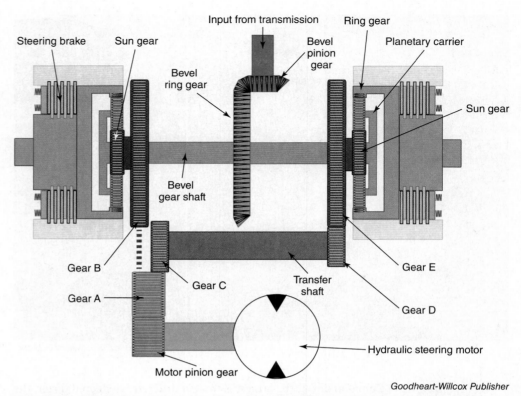

Figure 24-23. The Komatsu hydrostatic steering system is a differential steering system similar to the system equipped in a John Deere twin-rubber-track tractor. Note that on this illustration of the D275AX-5 dozer's system, the dashed line between Gears A and B indicates the two gears are in mesh.

Differences between Hydrostatic Transmission and Differential Steer

Hydrostatic-propelled dozers share similarities with differential-steered tractors. They both feature variable track control to deliver uninterrupted power to each track. Both systems can also counter rotate at various speeds.

The hydrostatic-propelled tractor uses dual-path controls, which means the right and left track are controlled independently by their own hydrostatic pump and motor. The steering differential has the left and right tracks tied together through a gear setup. When the operator commands the differential-steered tractor to steer, the differential simultaneously increases the speed of one track and decreases the speed of the other track. Late-model hydrostatic-propelled dozers appear to operate similarly to the differential steer tractors, but since they use two separate hydrostatic transmissions, they require an electronic control module to simultaneously increase the speed of one track while slowing the speed of the other track. The differential-steered tractor also uses a CVT, EVT, electric drive motor, or a powershift transmission. When equipped with a powershift transmission the operator will feel increases and decreases in machine speed as the transmission is upshifted and downshifted.

Differential Steering Advantages

Some machine owners prefer the powershift transmission over a hydrostatic transmission, and some tractors equipped with differential steering give owners that option. Differential-steered tractors provide excellent maneuverability.

In a differential steering system, the hydraulic steering motor is not operating at a high pressure when the machine is driving in a straight direction. The hydraulic steering motor only requires higher pressures when the tractor is steering. The differential steering system also has a longer service life compared to steering clutches and brakes that wear over time.

Hydrostatic Steering Terminology

This chapter has discussed different types of hydraulic controls as applied to track steering systems, specifically dual-path hydrostatic-propelled tractors and differential-steered tractors. The term *hydrostatic steering* could be used to describe both of these examples. In addition, manufacturers of off-highway equipment also use the term *hydrostatic steering* to describe a traditional tractor's hydraulic steering system. The three different usages of the phrase "hydrostatic steering" can be easily confused. **Figure 24-24** shows a comparison of hydrostatic steering terminologies based on three different steering applications.

Note

Caterpillar is unique in that they manufacture a tractor model line with each of the three different types of track steering systems:

- Hydrostatic-drive tractors (small dozers, D1–D4, and loaders, 953, 963, and 973K).
- Steering-clutch-and-brake tractors (large dozers, D10T2 and D11).
- Differential-steering tractors (midrange dozers, D6, D7, D8GC, D9T).

Criteria	Hydrostatic Steering (AKA Hydraulic Steering)	Dual-Path Hydrostatic Transmission and Steering	Differential Steering
Purpose of the "hydrostatic steering" system	Supply hydraulic power solely for the use of controlling a tractor's steering and is unrelated to the machine's propulsion system.	A hydraulic pump and a hydraulic motor are used to drive each side of the machine, which controls both the machine's steering and propulsion.	A hydraulic motor provides the steering input into the steering differential. It can be a closed-loop system, or a traditional implement pump that feeds a reversible hydraulic motor.
Components	• Hydraulic pump • Steering control valve • Steering cylinder	• On left side, variable-displacement, reversible hydraulic pump and a reversible hydraulic motor. • On right side, variable-displacement, reversible hydraulic pump and a reversible hydraulic motor.	Two inputs (transmission and a hydraulic motor) deliver power into a planetary gear system. The hydraulic motor can be driven by a dedicated closed-loop hydraulic pump or by the implement's hydraulic system.
Propulsion	Propulsion is completely unrelated to the hydrostatic steering. The machine can use a mechanical transmission, hydrostatic drive transmission, or an electric drive.	Dual-path hydrostatic transmissions. One pump and one motor on each side of the machine drive the tracks or drive wheels providing both propulsion and steering.	The transmission is solely responsible for the machine's propulsion.
Machine Examples	• Traditional agricultural tractor • Combine • Off-road haul truck • Motor grader • Forklift • Loader backhoe • Wheel loader • Tractor scraper	Twin-track tractor examples: • Dozers • Track loaders • Tracked skid steers Wheel-type tractor examples: • Skid steers • Ag swathers	• Dozers • Agricultural rubber-twin-track tractors Case 1650K calls their differential steering system "power steering." Komatsu calls it a "hydrostatic steering system (HSS)."

Goodheart-Willcox Publisher

Figure 24-24. The term *hydrostatic steering* can be used by manufacturers to describe three different steering applications in off-highway equipment: traditional hydraulic steering, dual-path hydrostatic drive steering, and differential steering.

Two-Speed Planetary Steering Systems

In 1947, International Harvester manufactured a machine with the first two-speed planetary steering system, the TD24 dozer. In 1982, Dresser purchased the construction machinery division of International Harvester. Today, the two-speed planetary dozer is manufactured by Dressta. Examples of Dressta dozers equipped with the two-speed planetary steering system include the following models: TD-14, TD-15M, TD-20M, TD-25, and TD-40.

Two-Speed Planetary Steering Operation

In a two-speed planetary steering system, the transmission delivers power to the bevel pinion and ring gear set. The bevel gear set delivers power to the right and left steering system. The operator steers the tractor by changing each track's speed with the tractor's steering system controls. Each track can be placed in one of the following four steering gear configurations:

- High speed.
- Low speed.
- Neutral.
- Brake.

Each steering system contains a combination planetary gear set. The bevel gear set drives a carrier assembly. The carrier assembly contains a set of cluster gears. Each cluster gear is a single gear assembly that has different-sized pinions that rotate all at the same speed and in the same direction. The gear set does not contain a ring gear. See **Figure 24-25**.

Three shafts are located inside the steering gear assembly. The outermost shaft is splined (coupled) to the low-speed sun gear, which is the sun gear with the medium-size diameter. A disc is coupled to this shaft and is held stationary when the operator shifts the track into low speed.

The middle shaft is splined to the high-speed sun gear, which is the sun gear with the smallest diameter. A disc is coupled to this shaft and is held stationary when the operator shifts the track into high speed.

The inside shaft is the output shaft that is driven by the sun gear with the largest diameter, called the pinion drive gear. The output shaft delivers power to the tractor's final drive, which consists of a pinion bull gear set. The inside shaft (output shaft) also contains a disc that can be held stationary by a steering brake.

Any time the transmission is in gear, the planetary carrier is driving forward or in reverse. Either the high-speed disc or the low-speed disc must be held in order for the planetary pinion drive gear to propel the output shaft. When either the high- or low-speed disc is held, it causes the carrier to react and drive the pinion drive gear, which propels the output shaft.

The original two-speed planetary steering system design used steering levers as controls. The position of the steering levers determines the following factors:

- If the high-speed disc is held (high).
- If the low-speed disc is held (low).
- If no speed discs are held (neutral).
- If no speed discs are held and the output shaft is held (brake steer).

Modern machines equipped with this system have a left-hand joystick that controls the tractor's propulsion and steering. The two-speed planetary steering system is not capable of counter rotation. When one track steers in high speed and the other in low speed, the machine is driving with two powered tracks, as opposed to a steering-clutch-and-brake tractor that must remove power from one track in order to steer. The dozer can also be steered by placing a track in neutral or applying the steering brake.

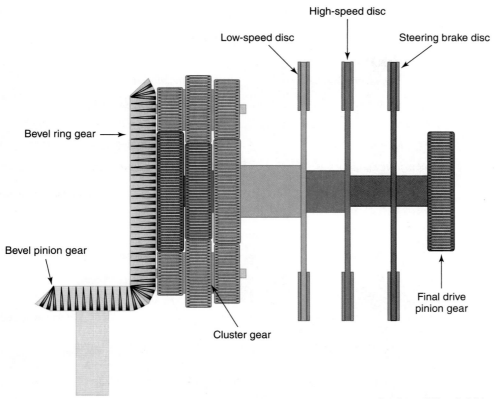

Figure 24-25. In a two-speed planetary steering system, the machine's transmission powers the ring and pinion bevel gear set. The low-speed disc holds the low-speed sun gear. The high-speed disc holds the high-speed sun gear. The steering brake disc holds the track stationary. Both the right- and left-track drives each contain a two-speed gearing system, both driven by the bevel pinion gear.

Goodheart-Willcox Publisher

Independent-Geared Lever-Operated Track Steering Systems

Older Case 1150B-E and 1450B track loaders and dozers use five separate control levers to control the machine speed, direction, and steering. See **Figure 24-26**. The tractors also have independent service brake pedals.

The tractor's propulsion and steering are achieved through the use of powershift-type clutches via countershafts and planetary gear sets, steering brakes, and bevel gears sets. **Figure 24-27** shows the order of power flow through the track steering system.

The operator can use different combinations of lever and brake pedal positions to operate the machine. The complexity of these controls makes it easy to see why late-model hydrostatic drive or differential-steered tractors use much simpler joystick controls.

Independent-Geared Lever-Operated Track Steering Operation

The low-range clutch and high-range clutch are located on the central shaft that receives power from the torque converter via the driveshaft. The two clutches control a single planetary gear set. The ring gear is the input. The planetary carrier is the output that drives the forward and reverse drive gear assembly. Refer to **Figure 24-28**.

The low-range clutch engages in the low-speed range, which results in the sun gear being held and causes the planetary carrier to drive the forward and reverse drive gear assembly in a torque multiplication mode. In high-speed range, the high-range clutch locks

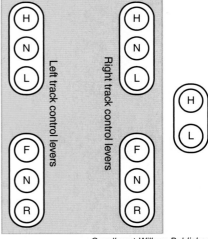

Goodheart-Willcox Publisher

Figure 24-26. This diagram of Case 1150B-E and 1450B track loader and dozer controls shows how the five individual track control levers are used to operate the tractor's forward and reverse propulsion, tractor speed, and steering.

Figure 24-27. The order of power flow through a Case 1150B-E and 1450B track loader and dozer.

1	Engine	
2	Torque converter	
3	Machine high or low clutch	
4	Left-hand forward or reverse clutch	Right-hand forward or reverse clutch
5	Left-hand low or high clutch	Right-hand low or high clutch
6	Left-hand brake and bevel gear shaft	Right-hand brake and bevel gear shaft
7	Bevel gear and final drive	Bevel gear and final drive
8	Left track drive sprocket	Right track drive sprocket

Goodheart-Willcox Publisher

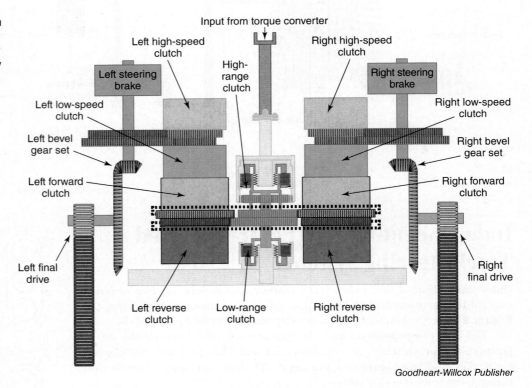

Figure 24-28. The powertrain of a Case 1150B-E and 1450B track loader and dozer uses countershafts, planetary gear sets, clutches, brakes, bevel gear sets, and final drives to control direction, speed, and steering.

Goodheart-Willcox Publisher

the planetary carrier and ring gear into a direct drive mode, resulting in the carrier driving the forward and reverse drive gear assembly in a direct drive 1:1 ratio.

The left and right tracks each have their own forward and reverse drive clutches and their own low-speed clutch and high-speed clutch. The forward and reverse drive gear assembly located on the central shaft has the responsibility of delivering power to the right and left track clutches. (The central shaft and left side of this system were shown in detail in **Figure 17-11**.)

How the forward and reverse drive gear on the central shaft meshes with the left and right track clutch shafts is unique. The forward and reverse drive gear assembly rotates counterclockwise and is in mesh with the reverse driven gear on the left track clutch shaft, causing it to rotate clockwise. The left track reverse driven gear is also in mesh with the right track clutch shaft reverse driven gear and causes it to rotate counterclockwise. When either of the left or right reverse clutches is engaged, it causes its clutch shaft to rotate. The clutch shaft delivers power to the track low- and high-speed clutch drums.

The forward and reverse drive gear assembly (rotating counterclockwise) is in mesh with the forward driven gear on the right track clutch shaft, causing the right forward driven gear to rotate clockwise (opposite of the right reverse driven gear). The right track forward driven gear is also in mesh with the left track forward driven gear and causes it to rotate counterclockwise (opposite of the left reverse driven gear). When either of the track forward clutches is engaged, it causes its clutch shaft to rotate and delivers power to the track low- and high-speed clutch drums.

Each left and right track clutch shaft has its own low- and high-speed clutches which receive power from the rotating drums. When the low- or high-speed clutch is engaged, it delivers power to a low- or high-speed drive gear that is responsible for driving a low- or high-speed driven gear on the pinion shaft. The right and left pinion shafts have a steering brake and a spiral bevel pinion gear. The spiral bevel pinion gear is in mesh with a spiral bevel ring gear. The spiral bevel ring gear delivers power to the pinion and bull gear final drive assembly.

Tractors Equipped with Four-Track Undercarriages

Machine manufactures produce three types of machines with four rubber tracks: articulated tractors, rigid frame four-wheel steer tractors, and MFWD tractors. None of these tractors use twin-track steering as described in this chapter.

Four-Rubber-Track, Articulated-Steering, Track-Type Tractors

In 1996, Case IH introduced a rubber-track tractor with four individual-track undercarriages, called the Quadtrac, **Figure 24-29**. The tractor is similar to other large four-wheel-drive agricultural tractors that use hydraulic articulated steering, which is explained in Chapter 25, *Wheeled Steering Systems*. Examples of other manufacturers who produce an articulated rubber-track tractor are John Deere, New Holland, and Versatile. John Deere manufactured a 764 HSD high-speed dozer with four rubber tracks and continues to manufacture an agricultural 9RX four-track tractor, **Figure 24-30**.

Articulated track-type tractors operate in a similar fashion to the traditional four-wheel-drive agricultural tractor. The tractors use traditional hydraulic articulated steering, have full-time four-track drive, are equipped with lockable differentials, and do not use any type of twin-track steering, such as differential steering, dual-path hydrostatic drive, or steering clutches and brakes. These tractors use four sets of tracks, all with the same height.

Four-Rubber-Track Rigid-Frame Four-Wheel Steer Track-Type Tractors

A few machines can be equipped with a rigid frame and four-rubber-track undercarriages, with all of the tracks having the same height. Ditch Witch produces a rigid-frame, four-track trencher that uses four-wheel steer. Chapter 25, *Wheeled Steering Systems*, explains four-wheel steer in more detail.

Goodheart-Willcox Publisher

Figure 24-29. One of the original Case IH Quadtrac tractors.

Phil McNew

Figure 24-30. The John Deere 9620RX rubber-track tractor and the Case IH Quadtrac.

Four-Rubber-Track, Rigid-Frame, MFWD Track-Type Tractor

Some agricultural MFWD row crop tractors can be purchased with four rubber tracks instead of four wheels. These tractors use a front-steerable axle like traditional row crop tractors. John Deere manufactures the 8RX tractor that comes from the factory with four rubber tracks. Two larger tracks are placed on the rear rigid axle. Two smaller tracks are placed on the front-steerable axle. See **Figure 24-31**.

Goodheart-Willcox Publisher

Figure 24-31. The John Deere 8RX tractor has two large tracks on the rear axle and two smaller tracks on the front steerable axle.

Summary

- Operators of twin-track tractors must drive the tracks at different speeds or in different directions to steer the tractor.
- The counter rotation steering method provides the tightest turning radius by propelling one track forward and the other track in reverse.
- The brake steer method achieves a tight turning radius by braking the inside track.
- The power turn steering method drives both tracks in the same direction but at different speeds.
- The neutral steer method has the operator shift the inside track into neutral.
- Dual-path hydrostatic-drive steering systems have a hydraulic pump and a hydraulic motor on each side of the tractor to provide propulsion and steering.
- Steering-clutch-and-brake systems require the steering clutches to be engaged and the steering brakes to be released in order to propel the machine in a straight line. A neutral steer is performed by releasing one steering clutch. A brake steer is performed by applying one steering brake.
- Differential steering systems use two inputs, a hydraulic steering motor and a CVT, electric motor(s), or powershift transmission, to drive a planetary gear set that delivers power to the left and right final drives.
- For differential-steered tractors, when the hydraulic steering motor is the only input, the tractor counter rotates. When the transmission is the only input, the tractor drives forward or in reverse in a straight path. When both the hydraulic steering motor and the transmission are inputs, the tractor moves (propels) and steers.
- Hydrostatic-driven and differential-steered dozers have better maneuverability compared to steering-clutch-and-brake dozers.
- A two-speed planetary steering system uses a set of cluster gears and brakes on each track that allow the operator to place each track in one of the following positions: neutral, brake applied, low speed, or high speed. The system is not capable of counter rotation.
- The term *hydrostatic steering* can describe traditional hydraulic steering for mobile equipment, the steering input for a differential-steered track-type tractor, or dual-path hydrostatic-propelled track-type tractors.
- The old independent-geared twin-track tractor requires the operator to manipulate multiple control levers instead of the joystick controls used in late-model machines.
- Case IH, John Deere, New Holland, and Versatile manufacture articulated tractors equipped with four-track undercarriages that are similar to traditional four-wheel-drive articulated tractors.

Technical Terms

brake steer
counter rotation
differential steering
hydrostatic steering system (HSS)
neutral steer
pivot turn
power steering
power turn
steering-clutch-and-brake system
steering tiller
turning radius
turn mode switch

Review Questions

Answer the following questions using the information provided in this chapter.

Know and Understand

1. Which of the following track steering methods creates the smallest turning radius?
   A. Brake steer.
   B. Counter rotation.
   C. Neutral steer.
   D. Power turn.

2. Which of the following track steering methods has the operator propel both tracks in the same direction, but one track is propelled faster than the other track?
   A. Brake steer.
   B. Counter rotation.
   C. Neutral steer.
   D. Power turn.

3. If the operator of a dozer steers the tracks by performing a brake steer, which direction will the tractor steer?
   A. Toward the stationary track.
   B. Toward the moving track.
   C. In the same direction as the ripper.
   D. None of the above.

4. Technician A states that steering-clutch-and-brake tractors are capable of counter rotating. Technician B states that Caterpillar's D10T2 and D11 dozers are equipped with a steering-clutch-and-brake system. Who is correct?
   A. Technician A.
   B. Technician B.
   C. Both A and B.
   D. Neither A nor B.

5. Which of the following is *not* an advantage of a hydrostatic-propelled track-type tractor?
   A. Hydraulic pumps are under constant high pressure when moving.
   B. Capable of counter rotation.
   C. Independent control of each track.
   D. Infinitely variable track speed control.

6. Dual-path hydrostatic-propelled dozers have all of the following advantages, EXCEPT:
   A. fewer power train components than steering-clutch-and-brake systems.
   B. no gear shifting.
   C. capable of counter rotation.
   D. excellent tolerance to fluid contamination within the system.

7. How must the operator of a steering-clutch-and-brake track-type tractor equipped with mechanical lever controls move the left and right steering levers to perform a sharp left steer?
   A. Move the left steer lever into *Neutral*, and move the right steer lever fully backward.
   B. Move both the left and right steer levers fully backward.
   C. Release the left steer lever to the forward position, and move the right steer lever into *Neutral*.
   D. Move the left steer lever fully backward, and release the right steer lever to the forward position.

8. Which statement is *incorrect* regarding the steering clutches in a steering-clutch-and-brake system?
   A. They can use multiple discs that are spring applied.
   B. They can use multiple discs that are hydraulically applied.
   C. They are engaged when the tractor is moving straight forward.
   D. They are applied with a band.

9. On a tractor equipped with a steering-clutch-and-brake system, which action will cause the torque converter to stall?
   A. Releasing the steering levers to the completely forward position and releasing the service brake pedal.
   B. Shifting the powershift transmission into *Neutral*.
   C. Releasing the steering levers to the completely forward position and applying the service brake pedal.
   D. All of the above.

10. Which of the following Caterpillar track-type tractor models are equipped with steering clutches and brakes?
    A. D1–D4.
    B. D6–D9.
    C. D7E.
    D. D10T2 and D11.

11. Which type of twin-track steering system uses a group of planetary gear sets and a hydraulic steering motor?
    A. Differential steer.
    B. Dual-path hydrostatic drive.
    C. Steering clutch and brake.
    D. Two-speed all gear.

12. Differential-steered Caterpillar dozers and AGCO differential-steered twin-track tractors have what commonality in the steering differential?
    A. Three sun gears connected together through a single shaft.
    B. Two built-in final drives.
    C. One planetary carrier that drives three individual final drives.
    D. Three bevel gears connected together through two shafts.

13. In a differential-steered Case dozer, the _____ are used to deliver power from the hydraulic steering motor to the steering differential.
    A. bevel gear shafts
    B. planetary carriers
    C. steering clutches
    D. sun gear shafts

14. Which manufacturer incorporates the use of a steering brake that is actuated by a turn mode switch on their largest differential-steered dozer, the D275AX-5?
    A. Case.
    B. Caterpillar.
    C. John Deere.
    D. Komatsu.

15. All of the following machines and systems can be considered to use some type of hydrostatic steering, *EXCEPT*:
    A. traditional combine harvester.
    B. differential-steered tractor.
    C. dual-path hydrostatic drive.
    D. two-speed planetary steering gear.

16. All of the following manufacturers have manufactured a two-speed planetary steering system, *EXCEPT*:
    A. Deere.
    B. Dresser.
    C. Dressta.
    D. International Harvester.

17. The Case 1150B-E and 1450B track-type tractors use what type of steering system?
    A. Differential steer.
    B. Dual-path hydrostatic drive.
    C. Independent-geared track steering.
    D. Two-speed planetary steering system.

18. All of the following manufacturers have produced a four-rubber-track tractor, *EXCEPT*:
    A. Case IH Agriculture.
    B. Case Construction.
    C. John Deere Agriculture.
    D. John Deere Construction.

Apply and Analyze

19. Dual-path hydrostatic-propelled dozers use _____ pumps and two motors.

20. When the steering control levers on a tractor equipped with a steering-clutch-and-brake system are released in the completely forward position, the steering clutches are _____ and the brakes are released.

21. When the steering control levers on a tractor equipped with a steering-clutch-and-brake system are pulled fully rearward, the steering clutches are released and the brakes are _____.

22. When the steering control levers on a tractor equipped with a steering-clutch-and-brake system are fully forward (released) and the service brake pedal is applied, the steering clutches are engaged and the brakes are _____.

23. When a Caterpillar differential-steered dozer's steering tiller is pushed forward, the machine will turn _____.

24. When a Caterpillar differential-steered dozer's steering tiller grip is twisted, the machine will move _____.
25. On a differential-steered track-type tractor, when the tractor is counter rotating, the differential is receiving input from _____.
26. On a differential-steered track-type tractor, when the tractor is driving straight ahead, the differential is receiving input from _____.
27. On a differential-steered track-type tractor, when the tractor is driving and steering simultaneously, the differential is receiving input from _____.
28. A John Deere twin-rubber-track differential-steered tractor has the transmission driving the pinion and ring bevel gear set, which drives the bevel gear shaft that powers the _____.

Critical Thinking

29. Describe all of the steering power flows through the Dressta two-speed dozer.
30. How does the Case independent-geared lever-operated dozer get power from the torque converter to provide input to the right and left track drives?

Chapter 25
Wheeled Steering Systems

Objectives

After studying this chapter, you will be able to:
- ✓ Describe the different types of steering systems used on off-highway wheeled machines.
- ✓ Explain the design and operation of steering control units (SCU).
- ✓ Explain the operation of steering amplifiers.
- ✓ Explain the operation of steering priority valves.
- ✓ Diagnose hydraulic steering system problems.
- ✓ Describe electronic steering applications and their advantages.
- ✓ Explain the operation of a Caterpillar Command Control Steering system.
- ✓ Summarize the steering features of motor graders.

Types of Wheeled Steering

Wheeled off-road machines vary in steering designs. The following are four types of wheeled steering systems used in mobile equipment:
- Dual-path hydrostatic transmission.
- Ackerman steering.
- Articulation steering.
- Combination steering.
- Multi-axle steering.

Dual-Path Hydrostatic Transmission

As explained in Chapter 13, *Hydrostatic Drives*, a dual-path hydrostatic transmission consists of a pair of hydrostatic transmissions that provide propulsion and steering for the machine. One hydraulic pump and one hydraulic motor power the right-side drive, and one hydraulic pump and one hydraulic motor power the left-side drive. Examples of wheeled machines that use dual-path hydrostatic transmissions for steering are skid steers, self-propelled windrowers, and zero turning radius (ZTR) lawn mowers.

One example of a machine that uses a dual-path hydrostatic drive that does not actually steer the machine is the all-wheel-drive Caterpillar M-series motor grader. The rear tandem wheels are powered by a powershift transmission, planetary final drives, and chain tandem drives. The Ackerman-style front steer axle has a dedicated HST to drive the right wheel and a HST to drive the left wheel. During a turn the ECM will slow the speed of the inside wheel, and increase the speed of the outside wheel, to compensate for the different distances the inside and outside front wheels must travel during the turn. In any turning arc, the outside wheel must travel a greater distance than the inside wheel. The difference in inside and outside wheel speeds do not cause the front axle to steer. Instead, the front axle is steered by hydraulic steering cylinders. The difference in front motor speeds compensates for the different distances the outside wheel and the inside wheel must travel during the turn, serving the same purpose as a differential. On most dual-path HSTs, the two pumps and motors are responsible for both propelling and steering.

Copyright Goodheart-Willcox Co., Inc.

Ackerman Steering

Ackerman steering is named after Rudolf Ackerman, who designed a steerable axle with a geometry that enables the inside wheel to turn at a greater steering angle than the outside wheel. The design allows the inside steering tire to follow a narrower turning arc than the outside steering tire. Without the Ackerman geometry, the machine would struggle to negotiate turns.

Linkage

In Ackerman steering systems, a mechanical linkage system connects the steering wheel to the steering knuckle through a series of links. The steering wheel splines directly to the steering gearbox or a steering shaft that connects the steering wheel to the gearbox.

The following components are commonly found in a steering linkage system:

- Steering wheel.
- Steering shaft.
- Universal joints.
- Steering gear.
- Pitman arm (not on rack-and-pinion systems).
- Drag link (not on rack-and-pinion systems).
- Tie rods.
- Steering knuckle.

Figure 25-1. This tractor steering system uses a parallelogram linkage with two tie rods. Each tie rod consists of a sleeve and two adjustable tie-rod ends.

Linkage system designs include parallelogram linkage and cross-steer linkage. *Parallelogram linkage* consists of a system of steering components positioned in the shape of a parallelogram. Most parallelogram linkages have two tie rods—each consisting of a sleeve with an inner tie-rod end and an outer tie-rod end. See **Figure 25-1**.

In a cross-steer linkage system, the Pitman arm transmits the steering motion of the gearbox to the drag link. See **Figure 25-2**. The *drag link* connects and transmits the rotary motion of the Pitman arm to one of the *steering knuckles*. A steering knuckle provides the pivoting action that allows the wheels to be steered left or right. Steering knuckles often pivot on pins, known as kingpins. A long tie-rod assembly connects the left-side steering knuckle to the right-side steering knuckle. As the drag link steers one knuckle, the tie rod steers the opposite knuckle.

Figure 25-2. A cross-steer linkage system has a long drag link that connects the Pitman arm to a steering arm. A long tie-rod assembly connects the two steering arms.

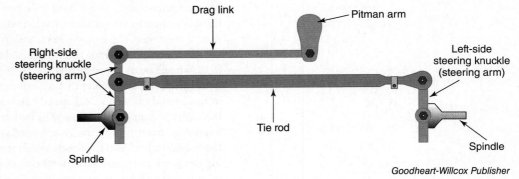

Two-Wheel and Four-Wheel Steering

Off-road machines can be equipped with front Ackerman steering (front-wheel steer) or rear Ackerman steering (rear-wheel steer). Front-wheel-steer machines include loader backhoes and traditional agricultural tractors. Rear-wheel-steer machines include combines, cotton harvesters, self-propelled forage harvesters, and some off-road forklifts.

Some rigid-frame machines are equipped with four-wheel steering. Examples include agricultural sprayers, telehandlers, large trenchers, and some four-wheel-drive agricultural tractors. Some four-wheel-steer machines can be configured to allow the operator to choose one of four different steering modes: front-wheel steer only, rear-wheel steer only, four-wheel steer (all-wheel steer), or crab steer (all wheels steered in the same direction and at the same angle). See **Figure 25-3**. Operators choose the four-wheel steer mode when operating at slow speeds and maneuvering in tight corners. Two-wheel front steer is used when travelling at fast forward speeds. Rear two-wheel steer is helpful when the machine is travelling long distances in reverse. Crab steering allows the machine to keep the load or implement straight ahead while steering towards the right or left.

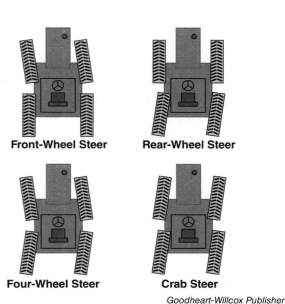

Figure 25-3. Four-wheel-steer tractors can be configured to have front-wheel steer, rear-wheel steer, four-wheel steer, and crab steer.

Most machines with Ackerman steering use an automotive-type power train with a drive axle and a steer axle. The most common configuration has the front axle as the steerable axle and the rear axle as the drive axle. As the wheels on the steerable axle pivot during a turn, the gears within the drive axle's differential rotate the left and right axle shafts at different speeds. The farther distance the outside drive wheel must travel during a turn requires it to spin faster than the inside wheel. The action of the drive axle's differential makes this possible. The front steerable axle can also be a drive axle equipped with a front differential.

The following steering designs are used in Ackerman steering:
- Manual steering.
- Power-assist steering.
- Steering control unit (SCU) steering. SCU steering is explained later in the chapter.

Manual Steering

The oldest type of wheeled steering is *manual steering*, which requires the strength of the operator to steer the wheels without any electric or hydraulic assistance. Some smaller, older machines use manual steering. It is difficult to steer heavily loaded machines with manual steering. At best, manual steering works on only the smallest off-highway machines, such as traditional lawn tractors. The advantage of manual steering systems is that they provide a mechanical connection from the steering wheel all the way to the wheels on the ground. See **Figure 25-4**.

A tractor with manual steering can use either a large steering wheel or a steering gearbox (steering gear) for steering torque multiplication. A ***steering gear*** is a gearbox that receives a rotating input from the steering wheel/

Figure 25-4. This old Ford tractor has a steering gearbox that drives a left Pitman arm and a right Pitman arm. Each Pitman arm connects to a drag link that connects to a steering arm that attaches to a steering axle spindle assembly that is not viewable due to the tire.

steering shaft. The steering gear multiplies the torque from the steering shaft input and translates it into lateral steering motion. Two types of steering gear designs are the recirculating ball steering gearbox and rack-and-pinion system. Both designs can be found in automotive and on-highway truck applications, but are not commonly used in late-model off-highway machines.

Note

Some steering gearboxes change the direction of the steering shaft by using only a bevel gear set.

Recirculating Ball Steering Gearbox

A *recirculating ball steering gearbox* has a worm shaft that is rotated by the steering wheel. The worm shaft spins inside a ball nut that contains a continuous row of ball bearings. As the worm shaft is spun, the ball nut slides up and down the worm shaft.

The teeth on the ball nut mesh with the sector gear's teeth. The sector gear is the output of the gearbox that directly splines to the Pitman arm. As the ball nut moves, it causes the Pitman arm to sweep to the left or right, effectively changing the steering wheel's rotational motion into a sweeping motion in the steering linkage. The term *recirculating ball* is based on the ball bearings continuously recirculating through an internal passageway (ball guides), which reduces friction. See **Figure 25-5**.

Rack-and-Pinion Steering

A manually actuated rack-and-pinion steering system has a pinion gear that is rotated by the steering column's shaft. The pinion gear meshes with a long horizontal rack that is attached to a right and left inner tie rod. A right and left outer tire rod connects the respective inner tie rod with the steering knuckle. As the operator rotates the steering wheel, the pinion gear forces the rack to move left or right, subsequently pulling or pushing the tie rods and steering knuckles to steer the tractor to the left or right.

Power-Assist Steering

In a *power-assist steering* system, a hydraulic pump supplies the hydraulic flow to a steering mechanism to help the operator steer a heavily loaded steer axle. The power assist can use an integral power steering gear or power rack-and-pinion steering.

Integral Power Steering Gear

An *integral power steering gear* contains an internal shaft, power piston, rotary valve, and sector gear. As the steering wheel is rotated, it causes the steering's internal shaft to rotate. The shaft is splined to the rotary valve and the power piston. The power piston is normally a recirculating ball nut design.

When the actuator's shaft is rotated, the shaft's splines manually actuate the rotary valve. The rotary valve acts like a directional control valve and directs hydraulic fluid pressure to the power piston. The power piston has teeth that mesh with the sector gear. See **Figure 25-6**.

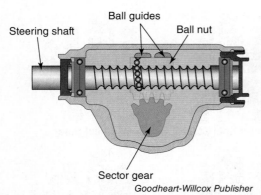

Figure 25-5. A manual recirculating ball gearbox contains a steering shaft, ball nut, and sector gear.

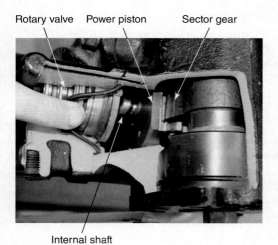

Goodheart-Willcox Publisher

Figure 25-6. An integral power steering gear contains an internal shaft, power piston, rotary valve, and sector gear. A—This cutaway shows the internal components of power-assisted steering used in automotive applications. B—The internal splines and three external teeth are visible on this power piston.

Note

If hydraulic fluid pressure is lost, the steering gear can still manually steer the tractor. However, much greater turning effort is required from the operator.

Chapter 25 | Wheeled Steering Systems

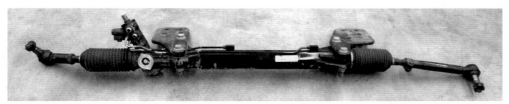

Figure 25-7. This power rack-and-pinion steering was removed from a Cascadia Freightliner on-highway truck.

Goodheart-Willcox Publisher

Power Rack-and-Pinion Steering

A power rack-and-pinion steering system operates similar to a manual rack-and-pinion steering system, except that a control valve is used to direct fluid to either side of the rack's power piston, which provides fluid pressure to assist in actuating the rack. Power rack-and-pinion steering systems are not commonly found in mobile machinery and are rarely used in large on-highway trucks. See **Figure 25-7**.

Articulation Steering

Articulation steering is found on many off-highway machines. In *articulation steering*, an articulation joint connects the front frame of the machine to the rear frame. Two hydraulic cylinders articulate (hydraulically pivot) the two frames so the tractor can steer to the left or to the right, **Figure 25-8**. The following machines use articulation steering:

- Agricultural four-wheel-drive tractors.
- Articulated dump trucks (ADTs).
- Scrapers.
- Motor graders.
- Large compactors.
- Some trenchers.
- Wheel dozers.
- Wheel loaders.
- John Deere high-speed rubber-track dozers.

Articulated machines use an SCU, which is covered later in this chapter.

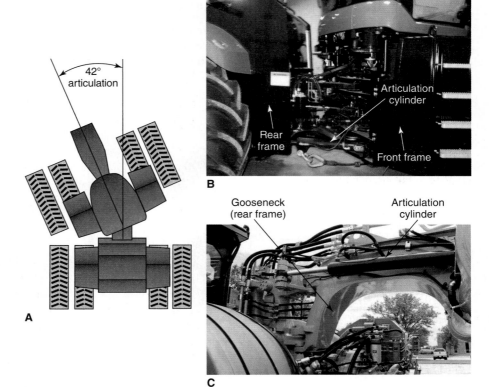

Figure 25-8. Off-highway machines can use articulation steering. A—Four-wheel-drive agricultural tractors often are equipped with articulation steering. B—This New Holland four-wheel-drive tractor uses articulation steering. C—Example of the rear frame and the left articulation cylinder for a scraper. D—Articulated wheel loader.

Goodheart-Willcox Publisher

Note
Although many telehandlers use four-wheel steer, Terex, Schaffer, Manitou, JCB, and VF Venieri are a few companies that have manufactured an articulated telehandler, often called a pivot-steer telehandler. Blaney also manufactures a similar machine. It too uses a telescoping boom and a pivot steering system, but that machine is labeled a pivot loader.

Combination Steering

A few off-road machines have *combination steering* that includes either articulation steering and Ackerman front-steer or two articulation joints. Examples are motor graders and some Case IH four-wheel-drive tractors.

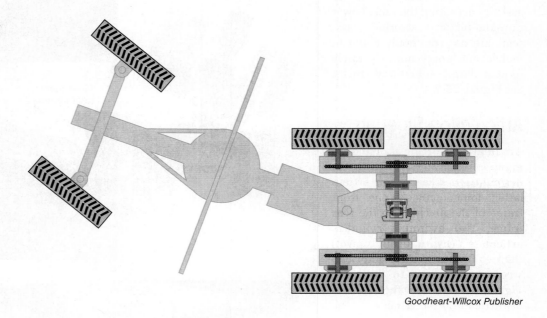

Figure 25-9. Motor graders often have articulation joint steering and front-axle steering to help turn in tight spaces.

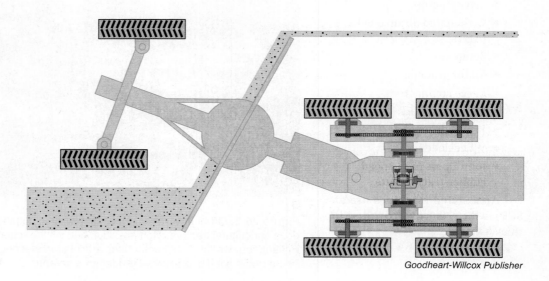

Figure 25-10. The articulation joint and front axle on a motor grader can be steered independently.

Motor Grader Steering

On motor graders, the operator can steer the articulation joint and the front axle together in the same direction or independently in opposite directions. The articulation joint often provides 20° or more of the machine's steering rotation. When the motor grader's articulation joint and front axle are steered together (combination) in the same direction, a motor grader can make a 180° turn in a distance approximately twice the overall length of the motor grader. See **Figure 25-9**. When the motor grader's articulation joint and the front axle are steered independently, the front axle can be positioned to avoid interfering with a windrow while the rear frame travels behind the blade, **Figure 25-10**. The operator can also avoid an object using this steering method. Motor grader steering is explained in detail later in this chapter.

Case IH 4WD Combination Steering

For several decades, small- and mid-frame Case IH four-wheel-drive Steiger tractors have offered combination steering, later called AccuSteer. Some Steigers, such as the 9350 tractor, used Ackerman steering (front steerable axle) combined with a traditional centrally located articulation joint. The 9310, 9330, and STX AccuSteer Steigers use a traditional centrally located articulation joint and a front-axle-style articulation joint. The front axle, like the rear axle, is a rigid non-steerable axle, but the front frame and axle assembly can pivot (articulate) like the center articulation joint.

AccuSteer allows a customer to choose between two options:

- Articulation-only steering.
- Combination articulation and front-axle articulation steering.

In the combination steer mode, the front axle is steered first. After the front articulation is steered 10°, the rear articulation is pivoted up to an additional 42°, providing a total of 52° of articulated steering. See **Figure 25-11**.

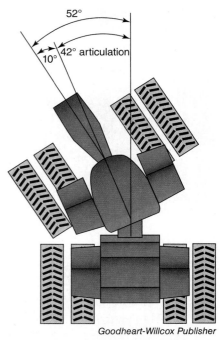

Goodheart-Willcox Publisher

Figure 25-11. A Case IH AccuSteer Steiger tractor has double articulation steering, providing 52° of total steering.

Note

Case IH AccuTurn is not the same as Case IH Steiger AccuSteer. AccuTurn is a feature that allows automated hands-free headland turning. It directs the tractor at the end of a row to turn around so it can plant or till directly adjacent to the previous planted row. This maneuver is completed completely hands-free, without operator assistance. AccuTurn is available on a variety of Case IH tractors, whereas Case IH AccuSteer is only available on certain Steiger four-wheel-drive tractors.

Case IH Steiger four-wheel-drive agricultural tractor combination steering is explained in further detail later in this chapter.

Multi-Axle Steering

Additional off highway machines can be equipped with multiple steering axles in a configuration other than the traditional twin axle four-wheel steer machine. Examples include all-terrain cranes, concrete pumper trucks, aviation rescue fire-fighting (ARFF) trucks, and military vehicles. Not all of these machines are equipped the same. For example, some ARFF trucks have one or two front steerable axles, and some are also equipped with rear axle steering. One significant benefit of rear steer is less tire wear.

All-terrain cranes and concrete pumper trucks can be equipped with eight axles, including up to four steerable axles. See **Figure 25-12A**. The multi-steer axle all-terrain cranes can be called all-wheel steer. The all-terrain crane operators can choose all-wheel steer or crab steer. Grove markets that their all-terrain crane is steer-by-wire. Some military vehicles can be configured with all eight axles being steerable axles. See **Figure 25-12B**.

A

GE_4530/Shutterstock.com

B

ID1974/Shutterstock.com

Figure 25-12. Multi-axle steerable machines. A—An example of a multi-terrain crane with seven axles, including two front and two rear steer axles. B—A military vehicle with eight steerable axles.

Steering control unit

Goodheart-Willcox Publisher

Figure 25-13. The internal components of an SCU are rotated by means of a steering wheel and steering shaft. The SCU provides true hydraulic steering with no mechanical link between the SCU and the steering cylinder.

Steering Control Units

For decades, mobile equipment manufacturers have used a unique 100% hydraulic power steering system in machines with both articulated steering and Ackerman steered axles. In this system, a manually operated *steering control unit (SCU)* is rotated by a steering wheel and steering shaft. See **Figure 25-13**. The hydraulic steering system has no direct mechanical connection between the SCU, the hydraulic pump, or the steering cylinder, unlike the mechanical linkages found in automotive power steering and manual steering systems. The only connection between the SCU and the steering cylinder are the hydraulic hoses that route oil from the SCU to the steering cylinder.

Flexibility and adaptability are two advantages of this style of hydraulic steering. The SCU is often located directly below the steering column, inside the steering console, or beneath the floor of the operator's cab. The hydraulic hoses provide the necessary flexibility to deliver oil to the steering cylinder(s), which may be located a great distance away from the SCU.

The downside of a traditional SCU is that it is not electronically controlled and is difficult to adapt to machines that use GPS-guided steering systems, also known as *auto-guidance steering systems*. However, kits are available for converting a traditional SCU system into an auto guidance steering system.

Steering control units can be called different names including the following:

- Hand metering unit (HMU) (Caterpillar's term).
- Hydrostatic steering.
- Steering hand pump.
- Steering control valve (SCV).
- Orbitrol valve.

> **Note**
>
> SCV is also a term used by John Deere to describe a selective control valve, which is a traditional directional control valve, not an SCU. When you encounter the term SCV, make sure you know which type of valve it is referencing.

In 1961, Lynn Charlson, of Char-Lynn, invented the first SCU. Char-Lynn was purchased by Eaton Corporation in 1970. In 2021, Danfoss purchased Eaton Hydraulics, which is the largest merger of mobile hydraulic manufacturers in modern history. Although there are multiple manufacturers of steering systems, Char-Lynn has held the majority of the market share for manufacturing SCUs for decades.

SCU Components, Design, and Operation

The SCU is responsible for completing two tasks related to steering a tractor. The SCU must control the direction of oil flow (right steer and left steer) and the quantity of oil flow (dictates how fast the wheels are turned). The primary components of an SCU include the gerotor assembly, drive shaft, control sleeve, centering pin, and control spool. See **Figure 25-14**. The SCU contains several components that are internally connected to one another. The manner in which the coupled components operate is described in the following sequence:

- The steering wheel's driveshaft is splined to the control spool.
- The control spool is coupled to the control sleeve through a centering pin and centering springs.
- The centering pin fits loosely inside the control spool and has a tight fit inside the control sleeve.

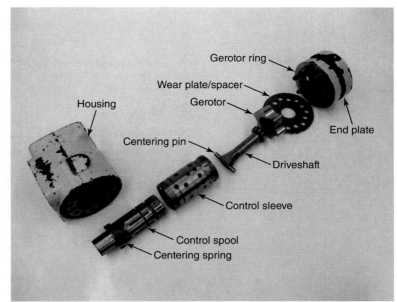

Goodheart-Willcox Publisher

Figure 25-14. A disassembled steering control unit. The main components within an SCU are the control spool and control sleeve assembly and the gerotor assembly.

- As the steering wheel turns the control spool, the control sleeve follows via the loose-fitting centering pin. When the steering wheel stops rotating, the centering spring centers the control spool/sleeve assembly.
- The driveshaft fits inside the control spool/sleeve assembly and has a yoke shape at the bottom of the shaft that slides over the centering pin.
- As the control spool/sleeve assembly spins, the centering pin causes the driveshaft to rotate.
- The driveshaft is splined to the internal gerotor that rotates inside the ring of the gerotor assembly.
- The SCU housing, control spool, and control sleeve work together to act as a rotary directional-control valve (DCV).
- The stationary components within an SCU are the gerotor's ring gear, the housing, the spacer/wear plate, and the end plate/cover plate.

Open-Center SCU

SCUs have different designs, such as open center, closed center, and load sensing. An open-center directional control valve (DCV) routes oil supplied from the hydraulic pump back to the reservoir at a low pressure when the DCV is in neutral. An open-center SCU is most likely used with either a fixed-displacement pump that is dedicated solely for steering or a fixed-displacement pump and a proportional flow divider valve. The divider valve proportions a percentage of pump flow to the open-center SCU. The remaining percentage of oil flow is used for other circuits, such as traditional hydraulic implements.

The open-center SCU design is not used with steering priority valves because these valves must receive a load-sensing signal from the SCU. Steering priority valves are explained later in this chapter.

Closed-Center Non-Load-Sensing SCU

When a closed-center DCV is in the neutral position, it blocks off the pump's oil supply, preventing it from returning to the reservoir. A closed-center non-load-sensing SCU, as shown in **Figure 25-15**, is used in a pressure-compensating hydraulic system. This particular SCU is rare because most SCUs are used in conjunction with a load-sensing steering priority valve.

Closed-Center Load-Sensing SCU

A traditional open-center or closed-center SCU has four ports—inlet, outlet, right steer, and left steer. Most SCUs are a closed-center, load-sensing design with a fifth hydraulic port, which is the load-sensing signal line. A load-sensing signal line is a small hydraulic line that is used to direct a hydraulic signal pressure (the pressure level required to perform the hydraulic function). The signal pressure is routed from the DCV or SCU and is sent to a device that requires a signal pressure to function, such as a steering priority valve, a loading-sensing variable-displacement pump, or an unloading valve (in the case of a fixed-displacement pump operating in a load-sensing hydraulic system). See **Figure 25-16**. This style of load-sensing SCU is used in multiple types of hydraulic systems:

- Open-center hydraulic system.
- Pressure-compensating hydraulic system.
- Load-sensing pressure-compensating (LSPC) hydraulic system.
- Flow-sharing hydraulic system.

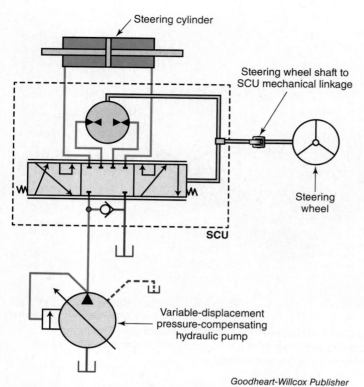

Figure 25-15. A schematic of a hydraulic steering system with a pressure-compensating hydraulic pump and a closed-center SCU without a load-sensing signal line.

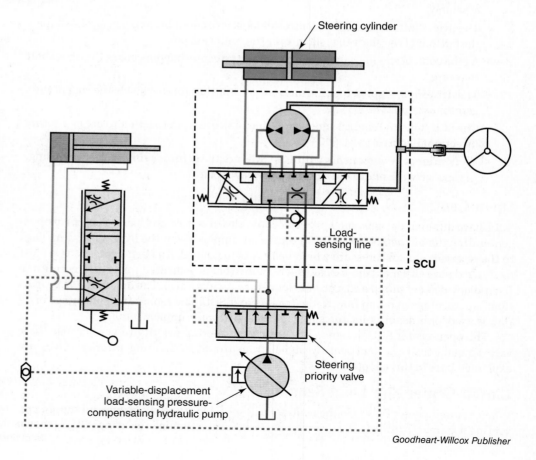

Figure 25-16. A closed-center load-sensing SCU is required if the machine is equipped with a steering priority valve. The signal line is also used for load-sensing hydraulic pumps. In this example, a load-sensing signal pressure is generated through the SCU and does not use a primary shuttle valve.

LSPC hydraulic systems are a style of load-sensing hydraulic system that uses a pressure compensator located prior to the DCV spool valve, also known as pre-spool compensation. Flow-sharing is a style of load-sensing hydraulic system that uses a pressure compensator located after the DCV spool valve, also known as post-spool compensation. Either system can use a variable-displacement pump with load-sensing signal pressure to its flow compensator valve, or a fixed-displacement pump with load-sensing signal pressure to its unloading valve.

In the traditional open-center hydraulic system and the traditional closed-center pressure-compensating hydraulic system, the SCU's load-sensing signal line is used only for the steering priority valve. However, in load-sensing hydraulic systems, whether pre-spool or post-spool, the SCU's signal line is used for the steering priority valve and the load-sensing hydraulic pump or unloading valve. Priority valves and load-sensing signal lines are detailed later in this chapter.

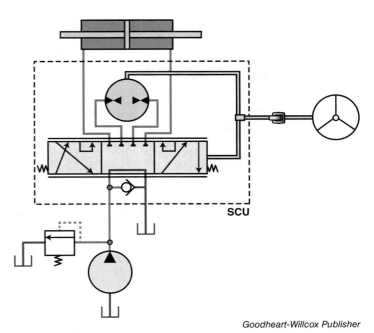

Goodheart-Willcox Publisher

Figure 25-17. A schematic of a hydraulic steering system in the neutral position. This system uses an open-center SCU in its design. When the operator rotates the steering wheel, both the control spool/sleeve assembly and the gerotor assembly rotate at the same speed and in the same direction as the steering wheel.

SCU Neutral Operation

If a tractor is driving straight forward, the SCU is in the neutral position. In **Figure 25-17**, the SCU is an open-center valve and the oil is routed back to the reservoir at a low pressure.

SCU Left or Right Steering Operation

When the operator rotates the steering wheel to the left or to the right, see **Figure 25-18**, the SCU routes the oil through the following path:

1. The control spool/sleeve assembly receives oil from the hydraulic pump.
2. The control spool/sleeve assembly directs the oil to the gerotor assembly.
3. The *gerotor assembly* meters the quantity of oil that is eventually sent to the steering cylinder. The quantity of oil is governed by how fast the steering wheel is rotated.
4. The control spool/sleeve assembly receives the metered oil from the gerotor assembly and directs it to the steering cylinder to steer the tractor to the left or to the right based on the operator's movement of the steering wheel.
5. The steering cylinder's return oil is routed back to the control spool/sleeve assembly.
6. The control spool/sleeve assembly directs the return oil to the reservoir.

SCU Emergency Steering Operation

As part of a truly 100% hydraulic steering system, an SCU has a special provision designed into the unit that enables the operator to steer the tractor in the

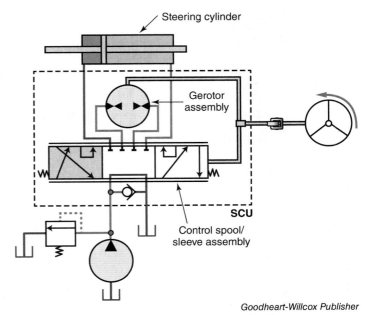

Goodheart-Willcox Publisher

Figure 25-18. This schematic shows a left turn. As the operator rotates the steering wheel to the left, the SCU control spool/sleeve assembly directs the oil in and out of the SCU to steer the tractor to the left. Note that for a right turn, the control spool/sleeve shifts to the far-right position and the oil follows the opposite flow path. Oil pressure forces the steering cylinder piston to the right end of its cylinder.

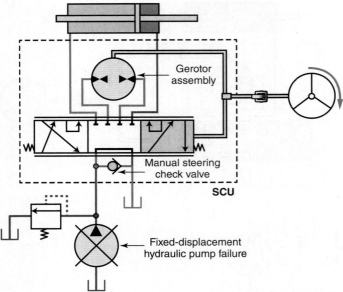

Figure 25-19. The manual steering check valve allows for emergency steering if the engine or hydraulic pump fails during operation. As the gerotor is spun, oil is drawn from the reservoir into the SCU to provide manual steering. This example shows manual steering to the right. Trace the path of the oil as it is extracted from the reservoir and travels through the check valve, into the control spool/sleeve assembly, and through the gerotor assembly before reaching the steering cylinder.

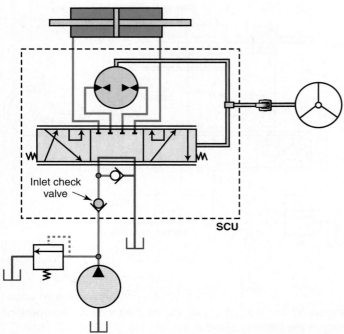

Figure 25-20. The inlet check valve prevents high-pressure steering oil from bleeding backward through the SCU when the steering pressure exceeds system pressure. This prevents steering kickback.

event that the engine dies or the hydraulic pump fails. This feature is called *emergency steering* or *manual steering mode*.

The *manual steering check valve* is located at the SCU's inlet and taps into the SCU inlet port and the SCU tank return port. If the operator attempts to steer when the hydraulic pump is unable to supply oil flow to the SCU, the manual steering check valve allows the SCU's gerotor to draw oil directly from the reservoir. The gerotor assembly uses this oil as the supply oil for emergency steering, **Figure 25-19**. As the operator manually rotates the steering wheel, the gerotor rotates within the gerotor ring, causing an expanding volume among the teeth near the internal gear's inlet port and a decreasing volume among the teeth near the internal gear's outlet port. The action of the gerotor during this mode causes it to act like a manual pump. For this reason, the SCU is sometimes called a steering hand pump. During the manual steering mode, because oil is unavailable from the hydraulic pump, the SCU requires a lot of effort from the operator just to turn the steering wheel. The SCU lacks the hydraulic horsepower normally received from the pump that is normally used to steer the cylinder(s).

Optional SCU Internal Valves

In addition to emergency steering, the following internal steering valve options are available:
- Inlet check valve.
- Inlet relief valve.
- Load-sensing relief valve.
- Cylinder port relief valves.
- Anti-cavitation check valve.

Inlet Check Valve

If external loads exerted a force on the steering cylinder, the steering pressure could exceed the main system pressure. The steering oil would then bleed backward through the SCU and cause *steering kickback* (jerkiness or recoil of the steering wheel). The *inlet check valve* prevents steering kickback. See **Figure 25-20**.

Inlet Relief Valve

Most hydraulic steering systems have a dedicated relief valve for the steering circuit. The inlet relief valve is designed to protect the steering circuit from rupturing due to over pressurization. It directs the oil flow to the reservoir when the steering cylinder reaches the end of its travel or encounters resistance, causing the steering pressure to increase. The inlet relief valve is commonly placed either inside the SCU, as shown in **Figure 25-21,** or inside the steering priority valve, which is discussed later in this chapter.

The SCU's inlet relief valve controls the steering relief pressure. Steering relief valve pressures are normally set below the main system pressure. For example, if the main system pressure is set at 3000 psi (207 bar), the steering relief valve pressure could be set as low as 2300 psi (159 bar).

If the hydraulic pump is used only for steering, the main system relief valve would be redundant and unnecessary for a system equipped with an SCU inlet relief valve. However, if the fixed-displacement pump's flow is divided by a proportional flow divider valve, the main system relief is required.

Load-Sensing Relief Valve

A steering load-sensing relief valve can also be used to control the steering relief pressure. The *load-sensing relief valve* establishes the steering relief pressure by dumping the steering signal pressure to the reservoir when the pressure exceeds its setting. Load-sensing hydraulic steering systems may use a load-sensing relief valve that is combined into the SCU or into the steering priority valve. See **Figure 25-22**.

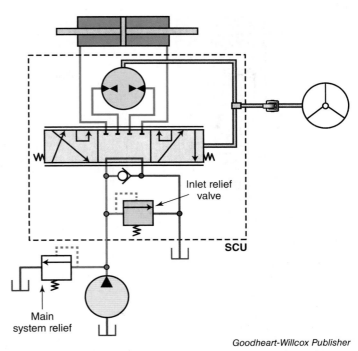

Goodheart-Willcox Publisher

Figure 25-21. An inlet relief valve can be placed inside the SCU. It is used as a hydraulic steering system relief valve.

Cylinder Port Relief Valves

When a DCV is in a neutral position, the circuitry located after the DCV has no protection from forces generated by external loads. Cylinder port relief valves protect the steering cylinder(s) and the steering hoses from over pressurization, which can rupture hoses, swell (balloon) cylinders, or rupture seals. See **Figure 25-23**.

Anti-Cavitation Check Valves

Anti-cavitation valves, also known as make-up valves, can be installed inside an SCU. See **Figure 25-24**. The valves prevent cylinder *cavitation* by drawing oil from the reservoir

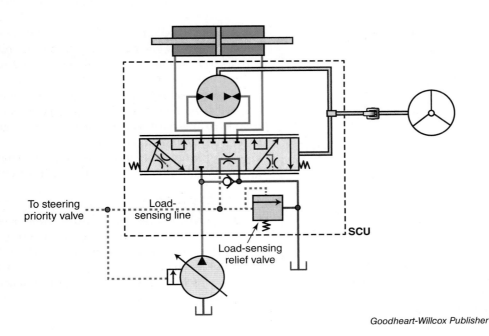

Figure 25-22. A steering load-sensing relief valve limits the hydraulic steering system pressure by dumping the steering signal pressure to the reservoir when the pressure exceeds its setting.

Goodheart-Willcox Publisher

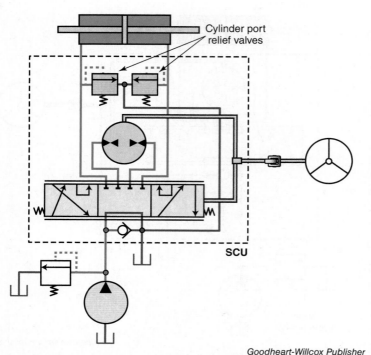

Figure 25-23. Cylinder port relief valves can be added inside the SCU. When the SCU is moved into a neutral position, the valves protect the steering working ports from pressure surges caused by external forces.

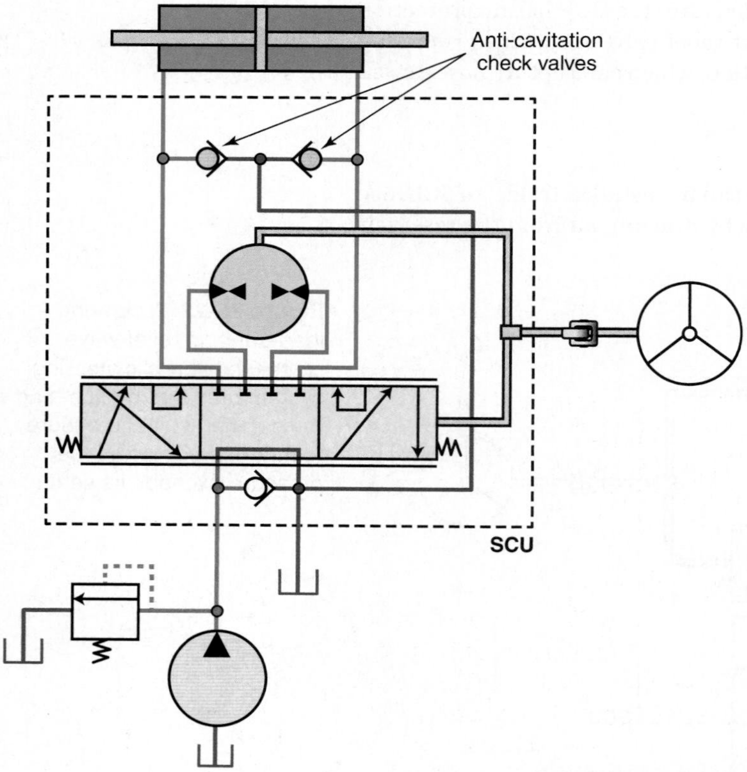

Figure 25-24. Anti-cavitation check valves can be installed inside an SCU. The valves draw make-up oil from the reservoir in the event that the steering cylinder(s) actuates fast enough that the pump alone is incapable of delivering all of the oil required.

whenever the cylinders actuate faster than the hydraulic pump can supply the required oil. Cavitation occurs within a hydraulic system anytime an expanding volume generates a low pressure area (such as a cylinder extending) and hydraulic fluid is unable to fill that low pressure area, resulting in the formation of air bubbles. When the bubbles are exposed to high pressure, they violently collapse, causing damage such as eroding metal surfaces.

SCU Features

SCUs are available in several configurations:
- Non-load-reaction.
- Load-reaction.
- Wide-angle.
- Cylinder damping.
- Q-Amp.

Non-Load-Reaction SCU

In a *non-load-reaction SCU*, the steering cylinder's left and right work ports are blocked when the SCU is in a neutral position. This configuration holds the steering cylinder(s) in a fixed position even when external loads are exerting forces on the steering cylinder(s). **Figure 25-15** through **Figure 25-24** show non-load-reaction SCUs.

Load-Reaction SCU

A *load-reaction SCU* allows the external loads placed on the steering cylinder(s) to actuate the hydraulic steering system. The SCU's neutral position has an open passage between the gerotor assembly and the steering cylinder work ports. See **Figure 25-25**. When the operator releases the steering wheel through the approximate midpoint of the turn, the forces exerted on the tractor's axle cause the steering cylinder(s) to return to the neutral position.

A single-acting differential cylinder cannot be used with a load-reaction SCU. The steering cylinder(s) area must be the same for left and right steer. A double-rod cylinder can be used. Two steering cylinders that have equal areas and are plumbed in parallel will also work, as shown in **Figure 25-25**.

Wide-Angle SCU

Large articulated tractors can use a *wide-angle SCU* that is specifically designed to smooth the bumpy steering motion of such vehicles. The SCU smooths the steering motion by increasing the amount of deflection between the SCU's control spool and sleeve during a turn.

Cylinder-Damping SCU

A *cylinder-damping SCU* is designed to smooth the steering motion on large articulated tractors. The cylinder damping is achieved by adjustable orifices that divert a small amount of fluid from the steering cylinder ports back to the reservoir whenever the steering wheel is turned.

Q-Amp SCU

Eaton offers an SCU that can magnify the volume of the oil flow that would normally be limited by the displacement of the SCU's gerotor. The SCU is called Q-Amp. Just like Q in a PQ curve, Q stands for fluid *quantity* and Amp stands for *amplification*. The increase in flow occurs when the steering shaft is rotated at a relatively high speed, such as 10 rpm. Internal variable orifices in the SCU divert a portion of the oil straight to the steering cylinder rather than being metered through the gerotor assembly. When the steering shaft is rotated at slower speeds—less than 10 rpm—the gerotor assembly meters all of the oil sent to the steering cylinders. The Q-Amp SCU can produce 60% more oil flow than the traditional gerotor is capable of delivering. The advantages of the Q-Amp SCU include the following:

- A smaller SCU can be used to deliver a larger amount of steering oil.
- When the vehicle is traveling at a high road speed, the SCU is not overly reactive as the operator slowly rotates the steering wheel.
- Fast turns are possible when the steering wheel is rotated quickly.

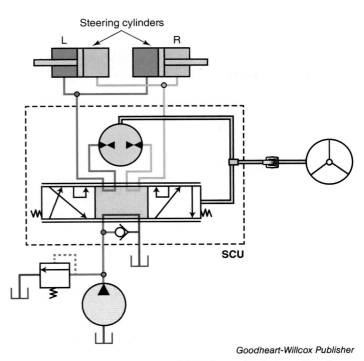

Goodheart-Willcox Publisher

Figure 25-25. A load-reaction SCU allows the external loads placed on the steering cylinder(s) to force the steering wheel back to a neutral position, which occurs when the operator releases the steering wheel midway through the turn.

Steering Amplifiers

Some off-road machines are so large that an SCU cannot supply an adequate amount of flow to the steering cylinders, so a *hydraulic steering amplifier* valve is used to multiply the requested oil flow from the SCU. For example, Danfoss hydraulic steering amplifiers are available with magnification factors ranging from four to twenty times. Mining trucks are examples of machines that use steering amplifiers. Hydraulic steering amplifiers are also used in large ships.

As the operator turns the steering wheel, the SCU acts similarly to a pilot controller and sends pilot oil to control the steering amplifier valve assembly rather than the steering cylinders. The SCU proportions the oil to the steering amplifier based on the speed at which the steering wheel is turning.

There is a directional valve inside the steering amplifier. It is pilot operated by the SCU oil and has two tasks. It takes the steering oil flow from the SCU and sends it to the flow-combiner valve that combines the SCU oil flow with oil flow from the steering pump and accumulator circuit, resulting in an amplified oil flow. The directional valve also sends the amplified oil in the correct direction, to steer right or to steer left. See **Figure 25-26**. The steering amplifier's flow-combiner valve and flow check valve work in conjunction with each other to amplify the quantity of oil flow that was sent by the SCU and send it to the amplifier's directional valve.

In this example, the cylinder-port relief valves and make-up valves have been moved from the SCU, and placed inside the steering amplifier. The steering priority valve has also been placed inside the steering amplifier.

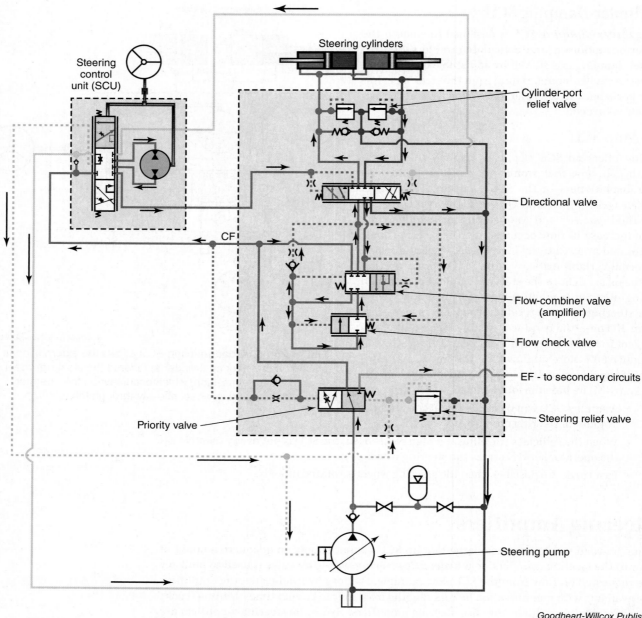

Figure 25-26. Hydraulic steering amplifiers are used on large machines that require more steering oil flow than an SCU is capable of delivering, such as mining trucks. It combines the SCU oil flow with oil flow from the steering pump and accumulator.

Traditional Steering Priority Valves

The majority of hydraulic steering systems use a steering priority valve. This valve ensures that steering demands are met before sending the remainder of the oil to the secondary implement circuits. See **Figure 25-27.**

A steering priority valve normally has five ports:
- Supply—receives a supply of oil from the hydraulic pump.
- Return—directs the oil to the reservoir.
- Load-sensing input—receives the steering signal (working) pressure.
- *Controlled flow (CF)*—directs the prioritized oil to the inlet of the SCU.
- *Excess flow (EF)*—directs the balance of the flow to the secondary circuits.

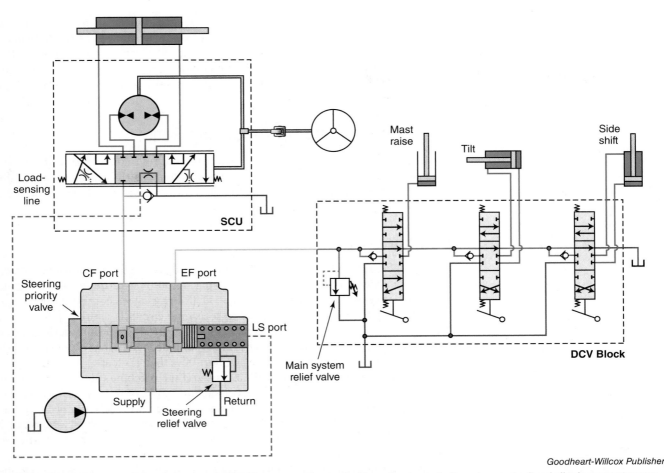

Goodheart-Willcox Publisher

Figure 25-27. A closed-center load-sensing SCU is commonly used with a steering priority valve and a fixed-displacement pump. The schematic resembles a Case 586e forklift hydraulic system. The main system relief valve is located downstream inside the forklift's DCV block. The steering relief is located in the priority valve, which dumps the steering load-sensing pressure.

The priority valve's control spool determines whether the pump's supply oil is sent to the SCU (controlled flow) or is sent downstream to the secondary circuits (excess flow). The control spool is operated by three variables—controlled flow pressure, load-sensing pressure, and bias-spring pressure. CF pressure opposes both load-sensing pressure and spring pressure.

Steering Priority Valve Operation—SCU in Neutral

If the closed-center SCU is in a neutral position (operator is not steering), the CF pressure will increase due to the blocked ports. The CF pressure is routed to the end of the control spool by means of an orifice and cross-drilled passageway through the center of the spool. With the SCU in a neutral position, there is no steering signal pressure.

The CF pressure builds until it can overcome the priority valve's bias spring. For this example, assume the bias spring equals 150 psi (10 bar) and the control spool's surface area equals one square inch. Once the CF pressure builds to 150 psi (10 bar), the control spool shifts, compressing the spring. All of the pump's supply oil is routed out of the EF port and sent to the forklift's mast DCV block. The forklift's mast DCVs are open centered and route the oil back to the reservoir at a low pressure, such as 150 psi (10 bar).

Steering Priority Valve Operation—SCU Steering Left or Right

As the operator rotates the steering wheel, the SCU develops a steering load-sensing signal pressure that is sent to the priority valve. The load-sensing pressure plus the bias-spring pressure ensures that the control spool meets the oil requirement of the SCU. After the SCU's oil demand is met, the CF pressure builds once again, assuming the fixed-displacement pump is producing excess oil. When the priority valve senses that the SCU's flow demand is met, the CF pressure overcomes the combined bias-spring pressure and signal pressure, resulting in the control spool directing the remaining balance of oil flow to the secondary circuits (EF port).

Steering Priority Valve Load-Sensing Relief

As mentioned earlier, the steering priority valve can contain a load-sensing relief valve (refer to **Figure 25-27**). The load-sensing relief valve controls the steering's maximum system pressure by dumping the load-sensing signal line to the reservoir when the signal pressure reaches the LS relief valve's setting.

Load-Sensing Line

Steering priority circuits often have one of two types of signal line configurations, either a static load-sensing signal or a dynamic load-sensing signal. The ***static load-sensing signal*** is the standard type of signal configuration. The priority valve's control spool receives a static pressure signal, meaning that the oil in the signal circuit is not flowing while the SCU is in a neutral position. Static load-sensing lines are used when the steering system has sufficient steering performance, the steering system does not lack response, and the steering load-sensing signal lines are relatively short (less than six feet).

The ***dynamic load-sensing signal*** circuit is used if the machine's signal line is longer than six feet, if the machine's steering circuit is less than stable, or if the steering operation is inconsistent. When the SCU is in a neutral position, the priority valve is generating a small signal pressure. The dynamic steering signal pressure is the result of a small amount of oil flowing back to the reservoir through an orifice. The dynamic load-sensing signal improves hydraulic steering responsiveness and helps in cold climates by warming up the circuit.

Engineers choose dynamic load sensing when the steering load-sensing line is longer than six feet or if the machine's steering is too sluggish (delayed) and needs to respond more quickly to the operator's commands. The dynamic flow of oil in the signal line generates a constant steering signal pressure, for example 50 to 150 psi, when the tractor is running and the SCU is in a neutral position.

Diagnosing Hydraulic Steering Systems

Hydraulic steering systems can become sluggish or difficult to steer. A short troubleshooting process is used to diagnose such problems. Before proceeding to SCU diagnostics, a technician should check the hydraulic pump, main system relief valve, and return line to eliminate them as the cause of the problem. Presuming the steering system pump, main system relief, and return line are operating properly, the technician can also check the priority valve pressures (EF, CF, and LS). However, it is usually easier to first check the SCU's internal leakage.

Be sure to follow the manufacturer's service literature for testing an SCU's internal leakage. An ***SCU internal leakage test*** assesses the amount of tolerated oil seepage between the parts operating inside the SCU. An example sequence consists of the following steps:

- Turn the steering wheel clockwise or counterclockwise until the wheels reach their end of travel.
- Ensure that the system has reached the operating temperature specified in the testing procedure.
- Use a torque wrench to apply the specified amount of torque to the steering wheel retaining bolt. A typical torque specification is 60 inch-pounds. Apply the torque in the same direction that the steering wheel was turned.

- With the wheels steered fully against their stops, apply the specified amount of torque to the steering wheel in the same direction the steering wheel was turned. The steering wheel should rotate slowly. Internal leakage in the SCU allows the steering wheel to rotate even though the wheels are locked.
- Record the number of steering wheel revolutions completed in one minute. The manufacturer's literature will specify the proper torque to apply, and a maximum acceptable rotational speed for the steering wheel, such as 4 rpm with 60 inch-pounds applied.
- Repeat the test procedure, turning the steering wheel and applying torque in the opposite direction.

If the rotational speed of the steering wheel exceeds the specification, the next step is to determine if the leakage is occurring in the steering cylinder or in the SCU. Most service literature recommends capping the steering cylinder's hydraulic hoses and repeating the tests. If the steering wheel's rotational speed is within specification with the cylinder hoses capped, the steering cylinder is internally leaking and needs to be rebuilt. If the speed is still too high with the steering cylinder hoses capped, the SCU has too much internal leakage and must be replaced.

If the SCU and steering cylinder(s) are found to be operating efficiently and the priority valve pressures meet requirements, inspect the priority valve. Determine if the spring is broken, the orifice is plugged, or the spool is seized in its bore.

Electronic Steering Features

Many mobile machines no longer use SCUs, but instead use electronic controls to steer the tractor, known as *steer-by-wire*. Specific manufacturers offer unique electronic steering features. Two examples are the John Deere ActiveCommand Steering and Caterpillar Quick Steer.

Steer-by-wire also provides additional benefits when integrated with global positioning satellite (GPS) systems, allowing manufacturers to offer automated guidance steering systems. A tractor equipped with such a system creates less overlap of chemicals when spraying farm fields, which saves fuel and reduces input costs. The guidance systems also reduce operator fatigue.

John Deere ActiveCommand Steering

The John Deere *ActiveCommand Steering system* is an optional feature on 8R tractors that uses electronic inputs, including four steering wheel position sensors housed in the steering module (located below the steering wheel), four wheel-angle sensors located at the front wheels (primary left, secondary left, primary right, and secondary right), and a gyrometer, which is used to sense the tractor's balance. The gyrometer input is used only at speeds above 12 mph. It is used at a reduced rate between 12 and 15 mph, and at a full rate above 15 mph. At higher speeds, resistance is added to the steering wheel, requiring the operator to increase steering wheel input over the level required at slower travel speeds.

Two steering wheel sensor inputs (sensors 1 and 2) are read by an ECM called the A-box; the other two steering wheel sensors (sensors 3 and 4) are read by an ECM called the B-box. The A-box reads the left primary and secondary wheel angle sensors. The B-box reads the right primary and secondary wheel angle sensors. The two ECMs communicate through "bridge connection" consisting of a dedicated harness connecting the two ECMs (A-box and B-box). If a problem occurs in one of the sensors, a fault code is activated, and the tractor's ECM quits reading the faulty sensor.

The A-box ECM is responsible for energizing the steering #1 enable solenoid, steering left #1 solenoid, and the steering right #1 solenoid. The B-box ECM energizes the steering #2 enable solenoid, steering left #2 solenoid, and steering right #2 solenoid. If the A-box or B-box encounters a problem, the good ECM will disable the faulty ECM.

The steering system has multiple features, such as variable ratio steering and variable effort steering. ***Variable ratio steering*** optimizes the number of turns the steering wheel must rotate based on the tractor's speed and the degree of sharpness of the turn being performed. ***Variable effort steering*** adjusts the turning effort required by the operator to best control the tractor under changing operating conditions, such as increased wheel resistance when turning at road speeds. A solenoid inside the wheel module (underneath the steering wheel), is used to apply a brake to apply resistance to the steering wheel to emulate the traditional resistance felt in an SCU steering system.

ActiveCommand Steering also eliminates steering wheel drift and prevents oversteering. Steering wheel correction is needed less often than with a conventional steering system.

An example of a John Deere 8R series ACS steering valve hydraulic schematic is shown in **Figure 25-28**. The steering system is comprised of four valve blocks bolted together: a steering priority valve, two sets of steering valve blocks (the ACS steering valve and the AutoTrac steering valve), and a parking brake pump-off manifold valve block. The two steering valve blocks are identical and provide redundancy in case a failure occurs in one of the steering valves. The schematic shows a right turn in progress. The position of each hydraulic valve during a right turn is indicated by yellow highlighting.

Both steering valve blocks are identical, having one nonproportional (on/off) steering enable solenoid, one steering enable spool valve, two steering solenoids (right and left) that are variable controlled with pulse-width modulation, and right and left steering spool valves. Both valve blocks are connected in parallel, receiving their supply of oil from the priority valve.

The tractor can be manually steered by the operator or steered hands-free with the assistance of GPS, known as the ***AutoTrac*** mode. During either of these modes, the A-box and B-box ECMs will actuate both of the steering valves. If a failure occurs in one of the steering control valves, the tractor will disable the operation of the bad valve and only use the good steering valve.

The transmission pump supplies pilot oil to the steering valve blocks. The pilot pressure operates the steering control spool valves located in each of the steering valve blocks, specifically the enable spool valve, right steer spool valve, and left steer spool valve. Whether the tractor is driving straight ahead or steering, the two ECMs energize the on/off steering enable solenoids (#1 and #2), allowing the tractor to be steered as commanded. With the enable solenoids energized, pilot pressure is routed to the right end of the steering enable spool valves, causing the valves to shift to the left. With the valves shifted to the left, the enable spool valves can route the oil from the steering priority valve to the left and right steering spool valves in each steering valve block.

When the tractor is commanded to steer right (either manually or with GPS-assisted AutoTrac), the ECMs pulse-width modulate the right steering solenoids (#1 and #2) in each steering valve block. The solenoids move a variable distance as commanded by the ECMs. As the steering solenoids are energized, they shift to the right (to the degree commanded by the controller). As they shift to the right, pilot pressure begins to build, causing the pilot pressure to shift the right steering spool valves to the right. As the steering spool valves shift to the right, they open a passage that allows oil from the priority valve and enable valves to flow to the steering cylinder(s). This causes the cylinder(s) to steer the tractor to the right. Oil exiting the steering cylinder(s) is routed back to the right steer spool valves, where it passes through the spool valves and is routed to the reservoir.

While the machine is steering to the right, the left steer solenoids #1 and #2 are de-energized, allowing their pilot oil to drain to the reservoir. As a result, the left steer spool valves are shifted to the left, which opens a passage that allows the steering working pressure to be sent to the signal network, which includes sending steering signal pressure to the steering priority spool valve and the pump flow compensator.

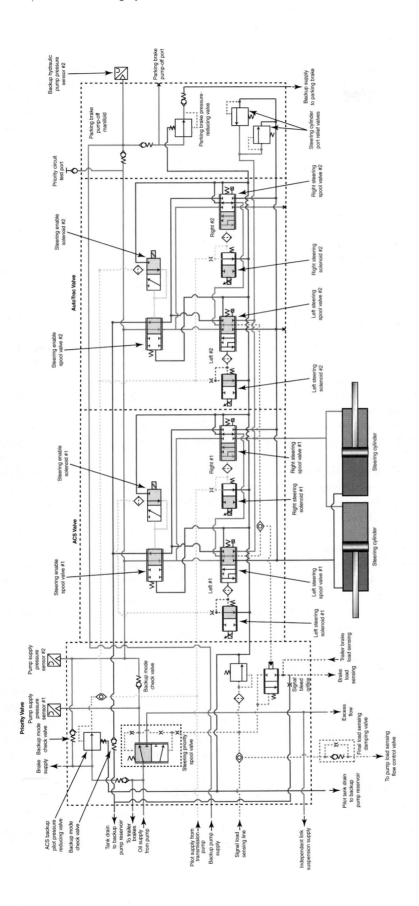

Figure 25-28. A hydraulic schematic of a John Deere 8R series tractor with ActiveCommand Steering that is steering to the right.

> **Pro Tip**
>
> 8R tractors with ACS have a secondary electrically driven hydraulic pump that operates if the engine is inoperable or the primary hydraulic pump fails. The secondary pump can operate in one of two different modes, a backup mode or when the tractor is placed in the tow mode. Whenever the tractor loses its primary oil supply and if the steering or brakes are actuated, the tractor enters the backup mode. The electrical motor will power the backup hydraulic pump, which supplies oil for steering and the service and parking brakes. The tow mode is entered by the operator removing a fuse and following a command sequence specified in the operator's manual.

If an 8R tractor is purchased without ACS, it will still have AutoTrac. One of the electronically controlled steering valves is not included, the ACS valve. A manually operated SCU is connected in parallel to the AutoTrac steering valve and steering cylinders. See **Figure 25-29.** The schematic is drawn with the tractor driving with hands-free steering to the right. The SCU is in a neutral position, and the AutoTrac valve is commanding the tractor to steer to the right.

Caterpillar Quick Steer

Caterpillar offers an electronic steering mode called *Quick Steer* on their small K and M series wheel loaders. The feature is best used when the wheel loader is cycling in short stops while loading trucks. With the machine in neutral and the parking brake off, the operator presses the appropriate button on the electronic monitor to put the steering system into Quick Steer mode.

On entering the Quick Steer mode, the machine's transmission is shifted into the first gear range. In this range, the loader has adjustable ground speed control and generates more flow for steering. Previously, the steering wheel had to be turned three revolutions to fully steer from stop to stop when the loader was stationary. The Quick Steer mode dramatically decreases the operator's turns of the steering wheel. It allows the operator to fully steer the loader from left to right (or vice versa) by rotating the steering wheel only from the 10 o'clock position to the 2 o'clock position.

In a Caterpillar 924K wheel loader equipped with Quick Steer, an ECM energizes a Quick Steer solenoid. When actuated, the solenoid hydraulically shifts a two-position Quick Steer spool valve. The spool valve is located in series between the SCU spool/sleeve and the SCU gerotor. When the Quick Steer spool valve is hydraulically actuated, it diverts the oil. The oil is prevented from passing through the SCU gerotor, causing the steering spool/sleeve to directly actuate the steering cylinders and resulting in an increased steering cylinder speed. See **Figure 25-30**.

Fendt VarioActive Steer

Fendt *VarioActive Steer* is an electronic steering mode that reduces the number of steering wheel revolutions required to steer full right to full left. Without VarioActive Steer engaged it takes close to four steering wheel revolutions to steer fully from one direction to the other. VarioActive Steer requires only two steering wheel revolutions to steer from fully in one direction to the other. The steering mode is entered using the tractor monitor's screen.

Case IH 4WD Combination Steering

The early Steiger tractors equipped with combination steering used a conventional SCU along with an electronically controlled double-selector valve. When the double-selector valve was de-energized, the valve directed oil to the articulation cylinders. When the steering electronic control module energized the double-selector valve, the valve directed oil to the front axle steering.

Chapter 25 Wheeled Steering Systems

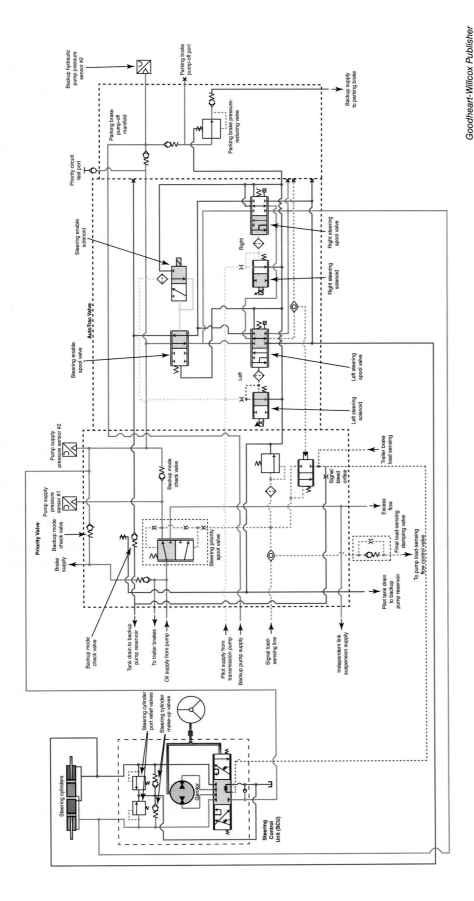

Figure 25-29. A hydraulic schematic of a John Deere 8R series tractor with AutoTrac and a manually operated SCU. The SCU is in a neutral position, and the AutoTrac valve is steering the tractor to the right.

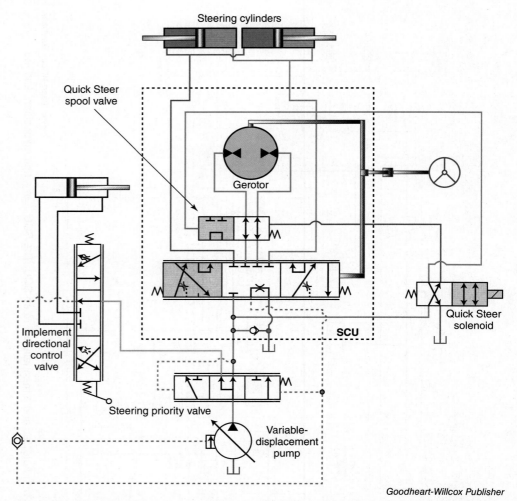

Figure 25-30. A simplified drawing of a Quick Steer solenoid and a Quick Steer spool valve.

Late-model STX Steiger AccuSteer tractors use electronic steering controls in conjunction with a traditional SCU to enable GPS auto guidance. The STX Steiger's steering controller monitors the following input data:

- Rear articulation axle potentiometer.
- Front articulation axle potentiometer.
- Steering control pressure sensor.
- Steering wheel motion sensor.
- Transmission output speed sensor.

Based on data received from the inputs, the steering controller can energize the following outputs:

- Front-steer left solenoid valve.
- Front-steer right solenoid valve.
- Left make-up solenoid valve.
- Right make-up solenoid valve.
- Selector bypass solenoid valve.

As shown in **Figure 25-31**, the steering priority valve delivers oil to the three control valves in parallel:

- Front-steering solenoid control valves.
- Make-up solenoid valves.
- SCU.

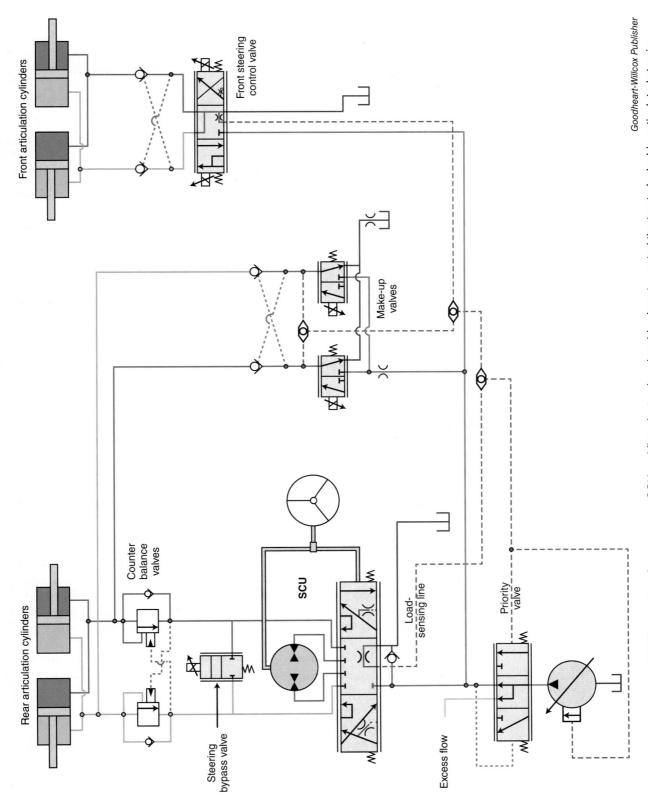

Figure 25-31. Case IH STX Steiger AccuSteer uses an SCU and five electronic solenoid valves to control the tractor's double-articulated steering. The SCU and make-up valves provide oil only to the rear articulation cylinders. The front articulation cylinders receive oil from the front steering control valve.

The SCU controls only the rear articulation steering cylinders. The bypass solenoid valve is a normally closed solenoid. When the bypass is de-energized, the SCU oil actuates the rear articulation cylinders. When the bypass valve is energized, a portion or all of the SCU oil can be bypassed and sent back to the reservoir instead of actuating the rear articulation cylinders. The make-up solenoids maintain the position of the rear articulation cylinders.

The front-steering solenoid valves directly control and maintain the position of the front articulation steering cylinders. The SCU, make-up solenoids, and bypass valve affect only the rear articulation cylinders.

Danfoss MultiAxis-Steer

Danfoss offers machine manufacturers a complete electro-hydraulically controlled four-wheel steering system that can be easily integrated to a machine. The system is called *MultiAxis-Steer*. It is fully electronically controlled, known as steer-by-wire. The system provides multiple steering modes: two-wheel steer, four-wheel steer, crab steer, dog steer, implement steer and custom steering modes. Examples of machines that can integrate this system are agricultural self-propelled sprayers, agricultural tractors, telehandlers, and wheel loaders.

The MultiAxis-Steer utilizes an electro-hydraulic steering control valve (EHi), an ECM labeled as PVED-CLS (which stands for *proportional valve digital-closed loop*), and several electronic inputs. Examples of electronic inputs include steering angle sensor (SASA), vehicle speed sensor, wheel angle sensor (WAS), man-machine interface (MMI), linear variable differential transformer (LVDT), and auto-guidance. A WAS is placed on the front axle, and sends master wheel angle input to the controller. A WAS is also placed on the rear axle and it is used as closed-loop feedback to the controller. Once the complete system is installed on the machine, the machine manufacturer inputs data into the controller, such as wheelbase length, distance between wheels, and maximum angles of the two steer axles.

Danfoss allows the operator to choose different steering modes, even while the machine is driving, based on the programming of two features: virtual axis position (VAP) and virtual axis angle (VAA). **Figure 25-32** illustrates multiple VAPs. When the VAP is positioned at the rear axle, the vehicle can have front two-wheel steer. When the VAP is positioned in the center of the machine, the vehicle can have four-wheel steer. When the VAP is extended beyond the rear axle, the vehicle can be configured with implement steer to improve steering while pulling trailing implements. With the VAP in the center, and when the VAA is adjusted while the machine is driving, the machine can be crab steered.

The operator chooses the VAP and VAA via an interface like the Danfoss HMR CAN rotary input sensor or control panel. The rear axle is controlled based on the position of the front axle. If the machine is travelling at a speed higher than the specified threshold, the rear axle is steered to a straight position and locked in place until the travel speed is reduced below the threshold.

In addition to being easily adaptable to a machine, the MultiAxis-Steer is SIL2 compliant. SIL is a certification standard, which stands for safety integrity level. SIL1-rated systems are the safest, and SIL4 systems present the highest risk.

The following are other advantages of the MultiAxis steering system listed by Danfoss:
- If used on sprayers, it can eliminate the extra tracks caused by the rear tire tracks enabling the rear tires to travel in the front tire tracks.
- If used on wheel loaders and telehandlers, the crab-steer function allows the machine to hug a fence line more closely when feeding livestock.
- If used on compactors, it can reduce the total number of passes needed, increasing efficiencies.

Caterpillar Command Control Steering

Some Caterpillar wheel loaders are equipped with Command Control Steering. Similar to Quick Steer, it reduces the operator's need to make multiple revolutions to steer the machine. The unique steering system does not use an SCU. A load-sensing hydraulic system

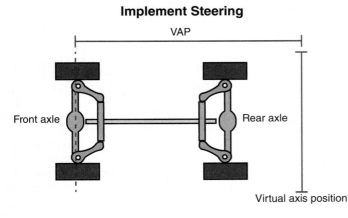
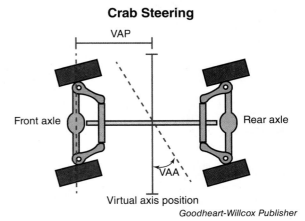

Figure 25-32. Danfoss offers machine manufacturers a complete fully-electronically controlled hydraulic steering system known as MultiAxis-Steer. The machine can adjust steering modes while driving.

supplies hydraulic fluid to a steering control valve that is actuated by a steering pilot valve. The steering pilot oil controls a steering control spool. Unlike SCUs that accept two to three complete revolutions of the steering wheel, the steering pilot valve limits the steering wheel to a rotation of 70° to the left and 70° to the right. The speed of the machine's steering is determined by the position of the steering wheel and is not controlled by the rotational speed of the steering wheel. See **Figure 25-33**.

A pressure-reducing valve directs pilot oil to the steering pilot valve and the implement control valves. As the operator actuates the steering wheel, the steering pilot valve directs pilot oil via a check valve and a neutralizer valve to the steering directional control valve spool. The steering directional control valve spool directs the steering pump fluid to actuate the steering cylinders. When the steering reaches the end of travel, a neutralizer valve (either the left or right valve) presses against its steering stop, causing the neutralizer valve to block pilot oil from acting on the steering directional spool valve.

Caterpillar Neutralizer Valves

Caterpillar wheel loader steering systems can be equipped with right and left *neutralizer valves*. When the loader reaches the end of its steering, a steering stop mechanically actuates the neutralizer valve, which acts like an electrical limit switch to hydraulically stop the steering. Caterpillar neutralizer valves can be used in traditional SCU steering systems or in the Caterpillar Command Control steering system.

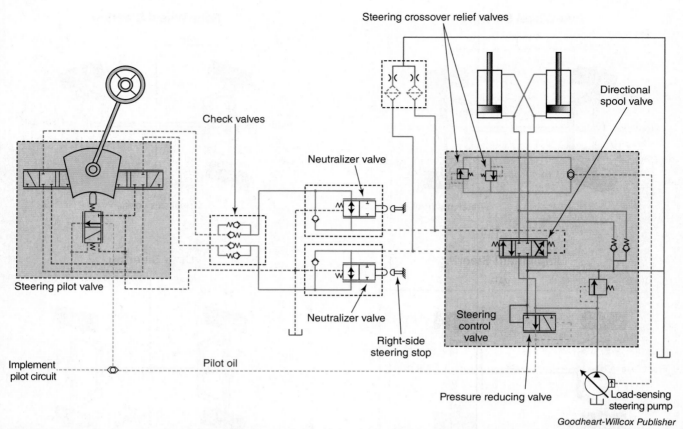

Figure 25-33. In Caterpillar Command Control Steering systems, a steering pilot valve actuates a steering directional control spool. The diagnostic pressure ports are not shown.

The neutralizer valves are located in the area of the articulation joint, **Figure 25-34**. In a Command Control Steering system, the left and right neutralizer valves are placed in series between the steering pilot valve and the directional spool valve. In an SCU steering system, the left and right neutralizer valves are placed in series between the SCU and the steering cylinder.

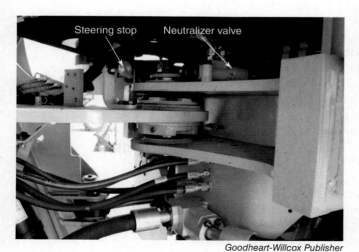

Figure 25-34. A steering neutralizer valve on a Caterpillar wheel loader.

Motor Graders

Motor graders have several unique features related to steering. They can be equipped with both articulation steering and front axle Ackerman steering. Two other features related to the front-steering axle are the differential speed of the front-wheel-drive graders and the wheel lean feature.

AWD and Steering

Motor graders can have six-wheel drive, known as all-wheel drive (AWD). As explained in Chapter 17, *Final Drives*, the rear tandems are driven with a powershift transmission that uses a differential to deliver power to two tandem chain drives. The front wheels are hydraulically driven. The Caterpillar M and non-suffix series motor grader has two separate hydrostatic drives that power the front drive wheels. While the machine is turning, the electronic control module (ECM) compensates so the front tires spin at different speeds due to the two different turning arcs. The ECM modulates the electronically controlled solenoid so the front outside tire spins at a faster speed than the front inside tire.

The front wheels of a motor grader have a hydraulic *wheel lean* function. This feature leans the tires in a parallel fashion to the left or right, up to 18° to either side. Wheel lean is helpful when the grader's blade is grading the soil, which can exert a force on the front axle causing it to drift sideways. The wheel lean compensates for this side thrust load, **Figure 25-35**.

Goodheart-Willcox Publisher

Figure 25-35. The wheel lean feature on motor graders offsets the side thrust on the front axle that is created when the blade is performing grade work.

Summary

- Four types of wheeled steering systems used in mobile equipment are dual-path hydrostatic transmission, Ackerman steering, articulation steering, and combination steering.
- The design of Ackerman steering allows the inside steered wheel to follow a narrower turning arc than the outside steered wheel.
- Steering designs used in Ackerman steering are manual steering, power-assist steering, and 100% hydraulic steering (SCU steering).
- In articulation steering, an articulation joint connects the front frame of the machine to the rear frame. Two hydraulic cylinders articulate (pivot) the two frames so the tractor steers to the left or to the right.
- A few off-road machines have combination steering that includes either articulation steering and Ackerman front-steer or two articulation joints.
- SCUs provide a true hydraulic steering system with no mechanical connection between the SCU and the steering cylinder.
- The purpose of the SCU's gerotor assembly is to meter the quantity of the steering oil.
- The purpose of the SCU's control spool/sleeve assembly is to control the direction of oil flow to a tractor's steering cylinder(s), which steers the tractor to the left or to the right.
- SCUs can be called hand metering units, hydrostatic steering, steering hand pumps, or steering control valves.
- A traditional open-center or closed-center SCU has four hydraulic ports—supply, return, left steer, and right steer.
- Most SCUs are closed-center and contain a fifth hydraulic port for the load-sensing signal line that is used in conjunction with a steering priority valve, and possibly an unloading valve or a variable-displacement pump's flow control valve.
- A load-sensing SCU is required if the tractor has a steering priority valve.
- An SCU contains a check valve at its inlet that connects the supply line to the return line, allowing the SCU's gerotor assembly to be used like a hand pump to supply emergency steering oil in the event of engine or hydraulic pump failure.
- A check valve located in series to the SCU's supply is used to prevent steering kickback.
- A steering inlet relief valve can be located inside the SCU or inside the steering priority valve and is used to limit the maximum steering pressure.
- The neutral position of a non-load reaction SCU blocks off the steering cylinder actuator ports.
- The neutral position of a load-reaction SCU has open actuator work ports.
- Wide-angle steering and cylinder damping are used in large articulated tractors to smooth bumpy steering motion.
- Q-Amp steering provides conventional gerotor metering if the steering wheel is turned slowly and increased flow if the steering wheel is turned quickly.
- A hydraulic steering amplifier is used to increase steering oil flow on large machines.
- The steering priority valve's controlled flow port supplies oil to the SCU.
- The steering priority valve's excess flow port supplies oil to the secondary circuits.
- A dynamic load-sensing signal line generates a low-pressure steering signal by constantly flowing oil in the steering signal line. Dynamic load sensing is used when the steering signal line is six feet or longer, or if the steering response is slow, less stable, or inconsistent.
- SCUs have an internal leakage amount specified in steering wheel rpms.

- Modern mobile hydraulic machines can be steered by a network of electronic controls, called *steer-by-wire*, instead of an SCU.
- John Deere's ActiveCommand Steering system provides variable ratio steering and variable effort steering. It eliminates steering wheel drift and prevents oversteering.
- Caterpillar Quick Steer provides an increase in steering oil flow when the loader is operating at a slow speed, enabling the operator to fully steer the loader with minimal movement of the steering wheel.
- Case IH 4WD combination steering (AccuSteer) provides both center articulation and front articulation steering.
- Caterpillar Command Control Steering is a steering system that uses load-sensing hydraulics and a pilot-actuated steering valve instead of an SCU. It controls the speed of the steering based on the position of the control valve and not the rotational speed of the steering wheel.
- Two neutralizer valves can be located in the articulation area of a machine. When one of the valves is pressed, as the machine reaches its end of steering, the valve prevents further steering.
- Motor graders can be equipped with both articulation steering and front axle Ackerman steering. The wheel lean feature compensates for side thrust load.

Technical Terms

Ackerman steering
ActiveCommand Steering system
articulation steering
auto-guidance steering systems
AutoTrac
cavitation
combination steering
controlled flow (CF)
cylinder-damping SCU
drag link
dynamic load-sensing signal
emergency steering
excess flow (EF)
gerotor assembly
inlet check valve
integral power steering gear
load-reaction SCU
load-sensing relief valve
manual steering check valve
MultiAxis-Steer
neutralizer valves
non-load reaction SCU
parallelogram linkage
power-assist steering
Quick Steer
recirculating ball steering gearbox
SCU internal leakage test
static load-sensing signal
steer-by-wire
steering control unit (SCU)
steering gear
steering kickback
steering knuckles
variable effort steering
variable ratio steering
VarioActive Steer
wheel lean
wide-angle SCU

Review Questions

Answer the following questions using the information provided in this chapter.

Know and Understand

1. Which steering system offers the machine engineers the advantages of flexibility and adaptability?
   A. Manual steering.
   B. Power-assist steering.
   C. Rack-and-pinion steering.
   D. Steering control units (SCUs).

2. Which SCU component is responsible for directing the oil in and out of the SCU?
   A. Driveshaft.
   B. Inlet check valve.
   C. Control spool/sleeve assembly.
   D. Gerotor assembly.

3. An SCU can be called by all of the following terms, *EXCEPT*:
   A. hand metering unit.
   B. steering hand-pump.
   C. steering control valve.
   D. selective control valve.

4. What two SCU components are coupled with a centering pin and centering springs?
   A. Gerotor and wear plate/spacer.
   B. Control spool and control sleeve.
   C. Driveshaft and gerotor.
   D. Control spool and end plate.

5. Which SCU configuration is designed to block both left and right steering cylinder ports when the SCU is in a neutral position?
   A. Load reaction.
   B. Load sensing.
   C. Non-load reaction.
   D. Q-Amp steering.

6. Which SCU configuration is designed to allow the external forces placed on a steering cylinder to return the SCU to a neutral position once the operator releases the steering wheel at mid-turn?
   A. Load reaction.
   B. Load sensing.
   C. Non-load reaction.
   D. Q-Amp steer.

7. Which SCU configuration smooths the steering operation of an articulated tractor by increasing the amount of deflection between the SCU's control spool and sleeve during a turn?
   A. Wide-angle steering.
   B. Cylinder damping.
   C. Load reaction.
   D. Non-load-reaction.

8. Within a steering priority valve, if the SCU is in a neutral position, through which port is the majority of the oil exiting the priority valve?
   A. Excess flow.
   B. Controlled flow.
   C. Load-sensing.
   D. Return.

9. A tractor has a fixed-displacement hydraulic pump, a steering priority valve, and a load-sensing SCU in its hydraulic steering system. Where is the signal pressure being sent?
   A. Hydraulic pump.
   B. Steering priority valve.
   C. Both A and B.
   D. Neither A nor B.

10. Engineers choose a dynamic load-sensing steering signal line for all of the following reasons, *EXCEPT*:
    A. signal line is 6′ or longer.
    B. steering is overly responsive.
    C. steering circuit is less than stable.
    D. steering is sluggish.

11. SCU leakage test results exceed specifications. The technician capped the hydraulic lines going to the steering cylinder and repeated the test. The second test results were within specifications. What is the problem?
    A. SCU.
    B. Steering cylinder.
    C. Both A and B.
    D. Neither A nor B.

12. A technician is reading a service manual to diagnose a possible leaking SCU. The service literature specification for the SCU is four revolutions per minute. If the SCU is bad, what type of results will the technician obtain during the test?
    A. One revolution per minute.
    B. Two revolutions per minute.
    C. Three revolutions per minute.
    D. Five revolutions per minute.

13. An 8R tractor equipped with Active Command Steering has how many steering wheel position sensors?
    A. One.
    B. Two.
    C. Three.
    D. Four.

14. Which term is used by John Deere to describe a tractor's GPS auto guidance system?
    A. ActiveCommand Steering.
    B. AccuSteer.
    C. AutoTrac.
    D. Quick Steer.

15. Which electronic steering option allows the operator to fully steer the loader from left to right (or vice versa) by rotating the steering wheel only from the 10 o'clock position to the 2 o'clock position?
    A. John Deere ActiveCommand Steering.
    B. Caterpillar Quick Steer.
    C. Auto guidance steering system.
    D. Case IH AccuSteer.

16. Which electronic steering option provides both articulation and front-axle steering?
    A. John Deere ActiveCommand Steering.
    B. Caterpillar Quick Steer.
    C. Auto guidance steering system.
    D. Case IH AccuSteer.

17. Which of the following Caterpillar steering systems determines the speed of steering by the position of the steering wheel and not the speed of the steering wheel?
    A. Ackerman steering.
    B. Combination steering.
    C. Command Control Steering.
    D. Dual-path hydrostatic steering.

18. Which machines can hydraulically lean their front steering wheels?
    A. Agricultural tractors.
    B. Motor graders.
    C. Skid steers.
    D. Wheel loaders.

Apply and Analyze

19. A(n) _____ steering system uses both articulation steering and front-axle steering.

20. A(n) _____ steering system pivots the machine's two frames with a pair of hydraulic cylinders to steer the machine.

21. A wheeled machine with _____ uses two pumps and motors, one pump and motor on each side of the machine, to propel and steer the machine.

22. The SCU component responsible for directing steering oil to the steering cylinder is the _____.

23. In the emergency steering mode, the _____ check valve allows the SCU to be operated as a small hand-pump to provide oil for steering.

24. The inlet check valve located in series to the SCU's supply is responsible for preventing _____.

25. The _____ are the SCU components responsible for keeping the control spool/sleeve assembly aligned while the SCU is in a neutral position.

26. The _____ is the SCU component responsible for metering the quantity of steering oil sent to the steering cylinder.

27. The _____ steering priority valve port supplies inlet flow to the SCU.

28. _____ is the John Deere electronic steering option that provides variable ratio steering and variable effort steering.

Critical Thinking

29. Describe the correct order of oil flow through an SCU during an active steering condition.

30. Describe the operation of a steering amplifier.

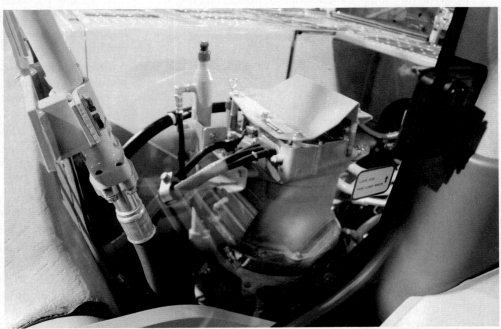

Goodheart-Willcox Publisher

The swing motor/generator of a Komatsu HB 365LC-3 hybrid excavator is shown here. The swing motor/generator functions as a motor during swing operations and acts as a generator to recapture energy during swing braking.

Chapter 26
Electric and Hybrid Drive Systems

Objectives

After studying this chapter, you will be able to:

- ✓ Explain how an alternator generates electricity.
- ✓ Describe the different types of traction motors.
- ✓ Explain the operation and unique features of electric-drive mining trucks.
- ✓ Explain the operation of electric-drive wheel loaders.
- ✓ Explain the operation of electric-drive dozers.
- ✓ List examples of electric and hybrid-electric excavators.
- ✓ Describe safe practices for working on machines with electric drive.

Off-highway equipment manufacturers have produced electric propulsion systems (electric-drive systems) for decades. Electric locomotives date back to the 1870s. General Electric (GE) was the first to manufacture an electric drive for mining haul trucks in 1963 with the introduction of the 772 electric drive. See **Figure 26-1**.

Electric drives are continuing to gain popularity. Some manufacturers use a diesel engine to drive one or more electric generators that produce(s) electricity primarily for propulsion. Examples include locomotives, haul trucks, dozers, and wheel loaders.

Some machines use high-capacity batteries in place of diesel engines to power the machine. The batteries are designed to deliver enough power to complete up to an eight-hour shift and then require recharging. Some hybrid machines use batteries or capacitors to capture energy during cyclical machine functions like shuttle shifting wheel loaders, swinging an excavator boom, or lowering an excavator boom. The time it takes to charge the batteries is one of the challenges for the industry. Manufacturers

Goodheart-Willcox Publisher

Figure 26-1. This GE 772E DC electric drive motor was used to propel an electric-drive haul truck.

are continuously investigating methods to decrease downtime and increase machine's availability time. Some solutions have technicians swapping out discharged batteries for charged batteries in place of discharged batteries. Another solution is for technicians to use fast mobile electrical charging stations that can be hauled to the job site. The weight of the batteries also poses challenges and requires manufactures to redesign the machine's weight distribution.

Robert LeTourneau, a key electric propulsion innovator in heavy equipment, developed diesel electric-drive wheel loaders, scrapers, and haul trucks. Joy Global purchased LeTourneau Technologies in 2011. Komatsu purchased Joy Global in 2017 and continues to manufacture electrically propelled machinery. In the 1970s, Euclid manufactured a unique electric-drive haul truck that was powered by a gas turbine.

Although construction equipment manufacturers have predominantly chosen powershift, mechanical drive, and hydrostatic drive propulsion systems, new electric-drive technology has been developed to help meet stringent emission requirements and increase fuel economy. A diesel-electric propulsion system uses a diesel engine to drive a generator. The *generator* converts the engine's mechanical rotational energy into electrical energy (electricity). The electricity is used to drive electric motors.

The electric propulsion system can also be used for braking, powering accessory drive systems, or for load testing the diesel engine. In addition to electric-drive mining trucks and locomotives, manufacturers produce numerous other types of electric-drive equipment. A few examples include wheel loaders, dozers, excavators, shovels, on-highway semi-trucks, mobile cranes, and backhoes.

Fully electric machines eliminate the diesel engine and are 100% emission free. The machines can be operated indoors or underground without the hazard of diesel exhaust fumes. Electric-drive machines can lower the time required to perform maintenance and reduce the total operating costs. They have reduced vibrations and can deliver instant torque rather than have a lag as a diesel engine ramps up to its rated speed. This allows them to provide more consistent performance. They also eliminate the difficult cold-starting issues that accompany diesel engines.

Electric-drive machines eliminate the engine oil and filter, diesel fuel and filters, air filters, mufflers, and engine after-treatment systems such as diesel exhaust fluid, EGR, and diesel particulate filter systems. Although engine coolant and radiators are no longer needed with the absence of an engine, electric drives still might use coolant to cool inverters, brake resistors, generators, and drive motors. Some electric motors and generators are air-cooled with a fan that is driven by a hydraulic motor requiring lots of hydraulic horsepower. Some generators and electric motors are cooled using their own dedicated hydraulic reservoir, coolers, and pumps.

Electric-drive machines can operate in environments that are off-limits to noisy diesel-powered equipment, such as urban locations like hospitals or schools. The in-cab noise can be reduced by four decibels. On average, the outside noise level of an electric machine can be ten decibels lower than for equivalent diesel-powered machines, and for every increase of three decibels, the overall volume is increased two times.

One target of the zero-emission machines is the rental equipment business. Some machines are battery-powered while others are powered by an AC electrical power cable. In the largest mining applications, an electric-powered shovel can only be used at a mine site that has affordable electrical power available and allowable for that mine site. The machine is less nimble than a diesel-powered shovel that can traverse throughout the site without being tethered to high-voltage power cables.

Alternators

The most common type of electrical generator is the alternating current (AC) generator, also known as an *alternator* or *traction alternator*. See **Figure 26-2**. The alternator is coupled to the diesel engine, which drives the alternator's rotor proportional to engine speed.

The alternator's ***rotor*** is an electromagnet consisting of a rotating, star-shaped iron core. Multiple coils of wire, called field poles, are placed around the rotor. See **Figure 26-3**. The alternating poles form north pole and south pole magnets when electricity is applied to the rotor. This is known as *exciting the field.*

The rotor spins inside a stationary housing called a ***stator***, which has multiple fixed windings. As the excited rotor passes by the stator's stationary windings, a voltage is induced in the windings. The voltage's waveform alternates from a positive voltage to a negative voltage. The alternating wave is known as an ***AC sine wave,*** **Figure 26-4.** One sine wave is called a ***cycle***. The machine's electronic control module (ECM) controls the electrical output of the alternator by controlling the rotor's field current. The field current is the amount of amperage applied to the rotor's field coils. A larger amount of field current causes the rotor to exhibit a larger amount of magnetism, resulting in more electrical current being induced in the stator's windings.

Goodheart-Willcox Publisher

Figure 26-2. This GE alternator is used to drive mining haul truck drive motors.

 Note

The number of pairs of poles determines the number of AC sine waves that are generated for every rotor revolution. For example, if the rotor has eight poles—four north poles and four south poles—it will produce four AC sine waves per rotor revolution.

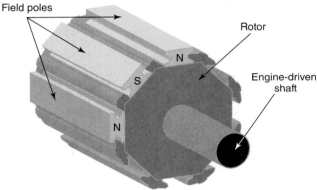

Goodheart-Willcox Publisher

Figure 26-3. An alternator contains a rotating iron rotor assembly that has multiple field poles. Every other field pole is wound in opposite directions to form alternating north and south electromagnetic poles. The stator windings are not shown.

Diodes are electronic devices that allow the flow of current in one direction only. Diodes are used to ***rectify*** AC sine waves; in other words, to convert the alternating current to direct current (DC) flow. Even if a machine uses AC electric drive motors, manufacturers commonly rectify the alternator's AC current to DC current. To operate AC electric drive motors, the DC current must then be inverted back to AC current. Manufacturers that use AC motors prefer rectifying the AC generator's current to DC, and then inverting the DC back to AC because of the difficulty in being able to accurately control the speed and torque in a straight AC generator to AC motor circuit. This process will be explained later in this chapter.

A diode acts in a similar manner to a hydraulic one-way check valve, allowing current to flow in one direction (forward) and blocking current flow in the opposite direction (reverse). A diode allows positive current to flow (known as *forward biasing*), but blocks a negative current trying to flow in a reverse direction (*reverse biasing*). A diode can block the negative portions of AC sine waves, leaving only the positive portions of the waves available to perform electrical work. See **Figure 26-5**. This is called *half wave rectification*. The challenge of half wave rectification is that the current pulses, making it impractical for performing electrical work.

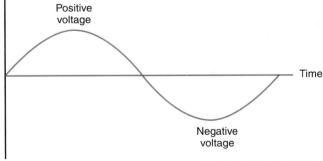

Goodheart-Willcox Publisher

Figure 26-4. A single-phase AC sine wave rises to a positive peak and then alternates to a negative peak. The single wave that starts from zero volts to maximum positive volts to maximum negative volts and back to zero volts is known as a *cycle*.

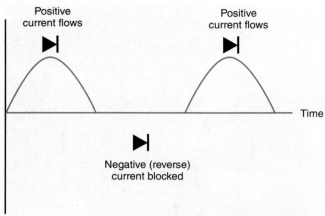

Figure 26-5. A diode can be used to rectify an AC sine wave by eliminating the negative current flow.

As the rotor's north and south poles alternatively pass by the stator, the current induced in the stator reverses direction. A ***bridge rectifier*** can be used to rectify reversed current flow into positive current. For single-phase AC, the bridge rectifier contains four diodes in a series block design.

In **Figure 26-6**, two of the diodes rectify the first half of a sine wave, delivering a half wave to drive the electric motor. Note that the rotor is depicted as a rotating horseshoe-shaped magnet to simplify the drawing. The other two diodes rectify the current that is flowing in a reverse direction. As the rotor's south pole passes by the stator, current is induced in the stator, which directs the current in the opposite direction to the bottom of the bridge rectifier. See **Figure 26-7**. The lower right diode directs the current to the electric motor. The electric motor directs the remaining current flow back to the bridge rectifier, where the upper left-hand diode directs the current back to the stator.

Note

Bridge rectifiers for single-phase AC consist of four diodes. Six diodes are used in a bridge rectifier to rectify three-phase AC.

As an electro-magnet with one north pole and one south pole makes a revolution within a single stator, two half waves are generated, with zero volts where the half waves meet. Manufacturers can alleviate this problem by designing alternators to produce three separate phases of current flow. Three different stator windings are used to produce ***three-phase AC***, **Figure 26-8**. The three AC phases are then rectified to produce

Figure 26-6. When the north pole passes by the stator, a positive current exits the stator and enters the top portion of the bridge rectifier. The top right diode directs the current to the electric motor to drive the motor. After the load, the current is directed back through the bridge rectifier through the bottom left diode, and back to the stator.

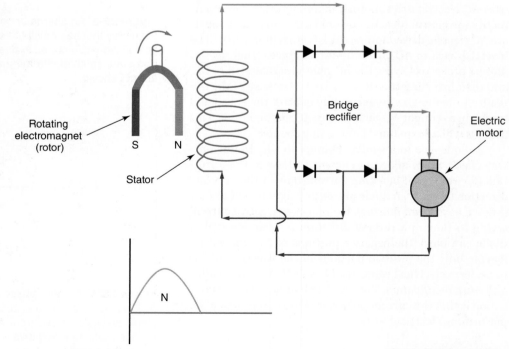

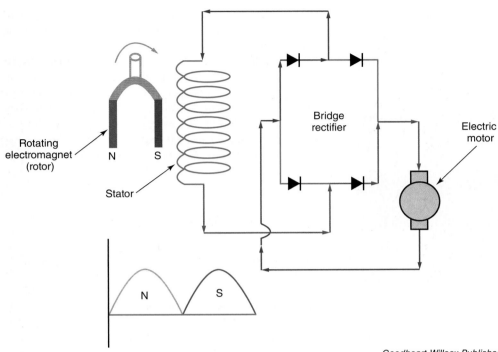

Figure 26-7. The graph shows the half of the sine wave produced as the north pole passed by the stator plus the half wave produced as the south pole passed by the stator. If a single diode is installed between the stator and motor, the part of the sine wave produced by the south pole would be blocked. However, both portions of the sine wave are usable if a bridge rectifier is placed in the circuit. The graph shows how a sine wave induced by a single north and south pole magnet can be rectified into two positive half waves.

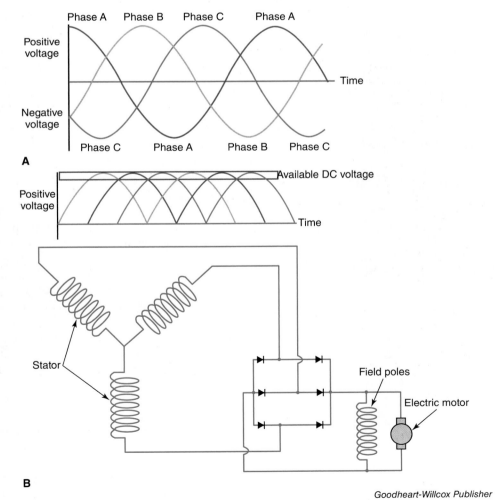

Figure 26-8. Traction drive systems often use alternators that produce three phases of current flow. A—Notice that the three phases alternate across a period of time. B—A six-diode bridge rectifier rectifies the three-phase AC into positive DC.

Figure 26-9. This traction alternator has a rotating rotor with multiple field windings. The rotor spins inside a stationary set of conductors called *stator windings*.

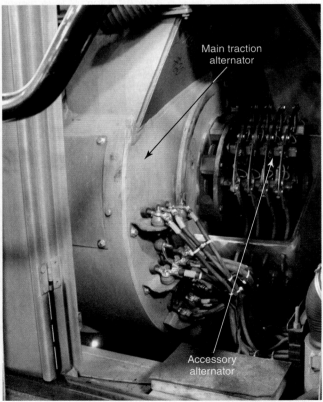

Figure 26-10. This locomotive alternator powers six traction motors and the locomotive's accessories. The large portion on the left drives the locomotive. The portion on the right provides electrical current for the accessories.

a DC current flow with a consistent positive voltage. The alternator in **Figure 26-9** provides a view of the rotor windings and stator windings.

The locomotive traction alternator in **Figure 26-10** generates electrical energy to drive six traction motors and accessories. The larger portion of the alternator is used to drive the traction motors. The smaller accessory alternator, located on the rear, excites the main alternator's field, charges the locomotive's batteries, and powers the locomotive's accessories, such as the three-phase radiator fan.

Traction Motors

Electric-drive systems can employ different motor designs. Motors are designed to be driven by AC or DC. Most electric motors normally deliver power to some type of gearing that increases their torque. On a locomotive, the traction motor often drives an external-tooth pinion gear that drives an external-tooth bull gear on the drive axle. In a mining truck, the electric motor commonly drives a planetary gear set's sun gear.

The following are examples of drive motors:
- DC motors.
  - Brush-type.
  - Brushless.
- AC traction motors.
  - AC induction.
  - AC brushless.
- Reluctance motors.
  - Switched reluctance motors.
  - Synchronous switched reluctance motors.

DC Traction Motors

DC traction motors are the oldest design and were the preferred motor used in locomotives for decades. DC motors are now rarely used in electric-drive systems. DC motors can have several different designs. Two distinct classifications are brush-type DC motors and brushless DC motors.

DC motor speed is controlled by varying the voltage to the motors. When the DC voltage is increased, the current flow in the windings is increased, resulting in a larger electromagnetic force that causes the motor's speed or torque to increase. The motor's direction is reversed by reversing the polarity (reversing the current flow).

Brush-Type DC Motors

The ***brush-type DC motor (BDCM)*** contains an ***armature*** (sometimes called a *DC rotor*) that consists of several loops of large copper conductors connected to copper commutator bars. See **Figure 26-11**. The ***brushes*** are electrical contacts made of carbon and are wear items, **Figure 26-12**.

Chapter 26 | Electric and Hybrid Drive Systems

Goodheart-Willcox Publisher

Figure 26-11. This GE armature is used in a DC electric drive motor for a mining truck.

Goodheart-Willcox Publisher

Figure 26-12. This brush is used in a locomotive's DC traction motor.

A *commutator* bar is a copper bar that serves as an electrical contact that transfers electrical current between one brush and one end of an armature's electrical conductor. A group of commutator bars are arranged around the circumference of a DC motor. See **Figure 26-13**. A commutator receives electrical current from a positive brush. Each commutator is connected to one of the armature's hard wire conductors. DC current is passed from the brush to the commutator through the hard wire conductor. The DC current is then directed through a commutator bar located 180° from the other commutator bar. The current is directed out of the armature through the negative brush. See **Figure 26-14**.

The current sent through the armature's conductors causes the conductors to form electromagnetic poles and emit electromagnetic lines of force. The armature is located inside a group of fixed windings known as a *stator*. The stator's windings (called *field windings*) are arranged around the inside circumference of the motor housing. Stator windings, when energized, become electromagnets, forming north and south pole magnets. The magnetic force of the armature and the stator causes the electromagnetic forces between the armature and the stator to attract, causing the armature to spin. The armature is splined to the motor's output shaft.

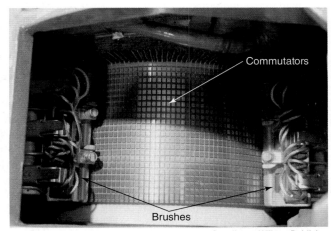

Goodheart-Willcox Publisher

Figure 26-13. The six brushes and commutators are viewed by removing the access cover on this DC traction motor.

Note

Two magnetic poles that are alike (two north poles or two south poles) repel each other. Opposite poles attract to one another. As current passes through the armature, the armature tries to move away from poles that are alike and move toward the opposite poles, causing the armature to rotate. As the armature spins, the commutators rotate past the brushes, allowing the current to flow through the next armature conductor. This sequencing action of current flowing through the next set of commutator bars is known as **commutation**. Commutation enables the motor's armature to generate the correct polarity of electromagnetic force, resulting in a continuous armature rotation.

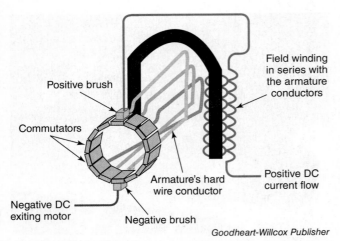

Figure 26-14. A BDCM contains an armature with copper commutator bars connected to hard wire conductors. Brushes direct current through the armature's wires. As the stator windings are energized, a magnetic field is created. This field forces the armature's current-carrying conductors to spin away from the magnetic field, causing the motor to rotate. The laminated steel stampings are not shown to better view the armature's wire conductors.

The brushes wear as they ride along the rotating commutator bars. For this reason, BDCMs require more service than other traction motors and are seldom used in modern electric-drive systems. Some locomotive manufacturers still offer BDCM drives in overseas applications.

The armature's core consists of numerous steel laminated stampings placed against one another. The stampings have slots to hold the hard wire conductors. The laminated stampings improve the magnetic field generated by the armature's hard wire conductors. Individual laminated stampings are used in place of a single iron core to improve the magnetic efficiency of the armature. A solid iron core armature would generate eddy currents, causing the motor to lose efficiency. Eddy currents are current flow resulting from counter-electromotive force that is induced in the armature's core and works against the motor's rotation.

BDCM designs are based on how the motor's windings are wired in conjunction with the armature's conductors. These designs include the following:

- Series.
- Shunt.
- Compound.
- Permanent magnet.

The series BDCM is the design used in traction motors. Series BDCMs provide maximum torque from a stop (excellent low-speed torque). In a shunt-type DC motor, the windings are connected in parallel with the armature. A compound BDCM has a series-parallel design. Permanent magnet motors do not have stator field windings. The shunt, compound, and permanent magnet DC motors are used in lower-horsepower applications. These applications include fan blowers, windshield wipers, and some small engine starter motors.

Brushless DC Motor

The rotor of a *brushless DC motor (BLDCM)* has a rotating permanent magnet surrounded by pairs of coil windings. The motor's electronic controls energize the coil windings, causing the rotor to spin. This is known as *electronic commutation*. As the coil windings are energized, the magnetic fields created cause the rotor's permanent magnets to spin the rotor. As previously mentioned, like magnetic poles repel and opposite poles attract.

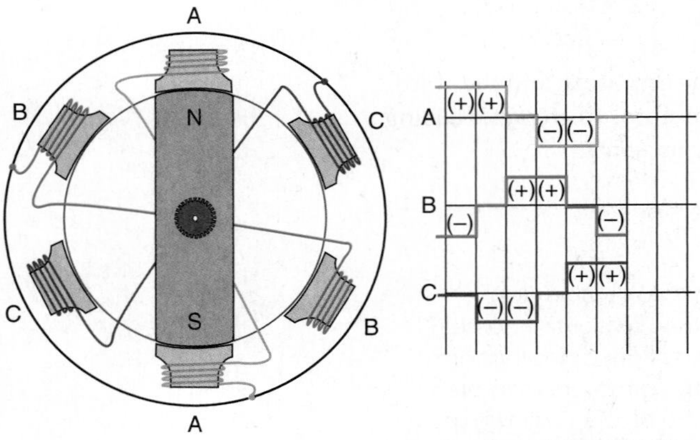

The electronic commutation of a BLDCM energizes a pair of coil windings that are 180 degrees opposite each, wound in opposite directions, and are connected in series to form an electromagnet. See **Figure 26-15**. When they are energized, one coil winding becomes the north pole magnet and the other coil winding becomes the south pole magnet. The electronic commutation energizes the coil windings in sequence with the correct polarity to cause the permanent magnet to be spun due to the pulling force or pushing force of electromagnetic coils. See **Figure 26-15**. BLDCMs are most commonly used in smaller-horsepower applications. BLDCM weigh less than BDCM for a given output, and do not experience sparking that can occur when brushes wear.

Figure 26-15. A BLDCM contains a magnet that is rotated when the windings are energized in the proper sequence. Electronic controls are used to correctly energize the windings with the correct polarity. The pairs of windings are labeled A, B, and C.

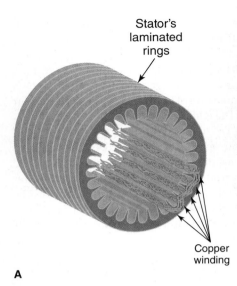

Figure 26-16. An AC induction motor contains a stationary stator. A—The stator has numerous laminated steel rings placed next to each other. Copper wire is placed in channels inside the body of the stator. B—This stator is from a Liebherr KDF 1501 AC induction motor. It is rated at 2000 hp (1500 kW) and operates at 1310 volts and 770 amperes.

AC Traction Motors

The brushes used in BDCMs wear over time, requiring the motors to be repaired. As a result, some manufacturers choose to use **AC traction motors** that are driven with AC drive systems. AC traction motors have two different designs—traditional AC induction traction motors and AC permanent magnet motors. Both types are found in diesel electric propulsion systems.

AC Induction Motor

One of the first types of AC traction motors employed in electric drives is the **AC induction motor**, which does not use commutators or brushes. An AC induction motor contains two primary components—a stationary stator and a rotating rotor. The stator is composed of numerous laminated stamped steel rings that are placed next to each other. Copper wire is wound and placed in channels inside the laminated stator body. The stator's windings have a clear resin coating. This coating insulates the conductors and prevents them from shorting with each other and the iron stator housing. See **Figure 26-16**. Multiple phases of AC electrical current (for example three-phase) is applied to the stator's copper windings, which results in a rotating magnetic field (rmf) that causes the rotor to rotate. The multiple phases of alternating current flowing through the field cause a uniform magnetic field to rotate. The speed of this rotating uniform magnetic field is known as synchronous speed.

A rotor, resembling an iron squirrel cage, is located inside the stator. See **Figure 26-17**. The ends of the rotor are made up of steel rings. Numerous bars, known as *conductor bars*,

Figure 26-17. An AC motor's rotor resembles a squirrel cage assembly that allows a current to be induced, causing it to rotate. The rotor's conductor bars can be parallel to the axis of the rotor, or arranged helically around the axis of the rotor. A—This rotor is from a Liebherr KDF 1501 AC induction motor. Note that the conductor bars are parallel to the rotor axis. B—A cutaway of an AC induction motor inside a belt drive used in a mining application. The conductor bars are arranged helically around the rotor axis.

connect the two rings, forming the cage. When alternating current is applied to the stator's windings, the rotating magnetic field (RMF) generates current in the conductor bars, which creates another magnetic field around the rotor. The attraction and repulsion between the stator's magnetic field and the rotor's magnetic field cause the rotor to rotate. Because the induction of current in the rotor causes the rotor to spin, the motor is known as an *AC induction motor*. The core of the rotor is made up of laminated steel discs, similar to those in the armature of a DC motor, to reduce the eddy currents.

Manufacturers commonly take the rectified DC current and invert it back into AC. Varying the speed and torque of an AC motor requires electronic controls to vary the AC *frequency* (number of cycles per second). This variable AC electric drive is often called **variable frequency drive (VFD).** A VFD is an electronic controller that varies the AC frequency sent to an AC motor. By varying the AC frequency, the VFD effectively controls the AC motor's output speed and torque. As the VFD reduces the AC frequency, the motor's speed is reduced. As the AC frequency is increased, the motor's speed increases. It is difficult to vary the frequency of three-phase AC for the purpose of controlling the speed of an AC motor. Therefore, manufacturers first rectify the alternator's AC into DC, which is easier to work with, and then invert it back into AC of the desired frequency.

> **Note**
>
> Advancements in modern electronics made variable frequency drives cost-effective and practical. Without these advancements, the majority of traction motors would still be BDCMs, which require less electronic sophistication to drive. Sophisticated AC drives require electronic controls that can rectify the AC to DC, invert the DC back into a variable AC output, and be able to handle high current flow (1000 or more amps) and high voltages (2600 or more volts).

An AC induction motor is also known as a *frequency motor* or **asynchronous motor.** The term *asynchronous* means the rotor spins at a speed slower than the speed of the induced magnetic field. The slower speed of the rotor is described as *rotor slip*. Locomotives; Caterpillar 794AC, 796AC, 798AC mining trucks; Liebherr T 236, T 264, T 274, and T 284 mining trucks; Komatsu 730E, 830E, 860E, 980E mining trucks; Hitachi EH3500AC, EH4000AC, and EH5000AC mining trucks; and Hitachi EX1900E-6 through EX8000E-6 electric-drive excavators use AC induction motors.

Permanent Magnet AC Motors

A **permanent magnet AC (PMAC) motor** is also called a *permanent magnet synchronous motor* and a *brushless AC motor*. See **Figure 26-18**. Like the brushless DC motor, a PMAC motor contains stationary coil windings in the stator and a rotating rotor with permanent magnets. However, the BLDC motor's coil windings are wound around individual stator iron poles and the stator in a PMAC motor is similar to a traditional AC induction motor stator with slots that hold the multiple phases of coil windings. Another difference is that a PMAC motor is driven with AC sine waves instead of DC trapezoidal square waves. One negative trait of BLDC motors occurs when the electronic controls commutate the coils, (shut off one coil and energize the next), the motor exhibits torque ripple, small decreases or increases in torque, due to the square wave DC coil energization. The PMAC motors use a smooth AC sign wave that eliminates the torque ripple associated with BLDCs.

Goodheart-Willcox Publisher

Figure 26-18. This Liebherr electric-drive travel motor is equipped with an AC permanent magnet drive motor that powers the planetary reduction final drive. The electric-drive motor is used in place of the traditional hydraulic motor. It is found in mining diesel electric drives, such as those on mobile crushers and mobile screens.

Compared to AC induction motors, PMAC motors in general offer a wider operating speed range, higher efficiency, and cooler operation that results in longer life of components. A PMAC motor also has more *power density* than an AC induction motor, meaning that it delivers more horsepower for a given motor size. The rotor also rotates at the same speed as the magnetic field making it a synchronous motor as compared to the asynchronous AC induction motor. Like other AC motors, a PMAC requires a VFD to operate. The John Deere 644K /644X wheel loader uses a PMAC motor.

BLDCMs and PMACs use rotors with permanent magnets. When the rotor is spinning, a backward *electromotive force (EMF)*, also known as *back electromotive force (BEMF)* and *counter-electromotive force (CEMF)*, is induced (generated) in the stator windings. EMF is voltage. When no current is applied to a BLDCM or PMAC, the permanent magnet rotor can be spun due to the machine's inertia. As the magnets spin inside the stator's housing, they cause a voltage to be produced. As a result, the motor acts as a generator and produces a voltage acting in the opposite direction as that normally used to drive the motor. If the motor is allowed to overspeed due to inertia, it can cause the voltage to increase to a damaging level.

If the motor is being electrically driven (due to current being applied to the motor), the motor still produces a CEMF. However, the voltage level of the CEMF is less than the voltage used to drive the motor. The motor's output torque is a function of the difference between the voltage being applied to the motor minus the CEMF created by the rotating motor.

Reluctance Motors

Another style of electric drive motors are reluctance motors. The two types include the switched reluctance and the synchronous reluctance motor.

Switched Reluctance Motors

A *switched reluctance motor (SRM)* has a rotor and a stator. The motor is simplistic and economical. The rotor does not have windings or permanent magnets. A SRM has an iron rotor with multiple poles. The rotor is made of many thin steel plates laminated together. The poles are magnetically permeable and have no conductors. The stator windings are separate individual coils, also known as concentrated windings, that can be replaced individually rather than requiring the entire stator winding to be replaced. The stator design can be called dual pole due to the opposing stator poles with concentrated coil windings. The only wear items on this type of motor are two bearings. Compared to AC motors, SRMs are heavier and noisier. SRMs also exhibit high torque ripple.

Although the motor is simplistic, the electronic controls are sophisticated and expensive to develop and produce. The control module energizes each pair of the stator windings independently. Reluctance (magnetic resistance) is developed between the aligned and unaligned rotor poles and stator poles. As the controller energizes pairs of stator windings (ahead of the rotor poles) in a synchronized fashion, the rotor is pulled, causing the stator poles to align with the rotor poles and resulting in rotation. The movement of the rotor results from the rotor taking the path of the least magnetic reluctance. Electronic controls synchronize the engagement of pairs of stator windings. This consecutive energizing of windings is called *sequential switching*. See **Figure 26-19**.

Switched reluctance (SR) is continuing to gain more popularity in the electric-drive sector. Caterpillar 988K XE wheel loaders, R1700XE LHD loaders, and D6XE dozers use SR. The John Deere 944K wheel loader uses four

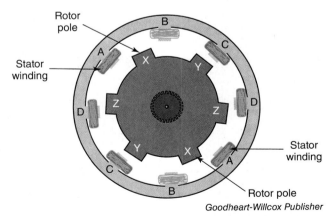

Goodheart-Willcox Publisher

Figure 26-19. In a switched reluctance motor, the stator contains windings that are energized in pairs, causing the rotor to turn along the path of least reluctance. The control module energizes the *A* stator windings to pull the rotor *X* poles counterclockwise. When the poles align, the control module next energizes the *B* stator windings, which pull the *Y* rotor poles into alignment with them.

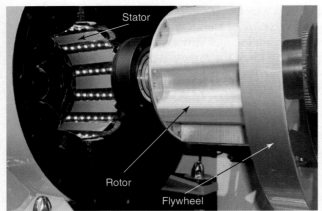

Figure 26-20. This Joy Global SRM has a rotor with a flywheel used to store energy. The stator and rotor are formed from numerous laminated steel plates.

switched reluctance motors. Joy Global/Komatsu Mining electric mining machines also use SRMs. The rotor within a Joy Global/Komatsu Mining SRM is made from the same steel stamping as the stator. The centers of the plates are cut out and laminated into a single rotor and the outside of the plates are laminated into the stator. The rotor assembly has no windings or permanent magnets. The stator uses independent wound coils as compared to traditional AC induction motors that have overlapping wire in the stator coils. The rotor in **Figure 26-20** has a flywheel used to store energy. This type of rotor is explained later in the chapter.

SRMs can also be used as generators during braking. During braking, the SR motors generate electrical current that is delivered to the SR generator, which acts like a motor assisting the engine to provide loader hydraulic functions. During braking, if the electrical energy generated by the motors becomes excessive, the controls will deliver electrical energy to the braking grids where it is dissipated as heat energy.

Controlling Switched Reluctance Motors

The torque of a switched reluctance motor is varied using insulated-gate bipolar transistors (IGBT) that act like on/off switches to apply DC electrical current to energize the motor stator's coils sequentially. The IGBTs are not variable, they are either switched fully open or fully closed. One IGBT is used to supply positive DC current from the positive DC bus to a stator coil. The DC bus is the electrical conductor that receives the rectified DC current. Another IGBT is used to direct the current exiting the stator coil to the negative DC bus. See **Figure 26-21**.

The Joy Global DC positive bus can have voltages as high as 700 DCV. One IGBT can handle 75 amps of current flow, therefore multiple IGBTs are connected in parallel to handle the amperage needed to drive a SR motor.

Figure 26-21. This simplified drawing shows that IGBTs are used as "on/off switches" to direct current through the SRM's stator coils.

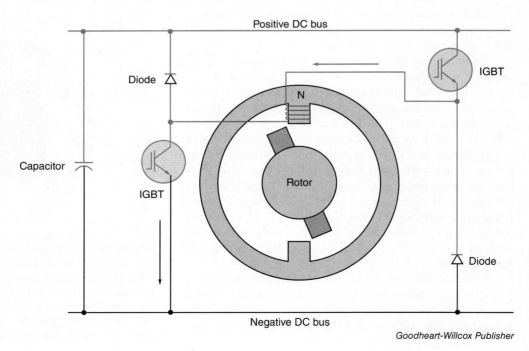

Synchronous Switched Reluctance Motors

The *synchronous switched reluctance motor (SynRM)* or (RSM) uses the same style stator that is used in traditional AC induction motors. It contains multiple phase stator windings distributed around the iron stator. The stator is sometimes described as a single salient pole with distributed windings. The rotor can be one of two types. Both rotors look similar from the end view. See **Figure 26-22**. One is made of stampings laminated axially with curved cavities (flux barriers) to provide magnetic reluctance and the cavities are surrounded by iron curvatures (known as flux carriers). The other type is assembled from radial laminations of steel and insulators rather than air pockets. The stator is energized with traditional AC inverters. The motor was first introduced in the 1920s.

A SynRM has less torque ripple than SRM, and the inverter and electronic controls are less sophisticated. The SynRM has excellent torque output and efficiency. The SRMs struggle to self-start (rotate), which requires the electronics to initially slowly start the RMF. The electronics slowly increases the frequency of the current to the coils. As the rotor begins to spin, the electronics synchronously increases the RMF to cause the attractive force to increase rotor speed. As the rotor experiences a load, it causes the rotor to slow behind the RMF. The electronics software continuously monitors the position of the rotor and adjusts the RMF as needed. SynRMs spin at synchronous speed and run cooler and produce more torque than AC induction motors. SynRM is used in public transportation locomotives. A variation of the SynRM with permanent magnets inside the rotor is used in automobiles.

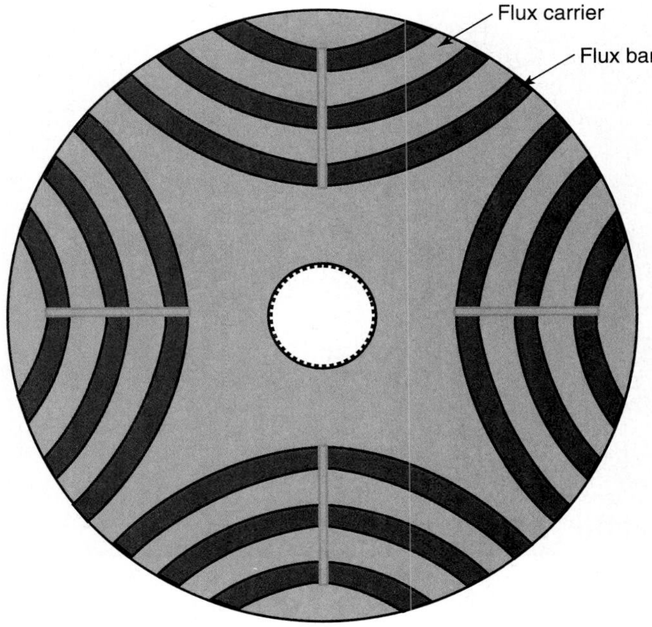

Figure 26-22. An end view of a synchronous reluctance motor's rotor.

Electric-Drive Mining Trucks

The mining industry has been using electric-drive equipment for decades and continues to use it today. Most electric-drive mining trucks use a pair of traditional AC induction motors in the rear axle. Each AC motor drives a planetary final drive, which drives a pair of wheels. The motors, when produced by a high-quality manufacturer and operated in the right environment with good maintenance practices, can achieve a very long life, such as 30,000 hours, and over 50,000 hours in some extreme cases.

Caterpillar AC Mining Truck

Caterpillar's 794 AC, 795F AC, 796 AC, and 798 AC mining trucks are electrically propelled, **Figure 26-23**. The 794 AC has a

Figure 26-23. A 794 AC Caterpillar electric-drive haul truck being shown at MineExpo.

Figure 26-24. An 795F AC generator is located at the rear of the truck and is driven by the engine through an isolation coupler (damper assembly) and drive shaft.

Figure 26-25. Electrical horsepower exhibits an inverse relationship between voltage and amperes. If one variable is increased, the other is decreased.

Figure 26-26. A Caterpillar 794 AC and 795F AC electric-drive truck's power control inverter cabinet is located beside the operator's cab.

320-ton payload capacity. The 795F AC truck has a 345-ton payload capacity. The 796 AC replaced the 795F AC and has a 360-ton capacity. The 798 AC has a 410-ton capacity. Caterpillar is the first manufacturer to offer a sole-sourced AC-powered truck. Caterpillar singularly designed and produced the electric-drive truck (795F AC) using several components from the mechanical 797 truck. Caterpillar dealers are the only entity needed to service the truck. The 794 AC is a product of Caterpillar purchasing Bucyrus, converting a MT 5300 Unit Rig truck to a 794 AC using the 795F AC electric drive train. All three 794 AC, 796 AC, and 798 AC truck use a C175-16 3500 hp tier-4 diesel-powered electric AC drive truck for low altitudes. The 798 AC high-altitude truck has a C175-20 4000 hp tier-2 engine and is used in South America.

A C175-16 diesel engine drives an AC generator at the rear of the 795F AC truck, **Figure 26-24**. The truck's electric drive operates at 2600 volts. Electrical power, measured in watts, is computed by multiplying the voltage times the current flow (amperes). For a given amount of wattage, an increase in voltage allows the engineers to decrease current flow. If the engineers designed the truck to operate on half the voltage, current would be doubled.

Amperage and voltage have an inverse relationship for a fixed amount of power (watts), **Figure 26-25**. For a given amount of electrical power, if a lower voltage is used, much larger electrical conductors would be required to accommodate the larger amount of current flow. High voltage and lower amperage reduces heat and increases component life.

The 2600 volts AC are rectified into 2600 volts DC. A control power inverter cabinet, **Figure 26-26**, converts the 2600 volts DC into a three-phase AC voltage that drives the right and left AC drive motors in the truck's axle housing. The control power inverter cabinet uses IGBT modules for controlling the truck speed and direction. IGBT modules are further discussed later in this chapter. The inverter cabinet directs the three-phase AC voltage to two AC induction traction motors in the truck's rear axle assembly. See **Figure 26-27**.

The Caterpillar AC trucks use a radial grid for braking. The *radial grid* converts the mechanical kinetic energy into heat energy, **Figure 26-28,** and acts like a brake resistor for slowing the truck. A *brake resistor* converts the electrical energy developed during a braking application into heat energy to be dissipated into the atmosphere. All three trucks have a hydraulic motor that drives a large fan that draws air into and through the electronic cabinet, blows it through the generator, and directs it to the two AC motors in the axle for cooling as needed. After startup, the air cooling is not always used or needed, depending on the climate.

Figure 26-27. A Caterpillar 795F AC truck has two AC induction motors in the truck's axle housing.

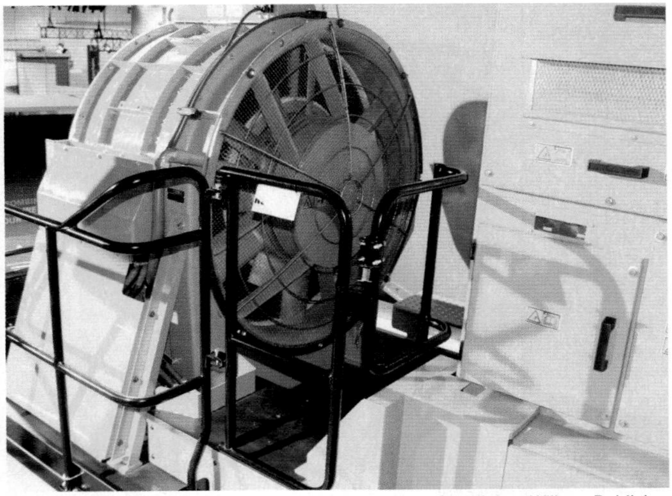

Figure 26-28. The radial grid acts like a brake resistor. It is located beside the control power inverter cabinet and has its own cooling fan.

Hitachi EH 4000 AC 3

Hitachi also manufactures electric-drive trucks. The EH 4000 AC 3 truck has a payload capacity of 243.6 tons and a rating of 2500 engine horsepower. See **Figure 26-29**. This truck has a single AC alternator and rectifies its output to DC. The DC is then inverted into AC and sent to two AC induction motors, one for the right drive and one for the left drive. The drive motors deliver power to a planetary final drive, which drives the left- or right-side dual wheels. A spring-applied caliper style parking brake is located at each drive motor. See **Figure 26-30**.

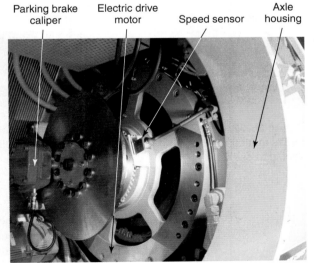

Figure 26-29. This Hitachi EH 4000 AC 3 electric-drive haul truck has a 243-ton payload capacity. Its resistor grid boxes dissipate energy during braking and are visible above the engine's exhaust, just right of the operator's cab.

Figure 26-30. The Hitachi EH 4000 AC 3 mining truck has spring-applied caliper-style parking brakes attached to each electric-drive motor. The brakes are accessible by removing the rear axle cover.

The truck is equipped with three advanced control features:

- **Slip/slide control.** This feature uses front wheel sensors, suspension and steering sensors to provide traction control and antilock brake control in slick conditions.
- **Pitch control.** This feature controls the AC drive motors to limit the amount of chassis bouncing that occurs in rough road conditions.
- **Side skid control.** This feature controls the two AC drive motors to improve turning when operating in slick road conditions.

Komatsu 930E

The Komatsu 930E truck shown in **Figure 26-31** has a 320-ton payload and uses the GE electric-drive propulsion system. The trucks have an AC generator and two AC motors. As in most electric-drive mining trucks, the AC is rectified into DC and then inverted back to AC so the frequency can be more accurately varied to control motor speed and torque.

The control cabinet, **Figure 26-32**, has three rows of IGBTs. IGBTs are used as braking choppers and AC inverters. A *braking chopper* is an electrical switch that limits the DC voltage by diverting the high voltage to braking resistors. The other two rows of IGBTs are dedicated to control the AC three-phase variable frequency to the AC induction motors. Located below the IGBTs are DC bus capacitors that are used to stabilize the DC bus voltage. In a VFD drive, the DC bus supplies the DC current that is inverted back into AC and controlled by the VFD. DC bus capacitors store excess charge and release the stored charge if voltage drops to maintain a stable DC bus voltage.

Goodheart-Willcox Publisher

Figure 26-31. The Komatsu 930E electric-drive truck uses GE's AC electric drive. The AC control cabinet is located beside the operator's cab.

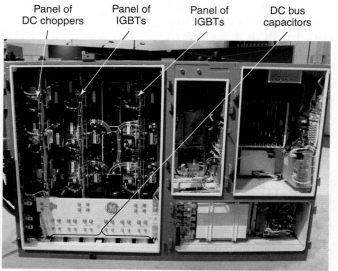

Goodheart-Willcox Publisher

Figure 26-32. This GE control cabinet contains the 12 air-cooled IGBTs that control the AC drive motors in the Komatsu 930E mining truck.

Komatsu Autonomous Truck

Komatsu has developed an *autonomous* electric-drive truck that can haul and deposit loads without an operator. The truck is designed to drive by itself and has no operator's cab. The Komatsu Autonomous Haulage System (AHS) operates using onboard electronic control systems, GPS technology, and an obstacle detection system. See **Figure 26-33**.

The autonomous truck offers the advantage of nonstop operation. The trucks reduce or eliminate operator risks at high altitudes or in other remote, dangerous locations, such as mine sites. The truck is also uniquely equipped with four independent electric drive motors, providing four-wheel drive and four-wheel steer. Unlike human-operated trucks, an autonomous truck can be driven straight to a load site and reversed straight to the dump site without having to turn around, because the truck is not dependent on the operator viewing the road.

A   **B**   *Goodheart-Willcox Publisher*

Figure 26-33. The Komatsu four-wheel drive, four-wheel steer truck does not have an operator's cab. A—A photo of the back of the truck. Without an operator's station and with four-wheel steer, the truck drives the same in forward and reverse. B—A close-up view of the steerable drive wheel.

Trolley Electric-Powered Mine Trucks

Mining trucks that must travel up long uphill grades can get a boost from an overhead electrical power supply. This system is called *trolley assist.* See **Figure 26-34**. When ascending a hill, trucks with trolley assist can extend current collectors into contact with overhead power lines. When electrical contact is made, the truck's engine speed is brought down to an idle, and the truck is driven by power from the overhead lines. The electrical power from the overhead lines enables a truck's electric propulsion to operate with increased horsepower, up to 90% more horsepower than could be achieved if the electric drive was powered by the diesel engine. At the end of the grade, the truck disconnects from the grid and the diesel engine takes over.

To reap the benefits of this system, the truck must have an electric drive and be equipped with trolley assist. A loaded truck with trolley assist can travel much faster up a grade than a diesel-electric truck without trolley assist. The engine is allowed to idle, so engine life is increased. The truck also spends less time under extreme load, so drive motor life is also extended. Many electric-drive mining truck manufacturers offer trolley assist as an option.

Electric-Drive Wheel Loaders

Wheel loaders are a common type of electric-drive machine due to their cyclical operation. A common application has the loader driving forward into a pile to load the bucket, backing up to reposition the loader, then driving forward to dump its bucket into a truck. Some loaders are electric drives while others are hybrids that recover and store the energy during the cyclical braking operation.

Joy Global SR/Komatsu Mining

Joy Global began offering switched reluctance (SR) electric-drive wheel loaders in 2002. Komatsu purchased Joy Global. SR electric-drive machines include underground mining wheel loaders, open pit wheel loaders, and shovels. WE1150-2, WE1350-3, WE1850-3 and WE2350-2 wheel loaders are electric drive. Their WE2350-2 wheel loader is the world's largest, with a 70-cubic-yard bucket.

Hitachi Construction Machinery (Europe) NV

Figure 26-34. An electric truck with trolley assist enables the haul truck to connect to overhead power lines, providing the truck high horsepower for long, steep grades.

SR Generator

Joy Global electric drives are unique in that they use an SR generator instead of a traditional AC generator. The single diesel-driven SR generator is controlled to send current flow to a pair of front end converters, which delivers power to the DC bus. The DC bus sends power to the load side converters, which deliver power to four independently controlled SR electric drive motors. The Joy Global SR drive uses a lower switching frequency than traditional AC induction drives, which increases component life.

Kinetic Energy Storage System

Joy Global underground mining hybrid loaders can be equipped with the hybrid **Kinetic Energy Storage System (KESS).** Instead of traditional high-capacity batteries or capacitors, the KESS uses a SRM rotor flywheel to recapture energy. As the brakes are applied, that energy drives the SRM rotor flywheel, which greatly increases the loader's ability to capture energy. This type of flywheel was shown in **Figure 26-20**.

John Deere 944K Electric-Drive Wheel Loader

The John Deere 944K electric-drive wheel loader's engine runs at three speed ranges, low (1200 RPM), mid (1500 rpm), and max 1800 RPM, to offer choices in fuel efficiency and reduction of emissions. The engine drives two 3-phase 480-volt AC permanent magnet generators that feed electrical power to the power electronics devices, which comprise six inverters that use IGBTs. A separate inverter is used for each of the four drive motors and each of the two generators. Based on operator commands and machine operating conditions, the power electronics independently control four electric switched-reluctance drive motors. Each of these motors drives a planetary final drive that propels a single drive wheel and tire. See **Figure 26-35.**

The loader can individually power one, two, three, or four wheels. The electric drive replaces the traditional torque converter, powershift transmission, front axle, and rear drive axle. Unlike a traditional loader power train, the 944K electric drive allows the machine's ground speed to be controlled independently from engine speed.

26.4.2.1 Dynamic Braking

John Deere originally called the 944K electric-drive wheel loader a hybrid electric loader. However, by most definitions, a hybrid has a storage device, such as a bank of batteries or capacitors, for capturing and later releasing energy. The 944K does not contain any high-voltage energy storage device. John Deere plans to label future models E-drive instead of hybrid.

Consider that a loader's operation often involves frequent forward and reverse cycling. When the 944K operator presses the brake pedal, the machine electronically controls the electric drive motors to brake. Instead of dispersing this energy as heat, which occurs in conventional braking applications, the loader is capable of ***dynamic braking***. This means the motors, acting as generators, convert the machine's mechanical energy back into electrical energy. This electrical energy is sent to the generators, which act like electric motors to help the engine maintain its speed rather than lugging.

The recaptured energy helps the engine perform other functions, such as hydraulically controlling the loader lift, bucket curl, and articulated steering. This energy savings, along with the efficiency gained from the electric drive, increases the loader's fuel efficiency.

If the machine is not performing any other useful work, such as operating the loader or bucket, during dynamic braking, the power electronics send the electrical energy from the electric motor to two liquid-cooled brake resistors at the rear of the machine. See **Figure 26-36**.

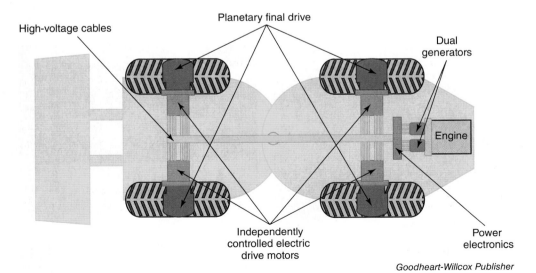

Goodheart-Willcox Publisher

Figure 26-35. The John Deere 944K electric-drive wheel loader has two engine-driven generators that send power to the power electronics, which are responsible for independently controlling four electric-drive motors.

The brake resistors convert the electrical energy developed during a braking application into heat energy that is dissipated into the atmosphere.

Locomotives, like electric-drive wheel loaders, use brake resistors, known as resistor grids, to dissipate heat. However, locomotives generate tremendous heat when they bring a long train of cars to a stop. As a result, their resistor grids must be designed to dissipate much more heat than the brake resistors in other types of electric drives. See **Figure 26-37**.

In addition to having dynamic braking, the 944K loader contains a traditional multi-disc parking and service brake assembly in each of the final drives. Each final drive brake is used for both service brakes and parking brakes. The service brake is hydraulically actuated and oil cooled. The parking brake is spring applied and hydraulically released.

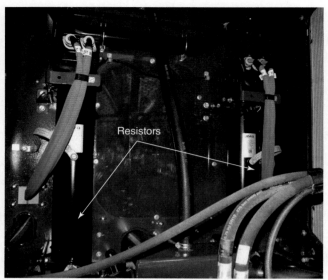

Figure 26-36. The John Deere 944K has two liquid-cooled brake resistors at the rear of the machine, accessible through the rear compartment access cover.

Electronic Traction Control

The 944K has *electronic traction control*, which monitors all four of the wheel motors to determine if any of the wheels begin to slip. When slip is detected, the electronic controls limit power to all four drive motors to reduce the machine's wheel slip. The loader has true four-wheel drive because it has no open differential.

The largest advantage of traction control is the reduction of costly tire wear. Loaders are notorious for spinning their tires while loading. The operator uses the machine's propulsion for driving into the face of the material being loaded. Some material is exceptionally harmful to slipping tires. An example is *shot rock*—rock that has been blasted or hydraulically hammered but has not yet been crushed. Although all tire slip increases tire wear, sharp shot rock dramatically decreases tire life during wheel slip.

Each generator drives two diagonally opposed wheels. One generator drives the front-right and rear-left wheels, and the other generator drives the front-left and rear-right wheels. This is strategy is based on a loader digging its bucket into a pile. The load is placed on the front axle and those tires have all the traction and the rear wheels will easily spin. The diagonal distribution of electrical current to the motors distributes the load diagonally.

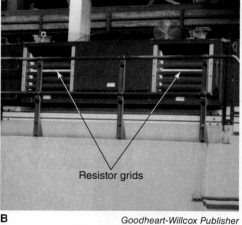

A B

Figure 26-37. Resistor grids are used on locomotives to convert braking energy into heat energy. A—These brake resistor grids are two of ten that can be used on one locomotive. B—Resistor grids are placed at the top of the locomotive.

Chapter 26 | Electric and Hybrid Drive Systems

Note

If a 944K's rear wheels lift off the ground when the machine drives into the face of a stock pile for loading, the machine will sense rear wheel slip and reduce power to all four drive wheels. Although you can see the wheels ratchet during the initial movement, they will not spin. Because the 944 has a separate inverter for each motor, it can actually control each motor individually if needed. The wheel speeds can be varied while steering so the outside wheels and inside wheels can turn at different speeds so the tires do not scrub like they do with a locked differential.

The loader has rim pull control with adjustable settings to limit the torque to the wheels to avoid slipping tires on soft and slippery applications.

944K Advantages

The 944K has no traditional axle or transmission fluids. It has a 13.5-liter engine compared to its competitor's mechanical-drive loader of the same size, which has an 18-liter engine. The electric motors are produced by Nidec, a Japanese locomotive company, and weigh approximately 1700 lb (780 kg) each. They have been engineered to limit heat buildup in the motors and use ceramic bearings. The motors have long service life and routinely last the life of the machine. A 944K can use up to 33% less fuel than a conventional loader of the same size and can be up to 8% more productive than a conventional loader. The inverters, including the internal IGBTs, also typically last the life of the machine. The loader has an 8-year, 20,000-hour hybrid-electric-drive warranty. At 15,000 hours, the loader's engine and electric-drive power cables should be replaced and the gearbox inspected. A similar service is performed at 30,000 hours, but the generators and motors typically do not need to be replaced or rebuilt at either interval.

John Deere 644K and 644X-Tier Electric-Drive Wheel Loader

John Deere produces an electric-drive wheel loader originally called the 644K, **Figure 26-38**, but now called the 644X E-Drive. The electric drive uses a constant-speed diesel engine to drive a brushless, 3-phase, 480-volt, permanent-magnet AC generator, which converts mechanical energy into electrical energy. The generator sends electrical power to an inverter. An early style inverter is shown in **Figure 26-39**. The current model wheel loader is the 644X-tier.

The inverter sends power to a brushless, 3-phase, 480-volt, permanent-magnet AC motor. The electric motor delivers power to a three-speed transmission, which delivers power to the front axle and rear axle. The mechanical three-speed transmission does not select neutral, forward, or reverse, which are achieved with the electric drive. The purpose of the transmission is simply to transfer power and change gear ratios. In

Philip McNew

Figure 26-38. A John Deere 644K electric-drive loader.

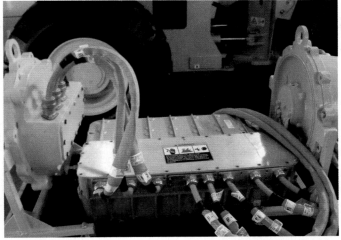

Goodheart-Willcox Publisher

Figure 26-39. An early version of the 644K inverter. The loader's inverter is liquid-cooled. The inverter controls the AC electric motor.

this application, the electric drive has replaced the torque converter. The electric drive allows the machine's ground speed to be controlled independently from the engine speed. See **Figure 26-40**.

The John Deere 644K electric drive is also labeled a hybrid electric loader, but like the 944K, it does not have an energy storage device. The 644K offers dynamic braking, allowing the electric drive to use braking energy to maintain engine speed. If the machine is not performing any other useful work during the dynamic braking, such as operating the loader or bucket, the inverter sends the electrical energy from the electric motor to a brake resistor. The brake resistor converts the electrical energy developed during a braking application into heat energy that is dissipated into the atmosphere. The brake resistors are also cooled with engine coolant. The integration of the brake resistor extends the life of the service brakes.

The inverters on the 644X and 944K are cooled by John Deere CoolGard II coolant with their own separate system, rather than the engine coolant. They have a dedicated electric water pump and their own cooler because they need to run at a lower temperature than the engine. The engine runs at a set speed and provides predictable hydraulics because of the fixed hydraulic pump speed. The engine speed of a mechanical-drive loader can be drawn down during operation. The 644K (644 X) loader performs similarly to a conventional 724 loader and costs less to ship than a 724 loader. It can save up to 5% in overall operating costs over its service life compared to the operating costs of a traditional 644 loader. If operating in a V-pattern when loading a truck, the loader can be up to 12% more fuel efficient than a conventional machine.

Hitachi ZW220HYB-5 Hybrid Wheel Loader

Hitachi was the first to mass-produce a hybrid electric wheel loader when they manufactured the ZW220HYB-5 in 2015 for the Japanese market. The loader has an engine-driven electric generator/motor assembly, two inverters, a capacitor, and two electric-drive motors. See **Figure 26-41**. The generator delivers electrical power to the left-side inverter, which sends power to the right-side inverter. The right-side inverter sends electrical power to the two drive motors. A capacitor assembly stores electrical energy that is captured when the machine is moving or during dynamic braking. The two electric-drive motors are responsible for driving the front and rear axles.

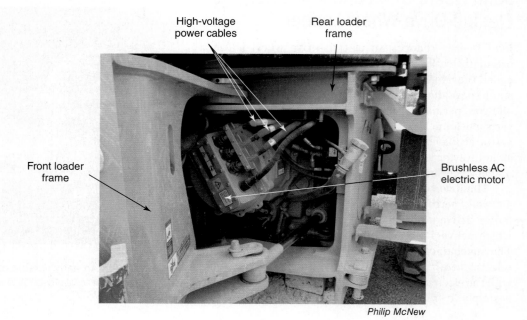

Figure 26-40. The John Deere 644K AC permanent magnet motor is visible at the articulation joint.

Philip McNew

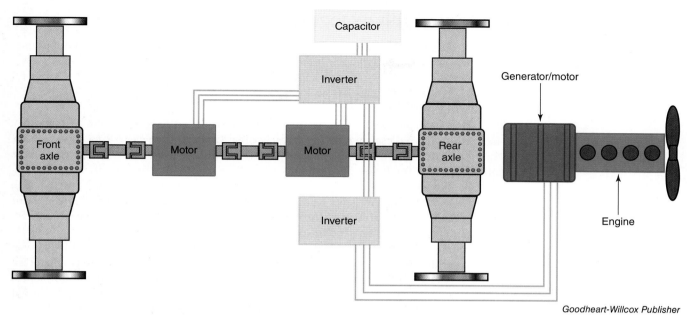

Figure 26-41. The Hitachi ZW220HYB-5 hybrid wheel loader uses an engine-driven generator that supplies power to two inverters. The inverters can direct energy to be stored in the capacitor or direct electrical energy to the two electric drive motors.

Volvo Battery Electric Wheel Loader

The Volvo L25 wheel loader is a battery electric wheel loader that uses a 48-volt, 39-kilowatt-hour battery pack to power the loader. The machine can be operated in continuous use from three and a half to six hours. The batteries can be fast charged in just two hours and do not have to be fully discharged (deep cycled) to be charged. They can be charged anytime. The machine includes an onboard 3-kilowatt charger that can fully charge the machine in 12 hours. A permanent-magnet AC motor powers the hydraulic system, including steering. An AC induction motor is coupled to the rear axle and used to propel the loader.

Caterpillar 988K XE Electric-Drive Wheel Loader

In 2017, Caterpillar introduced the 988K XE electric-drive wheel loader. In the machine, the mechanical transmission and torque converter have been replaced with an electric drive. The loader is an electric-drive machine that does not use any high-voltage, high-capacity battery or capacitor to store or recover electrical energy. Nor does it use any type of brake resistor.

The C18 engine drives a switched-reluctance generator. The generator sends power to an inverter (integrated power electronics). The inverter sends power to drive a switched-reluctance electrical motor, which sends torque to a drop box. The drop box delivers power to the front and rear axles. When the machine is slowing down, the motor acts like a generator and the generator acts like a motor.

Caterpillar R1700XE Battery Electric Underground Loader

Caterpillar produces an underground battery electric loader, the R1700XE, on a limited basis. It is labeled a load, haul, dump (LHD) loader. It uses SR technology for propulsion and hydraulics. Like the 988K XE wheel loader, this electric-drive loader also has traditional drive axles. The loader's batteries can be charged in 20 minutes using two Caterpillar MEC500 mobile equipment chargers. These portable chargers can be brought to the machine on the job site, much like a fuel truck, which can drive up to and refuel a machine.

Electric-Drive Dozers

Dozers are another machine that can be equipped with electric drives. Caterpillar has produced the D7e and D6XE dozers. It has also developed a D11XE prototype dozer.

Caterpillar D7E Electric-Drive Track-Type Dozer

The D7E is Caterpillar's first electric-drive dozer, introduced 2009. Much like a locomotive, the machine has a diesel engine that is responsible for driving the electric motors in series. The tractor does not use a high-capacity electrical storage device, such as a high-voltage battery or capacitor. It also does not have a torque converter or a powershift transmission. The power train consists of a C9.3 ACERT diesel engine coupled to a flywheel clutch, which drives the rotor of an AC generator. The electric drive operates at 480 V. The generator delivers the three-phase current through three high-voltage, high-current wires (conductors) into a power inverter. See **Figure 26-42**.

The power inverter controls the electrical output. It delivers both AC and DC power. The power inverter also sends high-voltage DC current to the accessory power converter, which is responsible for delivering DC electricity to the air-conditioning compressor, to traditional starting batteries to keep them charged, and to a hazardous voltage present lamp (indicator) that alerts the technician that high voltage is present.

The power inverter also propels the tractor by sending high-voltage and high-AC current through six wires to two infinitely variable electrical drive motors. The two drive motors drive a single bull gear that drives the differential steering's planetary carrier to provide the propulsion input into the track differential steering system. A dedicated steering hydraulic motor provides the steering input into the differential steering gear set to control the tractor's steering direction and steering speed. The axles deliver power to a right and left final drive assembly. See **Figure 26-43**.

Goodheart-Willcox Publisher

Figure 26-42. The power inverter receives high voltage, high current from the AC generator. The inverter then delivers high voltage, high current through six wires to the two individual electrical-drive motors. The motors provide an infinitely variable input to the machine's axle.

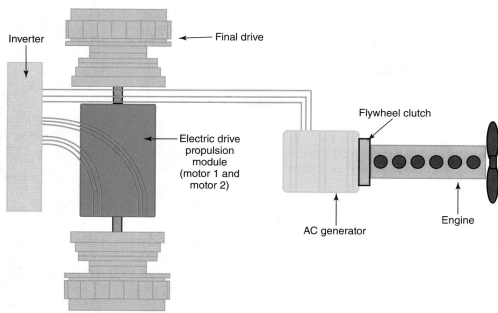

Figure 26-43. The D7E engine drives an AC generator that sends current to a power inverter. The inverter drives the tractor by controlling two AC drive motors in a propulsion module. The tractor uses hydraulic differential steering. The inverter also sends DC power to the accessory power converter to operate the air-conditioning compressor and charge the traditional starting batteries.

Goodheart-Willcox Publisher

The D7E tractor does not have a hybrid drive even though it has an electric drive. This means that the system does not store or recover high voltage or high current in any type of capacitor or battery.

D7E Electric Drive Advantages

The electric-drive system in the D7E dozer provides machine owners with the following advantages:

- An engine designed to operate at speeds within a narrow power band, which extends the life of the engine and enables the tractor to use up to 25% less fuel.
- It can lower operating costs.
- It has good straight-ahead dozing capacity and maneuverability in tight spaces (similar to hydrostatic drive and differential-steered dozers).
- It does not require gear shifting (similar to a hydrostatic drive).
- The electric-drive system has 60% fewer moving parts compared to a powershift transmission.
- The electric-drive powertrain components are expected to last up to 50% longer than traditional drive components (in similar conditions) and are configured in replaceable modules.
- It has no engine accessory drive belts, which improves machine serviceability.

The D7E is configured with a low oval-shaped undercarriage. In 2020, Caterpillar introduced the new D7 with elevated tracks and traditional mechanical powershift powertrain.

Caterpillar D6XE Electric-Drive Track-Type Dozer

In 2019, Caterpillar offered the first elevated-sprocket electric-drive dozer, the D6XE dozer. It is a differential steer tractor. The dozer does not use a high-capacity electrical storage device, such as a high-voltage battery or capacitor. Nor does it use any type of brake resistor.

Unlike the D7E, the D6XE uses a SR generator to drive one SR motor that drives a propulsion planetary carrier inside the differential steering gearbox, instead of two electrical motors. The SR generator sends three-phase, pulsed DC current to the power inverter, which directs three-phase, pulsed DC current to the SR motor. The pulsed DC current varies from –715 V to 715 V.

The D6XE has a 20,000-hour, 84-month electric power train warranty. The generator and motor bearings and seals must be replaced before 12,000 hours in order to maintain the warranty. The D6XE inverter is designed to last the life of the machine. However, if needed, unlike the D7E, the D6XE inverter is serviceable by Caterpillar technicians.

Caterpillar D11XE Electric-Drive Track-Type Dozer

Caterpillar is testing a prototype electric-drive in its largest dozer, the D11XE. It is the largest electric-drive dozer in the industry. The electric drive makes the dozer capable of a power turn, unlike traditional D11 dozers that use steering clutch and brakes. The D11XE also has no transmission gears to shift, as there are in a traditional D11.

Electric and Hybrid Electric Excavators

Manufacturers offer several types of electric-drive excavators: electric-powered compact excavators and loader backhoes that receive power from a battery or power cord, large mining electric-drive excavators and shovels that receive their power from an external AC electrical power supply, and diesel-powered excavators with hybrid electric slew drives.

Electric-Drive Compact Excavators and Loader Backhoes

Several manufacturers offer compact electric-drive excavators. Two manufacturers produce electric-drive loader backhoes.

Electric-Drive Mini and Compact Excavators

Numerous manufacturers produce prototypes or manufacture electric-powered mini excavators. Some are battery powered and some are powered with a plug-in power cord. Examples of prototype battery-powered compact excavators include the Yanmar CE SV17e compact excavator and Hitachi ZE19 and ZE85 models. The Hitachi machines are designed to operate for four or more hours with a one-hour rapid charge.

JCB manufactures the 19C-1E, E-TECH battery-powered mini excavator. When configured with the four-battery pack, it delivers up to five hours of operation, which is generally the amount of time a compact excavator is operated for a traditional day's work. The three-battery pack provides up to four hours of operation. The machine can be charged with 110 V AC, 230 V AC, or 415 V AC.

Volvo builds the ECR25 battery-powered electric-drive excavator. It uses three battery packs that provide 20 kilowatt-hours of energy. One *kilowatt hour* (kWh) equals the amount of power it takes to power an electric device that uses 1000 watts for one hour. The machine delivers four hours of work on a fully charged battery pack. A completely drained battery requires five hours to charge. It weighs 440 pounds (200 kilograms) more than the diesel-powered excavator. Volvo states that a traditional diesel-powered ECR25D excavator generates the noise equivalent of 8 to 10 ECR25 electric-powered machines. The battery is designed for 6000 hours of use, which is the equivalent of up to 10,000 hours on a diesel-powered machine due to the typical 40% idle time. The battery can handle up to 2000 charges and discharge cycles, which is the equivalent of one cycle per day for 200 days in a year times 10 years. The electric motor is designed to last the life of the machine.

Kato manufactures the 9VXB and 17VXB battery-powered compact excavators. They offer up to eight hours of operation on a full battery charge. The Kato 17VXE electric-powered mini excavator is powered by a three-phase, 50-amp, 480-volt AC power cable.

The Caterpillar 300.9D mini excavator can be powered by its own diesel engine or by a separate electric-powered hydraulic power unit (HPU). The HPU300 is a three-phase, 16-amp, 480-volt power source that supplies hydraulic power to the 300.9D mini excavator. When powered by the HPU, the mini excavator produces no exhaust emissions. The HPU300 has a front hitch bracket that allows the excavator's dozer blade to lift and carry the HPU when the mini excavator is being powered by its diesel engine.

Komatsu and Honda have partnered to produce the PC01E electric-drive micro excavator for overseas markets. The excavator is powered by Honda's mobile power pack (MPP). The PC01E is based on the diesel-powered PC01, which weighs only 661 pounds. The micro excavator can fit in the back of a truck. If the charge runs low, the drained MPP can be swapped with a charged MPP.

Load Backhoes

John Deere has partnered with a large utility company, National Grid, to test a prototype battery-powered loader backhoe, known as E-Power. The backhoe is labeled a 310-X tier. Deere is using X to denote hybrid and electric drive. The target is to provide eight to ten hours of operation on a fully charged battery. The battery charge time is also being tested, including a fast charge and recharge times when mobile quick chargers are used. The 310-X is approximately 25% quieter than the comparable 310L diesel-powered backhoe.

Case partnered with Moog, and in 2020 introduced the first fully electric-powered backhoe loader, the 580 EV, also known as project Zeus. Its 90-kilowatt-hour, 480-volt lithium-ion battery pack is designed to power the machine for eight hours in traditional backyard operations. A completely drained battery requires an eight-hour charge time. The battery pack is recharged with a three-phase, 220-volt charger.

Large Mining Electric-Drive Excavators and Shovels

Hitachi electric-drive excavators and Caterpillar electric-drive rope shovels receive their power from an external AC electrical power supply. The machines are environmentally friendly and run emission free because there is no diesel engine and exhaust system. They also have less daily maintenance. Because there is no diesel engine, there is no need to check the engine oil, air filters, or engine coolant.

Hitachi EX1900E-6 through EX800E-6 Electric-Drive Excavators

Hitachi EX1900E-6 through EX8000E-6 electric-drive excavators use a single three-phase AC induction motor that powers the machine's hydraulic system, HVAC, and electrical accessories. The electrical input to the motors ranges from 6600 to 6900 volts at 60 hertz. Their output ranges from 805 US horsepower (600 kilowatts) to 3218 US horsepower (2400 kilowatts).

Caterpillar Electric-Drive Rope Shovels

Caterpillar electric-drive rope shovels also use an external AC electrical power supply offering the same advantages as the AC electric-powered Hitachi excavators. Two machine model examples include 7395 and 7495. The shovel receives 7200 volts of power at 50/60 hertz. The high-voltage electrical power supply cord requires a carefully chosen approach for truck loading. The following are few examples and a few factors to consider:

- Single truck back-up loading.
  - Does not require overhead cable towers.
  - Good when working in close quarters
  - Requires minimum driver skill.
  - Requires extended spotter and truck exchange time.
  - Inability to load while the dozer is cleaning the work site
- Double truck back-up loading.
  - Requires twin overhead power cable towers.
  - More driver skill is necessary.
  - One truck can get into a loading position while the other truck is loaded.
  - Can load while the dozer is cleaning one of the work sites.
- Drive-by loading (undercarriages are parallel to wall and can move forward)
  - Good for belly-dump trucks and coal sites.

- Modified drive-by loading (undercarriages are parallel to wall and can move forward and the trucks back into loading site at an angle).
- Is good for end-dump trucks and popular for coal sites.
- Requires high driver skill.

Diesel-Powered Excavators with Electric Hybrid Slew Drives

Some examples of electric hybrid excavators are Hitachi ZH210-5, Kobelco SK210HLC, and Komatsu HB 215LC-A and HB 365LC-3. The hybrid technology is not used for propulsion, but for *slew*, which is swinging or rotating the excavator's house to the left or right. A generator and electric motor in conjunction with large liquid-cooled capacitors or batteries recover energy as the excavator swing comes to a stop. In hybrid excavators manufactured by Caterpillar and Volvo, hydraulic accumulators, rather than electrical capacitors, recover the energy. Depending on the design, hydraulic accumulators can be charged when slowing the slew (Caterpillar) or when lowering the boom (Volvo EC300E).

Electric/Hybrid Safety and Service

In electric propulsion and hybrid power systems, the high-voltage conductors are orange cables. The tremendous amount of voltage and current required to propel these large off-highway machines can easily cause a fatality. Technicians must be properly trained prior to servicing or repairing off-highway electric-drive systems.

Even if a machine does not have high-capacity batteries or capacitors, the system can still retain high voltage for a period of time after the machine has been shut down. Take the following precautions:

- Never perform any service or repair work on the electrical propulsion system while the machine is running or operating.
- After the machine has been shut down, follow the manufacturer's steps for de-energizing the electric-drive system.

On hybrid systems with high-energy capacitors or batteries, the system still contains a tremendous amount of energy stored in the capacitor or batteries, even when the machine is shut down. Take the following precautions:

- Never attempt to service or repair a charged battery or capacitor.
- Always follow the manufacturer's service literature for servicing and repairing hybrid and high-voltage systems.

Note

Manufacturers' diagnostic service tools and machine monitors can display high-voltage readings.

Some manufactures provide specific instructions for cleaning terminals with an alcohol and water solution any time the conductors are removed or installed from an electric-drive component. Some manufacturers specify always installing new hardware when reattaching the conductors to the component.

Gloves

Most electric-drive manufacturers require technicians who service their equipment to use special insulated rubber gloves designed for working with high voltages. Six different classes of gloves are available, based on the amount of voltage they can safely handle. See **Figure 26-44**. The gloves are often rolled from the cuff toward the fingers to see if they will hold air. If a hole or crack exists, voltage can leak through the glove and endanger the

Class	Max AC use voltage	Max DC use voltage (average)	Color of Label
00	500	650	Tan
0	1,000	1,500	Red
1	7.500	11,250	White
2	17,500	25,500	Yellow
3	26,500	39,750	Green
4	36,000	54,000	Oange

Goodheart-Willcox Publisher

Figure 26-44. Matrix of voltages and classes of insulated gloves.

technician. Leather gloves, specified by ASTM standards, are placed over the insulated rubber gloves to protect the rubber gloves.

Electric-drive machines should not be serviced or repaired any time high voltage is present. Prior to working on a high-voltage electric-drive system, technicians must follow the service literature for shutting down the system and depleting the high voltage. In an abundance of caution, a technician should double check that the system has had its high voltage properly depleted. The insulated gloves are worn while double checking for high voltages in electric-drive components.

Ground Fault, Low Isolation

The electric-drive's internal copper-core drive wires have a shielding that isolates the conductors from the machine's chassis ground. A cross-sectional drawing of a drive wire from a Caterpillar D6XE is shown in **Figure 26-45**. The internal conductor is made of copper and surrounded by wire insulation. The insulation is surrounded by Mylar tape. Braided stainless-steel grounded shielding surrounds the Mylar tape. The grounded shielding is surrounded by an orange nylon braided cover.

The stainless-steel shielding that surrounds the drive cables is grounded to the machine's chassis ground. The internal drive wires are isolated from the shielding and chassis ground. During start up on the D6XE, the electronics look for ground faults between the copper conductor and the grounded shielding. If no fault is detected, the machine will slowly step up voltage until it reaches the normal operating range.

During startup and normal operation, the electronics on the 644X and 944K Deere wheel loaders look for continuity from the internal core drive wires to the outside shielding. Normally there is 2,000,000 ohms of resistance. The machine is allowed to operate if the resistance is 500,000 ohms or above. However, if the resistance drops any lower, the machine will not operate, and a low isolation fault code will be set. A technician will have to repair the fault before the machine can operate.

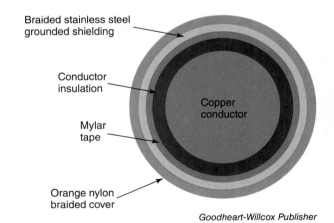

Goodheart-Willcox Publisher

Figure 26-45. A cross-sectional drawing of an electric drive cable.

Summary

- Alternators use a rotating electromagnet inside a stationary winding (stator) to develop three phases of AC flow.
- Bridge rectifiers are used to rectify AC to DC.
- DC traction motors have an armature consisting of wire conductors. Current passes through the field windings, carbon brushes, copper commutator bars, and armature conductors.
- Brushes wear over a period of time.
- AC traction motors do not require the use of commutator bars and brushes.
- Two types of AC motors are induction and permanent magnet.
- In switched reluctance motors, the stator coils are energized sequentially to pull the rotor, causing it to rotate.
- Electric-drive mining haul trucks use large resistor grids to dissipate the heat energy that occurs during braking.
- The John Deere 944K electric-drive wheel loader uses two AC generators to independently control four separate electric-drive wheel motors, providing independent drive, traction control, and dynamic braking.
- The John Deere 644K electric-drive wheel loader uses an engine-driven generator to drive an electric motor, which sends power into a three-speed transmission. The electric drive replaces the traditional torque converter.
- The Hitachi ZW220HYB-5 wheel loader uses a capacitor to recover energy.
- The Caterpillar D7E electric-drive dozer has an AC alternator that delivers power to an inverter, which sends electrical power to two electrical drive motors. The electrical drive motors send power to a bull gear in the differential steering assembly, which powers the left and right final drives.
- Hitachi, Kobelco, and Komatsu hybrid excavators use capacitors to recover energy.
- Electrical/hybrid machines can retain high voltage for a period of time after the machine has been shut down. Technicians working on these systems must be properly trained and take the necessary precautions.

Technical Terms

- AC induction motor
- AC sine wave
- AC traction motor
- alternator
- armature
- asynchronous motor
- autonomous
- back electromotive force (BEMF)
- brake resistor
- braking chopper
- bridge rectifier
- brushes
- brushless DC motor (BLDCM)
- brush-type DC motor (BDCM)
- commutation
- commutator
- counter-electromotive force (CEMF)
- cycle
- diode
- dynamic braking
- electromotive force (EMF)
- electronic commutation
- electronic traction control
- frequency
- generator
- kilowatt hour
- Kinetic Energy Storage System (KESS)
- permanent magnet AC (PMAC) motor
- pitch control
- power density
- radial grid
- rectify
- rotor
- shot rock
- side skid control
- slew
- slip/slide control
- stator
- switched reluctance motor (SRM)
- synchronous switched reluctance motor (SynRM)
- three-phase AC
- trolley assist
- variable frequency drive (VFD)

Review Questions

Answer the following questions using the information provided in this chapter.

Know and Understand

1. In an alternator, the _____.
   A. rotor is an electromagnet
   B. field poles are stationary
   C. stator rotates
   D. All of the above.

2. In an alternator, the _____ determines the number of sine waves generated for every rotor revolution.
   A. total number of field poles
   B. total number of pairs of poles
   C. number of stator windings
   D. strength of the field current

3. A three-phase AC bridge rectifier contains _____.
   A. two diodes
   B. three diodes
   C. four diodes
   D. six diodes

4. Modern electric propulsion systems *rarely* use _____.
   A. BDCM traction motors
   B. AC traction motors
   C. switched reluctance traction motors

5. Traction motor brushes are made of _____.
   A. aluminum
   B. carbon
   C. iron
   D. titanium

6. Commutators enable electricity to be transferred from the _____.
   A. field pole to the stator
   B. brush to the stator
   C. rotor to the armature's conductors
   D. brush to the armature's conductors

7. _____ causes an AC motor's rotor to rotate.
   A. An engine-driven pulley
   B. Magnetism generated by the field poles
   C. Induced magnetic force from the stator
   D. Magnetism generated from the brushes and commutators

8. Caterpillar's AC-powered electric truck operates at _____.
   A. 260 volts
   B. 600 volts
   C. 2600 volts
   D. 6000 volts

9. For a given amount of electrical power, if the voltage is increased, the amperage _____.
   A. decreases
   B. increases
   C. remains the same

10. Which of the following statements about trolley assist is *false*?
    A. Increases machine life.
    B. Allows faster travel speeds when loaded.
    C. Increases horsepower.
    D. Can be equipped on a mechanical drive truck.

11. John Deere's 944K wheel loader uses what type of high energy storage device to recover dynamic braking energy?
    A. None.
    B. Battery.
    C. Capacitor.
    D. Accumulator.

12. John Deere's 644K wheel loader uses what type of high-energy storage device to recover dynamic braking energy?
    A. None.
    B. Battery.
    C. Capacitor.
    D. Accumulator.

13. What does the John Deere 644K wheel loader use to assist with dynamic braking?
    A. Brake resistor.
    B. Hydraulic retarder.
    C. Radial grid.
    D. Wet service brakes.

14. Caterpillar's D7E accessory power converter sends DC power to all of the following, *EXCEPT*:
    A. air conditioning compressor.
    B. traditional starting batteries.
    C. hazardous voltage present lamp.
    D. traction motors.

15. Caterpillar's D7E propulsion system operates at _____.
    A. 12 volts
    B. 24 volts
    C. 480 volts
    D. 2600 volts

16. AC electrical propulsion systems normally _____.
    A. rectify AC to DC, then invert it back into AC
    B. use the AC generator to directly drive the AC motors
    C. use brushes and commutators

17. Hitachi, Kobelco, and Komatsu hybrid excavators use what type of high-energy storage device to recover energy when the swing is slowing to a stop?
    A. None.
    B. Battery.
    C. Capacitor.
    D. Accumulator.

18. Which of the following statements about electrical propulsion and hybrid power systems is *false*?
    A. Training is required in order to service or repair these systems.
    B. Repair work on the electrical propulsion system can be done while the machine is running.
    C. A machine that does not have a high-capacity battery can still exhibit high voltage after the machine has been shut down.
    D. High-voltage conductors are identified with orange cables.

Apply and Analyze

19. An alternating electrical wave of current flow is also known as an AC _____ wave.
20. One sine wave is also known as one _____.
21. In an alternator, the _____ form the north and south magnetic poles.
22. The process of converting AC electricity to DC electricity is known as _____.
23. A(n) _____ is an electronic device used to convert AC to DC.
24. A(n) _____ motor is simple and economical but requires complex and expensive electronic controls to drive the motor.
25. The Caterpillar AC electric truck uses a(n) _____ to dissipate heat during braking.
26. John Deere's 944K wheel loader uses a total of _____ electric-drive traction motor(s).
27. John Deere's 644K wheel loader uses a total of _____ electric-drive traction motor(s).
28. Caterpillar's hybrid excavator uses a(n) _____ to recover and store high energy when the swing is slowing to a stop.

Critical Thinking

29. Explain the electrical drive differences between a Caterpillar D7E dozer and a Caterpillar D6XE dozer.
30. What advantages do 100% electric-drive excavators offer over diesel-powered excavators?
31. What is a difference between a permanent magnet DC motor and a switched reluctance motor?

Appendix

This appendix includes tables that the student will find useful during the study of heavy equipment power trains and systems. The information is grouped into three categories: tool and fastener information, equipment sizes, and mechanical systems.

Fastener Information

The information in the following charts relates to tools and the classification of threaded fasteners. The tables in this section are referred to in Chapter 2, *Tools and Fasteners*.

Pressure Tap Part Number Chart					
SAE J1502 interchange	CNH	Caterpillar	John Deere	Parker	Eaton/Aeroquip/ Weatherhead
1/8"—27 NPT male tap	H434164	8T-3613	XPD323	PD323	FD90-1012-02-04
1/4"—18 NPT male tap	9845229	6V-3966	XPD343	PD343	FD90-1012-04-04
3/8"—24 ORB male tap	—	—	XPD331	PD331	FD90-1044-03-04
7/16"—20 ORB male tap	R55912	—	XPD345	PD341	FD90-1044-04-04
1/2"—20 ORB male tap	—	—	XPD351	PD351	FD90-1044-05-04
9/16"—18 ORB male tap	87026252	6V-3965	XPD361	PD361	FD90-1044-06-04
1/4"—18 NPT female tap	—	6V-3989	—	PD342	FD90-1034-04-04
1/8"—27 NPT female tap	—	6V-4142	—	PD322	FD90-1034-02-04
M14 × 1.5 ORB male tap	84320565	—	XPD367A	PD367A-6	FD90-1046-06-04
M18 × 1.5 ORB male tap	358968A1	—	—	PD3127-6	—
1/2" Tube ORF male tap	—	—	—	PD38BTL	—
7/16"—20 JIC (1/4" tube) male tap	R54805	—	—	PD34BTX	—
9/16"—18 JIC (3/8" tube) male tap	—	—	—	PD36BTX	—
3/4"—16 JIC (1/2" tube) male tap	—	—	—	PD38BTX	—
9/16"—18 ORF female tap	190117A1	—	—	PD34BTL-6	—
11/16"—16 ORF female tap	190119A1	—	—	PD36BTL-5	—
13/16"—16 ORF female tap	190316A1	—	—	PD38BTL-6	—
Dust cap	86502604	6V-0852	R77175	PDN-02C	FD90-1040-04
1/8"—27 NPT female coupler	—	6V-4143	—	PD222	FD90-1021-02-04
1/4"—18 NPT female coupler	73163493	6V-4144	RE219698	PD242	FD90-1021-04-04

Goodheart-Willcox Publisher

Pressure Tap Number Chart. This type of chart can be used to locate part numbers for pressure taps.

Metric Bolt Sizes			
Bolt Size (mm)	Standard Thread Pitch	Fine Thread Pitch	Extra Fine Thread Pitch
4	.70	—	—
5	.80	—	—
6	1.00	—	—
7	1.00	—	—
8	1.25	1.00	—
10	1.50	1.25	1.00
12	1.75	1.50	1.25
14	2.00	1.50	—
16	2.00	1.50	—
17	—	1.50	1.00
18	2.00	1.50	1.00
20	2.50	2.00	1.50
22	2.50	—	—
24	3.00	2.00	—
27	3.00	—	—
30	3.50	2.00	—
36	4.00	3.00	—
42	4.50	—	—
48	5.00	3.00	—
56	5.50	4.00	—
64	6.00	4.00	—
68	6.00	—	—

Goodheart-Willcox Publisher

Metric Bolt Sizes. Sizes and thread pitches for metric bolts.

US Customary Bolt Sizes

Bolt Size	Outside Diameter	Threads Per Inch (TPI) Unified National Coarse (UNC)	Threads Per Inch (TPI) Unified National Fine (UNF)
#8	0.1640"	32	36
#10	0.1900"	24	32
#12	0.2160"	24	28
1/4"	0.2500"	20	28
5/16"	0.3125"	18	24
3/8"	0.3750"	16	24
7/16"	0.4375"	14	20
1/2"	0.5000"	13	20
9/16"	0.5625"	12	18
5/8"	0.6250"	11	18
3/4"	0.7500"	10	16
7/8"	0.8750"	9	14
1"	1.0000"	8	12
1 1/8"	1.1250"	7	12
1 1/4"	1.2500"	7	12
1 1/2"	1.5000"	6	12

Goodheart-Willcox Publisher

US Customary Bolt Sizes. Common US customary bolt sizes and the number of threads per inch are listed in the matrix. For example, a 1/2" bolt is classified as (1/2" × 13) or (1/2" × 20). US customary bolts are also known as *standard* or *imperial*.

SAE Bolt Grade Strengths

SAE Grade	Tensile Strength	Yield Strength	Bolt Diameter Sizes
1	60,000 psi	36,000 psi	1/4"–1 1/2"
2	60,000 psi	36,000 psi	7/8"–1 1/2"
2	74,000 psi	57,000 psi	1/4"–3/4"
5	105,000 psi	81,000 psi	1-1/8"–1 1/2"
5	120,000 psi	92,000 psi	1/4"–1"
5.2	120,000 psi	92,000 psi	1/4"–1"
7	133,000 psi	115,000 psi	1/4"–1 1/2"
8	150,000 psi	130,000 psi	1/4"–1 1/2"
8.2	150,000 psi	130,000 psi	1/4"–1"

Goodheart-Willcox Publisher

SAE Bolt Grade Strength. SAE bolt strength ranges from 60,000 psi (grade 1) to 150,000 psi (grade 8.2).

Metric Bolt Grade Strengths

Metric Grade	Tensile Strength	Yield Strength	Bolt Diameter Sizes
4.6	60,000 psi (400 MPa)	34,800 psi (240 MPa)	M5–M36
5.8	75,400 psi (520 MPa)	57,000 psi (393 MPa)	M5–M24
8.8	116,000 psi (800 MPa)	92,800 psi (640 MPa)	M3–M16
8.8	120,350 psi (830 MPa)	95,700 psi (660 MPa)	M17–M36
9.8	130,500 psi (900 MPa)	104,000 psi (720 MPa)	M1.6–M16
10.9	150,800 psi (1040 MPa)	136,300 psi (940 MPa)	M6–M36
12.9	176,900 psi (1220 MPa)	159,500 psi (1100 MPa)	M1.6–M36

Goodheart-Willcox Publisher

Metric Bolt Grade Strengths. Metric grade bolts are listed in decimals. Their diameters are given in millimeters, signified by a "M" prefix. For example, a 10 mm bolt is marked M10.

Nut Grade Markings

SAE J995 prior to July 1999		SAE J995 after July 1999	ISO 898-2 Metric Nuts	
Grade 2	Grade 2	Grade 2	Class 5	
Grade 5	Grade 5	Grade 5	Class 6	
Grade 8	Grade 8	Grade 8	Class 8	
			Class 9	
			Class 10	
			Class 12	

Goodheart-Willcox Publisher

Nut Grade Markings. The markings on a nut indicate its grade. The grade of a nut should match the grade of the bolt to which it is attached.

Equipment Sizes

The tables in this section list the technical specifications of the various size classes of common construction and agricultural equipment. These tables are referred to in Chapter 3, *Construction Equipment Identification*, and Chapter 4, *Agricultural Equipment Identification*.

Excavator Sizes		
Class	**Weight**	**Engine Power**
Micro	0.27–1.5 metric tons	3.5–10 hp (2.6–8 kW)
Mini (compact)	1.75–10 metric tons	13–70 hp (9.6–52 kW)
Small (midsized)	12–20 metric tons	73–120 hp (54.5–89.5 kW)
Medium	22–36 metric tons	140–273 hp (104–203 kW)
Large	36–94 metric tons	273–542 hp (203–404 kW)
Hydraulic mining shovels	102–1270 metric ton	700–4500 hp (522–3360 kW)

Goodheart-Willcox Publisher

Excavator Sizes. Excavators are classified into different sizes based on weight and engine power.

Wheel Loader Sizes		
Class	**Engine Power**	**Bucket Heaped Capacity**
Compact	42–116 hp (31–86.5 kW)	0.75–1.5 yd^3
Small	124–168 hp (92.5–125 kW)	1.75–6.5 yd^3
Midsized	225–300 hp (168–223 kW)	375–12.0 yd^3
Large	316–541 hp (236–403 kW)	5.0–20.9 yd^3
Mining loaders	700–2300 hp (522–1715 kW)	7.0–70.0 yd^3

Goodheart-Willcox Publisher

Wheel Loader Sizes. Wheel loaders are classified into different sizes based on engine power and heaped capacity.

Track Loader Sizes			
Class	**Engine Power**	**Weight**	**Bucket Heaped Capacity**
Stand-on behind (32″–42″ wide)	20–36.9 hp (14.9–27.5 kW)	1660–3840 lbs	—
Compact rubber track	73–106 hp (54–79 kW)	8945–16,000 lbs	14.8–33.8 yd^3
Small steel track	110 hp (83 kW)	26,940 lbs	1.5–1.7 yd^3
Midsized steel track	140–160 hp (105–119 kW)	35,000–41,000 lbs	1.5–2.5 yd^3
Large steel track	173–202 hp (129–150.6 kW)	45,000–50,000 lbs	2.5–3.2 yd^3
Largest steel track	275 hp (205 kW)	63,900–85,000 lbs	*4.2 yd^3

*6.4 yd^3 for waste handling track loader

Goodheart-Willcox Publisher

Track Loader Sizes. Track loaders are classified into different sizes based on engine power, weight, and heaped capacity.

Dozer Sizes			
Class	Engine Power	Weight	Bucket Heaped Capacity
Small	80–104 hp (60–77 kW)	17,500–21,000 lbs	2–3 yd^3
Medium	13–238 hp (97–177 kW)	29,000–48,000 lbs	4–7 yd^3
Large	312–436 hp (233–325 kW)	87,000–106,000 lbs	7–17 yd^3
Mining	600–890 hp (447–663 kW)	154,000–254,195 lbs	18–45 yd^3

Goodheart-Willcox Publisher

Dozer Sizes. Dozers are classified into different sizes based on engine power, weight, and blade capacity.

Motor Grader Sizes				
Class	Engine Power	Weight	Blade Width	Blade Height
Compact	110–120 hp (82–89 kW)	12,800–17,000 lbs	10′	19″–21″
Small	130–230 hp (82–171 kW)	25,000–40,000 lbs	12′	19″–24″
Medium	210–285 hp (157–213 kW)	35,000–54,000 lbs	12′–14′	23.5″–27″
Large	300 hp (223 kW)	67,000 lbs	16′	31″
Mining	535 hp (399 kW)	161,695 lbs	24′	40″

Goodheart-Willcox Publisher

Motor Grader Sizes. Motor graders are classified into different sizes based on engine power, weight, and blade dimensions.

Tractor Scraper Sizes					
Class	Tractor Engine Power	Scraper Engine Power	Empty Operating Weight	Bowl Heaped Capacity	Bowl Struck Capacity	Rated Load
Small	215 hp (160 kW)	—	45,000 lb	14 yd^3	10 yd^3	30,000 lb
Medium	407 hp (304 kW)	290 hp (216 kW)	79,797–88,000 lb	24 yd^3	17 yd^3	57,610 lb
Large	500 hp (373 kW)	283 hp (211 kW)	100,000–118,000 lb	34 yd^3	24 yd^3	82,000 lb
Coal-mining	570–600 hp (425–447 kW)	290–451 hp (216–337 kW)	117,782–160,000 lb	50–73 yd^3	41–59 yd^3	76,000–110,000 lb

Goodheart-Willcox Publisher

Tractor Scraper Sizes. Tractor scrapers are classified into different sizes based on engine power, bowl capacity, and rated load.

Articulated Dump Trucks (ADTs)

Class	Rated Load Capacity	Bead Heaped Capacity	Bed Struck Capacity	Empty Operating Weight	Engine Power
Small	26.5–31 short tons	19.5–23 yd³	14–18 yd³	44,779–52,000 pounds	320–360 hp (237–268 kW)
Medium	37–45 short tons	27–32.7 yd³	21–24 yd³	64,000–80,000 pounds	450–504 hp (336–376 kW)
Large	60 short tons	45 yd³	35 yd³	93,644 pounds	577 hp (430 kW)

Rigid-Frame Haul Trucks

Class	Rated Load Capacity	Bead Heaped Capacity	Bed Struck Capacity	Empty Operating Weight	Engine Power
Small	40–50 short tons	33–41 yd³	21.5–30 yd³	54,000–56,000 lbs	510–600 hp (380–447 kW)
Medium	60–70 short tons	46–55 yd³	35–42 yd³	67,000–77,000 lbs	710–825 hp (529–615 kW)
Large	100 short tons	79 yd³	55 yd³	111,000 lbs	1025 hp (764 kW)

Mining Haul Trucks

Class	Rated Load Capacity	Bead Heaped Capacity	Bed Struck Capacity	Empty Operating Weight	Engine Power
Small	150 short tons	102 yd³	—	248,000 pounds	1450 hp (1081 kW)
Medium	200–250 short tons	137–230 yd³	—	292,000–350,000 pounds	1900–2650 hp (1416–1976 kW)
Large	350–496 short tons	280–350 yd³	—	558,000–793,664 pounds	3400–4600 hp (2536–3430 kW)

Goodheart-Willcox Publisher

Haul Truck Sizes. Haul trucks are classified into different sizes based on load capacity, weight, and engine power.

Skid Steer Sizes

Class	Engine Power	Machine Weight	Rated Operating Capacity
Stand-on behind	20–37 hp (14.9–27.5 kW)	1660–3840 lb	350–860 lb
Small	50–60 hp (36 kW)	5370–6200 lb	1370–1500 lb
Medium	70 hp (52 kW)	7000 lb	2000 lb
Large	80–110 hp (60–82 kW)	8600–10,200 lb	2500–3500 lb

Goodheart-Willcox Publisher

Skid Steer Sizes. Skid steers are classified into different sizes based on engine power, weight, and operating capacity. Because walk-behind skid steers must operate in tight areas, width is additional criteria for that class.

Compactor Sizes		
Class	Engine Power	Operating Weight
Mini walk-behind compactor rollers	4.8–9.7 hp (3.6–7.2 kW)	600–1600 lbs
Small vibratory compactors	22–36 hp (16.4–26.8 kW)	3300–5400 lbs
Midsized vibratory compactors	100–140 hp (74.5–104 kW)	18,000–28,000 lbs
Large vibratory compactors	150–175 hp (112–130 kW)	25,000–40,000 lbs
Large soil compactors	230–400 hp (171–298 kW)	40,000–80,000 lbs
Large landfill compactors	430–560 hp (320–417 kW)	90,000–120,000 lbs

Goodheart-Willcox Publisher

Compactor Sizes. Compactors are classified into different sizes based on engine size and operating weight.

Trencher Sizes				
Class	Engine Power	Weight	Digging Depth	Cutting Width
Mini walk-behind trenchers	11.7–22.3 hp (8.7–16.6 kW)	1020–1120 lbs	36″	4″–8″
Small trenchers	25–62 hp (18–46 kW)	3050–5610 lbs	42″–62″	4″–12″
Mid-to-large trenchers	100–120 hp (74.5–90 kW)	8700–8880 lbs	70″–97″	6″–24″
Pipeline trenchers	185–600 hp (138–447 kW)	24,000–205,000 lbs	96″–216″	24″–48″

Goodheart-Willcox Publisher

Trencher Sizes. Trenchers are classified into different sizes based on engine power, weight, and trench dimension.

Combine Classes 9, 10, 11–543 hp (405 kW) and Higher	
OEM Examples	Rated Engine Horsepower
Case IH 9250	625 hp (466 kW)
John Deere X9 1100	690 hp (514.5 kW)
Claas Lexion Terra Trac 8800	653 hp (480 kW)
New Holland CR10.90	700 hp (515 kW)
Fendt Ideal 10	790 hp (589 kW)
Combine Class 8–456–480 hp (340–358 kW)	
OEM Examples	Rated Engine Horsepower
Case IH AFX 8250	555 hp (413 kW)
Fendt Ideal 8	538 hp (401 kW)
Gleaner S98	471 hp (351 kW)
John Deere S780	540 hp (402 kW)
Claas Lexion Terra Trac 8700	577 hp (430 kW)
New Holland CR8.90	544 hp (400 kW)
Combine Class 7–370–408 hp (276–304 kW)	
OEM Examples	Rated Engine Horsepower
Case IH 7250	468 hp (349 kW)
Claas Lexion Terra Trac 7500	462 hp (340 kW)
Fendt Ideal 7	451 hp (336 kW)
Gleaner S97	451 hp (336 kW)
John Deere S770	449 hp (335 kW)
New Holland CR7.90	460 hp (338 kW)
Combine Class 6–313–360 hp (234–268 kW)	
OEM Examples	Rated Engine Horsepower
Case IH 6150	411 hp (306.5 kW)
Gleaner S96	398 hp (296 kW)
John Deere S760	382 hp (285 kW)
Claas Lexion Terra Trac 7400	408 hp (300 kW)

Goodheart-Willcox Publisher

Combine Harvester Sizes. Classes of combines are based on engine horsepower.

Mechanical Systems

This section provides information about various mechanical systems in heavy equipment. The tables in this section are referenced in Chapter 6, *Belt and Chain Drives*; Chapter 10, *Powershift and Automatic Transmission Theory*; Chapter 15, *Continuously Variable Transmissions*; and Chapter 22, *Tires, Rims, and Ballasting*.

ANSI Roller Chain Sizes			
Chain Size	Pitch	Distance between Inner Link Plates, or Roller Width	Roller Diameter
40	0.5″ (12.7 mm)	0.313″ [5/16″] (7.950 mm)	0.312″ (7.92 mm)
50	0.625″ (15.875 mm)	0.376″ [3/8″] (9.550 mm)	0.4″ (10.160 mm)
60	0.75″ (19.050 mm)	0.5″ [1/2″] (12.700 mm)	0.469″ (11.910 mm)
80	1.0″ (25.4 mm)	0.626″ [5/8″] (15.900 mm)	0.625″ (15.880 mm)
100	1.25″ (31.750 mm)	0.754″ [3/4″] (19.150 mm)	0.75″ (19.050 mm)
120	1.5″ (38.100 mm)	1.006″ [1″] (25.550 mm)	0.875″ (22.230 mm)
140	1.75″ (44.450 mm)	1.004″ [1″] (25.500 mm)	1″ (25.400 mm)
160	2″ (50.800 mm)	1.25″ [1 1/4″] (31.750 mm)	1.125″ (28.580 mm)
180	2 1/4″ (57.150 mm)	1.406″ [1 13/32″] (35.710 mm)	1.406″ (35.710 mm)
200	2 1/2″ (63.500 mm)	1.50″ [1 1/2″] (38.100 mm)	1.562″ (39.680 mm)
240	3″ (76.200 mm)	1.89″ [1 7/8″] (48.00 mm)	1.875″ (47.630 mm)

Goodheart-Willcox Publisher

ANSI Roller Chain Sizes. Examples of roller chain sizes and their respective pitch, width, and roller diameter.

Caterpillar TA22 Agricultural Transmission Clutch Apply Chart				
Gear	Reverse Shaft	Input Shaft	Intermediate Shaft	Output Shaft
Neutral	—	—	—	J
1F	—	C	—	G
2F	—	D	—	G
3F	—	—	E	G
4F	—	—	F	G
5F	—	C	—	B
6F	—	C	—	H
7F	—	D	—	B
8F	—	D	—	H
9F	—	—	E	B
10F	—	—	E	H
11F	—	—	F	B
12F	—	—	F	H
13F	—	C	—	J
14F	—	D	—	J
15F	—	—	E	J
16F	—	—	F	J
1R	A	—	—	J
2R	A	—	—	H
3R	A	—	—	B
4R	A	—	—	G

Goodheart-Willcox Publisher

Caterpillar Challenger MT 700, MT800, and MT 900 Clutch Apply Chart. The clutch apply chart for the Caterpillar TA22 powershift transmissions used in Challenger MT 700, MT800, and MT 900 series tractors. The powershift transmission has four countershafts and two planetary gear sets to provide 16 forward speeds and four reverse speeds. The letter indicates the clutch engaged for the listed transmission shaft. Note that clutches B and G on the output shaft are planetary gear clutches. The other clutches are countershaft gear clutches. Notice only two clutches are applied at one time to achieve power flow. The transmission is also used in Versatile Delta series articulated tractors.

Status of Members during Caterpillar CVT Operation

Steps	CVT Mode	Loader Travel Status	Low Forward, High Reverse Synchronizer	Low Reverse, High Forward Synchronizer	Auxiliary Synchronizer	Clutch A	Clutch B	Variator Pump Displacement
1	Low forward	Loader starts moving forward from a stationary position	Low forward	Low reverse	Neutral	Engaged	Released	Begins at max reverse displacement, then decreases moving toward neutral
2	Low forward	Increasing forward travel speed	Low forward	Neutral	Neutral	Engaged	Released	Decreases until yoke reaches neutral angle
3	Low forward	Nearly the fastest travel speed possible in low forward, approaching high forward	Low forward	High forward	Neutral	Engaged	Released	Increases positive displacement
4	Low forward to high forward shift	Making a synchronous shift	Low forward	High forward	Neutral	Engaged	Engaged	Maximum positive displacement
5	High forward	Loader traveling at the slowest rate in the high forward mode	Low forward	High forward	Neutral	Released	Engaged	Begins reducing displacement from the maximum positive yoke angle
6	High forward	Increasing forward travel speed	Neutral	High forward	Neutral	Released	Engaged	Decreases until yoke achieves neutral angle
7	High forward	Approaching fastest travel speed in high forward, near the auxiliary shift	Neutral	High forward	Auxiliary engaged	Released	Engaged	Yoke increasing its displacement in a reverse direction (negative yoke angle)
8	Auxiliary	Making a non-synchronous shift from HF to Aux	Neutral	High forward	Auxiliary engaged	Engaged	Released	This is the one point in the drive when the yoke makes a drastic angle shift from maximum negative to maximum positive angle.
9	Auxiliary	This is the slowest travel speed in the auxiliary mode (just leaving the high forward mode)	Neutral	High forward	Auxiliary engaged	Engaged	Released	Moving from a maximum positive yoke angle toward a neutral angle (decreasing)
10	Auxiliary	Maximum loader travel speed	Neutral	High forward	Auxiliary engaged	Engaged	Released	Yoke moves to maximum negative angle.

Goodheart-Willcox Publisher

Status of Members during Caterpillar CVT Operation. The Caterpillar 966K XE, 972M, XE, 966 XE, and 972 wheel loader CVT uses multiple synchronizers, clutches, and a hydrostatic transmission to provide its various speeds and change of direction. This matrix lists the status of the components based on the speed and direction of the loader.

Status of Members during Caterpillar CVT Operation (Continued)

Steps	CVT Mode	Loader Travel Status	Low Forward, High Reverse Synchronizer	Low Reverse, High Forward Synchronizer	Auxiliary Synchronizer	Clutch A	Clutch B	Variator Pump Displacement
NA	Neutral	Loader is stationary	Low forward	Low reverse	Neutral	Released	Released	Maximum negative displacement
1	Low reverse	Loader starts moving rearward from a stationary position.	Low forward	Low reverse	Neutral	Released	Engaged	Moving from maximum negative yoke angle toward neutral
2	Low reverse	Increasing reverse travel speed	Neutral	Low reverse	Neutral	Released	Engaged	Yoke reaches neutral angle.
3	Low reverse	Nearly the fastest travel speed possible in low reverse, approaching high reverse	High reverse	Low reverse	Neutral	Released	Engaged	Increases positive displacement
4	Low reverse to high reverse shift	Making a synchronous shift	High reverse	Low reverse	Neutral	Engaged	Engaged	Maximum positive displacement
5	High reverse	Loader traveling at the slowest rate in the high reverse mode	High reverse	Low reverse	Neutral	Engaged	Released	Begins reducing displacement from the maximum positive yoke angle
6	High reverse	Increasing reverse travel speed	High reverse	Neutral	Neutral	Engaged	Released	Yoke reaches neutral angle.
7	High reverse	Approaching fastest travel speed in high reverse	High reverse	Neutral	Neutral	Engaged	Released	Yoke increasing its displacement in a reverse direction (negative yoke angle)

Status of Members during Caterpillar CVT Operation. Continued.

Tire Load Index Values

Load Index Code	Load Capacity (Pounds)	Load Capacity (Kilograms)	Load Index Code	Load Capacity (Pounds)	Load Capacity (Kilograms)
90	1320	600	130	4180	2000
91	1360	615	131	4300	2060
92	1390	630	132	4400	2120
93	1430	650	133	4540	2180
94	1480	670	134	4680	2240
95	1520	690	135	4800	2300
96	1570	710	136	4940	2360
97	1610	730	137	5080	2430
98	1650	750	138	5200	2500
99	1710	775	139	5360	2575
100	1760	800	140	5520	2650
101	1820	825	141	5680	2725
102	1870	850	142	5840	2800
103	1930	875	143	6000	2900
104	1980	900	144	6150	3000
105	2040	925	145	6400	3075
106	2090	950	146	6600	3150
107	2150	975	147	6800	3250
108	2200	1000	148	6950	3350
109	2270	1030	149	7150	3450
110	2340	1060	150	7400	3550
111	2400	1090	151	7600	3450
112	2470	1120	152	7850	3550
113	2540	1150	153	8050	3650
114	2600	1180	154	8250	3750
115	2680	1215	155	8550	3875
116	2760	1250	156	8800	4000
117	2830	1285	157	9100	4125
118	2910	1320	158	9350	4250
119	3000	1360	159	9650	4375
120	3080	1400	160	9900	4500
121	3200	1450	161	10,200	4625
122	3300	1500	162	10,500	4750
123	3420	1550	163	10,700	4875
124	3520	1600	164	11,000	5000
125	3640	1650	165	11,400	5150
126	3740	1750	166	11,700	5300
127	3860	1800	167	12,000	5450
128	3960	1850	168	12,300	5600
129	4080	1900	169	12,800	5800

Goodheart-Willcox Publisher

Tire Load Index Values. A manufacturer's load index chart illustrates the amount of load a tire can handle in units of pounds and kilograms.

Tire Load Index Values (Continued)

Load Index Code	Load Capacity (Pounds)	Load Capacity (Kilograms)	Load Index Code	Load Capacity (Pounds)	Load Capacity (Kilograms)
170	13,200	6000	211	43,000	19,500
171	13,600	6150	212	44,100	20,000
172	13,900	6300	213	45,400	20,600
173	14,300	6500	214	46,700	21,200
174	14,800	6700	215	48,100	21,800
175	15,200	6900	216	49,400	22,400
176	15,700	7100	217	50,700	23,000
177	16,100	7300	218	52,000	23,600
178	16,500	7500	219	53,600	24,300
179	17,100	7750	220	55,100	25,000
180	17,600	8000	221	56,800	25,750
181	18,200	8250	222	58,400	26,500
182	18,700	8500	223	60,000	27,250
183	19,300	8750	224	61,500	28,000
184	19,800	9000	225	64,000	29,000
185	20,400	9250	226	66,000	30,000
186	20,900	9500	227	68,000	30,750
187	21,500	9750	228	69,500	31,500
188	22,000	10,000	229	71,500	32,500
189	22,700	10,300	230	74,000	33,500
190	23,400	10,600	231	76,000	34,500
191	24,000	10,900	232	78,500	35,500
192	24,700	11,200	233	80,500	36,500
193	25,400	11,500	234	82,500	37,500
194	26,000	11,800	235	85,600	38,760
195	26,800	12,150	236	88,000	40,000
196	27,600	12,500	237	91,000	41,250
197	28,300	12,850	238	93,500	42,500
198	29,100	13,200	239	96,500	43,750
199	30,000	13,600	240	99,000	45,000
200	30,900	14,000	241	102,000	46,250
201	32,000	14,500	242	104,500	47,500
202	33,100	15,000	243	107,500	48,750
203	34,200	15,500	244	110,000	50,000
204	35,300	16,000	245	113,500	51,500
205	36,400	16,500	246	117,000	53,000
206	37,500	17,000	247	120,000	54,500
207	38,600	17,600	248	123,500	56,000
208	39,700	18,000	249	128,000	58,000
209	40,800	18,500	250	132,500	60,000
210	41,900	19,000			

Tire Load Index Values. *Continued*

Glossary

A

abrasive wear. Wear to components caused by foreign particles. (17)

accumulator. A hydraulic energy storage device that provides a supply of hydraulic fluid to apply the brakes in the event of an engine or hydraulic pump failure. (19)

acetylene gas. A very unstable and highly flammable gas used in oxyacetylene welding. (1)

AC induction motor. An AC traction motor that contains two primary components—a stationary stator and a rotating rotor. It does not use commutators or brushes. (26)

Ackerman steering. A steerable axle with a geometry that enables the inside wheel to turn at a greater steering angle than the outside wheel. (25)

AC sine wave. The wave form created when alternating current flow is graphed against time. (26)

active air ride seat. An air suspension seat design equipped with multiple sensors (position, pressure, displacement, and temperature) to provide input to an electronic control module that controls a dampening assembly connected to the seat. (21)

ActiveCommand Steering system. John Deere steering system that uses electronic inputs to sense the tractor's balance. The steering system has variable ratio steering and variable effort steering. (25)

AC traction motor. A motor that is driven with a variable frequency AC sign wave, also referred to as an *asynchronous motor* because the rotor spins slower than the induced rotating magnetic field. (26)

actuating disc. Part of a ball ramp disc brake that has a flat surface on one side that mates with a brake friction disc. The other side has pockets for steel balls. (19)

adaptive learning. The process in which a transmission ECM determines if a clutch engages too slowly or too quickly and then commands a solenoid to respond more quickly or more slowly. (11)

adhesive wear. Wear that occurs when friction between two components results in the two surfaces welding to each other. As the components attempt to move, material lifts off both components, tearing off or transferring material. (17)

Advanced Stone Protection (ASP). A system that uses acoustic sensors to detect the presence of stones. (4)

agricultural chain. A chain commonly found on agricultural machines and used to move material. Also referred to as a *conveyer chain*. (6)

air-assisted scraper. A towable scraper in which air is directed through the floor to loosen the dirt, helping the scraper load more dirt. (3)

air bag. A type of suspension spring that provides support using a rubber bladder inflated with compressed air. Also referred to as an *air spring*. (21)

air brake actuator. Part of the foundation brakes that converts air pressure into a linear force. Also referred to as an *air chamber* or *brake chamber*. (20)

air brake supply system. Part of an air brake system that contains a compressor, governor, air dryer, and supply tank. (20)

air chamber. Part of the foundation brakes that converts air pressure into a linear force. Also referred to as an *air brake actuator* or *brake chamber*. (20)

air compressor. Part of the air brake supply system that generates compressed air to apply the air brakes. (20)

AirCushion. A type of John Deere automatic air suspension equipped with an electric-powered air compressor, two air springs, and a walking beam to improve the ride quality of 8RT and 9RT twin-rubber-track tractors. (21)

air disc brakes. Air-operated external disc brakes that use a caliper with pistons that press brake pads into a rotating disc rotor. (20)

air dryer. Part of the air brake supply system that removes contaminants and moisture from the air. (20)

air gap. The distance the armature must move in an electromagnetic clutch for a pulley to couple to a rotor. (8)

airless tires. Non-pneumatic tires that have pockets (holes), arranged in honeycomb or spoke-shaped patterns within the tire sidewall and body, that allow the rubber webbing to cushion and support the load placed on the tire. (22)

air-over-hydraulic brakes. A brake system that incorporates the use of both air and hydraulics to actuate the service brakes. (20)

air ride seat. An operator seat equipped with an air bag, electric-powered compressor, and hydraulic shock absorber to adjust seat height and cushioning and dampen overall movement. Also referred to as an *air suspension seat*. (21)

air seeder. A planting implement for small grains that integrates a chisel plow and a large seed cart that contains both the seed and chemicals, which are metered and distributed by air pressure. (4)

air spring. A type of suspension spring that provides support using a rubber bladder inflated with compressed air. Also referred to as an *air bag*. (21)

Note: The number in parentheses following each definition indicates the chapter in which the term can be found.

air supply tank. Part of the air brake supply system located directly after the compressor or after the air dryer. Stores the compressed air used to apply the brakes. Also referred to as *supply reservoir* or *wet tank*. (20)

air suspension seat. An operator seat equipped with an air bag, electric-powered compressor, and hydraulic shock absorber to adjust seat height and cushioning and dampen overall movement. Also referred to as an *air ride seat*. (21)

air wedge brake. A type of brake that uses a push rod to extend a wedge with two rollers against two angled plungers to force the brake shoes against the drum. (20)

alcohol evaporator. Part of the air brake supply system that injects small amounts of alcohol into the air system to reduce air line icing. (20)

alternator. An electrical generator that produces alternating current (AC). (26)

amboid gear set. A ring-and-pinion gear set in which the pinion gear is positioned above the center axis of the ring gear. (7)

angle of repose. The slope of the bucket's dirt in relationship to the bucket's top plane. (3)

angular load. A combination of radial and thrust loads. (7)

annulus gear. A large internal-toothed gear. In a planetary gear set, the annulus gear surrounds the entire set and meshes with the planetary pinion gears. Also referred to as a *ring gear*. (7)

antifriction bearing. A bearing consisting of two steel rings called races that surround rolling elements, such as balls, rollers, or needles. (7)

anti-slip regulation (ASR). A traction control system used on four-wheel-drive ProDrive-equipped harvesters. The system monitors wheel slip and controls the hydraulic motors to provide maximum traction. (15)

A-post display. A monitor mounted in the cab pillar of a machine that gives the operator real-time information about the machine's operation. Also referred to as a *corner post display*. (15)

apron. A movable metal bulkhead on a scraper that is opened to load and unload the scraper and serves as the front wall of the scraper bowl when it is lowered. (3)

armature. Brush-type DC motor component that consists of several loops of large copper conductors connected to copper commutator bars. (26)

articulated agricultural tractor. A tractor with full-time four-wheel drive, equal-size front and rear tires, and an articulated frame. (4)

articulated dump truck (ADT). A truck with a dump bed and an articulated frame. Typically equipped with two rear drive axles with single wheels and often equipped with a front drive axle, resulting in six wheels that are all powered. (3)

articulation joint. A joint on an articulated machine that allows the front and rear frames to pivot in a horizontal plane and steer the machine to the left or right. (21)

articulation steering. Steering system in which an articulation joint connects the front frame of the machine to the rear frame. Two hydraulic cylinders articulate the two frames so the tractor can steer to the left or to the right. (25)

asbestos. A cancer-causing material used in older friction material. (8)

aspect ratio. The ratio of a tire's height to its cross-sectional width. (22)

asperities. Microscopic high spots on a component's surface. (17)

asynchronous motor. An AC induction motor. The term *asynchronous* refers to the fact that the rotor spins at a speed slower than the speed of the induced rotating magnetic field. (26)

auto field operation. A transmission sub-mode of operation designed to maintain the engine speed while operating in the field. (10)

auto-guidance steering system. A steering system that uses electronic controls coupled with a global positioning satellite (GPS) system to program and direct the machine's path of movement. (25)

auto-guidance system. A feature on many late-model agricultural machines that uses electronic global positioning satellite (GPS) systems to automatically steer the machine as it is working in the field. (4)

auto-kickdown mode. An electronic transmission control mode that automatically downshifts the transmission when the load is too heavy. (11)

automated external defibrillator (AED). A portable electronic device used to diagnose and treat a person through defibrillation, the application of electrical therapy to help the heart reestablish an effective rhythm. (1)

automatic ride control. A feature that employs one or more accumulators to dampen implement movement. It engages the machine's ride control at a specific travel speed programmed into the electronic control module. (21)

automatic slack adjuster (ASA). A type of slack adjuster that will automatically adjust the slack as brake shoes wear. (20)

automatic traction control (ATC). A traction control system in which an ECM monitors wheel slip and modulates solenoid valves to vary the apply pressures to the cross-axle differential locks and inter-axle differential locks, thus reducing wheel slip. (16)

automatic transmission. A transmission that automatically upshifts and downshifts based on the machine's travel speed and load. (10)

autonomous. A term referring to a machine that can operate by itself. (26)

auto road operation. A transmission sub-mode of operation that enables the operator to control the speed of the tractor using the engine's throttle. (10)

Autoshift. An electronic control system in older Caterpillar dozers that allows the operator to choose the specific speeds for forward and reverse any time the dozer is making a directional shift. (11)

auto shift control. A control that automatically shifts the transmission based on engine load, tractor speed, and the current gear speed. (10)

B

back electromotive force (BEMF). The induced voltage produced by a rotating permanent magnet motor that acts in an opposite direction of the current flow that is driving the motor. Also referred to as *counter-electromotive force (CEMF)*. (26)

backfire. A flame that travels back through an oxyacetylene torch tip and results in a single loud pop that extinguishes the torch's flame. (1)

backlash. The amount of clearance (play) between meshing gear teeth. (7)

ball-and-cage joint. A U-joint containing an inner race with six channels in which six balls ride. A cage retains six ball bearings. As the inner race is driven, it causes the balls to drive, which propel the outer bearing race. Also referred to as a *Rzeppa U-joint*. (18)

ballasting. Adjusting the amount and position of weight that is added to a tractor. (22)

ball bearing. An antifriction bearing consisting of several spherical rolling elements between inner and outer races. (7)

ball ramp disc brake. A brake that uses an actuating assembly composed of two actuating discs with teardrop pockets, steel balls, and springs. As the discs are actuated, the steel balls roll up the ramps in the teardrop pockets to apply pressure to the brake discs. (19)

band. A flat strap lined with friction material, shaped in a circular fashion that wraps around a drum. When engaged, it holds the drum stationary. (8)

band brake. A brake that consists of a flexible band around the circumference of a drum. (19)

banded V-belt. Multiple V-belts joined together by a top layer of material (tie-band). The belt is used on pulsating drives to reduce slip that would occur with a matched set of multiple single-drive belts. Also referred to as *joined belt*. (6)

bar axle. An axle in which a solid bar serves as the drive axle. The solid bar protrudes from the inboard planetary final drive, driven by the planetary carrier. This axle allows the drive wheels to be repositioned by sliding them inward or outward as needed. (18)

basket. A circular housing that houses the internal components of a multiple disc clutch. Also referred to as *carrier*, *cylinder*, or *drum*. (8)

basket hitch. A lifting hitch formed by trapping the load and placing both ends of the sling into the hoist's hook or collector ring. (5)

beach marks. Curved lines that radiate from one or more origin points of a slow occurring fatigue fracture. (17)

bead. A group of high-strength steel wires encased in rubber at the edges of a tire that form a firm circular base for attaching the tire to the rim of the wheel. (22)

bearing. A part that supports and allows rotation of a shaft or gear while minimizing drag, friction, resistance, heat, and wear. (7)

bearing plate U-joint. A U-joint with bearing cups that are fitted with plates on the outside surface that are bolted to the yoke. (18)

bearing strap U-joint. A U-joint with two metal straps that are bolted around the circumference of the bearing cup to retain the U-joint in the yoke. (18)

Belleville washer. A cone-shaped disc that flattens when pressed. It is used to apply the clutch apply pressure in a spring-applied clutch or brake, or used to retract the piston in an oil-applied clutch spring-released clutch. (8)

belt length. The overall length of the belt if it is cut and measured from one end to the other end. (6)

Bendix Weiss U-joint. A CV joint with two ball yokes. Four balls transmit power to the two yokes. A fifth smaller ball is a spacer between the two shafts. (18)

bevel gears. Gears that have straight cut teeth or helical cut teeth on the angled face of the gear. They are used to change the direction of power flow at a 90° angle. (7)

bias-ply tire. A thick sidewall tire constructed with layers of nylon, polyester, or rayon cords that are angled at 45° to the tire's centerline and extend in opposite directions across the tire from bead to bead. (22)

bladder accumulator. A type of gas accumulator that consists of a synthetic rubber bladder, also called a bag, inside a metal housing. (19)

blade tilt. The sloping down of a dozer blade to the left or to the right from a level, horizontal position. (3)

blocker ring. The part of a block synchronizer that contacts a speed gear and equalizes the speeds of the transmission shaft and speed gear so the sleeve can engage the gear. Also referred to as a *synchronizer ring*. (7)

blocking. The actions taken or methods used to prevent a component or machine from moving. (5)

block synchronizer. A type of synchronizer consisting of a hub that is splined to the transmission shaft, a movable sleeve, a blocker ring, and inserts. Also referred to as a *cone synchronizer*. (7)

body-up reverse neutralizer. A feature that prevents an operator from backing up a truck while the bed is raised. (3)

bogie. An assembly of four small rollers installed on a pivoting frame that is mounted in an undercarriage. The assembly is commonly equipped on agricultural rubber-track tractors in place of traditional undercarriage track rollers. (23)

bolt. A fastener with external threads on part of its shaft and a head on the opposite end. (2)

boost-assist master cylinder. A master cylinder that uses a boost piston to increase brake pressure. (19)

bottom-of-travel switch. A normally closed switch in inching pedal assemblies that acts as a neutralizer when actuated to the open position. (11)

Bourdon-tube pressure gauge. The most common type of mechanical pressure gauge used by hydraulic technicians. (2)

box-end wrench. A nonadjustable wrench that is closed at each end and uses points around the inside diameter of its jaws to grip hexagonal bolt heads and nuts. (2)

brake. A band or clutch used to hold a shaft, gear, or component stationary. (8)

brake chamber. Part of the foundation brakes that converts air pressure into a linear force. Also referred to as an *air brake actuator* or *air chamber*. (20)

brake resistor. Component in an electric drive machine that converts the electrical energy developed during a braking application into heat energy to be dissipated into the atmosphere. Often is cooled by the engine's coolant. (26)

brake steer. A track steering method in which the operator propels one track in either forward or reverse, while holding the other track stationary by applying a brake. Also referred to as the *pivot turn*. (24)

braking chopper. An electrical switch that limits the DC voltage by diverting the high voltage to braking resistors. (26)

breaker. A rubberized layer of cords that is placed between a bias-ply tire's carcass and tread for increased protection. Some bias-ply tires incorporate the use of steel breakers. (22)

breaker bar. A stronger handle that should be used in place of a ratchet when loosening stubborn fasteners. Also referred to as a *flex handle*. (2)

breakout force. The amount of force exerted at the bucket as it is digging. (3)

bridge rectifier. Device containing diodes, used to rectify AC current flow into DC current flow. (26)

bridle. A lifting sling formed with multiple legs. (5)

brinelling. A surface fatigue failure in bearings that appears with indentations caused by the needle, roller, or ball bearings. (17)

brittle fracture. A fast type of fracture. The material does not elongate, but breaks apart quickly. (17)

broadcast spreader. A paddle-type spinner to sling seed, chemicals, and fertilizer in a circular pattern. (4)

brushes. DC motor components made of carbon. Direct current is passed from the positive brush to the commutator. The current is directed out of the armature through the negative brush. (26)

brushless DC motor (BLDCM). A DC motor with a rotating permanent magnet surrounded by a group of windings. The motor's electronic controls energize the windings, causing the rotor to spin. (26)

brush-type DC motor (BDCM). A motor that uses brushes to send direct current through an armature's copper commutator bars. Speed is controlled by varying the DC voltage. (26)

bull-and-pinion. A gear set consisting of a small external tooth gear that drives a large external tooth gear. (7)

bull gear. A large spur gear driven by a pinion. (7)

bull gear. The large external-toothed ring gear in a bull-and-pinion gear final drive. (17)

burst disk. A safety device installed in compressed gas cylinders that will rupture at a specified pressure to allow the controlled release of the gas. Also referred to as *a rupture disk*. (1)

bushing turn. The process of removing the track chain from a tractor, putting the track on a track link press, pressing the bushings out of the links, and turning the bushings 180° before reinserting them into the links in order to gain additional track life from the opposite side of the bushings. (23)

C

cage. In an antifriction bearing, a part that does not carry any load, but is used to evenly space the rolling elements. (7)

calibration mode. A mode on electronically controlled transmissions that allows the ECM to learn transmission clutch values, such as clutch fill and clutch hold. (11)

caliper. The cast housing in a hydraulically actuated caliper disc brake that contains one or more pistons that apply pressure to the brake pads to slow or stop a rotating rotor. (19)

camber. The tire angle formed between the vertical axis of the tire and the vertical axis of the tractor when viewed from the front or back of the axle. (21)

cam shifter. A shifter that consists of a cam plate with slots that cause the rollers of a shift fork to move along a set path as the cam plate is rotated. (7)

carcass. The rubber body of a tire that contains the tire's plies and beads. (22)

Cardan U-joint. A U-joint that rotates at nonuniform velocities. Also referred to as a *Hooke U-joint* and *conventional U-joint*. (18)

cardiopulmonary resuscitation (CPR). An emergency procedure that combines chest compression with artificial ventilation to preserve heart and lung function until further measures are taken to restore heart and lung function to the victim. (1)

carriage assembly. The part of a forklift that lifts and lowers materials. It is a metal frame with movable forks that is raised and lowered by a chain drive. (3)

carrier. A circular housing that houses the internal components of a multiple disc clutch. Also referred to as *basket*, *cylinder*, or *drum*. (8)

carrier roller. A roller with a center flange, often mounted on a pedestal, that is positioned on top of the track frame to help support and align the top half of the track chain. (23)

carrier undercarriage. A type of undercarriage designed primarily to propel the machine and to support the machine with a stable footing. (23)

casing. The leftover rubber shell of a worn tire that is used as the base when manufacturing a retread tire. (22)

cavitate. Restricting the oil into a pump's inlet causing the fluid to vaporize, resulting in the formation of air bubbles. As the fluid vaporizes and when exposed to pressure, the bubbles implode, causing shock waves that erode and pit the metal in the pump. (13)

cavitation. Phenomenon that occurs within a hydraulic system anytime an expanding volume generates a low pressure area (such as a cylinder extending) and hydraulic fluid is unable to fill that low pressure area, resulting in the formation of air bubbles. When the bubbles are exposed to high pressure, they violently collapse and cause damage, such as eroding metal surfaces. (25)

center axle through drive. A gear shaft assembly located in the center axle of an ADT that is responsible for directly transferring power through the center axle back to the rear axle. (16)

center cam. No-spin differential component located at the center of the spider that meshes with the left and right holdout rings and driven clutches. (16)

center of gravity (COG). The balancing point of the load. (5)

center support bearing. Bearing that is commonly found in articulated vehicles to support the driveshafts between the two articulating frames. It is also used in longer driveshaft applications to support the driveshaft and limit its vertical movement. Also referred to as *hanger bearing* and *midship support bearing*. (18)

center tread idler. The idler used on a Caterpillar SystemOne undercarriage that has twice as much material as a standard idler and is designed to ride between the rails, making contact with only the bushings. (23)

centrifugal clutch. A clutch that uses a governor flyweight assembly to engage a drive based on a set speed. (8)

chain breaker. A tool used to drive the pins out of the links and split the chain. (6)

chain elongation. The lengthening of a chain over a period of time due to wear. (6)

charge relief. The higher-pressure charge relief valve inside a closed-loop HST, normally located inside the pump. It provides charge pressure relief when the transmission is in a neutral position. Also referred to as *neutral relief valve*. (13)

check valve. A safety valve used in welding to prevent the reverse flow of gas. (1)

chisel. A long-bladed hand tool with a beveled cutting edge that is struck with a hammer or mallet to cut or chip metal surfaces. (2)

chisel plow. A tillage tool that looks similar to a field cultivator but has V-shaped or twisted spikes instead of shovels. It works deeper below the surface than a field cultivator and is used for primary tillage. (4)

choker hitch. A lifting hitch formed with one end of the sling passing under the load and through a fitting or eyelet on the other end of the sling. (5)

clamp master link. The master link used on a Caterpillar SystemOne undercarriage that has a center strut that strengthens the master link. (23)

clean grain auger. The auger at the bottom of a combine harvester that collects the clean grain and delivers it to the clean grain elevator. (4)

clearance-sensing ASA. A type of automatic slack adjuster that uses a gear-clutch mechanism that automatically compensates for wear. (20)

clevis pin. A fastener with a straight shank and shouldered head with a drilled hole through the end of the clevis. (2)

closed-loop HST. A hydrostatic transmission in which the pump is designed to drive a closed-loop motor. The motor's return oil is sent directly back to the pump's inlet instead of the reservoir. (13)

cluster gear. A large and small pinion gear coupled together and used in a compound planetary gear set. (15)

clutch apply chart. Chart that shows which clutches are applied in specific gear settings. These charts help technicians determine whether the transmission has a common planetary control that is malfunctioning. (9)

clutch disc. A steel disc that is lined with friction material. (8)

clutch hold value. A value that represents the current that the ECM driver uses to hold the clutch. (11)

clutch pedal. A foot-operated pedal used to disengage engine power from the drive train. Also referred to as an *inching pedal*. (10)

clutch release mechanism. The spring(s) used to hold the clutch piston in a retracted position when the clutch's apply oil is drained. (8)

cogged V-belt. A belt with corrugations along the bottom that resemble ribs. Also referred to as a *corrugated V-belt*. (6)

collar shift transmission. A nonsynchronized transmission in which output and countershaft gears are in constant mesh and are coupled to and decoupled from the output shaft by a splined shift collar. (7)

collector ring. A ring used at the top of a chain sling to collect multiple legs. (5)

combination fracture. A failure that occurs to a component as the result of a combination of fatigue fracture, impact fracture, ductile fracture, or brittle fracture. (17)

combination steering. Steering system that includes articulation steering and Ackerman front-steer or two articulation joints. (25)

combination tool. A single implement consisting of multiple tillage tools. The combination tool may include any combination of disc harrows, field cultivators, chisel plows, finish harrows, and spike-toothed harrows. (4)

combination wet disc brake. A brake used for both the parking brake and the service brake. (19)

combination wrench. A wrench with one open end and one closed end. Both ends are typically the same size. (2)

commissioning. Term used by the International Fluid Power Society referring to the process of returning a hydraulic system to service following pump or motor replacement. (14)

commutation. The sequencing action of current flowing through commutator bars. Commutation enables the motor's armature to generate the correct polarity of electromagnetic force, resulting in a continuous armature rotation. (26)

commutator. A copper bar that serves as an electrical contact that transfers electrical current between one brush and one end of an armature's electrical conductor. (26)

compensating port. An opening in the master cylinder that enables fluid in the outlet port to flow back into the reservoir as fluid temperature rises and fluid expands. (19)

compensation cylinder. Hydraulic cylinder(s) that is responsible for hydraulically self-leveling the carriage (forks or front attachment) on a telehandler. (3)

compensation valve. A valve used with dual master cylinders to connect the two separate pressures from the master cylinders in order to equalize the brake pressures. (19)

compound planetary gear set. Gear set in which two simple planetary gear sets are combined into one assembly and one of the planetary members is shared. (9)

concaves. A set of wire shells located under the combine's cylinder or rotor, through which threshed grain falls into the conveyance augers or onto a grain pan. (4)

cone. The inside race of a tapered roller bearing. (7)

cone synchronizer. A type of synchronizer consisting of a hub that is splined to the transmission shaft, a movable sleeve, a blocker ring, and inserts. Also referred to as a *block synchronizer.* (7)

Conrad bearing. A type of ball bearing that does not have a notch for loading balls. (7)

constant velocity U-joint (CV joint). A universal joint that enables a driveshaft to spin at a uniform speed. (18)

continuously variable transmission (CVT). A transmission that provides an infinite range of speed control rather than a limited number of discrete gear ratios. Also referred to as an *infinitely variable transmission* or a *stepless transmission.* (15)

continuous-running PTO. A power take-off that is engaged and disengaged with a two-section, or two-stage, clutch. (4)

controlled flow (CF). Hydraulic pump oil flow that is delivered to the hydraulic steering control unit via the steering priority valve. (25)

control oil. Oil flow that is used to actuate a servo piston. Also known as *servo oil.* (13)

conventional combine. A combine that uses a rotating cylinder, positioned parallel to the header and feeder drum, as the primary means of threshing the crop. (4)

conventional U-joint. A U-joint that rotates at nonuniform velocities. Also referred to as a *Cardan U-joint* and *Hooke U-joint.* (18)

conveyor chain. A chain commonly found on agricultural machines and used to move material. Also referred to as an *agricultural chain.* (6)

corner post display. A monitor mounted in the cab pillar of a machine that gives the operator real-time information about the machine's operation. Also referred to as an *A-post display.* (15)

corn header. A combine harvester header with individual row units that strip ears of corn from their stalks. (4)

corrosion wear. Wear that occurs when oil chemically reacts to a foreign substance, causing the components to corrode. (17)

corrugated V-belt. A belt with corrugations along the bottom that resemble ribs. Also referred to as a *cogged V-belt.* (6)

cotter pin. A flexible teardrop-shaped pin with two legs that is used with a bolt or nut to prevent the fastener from loosening. (2)

counter-electromotive force (CEMF). The induced voltage that is produced by a rotating permanent magnet motor that acts in the opposite direction of the current flow applied to the motor. Also referred to as *back electromotive force (BEMF).* (26)

counter rotation. A track steering method in which the operator propels one track forward and propels the other track in reverse. (24)

countershaft powershift transmission. A type of transmission that resembles a traditional manual transmission except the countershaft powershift uses multiple disc clutches that are applied with oil pressure or springs to lock the gears to the countershaft and distribute power flow. (10)

coupling clutch. A rotating clutch that couples two components in a planetary transmission. (8)

coupling gear. The component in a Caterpillar planetary powershift transmission that connects the #1 clutch to the #1 planetary carrier, allowing the carrier to be held in reverse. (9)

coupling phase. A phase of torque converter operation denoted by the impeller and turbine rotating at nearly the same speed, and the stator freewheeling. During this phase, vortex oil flow in the torque converter diminishes and rotary oil flow increases. (12)

crab steer. A steering technique that allows a machine to stay oriented straight ahead while moving at an angle. It is accomplished by angling front and rear wheels in the same direction. (3)

crawler tractor. A type of undercarriage designed for strong pulling power in addition to propelling and supporting the machine. Also referred to as a *tractor undercarriage.* (23)

creeper control. Feature found in off-highway equipment. It allows a machine to travel at very slow speeds while the engine runs at high rpms, allowing for a large amount of implement-hydraulic flow to operate attachments, such as brooms, brush cutters, and snowblowers. (13)

creeper transmission. A transmission equipped with optional gearing that enables travel at very slow speeds, such as at a rate of feet per minute instead of miles per hour, while being able to maintain a high engine speed for maximum hydraulic flow or PTO speed. (10)

cribbing. A means of temporarily supporting a load with hardwood or engineered plastic blocks stacked in alternating directions. (5)

cross. U-joint component that is cross-shaped and has four trunnions (cross ends). Also referred to as a *spider*. (18)

cross-axle differential lock. A locking differential that is used to lock the right and left drive wheels. Also referred to as a *locking inter-axle differential*. (16)

cross shaft. A perpendicular shaft on which spider gears are mounted in a differential with four spider gears. (16)

crowfoot socket. An open-end socket that resembles an open-end or flare-nut wrench. Its non-cylindrical design allows access to otherwise hard-to-reach areas. (2)

crown. The center portion of the tire's tread that contacts the ground surface. (22)

crush sleeve. A sleeve designed to crush (collapse) as the pinion nut is tightened to set the correct pinion bearing preload. (16)

C tire. Tire classified for installation on compactors. (22)

cup. The outside race of a tapered roller bearing. (7)

cushioned hitch. A scraper's hitch design that improves the life of the hitch by reducing shock loads placed on the hitch. (3)

Customer Support System (CSS). A term used by Komatsu for the online software that provides its service literature. (11)

cut-and-blow forage harvester. A type of forage harvester with a fan after the cutter head that blows silage through a spout, into a wagon. (4)

cut-and-throw forage harvester. A type of forage harvester with a high-speed cutter head that cuts the silage and throws it through a spout, into a wagon. (4)

cut-in pressure. In a hydraulic brake system, a pressure at which the accumulator is not fully charged and is capable of receiving pump flow. (19)

cut-in pressure. In an air-brake system's governor, it is the minimum pressure that allows the compressor to build system pressure, typically 25 psi less than the cut-out pressure. (20)

cut-out pressure. In a hydraulic brake system, the pressure at which an accumulator is fully charged and fluid is prevented from further charging the accumulator. (19)

cut-out pressure. The maximum pressure of an air brake system. When the cut-out pressure is reached, the governor signals the compressor to cut-out, or unload. (20)

cycle. One sine wave. (26)

cycle time. The amount of time it takes for a bucket, stick, or boom to actuate—extend, retract, or both fully extend and retract—or for the excavator's house to slew. (3)

cycling transmission. A transmission used in machines that operate for short distances and frequently cycle between forward and reverse propulsion. (10)

cylinder. A circular housing that houses the internal components of a multiple disc clutch. Also referred to as *basket*, *carrier*, or *drum*. (8)

cylinder damping SCU. Steering control unit configured to smooth steering motion by using adjustable orifices that divert some of the fluid from the steering cylinder ports back to the reservoir whenever the steering wheel is turned. (25)

cylindrical roller bearing. A type of antifriction bearing that has multiple cylindrical rollers between inner and outer races. Also referred to as a *straight roller bearing*. (7)

D

dampening springs. Springs used to help prevent vibrations produced by engine pulsations from being transferred to the rest of the driveline. (8)

dead axle. An axle that supports the weight of the machine but does not propel the drive wheels. Also referred to as a *lazy axle*. (18)

dead end. The wire rope end (opposite the live end) that does not support the weight of the load. Also referred to as the *dead line* or *free end*. (5)

dead line. The wire rope end (opposite the live end) that does not support the weight of the load. Also referred to as the *dead end* or *free end*. (5)

deep-cycle battery. A battery that is designed to be deeply discharged and recharged frequently. (3)

delivery port check valve. A check valve that allows dry air to exit the air dryer. (20)

depth gauge. A measurement tool with sliding probes that measure the linear distance below a specific point. (23)

detachable link chain. A chain used for low-speed and low-torque applications. It is designed to be easily disassembled and assembled and can be driven in only one direction. (6)

detent mechanism. A mechanism consisting of a ball, a spring, and a notch in the shift rail that work together to prevent the transmission from popping out of gear. (7)

diagnostic pressure tap. A device placed in fluid pressure systems to create a port through which safe and easy pressure measurements may be taken. (2)

Diagnostic Technical Assistance Center (DTAC). A term used by John Deere for its manufacturer solutions group. (11)

dial calipers. A precision tool used to take inside, outside, and depth measurements. (2)

dial indicator. A precision measuring device used to make a variety of measurements, including runout and shaft endplay. (2)

diaphragm accumulator. A type of gas accumulator that is a small vessel containing a flexible synthetic rubber diaphragm. (19)

differential. A drive mechanism designed to allow two different wheels (or two different axles) to spin at different speeds as a machine turns a corner. (16)

differential carrier. The structure that holds the differential case in proper alignment with the pinion gear. (16)

differential case. The housing that contains or supports the components of the differential assembly (side gears, ring gear, spider gears, cross shaft, and bearings). It can be installed and removed from the axle's housing as an assembly. (16)

differential pressure gauge. A measuring tool that allows a technician to simultaneously measure the difference between two pressures. (2)

differential steering. A twin-track steering system comprised of planetary gear sets, external-tooth gears, a reversible hydraulic motor, and bevel gear sets. It receives input from the transmission and the hydraulic steering motor, either individually or together, to deliver power to the right and left final drives. (24)

diode. Electronic device that allows the flow of current in one direction only; used to convert AC to DC. (26)

DirectDrive (dual clutch) transmission. A John Deere manual transmission that has two automated clutches that engage and disengage two separate input shafts. One clutch is engaged to select even-numbered gears and the other clutch is engaged to select odd-numbered gears. (7)

direct-drive powershift transmission. A type of powershift transmission without a torque converter. Input is supplied directly by the engine via a flywheel-driven dampener assembly. (11)

directional clutch. A clutch inside a transmission that, when engaged, causes the machine to move in a specific direction, either forward or reverse. (10)

discharge valve. A check valve that allows compressed air out of the air compressor during the compression cycle. (20)

disc harrow. A tillage tool consisting of a group of discs. (4)

disengaged clutch clearance. The distance between the snap ring and pressure plate in an oil-applied clutch. It is the same distance the piston must move to engage the oil-applied clutch. (8)

dog clutch. A term used to describe a no-spin spider and a driven clutch assembly. (16)

double. Two seeds being planted at the same location due to planter malfunction. (4)

double Cardan U-joint. A CV joint that consists of two conventional U-joints inside a center yoke. (18)

double-choker hitch. A lifting hitch with two separate slings choked on the load. (5)

double-reduction final drive. A final drive consisting of two reduction gear sets. (17)

double V-belt. A belt with the appearance of two V-belts placed onto each other to form a hexagon shape. Both sides of the belt are used to drive pulleys. (6)

double-wrap basket hitch. A basket hitch formed by fully wrapping a single sling around the load. (5)

double-wrap choker hitch. A hitch formed by wrapping the sling completely around the load once before forming the choker hitch. (5)

down force. Force applied to a planter's row units to help them cut through crop residue and cut seed trenches in difficult soils. (4)

drag link. Component of a cross-steer linkage system that connects and transmits the rotary motion of the Pitman arm to one of the steering knuckles. (25)

drain valve. A manually operated or automatic valve used to remove moisture from an air brake system. (20)

draper belt. Long rubber belt used in some straight cut headers to deliver the cut crop to feeder house. (4)

drill. A seeding implement that meters seed by volume. Drills plant in much narrower rows than a row crop planter. (4)

driven clutch. The component in a no-spin differential that is driven by the spider and is responsible for driving the side gear. (16)

driver key. A multi-piece wheel component that fits in the pocket of the bead seat band (or outer flange on three-piece rims) and the pocket of the rim base to retain the locking ring. (22)

drive segments. The five segments that make up a split drive sprocket. See also *drive sprocket*. (23)

driveshaft. A hardened steel tube used to connect two drivetrain components. Also referred to as a *propeller shaft*. (18)

drive sprocket. A toothed wheel (driven by a final drive) in an undercarriage that propels the track chain's bushings to move the machine forward or backward. (23)

drive tumbler. A drive sprocket used on a crane or excavator undercarriage. On cranes it may be driven by a chain that is powered by a hydraulic motor or gearing. (23)

drop axle. Design in which the final drive drops the power coming from the differential to a lower axis on the machine, often used on low-track crawlers. (17)

drop box. A power train component that connects the engine's crankshaft to the speed transmission. (10)

drop-center rim. A single-piece rim with a recessed (dropped) center that aids in removal and installation of the tire. (22)

drum. A circular housing that houses the internal components of a multiple disc clutch. Also referred to as *basket*, *carrier*, or *cylinder*. (8)

drum brake. A brake that contains two brake shoes that are wedged against a rotating drum. (19)

dry disc brake assembly. A type of brake assembly with one or more discs that are air cooled, and are not lubricated with hydraulic fluid. (19)

dry tank. An air tank located after the supply tank in an air brake system. (20)

dual brake circuit. An air brake system that contains two separate circuits: a primary brake circuit and a secondary circuit. Each circuit has a dedicated air tank. (20)

dual master cylinders. Two separate single master cylinders. The right master cylinder operates the right side brake, and the left master cylinder operates the left side brake. (19)

dual-path hydrostatic transmission (HST). Hydrostatic drive system consisting of two separate hydrostatic transmissions for the purpose of propelling the machine and steering the machine. One pump and one motor drive the left side of the machine, and one pump and one motor drive the right side of the machine. (13)

dual service/spring brake chamber. A type of air brake actuator that has a front chamber used for the service brake and a rear chamber used for the parking brake. (20)

ductile fracture. A slow fracture that results in plastic deformation of the material. (17)

Duo-Cone seal. A seal that retains fluids and prevents contamination between a rotating component and a static component. The seal consists of two metal face seal rings that are held compressed against each other by means of two rubber torics. (18)

duo-servo drum brake. A type of drum brake with both shoes attached to the backing plate by pins or hold-down clips and the tops of the shoes held against an anchor pin with a return spring. The bottom of one shoe pushes the bottom of the other shoe into the brake drum, causing both shoes to self-energize during brake application. (19)

dynamic braking. Occurs in electric propelled machines when reducing propulsion speed, the electrical motor converts the machine's mechanical energy (momentum) back into electrical energy. That energy drives the generators, which act like electrical motors. In turn, the motors attempt to drive the diesel engine and thereby slow (brakes) the machine's momentum. (26)

dynamic load-sensing signal. A steering signal with a small amount of flow and pressure used to develop a signal pressure while the steering valve is in a neutral position. This makes the system more responsive than a static steering signal system. (25)

dynamic retarder. A device designed to help slow a vehicle. It consists of a driven rotor that spins inside stationary housing, which will be flooded with oil to provide resistance. (12)

E

ejector. A hydraulically driven movable wall that is moved forward to empty a scraper bowl and moved back when the scraper is loading. (3)

elastomer suspension cylinder. A type of suspension cylinder that has rubber (elastomer) springs for support instead of oil or nitrogen. (21)

electric clutch. A clutch containing a rotor, field coil, armature, clutch plate and pulley (hub) that uses electricity to linearly move the armature and couple the rotor to the pulley via the clutch plate. Also referred to as an *electromagnetic clutch* or *electro-mechanical clutch*. (8)

electrocution. Injury or death by electrical shock. (1)

electromagnetic clutch. A clutch containing a rotor, field coil, armature, clutch plate and pulley (hub) that uses electricity to linearly move the armature and couple the rotor to the pulley via the clutch plate. Also referred to as an *electric clutch* or *electro-mechanical clutch*. (8)

electromagnetic particle clutch. An electric clutch with a small cavity between the input and output members that contains a magnetic powder. As electrical current is applied to the field coil, the resulting magnetic field causes the particles to bind together causing the input and output members to engage. (8)

electro-mechanical clutch. A clutch containing a rotor, field coil, armature, clutch plate and pulley (hub) that uses electricity to linearly move the armature and couple the rotor to the pulley via the clutch plate. Also referred to as an *electric clutch* or *electromagnetic clutch*. (8)

electromotive force (EMF). The electrical force that results in electrical current. (26)

electronic anti-stall control. Normally closed two-position destroke solenoid that is modulated by an ECM in order to limit stalling of the pump. (13)

electronic clutch pressure control (ECPC). A term used by Caterpillar to refer to electronic clutch modulation that optimizes the transmission's clutch life and shift feel by using individual electronic proportional solenoid valves to engage the clutches. (11)

electronic commutation. Using electronic controls to sequentially energize a motor's field coils for the purpose of controlling the motor's speed. (26)

electronic service tool. A program installed on a computer or other digital device that allows a technician to communicate with the machine's ECMs, read error codes, and perform calibrations. (11)

Electronic Technical Information Manuals (eTim). A term used by Case New Holland for the online software that provides its service literature. (11)

electronic traction control. Feature that monitors all four drive motors to determine if any of the wheels begin to slip. If slip is detected, the electronic controls limit power to all four drive motors to reduce the machine's wheel slip. (26)

elevated-sprocket undercarriage. An undercarriage with drive sprockets that are raised and positioned slightly forward from the second idlers at the rear of the tracks. Also referred to as a *high-track undercarriage*. (23)

elevating scraper. Scraper that is loaded by a hydraulically operated elevator assembly consisting of steel flighting mounted on chains. (3)

elevator. A housing on a combine that contains chain-driven rubber paddles that move grain from an auger to a new location. (4)

embolism. An air bubble in a person's bloodstream that can block a blood vessel and cause a stroke or heart attack and result in death. (1)

emergency steering. An SCU feature that enables the operator to steer the tractor if the engine dies or the hydraulic pump fails. (25)

endless round belt. A belt with a round cross-section that is used for driving a pulley. It is used in serpentine drives and other drives that require the belt to turn a quarter turn or twist. Also referred to as a *round belt*. (6)

endplay. The distance a shaft or gear can move after bearing installation. (7)

Enhanced AutoShift (EAS). A Caterpillar electronic transmission control mode that can be chosen by the operators of late-model machines to make the powershift transmission behave like an automatic transmission. (11)

epicyclical. Having parts that move around the circumference of another part. The term sometimes used to describe a planetary gear. (9)

epicyclic gear set. A gear set consisting of a central gear called a sun gear, planetary gears, planetary gear carrier, and a ring gear. Different gear ratios can be achieved by holding and driving different combinations of members in the gear set. Also referred to as a *planetary gear set*. (7)

equalizer bar. The oscillating bar installed on machines with pivoting undercarriages. It attaches to the front of the left- and right-side track frames and also attaches to a center pin in the center of the machine's frame. (23)

error codes. Diagnostic reference numbers tied to specific faults. (11)

E tire. Tire equipped on earthmoving machines, such as rigid frame haul trucks and scrapers. (22)

excavator final drive. Final drive that receives its input from a hydraulic motor and has two or three planetary gear sets. (17)

excess flow (EF). Hydraulic pump oil flow that is delivered to the secondary circuit(s) via the steering priority valve after the steering control unit's oil demand is met. (25)

exercising the tracks. Driving a track-type machine at regular intervals to prevent it from sitting stationary for a long period, which could damage undercarriage components. (23)

F

falling object protective structure (FOPS). A safety structure attached to a machine to protect the operator from falling objects. (1)

fall protection. The use of a full-body harness and shock-absorbing lanyard that is securely attached to an anchor point to prevent the technician from falling to the ground. (1)

false brinelling. A type of bearing surface fatigue in which the indentations show no evidence of the precision grinding marks. It is often caused by excessive vibration that occurs when the bearing is stationary. (17)

fast adapt. Adaptive learning that takes place after the transmission has been rebuilt or a major component has been replaced. The technician enters the calibration mode and forces the ECM to learn the volume of oil and amount of time for clutches to engage. (11)

fatigue fracture. A fracture that occurs due to numerous cyclical stresses over time. (17)

feeder house. The interface between the header and the rest of the combine. It serves as the mounting point for the header and feeds crop from the header into the thresher. (4)

feeler gauge. A measuring device with a series of thin blades that are stacked to measure small distances between two components. (2)

feller buncher. A specialized forestry excavator with a cutting head attachment and tree trunk clamp at the end of the boom. It fells and bunches the trees. (3)

fertilizer. A chemical that promotes plant growth. (4)

field cultivator. A tillage tool that uses multiple rows of shovels to break up the soil. It operates at shallower depths than moldboard plows or disc harrows and is used to prepare the soil for planting. (4)

filler wire rope. A wire rope constructed with two layers of wire with the same diameter around a center and smaller diameter wires filling in spaces between the outer and inner layers. The inner layer has half as many wires as the outer layer. (5)

filling-notch bearing. A ball bearing that has a notch or slot in the outside and inside races, which is used to load the balls into the bearing assembly. (7)

fill valve. Part of a hydraulic piston track tensioner that serves as a hydraulic inlet port for filling the tensioner's piston with grease to increase track tension. (23)

final drive. A gearing system that multiplies the machine's torque and reduces the speed one final time prior to the drive wheel or track drive sprocket being driven. (17)

finish harrow. An implement that has a rotary bladed spool that pulverizes the soil. It is used to prepare a seedbed for a planter. (4)

first aid. Medical assistance given to a sick or injured person until full medical treatment is available. (1)

first order vibration. A vibration that occurs once during every revolution of the shaft. Out-of-balance driveshafts and out-of-round driveshafts exhibit this type of vibration. (18)

fist grip clip. A variation of a *wire rope clip* that uses a second saddle instead of a U-bolt. (5)

fixed bale chamber. A round hay baler chamber design that can produce only a fixed-diameter bale with a star-shaped center. (4)

fixed caliper. A type of caliper that uses two hydraulic pistons that are opposed to one another. (19)

fixed-displacement motor. A motor that will operate at a constant speed for a given amount of input flow. (13)

fixed-displacement pump. A hydraulic pump that produces a fixed volume of oil for each revolution. (11)

fixed pin synchronizer. A type of pin synchronizer used in the rear auxiliary transmission in semi-trucks to provide high and low ranges for the truck's main transmission. (7)

fixed stator. A type of torque converter stator that is locked to the stator shaft and cannot rotate in either direction. (12)

flap. A piece of rubber that is installed around a tube tire's inside circumference, between the inner tube and the rim, to protect the tube from damage. (22)

flare-nut wrench. A wrench designed for loosening and tightening fittings that attach lines, such as fuel lines. Also referred to as a *line wrench*. (2)

flashback. A flame that travels back through an oxyacetylene torch head. It can cause the hoses and cylinder to explode. (1)

flashback arrestor. A safety device used on welding hoses to quench a flame, preventing a flame from traveling back into the hose. (1)

flat belt. A belt that has the appearance of a continuous flat strap. The belt is driven with a crowned pulley and was used on old farm tractors to power implements, and is currently used in mining conveyor systems. (6)

Flemish eye. A wire rope end termination formed by initially unwinding the end of a wire rope into two equal parts, forming an eyelet and weaving the two parts back into the wire rope. (5)

flex handle. A stronger handle that should be used in place of a ratchet when loosening stubborn fasteners. Also referred to as a *breaker bar*. (2)

floating caliper. A type of caliper that uses one or more pistons located only on the inboard side of the caliper assembly. Also referred to as a *sliding caliper*. (19)

flow sharing. A type of hydraulic system that, when the pump cannot satisfy the system demands, will proportion the available hydraulic flow to each of the commanded hydraulic functions based on the amount of oil requested for each function. (3)

fluid coupling. A straight-finned hydrodynamic fluid drive consisting of only an impeller and turbine and incapable of torque multiplication. (12)

fluid injection. When pressurized fluid penetrates the skin and injures or kills the affected person. (1)

fluid reservoir. A reservoir used to supply fluid to the master cylinder piston. (19)

flushing relief valve. Lower-pressure charge relief valve inside a closed-loop HST, normally located inside the motor. It provides charge pressure relief when the transmission is in forward or reverse. Also referred to as a *replenishing relief valve* or a *hot-oil purge relief valve*. (13)

flushing valve. Shuttle valve used to flush the oil from the motor and pump. This aids in cooling the transmission. Also known as a *replenishing shuttle valve* or *hot-oil purge shuttle valve*. (13)

flywheel clutch. A flywheel-mounted clutch that can couple or disengage engine power from the transmission input shaft (traction clutch) or the PTO input shaft (PTO clutch). (8)

FNR lever. Lever that is connected to the pump's control valve and is used to control the speed and the direction of the HST. Also known as a *propulsion lever* or *forward and reverse lever*. (13)

FNR valve. A valve used in a speed-sensing hydrostatic drive with three fixed positions: neutral, forward, and reverse. (13)

foam-filled tire. Tire that has a chemical solution pumped into the tire that expands to fill the interior of the tire with a solid foam material. (22)

foldback eye. A wire rope end termination formed by turning back the dead-end wire and securing it in a fitting. Also referred to as a *turnback eye*. (5)

forage harvester. A machine designed to cut and chop complete plants into silage and blow the silage into a wagon. (4)

forward and reverse lever. Lever that is connected to the pump's control valve and is used to control the speed and the direction of the HST. Also known as a *propulsion lever* or *FNR lever*. (13)

foundation brakes. The mechanical components of an air brake system that apply friction material to a rotating wheel assembly. (20)

four-in-one bucket. A bucket with a front face that can be opened and closed hydraulically. Also known as a *clam shell bucket* or *multipurpose bucket*. (3)

four-wheel steer. A term applied to rigid frame machines that use both front and rear steer axles. (18)

free end. The wire rope end (opposite the *live end*) that does not support the weight of the load. Also referred to as the *dead end* or *dead line*. (5)

free play adjustment. Adjustment that allows the clutch pedal to move a short distance (for example, 1″) before the release mechanism begins to actuate the pressure plate. (8)

frequency. The number of cycles per second. (26)

fretting corrosion. A surface wear failure that occurs between two close-tolerance stationary components that have moved. The movement causes the microscopic asperities to wear, often leaving a brownish-red oxide debris on the component's surface. (17)

friction bearing. A type of bearing that has no moving parts and is usually made from a relatively soft material. (7)

friction clutch. A clutch that uses a friction-lined plate to transfer torque. (8)

friction discs. The discs (in a multiple disc clutch) that are lined with friction material and are stacked in an alternating fashion between the separator plates. The discs normally have internal splines and align with a shaft or hub. (8)

F tire. Non-cleated tire (no drive lugs) with flat ribs and installed on the front, non-drive axle of agricultural two-wheel-drive tractors. (22)

full-floating axle. An axle shaft responsible only for driving the wheel. The weight of the machine is exerted on the axle housing instead of the axle shaft. (18)

full-round yoke. Consists of two round lobes that attach to two of the U-joint bearing cups. This yoke is used with either a retaining-ring-type U-joint or a bearing-plate-type U-joint. (18)

full suspension switch. An electric switch in a Volvo ADT that allows the operator to fully extend the hydraulic suspension cylinders. (21)

fusible plug. A cylinder safety device filled with a metal alloy that will melt at a specific temperature to allow the controlled release of the gas. (1)

G

gang. A group of discs on a disc harrow. (4)

gas accumulator. A type of accumulator that is filled with a gas to a precharge pressure. (19)

gauge wheels. The two wheels outside the pair of trench-cutting discs on a row crop planter. They determine the depth of the trench cut by the discs. (4)

generator. A component that converts the engine's mechanical rotational energy into electrical energy (electricity). (26)

gerotor assembly. The gear mechanism inside a steering control unit that meters the quantity of oil used to actuate a tractor's steering cylinder. (25)

governor. Part of the air brake supply system that senses and regulates the air system's pressure. (20)

grade. A measurement of a bolt's tensile and yield strength. (2)

graduating rubber spring. Part of the treadle valve that provides brake pedal resistance so the operator can feel and control the amount of brake pressure being applied. (20)

ground-drive pump (GDP). A pump driven when the machine is moving that provides oil for tractor steering and brakes in the event of a loss of hydraulic flow. (10)

ground pressure. The amount of pressure a machine's weight exerts on the soil. (23)

ground-speed PTO. A power take-off that is powered only when the tractor is moving forward or reverse. (4)

ground speed radar. An electronic device used in late-model tractors to measure the tractor's true ground speed. (22)

grouser. A vertical steel lug (or lugs, depending on the design) that protrudes from track shoes. (23)

G tire. Tire classified for installation on motor graders. (22)

guides and guards. Metal housings that help direct the track chain, prevent debris from packing between the rollers and the links, and protect the rollers, idlers, and sprockets from impact damage. (23)

H

half-round yoke. A yoke that has two partial lobes that act as cradles for the U-joint bearing cups. The U-joint is secured in the yoke by two straps or two U-bolts. (18)

hanger bearing. Bearing that is commonly found in articulated vehicles to support the driveshafts between the two articulating frames. It is also used in longer driveshaft applications to support the driveshaft and limit its vertical movement. Also referred to as *center support bearing* and *midship support bearing*. (18)

hard bar. A rigid metal tube that connects to the front of both the left and right undercarriage track frames and attaches to the tractor's frame. (23)

harvester final drive. A gearbox mounted on the outside ends of a rigid axle frame. It receives power from a transmission gearbox through a driveshaft and coupler. Can use bull-and-pinion gears, or both bull-and-pinion and planetary gears. (17)

hauling transmission. A transmission used in machines that haul payloads of dirt long distances at higher travel speeds. (10)

hay baler. Implement that is used to pick up a swath of hay and pack it into a bale. (4)

heaped capacity. A volume equal to the struck capacity plus the additional material that can be heaped on top of the bucket when the top plane of the bucket is parallel to the ground. (3)

heavy-duty extended life (HDXL) with Duralink track. A track design with taller, crowned links to achieve a longer life. (23)

heavy-duty track. A sealed and lubricated track with links, bushings, and pins that are enhanced for higher stress and impact loading environments. (23)

heel. The portion of a ring gear's tooth closest to the outer edge of the gear. (16)

helical gear. Gear with external teeth that are cut at an angle to the gear's centerline. (7)

herringbone gear. A gear with a chevron-shaped tooth pattern. (7)

high-clearance tractors. A variation of utility and row crop tractors that are designed to have extra ground clearance. (4)

high-pressure washer. A cleaning device that uses a pump to build water pressure to spray a stream of water at high velocities. (2)

high-track undercarriage. An undercarriage with drive sprockets that are raised and positioned slightly forward from second idlers at the rear of the tracks. Also referred to as an *elevated-sprocket undercarriage*. (23)

hitch. The manner in which a sling is configured for lifting. (5)

hoist. Another term for a haul truck dump bed. (3)

holding clutch. A multiple disc clutch used to hold a shaft or gear stationary. The clutch normally uses a housing, such as a transmission case, instead of a rotating drum. Also referred to as a *stationary clutch* or *brake*. (8)

hold-off brake chamber. A type of air brake actuator that contains a heavy-duty spring that applies the parking brake when no air is supplied to the chamber. Also referred to as a *single parking brake chamber*. (20)

holdout ring. No-spin differential component that disengages from the center cam inside the spider, causing the driven clutch to disengage from the spider. (16)

Hooke U-joint. A U-joint that rotates at nonuniform velocities. Also referred to as a *Cardan U-joint* and *conventional U-joint*. (18)

hopper level sensor. Sensor that alerts the operator when the planter's seed hoppers become low and need to be refilled. (4)

horizontal directional drill. A drilling rig that is capable of steering the drill head in a different direction while drilling. (3)

horizontal oil level mark. A line across the face of the final drive housing that must be oriented in a specific position before the final drive fill level is checked. (17)

horizontal sling angle. The angle formed between the horizontal plane of the component being lifted and the lifting sling. (5)

hot-oil purge valve. Shuttle valve that is used to flush the oil from the motor and pump. This aids in cooling the transmission. Also known as a *replenishing valve* or *flushing valve*. (13)

housing. The stationary case (such as a transmission case) used in a holding clutch for holding a shaft or gear stationary. It is used in place of a rotating drum. (8)

hub. Either the input or output member of a multiple disc clutch. It contains external splines and often aligns with the internal splines on the friction discs. (8)

HydraCushion. A type of John Deere suspension system equipped with an axle swing arm, two double-acting hydraulic suspension cylinders, and three large, gas-filled accumulators to improve ride quality and reduce power hop on 9R articulated tractors. (21)

hydraulically actuated caliper disc brake. A type of brake that uses a caliper to squeeze two brake pads against a single rotating disc known as a rotor. (19)

hydraulic brake. A brake that uses fluid pressure to apply a piston, which actuates the brake mechanism. (19)

hydraulic slack adjuster. A device used to adjust for the brake slack that occurs due to brake wear. (19)

hydraulic steering amplifier. A steering valve used to multiply the oil flow requested from the SCU and sent to the steering cylinders. (25)

hydrodynamic drive. A fluid drive system that operates at a relatively low fluid pressure and relies on the fluid's mass and velocity for transmitting power. It contains an impeller and a turbine and is used to transmit power through a rotating shaft. (12)

hydrophobic. A term used to describe fluid that will repel water. (19)

hydro-pneumatic suspension cylinder. Cylinder equipped with a double-acting rod and filled with a volume of oil and nitrogen gas that work together to absorb chassis shock loads. Also referred to as *nitrogen over oil cylinder*, *oil-pneumatic hydraulic cylinder*, or *shock absorber*, or *strut*. (21)

hydrostatic. The ability to transfer power via fluids at rest or under pressure. (13)

hydrostatic drive. A high-pressure fluid drive system that uses a hydraulic pump to drive a hydraulic motor. The pump and motor have tight sealing surfaces and are designed not to slip. Also referred to as a *hydrostatic transmission (HST)*. (13)

hydrostatic steering system (HSS). The term that machine manufacturer Komatsu uses to describe a differential-steered tractor. (24)

hydrostatic transmission (HST). A high-pressure fluid drive system that uses a hydraulic pump to drive a hydraulic motor. The pump and motor have tight sealing surfaces and are designed not to slip. Also referred to as a *hydrostatic drive*. (13)

hygroscopic. A term used to describe fluid that will absorb water. (19)

hypoid gear set. A ring-and-pinion gear set in which the pinion gear is positioned below the center axis of the ring gear. (7)

I

idler. A center-ribbed wheel with outer edges along which the track chain runs to stay aligned in the undercarriage. (23)

idler flange. The raised center circumference of an undercarriage idler that rides between the track links. Also referred to as an *idler rib*. (23)

idler rib. The raised center circumference of an undercarriage idler that rides between the track links. Also referred to as an *idler flange*. (23)

impact fracture. A fracture caused by an abrupt, extreme force (shock load). (17)

impeller. The engine-driven input member of a hydrodynamic fluid drive. Sometimes called a pump. (12)

impeller clutch. A multiple disc clutch on some torque converters that can modulate impeller engagement to reduce the converter's output torque. (12)

implement lock. A machine safety device that prevents the implement from moving. (1)

inboard bull-and-pinion final drive. Inboard final drive in which the pinion gear receives its power directly from the differential side gear. The service brakes normally are integrated on the pinion gear shaft. The pinion gear drives the bull gear. The center of the bull gear is directly splined to the axle shaft, which propels the drive wheel. (17)

inboard final drive. Final drive that is integrated in the axle housing and is located right next to the differential. (17)

inboard hydraulically actuated wet disc brake. A type of brake located between the rear differential and the inboard planetary final drive consisting of a hydraulically actuated piston, a friction disc, and a backing plate. (19)

inboard single-reduction planetary final drive. Inboard final drive that attaches directly to the axle's differential housing. At each side, one of the differential's side gears drives the planetary sun gear. The ring gear is bolted stationary to the axle housing, and the planetary carrier is the output member that attaches directly to the axle's driveshaft. (17)

inching pedal. A foot-operated pedal used to disengage engine power from the drivetrain. Also referred to as a *clutch pedal*. (10)

inching pedal potentiometer. A variable resistor that provides a variable voltage as an input to the ECM, indicating the position of the inching pedal. (11)

inching valve. A foot-operated valve used on some HSTs to perform functions similar to a transmission clutch. (13)

included angle. The angle formed by the sling directly beneath the hook. The maximum allowable included angle for a hook is 90°. (5)

Independent Link Suspension (ILS). A type of John Deere suspension system equipped with dual control arms, two double-acting hydraulic suspension cylinders, and four gas-filled accumulators to improve ride quality and reduce power hop on an 8R MFWD tractor. (21)

independent PTO. A power take-off that is engaged and disengaged by a clutch that is independent of the transmission. (4)

Individual Clutch Modulation (ICM). A term used by Caterpillar that indicates the use of multiple modulating valves. (11)

infinitely variable transmission. A transmission that provides an infinite range of speed control rather than a limited number of discrete gear ratios. Also referred to as a *continuously variable transmission (CVT)* or *stepless transmission*. (15)

inlet check valve. A valve in the primary air tank and in the secondary air tank that prevents a circuit with good air pressure from bleeding to a circuit that is leaking. (20)

inlet check valve. The check valve located in series to a steering control unit that is used to prevent steering wheel kickback that can occur when a load is exerted on the steering cylinder, causing the pressure to rise above the system pressure. (25)

inlet/exhaust valve. A two-way valve in the governor that either allows system pressure to pass through the governor to the unloader valve when the governor reaches cut-out pressure (inlet), or blocks system pressure while allowing the unloader valve to exhaust the air in the unloader valve to the atmosphere when the governor reaches cut-in pressure (exhaust). (20)

inlet valve. A check valve that allows air into the cylinder of an air compressor and blocks air from exiting the cylinder. (20)

inner liner. A rubberized layer that surrounds the entire inside of a tire to maintain tire air pressure. (22)

inner tube tire. A tire design that incorporates an inflatable rubber tube inside the body of the tire. (22)

in-phase. A term describing a two-piece driveshaft that is installed with the front and rear yokes in alignment with each other. (18)

input clutch. A multiple disc clutch used to drive a shaft or a gear. (8)

inserts. The parts of a block synchronizer that push the blocker ring into contact with the speed gear. They ride in slots in the synchronizer hub. Also referred to as *keys*. (7)

integral HST. A hydrostatic drive system in which the pump and motor are directly connected to each other, eliminating the need for high pressure drive lines. (13)

integral power steering gear. Power-assist steering design that contains an internal shaft, power piston, rotary valve, and sector gear. The rotating steering wheel causes the steering's internal shaft to rotate. The shaft is splined to the rotary valve and the power piston. (25)

IntelliAxle. A John Deere rear suspension system equipped on swathers that has a double-acting hydraulic cylinder attached to each caster wheel for suspension cushioning and steering assistance. (21)

inter-axle differential (IAD). A type of differential that is designed to allow a speed difference between two drive axles. In an ADT, it is normally constructed using a compound planetary gear set. (16)

interlock mechanism. A mechanism that prevents two gears from being selected at one time. (7)

internal pressure override (IPOR) valve. A valve placed in series between a charge pump and the inlet to the hydrostatic drive pump's servo control valve. It is used to protect the machine from high pressure overloads for extended periods of time. (13)

internal purge air dryer. A type of air dryer that uses the air inside the dryer to regenerate the desiccant material. (20)

internal wet disc brake. A type of brake that contains friction discs, steel separator plates, a piston, and a spring mechanism. (19)

inverse shuttle valve. A valve that senses the front and rear service brake accumulator pressures and will block flow to both brake circuits when the accumulators are fully charged. (19)

inverted basket hitch. An unstable lifting hitch made with a single sling that is loosely placed over a hook with the two ends of the sling attached to the load. (5)

I tire. Tire with flat ribs that is commonly installed on trailing implements. (22)

J

job hazard analysis (JHA). A review conducted before starting a job to determine potential risks related to the job, tools, and surrounding environment in order to reduce risks. (1)

joined belt. Multiple V-belts joined together by a top layer of material (tie-band). The belt is used on pulsating drives to reduce slip that would occur with a matched set of multiple single-drive belts. Also referred to as a *banded V-belt*. (6)

jounce. The compressing action of a suspension spring or hydraulic suspension cylinder. (21)

K

Kinetic Energy Storage System (KESS). An energy storage system used in Joy Global unground mining hybrid loaders that uses a SRM rotor flywheel to recapture energy. (26)

keys. The parts of a block synchronizer that push the blocker ring into contact with the speed gear. They ride in slots in the synchronizer hub. Also referred to as *inserts*. (7)

kidney loop filtration cart. An external hydraulic pump and filtration system that pulls oil out of a reservoir, filters the oil using a high efficiency filter, and returns the oil back to the machine's reservoir. (12)

kilowatt hour. The amount of power it takes to power an electric device that uses 1000 watts for one hour. (26)

Kinetic Energy Storage System (KESS). System on a hybrid electrical loader that increases the ability to capture energy. As the brakes are applied, that energy drives an SRM rotor flywheel, which increases the loader's ability to capture energy. (26)

kingpin. The shaft or strut that functions as the pivot point for a steered wheel. (21)

knuckleboom loader. A specialized forestry excavator that acts as a crane for moving and loading felled trees. (3)

L

laminated tire. A tire made by attaching similarly sized rectangular rubber scraps to a rim and fusing the rubber together with extreme heat and pressure. (22)

Lang-lay wire rope. A wire rope with the wire and strands rotated in the same direction. (5)

lay length. The parallel distance (measured on the length of the wire rope) that it takes for one strand to make an entire helical revolution around the wire rope. Also referred to as *rope lay*. (5)

lazy axle. An axle that supports the weight of the machine but does not propel the drive wheels. Also referred to as a *dead axle*. (18)

leading-trailing shoe drum brake. A type of drum brake that has the bottoms of the brake shoes anchored to the backing plate by a pin or an anchor strap. Only one shoe slows the drum based on the direction of machine travel. Also referred to as *nonservo drum brake*. (19)

leaf spring suspension. A type of suspension system designed with one or more flat strips of rectangular steel (leaves) shaped into an arch to support the machine's weight and absorb shock loads. (21)

left-hand-lay wire rope. A wire rope with the strands rotated in a counterclockwise pattern. (5)

Liebherr Information and Documentation Services (LIDOS). A term used by Liebherr for the online software that provides its service literature. (11)

lightbar. A display panel with a series of indicator lights to the left and right of center that tell the operator which way to steer. (4)

limited-slip differential. A differential with two multiple disc clutch packs, one for each side gear. When driving straight ahead, this differential drives both side gears at equal speeds. When one drive wheel loses traction and begins to spin, the drive wheel with the most resistance/traction receives all the power. (16)

liners. Steel plates that help protect the walls or floors of a haul truck bed. (3)

line wrench. A wrench designed for loosening and tightening fittings that attach lines, such as fuel lines. Also referred to as a *flare-nut wrench*. (2)

link height. The distance between the top rail and the bottom rail on an individual track link. (23)

link pitch. The distance from the center of the bushing hole to the center of the pin hole on a single track link. (23)

live axle. A drive axle that receives power from the transmission and transfers the power to the drive wheels to propel the machine. (18)

live end. The wire rope end suspending the weight of the load. Also referred to as the *live line*. (5)

live line. The wire rope end suspending the weight of the load. Also referred to as the *live end*. (5)

live PTO. A PTO that can continue to operate even if the tractor quits moving. (4)

load angle factor. A crushing load multiplier that is applied to the sling's overall load. (5)

load index values. Numerical values given to tires to indicate the weight load—in pounds or kilograms—a tire can handle. (22)

Load Match. A feature on John Deere compact utility tractors equipped with an electronically controlled HST that, when activated, prevents stalling of the engine when the tractor is working under heavy loads. (13)

load reaction SCU. Steering control unit that allows the external loads placed on the steering cylinder(s) to actuate the hydraulic steering system. (25)

load-sensing relief valve. SCU component that establishes the steering relief pressure by dumping the steering signal pressure to the reservoir when the pressure exceeds its setting. (25)

locking differential. A type of differential that locks one of the side gears to the differential housing, causing the side gear to rotate at the same speed as the differential case. The differential drives both axles at the same speed, regardless of the resistances exerted on the two drive wheels. (16)

locking inter-axle differential. A differential that is designed to lock two drive axles to drive at the same speed. Also referred to as *cross-axle differential lock*. (16)

lockup clutch. A component on some torque converters that can mechanically couple the turbine with the impeller or torque converter's shell so that the turbine will spin at exactly the same speed as the engine. (12)

lockup phase. In torque converters equipped with a lockup clutch, a phase of operation that provides direct drive through the torque converter, eliminating the inefficiency of the coupling phase and providing an increase in machine speed. (12)

longwall mining. A subterranean mining technique that removes sediments along a long wall while using supports to control collapse of the mine roof. (3)

low ground pressure (LGP) undercarriage. An undercarriage with an extended length and wider tracks (compared to a standard undercarriage) to increase soil contact for better flotation. (23)

low-pressure switch. A switch that illuminates a warning light when the pressure in an air brake system drops below a set pressure. (20)

low sidewall (LSW) tire. A type of Goodyear tire that features a shorter sidewall to improve the tire's stability and reduce power hop and road lope. (22)

LS tire. Tire classified for installation on log-skidders. (22)

L tire. Tire classified for installation on construction wheel loaders. (22)

lube passage. A rifle-drilled passageway through the center of a shaft that allows oil to be directed to clutches for lubrication. (8)

M

magnetorheological fluids. Types of oil with microscopic iron particles suspended within the fluid that can change fluid viscosity (fluid thickness) by the application of a magnetic field. (21)

main pressure. The main transmission operating pressure that is used to apply the transmission's clutches. Also referred to as *regulated pressure, transmission charge pressure*, and *transmission operating pressure*. (11)

manual bypass valve. Valve that allows the fluid from one leg of an HST drive loop to be bypassed into the other leg of the loop. (13)

manual slack adjuster. A type of slack adjuster that must be manually adjusted by rotating the hex head on the worm shaft. (20)

manual steering check valve. Valve that allows the SCU's gerotor to draw oil directly from the reservoir if the operator attempts to steer when the hydraulic pump is unable to supply oil flow to the SCU. (25)

manual transmission. A transmission design that requires the operator to press a clutch pedal and move a gearshift lever to change gears. (7)

manufacturer solutions group. A group of personnel assigned to assist technicians with difficult technical problems. (11)

margin pressure. The difference between pump outlet pressure and signal pressure when a hydraulic actuator is moving in a load-sensing hydraulic system. (2)

marker. A guide for row positioning. (4)

master clutch. A modulated clutch used to disengage engine power from the drivetrain. (10)

master cylinder. Part of a hydraulic brake system that consists of an actuator assembly with a small hydraulic piston that acts as the input cylinder to the brake's hydraulic system. (19)

master link. The top ring of a chain sling that collects or holds the bridle legs. (5)

master link. A link that provides for easy chain removal. It has a spring clip that slides into two recessed slots on the master link pins. (6)

master link. A link used to couple and uncouple track chains. It is a two-piece mating link with notched teeth. The two halves of the link are unbolted to split the chain or bolted together to join the ends of the chain. (23)

master pin. A pin used to couple and uncouple track chains. The master pin has a slightly smaller diameter than the rest of the track's pins and a designation mark that identifies it. (23)

mechanically actuated brake. A brake that uses mechanical linkage actuated by the operator to apply the brakes. (19)

mechanical schematic symbols. Shapes used in a diagram that explains power flow through a mechanical transmission. (8)

MFWD clutch. The clutch pack inside a transmission used to engage front wheel drive to achieve four-wheel drive. (10)

micrometer. A precision instrument that can measure to one ten-thousandth of an inch (0.0001″) or one thousandth of a millimeter (0.001 mm). Different types are used to measure dimensions, diameters, thicknesses, and depth of parts and openings. (2)

midship support bearing. Bearing that is commonly found in articulated vehicles to support the driveshafts between the two articulating frames. It is also used in longer driveshaft applications to support the driveshaft and limit its vertical movement. Also referred to as *center support bearing* and *hanger bearing*. (18)

Mine Safety and Health Administration (MSHA). The United States federal agency responsible for ensuring mine site safety. (1)

modulation. The varying of transmission clutch pressure to optimize shift feel and clutch life. (11)

moldboard plow. A tillage tool that slices into and inverts the soil, burying the crop residue on the first pass. (4)

moused (mousing). Using a wire to prevent the pin in a screw-pin shackle from loosening. (5)

mower. An implement designed to cut and lay down hay crops to dry in the field. (4)

mower conditioner (MOCO). A type of mower that contains an impeller or two long, horizontal rollers that are designed to crimp or crush the hay crop. (4)

MultiAxis-Steer. A Danfoss-manufactured complete electro-hydraulically controlled four-wheel steering system that can be easily integrated to a machine. (25)

multifunction handle (MFH). A handle that functions as a propulsion lever and has several buttons that allow the operator to control transmission functions. (15)

multi-function valve. A single cartridge valve within a Danfoss series-90 HST that includes an integrated make-up check valve, a bypass valve, pressure limiter valve, and cross-over relief valve. (13)

multi-grouser. A track shoe design that has two or more short grouser bars on each track shoe for increased maneuverability. (23)

multi-hybrid planter. A type of planter that has the ability to change the type of seed it is planting based on the location in the field. (4)

multi-piece rim. A rim design in which two to seven separate components form the rim assembly. Also referred to as a *split rim*. (22)

multiple disc clutch. A clutch mechanism comprised of alternating friction discs and separator plates that, when squeezed together, will couple a drum (or case) to a hub or a shaft. (8)

multiple strand chain. A chain that has longer pins that connect additional strands. One sprocket assembly consists of several sprockets to mate with the multiple strands. (6)

multiplication phase. A phase in torque converter operation in which oil expelled from the turbine is redirected via the stator to assist in rotation of the impeller. It is denoted by a large difference in the speed of the impeller and turbine, a locked stator, and vortex oil flow. (12)

multi-ribbed belt. A V-ribbed belt that snakes around all of the pulleys in the drive. Also referred to as a *multi-V belt* or *serpentine belt*. (6)

multi-V belt. A V-ribbed belt that snakes around all of the pulleys in the drive. Also referred to as *serpentine belt* or *multi-ribbed belt*. (6)

Multi-Velocity Program (MVP). An optional electronic control in Caterpillar D5N, D6N, and D6T dozers that allows an operator to choose one of five different operating ranges for forward and reverse. While operating in the specific range, the engine automatically varies to provide the best production results and fuel efficiency. (11)

N

needle roller bearing. A type of antifriction bearing that has thin cylindrical rollers, known as needles, between an inner and an outer race. (7)

neutralizer valves. Valves in the articulation area of a machine that stop the steering when they are mechanically pressed (as the machine reaches its end of travel). (25)

neutral relief valve. Charge relief in a closed-loop HST that is normally located in the pump. It is so named because it provides charge pressure relief when the transmission is in neutral. Also referred to as the *charge relief*. (13)

neutral safety switch. Electrical switch that prevents the engine from starting anytime the propulsion lever is outside of neutral. (13)

neutral steer. A track steering method in which the operator powers one track and shifts the other track into neutral. (24)

nitrogen-over-oil cylinder. See *hydro-pneumatic suspension cylinder*. (21)

nonintegrated final drive. A final drive that is not part of the front or rear axle housing. (17)

non-load reaction SCU. Steering control unit configured to hold the steering cylinder(s) in a fixed position even when external loads are exerting forces on the steering cylinder(s). (25)

non-pneumatic tire. Tire designed not to be filled with compressed air, high-purity nitrogen, or foam. (22)

non-positive-displacement pump. A pump that works on the principle of centrifugal force and reduces output as resistance to flow increases. (4)

nonrecordable accident. An accident that requires only superficial care, such as a temporary bandage. (1)

nonservo drum brake. A type of drum brake that has the bottoms of the brake shoes anchored to the backing plate by a pin or an anchor strap. Only one shoe slows the drum based on the direction of machine travel. Also referred to as *leading-trailing shoe drum brake*. (19)

normally applied clutch. A clutch that uses spring tension to keep the multiple disc clutch in an engaged position. Oil pressure or mechanical force must be used to release the normally applied clutch. (8)

no-spin differential. A differential that delivers power to both drive wheels when the machine is driven in a straight direction. When the machine turns, the outside wheel is allowed to slip, free-wheel, and spin faster, while the inside drive wheel is positively driven by the differential. (16)

no-spin spider. Component of a no-spin differential that consists of a circular ring with four steel dowels protruding from it. The dowels fit in place of the standard differential cross shaft and the no-spin spider replaces the spider pinions found in a standard differential. Dog teeth on both sides of the spider mate with the right- and left-side driven clutches. (16)

nut. A hex-shaped fastener with internal threads designed to mate with a bolt or stud. A nut is used to clamp, attach, and fasten components. (2)

O

Occupational Safety and Health Administration (OSHA). The United States federal agency that is responsible for ensuring that employees have a safe work environment. (1)

off-line kidney-loop filtration. A bypass filtration system that is separate from the main hydraulic system. It is designed to filter the fluid inside the reservoir and allows a very efficient filter to be used without interfering with system operation. (13)

offset. The distance of a rim's centerline to the outside disc surface. (22)

offset chain. A chain that uses one pair of links for each section of chain. One end of the link has a large hole that receives the bushing. The other end has a small hole that receives the pin. The roller is placed around the bushing, and then the pin is inserted through the bushing and neighboring links. (6)

offset wrench. A box-end wrench that is angled. The angle prevents the wrench from interfering with hardware in the same plane as the nut being attached. (2)

off-the-road (OTR) tires. Tires (often used on heavy equipment) that are designed to operate on unpaved surfaces, such as gravel, dirt, and turf. (22)

oil-applied clutch. A multiple disc clutch that uses oil pressure to squeeze together the alternating clutch discs and separator plates for the purpose of coupling a hub (shaft) to a drum (housing). (8)

oil-pneumatic hydraulic cylinder. See *hydro-pneumatic suspension cylinder*. (21)

one-way clutch. A mechanical clutch that is designed to lock when turned in one direction and unlock, or overrun, when turned in the opposite direction. Also referred to as an *overrunning clutch*. (8)

one-way roller clutch. A type of one-way clutch that uses two races, one of which has cam-shaped pockets that each contain a roller and spring. (8)

on/off solenoid. A solenoid that either fully blocks oil flow or allows oil flow through the valve when system voltage is applied to the coil. (11)

open bowl scraper. A scraper with a bowl that has rigid sides, an apron that forms a movable front wall, and an ejector that serves as a movable back wall. (3)

open-center grouser. A track shoe design with an opening in the center of the shoe that allows the drive sprocket to push material out of the shoe. (23)

open differential. A non-locking differential. (16)

open-end wrench. A nonadjustable wrench that is open at each end and used for loosening, turning, and tightening fasteners. Open-end wrenches are usually double ended with different-sized openings. (2)

open-loop HST. Hydrostatic transmission in which the return oil is routed directly to the reservoir, and the pump does not receive any oil returning from the motor. (13)

operating levers. The levers in a John Deere PermaClutch that are actuated with a hydraulic piston. When hydraulically actuated, the levers apply force to engage the clutch. (8)

operating load. The load that a loader can safely handle; defined by Society of Automotive Engineers (SAE) to be 50% of the static tipping load for a wheel loader and 35% of the static tipping load for a track loader. (3)

operating pressure. The range between the minimum, or cut-in, pressure and the maximum, or cut-out, pressure of an air brake system. Also referred to as *system pressure*. (20)

oscillating axle. A suspension system design that has an oscillating tube in the center of the axle housing to allow the axle to oscillate and maintain four-wheel ground contact as the machine travels over rough terrain. (21)

oscillation joint. A type of hitch installed on an articulated dump truck that allows the front and rear frames to rotate (twist/oscillate) clockwise and counterclockwise in a vertical plane, independently from one another. (21)

outboard double-reduction planetary final drive. A final drive mounted to a traditional drive axle housing consisting of two planetary gear sets. The differential axle shaft drives the first sun gear. The two ring gears are fixed. The first reduction carrier drives the second reduction sun gear. The second reduction carrier drives the wheel hub. (17)

outboard final drive. A final drive that is integrated in the axle housing and is located near the drive wheels or track drive sprockets. (17)

outboard hydraulically actuated internal wet disc service brake. A type of brake used on machines with outboard planetary final drives. (19)

outboard single-reduction planetary final drive. A final drive mounted to a traditional drive axle housing and consisting of one planetary gear set with a fixed ring gear. The differential axle shaft drives the sun gear. The reduction carrier drives the wheel hub. (17)

out-of-phase. A term describing two driveshafts that are separated and installed with the yokes not aligned with each other. (18)

output clutch. A clutch within a countershaft transmission that when engaged transfers power from a countershaft to the transmission's output shaft. (10)

overload failure. A failure that occurs when an operator repeatedly exceeds the machine's operating capacity. (17)

overrunning clutch. A mechanical clutch that is designed to lock when turned in one direction and unlock, or overrun, when turned in the opposite direction. Also referred to as a *one-way clutch*. (8)

overrunning stator. A stator with an overrunning clutch that holds the stator stationary during the torque-multiplication phase of torque converter operation and allows it to overrun during the coupling phase. (12)

oxyacetylene torch. A welding tool that is used with a combination of oxygen and acetylene gases to weld, heat, and cut metal. (2)

P

pads. Hardened metal sheets configured with different designs that are bolted to the track links to support the tractor, protect the chain, and provide traction to propel the tractor. Also referred to as *track shoes*. (23)

panhard rod. A suspension component attached to the top of an MFWD tractor front axle and to the tractor's frame to help locate and stabilize the front axle, limiting axle lateral movement. (21)

parallel linkage. Bucket curl linkage that is positioned in parallel to the lift arms. (3)

parallelogram linkage. Linkage that forms a parallelogram, usually having four tie rods—two inner tie rods and two outer tie rods. (25)

parallel rake. A rake design that has several rotating bars with tines. The rotating bars form a rotating reel. (4)

parking brake. A static brake used to hold a stationary machine in a parked position. (19)

parking brake control valve. A manual valve used to apply and release the parking brake. (20)

parts washer. A cleaning device (for smaller components) with a basin similar to a large sink, a lid, a drain, and an electric pump. Parts are held in the basin and scrubbed with a brush as the solvent or aqueous solution flows over the part. (2)

Pascal's law. A law that states that when a force is placed on a liquid in a container, pressure develops that acts equally in all directions. (19)

passive air ride seat. An air suspension seat design that does not feature any type of sensors, mechanical feedback lever, or electronic controls to vary the seat's suspension based on the terrain and machine operation. (21)

PAT dozer. A dozer with a blade that can hydraulically lift, angle, and tilt. Short for Power Angle Tilt dozer. See also *six-way blade*. (3)

pawl. A thin piece of curved metal that falls into the spaces between the teeth of a ratchet wheel or bar. (2)

PermaClutch. A John Deere flywheel clutch containing a traction clutch and PTO clutch. It requires a hydraulically applied piston to apply pressure to engage the clutch. The clutch is cooled and lubricated with oil. (8)

permanent magnet AC (PMAC) motor. An AC traction motor that has a rotating permanent magnet surrounded by a group of windings. The motor is driven with an AC sine wave. (26)

personal protective equipment (PPE). Equipment and clothing designed to protect employees from potential injuries or illnesses. (1)

pillow-block bearing. A bearing that has an integral housing that is designed to be bolted to a mounting surface. It is sometimes used in heavy equipment to support external shafts with high radial loads. (7)

pilot bearing. A bearing located in the center of the flywheel or crankshaft that is used to support the end of the clutch shaft or the transmission input shaft. (8)

pinion cone variation number. A number, located on the end of the pinion gear, that is used to adjust the pinion gear depth. The number, preceded by a positive (+) or negative (–) symbol, equals the number of variance in thousandths of an inch that the pinion gear's face protrudes too far towards the ring gear's center axis (+) or away from the ring gear's center of axis (-). (16)

pinion gear. A small gear that drives a larger gear. Often used as the input gear in a differential. The pinion teeth can be external spur, helical, or bevel design. (7)

pin synchronizer. A type of synchronizer that uses pins to move a collar into contact with a range gear to equalize speeds of the transmission shaft and range gear. (7)

pintle chain drive. A chain drive formed with tapered C-shaped link assemblies. One link assembly has two side links and a barrel. The opposite end of the link has two small holes that receive a pin. Pins are used to join each link with the adjacent links. (6)

piston accumulator. A type of gas accumulator that consists of a cylinder, with an internal piston and seals, and a gas fill valve. (19)

piston free travel. The clutch disengaged clearance on hydraulically applied clutches. (11)

pitch. The angle of the dozer blade as it is rotated around a horizontal axis, similar to a loader bucket's dump and curling motion. (3)

pitch. The distance on a chain from the center of one pin to the center of the next pin. (6)

pitch circumference. A sprocket's pitch multiplied by the number of teeth on the sprocket. (6)

pitch circumference. The distance along an imaginary circle that passes through the center of each tooth on a gear. (7)

pitch control. Advanced electric drive truck control feature in which the AC drive motors are controlled to limit chassis bouncing in rough road conditions. (26)

pitch diameter. The distance across a sprocket's pitch circumference. (6)

Pitman arm drive. A drive system consisting of a linkage connected to a pulley cam assembly. Since the linkage is connected off center, it reciprocates as the pulley turns. This reciprocating motion is translated through the pivot arm, causing the sickle bar knife to reciprocate. (4)

pivot shaft. A shaft on an elevated sprocket machine that extends across the entire machine and attaches to the rear of the left- and right-side track frames to allow the undercarriages to pivot. (23)

pivot turn. A track steering method in which the operator propels one track in either forward or reverse, while holding the other track stationary by applying a brake. Also referred to as the *brake steer*. (24)

planetary carrier. A frame that planetary gears are attached to that keeps the gears in position relative to each other. (9)

planetary gear set. A gear set consisting of a central gear called a sun gear, planetary gears, planetary gear carrier, and a ring gear. Different gear ratios can be achieved by holding and driving different combinations of members in the gear set. Also referred to as an *epicyclic gear set*. (7)

planetary pinion gears. Gears in a planetary gear set that mesh with and revolve around the sun gear. (9)

plies. Layers of fabric-type cords that are bonded into the rubber of a tire carcass to give strength to the tire. (22)

plunger. The part of a hay baler that packs hay into a firm bale. (4)

ply rating. A numerical value given to a bias-ply tire to designate the tire's strength and load-carrying capacity. (22)

pneumatic system. A system that uses compressed air for tools and some suspension or brake systems. (1)

pneumatic tire. A tire that is inflated with compressed air or high-purity nitrogen to support a weighted load. (22)

pop-off valve. A valve in an air brake system that vents system pressure when it reaches 150 psi. Also referred to as a *safety pressure relief valve*. (20)

positive-displacement pump. A pump that is designed to displace a set amount of fluid per cycle or revolution, regardless of resistance. (4)

positive pin retention track. A track design that uses a snap ring at the end of the track pins to positively retain the pin within the chain link, reducing pin endplay. (23)

power-assist steering. A steering system in which a hydraulic pump supplies the hydraulic energy to a steering mechanism to help the operator steer a heavily loaded steer axle, using an integral power steering gear or a power rack-and-pinion steering. (25)

power density. Amount of horsepower that can be delivered for a given motor size. (26)

power hop. A condition that causes a tractor's front axle (or whole tractor in the severest cases) to bounce when operating with high draft loads in the field. (21)

power neutral. A CVT transmission mode in which the tractor is held stationary by means of a hydraulic motor that counteracts the mechanical input from the engine. Also referred to as *power zero (pz)*. (15)

powershift transmission. A transmission that relies on the command of the operator to upshift and downshift. It is different from a manual transmission in that it does not require the operator to depress the clutch pedal to interrupt engine power for completing a shift. (10)

power steering. The term that machine manufacturer Case New Holland uses to describe differential-steering. (24)

power take-off (PTO). A splined stub shaft that protrudes from the front, middle, or rear of the tractor and can be connected to a driveshaft to power an implement, such as a baler, mower, or post-hole digger. (1)

power turn. A track steering method in which the operator propels both tracks in the same direction, but one track is propelled faster than the other track. (24)

power zero (pz). A CVT transmission mode in which the tractor is held stationary by means of a hydraulic motor that counteracts the mechanical input from the engine. Also referred to as *power neutral*. (15)

pre-charge chamber. A chamber in an inline square hay baler where hay is delivered by the pickup and a flake is formed. (4)

precharge pressure. The level of gas pressure in a gas accumulator when no liquid is present. (19)

precision chain. A chain that delivers power rather than conveys material. (6)

preload. The amount of static force applied to the bearing rollers after they have been adjusted for zero endplay. (7)

preselect gear. A feature on some powershift transmissions that allows the operator to preselect a specific initial forward or reverse gear from a stop (neutral position). (10)

preselect mode. An adjustable preprogrammed setting that determines the maximum travel speed that the tractor can achieve within that preprogrammed travel-range setting. Multiple preselect modes allow the operator to switch between different transmission settings for different situations with the push of a button. (15)

press-fit bearing race. A bearing race that has an interference fit with the mating shaft or housing. Pressure must be applied to install or remove the race. (7)

pressure cutoff valve. A Bosch Rexroth valve similar to an internal pressure override valve. It is located inside the HST pump and provides protection from high-pressure overloads in an HST. This valve controls the system's high pressure during gradual pressure buildup rather than from pressure spikes by dumping the servo supply oil to the reservoir. (13)

pressure passage. A rifle drilled passage through the center of a shaft that is often plugged with a steel check ball. The passage is used to direct oil to a clutch for the purpose of applying it. (8)

pressure-protection valve. A valve that isolates the main air brake system from accessory circuits. (20)

pressure-reducing valve. A normally open pressure control valve that senses and reduces pressure downstream. (11)

pressure-release solenoid. A solenoid valve used in an HST, that performs the same basic function as a manual bypass valve. (13)

primary air brake circuit. The circuit that applies rear tandem service brakes and trailer service brakes in on-highway trucks. (20)

primary shoe. The front shoe in a duo-servo drum brake. (19)

primary tillage. Aggressive working of soil at depths ranging anywhere from 6″ to 2′. (4)

proof testing. A testing method in which weight is added to the sling until it is slightly above the working load limit (WLL) or until it fails. (5)

propeller shaft. A hardened steel tube used to connect two drivetrain components. Also referred to as a *driveshaft*. (18)

proportional solenoid control valve. A solenoid valve controlled by an ECM that varies the current to the coil and allows the solenoid to vary the degree to which it opens or closes. Also referred to as a *variable solenoid control valve*. (11)

proportionator pump/motor. In a ProDrive-equipped combine, a single pump and motor assembly that acts like a scavenge pump, drawing lube oil out of the transmission and returning it to the engine separator gear case. (15)

propulsion lever. Lever that is connected to the pump's control valve and is used to control the speed and the direction of the HST. Also known as a *forward and reverse lever* or *FNR lever*. (13)

pull belt. CVT flexible steel belt that drives by pulling the output pulley sheave. (15)

pulley pitch diameter. The diameter of the pulley where the V-belt's core cords ride on the pulley. (6)

pulser lever. Control lever used to upshift and downshift a powershift transmission. (10)

pulse width modulation (PWM). Varying the off time and on time of the full system voltage for a given cycle. (11)

punch. A brass or hardened steel tool designed to be struck by a hammer or mallet to drive pins or other machine components into or out of an opening. (2)

purge valve. Part of the air dryer that expels heavy contaminants from the dryer during the purge cycle. (20)

purge volume. The area in an air dryer between the desiccant cartridge and the dryer's outer shell. (20)

push arms. Arms that connect a tractor to a dozer blade and can be lifted and lowered. (3)

push belt. A CVT flexible steel belt that drives by pushing the belt to the output pulley. (15)

pusher axle. A dead axle located ahead of a drive axle. (18)

Q

quick coupler. A device that allows an excavator operator to quickly uncouple the bucket and change attachments. (3)

quick-hitch. A coupling adapter installed on a three-point hitch to make connecting the tractor to an implement much quicker. (4)

quick-release valve. A valve that quickly releases service brake pressure when the treadle valve is released. (20)

Quick Steer. Caterpillar steering mode found on Caterpillar's small K series wheel loaders that allows for a higher rate of steering flow when traveling at slower ground speeds, dramatically decreasing the operator's turns of the steering wheel and allowing the operator to fully steer the loader from left to right (or vice versa) by rotating the steering wheel only from the 10 o'clock position to the 2 o'clock position. (25)

R

race. A metal ring that provides a smooth rolling surface for the rolling elements in a friction bearing. (7)

rack-and-pinion gear set. A gear set that uses an external-toothed spur or helical gear to drive an external-toothed rack. The gear set changes rotary motion into linear motion. (7)

radial grid. Component that acts like a brake resistor for retarding the truck. Converts mechanical kinetic energy into heat energy. (26)

radial lift. A skid steer lift arm design that causes the lift arms to move in a pronounced arc. (3)

radial load. A load that is applied perpendicular (at a right angle) to the shaft. (7)

radial-ply tire. A thin sidewall tire constructed with steel cords that are perpendicular to the tire tread and extend across the tire from bead to bead. (22)

rake. An implement that is designed to pick up and windrow hay. (4)

range transmission. A gear box that provides multiple speed ranges. (7)

ratchet. A device in which a toothed bar or wheel is engaged by a pawl to permit motion in one direction only. (2)

ratchet drive size. The size of the square that fits the opening on sockets or other attachments. (2)

ratcheting wrench. A wrench with a built-in ratcheting mechanism. (2)

ratchet marks. Short, abrupt lines that occur at the origin of a failure. These lines indicate a very high stress load. (17)

ratio valve. A valve that limits the maximum pressure to the torque converter, which is necessary for starting an engine in cold climates. When the torque converter's flow demands are met, pressure rises and the ratio valve directs the remaining oil to the reservoir. (11)

reactor. The component in a torque converter that acts like a recycler by redirecting the oil exiting the turbine back to the impeller. The reactor changes the trajectory of returning oil so that it assists rotation of the impeller, thereby multiplying input torque. Also referred to as a *stator*. (12)

reap. To cut and windrow a crop. (4)

rebound. The extending action of a suspension spring or hydraulic suspension cylinder. (21)

recap. A tire that is created with a process that removes the worn tire tread from a used tire casing and adheres a new rubber tire tread onto the old tire casing. Also referred to as a *retread tire*. (22)

recirculating ball steering gearbox. A steering gearbox consisting of a worm shaft, a ball nut, recirculating balls, and a sector gear. (25)

recoil mechanism. A spring, nitrogen-filled accumulator, or series of cone plates connected to the front idler that allows the idler to retract within the track frame. (23)

recordable accident. An accident that results in a fatality or an injury that causes an employee to lose consciousness or miss future work days, limits the employee's work abilities, or requires the care of a physician. Also referred to as a *recordable incident*. (1)

recordable incident. An accident that results in a fatality or an injury that causes an employee to lose consciousness or miss future work days, limits the employee's work abilities, or requires the care of a physician. Also referred to as a *recordable accident*. (1)

rectify. To convert alternating current to direct current. (26)

reeving. An improper lifting technique in which a single sling is attached to a load by threading the sling through both lifting eyes and attaching both ends of the sling to the lifting hook. (5)

reflashing. Reprogramming a machine's computer by installing new software to fix, modify, or improve the machine's performance. (11)

regular-lay wire rope. A wire rope with the wires rotating in the *opposite* direction of the strands. (5)

regulated pressure. The main transmission operating pressure that is used to apply the transmission's clutches. Also referred to as *main pressure*, *transmission charge pressure*, and *transmission operating pressure*. (11)

relay valve. A valve used to increase the reaction time for applying and releasing service brakes. (20)

release bearing. A bearing that is used to transmit the force from a sliding collar to the pressure plate's release levers (or fingers) to release a clutch. Also referred to as a *throw out bearing*. (8)

relief valve. Part of a hydraulic piston track tensioner that will release grease and loosen track tension when opened. (23)

replenishing port. An opening in the master cylinder that allows fluid from the reservoir to fill the void behind the master cylinder piston. (19)

replenishing valve. Shuttle valve that is used to flush the oil from the motor and pump. This aids in cooling the transmission. Also known as a *hot-oil purge valve* or *flushing valve*. (13)

retarder. A device that helps slow a machine and reduces brake wear. (3)

retread tire. A tire that is created with a process that removes the worn tire tread from a used tire casing and adheres a new rubber tire tread onto the old tire casing. Also referred to as a *recap*. (22)

return springs. The springs inside a multiple disc clutch that are used to return the piston to a disengaged position when the apply oil is drained. (8)

reverse idler. A gear that is positioned between a gear on a transmission's countershaft and a gear on the transmission's output shaft in order to reverse the rotation of the output shaft. (7)

reversible pump. A pump that can internally reverse the direction of the pump's flow. (13)

ride control. One or more accumulators plumbed in parallel to a machine's implement lift cylinder to dampen the implement's movement during transport. (21)

right-hand-lay wire rope. A wire rope with the strands rotated in a clockwise direction. (5)

rigid axle. A non-steerable axle in which the wheels cannot pivot on the axle housing. Commonly used on articulated steering machines. (18)

rigid-frame haul trucks. Trucks with a solid, unarticulated frame and, typically, a single set of dual rear wheels and a pair of front-steer wheels. (3)

rim. The flanged portion of the wheel against which the tire mounts. (22)

rimpull. The amount of torque delivered at the ground, through the tire or track shoe. (12)

ring-and-pinion gear set. A gear set consisting of a small external-toothed gear that is meshed with a large internal-toothed gear. (7)

ring gear. A large internal-toothed gear. Also referred to as an *annulus gear*. (7)

ring gear. In a planetary gear set, the ring gear surrounds the entire set and meshes with the planetary pinion gears. (9)

ring gear. A gear, attached to the differential housing, that is driven by a spiral bevel pinion gear or an external toothed spur or helical gear. The spiral bevel pinion and ring gear provide a torque multiplication and change the power flow at a 90° angle. (16)

road lope. Severe tractor suspension bouncing and swaying that occurs when transporting a tractor on pavement at increased speeds. (21)

rock drill. A drill designed for drilling through rock formations so explosives can be inserted to blast the rock. (3)

rock ejectors. Devices that prevent large rocks from getting wedged between dual tires. (3)

rod weeder. An implement that has a long rod that is dragged under the soil's surface to uproot vegetation. (4)

roller chain. A chain assembly formed by links, pins, bushings, and rollers. Rollers surround the bushings and provide the drive surface that meshes with the sprockets. (6)

rollerless chain. A chain that has the same appearance as a traditional roller chain, except it is missing the rollers. The two neighboring link assemblies are joined through the traditional outside links and pins. (6)

roll out bucket. A bucket designed to increase the loader's dump height by rolling forward, away from the loader's frame. (3)

roll over protective structure (ROPS). A safety device attached to mobile machines to prevent the machine from crushing the operator in the event of a roll over. (1)

roll pin. A small fastener designed as a short tube with a slot that runs the length of the pin. As the pin is inserted into a component, the pin compresses to hold itself in place. (2)

room and pillar. A subterranean mining technique that leaves behind a grid of square earthen pillars to support the roof. (3)

root. The bottom cavity or space between the base of one gear tooth and the base of the adjacent tooth. (7, 16)

rope lay. The parallel distance (measured on the length of the wire rope) that it takes for one strand to make an entire helical revolution around the wire rope. Also referred to as *lay length*. (5)

rotary combine. A type of combine that uses a rotating rotor(s) as the primary means of threshing the crop. Threshing occurs as the crop makes multiple passes around the rotor. (4)

rotary encoder. A rotary speed dial on the multifunction handle that is used to set the maximum travel speed for a preselect mode. (15)

rotary hoe. An implement consisting of several spiked wheels that are drug across the soil to reduce weeds, improve the plant emergence, and minimize soil crusting. (4)

rotary oil flow. Torque converter oil flow around the outer circumference of the torque converter. It occurs due to centrifugal force during the coupling phase and causes an overrunning stator to freewheel. (12)

rotary rake. A rake design that has one or more rotating vertical spindles, known as rotors. Each rotor has several horizontal tine-arms extending outward from it. A number of spring-loaded tines are attached to the end of each tine-arm. (4)

rotary tiller. A tillage tool that has a rotating shaft with multiple knives that pulverize the soil. It is typically mounted on a three-point hitch and driven by a PTO. (4)

rotating clutch. A multiple disc clutch that is used to drive a shaft or gear (input clutch), or couple two components (coupling clutch). (8)

rotochamber. A type of air brake actuator that uses a lobed diaphragm designed to roll as it actuates. (20)

rotor. A rotating disc that is part of a caliper disc brake. (19)

rotor. In an alternator, an electromagnet consisting of a rotating, star-shaped iron core with multiple coils, called field poles, around its outer perimeter. (26)

round belt. A belt with a round cross-section that is used for driving a pulley. It is used in serpentine drives and other drives that require the belt to turn a quarter turn or twist. Also referred to as an *endless round belt*. (6)

row crop planter. A type of seeder that controls the planting based on a seed population of a set number of seeds per acre or seeds per hectare. A seeder typically plants with wider spacing between rows than a drill does. (4)

row crop tractor. A type of tractor that has wheels that can be spaced to match the spacing between crop rows. (4)

R tires. Tires (often having drive lugs, known as cleats, in chevron shapes) that are most commonly installed on drive wheels but can also be installed on non-drive wheels. They are popularly found on agricultural machines but can also be found on construction machines. (22)

rubber track pads. Steel pads with rubber bonded to the face of the pad. (23)

runout. How much a shaft moves radially from its center axis. (18)

rupture disk. A safety device installed in compressed gas cylinders that will rupture at a specified pressure to allow the controlled release of the gas. Also referred to as a *burst disk*. (1)

Rzeppa U-joint. A U-joint containing an inner race with six channels in which six balls ride. A cage retains six ball bearings. As the inner race is driven, it causes the balls to drive, which propel the outer bearing race. Also referred to as a *ball-and-cage joint*. (18)

S

safety data sheets (SDS). Printed materials that provide end users important information regarding products. (1)

safety pressure relief valve. A valve in an air brake system that vents system pressure when it reaches 150 psi. Also referred to as a *pop-off valve*. (20)

scavenge pump. A pump that transfers fluid from one reservoir to another reservoir. (11)

screw. A threaded fastener that is tightened by applying torque to the head of the fastener. (2)

SCU internal leakage test. Procedure to assess the amount of tolerated oil seepage between the parts operating inside a steering control unit. (25)

sealed and lubricated track (SALT) undercarriage. An undercarriage designed with internally lubricated pins and bushings that are plugged and sealed to keep contaminants out and lubrication in for extended track service life. (23)

sealed bearing. A type of antifriction bearing that has seals on the outside of the races to keep contamination out of the bearing and keep lubrication in. (7)

Seale wire rope. A wire rope constructed with larger diameter wires in the outer edge of the rope to provide better resistance to abrasion. (5)

secondary air brake circuit. The circuit that applies front service brakes and may be used to supply air to the trailer service brakes via the trailer lever control valve in on-highway trucks. (20)

secondary brake pedal. A left-hand pedal that allows the operator to apply the parking brake in the event that the service brake is inoperative. (19)

secondary shoe. The rear shoe in a duo-servo drum brake. (19)

secondary tillage. Working of the soil at depths of less than 6″. (4)

second order vibration. A vibration that occurs twice during every revolution of the component. This vibration can be caused by driveshaft angularity problems. (18)

seed tube sensors. Sensors that alert the operator if a seed tube on a planter becomes plugged and no longer allows seed to drop into the trench. (4)

self-cleaning grouser. A track shoe design that enables dirt and mud to fall freely out of the tracks when the shoes separate slightly as the tracks rotate around the drive sprocket and idler. (23)

self-energizing. The effect of a brake band, that when applied to one side causes further force on the other side to apply the band. (8)

self-energizing braking. A braking action in which the rotation of the brake drum increases the braking force applied by the brake shoe. In nonservo drum brakes, one shoe is self-energized. In duo-servo drum brakes, both shoes are self-energized. (19)

self-leveling. A feature that holds the work tool at the desired pitch as the loader frame is lifted and lowered. (3)

self-leveling. A suspension system feature that has an electronic control module automatically adjust the height of the suspension cylinders as operating conditions change. (21)

semi-active seat suspensions. Operator seat designs that use hydraulic cylinders, commonly filled with magnetorheological fluids, to change the seat's suspension stiffness and dampen unwanted seat movement. (21)

semi-floating axle. Type of axle in which the weight of the machine is exerted onto the axle bearings. The weight is then transferred to the axle shaft and onto the drive wheel. (18)

separator plates. The flat steel plates, without friction material, located in an alternating fashion between the friction discs inside a multiple disc clutch. (8)

serpentine belt. A V-ribbed belt that snakes around all of the pulleys in the drive. Also referred to as a *multi-V belt* or *multi-ribbed belt*. (6)

Service Advisor. A term used by John Deere for the online software that provides its service literature, or to describe the electronic service tool that communicates with the machine's ECMs. (11)

service brake. A dynamic brake used to bring a moving machine to a stop. (19)

service brake control valve. A valve that directs pump flow to apply the brakes when the operator depresses the brake pedal. (19)

service code. Agricultural and construction tire classifications that are based on function and type of machine. (22)

Service Information System (SIS). A term used by Caterpillar for the online software that provides its service literature. (11)

servo oil. Oil flow that is used to actuate an HST servo piston. Also known as *control oil*. (13)

shackle. A U-shaped device that uses a pin or bolt to attach slings to the loads and hooks. (5)

shear bar. A bar in a forage harvester that is positioned so the cutter head cuts the crop as the cutting head blades rotate by the bar. (4)

shear bolt. A low-grade bolt designed to sacrificially break when a PTO shaft becomes overloaded due to an implement hitting an obstruction. (8)

shear hub. A hub on a rotary mower that is designed to shear in the event a knife hits an obstruction, protecting the mower's gearbox. (4)

shear pin. A low-grade sacrificial bolt that couples the PTO driveshaft to an implement. (4)

shear pin. A low-grade bolt that is designed to fail when the implement encounters an obstruction. (18)

sheave. Another name for a pulley. (6)

shift collar. A sliding collar that is splined to the output shaft and has external splines that can mesh with the internal splines of an output gear to couple the gear to the output shaft. Also referred to as a *sliding clutch*. (7)

shift fork. A C-shaped lever that mates with a groove in a sliding gear, collar, or synchronizer sleeve. The fork is used to slide the sliding gear, shift collar, or a synchronizer sleeve to shift transmission gears. (7)

shift gear. In one type of collar shift transmission, a gear next to a driven range gear and splined to the transmission shaft and is used in combination with a shift collar to couple the range gear to the shaft. (7)

shift rail. A rod that is used to actuate a shift fork. (7)

shock absorber. See *hydro-pneumatic suspension cylinder*. (21)

shot rock. Rock that has been blasted or hydraulically hammered but has not yet been crushed. (26)

shoulder. The portion of a tire where the tread joins the sidewall, including the outer edge of the tread. (22)

shovel. A pointed tool on a field cultivator that breaks up the soil. Also referred to as a *sweep*. (4)

shuttle shift. A feature on some powershift transmissions that allows the operator to quickly shift between forward and reverse without moving the gearshift lever. (10)

sickle bar knife drive. A device consisting of triangular knife sections riveted or bolted to a cutter bar, which is cycled back and forth through cutter guards to cut crops. (4)

side dressing. Applying fertilizer between the rows of crops so that the crop has access to the nutrients at its side. (4)

side dump bucket. A bucket with one open side, which allows it to dump when it is tipped to that side. (3)

side gears. Gears inside a differential that drive the wheels and are attached to the axle shafts. (16)

side skid control. Advanced electric drive truck control feature that controls the two AC drive motors to improve turning when operating in slick road conditions. (26)

sidewall. The flexible rubber exterior that covers the sides of a tire from the bead to the tire's shoulder. (22)

sieves. Devices in a combine harvester that resemble finned combs and allow air from the cleaning fan to blow through the grain to separate the chaff. (4)

signal pressure. Pressure sent by the governor to the compressor's unloader valves when the cut-out pressure is reached. (20)

silage. A crop that is harvested by chopping the entire plant at high moisture levels into small pieces so it can be fed to livestock. (4)

silent chain. A quiet chain that consists of several links assembled with pins, but no rollers. (6)

simple planetary gear set. Gear set consisting of a sun gear, planetary pinion gears, a planetary carrier, and a ring gear. (9)

simple wire rope. A wire rope constructed with wires that are all the same size. (5)

single master cylinder. A master cylinder with just one piston. (19)

single parking brake chamber. A type of air brake actuator that contains a heavy-duty spring that applies the parking brake when no air is supplied to the chamber. Also referred to as a *hold-off brake chamber*. (20)

single-path HST. Hydrostatic drive system consisting of one variable-displacement reversible hydraulic pump and one hydraulic motor. (13)

single-piece rim. A rim design that consists of one complete wheel that cannot be taken apart into multiple pieces. (22)

single-reduction final drive. A final drive that achieves gear reduction through a single gear set. (17)

single service brake chamber. A type of air brake actuator that contains a pressure plate, non-pressure plate, rubber diaphragm, push plate, rod assembly, return spring, and a clamp ring that couples together the assembly. (20)

single-speed motor. A fixed-displacement motor. (13)

singulation. The placing of single seeds, with consistent spacing between seeds. (4)

singulator. The device on a row crop planter that ensures only one seed is pulled onto each of the seed disc's individual seed holes. (4)

six-way blade. A blade on a PAT dozer, which can lift, angle, and tilt. (3)

skid steer. A compact self-propelled loader with independently controlled drive wheels on each side of the machine. (3)

skip. Failure of a planter to deposit a seed at the designated planting interval. (4)

skip loader. A unique type of wheel loader built on a loader backhoe chassis that resembles an agricultural utility tractor with a rear box blade. It is also referred to as a *landscape tractor*. (3)

skip shifting. A feature on some powershift transmissions that allows the operator to skip gears while upshifting, reducing the overall lag time associated with shifting. (10)

skiving. The process of removing a narrow layer of rubber from each end of a flat belt for the purpose of attaching splices onto each end of the belt. (6)

slack adjuster. Part of cam brakes that transforms the brake chamber's linear force into a rotational torque applied to the camshaft and eliminates brake slack caused by brake shoe wear. (20)

slave cylinder. The actuating piston that applies the service brakes. Also referred to as a *wheel cylinder*. (19)

sleeve. The part of a block synchronizer that slides over the dog teeth on the speed gear and blocker ring and the external splines on the synchronizer hub to couple the speed gear to the transmission shaft. (7)

slew. The crane or excavator control that forces the machine to swing the house to the right or left. (3, 26)

sliding caliper. A type of caliper that uses one or more pistons located only on the inboard side of the caliper assembly. Also referred to as a *floating caliper*. (19)

sliding clutch. A sliding collar that is splined to the output shaft and has external splines that can mesh with the internal splines of an output gear to couple the gear to the output shaft. Also referred to as a *shift collar*. (7)

sliding gear transmission. A type of transmission in which splined gears slide into and out of mesh with mating gears on a parallel shaft when the operator moves the gearshift lever. (7)

slip clutch. A clutch that is designed to slip when it encounters an excessive load, such as when a PTO shaft becomes overloaded due to an implement hitting an obstruction. (8)

slip fit race. A bearing race with a small space between it and the shaft or housing. (7)

slip/slide control. Advanced electric drive truck control feature that uses front wheel sensors and suspension and steering sensors to provide traction control and anti-lock brake control in slick conditions. (26)

slip splines. A sliding joint on a two-piece driveshaft consisting of external splines on one end of a driveshaft and internal splines on the mating driveshaft. Allows the overall length of the driveshaft assembly to change as the axle assembly raises and lowers. (18)

slow adapt. Adaptive learning that takes place during normal operation. The ECM determines the volume of oil needed to engage the clutch and adjusts the ECM's software. (11)

SmartRide. A suspension system, used in late-model Challenger MT700 and MT800 series tractors and Fendt 900 and 1100 Vario MT series tractors, that allows the track rollers to pivot. (21)

snap ring U-joint. U-joint that has a retaining ring that holds the bearing cup inside the yoke. The retaining ring can be located either inboard or outboard of the bearing cup. (18)

socket. A short steel tube that attaches to the ratchet handle's quick-release, square-shaped drive coupler. It is used to more easily turn fasteners. (2)

socket drive size. The size of the square that fits the drive size on the ratchet handle. (2)

solenoid control valve. A valve that consists of a coil wire wrapped around an iron core. Electrical current applied to the coil creates a magnetic field and causes an internal core plunger rod to open or close a fluid passage. (11)

solid tires. Non-pneumatic tires of which the entire tire body and interior are manufactured completely of solid rubber. (22)

spalling. A surface contact stress fatigue in which the mating surface of a component chips, breaks, or flakes off in pieces. (17)

speed handle. A wrench consisting of a rod with a C-shaped center section, a rotating handle on one end, and a long shaft on the other end. The long shaft ends in a square head drive that attaches to sockets. (2)

speed matching. A feature on some powershift transmissions that matches the transmission gear to the tractor's travel speed. (10)

speed rating. A tire sidewall marking that designates the top speed that the manufacturer has designed the tire to travel. (22)

speed-sensing signal pressure. A control (signal) pressure that is proportional to engine speed used to control the HST servos in the variable-displacement pump and motor. (13)

speed-sensing valve. A valve that is manually operated by the engine's governor to produce an HST signal control pressure that is proportional to engine speed. (13)

spelter socket. A wire rope end termination that uses a polyester-based compound resin to secure the wire rope in the socket. (5)

spherical roller bearing. A type of antifriction bearing that has barrel-shaped rollers between inner and outer races. (7)

spider. U-joint component that is cross-shaped and has four trunnions (cross ends). Also referred to as a *cross*. (18)

spider gears. Small bevel gears that transfer power from the case to the side gears in a differential. (16)

spider key. The long tooth on a no-spin spider. (16)

spike-toothed harrow. A tillage tool that has a mat of spikes that are dragged across the soil's surface to smooth it. (4)

spiral bevel gear set. A gear set consisting of a small bevel gear with helical teeth that mesh with the helical teeth on large beveled ring gear. (7)

splice. A group of metal loops that are pressed onto each end of a baler's belt using a belt splice press tool. (6)

split guide rings. Trough-like rings in the impeller and turbine that form a donut-shaped void between the two members in an assembled torque converter. The split guide rings help direct oil as it cycles through the torque converter. (12)

split HST. Hydrostatic drive system in which an engine-pump is separate from the hydraulic motor and connected to it by hoses or tubing. (13)

split pin synchronizer. A type of pin synchronizer in which the hourglass-shaped pins are split longitudinally and in which friction rings on the synchronizer engage friction cups that are splined to the speed gear. (7)

split rim. A rim design in which two to seven separate components form the rim assembly. Also referred to as a *multi-piece rim*. (22)

splitting stands. Support equipment used to split and support the front and rear ends of a tractor to provide access to components. (5)

spotter. A person with a radio, or other means of contact, that is in direct communication with a machine operator to help guide and prevent the operator from contacting power lines or nearby structures. (1)

sprag clutch. A type of one-way clutch that uses dog-bone–shaped pieces that stand up and lock the clutch when it is rotated in one direction and roll over and allow the clutch to overrun when it is turned in the opposite direction. (8)

sprayer. An implement designed to apply liquid chemicals to a field. (4)

spring-applied oil-released clutch. A multiple disc clutch that uses spring pressure to apply the piston to squeeze the friction discs and separator plates together for the purpose of coupling a hub (shaft) to a drum (housing). Oil pressure is required to overcome the spring tension to release the clutch. (8)

spring-applied parking brake. A type of parking brake that requires hydraulic pressure to release the parking brake. (19)

sprocket. A toothed wheel that meshes with a chain. (6)

sprocket pitch. The distance from the center of one bushing on the drive sprocket to the center of the adjacent bushing on the sprocket. (23)

spur gear. An external-toothed gear that has straight cut teeth. (7)

stabilizers. Outriggers on a machine such as a crane or backhoe that help the machine maintain a stable footing during operation. (3)

stall mode. A mode of torque converter operation denoted by a stationary turbine and the impeller rotating at high speed. (12)

standard grouser. A track shoe design that has a single tall grouser that digs deep into the soil. (23)

stand pipe. Check valve in a U-joint that prevents lubricant from draining out of the trunnion that is pointing upward. (18)

star rating. A one-, two-, or three-star designation for a radial-ply tire that describes the minimum amount of tire pressure that is required for a given tire load on a construction tire, or the maximum allowable load an agricultural tire can handle based on a specific air pressure. (22)

static load-sensing signal. Steering signal that has a pressure based on the steering cylinder's working pressure. Unlike the dynamic load-sensing signal, it does not have flow when the SCU is in a neutral position. (25)

static tipping load. The amount of force that a loader must exert to lift the machine's rear tires or rear tracks off the ground. (3)

stationary clutch. A multiple disc clutch used to hold a shaft or gear stationary. The clutch normally uses a housing, such as a transmission case, instead of a rotating drum. Also referred to as a *holding clutch* or a *brake*. (8)

stator. The component in a torque converter that acts like a recycler by redirecting the oil exiting the turbine back to the impeller. The stator changes the trajectory of returning oil so that it assists rotation of the impeller, thereby multiplying input torque. Also referred to as a *reactor*. (12)

stator. The stationary windings inside an alternator or electrical motor. (26)

steer axle. An axle, used for steering, that allows the wheels to pivot. (18)

steer-by-wire. Electronically controlled steering consisting of steering wheel speed sensors, wheel angle sensors, and solenoids that are used to actuate a steering control valve. (25)

steering clutch and brake system. A twin-track steering system that uses a lever (or an electronic joystick) for each track to control the application and disengagement of steering clutches and steering brakes that propel and steer each track of the machine. (24)

steering control unit (SCU). Manually operated group of hydraulic components rotated by a steering wheel and used to control the actuation of a hydraulic steering system. (25)

steering gear. A gearbox that provides steering torque multiplication in a manual steering system. A gearbox receives a rotating input from the steering wheel/steering shaft. The steering gear multiplies the torque and translates it into a lateral steering motion. (25)

steering kickback. Jerkiness or recoil in a steering wheel caused by steering oil bleeding backward through the steering control unit when external forces are exerted on the steering cylinder(s). (25)

steering knuckles. Steering linkage system components that provide the pivoting action that allows the wheels to be steered left or right. (25)

steering tiller. Caterpillar's term for the single joystick control on their differential-steered twin-track tractors that operates the machine's direction, speed, and steering. (24)

stepless transmission. A transmission that provides an infinite range of speed control rather than a limited number of discrete gear ratios. Also referred to as a *continuously variable transmission (CVT)* or *infinitely variable transmission*. (15)

straight cut header. A combine harvester header used for harvesting grain and equipped with a rotating reel, long auger or rubber draper belts, and a reciprocating knife. (4)

straight roller bearing. A type of antifriction bearing that has multiple cylindrical rollers between inner and outer races. Also referred to as a *cylindrical roller bearing*. (7)

straw walkers. Parts of a conventional combine harvester (cylinder-type combine) that move forward and rearward in order to move straw to the rear of the machine, where it is discharged. (4)

stress riser. The initial high area of concentrated stress from which fractures propagate. (17)

stretch fit V-ribbed belt. A V-ribbed belt that does not use a tensioner and must be stretched over the pulleys during installation. (6)

stripper header. A combine harvester header with a long rotating rotor that has comb-like teeth that strip the grain off the crop's stems. (4)

stroke-sensing ASA. A type of automatic slack adjuster that has a push rod with an actuator piston and actuator that actuates when brakes are applied and causes the worm shaft to adjust for wear. (20)

struck capacity. The volume of material that it takes to fill a bucket even with its cutting and rear edges when the bucket is in a level position. (3)

strut. Suspension cylinder that forms a structural connection between the machine's frame and wheel, with the cylinder also acting as the steering pivot shafts. See also *hydro-pneumatic suspension cylinder*. (21)

stud. A headless rod with threads on both ends. (2)

stuffer fork. The part of a hay baler that moves a flake from the pre-charge chamber into the bale chamber. (4)

sub-soiler. A tillage tool that has heavy-duty rippers mounted solidly to a tool bar. It cuts deeper than field cultivators and chisel plows. Also referred to as a *V-ripper*. (4)

suction screen. A wire mesh filtration device used to filter large particles from entering the inlet of the transmission pump. Also referred to as a *suction strainer*. (11)

suction strainer. A wire mesh filtration device used to filter large particles from entering the inlet of the transmission pump. Also referred to as a *suction screen*. (11)

suitcase weight. Cast iron weights often attached to the front, rear, or side of a tractor to correctly ballast the tractor's weight. (22)

sun gear. The central gear in a planetary gear set that is surrounded by the planetary pinions. (9)

supply reservoir. Part of the air brake supply system located directly after the compressor or after the air dryer. Stores the compressed air used to apply the brakes. Also referred to as *air supply tank* or *wet tank*. (20)

surface mining. A mining technique that removes surface soil to expose deposits of the material being mined. (3)

suspension systems. Groups or assemblies of components installed on a machine to absorb shocks as the machine travels over rough terrain and as the machine handles loads while performing work. (21)

swaged socket. A wire rope end termination formed by placing the fitting and wire rope into a die that is hydraulically pressed through a tapered ring. (5)

swath control. A feature that electronically activates and deactivates rows or groups of rows on a planter to avoid planting on top of previously planted rows. (4)

swather. An implement or self-propelled mower designed to mow a crop and gather the cut crop into a single swath, or windrow. Also referred to as a *windrower*. (4)

sweep. A pointed tool on a field cultivator that breaks up the soil. Also referred to as a *shovel*. (4)

swept area. The amount of brake drum or rotor area that sweeps past the friction material during a single revolution of the tire. (19)

swing arm. A suspension component that is attached to the front frame of a large articulated tractor and allows the front axle to move up and down. (21)

switched reluctance motor (SRM). An electrical motor without brushes that uses DC current to energize pairs of stator windings, causing a magnetic reluctance to pull the rotor, causing it to rotate. (26)

synchromesh transmission. A type of transmission with gears that are in constant mesh. Synchronizers match engine speed (input shaft) to the transmission speed (output shaft) to prevent grinding gears and gears clashing while gears are shifted. Allows gears to be changed while the equipment is moving. Also referred to as a *synchronized transmission*. (7)

synchronized transmission. A type of transmission with gears that are in constant mesh. Synchronizers match engine speed (input shaft) to the transmission speed (output shaft) to prevent grinding gears and gears clashing while gears are shifted. Allows gears to be changed while the equipment is moving. Also referred to as a *synchromesh transmission*. (7)

synchronizer cone. In a split pin synchronizer, a cone-shaped disc that is splined to the speed gear and engages the synchronizer's friction ring. Also referred to as a *synchronizer cup*. (7)

synchronizer cup. In a split pin synchronizer, a cone-shaped disc that is splined to the speed gear and engages the synchronizer's friction ring. Also referred to as a *synchronizer cone*. (7)

synchronizer hub. The part of a synchronizer that is splined to the transmission shaft. (7)

synchronizer ring. The part of a block synchronizer that contacts a speed gear and equalizes the speeds of the transmission shaft and speed gear so the sleeve can engage the gear. Also referred to as a *blocker ring*. (7)

synchronous belt. A flat belt with cogs or teeth, used as a traditional belt drive or for applications that require two or more pulleys to maintain precise synchronization with each other. Also referred to as a *timing belt*. (6)

synchronous switched reluctance motor (SynRM). A type of switched reluctance motor that has the same style stator used in traditional AC induction motors. It contains multiple phase stator windings distributed around the iron stator. The rotor is comprised of iron stampings (flux carriers) that contain air pocket cavities or insulators (flux barriers). (26)

SystemOne undercarriage. A Caterpillar undercarriage that is designed to provide longer life, while reducing maintenance and downtime related to servicing undercarriages. It also reduces operating costs. (23)

system pressure. The range between the minimum, or cut-in, pressure and the maximum, or cut-out, pressure of an air brake system. Also referred to as *operating pressure*. (20)

T

tag axle. A dead axle located behind a drive axle. (18)

tailings auger. The part of a combine harvester that delivers unthreshed crop to the tailings elevator, which sends the crop back to the cylinder or rotor to be rethreshed. (4)

tandem. The type of chain drive found on each side of a motor grader and consisting of two drive sprockets, two chains, and two driven sprockets. (3)

tandem chain drive. A drive system consisting of a pair of drive sprockets that drive a pair of chains and driven sprockets that are splined to a drive wheel. Each side of the machine has two chains used to propel a set of tandem wheels. (17)

tandem master cylinder. A master cylinder that contains two pistons that are located one in front of the other. (19)

tapered roller bearing. A type of antifriction bearing consisting of multiple tapered (cone-shaped) rollers housed in a cage and positioned between two tapered races. (7)

Technical Service Group (TSG). A term used by CNH for its manufacturers solutions group. (11)

tedder. An implement that has several small PTO-driven rotors with tines designed to pick up hay and turn it over. (4)

telehandler. A machine with a telescoping boom that is designed to lift and move materials using an attachment on the end of the boom, such as pallet forks or a hay grapple. Also called a rough terrain forklift. (3)

tensile strength. The amount of stress that can be placed on a component before it stretches and breaks. (17)

termination efficiency. The value factored into a rope's WLL based on the termination method. (5)

three-phase AC. Three separate phases of alternating current flow produced by three different stator windings. (26)

three-point hitch. A device that uses three arms to attach an implement to a tractor and is used to lift and lower the implement. (4)

three points of contact. The consistent contact of two hands and one foot or two feet and one hand with a ladder to minimize the risks associated with entering and exiting a machine, such as slipping on slick steps or having a boot become lodged in the crevice of a step. (1)

three-stage torque converter. A torque converter design with three layers of turbine blades and a stator with two layers capable of multiplying engine torque up to five times. (12)

thresh. To rub a crop so the grain breaks out of its hulls. (4)

throw out bearing. A bearing that is used to transmit the force from a sliding collar to the pressure plate's release levers (or fingers) to release the clutch. Also referred to as a *release bearing*. (8)

thrust bearing. A type of antifriction bearing that has rollers that are arranged so their axes are perpendicular to the shaft. It is designed to handle thrust loads. (7)

thrust load. A force that is applied parallel to the shaft, or axial to the shaft. Also referred to as an *axial load*. (7)

tilling. The act of cutting, agitating, stirring, and overturning the soil in preparation for seeding and to control weeds. (4)

timing belt. A flat belt with cogs or teeth, used as a traditional belt drive or for applications that require two or more pulleys to maintain precise synchronization with each other. Also referred to as *synchronous belt*. (6)

timing chain. A chain used to synchronize the rotation of two or more shafts. (6)

toe. The portion of a ring gear's tooth closest to the inside diameter of the ring gear. (16)

ton-kilometer per hour (TKPH). A calculated value that designates the workload capacity of earthmoving tires without the tires overheating. (22)

ton-mile per hour (TMPH). A calculated value that designates the workload capacity of earthmoving tires without the tires overheating. (22)

top land. The outer edge surface of a ring gear's tooth. (16)

top-of-travel switch. A switch in inching pedal assemblies used for certain transmission features, such as speed matching. (11)

toric. A large rubber ring that applies pressure to one of the duo-cone metal seals. The term toric is derived from its torus shape. (18)

toroidal drive. A CVT that uses one or two sets of cone-shaped rollers that pivot between two toroidal-shaped discs. (15)

torque. The strength of force being applied through an arcing motion. (2)

torque arm. A hollow cast suspension component that attaches the front axle of a MFWD tractor to the tractor's frame, has the MFWD driveshaft extend through its center, and allows the front axle to oscillate. (21)

torque converter. A hydrodynamic fluid drive consisting of an impeller, a turbine, and a stator. It multiplies input torque and acts as an automatic clutch by decoupling the engine from the transmission when the equipment is at rest. (12)

torque divider. A torque converter design that includes an internal planetary gear set. The planetary gear set provides approximately 25% of the total torque multiplication. (12)

torque-multiplication phase. A phase of torque converter operation that occurs when there is a large difference between impeller and turbine speeds and the stator is locked. During this phase, oil expelled from the turbine is redirected by the stator to assist in rotating the impeller, assisting input torque. (12)

torque multiplier. A tool containing gearing and used by technicians to multiply torque. (2)

torque wrench. A calibrated tool used to apply a specified amount of torque to a fastener through an attached socket. (2)

torsional stress fracture. A fracture exhibiting a 45° break that indicates the torsional twisting that occurred during the failure. (17)

tow valve. A valve that enables the operator to release the parking brake on a machine with a dead engine by pressing the left service brake pedal. (19)

track chain. A heavy-duty chain loop built with interconnected track links, pins, and bushings that is the driven component of a track undercarriage system. (23)

track frame. A large housing that provides a mounting place for the undercarriage components, such as the track rollers, carrier rollers, a track tensioner/recoil mechanism, a front idler, and a rear idler (on elevated sprocket undercarriages). (23)

track pitch. The distance from the center of one pin to the center of the next pin in a track chain. (23)

track rollers. Single- or double-flanged rollers installed below the track frame, between the bottom of the track frame and the track links, to evenly transfer the load of the machine to the chain's links. (23)

track shoes. Hardened metal plates, available in various designs, that are bolted to the track links to support the tractor, protect the chain, and provide traction to propel the tractor. Also referred to as *pads*. (23)

track tensioner. A mechanism (hydraulic piston, threaded sleeve and lock collar, or hydraulic system) that sets and maintains track tension in an undercarriage. (23)

traction clutch. A clutch used to couple or disengage engine power from the transmission's input shaft. Also referred to as a *transmission clutch*. (8)

tractor undercarriage. A type of undercarriage designed for strong pulling power in addition to propelling and supporting the machine. Also referred to as a *crawler tractor*. (23)

transmission charge pressure. The main transmission operating pressure that is used to apply the transmission's clutches. Also referred to as *main pressure*, *regulated pressure*, and *transmission operating pressure*. (11)

transmission clutch. A clutch used to couple or disengage engine power from the transmission's input shaft. Also referred to as a *traction clutch*. (8)

transmission-driven PTO. A power take-off that is engaged and disengaged using the transmission clutch. (4)

transmission operating pressure. The main transmission operating pressure that is used to apply the transmission's clutches. Also referred to as *main pressure*, *regulated pressure*, and *transmission charge pressure*. (11)

tread. The outer surface of a tire that contacts the soil or pavement. It is manufactured with a distinct pattern, design, and depth. (22)

treadle valve. A foot-operated service brake control valve used to apply the service brakes. (20)

tread wear rate. A measurement of the amount of rubber tread worn off the tire per hour of use. (22)

Triple Link Suspension (TLS). A type of John Deere suspension system equipped with a panhard rod, two double-acting hydraulic suspension cylinders, a torque arm, and three gas-filled accumulators to improve ride quality and reduce power hop on a 6R and 7R MFWD tractor. (21)

trolley assist. A system providing additional electric power to a truck through overhead power lines. (3)

trolley assist. A system by which an electric drive truck can connect to overhead power lines while moving. The overhead power lines provide the truck with increased horsepower. After the truck completes the ascent, it disconnects from the overhead power lines and resumes normal operation. (26)

true brinelling. A type of surface fatigue in which permanent indentations in the bearing race occur without the loss of metal. Fine lines (precision grinding marks) are visible in the brinelling indentations. An impact or shock overload causes the bearings to indent the race. (17)

truing a shaft. When fabricating a drive shaft, using a dial indicator to measure shaft runout while using a punch to reposition the driveshaft's fittings to ensure the driveshaft has little runout. (18)

turbine. The output member of the hydrodynamic fluid drive. (12)

turnback eye. A wire rope end termination formed by turning back the dead-end wire and securing it in a fitting. Also referred to as a *foldback eye*. (5)

turning brakes. Brakes that can be used individually to help the machine make a tight turn. (19)

turning radius. The radius of the circumference formed by a tractor when it is turning. The circumference is measured from the widest point of the machine's undercarriage while steering. (24)

turn mode switch. The switch used to engage the left or right turning brake on a Komatsu D275AX-5 dozer. (24)

Tweel. The name given to tire manufacturer Michelin's combination wheel and airless tire assembly. (22)

two-speed motor. Motor that can be actuated between two fixed displacements, a high-speed small displacement and a low-speed large displacement. (13)

U

ultimate strength. The average load that causes the sling to fail. (5)

ultrasonic measurement tool. A handheld electronic tool that uses a probe for measuring the thickness of track components to determine the amount of track component wear. (23)

undercarriage. The parts of a track-type machine that support and propel the machine. An undercarriage consists of side frames that house and support drive sprockets and idlers, carrier rollers and track rollers, track tensioners and recoil mechanisms, and track chains and shoe assemblies. (23)

universal joint (U-joint). A component that connects two shafts together and makes it possible for one shaft to transfer rotation to the other through a range of angles. It typically consists of two Y-shaped yokes connected to bearing cups on a cross. Needle rollers inside the cups allow the cross to pivot within the yokes as the shafts rotate. (18)

unloading valve. A valve designed to dump excess pump flow to the tank at a pressure set by the valve's spring pressure value. (19)

unsafe act. Careless action by a worker in which the worker willingly chooses to take risks that are unnecessary. (1)

V

valve stem. A pneumatic valve that contains a removable core used to inflate a tire with compressed air or nitrogen. (22)

variable bale chamber. A round hay baler chamber design that gives farmers the option of changing the bale's diameter while being able to maintain a consistent bale density. (4)

variable-capacity torque converter. A two-member impeller torque converter that can reduce wheel slippage by disengaging the outer impeller to reduce the torque converter's torque multiplication. (12)

variable diameter pulley (VDP) system. Two pulleys, a speed pulley and a tension pulley, that change their width to provide a change of speed. (6)

variable-displacement motor. A hydraulic motor that can vary the effective volume of its chambers to adjust the motor's output speed and torque. (13)

variable-displacement pump. A hydraulic pump that can vary the volume of fluid it pumps, such decreasing the system flow when flow is not needed. (11)

variable effort steering. Steering system that adjusts the turning effort required by the operator to best control the tractor under changing operating conditions, such as requiring more steering wheel turning effort when traveling at road speeds. (25)

variable frequency drive (VFD). The electronic controls employed by manufacturers used to vary the AC frequency sent to the AC motor, which effectively controls the AC motor's output speed and torque. (26)

variable rate planter. A planter in which seed distribution rates can be adjusted. (4)

variable ratio steering. Steering system that optimizes the number of turns the steering wheel must rotate, depending on the tractor's speed and the degree of sharpness of the turn being performed. (25)

variable solenoid control valve. A solenoid valve controlled by an ECM that varies the current to the coil and allows the solenoid to vary the degree to which it opens or closes. Also referred to as a *proportional solenoid control valve*. (11)

variable speed belt. A wide belt with a thin cross-sectional depth. It is used with two variable sheave pulleys that vary the belt's speed. (6)

variator. The variable input in a CVT. Can be a hydraulic HST drive, or a mechanical drive such as a toroidal drive or a variable steel-belt drive. (15)

VarioActive Steer. An electronic steering mode used by Fendt that reduces the number of steering wheel revolutions required to steer full right to full left. (25)

V-belt. A belt with a V wedge–shaped cross section. The belt drives the pulley using a wedging action. (6)

V-belt gauge. A V-shaped plastic gauge into which a V-belt is placed. The size of the belt is determined by viewing the top edge of the belt on the size of the gauge. (6)

vertical lift. A skid steer lift arm design that arranges the hydraulic cylinders and lift arms so the arms move in a very shallow arc. (3)

vibratory compactor. A compactor with a drum that vibrates to help it compact material. (3)

viscosity. The thickness of the oil, also known as the resistance to fluid flow, measured in units of Saybolt Universal Seconds (SUS) or centistokes (cSt). (11)

vortex oil flow. A short cyclical flow of oil that occurs during the torque-multiplication phase of torque converter operation. It is denoted by oil rapidly recirculating from the impeller, to the turbine, to the stator, and back to the impeller. (12)

V-ribbed belt. A thin flat belt formed with multiple V-shaped ribs, tensile cords, and a belt cover. This belt drives based on friction. (6)

V-ripper. A tillage tool that has heavy-duty rippers mounted solidly to a tool bar. It cuts deeper than field cultivators and chisel plows. Also referred to as a *subsoiler*. (4)

W

walking beam suspension. A type of suspension system consisting of an equalizer beam with a center pivot attached by elastomer springs to both the front tandem axle and rear tandem axle on one side of the rear truck frame of an ADT. (21)

Warrington wire rope. A wire rope constructed with large and small diameter wires in an alternating pattern. (5)

washer. A thin piece of metal that is placed under a bolt head or nut to distribute the clamping force of the fastener. (2)

wedges. Tapered cast iron parts that are designed to fit snuggly inside a hub's hole to secure a wheel to an axle. (22)

wedge socket. A wire rope end termination that is used to secure the end of a crane's wire rope to the crane's load block or overhaul ball. (5)

welding. The process of using heat to fuse two pieces of metal. (1)

wet tank. Part of the air brake supply system located directly after the compressor or after the air dryer. Stores the compressed air used to apply the brakes. Also referred to as *air supply tank* or *supply reservoir*. (20)

wheel. A rotating assembly that couples the drive hub (or non-driving hub) to the tire. (22)

wheel chock. Wedge-shaped block inserted in the front and in the rear of a machine's tire to prevent the machine from moving while it is being unloaded or repaired or when it is not in service. (1)

wheel cylinder. The actuating piston that applies the service brakes. Also referred to as a *slave cylinder*. (19)

wheel lean. Motor grader feature that leans the tires in a parallel fashion to the left or right, up to 18° to either side. It is used to compensate for side thrust load. (25)

wheel rake. A type of rake that has several thin ground-driven wheels with tines around the circumference of each wheel. (4)

wheel weights. Cast iron weights bolted to a tractor's wheels to correctly ballast a tractor's weight. (22)

wide-angle SCU. Steering control unit configured to smooth steering motion by increasing the amount of deflection between the SCU's control spool and sleeve during a turn. (25)

wide-sweep blade plow. A tillage tool that is similar to a field cultivator but uses a single row of very wide sweeps. (4)

windrower. An implement or self-propelled mower designed to mow a crop and gather the cut crop into a single swath, or windrow. Also referred to as a *swather*. (4)

windrow pickup header. A combine harvester header or forage harvester header designed to pick up and feed previously cut and windrowed crops into the combine or forage harvester. (4)

wing-type U-joint. U-joint that has bearing cups that are bolted directly to a yoke. (18)

wing-type yoke. A yoke that has four threaded bolts used to connect it to a wing-type U-joint. (18)

winnow. Blow chaff away from the grain. (4)

wire rope clip. A clamp used to fix the loose end of the loop back to the wire rope and create an eyelet. (5)

wire rope lay. The way the wires are laid to form strands and the direction the strands are laid around the core. (5)

wire rope sling. As defined by OSHA, "an assembly that connects the load to the material handling equipment." (5)

working load limit (WLL). The maximum load a sling can safely handle while lifting. (5)

worm drive gear set. A gear set in which a gear with helical-cut teeth, resembling the threads on a large screw, drives another gear. (7)

Y

yield mapping. A method of evaluating yield by superimposing yield monitor data on a colored grid-like map. This allows producers to see areas of the field that have high and low yields. (4)

yield monitoring. A method of calculating the momentary yield by using a sensor to weigh the harvested crop. A controller uses the yield measuring sensor data along with the machine travel data to calculate and display the momentary yield (in bushel per acre). (4)

yield strength. The point at which a component deforms plastically (does not return to its original condition after deformation). (17)

yoke. A Y-shaped component located in a driveline used to connect to a U-joint. (18)

Z

Z-bar linkage. Bucket curl linkage that is arranged in the shape of a *Z*. (3)

Index

540-rpm implements, 140

A

ABC fire extinguishers, 4, 5
ABDS (accessory belt drive system), 197
abrasive wear, 607
AC (alternating current)
    generator. *See* alternator
    induction motors, 891–892
    sine wave, 885
    traction motors, 891–893
    wheel motor driven by, 87
Accelerated Pre-Separation (APS) system, 135
acceleration response control (ARC), 524
accessory belt drive system (ABDS), 197
accidents, workplace, 14–15
accumulators
    bladder, 671–672
    charge with nitrogen, 674–676
    components and functions, 669–671
    diaphragm, 672
    hydraulic fluid, 654, 671–672
    nitrogen gas, 670–671
    piston, 672
    precharge pressure, 670, 673–675
    recoil mechanism, 803
    repair, 676–677
    safety, 672–673
    service and repair, 673
accumulator charging spool valve, 678
accumulator effect valve, 474
accumulator isolation check valve, 678, 679
AccuSteer, 855
acetylene gas
    cylinder colors and icons, 25
    defined, 22
    safety and use, 22–23
acetylene hose, 48
acetylene regulator, 48
Ackerman steering
    defined, 850
    front steerable axle, 855
    linkage components, 850
    manual, 851–852
    power-assist, 852–854
    two-wheel and four-wheel, 851
acorn nuts, 55
ACR (acceleration response control), 524
active air ride seat, 741
ActiveCommand steering system, 867–870
actuating discs, 659–661

Adaptive Hydro-Pneumatic Cab Suspension (HCS) Plus, 760
adaptive learning, 389–390
adhesive wear final drive, 607
ADT. *See* articulated dump trucks
Advanced Stone Protection (ASP) system, 134
AED (automated external defibrillator), 7
AGCO
    air brake systems, 689
    Challenger single rotor combine, 135
    differential-steered twin-track tractors, 832, 833
    Gleaner brand combine, 135–136
    Hesston hay baler, 124
    RoGator sprayers, 150
    White brand row crop planter, 113
agricultural chain, 200
agricultural equipment
    axles and, 568, 615
    chains, 200–202
    combines, 130–138
    differential locks, 570
    fertilizer applications, 151–154
    forage harvesters, 127–130
    harvester final drives, 595–597
    hay balers, 123–127
    machine and implement types, 107
    mowers, 120
    planting and seeding, 112–119
    PTO driveshafts, 637
    rakes, 122–123
    split jack stands, 161
    sprayers, 147–150, 442
    tillage tools, 108–112
    tires, 790–792
    tractors, 138–146, 517, 791
    wheel slip and, 780–781
    wheel-spacing, 787–790
    See also entries for names of models and manufacturers
agricultural power take-offs, 13
agricultural suspension systems, 741–759
    air, 757–759
    four-wheel drive agricultural tractors, 752–754
    GlideRider, Massey Ferguson, 742
    MFWD equipment, 742–752
    oscillating axle with stops, 741–742
    torsion suspension, John Deere, 742
    twin-rubber tractors, 754–757
agricultural tire classifications, 776–778
agricultural utility tractor, 74
air-assisted scraper, 84
air bag, 757–758

air bag suspension, 150
air brake actuator
    cam brakes, 720
    defined, 655, 715
    defining components, 715
    dual service/spring brake chamber, 717–719
    rotochamber, 717
    single service brake chamber, 716
    single wedge, 723
    slack adjusters, 720–721
    twin wedge, 723–724
air brakes
    advantages, 726
    air dryer, 695–698
    air-over-hydraulic, 722, 724–725
    dual brake circuit, 702–707
    foundation, 714–724
    hydraulic brakes v., 726
    moisture,
    single-circuit, off-highway, 708–710
    supply system components, 689–702, 690
    valves, 710–714
air brake valves
    parking brake control, 713–714
    pressure-protection, 714
    quick release, 713
    relay, 710–712
air chamber, 715. *See also* air brake actuator
air chisels, 40
air compressor
    compression cycle, 691–692
    defining functions, 690
    inlet/exhaust valve, 693
    intake cycle, 690–691
    unloader valve and governor, 692–695
air conditioning compressors, 275
aircraft sprayers, 147
AirCushion suspension, 755–757
air disc brakes, 722
air drills, 40
air dryer, 695–698
air gap, 274, 276
airless tires, 774
air-over-hydraulic brake systems, 722, 724–725
air pressure, tires and, 781
air pressure brakes, 644
air ratchets, 40
air ride seat, 740
air seeders, 119
air spring, 757–758

air supply system components
    air compressor, 690–695
    air dryer, 695–698
    alcohol evaporators, 698
    compressor, 690–695
    drain valves, 698–700
    low-pressure switches, 702
    safety valves, 700–701
    supply tank, 689–690, 698, 699
air suspension seats, 740–741
air suspension systems
    Case IH swather, 757–758
    coil spring/shock absorber, 759
    diagnosis, 759
    rotary mower, John Deere, 758
    sprayers, 758–759
air wedge brakes
    defining functions, 722
    single, 723
    twin wedge, 723–724
alcohol evaporator, 698
Allison automatic transmissions
    1000 and 2000 series, 311–312
    1000–3000 series transmission, 287
    1000–4000 series automatic, 310–311
    3000 series, 312, 401, 402
    3700 and 4700 series, 318–320
    4000 series, 312
    6625 Off-Road Series (ORS), 320–321
    applications, 309
    countershaft-style, 309
    dynamic retarder, 429, 430
    forward gears, 314–318
    overdrive gears, 316–318
    powerflow, 1000-4000 series, 312–314
    reverse gears, 318–319
    TC10 series, 327–329
    transmission fluids, 372
    twin-turbine torque converters, 421
    variable stators, 424
all-terrain (AT) crane, 162
all-terrain vehicles, 855
all-wheel steer loader, 91
alternate-lay wire rope, 166
alternating current (AC). See AC (alternating current)
alternator, 193, 197, 884–888
amboid gear set, 218, 219
American National Standards Institute (ANSI), 202
angle, dozer blade, 78
angle of repose, 74
angular load, bearings, 243
angular loading, shouldered eyebolts, 175
anhydrous ammonia, 152, 153

annulus (ring) gear, 217
ANSI roller chain sizes, 924
anti-cavitation check valve, SCUs, 861–862
anti-extrusion valve, 671
antifriction bearings, 242
    service/repair, 246–248
    types, 243–246
anti-slip regulation (ASR), 555
antler rack, 80
A-post display, 522, 523
applicators, 147–150
applicator sprayer pump, 149
apron, 82
apron, scraper, 82
APS. See Accelerated Pre-Separation system (APS); Advanced Stone Protection (APS)
APS SYNFLOW HYBRID, 135
aramid belts, 191
ARC (acceleration response control), 524
ARFF trucks, 855
armature, 274
articulated agricultural tractor, 144, 146
articulated chain trencher, 99
articulated dump trucks (ATDs), 86, 87, 88, 567, 573, 574
    charging suspension, 738
    hydro-pneumatic suspension cylinders, 736–738
    sizes, 921
    suspension systems, 737–738
articulated joint, 736
articulated rubber-track tractors, 594
articulated steering track-type tractors, 843
articulation steering, 81, 853–854
articulation steering locks, 17
ASA (automatic slack adjuster), 721
asbestos, 255
aspect ratio, 775–776
asperities, 610
asphalt compactors, 96
asynchronous motor, 892
ATC (automatic traction control), 570–571
ATD (articulated dump trucks). See articulated dump trucks (ATDs)
Auburn Gear Power Wheel, 662
auger, 83
auto field operation, 333
auto-guidance systems, 128, 138, 146, 856
auto-kickdown mode, 391
automated external defibrillator (AED), 7
automatic belt tensioner, 197
automatic differential lock, 570

automatic drain valves, 700
automatic ride control, 740
automatic slack adjuster (ASA), 721
automatic synchronized transmission, 234–240
automatic traction control (ATC), 570–571
automatic transmissions
    Allison 1000–4000, 310–311
    characteristics, 330
    clutch apply charts, 392
    defined, 329
    See also planetary gear set theory; planetary transmissions
autonomous electric drive, 898, 899
Autonomous Haulage System (AHS), 898, 899
AutoPower, 531–539
auto road operation, 333
Autoshift, 391
auto-shift control, 333, 334
AutoTrac steering valves, 870
average speed per shift, tires and, 783
AWD motor graders and steering, 876–877
axial load, bearings, 243
axial rotor combine, 136
axles
    bar and flange-mount, 615–616
    dead and live, 618
    differentials and, 565–567.
    full and semi-floating, 617
    heavy equipment and, 567–568
    steer and rigid, 618–619

B

back electromotive force (BEMF), 893
backfilling, 78
backfire, 23
backhoe
    boom, 67
    control pattern, 65
    See also excavators
backlash, 248, 580–581
balance driveshaft, 636
bale monitors, in-cab, 126–127
baler belts, 190
bale spears, 142
ball-and-cage joint, 626
ballasting, 790
ballasting tires
    calculators and weight ratios, 791–792
    liquid, 790–791

tools, calculate weight ratios, 791–792
weight-to-horsepower ratios, 792
weight types, 790
ball bearings, 244
ball-peen hammer, 39
ball ramp disc brakes, 659–661
band brakes, 669
banded V-belts, 193
bands, clutch, 276–278
bar axle
    mount type, 615–616
    wheel spacing, adjust, 788–790
barring tool, 427
basket, clutch, 263
basket hitch, 167, 178, 179
bath lubrication, 209
battery-powered compact excavators, 908–909
BDCM (brush-type DC motors), 888–890
beach marks, 607–608
bead, tire, 767
beam torque wrench, 36
bearing cup, 622, 624
bearing plate U-joints, 622
bearings, 242–248
    adjusting, endplay/backlash, 248
    antifriction, 242, 243–246
    ball, 244
    contamination, 248
    friction, 242–243
    functions, 242
    installing, 247
    load, types, 243
    pilot, 255
    press fit bearing races, 246
    release (throw out), 255
    removal, 247
    roller, 245–246
    service and repair, 246–248
    slip fit race, 246
bearing strap U-joints, 622
bed liners, haul trucks, 86
BelAZ dump truck, 87
Bell, articulated dump trucks, 87
Belleville washers, 266, 268
bell housing, 413
belt deflection, 193
belt drive ratios, 200
belt lacings, 190
belt length, 191
belt pulleys, 199
belts, 188–199
    classifications, 189–199
    draper straight cut headers, 131

    flat belts, 189
    installing, zip-tie method, 198
    joined, 192
    operating temperature, 192
    pulley alignment, 199
    round baler belts, 189–190
    round belt, 189
    tension, 193
    timing belts, 190
    variable speed, 194
    V-belts, 190–192
    V-ribbed, 197, 198
BEMF (back electromotive force), 893
Bendix AD-9 air dryer, 696–698, 698
Bendix air compressor, 691–692, 693
Bendix-Weiss U-joint, 626
bent or cracked wheels, tires and, 781
bevel gears, 217–218
    amboid, 218, 219
    hypoid, 218, 219
    spiral, 218
bias-ply tires, 768, 776
bi-directional chains, 207
black sockets, 35
bladder accumulator, 671–672
blades
    dozer, 77–78
    See also motor graders
blade side shift, 80
BLDCM (brushless DC motor), 890
bleed port plug, 490
bleed vent, 490
blind spots, 20
block chain drive, 201
block (cone) synchronizers, 228–231
    components and function, 228–230
    inspecting, 231
    operation sequence, 230
    spring-detented, 230–231
blocker ring, 230, 231
block graph, 487
blocking, 161, 162
Bobcat
    all-wheel steer loader, 91
    compact track loader, suspensions, 739
    T650, 75
body-up reverse neutralizer, 86
bogies, undercarriages, 804
bolt grade strengths, 917–918
bolt-on sprocket, 802
bolts
    grades and grade markings, 54
    parts of, 54
    sizes, 916–917

    threads, 54–55
    uses and types, 53
boom and pillar mining, 69–70
boom bold, 148–149
boom control, 64, 148
boom lateral tilt, 148
booms, 66
boom swing, 68
boost-assist master cylinder, 651–652, 683
Bosch Rexroth
    A4VG hydrostatic transmissions pump, 449, 490
    DA controlled hydrostatic transmission, 450–453, 453
    DA valve plate adjustment, 486–487
    hydrostatic transmission pump, 480–481
    null adjustment, 485
    piston pump, 446
    pressure cutoff valve, 470–471
    single-servo hydrostatic transmission pump, 482
    See also Rexroth entries
bottom-of-travel switch, 382
Bourdon-tube pressure gauge, 43–44, 395
bowl, 82
box blade, 74
box-end wrenches, 33
brake chamber, 715
brake pedal operation, 648–649
brake pressure, 647
brake resistor, 896, 902
brakes
    air (See air brakes)
    components, 643
    electromagnetic, 276
    holding clutches, 264
    hydraulic, 645–653
    off-highway vehicles, 655–669
    ProDrive, 555
    types, 643–645
brake steer, 824
braking chopper, 898
breaker, 768
breaker bar wrench, 34
breakout force, 64
bridge crane, 162, 163
bridge rectifier, 886
bridle, 172–173
bridle lifts, 182
bridle and swivel, 100
brinelling, 610, 611
brittle fracture, 608
broadcast spreaders, 119
brushes, DC motor, 888, 889

brushless DC motor (BLDCM), 890, 893
brush-type DC motors (BDCM), 888–890
buckets, 142
    curl control, 71
    extend or curl control, 64
    excavator, 65
    wheel loader, 72
bull-and-pinion final drive, 590, 595, 597
bull-and-pinion gear set, 217
bullet rotor, combine, 135
bull gear, 217, 589
burns, 9
burst disk, 24
bushings, 201–202, 810–811
bushing turn, 810
button-head screw, 56
bypass filtration, 458, 460
bypass switch, 429

C

cab air suspension, 760
cable-laying operations, 99–100
cab suspension systems, 759–760
cage, ball bearing, 244
calibration
    electronic controls, 382–384
    micrometers and, 49
calibration mode, 382, 516
caliper, 661
caliper disc parking brake, 661
camber, 746–747
cam brakes, 720
cam shifter, 242
Camso SolidAir tires, 774
cam twists, 194
CAN (controller area network), 516
cap screw, 683
carcass, 767
Cardan
    differential, 577
    U-joint, 622, 628
cardiopulmonary resuscitation (CPR) training, 7
carriage assembly, forklift, 92
carrier, 263
carrier roller, 804–805
carrier undercarriage, 799
carry dozer (CD) blade, 78
Case Astra ADT, 574, 575
Case/Case IH
    1620/1640 three-speed sliding gear, 242
    2188 combine, 195
    4WD combination steering, 855, 870, 872, 874

90XT skid steer, 605
Advanced Seed Meter (ASM), 114
AFX rotor, 135
air tank, brakes, 699
brake assembly, 645, 659
bull-and-pinion final drives, 597
caliper brakes, 663
collar-shift combine transmissions, 228
combination steering, 854
combination wet disc brake, 666–669
diagnose weak hydrostatic transmission, 500
dual-path hydrostatic transmission models, 826
Early Riser positive air drum seeder, 114
final drives, 596, 597
four-track tractors, 843
four-wheel drive, 618
friction discs, 662
hydraulically actuated ball ramp disc brakes, 660–661
independent-geared lever-operated track steering system, 841–842
inline integral hydrostatic transmission, 443
IPOR valves, 470
Magnum countershaft powershift transmission, 267, 332–340, 383
master cylinder, 650
Moog Zeus electric/hybrid backhoe loader, 909
motor graders, 81
MX Magnum, 385
parking brake, 644
pressure cut off valve, 470
Quadtrac undercarriage, 818
reducer gear, 597
single-reduction planetary final drive, 602
single rotor combine, 135
sliding gear transmission, 241
sprayer, 151
spring-applied caliper parking brake, 665
Steiger 12-speed powershift, 394, 399
Steiger 4WD tractors, 232
Steiger axles, 592, 594
Steiger Synchroshift transmission, 264, 393–394
swather suspension, 757–758
tapered wedges, 789

three-speed sliding-gear combine transmission, 221–223
Titan Floater air brakes, 689
towing safety, 681–682
transmission hydraulic circuit, E/F series wheel loaders, 373–375
transmission oil flow, 426
case drain filtrations, 458
Case IH combine feeder and rotor CVT, 513–517
Case IH Magnum and New Holland T8 CVT, 518–527
Case IH Magnum countershaft powershift transmission, 332–340
Case New Holland, 521–525, 835
casings, tire, 779
castle nut, 55, 665
Caterpillar
    160M motor grader, 384
    24M mining motor, 308–309
    242D backhoe, split pin synchronizer, 234
    637G pressure sensor, 698
    673K tractor transmissions, 84
    6x6 ATDs, 574
    740B ADT, 737
    777G dump truck, 17
    938G wheel loader, 414
    988K XE wheel loader, 905
    AC mining trucks, 895–897
    advanced ride management (CARM), 741
    all-wheel drive dual-path hydrostatic transmission, 849
    articulating dump trucks, 87
    Bendix compressor, 693
    brake systems, 703
    CAT ET service tool, 395
    chain tension skid steers, 604
    Challenger combine, 135
    clamp master link, 816
    Command Control Steering, 874–876
    continuously variable transmissions operation, 926–927
    cooling internal wet disc brakes, 679–680
    countershaft power transmission, 340–348
    coupling gear, 300
    CX automatic transmission, 381
    D11XE electronic-drive track-type dozer, 908
    D5B track-type tractor, 255
    D5H dozer, 331

D6B single-reduction final drive, 594, 595
D6R dozer, 264, 298–300, 300–303
D6R transmission oil flow, 375–376
D6XE track-type dozer, advantages, 907–908
D7E, 77
D7E electric-drive track-type dozer, advantages, 906–907
differential steer, 436, 832, 834
DOT fluids, 652–653
double reduction final drive assembly, 289
D series, 464, 575
DSH dozer, 77, 632
dual circuit air brakes, 705, 707
dual-path hydrostatic transmission models, 826
dynamic retarders, 429
ECPC transmission control, 387
electric drive dozers, 834
electronic controls, 390
elevated sprocket dozer, 76
elevated-track dozer, three-speed transmissions, 295–297
final drives, 600
Flexport tires, 774
fluid levels, final drive, 603
four-speed planetary transmission, wheel-type, 304–306
haul truck air brakes, 708–709
H, M, and non-suffix series motor grader, 341–348
hybrid mini excavator, 908
hydraulic accumulator, 910
hydro-pneumatic suspension cylinders, 732, 734
ICM system, 386
impeller clutch, 419
integrated tool carrier (IT), 72
internal conductor, 911
K series, 464, 556, 558
manual slack adjuster, 720–721
master cylinder, 652
mining trucks, 86, 331
motor graders, 81
M series motor grader, 331
multiple disc stator clutch, 417
neutralizer valves, 875–876
NVH tools, 631
outboard-double-reduction planetary final drive, 590–591
planetary transmission, 264, 265
pressure-protection valves, 714

Quick Steer, 870, 872
radial grid braking, 896
rigid haul trucks, 87
rope shovel, 909–910
rotating clutch, 264
rotochamber, 717
R1700XE battery electric underground loader, 905
scraper/haul track planetary transmissions, 309, 386
Service Information System (SIS), service literature, 391
single inboard brake disc, 665
steering-clutch-and-brake dozer, 828–832
suspension systems, 738
switched reluctance (SR) drive, 893
SystemOne Undercarriage, 816
TA22 agricultural transmission clutch apply chart, 925
Tetra Gauge, 44, 395
three-speed planetary transmission, wheel-type, 304
TMPH tire monitoring, 783
torque converters, 413
towing safety, 682
track chains, 801
track steering systems, 839
transmission fluids, 372
transmission removal/installation, 404–405
transmission yoke, 624
treadle valves, 709–710
undercarriage classifications, 798–799
upstroke pumps, ECM, 465
variable-capacity torque converter, 419–420, 420
variable-displacement motors/pumps, 443, 464–465
wheel loader, 540–548
cavitation, 445, 861–862
CEMF (counter-electromotive force), 893
center axle through drive, 573
center cam, 572
centering spring, 485
center link pin diameter, 139
center of gravity (COG), 178, 180–182
center punch, 38
center support bearings, 626
center tread idler, 816
center valve, 651–652
centistokes (cSt), 371
centrifugal clutches, 279
centrifugal pump, sprayer, 153

centrifugal sprayer, 147
chain breaker, 208
chain drive
    checking for wear, 206–208
    defined, 81
    pull belt, 512
    roller, 201–202
    tandem, 597–600
    See also Chains
chain-driven bars, 83
chain-driven trenchers, 97–98
chain elongation, 206–207
chain hoists, 162–163
chain links, 201
chain pitch, 202
chains
    block, 205–206
    conveyor, 200
    detachable link, 205
    elongation, 206–207
    installing, 208
    lubricating, 209–210
    offset, 203
    pintle, 205
    ratio formula, sprocket, 210
    rollerless, 204
    silent, 204
    specifications, wear, 207–208
    types, 200
    worn offset chain drive, 204
    See also chain drive
chains, tire, 772, 773
chain slings
    advantages/disadvantages, 173
    inspecting, 173
    ratings and grades, 173
    uses, 172
chain wear gauge, 208
Challenger
    idlers, 817
    MT 800 series, 482, 484
    rubber-track tractor, 146
    suitcase weights, tires, 790, 791
    suspension systems, 757
    twin-track, 818
    undercarriages, 800
change relief valve, 455
charge cycle, air dryer, 696–697
charge pressure, 497–498, 504
charge pressure filtration, 458
charge pump, 444–445, 497, 499
Charlson, Lynn, 856
Char-Lynn, 856
check balls, clutch, 265–266
check valve, 23

chisel plow, 108
chisel plow cultivator, 110
chisels, 38
choker hitch, 167, 178, 179
chopper, 129
chords, 162
chrome sockets, 34, 35
circle drawbar, 79
circle gear box, 80
circle lift cylinders, 80
Claas Lexion combine, 135, 923
clamp master link, 816
clam shell bucket, 72
clean grain auger, 134, 135, 136, 137
clean grain elevator, 137
cleaning, compressed air, 11
cleaning equipment, 46–48
cleaning fan, 194
cleanliness, shop, 430–431
clearance-sensing ASA, 721
clevis pin, 57
clicker torque wrench, 36
closed-center load-sensing SCU, 858–859
closed-center non-load-sensing SCU, 857–858
closed-loop filtration, 458
closed-loop hydrostatic transmissions, 437, 438, 504
closing the farrow, planting, 116
clothing, safety equipment and, 2
cluster gear, 520
clutch disc, 254
clutches
    bands, 277–278
    Belleville washers, 266, 268
    centrifugal, 279
    dual-stage PTO, 257
    electromagnetic, 274–276
    mechanical schematics, 279
    multiple-disc, 263–270
    one-way, 270–274
    planetary controls. *See* planetary gear set
    pressure specifications, 391, 394
    release mechanism, 255
    ring-to-frame, 514–517
    rotating, 264
    slip, 278–279
    sprag, 271, 273, 417
    stationary, 264
    traction, 253–262
clutch fill value, 383
clutch hold values, 383
clutch pack, 265, 402, 403
clutch pedal, 329, 344

CNH. *See* Case New Holland
coal body haul truck, 86
coasting, 479, 502
COG. *See* center of gravity (COG)
cogged V-belt, 192
coiled spring suspension, 150, 151
coiled wire electromagnetic clutches, 274–276
coil spring and shock absorber suspension, 759
collar shift transmissions
    Case HI, 228
    defined, 224
    John Deere 4200 compact utility tractor, 225–227, 268–270
    range driven gear, 225–226
    shift collar, 225
    shift gear, 225
    sliding clutch, 225
    unsynchronized, 226
collector rings, 173
combination brake, 644
combination fracture, final drive, 609
combination master cylinder, 651
combination planetary countershaft powershift transmission
    applications, 348
    forward powerflow, 349
    John Deere 8R 16-speed powershift transmission, 352–367
    reverse powerflow, 350
combination steering, 854–855
combination tool, 108, 111
combination wet disc brake, 667–669
combination wrench, 33, 34
combine cleaning fan VDP, 194
combine harvester, 130–138
    conventional, 138, 155
    corn header, 132, 133
    feeder houses, 133–134
    row crop header, 132–133
    separating, 136
    sizes, 138
    straight cut header (table), 130–131
    stripper header, 132, 133
    threshing classifications, 134–136
    windrow pickup header, 131
    yield monitoring/mapping, 138
combine rotor VDP, 194–196
combines
    classes, 923
    cleaning fan, VDP, 194
    rotor VDP, 194
combustibles
    fire and, 5, 6

    storing safely, 22, 24
Command Control Steering, 874–876
commercial motor vehicle (CMV), 15
commissioning, 489–490
commutation, 889
commutator bar, 889
compact excavators, 68
compact high-lift forklifts, 94
compactors
    application and type, 96
    capacity, 97
    controls, 95–96
    high energy pull-type, 97
    pneumatic, 95–96
    propulsion types, 95
    roller designs, 97
    sizes, 922
    uses, 95
    vibratory, 96
    wheel and drum, 96
compact rubber-track loader, final drive, 601
compact sprayers, 147
compact track loaders, 75, 739
compact utility tractor, 144–145
compensating port, master cylinder, 648
compensation cylinder, 92
compensation valve dual master cylinder, 651
component operation, Caterpillar continuously variable transmissions, 542
compound BDCM, 890
compound planetary gear set, 289, 531–532, 550–551
compressed air systems, 9–11
compression cycle, 691–692
compression section, 191
concaves, 134
conduction bars, 891
cone, 245
cone clutch, 721
cone synchronizers, 228–231
Conrad ball bearing, 244
constant velocity (CV), 627
constant velocity (CV) U-joints, 625–626
construction equipment
    compactors, 95–97, 922
    countershaft powershift transmission, 340–348
    dozers, 79, 594, 779, 920, 834, 906–908
    excavators, 64–70
    forklifts, 92–94
    haul trucks, 85–89, 309, 732–734
    horizontal drills, 101

Index

motor graders, 79–81, 599, 603, 605, 855, 920
overview, 63
retreads, 779
scrapers, 82–85, 309, 738–739
skid steers, 89–91, 603–605
suspension systems, 732–742
telehandlers, 92–93, 618–619
tire classifications, 778–779
tire life, 782
track loaders, 75–76, 919
trenchers, 97–99, 922
types and uses, 63
wheel loaders, 70–74, 899–905, 919
wheel slip and, 780
See also entries for individual topics
contamination control, 430–431
continuously variable transmissions (CVT), 145, 511
   agricultural tractors, 517
   Case IH combine feeder and rotor, 513–517
   Case IH Magnum, 332, 518–528
   Caterpillar wheel loader, 540–545
   Fendt Vario, 528–531
   hydrostatic drives (HST), 555–558
   John Deere IVT, 531–539
   John Deere two-speed automatic ProDrive, 545, 549–555
   mechanical input, hydraulic variator, 512–513
   mechanical variators, 511–512
   New Holland T8, 518–528
continuous PTO clutches, 257
continuous-running power take-off (PTO), 141
controlled flow (CF), 864
controller area network (CAN), 516
control port, air dryer, 696
conventional 10-wheel (end-dump) truck, 87, 88
conventional combine, 138, 155
conventional cylinder threshing, combine, 134–135
conventional single-disc traction clutch, removal/installation, 262
conventional U-joint, 622–629, 633–634
conventional V-belts, 191
conveyance cross auger, 134
conveyor chain, 200
cooler, 153
cooling, filtering hydrostatic transmissions, 458–461
cooling internal wet disc brakes, 679–681
corner post display, 522, 523

corn headers, 128, 132, 133
corrosion wear, final drive, 607, 609
cotter pin, 57
coulter blade, 151
counter-electromotive force (CEMF), 893
counter rotation, 823
countershaft powershift transmission
   Caterpillar H, M, K, and non-suffix series motor grader, 340–348
   combination planetary, 348–367
   John Deere 8R 16-speed, 597–600
   John Deere e23, 348, 351
   power flow, 336–338
countershaft-style transmission, 309, 332–340
counterweights, 69
couples, excavator, 66
coupling clutches, 264
coupling gear, 300
coupling phase, 418
crab steer,] 93, 851
cranes, 162
crawler dump beds, 87
crawler three-stage converters, 423
crawler tractor, 799
crawler undercarriage, 798
creeper control, 465
creeper first forward, 341
creeper powerflow, 340
creeper transmissions, 332
creeping, 496
crimping rollers, 121
crop dusters, 147, 148
Crosby Sling Saver, 176
cross, U-joint, 622, 625
cross-axle differential lock, 568
cross-over relief valves, 456
cross shaft, 564
crowfoot socket, 35, 36
crown, 768
crowned pulley, 189
crushing rollers, 121
crush sleeve, 578
CSS (Customer Support System), 391
cup, 245
curve vanes, 414
cushion blade, 78
cushioned hitch, 85, 738–739
Customer Support System (CSS), 391
cut-and-blow forage harvester, 129
cut-and-throw forage harvester, 129
cut-in pressure, 678, 692–695
cut-out pressure, 678, 692–695
cut-out valve, 698

cutter head knives, 129
cutter knives, 121
cutting crops, 120
cutting edge, 82
CV joints (constant velocity U-joints), 626
CVT. *See* continuously variable transmissions (CVT)
CVT CORP, 512
cycle, sine wave, 885
cycle times, 64, 71–72
cycling transmission, 330–331
cylinder, clutch, 263
cylinder-dampening steering control unit (SCU,) 863
cylinder port relief valve, 861
cylinders, gas, 24–25
cylindrical roller bearing, 245

D

dampening springs, clutch, 254
Danfoss
   Eaton Hydraulics and, 856
   MultiAxis-Steer, 874
   Series-20, 466, 504
   Series-90, 447, 458, 468, 470–473, 485–486
DAYCO, 197, 198
DC traction motors, 888–890
DCV. *See* directional control valve (DCV)
dead axle, 618
dead-blow hammer, 39
dead end, 168–169
dead line, 168–169
deceleration, 455–456
deep-cycle batteries, 94
Deere. *See* John Deere
defibrillator, 7
delivery port check valve, 696, 697
delta zero, 45
Delta Zero pressure meter, 395
depth gauge, 811–812
depth micrometer, 49, 51–52, 485
depth rod, 52
desiccant cartridge, 697, 698
detachable link chain, 201, 205
detent mechanism, 241–242
diagnostic pressure taps, 41–42
diagnostic receptacles (DR), 42
diagnostic software, transmission, 382
Diagnostic Technical Assistance Center (DTAC), 397
diagnostic troubleshooting flowcharts, 391
diagonal cutters, 39
dial calipers, 52, 207

dial indicator, 53
dial torque wrench, 36, 578
diameter formula, 203
diaphragm accumulator, 672
die grinders, 40
differential carrier, 564, 565
differential case, 563, 564, 579
differential pressure gauge, 44–46
differentials
    adjusting, 577–583 (*See also* differentials, adjusting)
    components, 563–564
    diagnosis, 576–577
    functions, 563
    heavy equipment and, 567–568
    inter-axle (IAD), 573–576
    limited-slip differential, 571
    location, 568
    locking, 568–571
    non-locking, 568
    no-spin, 571–572
    operation, 565–567
    repairing, 577–583
differential steering
    advantages, 838
    AGCO, 833
    applications, 839
    Case New Holland, 835
    Caterpillar electric-drive dozers, 834
    components, 76, 77, 832
    Fiat power steering, 835
    hydrostatic steer v., 838
    John Deere twin-rubber-track, 836, 837
    Komatsu hydrostatic steering system (HSS), 836–837
    multiple planetary gear sets and, 289
    operation, 832–833
digital volt ohm meters (DVOMs), 388
diodes, 885–886
direct drive, 220, 286
DirectDrive dual clutch transmission, 234–240
    components and function, 234–235
    dual path module (gear), 236–238
    powertrain power flow, 235
    PowrReverser module, 235–236
    range transmission, 238–240
    speed control lever, 240
direct drive clutch, 551–552
direct-drive powershift transmission, 329–330, 381
directional clutch, 330, 401
directional control valve (DCV), 483, 502, 504, 857

direct mount torque converter, 413
disc brakes
    air, 722
    ball ramp, 659–661
    external caliper, 661–664
    internal wet, 664–669, 679–680, 683
discharge valve, 691, 692
disc harrow, 108–109
disengaged clutch clearance, 267, 399
displacement limiter adjustment, 447
distortion, 609
Ditch Witch SK 750, 76
dog clutch, 572
dog teeth, 223, 225–226, 228, 230–231, 568–569, 572–573
Doosan, 573, 574, 738
DOT master cylinder fluid types, 652–653
double-acting lift cylinders, 94
double Cardan U-joint, 625
double-choker hitch, 178
double-eye web sling, 171
double-flange rollers, 804
double helical gear, 217
double-reduction final drives, 289, 588, 594–596
doubles, seed placement, 115
double V-belt, 192, 193
double-wrap hitches, 178, 179
down force, 115
dozers
    blades, 76–80, 95
    electric drive, 834, 906–908
    loader and wheel tires, 779
    single-reduction drop-axle bull-and-pinion final drives, 594
    sizes, 79, 920
    track-type, 79
    wheel, 79
draft arm, 82, 139
drag link, 850
drain-case flow, 492–495
drain cock, 699
drain valves
    automatic, 700
    functions, 698
    manual, 699, 703
draper belts, 130, 131
drawbar horsepower, 139
drawbar planter frame, 116, 117
drawbars, 79, 84, 143
drawn planters, 116–117
drift punch, 38
drill bit, 99–100, 101
drill fluid, 99–101
drill-head carriage assembly, 217

drill rod, 99–100, 101
drills, 40, 99–101, 117–119
drip lubrication, 209
drive clutches, 572
drivelines
    angles, 626–628
    components, 621
    driveshaft assembly, 622–626
    inspecting, 629
    phases, 628–629
    power-take-off (PTO) driveshaft, 636–638
    repair and maintenance, 629–636
    service brake, 656
    slopes, 632–633
drive-loop legs, 469
driven clutches, 572
drive-pin punch, 38
drive pressure, 499–501
drive pulley, 198
drive segments, 802
driveshaft
    assembly, 622–626
    balancing, 636
    fabrication, 634–635
    fasteners, 199
    lubrication, 636
    securing, 142
    straightening, 635–636
drive sprocket, 201, 802
drive tumbler, 802
droop control, 524
drop-axle, 594
drop box, 334
drop-center rim, 784
drum, clutch, 263
drum brakes, 647, 656–659
dry disc brake assembly, 659–661
dry tank, 698
DTAC (Diagnostic Technical Assistance Center), 397
dual brake circuit, 702–707
dual-disc traction clutch, 255–256
dual lift arms, 91
dual master cylinder, 650, 651
dual-path hydrostatic transmission (HST), 75, 76, 437, 441–439–442
    advantages, 825–826
    diagnosing skid steer, 502–503
    disadvantages, 826
    examples, 826
    steering applications, 839
    track-type tractors and, 825
    wheeled steering and, 849
dual path module transmission, 236–238

dual service/spring brake chamber, 717–719
dual-servo trunnion bearings, 484–485
dual-stage PTO clutches, 257
dual steel disc openers, 113
dual strut independent air suspension, 758
ductile fracture, 608, 609
dump bed, 85, 86
dump bed locks, 17
dump cycle time, 72
dump trucks, 87–89. *See also* haul trucks
Duo-Cone seals, 619–621, 805
duo-servo drum brakes, 657–658
duty cycle scale, 388
dynamic brakes, 643
dynamic braking wheel loader, 901
dynamic load-sensing signal, 866
dynamic retarders, 429–430
dynamometer, 629–630

E

earmuffs, 3
earplugs, 3
earthmoving equipment tires, 770, 779
Eaton Corporation, 856
Eaton Heavy Duty Hydrostatic Transmission Pump and Main Motor Seizing Guide (No. 3-409), 445
Eaton Series-1
    diagnose weak hydrostatic transmission, 500
    dual-servo pump, 485
    dual-servo pump and motor, 464
    dual-servo trunnion, 484–485
    fixed-displacement motor, 466
    hydrostatic transmission pumps, 445, 446, 448, 449, 453, 456–457, 459, 501
    internal pressure override (IPOR) valve, 468–470
    test ports, 501
    two-speed axial piston motor, 461–463
    two-speed hydrostatic transmission motor, 465
    valve assembly numbers, 496
Eaton Ultra-Shift centrifugal clutch, 278
eccentric rod, 484
eccentric screw pin, 486
ECM. *See* electronic control module (ECM)
ejector, 82, 83
ela belts, 198

elastomer suspension cylinder, 736
electrical fires, 4–5
electrical shock, welding and, 25
electrical transmission controls, 380–381
electric clutch, 273. *See* electromagnetic clutches
electric control module (ECM), 331
electric drive dozers, 77, 906–908
electric-drive systems. *See* electric propulsion systems
electric-drive wheel loaders
    applications, 899
    Caterpillar 988K XE, 905
    Hitachi ZW220HYB-5, 904–905
    John Deere 644k/644k-tier electric drive wheel loader, 903–904
    John Deere 944K, 901–903
    Joy Global SR generators, 899–900
    Volvo L25, 905
electric/hybrid excavators, 908–911
electric-powered hydraulic power unit (HPU), 908
electric propulsion systems
    alternators, 884–888
    electric-drive wheel loaders, 899–905
    electric/hybrid excavators, 908–910
    ground-fault/low isolation, 911
    history and applications, 883–884
    mining trucks, 895–899
    reluctance motors, 893–895
    safety, high-voltage and, 910–911
    traction motors, 888–895
    wheel loaders, 900–905
electric tools, 40
electrocution, 25
electrohydraulically controlled transmissions, 331
electro-hydraulic steering control valve (EHi), 874
electromagnetic brakes, 276
electromagnetic clutches, 273–276
electromotive force (EMF), 893
electronically controlled displacement, 473
electronically controlled hydrostatic transmission pumps, 448
electronic anti-stall control, 473
electronic clutch pressure control (ECPC), 387
electronic commutation, 890
electronic control, 381
    advantages, 382–389
    locking differentials, 570
electronic control module (ECM), 115, 382
    adaptive learning, 389–390

    alternator output, 885
    AWD motor grader steering, 877
    bypass switch, 429
    electronic anti-stall control valve, 473
    impeller clutch, 419
    pressure-release solenoids and, 468–470
    reflashing, 390
    swashplate angle sensor, 449
    upstroke pumps, 464–465
    virtual operating speed, 390–391
Electronic Data Link (EDL), 395
electronic fingertip control (FTF) sensors, 828–829
electronic self-leveling, 738
electronic service tools, 383, 395
electronic steering systems, 867–874
Electronic Technical Information Manuals (eTim), 391, 395, 396
Electronic Technician (ET), 395
electronic traction control, 902–903
elevated drives, 594
elevated-sprocket undercarriages, 799–800, 803
elevating scraper, 83
elevator, 136
embolism, 11
emergency preparedness
    case study, 4
    combustibles, 6
    fire safety, 4–6
    first aid, 7
    safety data sheets (SDS), 6–7
    See also entering machines safely; first aid
emergency steering, 859–860
EMF (electromotive force), 893
endless round belt, 189
endless sling (EN), 171–172
endplay, 248, 605
end terminations, 168–170
engine-to-ring (ETR) clutch, 514–517
Enhanced AutoShift (EAS), 391
entering machines safely, 15–16
epicyclic gear set, 217, 285
equalizer bar, 800
equalizer valve, 650, 651
error codes, 391, 394
ethylene propylene diene monomer (EPDM), 197
E tires, 779
ETR (engine-to-ring) clutch, 514–517
Euclid electronic-drive, 884
excavators, 64–70

attachments, 65
backhoe control pattern, 65
compactor attachment, 95
controls, 64–65
counterweights, 69
couples, 66
diesel-powered electronic slew drive, 910
electric/hybrid, 908–910
final drives, 600–601
forestry, 69
hydrostatic transmissions and, 441
ISO control patterns, 65
large mining, electric-drive, 909–910
mining, 69–70
mounting, 68
sizes, 68, 919
types and uses, 66–67
excess flow (EF), 864
exciting the field, 885
exercising the tracks, 816
exhaust diaphragm, 705
exhaust ventilation, 13
exiting machines safely, 15–16
extended-to-rear (XR) undercarriages, 798
external caliper disc brakes, 661, 663
external toothed gear, 216, 575
external vanes, 415
eyebolts, 175
eye hook, 174
eye injuries, 10
eye protection, 2
eyewash station, 7

F

face protection, 2
fall-arrest system, 14
falling object protective structure (FOPS), 19
fall protection, 14
false brinelling, 610, 611
fast adapt, 390
fasteners, 53–57
    bolts, 53–55
    metric bolt grade strengths, 918
    metric bolt sizes, 916
    nut grade markings, 918
    nuts, 55
    pressure tap part number chart, 915
    retainers, 57
    SAE bolt grade strengths, 917
    screws, 56
    studs, 55
    Torx, 56–57
    types and uses, 53
    US customary bolt sizes, 917
    washers, 55–56
fastener sockets, 37
fatigue fracture, 607
Federal Motor Vehicle Safety Standards (FMVSS) 121, 721
feedback link, 448–449
feeder houses, 133–134
feeler gauge, 53, 231
feller buncher excavator platform, 69
Fendt Vario CVT, 528–531, 757, 870
fertilizer
    anhydrous ammonia, 152, 153
    applications, 151–152
    chemicals, 147
    defined, 151
    dry, 154
    liquid manure, 153–154
Fiat power steering, 835
field cultivator, 108, 110
field speed, 461
field windings, 889
filler wire rope, 164–165
filling-notch ball bearing, 244
fill valve, 803
filter bypass valves, 458
final drives, 564
    compact rubber-track loader, 601
    elevated-sprocket track-type tractor double-reduction, 594–595
    elevated-sprocket track-type tractor single-reduction, 594
    excavator, 600–601
    failure analysis, 606–611
    functions/types, 587
    gear reductions, 588
    harvesters, 595–597
    integrated into axle housing, 588–594
    location, 588
    not integrated into axle housing, 594–597
    repair, 604–605
    service, 602–604
    tandem chain, 597–600
finish harrow, 108, 112
fire alarm switches, 4
fire classifications, 5
fire extinguishers, 4–5
fire suppression, 4–5
first aid, 7. *See also* emergency preparedness; safety procedures
first-aid stations, 7
first-order vibration, 629–630
fist grip clip, 169
fixed bale chamber, 125
fixed caliper, 662, 664
fixed-displacement hydraulic pump, 677
fixed-displacement motor, 442–443
fixed-displacement pump, 372, 436, 653
fixed engine speed, 524
fixed pin synchronizers, 231–233
fixed stator, 416–417
flange-mount axle, 615–616
flange-mounted wheel spacing adjustments, 788
flange nut, 55
flap, 766
flapper valve, 652
flare-nut wrench, 33
flashback, 23
flashback arrestor, 24
flat belts, 189
flat cam brake, 720
flat eye (FE) web sling, 171
Flemish eye, 170
flex handle wrench, 34
floating caliper, 663
floor jacks, 160
flowcharts, troubleshooting, 394, 395
flow rate, applicator, 149
flow sharing, 80
fluid coupling, 411–412. *See also* hydrodynamic drives
fluid hazards, 7–9
fluid injection, 7–10, 44
fluid loss, 372
fluid pressure testing equipment, 41–46
fluid reservoir, 650
flushing relief valve, 455
flushing shuttle valve, 455, 456
flywheel clutch, 224, 253–262. *See also* traction clutches
FMVSS (Federal Motor Vehicle Safety Standards), 121, 708, 721
FNR (forward and reverse) lever, 446–448
FNR (forward and reverse) valve, 450
FNRP (forward, neutral, reverse, park) lever, 523
foam-filled tires, 772
foldable ROPS, 18
foldback eye, 170
foot protection, 3
FOPS (falling object protective structure), 19
forage harvesters, 127–130
force measurement, 71
Ford tractor, 851
forestry excavation equipment, 69

forklifts, 92–94
forward, neutral, reverse, park (FNRP) lever, 523
forward neutral reverse (FNR) valve, 450
foundation brakes
    actuators, 715–719
    air disc brakes, 722
    air wedge brakes, 723
    cam brakes, 720
    components, 714–715
    slack adjusters, 720–721
    wedge, 722–724
four-track steering systems, 843–844
four-speed planetary transmission, 304–307
four-track undercarriage tractors, 843
four-wheel drive tractors, 145–146, 550, 554
four-wheel steer, 618–619, 851
fpm (feet per minute), 191
fractures, 607–609
free end, 168–169
free play adjustment, 259–260
freewheeling, 479, 502
frequency, 892
fretting corrosion, 610, 611
friction bearings, 242–243
friction belts, 190–192
friction clutch, 253, 263
friction discs, 263
friction plate synchronizers, 234
front-end loaders, 142–143. *See also* wheel loaders
front suspension, 732–733
front-wheel steer, 81, 86–87, 851
F tires, 778
fulcrum pivot pin, 705
full-floating axle, 617
full-round yoke, 622
full suspension switch, 738
fully articulated static tipping load, 73
fungicides, 147
Funk transmission, 382–383, 390, 413
furrow, 113–116
fusible plug, 24

G

gang of discs, 108–109
gantry crane, 162, 163
gas accumulators, 670–677
    bladder accumulator, 671–672
    defined, 670
    diaphragm accumulator, 672
    piston accumulator, 672

    safety, 672–673
    servicing and repairing, 673–677
Gates, belt manufacturer, 197–199
gauge wheels, 113
gear ratios
    direct drive, 220
    overdrive, 220
    simple planetary gear set, 287–288
    torque multiplication/speed reduction, 219–220
gear reductions, 588
gears
    advantages/disadvantages, 221
    bevel, 217–218
    bull-and-pinion, 217
    external and internal toothed, 217
    functions, 215
    helical, 216
    herringbone, 217
    manual transmissions, 221–240
    planetary gear sets, 217
    rack-and-pinion gear set, 217, 218
    spiral bevel, 218, 219
    spur, 216, 218
    tooth contact pattern, 581–583
    worm drive, 218–219
gear sets, 289–291, 293–295
General Electric (GE)
    electric drive, 883, 898
    wheel motor, 87
generators, 884–888
gerotor assembly, 859, 860
glasses, safety, 2
global positioning satellite (GPS), 128
    fertilizer application and, 151
    forage harvester and, 128
    planters and, 115–116
    sprayers, 151
    yield monitoring, 138
gloves, safety, 2, 21, 46, 47, 152, 910–911
goggles, safety, 48, 152
Goodyear tires, 778
governor, 692–695
GPS. *See* global positioning satellite (GPS)
graders. *See* motor graders
graduating rubber spring, 703–706
grain drills, 118–119
grain tank, 137
grapple, 65
grease fitting, 209
grinders, 40
grommet sling, 171
ground-drive pump (GDP), 354–355
ground fault, 911
ground pressure, 797

ground speed matching, 331
ground-speed power take-off (PTO), 141–142
ground speed radar, 780
grousers, 805–807
G tires, 779
guards, machine, 12, 805
guides, undercarriage, 805

H

half-round yoke, 622
hammer excavator, 65
hammers, 39
handheld locators, 101
handle-bar torque multiplier, 36–37
hand tools
    chisels, 38
    defined, 32
    fastener sockets, 37
    hammers, 39
    mallets, 39
    punches, 37–38
    safety and, 31–32
    screwdrivers, 37
    wrenches, 32–37
    See also entries for individual tools
hanger bearings, 626
hard bar, 799–800
hard hats, 3, 46
harvester final drives, 595–597
harvesters, forage, 127–130
hauling transmission, 330, 331
haul trucks
    applications, 86
    articulated dump trucks (ADTs), 87
    bed types, 86
    capacity, 87
    controls, 86
    crawler dump beds, 87
    defined, 85
    dump bed (hoist), 85
    front suspension, 732–733
    loader capacity, 74
    on-highway dump trucks, 87–89
    planetary transmissions, 309
    rear suspension, rigid, 733–734
    rigid-frame, 86–87, 732–733
    single-circuit air brakes and, 708–709
hay balers
    controls, 127
    defined, 123
    in-cab bale monitors, 126–127
    inline square, 124–125
    knotting assembly, 124

lubrication systems, 127
pickup, 125
round, 125–126
square, small/large, 123–124
stuffing fork, 125
twin v. net wrap, 126
hay grapple, 94
hazardous waste disposal, 47
headland management system, 570
headphones, 2
heaped capacity, 74
hearing protection, 2–3
heat stain, 606
Heavy Duty Hydrostatic Transmission Pump and Motor Sizing Guide, 445
heel ring gear tooth, 582
helical cut teeth, 218
helical gears, 216, 218
herbicides, 147
herringbone gears, 217
high-clearance tractor, 145, 146
high-efficiency particulate air (HEPA), 431
high-pressure relief valves, 456–457
high-pressure washers, 46
high-track undercarriages, 803
Hitachi
    AC traction drive, 892
    ECX1900E-6 through EX800-6 electronic-drive excavators, 909–910
    EH 4000 AC 3, 897–898
    electric-drive excavators, 908
    Hitachi ZW220HYB-5 wheel loader, 904–905
    rigid haul trucks, 87
    ZH210-5 hybrid excavator, 910
hitches, 178
hoist, 86
hoist hydraulic control valve, 679–680
hoist swivel rings, 175, 176
holding clutches, 264
hold-off brake chamber, 717
holdout ring, 572
Honda, PC01E electric-drive micro excavator, 909
Honey Bee combine header, 190
Hooke U-joint, 622, 628
hooks, 174–175, 181
hopper level sensors, 116
horizontal directional drills (HDDs), 99–101
horizontal I-beam crane, 162
horizontal oil level mark, 603
horizontal sling angle, 176–177

horsepower
    ballasting tires, 792
    drawbar, 139
    utility tractor, 145
horsepower management, 90
hot-oil purge valve, 455, 495–496
house swing excavator control, 64
housing, 264
HPU (hydraulic power unit), 908
HST. *See* hydrostatic transmission (HST)
H steer pattern, 90
hub, clutch, 263
hybrid compact excavator, 908
hybrid seed, 116
HydraCushion, 743, 752–755
hydraulically actuated caliper disc brakes, 681–682
hydraulically actuated ejector, 82
hydraulic box blade, 74
hydraulic brakes, 645
    accumulators, 669–677
    adjusting internal wet disc, 683
    air brakes v., 726
    air-over, 722, 724–725
    ball ramp disc, 659–661
    band, 669
    circuits and operation, 677–679
    control valves, service brake, 653–654
    cooling internal wet disc brakes, 679–680
    external caliper disc, 661–664
    hydraulically actuated caliper disc, 681–682
    internal wet disc, 664–669
    master cylinder, 647–653
    off-highway vehicles, 655–669
    slack adjusters, 655–656
    swept area, 669
    towing and, 681–682
hydraulic clutch control valve, 381
hydraulic cutterhead, 69
hydraulic external caliper disc brakes, 681–683
hydraulic fluid, 671–672
hydraulic front-loading shovels, 69
hydraulic input, Case IH, 515–516
hydraulic oil, 6
hydraulic power unit (HPU), 908
hydraulic press, 41, 398, 610, 809, 811
hydraulic pressure brakes, 644
hydraulic principles, 645–647
hydraulic pumps, 84, 353, 373, 826
hydraulic retarder, 429
hydraulic shock absorber, 151

hydraulic shovel, 70
hydraulic slack adjuster, 655, 656
hydraulic sprayers, 147
hydraulic steering, 863–869
hydraulic suspension, 150
hydraulic system pressure, 8–9
hydraulic telescoping crane, 162, 163
hydraulic tools, 41
hydraulic variator, 512–513
hydrodynamic drives, 411–434
    components and function, 411–412
    contamination control, 430–431
    defined, 435
    dynamic retarders, 429–430
    one-way clutches, 270
    torque converters, 412–430 (*See also* torque converters)
hydrophobic, 653
hydro-pneumatic accumulators. *See* gas accumulators
hydro-pneumatic suspension cylinders, 732–736
hydroscopic fluids, 652
hydrostatic, 435
hydrostatic drive diagnostics
    charge pressure, too high, 504
    coasting/freewheeling, 502
    creeping transmission, 496
    directional control valve effects, 504
    dual-path skid steer, 502–503
    low/sluggish power, 496–501
    not moving, either direction, 502–503
    overheating, 491–496
    plugged directional control valve orifice, 502
    pump isolation, 504
    weak power, 500
hydrostatic drives
    accumulator effect, 474
    advantages, 443
    Caterpillar K series wheel loader, 556, 558
    charge pump, 444–445
    components and function, 411–412, 435
    deceleration, 455–456
    diagnostics (*See* hydrostatic drive diagnostics)
    disadvantages, 443
    dual-path, 437, 441–439–442
    electronically controlled displacement valve, 473
    electronically controlled pumps, 448
    electronic anti-stall control valve, 473
    electronic pressure-release solenoid, 468

excavators and, 441
feedback link, 448–449
filtering and cooling, 458–461
fixed-displacement motor, 436, 442–443
flushing valves, 455
high-pressure relieve valve, 456–457
inching valve, 467
infinitely variable/variable-displacement pumps/motors, 464
integral, 443, 444
internal pressure override (IPOR) valve, 468–470
John Deere dual motor ProDrive, 556, 557
manual bypass valve, 468
manually controlled pumps, 446–448
multi-function valves, 457–458
oil flow, 453–454
oil flowchart, 457
open and closed loop, 436–437
over-speed limit valve, 473–474
piston pump frames, 445–446
pressure cutoff valve, 470–471
pump controls, 446–449
reverse propulsion, 458
servo oil, 453–454
single-path, 437, 441–442
single-speed, fixed-displacement motor, 466
speed-sensing valve, 450–453
swathers, 439–441
two-speed motor, uses, 461–464
variable-displacement motors, 436, 442–443
hydrostatic drive service
centering adjustments, 479–480
coasting/freewheel, 479
manual bypass valve, opening, 485–486
safety, 479
shaft run-out, 488–489
single-servo cradle bearing pumps, 480–484
startup, 489–490
test stand, using, 504–505
uncommanded actuation, 484
valve plate adjustment, 486–487
hydrostatic module, 532, 533
hydrostatic steering motor, 838–839
hydrostatic steering systems (HSS), 836–837
hydrostatic steering terminology, 839
hydrostatic transmission (HST), 150

directional control valve, 446–448
differential steer, hydrostatic steer v., 838
inching valve, combination master cylinder, 651
John Deere, 270
test stand, using, 504–505
See also hydrostatic drives
Hydrotechnik fittings, 42
hygroscopic, 652
hypoid gear set, 218, 219

I

IAD (inter-axle differential), 573–576
I-beam crane, 162
ICM (Individual Clutch Modulation), 385, 386
idler flange, 803
idler rib, 803
idlers
functions and components, 802–803
inspection, 811–812
IGBT (insulated-gate bipolar transistors), 894
impact compactors, 97
impact fracture, 608
impact sockets, 34
impact wrenches, 40
impeller
functions, 411, 412, 414
multiple, 419–420
impeller clutch, 419
implement locks, 16, 17
implement pump, 512
implements
driveshafts, 142
loaders, 142–143
PTOs and, 140
improved plow steel (IPS), 166
inboard final drives, 588–589
inboard hydraulically actuated wet disc brakes, 665
inch dial calipers, 52
inching pedal, 329
inching pedal potentiometer, 381–382
inching valve, 467, 651
included angle, 175
independent-geared lever-operated steering, 841–843
Independent Link Suspension (ILS), 746–752
independent power take-off (PTO(, 141, 257
independent wire rope core (IWRC), 164,

165
Individual Clutch Modulation (ICM), 385, 386
industrial loader, 74
inertia compactor, 97–98
infinitely variable motors, 464
infinitely variable pumps, 465
infinitely variable transmission (IVT), 511
John Deere AutoPower, 531–539
See also continuously variable transmissions (CVT)
injuries. *See* safety practices
inlet check valves, 703, 860
inlet/exhaust valve, 693
inlet relief valve, 860–861
inlet valve, 690–691
inline square baler, 124–125
inline vertical hitches, 178
inner liner, 767
inner tube tire, 766
in-phase driveline, 628–629
input clutches, 264, 330
input pulley, 194, 200
insecticides, 147
inserts, 229, 230
inside micrometer, 49, 51
insulated-gate bipolar transistors (IGBT), 894
intake cycle, 690–691
integral differential carrier, 564
integral hydrostatic transmission, 519
integrated final drives, 588–594
IntelliAxle, 742
inter-axle differential (IAD), 573–576
interlock mechanism, 242
internal pressure override (IPOR) valve, 468–470
internal purge air dryer, 698
internal wet disc brakes, 664–669, 683
International Fluid Power Society (IFPS), 9, 489
International Standards Organization (ISO), 371
inverse shuttle valve, 678, 679
inverted basket hitch, 178, 180
involute curvature, 216
IPOR (internal pressure override) valve, 468–470
ISO excavator control pattern, 65
ISO skid steer pattern, 90
I tires, 778
IVT. *See* infinitely variable transmission

J

jacks, 160
jam nut, 55, 683
jaw clutch differential lock, 568–569, 572
JCB manufacturers, 71
    E-TECH battery-power mini excavator, 908
    FastTrac agricultural equipment, 618–619
    single lift arm skid steer loaders, 91
jib crane, 162, 163
job hazard analysis (JHA), 14
John Deere
    16-speed transmission, 8R16, 330, 352–367
    19-speed planetary transmissions, 322
    4005 compact utility tractor, 141
    4200, 4300, 4400, transmission types, 268–270
    4200 compact utility tractor, collar shift, 225–227
    644k/644k-tier electric drive wheel loaders, 903–904
    6R series, 570
    764 articulated four-track dozer, 79
    8000 Series MFWD tractor, 590
    8R IVT ProDrive (AutoPower), 531–539
    8R series tractor, 352
    944K electric drive wheel loader, 901–903
    9600 final drive, 597
    9770 STS combine, 196
    9R series, 589
    ActiveCommand steering system, 867–870
    Adaptive Hydro-pneumatic Cab Suspension (HCS) Plus, 760
    air brake systems, 689
    Allison twin-turbine torque converters, 421
    applicator, 153
    articulated dump trucks, 87
    axial rotor combine, 135
    axle assembly, 8000 series, 568, 580
    bullet rotor, 135
    Central Commodity System (CCS), 114
    combine classes, 138
    compact utility, 145
    CTS (Cylinder Tine Separation), 135
    DBI 20 48row30 planter, 113
    Diagnostic Technical Assistance Center (DTAC), 397
    differential steer, 436
    differential-steered twin-track tractors, 832
    DirectDrive transmission, 234–240
    drive wheel, 818
    drop axles, 594
    drum brakes, 657, 658
    dual motor ProDrive, 556, 557
    dual-path hydrostatic transmission models, 826
    dynamic braking, ED wheel loader, 901
    e23 countershaft transmission, 348, 351, 392
    EDL electronic data link, 395
    E-II series ADTs, 737
    electric drive, 944K, 567
    Electronic Data Link (EDL), 395
    ExactEmerge planters, 114, 115
    external caliper brakes, 663
    final drives, 597
    four-track undercarriages, 843–844
    friction plate synchronizers, 234
    front-wheel steering sprayer, 150
    HydraCushion suspension, 752–754
    Independent Link Suspension (ILS), 746–752
    integral hydrostatic transmission, 443
    IntelliAxle suspension, 742
    IPOR valves, 470
    load backhoe, 909
    LoadMatch, 473
    MFWD clutch, 8000 series, 270, 403
    motor graders, 81
    Nidec electric motors, 903
    PermaClutch, 258–259, 262
    Powershift transmission, 259
    pressure taps, 42
    Quad-Range transmissions, 235, 259
    rack-and-pinion gear sets, 217
    rear bar axle, 616
    right hand service brake, 665
    rotary mower suspension, 758
    rubber belt VDPs, 511–512
    S670, 597
    selective control valve, 856
    self-propelled forage harvester, 129
    Service Advisor, literature, 391, 395
    sprayer models, 147
    sprayer suspension, 758–759
    spring-applied parking brakes, 666
    steer shaft, 618
    STS axial rotor combine, 136
    STS combine, 195, 196
    suspensions, comparing, 738
    switched reluctance (SR) drive, 893
    Triple Link Suspensions, 742–746
    twin-rubber-track differential steering, 836, 837
    two-speed automatic ProDrive CVT, 545, 549–555
    variable-displacement motors/pumps, 443
    VDP system, 196
    walking beam suspension, AirCushion, 755–756, 757
    8R series tractor,
    wedges, wheel spacing, 789, 790
    X9 1000 and 1100 combines, 136, 464
joined belts, 192, 193
Joint Industry Council 37° flare fitting (JIC), 42
jounce, 733
Joy Global, 884
    Kinetic Energy Storage System (KESS), 900
    SR generator, 900
    switched reluctance (SR) drive, 894
    switched reluctance (SR) electric-drive wheel loader, 899–900
    See also Komatsu
joystick controls, 64, 65, 834

K

Kato battery-powered compact excavators, 908
Kawasaki wheel loader, 701
Kent-Moore (OTC) tools, 631
KESS (Kinetic Energy Storage System), 900
keys, synchronizer hub, 229, 230
keyway, 480
kidney-loop filtration cart, 429, 430
Kinetic Energy Storage System (KESS), 900
kingpin, 732–733
knife, fertilizer, 151
knife driver, 120–121
knuckle boom loaders, 69
Kobelco hybrid excavator, 910
Komatsu
    930E AC motor, 898
    AC traction drive, 892
    articulated dump trucks, 87
    Autonomous Haulage System (AHS), 898, 899
    Customer Support System (CSS), 391
    differential-steered twin-track tractors, 832

drop-axle dozers, 594
HB 215LC-A hybrid excavator, 910
hydro-pneumatic suspension cylinders, 732
hydrostatic steering system, 838
Joy Global and, 884
motor graders, 81
onboard monitors, 395
PC01E electric-drive micro excavator, 909
rigid haul trucks, 87
steering-clutch-and-brake systems, 832
suspensions, comparing, 738
switched reluctance (SR) drive, 894
See also Joy Global

L

lacings, 162
laminated tires, 774
landfill compactor, 96
Landoll grain drill, 118
landscape tractor, 74
Lang-lay wire rope, 166
lattice boom crane, 162, 163
lay length, 167
lazy axle, 618
leading-trailing shoe drum brake, 657
leaf spring suspension, 731, 732
LeeBoy motor graders, 81
left-hand-lay wire rope, 166
LeTourneau, Robert, 884
LGP (low-ground-pressure undercarriages), 798
Liebherr
    AC traction drive, 892
    drop axle dozers, 594
    dual-path hydrostatic transmission models, 826
    electric-drive travel motor, 892
    hydro-pneumatic suspension cylinders, 732
    Information and Documentation Services (LIDOS), 391
    LTM 1750-9.1 mobile crane, 567
    rigid haul trucks, 87
lift capacity, 74
lift cylinder, 93–94
lifting
    dozer blade and, 77
    equipment (*See* lifting equipment)
    by hand, 159
    principles, 176–182
    planning a lift, 182
    skid steer loaders and, 91

lifting triangle, 177
lightbar, 151
light-duty V-belt, 192
lights, 40
limited-slip differential, 571
linear variable differential transformer (LVDT), 874
liners, 86
line wrench, 33
linkage systems, 850
link height, 801
link pitch, 813
lip piston seal, 266, 267
liquid ballast, 790–791
liquid chemicals, 147
liquid fertilizer, 151
liquid manure, 153–154
live axle, 618
live end, 168, 169
live line, 168, 169
load angle factor, 177
loaders. *See* wheel loaders
load factor, 180
load index values, 770
load-reaction steering control unit (SCU), 862
loads
    average tire, 783
    bearings, 243
    machine weight and, 68
load-sensing line steering priority valve, 866
load-sensing relief valve, 861, 866
load-sensing steering control unit (SCU), 858–859
lockable differential, 81, 554–555
locking differentials, 568–571
locking inter-axle differential, 568
lock-out, tag-out (LOTO), 12
lockout lever, 17
lock pins, 17
lock screw, 484
lockup clutch, 418–419
lockup phase, 418
locomotives
    AC induction drives, 892
    BDCM drives, 889, 890
    brake resistors, 902
    DC traction motor, 888, 889
    Nidec, 903
long-reach excavator, 67
longwall mining, 70, 71
loss-time accident, 15
low ground-pressure (LGP) undercarriage, 798

low-pressure flushing relief, 455
low-pressure switches, 702
low sidewall (LSW) tires, 778
LS tires, 779
LSW (low sidewall tires), 778
L tires, construction machinery, 779
lube passage, 265
lubrication
    chains, 209–210
    clutch, 258, 265
    driveshaft, 636
LVDT (linear variable differential transformer), 874

M

machine safety
    articulation steering locks, 17
    blind spots, 20
    capacities and limitations, 19
    compartments, stay out of, 19
    dump bed locks, 17
    entry and exit, 15–16
    falling object protective structure (FOPS), 19
    overhead power lines, 20
    riders, seating and, 21
    rollover protective structure (ROPS), 17–18, 76
    seat belts, 19
    starting procedure, 20
    towing large implements, 21
    unsafe acts and, 15
    wheel chocks and, 15
magnetic poles, 889
magnetorheological fluids, 740–741
main piston pumps, 444–446
main pressure, 376–377
make-up valve, 501, 504
mallets, 39
manifolds, 153
Manitou steering, 854
man-machine interface (MMI), 874
manual, manufacturer, 11
manual bypass valve, 468, 485–486
manual displacement control (MDC), 485
manual slack adjuster, 720–721
manual steering, 851–852
    check valves, SCU, 860
    control unit (SCU), 856
    mode, SCUs, 860
manual transmissions
    bearings, 242–248
    collar shift, 224–228
    defined, 221

detent mechanism, 241–242
interlock mechanism, 242
shift forks, 241–242
shift rails, 241–242
sliding gear, 221–224
synchronized, 228–240
unsynchronized, 226
manufacturer solutions groups, 396–397
margin pressure, 44
markers, 116, 117, 151
Massey Ferguson
axial rotor combine, 135
Delta combine, 135
GlideRider suspension, 742
master clutch, 330
master cylinder brakes, 647–653
boost-assist, 651–652
combination, hydrostatic transmission inching valve and, 651
compensation valve, 651
fluid types, 652–653
operations, 648–649
types of, 649–651
master links
chain, 173, 175, 178, 809
fitting, 173
master pin, 809
MDC (manual displacement control), 485
measuring tools, 49–53
dial calipers, 52
dial indicator, 53
feeler gauge, 53
micrometers, 49–52
mechanical controls, powershift transmissions, 378–380
mechanical diode, 271, 274
mechanical front-wheel-drive (MFWD) clutches, 268, 269
Case IH Magnum, 334
inboard hydraulically actuated wet disc, 665
Independent Link Suspension (ILS) John Deere, 746–752
John Deere 8R, 356
oscillating axle stops, 741–742
Triple Link Suspensions, John Deere, 743–746
mechanical front-wheel-drive (MFWD) tractors, 83, 145
mechanical input hydraulic variator (CVT), 512–513
mechanical leverage, brakes, 644
mechanical locking differentials, 568–569
mechanically actuated brakes, 644–645

mechanically applied external caliper disc brakes, 663
mechanical schematics
clutches, 279
diagnostics, 394–396
mechanical schematic symbols, 278
mechanical tachometer, 630
mechanical turnbuckle, 77
mechanical variators, continuously variable transmissions, 511–512
media blasting equipment, 47–48
metal cutting, 48–49
metric bolt grade strengths, 918
metric bolt sizes, 916
metric micrometer, 49–50
MFH (multifunction handle), 521–522
MFWD clutches. See mechanical front-wheel-drive (MFWD) clutches
Michelin Tweel tires, 774
MICO Quadrigage, 44, 395
micrometers, 49–52
midship support bearings, 626
Miner Elastomer Products Corporation, 736
Mine Safety and Health Administration (MSHA), 2, 15
mini excavators, 68
Minimess test points, 42
mining equipment
Caterpillar AC mining truck, 895–897
elastomer-style suspension cylinder, 736
electric-drive excavators/rope shovels, 909–910
haul trucks, 85–86, 88, 921
Hitatchi EH 4000 AC 3, 897–898
hydro-pneumatic suspension cylinders, 733
Komatsu 930E GE electronic dive, 898
Komatsu autonomous truck, 898, 899
motor graders, 81
servicing, 734–736
suspension systems, 732–742
towing safety, 682
trolley assist electric-powered, 899
wheel loaders, 72
mini skid steers, 75
MIN/MAX pressure, 45, 395
MOCO (mower conditioner) swather, 121
modulation, electronic control, 384–389
moldboard plow, 108
moldboard-shaped blade, 79
motor graders, 599
applications, 80
AWD and steering, 876–877

chain drive fill levels, 603
combination steering, 855
controls, 80, 81
drive types/steering, 81
hydraulic functions, 80
replace drive chains, 605
sizes, 81, 920
types and uses, 79
mousing wires, 174
mower conditioner (MOCO) swather, 121, 122
mowers, 120
MT 845 Challenger, 387
multi-axle steering, 855
multifunction handle (MFH), 521–522
multi-function valves, 457–458, 485–486
multi-groove adjustable pliers, 39
multi-grouser, 805, 806
multi-hybrid planter, 116
multilink chain, 204
multi-piece rims, 784–787
multiple-disc clutches, 263–270
advantages, 265
Case IH Magnum transmission, 267
Case IH Steiger Synchroshift transmission, 264
components, 263
differential lock, 570
off-highway countershaft, 327–329
oil applied engagement, 265–268
operation, 265
service and repair, 397–403
spring-applied manually released, 268–270
spring-applied oil-released engagement, 268
stator clutch, 417
uses, 264
multiple planetary gear sets, types, 289
multiple strand roller chain, 203
multi-rib belt, 197
multi-V belt, 197
Multi Velocity Program (MVP), 391
mushroomed head punch, 38

N

narrow V-belt, 192
National Fire Protection Agency (NFPA), 25
national pipe taper (NPT), 42
needle nose pliers, 39
needle roller bearing, 245, 623
neoprene, 197
neutralizer valves, 875–876
neutral relief value, 455

neutral safety switch, 448
neutral steer, 824
New Holland
    Advanced Stone Protection (ASP), 134
    chain drive continuously variable transmissions s, 512
    inline integral hydrostatic transmission, 443
    parking brake adjustment, 682
    round balers, 126
    telehandler brakes, 683
    twin axial rotors, 136
nitrogen
    accumulator brake, 670–671, 674–676, 803
    anhydrous ammonia, 152
    hydro-pneumatic suspension cylinders, 735–736
    service tool, 673–674, 675
nitrogen-filled tires, 771, 772
nitrogen-over-oil cylinders, 732–736
noise, hearing protection and, 2–3, 22
noise reduction ratings (NRR), 3
non-caged needle rollers, 245
non-directional corn header, 128
non-elevated undercarriages, 800
non-filling-notch ball bearings, 244
non-load reaction steering control unit, 862
non-load-sensing steering control unit, 857–858
non-locking differential, 568
non-loss time accident, 15
non-pneumatic tires, 773–774
non-positive-displacement pump, 147
nonrecordable accident, 15
nonservo drum brake, 657, 658
non-shouldered machinery eyebolt, 175
nonuniform velocity, 628
normally applied clutches, 268
no-spin differentials, 571–572
no-spin spider, 571–572
nozzles, 149, 151
nurse cart, 152
nut grade markings, 918
nuts, 55

O

Occupational Safety and Health Administration (OSHA)
    compressed air and, 11
    defined, 2
    gas cylinders and, 25
    personal protective equipment regulations, 2–3
    seat belts and, 19
    sling angles and, 177
    tracking accidents and incidents, 14–15
    wheel chocks and, 15
off-highway
    air brakes and, 689
    axles, 567, 590, 615. *See also* axles
    brake systems, 655–669, 703
    clutch pedals, 329
    continuously variable transmissions, 545
    countershaft multiple disc clutch transmissions, 327–329
    fixed-displacement motor, 446
    hydrostatic/hydrodynamic drives and (*See* hydrostatic drives; hydrodynamic drives)
    rubber track type, 817
    single-circuit air brakes, 708–710
    tandem chain drives, 597–600
    transmission applications, 329–331
    variable-displacement motors and, 442, 443
off-line kidney-loop filtration, 458, 461
offset chain, 203, 204
offset wheel spacing, 787–788
offset wrench, 33, 34
off-the-road (OTR) tires, 765
Ohm's law, 388
oil-applied clutches, 265–268
oil contamination, 498–499
oil cooled PTO clutch, 258–259
oil cooler, 491–492
oil flow
    hydrostatic transmissions, 453–454, 457
    rotary, 418
    torque converters, 426
    transmission, 376–377
    vortex, 416
oil-pneumatic hydraulic cylinders, 732–736
one-way check valve, 691, 692
one-way clutches, 270–274
on/off solenoids, 385
open bowl scraper, 82, 83
open-center grouser, 805, 806
open-center steering control unit, 857
open differential, 568
open-end wrenches, 32
opening the furrow, 113
open-loop hydrostatic transmission, 436–437
operating levers, 258–262

operating load, 73
operating pressure, 376–377, 692–695
original equipment manufacturer (OEM), 42
O-ring belt, 189
O-ring boss (ORB), 42
O-ring face seal (ORF), 42
O-ring piston seal, 266, 267
oscillating axle, 741–742
oscillation joint, 736
oscilloscope, 631
OSHA. *See* Occupational Safety and Health Administration (OSHA)
OTC electronic digital pressure meter, 44–46
OTG (output transfer gears), 573
outboard final drives, 590–593
outboard hydraulically actuated internal wet disc service brakes, 666
out-of-phase driveline, 628–629
output clutch, 330
output force, 645–647
output pulley, 194, 196, 200
output shaft, 228
output transfer gears (OTG), 573
outside micrometer, 49
overdrive, 220, 316–318, 341
overhead power lines, 20
overhead shop doors, 13
overheating, 491–496
    drain-case flow, 492–495
    hot-oil purge valve, 495–496
    hydraulic oil, 493
    oil temperature, 491–492
overload failure, 607
overrunning clutch, 270–274
overrunning stator, 417
over-speed limit valve, 473–474
oxyacetylene torch, 48, 49
oxyacetylene welding, 22–25
    acetylene gas, 22–23
    electrical shock, 25
    oxygen, 23
    safety, 23–25
oxygen cylinders
    colors and icons of, 25
    storage and transport, 24
    See also cylinders, welding safety and
oxygen hose, 48
oxygen regulator, 48

P

padfoot roller design, 97
pads, rubber track, 805–808

panhard rod, 744
parallel linkage, 72, 73, 76
parallelogram linkage, 850
parallel rake, 122, 123
parallel yokes, 627
parking brake, 644
    adjusting New Holland, 682
    control valve, 713–714
    dual service/spring chamber, 717–719
    rotor, 663
    single chamber, 716–717
parts washer, 46–47
Pascal's law, 645
passive air ride seat, 740
passive deceleration, 517
PAT (Power Angle Tilt) dozer, 78
pawl wrench, 33
payload rigid frame dump truck, 87
PermaClutch, 258–259, 262
permanent magnet
    brushless DC motor, 890, 893
    synchronous motor, 892–893
permanent magnet (PMAC) motor, 892–893
personal protective equipment (PPE)
    anhydrous ammonia and, 152
    bearing installation, 248
    changing multi-rim tires and, 786
    electrocution safety and, 25
    eye protection, 2, 11, 31, 46, 48
    face guards/shields, 12, 46, 48
    foot protection, 3
    hard hats, 3
    hearing protection, 2–3
    high-pressure washers and, 46
    media blasting equipment and, 48
    OSHA and, 2
    oxyacetylene gas safety and, 25
    parts washers and, 47
    welding and, 21–22
    See also safety practices
pesticides, 147
PGA (power guide axle), 497
photo tachometer, 630
pillow-block bearing, 246, 247
pilot bearing, 255
pilot control pressure, 453
pinion backlash, 580–581
pinion bearing, 578
pinion cone variation number, 579
pinion depth, setting, 578–589
pinion gear, 217, 578
pinion-to-ring gear, 564
pin punch, 38
pins, loader frame, 73

pin synchronizers, 231–234
pintle chain, 201, 205
pipeline trenchers, 98
piston accumulator, 672
piston free travel, clutch, 267, 399–400
piston pumps, 444–446
piston seal, 266, 267, 663
piston travel, 399–400
pitch
    chain, 202
    dozer blade, 77
    elongation and, 207
pitch circle, 216
pitch circumference, 203, 216
pitch control, 898
pitch diameter, 203
pitch measurement, 813
Pitman arm drive, 120, 850
pivot shaft, 800
pivot turn, 824
plain bearing, 242–243
planetary carrier, 283, 284, 520, 599
planetary controls. *See* clutches
planetary final drives, 597
planetary-gear automatic transmission, 87
planetary gear sets, 217
    Allison Transmissions, 309–321
    clutch apply chart, 302, 306, 307, 309
    compound, 289
    engagement, 287
    epicyclical, 285
    fracking transmissions, 321
    inter-axle differentials, 575–576
    John Deere 19-speed, 322
    multiple, 289
    power flows, 290
    ProDrive continuously variable transmissions, 554–555
    Ptolemy and, 285
    reverse gears, 300–303
    simple, 283–289, 295
    torque converters and, 424–425
planetary inputs, 540
planetary outputs, 540, 542
planetary pinion gears, 283, 284
planetary transmissions
    Allison Transmissions, 309–321
    Caterpillar, 264, 265, 295–300, 304–309, 308–309
    fracking, 321
    John Deer, 322
    overdrive, 316–318
planters and seeders
    air seeders, 119

    broadcast spreader, 119
    chemicals and, 116
    classifications, 117
    closing the farrow, 116
    clutches, 116
    drills, 118–119
    firming the seedbed, 116
    frames, 116
    markers, 116, 117
    metering, 114–115
    multi-hybrid planter, 116
    opening the furrow, 113
    placing seed, 115–116
    planter sensors, 116
    row crop, 113
    seeder categories, 113
    seeding, 115
    sensors, 116
    vacuum planters, 114
planter sensors, 116
plate synchronizers, 235
pliers, 39
plies, 767
plow steel, 166
plumb bob, 181
plunger, 123
ply rating 769
PM (preventive maintenance) inspection, 673
PMAC (permanent magnet AC) motor, 892–893
PMW. *See* pulse width modulation (PWM)
pneumatic accumulators, 670, 738
pneumatic compactors, 95–96
pneumatic hazards, 10–11
pneumatic systems, 9
pneumatic tires, 96
    bias-ply, 768
    chains, 772, 773
    construction components, 766–768
    foam-filled, 772
    load index values, 770
    nitrogen-filled, 771, 772
    ply rating, 769
    radial-ply, 769
    speed rating, 771
    star rating, 769–770
pneumatic tools, 40
pop-off valve, 700
POR (pressure override valve), 470
positive-displacement pump, 147, 153
positive traction differential, 571
Power Angle Tilt (PAT) dozer, 78
power-assist steering, 852–854
power density, 893

power divider. *See* inter-axle differential
power guide axle (PGA), 497
power hop, 743
power lines, 20
power neutral, 526
power rack-and-pinion steering, 853
powershift transmissions, 259, 329
    as automatic, 330
    characteristics, 330
    clutch apply charts, 392
    clutch pedal controls, 381–382
    electrohydraulically controlled, 331
    hydraulically controlled, 331
    inter-axle differentials, 576
    multiple planetary gear sets, 289
    preselect gear, 331
    skip shifting, 331
    speed matching, 331
power shuttle transmission, 427
power steering, 835
power systems, 187–199
power take-offs (PTOs), 13, 112
    Case IH, 513–514
    CNH continuously variable transmissions mode, 525
    continuous-drive, 141
    coupler, 638
    drawbar adjustment, 144
    driveshaft, 636–638
    functions, 140
    ground-speed, 141–142
    independent, 141
    mid-mounted, 140
    pump shaft, Deere 8R IVT, 534
    shaft guards, 142
    traction clutch assemblies and, 256–259
    transmission-drive, 141
power tools, 40–41
powertrain power flow, 235
power turn, 824
power zero (pz), 526
PowerZero mode, 534
PowrReverser module, 235–236
pre-charge chamber, 124–125
precharge pressure, 673–675, 670
precision chains, 201
preload, 248, 578, 605
preselect gear, 331
preselect mode, 521, 522–523
press fit bearing races, 246
pressure cutoff valve, 470–471
pressure differential valve, 380
pressure fluid. *See* fluid pressure testing equipment

pressure gauges, 42–43
pressure-limiter valve, 471–473
pressure override (POR) valve, 470
pressure passages, clutch, 265
pressure-protection valve, air brake, 714
pressure-release solenoid, 468
pressure sensor, 11, 698, 707
pressure tap part number chart, 915
pressure transducers, 45
press wheel grain drill, 118–119
prevailing torque nut, 55
prevention maintenance (PM) inspection, 673
primary air brake circuit, 702, 703
primary shoe, 658
primary tillage, 108
priority valve, 678
ProDrive transmission, 545, 549–557
propeller shaft, 622. *See also* driveshaft assembly
proportional solenoid control valve, 387
proportionator pump/motor, 550
propulsion, sprayer, 150
propulsion levers, hydrostatic transmission pumps, 446–448
Pro-test test-points, 42
protractor, driveline angles and, 632–633
Ptolemy, 285
PTOs. *See* power take-offs (PTOs)
pull belts, 512
pulley cap screw
pulley pitch diameter, 199
pulleys
    alignment, belt, 199
    crowned, 189
    drive/driven, 194
    fasteners, drive shaft, 199
    flat belts and, 189
    pitch diameter, 199
    speed, design, 194
    tension, 194
    V-ribbed belts and, 190–191, 197
pulser lever, 351
pulse-width modulation (PWM), 333, 388
punches, 37–38
punctures, tires and, 781
purge cycle, 697–698
purge valve, 697
purge volume, 697
push arms, 77
push belts, 511–512
pusher axle, 618
PVED-CLS (proportional valve digital-closed loop), 874
PWM (pulse-width modulation), 333, 388

Q

Q-Amp steering control unit, 863
Quad Range transmission, 259
Quadrigage, 44, 396
quick couplers, 41, 42, 45, 66, 73
quick-hitch, 139, 140
quick-release valve, 713
Quick Steer, 867, 870, 872

R

Raba 694 axle outboard-double-reduction planetary, 592–593
races, 242, 246
rack-and-pinion gear set, 217, 218
rack-and-pinion steering, 851
rack back cycle time, 72
radial grid braking, 896
radial lift, 91
radial load, 243
radial-ply tires, 769
raise cycle time, 71
rakes, 122–123
range transmission, 225–226, 238–240
ratchet drive size, 34, 35
ratcheting differential. *See* no-spin differentials
ratcheting wrench, 33, 34
ratchet marks, 608, 609
ratio valve, 380
reactor, 412. *See also* stator
reaping, 130–133
rear suspension, 733–734
rear visibility, 20
rear-wheel steer, 851
rebound, 733
recap tires, 779
recirculated ball steering gearboxes, 851
recordable accidents/incidents, 14–15
red cab lock, 17
reducer gear, 597
reflashing, 390
regular-lay wire rope, 166
regulated pressure, 376–377
relay valve, 710–712
release adjustment, clutch pedal, 260–262
release bearing, 254, 255, 2-57
relief pressure, 11, 499
relief valve, 803
replenishing port, 648
replenishing valve, 455
resistor grids, 902
respirators, 22, 48, 152
retainers, 57

retarders
- haul trucks, 86, 709
- *See also* dynamic retarders

retread tires, 779
return springs, 266
reverse
- Case IH combine feeder, 516–517
- CNH CVY, 523
- Fendt Vario CVT, 530–531
- planetary configuration, 285

reversed eye (RE) web sling, 171
reverse gears
- Allison Transmissions, 318–319
- Case IH Magnum powerflow, 339–340
- Caterpillar CVTs, 545, 547–548
- Deere 8R IVT, 537–538

reverse idler, 216
reverse-lay wire rope, 166
reverser, 427
reverse speed
- Caterpillar H, M, and non-suffix series, 348
- John Deere 8R, 357, 367

reversible pumps, 435–436
Rexroth MCR hydraulic drive motor, 597, 644. *See also* Bosch Rexroth
ride control, 739–740
rider safety, 21
right-handed threads, 55
right-hand-lay wire rope, 166
rigid axle, 618–619, 741
rigid-frame haul trucks, 86–87, 921
rigid-frame tractor, 844
rimpull, 419
rims, 783
ring-and-pinion gear set, 217
ring gear, 217, 283, 284, 563, 564
ring gear backlash, 580–581
ring-to-frame (RTF) clutch, 514–517
ring unit (RU) hydrostatic unit, 532
ripper dozers, 79
RMF (rotating magnetic field), 892
road lope, 752
rock drills, 99, 101
rock ejectors, 86–87
rockshaft, 139
rock traps, 133–135
rock wheel trencher, 98
rod weeder, 108, 111
roller bearings, 245–246
roller chains, 201–203, 210
roller design, compactor, 97
roller frame, 799
roller inspection, 812–813

rollerless chain, 201, 204
rollers
- mower conditioner (MOCO), 121, 122
- *See also* compactors

roll out bucket, 72
rollover protective structure (ROPS), 17–18, 76
roll pin, 57
room and pillar mining, 70, 71
root, 216
root ring gear tooth, 582
rope diameter, inspect, 167
rope lay, 167
rope-shovels, 85, 909–910
ROPS (rollover protective structure), 17–18, 76
rotary combines, 135
rotary disc, 120
rotary encoder, 521
rotary hoe, 108, 111
rotary mowers, 120, 121, 758
rotary oil flow, 418
rotary rake, 122, 123
rotary tiller, 108, 112
rotating bucket, 65
rotating clutches, 264
rotating concave rotary combine, 136
rotating magnetic field (RMF), 892
rotochamber, 717
rotor
- alternator, 885
- brake assembly, 656, 661
- continuously variable transmissions, 513–517

rotor slip, 892
rough-terrain crane, 162
rough-terrain forklift, 92
round baler belts, 189–190
round balers, 125–126
round belt, 189
row crop header, 132–133
row crop planter, 113
row crop tractor, 145
RSM (synchronous switched reluctance motor), 895
RTF (ring-to-frame) clutch, 514–517
R tires, 776–777
rubber track drives, 817–818
rubber-track loaders, 75, 76, 91, 601, 602
rubber track pads, 805, 807
rubber track systems, 817–818
rubber-track tractors, 146
runout, 635
rupture disk, 24
Rzeppa U-joint, 626

S

saddles, 169
SAE (Society of Automotive Engineers)
- bolt grade strengths, 54, 917
- breakout force, 71
- chain lubricants, 209
- communication protocols, 396
- operating load, 73
- seat belt regulations, 19
- viscosity ranges, 372

safety-bolt shackles, 174
safety data sheets (SDS), 6–7
safety glasses, 2, 11, 31, 46, 48
safety guards, 12, 13
Safety in Welding, Cutting, and Allied Processes (ANSIZ49.1), 21
safety poppet valve, 671
safety practices, 11–13
- accumulator, 672–673
- anhydrous ammonia and, 152
- asbestos, 255
- bar axle tractors and, 616
- blocking and cribbing, 161
- brake fluids, 653
- center of gravity and, 181
- chain slings and, 173
- changing multi-rim tires, 786
- charging accumulator, 675, 676
- check valve, 23
- clutches, sudden lunge, 332
- compressed air, 701
- cylinders, 24–25
- electric/hybrid service and, 910–911
- emergency preparedness, 3–7
- exhaust ventilation, 13
- eye injury, radar faces and, 780
- fall protection, 14
- flashback arrestor, 24
- fluid hazards, 7–9
- fluid injection injury, 377
- hand tools, 31–32
- hooks, slings and, 175
- hydro-pneumatic suspension cylinders, 735
- hydrostatic drive adjustments, 479
- idler repair, 812
- jacks and jack stands, 161
- jack stands, clutch release, 335
- job hazard analysis (JHA), 14
- lifting, 159, 182
- lock-out/tag-out (LOTO) procedure, 12
- lubricating chains and, 209
- machine operation safety, 19–21

machine safety, 15–19
manufacturer's manuals, 11
nitrogen gas, 772
on-site work and, 11–12
overhead shop doors, 13
oxygen/acetylene, 22–23
personal protective equipment (PPE) and, 2–3
pneumatic hazards/injuries, 9–11
power take-off (PTO), 13
reversible pumps, 436
self-propelled forage harvester, 129
shields/guards, 12
stalling torque converters, 429
tires and, 765–766
towing, brakes and, 681
tracking accidents, 14–15
undercarriage maintenance and, 808, 809
welding and, 21–22
wire rope clips, 170
wire rope inspection, 166–167
wire rope slings and, 168
wood blocking, 162
safety shields, 12
safety valves, 700–701
safety vests, 2
SALT (sealed and lubricated track) undercarriages, 801
Saur-Danfoss, 446
S-cam brake, 720
scarifier, 80
scavenge pumps, 355, 373
Schrader valve, 675, 768
scrapers
    applications, 83
    capacity, 85
    controls, 82–83
    cushioned hitch suspension, 738–739
    defined, 82
    open bowl/elevating, 83
    planetary transmission, Caterpillar, 309
    tractor/towed, 83
screwdrivers, 37
screw-pin shackle, 174
screws, 56
SCU. See steering control units
sealed and lubricated track (SALT) undercarriages, 801
Seale wire rope, 164, 165
seat belts, 19
secondary air brake circuit, 702, 703
secondary brake pedal, 669
secondary shoe, 658
secondary tillage, 108

second order vibrations, 630
seeders. See planters and seeders
seed tube sensors, 116
selective control valve (SCV), 744–745
selective shims, 401–402
self-cleaning grouser, 805, 807
self-energizing braking, 657
self-energizing effect, clutch bands, 276
self-leveling controls, 73, 738
semi-active seat suspensions, 740
semi-automatic transmission. See powershift transmissions
semi-floating axle, 617
separator plates, clutch, 263
sequential switching, 893
series BDCM, 890
serpentine belt, 197, 198
Service Advisor, 391, 395
service brakes
    application, 643
    control valves, 653–654
    ProDrive, 555
    rotor, 663
service brake valve, 678
service code classifications, 776–778
Service Information System (SIS), 391
servo caps, 447, 485
servo oil, 450, 453–454
servo valves, 448
set screw, 56, 485
shackles, 175
shaft endplay, 401
shaft guards, 142
shaft run-out, 488–489
shear bar, 129
shear bolt, 278
shear excavator, 65
shear hub, 120
shear pin, 142, 143, 637
sheaves. See pulleys
sheep's foot roller, 97
shields, drive belt, 196
shift collar, 225
shift detent, 241
shift fork, 229, 241–242
shift fork groove, 228
shift gear, 225
shift rails, 241–242
shock absorbers, 732–736
shock load, 608
shot rock, 902
shoulder, pneumatic tires, 768
shouldered eyebolts, 175
shovels, 110
shuttle shift, 333

shuttle valve, 455
sickle bar knife drive, 120
sickle bar mowers, 120
sickle bars, 120
side bearing preload, 580
side dressing, 152
side dump buckets, 72
side gear, 563, 564
side shields, 2
sidewalls, 767
sieves, 134
sieves, thresher, 136
signal oil, 450
signal pressure, 450, 693
silage, 12, 127
silent chain drive, 201, 204, 512
simple planetary gear set
    components, 283, 284
    configuration, determine, 284
    configurations, 285–287
    diagram, 295
    direct drive, 286
    forward overdrive, 285–286
    forward torque multiplication, 285
    gear ratios, 287–288
    member size relationship, 284
    neutral, 287
    principles, 284
    reverse, 285
    state of planetary members, 284
simple wire rope, 164, 165
single-circuit air brakes, 708–710
single circuit treadle valve, 709–710
single-disc traction clutch, 254–255
single-drive pressure test port, 481–482
single drive segment, 802
single engine scraper, 84
single flange roller, 804
single lift arm skid steer loaders, 91
single master cylinder, 647, 649–650
single parking brake chamber, 716–717
single-path hydrostatic transmission (HST), 437, 441–442
single piece rim, 784
single-reduction drop-axle bull-and-pinion final drives, 594
single-reduction final drives, 594
single roller compactors, 96
single rotor rotary, threshing combine, 135–136
single service brake chamber, actuator, 716
single-servo cradle bearing pumps, adjustments, 480–484
single-speed motor, 442–443

single standard grouser, 805, 806
single-strand roller chain, 203
single wedge actuator, 723
singulation, 114–115
sirometer, 630
six-way blade, 78
skid steer, 75, 921
skid steer loaders (SSLs)
    adjusting chain tension, 604
    applications and attachments, 91
    capacity, drive and lift, 91
    chain drive fill levels, 603
    chain drive housing breaker, 603
    controls, 89–90
    defined, 89
    drive types, 91
    ISO steer pattern, 90
    replace drive chains, 604–605
    steer H pattern, 90
    tensioning drive chains, 604
    weight distribution, loaded/unloaded, 89
skip loader, 74
skips, seed placement, 115
skip shifting, 331
slack adjusters, 655–656, 720–721
slave cylinder, 647
sledgehammer, 39
sleeve, 228
slew, 64
slew drives, 910
slide cutting, 78
slider, 228
slide skid control, 898
sliding caliper, 663
sliding clutch, 225
sliding collar, 233
sliding gear clutch, 232
sliding gear transmission, 221–224
sliding mast forklift frame, 92
sling hitches, 178
ings
    chain, 172–173
    fittings, 173–176
    polyester endless round, 172
    positioning, load capacity and, 177
    thetic, 171–172
    rope, 170–171
    s, 175
    142, 143, 278–279
    246
    s, 39
    ol, 898

slot-filled ball bearing, 244
slow adapt, 389
slugging, 199
snap rings, 57
snap-ring U-joints, 624
Society of Automotive Engineers (SAE), 19, 73, 209, 372
socket-cap head, 56
socket driver, 33–35
socket wrench, 33
soft-faced mallet, 39
soil compactor, 96
soldering irons, 40
solenoid apply charts, 391, 393–394
solenoid-controlled hydrostatic transmissions, 448
solenoid control valve, 385
solenoid testing, 388–389
solid tires, 774
solvent tanks, 431
spalling, 609
speed
    belt in fpm, 191
    power-take offs (PTOs) and, 140
    torque and, 188
speed control lever, 240
speed gear ring, 230
speed handle wrench, 34
speed matching, 331
speed rating, tires, 771
speed selector spool, 380
speed-sensing signal pressure, 450, 453
speed-sensing valve, 450–453
spelter socket, 170
spherical roller bearing, 246
spider, U-joint, 622, 625
spider gear, 422, 563, 564
spider key, 572
spike-toothed harrow, 108, 110, 111
spiral bevel gear sets, 218, 219
spiral bevel pinion gears, 579
splice, 190
splined driveshaft, 140
splined stub shaft, 140
split guide rings, 414
split hydrostatic transmission, 443, 444
split pin synchronizer, 233–234
split rim, 784
split splines, 622
split throttle, 523–524
splitting stands, 161
splitting the track, 809
spool valve, 387, 652
spotter, 20
sprag clutch, 271, 273, 417

sprayers, 147–150
    agitation, 150
    air suspension, 758–759
    applicator sprayer pump, 149
    boom controls, 148–149
    centrifugal, 147
    functions, 147–148
    hydraulic, 147
    hydrostatic transmissions and, 442
    markers, 151
    nozzles, 149
    propulsion, 150
    pull-type, 151
    pumps, 147
    self-propelled, 151
    sizes, 151
    steering, 150
    suspension, 150
    types, 147–148
spreader bar, 179
spring-applied caliper parking brake, 665
spring-applied external caliper disc brakes, 664
spring-applied manually released multiple-disc clutches, 268–270, 271
spring-applied oil-released multiple-disc clutch engagement, 268
spring-applied parking brake, 666
spring-applied parking brake caliper, 663
spring-detented block synchronizers, 230–231
spring-loaded collar, 142
spring-loaded pins, 142
spring-loaded poppet valve, 671
spring scale, 578
sprocket pitch, 813
sprockets
    defined, 202
    diameter formula, 203
    drive, 201
    pitch circumference, 203
    ratio formula, chain, 210
    wear, inspect for, 206, 811–812
spur gears, 121, 216, 218
square baler operation, 123–124
square-cut piston seal, 266, 267
square-cut seal, 663
SRM (switched reluctance motor), 893–894
SSLs. *See* skid steer loaders (SSLs)
stabilizers, 98
stall mode, 417
stall testing, 428
standard differential, 568

standard grouser, 805, 806
standard undercarriage, 798
stand pipes, 625
star rating, 769–770
star washers, 55–56
static load-sensing signal, 866
static tipping load, 73
stationary clutches, 264
stationary planetary carrier, 600
stators
    alternator, 885
    functions, 412, 413
    overrunning, 417
stator windings, 889
steer axle, 618–619
steer-by-wire systems, 867, 874
steering
    electronic, 146
    forklift, 92
    gear sets and, 217
    sprayer, 150
    telehandler rough-terrain forklift, 93
    wheeled. *See* wheeled steering
steering amplifiers, 863–864
steering clutch and brake, 76, 77
steering-clutch-and-brake systems
    components, 827
    controls, 828–830
    differential steering systems, 832–837
    disadvantages, 832
    examples, 832
    functions, 828
    operation, 830–832
steering control units (SCUs)
    anti-cavitation check, 861–862
    auto-guidance steering systems, 856
    closed-center load-sensing SCU, 858–859
    closed-center non-load-sensing, 857–858
    components, 857
    cylinder-dampening, 863
    cylinder port relief, 861
    emergency steering, 859–860
    functions, 856
    inlet check valve, 860
    inlet relief valve, 860–861
    internal leakage test, 866–867
    left or right operation, 859
    load-reaction, 862
    load-sensing relief valve, 861
    neutral operation, 859
    non-load reaction, 862
    open-center, 857

Q-Amp, 863
wide-angle, 862
steering control valve (SCV), 856. *See also* steering control units (SCU)
steering drift compensation, 90
steering gear, 851
steering kickback, 860
steering knuckles, 850
steering priority valves, 864–866
steering tiller, 834
steering wheel, 80
steer pattern, 90
Steiger tractors, steering systems, 855, 872–873
stepless transmission, 511. *See also* continuously variable transmissions
straight cut header, 121, 130–131
straight dozing, 78
straightening driveshaft, 635–636
straight roller bearings, 245
straight static tipping load, 73
strands, wire rope construction, 164, 165
strand wire arrangement, 164–165
straw walker, 134, 138
stress riser, 607
stretch, chain elongation, 206–207
Stretch-Fit V-ribbed belt, 198
stripper header, 132, 133
stroke-sensing automatic slack adjuster, 721
struck capacity, 74
struts, 732–733
studs, 55
stuffer fork, hay baler, 125
sub-compact tractor, 144
sub-soiler, 110
suction filtration, 458
suction screens, 373
suction strainers, 373
suitcase weight, 790, 791
sun gear, 283, 284. *See also* planetary gear sets
sun gear shaft, 520
supply reservoir, 689
supply tank, 698, 699
surface mining, 70, 71
suspended front axle (SFA), 743–746
suspension systems
    agricultural, 741–759
    cab, 759–760
    construction and mining, 732–742
    defined, 731
swaged socket, 170
swashplate angles

    examples, 442–443, 445, 446, 485
    hydrostatic transmissions and, 448–449
swath control, 115
swathers, 120–121
    air suspension, 757–758
    hydrostatic transmissions, 439–441
    mower conditioner (MOCO), 121, 122
    rear caster wheels, 92
    self-propelled, 121
sweeps, 110
swept area, 669
swing arm, 753
switched reluctance motor (SRM), 893–894
swivel hooks, 174
synchromesh transmissions. *See* synchronized transmissions
synchronized transmissions
    block synchronizers, 228–231
    components and function, 228
    defined, 228
    dual clutch, 234–240
    friction plate, 234
    pin synchronizers, 231–234
    three-speed transmission, 226
synchronizer assembly, 533–534
synchronizer cones, 233
synchronizer cups, 233
synchronizer hub, 228
synchronizer rings, 230
synchronizer shaft, 533–534
synchronizer sleeve, 233
synthetic slings, 171–172
SystemOne Undercarriage, 801, 816
system pressure, 692–695

T

tachometers, 140, 141
tag and lock-out procedures, 12
tag axle, 618
tailings auger, 134, 136, 137
tailings elevator, 137
tandem chain final drives, 597–600
tandem fill levels, 603
tandem master cylinder, 650, 651
tandem-roller compactor, 96
tandems, 81
tandem scraper, 84
tandem service brakes, 654
tapered roller bearing, 245
tapered wedges, wheel spacing, 789
Technical Service Group (TSG), 396
tedders, 123

telehandler, 92–93, 618–619
tensile strength, 54, 607
termination efficiency, 168
test stand, 504–505
three-phase alternating current (AC), 886–888
three-point hitch, 139, 140
three-stage torque converter, 421–423
thresh, 130
threshing, 134–138
throttle controls, 423, 539, 760
throughput, 138
throw out bearing, 254, 255, 257
thrust bearing, 246
thrust load, 243
thumb excavator, 65
tie-band, 193
tillage tools, 108–112
tilling, 108
tilt, blade control, 77–78
timing belts, 190
timing chain, 200, 204
tire cage, 766
tire chains, 772, 773
tire load index values, 928–929
tires
    agricultural service classifications, 776–778
    airless, 774
    changing multi-piece rim, 785–787
    construction tire classifications, 778–779
    hydraulic dismounting tool, 786
    inspecting, 781–782
    maximizing life of, 782
    non-pneumatic types, 773–774
    off-the-road (OTR), 765
    pneumatic, 766–773
    safely and, 765–766
    sizes, aspect ratio and, 775–776
    solid, 774
    spelling differences, 765
    tire cage, 766
    ton-mile per hour value, 782–783
    tread wear rate, 782
    weighting and ballasting agricultural, 790–792
    wheels and rims and, 783–785
    wheel slip, 780–781
    wheel spacing, adjusting, 787–790
TKPH (ton-kilometer per hour), 782–783
TLS (Triple Link Suspension) Plus, 743–746
TMPH (ton-mile per hour), 782–783
toe ring gear tooth, 582

ton-kilometer per hour (TKPH), 782–783
ton-mile per hour (TMPH), 782–783
tools
    cleaning, 46–47
    electric, 40
    electronic service, 402, 404
    fluid pressure testing equipment, 41–46
    hand, 32–39
    hydraulic, 41
    measuring, 49–53
    media blasting, 47–48
    metal cutting, 48–49
    pneumatic, 40
    power, 40–41
    pressure testing, 41–46
    welding, 48–49
tooth contact pattern, 581–583
top land, 582
top-of-travel switch, 382
toric ring, 625
torque, 32–33, 187–188
torque arm, 743
torque converters, 412–430
    component removal and cleaning, 431
    components and functions, 412, 415, 428
    contamination control, 430–431
    coupling phase, 418
    design variations, 418–425
    external vanes, 414
    failure of, 428–429
    impeller clutch, 419
    inspecting, 428
    installation, 426–427
    lockup clutches, 418–419
    mounting designs, 412–414
    multiple impellers, 419–420
    multiple stage, 421–423
    multiple turbines, 420–421
    oil flow, 426
    operation, 414–416
    split guide rings, 414
    testing, 427–428
    torque divider, 424–425
    torque multiplication phase, 416–417
    twin-turbine, 420
    variable-pitch stator, 423–424
torque converter stator clutches, 270, 273
torque divider, 417, 424–425
torque multiplication, 220, 416–417
torque multipliers, 36–37
torque-sensing, 194–196
torsional stress fracture, 608, 609
Torx fasteners, 56–57

towing safety, 21, 681–682
tow valve, 681
TPI (threads per inch), 54
TPMS (Truck Payload Measurement System), 734
track chains, 800, 801
track links, 800–801
track loaders, 75–76, 919
    applications, 75
    Bobcat leaf spring suspension, 729
    controls, 75
    gear sets, 217
    rubber, 76, 91
    sizes, 76, 919
    types, 75–76
track pins, 800–801
track pitch, 813
track rollers, 799, 804
track sag, 808
track shoes, 806–807, 813
track steering systems
    dual-path hydrostatic transmissions, 825–826
    hydraulic steering motor, 837–838
    hydrostatic steering terminology, 839
    steering-clutch-and-brake systems, 827–832
track tension, 807–808
track tensioner, 803
track wear gauge, 813–814
traction alternator, 884–888
traction clutches
    conventional single-disc, 262
    dual-disc, 255–256
    free play adjustment, pedal, 259–260
    function, 253
    mounting, 254
    PermaClutch, 258–259
    PTO assemblies, 256–259
    release/operating lever adjustment, 260–262
    removal/installation, 262
    service/adjustment, 259–263
    single-disc, 254–255
traction control, 902–903
traction motors
    AC, 891–898
    DC, 888–890
    function, 888
tractors, 138–146
    auto-guided systems, 146
    compact utility, 144–145
    differential steer, 436
    drawbars, 143
    four-track undercarriages, 843–844

four-wheel-drive, 145–146
   high-clearance, 145, 146
   implements, 140
   loaders, 142–143
   power take-offs (PTOs), 140–142
   quick-hitch, 140
   row crop, 145
   sub-compact, 144
   three-point hitch, 139, 140
   twin-rubber-track, 146
   two-wheel drive, 145
   uses, 138
   utility, 145
tractor scraper sizes, 920
tractor-trailers, 88–89
tractor undercarriage, 799
traditional track dozers, 79
traditional track loaders, 75, 76
transmission charge pressure, 376–377
transmission clutch, 141, 253–262. *See also* traction clutch
transmission controls
   diagnosing low power, 496–497
   electrical, 380–381
   electronic, 381
   mechanical, 378–380
transmission control valve, 379–380
transmission diagnostics
   clutch pressure specifications, 394
   diagnose automatic powershift using clutch apply charts, 391
   electronic service tools, 395
   error codes, 394
   flowcharts, troubleshooting, 394, 395
   manufacturer solutions, 395–397, 396–397
   power flows/mechanical schemas, 394–396
   pressure gauges, 395
   service literature, 391
   solenoid apply chart, 393–394
transmissions
   contamination control, 430–431
   cycling, 330–331
   diagnostic software, 382
   electrohydraulically controlled, 331
   filtration, 373
   fluids, 371–372
   hauling, 330, 331
   hydraulically controlled, 331
   hydraulic circuit, 373–375
   input shaft speed, 384
   off-highway applications, 329–331
   oil flow, 375–376
   operating pressure, 376–377
   pumps, 372–373
   removal/installation, 403–405
   scavenge pump, 373
   transmission-driven power take-off, 141
transmission speed, 228
tread, 767, 781
treadle valve, 703–707, 709–710
trenchers
   applications, 98–99
   controls, 98
   sizes, 99, 922
   types and uses, 97–98
trench openers, planting/seeding, 113
trial-and-error method, determining center of gravity, 181
triangle and choker (TC) sling, 171
triangle and triangle (TT) web sling, 171
Triple Link Suspension (TLS), 743–746
trolley assist, 87, 899, 900
truck-mounted crane, 162, 163
truck-mounted forklift, 94
Truck Payload Measurement System (TPMS), 734
true brinelling, 610, 611
truing a shaft, 635
TSG (Technical Service Group), 396
turbine, 411, 412. *See also* hydrodynamic drives
turnback eye, 170
turnbuckle, 77
turning brakes, 643
turning radius, 81
turn mode switch, 837
Tweel, 774
twin countershaft (TC), 327–329
twin-rubber-track tractors, 146
   differential steering, John Deer, 836, 837
   twin-coil and hydraulic shock absorbers, 757
   walking beam AirCushion suspension systems, 754–757
twin-track turning radiuses
   brake steer, 824
   counter rotation, 823
   dual-path hydrostatic transmission and, 825–826
   functions, 823
   neutral steer, 824
   power turn, 824
   steering-clutch-and-brake systems, 827–832
twin-turbine torque converter, 420
twin wedge actuator, 723–724

two-speed motors, 442–443
   hydrostatic transmissions, 461–464
two-speed planetary steering system
   applications, 840
   operation, 840–841
two-wheel steer, 93, 851

U

U-bolt, 169
U-joint. *See* universal joint
ultimate strength, 168
ultrasonic measurement tool, 810
uncommanded actuation, hydraulic pump, 484
undercarriages
   bogies, 804
   bushing inspection, 810–811
   carrier rollers, 804–805
   Caterpillar classifications, 798–799
   center tread idler, 816
   clamp master link, 816
   cost, disadvantages, 798
   drive sprockets, 802, 811
   elevated-sprockets, 799–800, 803
   equalizer bar, 800
   exercising the tracks, 816
   four-track tractors, 843–844
   functions, 797–798
   ground pressure, 797
   guides and guards, 805
   hard bar, 799–800
   heavy-duty extended life (HDXL), 801
   heavy-duty track, 801
   high-track, 803
   idler, 802–803, 811–812
   inspection tools, 809–810
   length, track width and, 797
   manufacturer's handbooks, 798
   operating tips, 816
   pitch measurement, 813
   pivot shaft, 800
   positive-pin-retention track, 801
   recoil mechanism, 803
   roller inspection, 812–813
   rubber track systems, 817–818
   sealed and lubricated track (SALT), 801
   splitting the track, 809
   SystemOne, Caterpillar, 801, 816
   track chains, 800–801, 801
   track frame, 799
   track link inspection, 810
   track links, 800–801
   track pins, 800–801

track rollers, 804
track shoes, 805–807, 813
track tension, 803, 807–808
track wear, 813–815
unit planter, 116
universal blade, 78
universal control module (UCM), 520
universal joint, 622–625
unloader valve, 679
unsafe acts, 15
unsynchronized manual transmission, 226
upstroke pumps, 464–465
US customary bolt sizes, 917
U.S. Department of Transportation (DOT), 25, 652–653
US National Farm Progress show, 589
utility tractor, 145

V

vacuum planters, 114
valve plate adjustment, 486–487
valves
    accumulator charging spool, 678
    accumulator effect, 474
    accumulator isolation check, 678
    air brake, 710–714
    anti-cavitation check, 861–862
    AutoTrac, 870, 871
    Caterpillar H, M, and non-suffix series, 344
    center, 651–652
    compensation, 651
    cylinder port relief, 861
    directional control, 857
    discharge, 691, 692
    drain, 698–700
    dual circuit treadle, 703–707
    electronically controlled displacement valve, 473
    electronic anti-stall control, 473
    electronic pressure-release solenoid, 468
    equalizer, 650, 651
    fill, tensioner piston, 803
    flapper, 652
    flushing, 455
    high-pressure relief, 456–457
    hot-oil purge, 455, 495–496
    inching, 467, 651
    inlet, 690–691
    inlet check, 703, 860
    inlet/exhaust, 693
    internal pressure override (IPOR) valve, 468–470
    inverse shuttle, 678
    load-sensing relief, 861
    make-up, 501, 504
    manual bypass, 468, 485–486
    multi-function, 457–458
    over-speed limit valve, 473–474
    pressure-limiter valve, 471–473
    pressure override (POR), 470
    pressure settings, 455
    priority, 678
    purge, 697
    relief, tensioner piston, 803
    safety poppet, 671
    safety pressure relief (pop-off), 701
    Schrader, 675, 768
    selective control, 856
    service brake, 678
    service brake control, 653–654
    single circuit treadle, 709–710
    spool, 652
    steering priority, SCUs, 864–866
    tow, 681
    unloading, 679
valve stem, 768
variable-capacity torque converter, 419–420
variable-diameter pulley (VDP) system, 194–196, 511–512
variable-displacement hydrostatic transmissions, 550
variable-displacement motors, 436
    components and application, 464–465
    hydrostatic transmissions and, 442–443
variable-displacement pumps, 372, 464–465
    brakes, 653
    hydrostatic transmissions and, 442–443
variable frequency drive (VFD), 892
variable-pitch stator, 423–424
variable rate planter, 115
variable solenoid control valve, 387
variable speed belt, 194–196
variator, 511
VarioActive steering, Fendt, 870
V-belt gauge, 191–192, 192
V-belts, 190–193, 199
VDP. See variable-diameter pulley (VDP)
Venieri articulated telehandler, 854
ventilation systems, 13
Versatile
    axial rotor combine, 135
    RCR rotating concave rotary, 136
    SX 240, 275, and 280, 412
vertical auger, 83
vertical hitch, 167
vertical inline hitch, 178
vertical lift, 91
VFD (variable frequency drive), 892
vibrating reed tachometer, 630
vibrations, driveline, 629–632
    analyzing, 629–631
    case study, 632
    measuring, 631
    out-of-balance shafts, 631
vibratory compactor, 96
vibratory plow trencher, 98
viscosity, 371–372
Volvo
    articulated dump trucks, 87
    ECR25 battery-powered electric-drive excavator, 908
    hydraulic accumulator, 910
    L25 battery electric wheel loader, 90512
    motor graders, 81
    rigid haul trucks, 87
    single lift arm skid steer loaders, 91
    suspensions, comparing, 738
vortex oil flow, 416
VPM (vibrations per minute), 96
V-ribbed belts, 197–198
V-ripper, 108, 110
V-shaped drawbar, 80
V-trenching, 78

W

walk-behind trenchers, 97, 99
walking beam suspension, 737
    John Deere 8RT AirCushion, 755–756, 757
    laminated blocks, 738
    rubber, 738
Warrington wire rope, 164, 165
WAS (wheel angle sensor), 874
washers, 55–56, 175
water pump, 197
wear
    chains/sprockets, 206–208
    final drives, 606–607
    tire, 781
web slings, 171–172
wedge brakes, 722–724
wedges, wheel spacing and, 788–789
wedge socket, 169
weighting and ballasting tires, 790–792
welding equipment, 48–49

welding safety, 21–25
wet clutches, 253
wet disc brake, 683, 644
wet parking brake, 555
wet tank, 689, 698, 699
wheel and drum compactors, 96
wheel angle sensor (WAS), 874
wheel chocks, 15
wheel cylinder, 647
wheel dozers, 79
wheeled excavators, 66
wheeled steering systems
    Ackerman, 850–853
    articulation, 853–854
    Caterpillar Command Control Steering, 874–876
    combination, 854–855
    control units, 856–863
    dual-path hydrostatic transmission, 849
    motor graders, 876–877
    multi-axle, 855
    steering amplifiers, 863–864
wheel lean, 877
wheel loaders
    applications/attachments, 72
    capacity, 74
    controls, 71
    cycle time, 71–72
    electric drive, 899–905
    frame designs, 72–74
    sizes, 919
    types and uses, 70–71
    uses, 72

wheel rake, 122
wheels
    defined, 783
    material wedged in, 781
    rims, 783–785
wheel slip, 780–781
wheel spacing
    bar axle, 788–790
    flange-mounted, 788
    offset, 787–788
    wedge removal and, 790
wheel weights, tires, 790
wide-angle steering control unit, 862
wide-sweep blade plow, 108, 112
windrower swather, 121
windrow pickup header, 128, 131
wing nut, 55
wing-type U-joints, 624, 625
wing-type yoke, 622
winnow, 130
wire rope
    abrasions and flexibility, 165
    classification, 165
    cores, 164, 165
    end terminations, 168–170
    inspecting, 166–167
    lay of, 166
    slings, 167–168, 175
    steel, grades, 166
wire rope clips, 168–170
wire rope lay, 166
wire rope shovels, 69, 70
wire rope slings, 167, 170–171
wire rope strands, 164, 165

WLL. *See* working load limit (WLL)
Woodruff key, 199
working load limit (WLL)
    chain slings, 173
    eyebolts and, 175
    hitches, 178
    shackle, slings, 174
    wire rope slings, 168
worm drive gear, 218–219
worm gear drive, 720
wrenches, 32–33

X

XL (extra-long track) undercarriages, 798
XR (extended-to-rear) undercarriages, 798
XW (extra-wide track) undercarriage, 798

Y

yield mapping, 138
yield monitoring, 138
yield strength
    bolts and, 54
    final drive fractures, 607
yokes, 622–623, 628–629

Z

Z-bar linkage, 72, 73, 76
zerk, 209
zero-emission machines, 884
zero turning radius (ZTR) lawn mowers, 849
zip-tie method, 198